Numerical Methods in Thermal Problems
Volume V, Part 2

Edited by

R. W. Lewis and K. Morgan
Civil Engineering Department, University College of Swansea, Wales

and W. G. Habashi
Mechanical Engineering Department, Concordia University, Montreal, Canada

Proceedings of the Fifth International Conference held in Montreal, Canada on June 29th–July 3rd, 1987

PINERIDGE PRESS

Swansea, U.K.

First Published 1987 by
Pineridge Press Limited
54, Newton Road, Mumbles, Swansea, U.K.

ISBN 0-906674-65-4

Copyright © 1987 the Contributors named in the list of contents.

International Conference on 'Numerical Methods in
 Thermal Problems' *(5th: 1987: Montreal, Canada)*
Numerical methods in thermal problems:
problems of the fifth international conference.
 1. Heat — Transmission — Mathematics
 2. Numerical calculations
 I. Lewis, R. W.   II. Morgan, K. (Kenneth)
 536'.2'001511   QC320.2

ISBN 0-906674-65-4

All rights reserved.
No part of this publication may be reproduced, stored in a retrieval system or transmitted, in any form or by any means, electronic, mechanical, photocopying or otherwise, without prior permission of the publishers Pineridge Press Ltd.

Printed and bound in Great Britain by
Dotesios (Printers) Ltd., Bradford-on-Avon, Wiltshire

PREFACE

This proceedings contains the papers presented at the Fifth International Conference on Numerical Methods in Thermal Problems held in Montreal during the period June 29th - July 3rd, 1987. Over two hundred and thirty abstracts were submitted to the organising committee and their high standard made it extremely difficult to leave out so many, but constraints of space and time demanded that over a third be rejected. The organisers would, however, like to thank the authors concerned for sending in their manuscripts for consideration.

The ever continuing interest in the application of numerical methods for the solution of thermal based problems is aptly demonstrated in these proceedings. The papers, contributed by authors from many countries, have been separated into nine sections, which cover the main areas of interest, e.g. 'heat conduction', 'free and forced convection', 'heat and mass transfer', etc.. In general, the manuscripts have been placed into the appropriate field of interest apart from the more diversified groups of 'novel computational techniques' and 'industrial and scientific applications'.

The proceedings are printed from direct lithographs of the authors' manuscripts and the editors do not accept responsibility for any erroneous comments or opinions expressed herein. As in previous conferences of this highly successful series, a planned 'state-of-the-art' archival text will be produced after the conference.

Last, but by no means least, the conference organising committee wishes to acknowledge the collaboration of the international journals *Numerical Methods in Engineering* and *Communications in Applied Numerical Methods* and also the support of *PRATT AND WHITNEY CANADA, CONCORDIA UNIVERSITY* and the *Natural Sciences and Engineering Research Councils of Canada.*

R.W. LEWIS, K. MORGAN and W.G. HABASHI

May, 1987.

CONTENTS

PREFACE

VOLUME I

SECTION 1   Phase Change Problems

1. J.M. SULLIVAN, Jr. and D. R. LYNCH   5
   "Grid generation for dendritic growth simulations
   on deforming elements"

2. M.P. KANOUFF   15
   "Weld pool modeling"

3. D. LOYD and B. KARLSSON   27
   "Heat transfer at cabinet walls containing phase
   change material"

3. J. CALDWELL   39
   "Numerical solution of one-dimensional
   melting/solidification model problems

5. R.O. STAFFORD, J.W. KLAHS, D.F. PINELLA   62
   "Numerical Methods for solidification simulation"

6. M. REZAYAT and T.E. BURTON   84
   "Combined boundary-element and finite-difference
   simulation of cooling and solidification in
   injection moulding"

7. M.J. BURTON and R.J. BOWEN   96
   "Modelling ice plug formation in cryogenic
   pipe freezing"

8. R. HAMAR, and S. THIBAULT   108
    "Modelization of thermal transfer in D.T.A. cell"

9. N. KITAHARA, H. YANO and A. KIEDA   120
    "Approximate Solutions with heat polynomials
    for Stefan problem"

10. B. CARRUPT. J.-L.DESBIOLLES and M. RAPPAZ   132
    "FDM simulation of the surface treatment of
    materials by laser"

11. C.A. Van der STAR   143
    "The application of the enthalpy method to
    phase change problems in annuli"

12. M. CARMIGNANI, A. MASNATA, V. RUISI, A. TORTORICI   154
    "A mathematical model of thermal cycles in welding"

13. Z. ABDULLAH and M. SALCUDEAN   164
    "Computation of casting solidification in the
    presence of natural convection"

14. S. LIN, C.K. KWOK, W.L.DAI   176
    "A study on the short-time starting temperature
    profile for numerical solution of phase change
    problems"

15. A.F.A. HOADLEY, T.J. SMITH and D.M. SCOTT   185
    "The incorporation of natural convection effects
    in solidification simulation"

16. A.K. PANI and P.C. DAS   198
    "Finite element approximation to a class of
    one dimensional ablation problems"

SECTION 2   Heat Conduction

1. BENGT SUNDEN   207
    "Numerical prediction of transient heat conduction
    in a multi-layered solid with time-varying
    surface conditions"

2. J.L. WEARING, C. PATTERSON, M.A. SHEIKH and
    A.G. ADBUL RAHMAN   219
    "A regular indirect boundary element method for
    heat conduction"

3. M.S. Ingber and A.K. MITRA 231
    "Solution of the transient heat conduction problem
    in zoned-homogeneous media by the boundary element
    method"

4. S.M. CARTER, R.F. BARRON, R.O. WARRINGTON and
    R.P. KOBS 241
    "The boundary element method applied to
    cryosurgical probe tip design"

5. C.-C WONG and R. K-T. WONG 253
    "Numerical methods for solving the network model of
    three-dimensional diffusion problems"

6. D. P. UPDIKE and A. KALNINS 265
    "Heat conduction in shells of revolution"

7. D.A. KOUREMENOS, K.A. ANTONOPOULOS 276
    "Numerical simulation of the thermal problem
    in hyperthermia treatments"

8. LIU GAO-LIAN and ZHANG DAO-FANG 284
    "Numerical methods for solving inverse problem of
    heat conduction with unknown boundary based on
    variational principles with variable domain"

SECTION 3

Natural and/or Forced Convection

1. K.H. WINTERS 299
    "Oscillatory convection in crystal melts:
    The horizontal Bridgman process"

2. A. HAHRMANN and W. NITSCHE 311
    "Comparative numerical and experimental
    investigation on transient temperatures in convec-
    tively heated non-homogeneous structures"

3. T. FUSEGI and B. FAROUK 321
    "Turbulent natural convection-radiation inter-
    actions of a non-gray gas in a square cavity"

4. C. NONINO and S. DEL GIUDICE 332
    "Turbulent forced convection in two-dimensional
    recirculating flows"

5. Y. LE PEUTREC and G. LAURIAT 344
    "Practical evaluation of improved upwind finite
    difference schemes for natural convection in
    enclosures"

6. P.H. OOSTHUIZEN and J.T. PAUL 356
"Natural convective heat transfer across a cavity with elliptical ends"

7. DANIEL DICKER 368
"A mathematical study of a radial-wall heat exchanger"

8. I.M. RUSTUM and H.M. SOLIMAN 380
"Developing heat transfer in internally finned tubes"

9. E. ZIMMERMAN and S. ACHARYA 392
"Natural convection in an enclosure with a vertical baffle"

10. T. ITO and M. YAMAGUCHI 404
A numerical study of turbulent forced-convection heat transfer to supercritical helium"

11. C.T. NGUYEN and N. GALANIS 414
"Combined forced and free convection for the developing laminar flow in horizontal tubes under uniform heat flux"

12. LAI-CHEN CHIEN 426
"Forced convection heat transfer from the flow around an impulsively started sphere"

13. T. KOBAYASHI and Y. MORINISHI 437
"A numerical experiment of incompressible turbulent swirling flow in rectangular straight pipe"

14. P. ANDRE, J. BATINE and R. CREFF 448
"Study of the thermal fluid field for pulsed flows with compressible fluids"

15. J.C. DUH and WEN-JEI YANG 459
"Effects of Prandtl number on transport phenomena in evaporating sessile drops"

16. J. MAQUET, G. GOUESBET, A. BERLEMONT 472
"A computer code for natural convection in an enclosed cavity with a free surface"

17. K. NOTO and R. MATSUMOTO 484
"Breakdown of the Karman vortex street due to natural convection (case from an elliptical cylinder whose major axis oriented at right angle to main stream)

18. N. TOSAKA and H. FUKUSHIMA — 500
"Numerical simulations of laminar natural convection problems by the integral equation method"

19. R.W. KNIGHT and M.E. CRAWFORD — 512
"Simulation of convective heat transfer in pipes and channels with periodically varying cross-sectional area"

20. D. KUHN and P.H. OOSTHUIZEN — 524
"Transient three-dimensional natural convective flow in a rectangular enclosure with two heated elements on a vertical wall"

21. S. LAVOIE, T.H. NGUYEN and C.A. LABERGE — 536
"Heat transfer by natural convection between two concentric cylinders"

22. P. GUANG MAO and T. HUNG NGUYEN — 547
"Higher-order accurate numerical solution to the problem of natural convection in a rectangular cavity"

23. M.A. KALAM and R. KUMAR — 559
"Numerical study of laminar natural convection in vertical annuli"

24. K. NOTO and R. MATSUMOTO — 571
"Three-dimensional natural convection heat transfer from a single plate"

25. A.N. THORNHILL and E.K. GLAKPE — 586
"Turbulent natural convection in boundary-fitted coordinates"

26. T. FUJII, S. KOYAMA and K. SHINZATO — 597
"Forced convection heat transfer inside a locally heated tube - numerical analysis as a conjugated problem"

27. S. KOSHIZUKA, Y. OKA and Y. TOGO — 609
"An evaluation of three filtering methods applied to three higher-order difference schemes of convection"

28. P. ROSA and F. PIRONTI — 621
Heat transfer natural convection steady state simulation by finite elements between vertical enclosed concentric cylinders"

29. A.S. BARTOSIK, R. SOBOTINSKI and A.J. WANIK — 632
"Numerical prediction of heat transfer in fully developed pulsating turbulent flow"

30. A. MOJTABI, D. QUAZAR and M.C. CHARRIER-MOJTABI 644
"An efficient finite element code for 2D steady state porous annular layer"

SECTION 4

Heat and Mass Transfer

1. L. IMRE, A. BITAI, Cs. HORVATH, S. SZENTGYORGYI & L. Banhidi 657
"Thermal analysis of human body - clothing, environment system"

2. J. A. TINKER 669
"Modelling the thermal conductivity of multiphase materials containing moisture"

3. G.J. ANDERS, H.S. RADHAKRISHNA, J.A. ROIZ 681
"Numerical solutions to the heat transfer problem in the vicinity of underground power cables"

4. V.R. VOLLER 693
"A numerical method for analysis of solidification in heat and mass transfer systems"

5. R. KOHONEN, T. OJANEN 705
"Non-steady-state coupled diffusion and convection heat and mass transfer in porous media"

6. L. ROBILLARD, H. WANG CHONG and P. VASSEUR 717
"Multiple steady-states in a confined porous medium with localized heating from below"

7. T. KODAMA and S. KOTAKE 728
"Coarse-find mesh method for locally complex flows of heat and mass transfer"

8. M. NOVAKOVIC, A. VEHAUC and Z. KOSTIC 738
"Steady alternating state (SAS) numerical method for determination of heat and mass transfer between fluid streams"

9. A.M. CRAWFORD, H.S. RADHAKRISHNA and K.C. LAU 749
"Application of the integrated finite difference technique to heat transfer through unsaturated particulate media"

10. G.A. CLUTE and A.M. CRAWFORD 761
    "SFM: A symmetrical four-well ates model"

11. X. ZHANG, T. HUNG NGUYEN, R. KAHAWITA and PU WANG 773
    "Spectral and spectral-finite difference methods in wavenumber prediction of penetrative convection"

12. F.C. LAI, F.A. KULACKI and V. PRASAD 784
    "Numerical study of mixed convection in porous media"

13. N. KLADIAS and V. PRASAD 797
    "Numerical study for inertia and viscous diffusion effects on Bernard convection in porous media"

14. D.A. KOUREMENOS, K.A. ANTONOPOULOS 811
    "Finite-difference solution of the transient bioheat transfer equation during local hyperthermia in inhomogeneous tissues containing arteries and veins"

15. B. ZAPPOLI, C. MIGNON and N. MATHE 821
    "A pseudo compressible method for computing cavity flows with surface reaction"

16. D.A. KOUREMENOS, J.G. KOULIAS AND K. A. ANTONOPOULOS
    "Heat and mass transfer in vertical annular 833
    two-phase counter-flow "

SECTION 5

Fire and/or Combustion Simulation

1. T. NAKAMURA, T. OMORI, K. YASUSAWA, I. NAKAMACHI 845
   and H. TANIGUCHI
   "Radiative heat transfer analysis in a forge furnace"

2. T.K. PHUOC and P. DURBETAKI 857
   "Modeling mechanism of ignition phase transition"

3. A. K. GUPTA and D.G. LILLEY 869
   "The role of diagnostics for improved simulation of practical flowfields"

4. L. POST 884
   A mathematical model of the combustion-chamber in a glass-furnace"

5. T. SANO and S. KOTAKE 896
   "A rational algorithm for chemical kinetics; calculation of combustion flows"

6. NEVIN SELCUK 907
   "Finite difference solution of three dimensional flux equations for radiative transfer in furnaces

7. N. LARAQUI and J. BRANSIER 918
   C. VOVELLE, J.L. DELFAU and M. REUILLON
   Modelling of the thermal degradation of vertical PMMA slabs"

VOLUME II

SECTION 6

Nuclear Research and Technology

1. C.A. ESTRADA-GASCA and M.H. COBBLE 935
   "Transient heat conduction in the WIPP problem"

2. D.F. FLETCHER and A. THYAGARAJA 945
   "Numerical simulation of two-dimensional transient multiphase mixing"

3. D.R. CROFT and T.W. EATON 957
   "Thermal analysis of targets used for anti-proton production at CERN"

4. J.C. KITELEY, M.B. CARVER and A. TAHIR 968
   "Numerical modelling of phase distribution in interconnected subchannels"

SECTION 7

Thermal and/or Mechanical Loads

1. R.C. GULARTE, J.A. NEMES, F.R. STONESIFER and
   C.I. CHANG 983
   "Failure of mechanically loaded laminated composites subjected to intense localized heating"

2. L. KARLSSON 994
   "Deformations in butt-welding of large plates with special reference to the temperature dependence of the yield limit of the material"

3. M.A. FANELLI 1006
    "The numerical solution of 3-D thermoelastic problems using only laplace equation discretized formulations"

4. D. LAPOINTE, C.A. LABERGE, R. BALDUR 1026
    "Plate-cylinder intersections subjected to thermal loads"

5. Z. P. MOURELATOS 1038
    Calculation of thermal deformation of a piston and cylinder of a low-heat-rejection engine"

6. P. AGATONOVIC and F. KOCH 1053
    "The thermal analysis and optimization in thermal isolation of engine components"

7. A. CAUVIN 1065
    "Structural effects of thermal variations on r.c. frames and grids in the cracked stage"

8. A. PIETRZYK and P. WITAKOWSKI 1077
    "A study of thermal stresses development in maturing concrete"

9. H-Q. YANG, CHU CHEN and J. WANG 1089
    "Numerical simulations of residual stress and distortion during welding pipes"

10. H.C. FU, S.F. NG and M.S. CHEUNG 1101
    "Continuous composite bridge subjected to negative thermal gradients"

11. W. L. KO and T. OLONA 1112
    "Effect of element size on the solution accuracies of finite-element heat transfer and thermal stress analyses of space shuttle orbiter"

12. A.P. CHAKRAVARTI, L.M. MALIK, A.S. RAO and J.A. GOLDAK 1131
    "Prediction of distortion in overlayed repair welds"

13. M.C. SINGH and P.K. CHAN 1144
    "Finite element solution for a round cornered square plate subjected to thermal effects over a central region"

SECTION 8

Novel Computational Methods

1. H.M. GAUER 1155
   "Analytical and computational techniques in the treatment of heat transport problems"

2. H.Q. YANG, K.T. YAND and J.R. LLOYD 1169
   "Three-dimensional finite-difference calculations of buoyant flow in arbitrary parallelepiped enclosures using curvilinear non-orthogonal coordinates"

3. C. HABERLAND and A. HAHRMANN 1182
   "An unconditionally stable one-step formula for the determination of transient temperatures in Aero/space structures"

4. M.A. KEAVEY 1193
   "An isoparametric boundary solution for thermal radiation"

5. P.H. GASKELL and N.G. WRIGHT 1204
   "A multigrid algorithm for the investigation of thermal, recirculating fluid flow problems"

6. M. SHOUCRI, P. BERTRAND and M. FEIX 1216
   "The application of shape preserving splines for the numerical solution of differential equations"

7. J.W. GRAHAM and S.N. SARWAL 1222
   "Sampled data model for a transient solution to a nonlinear heat transfer system"

8. K.K. TAMMA and S.B. RAILKAR 1231
   "Recent advances in the development of transfinite element formulations for nonlinear/linear transient thermal problems"

9. GANQUAN XIE 1243
   "TCR method for determining the thermal transmittance of 2-D body"

10. L.G. MARGOLIN and A.E. TARWATER 1252
    "A diffusion operator for Lagrangian meshes"

11. K. KUDO, H. TANIGUCHI, W.-J. YAND,
    H. HAYASAKA, T. FUKUCHI and I. NAKAMACHI 1264
    "Monte Carlo method for radiative heat transfer
    analysis of general gas-particle enclosures"

12. J.H. CHIN and D.R. FRANK 1276
    "Modeling of severe thermal gradients near fluid
    passages using special finite elements"

13. A.L.G.A. COUTINHO, J.L.D. ALVES, L. LANDAU and
    L. C. WROBEL 1287
    "Element-by-element solution of steady state
    heat conduction by Lanczos algorithm"

14. C. LACROIX and P.L. VIOLLET 1299
    "Prediction of the heat transfer between fluid
    and structure in transient flows using local
    wall models"

15. M.B. HSU 1307
    "Simulation of gaps and channels in conductive
    solid heat transfer"

16. M.Y. NUMAN 1315
    "Evaluation of the thermal response factors and
    transient conduction heat flow of multi-layered
    slabs by hopscotch finite difference algorithm"

17. V.P. MANNO and J.A. HAUBOLD 1326
    "A finite difference convective algorithm
    with reduced dispersion"

18. J.C. HUNTER, H.M. TSAI, J.F. LOCKETT, P.R. VOKE,
    M.W. COLLINS AND D.C. LESLIE 1338
    "A comparison of large-eddy simulation predictions
    of a thermal layer with holographic interferograms"

19. L.C. WROBEL 1348
    "A novel boundary element formulation for
    nonlinear transient heat conduction"

20. E. WEILAND and J.P. BABARY 1358
    "A solution to the IHCP applied to ceramic
    product firing"

21. W.S. DUNBAR and A.D. WOODBURY 1368
    "Semi-analytic solution of the heat convection
    equation"

22. B. DORRI and V. KADAMBI 1377
    "Boundary integral analysis with orthogonal basis functions"

23. P. HANLEY and W.L. HARRIS 1387
    "Chebychev polynomial solution of the heat equation"

24. N. DEVANATHAN, Y. YANG and B. SZABO 1399
    "Solution of nonlinear thermal problems by the p-version of the finite element method"

25. M.C. IMGRUND 1414
    "Using the approximate optimization algorithm in the ANSYS program for the solution of optimum convective surfaces and other thermal design problems"

26. A. S. L. SHIEH 1425
    "A domain partitioned multigrid method for the solution of the passive scalar advection-diffusion equation"

27. M. GADBOIS and P.A. TANGUY 1434
    "A 3-D finite element investigation of Non-Newtonian thermal fluid flow"

SECTION 9

Industrial and Scientific Applications

1. S.Y. SHIM, P.J. MILLS and S. MEYER 1437
   "SPORTS-M: a thermalhydraulics computer model for pool-type reactors"

2. J.F. STELZER and R. WELZEL 1449
   "Some exceptional nonlinear temperature field problems and their solution by the finite element method"

3. M.G. DAVIES 1460
   "The reduction of room radiant and convective exchange to star based systems"

4. H. YANO, Y. TANAKA and A. KIEDA 1472
   "Heat transfer characteristics of split-film sensor"

5. J.A. VISSER and E.H. MATHEWS 1484
    "Numerical prediction of temperature distributions
    in a steel bar during hot rolling"

6. J.D. JONES 1495
    "Application of the Galerkin method to analysis
    of heat transfer in a low-heat-rejection diesel
    engine"

7. M. BOUZIDI and P. DUHAMEL 1506
    "An algorithm for solving the eigenvalue problems
    associated with heat diffusion in composite slabs"

8. L. CLAES 1518
    "Determination of temperature gradients in
    a circular concrete shaft"

9. A. BUIZZA, M. CARNIMEO and M. SYLOS LABINI 1530
    "Finite element analysis of the thermal field
    induced by a buries power cable"

10. Y. CAYTAN 1542
    "Numerical simulation of the operation of
    natural draft cooling towers"

11. P. CIRESE 1552
    "Microcomputer software for preliminary analysis
    and design of cooling of electronic equipments

12. A. DEL PUGLIA, F. FONTANELLA, D. QUILIGHINI, 1561
    B. TESI, G. ZONFRILLO and P. LOMBARDI
    "A mathematical model of thermal field
    in semicontinuous copper casting"

13. J.A. THURLBY, E.A. KOWALCZYK, M.J. CUMMING and 1573
    G.J. THORNTON
    "A solution method for models of thermal
    processes in packed beds of solids"

14. Y. JALURIA 1585
    "Numerical simulation of the transient
    processes in a heat treatment furnace"

15. M.S. KHANNICHE and A. HOOPER 1597
    "Thermal modelling of high power semiconductors"

16. T.E. TEZDUYAR, H.A. DEANS and J. MARBLE 1609
    "Finite element/finite difference analysis
    of a deep-well oxidation process"

17. A. CARAPANAYOTIS and Salcudean　　　　　　　　　1622
    "The effect of inlet velocity orientation
    on the scavenging efficiency in a uniflow
    scavenged engine"

18. J. PERSSON and A. SOLMAR　　　　　　　　　　　　1634
    "Design of a latent heat storage using numerical
    simulation and experimental results.
    An industrial case study"

19. W. SHYY and J.T. DAKIN　　　　　　　　　　　　　1647
    "Three-dimensional natural convection in
    a high-pressure mercury discharge lamp"

20. M. TIROVIC and D. STEVANOVIC　　　　　　　　　　1653
    "Temperature field of vehicles' clutches
    and brakes"

21. N. OZOE and T. MATSUI　　　　　　　　　　　　　　1654
    "Numerical computation of Czochralski bulk flow
    for metallic silicon"

22. P. PROULX, J. MOSTAGHIMI and M.I. BOULOS　　　1676
    "Numerical simulation of the DC plasma jet
    two-phase reactor"

23. P. DURBETAKI　　　　　　　　　　　　　　　　　　1688
    "Solid particle injection in a laminar
    boundary layer"

24. R.N. SMITH and J.R. WARNOT, Jr.　　　　　　　　1700
    "A numerical study of freezing in a finned
    enclosure with a convective boundary condition"

25. G. JOLY, J.P. KERNEVEZ, R. LUCE　　　　　　　　1712
    "Heat exchange in the presence of nonlinearities:
    Comparison between a one-dimensional model
    and a two-dimensional model"

26. Y.-H. CHOI, S. VENKATESWARAN, C.L. MERKLE and　1722
    C.W. LARSON
    "Numerical solution of low Reynolds number
    flows with heat addition"

27. N.L. KALTHIA and M.G. TIMOL　　　　　　　　　　1734
    "Heat transfer in forced convection boundary
    layer flows of viscoinelastic fluids

28. M.G. TIMOL and N.L. KALTHIA　　　　　　　　　　1746
    "Numerical study in heat transfer of
    Non-Newtonian flows past a vertical plate"

29. M. WEINSTEIN and P.G. DANIELS — 1758
"On the nonlinear instability of the buoyancy-layer flow in a heated vertical slot"

30. CHEN XIANGFU — 1767
"Analysis of F.E.M. and practical research on tall cold storage building"

31. K. WENQI, ZHU JUNJIE, GU SONG and DI YUNZHEN — 1776
"Three dimensional thermal field and its boundary conditions"

32. V.D. MURTY — 1781
"A numerical investigation of double diffusive convection using the finite element method"

33. R.L. AKAU, G.W. KRUTZ and R.J. SCHOENHALS — 1794
"Experimental and analytical study of an arc welding process"

34. S. ROGERS and L. KATGERMAN — 1806
"Heat transfer and solidification of a stream of molten metal during atomisation by an impinging gas jet"

35. JAY I, FRANKEL, TAI-PING WANG and GARY W. HOWELL — 1817
"The use of the finite integral tranform technique for thermal analysis in micro-electric chip modules"

36. A.P.S. SELVADURAI and M.C. AU — 1829
"Numerical modelling of heat and moisture flow in a porous solid"

37. G.D. RICHARDS — 1839
"Computer simulation of forest fires"

38. P. BISWAS — 1847
"Nonlinear vibrations of a skew cylindrical panel at elevated temperature"

39. M.A. MUNTASSER and J.C. MULLIGAN — 1855
"Local non-similarity solutions for free convection problems using accelerated successive replacement method"

40. G. SUBHASH, S. RAHAMATULLA and S. VASUDEVA RAO — 1868
"Transient thermal behaviour of steam chest during turbine startups"

41. G. CESINI, G. LUCARINI, M. PARONCINI and R. RICCI   1877
    "Numerical and experimental study of natural
    convection in parallelogrammic enclosures"

42. N.E. BIXLER and C.R. CARRIGAN,   1890
    "Finite Element Analysis of Heat Transport in
    a Hydrothermal Zone"

43. C.A. BREBBIA and L.C. WROBEL   1902
    "Further Advances in the use of Boundary
    Elements for Modelling Thermal Problems"

VOLUME II

SECTION 6

Nuclear Research and Technology

TRANSIENT HEAT CONDUCTION IN THE WIPP[†] PROBLEM

By
C. A. Estrada-Gasca[‡] and M. H. Cobble
New Mexico State University
Las Cruces, New Mexico 88003

Abstract

Comparison is made of the exact solutions, and the finite element solutions for temperature histories at a nuclear waste site. A time dependent heat source is considered. The resulting errors of the finite element method are found for a class of nonlinear diffusion problems connected with the disposal of nuclear wastes.

Introduction

Waste isolation pilot plant (WIPP) is a nuclear waste repository (NWR) that is being developed by the U.S. Department of Energy as a research and development facility to demonstrate the safe disposal of radioactive waste [7]. 6×10^6 cubic feet (1.829×10^6 cubic meters) of transuranic nuclear waste resulting from the defense activities of the United States are planned to be housed at WIPP [11].

Among the general considerations that have to be made for the safety repository of nuclear waste in a geological formation are the depth of burial, the properties and dimensions of the host rock, the tectonic stability and the hydraulic properties of the host rock [10]. From the thermal point of view, the first two considerations are the important ones. Moreover, the temperature in the near-field as well as in the far-field of a repository is probably the most important single parameter concerning the safe disposal of high-level waste. Therefore, the determination of temperature change is essential in the design and environmental impact of the facility. This determination implies a need to find a solution to the heat conduction equation.

The general problem of a NWR is a time dependent, nonlinear, non-homogeneous, multidimensional heat conduction problem. To predict the thermal field and the heat transfer rates from a NWR, several computer codes have been applied using finite difference and finite element methods [3], [8]. However, just a few analytical approaches to a NWR thermal

---

[†] Waste Isolation Pilot Plant.
[‡] At present time working for Instituto de Investigaciones Electricas, Departamento de Combustibles Fosiles, Apdo. Postal 475, Cuernavaca, Morelos, 62000, Mexico.

problem have been made. Sweet and McCreight [9], based on reference [1], solve exactly the simplified one-dimensional semi-infinite linear WIPP problem. The purpose of this study is to solve a class of "WIPP" problems exactly, and then numerically using a finite element method. The comparison between the solutions is made and the errors of the numerical technique are presented.

Controlling Equations

The temperature distribution in a nuclear waste site is subject to the following partial differential equation

$$\frac{\partial}{\partial x}\left[k(T)\frac{\partial T}{\partial x}\right] = C(T)\frac{\partial T}{\partial t} \quad (1)$$

where

$$T = T(x,t)$$
$$k(T) = k_o F(T) \quad (2)$$

and from reference [2],

$$C(T) = \rho(T)\, c_p(T) = C_o F(T) \quad (3)$$

Boundary and initial conditions, $T(x,t)$:

1. $k(T)\dfrac{\partial T}{\partial x}(0,t) = -\ddot{q}(t)$, (Subterranean Site)

2. $T(\ell,t) = T_\ell$, (Surface)

3. $T(x,0) = T_o$

The equations (1) - (3) and the listed boundary and initial conditions define a class of "WIPP" problems that are solved exactly, and numerically.

Transient Solution

The transient problem is subject to a nuclear heat source of the form

$$\ddot{q}(t) = \ddot{q}_o e^{-\lambda t} \quad (4)$$

see reference [4], and from reference [9]

$$k(T) = k_o F(T) = k_o (T_a/T)^p \quad (5)$$

so

$$C(T) = C_o F(T) = \rho_o c_{p_o} (T_a/T)^p \tag{7}$$

The transformation

$$V(T) = \int_{T_r}^{T} k(T)dT \tag{8}$$

which is the Kirchhoff transformation, is used to solve analytically the nonlinear problem. The exact solution is given by the following equation

$$T(x,t) = T_r \left[1 - \frac{p-1}{k_o T_r} V(x,t) \left(\frac{T_r}{T_a}\right)^p \right]^{1/(1-p)} \tag{8}$$

where V is the transformed temperature given by

$$V(x,t) = V_\ell + \ddot{q}_o \frac{\sqrt{\alpha_o} \sin\sqrt{\lambda/\alpha_o}\,(\ell - x)}{\sqrt{\lambda} \cos\sqrt{\lambda/\alpha_o}\,\ell} e^{-\lambda t}$$

$$+ \frac{2}{\ell} \sum_{n=1}^{\infty} \frac{e^{-\alpha_o \gamma_n^2 t}}{(-1)^{n+1}} \left[\frac{V_o - V_\ell}{\gamma_n} \cos \gamma_n x + \ddot{q}_o \frac{\alpha_o \sin \gamma_n (\ell-x)}{(\lambda - \alpha_o \gamma_n^2)} \right] \tag{9}$$

where

$$\gamma_n = \frac{(2n-1)\pi}{2\ell}, \quad n = 1, 2, 3, \ldots \tag{9a}$$

and

$$V_\ell = \frac{k_o T_r}{p-1} \left[1 - \left(\frac{T_r}{T_\ell}\right)^{p-1} \right] \left(\frac{T_a}{T_r}\right)^p \tag{10}$$

and

$$V_o = \frac{k_o T_r}{p-1} \left[1 - \left(\frac{T_r}{T_o}\right)^{p-1} \right] \left(\frac{T_a}{T_r}\right)^p \tag{11}$$

Steady State Solution

Let us assume for simplicity that $T_r = T_a = T_\ell$ and $T_o = T_\ell$, then $V_\ell = 0$. The steady state situation occurs when $t \to \infty$, so $\ddot{q}(t) \to 0$, and from equation (9) $V(x,\infty) = 0$. Substitution into (8) gives $T(x,\infty) = T_r = T_l = T_o$, which is the expected value.

Numerical Solution

The thermal problem was also solved numerically using the finite element method. The main idea behind this method can be presented by the method of weighting residuals. In this method, the solution is replaced by a finite linear combination of basis functions, which are usually polynomials. When this expansion is substituted into the partial differential equation, a residual term appears. The objective is to select the undetermined coefficients of the expansion such that the residual is minimized in some sense. The sense chosen by the method of weighted residuals consists of integrating over the domain of the solution function, the product of the residual and some specified weighting functions and setting this integral to zero. In linear algebra, it is said that the residuals are orthogonal to the weighting functions. This produces a system of equations which can be solved for the unkown coefficients. When this method is applied to the heat conduction equation (a parabolic partial differential equation), the integration is made over the spatial coordinate and a system of ordinary differential equations in time is obtained. This system is solved using a Crank-Nicolson scheme (an implicit finite difference method [6]. The basis functions used were polynomials of second order (parabolas). The number of nodes was 30 which corresponds to 29 elements for the x-coordinate.

Results

Table 1 gives a listing of the parameters used to plot the solutions. Two thermal load models were used.
One corresponds to high level waste (HLW) with 12 w/m^2 as the initial thermal load, and 32.9 years as the half life. The other corresponds to transuranic nuclear waste (TRU) with 0.7 w/m^2 and 21661 years as the initial thermal load and half life respectively.

Figure 1 shows the transient thermal response at several points above WIPP with an areal thermal load of 12 w/m^2 (HLW) located 660 meters deep in salt. It is clear that ΔT (temperature rise) gets its maximum values for x=0. Similar results are obtained when TRU nuclear waste is used. Figure 2 shows a plot of the exact and finite element temperature rise solutions. These solution represent the transient thermal response at the plane where the nuclear waste repository is located (x=0). The parameters used correspond to HLW in salt. The exact ΔT(max) was 94.6 °C, and occured 41 years

Table 1
WIPP Parameters
Nuclear Waste in Salt

$k_o = 5.46$ watts/m-°K	$p = 1.21$
$\alpha_o = 74.92$ m²/yr	$C_o = 2298$ kJ/m³-oK
$l = 660$ m	$\ddot{q}_{o_1} = 12$ watts/m² (HLW)
$T_o = 300$ °K	$1/\lambda_1{}^{\ddagger} = 47.4$ years (HLW)
$T_l = 300$ °K	$\ddot{q}_{o_2} = 0.7$ watts/m² (TRU)
$T_a = 300$ °K	$1/\lambda_2 = 31250$ years (TRU)

$\ddagger$ with $\lambda_i = \ln [2/(t_{\frac{1}{2}})_i]$.

after emplacement of the nuclear waste. Figure 3 plots the differences between the analytical and numerical solutions. The error of the finite element method shows a strong oscillation at earlier times, which tends to diminish but the difference increases until a maximum of 6.37 °C occurs at 35 years, then the differences decrease and tend to zero. At 41 years, where where the maximum occurs, the error of the finite element solution with respect to the analytical one was 6.7 % (a difference of 6.36 °C).

Similar results are obtained when the TRU thermal load is considered. Figures 4 and 5 show the results. The analytical $\Delta T(max)$ for this case was 79.36 °C and occurred 6100 years after emplacement. The numerical solution for this time was 77.5 °C, i.e. an error of 2.4 %. The deviation of the numerical solution from the exact one has the same general behavior as that of the HLW. The maximum error of the finite element solution was 3.2 % (a difference of 2.9 °C) and occurred at 4000 years. It is clear that the error of the numerical solution depends strongly on the elements considered as well as on the time steps used. For smaller elements and time steps the numerical error is smaller.

Conclusion

The errors, resulting from the utilization of a finite element method to solve the nonlinear diffusion equation for a class of WIPP problems, were determined.

For the two types of waste analyzed, the maximum errors were located close to the maximum of the temperature rise. In both cases, when time goes to infinity, the error goes to zero. Finally it is mentioned that reference [5] treated the same problem, but with a constant source.

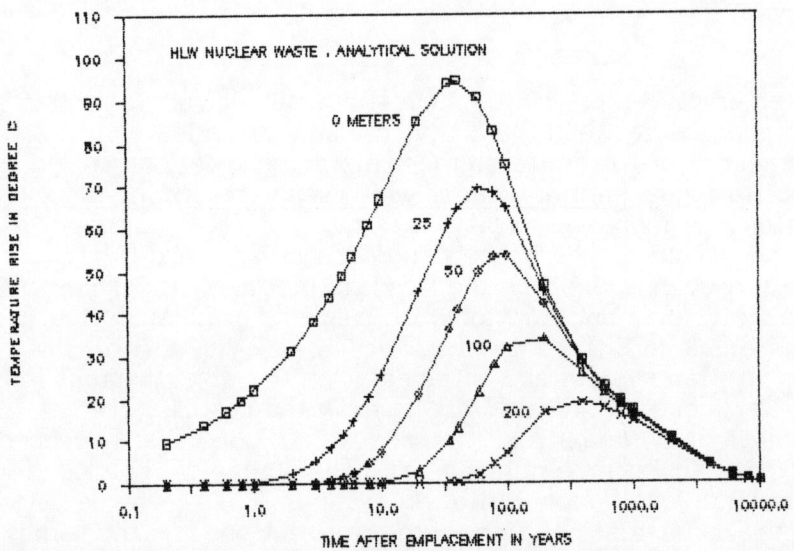

Figure 1. Transient thermal response at several points above a NWR. Labels on curves indicate height above NWR.

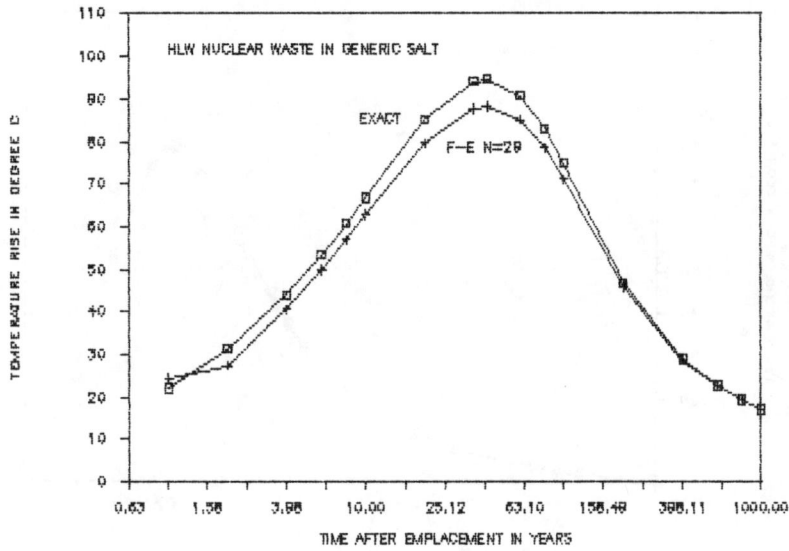

Figure 2. Comparison of the analytical and finite element solutions at x=0, using HLW in salt.

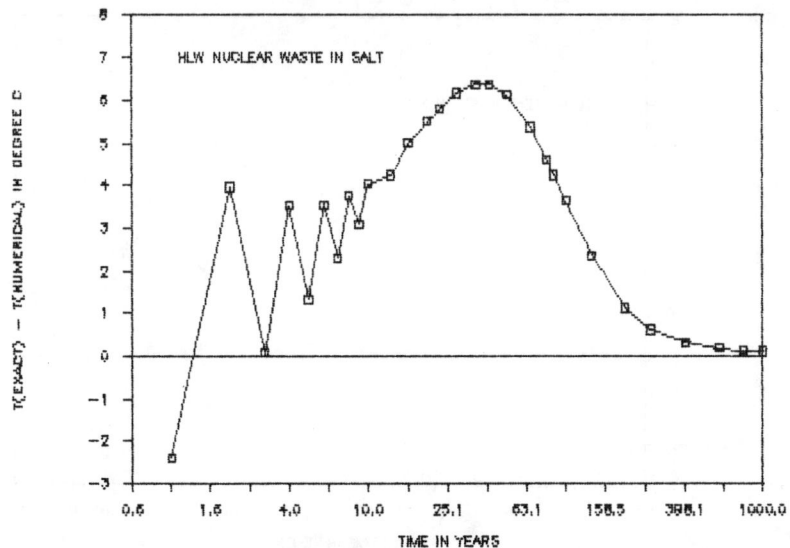

Figure 3. Difference of the exact solution with the finite element solution of Figure 2

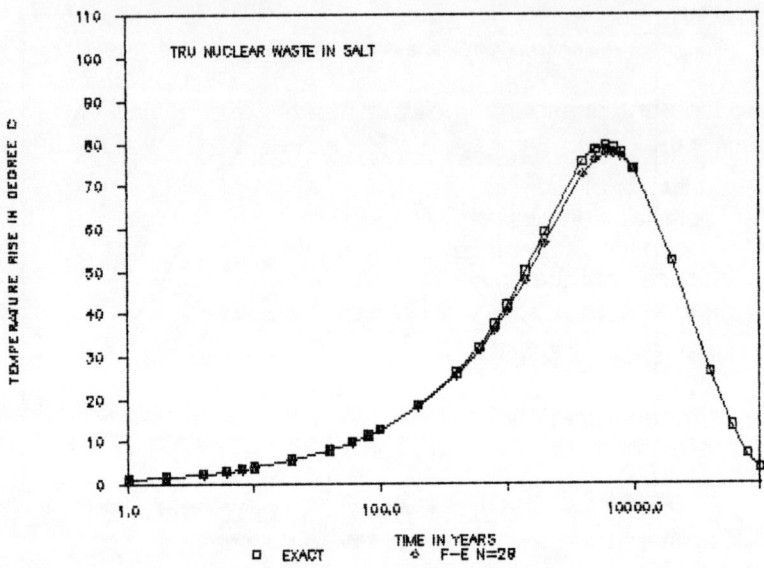

Figure 4. Comparison of the analytical and finite element solutions at x=0, using TRU in salt.

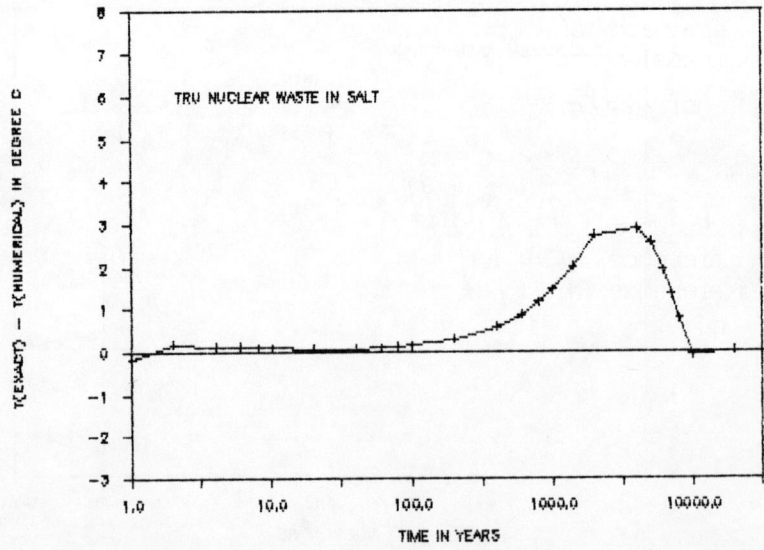

Figure 5. Difference of the exact solution with the finite element solution of Figure 4.

Nomenclature

c_p	Constant pressure specific heat, $j/Kg\text{-}°K$
C	Density-specific heat product, $J/m^3\text{-}°K$
$F(T)$	Function of T
k	Thermal conductivity, $w/m\text{-}°K$
ℓ	Thickness of geological layer, m
n	Natural number
p	exponent in a polynomial expression
q	Heat flux, w/m^2
t	Time, years
T	Temperature, $°K$
T_a	Reference temperature for k, $°K$
T_ℓ	Temperature at $x = \ell$, $°K$
T_o	Temperature at $t = 0$, $°K$
T_r	Reference temperature for $V(T)$, $°K$
V	Transformed temperature, w/m
x	Spatial coordinate, m

Greek Symbols

α	Thermal diffusivity, m^2/yrs
λ	Decay constant, yrs^{-1}
γ_n	Expansion coefficient, m^{-1}
ρ	Density kg/m^3

Subindex

n	Index, $n = 1, 2, 3, \ldots$
o	Reference
ℓ	Reference value at $x=\ell$

References

[1] Carslaw, H.S. and Jaeger J.C., 1959. Conduction of Heat in Solids, 2nd. Edition, Oxford at the Clearendon Press, pp. 10-11.

[2] Cobble, M.H., 1967. "Nonlinear Heat Transfer of Solids in Orthogonal Coordinate Systems", International Journal of Non-Linear Mechanics, Vol. 2, Pergamon Press Limited, pp. 417-426.

[3] Eaton, R.R., 1984. Three Dimensional Thermal Analysis for the WIPP in Situ Test Room A2 Heater Configuration, SAND84-2220. Sandia National Laboratories.

[4] Estrada-Gasca, C.A. 1986. Analytical Methods of Heat Transfer Compared with Numerical Methods as Related to Nuclear Waste Repositories. Doctoral Dissertation, New Mexico State University.

[5] Estrada-Gasca, C.A. and M.H. Cobble, 1986. The WIPP Problem. First World Congress in Computational Mechanics. September 22-26, 1986. Austin, Texas.

[6] Lapidus, L. and G.F. Pinder, 1982. Numerical Solution of Partial Differential Equations in Science and Engineering. John Wiley & Sons.

[7] Matalucci, R.V., et al., 1982. WIPP Research and Development Program: In Situ Testing Plan. SAND81-2628. Sandia National Laboratories.

[8] Maxwell, D.E., K.K. Wahi and B. Diel, 1980. The Thermomechanical Response of WIPP Repositories. SAND79-711, Sandia National Laboratories.

[9] Sweet, J.N. and J.E. McCreight, 1980. Thermal Properties Measurement on Rocksalt Samples from the Site of the Proposed Waste Isolation Pilot Plant. SAND80-0799. Sandia National Laboratories.

[10] _____, 1976. The Selection and Evaluation of Thermal Criteria for Geologic Waste Isolation Facility in Salt, Y/OWI/SUB-76/07220, prepared by Science Applications, Inc. For Office of Waste Isolation, Union Carbide Corporation, Oak Ridge, TN.

[11] _____, 1980. Waste Isolation Pilot Plant, Final Environmental Impact Statement. U.S. Department of Energy. Volume 1.

NUMERICAL SIMULATION OF TWO-DIMENSIONAL TRANSIENT MULTIPHASE MIXING

D. F. Fletcher and A. Thyagaraja

Culham Laboratory, Abingdon, Oxfordshire, OX14 3DB, U.K.

SUMMARY

In this paper we describe a finite-difference model of multiphase mixing. The model is applicable to the study of a hot fluid poured into a cold vaporisable fluid. We represent this situation using the continuum equations of multiphase flow and give a description of the equations to be solved and the boundary conditions used. The outline of the numerical solution scheme is described and the new features we have developed are highlighted. Emphasis is placed on the need to ensure qualitative consistency of the finite difference scheme with the properties of the differential equations. Finally, we present some sample calculations showing that the model gives good general agreement with experimental data.

1. INTRODUCTION

The calculation of multiphase flow is of considerable interest in the nuclear and process industries. A particular application, the modelling of transient buoyancy-driven multiphase mixing which occurs on relatively slow timescales (~ 1 second), is the subject of this paper. For example, we are concerned with the modelling of the behaviour of a jet of hot melt poured into a pool of cold vaporisable liquid (usually water). This situation may occur in certain accidents within the metal casting industry or may arise in the progression of hypothetical accidents in the nuclear industry if core material melts and pours into water [1].

This problem is particularly difficult to model since the melt may be at a very high temperature (~ 3500 K) so that the heat transfer rate to the cold liquid (water in our application) could be of the order of 5 MW/m^2 from thermal radiation alone, leading to very high vapour production rates. A recent review [2] of the available models of this process has identified the need to develop a transient, 3-component (melt (M), water (W), steam (S)) multiphase flow model. To be a useful tool the model needs to have the following properties:

(a) The resulting computer code must be stable, robust and ensure that quantities such as volume fractions always remain positive.

(b) It must be able to work with virtually arbitrary constitutive relations for interphase drag, heat-transfer rates and melt particle fragmentation.

(c) There is no great need for a high order accuracy in the finite difference scheme since many of the constitutive relations are only known approximately and the experimental data are global measurements with substantial uncertainties.

(d) It must be relatively cheap to use so that the effect of uncertain parameters can be scoped.

In the remainder of this paper we describe the model, finite difference equations and solution procedure developed to satisfy the above aims, together with the simulation of a particular experiment. Due to the short length of this paper we draw heavily on our earlier work described in references 3,4 and 5.

2. FORMULATION OF THE MATHEMATICAL MODEL

We assume that mixing takes place axisymmetrically in a right-circular cylinder of radius R, height H, open at the top. The flow velocities are small enough to allow the incompressible approximation, so that the mass densities ρ_M, ρ_W and ρ_S may be taken as constants. This results in a set of equations of the form

$$\frac{\partial}{\partial t}(\alpha_i \phi_i) + \frac{1}{r}\frac{\partial}{\partial r}(rU_i \alpha_i \phi_i) + \frac{\partial}{\partial z}(V_i \alpha_i \phi_i) = S_{\phi i} \qquad (1)$$

where α_i is the volume fraction, U_i is the horizontal velocity (in the r-direction) and V_i is the vertical velocity (in the z-direction) of species i (i = M, W, S). Table 1 below lists the variables ϕ_i and the source terms for the water species equations, plus the melt enthalpy and melt length-scale equations.

Equation	ϕ	source term
Conservation of water mass	1	$\dot{m}_W/\rho_W$
Conservation of radial momentum	$\rho_W U_W$	$-\alpha_W \dfrac{\partial \overline{p}}{\partial r} - \alpha_W \dfrac{\partial \chi}{\partial r}$ $+F^r_{WM} + F^r_{WS} + F^r_{W\dot{m}}$
Conservation of axial momentum	$\rho_W V_W$	$-\alpha_W \dfrac{\partial \overline{p}}{\partial z} + g\alpha_W \alpha_M (\rho_M - \rho_W)$ $+g\alpha_W \alpha_S (\rho_S - \rho_W)$ $+F^z_{WM} + F^z_{WS} + F^z_{W\dot{m}}$
Melt enthalpy	$\rho_M H_M$	$-\dot{m}_S h_{fg}$
Melt length-scale	L_M	$-\alpha_M (L_M - L_{crit})/\tau_L$

Table 1 : Coefficients and source terms in the conservation equations.

In addition to the above equations the volume fractions must satisfy the constraint

$$\alpha_M + \alpha_W + \alpha_S = 1 \qquad (2)$$

which implies the following elliptic constraint,

$$\frac{1}{r}\frac{\partial}{\partial r}\left(r(\alpha_M U_M + \alpha_W U_W + \alpha_S U_S)\right)$$
$$+ \frac{\partial}{\partial z}(\alpha_M V_M + \alpha_W V_W + \alpha_S V_S) = \frac{\dot{m}_S}{\rho_S} - \frac{\dot{m}_W}{\rho_W} \qquad (3)$$

The above equation allows the common pressure to be determined, as will be illustrated in section 3.

2.1 Form of the source term

In this section we describe the source terms, together with the necessary constitutive relations needed to close the problem. We assume that $\dot{m}_M = 0$ so that there are no internal sources or sinks of melt in the solution domain. Melt is either present initially in the solution domain or is injected at a boundary. We assume that there is no condensation of steam (i.e. $\dot{m}_S \geqslant 0$), since in all problems of interest to us the water is saturated. In addition, we have $\dot{m}_W + \dot{m}_S = 0$ and set

$$\dot{m}_W = -\alpha_W \alpha_M 6h(T_M - T_W)/L_M h_{fg} \qquad (4)$$

where we have assumed the melt to be in the form of spheres of diameter L_M. T_M and T_W are the melt and water temperatures, h_{fg} is the latent heat of vaporisation and h is the appropriate heat transfer coefficient (usually radiation plus film boiling).

In the momentum equations we work with a reduced pressure, defined by

$$\overline{p} = p - \chi = p - \int_z^H g(\alpha_M \rho_M + \alpha_W \rho_W + \alpha_S \rho_S) dz \qquad (5)$$

Thus the usual axial momentum source term $-\alpha_W \frac{\partial p}{\partial z} - \alpha_W \rho_W g$ takes the form given in table 1 and $\frac{\partial \chi}{\partial r}$ terms appear in the radial momentum equations. The terms F_{WM}^r and F_{WM}^z represent interphase drag between the melt and water in the r and z directions respectively. These terms were modelled using the drag law proposed by Harlow and Amsden [6] and their exact form is given in [5]. The important point to note is that

$$F_{WM}^r \equiv D_{WM}^r(\alpha_M, \alpha_W, L_M, L_W, U_M, U_W, V_M, V_W)(U_M - U_W) \qquad (6)$$

so that provided $D_{WM}^r = D_{MW}^r$ Newton's third law is satisfied. The terms $F_{W\dot{m}}^z$ etc. represent evaporation reaction forces and are not present in the melt equation. These take the form

$$F_{W\dot{m}}^r = -\dot{m}_S U_W, \quad F_{S\dot{m}}^r = \dot{m}_S U_W \qquad (7)$$

In the enthalpy equation the kinetic energy of the melt and terms arising from pressure and drag work have been neglected, since they are small compared to the thermal energy terms. The source term ensures that the melt cools by an amount consistent with the heat used to produce vapour. The melt temperature is determined from the enthalpy by using a suitable caloric equation of state. In the melt length-scale equation the source term causes the melt length-scale to reduce to L_{crit} with a fragmentation rate $1/\tau_L$.

2.2 Boundary and initial conditions

We assume that $U_i = 0$ on $r = 0$, $U_i = 0$ on $r = R$, $V_i = 0$ on $z = 0$. At the top ($z = H$) we set $U_i = 0$ and $\frac{\partial V_W}{\partial z} = \frac{\partial V_M}{\partial z} = 0$. A uniform steam velocity outlet profile is set with the flow rate determined from the volume integral of equation (3). At $t = 0$ an initial velocity and volume fraction field is specified together with the length-scale and temperature of each species. Any melt injected into the solution domain is given a specified velocity, length-scale and enthalpy.

3. SOLUTION SCHEME

Only the outline of the solution scheme is presented here, full details are given in references 3,4 and 5. The equations given in the previous section were finite differenced on a staggered grid. All convective terms were upstream differenced for stability. The solution procedure is as follows:

(a) Time advance the α_M equation to get $\alpha_M(t+\Delta t)$ using the melt velocity and α_M fields at time t.

(b) Similarly time advance L_M and H_M and determine T_M using the caloric equation.

(c) Time advance the α_W equation treating the source term implicitly.

(d) Determine $\alpha_S = 1 - \alpha_W - \alpha_M$

(e) Determine new velocity fields using the pressure field at time t.

(f) Substitute the new velocity and volume fraction fields into the finite differenced form of equation (3) to determine the local continuity error.

(g) If this error is too large update the pressures using Newton's method, go back to step (e) and repeat the procedure using the modified pressure field. Iterate steps (e)→(g) until the local continuity errors (suitably normalised) are below the desired accuracy level.

Typically after 2 iterations the above procedure converges and the code steps forward in time again. The chosen method of time advancing the α's ensures that they remain positive and if the continuity error is reduced to a suitably small value they remain less than unity [3]. For stability we require the Courant number $\left(\text{Max}\left(\frac{V_i \Delta t}{\Delta z}, \frac{U_i \Delta t}{\Delta r}\right)\right)$ to be less than unity and for accuracy we require the product of any rate parameter and the time-step to be less than unity e.g. $\Delta t/\tau_L \ll 1$.

In the momentum equations the velocities of all species are coupled at each grid location due to the drag terms. Thus at each grid point we invert a 3×3 matrix to obtain the new velocities of all 3 species simultaneously. This procedure ensures that even with extremely large drag forces Newton's third law is exactly satisfied. A semi-implicit scheme is used for accuracy.

The pressure correction is carried out by first correcting the pressure level in vertical slabs. The pressure is then corrected locally within each slab. Thus we avoid the need to solve Poisson's equation directly but instead use a TDMA solution for the blocks and then a TDMA solution within each block.

4. RESULTS

The code has been extensively tested on model problems and has been found to give grid independent solutions and to predict solutions which depend continuously on the initial data. It has been used to simulate the one-dimensional mixing of heated ball-bearings with water [4] and has been used to model a two-dimensional mixing experiment carried out at Argonne [5]. In this section we present a further comparison with the Argonne experiment (CWTI-9) in which the melt length-scale is evolved with time, unlike in our earlier work where we assumed pre-fragmented melt of a fixed size. The experimental geometry is shown in Figure 1 below.

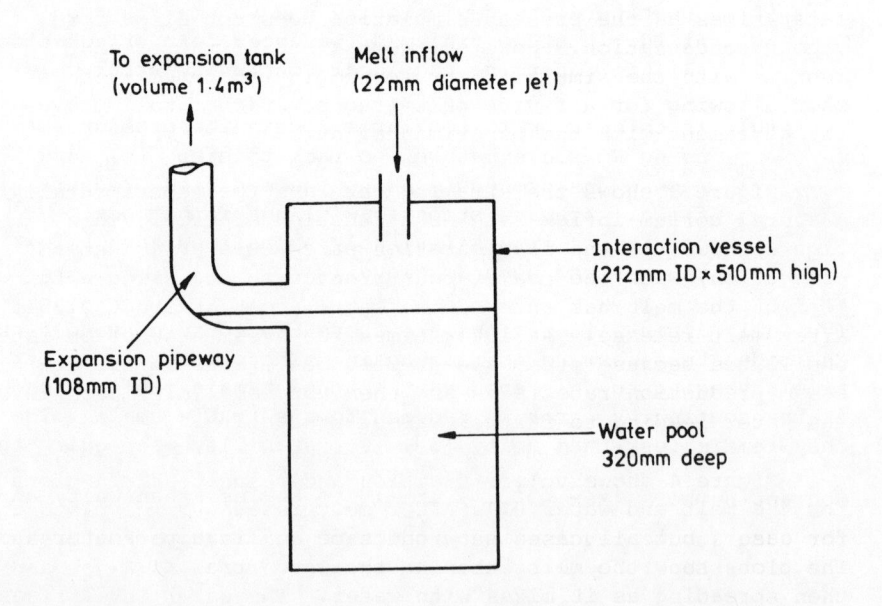

Fig 1.   Illustration of the experimental geometry.

A full description of the experiment and chosen constitutive relations is beyond the scope of this paper. The experiment is reported in reference 7 and we use the same constitutive relations as those used in our previous calculations (see section 6.1 of reference 5). A 10×10 grid and a time-step of 5×10^{-6}s was used. Briefly, the experiment consisted of the injection of molten corium into a vessel partially filled with water. The vessel was closed except for a pipeway on the side which allowed the steam produced to be collected in a tank (see figure 1). In our calculations the vessel is assumed to be open at the top (we cannot model 3-D effects) and we determine how much steam is produced. In this way we can use our data to generate a pressure transient similar to that measured in the expansion vessel by the experimenters.

A comparison of the calculated pressurization results for three different assumptions about the melt fragmentation parameters, (case 1 $\equiv L_{crit}$= 1mm, τ_L = 0.02s, case 2 $\equiv L_{crit}$= 2mm, τ_L = 0.0033s, case 3 $\equiv L_{crit}$= 2mm, τ_L = 0.02s), is shown in figure 2. The initial melt length-scale was set equal to the pour diameter. The plot shows that the experimental data is well-fitted for this range of parameters. These are consistent with the experimentally determined particle size distribution and break-up rates resulting from hydrodynamic instabilities. The agreement is less good at later times as the present simulation does not allow for vapour condensation. However, comparison of the present results with the simulation presented in reference 5 shows that allowing for a finite melt fragmentation rate improves the agreement with the experimental data.

Figure 3 shows the steaming rate and the experimentally measured corium inflow rate as a function of time. This figure shows how the fragmentation parameters affect steam production, with the peak vapour production occurring when most of the melt has entered the water pool, at about 0.25s after melt release. At later times the steam production rate oscillates because steam sweeps melt out of the water, the steam production rate falls and then the melt falls back into the water leading to further vapour production.

Figure 4 shows volume fraction and velocity field plots for the melt and water 0.1s after melt release. The plots are for case 1 but all cases show the same qualitative features. The plots show the melt entering the water pool as a jet and then spreading as it mixes with water. The water level rises due to the entry of melt and steam production. The water velocity plot shows the level swell rate to be about 0.5 m/s and also shows that some water is carried along with the leading edge of the jet. At later times vapour production becomes very rapid and the water pool 'froths up' to fill the

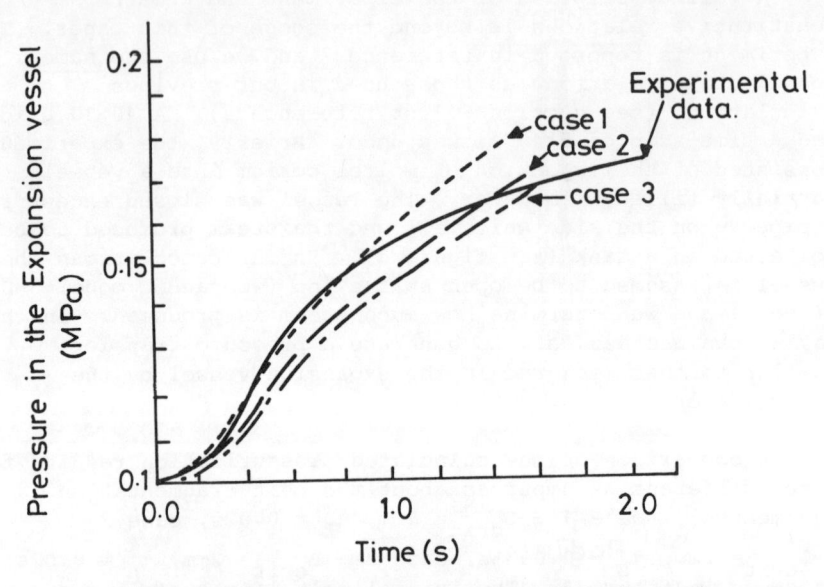

Fig.2 A comparison of the predicted and measured vessel pressurisation data.

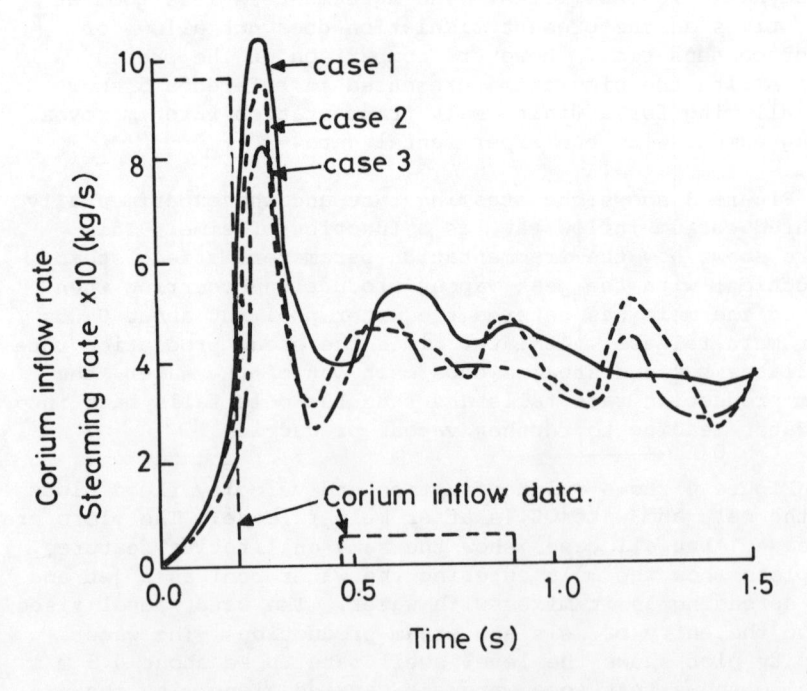

Fig.3. The transient steaming rate.

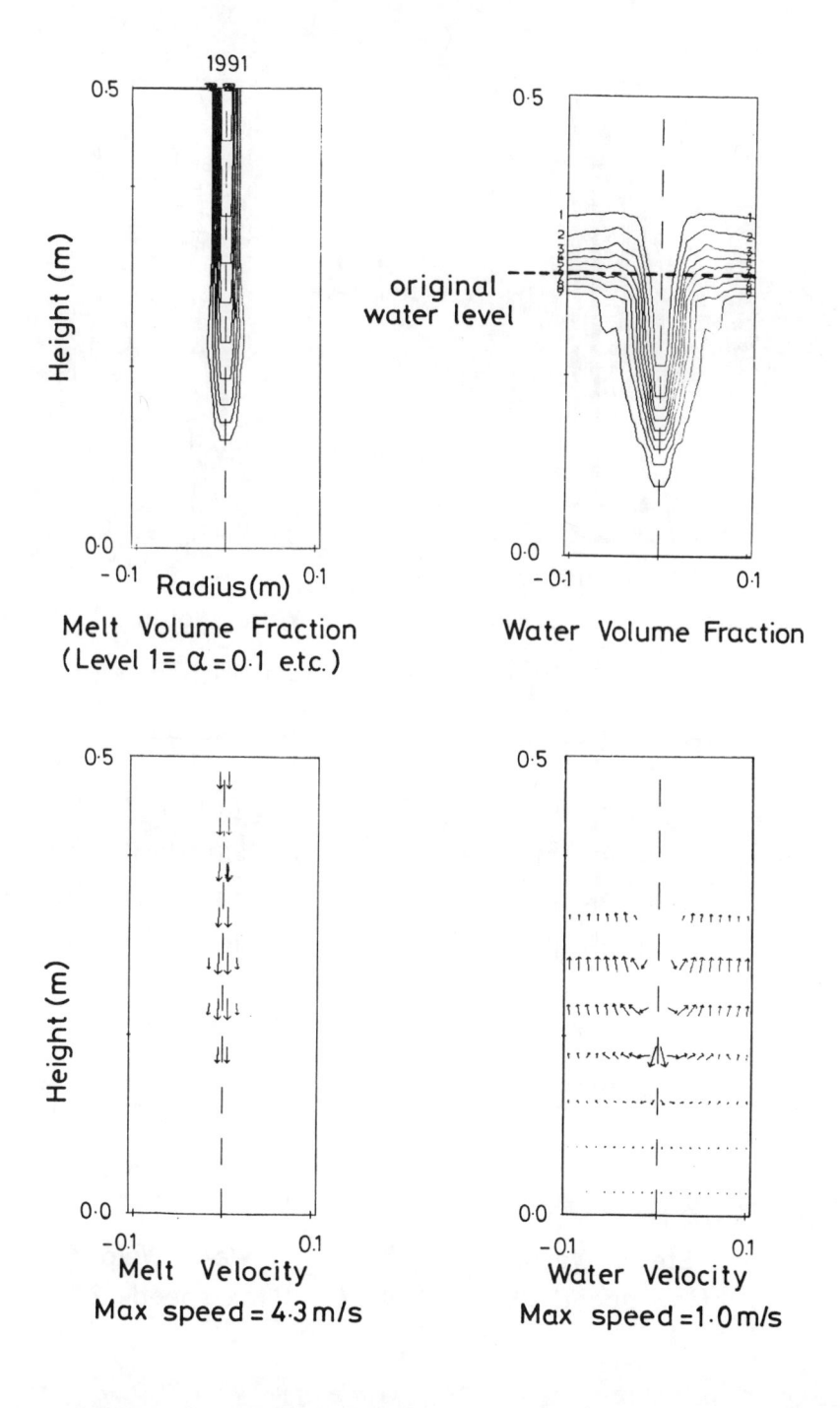

Fig. 4: Volume fraction & velocity fields at t=0.1s.

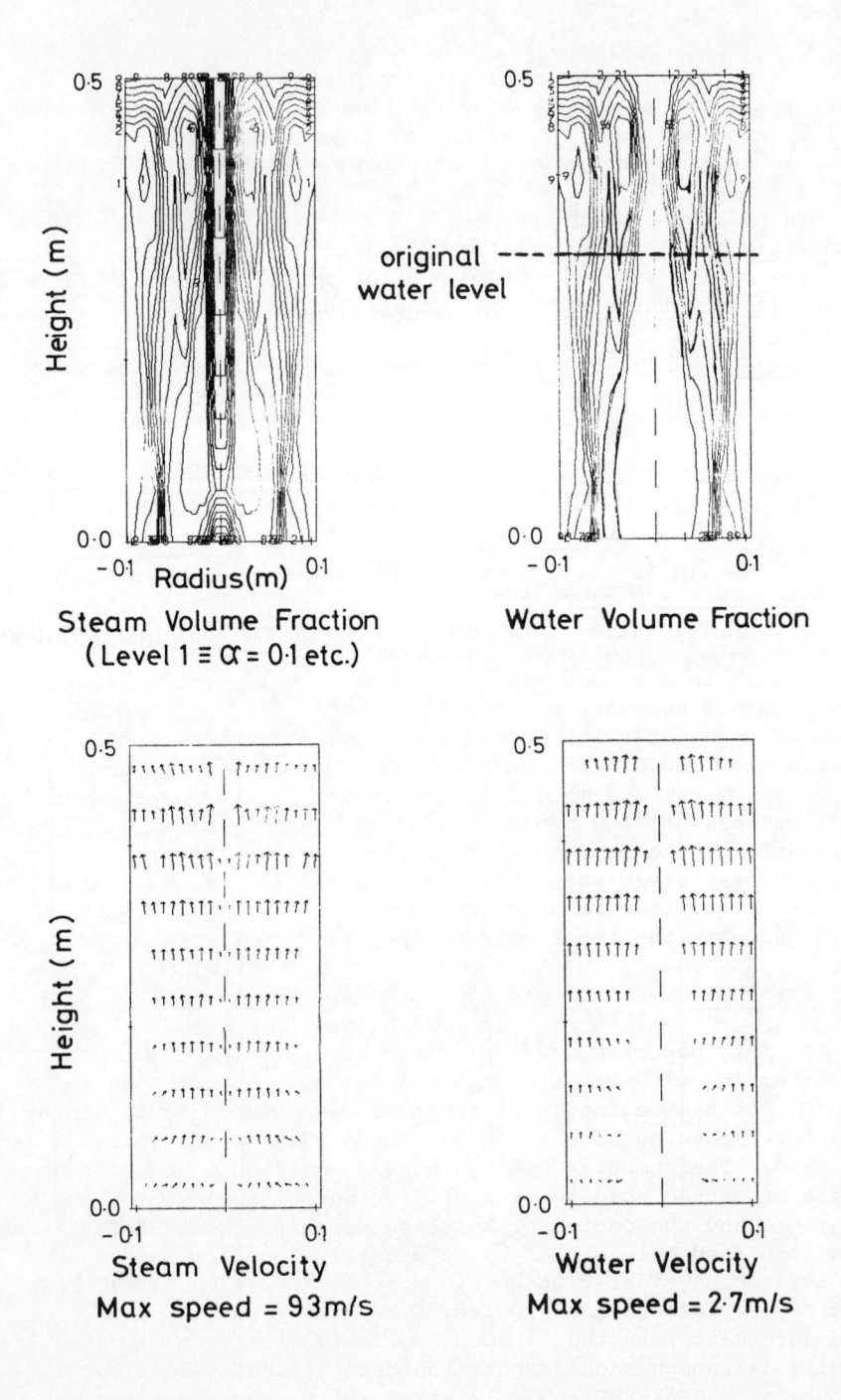

Fig. 5: Volume fraction & velocity fields at t = 0.2s.

vessel. Figure 5 shows volume fraction and velocity plots for water and steam at a time of 0.2s. The figure shows that most of the water has been pushed to the side of the vessel and steam is exiting the system with a peak velocity of 93 m/s.

In the experiment 13% of the corium and 36% of the water were estimated (by the experimenters) to have been expelled from the mixing vessel. Table 2 below shows the calculated values for the three different simulations performed.

Case	% melt expelled	% water expelled
1	26	93
2	17	73
3	12	79

Table 2: Melt and Water 'Sweep-out' Data

The data in the above table shows that the melt 'sweep-out' is most accurately modelled for a final particle size of 2mm. However, the model predicts that approximately twice as much water was 'swept-out' as observed in the experiments. This difference could be due to differences in the experimental and modelling geometries (in the experiment the steam has to flow around a bend so this may cause both water and melt to be deposited by the flow) or it may be due to the chosen steam-water drag law. However, the simulation shows good qualitative agreement with the experimental data and highlights the importance of the 'sweep-out' phenomena.

5. DISCUSSION

In this paper we have described a multiphase flow model of the mixing of a hot fluid jet with a cold vaporisable liquid. At the beginning of the paper we set out four guiding aims for producing such a model. These have been largely realised. The computer code is stable and robust due to the choice of method used to finite difference the conservation equations and the choice of solution scheme. The constitutive relations need only satisfy very general properties which are required on physical grounds alone, for example, the drag laws must satisfy Newton's third law. The code gives generally good agreement with the limited available experimental data. Finally, although each computation uses typically 3500 cpu seconds on a CRAY-XMP, this is still relatively cheap and far less costly than experimental studies of mixing. Thus the code can be used to determine the important parameters and effects which can then be examined in detailed experimental studies. To the best of our knowledge the code described in this paper is unique in its ability to model experiments of this type.

ACKNOWLEDGEMENT

The authors would like to thank Mrs G. Lane for making such a good job of typing this manuscript.

REFERENCES

1. CRONENBERG, A.W. - Recent Developments in the Understanding of Energetic Molten Fuel-Coolant Interactions. Nuclear Safety, 21, 319-337 (1980)

2. FLETCHER, D.F. - A Review of Coarse Mixing Models. Culham Laboratory report CLM-R251 (1985)

3. THYAGARAJA, A., FLETCHER, D.F. and COOK, I. - One Dimensional Calculations of Two-phase Mixing Flows. Int. J. Numer. Methods Eng., 24, 459-469 (1987)

4. FLETCHER, D.F. and THYAGARAJA, A. - Numerical Simulation of One-dimensional Multiphase Mixing. Culham Laboratory preprint CLM-P776 (1986)

5. THYAGARAJA, A. and FLETCHER, D.F. - Buoyancy-driven, Transient, Two-dimensional Thermo-hydrodynamics of a Melt-water-steam Mixture. Culham Laboratory preprint CLM-P790 (1986)

6. HARLOW, F.H. and AMSDEN, A.A. - Flow of Interpenetrating Material Phases. J. Comp. Phys., 18, 440-465 (1975)

7. SPENCER, B.W., MCUMBER, L., GREGORASH, D., AESCHLIMANN, R. and SIENICKI, J.J. - Corium Quench in Deep Pool Mixing Experiments. Paper presented at the National Heat Transfer Conference, Denver, Colorado, U.S.A. August (1985)

THERMAL ANALYSIS OF TARGETS USED FOR ANTI-PROTON PRODUCTION AT CERN.

D R Croft, Sheffield City Polytechnic, Sheffield, U.K.
T W Eaton, CERN, Geneva, Switzerland

Summary

A target is a device used to convert protons to antiprotons; these particles are used extensively in nuclear physics research at CERN.

Due to the very low conversion rate of protons to antiprotons there is considerable energy degradation within the core of the target. This demands an elaborate cooling system to ensure the structural integrity of the target/cooling system assembly. It is important then to be able to assess the thermal loading on the components and the whole assembly prior to installation.

The thermal analysis of the assembly is made complicated by a number of factors; the pulsing spatial variation of the internal heat generation; the directional variation and temperature dependence of the thermal properties of the materials involved; the geometry of the components.

This paper presents the numerical analysis of the dynamic and steady-state thermal performance of such a target.

1. PROBLEM STATEMENT

In Nuclear Physics research, small particles called antiprotons are the "raw material" from which hosts of other significant particles are produced.

These particles seem reluctant to come into being as it takes 10^7 protons to produce 1 antiproton. This "conversion" process is achieved by directing high energy proton beams at iridium metal targets. The protons have an energy of 24 GeV and are confined in a Gaussian beam distribution of $4\sigma = 2.0$mm; there are 10^{13} protons released in a pulse lasting 0.5ms and this pulse repeats every 2.4 s.

This pulsing of energy release generates instantaneous rises in temperature of about 2000°C in the iridium on top of high "steady-state" temperatures which exist in the target for most of the cycle. Since it is not possible to cool the highly radioactive iridium rod directly, then a cooling system has to be designed to remove the internal heat generation. The general layout of the iridium rod and the cooling assembly is shown in Fig.1.

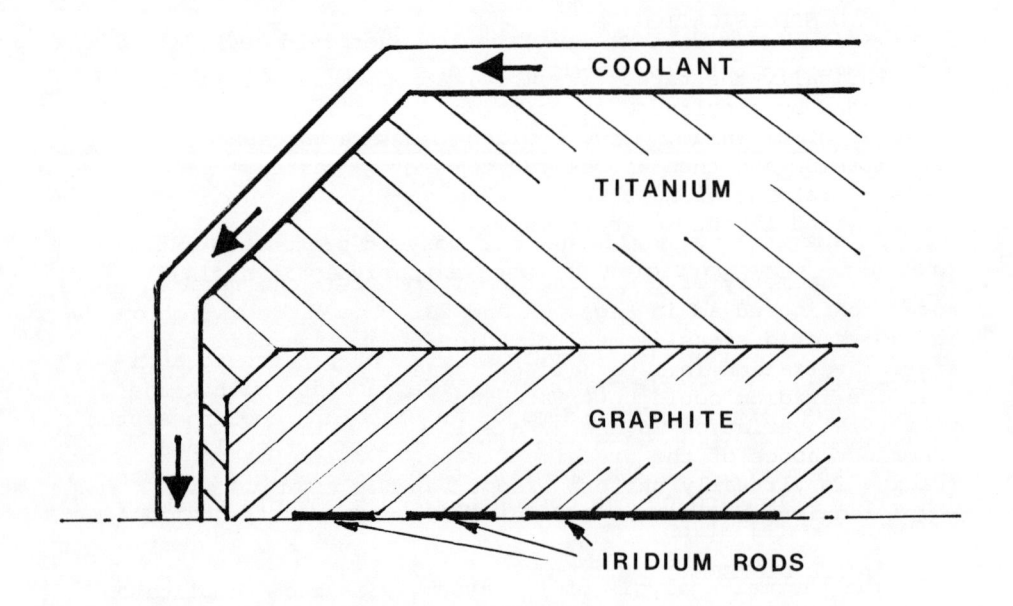

FIG 1. GENERAL LAYOUT OF TARGET

Basically the iridium is surrounded by polytropic graphite which has a radial thermal conductivity of the same order as copper, and a further sleeve of titanium through which the cooling channels are bored.

From a mechanical engineering viewpoint, the important consideration is the long-term integrity of the target container, when subjected perhaps to 10^6 pulses of protons, ie, one month of beam pulsing. This long-term integrity is principally determined by

a) the steady-state temperature distribution which builds up over the whole target assembly after about 30 minutes of pulsing. This will determine the basic stress levels in the structure.

b) the instantaneous temperature rises in the iridium which will produce fatigue stresses along the centre line of the structure.

This paper describes the numerical analysis of the behaviour of a model of the target assembly. Although the example used here is an iridium/graphite/titanium combination it is obviously a straightforward matter to adapt the model to other materials and geometric configurations.

2. GRID SPECIFICATION

Basically the target assembly is a thin rod of iridium surrounded by a cylinder of polytropic graphite and a further cylinder of titanium. The cooling channels bore axially along the titanium and then across the top of the target; the coolant returns along channels also bored into the titanium. A simplified layout is shown in Fig.1.

The geometry of the target requires axial and raidal nodes configured as in Figs. 2A and 2B. The large variation in radial grid sizes is entirely dictated by the peculiar spatial variation in internal heat generation terms. Otherwise the iridium could be treated as a point source of energy with axial variation in energy dissipation allocated to the innermost nodes of the graphite sleeve. The axial grid spacing is virtually uniform except for the titanium cap which requires a slightly smaller grid.

3. FINITE DIFFERENCE EQUATIONS

3.1. There are four groups of typical F.D.E's for;

(i) Internal nodes in the graphite, iridium, and titanium, which should allow for spatial variations in the internal heat generation terms, temperature and directional variations in thermal conductivities, variable axial and radial grid sizes (see Fig 3A).

(ii) Interface nodes at the common surfaces at various axial and radial locations involving graphite and iridium, titanium and iridium, and titanium and graphite. Some of these nodes will be associated with internal heat generation terms and most will involve a change of radial/axial grid size in moving from one material to the other. Sometimes the interface node will be a "corner" of, for example, iridium poking into a graphite environment, again with radial and axial grid size changes, internal heat generation terms and variable thermal conductivity. (see Figs. 3A and 3B)

(iii) Boundary nodes, at the surfaces between the cooling water and the titaniumm.

(iv) The boundary nodes at the opposite end of the target assembly to the cooland system.

3.2 Typical F.D.E's

3.2.1. Internal node in a uniform grid (Fig.3A)

The governing partial differential equation is:

$$\frac{1}{r}\frac{\partial}{\partial r}(rk\frac{\partial T}{\partial r}) + \frac{\partial}{\partial z}(k\frac{\partial T}{\partial z}) + H = 0$$

For node (I.J),

$$T(I,J) = A\left\{T(I+1,J) * \left[\frac{k_{I,J}}{2R(I)DR2} + \frac{k_{I+1,J}}{(DR2)^2}\right]\right.$$

$$+ T(I-1,J) * \left[\frac{k_{I-1,J}}{(DR2)^2} - \frac{k_{I,J}}{2R(I)DR2}\right]$$

$$+ T(I,J+1) * \left[\frac{k_{I,J+1}}{(DZ2)^2}\right]$$

$$\left. + T(I,J+1) * \left[\frac{k_{I,J-1}}{(DZ2)^2}\right] + H(I,J)\right\}$$

where $1/A = \frac{k_{I+1,J}}{(DR2)^2} + \frac{k_{I-1,J}}{(DR2)^2} + \frac{k_{I,J+1}}{(DZ2)^2} + \frac{k_{I,J-1}}{(DZ2)^2}$

For the polytropic graphite material the radial and axial thermal conductivities are radically different in value as specified in section 3.3. For titanium, the values are temperature dependent only. The internal heat generation term H(I,J) only applies to the iridium and the titanium end cap (see Fig.4.)

3.2.2. All the other nodes except those at the boundary are permutations of the equation in 3.2.1. They have to include the effects of non-uniform grids and material changes at the various interfaces and positions throughout the field as shown in Figs 2A and 2B.

3.2.3 The boundary node is best derived using the control volume energy balance approach [4]

3.3. Thermal properties [5], [6]

The thermal properties of polytropic graphite, iridium and titanium are:

radial : $K_r = (440 - 0.3 * T)$ W/mK
axial : $K_a = (0.005 - 0.01) k_r$ W/mK $\}$ For poly graphite

$K = (149.25 - 0.625*T)$ w/mK   for Iridium

$K = 4.306 + 0.0354*T - 3.472 \times 10^{-5}*T^2$ W/mK for Titanium

where T is the temperature in °C.

The spatial variation of internal heat generation within the iridium is derived as described in ref [3] and shown in Fig.4.

For iridium, the specific heat = 134 J/kgK
the density = 22400 kg/m³

The coolant heat transfer coefficient was evaluated at between 2500 - 5000 w/m²k depending upon geometry and flow rate. With water flow rates in the region of 5 kg/min then the rise in coolant temperature was ignored, since it rarely exceeded 2°C.

3.4. Instantaneous Temperature Rise

Ref [3] can be used to evaluate the distribution of energy (eV/cm³/proton) within the proton beam.

For example, for the configuration described in this paper the energy level is 3.6 GeV/cm³/proton at node (23,6) in the iridium. Each pulse contains 10^{13} protons.

Instantaneous temperature rise/pulse

$$= \frac{(3.6 \times 10^9) \times 10^{13} \times \text{eV/J}}{\text{density} \times \text{specific heat}}$$

$$= \frac{(3.6 \times 10^9) \times 10^{13} \times 1.6 \times 10^{-19}}{22.4 \times 0.134} = \underline{1919\,°C}$$

Repeating this calculation over the iridium segments and the titanium front window gives temperature pulses in the range 250°C to 3500°C

3.5. Surface heat transfer coefficient

Using the correlations of ref [5], the surface heat transfer coefficient lies in the range 3000 - 5000 w/m²K.

4. SOLUTION PROCEDURE

The simplest procedure is to use the Gauss-Seidel technique with over-relaxation. For a grid of 41 x 19 subdivision the optimum over-relaxation factor ω is 1.77

However, a much lower value is preferred (1.2 - 1.35) because of the irregular grid and non-linear coefficient in the thermal conductivity terms.

5. COMPUTED RESULTS AND DISCUSSION

Fig. 5 shows some steady-state profiles within the target assembly. There is nothing particularly suprising from a thermal viewpoint. The effect of the directional properties of the polytropic graphite can be seen to minimise axial heat flow and provide low resistance for the heat to flow radially outwards from the iridium segments. Using a natural graphite would significantly increase the steady-state temperature of the titanium front window.

The most important consideration is whether or not the instan(can)eous and steady-state temperatures will cause stresses sufficient in magnitude to cause cracking of the material, or, if cracking is unavoidable, then as long as the cracks do not permeate the titanium shell, there will be no route for the iridium products to leak into the water.

Calculation of the type in ref [1] shows that using this thermal data will give alternating stress levels in the titanium front window of around 300 MN/m^2. Using the fatigue data [6], suggests that for anodised IMI 318 titanium alloy a fatigue life of about 10^6 cycles is well within the maximum possible. For a pulse cycle time of 2.4 s, this means that the target assembly will maintain its structural integrity for around 40 days, considerably longer than previous designs.

6. REFERENCES

1. Eaton, T.W. et al, Conducting Targets for anti-proton production of ACOL. High Energy Accelerator Conf. Vancouver 1985.

2. Eaton, T.W. et al, Recent Work on the Production of Anti-protons from Pulsed Current Targets. (To be submitted to the Journal, Nuclear Instruments and Methods) 1986.

3. Eaton, T.W et al, Calculations of Beam Energy Deposition, Adiabatic Temperature Rise and Thermal Stress in the Windows of High Density Passive Targets. CERN/PS/AA/ Note 85 - 14 November 1985

4. Croft, D.R and Stone, J.A.R., Heat Transfer Calculations using Finite Difference Equations. Pavic Press 1986

5. Engineering Sciences Data: 67016. Heat Transfer Correlations for fully developed Turbulent Flow - their scope and limitations. Engineering Sciences Data Unit, London 1977

6. Data from Fairey Hydraulics Ltd., U.K.

FIG. 2A. RADIAL NODE CONFIGURATION

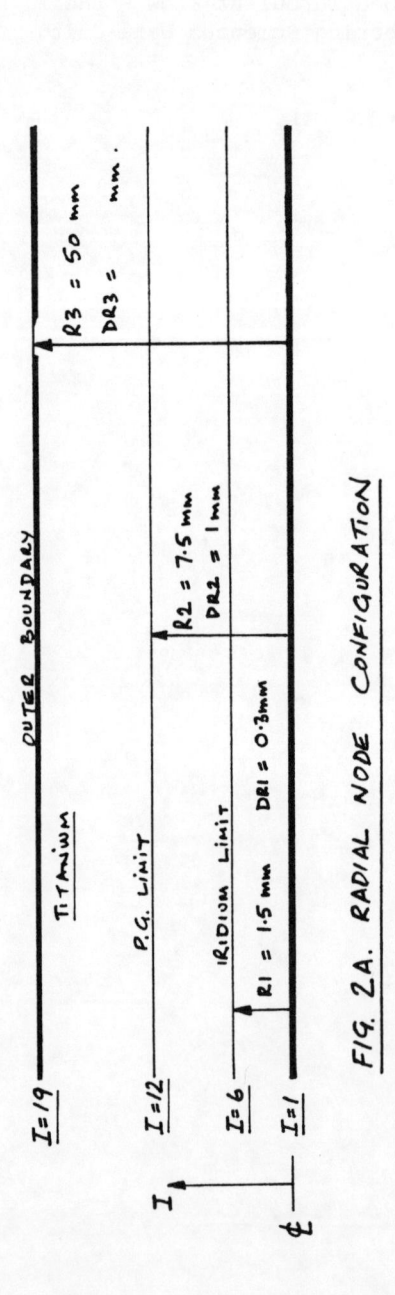

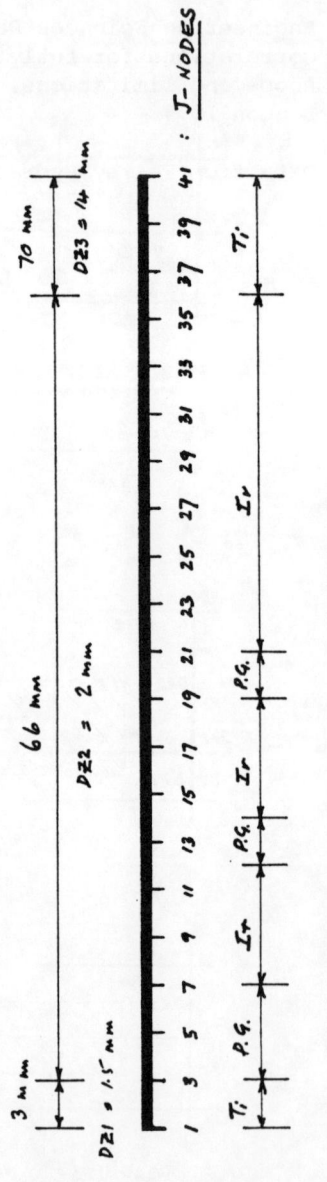

FIG. 2B. AXIAL NODE CONFIGURATION

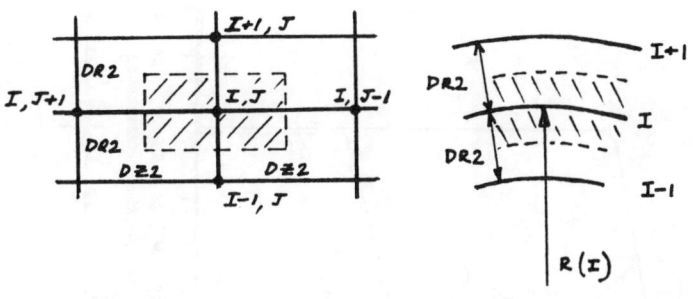

FIG. 3A. INTERNAL NODE IN ONE MATERIAL

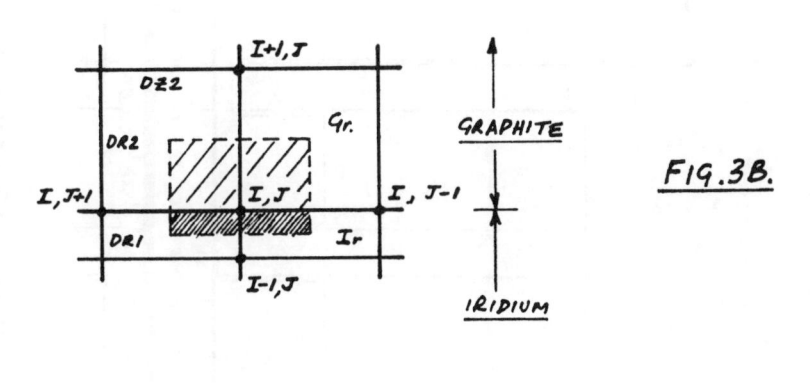

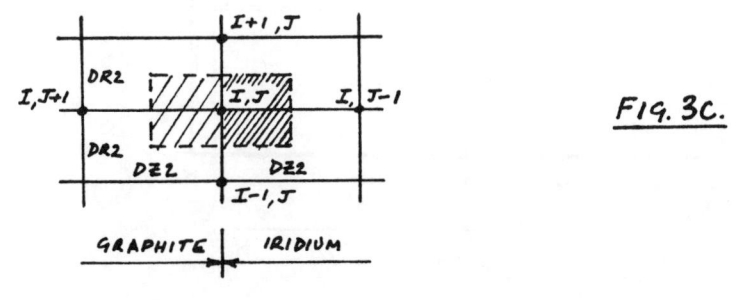

FIG. 3B.

FIGS. 3B. AND 3C. INTERFACE NODES

FIG. 3C.

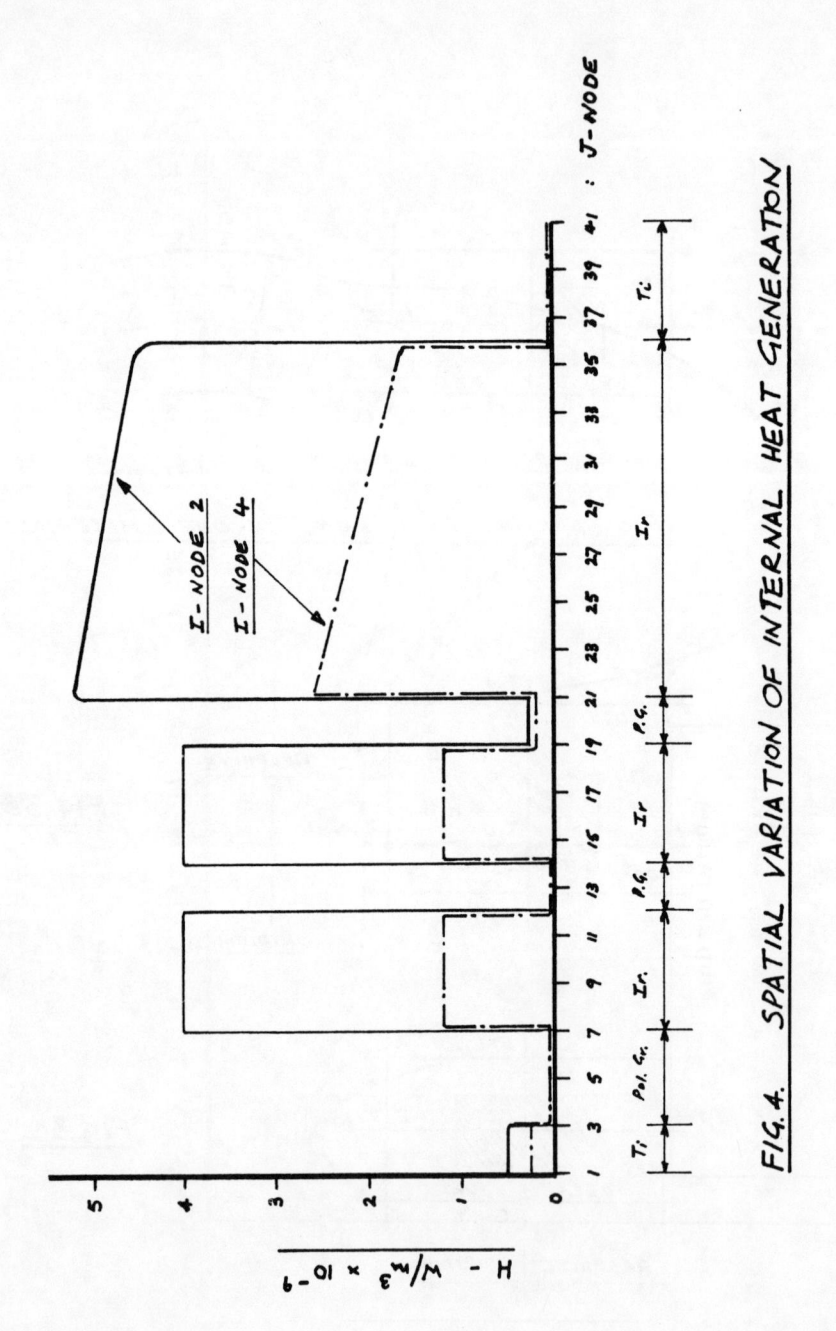

FIG. 4. SPATIAL VARIATION OF INTERNAL HEAT GENERATION

FIG. 5. STEADY-STATE TEMPERATURE PROFILES

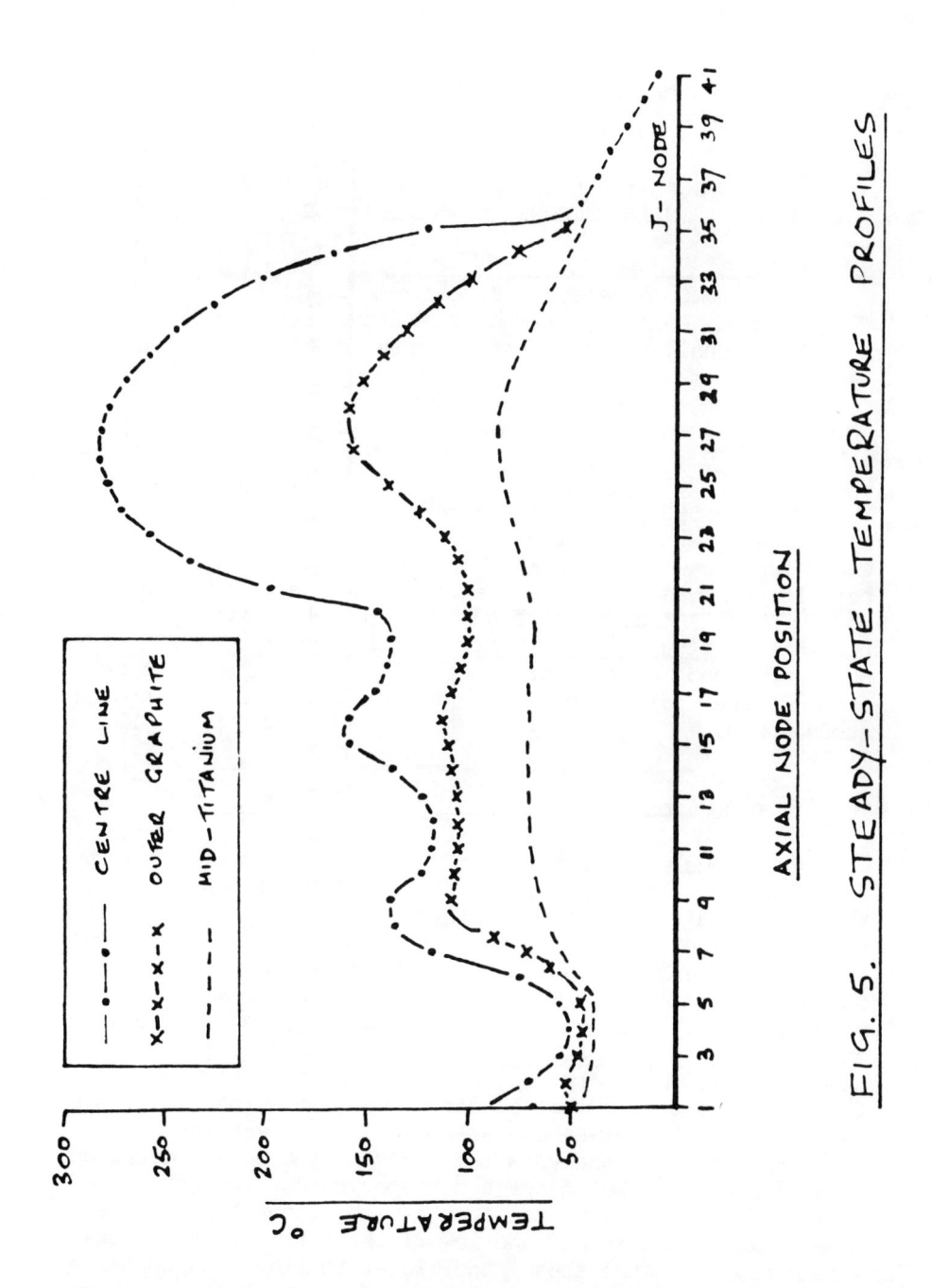

NUMERICAL MODELLING OF PHASE DISTRIBUTION IN INTERCONNECTED SUBCHANNELS

J.C. Kiteley, M.B. Carver, A. Tahir
Chalk River Nuclear Laboratories
Chalk River, Ontario, Canada. K0J·1J0

SUMMARY

The ASSERT subchannel code has been developed specifically to model flow and phase distributions within CANDU fuel bundles. ASSERT uses a drift-flux model which permits the phases to have unequal velocities, and can thus model phase separation tendencies which may occur in horizontal flow. The basic principles of ASSERT are outlined, and computed results are compared against data from three air-water experiments in interconnected subchannels.

1. INTRODUCTION

As a nuclear reactor system relies on fluid circuits for energy transport, mathematical modelling of thermalhydraulic phenomena plays an important role in reactor design and development, and methods of improving the accuracy and efficiency of thermalhydraulic computations are sought continually. Throughout most of the reactor piping network, the fluid behaviour may be adequately described by one-dimensional models. However, in the reactor fuel channel, flow must distribute itself amongst the intricate flow passages of the fuel bundle. One-dimensional analysis is adequate to simulate overall or bulk energy transfer, but multi-dimensional analysis is necessary to model detailed local distribution of flows and temperatures. In the CANDU fuel bundle, the flow passages are horizontally oriented. Under boiling conditions and low flows there is, therefore, a tendency for the steam and water to flow in opposite directions with respect to gravity. A model of two-phase flow, which permits the water and steam phases to flow with different velocities, is required to simulate this gravity induced separation tendency.

The Chalk River Nuclear Laboratories (CRNL) are involved in an extensive program of research and development of non-equilibrium two-phase flow modelling and multi-dimensional computational analysis of flow in fuel bundles and related geometries. Development of the ASSERT subchannel code, which incorporates advanced techniques, is well underway and models and numerical methods used are discussed in a number of publications [1,2,3]. Parallel phenomenological investigation is required to ensure that the computer codes incorporate the mechanism necessary to simulate experimentally observed trends. The development program, therefore, contains a number of coordinated experimental and analytical projects, each of which acts as an auxiliary building block, providing information essential for the central project. This paper concentrates on one of these fundamental projects, which specifically examines the effect of phase distribution in interconnected subchannels. Experiments have been conducted on air and water flowing in vertical and horizontal interconnected channels at Ecole Polytechnique [4,5] and Ontario Hydro Research [6]. The experiments were designed so that the separate effects of gravity, turbulent mixing and pressure gradient could be examined.

In this paper numerical simulations of these experiments generated by the ASSERT code are presented, and computed velocities and phase distributions are compared to experimental data for three sets of air-water experiments.

2. THERMALHYDRAULIC MODEL

The thermalhydraulic model equations used in ASSERT-4 are derived from the two-fluid formulation. The two-fluid equations are combined to obtain the ASSERT model equations. The transportive form is obtained from the conservative form merely by subtracting the identity expressed by the mass equations. ASSERT has options to solve either the two-fluid equations using either the drift-flux or the homogeneous mixture model.

2.1 Conservation Equations

Mixture mass (conservative form)

$$\frac{\partial \rho}{\partial t} + \vec{\nabla} \cdot (\rho \vec{V}) = 0 \qquad (1)$$

where: $\rho = (\alpha\rho)_g + (\alpha\rho)_f = \alpha_g^+ \rho_g + \alpha_f^+ \rho_f$

$(\rho\vec{V}) = (\alpha\rho\vec{V})_g + (\alpha\rho\vec{V})_f = \rho\vec{V}$

Mixture momentum (conservative form)

$$\frac{\partial(\rho\vec{V})}{\partial t} + \vec{\nabla} \cdot \left(\rho\vec{V}\vec{V} + \frac{(\alpha\rho)_g(\alpha\rho)_f}{\rho} \vec{V}_r \vec{V}_r^i\right) + \vec{\nabla}P = -\vec{F}_w^i + \rho\vec{g} \qquad (2)$$

Mixture energy (transportive form)

$$\rho \frac{\partial h}{\partial t} + \rho\vec{V} \cdot \vec{\nabla}h + \vec{\nabla} \cdot \left(\frac{(\alpha\rho)_g(\alpha\rho)_f}{\rho}(h_g - h_f)\vec{V}_r^i\right) =$$

$$q_w'''^i - \vec{\nabla} \cdot ((\alpha\vec{q}'')_g + (\alpha\vec{q}'')_f)^i \qquad (3)$$

Phasic energy (transportive form)

— liquid:

$$(\alpha\rho)_f \frac{\partial h_f}{\partial t} + (\alpha\rho\vec{V})_f \cdot \vec{\nabla}h_f = q_{wf}'''^i - \vec{\nabla} \cdot (\alpha\vec{q}'')_f^i + q_{if}'''^i \qquad (4)$$

— vapour:

$$(\alpha\rho)_g \frac{\partial h_g}{\partial t} + (\alpha\rho\vec{V})_g \cdot \vec{\nabla}h_g = q_{wg}'''^i - \vec{\nabla} \cdot (\alpha\vec{q}'')_g^i + q_{ig}'''^i \qquad (5)$$

$\alpha_g + \alpha_f = 1$, for simplicity we use α to denote α_g

+ — variables to be defined by state relationships, and
i — variables to be defined by constitutive relationships

2.2 Subchannel Equations

ASSERT uses the subchannel approach similar to that used in the COBRA-IV computer code [7]. Subchannels are defined as the flow areas between rods, bounded by the rods themselves and imaginary lines linking adjacent rod centres. Subchannels are divided axially into a number of control volumes which communicate axially with neighbours in the same subchannel and laterally across fictitious boundaries (gaps) with control volumes in neighboring subchannels. The relationship between the reactor core, a fuel channel and the definition of particular subchannels is given in Figure 1.

The development of finite difference analogs to express the Equations (1-5) with respect to subchannel control volumes follows the approach used in COBRA, but unlike COBRA, the transverse gravity terms are retained, making it possible to use ASSERT-4 to model the effect of gravity on horizontal two-phase flows even if the homogeneous option is used. Spatial differenced versions of the model equations are derived by applying the conservation equations to a representative control volume taken from subchannel i(k) which shares gap k with an adjacent control volume in subchannel j(k) between axial nodes j-1 and j. Details are given in the ASSERT User's Manual [8].

2.3 Closure Relationships

The required closure relationships, as indicated above, are the equations of state and constitutive relationships relating relative velocity, fluid friction, wall heat transfer, interfacial heat transfer and thermal mixing to primary variables, phasic flow velocities, densities, enthalpies and pressure.

The relative velocity is the heart of the successful application of the ASSERT model to horizontal bundles and channels. It comprises several effects including:

i) relative velocity due to cross section averaging
ii) local relative velocity due to gravity separation
iii) turbulent diffusion of void, both between neighbouring channels and towards a preferred phase distribution pattern.

3. VALIDATION STUDIES USING AIR AND WATER

The early part of ASSERT development concentrated on the development of a suitable thermalhydraulic model [1] and then on validation, which, of course, involves continuous development.

3.1 The Square Channel Experiments of Tapucu

The work of Tapucu on exchange of air-water mixtures flowing in two parallel square communicating channels was used for initial testing of the ASSERT code. Details of these experiments and the experimental technique are reported in Reference [4]. Channel dimensions are given in Figure 2a. The experiments were run at the same initial nominal mass flux of 3100 kg $m^{-2} s^{-1}$ in both subchannels but for different initial voids and different orientations. The key parameters: pressure, void fractions, and liquid and gas flow rates in both channels along the interconnected region were measured at several axial locations.

A sufficiently general model should be able to reproduce trends quantitatively throughout the entire spectrum of experimental conditions. This is not an unrealistic demand in this case as the data base and range of parameters are quite limited. The current research has modelled both void fraction and mass flux for the entire range of 17 experiments (7 vertical flow and 10 horizontal flow), however only two representative cases are shown here.

In the vertical orientation, only two mechanisms are active, the diversion cross-flow and the turbulent exchange. The experiments are well documented and good agreement with measured pressure drops was first obtained by a single-phase friction factor and the modified homogeneous two-phase multiplier. Together with a good estimate of form loss, these closely determine the pressure driven diversion cross flow. Pressure driven cross flow induces a co-current flow of air and water to flow from the higher pressure channel. In the experiments, pressure is quickly equalized in the slot, but the high void channel requires a higher initial pressure to overcome the two-phase pressure drop. The initial tendency in the vertical experiments is therefore for the recipient or low void channel to gain air and water from the donor. However, when the void fraction in the donor is high, the tendency towards turbulent exchange increases. This results in some counter-current flow in which some liquid returns from the recipient to the donor and is replaced by air. This tendency is readily simulated by the diffusion model. Increased diffusion augments the tendency towards counter-current exchange. Computed profiles of pressure, void fraction, and liquid and vapour mass flow for the vertical flow cases are compared to measurements in [9].

In the horizontal orientation, with one channel above the other cross flow is now driven by gravity as well as pressure and diffusion. In simulating the experiments it was postulated that although the internal distributions would be different from the vertical case, the turbulent exchange would be of the same magnitude. Attention was therefore turned to the formulation of the relative velocity due to gravity. Two orientations were examined in the experiments, with the donor or high void fraction channel above and below the recipient or low void channel, denoted by H_R^D and H_D^R respectively. In the former case, gravity drift does not cause exchange, and diversion cross-flow and turbulent exchange dominate. In the latter gravity drift is significant, but the initial interchange is a pressure driven diversion cross-flow during which both air and water flow from the donor into the recipient above. Eventually gravity forces tend to become dominant and counter-current lateral flow is set up.

Figures 3 and 4 show typical comparisons of void fraction and mass flux profiles for the horizontal flow case H_R^D -1 and its inverse H_D^R -1. Note that in the void fraction profiles, donor void fraction initially increases somewhat, although air transfer is taking place to the receiver. This is typical of the experiments and is due to the fact that at the low experimental pressures (nominally 0.15 MN/m^2) the air expands significantly as it descends the pressure, gradient. This particular pair of experiments was chosen because of the interesting behaviour of mass flux. This clearly illustrates that the initial tendency towards pressure driven co-current exchange is eventually overcome by gravity driven counter-current exchange.

3.2 The Rod-Bundle Subchannel Experiments of Tapucu

Similar experiments have recently been completed by Tapucu [5] in which the channels were fabricated to a form which simulates the shape of neighbouring subchannels in a rod bundle as shown in Figure 2b. In this case, the channels were run in five orientations, the vertical and horizontal positions mentioned above, plus horizontal equal elevation and inclined orientations. The same nominal mass flux was used. All 24 of these experiments (7 vertical flow and 17 horizontal flow) were simulated using ASSERT, but only two representative cases can be shown here.

In this study the pressure drop was matched using a two-phase friction factor multiplier derived from the experimental data of two-phase multiplier versus void fraction and mass flux. This provided a good fit of the ASSERT code results to the experimental pressure drop.

Computed and experimental results are shown for a horizontal case H_R^D -4 in Figure 5. Again the cross-over tendency is apparent in the mass flux profile and is simulated quite well by the program. Although the code does not predict as much initial variation as the experiment, the cross-over occurs at the same location. A horizontal flow case with the two subchannels at equal elevation (side-by-side) is shown in Figure 6. In this case the buoyancy effect will not be present and mass exchange is primarily due to turbulent mixing, as in a vertical flow case, and, in fact, the results are very close to the equivalent vertical flow case.

3.3 The Rod-Bundle Subchannel Experiments of Shoukri

The Shoukri experiments provide a more detailed investigation of mass exchanges between two laterally interconnected horizontal air-water flows in that a wider range of mass flux and void fraction were used. The experimental work was completed by Shoukri et al [6] in 1982 at the Ontario Hydro Research Division. These experiments were run for low pressure, air-water, adiabatic conditions.

The experimental test section (air-water loop) consisted of two identical interconnected horizontal subchannels. The two subchannels were aligned one above the other in order to include the effect of gravity in the redistribution process. A simplified diagram of the two subchannel geometry is shown in Figure 2c. There were 92 horizontal flow experiments, 12 of which were modelled in this validation study. The experiments were performed for 3 different mass fluxes (950, 1360, 1560 kg·m^{-2}·s^{-1}) and for average volumetric flow quality up to 0.80. The experiments were controlled so that the exchange of air and water between the two subchannels would take place in the absence of an imposed transverse pressure gradient. Thus the diversion crossflow effects were eliminated and the mass exchange was due only to gravity separation and turbulent mixing.

A comparison of the computed and measured profiles of void fraction and mass flux are shown in Figures 7 and 8 for 2 values of mass flux. The ASSERT code qualitatively predicted the correct trends for lateral mass exchange. The tendency of the two-phase mixture to stratify increased with decreasing mass flux. The ASSERT code predictions show that at low mass flux (Figure 7) the gravity separation (buoyancy) process is dominant. For higher mass flux (Figure 8), the turbulent void diffusion mechanism becomes more significant and the rate of vapour-liquid exchange is slower. In these experiments the ratio between the flowrates in each subchannel was reported individually for air and water similar ratios for air flow ratios. As the ratio for the vapour flowrate is often quite large, the computed variation of this ratio compared to the experimental variation in Figures 7 and 8, rather than showing the actual mass flow.

CONCLUSIONS

The first phase of the ASSERT advanced subchannel code development illustrated that the code is capable of computing flow and phase distribution effects in horizontal channels and fuel bundles. This study has extended the validation data base of the code by simulating air-water experiments for three different types of interconnected subchannels. It has been shown that the code is able to predict phase exchange including co-current and counter-current flow in considerable detail. Current work is now concentrating on extending the modelling of flow regimes to incorporate both subcooled boiling and post-dryout heat transfer.

ACKNOWLEDGEMENTS

This work was partly funded under the AECL/CANDU Owners Group (COG) Cooperative agreement. The authors wish to acknowledge the support and interest of R.E. Pauls and W.I. Midvidy of Ontario Hydro, and S. Alikhan of New Brunswick Power.

REFERENCES

[1] M.B. Carver, A. Tahir, D.S. Rowe, A. Tapucu and S.Y. Ahmad, "Computational Analyses of Two-Phase Flow in Horizontal Rod Bundles", Nuclear Engineering & Design 66 101-123, 1984.

[2] M. Shoukri, A. Tahir and M.B. Carver, "Numerical Simulation of Two-Phase Flow in Horizontal Interconnected Subchannels", CNS/ANS International Conference on Numerical Methods in Nuclear Engineering, Montreal, 1983.

[3] A. Tahir and M.B. Carver, "ASSERT and COBRA Prediction of Flow Distribution in Vertical Bundles", Proceedings CNS/ANS International Conference on Numerical Methods in Nuclear Engineering, 371-387, Montreal, 1983.

[4] Tapucu, A. and Gencay, S., "Experimental Investigation of Mass Exchange Between Two Laterally Interconnected Two-Phase Flows", Ecole Polytechnique de Montreal, Reports IGN-354 and IGN-394, 1980.

[5] Tapucu, A., Geckinli, M., and Troche, N., "Experimental Investigation of Mass Exchange Between Two Laterally Connected Two-Phase Flows (Subchannel Geometry)", Ecole Polytechnique de Montreal, Reports IGN-556 and 557, 1984.

[6] Shoukri, M., Tawfik, H. and Chan, A., "Two-Phase Redistribution in Horizontal Subchannel Flow", Second International Topical Meeting on Nuclear Reactor Thermal-hydraulics, Santa Barbara CA, January 1983.

[7] Stewart, C.W., Wheeler, C.L., Cena, R.J., McMonagle, C.A., Cuta, J.M., and Trent, D.S., "COBRA-IV - The Model and Method", Battelle Pacific Northwest Laboratories, BNWL-2214, 1977 July.

[8] Judd, R.A., Tahir, A., Kiteley, J.C., Carver, M.B., Rowe, D.S., Stewart, D.G., and Thibeault, P.R., "ASSERT-4 (Version 1) User's Manual", Atomic Energy of Canada Limited, AECL-8573, 1984.

[9] Carver, M.B., Tahir, A., Kiteley, J.C., Judd, R.A., "Computation of Two-Phase Distribution and CHF in Horizontal Bundles Using the ASSERT Subchannel Code", CNS 6th Annual Conference, p1808-1816, Ottawa, 1984.

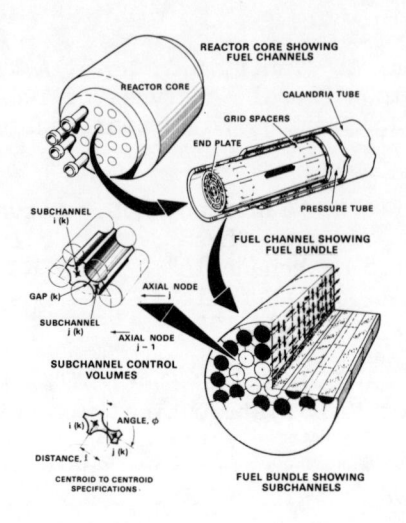

FIGURE 1   Formulation of Subchannel Control Volumes

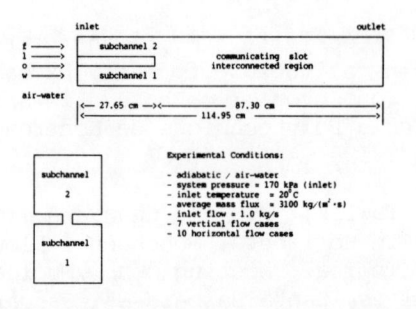

a)   Tapucu Square Channel

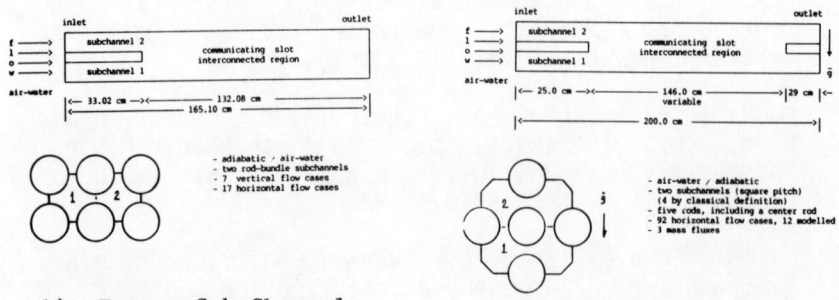

b)   Tapucu Sub-Channel

c)   Shoukri Sub-Channel

FIGURE 2   Geometry of the Air-Water Experiments

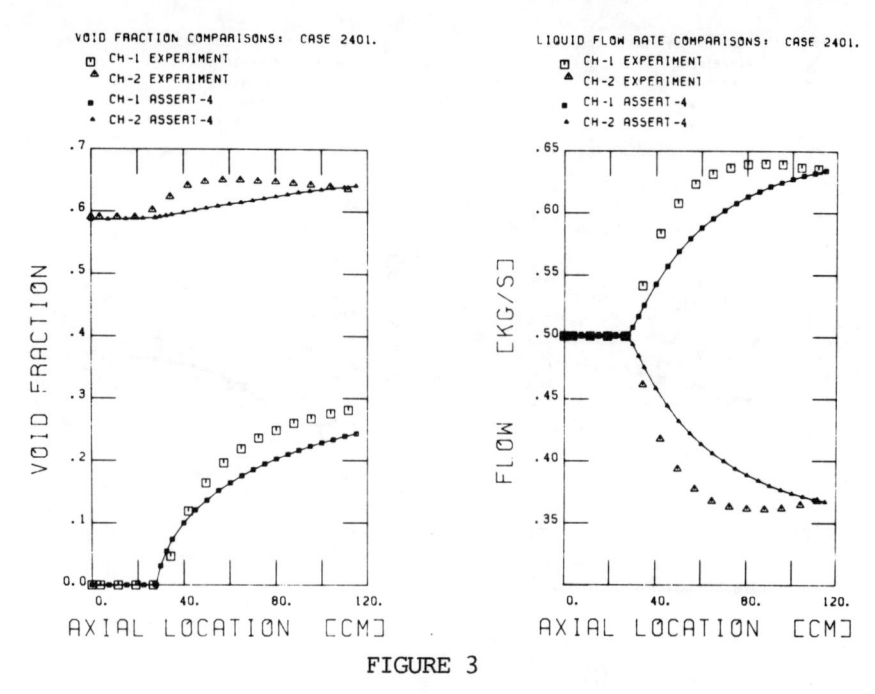

FIGURE 3

Computed and Measured Profiles of Void Fraction and Mass Flux: Tapucu Horizontal Case H_R^D-1 (Square Geometry)

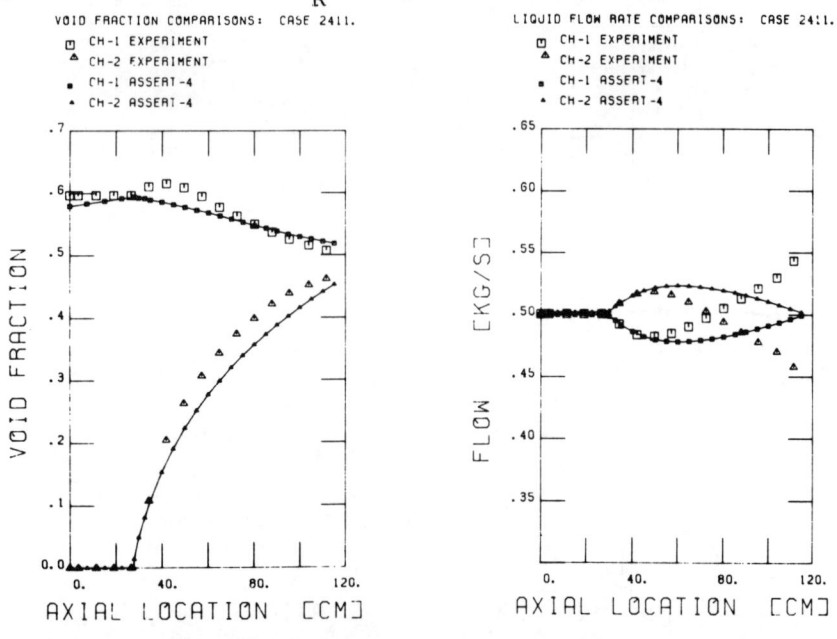

FIGURE 4

Computed and Measured Profiles of Void Fraction and Mass Flux: Tapucu Horizontal Case H_D^R-1 (Square Geometry)

FIGURE 5

Computed and Measured Profiles of Void Fraction and Mass Flux: Tapucu Horizontal Case $H_R^D=4$ (Bundle Geometry)

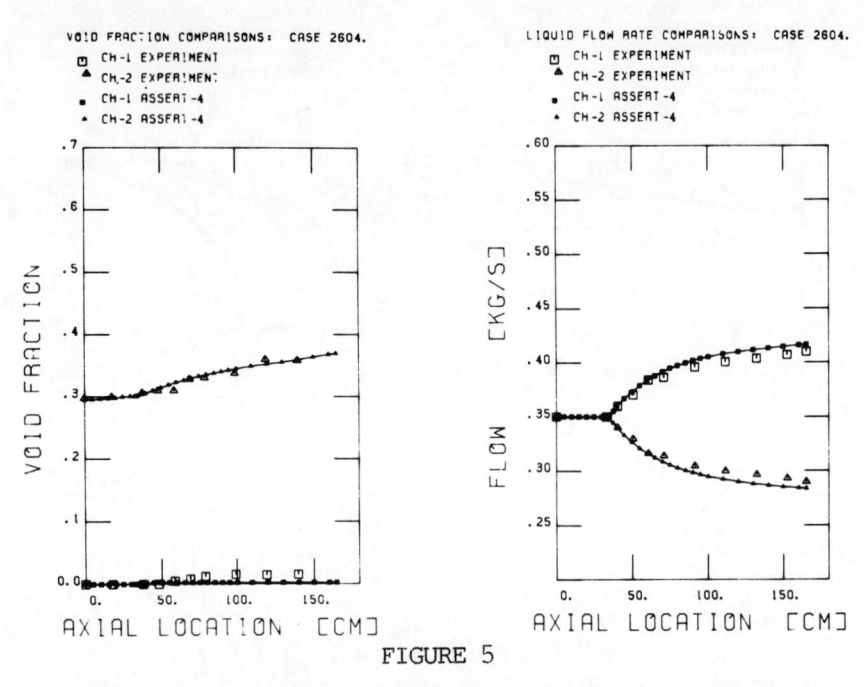

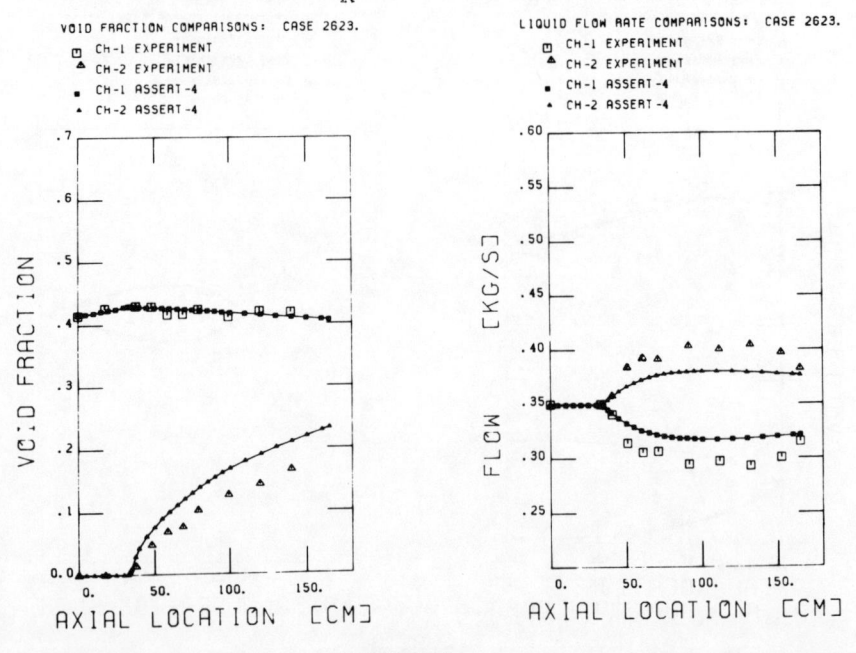

FIGURE 6

Computed and Measured Profiles of Void Fraction and Mass Flux: Tapucu Horizontal Case H D=R (Bundle Geometry)

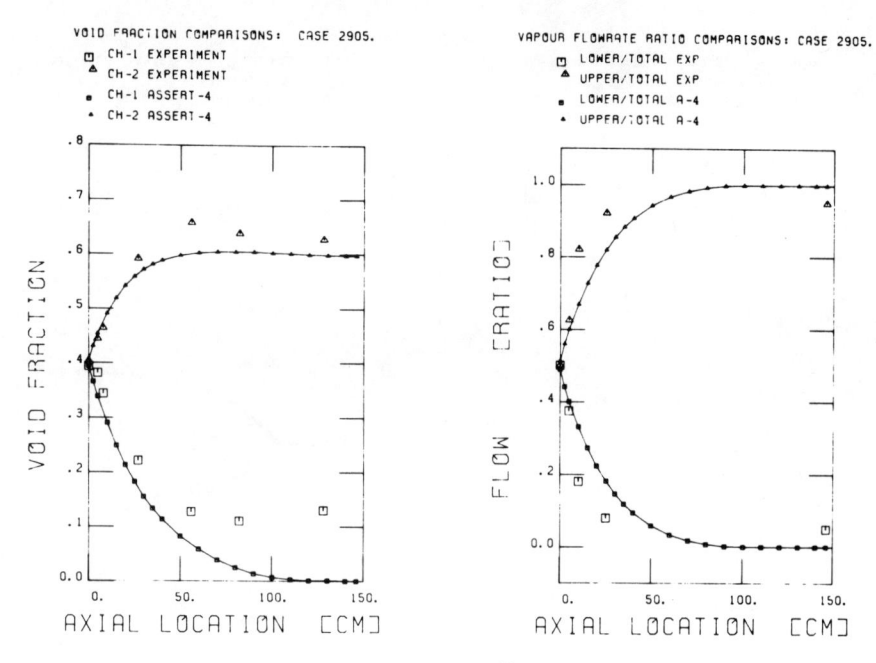

FIGURE 7

Computed and Measured Profiles of Void Fraction and Mass Flux: Shoukri Case B0105 – Mass Flux = 950 kg·m^{-2}·s^{-1}

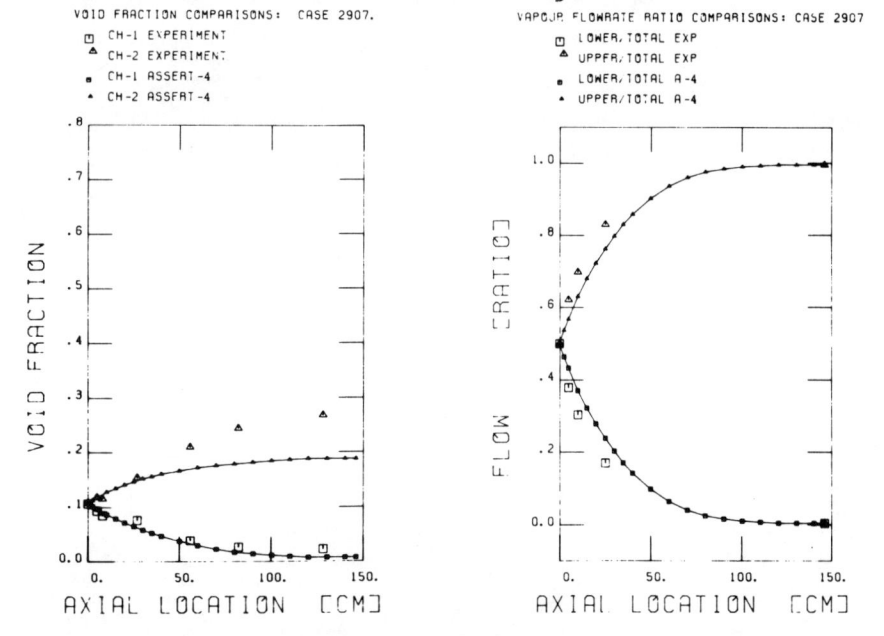

FIGURE 8

Computed and Measured Profiles of Void Fraction and Mass Flux: Shoukri Case C0102 – Mass Flux = 1560 kg·m^{-2}·s^{-1}

SECTION 7

Thermal and/or Mechanical Loads

Failure of Mechanically Loaded Laminated Composites Subjected to Intense Localized Heating

by

R. C. GULARTE, J. A. NEMES, F. R. STONESIFER
AND C. I. CHANG

*Structural Integrity Branch
Naval Research Laboratory
Washington DC 20375-5000*

ABSTRACT

A numerical procedure is presented for predicting failure in laminate composites subjected to simultaneous intense localized heating and applied mechanical loads. The method consists of a nonlinear two-dimensional, axisymmetric finite difference thermal analyzer which considers the effects of surface ablation, radiation and convection losses and temperature dependent thermophysical properties. The thermal results are used in conjunction with a flat plate finite element code to predict failure of the laminate composite. Predictions based on the numerical analyses are compared with experimental results obtained from graphite epoxy coupons spot-irradiated at various intensities.

INTRODUCTION

The objective of the approach presented herein was to develop and experimentally verify a numerical method for predicting the response of mechanically loaded and intensely heated composite structures. Due to the anticipated and subsequently verified nonlinearity of the thermal and mechanical responses, numerical methods were used. A thermal/mechanical methodology was developed in which the thermal analysis was performed using finite difference techniques and the mechanical response was determined using finite element methods. Experimental verification consisted of thermal tests of non-loaded specimens for thermal property determination and combined thermal and mechanical testing for the thermal/mechanical methodology verification.

The approach presented consists of the integration of a high temperature material property data base with a nonlinear thermal analysis capability to produce a temperature distribution within an intensely locally heated structure which, in turn, was used in the mechanical code to predict failure. It should be emphasized that the subject of the failure of laminate composites is controversial even under ambient conditions. The analysis and experimental verification concentrated on the response of laser spot-irradiated mechanically loaded organic matrix composite structures. Adverse effects experienced under laser irradiation includes ablation (of the matrix and fibers), degradation of mechanical properties, severe thermal stresses (due to high thermal gradients) that result from spot-irradiation and stress intensification (resulting from localized burnthrough). The thermal degradation of material properties results in material nonlinearities, and ablation results in geometric nonlinearities in the mechanical analysis.

THERMAL MODELING

The purpose of the thermal analysis was to determine the transient spatial temperature distributions within the structure. These results were then used as input into the stress and failure calculations which were used to determine the structures overall survivability. To illustrate the fundamental aspects of the thermal formulation consider a one-dimensional problem in which a plate is subjected to a uniform constant-intensity irradiation on one surface. The surface is assumed to be ablated when the surface temperature reaches the ablation temperature, T_s. The recession rate of the ablating surface is denoted by V and z was adopted as the convective coordinate which moves with the ablating front. The Fourier heat conduction formulation of this problem is given by:

$$\frac{\partial}{\partial t}\left[K\frac{\partial T}{\partial z}\right] + \rho C_p V \frac{\partial T}{\partial z} = \rho C_p \frac{\partial T}{\partial t} \tag{1}$$

where T, K, ρ and C represent the temperature, thermal conductivity, density and heat capacity. The two-dimensional axisymmetric lumped mass finite difference model chosen is illustrated in Fig. 1. The energy balance can be expressed as:

$$\rho H_s V = \alpha I_o + K \frac{\partial T}{\partial x} + I_r + I_c. \tag{2}$$

Where additionally H_s represents the ablation energy, α is the absorbitivity, I_o represents the incident heat flux from the laser, and I_r and I_c denote radiation and convective heat losses. The boundary condition on the irradiated surface ($z = 0$) is:

$$T(0,t) = T_s = \text{Constant} \tag{3}$$

and the opposite side of the plate with thickness, $L(z = L - z_b)$ is assumed insulated, or

$$\frac{\partial T}{\partial z} = 0. \tag{4}$$

The finite difference formulation and the computational algorithm for the solution of the above equations are contained in Ref. [1].

To illustrate the specific aspects of the thermal formulation, consider the surface element i, shown crosshatched in Fig. 1. The basic difference equations are derived by considering surface element i subjected to incident, I_o, reirradiated, I_r and convected, I_c, heat fluxes. Additionally, over a time increment Δt, element i also absorbs heat ΔQ_K through conduction from the surrounding elements shown as 1, 2 and 3. The net heat conducted is approximated by:

$$\Delta Q_K = \sum_{i=1}^{3} K_{ij}(T_j - T_i)\Delta t \tag{5}$$

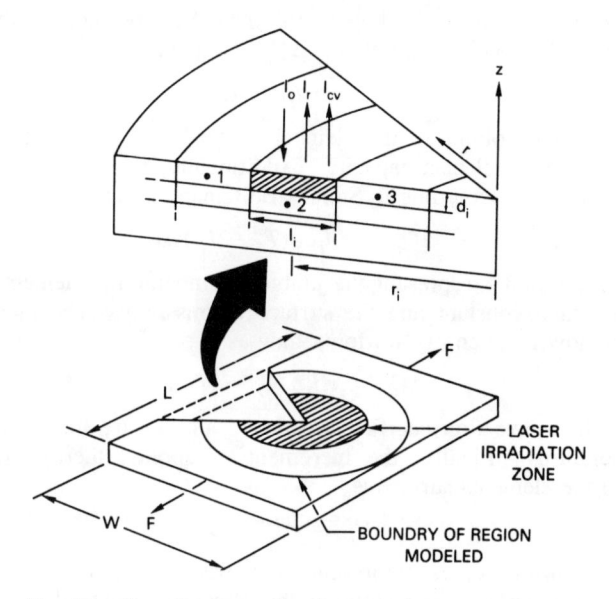

Fig. 1 — Two-dimensional axisymmetric finite difference heat transfer model

where T_i and T_j are the temperatures associated with element i and j at the beginning of the time increment and K_{ij} is the effective heat conductivity between elements i and j. The effective heat conductivity for the radial configuration shown in Fig. 1 is obtained by using a finite difference approximation of the Fourier equation and using continuity of flux across the interface area, A_{ij}, between elements i and j resulting in:

$$K_{ij} = \frac{2A_{ij}}{l_i/k_{ri} + l_j/k_{rj}} \qquad (6)$$

where k_{ri} and k_{rj} are the effective radial conductivities associated with elements i and j which were obtained from the following expression:

$$\begin{Bmatrix} k_x \\ k_y \\ k_{xy} \end{Bmatrix} = \begin{bmatrix} \cos^2\phi & \sin^2\phi & -2\sin\phi\cos\phi \\ \sin^2\phi & \cos^2\phi & -2\sin\phi\cos\phi \\ \sin\phi\cos\phi & -\sin\phi\cos\phi & 0 \end{bmatrix} \begin{Bmatrix} k_f \\ k_t \\ 0 \end{Bmatrix} \qquad (7)$$

where k_t and k_f are the transverse and fiber direction conductivity for an individual lamina, k_x and k_y denote the conductivity in the width and longitudinal directions and ϕ is the clockwise angle between the fiber and width direction. Since, as will be seen in the experimental verification part of this paper, heat conduction in the x-direction into the small unheated region adjacent to the spot irradiation was believed to be most important in defining the failure response, the values of k_x for all plys were averaged and this value was used for the radial conductivity. For conduction in the thickness direction expression (6) is valid except that the dimensions l_i and l_j are replaced by d_i and d_j. Additionally, k_{ri} and k_{rj} become corresponding values for the transverse direction. The thermal model addressed was essentially a disk with a diameter equal to the width of the specimen irradiated, as shown in Fig. 1, and insulated on the boundary, shown in the figure, and on the rear surface.

With respect to the remaining terms in the energy balance, the conduction associated with the specific heat effect, ΔQ_p, is given by:

$$\Delta Q_p = r_i l_i d_i \rho_i C_i \Delta t \tag{8}$$

where $r_i l_i d_i$ is the volume of the element subtended by an angle equal to unity and C_i is the elemental heat capacity. Additionally, the net heat increment due to surface irradiation according to Stefans-Boltzmann's Law is:

$$\Delta Q_r = r_i l_i \sigma \epsilon (T_o^4 - T_i^4) \Delta t \tag{9}$$

where T_o, T_i, σ and ϵ represent the ambient temperature, element temperature, Stefans-Boltzmann constant and the surface emmissivity. The incremental convection heat flow is given by Newton's Law as

$$\Delta Q_c = r_i l_i h (T_r - T_i) \Delta t \tag{10}$$

where h is the convection coefficient, and T_r and T_i are the recovery and element temperatures. Finally, the increment of applied thermal (laser) energy absorbed on the elements surface is:

$$\Delta Q_o = r_i l_i \alpha I_o \Delta t \tag{11}$$

where the absorptivity, α, represents the fraction of incident laser energy absorbed by the surface. Substituting the above incremental energies into the energy balance, expression (2), the following expression for the incremental change in temperature is given by:

$$\Delta T_i = \frac{\left[\sum_{j=1}^{4} K_{ij}(T_j - T_i) + r_i l_i \alpha I_0 + r_i l_i \sigma \epsilon (T_o^4 - T_i^4) + r_i l_i h (T_r - T_i) \right] \Delta t}{r_i l_i d_i \rho_i C_i} \tag{12}$$

Since all of the quantities on the right hand side of the above expression are known at the end of the previous integration interval, this formulation constitutes an explicit foreward difference method. That is, the temperature for each element in the mesh (typically 40 nodes in the radial direction and from 16 to 64 nodes in the thickness direction) is computed at successive time increments to generate the complete transient thermal response of the model. Additionally, the formulation can accommodate temperature-dependent properties. The ablation phenomena is accommodated by introducing a large heat capacity value at the prescribed ablation temperature. When a given surface element ablates, the incident radiation, I_o, is transferred to the appropriate adjacent element in the depthwise z-direction.

STRUCTURAL MODELING

For the analysis of composite structures a finite element approach using Mindlin plate theory has been developed [2]. In the Mindlin formulation, deflections due to both bending and transverse shear are included. With respect to the x-, y-, z- coordinate system shown in Fig. 2, the displacement field for each element is given by:

$$\left. \begin{array}{l} u(x,y,z,t) = u_o(x,y,t) - z\psi_x(x,y,t) \\ v(x,y,z,t) = v_o(x,y,t) - z\psi_y(x,y,t) \\ w(x,y,z,t) = w_o(x,y,t) \end{array} \right\} \tag{13}$$

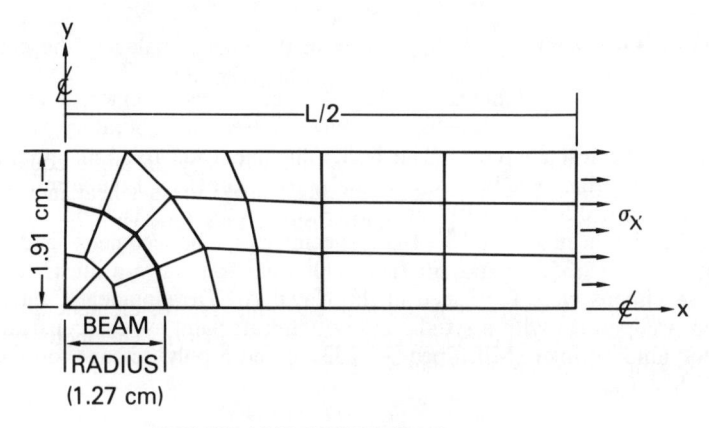

Fig. 2 — Finite element model of mechanically loaded and spot-irradiation specimen

BOUNDARY CONDITIONS

$X = L/2 : \psi_x = \psi_y = w = 0$

$\sigma_x \cong 0.25 \text{ UTS}$

$X = 0 : \psi_y = u = 0$

$y = 0 : \psi_x = v = 0$

where u_o, v_o and w_o are the midplane displacements in the x-, y- and z-directions and ψ_x and ψ_y are rotations normal to the x-z and y-x planes.

For the composite material being considered in this analysis, it is appropriate to consider each lamina as an orthotropic solid in which the stress-strain relations are:

$$\begin{Bmatrix} \sigma_{xx} \\ \sigma_{yy} \\ \sigma_{xy} \\ \sigma_{yz} \\ \sigma_{xz} \end{Bmatrix} = \begin{bmatrix} \bar{Q}_{11} & \bar{Q}_{12} & \bar{Q}_{16} & 0 & 0 \\ \bar{Q}_{12} & \bar{Q}_{22} & \bar{Q}_{26} & 0 & 0 \\ \bar{Q}_{16} & \bar{Q}_{26} & \bar{Q}_{66} & 0 & 0 \\ 0 & 0 & 0 & \bar{Q}_{44} & \bar{Q}_{45} \\ 0 & 0 & 0 & \bar{Q}_{45} & \bar{Q}_{55} \end{bmatrix} \begin{Bmatrix} \epsilon_{xx} - \alpha_x \Delta T \\ \epsilon_{yy} - \alpha_y \Delta T \\ \gamma_{xy} - \alpha_{xy} \Delta T \\ \gamma_{yz} \\ \gamma_{xz} \end{Bmatrix} \quad (14)$$

where $\bar{Q}_{ij}$ ($i,j = 1,2,6$) are the transformed reduced in-plane stiffness components and $\bar{Q}_{ij}$ ($i,j = 4,5$) are the reduced transverse shear stiffnesses for a state of plane stress [3]. The $\alpha \Delta T$s are the transformed thermal strains. The above expression in conjunction with the strain-displacement relationship and boundary conditions is used to formulate the potential energy of the finite element mesh. Minimization of the functional with respect to nodal displacements yields the matrix equilibrium equation:

$$[K]\{\delta\} = \{F^e\} + \{F^T\} \quad (15)$$

where $\{\delta\}$ refers to the overall system displacement vector, $[K]$ is the system stiffness matrix and $\{F^e\}$ and $\{F^T\}$ represent the mechanical and thermal statically equivalent nodal point loads vector.

THERMAL TESTING

To investigate rapid intense heating in a controlled manner, a 15 kW continuous wave, 10.6 μ CO_2 laser at the Naval Research Laboratory was used to spot-irradiate the test coupons. The beam diameter was fixed at approximately 1.27 cm with profiles which were essentially axisymmetric. Five nominal intensities ranging from 0.5 to 2.5 kW/cm^2 were employed. A Mach 0.3 airflow was passed over the heated surface of all specimens. The geometry and stacking sequence of the AS/3501-6 graphite/epoxy thermal coupons are illustrated in Fig. 3. The specimens were fabricated at the Naval Air Development Center. Each specimen was painted with a standard Navy aircraft paint system consisting of an epoxy polyimide primer (Mil. Spec. -P-23377) and a polyurethane top coat (Mil. Spec. -C-83286).

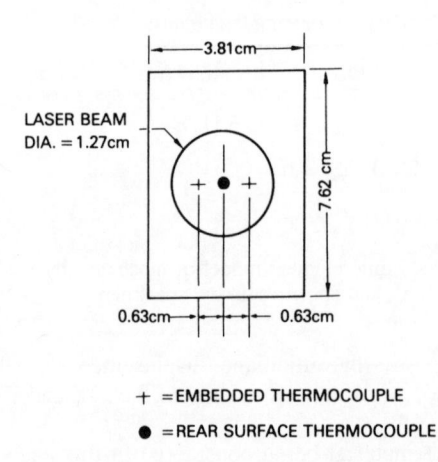

SPEC. THICKNESS (mm.)	STACKING SEQ.
3.18 (24 PLY)	[±45/0$_2$/±45/0$_2$/±45/0/90]$_s$
6.35 (48 PLY)	[±45/0$_2$/±45/0$_2$/±45/0/90]$_{2s}$
12.7 (96 PLY)	[±45/0$_2$/±45/0$_2$/±45/0/90]$_{4s}$

Fig. 3 — Configuration and stacking sequence for thermal test specimen

The temperature dependent transverse conductivity, specific heat, density, ablation temperature and ablation energy for graphite/epoxy, shown in Fig. 4, were established in a previous investigation [1,4]. In this study, the laser testing on the unloaded thermal specimens was conducted to derive two key material parameters for the general two-dimensional axisymmetric thermal analysis—the fiber direction conductivity and the absorptivity of the painted composite surface. These quantities were determined using a combined analytical and experimental method in which the temperatures obtained from the two embedded and rear surface thermocouples were compared with analytical results using assumed fiber conductivity and surface absorptivity. The assumed parameters were systematically varied until consistently good agreement was achieved between computed and measured thermal responses for the three thicknesses of materials for the five different intensities.

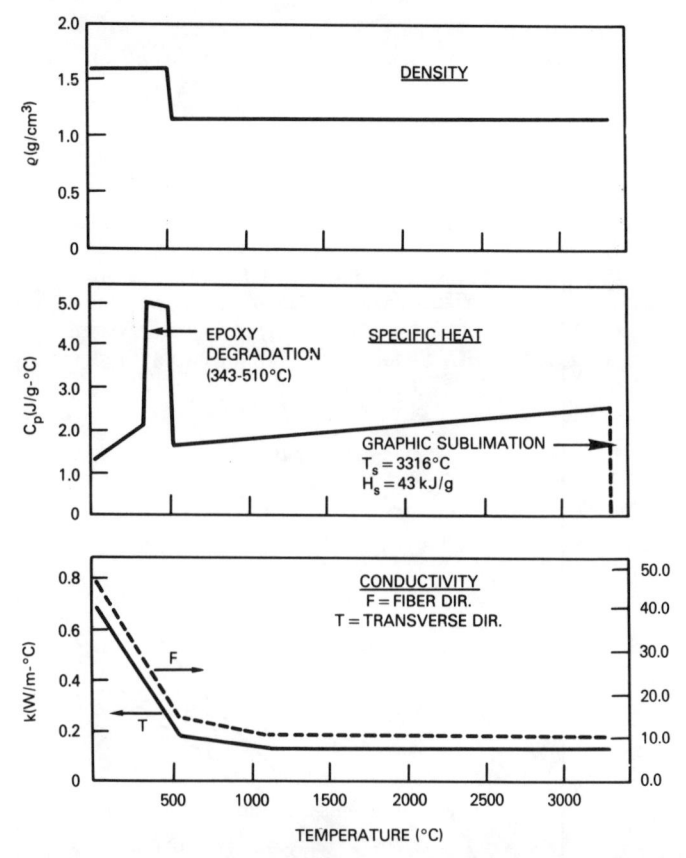

Fig. 4 — Temperature dependence of AS/3501-6 graphite epoxy

Figure 5 compares the resulting analytically derived and experimental temperature vs. time histories for the 24-ply laminates exposed to 1.5 kW/cm² irradiation. As can be seen from this figure the agreement between theory and experiment is quite good in that the complex time-dependent resin decomposition is accounted for in the thermal model by incorporating their effects into a heat capacity vs. temperature relationship. The fiber direction conductivity vs. temperature, also derived from combined experimental/analytical method, is shown as the dashed curve in the lower portion of Fig. 4. The shape of the curve was suggested by the known relationship between transverse conductivity and temperature. It was found that by scaling the transverse conductivity by a factor of 17 a fiber conductivity vs. temperature curve giving good correlation with the experimental data was obtained. Actually, this factor is quite close to the ratio of fiber to transverse conductivities at room temperature [4,5]. Additionally, the thermal data indicated that the absorptivity of the painted surface was approximately 0.5 for intensity values ranging from 0.5-2.5 kW/cm².

THERMAL/MECHANICAL TESTING

To assess the accuracy of the thermal/mechanical analytical method for predicting the response of composite structures subjected to an intense localized thermal environment, the composite test specimens shown in Fig. 6 were subjected to laser irradiation while under sustained uniaxial tension.

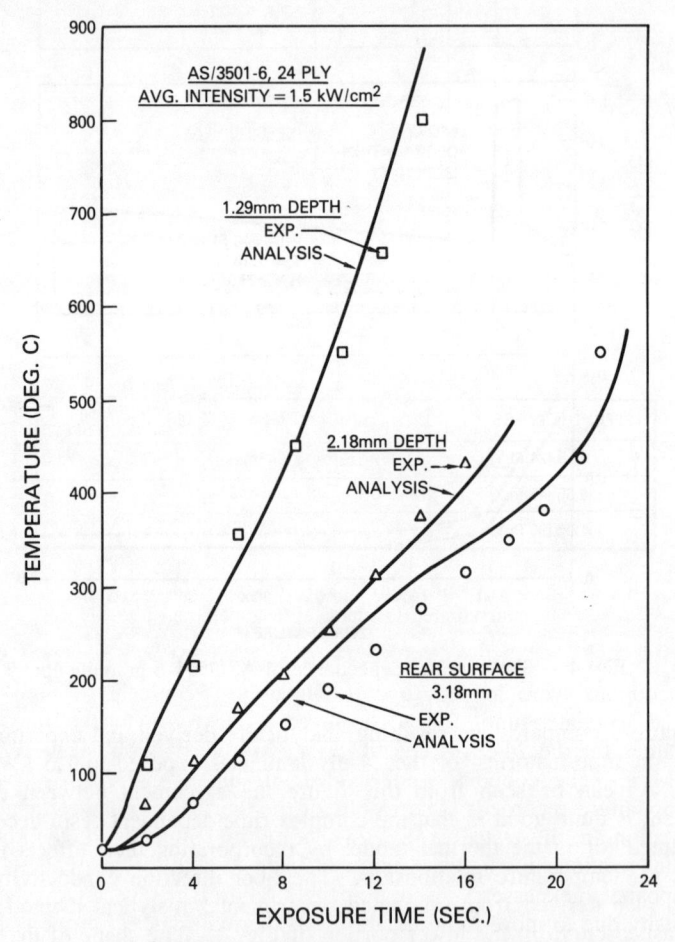

Fig. 5 — Comparison of theoretical and experimental thermal responses for 24-ply graphite epoxy

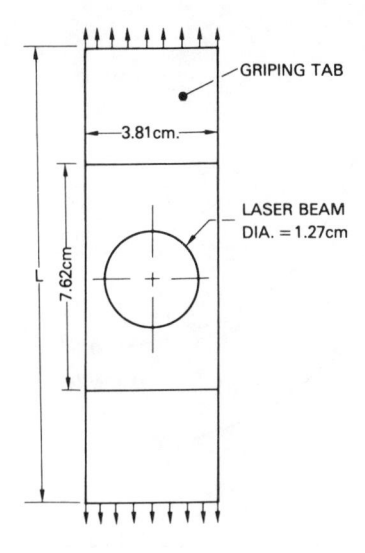

SPEC. THICKNESS (mm.)	STACKING SEQ.	LENGTH (cm.), L (TENSION SPEC.)
3.18 (24 PLY)	$[\pm 45/0_2/\pm 45/0_2/\pm 45/0/90]_s$	25.4
6.35 (48 PLY)	$[\pm 45/0_2/\pm 45/0_2/\pm 45/0/90]_{2s}$	30.4
12.7 (96 PLY)	$[\pm 45/0_2/\pm 45/0_2/\pm 45/0/90]_{4s}$	38.1

Fig. 6 — Configuration and stacking sequence for irradiated mechanical test specimen

The coupons were loaded to approximately 25 percent of their ultimate strength and irradiated until failure occurred. The room temperature ultimate strength values for the 24-, 48- and 96-ply composite material were 1081, 1076 and 898 MPa. The coupons were loaded, irradiated and the exposure time required for failure recorded. The results of the thermal/mechanical analysis and experimentation are illustrated in Fig. 7. The figure compares the theoretical and experimental relationships between nominal intensity and failure time, t_f, for the three composite thicknesses. As anticipated, failure time varies inversely with beam intensity due to more rapid ablation and thermal material property degradation at higher laser intensities. Additionally, both the analytical and experimental results indicate that, for a given intensity, longer exposure times are required to fail the thicker laminates due to the requirement that more extensive through-thickness damage in the irradiated zone is required to elevate the stress level to a sufficiently high level for failure to initiate. The best agreement between analysis and experimental data was obtained with the 48-ply laminate at higher intensities. At the intermediate laser intensities the calculated failure times consistently underestimated the measured failure times for all thicknesses.

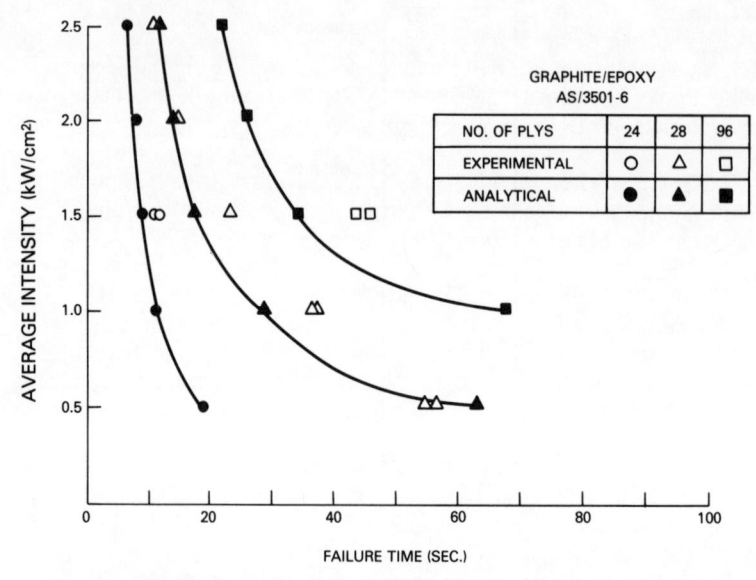

Fig. 7 — Comparison of analytical and experimental failure times for mechanically loaded spot-irradiated graphite epoxy specimens

CONCLUSIONS

A combined thermal and mechanical model has been successfully employed to predict the radiation time to cause failure in spot-irradiated laminated graphite/epoxy at nominal beam intensities from 0.5-2.5 kW/cm^2. The lumped-mass, finite difference thermal analysis in which resin decomposition is included in the heat capacity vs. temperature relationship provides good correlation with experimentally-determined transient temperature profiles. The fiber direction conductivity of graphite/epoxy as a function of temperature was determined using a combined analytical/experimental method. By similar means, the absorptivity of the paint used was found to be essentially constant for the range of intensities studied. The concept of using a two-dimensional axisymmetric heat transfer model with an average radial conductivity has been found to give good results; although, in the final analysis the resulting failure times were shorter.

Further examination of the thermal analyses revealed that depending on the intensity level, material failure is governed by two types of thermal effects. That is, material removal due to ablation was the dominant factor contributing to failure at high irradiation levels. Degradation of material properties due to thermal diffusion was the major cause for failure at lower intensities. No ablation was observed for nominal intensities of 0.5 kW/cm^2 where a maximum surface temperature of 2870° C was calculated.

In summary, disparities in analyses may be attributed to slight inaccuracies in the thermal calculations which in turn may have led to errors in defining the temperature-dependent mechanical properties. The lack of experimentally-determined properties at elevated temperatures could also have contributed to uncertainties in the analysis. The use of classical plate theory to model the complex stress state in the vicinity of ablated or degraded material could be an oversimplification.

REFERENCES

1. Griffis, C.A., Masumura, R.A. and Chang, C.I., "Thermal Response of Graphite Epoxy Composite Subjected to Rapid Heating," Journal of Composite Materials, Vol. 15, p. 427 (September 1981).

2. Sun, C.T., Chen, J.K. and Chang, C.I., "Failure of a Graphite/Epoxy Laminate Subjected to Combined Thermal and Mechanical Loading," Journal of Composite Materials, Vol. 19, p. 408 (September 1985).

3. Jones, R.M., "Mechanics of Composite Materials," Scripta Book Company, Washington, D.C. (1975).

4. Menousek, J.F. and Monin, D.L., "Laser Thermal Modeling of Graphite Epoxy," NWC Technical Memorandum Report 3834, Naval Weapons Center (China Lake, CA) (June 1979).

5. Ender, T., Clark, H.T. and Grimm, T.C., "Response of Advanced Fiber Composite Materials to Laser Heating," Final Report, NWC Contract No. N00123-74-C-1482, Naval Weapons Center (China Lake, CA) (October 1979).

DEFORMATIONS IN BUTT-WELDING OF LARGE PLATES WITH SPECIAL REFERENCE TO THE TEMPERATURE DEPENDENCE OF THE YIELD LIMIT OF THE MATERIAL

Lennart Karlsson

Department of Mechanical Engineering
Luleå University of Technology
S-951 87 Luleå, Sweden

ABSTRACT

Simulation of automatic butt-welding of large plates was investigated. The plates were tack-welded before the butt-welding. The simulation includes the tack-welding, the butt-welding and the cooling to room temperature. The simulation should lead to an understanding of how the temperature dependence of the yield limit influence the change in gap width in front of the moving arc. The finite element method was used in the calculations of the temperature fields and the pertinent deformations. Plane stress conditions were assumed. Two verifying experiments were performed using two different values of the heat input. One calculation was performed for each of the experiments. In these two calculations the yield limit of the material was made dependent of the temperature history of the material. In a third calculation the yield limit was dependent on the temperature only. Overall good agreement between calculated and measured values of the change in gap width was obtained. The simulation where the yield limit was dependent on the temperature history gave a slightly larger increase in gap width than the simulation where the yield limit was temperature dependent only.

1.   INTRODUCTION

Automatic butt-welding of large plates was simulated. The two plates were tack-welded before the butt-welding. The simulations included the tack-welding, the butt-welding and the cooling to room temperature. The main object of this investigation was to study the change in gap width in front of the moving arc in root-bead butt-welding and one-pass butt-welding of a low-alloy steel. Special emphasis was given to the temperature dependence of the yield limit. In automatic butt-welding it is important to know how the gap width, Δg, between

DEFORMATIONS IN BUTT-WELDING
OF LARGE PLATES

the two tack-welded plates changes during the welding process. This change in gap width depends on the welding parameters, the tack-welding, the restraint of the plates and the size of the plates. The most important welding parameters are the arc energy per unit length of the weld and the welding speed. Also the thermal stresses during the cooling process and the resulting residual stresses were studied.

Two different experiments were performed and are reported here. In both experiments two plates of the size 0.5×1.0×0.01 m^3 were first tack-welded in the order 1, 2, 3, and 4 and then butt-welded, see Fig. 1. Each tack-weld had the length 30 mm. The tack-welds and the butt-welds were made in an automatic gas-metal arc-welding process. The butt-welding started at $x_2 = 0$, see Fig. 1, and time t = 0. The welding arc moved along the x_2-axis with a constant velocity v = 0.005 m/s. In the first experiment root-bead butt-welding was performed using a V-shaped groove and the gross heat input Q = 0.8 MJ/m. In the second experiment one-pass butt-welding was performed using an Y-shaped groove and the gross heat input Q = 1.6 MJ/m. The temperature was measured during the butt-welding at x_1 = 10 mm and x_2 = 500 mm. The arc efficiency η was taken as η = 0.75 in case of V-shaped groove (Q = 0.8 MJ/m) and in case of an Y-shaped groove (Q = 1.6 MJ/m) the arc efficiency η was taken as η = 0.72.

The change in gap width during the butt-welding was measured with a video camera which was attached to the welding device, see [1]. The camera recorded the gap 0.12 m in front of the arc. The change in gap width was determined by taking the difference between the gap during the butt-welding and the gap before the butt-welding.

The welding procedure described above was simulated by a mathematical model as is shortly described below. A detailed description and investigation of this model is given in [1] and [2].

One simulation was performed for the first experiment (V-shaped groove) and two simulations for the second experiment (Y-shaped groove). The two simulations for the second experiment differed from each other with respect to the modelling of the temperature-dependence of the yield limit. The influence of different values of the mechanical material properties on the change in gap width was studied in [3].

DEFORMATIONS IN BUTT-WELDING OF LARGE PLATES

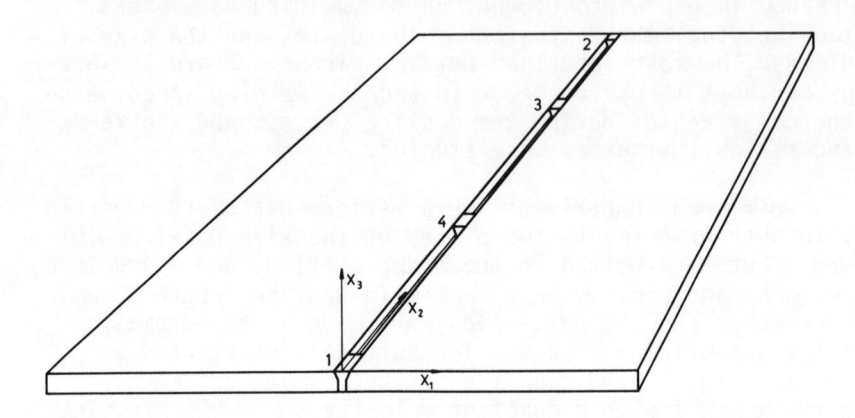

Fig. 1. Butt-welded plate to be analysed. Heat source starts at $x_2 = 0$ and moves along x_2-axis with constant velocity.

2. THERMAL ANALYSIS

The material investigated is a low-alloy steel with a yiel stress of 290 MPa. The filler material used was ESAB 1.2/12.51. The temperature-dependent thermal properties were taken from [1]. Uniform properties were assumed through the plate thickness.

The temperature field and the mechanical field were assumed to be thermodynamically uncoupled, see [1]. The heat content (enthalpy) was used as the dependent variable in the heat conduction equation. The finite element method (FEM) was used in the analysis. Owing to symmetry only one half of the plate need be analysed.

2.1 Temperature Calculations in Tack-Welding

In the calculation of temperature fields due to tack-welding the $x_1 x_2$-plane was divided into 5600 lowest-order triangular elements. Convective surface heat transfer was simulated by heat sinks according to [1]. The calculation started with the heat input simulating tack-welds 1 and 2. The heat input was applied during a few seconds. After 60 s the heat input simulating tack-weld 3 was applied and the heat input for tack-weld 4 was applied after another 60 s.

2.2 Temperature Calculations in Butt-Welding

The welding speed v used was high enough to allow the heat flow along the weld to be neglected as compared to the heat flow transverse to the weld. Two-dimensional temperature

DEFORMATIONS IN BUTT-WELDING
OF LARGE PLATES

fields (in the $x_1 x_2$-plane) were used in the analysis. The temperature fields were calculated from one-dimensional temperature fields according to [1] and [4]. This modelling of the temperature field was in [1] shown to be very accurate. Calculated and measured temperatures are given in Fig. 8.

3. MECHANICAL ANALYSIS

The material was assumed to be thermo-elastoplastic with temperature-dependent material properties. Volume changes due to phase transformations were accounted for by use of the thermal dilatation ε^T as given in Fig. 2 and Fig. 3. The curves in Fig. 2 was based on $\Delta t_{8/5} = 6$ s (Q = 0.8 MJ/m) which is the time for cooling from 800°C to 500°C. This cooling time was about the same for the entire heat-affected zone where the phase transformations occur. During heating the top curve in the dilatation diagram was followed. Depending on the maximum temperatures reached, different curves were followed during cooling. The curves in Fig. 3 was based on $\Delta t_{8/5} = 22$ s (Q = 1.6 MJ/m). The dilatation-temperature diagram was obtained from a steel manufacturer [5].

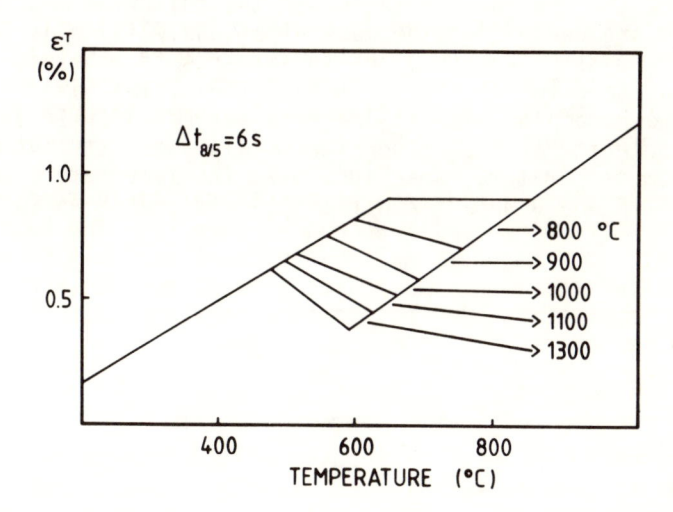

Fig. 2. Thermal dilatation ε^T for the low-alloy steel used with heat input Q = 0.8 MJ/m.

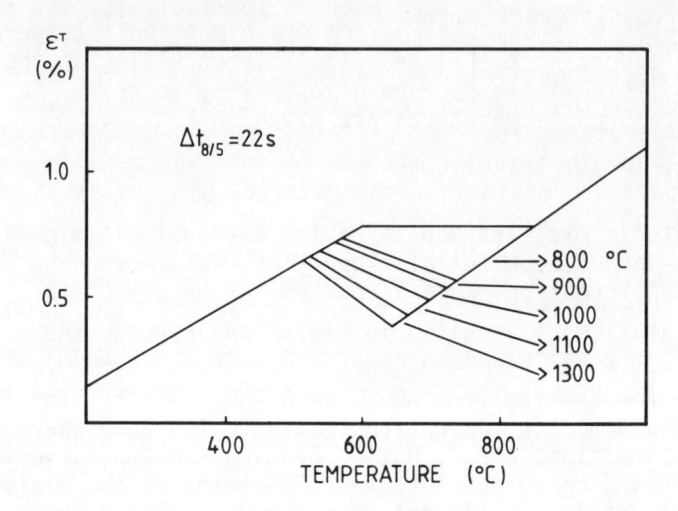

Fig. 3. Thermal dilatation ε^T for the low-alloy steel used with heat input Q = 1.6 MJ/m.

Beside the thermal dilatation ε^T the following material parameters are needed for the mechanical analysis, elastic modulus E, Poisson's ratio ν, yield stresses for the base material σ_{yb} and for the filler material σ_{yf}, the hardening moduli. The hardening moduli were here set to zero. In order to compensate for that a rather large value was attributed to σ_{yf} at room temperature, see Figs. 5-7. The temperature dependence of E and ν were taken from [1] and are shown in Fig. 4.

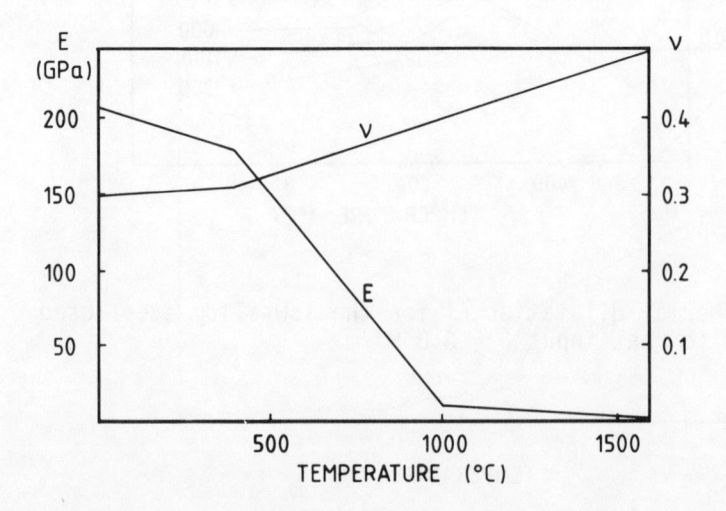

Fig. 4. Assumed temperature variation of E and ν (from [1]).

DEFORMATIONS IN BUTT-WELDING OF LARGE PLATES

The five different curves for σ_{yb} in Fig. 5 and in Fig. 6 were obtained by use of the thermal dilatation curves in Fig. 2 and Fig. 3, respectively. During heating the curve furthest to the right of the five curves was followed. Depending on the maximum temperatures reached, different curves were followed during cooling. The yield limit was in this way made dependent on the temperature history of the material. The value at room temperature of σ_{yb} was obtained from a steel manufaturer [5]. The slope of the curves in the temperature ranges up to about 500°C and above 800°C were taken from [1]. The slopes of the curves and the break points in Fig. 7 were also taken from [1].

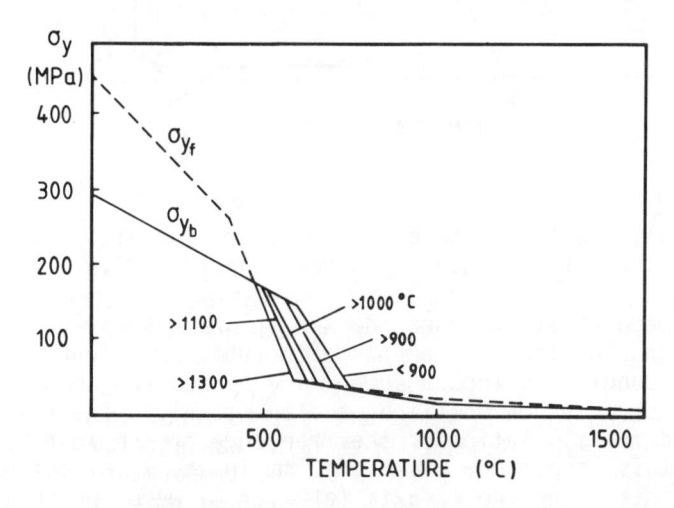

Fig. 5. Temperature dependence of the yield limit for the material used with heat input Q = 0.8 MJ/m.

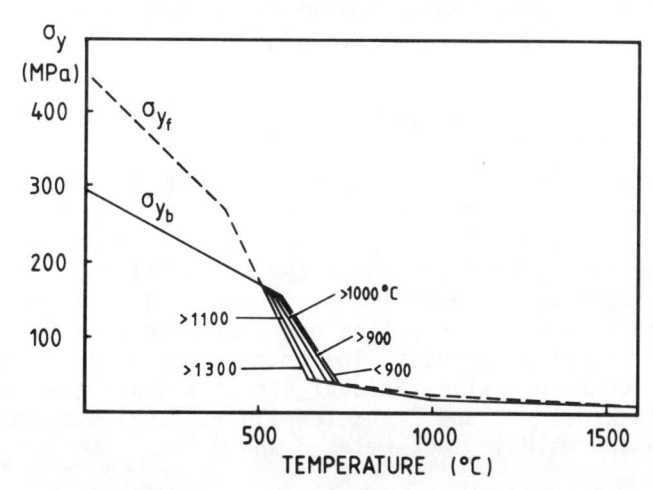

Fig. 6. Temperature dependence of the yield limit for the material used with heat input Q = 1.6 MJ/m.

DEFORMATIONS IN BUTT-WELDING
OF LARGE PLATES

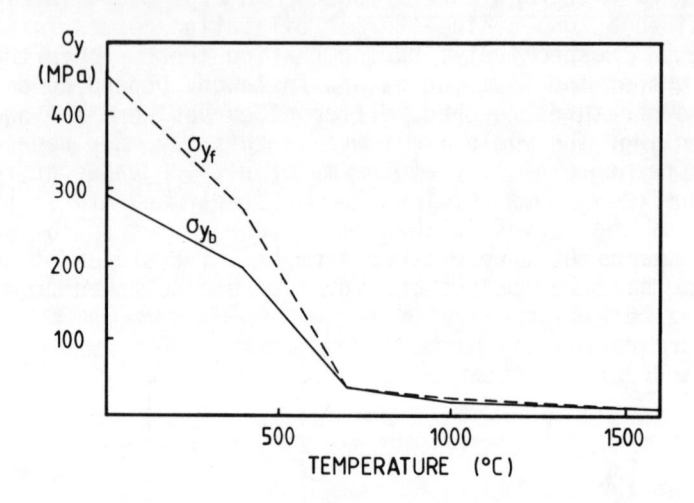

Fig. 7. Simplified temperature dependence of the yield limit for the material used with heat input Q = 1.6 MJ/m.

In the mechanical analyses the $x_1 x_2$-plane was divided into 700 triangular finite elements with cubic base functions. Plane stress conditions were assumed. The system had about 4200 degrees of freedom. One line of elements along the x_2-axis was used for simulation of the free edge of the weld. The elements in this line had a width of 2 mm in the x_1-direction. The nodal points along the x_2-axis (at x_1 = 0) were locked in the x_1-direction. In order to simulate the free edge these elements were given a very low value of the elastic modulus in front of the moving arc. Behind the moving arc the elements were given values of the temperature-dependent elastic modulus according to Fig. 4.

3.1 Mechanical Analysis of Tack-Welding

The mechanical analysis of the tack-welding started with the simulation of the welding of tack-weld 3 when the temperature in tack-welds 1 and 2 was about 250°C. The reason for this was to avoid ill-conditioned systems of equations. The analyses continued with simulation of tack-weld 4. The analyses were finished when the plates had cooled to room temperature after about 20 minutes. The order 1, 2, 3, 4 of the tack-welds gives compressive residual stresses transverse to the weld in tack-welds 1 and 2 and tensile residual stresses transverse to the weld in tack-welds 3 and 4.

DEFORMATIONS IN BUTT-WELDING
OF LARGE PLATES

3.2 Mechanical Analysis of Butt-Welding

In the simulations of the butt-welding the residual stress fields after the tack-welding were used as initial stress fields. The mechanical analyses of the butt-welding were performed for times t = 0 to t = 40000 s using about 100 time (load) increments for the welding process and about 20 time (load) increments for the cooling process.

The change in gap width was calculated 0.12 m in front of the moving heat source. It was taken as twice the difference between calculated transverse displacement 2 mm from the line of symmetry during the butt-welding and the corresponding displacement at the start of the butt-welding.

4. RESULTS

4.1 Temperature Fields in Butt-Welding

Calculated and measured temperatures at x_1 = 10 mm and x_2 = 510 mm for root-bead butt-welding (Q = 0.8 MJ/m) and one-pass butt-welding (Q = 1.6 MJ/m) are shown in Fig. 8.

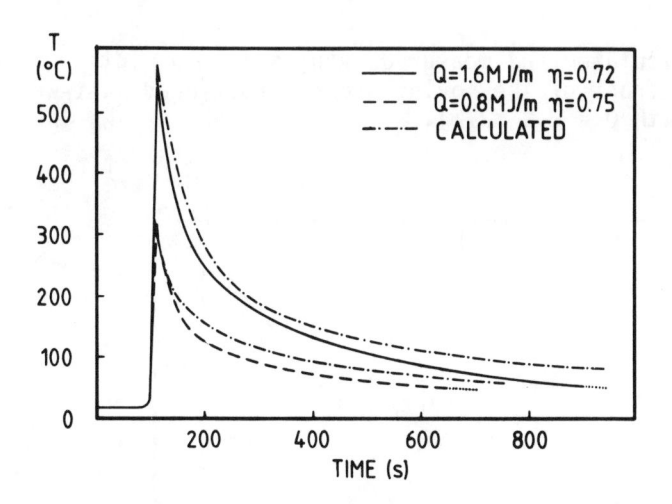

Fig. 8. Calculated and measured temperatures at (x_1, x_2) = (10,510) mm for root-bead butt-welding and one-pass butt-welding.

4.2 Change in gap width

Calculated and measured values for the change in gap width 0.12 m in front of the moving arc are shown in Fig. 9 for root-bead butt-welding (with Q = 0.8 MJ/m and) with yield

DEFORMATIONS IN BUTT-WELDING
OF LARGE PLATES

limit according to Fig. 5. In case of one-pass butt-welding (with Q = 1.6 MJ/m and) with yield limit according to Fig. 6 the corresponding results are given in Fig. 10.

The measured values fall within length of vertical lines in Figs. 9 and 10.

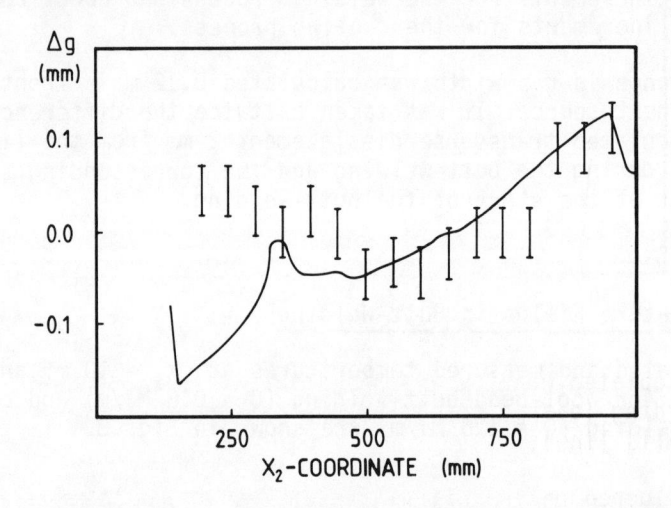

Fig. 9. Calculated and measured changes in gap width 0.12 m in front of the moving arc for root-bead butt-welding (with Q = 0.8 MJ/m).

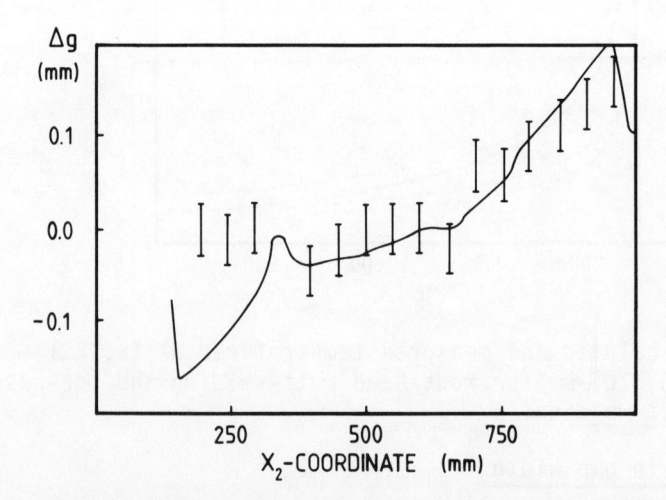

Fig. 10. Calculated and measured changes in gap width 0.12 m in front of the moving arc for one-pass butt-welding (with Q = 1.6 MJ/m).

DEFORMATIONS IN BUTT-WELDING
OF LARGE PLATES

The calculated changes in gap width for root-bead butt-welding and one-pass butt-welding are compared in Fig. 11.

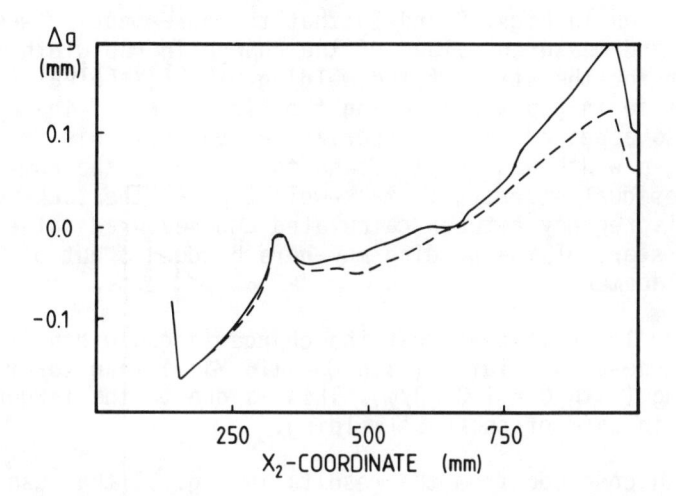

Fig. 11. Calculated changes in gap width for root-bead butt-welding (dashed line) and one-pass butt-welding (solid line).

The influence on the calculated change in gap width from the two different models of the temperature dependence of the yield limit according to Figs. 6 and 7 is shown in Fig. 12. The solid line show the change in gap width in the case where the simplified temperature dependence of the yield limit according to Fig. 7 was used.

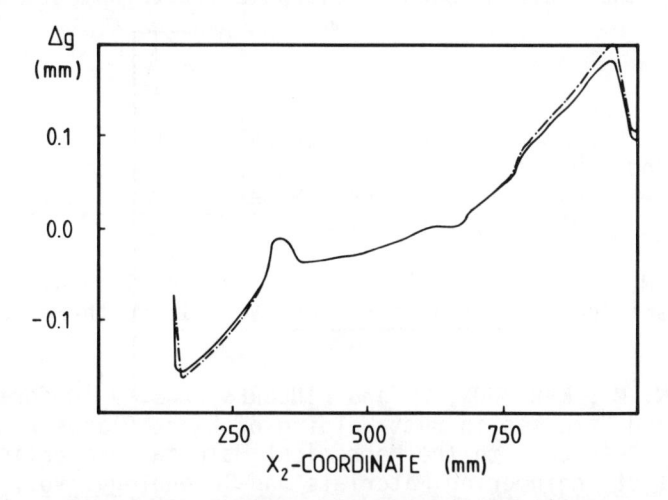

Fig. 12. Calculated change in gap width for one-pass butt-welding for the two different models of the temperature dependence for the yield limit. Solid line calculated with simplified temperature dependence according to Fig. 7.

5. DISCUSSION AND CONCLUSIONS

It is seen in Figs. 9 and 10 that the agreement between calculated and measured values of the change in gap width is good except for the start of the welding. In [1] the calculated decrease in gap width during the first part of the welding also appeared in the experimental values. This decrease in gap width was in [1] found to be due to the compressive residual stresses in tack-weld 1 after the tack-welding. The discrepancy between calculated and measured values during the start of the welding may here be due to out-of-plane displacements.

In Fig. 11 it is seen that the change in gap width is larger for one-pass welding (with Q = 1.6 MJ/m) than for root-bead welding (with Q = 1.8 MJ/m). This is due to the larger heat input in case of one-pass welding.

One can conclude from the results in Fig. 12 that use of the detailed modelling of the yield limit (according to Fig. 6) gives a slightly better result than use of the simplified modelling of the yield limit (according to Fig. 7) as compared to measured values shown in Fig. 10.

ACKNOWLEDGEMENTS

The experiments were performed at the Swedish Institute of Production Engineering Research, Göteborg, Sweden in collaboration with Dr Gunnar Lindén. The project was financially supported by the Swedish Board for Technical Development (STU).

REFERENCES

1. JONSSON, M., KARLSSON, L. and LINDGREN, L-E. - 'Deformations and Stresses in Butt-Welding of Large plates', in R.W. Lewis and K. Morgan (eds.) Numerical Methods in Heat Transfer - Vol. III, Wiley, 1985.

2. KARLSSON, L. - 'Thermal Stresses in Welding', in R. Hetnarski (ed.), Thermal Stresses, Vol. I, North-Holland, 1986.

3. JONSSON, M., KARLSSON, L. and LINDGREN, L-E. - 'Deformations and Stresses in Butt-Welding of Large Plates with Special Reference to the Mechanical Material Properties', Journal of Engineering Materials and Technology, Vol. 107, pp. 265-270, 1985.

4. ANDERSSON, B. and KARLSSON, L. - 'Thermal Stresses in Large Butt-Welded Plates', J. of Thermal Stresses, Vol. 4, Nos. 3-4, pp. 491-500, 1981.

5. Svenskt Stål AB, Box 1000, 613 01 Oxelösund, Sweden (Mr Lars Höglund).

THE NUMERICAL SOLUTION OF 3-D THERMOELASTIC PROBLEMS USING ONLY LAPLACE EQUATION DISCRETIZED FORMULATIONS

Michele A. Fanelli

ENEL, Italian National Power Agency

FOREWORD

The application of potential theory to the solution of Theory of Elasticity problems harks back to an old tradition, witness the Boussinesq treatment of the half-space point loading.

In a preceding study by the Author, published on "MECCANICA" [1] it has been shown how <u>plane</u> thermoelastic problems can be solved by using only numerical methods suited to treat the Laplace equation $\nabla^2 \varphi = 0$, hence by introducing only one d.o.f. for every node of the discretization.

In the present paper it is proposed to show how similar techniques can be used in a <u>three-dimensional</u> case. The reduction achieved in the number of nodal unknowns is from 3 d.o.f. to 1; thus it is hoped that - notwithstanding the greater complexity of the present treatment - a computational advantage can be gained over the direct approach, at least for discretization meshes of a few thousands of nodes (which is not unusual for 3-D domains).

The basic formulation draws from fairly classical works [3, 4, 5, 6, 7] and differs from more recent ones (e.g. GOODIER as quoted in [10]) insofar as the particular solutions of the thermoelastic problems do not entail, in the present approach, any integration, but rather are expressed directly in terms of the thermal field.

1. FORMULATION OF THE ELASTIC PROBLEM IN THE ABSENCE OF BODY FORCES AND OF THERMAL VARIATIONS

Let

(1.1) $\quad \xi(x, y, z); \quad \eta(x, y, z); \quad \zeta(x, y, z)$

be the three displacement components in the generic point $P(x, y, z)$ inside the domain Ω where we are considering a thermal distribution:

(1.2) $\quad \vartheta(x, y, z)$

Both the scalar field ϑ and the vector field ξ, η, ζ can be time-dependent; if so, the timewise variations are supposed to be sufficiently slow, so that accelerations $\frac{\partial^2 \xi}{\partial t^2}, \frac{\partial^2 \eta}{\partial t^2}, \frac{\partial^2 \zeta}{\partial t^2}$, and thus inertial forces, can be neglected.

Let

(1.3) $\quad e = \dfrac{\partial \xi}{\partial x} + \dfrac{\partial \eta}{\partial y} + \dfrac{\partial \zeta}{\partial z}$

be the "cubic dilatation" evidently a function of the point:

(1.4) $\quad e = e(x, y, z) \quad$ (also a function of t if conditions are not steady-state)

One can show that e is a harmonic function:

$$\nabla^2 e = \frac{\partial^2 e}{\partial x^2} + \frac{\partial^2 e}{\partial y^2} + \frac{\partial^2 e}{\partial z^2} = 0 \quad \text{(with nil thermal variations).}$$

The equilibrium equations, in the case of zero body force and zero thermal variations, can be written in the following way:

(1.5)
$$\begin{cases} \dfrac{1-\nu}{1-2\nu} \dfrac{\partial e}{\partial x} - \left(\dfrac{\partial \omega_z}{\partial y} - \dfrac{\partial \omega_y}{\partial z}\right) = 0 \\[2mm] \dfrac{1-\nu}{1-2\nu} \dfrac{\partial e}{\partial y} - \left(\dfrac{\partial \omega_x}{\partial z} - \dfrac{\partial \omega_z}{\partial x}\right) = 0 \\[2mm] \dfrac{1-\nu}{1-2\nu} \dfrac{\partial e}{\partial z} - \left(\dfrac{\partial \omega_y}{\partial x} - \dfrac{\partial \omega_x}{\partial y}\right) = 0 \end{cases}$$

where the local rotations $\omega_x, \omega_y, \omega_z$ are thus defined:

$$(1.6) \quad \begin{cases} \omega_x = \frac{1}{2}\left(\frac{\partial \zeta}{\partial y} - \frac{\partial \eta}{\partial z}\right) \\ \omega_y = \frac{1}{2}\left(\frac{\partial \xi}{\partial z} - \frac{\partial \zeta}{\partial x}\right) \\ \omega_z = \frac{1}{2}\left(\frac{\partial \eta}{\partial x} - \frac{\partial \xi}{\partial y}\right) \end{cases}$$

Restricting ourselves for the moment to <u>nil body forces</u> and <u>nil thermal variations</u>, equilibrium equations (1.5) are identically satisfied by assuming two indipendent harmonic functions:

(1.7) $\quad \varphi(x, y, z), \quad \nabla^2 \varphi = 0 \quad$ and

(1.8) $\quad \chi(x, y, z), \quad \nabla^2 \chi = 0, \quad$ and posing:

(1.9) $\quad e = -\frac{1-2\nu}{1-\nu}\left(\chi + x\frac{\partial \chi}{\partial x} + y\frac{\partial \chi}{\partial y} + z\frac{\partial \chi}{\partial z}\right),$

a harmonic function of (x, y, z);

$$(1.10) \quad \begin{cases} \omega_x = \frac{\partial \varphi}{\partial x} + y\frac{\partial \chi}{\partial z} - z\frac{\partial \chi}{\partial y}, \\ \omega_y = \frac{\partial \varphi}{\partial y} + z\frac{\partial \chi}{\partial x} - x\frac{\partial \chi}{\partial z}, \\ \omega_z = \frac{\partial \varphi}{\partial z} + x\frac{\partial \chi}{\partial y} - y\frac{\partial \chi}{\partial x} \end{cases}$$

(see [6]; $\omega_x, \omega_y, \omega_z$ are also harmonic functions)

However, since boundary conditions are usually expressed in terms of ξ, η, ζ and/or of "surface tractions" X_n, Y_n, Z_n (force per unit surface), it is necessary to take into account the following relationships:

$$(1.11) \begin{cases} \nabla^2 \left(\xi + \dfrac{x \cdot e}{1 - 2\nu} \right) = 0 \\ \nabla^2 \left(\eta + \dfrac{y \cdot e}{1 - 2\nu} \right) = 0 \\ \nabla^2 \left(\zeta + \dfrac{z \cdot e}{1 - 2\nu} \right) = 0 \end{cases}$$

(see [6]); these show that, having chosen three suitable harmonic functions ψ_1, ψ_2, ψ_3:

$$(1.12) \begin{cases} \psi_1(x,y,z) \;;\; \nabla^2 \psi_1 = 0 \\ \psi_2(x,y,z) \;;\; \nabla^2 \psi_2 = 0 \\ \psi_3(x,y,z) \;;\; \nabla^2 \psi_3 = 0 \end{cases}$$

one can write (see [7]):

$$(1.13) \begin{cases} \xi = -\dfrac{x \cdot e}{1 - 2\nu} + \psi_1 \\ \eta = -\dfrac{y \cdot e}{1 - 2\nu} + \psi_2 \\ \zeta = -\dfrac{z \cdot e}{1 - 2\nu} + \psi_3 \end{cases}$$

It can be shown that $e = \dfrac{\partial \xi}{\partial x} + \dfrac{\partial \eta}{\partial y} + \dfrac{\partial \zeta}{\partial z}$ is a harmonic function.

The four harmonic functions $\chi, \psi_1, \psi_2, \psi_3$ are of course not independent, since at every point the following relationship must hold:

(1.14) $\dfrac{\partial \xi}{\partial x} + \dfrac{\partial \eta}{\partial y} + \dfrac{\partial \zeta}{\partial z} = e$, which determines a differential link between ψ_1, ψ_2, ψ_3 and χ.

It is to be noted that, since the left-hand as well as the right-hand member of (1.14) are harmonic functions of (x,y,z), it is sufficient to impose (1.14) on the frontier S of Ω in order to satisfy it over all of the domain Ω.

As for the "surface tractions" X_n, Y_n, Z_n, the following relationships hold (see [6]):

$$(1.15) \begin{cases} \dfrac{\partial \xi}{\partial n} = \dfrac{1}{2G} X_n - \dfrac{\nu}{1-2\nu} e \cos(x,n) + \omega_y \cos(z,n) - \omega_z \cos(y,n) \\[2mm] \dfrac{\partial \eta}{\partial n} = \dfrac{1}{2G} Y_n - \dfrac{\nu}{1-2\nu} e \cos(y,n) + \omega_z \cos(x,n) - \omega_x \cos(z,n) \\[2mm] \dfrac{\partial \zeta}{\partial n} = \dfrac{1}{2G} Z_n - \dfrac{\nu}{1-2\nu} e \cos(z,n) + \omega_x \cos(y,n) - \omega_y \cos(x,n) \end{cases}$$

where $\vec{n}$ is the normal to the surface element, dS.

2. BUILDING UP GENERAL SOLUTIONS WITH NIL BODY FORCES AND NIL THERMAL VARIATIONS

Let us now see how we can devise a numerical procedure for determining the harmonic functions:

$$(2.1) \quad \begin{cases} \chi(x,y,z) ; \\ \psi_1(x,y,z) ; \quad \psi_2(x,y,z) ; \quad \psi_3(x,y,z). \end{cases}$$

We suppose the Ω domain has been discretizes into a F.E. mesh. Its surface S will be discretized too; the nodes belonging to S are indicated with an index i variable between 1 and N:

$$(2.2) \quad 1 \leq i \leq N$$

Let us build up N "unit basic solutions" of the LAPLACE equation:

$$(2.3) \quad \nabla^2 \Phi = 0 ; \quad \text{they are denoted } \Phi_1, \Phi_2, \ldots \Phi_i, \ldots \Phi_N,$$
$$\Phi_i = \Phi_i(x,y,z), \quad 1 \leq i \leq N,$$

each with the boundary conditions:

$$(2.4) \quad \Phi = \Phi_i = F_i \quad \text{over} \quad S,$$

being F_i the shape function pertaining to the node $\underline{i}$ ($F_i \equiv 1$

in node i, $F_i \equiv 0$ in every other node):

(2.5) $\begin{cases} F_i(x_i, y_i, z_i) = 1 \\ F_i(x_k, y_k, z_k) \equiv 0 \end{cases}$ for every node k belonging to S with $k \neq i$.

Otherwise, the generic F_i should satisfy continuity requirements, possibly down to the second-order derivatives (surface cubic splines would be adequate).

Those solutions $\Phi_i(x, y, z)$ of (2.3) will have to allow a "good" estimation of partial derivatives, down to second--order ones, inside Ω.

It is obvious that - barring special situations - the harmonic functions (2.1) constituting the unknowns of our problem can be expressed as follows:

(2.6) $\begin{cases} \chi = \sum_{i=1}^{N} \chi_i \cdot \Phi_i(x,y,z) & \text{with } \chi_i = \text{unknown value assumed by functions } \chi \text{ in the node } \underline{i} \text{ of S;} \\ \psi_1 = \sum_{i=1}^{N} \psi_{1i} \cdot \Phi_i(x,y,z) \\ \psi_2 = \sum_{i=1}^{N} \psi_{2i} \cdot \Phi_i(x,y,z), & \psi_{1i} = \text{unknown value assumed by function } \psi_1 \text{ in the node } \underline{i} \text{ of S; etc.} \\ \psi_3 = \sum_{i=1}^{N} \psi_{3i} \cdot \Phi_i(x,y,z), \end{cases}$

(it will also be $\varphi = \sum_{i=1}^{N} \varphi_i \cdot \Phi_i(x, y, z)$ with φ_i = unknown value assumed by function φ in the node $\underline{i}$ of S; but we will have no further use for this relationship).

Based on (1.9) we will then get:

(2.7) $e = -\frac{1-2\nu}{1-\nu} \sum_{i=1}^{N} (\Phi_i + x \frac{\partial \Phi_i}{\partial x} + y \frac{\partial \Phi_i}{\partial y} + z \frac{\partial \Phi_i}{\partial z}) \chi_i$

and based on (1.13):

$$(2.8) \begin{cases} \xi = -\dfrac{1}{1-2\nu} x \cdot e + \sum_{i=1}^{N} \psi_{1i}\, \Phi_i \\[2mm] \eta = -\dfrac{1}{1-2\nu} y \cdot e + \sum_{i=1}^{N} \psi_{2i}\, \Phi_i \\[2mm] \zeta = -\dfrac{1}{1-2\nu} z \cdot e + \sum_{i=1}^{N} \psi_{3i}\, \Phi_i \end{cases}$$ where $\underline{e}$ is given by (2.7)

Condition (1.14) at the generic node P_k on S will be expressed as follows:

$$(2.9)\quad \begin{cases} \dfrac{\partial \xi}{\partial x} + \dfrac{\partial \eta}{\partial y} + \dfrac{\partial \zeta}{\partial z} = -\dfrac{3}{1-2\nu} e - \dfrac{1}{1-2\nu}\left(x \dfrac{\partial e}{\partial x} + y \dfrac{\partial e}{\partial y} + z \dfrac{\partial e}{\partial z}\right) + \\[2mm] \qquad + \sum_{i=1}^{N} \left(\psi_{1i} \dfrac{\partial \Phi_i}{\partial x} + \psi_{2i} \dfrac{\partial \Phi_i}{\partial y} + \psi_{3i} \dfrac{\partial \Phi_i}{\partial z}\right) = e, \end{cases}$$

with $\underline{e}$ given by (2.7). Working out the formulae one gets:

$$(2.10)\quad \begin{cases} \sum_{i=1}^{N} \chi_i \left[-(5-4\nu)\Phi_i - (3-4\nu)\left(x \dfrac{\partial \Phi_i}{\partial x} + y \dfrac{\partial \Phi_i}{\partial y} + z \dfrac{\partial \Phi_i}{\partial z}\right) + \right. \\[2mm] \qquad + x^2 \dfrac{\partial^2 \Phi_i}{\partial x^2} + y^2 \dfrac{\partial^2 \Phi_i}{\partial y^2} + z^2 \dfrac{\partial^2 \Phi_i}{\partial z^2} + 2\left(xy \dfrac{\partial^2 \Phi_i}{\partial x \partial y} + xz \dfrac{\partial^2 \Phi_i}{\partial x \partial z} + \right. \\[2mm] \qquad \left. \left. + yz \dfrac{\partial^2 \Phi_i}{\partial y \partial z}\right)\right] + \sum_{i=1}^{N}\left(\psi_{1i}\dfrac{\partial \Phi_i}{\partial x} + \psi_{2i}\dfrac{\partial \Phi_i}{\partial y} + \psi_{3i}\dfrac{\partial \Phi_i}{\partial z}\right) = 0, \end{cases}$$

to be imposed on every node P_k on S.

Eqs. (2.10), written successively for $k = 1, 2, \ldots i, \ldots N$, yield N linear conditions for the 4N unknowns

(2.11) $\chi_i, \psi_{1i}, \psi_{2i}, \psi_{3i}$ $(1 \leq i \leq N)$,

so that the actually independent unknowns number only 3 N. Eqs. (2.10) can be written down concisely as follows:

(2.12) $[C] \begin{Bmatrix} \chi \\ \psi \\ \cdots \end{Bmatrix} = \{0\}$, where $[C]$ is a N x 4 N matrix

whose structure is depicked hereunder.

(2.13)

$$\text{UNKN.} \rightarrow \chi_1 \; \psi_{11} \; \psi_{21} \; \psi_{31} \; \cdots \; \chi_i \; \psi_{1i} \; \psi_{2i} \; \psi_{3i} \; \cdots \; \chi_N \; \psi_{1N} \; \psi_{2N} \; \psi_{3N}$$

$$[C] = \begin{bmatrix} c_{11} & \frac{\partial \Phi_1}{\partial x}\bigg|_1 & \frac{\partial \Phi_1}{\partial y}\bigg|_1 & \frac{\partial \Phi_1}{\partial z}\bigg|_1 & \cdots & c_{1i} & \frac{\partial \Phi_i}{\partial x}\bigg|_1 & \frac{\partial \Phi_i}{\partial y}\bigg|_1 & \frac{\partial \Phi_i}{\partial z}\bigg|_1 & \cdots & c_{1N} & \frac{\partial \Phi_N}{\partial x}\bigg|_1 & \frac{\partial \Phi_N}{\partial y}\bigg|_1 & \frac{\partial \Phi_N}{\partial z}\bigg|_1 \\ \cdots & & & & & & & & & & & & & \\ c_{k1} & \frac{\partial \Phi_1}{\partial x}\bigg|_k & \frac{\partial \Phi_1}{\partial y}\bigg|_k & \frac{\partial \Phi_1}{\partial z}\bigg|_k & \cdots & c_{ki} & \frac{\partial \Phi_i}{\partial x}\bigg|_k & \frac{\partial \Phi_i}{\partial y}\bigg|_k & \frac{\partial \Phi_i}{\partial z}\bigg|_k & \cdots & c_{kN} & \frac{\partial \Phi_N}{\partial x}\bigg|_k & \frac{\partial \Phi_N}{\partial y}\bigg|_k & \frac{\partial \Phi_N}{\partial z}\bigg|_k \\ \cdots & & & & & & & & & & & & & \\ c_{N1} & \frac{\partial \Phi_1}{\partial x}\bigg|_N & \frac{\partial \Phi_1}{\partial y}\bigg|_N & \frac{\partial \Phi_1}{\partial z}\bigg|_N & \cdots & c_{Ni} & \frac{\partial \Phi_i}{\partial x}\bigg|_N & \frac{\partial \Phi_i}{\partial y}\bigg|_N & \frac{\partial \Phi_i}{\partial z}\bigg|_N & \cdots & c_{NN} & \frac{\partial \Phi_N}{\partial x}\bigg|_N & \frac{\partial \Phi_N}{\partial y}\bigg|_N & \frac{\partial \Phi_N}{\partial z}\bigg|_N \end{bmatrix}$$

where the generic coefficient C_{ki} of unknown χ_i, in the equation pertaining to the surface node P_k, has the structure:

(2.14)
$$\begin{cases} C_{ki} = -(5-4\nu) \Phi_i (P_k) - (3-4\nu)(x \frac{\partial \Phi_i}{\partial x} + y \frac{\partial \Phi_i}{\partial y} + z \frac{\partial \Phi_i}{\partial z})_{P_k} + \\ \quad + (x^2 \frac{\partial^2 \Phi_i}{\partial x^2} + y^2 \frac{\partial^2 \Phi_i}{\partial y^2} + z^2 \frac{\partial^2 \Phi_i}{\partial z^2})_{P_k} + 2(xy \frac{\partial^2 \Phi_i}{\partial x \partial y} + xz \frac{\partial^2 \Phi_i}{\partial y \partial z} + \\ \quad + yz \frac{\partial^2 \Phi_i}{\partial y \partial z})_{P_k} , \end{cases}$$

(see 2.10), and the vector of the unknowns is ordered according to the same order of the N surface nodes:

$$
(2.15) \quad \left\{\begin{array}{c} \chi \\ \psi \\ \cdots \end{array}\right\} = \left\{\begin{array}{c} \chi_1 \\ \psi_{11} \\ \psi_{21} \\ \psi_{31} \\ \cdot \\ \cdot \\ \cdot \\ \chi_N \\ \psi_{1N} \\ \psi_{2N} \\ \psi_{3N} \end{array}\right\}
$$

We have now to impose the actual boundary conditions, relating in general to displacement unknowns ξ, η, ζ over a portion (S_u) of surface S, and to "surface tractions" X_n, Y_n, Z_n over the remaining portion ($S_T = S - S_u$) of S (*).

The "displacement" conditions (which we suppose as given in terms of cartesian components ξ, η, ζ) are immediately obtained considering (2.8). Let $\xi_{ak}, \eta_{ak}, \zeta_{ak}$, be the <u>assigned</u> components of displacements at node <u>k</u> ; it will be:

$$
(2.16) \quad \begin{cases} \xi_{ak} = \frac{1}{1-\nu} \sum_{i=1}^{N} (x\Phi_i + x^2 \frac{\partial \Phi_i}{\partial x} + xy \frac{\partial \Phi_i}{\partial y} + xz \frac{\partial \Phi_i}{\partial z})_{P_k} \cdot \chi_i + \\ \qquad + \sum_{i=1}^{N} \psi_{1i} \Phi_i (P_k) \\ \\ \eta_{ak} = \frac{1}{1-\nu} \sum_{i=1}^{N} (y\Phi_i + xy \frac{\partial \Phi_i}{\partial x} + y^2 \frac{\partial \Phi_i}{\partial y} + yz \frac{\partial \Phi_i}{\partial z})_{P_k} \cdot \chi_i + \\ \qquad + \sum_{i=1}^{E} \psi_{2i} \Phi_i (P_k) \\ \\ \zeta_{ak} = \frac{1}{1-\nu} \sum_{i=1}^{N} (z\Phi_i + xz \frac{\partial \Phi_i}{\partial x} + yz \frac{\partial \Phi_i}{\partial y} + z^2 \frac{\partial \Phi_i}{\partial z})_{P_k} \cdot \chi_i + \\ \qquad + \sum_{i=1}^{N} \psi_{3i} \Phi_i (P_k) \end{cases}
$$

(*) Obviously, if $S_T = S$, i.e. if boundary conditions are expressed only in terms of "surface tractions", the system of X_n, Y_n, Z_n must be in static equilibrium.

The boundary conditions pertaining to "surface tractions" X_n, Y_n, Z_n are more cumbersome. Going back to definitions (1.6) of $\omega_x, \omega_y, \omega_z$ as well as to (1.13) and to expression (1.9) of $\underline{e}$, equations (1.15) yield:

(2.17)
$$\frac{1}{2G} X_n = \frac{\partial \psi_1}{\partial n} + (\chi + x \frac{\partial \chi}{\partial n} + y \frac{\partial \chi}{\partial y} + z \frac{\partial \chi}{\partial z}) \cdot \cos(x,n) +$$

$$+ \frac{x}{1-\nu} [2(\frac{\partial \chi}{\partial x} \cos(x,n) + \frac{\partial \chi}{\partial y} \cos(y,n) + \frac{\partial \chi}{\partial z} \cos(z,n)) +$$

$$+ x(\frac{\partial^2 \chi}{\partial x^2} \cos(x,n) + \frac{\partial^2 \chi}{\partial x \partial y} \cos(y,n) + \frac{\partial^2 \chi}{\partial x \partial z} \cos(z,n)) +$$

$$y(\frac{\partial^2 \chi}{\partial x \partial y} \cos(x,n) + \frac{\partial^2 \chi}{\partial y^2} \cos(y,n) + \frac{\partial^2 \chi}{\partial y \partial z} \cos(z,n)) +$$

$$+ z(\frac{\partial^2 \chi}{\partial x \partial z} \cos(x,n) + \frac{\partial^2 \chi}{\partial y \partial z} \cos(y,n) + \frac{\partial^2 \chi}{\partial z^2} \cos(z,n))] +$$

$$+ \frac{1}{2} \Big\{ (\frac{\partial \psi_1}{\partial z} - \frac{\partial \psi_3}{\partial x}) \cos(z,n) + (\frac{\partial \psi_1}{\partial y} - \frac{\partial \psi_2}{\partial x}) \cos(y,n) +$$

$$+ \frac{x}{1-\nu} [(2 \frac{\partial \chi}{\partial z} + x \frac{\partial^2 \chi}{\partial x \partial z} + y \frac{\partial^2 \chi}{\partial y \partial z} + z \frac{\partial^2 \chi}{\partial z^2}) \cos(z,n) +$$

$$+ (2 \frac{\partial \chi}{\partial y} + x \frac{\partial^2 \chi}{\partial x \partial y} + y \frac{\partial^2 \chi}{\partial y^2} + z \frac{\partial^2 \chi}{\partial y \partial z}) \cos(y,n)] -$$

$$- \frac{z \cos(z,n) + y \cos(y,n)}{1-\nu} (2 \frac{\partial \chi}{\partial x} + x \frac{\partial^2 \chi}{\partial x^2} + y \frac{\partial^2 \chi}{\partial x \partial y} +$$

$$+ z \frac{\partial^2 \chi}{\partial x \partial z}) \Big\},$$

together with analogous expressions for Y_n and Z_n (these are easily obtained by permutation of axes and indices of the ψ's).

By taking into account (2.6) we get eventually for every node P_k of S where a surface traction, e.g. X_n, is assigned:

(2.18)

$$\frac{1}{2G} X_{an}(P_k) = \sum_{i=1}^{N} \psi_{1i} \left(\frac{\partial \Phi_i}{\partial x}\bigg|_{P_k} \cdot \cos(x,n)_k + \frac{\partial \Phi_i}{\partial y}\bigg|_{P_k} \cdot \cos(y,n)_k + \right.$$

$$\left. + \frac{\partial \Phi_i}{\partial z}\bigg|_{P_k} \cdot \cos(z,n)_k \right) + \frac{x_k}{1-\nu} \left[2 \sum_{i=1}^{N} \chi_i \left(\frac{\partial \Phi_i}{\partial x}\bigg|_{P_k} \cdot \right. \right.$$

$$\left. \cdot \cos(x,n)_k + \frac{\partial \Phi_i}{\partial y}\bigg|_{P_k} \cdot \cos(y,n)_k + \frac{\partial \Phi_i}{\partial z}\bigg|_{P_k} \cos(z,n)_k \right) +$$

$$+ x_k \sum_{i=1}^{N} \chi_i \left(\frac{\partial^2 \Phi_i}{\partial x^2}\bigg|_{P_k} \cdot \cos(x,n)_k + \frac{\partial^2 \Phi_i}{\partial x \partial y}\bigg|_{P_k} \cdot \cos(y,n)_k + \right.$$

$$\left. + \frac{\partial^2 \Phi_i}{\partial x \partial z}\bigg|_{P_k} \cdot \cos(z,n)_k \right) + y_k \sum_{i=1}^{N} \chi_i \left(\frac{\partial^2 \Phi_i}{\partial x \partial y}\bigg|_{P_k} \cdot \cos(x,n)_k + \right.$$

$$\left. + \frac{\partial^2 \Phi_i}{\partial y^2}\bigg|_{P_k} \cdot \cos(y,n)_k + \frac{\partial^2 \Phi_i}{\partial y \partial z}\bigg|_{P_k} \cdot \cos(z,n)_k \right) +$$

$$+ z_n \sum_{i=1}^{N} \chi_i \left(\frac{\partial^2 \Phi_i}{\partial x \partial z}\bigg|_{P_k} \cdot \cos(x,n)_k + \frac{\partial^2 \Phi_i}{\partial y \partial z}\bigg|_{P_k} \cdot \cos(y,n)_k + \right.$$

$$\left. + \frac{\partial^2 \Phi_i}{\partial z^2}\bigg|_{P_k} \cdot \cos(z,n)_k \right) + \frac{1}{2} \left\{ \sum_{i=1}^{N} \left[\left(\psi_{1i} \frac{\partial \Phi_i}{\partial z}\bigg|_{P_k} - \right. \right. \right.$$

$$\left. - \psi_{3i} \frac{\partial \Phi_i}{\partial x}\bigg|_{P_k} \right) \cos(z,n)_k + \left(\psi_{1i} \frac{\partial \Phi_i}{\partial y}\bigg|_{P_k} - \psi_{2i} \frac{\partial \Phi_i}{\partial x}\bigg|_{P_k} \right) \cdot$$

$$\cdot \cos(y,n)_k \right] + \frac{x_k}{1-\nu} \left[\sum_{i=1}^{N} \chi_i \left(2 \frac{\partial \Phi_i}{\partial z}\bigg|_{P_k} + x_k \frac{\partial^2 \Phi_i}{\partial x \partial z}\bigg|_{P_k} + \right. \right.$$

$$\left. + y_k \frac{\partial^2 \Phi_i}{\partial y \partial z}\bigg|_{P_k} + z_k \frac{\partial^2 \Phi_i}{\partial z^2}\bigg|_{P_k} \right) \cos(z,n)_k + \sum_{i=1}^{N} \chi_i \left(2 \frac{\partial \Phi_i}{\partial y}\bigg|_{P_k} + \right.$$

$$\left. + x_k \frac{\partial^2 \Phi_i}{\partial x \partial y}\bigg|_{P_k} + y_k \frac{\partial^2 \Phi_i}{\partial y^2}\bigg|_{P_k} + z_k \frac{\partial^2 \Phi_i}{\partial y \partial z}\bigg|_{P_k} \right) \cos(y,n)_k \right] -$$

$$- \frac{z_k \cos(z,n)_k + y_k \cos(y,n)_k}{1-\nu} \cdot \sum_{i=1}^{N} \chi_i \left(2 \frac{\partial \Phi_i}{\partial x}\bigg|_{P_k} + \right.$$

$$\left. + x_k \frac{\partial^2 \Phi_i}{\partial x^2}\bigg|_{P_k} + y_k \frac{\partial^2 \Phi_i}{\partial x \partial y}\bigg|_{P_k} + z_k \frac{\partial^2 \Phi_i}{\partial x \partial z}\bigg|_{P_k} \right) \right\},$$

together with analogous expressions for Y_n, Z_n (these are easily obtained by permutation of axes and indices of the ψ's).

If, instead of referring to Cartesian axes, the conditions at P_k on S should be expressed in another reference system (e.g. an "intrinsic" one, the three local orthogonal directions being the perpendicular to S in P_k and two mutually orthogonal axes tangent to S in P_k), then expressions (2.16),(2.17),(2.18) would undergo formal transformations, according to linear combinations whose coefficients are the director cosines of the local triedron (in P_k) with respect to the Cartesian global axes x,y,z.

Obviously, the total number of eqs. (2.16) and (2.18) will be in any case 3 N [notice that (2.16) and (2.18) are mutually and orderly exclusive], so that the mathematical problem is formally closed (*).

3. PARTICULAR SOLUTIONS IN PRESENCE OF THERMAL VARIATIONS

In case we have a thermal distribution

(3.1) $\vartheta(x, y, z)$ known (and possibly time-dependent with the restriction of <u>slow</u> time-wise variations as previously specified), the indefinite equilibrium equations in terms of the vector field ξ, η, ζ can be written down as follows:

(*) <u>Stress</u> are obtained through formulae:

$$\sigma_x = 2G[\chi + x\frac{\partial \chi}{\partial x} + y\frac{\partial \chi}{\partial y} + z\frac{\partial \chi}{\partial z} + \frac{x}{1-\nu}(2\frac{\partial \chi}{\partial x} + x\frac{\partial^2 \chi}{\partial x^2} + y\frac{\partial^2 \chi}{\partial x \partial y} + z\frac{\partial^2 \chi}{\partial x \partial z}) + \frac{\partial \psi_1}{\partial x}]$$

and analogous ones for σ_y, σ_z;

$$\tau_{xy} = G[\frac{\partial \psi_1}{\partial y} + \frac{\partial \psi_2}{\partial x} + \frac{x}{1-\nu}(2\frac{\partial \chi}{\partial y} + x\frac{\partial^2 \chi}{\partial x \partial y} + y\frac{\partial^2 \chi}{\partial y^2} + z\frac{\partial^2 \chi}{\partial y \partial z}) + \frac{y}{1-\nu}(2\frac{\partial \chi}{\partial x} + x\frac{\partial^2 \chi}{\partial x^2} + y\frac{\partial^2 \chi}{\partial x \partial y} + z\frac{\partial^2 \chi}{\partial x \partial z})],$$

and analogous ones for τ_{yz}, τ_{xz}, where account is taken of (2.6).

$$(3.2)\begin{cases} 2\dfrac{1-\nu}{1-2\nu}\dfrac{\partial}{\partial x}(\dfrac{\partial\xi}{\partial x}+\dfrac{\partial\eta}{\partial y}+\dfrac{\partial\zeta}{\partial z}) - \dfrac{\partial}{\partial y}(\dfrac{\partial\eta}{\partial x}-\dfrac{\partial\xi}{\partial y}) + \dfrac{\partial}{\partial z}(\dfrac{\partial\xi}{\partial z}-\dfrac{\partial\zeta}{\partial x}) - \\[4pt] \quad - 2\dfrac{1+\nu}{1-2\nu}\alpha\dfrac{\partial\vartheta}{\partial x} = 0 \\[8pt] 2\dfrac{1-\nu}{1-2\nu}\dfrac{\partial}{\partial y}(\dfrac{\partial\xi}{\partial x}+\dfrac{\partial\eta}{\partial y}+\dfrac{\partial\zeta}{\partial z}) - \dfrac{\partial}{\partial z}(\dfrac{\partial\zeta}{\partial y}-\dfrac{\partial\eta}{\partial z}) + \dfrac{\partial}{\partial x}(\dfrac{\partial\eta}{\partial x}-\dfrac{\partial\xi}{\partial y}) - \\[4pt] \quad -2\dfrac{1+\nu}{1-2\nu}\alpha\dfrac{\partial\vartheta}{\partial y} = 0 \\[8pt] 2\dfrac{1-\nu}{1-2\nu}\dfrac{\partial}{\partial z}(\dfrac{\partial\xi}{\partial x}+\dfrac{\partial\eta}{\partial y}+\dfrac{\partial\zeta}{\partial z}) - \dfrac{\partial}{\partial x}(\dfrac{\partial\xi}{\partial z}-\dfrac{\partial\eta}{\partial x}) + \dfrac{\partial}{\partial y}(\dfrac{\partial\zeta}{\partial y}-\dfrac{\partial\eta}{\partial z}) - \\[4pt] \quad - 2\dfrac{1+\nu}{1-2\nu}\alpha\dfrac{\partial\vartheta}{\partial z} = 0 \end{cases}$$

We propose now to look for <u>particular solutions</u> of (3.2) in some remarkable cases, to wit:

A) when the distribution $\vartheta(x, y, z)$ is time-independent:

$$(3.3)\qquad \nabla^2\vartheta = \dfrac{\partial^2\vartheta}{\partial x^2} + \dfrac{\partial^2\vartheta}{\partial y^2} + \dfrac{\partial^2\vartheta}{\partial z^2} = 0 ; \qquad \text{(steady-state).}$$

If the boundary conditions are given by prescribing ϑ on nodes P_k of S, then the solution of (3.3) inside Ω can be written down explicitly using the "unit basic solutions" Φ_i (2.3, 2.4, 2.5):

$$\vartheta = \sum_{k=1}^{N} \vartheta_k \cdot \Phi_k (x, y, z).$$

B) when the distribution $\vartheta(x, y, z, t)$ corresponds to the mth "natural cooling mode":

$$(3.4)\qquad a\nabla^2\vartheta = - K_m \vartheta \qquad (a = \text{thermal diffusivity,}$$

$$\dfrac{1}{K_m} = \text{time-constant of the } m^{th} \text{ N.C.M.})$$

C) when the distribution $\vartheta(x, y, z, t)$ is periodic in time with period $T = \dfrac{2\pi}{\omega}$:

(3.5) $$\begin{cases} a \nabla^2 \vartheta = \dfrac{\partial \vartheta}{\partial t}, \quad \text{with } \vartheta = \vartheta_S \sin \omega t + \vartheta_C \cos \omega t, \\ \vartheta_S = \vartheta_S(x, y, z) \text{ and } \vartheta_C = \vartheta_C(x, y, z) \end{cases}$$

Obviously the boundary conditions (prescribed temperatures on S, e.g.) too are in this case sinusoidal in time with period T.

Fanelli and Giuseppetti in [2] ("The Computation of Time-dependent Thermal Structural Effects ...", on "Numerical Methods in Heat Transfer, Vol. II, J. Wiley & Sons, 1983) showed that these three remarkable cases can generate, by suitable linear superposition, any practical case of interest.

We will now therefore look for particular solutions of the thermoelastic equations (3.2) with ϑ given either by A), B) or C); it can be shown that these particular solutions give a system of surface traction which is self-equilibrated.

Ad A) - Let us build, starting from the scalar field $\vartheta(x,y,z)$ satisfying (3.3), the vectorial field thus defined:

(3.6) $$\begin{cases} \vec{q} = (q_x, q_y, q_z), \\ \operatorname{curl} \vec{q} = \left(\dfrac{\partial q_z}{\partial y} - \dfrac{\partial q_y}{\partial z}, \dfrac{\partial q_x}{\partial z} - \dfrac{\partial q_z}{\partial x}, \dfrac{\partial q_y}{\partial x} - \dfrac{\partial q_x}{\partial y} \right) = \\ \qquad = \operatorname{grad} \vartheta = \left(\dfrac{\partial \vartheta}{\partial x}, \dfrac{\partial \vartheta}{\partial y}, \dfrac{\partial \vartheta}{\partial z} \right) \ (*), \\ \operatorname{div} \vec{q} = \dfrac{\partial q_x}{\partial x} + \dfrac{\partial q_y}{\partial y} + \dfrac{\partial q_z}{\partial z} = 0. \end{cases}$$

(*) Curl $\vec{q}$ can be identified, apart from the sign and from a factor containing the thermal conductibility coefficient, with the vector "heat flow density" associated with the scalar field $\vartheta(x, y, z)$; $\vec{q}$ is then the "vector potential" of such flow density.

It can be shown that $\nabla^2 q_x = \nabla^2 q_y = \nabla^2 q_z = 0$. By using the functions $\Phi_i(x, y, z)$, see (2.4), let us pose

$$(*) \begin{cases} q_x = \sum_{i=1}^{N} q_{xi} \cdot \Phi_i, \\ q_y = \sum_{i=1}^{N} q_{yi} \cdot \Phi_i, \\ q_z = \sum_{i=1}^{N} q_{zi} \cdot \Phi_i, \end{cases}$$

and let us impose the relationships:

$$\frac{\partial q_z}{\partial y} - \frac{\partial q_y}{\partial z} = \frac{\partial \vartheta}{\partial x} = \sum_{i=1}^{N} \vartheta_i \frac{\partial \Phi_i}{\partial x}\bigg|_{P_k} = \sum_{i=1}^{N} \left(q_{zi} \frac{\partial \Phi_i}{\partial y}\bigg|_{P_k} - q_{yi} \frac{\partial \Phi_i}{\partial z}\bigg|_{P_k} \right)$$

$$\frac{\partial q_x}{\partial z} - \frac{\partial q_z}{\partial x} = \frac{\partial \vartheta}{\partial y} = \sum_{i=1}^{N} \vartheta_i \frac{\partial \Phi_i}{\partial y}\bigg|_{P_k} = \ldots\ldots$$

$$\frac{\partial q_y}{\partial x} - \frac{\partial q_x}{\partial y} = \frac{\partial \vartheta}{\partial z} = \sum_{i=1}^{N} \vartheta_i \frac{\partial \Phi_i}{\partial z}\bigg|_{P_k} = \ldots\ldots$$

at every node P_k of surface S. These relationships allow to determine q_{xi}, q_{yi}, q_{zi} at each node P_i of S and hence, through (*), over all of Ω.

By posing now:

$$(3.7) \begin{cases} \xi_b = \dfrac{2(1+\nu)}{7-8\nu}\,(x\vartheta + y q_z - z q_y), \\ \eta_b = \dfrac{2(1+\nu)\,a}{7-8\nu}\,(y\vartheta + z q_x - x q_z), \\ \zeta_b = \dfrac{2(1+\nu)\,a}{7-8\nu}\,(z\vartheta + x q_y - y q_x), \end{cases}$$

The elastic equilibrium eqs. (3.2) are identically satisfied. This (3.7) is thus the particular solution sought after.

Obviously, this particular solution will originate displacements and/or "surface tractions" not corresponding to prescribed boundary conditions. For instance, surface tractions corresponding to (3.7) are given by:

$$(3.8) \quad \sigma_x = \frac{2 E a}{7 - 8\nu} \left(x \frac{\partial \vartheta}{\partial x} + y \frac{\partial q_z}{\partial x} - z \frac{\partial q_y}{\partial x} - \frac{5}{2} \vartheta \right),$$

and analogous ones for σ_y, σ_z;

$$(3.9) \quad \tau_{xy} = \frac{E a}{7 - 8\nu} \left[x \left(\frac{\partial \vartheta}{\partial y} - \frac{\partial q_z}{\partial x} \right) + y \left(\frac{\partial q_z}{\partial y} + \frac{\partial \vartheta}{\partial x} \right) + z \left(\frac{\partial q_x}{\partial x} - \frac{\partial q_y}{\partial y} \right) \right],$$

and analogous ones for τ_{yz}, τ_{xz}; from these "surface tractions" are derived according to the well-known Cauchy relationships.

Ad B) - By posing:

$$(3.10) \begin{cases} \xi_b = \dfrac{1+\nu}{1-\nu} \dfrac{a}{M_m} \dfrac{\partial \vartheta_m}{\partial x}, & M_m = -\dfrac{k_m}{a}, \\[1em] \eta_b = \dfrac{1+\nu}{1-\nu} \dfrac{a}{M_m} \dfrac{\partial \vartheta_m}{\partial y}, \\[1em] \zeta_b = \dfrac{1+\nu}{1-\nu} \dfrac{a}{M_m} \dfrac{\partial \vartheta_m}{\partial z}, \end{cases}$$

and recalling that $\nabla^2 \vartheta_m = M_m \vartheta_m$, eqs. (3.2) are identically satisfied.

In this case as well, displacements and/or tractions" will be originated that will not correspond to prescribed boundary conditions, in particular:

$$(3.11) \quad \sigma_x = \frac{E a}{M_m} \frac{1}{1-\nu} \left[\frac{\partial^2 \vartheta_m}{\partial x^2} - M_m \vartheta_m \right], \quad = - \frac{E a}{M_m (1-\nu)} \left(\frac{\partial^2 \vartheta_m}{\partial y^2} + \frac{\partial^2 \vartheta_m}{\partial z^2} \right),$$

and analogous ones for σ_y, σ_z;

(3.12) $\quad \tau_{xy} = \dfrac{E\,\alpha}{(1-\nu)\,M_m}\,\dfrac{\partial^2 \vartheta_m}{\partial x\,\partial y}$, and analogous ones for τ_{yz}, τ_{xz}, from which the "surface tractions" X_{nb}, Y_{nb}, Z_{nb} through the Cauchy relationships.

In this case the Maxwell stress functions: $A_1(x, y, z)$, $A_2(x, y, z)$, $A_3(x, y, z)$ such that (LOVE, op. cit., Chapt. II):

$$(3.13)\begin{cases} \sigma_x = \dfrac{\partial^2 A_3}{\partial y^2} + \dfrac{\partial^2 A_2}{\partial z^2}, \quad \sigma_y = \dfrac{\partial^2 A_1}{\partial z^2} + \dfrac{\partial^2 A_3}{\partial x^2}, \quad \sigma_z = \dfrac{\partial^2 A_2}{\partial x^2} + \dfrac{\partial^2 A_1}{\partial y^2} \\[6pt] \tau_{xy} = -\dfrac{\partial^2 A_3}{\partial x\,\partial y}, \quad \tau_{yz} = -\dfrac{\partial^2 A_1}{\partial y\,\partial z}, \quad \tau_{xz} = -\dfrac{\partial^2 A_2}{\partial x\,\partial z}, \end{cases}$$

are simply:

$$(3.14) \quad A_1 = A_2 = A_3 = -\dfrac{E\,\alpha}{(1-\nu)\,M_m}\,\vartheta_m .$$

NOTE: If all that is of interest are the <u>variations</u> of stress--strain effects occurring during the cooling (or heating) process, with respect to an origin situation in which $\vartheta(x, y, z, t=0) = \sum_m \beta_m \vartheta_{m0}$ [in this case for $t > 0$ $\vartheta(x, y, z, t) = \sum_m \beta_m \vartheta_{m0}\, e^{-K_m t}$, it is necessary to use, in place of (3.10):

$$\xi_b = \dfrac{1+\nu}{1-\nu}\,\dfrac{\alpha}{M_m}\,\left(\dfrac{\partial \vartheta_m}{\partial x} - \dfrac{\partial \vartheta_{m0}}{\partial x}\right)$$

and analogous ones; in place of (3.11), (3.12)

le $\quad \sigma_x = -\dfrac{E\,\alpha}{(1-\nu)M_m}\left(\dfrac{\partial^2 \vartheta_m}{\partial y^2} - \dfrac{\partial^2 \vartheta_{m0}}{\partial y^2} + \dfrac{\partial^2 \vartheta_m}{\partial z^2} - \dfrac{\partial^2 \vartheta_{m0}}{\partial z^2}\right)\ldots\quad$ and

$\tau_{xy} = \dfrac{E\,\alpha}{(1-\nu)M_m}\left(\dfrac{\partial^2 \vartheta_m}{\partial x\,\partial y} - \dfrac{\partial^2 \vartheta_{m0}}{\partial x\,\partial y}\right)\ldots;\quad$ in place of (3.14), lastly

$$A_1 = A_2 = A_3 = -\dfrac{E\,\alpha}{(1-\nu)M_m}\,(\vartheta_m - \vartheta_{m0}).$$

Every effect will then be given by a sum (with respect to index m) of products containing the factor $(e^{-K_m t} - 1)$ and factors function only of (x, y, z).

Ad C) — Posing:

$$(3.15) \begin{cases} \xi_b = \dfrac{1+\nu}{1-\nu} \dfrac{a\,a}{\omega} \left(\dfrac{\partial \vartheta_c}{\partial x} \cdot \sin\omega t - \dfrac{\partial \vartheta_s}{\partial z} \cos\omega t\right), \\[6pt] \eta_b = \dfrac{1+\nu}{1-\nu} \dfrac{a\,a}{\omega} \left(\dfrac{\partial \vartheta_c}{\partial y} \cdot \sin\omega t - \dfrac{\partial \vartheta_s}{\partial y} \cos\omega t\right), \\[6pt] \zeta_b = \dfrac{1+\nu}{1-\nu} \dfrac{a\,a}{\omega} \left(\dfrac{\partial \vartheta_c}{\partial z} \cdot \sin\omega t - \dfrac{\partial \vartheta_s}{\partial z} \cos\omega t\right), \end{cases}$$

and, recalling that $\vartheta = \vartheta_S \sin\omega t + \vartheta_c \cos\omega t$, with $\vartheta_S = \vartheta_S(x,y,z)$, $\vartheta_c = \vartheta_c(x,y,z)$ determined by the boundary conditions; also that

$$\nabla^2 \vartheta_S = -\dfrac{\omega}{a}\vartheta_c, \qquad \nabla^2 \vartheta_c = \dfrac{\omega}{a}\vartheta_S,$$

one sees that (3.2) are identically satisfied.

It is obvious that also this particular solution will give rise to displacements and/or "surface tractions" generally incompatible with prescribed boundary conditions; for the surface tractions corresponding to (3.15) one gets:

$$(3.16) \begin{cases} \sigma_x = \dfrac{E\,a\,a}{(1-\nu)\omega}\left[\left(\dfrac{\partial^2 \vartheta_c}{\partial x^2} - \dfrac{\omega}{a}\vartheta_S\right)\sin\omega t - \left(\dfrac{\partial^2 \vartheta_S}{\partial x^2} + \dfrac{\omega}{a}\vartheta_c\right)\cos\omega t\right] = \\[8pt] = \dfrac{E\,a\,a}{(1-\nu)\omega}\left[-\left(\dfrac{\partial^2 \vartheta_c}{\partial y^2} + \dfrac{\partial^2 \vartheta_c}{\partial z^2}\right)\sin\omega t + \left(\dfrac{\partial^2 \vartheta_S}{\partial y^2} + \dfrac{\partial^2 \vartheta_S}{\partial z^2}\right)\cos\omega t\right], \end{cases}$$

and analogous ones for σ_y, σ_z;

$$(3.17) \quad \tau_{xy} = \dfrac{E\,a\,a}{(1-\nu)\omega}\left[\dfrac{\partial^2 \vartheta_c}{\partial x\,\partial y}\sin\omega t - \dfrac{\partial^2 \vartheta_S}{\partial x\,\partial y}\cos\omega t\right],$$

and analogous ones for τ_{yz}, τ_{xz}, from which surface tractions X_{nb}, Y_{nb}, Z_{nb} according to Cauchy.

In this case the Maxwell stress functions $A_1(x,y,z)$, $A_2(x,y,z)$, $A_3(x,y,z)$ are simply:

$$(3.18) \quad A_1 = A_2 = A_3 = -\dfrac{E\,a\,a}{(1-\nu)\omega}(\vartheta_c \sin\omega t - \vartheta_S \cos\omega t).$$

4. BUILDING UP THE DESIRED SOLUTION OF THE THERMOELASTIC PROBLEM AS A LINEAR COMBINATION OF PARTICULAR SOLUTIONS DEFINED SUB 3) AND OF GENERAL SOLUTIONS WITH NIL BODY FORCES

Let now $\bar{\xi}_b, \bar{\eta}_b, \bar{\zeta}_b; \bar{\sigma}_{xb}, \bar{\sigma}_{yb}, \bar{\sigma}_{zb}, \bar{\tau}_{xyb}, \bar{\tau}_{yzb}, \bar{\tau}_{xzb}; \bar{X}_{nb}, \bar{Y}_{nb}, \bar{Z}_{nb}$ be the displacements, stresses, surface tractions generated by superposition of particular solutions of (3.2); if $\bar{\bar{\xi}}_k \ldots \bar{\bar{Z}}_{nk}$ are the corresponding quantities prescribed on the generic node P_k of surface S in the particular problem in hand (*), we will have to look for a solution of the type treated under 2) by imposing at every node P_k of S:

$$(4.1) \quad \left. \begin{array}{l} \xi_{ak} = \bar{\bar{\xi}}_k - \bar{\xi}_{bk} \\ \eta_{ak} = \bar{\bar{\eta}}_k - \bar{\eta}_{bk} \\ \zeta_{ak} = \bar{\bar{\zeta}}_k - \bar{\zeta}_{bk} \end{array} \right\} \quad \text{or} \quad \left\{ \begin{array}{l} X_{ank} = \bar{\bar{X}}_{nk} - \bar{X}_{nbk} \\ Y_{ank} = \bar{\bar{Y}}_{nk} - \bar{Y}_{nbk} \\ Z_{ank} = \bar{\bar{Z}}_{nk} - \bar{Z}_{nbk} \end{array} \right.$$

to be introduced in eqs. (2.16), (2.18) respectively; those, together with the condition eqs. (2.10), formally close the mathematical problem.

This solution of course will be added to the linear combination of particular solutions alluded to at the beginning of this paragraph.

With this, a formal solution of the general thermoelastic problem has been achieved by using only the F.E. formulation of problems connected with the ∇^2 operator; algebraic systems stemming from this discretized formulation either contain only one variable per node (if extended to all the nodes of Ω), or relate only to surface nodes P_k (S); in this last event, they are at most of the order 4 N if $1 \leq k \leq N$. The only matrices to be used are those usually called "conductivity matrix" and "thermal capacity matrix".

The method above outlined could be convenient for domains Ω of simple shape and containing a few thousands of discretization nodes.

(*) Note that, if $\bar{\bar{X}}_{nk}, \bar{\bar{Y}}_{nk}, \bar{\bar{Z}}_{nk}$ are prescribed over all of S, they must from an equilibrated system of forces.

BIBLIOGRAPHIC LIST

1. FANELLI, M. - Thermal stresses in Plane Elasticity: Numerical Solutions based only on Harmonic Functions, Meccanica **4**, (1986)

2. FANELLI, M., GIUSEPPETTI, G. - The Computation of Time--dependent Thermal Structural Effects by Means of Influence Functions Pertaining to Unit Temperature Distributions: A Unified Approach Force from Restrictive Hypotheses, Numerical Methods in Heat Transfer, Edited by R.W. Lewis and K. Morgan, B.A. Schrefler, J. Wiley & Sons, (1983)

3. BORCHARDT, C.W. - Ges. Weske, Berlin (1988), p. 245

4. CERRUTI V., Roma, Accad. Linc. Rend. (Ser. 4), t. 2, (1886)

5. LAMB, H. - Hydrodynamics, Cap. XI

6. LOVE, A.E.H. - A Treatise on the Mathematical Theory of Elasticity, (1926), cpp. II, I, etc.

7. TEDONE - Amm. di Mat. (ser. 3), t. 8, (1903), p. 129; (Ser. 3), t. 10, 1904, p. 13 Roma, Accad. Linc. Rend. (Ser. 5), t. 14, (1905), pp. 76 e 316.

8. BELLUZZI O. - Scienza delle Costruzioni -- Ed. Zanichelli, Bologna, (1951), Cap. XXV

9. BOLEY, B., WEINER, J. - Theory of Thermal Stresses, J. Wiley & Sons, (1966)

10. HETNARSKI, R.B. - Thermal stresses I - North Holland, (1986)

PLATE-CYLINDER INTERSECTIONS

SUBJECTED TO THERMAL LOADS

D. Lapointe, C.A. Laberge, R. Baldur

Ecole Polytechnique

Montréal, Québec, Canada

ABSTRACT

The present paper is an extension of work on stresses in corner radii already described by the authors in previously published reference [1]. Whereas the original study concerned itself with pressure effects only, the report included here deals specifically with thermal loads. As before, the results are limited to inside corner radii between cylinders and flat head closures. Similarly, the analysis is based on a systematic series of finite element calculations with the significant parameters covering the field of useful design boundaries. The 334 elements containing some 1800 degrees of freedom ensure a realistic determination of local stresses and a large number of complete solutions enables the presentation of smooth design curves. The results are condensed into a rapid method for the determination of fatigue dependent stresses in pressure vessels subjected to a significant, variable thermal load. The paper takes into account the influence of the film coefficient, temporal temperature variations, and material properties. A set of coefficients provides a convenient method of stress evaluation suitable for design purposes.

NOMENCLATURE

B_i = Biot number = $\dfrac{hb}{k}$

b = half thickness of the cylinder = $\dfrac{t_s}{2}$

d = thermal diffusivity
E = Young's modulus

f = thermal stress factor correction
h = surface heat transfer coefficient
k = coefficient of thermal conductivity
L = cylinder length

$$N = \text{Fourier modulus} = \frac{d\,\Delta t}{b^2}$$

r = corner radius
R_m = mean cylinder radius
S_t = Thermal Stress Factor (TSF)
T = plate thickness
t_s = cylinder thickness

α = coefficient of thermal expansion

$$\beta = \sqrt[4]{3(1-\nu^2)/t_s^2\,R_m^2}$$

ΔT = temperature change
Δt = time interval for the change ΔT to occur
ν = Poisson's ratio
σ_L = longitudinal (meridional) stress

σ_T = tangential (hoop) stress

σ_{max} = maximum principal stress (absolute value)

Note: in all cases, units are the same than the ones used in the ASME Boiler and Pressure Vessel Code [2].

1. INTRODUCTION

The original work by the authors on stress concentration factors at cylinder-plate junctions was limited to pressure effects [1]. That paper contained a method suitable for design purposes and was concerned mainly with maximum stresses and the stress attenuation along the cylinder which is a determining factor for the hub length requirements.

The present publication reports on the results of the continuation of the study in which the effects of thermal loads were considered. In the vicinity of pressure vessel closures, the thermally induced stresses depend on a large number of parameters, the principal ones being the time-dependent temperature variations, the film coefficient of heat transfer, material properties and the vessel geometry. The latter is taken to consist principally of a cylinder, flat head closure and the transition corner radius. The ASME Boilers & Pressure Vessel Code [2] requires the evaluation of thermally induced stresses for calculations involving fatigue effects. Although the final analysis has to be based on detailed analytical method, the process at the initial design stage could benefit from a less time-consuming technique. The work by Heisler [3] deals with slabs and cylinders individually, and Lapointe [4] treats the cylinder plate junctions under thermal shock only. Neither of the above references is suitable for quick design estimates.

The method presented in this paper is intended specifically to serve as a tool for design purposes. It gives quick and realistic estimates of maximum stresses to be expected near cylinder head closures. The authors were not able to condense the results into a single set of parameters which would take into account all the parameters mentioned previously. The method consists of two stages. In the first, a somewhat conservative result is obtained by assuming a "standard" set of conditions of a thermal shock and an infinite film coefficient (surface temperature equal to fluid temperature). If further refinements are required, correction factors (all less than unity) can be evaluated in order to obtain a more realistic result including the temporal temperature variations, material properties and the film coefficient.

In the case of thermal stresses the term "stress concentration factor" no longer applies. Instead, results are presented in the form of "thermal stress factors". The latter express maximum stresses relative to the stress developed in the cylindrical portion of the vessel if it were subjected to a linear through-thickness temperature variation.

The method of analysis is based on the finite element technique. Each geometry is represented by a mesh of 334 triangular, axisymmetric refined elements with each node having 9 degrees of freedom [5]. These elements ensure that all surface stresses are obtained directly without extrapolation; an important feature for stress concentration work. The distribution of the nodes across the cylinder thickness is logarithmic for a good representation of the temperature distribution. Some 250 complete solutions were obtained to determine the effects of the different sets of parameters. Each solution contained the results for the temperature, displacement and all stress components of the 334 elements (211 nodes) at all time intervals from the start of the transient up to a steady state condition. It is estimated that a total of 1.5 million stresses were considered in order to produce the results contained in this paper.

2. GEOMETRY

Fig.1 shows the basic geometry and the main parameters of the cylinder-plate intersection. The straight portion of the cylinder was made at least equal to the attenuation length suggested by the ASME Code [2] or other references [6] and given by:

$$L = \pi / \beta \quad \text{where} \quad \beta = \sqrt[4]{\frac{3(1-\nu^2)}{t_s^2 R_m^2}}$$

which, for steel, with Poisson's ratio of 0.3 becomes:

$$L = 2.5 \sqrt{t_s R_m}$$

The complete intersection was modelled by means of 334 triangular axisymmetric elements formed by 211 nodes. The element density was increased at the inside corner in order to improve the accuracy of the surface stresses in that region. Results were obtained for a temperature change at the inside surface, i.e. the boundary ABCD of Fig. 1.

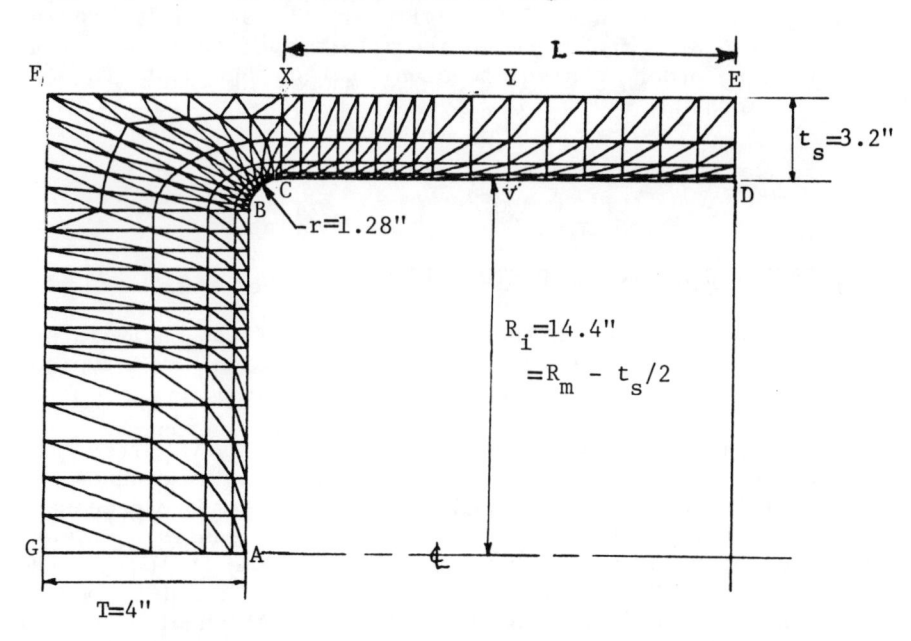

Fig.1 Geometry of a typical case

The elastic boundary conditions applied to the model were:

a) elimination of rotations and radial displacements on the axis of symmetry GA;

b) elimination of rotations at the cylinder end DE in order to simulate an infinite length;

c) elimination of axial displacement at the cylinder end DE in order to apply the correct membrane stresses in the cylinder caused by the radial variation of temperature.

The thermal boundary conditions were as follows:

d) elimination of radial variation of temperature on the axis of symmetry GA;

e) elimination of axial variation of the temperature at the cylinder end DE;

f) elimination of the radial variation of the temperature on the outside surface of the cylinder (EF) and of the axial variation of the temperature on the outside surface of the plate (FG) in order to simulate a thermal insulation of the outside surface;

g) elimination of the axial variation of the temperature on the inside surface of the cylinder (CD) and of the radial variation of temperature on the inside surface of the plate (AB) in order to simulate a uniform surface heat transfer conditions.

3. STRESS VARIATION AND DISTRIBUTION

The manner in which the final results of the analysis are presented was decided after the examination of typical results. The important aspects were as follows.

The maximum stresses occur some time after the application of the thermal load. The actual delay depends on the geometry and other factors and it was therefore necessary to ensure that the transient finite element analysis always extended over a sufficient period of time to include the critical case. Fig. 2 shows a typical case taken at a node on the inside corner radius after the application of a thermal shock (time 0 on the graph).

Fig. 3 shows a typical spacial variation of stresses at the instant of time when maximum conditions occur. The graphs indicate that the critical conditions take place (typically) at the inside corner radius and that the longitudinal stresses predominate. It is further shown, that, away from the corner, both longitudinal and tangential stresses attenuate to roughly the same values.

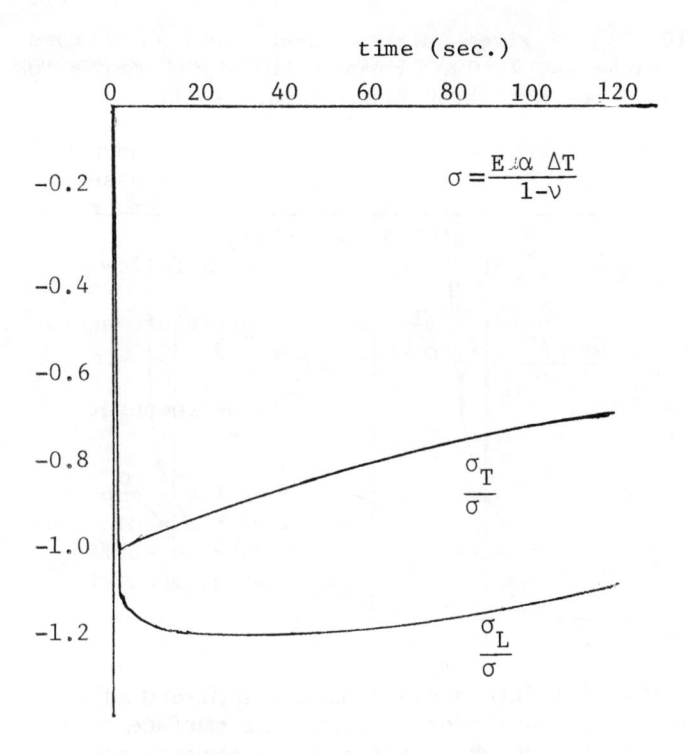

Fig. 2 Stress variation with time at the node where maximum stress occurs for a typical case

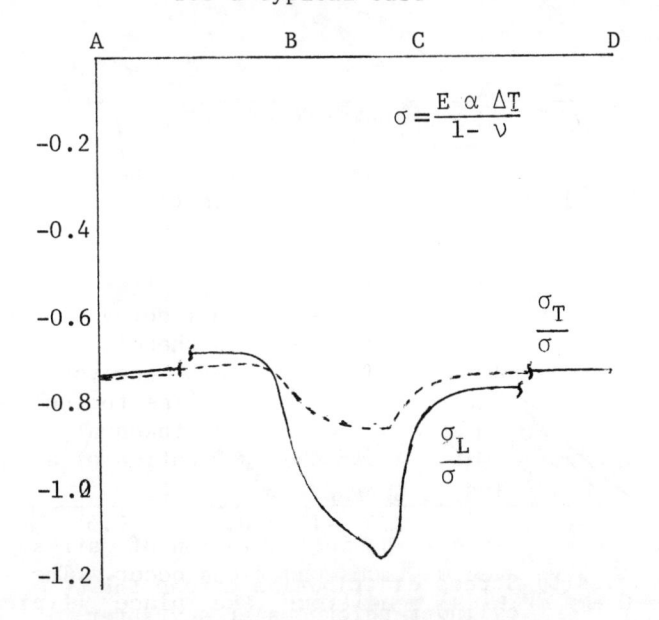

$$\sigma = \frac{E \alpha \Delta T}{1-\nu}$$

$\frac{\sigma_T}{\sigma}$

$\frac{\sigma_L}{\sigma}$

Fig. 3 Surface stress distribution at the time at which the maximum stress occurs for a typical case

Finally, Fig. 4 gives stress distributions across the cylinder thickness. The through-thickness variation is non-linear and does not vary in character as the stress levels change with time.

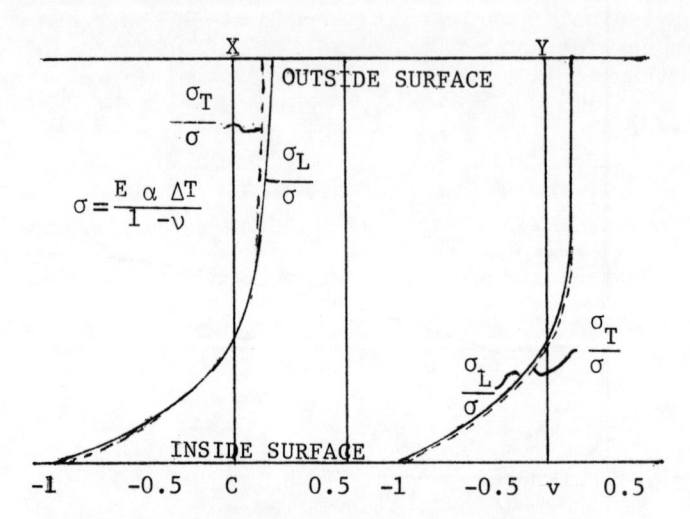

Fig. 4(a) Stress distribution across the cylinder thickness at the time at which the maximum stress occurs for a typical case

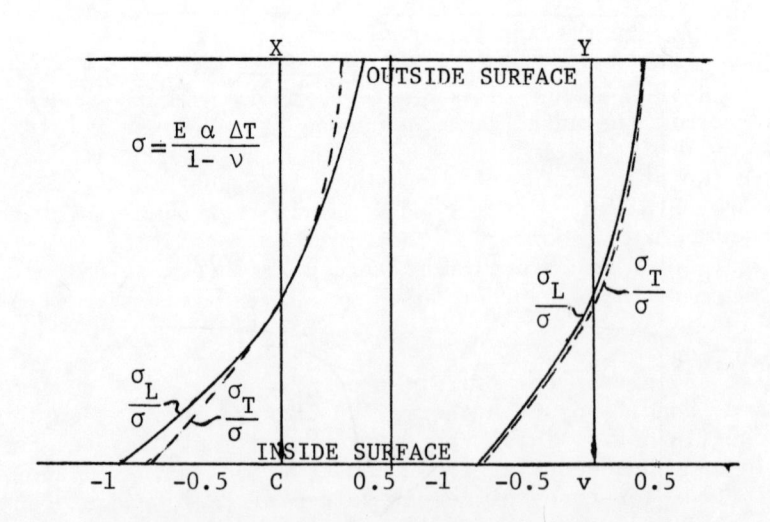

Fig. 4(b) Stress distribution across the cylinder thickness at a greater time than in Fig. 4(a) for the same typical case

On the basis of the foregoing it was decided to simplify the presentation of the results by considering only the maximum stresses at the surface location and at the instant that that value occured. In all cases this corresponded to the longitudinal (meridional) stress at the corner radius on the cylinder side and occured some time after the initiation of the transient. From the design point of view the exact location and time are not significant and this information is not contained in the paper. Furthermore, the stress distributions across the body of the vessel are omitted from the results. These are presented in a non-dimensional form relative to a cylinder stress subjected to a linear temperature gradient. The coefficient used in the paper, the Thermal Stress Factor (TSF) is defined as:

$$S_t = \frac{\sigma \max}{\sigma}$$

where σ_{max} = maximum absolute value of the principal stress anywhere in the body at any time after the thermal load is applied.

$\sigma = \frac{E \alpha \Delta T}{1-\nu}$ = maximum thermal stress in a cylinder based on a linear variation of the temperature across its thickness.

ΔT = total variation of temperature on the inside surface during the transient.

4. ANALYTICAL RESULTS AND PROPOSED METHOD

The large number of parameters involved in the analysis did not permit the authors to condense the results into a single set of design graphs. Instead, the present method is based on two stages. In the first one, somewhat conservative values are obtained. The second stage is then used if further refinements are required. The latter consists of the determination of correction factors which are applied to the first stage results in order to obtain a more realistic stress evaluation.

Fig. 5 gives the first stage results obtained for "standard" conditions of a thermal shock with infinite surface film coefficient. This represents the most severe case possible and consequently gives conservative values. Graphs are plotted for three different values of R_m/t_s and a double interpolation is required for a specific geometry. However, it was found that for given values of r/t_s and t_s/T, the TSF varies almost linearly with R_m/t_s. Interpolation between the three sets of graphs should therefore be quite reliable. The range of the parameters covers most of the design proportions found in practice.

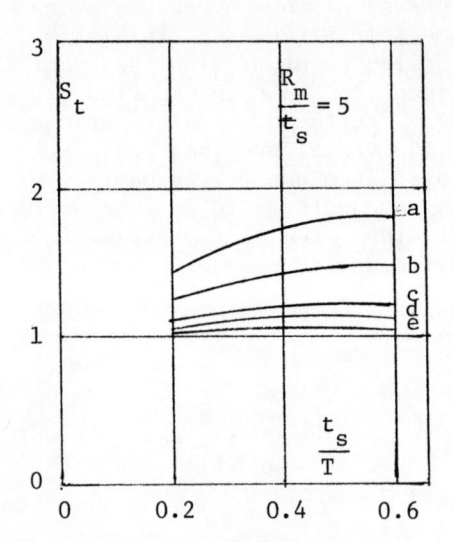

Fig. 5(a) Thermal Stress Factors

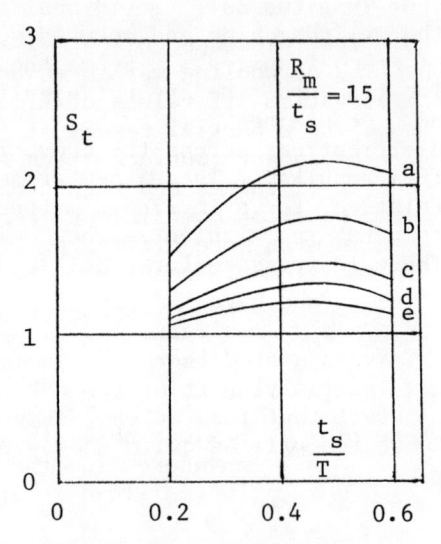

Fig. 5(b) Thermal Stress Factors

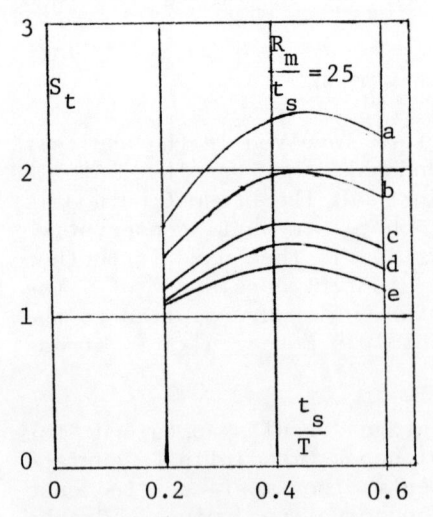

Fig. 5(c) Thermal Stress Factors

Legend for Fig. 5
a : $\dfrac{r}{t_s} = 0.1$
b : $\dfrac{r}{t_s} = 0.2$
c : $\dfrac{r}{t_s} = 0.4$
d : $\dfrac{r}{t_s} = 0.6$
e : $\dfrac{r}{t_s} = 1.0$

The calculations have shown that the TSF is independent of material properties provided the conditions of thermal shock and infinite film coefficient are respected. The values obtained from Fig.5 can therefore be used for preliminary design purposes without further refinements.

The presence of a large film coefficient and a slow heating or cooling may in some cases result in considerable conservatism if values of Fig. 5 are used. Fig. 6 provides a means for a partial remedy of the situation. The effects of time-dependent surface temperature variations give a wide scatter of values and when combined with the influence of the surface film coefficient render the condensation of the results even more difficult. The authors have used great many finite element results with different combinations of those factors and have finally arrived at a graph which they feel is reasonably realistic and always conservative. As a refinement it is suggested therefore to improve the thermal stress factor by multiplying it by the correction obtained from Fig. 6. It is felt that for quick design calculations the values so obtained will be quite adequate.

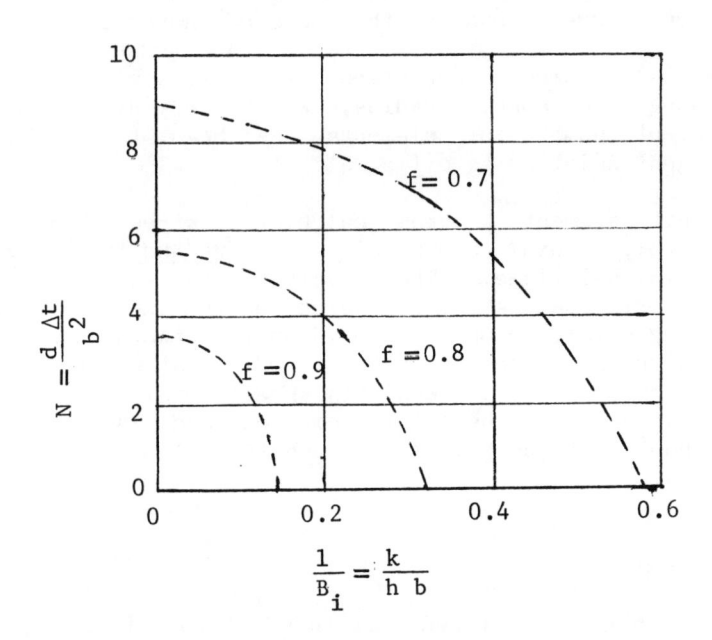

Fig. 6 Thermal stress factor correction
.due to the presence of a film
coefficient and a temperature-time
variation

5. DISCUSSION OF RESULTS

The results contained in the paper deal with a problem which is difficult to confirm experimentally. A test which would simulate the conditions of a thermal shock would require large quantities of heat at the surface of a plate-cylinder intersection. Furthermore, stresses would have to be measured at the surface of the corner radius and, unless large vessel dimensions were used, the concentrated nature of the maximum stress area would make the measurements difficult and probably unreliable. In view of the above comments it is not surprising that the authors have not been able to find any reference in the experimental literature on the subject matter being treated. It appears that, at least at present, the theoretical results provide the only source of information.

The authors feel that the data presented herein is valid and have come to this conclusion on the basis of the following observations:

(a) Based on previous experience, the finite elements give excellent results for the stresses at the center of the plate and at the cylinder away from the corner radii where theorical values can be obtained from the distribution of the temperature across the thickness of the cylinder.

(b) There is a smooth variation of stresses at all points including the corner radius, which indicates that a sufficient number of elements has been taken for a convergent solution (see fig. 3).

(c) The finite element program calculates stresses in the tangential, radial and axial directions and the shear in the meridional plane. These values are then used to derive principal stresses which, on the inside surface, have tangential, normal and meridional orientations. Of the foregoing stresses, the normal must be zero. The results obtained by the analysis showed consistently very small values for the latter component and, in all cases, less than 1% of the calculated longitudinal and tangential stresses.

6. CONCLUSIONS

The thermal stress factors for plate-cylinder intersections contained in this paper cover the most common range of pressure vessels. The suggested design procedure comprises two stages. In the first, a linear double interpolation produces conservative stress values on the assumption of a thermal shock and an infinite film coefficient. A further, also conservative, correction factor can then be determined in order to allow for the time-dependent temperature variations and a surface heat transfer film coefficient.

The paper shows that the effect of the corner radius depends on many parameters and that the practice of its determination on the basis of cylinder thickness alone is insufficient and, in some cases may lead to an unconservative design.

This paper concludes the present phase of studies on corner radii. The results contained herein, combined with those given in reference [1] will provide complete design information for any operating conditions to which a pressure vessel may be subjected.

REFERENCES

1. LABERGE, C.A., and BALDUR, R., "The Effect of Corner Radius on Plate-Cylinder Intersections", Journal of Pressure Vessel Technology, Volume 102, pp. 79-83, February 1980.

2. ASME Boiler & Pressure Vessel Code

3. HEISLER, M.P., "Transient Thermal Stresses in Slabs and Circular Pressure Vessels", Journal of Applied Mechanics, 1953.

4. LAPOINTE, D., "Concentration de contraintes dans le congé à l'intersection d'une plaque et d'un cylindre sous l'effet d'un choc thermique", Ecole Polytechnique, Montréal, 1984.

5. LAGUE, G., and BALDUR, R., "An Axisymmetric Finite Element in Natural Coordinates", ASME Paper No. 75-WA/PVP-24, 1975.

6. HARVEY, J.F., "Theory and Design of Modern Pressure Vessels", Van Nostrand Reinhold Co., 1974.

CALCULATION OF THERMAL DEFORMATION OF A PISTON AND CYLINDER OF A LOW-HEAT-REJECTION ENGINE

Zissimos P. Mourelatos

Fluid Mechanics Department
General Motors Research Laboratories
Warren, Michigan 48090-9055, USA

ABSTRACT

A method is presented for calculating the thermal deformations of the piston and cylinder of a Low-Heat-Rejection (LHR) engine very accurately. First, the temperature field in the piston and cylinder is calculated and then the MSC/NASTRAN finite-element code is used to find the corresponding thermal deformations.

The analysis accounts for the piston motion and temperature-dependent material properties. The piston motion is simulated by solving the time-dependent heat-conduction problem for a thin layer of the piston crown, cylinder liner, and firedeck. In the remaining piston and cylinder domain, the steady-state heat-conduction problem is solved by the finite-element method.

For the calculation of the temperature field, two independent analyses are performed, depending on whether the heat-transfer coefficients are known or not. When the heat-transfer coefficients are known, they can be time dependent and temperature dependent. When they are unknown, a collocation method which uses an optimization technique is used to calculate a set of equivalent time-dependent heat-transfer coefficients in order for the corresponding temperature field to match measured temperatures at preselected collocation points.

From a number of case studies, it is concluded that the developed method calculates the thermal deformation of the piston and cylinder very accurately.

1. INTRODUCTION

The piston assembly has long been recognized as a major source of engine mechanical friction. It is an area where significant reduction in engine friction can be achieved through design modifications. Both the piston ring and the piston skirt friction are considerable. For an accurate calculation of the piston skirt friction, an elastohydrodynamic lubrication analysis is needed [1].[1] Such a study requires an accurate estimation of the piston skirt and cylinder bore thermal expansion. A 250 micron radial expansion of a piston skirt of an automotive engine has been reported [2].

Estimating the thermal deformation of the piston and the cylinder with reasonable accuracy first requires an accurate calculation of the associated temperature field. However, such a calculation is a complicated problem because of:

1. the piston motion,
2. the temperature-dependent material properties,
3. the inability to accurately estimate the heat-transfer coefficients, and
4. the inability to accurately estimate the radiation heat flux.

In this paper, a method is presented to calculate accurately the thermal deformations of a piston-connecting rod and cylinder cap assemblies of a LHR engine. The analysis accounts for the piston motion and possible temperature-dependent material properties. Two independent analyses are performed depending on whether the heat-transfer coefficients are known "a priori" or not. In the first case, the heat-transfer coefficients can be time dependent and temperature dependent if their distribution in time and temperature dependency are known. In the second case, when the heat-transfer coefficients are unknown, a collocation method which uses an optimization technique is utilized to calculate the distribution in time of a set of equivalent heat-transfer coefficients in order for the corresponding temperature field to match measured temperatures at preselected collocation points.

2. OVERVIEW OF ANALYSIS APPROACH

The periodic fluctuations in piston position, gas temperature, and heat-transfer coefficients in the combustion chamber have a small effect on metal temperatures. The temperature variation is about 5 to 10°C at 1 mm below the surface of the engine metal [3]. However, the piston motion effect is important in accurately estimating the temperature profile along the piston crown and cylinder liner and it

---

1. Numbers in brackets designate references at end of paper.

becomes more profound when a ceramic coating with low thermal conductivity is used. The piston motion is simulated by solving the time-dependent heat-conduction problem for a thin layer of the piston crown, cylinder liner, and firedeck. An explicit finite-difference scheme in time is used. In the remaining domain of the piston and cylinder, the steady-state heat-conduction problem is solved by the finite-element method. Compatibility of the two models is ensured at the common interface. Temperature-dependent material properties (thermal conductivity, specific heat, Young's modulus, and thermal expansion coefficient) can be used in the analysis.

The heat-transfer coefficients describing the convective boundary conditions are very difficult to estimate accurately. Furthermore, the temperature field is greatly dependent on the heat-transfer coefficients. Every researcher, to our knowledge, has used either a specified time-varying or a specified time-average approximation ([3] and [2], respectively) of these heat-transfer coefficients, known from the beginning. Time-dependent and temperature-dependent heat-transfer coefficients can be used in this analysis when they are known "a priori." The temperature dependency gives a spatial distribution to the heat-transfer coefficients, but it makes the analysis nonlinear. However, the heat-transfer coefficients cannot always be estimated accurately for the purpose of this research. The thermal radial deformation of the piston skirt and cylinder liner must be calculated with an accuracy of a few microns. But the thermal deformations correspond to a temperature field which is very sensitive to the heat-transfer coefficients. Therefore, the heat-transfer coefficients must be estimated or calculated very accurately. To circumvent this difficulty, a collocation method which uses an optimization algorithm was developed. This method calculates the distribution in time of a set of equivalent heat-transfer coefficients in order for the corresponding temperature field to match measured temperatures at preselected collocation points. Since this method calculates the temperature field based on measured temperatures at specific points, it accounts for radiation effects through the calculated equivalent heat-transfer coefficients. The optimization algorithm, having as independent variables a set of time-dependent equivalent heat-transfer coefficients, performs a number of iterations systematically. At each iteration a distribution of each equivalent heat-transfer coefficient is assumed and the corresponding temperature field is calculated taking into account (1) the time dependency introduced by the piston motion, and (2) the nonlinear effect of the temperature dependency of the specific heat and thermal conductivity of the subject materials. The temperatures at a number of points which are spread over the whole cylinder cap-piston-connecting rod domain, are compared with measured temperatures at the same points. If there is not agreement within a tolerance, the optimization search updates the initially assumed equivalent heat-transfer coefficients and proceeds to the next iteration.

Once the temperature field in the piston and cylinder cap assemblies has been calculated, the MSC/NASTRAN finite-element code [4] is used to find the corresponding thermal deformations. The piston skirt and the cylinder bore can be manufactured with an initial profile exactly opposite to the calculated thermal deformation profile at the operating conditions. This will compensate for the expected thermal deformation, eliminating the important thermal effects in controlling the piston skirt-cylinder bore radial clearance.

3. DESCRIPTION OF THE MODEL

The heat-transfer analysis of the piston and cylinder cap is generally a three-dimensional problem. In this study, a single-cylinder engine with a cardanic mechanism, which replaces the usual crank shaft connecting rod arrangement [5], is analyzed. The piston-connecting rod-cylinder cap assembly of this experimental engine is axisymmetric. Therefore, an axisymmetric heat-transfer analysis is conducted which considerably reduces the required computational effort.

The model consists of two submodels; a finite-difference submodel for a thin layer of the cylinder liner, firedeck, and piston crown and a finite-element submodel for the rest of the cylinder and piston-connecting rod. The finite-difference submodel is one-dimensional, time dependent and it accounts for the cyclic piston motion. This motion is important in determining the actual steady-state bulk temperature distribution at the common interface of the two submodels. The finite-element submodel is two-dimensional, axisymmetric, steady-state and it determines the steady state temperature field of the model. The two submodels supply boundary conditions to each other through their common interface.

3.1 Finite-Element Submodel

The temperature field in the piston and cylinder cap is governed by the following axisymmetric heat-conduction boundary-value problem.

$$\frac{\partial}{\partial x_i} [rk_{ij} \frac{\partial T}{\partial x_j}] = 0 \qquad \text{in } \Omega$$

$$T = T_o \quad \text{on } \Gamma_1$$

$$k_{ij} \frac{\partial T}{\partial x_j} \eta_i = q \quad \text{on } \Gamma_2 \qquad (1)$$

$$k_{ij} \frac{\partial T}{\partial x_j} \eta_i = -h(T-T_\infty) \quad \text{on } \Gamma_3,$$

where k_{ij} is the material thermal conductivity ($k_{11}=k_{22}=k$ and $k_{12}=k_{21}=0$), η_i is the unit vector normal to the boundary in the ith direction, T is the temperature field, $x_1=r$ is the axial direction, and $x_2=z$ is the vertical direction.

The boundary Γ of the domain Ω consists of three disjoint segments Γ_1, Γ_2, and Γ_3. A constant temperature, T_o, is specified on Γ_1, a specified heat flux, q, is specified on Γ_2, and a convective boundary condition with a heat-transfer coefficient h and an ambient temperature T_∞, is specified on Γ_3. Summation is taken with respect to the index repeated exactly twice in a term. The boundary value problem of equation (1) is solved with the Finite-Element Method (FEM). Galerkin's method, which is a weighted residual method, is used to find the corresponding variational form. Linear triangular elements are used for the discretization of both the piston and cylinder cap domains (Figures 1 and 2). Because of symmetry, only half of the model is analyzed.

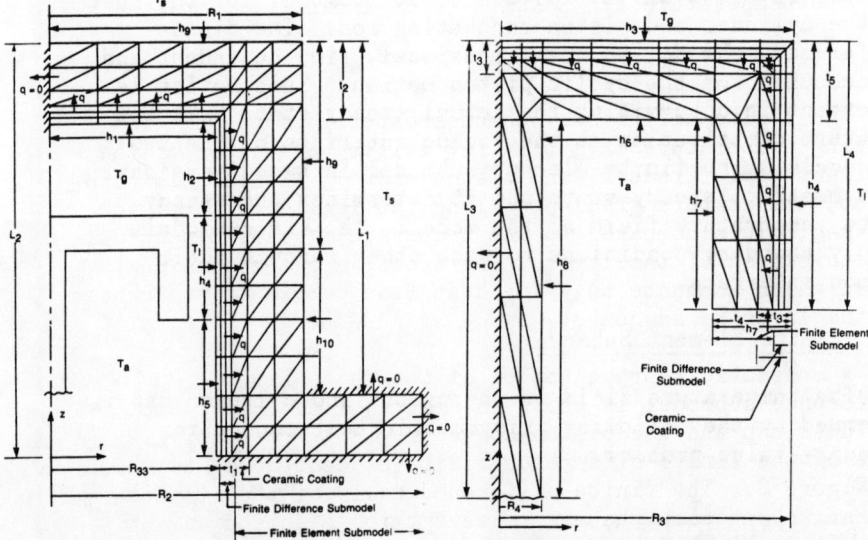

Figure 1. Transverse Section of the Cylinder Cap Model

Figure 2. Transverse Section of the Piston-Connecting Rod Model

A transverse section of the cylinder cap model is shown in Figure 1. Two different materials may be used. A thin layer along the cylinder liner and the firedeck may be of ceramic material. The rest of the model may be of a different material. An automatic mesh generation based on given geometry parameters is incorporated in the program. The heat-transfer coefficients for the convective type of boundary conditions and the regions over which these boundary conditions are applied are shown in Figure 1. Only a part of the outer surface of the cylinder cap exchanges heat with the surrounding. The remaining outer surface is assumed to be insulated (q=0). At the common interface of the two submodels there is a heat flux in the finite-element model. This heat flux is treated as a boundary condition by the finite-element model and it is due to the temperature difference between each cell of the last layer of the finite-difference model and the adjacent finite-element of the finite-element model.

A transverse section of the piston-connecting rod submodel is shown in Figure 2. Again, there is the capability of using different material for a thin layer of the piston crown from the material used for the rest of the model. These materials may have temperature-dependent thermal conductivity and specific heat values. An automatic mesh generation similar to that of the cylinder cap model is used. The boundary conditions are similar to those of the cylinder cap model and are indicated in Figure 2. The analysis does not include the whole connecting rod structure, but it extends up to a specific cross section which has a uniform temperature. Both the location of the cross section and its uniform temperature are assumed to be known.

3.2 Finite-Difference Submodel

The finite-difference submodel determines the time-varying temperature in a thin few millimeters layer of the cylinder cap, firedeck, and piston crown. Within this thin layer, temperature gradients along the length of the layer are negligible compared to the temperature gradients across the thickness of the layer. Therefore, a one-dimensional model which accounts only for heat flux across the thickness of the layer is adequate.

A schematic representation of the whole finite-difference submodel is shown in Figures 1 and 2 for the cylinder cap and the piston-connecting rod, respectively. A part of the finite-difference finite-element model is shown in Figure 3. The finite-difference cells of the first layer exchange heat with the environment through an equivalent convective type of boundary condition, which simulates the actual heat exchange. The ambient temperatures T_g and T_1 in the combustion chamber and in the piston skirt-cylinder

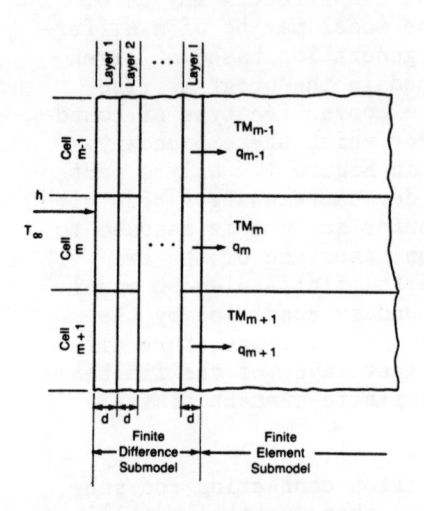

Figure 3. Finite Difference-Finite Element Model

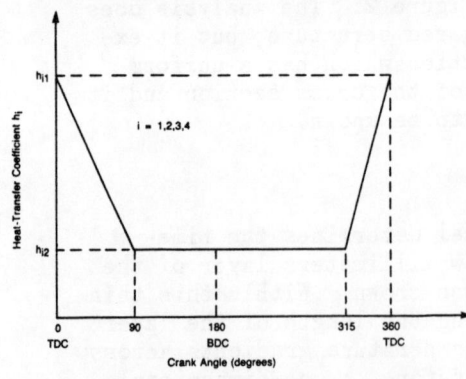

Figure 4. Approximation of Heat-Transfer Coefficients in the Combustion Chamber vs. Crank Angle

liner gap, respectively (Figures 1 and 2) are assumed to have a known distribution with respect to time. The remaining reference temperatures T_a and T_s are assumed to be known and time dependent.

The heat-transfer coefficients in the combustion chamber, h_1, h_2, and h_3 (Figures 1 and 2), have a peak value around the piston top dead center (TDC) position and then they level off at a considerably lower value for the rest of the cycle [2]. In this analysis, the time-dependent heat transfer coefficients h_1, h_2, h_3, and h_4 are approximated by a piecewise linear distribution in time as shown in Figure 4, when their actual distribution is not known "a priori." A two-stroke engine cycle is used. Two constants are required to describe each one of these time-dependent heat-transfer coefficients. One represents the peak value at the TDC position and the other the lower value at the bottom dead center (BDC) and during the most part of the remaining cycle. The remaining heat-transfer coefficients (Figure 1 and 2) are assumed to be time independent throughout the engine cycle.

An explicit forward-difference scheme is used to calculate the time-dependent temperature field in the finite-difference submodel. This scheme expresses the conservation of energy for each cell. The set of finite-difference equations which give the temperature field at the time t+1, as a function of the temperature field at the time t, for the 1 layer of cell m (Fig. 3) is as follows:

$$T_1^{t+1} = T_1^t + \frac{A_{1m}}{m_{1m}c_{p_{1m}}} [h(t)(T_\infty - T_1^t) - \frac{k}{d}(T_1^t - T_2^t)] \delta t$$

$$T_2^{t+1} = T_2^t + \frac{kA_{2m}}{dm_{2m}c_{p_{2m}}} [T_1^t - 2T_2^t + T_3^t] \delta t \qquad (2)$$

$$\vdots$$

$$T_l^{t+1} = T_l^t + \frac{kA_{lm}}{dm_{lm}c_{p_{lm}}} [T_{l-1}^t - 2T_l^t + TM_m] \delta t$$

where k is the thermal conductivity of the material, A_{im} is the average area of cell m of layer i, m_{im} is the mass of cell m of layer i, d is the thickness of each layer, c_p is the specific heat of cell m of layer i, h(t) is the heat-transfer coefficient at time t, T is the reference temperature, TM_m is the bulk temperature of the finite element adjacent to layer of cell m, and δt is the time increment.

This finite-difference scheme is convergent when the time increment δt satisfies the following relationship [6]

$$1 - \frac{2kA_{im}\delta t}{dm_{im}c_{p_{im}}} \geq 0 \qquad (3)$$

for all i and m when h<k. The engine cycle is divided into eight time increments, the maximum of which satisfies equation (3).

The heat flux in cell m of the finite-element model at the time t+1 is as follows:

$$q_m^{t+1} = \frac{kA_{lm}}{d} [T_l^{t+1} - TM_m] \qquad (4)$$

An average value of this heat flux over the engine cycle and for each cell is used as a boundary condition to the finite-element model.

3.3 Interaction Between the Finite-Difference Submodel and the Finite-Element Submodel

For every set of heat-transfer coefficients the equilibrium between the finite-difference submodel and the finite-element submodel is found by the following iterative procedure (Fig. 5):

1. An initial steady-stage bulk temperature field TM^o (Fig. 3) is assumed.

2. The finite-difference submodel runs for a number of engine cycles until cyclic equilibrium is achieved for the time-dependent temperature of each layer and each cell. The bulk temperature field TM^k of the previous iteration is used by the finite-difference scheme as a boundary condition (Fig. 3).

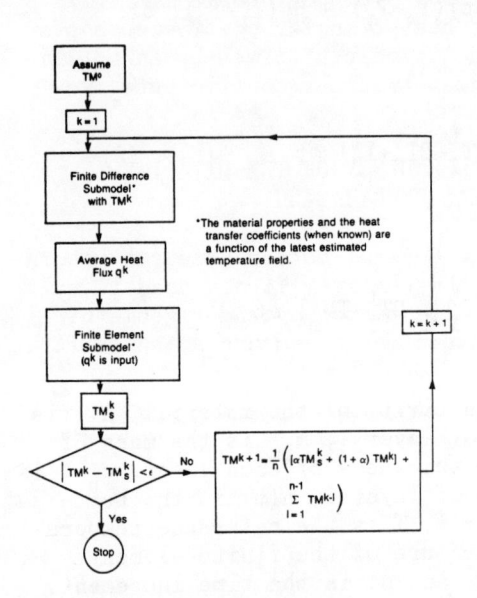

Figure 5. Equilibrium Between the Finite Element Submodel and Finite Difference Submodel

3. An average heat flux q^k over the engine cycle is calculated using equation (4) for each cell.

4. The finite-element submodel runs with the specified heat flux distribution q^k from Step 3. This run gives a new steady-state bulk temperature field TM_s^k.

5. If $|TM^k - TM_s^k|$ is greater than a given tolerance, a new estimation of TM for the next iteration is found using a combination of the underrelaxation technique and the averaging of n previous estimates as follows:

$$TM^{k+1} = \frac{1}{n} \{[\alpha \; TM_s^k + (1-\alpha) \; TM^k] + \sum_{l=1}^{n-1} TM^{k-l}\} \qquad (5)$$

where $\alpha < 1$. If k-1 < 0 then k-1 = 0.

Steps 2 to 5 are repeated until the steady-state bulk temperature field TM converges within a tolerance of 1 to 3 degrees C. In Step 5, the underrelaxation technique does not allow TM to change considerably from one iteration to the next iteration, while the averaging of the previous estimates does not let TM fluctuate about a mean value. Less than 10 iterations are required for the above scheme to converge.

The above iterative scheme is used to calculate the temperature field in the piston-cylinder domain and it is used only one time when the heat-transfer coefficients are known. However, it is used a number of times by an optimization technique when the heat-transfer coefficients are unknown.

4. EFFECT OF TEMPERATURE DEPENDENCY OF THE MATERIAL PROPERTIES ON THE THERMAL DEFORMATION

The latest design of the General motors Research Laboratories experimental cardanic mechanism [7] was used to validate the present analysis. The piston-cylinder geometrical characteristics of the cardanic mechanism are shown in Table 1.

Table 1. Cardanic Mechanism Parameters and Piston-Cylinder Geometry

Parameters		
Bore	84.924 mm	
Stroke	87 mm	
Velocity	2000 rpm	
Connecting Rod Length	310.957 mm	
Piston Skirt — Cylinder Bore Radial Clearance	38 microns	
Piston—Cylinder Geometry*		
L_1 = 210 mm	t_1 = 2.6 mm	R_1 = 57.15 mm
L_2 = 225	t_2 = 6.3	R_2 = 76.2
L_3 = 261	t_3 = 2.6	R_3 = 42.462
L_4 = 87	t_4 = 5.362	R_{33} = 42.5
	t_5 = 10.0	R_4 = 12.5

*see Figures 1 and 2

For an accurate calculation of the temperature field and consequently the radial thermal deformation of the piston and cylinder, the temperature dependency of the material properties must be included in the analysis. However, this makes the problem nonlinear.

Figures 6 and 7 demonstrate the effect of the temperature-dependent thermal conductivity and specific heat on the

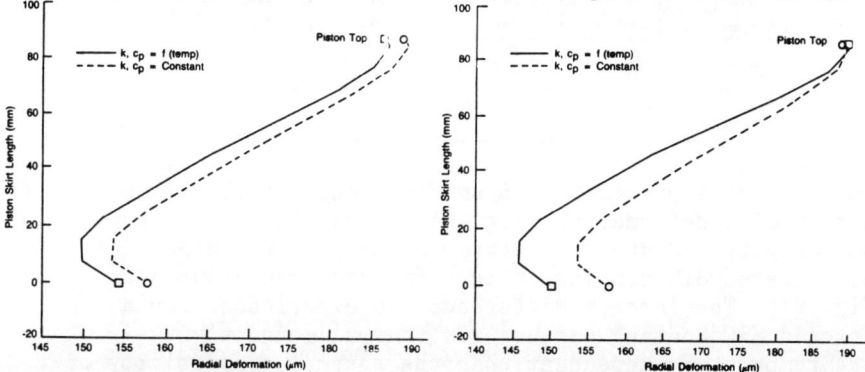

Figure 6. Radial Deformation of Piston Skirt with a 2.6 mm Coating of Alfa-SiC

Figure 7. Radial Deformation of Piston Skirt with a 2.6 mm Coating of Sintered-SiC

radial deformation of the cardanic mechanism piston skirt.
A 2.6 mm ceramic coating of alpha SiC, and sintered SiC was
used. The Young's modulus and thermal expansion coefficient
of the ceramic coating were always temperature dependent.
The baseline heat-transfer coefficients and ambient temperatures of Table 2 were used. Only the h_1, h_2, h_3, and h_4

Table 2. Baseline Heat-Transfer Coefficients and Ambient Temperatures

Baseline Heat-Transfer Coefficients		Baseline Ambient Temperatures	
$h_{11} = 6.0 \cdot 10^{-4}$ W/mm²·K	$h_5 = 4.5 \cdot 10^{-4}$ W/mm²·K	$T_g =$	1500°C at TDC
$h_{12} = 0.8 \cdot 10^{-4}$	$h_6 = 6.0 \cdot 10^{-4}$	=	500°C at BDC
$h_{21} = 6.0 \cdot 10^{-4}$	$h_7 = 5.8 \cdot 10^{-4}$	$T_l =$	1125°C at TDC
$h_{22} = 0.8 \cdot 10^{-4}$	$h_8 = 5.0 \cdot 10^{-4}$	=	375°C at BDC
$h_{31} = 6.0 \cdot 10^{-4}$	$h_9 = 5.5 \cdot 10^{-4}$	$T_a =$	200°C
$h_{32} = 0.8 \cdot 10^{-4}$	$h_{10} = 4.0 \cdot 10^{-4}$	$T_s =$	40°C
$h_{41} = 4.5 \cdot 10^{-4}$			
$h_{42} = 0.6 \cdot 10^{-4}$			

heat-transfer coefficients and the T_g and T_l ambient temperatures were time dependent. The calculation of the temperature field and the corresponding thermal deformation was based on the material properties of Table 3. For an alpha

Table 3. Material Properties

MATERIAL		PARTIALLY STABILIZED ZIRCONIA		ALFA SILICON CARBIDE		SINTERED SILICON CARBIDE		CAST IRON	
Thermal Conductivity	$\frac{W}{m°C}$	25°C	2.92	20°C	87.0	20°C	104.0	54.4	
		1000°C	2.51	600°C	49.0	600°C	39.0		
Density	$\frac{kgr}{m^3}$	5910.0		3140.0		2980.0		7200.0	
Specific Heat	$\frac{J}{kgr°C}$	25°C	502.4	20°C	669.0	20°C	710.0	480.0	
		1000°C	628.0	600°C	1120.0	600°C	1087.0		
Thermal Expansion Coefficient	$°C^{-1} \cdot 10^6$	200°C	8.0	20°C	2.5	20°C	2.5	200°C	10.0
		800°C	10.0	1000°C	4.2	1000°C	4.2	700°C	13.57
Young's Modulus	$10^4 \cdot \frac{Nt}{mm^2}$	25°C	20.3	20°C	4.03	20°C	3.87	20.7	
		1000°C	16.5	900°C	3.91	900°C	4.08		
Poisson Ratio	—	0.23		0.14		0.13		0.3	

SiC coating, approximately 5 μm difference in the piston
skirt radial deformation (Fig. 6) occurs when the thermal
conductivity and specific heat were temperature dependent.
For sintered SiC coating, this difference was about 8 μm
(Fig. 7). The largest difference was experienced for a
sintered SiC coating because its thermal conductivity is
more temperature dependent than the thermal conductivity of
alpha SiC (Table 3). For a PSZ coating, there is no difference at all because its thermal conductivity and specific
heat are practically temperature independent (Table 3) for

the temperature range of the specific example. Therefore, the more the material properties are temperature dependent, the more important is the effect of their temperature dependency in calculating the piston and cylinder thermal deformation accurately.

5. DESCRIPTION OF THE METHOD

Whenever a good estimation of the heat-transfer coefficients is available, the direct algorithm of Figure 5 is used to calculate the temperature field in the piston and cylinder. Once the temperature field is calculated, the corresponding thermal deformations are calculated using the MSC/NASTRAN code [4]. However, the heat-transfer coefficients are difficult to estimate accurately. Furthermore, even if an accurate estimation of these heat-transfer coefficients is available, the inability to calculate the radiation effects accurately does not allow an accurate calculation of the temperature field. T. Morel and R. Keribar [8] state that the heat transfer due to radiation may be up to 50% of the total heat transfer in a LHR engine.

A collocation method which uses an unconstrained optimization algorithm was developed to compensate for the difficulties described above. This method calculates the distribution in time of a set of equivalent heat-transfer coefficients in order for the corresponding temperature field to match measured temperatures at preselected collocation points. The collocation points are distributed over the whole piston-cylinder domain. Since this method bases the calculation of the temperature field on measured temperatures, it accounts for radiation effects through the calculated equivalent heat-transfer coefficients. The optimization search uses the following set of equivalent heat-transfer coefficients as independent variables (see Figures 1, 2, and 4):

h_{11}, h_{12}, h_{21}, h_{22}, h_{31}, h_{32}, h_{41}, h_{42},

h_5, h_6, h_7, h_8, h_9, and h_{10}.

There is not a unique solution for these equivalent heat-transfer coefficients. However, any solution gives a temperature field which is very accurate for practical purposes. The ambient temperatures are assumed known in this analysis since it is easy to estimate (or measure) them. Nevertheless, the present method can be easily extended to include the ambient temperatures in the set of independent variables.

For a given set of measured temperatures at specified points of the piston and the cylinder, the optimization search uses the algorithm of Figure 5 to find the whole temperature field as follows:

1. The optimization search specifies a set of equivalent heat-transfer coefficients (independent variables).

2. The algorithm of Figure 5 gives a temperature field which corresponds to the set of equivalent heat-transfer coefficients of Step 1.

3. The objective function which must be minimized is formed. This objective function is the summation over all the collocation points of the absolute differences of the measured temperatures from the corresponding temperatures from Step 2.

If the objective function of Step 3 is not minimum, the optimization search updates the assumed set of equivalent heat-transfer coefficients and goes to Step 2. From the definition of the objective function, its minimum value is theoretically zero. However, the optimization search stops when either the objective function value is very small or the search cannot reduce it anymore. The temperature field of the latest Step 2 is used to calculate the thermal deformation of the piston and cylinder. The flexible polyhedron search as described by D. M. Himmelblau [9] is utilized in this analysis.

5.1 Selected Results

Since temperature measurements were not available, a simulation of measured temperatures was performed. The algorithm of Figure 5 was run using the baseline heat-transfer coefficients of Table 2 and a set of ambient temperatures T_∞^*. The ambient temperatures T_∞^* were different from the set of ambient temperatures T_∞ of Table 2. The temperatures at preselected collocation points given by the above run were used as hypothetical temperature measurements. The baseline ambient temperatures T_∞ of Table 2 were used in the optimization search. A difference between T_∞^* and T_∞ was introduced in order to test how well the collocation method can match a set of given measured temperatures obtained under conditions which the collocation method cannot account for. This provides a very strong test on the capability of the collocation method to match an arbitrary set of measured temperatures. However, a realistic set of measured temperatures is necessary for the collocation method to give a solution.

Figures 8 through 11 compare the radial deformation which corresponds to the hypothetically measured temperature field with the radial deformation which corresponds to the temperature field found by the collocation method. The abscissa gives the radial deformation of the piston skirt or cylinder liner in microns and the ordinate gives the length of the piston skirt or cylinder liner in millimeters. Thirteen and nineteen collocation points were used for the piston model and the cylinder model, respectively. In Figures 8 through 11, a considerable interference between the piston and cylinder liner occurs when the piston and liner are manufactured with the "hot" radial clearance. Therefore, the piston and cylinder must be manufactured with an initial profile opposite to the calculated thermal deformation profile at the operating conditions.

In Figures 8 and 9, a 2.6 mm coating of PSZ was used and the ambient temperatures T_∞^* for the simulated temperature

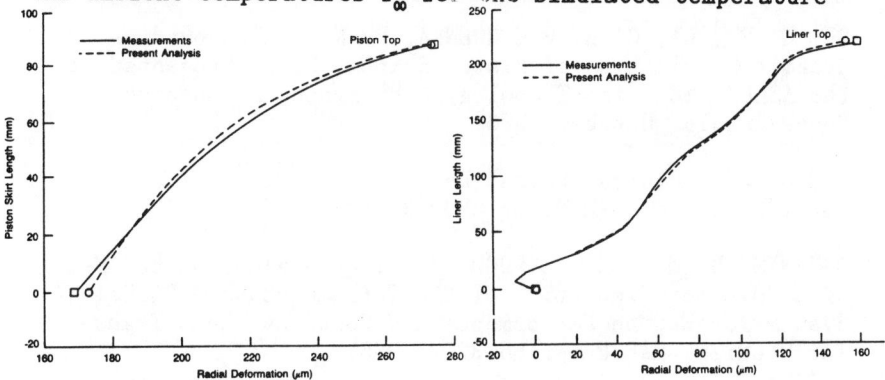

Figure 8. Radial Deformation of Piston Skirt with a 2.6 mm Coating of PSZ for $T_\infty^* = 1.2 \cdot T_\infty$

Figure 9. Radial Deformation of Cylinder Liner with a 2.6 mm Coating of PSZ for $T_\infty^* = 1.2 \cdot T_\infty$

measurements were 20% higher than the ambient temperatures T_∞ used in the collocation method. The collocation method calculated the radial deformation of the piston skirt and the radial deformation of the cylinder liner with an accuracy of 5 microns and 2 microns, respectively. In Figures 10 and 11, cast iron was used for the whole piston and cylinder

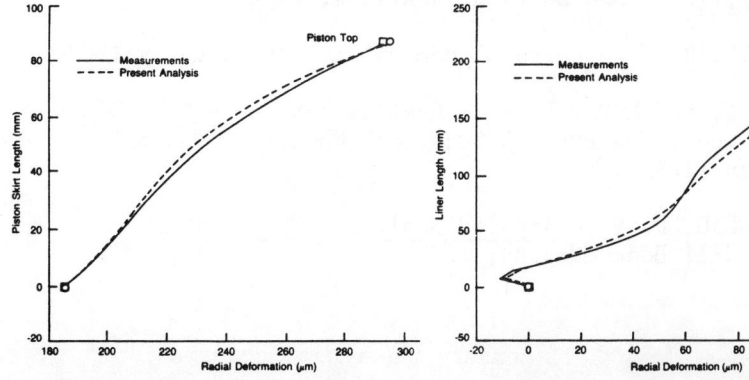

Figure 10. Radial Deformation of Cast Iron Piston Skirt for $T_\infty^* = 1.2 \cdot T_\infty$

Figure 11. Radial Deformation of Cast Iron Cylinder Liner for $T_\infty^* = 1.2 \cdot T_\infty$

assemblies and again T_∞^* was 20% higher than T_∞. The collocation method calculated the radial deformation of both the piston skirt and cylinder liner with a 5 micron accuracy.

From the above case studies, it is concluded that the developed collocation method gives very satisfactory results even under severe conditions introduced by uncertainties in the provided measured temperatures.

6. REFERENCES

1. OH, K. P., LI, C. H. and GOENKA, P. K. - Elastohydrodynamic Lubrication of Piston Skirts, Paper Presented at the ASLE/ASME Joint Tribology Conference, Pittsburgh, Pennsylvania, October 1986.

2. LI, C. H. - Piston Thermal Deformation and Friction Considerations, SAE Paper 820086 (1982).

3. ASSANIS, D. N., and HEYWOOD, J. B. - Development and Use of a Computer Simulation of the Turbocompounded Diesel System for Engine Performance and Component Heat Transfer Studies, SAE Paper 860329 (1986).

4. MACNEAL, R. H. - The Nastran Theoretical Manual, NASA SP-221 (01), 1972.

5. ISHIRA, K., and MATSUDA, T. - Theories and Experiments on Perfectly Balanced, Vibrationless, Geared Devices to Convert Linear Reciprocating Motion to Rotary Motion or Vice Versa, ASME Paper 72-PTG-12 (1972).

6. CHAPMAN, A. J. - Heat Transfer, MacMillan Publishing Company, Inc., 3rd Edition, New York, 1974.

7. TRIPPETT, R. J. - personal communications, January 1986.

8. MOREL, T, and KERIBAR, R. - Heat Radiation in D.I. Diesel Engines, SAE Paper 860445, SAE Congress, Detroit, Michigan (1986).

9. HIMMELBLAU, D. M. - Applied Nonlinear Programming, McGraw-Hill Book Company, 1972.

THE THERMAL ANALYSIS AND OPTIMIZATION IN THERMAL ISOLATION OF
ENGINE COMPONENTS

Petar Agatonovic and Franz Koch

MAN Technologie GmbH
Munich, Germany

SUMMARY

Due to necessary external cooling of current metallic engine components a great part of the energy heat input is lost. Hence, limited cooled or uncooled engines have been proposed with components from the materials capable to withstand the high temperature conditions as, for example, ceramic materials.

The shortcoming of the monolithic ceramic materials in their stress concentration and small flow presence sensitivity may be overcome with the ceramic composite materials. The efficient use of this kind of material, treated in the study, requires that the anisotropic material behaviour must be considered and the input data on thermal and mechanical loads properly involved. With the help of design method based on two- and three-dimensional heat transfer and stress analysis, and some useful simplifications, the optimal component form can be evaluated.

In the present paper, the design methodology applicable to ceramic composite components is described and some recent development results in engine piston and cylinder head isolation are presented. Some important factors influencing the heat transfer analysis on isolated engine components are discussed.

1. INTRODUCTION

Material limitation and lubrication requirements in a conventional internal combustion engine, require external cooling of cylinder components. Because of this, at the current level of the engine technology, about one third of the supplied energy is rejected in the form of waste heat in the cooling system. Additional power output leading to a more energy-efficient engine can be achieved if the waste heat utilisation can be improved. To this end limited cooled or "adiabatic" engine have been proposed with combustion chamber isolation to allow operation with reduced heat losses. The insulated components include piston, cylinder head and liner, and exhaust ports and valves. The towards combustion chamber turned walls of the components must be from materials capable to withstand the high temperature conditions.

The mechanical and physical properties of ceramics as, for example, the high temperature strength and low thermal expansion, offer an advantage over other materials for the use in limited cooled engines. The successful application of ceramic, results in lower friction, weight, and inertia of engine components and allow higher engine operating temperatures with all of these contributing to the significant energy-saving potential. However, the conventional monolithic ceramic materials manifest a wide scatter in strength properties and since they do not yield, they are not capable of redistributing high local stresses as typical other design materials can do. Because of this, they are extremely sensitive to stress concentration and lose strength due to the presence of very small flaw. This shortcoming of the monolithic ceramic materials can be overcome with the ceramic composite materials which could combine the advantages of both the monolithic ceramics and cancel their weaknesses. This idea has been realised with the class of materials called CERASEP [1]. The recent developments of this kind of material with a Silicon-carbide matrix and Silicon-carbide fibre reinforcement for the application in MAN Diesel engine show considerable improvement in the fracture toughness and thermal fatigue resistance. The SiC/SiC CERASEP exhibit toughness over 25 Mpa$\sqrt{m}$ (Table 1), which even increase at high temperature. This is related to a fibre pull-out

TABLE 1 Monolithic ceramics and ceramic-composites [2]

Material	Flexural strength (Mpa)	Fracture toughness (Mpa$\sqrt{m}$)
Al_2O_3	550	5
Al_2O_3 SiC whiskers	800	9
Al_2O_3 SiC coated fibres	-	10.5
SiC	500	4
SiC/SiC CERASEP	750	25
ZrO_2	200	5
ZrO_2/BN-coated SiC fibres	450	22
RB Si_3N_4	260	3
RB Si_3N_4/SiC whiskers	900	20

phenomenon which leads to high fracture energy. Other important advantages of composite ceramics over conventional ceramic materials, especially for engine application, is the possibility to control their heat and strength properties in different directions. Results obtained on SiC/SiC CERASEP show that the fibre orientation effects the thermal diffusivity values. Due to the matrix porosity thermal diffusivity perpendicular to the fibre orientation is substantially reduced allowing very effective isolation in the same direction [1].

2. HEAT TRANSFER AND ISOLATION OF ENGINE COMPONENTS

Rapidly varying gas pressure and temperature and local velocity change makes the processes inside the engine cylinder very complex. The heat flow path between the working gas and the cylinder components varies throughout the engine cycle and spatially. It can be treated in two ways: as the overall effect which provides a steady, or mean, rate of heat loss to the walls or with the instantaneous rates varying from small negative values during intake to much higher, positive values during compression, combustion, expansion and exhaustion phases of the cycle. Integrating instantaneous rates with respect to time, gives the steady, or mean, change as mentioned above. For the purpose of thermal stress analysis the first approach is, in general, sufficient. The fluctuation on temperature at the surface is rapidly reduced at a very small distance from the surface and for most practical purposes the component temperatures may be taken to be the steady state values.

The surface temperature variation during the cycle, should not be confused with engine start-up and shut-down transients which diminish under steady-state engine operation conditions and require separate treatment.

The heat flow within the combustion chamber is accomplished by three different mechanisms: convection, conduction and radiation. The convective heat transfer rate ($\dot{q}_c$) at gas-to-wall interface can be expressed as

$$\dot{q}_c = h \cdot (T_G - T_W) \qquad (1)$$

For computing purposes the convective heat transfer coefficient (h) must be known. Unfortunately, it is not possible to predict this parameter and its spatial variation, as the necessary gas transport properties and their change can also not be predicted. For this reason, empirical values and relationships must be used.

Radiation heat transfer in the engine was frequently estimated as not significant enough to be included in the analysis. Moreover, because of the compexity of the measurements the heat losses through convention and radiation have seldom been measured separately. However, the general relationship:

$$\dot{q}_r = K_r \cdot A \cdot (T_G^4 - T_W^4) \qquad (2)$$

shows that in case of insulated engine due to the higher temperature conditions this part of heat flow can increase. Therefore, the separate consideration of the two form of heat transfer from the combustion chamber gases to the walls for limited cooled engine may be expected to become increasingly important.

The third mechanism of heat transfer, due to the conduction, effects the heat losses from gas to the walls in controlling surface wall temperature (T_W in (1) and (2)). To obtain component temperature distribution for surfaces with large temperature variation and with complicated internal heat transfer path, two-dimensional or three-dimensional heat flow analyses should be applied. Simplified computations using one-dimensional models and heat transfer coefficient formulas (Woschni, Annand) gives uniform component temperatures and are not adequate for detailed thermal stress calculations. On the other hand, the complexity of continuously changing boundary conditions and geometry of the system within the engine force simplifications. In summarizing the discussion subjects outlined above, the estimation of component temperature distribution is based on different simplifications with following general assumptions:

(1) The heat transfer surface temperatures are constant with time (cycle surface temperatures variation are not considered).
(2) The spatial variation in gas temperature is ignored. The spatial variation in heat transfer rates at surfaces are considered through the heat transfer coefficient h.
(3) Separate consideration of convective and radiative gas-to-wall heat transfer is ignored. Heat transfer coefficient h includes the radiative portion of the heat transfer.

Additionally, in solving practical problems further, different simplifications concerning component geometry idealization and heat transfer boundary conditions may be also necessary.

3. HEAT TRANSFER ANALYSIS

The thermal analysis method developed and used during the present examination have envolved into a very useful procedure for design and optimization of engine components. The method is illustrated with the flow-chart in Fig. 1. In the following paragraphs some important details will be presented and discussed using the computing results according to this method.

All computings were carried out with the FE-elemente Code MARC using higher order isoparametric elements. The nonlinearities of the problem were included through temperature dependent properties and nonlinear boundary conditions. The thermal conductivity was implemented anisotropic with the preferred orientation according to the composite material construction

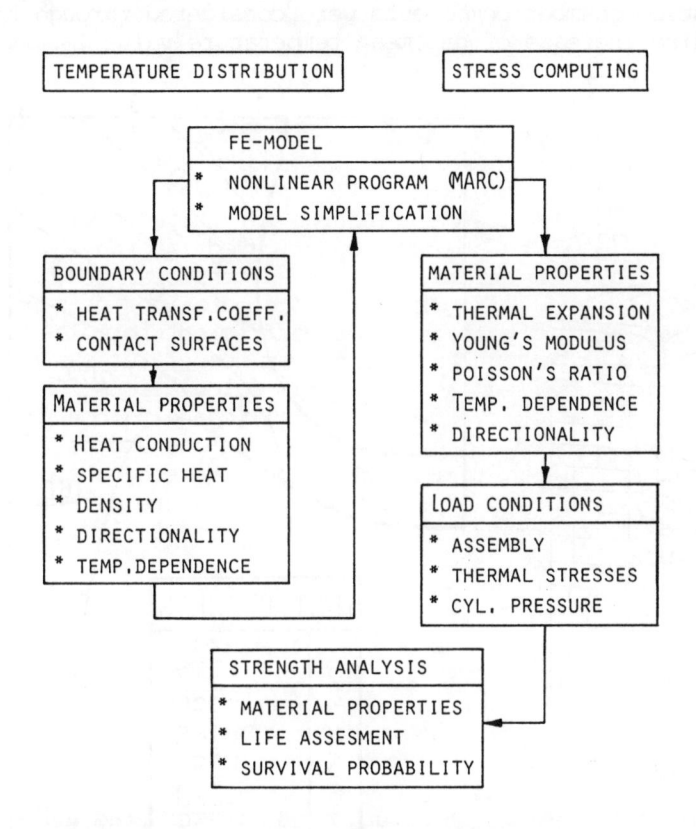

Figure 1: Design of ceramic engine components using special user subroutines.

3.1 Thermal analysis of the piston

By the great number of analysed design and material variants, to keep the expenditure of the examination within limits, it was necessary to reduce finite element models of the piston so far as possible using axial symmetric presentation and simple mesh structure. This was succeeded with the mesh design shown in Fig. 2. Presented is one variant of the piston with ceramic composite insulation parts bolted to the piston crown.

The convective heat tranfer coefficient values along external surfaces were evaluated based on the measurements on a serial engine. For this purpose, separate analysis was performed using iterative finite element computations. The full computings were repeated varying the h-coefficient values until adequate agreement between the measured and calculated temperature distribution was achieved (see Fig. 3). The evaluated values of heat transfer coefficient were used in the analysis of isolated components. The higher temperature level for gas and combustion chamber components were considered through the corresponding increase of gas mean temperature value T_G. It is

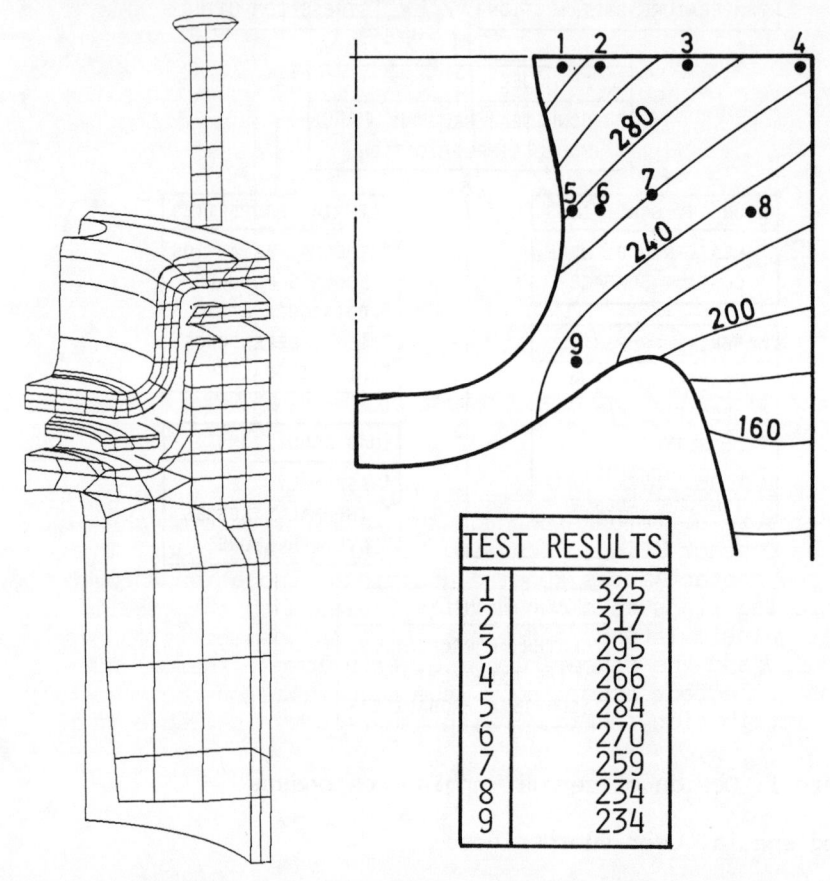

Figure 2: FE-Mesh for piston

Figure 3: Heat transfer coefficient evaluation

TEST	RESULTS
1	325
2	317
3	295
4	266
5	284
6	270
7	259
8	234
9	234

clear that heat transfer coefficient may be different under conditions of isolated engine. However, the proper values can be evaluated only experimentally by the measurements on developed components. For preliminary design, therefore, these assumptions cannot be avoided.

The first series of computings with ceramic insert fixed due to casting, have shown that direct exposition of a ceramic part to the piston crown causes to high temperature gradients through the walls, leading to high thermal loads. This can be avoided with additional isolation separating "hot" ceramic components from "cold" aluminium body of the piston, for example, with the air gap. Unfortunately, this kind of isolation is not capable of carrying or transferring the mechanical loads, so that limited contact areas are not to be avoided (compare Fig. 2). The heat flow within the piston is very strongly influenced through the disposition and the heat

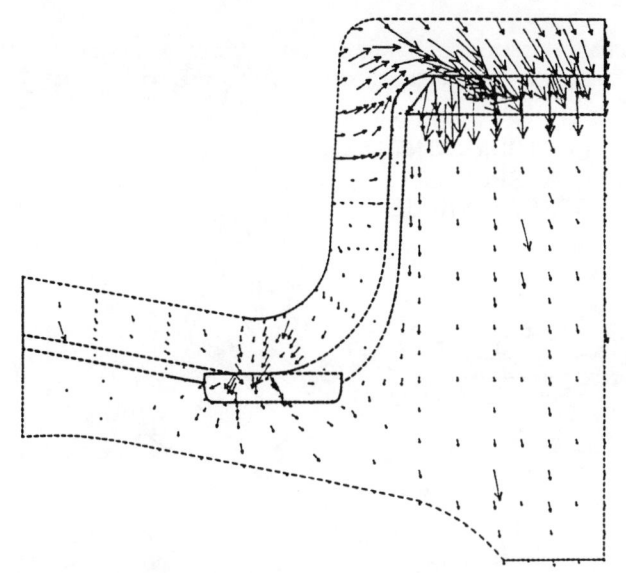

Figure 4: Heat flow within the piston

transfer conditions in these areas, as the Fig. 4 shows. The temperature difference through the wall increase at the place of the contact (Fig. 5) causing higher thermal loads. Therefore, the contact areas require additional isolation through the separating rings as shown in Fig. 2. Based on the results of numerical simulation, different solution possibilities can be compared and the favourable, i.e. with lowest thermal gradients selected. Accordingly, the comparence of the temperature distribution in Fig. 5 with the results available in the

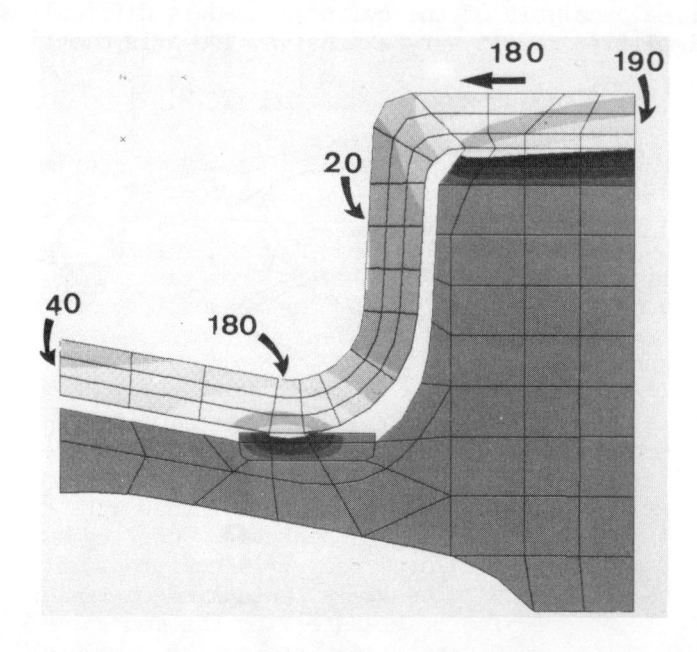

Figure 5: Temperature distribution and temperature differences

literature [5] show significantly lower temperature differences in all directions. The corresponding stress distribution for the case of maximum thermal and pressure loads based on material linear behaviour (Fig. 6) is characterised through smooth stress transitions and reasonable maximum values.

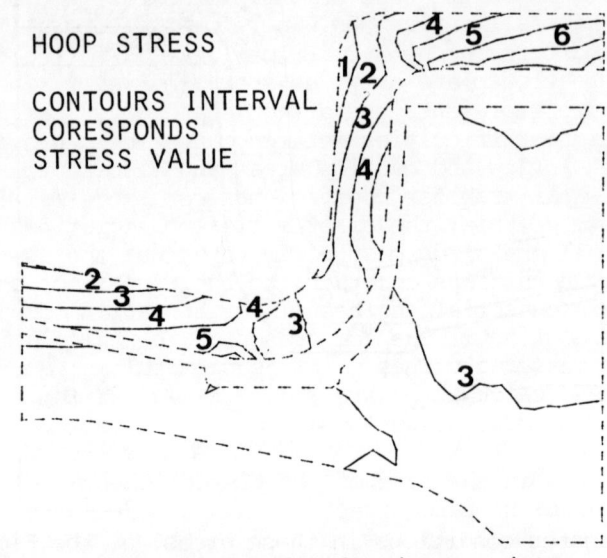

HOOP STRESS

CONTOURS INTERVAL CORESPONDS STRESS VALUE

Figure 6: Stress distribution at piston

Thus, the results in Figs. 5 and 6 make clear that through the optimization of design the isolation capacity of ceramic parts can be increased and the thermal loads reduced.

3.2 Thermal analysis of cylinder head

Numerical analysis of the cylinder head is difficult and also on simplified models very expensive. The axisymmetric mo-

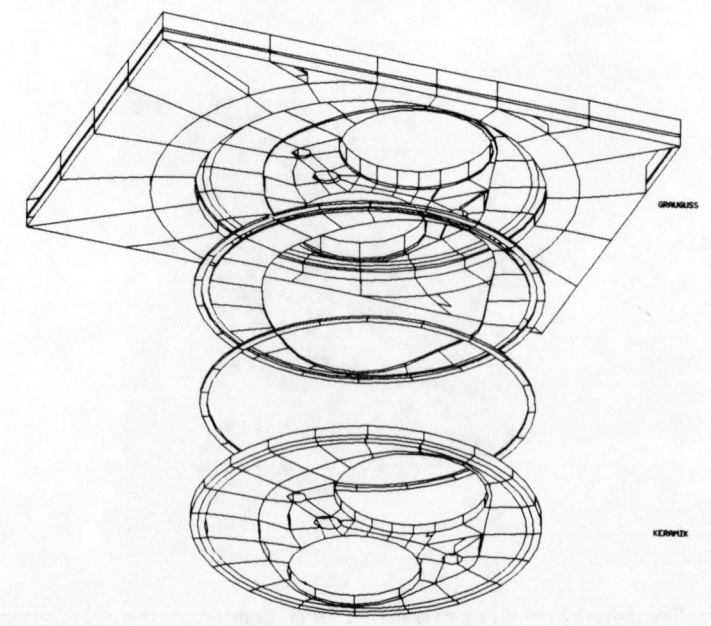

Figure 7: Cylinder head plate idealization

dels, as in the case of the piston, are not detailed enough to examine all influences. Therefore, a compromise solution was found through the 3-D FE-Idealization of the cylinder head foundation plate only, as shown in Fig. 7. The model in this figure represents the design state after optimization of the design. Difficulties were encountered in properly defining heat inflow and outflow areas and also in the evaluation of the heat transfer coefficients for the simplified model. Because of this, the first series of computings were carried out on all grey iron cast plate to adjust the heat transfer boundary conditions according to the experience and the measurements on serial engines. These iterative computings were carried out until the comparison of the calculation results and the measured values exibit successful agreement. The results of end calculation are shown in Fig. 8. The so evaluated boundary conditions were used for all further calculations.

As in the case of the piston, the calculation with the insulation plate direct contact to metal exhibits to high thermal loads in the form of the temperature differences through the wall and between different parts of the plate. Therefore, an attempt was made to reduce these thermal loads through the appropriate introduction of air gaps between the insulation

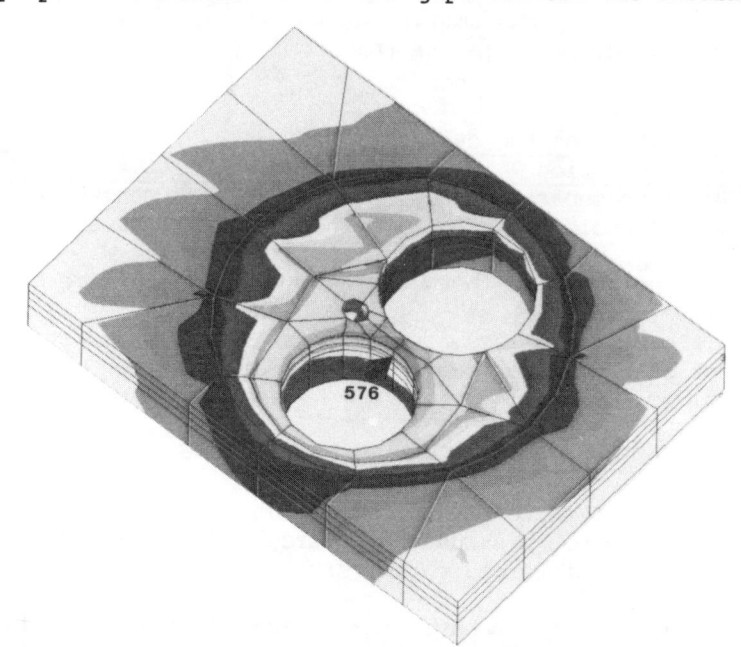

Figure 8: Temperature distribution at metalic cylinder head

plate and the metal. The interesting results of this examination (compare Table 2) is: the lowest thermal loads appear with the air gaps far outside the valve areas. Because of this, this solution was selected for further examination leading to the design presented in Fig. 7.

In addition, different kinds of insulation plate fixtures were examined. It was found that the use of the press fit should be avoided, and for the purposes of free thermal dilatation and better thermal isolation the ceramic insert should be built-in with a circumferential gap.

The stress calculation for combined thermal and pressure loads based on linear material behaviour shows peak stress in the area surrounding the intake valve opening (Fig. 9). The bridge between the valves in he middle of the cylinder exibits lower stresses due to controlled outflow of superfluous heat input at this place.

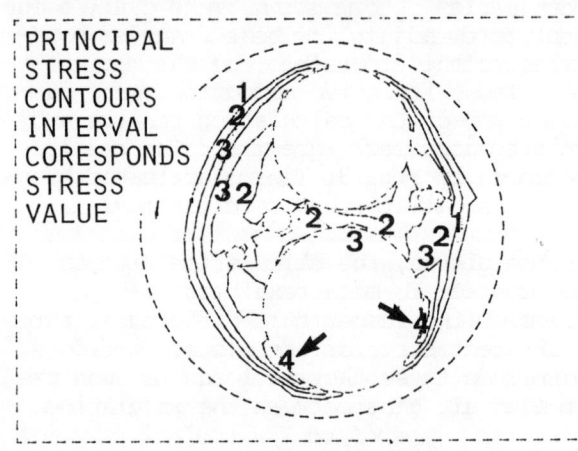

PRINCIPAL STRESS CONTOURS INTERVAL CORESPONDS STRESS VALUE

Figure 9: Stress distribution at the isolated cylinder head

TABLE 2: Insulation of the cylinder head

Variant	Description	Results
1	Thorough separation due to air gap at whole back face area	To high stress at bridge between the valves
2	Air gap at bridge between the valves only	Nearly the same as above
3	Air gap at all face areas but not at bridge between the valves (see Fig. 7)	Well-balanced stress distribution with mode-rate peaks

3.3 Specific influences

Some important factors influencing the heat transfer on isolated engine components must be particularly considered in the case of fibre reinforced ceramics:

(a) Anisotropic material behaviour - As stated before, the fibre orientation effects the heat conduction behaviour of material used. The heat conduction through the thickness, is a few times lower causing excellent isolation behaviour in the same direction. In return for that, high heat conduction in fibre direction, effects the thermal load release in, for example, in case of piston insert (Fig. 2) in hoop and radial direction. As a result convenient property distribution is achieved. The efficient use of these advantages requires exact heat transfer and thermal stress analysis with full consideration of anisotropic material properties, caused through the composite construction.

(b) Heat radiation across internal gaps - Increase of ceramic parts temperature can cause additional heat transfer within the air gaps due to radiation. This kind of radiation transfer remains within the structure boundaries and can be accounted for as additional conductivity $_r$, as shown in [6]. The estimated values of the heat conductivity increase in this way for the air gaps between the ceramic and metal parts is about 40 %.

(c) Heat transfer across an interface - At the interface between two objects in contact, an additional resistance to the heat flow is produced. The solid-to-solid conduction of the interface depends on the actual area of contact and, therefore, on surface roughnesses and the pressure of the contact. It was shown [7] that the pressure dependence is similar as in the case of electric resistance following the relationship of the type $R = C \cdot p^{-n}$. The contact resistance was measured using the laser flash method. Typical results of the measurements are presented in Fig. 10. It can be seen that the heat conductivity at the interface depends on the contact pressure approaching some limiting values after pressure increase. The measured curve can be approximated well with the above relationship. On the other hand, the further measurements have

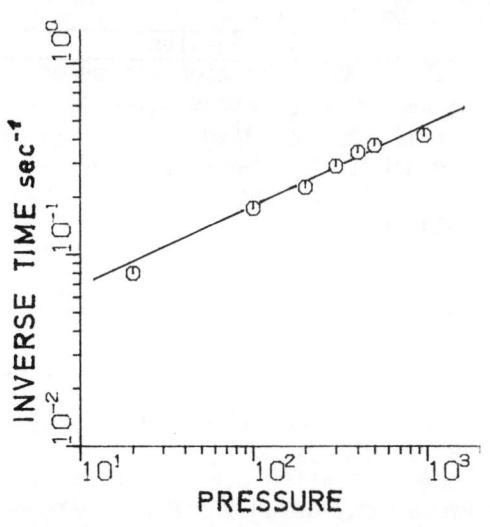

Figure 10: Results of heat transfer measurements

shown that the repeated loading can cause reduction in the contact resistance, probably due to the embedding and diffusion within the interface.

For the proper evaluation of the maximum thermal stresses in ceramic components, the contact conditions are mostly important and should always be considered.

(d) Surface heat fluctuation - Calculation simplification due to nonconsideration of the wall surface temperature swing during a cycle should not be disregarded. Using the materials with low thermal conductivity and small thermal diffusivity, such as ceramic materials, the surface temperature becomes not only higher, but its amplitude also becomes greater. Under these conditions the corresponding surface thermal stresses can achieve significant values. The effective consideration of this effect can be complicated by the presence of deposits.

4. CONCLUDING COMMENT

The paper summarized the approach to thermal analysis and optimisation of isolated engine components using composite ceramics. Numerical analysis methods can be very efficient and useful tools in the evaluation of proper design when some important factors influencing the heat transfer on isolated engine components, as:

- Anisotropy of material thermal and strength properties
- Heat transfer across the interfaces and internal gaps,
- surface heat spatial and temporal fluctuations,

are properly taken into account. Since the amount required for the production and testing of numerous component solutions is very high significant cost and schedule savings can be achieved by using this approach.

5. REFERENCES

[1] DAUCHIER, M.; G. BERNART and C. BONNET - Properties of Silicon Carbide Based Ceramic-Ceramic Composite, 30th National SAMPE Symposium, March 19-21, 1985, p. 1519.
[2] KLEIN, A.J. - Ceramic-Matrix Composites, Advanced Materials & Processes 9/(1986), p. 26.
[3] BENSON, R.S. and N.D. WHITEHOUSE - Internal combustion engines, Pergamon Press, Oxford, 1979.
[4] ASSANIS, D.N. at all - A Computer Simulation of the Turbocharged Turbocomponent Diesel Engine System, National Aeronautics and Space Administration, NASA CR-174971, August 1985.
[5] KAMO, R. and W. BRYZIK -Cummins/TACOM Advanced Adiabatic Engine, SAE Paper 840428 (1984)
[6] STELZER, J.F. - Experience in Dealing with Non-Linearities, Expecially Heat Radiation, using the Finite Element Method, Numerical Methods in Thermal Problems, Ed. R.W. Lewis and K. Morgan, Pineridge Press, 1985, p. 301.
[7] PRESTON, S.D. - Thermal contact resistance at the pellet/cladding interface in the fast reactor boron carbide control rod pins, High Temperatures-High Pressures, 17, 253-60 (1985).

Structural effects of thermal variations on r.c. frames and grids in the cracked stage.

ALDO CAUVIN
Professor of Structural Engineering.
University of Pavia
Pavia Italy

Summary

As known variation of temperature across the depth of structural elements can produce high flexural and axial stresses in statically indeterminate structures.
However if these stresses are computed according to linear elastic theory, unacceptably high and unrealistic values are obtained, because the stiffness reduction due to cracking and, in case, plastic behaviour, produced both by external loads and thermal actions, is not considered.
It is therefore necessary to adopt an incremental method of analysis which permits to simulate the effect of flexural, shear and torsional cracking, in a efficient and sufficiently simple way, so that the influence of the gradually decreasing structural stiffness on distribution of thermal stresses can be realistically evaluated.
In this paper a computer method for the analysis of r.c. monodimensional structures is used which permits to simulate the loadig process and its consequences on stresses in the following way:
-Permanent loads are applied incrementally to the structure and the cracking pattern due to this kind of load established
-Thermal actions are applied taking into account the reduction of stiffness taking place in the preceding steps
-Variable loads are applied and the collapse of the structure simulated, both by increasing thermal actions and external forces.
The problem is not only of academic interest as these problems do occur frequently in practice and are not easily solved by conventional means.
Therefore a realistic example of calculation has been briefly described.

1.Introduction

Stresses produced by thermal variations in r.c. statically indetermined structures can be computed using linear elastic analysis. However in this case

unrealistically high values of stresses are obtained because the favourable effect of cracking and plastic behaviour of the structure is disregarded.
On the other hand if these phenomena are not taken into account in a rational way, the opposite (and potentially dangerous) result can be obtained, that is these effects can be underestimated.
This is true particularly in the case of cracking:overestimation of the reduction of stiffness due to this phenomenon leads to an evaluation on the unsafe side of action effect distribution at service load level and therefore to unreliable verifications.
At ultimate load level a safe evaluation of the influence of plastic behaviour on thermal effects is obtained by adopting high values of concrete and steel resistances.This in turn leads to an overevaluation of the yielding moment M_{yu} which is itself an unfavourable effect especially if the ultimate limit state is obtained by increasing external loads.It is therefore not easy at first glance to establish which values of material design resistances lead to the safest results.
In any case it is possible to control the situation at ultimate limit state by adopting ductile critical sections to the extent that for particularly ductile structures thermal effects(as well as other types of imposed deformations) can be ignored.
On the other hand resistances which control the yielding moments and especially the yielding resistance of reinforcing steel can be determined realistically, because its statistical distribution has a small standard deviation, while the tensile resistance of concrete ,which controls the cracked behaviour, has a sparse distribution(corresponding to a standard deviation varying between 15% and 18% of the mean value according to CEB Model Code[1]).In addition cracking is a more complicated and elusive phenomenon than yielding of the section.
As a consequence cracking is difficult to simulate and predict:it is therefore important that the simulation be performed so that the results be on the safe side.
On this basic problem our attention will be concentrated in this paper.

2.Nonlinear analysis in presence of thermal actions

The program is based on an incremental-iterative procedure,in the sense that loads and actions are increased step by step according to a given non proportional monotonically increasing law and the structure is analyzed for each loading increment,using the stiffness relative to the preceding loading step.An iterative correction can be

performed within each step, using the well known
Newton-Raphson algoritm.
The load history is represented in fig.1 and can be
described as follows:
<u>First</u>: permanent loads are applied. in n_1 loading
steps
<u>Second</u>: the effects of creep due to permanent loads
are, in case, simulated. in a given number(n_2) of "time"
steps.

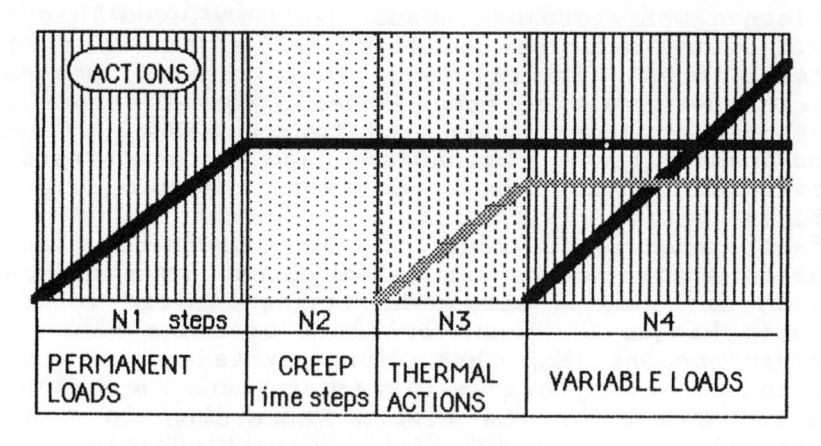

Fig.1-non proportional load history

<u>Third</u>: the action due to thermal effects are applied
in n_3 loading steps.
A linear or nonlinear thermal field can be defined
across the section of each element(see fig.2).
In the case of nonlinear stress field, it is
decomposed in linear and nonlinear components. While
the linear component is responsible for the
hyperstatic effects, the nonlinear one generates
within each element a self equilibrating stress field
which must be considered when computing the total
stress in a given fiber of a given element[11].

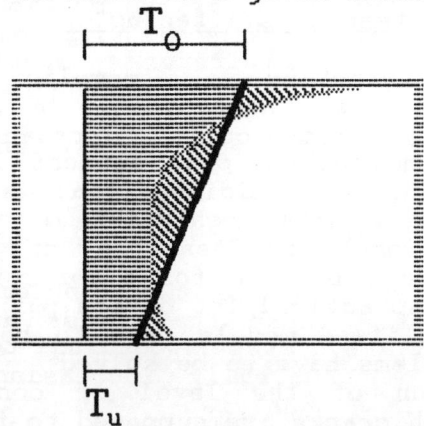

Fig.2-Plane and non plane thermal field

Fourth: variable loads are applied in n_4 loading steps eventually until collapse of the structure.
The following nonlinear effects are simulated in a simplified way:

- Flexural cracking: this phenomenon has a very strong effect on stresses produced by thermal variation and will be discussed in the following paragraphs.
- Torsional and shear cracking: If plane grids (or beams curved in the horizontal plane) are studied the influence of torsion needs be simulated. This is perfomed by reducing the torsional stiffness of the element when flexural and shear-torsion diagonal cracks are likely to occur. When diagonal cracks are present, the element tends to behave according to a tube-space truss mechanism and its torsional stiffness can be defined accordingly. See ref. [2], [4], [8], [12].
- Plastic behaviour: Structural behaviour in the elasto-plastic stage can be simulated either in a discrete way by introducing strain hardening limited rotation plastic hinges in those critical sections where the yieldig moment M_{yu} has been exceeded or in a continuous way by gradually reducing the "equivalent" elastic module of the element according to a semi empirical expression [2]
- Creep : it is simulated by introducing, after the incremental application of permanent and semi-permanent loads, a number of "time steps" in which the load is kept constant, while the moment-curvature laws are shifted by a quantity $K(\phi)$, ϕ being the creep coefficient and K the curvature of the element [2]
- Second order effects: The program takes into account the influence of second order effects using well known analytical procedures. See ref. [2]

3. Flexural cracking in monodimensional elements of concrete
Simulation of "tension stiffening"

Flexural cracking is a complicated phenomenon which could, at least theoretically be simulated with accuracy by a "microscopic" approach, that is by dividing each monodimensional element in an adequate number of nonlinear bidimensional elements, each simulating the material behaviour in a very limited area. This approach is feasible only for simple structures as a subtitute for experimental tests, but is obviously unpractical for design purposes.
Therefore simplified models have to be used and the following problems have to be solved:
- The evaluation of the level of concrete tensile stress at which cracks are supposed to form.
- The evaluation of the cracked element stiffness

in tension between cracks(tension stiffening effect)
-The evaluation of the stress level at which "stabilized" cracking may be supposed to have taken place.
The first and the third problems will be treated in paragraph 4.
For the simulation of the "tension stiffening" effect, a number of methods have been proposed[10][13] among which one proposed by the author which is extensively described in [2].
Comparisons of this method with experimental tests have given satisfactory results[13] and therefore it was adopted for this work. It can be briefly described as follows:
Each element belonging to the frame is divided in a given number of short elements. For each element the moment of inertia of the cracked section is introduced when, within the element, the adopted tensile strength of concrete in tension is exceeded. In computing the moment of inertia of the cracked section the influence of "tension stiffening" is simulated by introducing a "virtual" additional

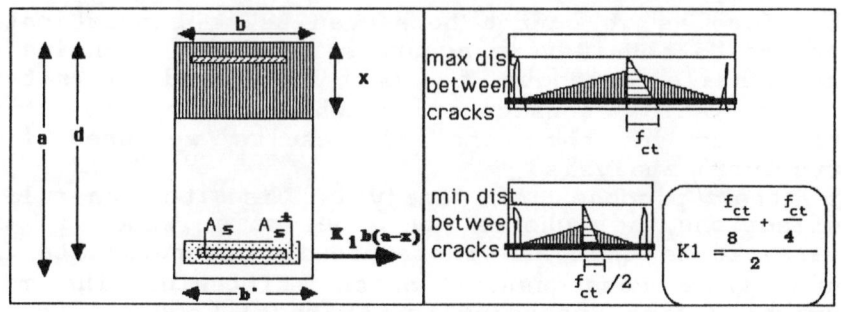

Fig.3-Simulation of "Tension Stiffening"

steel area A_s^* which can resist a constant force $K_1 b(a-x)$, K_1 being the mean tensional stress between cracks which will be subsequently called "tension stiffening coefficient" (see fig.3)
A_s^* decreases as loads increase, thus simulating the decreasing influence of " tension stiffening" with increasing loads(which can be verified experimentally).
The tension stiffening coefficient can be computed in function of the tensile resistance of concrete, adopting the simplified assumption of uniform distribution of adherence stresses along tensile reinforcement between one crack and the following. If this assumption is made tensile stresses in concrete vary linearly along the element axis. If again the assumption is made that tensile stresses also vary linarly along the section depth, the mean value of these stresses is equal to $f_{ct}/4$ in the

possible distance(upper limit situation) and equal to $f_{ct}/8$ in the lower limit situation(minimum possible distance between cracks).If again a mean value is assumed between these two extremes a value of the tension stiffening coefficient equal to $K1=3/16.f_{ct}$ is obtained.

4.Evaluation of concrete strength in tension.Lag in formation of stabilized cracking

Every simulation of the cracking formation process must necessarily be based on an assumed value of concrete tensile strength.On the other hand ,as already said,it is well known that this parameter can only be determined with considerable uncertainty given the great dispersion of test results.
We cannot ,in evaluating structural behaviour in the cracked stage,reach the same degree of accuracy that we can obtain in the calculation of yielding moments in critical sections,which are in most cases mainly influenced by steel strength f_{sy},and to a lesser degree by concrete compressive strength f_{ck}.
Therefore as we cannot hope,even using sophisticated methods,to obtain very accurate results,the choice of the tensile strength f_{ct} must be based on safety probabilistic considerations,which are in turn influenced by the kind of results we need from structural analysis.
If the purpose of analysis is to calculate deflections,the choice on the safe side is to overevaluate them and therefore to underevaluate f_{ct} (and ,as a consequence,tension stiffening).The most logical choice according to CEB Model Code philosophy [1] is to adopt a characteristic value with a 5% probability of not being exceeded ($f_{ck.05}$)
If the purpose is to evaluate redistribution of moments due to cracking,it is not easy (and probably not even possible)to establish which value of f_{ct} would yield the safest distribution of action effects.In this case it is most reasonable to look for the "most probable" result and therefore choose a mean value of resistance(f_{ctm})
If at last we need to evaluate thermal effects,the choice on the safe side is of course to overevaluate them and therefore to overevaluate f_{ct}(and tension stiffening).
A value of f_{ct} with a 95% probability of not being exceeded can be adopted($f_{ct0.95}$).
This may however not be enough.
The method of simulating stiffness reduction due to cracking which was described in the previous paragraph(as well as most other methods of this

kind)are based on the assumption that,as soon that,in a given element,the tensile strength of concrete is reached in the most stressed fiber,fully opened "stabilized" cracks form in the element and its stiffness can be reduced accordingly.
However this assumption does not correspond to results of experimental tests as it has been observed by Macchi and Sangalli[10].In fact there is a "lag" between the reaching of the limit value of f_{ct} and the formation of stabilized cracking.
If a p coefficient is introduced such that the current stiffness of an element $(EJ)_{cur}$ is given by the following expression

$$(EJ)_{cur}=p.(EJ)_{II}+(1-p)(EJ)_{I} \qquad (1)$$

$(EJ)_{II}$ being the stiffness of the cracked element(taking into account tension stiffening) and $(EJ)_{I}$ the stiffness of the uncracked element,it may be observed in tests such as the one represented in

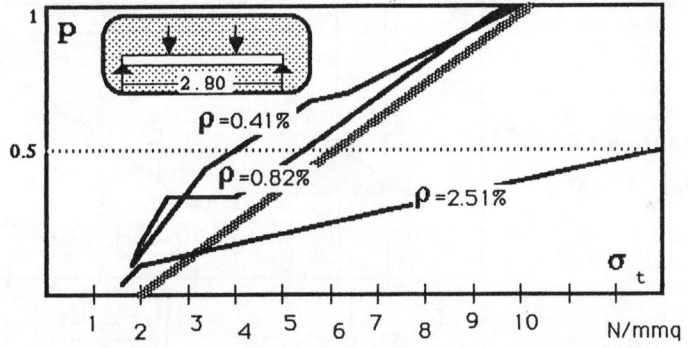

Fig.4-Process of stabilized cracking formation

fig.4 that the coefficient p varies,with increasing of tensile stress of concrete in the most stressed fiber according to a law which is also a function of reinforcement ratio ρ and which may ,as a first approximation,be taken as linear(fig.4)
It may be concluded that a situation of fully open and stabilized cracks takes place gradually in the element and is fully developed for values of maximum tensile stress in concrete that are much higher then the tensile strength f_{ct}.
This fact may be explained intuitively:in fact when fct is reached in the most stressed fiber a crack must begin;however an increase in stress in the adjacent fibers is required for the extension of this crack toward the neutral axis;this can happen only by furtherly increasing external actions.
This phenomenon,for the reason which have been previously said should be taken into account when the effects of thermal actions need be computed.

This has been performed in two ways in the computer program which has been previously described :
-In the most "accurate" way when the assumed value of f_{ct} is reached in a given element a reduced value of element stiffness is introduced according to expression (1) in which the value of p is assumed according to the following linear law

$$p = \frac{1}{(\sigma_{t1}(p)-f_{ct})} (\sigma_t - f_{ct}) \qquad (2)$$

where $\sigma_{t1}(p)$ is the value of tensile stress in concrete(assuming that the element is uncracked)corresponding to the full formation of stabilized cracking.This parameter can only be determined experimentally.
-In a more approximate way the stiffness $(EJ)_{II}$ is introduced when a tensile stress is reached equal to $(f_{ct} + \sigma_{t1}(p))/2$

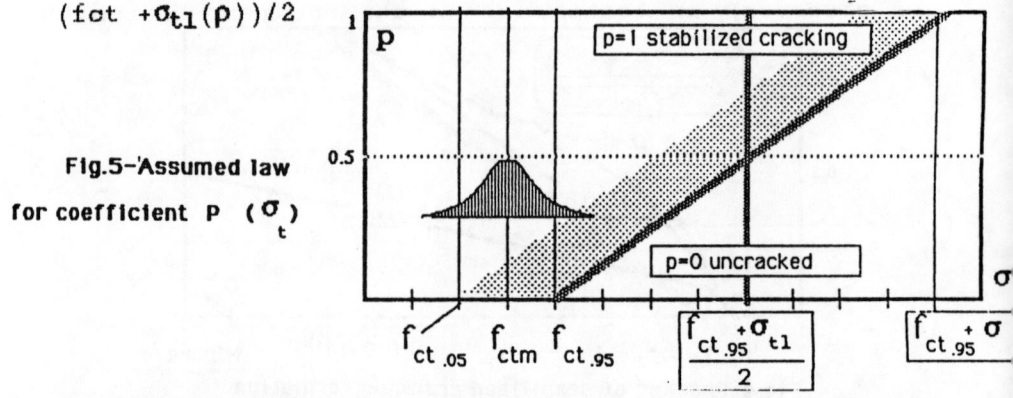

Fig.5-Assumed law for coefficient P (σ_t)

In fig.5 the laws of variation of p are represented in the "accurate" and simplified cases and for different possible choices of the limit value of f_{ct}.

5. Example

The frame of fig.6 has been analyzed both linearly and non linearly and for different values of f_{ct} and corresponding values of K_1(tension stiffening coefficient).The 6 cases which were considered are summarized in the table of fig.7.
The lag in formation of stabilized cracking was simulated in case n.5.
The load history is represented in fig.6:vertical loads were applied first to the beams in 10 steps so that cracking might take place due to external loads.An uniform thermal field of +40°C was then applied to the central column(again in 10 steps),so that moments of the same order as those produced by external loads(and of the same sign in the central

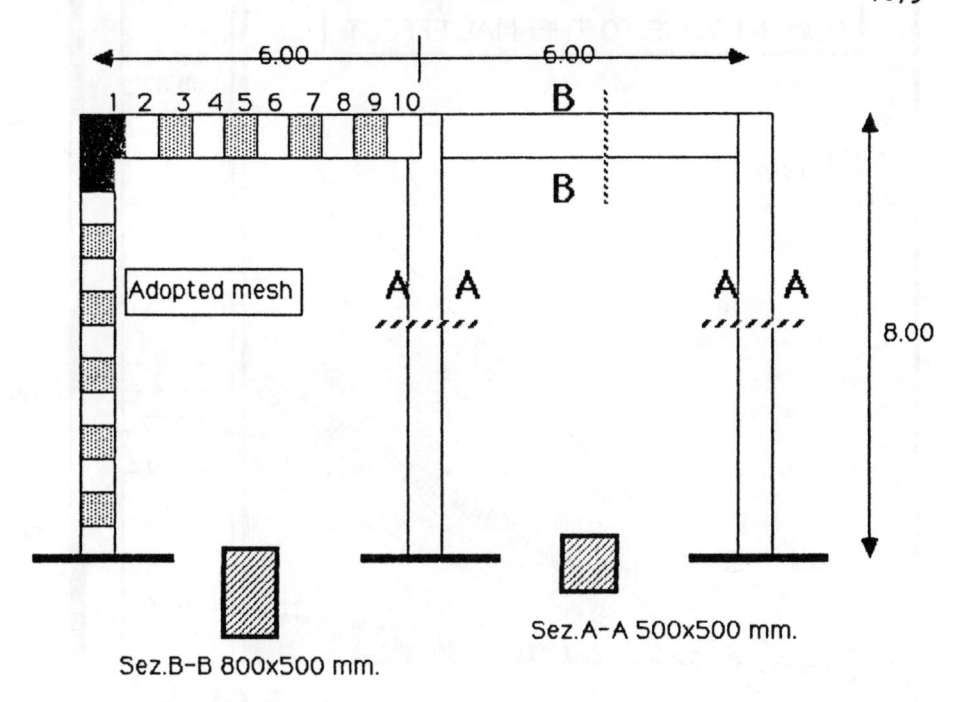

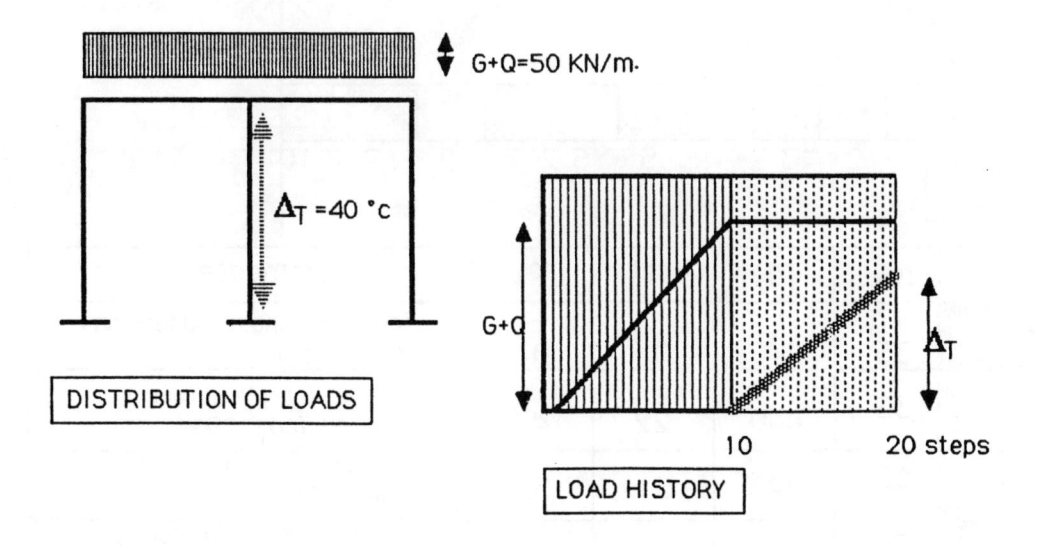

Fig.6-Considered example

MOMENTS DUE TO THERMAL EFFECT

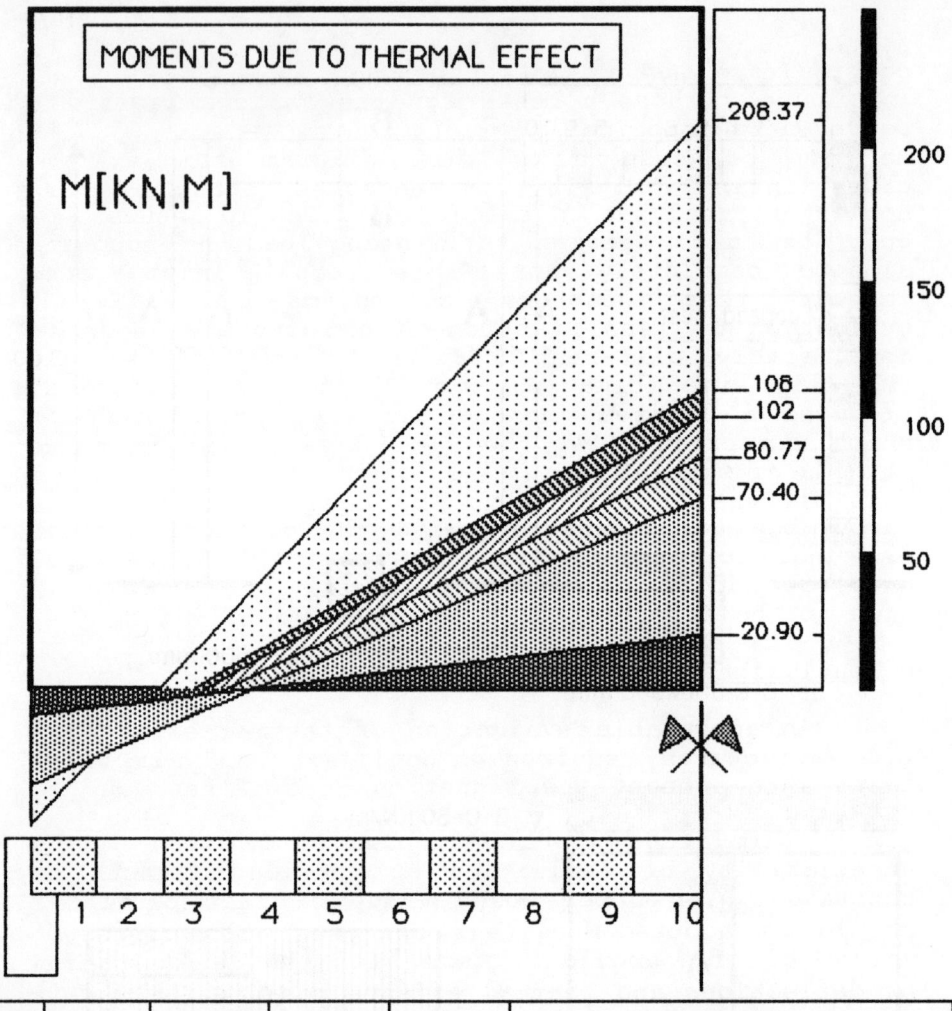

M[KN.M]

208.37
108
102
80.77
70.40
20.90

1 2 3 4 5 6 7 8 9 10

Case	fck	fsy	fct	K1	Comments
1	35	440	0	0	Tensile strength and tension stiffening disregarded
2	35	440	22	4.12	Tensile strength of concrete assumed=fctk0.05
3	35	440	31	5.81	Tensile strength of concrete assumed=fctm
4	35	440	40	7.50	Tensile strength of concrete assumed=fctk0.95
5	35	440	40	7.50	Lag in formation of stabilized cracking taken into account
6	35	440	Linear case		

Fig.7-Thermal effect in beams of example frame for the given cases

section)were induced in the beams.
In figure 7 those moments for the various cases are represented graphically
As may be seen from this diagram the extreme cases(linear analysis of uncracked structure and nonlinear analysis disregarding tensile stress of concrete and tension stiffening)lead,on opposite side,to completely unreliable results.Taking into account these factors according to different evaluation of tensile strength of concrete leads to considerably different results.Taking into account the lag in formation of "stabilized" cracking produced a limited increase in the thermal moments which was practically the same for the two proposed methods of simulation.
This increase could however be much greater in less favourable cases(case in which the critical section are not cracked after application of external load,case in which thermal effects are opposite in sign to the effect of external loads)

6.Conclusions

An acceptable evaluation of thermal effects in r.c. structures requires a nonlinear analysis which takes into account the reduction in stiffness due to cracking.
-In simulating this phenomenon,given the unreliability of evaluation of tensile strength of concrete f_{ct},an upper bound value of this parameter should be adopted.Nonlinear analyses with different values of f_{ct} should be done for the determination of deflections and thermal effects,because the safety requirements in the two cases have opposite signs.
-Taking into account tension stiffening is an essential requirement in thermal analysis:by no means it can be considered a "secondary" effect as in other kind of nonlinear analysis.
-The phenomenon of gradual formation of stabilized cracking should also be taken into account as it can have considerable influence on the results.
However we don't think a very sophisticated method of simulation is necessary for this effect.

7.References:

1. CEB-FIP "Model Code" Volume 2,Chapter 11;CEB Bulletin N.124/125 E,April 1978
2. CAUVIN A,Nonlinear Computer Program:idealization of constitutive laws.Second order effects and creep.Behaviour in torsion;Lessons 8,9,10 of Course:"Nonlinear analysis and design of r.c. and

prestressed structures".CEB,Universita' di Pavia,Institut fur Bautechnik,Berlin,Pavia,September 1981

3. CAUVIN A,Analisi non lineare di telai piani in CA.Giornale del Genio Civile;fascicolo 1,2,3-1978

4. CAUVIN A,Analisi non lineare di graticci piani in CA.Giornale del Genio Civile;fascicolo 4,5,6-1983

5. CAUVIN A,Practical Applications of an Incremental Nonlinear Analysis Program to Design of Tall r.c. Frames and Coupled Shear Walls.Presented at:Third International Conference on Tall Buildings,Chicago,January 6,1986 .

6. PARK R.,PAULAY T.,Reinforced Concrete Structures.John Wiley &Sons,1975 Edition.

7. ACI COMMITTEE 318,Building Code requirements for Reinforced Concrete(ACI 318-83).ACI Detroit,1983

8. CAUVIN A.
Redistribution of torsional moments in curved r.c. beams. Proceedings of the Second International Conference on Numerical Methods for Nonlinear Problems.Barcelona,Spain 1984

9.CAUVIN A
Nonlinear Analysis of Coupled Shear Walls in Tall Buildings.Proceedings of the Third International Conference on Numerical Methods for Nonlinear Problems.Dubrovnik,Jugoslavia 1986.

10. MACCHI G.,SANGALLI D.
Effect of crack formation process and Tension Stiffening on thermal stress relaxation in reinforced concrete.containments.Proceedings of the 5th International Conference on Structural Mechanics in Reactor Technology.Berlin,August 1979.

11.CEB Bulletin d'Information n.167:Thermal effects in concrete structures.January 1985

12.CAUVIN A
Simulation of cracked behaviour due to flexure,shear and torsion in nonlinear analysis of monodimensional statically indeterminate structures.
Proceeding of the conference:Fundamental developments in design models.Karlsruhe,19,21 November 1986.

13.MOOSECKER W
Deformational behaviour of reinforced and prestressed concrete elements.Lesson 5 of Course:"Nonlinear analysis and design of r.c. and prestressed structures".CEB,Universita' di Pavia,Institut fur Bautechnik,Berlin,Pavia,September 1981

8.Aknowledgements

The Author is grateful to Prof.Giorgio Macchi for his suggestions and advice. This work was performed with the partial financial support of the Italian Ministry of Education
(MPI)

A STUDY OF THERMAL STRESSES DEVELOPMENT IN MATURING CONCRETE

A. Pietrzyk
Warsaw University of Technology, Warsaw, Poland
P. Witakowski
Institute of Fundamental Technological Research, Warsaw, Poland

SUMMARY

The thermodynamic theory of simply maturing medium is proposed for studying the thermal stresses development in maturing concrete casted in large blokcs. The theory, based on the I and II principles of thermodynamics leads to two constitutive equations similar to those of the theory of thermal stresses. However, the material constants are replaced by the functions depending on the thermodynamic history of the material. The heat sources function appearing in the heat conduction equation is determined by the third constitutive equation, which expresses it by the set of functions W determined experimentally in isothermic conditions. The heat conduction equation was solved iteratively by finite element method. New values of heat sources output were calculated at each time step for each element. The mechanical equilibrium equation was then solved for each time step with temperature gain at this step as load.

1. INTRODUCTION

Since the cement hydration is an exothermic process, concrete temperature increases after casting. This, in case of massive structures, can induce thermal stresses several times greater than the future service stresses and lead to cracks development disqualifying the structure. The danger of damage of the structure increases with it's dimensions, because a growth of size makes the hydration heat abstraction difficult and leads in consequence to higher temperature rise and higher thermal stresses. Thus, the whole designing logic should be reversed in case of massive structures. The design engineer is used to the fact, that in case the admissible stresses are exceeded in a given cross section of a structure, he has to design a bigger one. Here we have just the opposite situation. There are numerous works from several countries devoted to the problem of thermal stresses in concrete massives (see [1]-[12])

for thermal stresses caused by hydration heat has been already for fifty years the greatest problem in the construction of massive hydrotechnic structures and especially in the construction of concrete gravity dams. Thermal stresses, which develop usually within the first 100 hours after casting cause a greater danger for the structure to remain a monolith than the future service stresses. So, the whole technology of building a structure should be adapted to avoid this danger. This problem has enormous practical and economical meaning because cracks can lead to the collapse of the whole structure with the terrible effect, while the sealing of a structure after construction is extremely expensive. The most important meaning has this phenomenon in the nuclear reactor buildings. In this case absolute tightness is required due to the danger of radioactive contamination of the environment.

The design of a structure with respect to thermal stresses caused by hydration heat of cement is much more complicated than the classical structural design. This complications are enlarged by the fact, that several phenomena connected with the hydration are not yet finally explained. Thus, the phenomenological approach is necessary. It seems obvious, that two effects neglected in classical design methods has to be considered:
- the fact, that concrete properties are time-dependent, because the maximum thermal stresses occur in the early stages of concrete maturing, when material properties are changing very quickly,
- the fact, that concrete properties are heterogeneous, because the concrete maturing rate depends on the temperature, which develops in each point in a different way.

The fact that concrete maturing rate depends on temperature, makes the whole process a coupled one - hydration heat generation causes the tempeature to rise which in turn causes (usually) the rate of heat generation to increase. This causes the necessity of simultaneous calculations of mechanical (displacements, strains and stresses) and thermal (temperature and heat generation) parameters.

2. SOURCES FUNCTION

The intensity of hydration heat generation is determined by the heat sources' power density. It's dependence upon time we call the sources' function and designate by W. Sources' function depends on 8 factors:
- mineralogical composition of clinker,
- degree of grind,
- gypsum contents,
- blast-furnace and puzzolanic additions contents,
- stale degree,
- special additions contents,
- cement-water ratio,
- temperature.

The main task of the technology of cement and concrete for massive structures is to choose these factors so as to make the sources' function's values as low as possible. The example of the sources' function is shown in fig. 1. Its worth noting, that the description of the dependence of sources' function on temperature by the so-called temperature's function - introduced by Rastrup in [13] and commonly used - at closer inspection turned out to be insufficiently accurate for designing requirements. This dependence is the specific property of each cement and has to be determined experimentally by a sequence of isothermic tests. All investigations confirm the well known fact, that the choice of cement and composition of concrete mix are fundamental for the occuring values of thermal stresses.

3. THEORETICAL FOUNDATIONS

The thermodynamic theory of simply maturing medium constitutes the theoretical foundation for the present work. It is based on the following assumptions:
- the medium consists of n phases, each of which has a given internal potential;
- the process of maturing consists in internal phase transitions in which the total potential decreases and it's changes manifest themselves through internal heat sources;
- the rate of phase transitions depends on temperature;
- regardless the temperature, the maturing process leads the medium through the same sequence of structures;
- the advancement of maturing process is indicated by heat generated by internal sources;
- the set (ε_{ij}, g_α, T), where:
  ε_{ij} - strains;
  g_α, ($\alpha = 1, 2, 3, \ldots, n$) - phase composition;
  T - absolute temperature;
constitutes a closed set of state parameters;
- the changes of ε_{ij} and T are small.

According to the hypothesis of simple maturing, the change of temperature causes the change of concrete maturing rate, exclusively. Several investigations prove that in the temperatures range interesting for concrete structures this hypothesis is fulfilled with sufficient accuracy.

4. BASIC EQUATIONS

On the assumptions mentioned above the classical geometrical and equilibrium equations hold. From the I and II principles of thermodynamics we obtain two constitutive equations:

(1) $$\sigma_{ij} = 2\mu(g_\alpha)\varepsilon_{ij} + (\lambda(g_\alpha)\varepsilon_{kk} - \gamma(g_\alpha)\Theta)\delta_{ij}$$

(2) $$\eta(g_\alpha)\Theta_{,ii} - c_\varepsilon(g_\alpha)\dot{\Theta} = -W$$

analogous to the equations of the theory of thermal stresses. The hypothesis of simple maturing leads to the third constitutive equation:

(3) $$W = \int_0^t \dot{W}_T(Q_T^{-1}(Q(t)))dt$$

where: Q - hydration heat density, Θ - temperature gain over the initial values, μ, λ, γ, η, c_ε - material coefficients, the subscript T denotes values obtained from isothermic tests.

5. ADDITIONAL INFORMATION

The theory admits all the classical boundary and initial conditions. However, for more accurate description of concrete maturing rather complicated initial conditions were introduced. The process was assumed to have two characteristic times t1 and t2. The first one meant the time when the casting was finished and the second one meant the end of setting. Up to the time t1 concrete was treated as ideal fluid, between the times t1 and t2 as viscous fluid and after the time t2 as a solid body with increasing stiffeness. Therefore the initial conditions were set for equation (3) at the time 0, for equation (2) at the time t1 and for equation (1) at the time t2. The theory admits any changes of material properties with respect to time. They are assumed to be obtained experimentally for a given concrete. According to the theory it is sufficient to determine them at one temperature, e.g. at 20° C. Dividing the whole process into short intervals, the changes of all interesting quantities can be found within each of them according to the theory presented above. Their actual values can be found by summing up the changes from the beginning of the process.

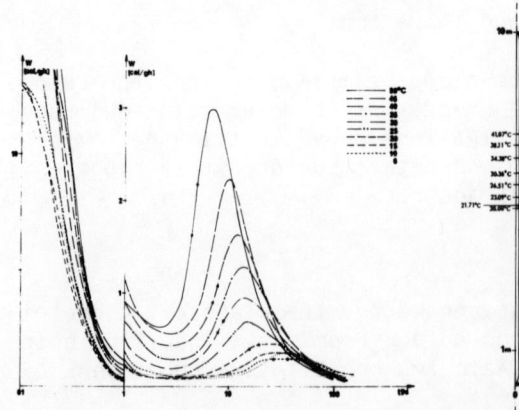

Figure 1. Sources function of clinker from Wiek cement plant; specific area 1700 cm^2/g.

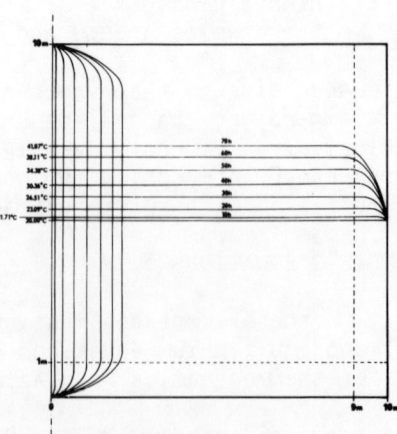

Figure 2. Temperature disbution in a 10 m high concrete cylinder.

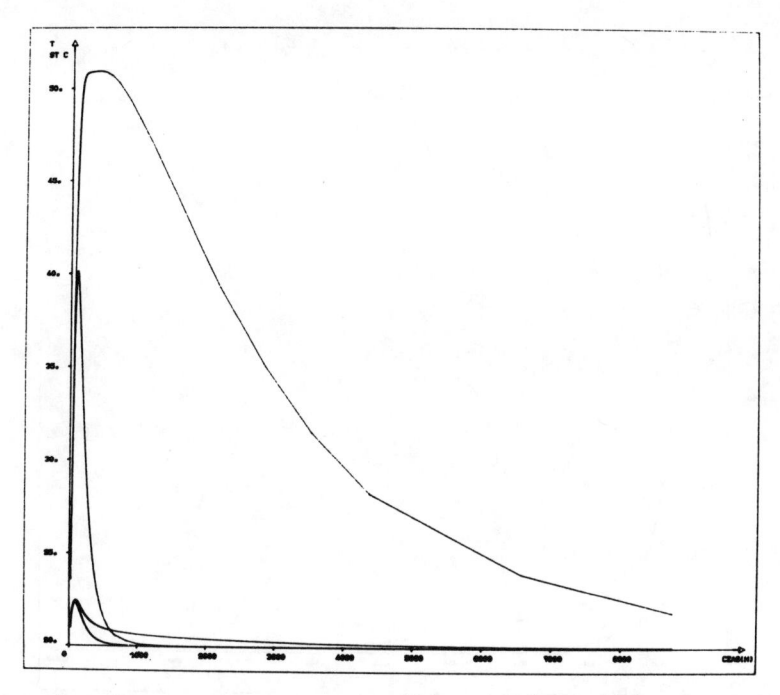

Figure 3. Temperature T development in selected points in 2 m and 10 m high blocks up to 8000 hours after casting.

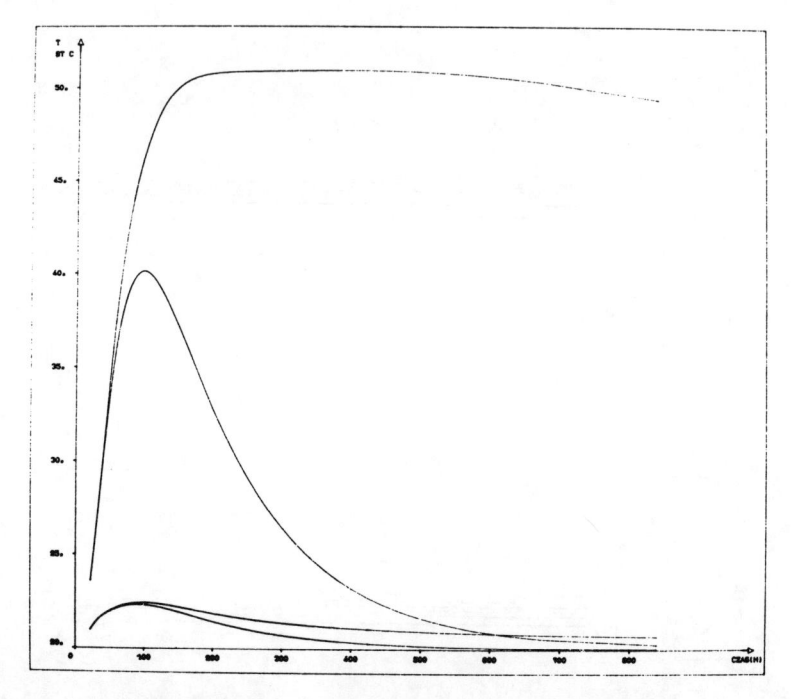

Figure 4. Temperature T development in selected points in 2 m and 10 m high blocks up to 800 hours after casting.

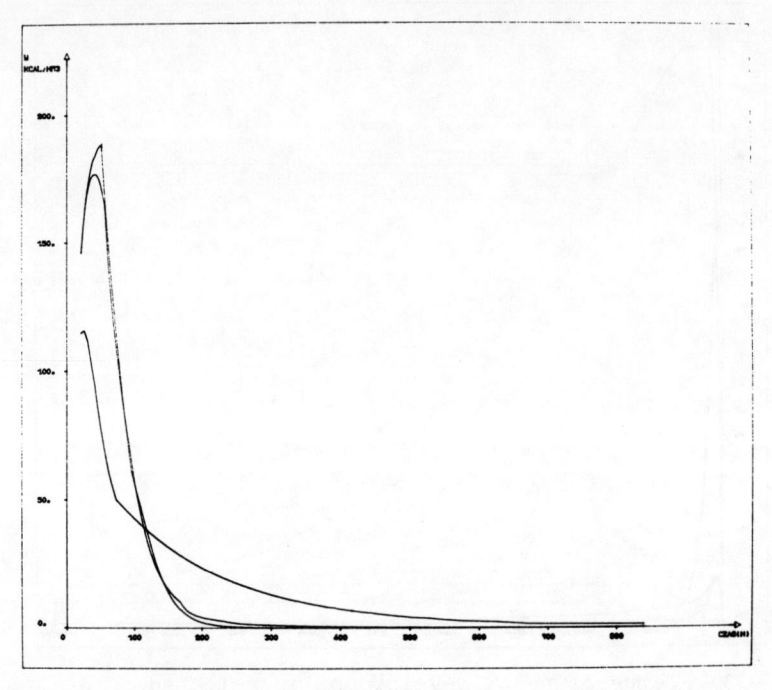

Figure 5. Sources function W development in selected points in 2 m and 10 m high blocks up to 800 hours after casting.

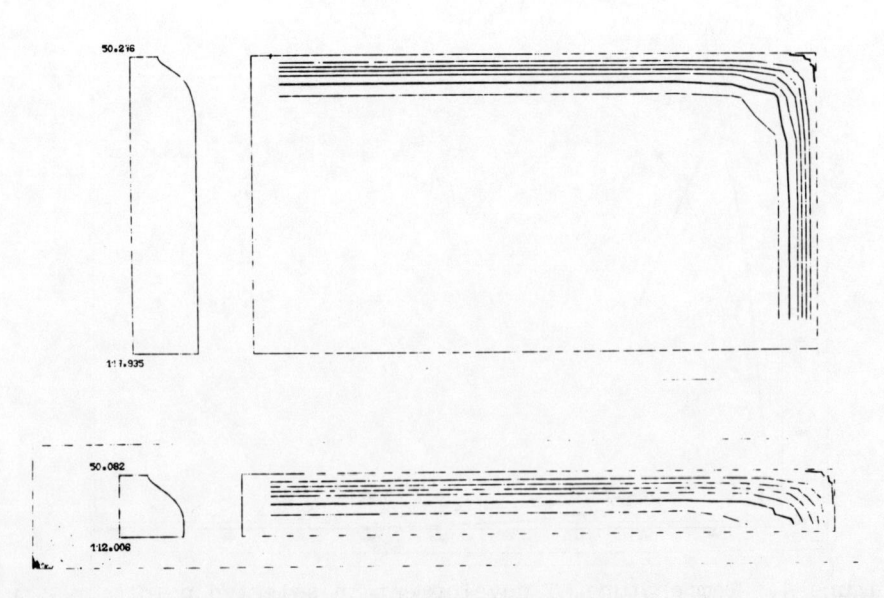

Figure 6. Contour map of sources function W in 2 m and 10 m high blocks 3 days after casting.

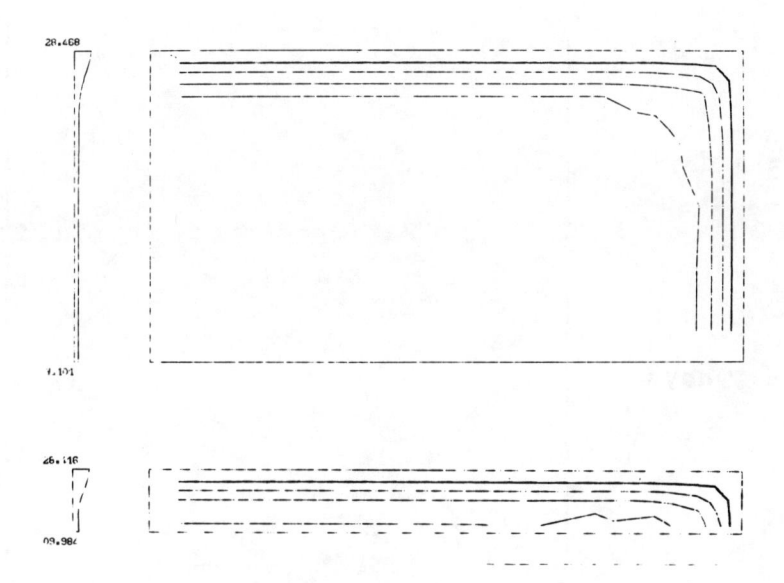

Figure 7. Contour map of sources function W in 2 m and 10 m high blocks 1 week after casting.

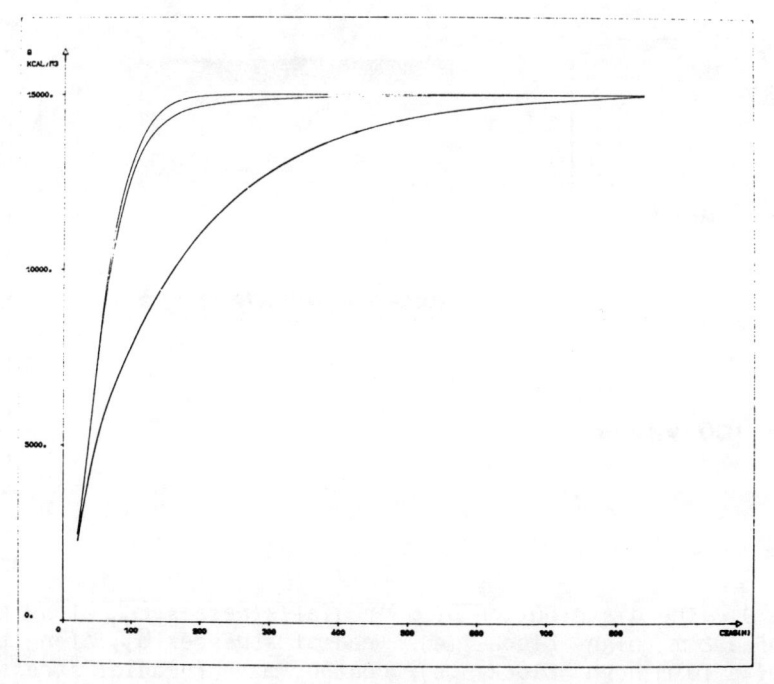

Figure 8. Total heat generated Q development in selected points in 2 m and 10 m high blocks up to 800 hours after casting.

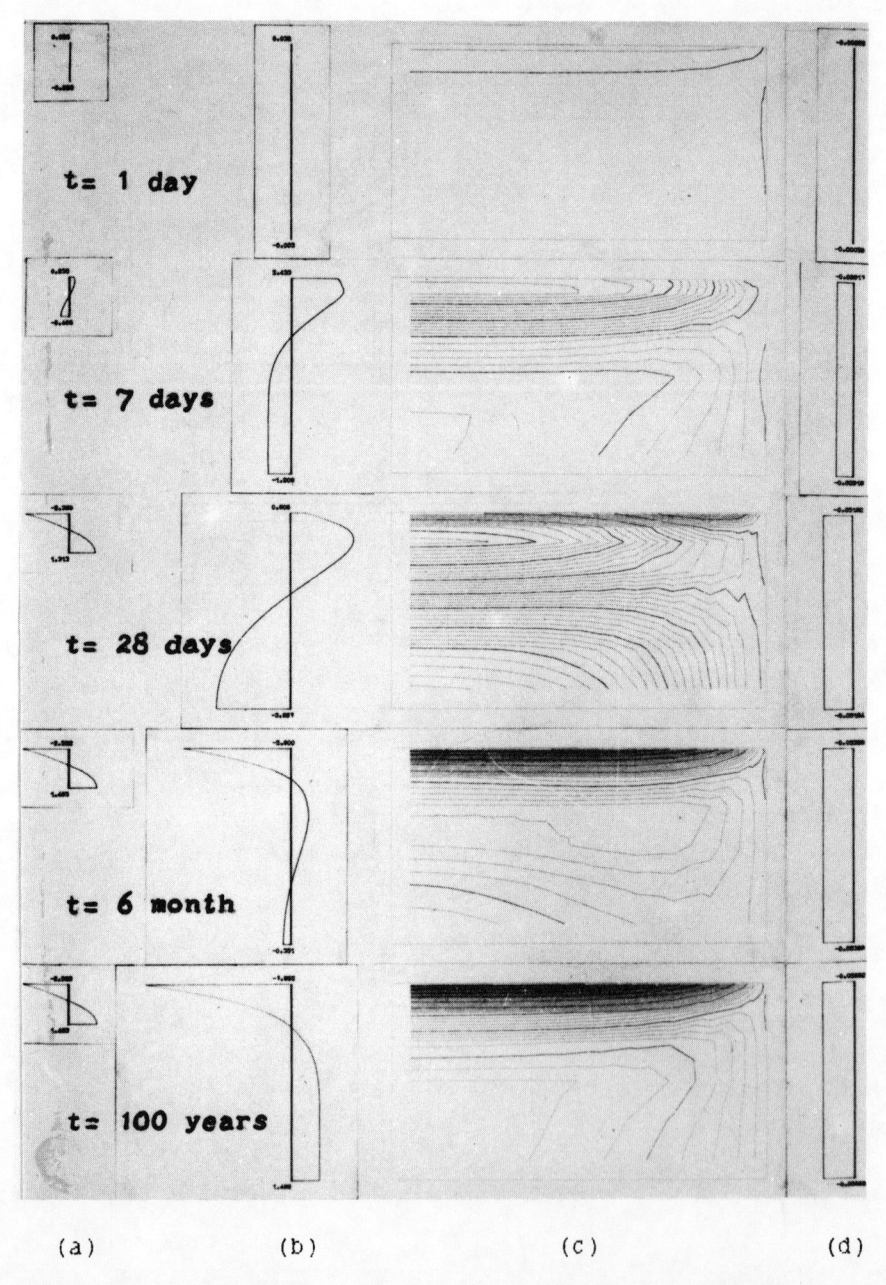

Figure 9. The distribution of : radial stresses σ_{rr} along the axis of a 2 m high block (a); radial stresses σ_{rr} along the axis of a 10 m high block (b); contour map of radial stresses σ_{rr} in a 10 m high block (c); radial strains ε_{rr} along the axis of a 10 m high block (d).

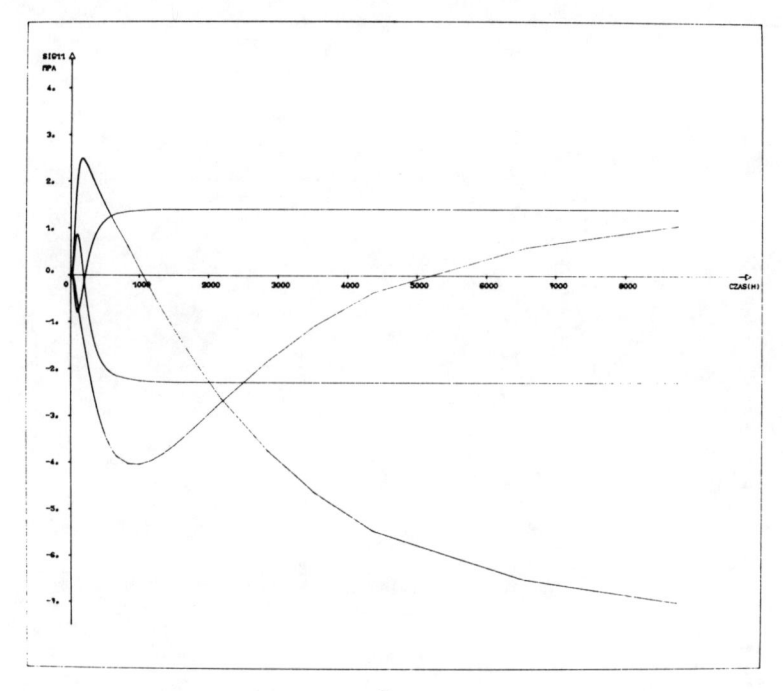

Figure 10. Radial stresses σ_{rr} development in selected points in the 2 m and 10 m high blocks up to 8000 hours after casting.

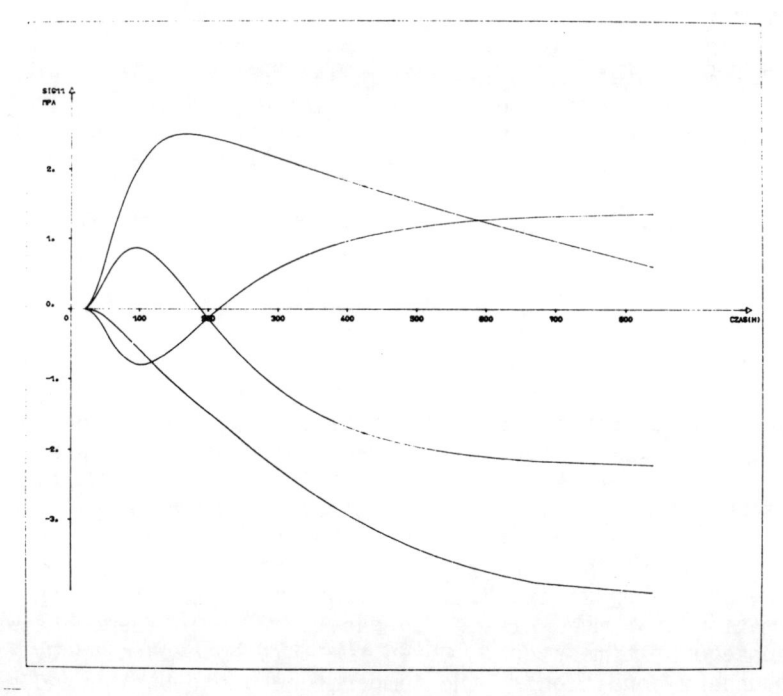

Figure 11. Radial stresses σ_{rr} development in selected points in the 2 m and 10 m high blocks up to 800 hours after casting.

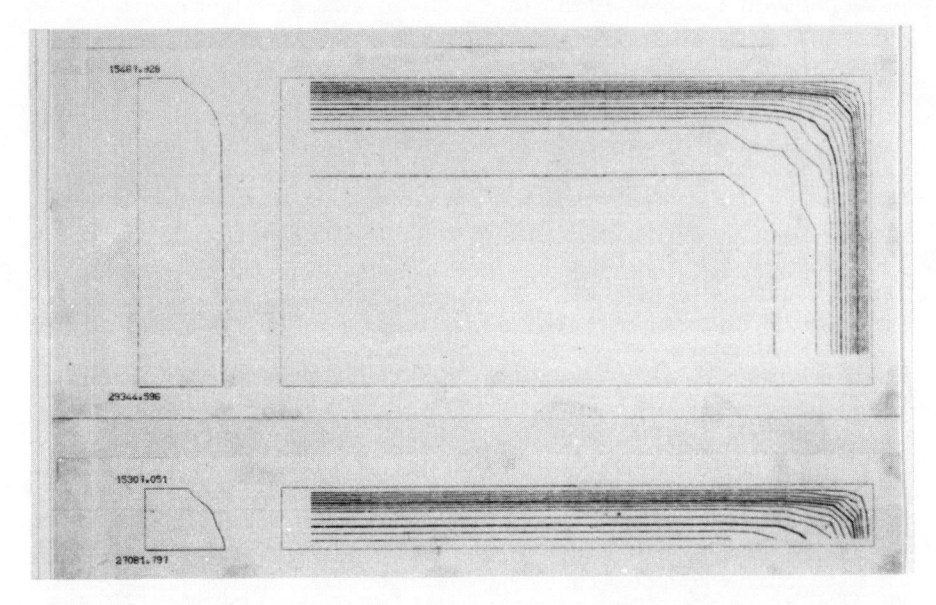

Figure 12. Comparison of Young modulus distributions in the 10 m and 2 m high blocks in the age of 1 week.

6. COMPUTATIONS

The computational algorithm based on the finite element method was developed for performing calculations according to the theory presented above. For transient heat conduction equation the following iterative scheme was obtained (see [14]):

(4) $\quad (C_n/\Delta t_n + K_n \theta_n)T_{n+1} = (C_n/\Delta t_n - K_n(1-\theta_n))T_n + F_{n+1}\theta_n + F_n(1-\theta_n)$

The subscript n denotes the number of time step. Matrices C and K depend on the specific heat and thermal conductivity, respectively. Due to the lack of relevant data, they were assumed to be constant during calculations, although the algorithm is prepared to accept their changes at every time step. The heat sources output for two subsequent time steps were denoted by the F_n and F_{n+1} vectors. They were calculated at each time step. The set of experimentally obtained functions $W_T(t)$ was used to update the values of F vectors according to the results obtained at the earlier time steps. Specially developed extrapolation algorithm employing the concept of reduced time was utilized for approximate F_{n+1} calculations. θ_n defined the approximation in time, giving the partition of influence between two subsequent time steps. Given the initial condition To and the boundary conditions, the temperature distribution and heat generated in each of the elements was established for every time step. The mechanical equilibrium equation was then solved with temperature gain at the actual step as load. Young modulus, Poisson ratio and the strength of concrete were calculated for all elements at the given time step according to the heat produced in each element up to the moment considered. The stresses obtained for each time step were then summed up to form a stress-time relationship.

As an example thermal stresses in maturing concrete cylinder were calculated for several thicknesses. The results for two cylinders are given here, both having the same diameter of 10 m and the height of 2 m and 10 m, respectively. Both cylinders were assumed to be made of the same concrete. In fig.2 the temperature distribution development in the bigger block is given. It can be seen that the temperature gradients occur in the relatively narrow zone along the surface of the cylinder, at least at the early stages of the process. In this zone the highest stresses would develop. The temperature distribution in time in two points: one situated in the center of the top surface and the other one in the center of the block for both 2 m and 10 m high blocks are drawn in figures 3 and 4 with various scales at the time axis. It is evident that in the bigger block the temperature in the center reaches higher values and remains elevated for considerably longer time than in the smaller one. In fig. 5 the sources function distribution in time is shown in the similar manner up to 800 hours after casting. This function also reaches higher values in the case of the bigger block as compared to the smaller one. However, concrete maturing runs faster in this case and the sources function decreases more rapidly than for the smaller block. In figures 6 and 7 spatial distribution of sources function is presented in the form of contour maps for both heights of the blocks considered 3 days and 1 week after casting, respectively. At the early stages of maturing the heat generation is bigger in the central regions of the blocks, while at the later stages it takes place rather in the outer regions. Since heat abstraction is easier in this regions and heat generation rates are smaller, it can be stated that the most dangerous are the early stages of maturing up to about 100 hours after casting. In fig. 8 the relation between the total heat generated and time is shown up to 800 hours. The maturing runs almost identically in the outer regions of both blocks, while significant differences can be noted between the central regions of both blocks. In the bigger block concrete matures faster due to higher temperature. The σ_{rr} stresses are shown in fig.9 for five selected times in the form of both a contour map and a diagramm of the σ_{rr} stresses distribution along the axis of the cylinder. The 2 m and 10 m high cylinders are compared. At the far right-hand side of this picture the strain distribution is shown. As follows from this figure no relationship can be observed to exist between the summed up stresses and strains. The development of radial stresses σ_{rr} is depicted in figures 10 and 11 up to 8000 hours and 800 hours, respectively. What may seem to be a paradox, according to classical design, the bigger block would undergo damage (cracking) while the smaller one would remain intact. The distribution of Young modulus values in the cylinder is shown in fig. 12 for both heights of the block. It is evident that the Young modulus is varying with the position in the block. It is also varying with respect to time in the manner similar to that of the total heat generated Q dependence on time.

7. REFERENCES

1. HOFFMANN E. - Studies on stress state in concrete gravity dams aided by building site measurements, Dissertation, Karlsruhe, 1933, (in German).
2. GLOVER R. - Calculation of temperature distribution in a succesion of lift to release of chemical heat, Journal of the Amer. Conc. Inst., 9, (1937).
3. HIRSCHFELD K. - Temperature distributions in concrete, Springer, Berlin, 1948, (in German).
4. HOUGHTON D.L. - Field study of interior temperatures in concrete, Journal of the Power Division Proceedings of the American Society of Civil Engineers, (1959).
5. HANSEN T.C. - Surface cracking of mass concrete structures at early form removal, Bulletin RILEM No.28, Paris, 1965.
6. ALIEKSANDROWSKIJ S.W. - Design of concrete and reinforced concrete structures with respect to temperature and humidity (accounting for creep), Strojizdat, Moscow, 1966, (in Russian).
7. ZAFOROZIEC I.D., OKOROKOW S.D., PARIJSKIJ A.A. - Heat generation in concrete, Literature on Building Publishers, Leningrad - Moscow, 1966, (in Russian).
8. Studies of stress state in tall concrete dams (theme G-P-2, stage II), COMECOM, Steady Commision for Electric Energy, Dresden, 1970, (in Russian).
9. WITAKOWSKI P. - Analysis of thermal stresses in concrete massives, Transport and Communication Publishers, Warsaw, 1977, (in Polish).
10. ROELFSTRA P.E., WITTMANN F.H. - Modelling the time dependent behaviour of concrete, Tr. 6th Int. Conf. SMiRT., Paris, 1981.
11. ZEITLER W. - Studies on temperature and stress state in concrete elements due to hydration, Dissertation, Technische Hochschule Darmstadt, Darmstadt, 1983, (in German).
12. YAMAZAKI M., MIYASHITA T., MORIKAWA H., HAYAMI Y., - An analytical study on the thermal stress of mass concrete, Kajima Institute of Construction Technology, Report No. 57, Tokyo, 1985.
13. RASTRUP E. - Heat of hydration in concrete, Magazine of Concrete Research, 6, (1954).
14. ZIENKIEWICZ O.C. - Finite element method in engineering practise, 3rd ed., McGraw - Hill, London, 1977.

NUMERICAL SIMULATIONS OF RESIDUAL STRESS
AND DISTORTION DURING WELDING PIPES

Hong-Qing Yang[I], Chu Chen[II] and Jianhua Wang[II]

SUMMARY

A great amount of effort has been devoted to study the residual stress resulting from welding process in recent years. This is because the welding thermal stresses and the resulting residual stresses have important influence on the strength of welded structures, and causes brittle fracture, buckling and weld cracking. In this paper, the constitutive relations of thermo- elastic-plastic material behaviour with temperature dependent physical properties are derived, and a finite element method based numerical scheme for the determination of the thermally induced elastic-plastic stress and distortion is described for axisymmetric pipe welding. The detailed generating process of thermal stresses is tracked. The effect of the pipe wall thickness and welding parameters are also addressed.

1. INTRODUCTION

Welding has been widely used in the fabrication of building structures, ships, pressure vessels to name a few. However, the inherent problems of the welding procedure still need to be solved to get reliable joints. One of the major problems is stress, which causes distortions during welding, and persists as residual stress, consequently influencing the buckling strength and brittle fracture strength of the welded structures.

To better understand the formation of the welding residual stresses, it is important to obtain the entire history of the

---

I. Department of Aerospace and Mechanical Engineering
   University of Notre Dame, Notre Dame, IN 46556, U.S.A.
II. Department of Material Science and Engineering
    Shanghai Jiao Tong University, Shanghai,
    People's Republic of China

stresses and strains. Early studies were more or less experimental or qualitative if analytical. Fortunately, the application of computers have made quantitative analysis possible through numerical analysis. In 1971, Ueda and Yamakawa [1] developed a method of thermo- elastic-plastic analysis of the thermal stresses during welding based on the finite element method, and from then on research in this field has progressed rapidly [2-7]. The present study is motivated from the investigation of stress relaxation in welded pipes by non-uniform post-welded heat treatment, a method which was successful in the flat plate case. This paper represents our continual efforts on the applications of numerical method in welding processes [8], including heat transfer during welding [9-10], welding residual analysis in flat plate [11].

The constitutive equations describing the thermo-elastic-plastic material behaviour with temperature dependent physical properties are derived in the tensor form so that they can be applied to more general coordinates used. The example is studied to track the formation of residual stresses and distortion during welding pipes.

2. THERMAL ELASTIC-PLASTIC THEORY

2.1. Constitutive Equations

The welding process stands out as an extremely complex material process. Temperature varies from ambient to above melting point and as a result, temperature dependent physical properties nonlinearises the problem over and above the already existing nonlinearity of plastic deformation in the heated affected zone (HAZ). The problem can be made tractable by linearization and by using increments of displacement, strain and stress.

The total strain is decomposed into elastic strain increment $d\varepsilon^e_{kl}$, plastic strain increment $d\varepsilon^p_{kl}$, and thermal strain increment $d\varepsilon^T_{kl}$,

$$d\varepsilon_{kl} = d\varepsilon^e_{kl} + \alpha \, d\varepsilon^p_{kl} + d\varepsilon^T_{kl} \tag{1}$$

where α is a coefficient for the purpose of finite element analysis. It will be discussed later.

According to elastic theory, stress σ_{ij} is related to elastic strain through generalized Hooke's law,

$$\sigma_{ij} = D^e_{ijkl} \, \varepsilon^e_{kl} \tag{2}$$

D^e_{ijkl} is the elastic moduli of the material,

$$D^e_{ijkl} = E/(1+\mu) \, (\delta_{ik} \delta_{jl} + \mu/(1-2\mu) \, \delta_{ij} \delta_{kl}) \tag{3}$$

E is the Young's modulus, μ the Poisson's ratio.

Recalling that physical properties are temperature dependent, the incremental elastic strain is given as,

$$d\varepsilon^e_{kl} = (D^e_{ijkl})^{-1} \sigma_{ij} + \partial(D^e_{ijkl})^{-1}/\partial T \, \sigma_{ij} \, dT \tag{4}$$

Taking change of thermal expansion coefficient β with temperature into account, the thermal strain increment is found to be

$$d\varepsilon^T_{kl} = (\beta + \partial\beta/\partial T \, dT)\delta_{kl} \, dT \tag{5}$$

To find plastic strain increment, the yielding condition of Von Mises is defined by the yield function

$$G = f - f_0 \tag{6}$$

where
$$f = 3/2 \, S_{ij}S_{ij} \tag{7}$$

S_{ij} is the stress deviator, given as

$$S_{ij} = \sigma_{ij} - 1/3 \, \sigma_{mm} \delta_{ij} \tag{8}$$

and f_0 is the yield stress which is a function of temperature, and plastic strain ε^p_{kl} history.

In plastic state the condition of $G = 0$ is always satisfied, while

$$dG = \partial f/\partial\sigma_{ij} \, d\sigma_{ij} - (\partial f_0/\partial\varepsilon^p_{kl} \, d\varepsilon^p_{kl} + \partial f_0/\partial T \, dT) = 0. \tag{9}$$

According to plastic flow theory,

$$d\varepsilon^p_{kl} = \lambda \, \partial f/\partial\sigma_{kl} \tag{10}$$

λ is a constant of proportionality, and from (1), (4), (9) and (10) it can be shown,

$$\lambda = [\partial f/\partial\sigma_{ij} \cdot D^e_{ijkl} \cdot d\varepsilon_{kl} - \partial f/\partial\sigma_{ij} \cdot D^e_{ijkl} \cdot (d\varepsilon^T_{kl} - \partial f_0/\partial T \, dT)]/S \tag{11}$$

$$S = \partial f_0/\partial\varepsilon^p_{kl} \cdot \partial f/\partial\sigma_{kl} + \partial f/\partial\sigma_{ij} \cdot D^e_{ijkl} \cdot \partial f/\partial\sigma_{kl} \tag{12}$$

The value of λ gives the criterion for the loading process,

$$\lambda > 0 \quad \text{while loading} \tag{13}$$

$$\lambda = 0 \quad \text{while neutral} \tag{14}$$

$$\lambda < 0 \quad \text{while unloading} \tag{15}$$

If a modified plastic moduli is defined as,

$$D^p_{ijkl} = D^e_{ijkl} \cdot \partial f/\partial \tilde{\sigma}_{ij} \cdot D^e_{ijkl} \partial f/\partial \tilde{\sigma}_{kl} /S \tag{16}$$

and the elastic-plastic moduli can be written as,

$$D^{ep} = D^e_{ijkl} - \alpha D^p_{ijkl} \tag{17}$$

combining temperature related stresses into H^{ep}_{ij},

$$H^{ep}_{ij} = D^{ep}_{ijkl} (\beta + \partial \beta/\partial T \cdot dT) \delta_{kl} dT + D^e_{ijkl} \cdot \partial (D^e_{ijkl})^{-1}/\partial T \tilde{\sigma}_{ij}$$

$$+ \alpha/S \cdot D^e_{ijkl} \cdot \partial f/\partial \tilde{\sigma}_{kl} \cdot \partial f_\theta/\partial T \tag{18}$$

a general relation is obtained,

$$d\sigma_{ij} = D^{ep}_{ijkl} d\varepsilon_{kl} - H^{ep}_{ij} dT \tag{19}$$

As seen from (1), α is the coefficient of plastic strain increment, the value of α, $0 \leqslant \alpha \leqslant 1$, represents the state of the material. In elastic range, $\alpha = 0$. In yielding state, if $\lambda > 0$, then $\alpha = 1$; if $\lambda < 0$, $\alpha = 0$. In process of change from elastic state to plastic, or vice versa, an appropriate value of α between 0 and 1 gives the real state of the material accurately.

In all, the constitutive equation is

$$d\tilde{\sigma}_{ij} = D_{ijkl} d\varepsilon_{kl} \tag{20}$$

in elastic range and unloading with $\lambda < 0$,

$$D_{ijkl} = D^e_{ijkl}, \text{ and } \alpha = 0 \tag{21}$$

in plastic range and $\lambda > 0$,

$$D_{ijkl} = D^{ep}_{ijkl} \text{ and } \alpha = 1 \tag{22}$$

2.2 Principle of Virtual Work

At time t, assuming equilibrium state of body and known stress $\tilde{\sigma}_{ij}$, any increment in external surface force dP_i on S_p, du_i on S_u, and internal body force df_i, can be related by the following,

$$d\tilde{\sigma}_{ij,j} + df_i = 0 \tag{23}$$

The strain displacement relation is

$$d\varepsilon_{ij} = 1/2 (du_{i,j} + du_{j,i}) \tag{24}$$

and stress strain relation (20). Boundary conditions are

$$d\tilde{\sigma}_{ij}\, \underline{n}_j = d\underline{p}_i \qquad \text{on } S_p \qquad (25)$$

$$du_i = d\underline{u}_i \qquad \text{on } S_u \qquad (26)$$

If the system is given a virtual displacement du_i, the work done by external force and internal body force is

$$\delta W = \int_V df_i\, \delta du_i\, dV + \int_S d\underline{p}_i\, \delta du_i\, dS \qquad (27)$$

while the virtual internal energy increment is

$$\delta U = \int_V d\tilde{\sigma}_{ij}\, \delta d\varepsilon_{ij}\, dV \qquad (28)$$

According to the principle of virtual work,

$$\delta W = \delta U \qquad (29)$$

therefore,

$$\int_V \delta d\varepsilon_{ij}\, d\tilde{\sigma}_{ij}\, dV = \int_V df_i\, \delta du_i\, dV + \int_S d\underline{p}_i\, \delta du_i\, dS \qquad (30)$$

2.3 Finite Element Equations

As a result of the symmetric property of stress and strain,

$$d\tilde{\sigma}_{ij} = d\tilde{\sigma}_{ji} \qquad (31)$$

$$d\varepsilon_{ij} = d\varepsilon_{ji} \qquad (32)$$

Previous equations can now be rewritten in the matrix form,

$$\{d\tilde{\sigma}\} = [D]\, \{d\varepsilon\} - \{H\}\, dT \qquad (33)$$

Let L be the differential operator for the strain displacement relation. The equation (24) will become,

$$\{d\varepsilon\} = L\{du\} \qquad (34)$$

Through discretization, the displacement is expressed in terms of its value at each node, or

$$\{du\} = \Sigma[N]\, \{du\}^e \qquad (35)$$

Here [N] is the shape function, the superscript e refers to the element. Defining the matrix [B] as,

$$[B] = L[N] \qquad (36)$$

equation (30) becomes,

$$\{\delta du\}^T \left(\int_V [B]^T \{d\hat{\sigma}\} dV - \int_V [N]^T \{df\} dV - \int_S [N]^T \{dp\} dS \right) = 0. \tag{37}$$

This is for any virtual displacement $\{\delta du\}$. By substituting (33) into (37), following is obtained,

$$\left(\int_V [B]^T [D] [B] \, dV \right) \{du\} - \left(\int_V [B]^T \{H\} dV \right) dT^e -$$

$$\int_V [N]^T \{df\} \, dV - \int_S [N]^T \{dp\} dS = 0 \tag{38}$$

For simplification, let

$$[K] \{du\} = \{dF\} \tag{39}$$

where, the stiffness matrix [K] is

$$[K] = \Sigma \int_{V^e} [B]^T [D] [B] dV^e \tag{40}$$

$$\{dF\} = \Sigma \left[\int_{V^e} [B]^T \{H\} dV^e \, dT^e + \int_{V^e} [N]^T \{df\} \, dV^e \right.$$

$$\left. + \int_{S^e} [N]^T \{dp\} \, dS^e \right] \tag{41}$$

For welding, $\{df\}=0$, $\{dp\}=0$, which simplifies the above to the following,

$$\{dF\} = \Sigma \int_{V^e} [B]^T \{H\} \, dV^e \, dT^e \tag{42}$$

The solution of thermo-elastic-plastic problems is essentially done by linearizing the non-linear stress-strain relation in an incremental way. In general, the external force $\{dp\}$ does not exist. The force term in (39) are basically due to the temperature variation ΔT from time t to t+Δt. To account for the contribution of the ΔT accurately, the ΔT is divided into several increments (say n) according to its magnitude. With this dT = $\Delta T/n$, the right hand side of (39) is formed through expression (42). According to the status of the element the stiffness matrix [K] can also be formed by (40). The displacement $\{du\}$ is solved through a system of linear equations (39) by Gaussian elimination or matrix inversion. Once $\{du\}$ is obtained, the strain is found from (34) and in turn the stress can be found through (33). Plastic strain can be calculated through expression (10). For further calculations, another temperature difference $\Delta T/n$ is added. A similar procedure is followed until the whole load ΔT is exerted on the system. Marching in time, the new temperature variation from t+Δt to t+2Δt is evaluated either numerically or analytically. The procedure is repeated until ambient temperature is reached. The calculated residual stress are thus retained.

RESULTS AND DISCUSSIONS

In the problem considered, thermal stresses induced by

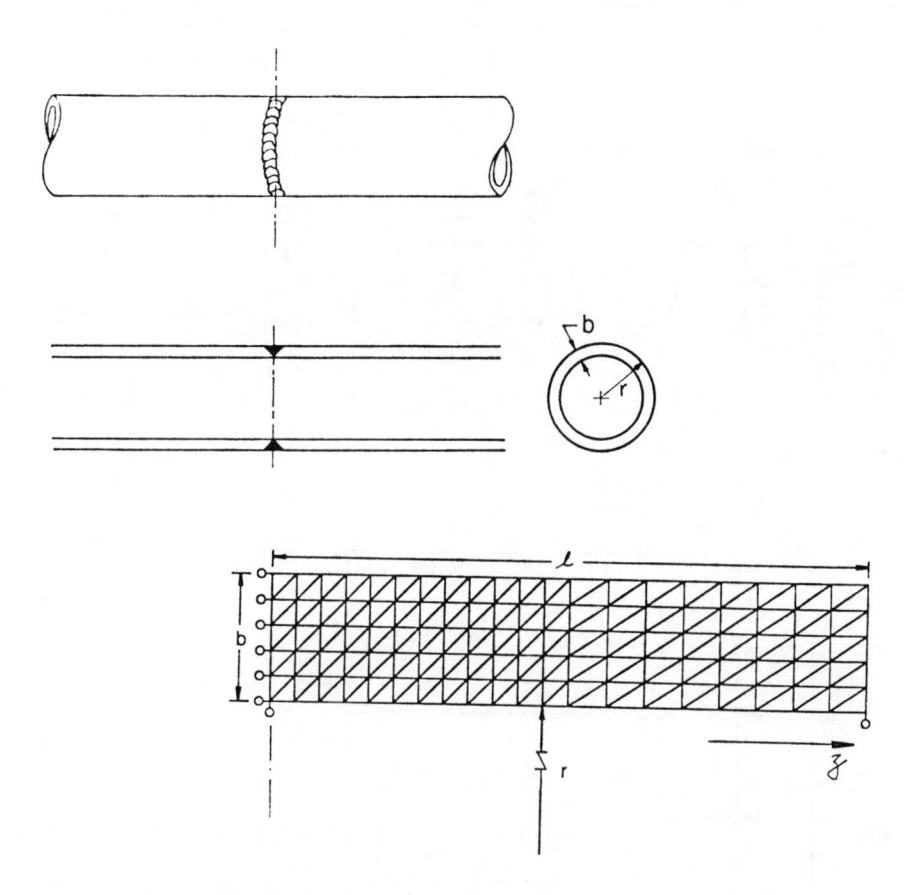

Fig.1 Welding Structure and Finite Element Meshes

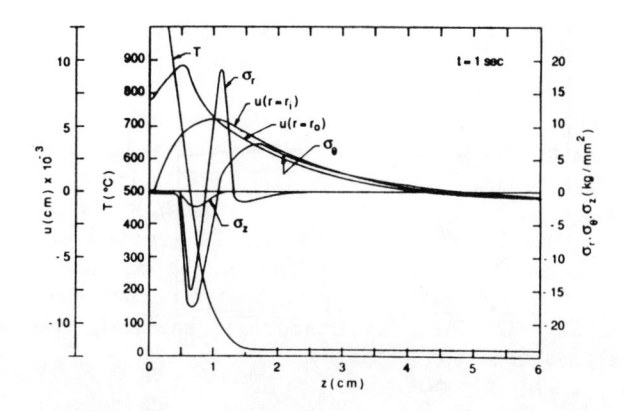

Fig. 2 (a) Temperature, Stresses and Distortions at 1 Second

(b) t= 10 seecond

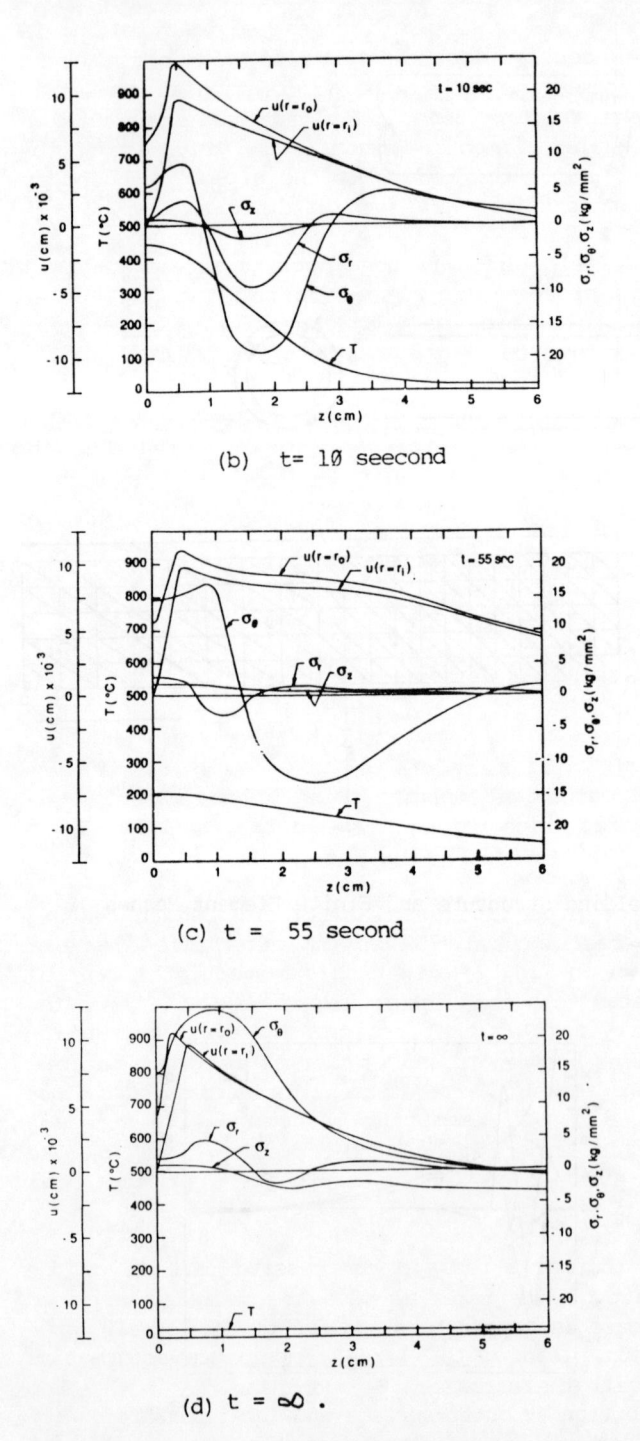

(c) t = 55 second

(d) t = ∞ .

Fig. 2 Temperature, Stresses and Distortions at Various Time

welding pipes is calculated. The general geometry and the finite element mesh is shown in Fig. 1. It is assumed that welding is performed instantaneously around the circumference, hence deformation is axisymmetric. Since, the deformation is symmetric with respect to the welding line, only half the structure is discretized by triangular elements. Here h is the thickness of the pipe, r the radius, and l the length of the pipe. The boundary conditions of displacement are also given in the figure.

The temperature field which is the input to stress anaylsis during the welding process was obtained from the analytical solution assuming thin pipe wall thickness. The non-linear conduction problem related to welding has been studied in [10].

The yield stress and Young's modulus approach zero once temperature is higher than a certain value T_m ($T > T_m$). In our model, the following relations of the physical properties were taken,

$$f_\emptyset = \tilde{\sigma}_s = 2500 \; (T_m - T)/T_m + 10 \; (kg/cm^2) \quad \text{if } T < T_m$$

$$= 10 \quad \text{if } T > T_m \quad (43)$$

$$E = 2.0 \times 10^6 \; (T_m - T)/T_m + 5.0 \times 10^4 \; (kg/cm^2) \quad \text{if } T < T_m$$

$$= 5.0 \times 10^4 \quad \text{if } T > T_m \quad (44)$$

here T_m, so called mechanical melting point of the material, is 700°C. For temperatures higher than T_m, the thermal load ΔT was set to zero in the code used.

The results for pipe thickness b=0.4 cm, length l=24cm, radius r=10 cm and heat input Q_v=750 cal/cm after time elapse of 1 second, are shown in Fig. 2 (a). At this moment, the central region has been heated to a temperature higher than T_m, hence the stresses are nearly zero in that region. However, the region where temperature lies between T_m and $T_\emptyset$ thermal expansion occures. $T_\emptyset$ is the initial and ambient temperature. This expansion can not uniformly follow $\beta(T-T_\emptyset)$, resulting in compressive stresses (where ever $\beta(T-T_\emptyset)$ is larger then real deformation), and tensile stresses (where ever $\beta(T-T_\emptyset)$ is smaller than the real deformation).

Actually the thermal stress in compressive part is of the order of yield limit of the material due to decrease in yield limit $\tilde{\sigma}_s$ with increase in temperature and due to the locally large temperature variation, leading to large temperature deformation (large $\beta(T-T_\emptyset)$). In all the stresses, the dominant one is the hoop stress $\tilde{\sigma}_\theta$. The distortion at outer surface and inner surface of the cylinder are shown in figure (2). The distortion in the central region follows a different pattern.

After 10 seconds (Fig.2(b)), the temperature in the central region drops to the range below T_m and as such the material recovers its ability to deform. However the effective stress is following the yield limit with the change in temperature. The region of compressive stress moves away from the center. The stress distribution at t=55 second (Fig. 2(b)) is similar to the one at 10 second, except that as temperature decreases the stresses are higher in the central region. The residual stresses are obtained upto the time when temperature is back to the ambient. (Fig. 2(d)).

Qualitatively, the residual stress $\tilde{\sigma}_\theta$ at the center are shown in Fig. 3. The highest residual stress and temperature reside in this region. It can be divided into four time regions. In time region I temperature increase results in compressive stress. In region II the stress of the material is in the plastic region, therefore the stress is equal to $\tilde{\sigma}_s$, which follows the temperature variation. In region III, $T > T_m$, stress is almost zero. In region IV, the stress is recovered, while still following $\tilde{\sigma}_s$ to the final residual one.

Fig.4 presents the residual hoop stress for the case when heat input per unit area Q_v/b is fixed, and thickness of the pipe b changed. It is seen that the width w within which the stress is of the order of yield limit $\tilde{\sigma}_s$ is almost invariant with respect to b. The distortions are however very different. Apparently, the thicker the pipe, the stiffer the system, and therefore smaller the distortion. In the non-uniform post-welded heat treatment, the width w where the residual stress is around yield limit is a very important factor in determining the non-uniform heating. With the present study, general approach of anaylzing the residual stress and its distribution can be obtained and proper parameters of post-welded heat treatment can be chosen. Further work is being undertaken.

CONCLUDING REMARKS

The constitutive relations of thermo-elastic-plastic material behaviour is described in tensor notation with consideration of material from elastic state to plastic state by using proper coefficenet α. The application of tensor transformation enables one to analyse non-Cartesian coordinates as well. The thermal stress and distortion induced by welding pipes are tracked and the region where the residual stress is the order of yield limit is found to be insensitive to the pipe thickness for the case of fixed Q_v/b. The distortions are dependent on b and becomes smaller as b increases due to increased stiffness of the structure.

REFERENCES

1. UEDA, Y., and YAMAKAWA, T. - Analysis of Thermal-

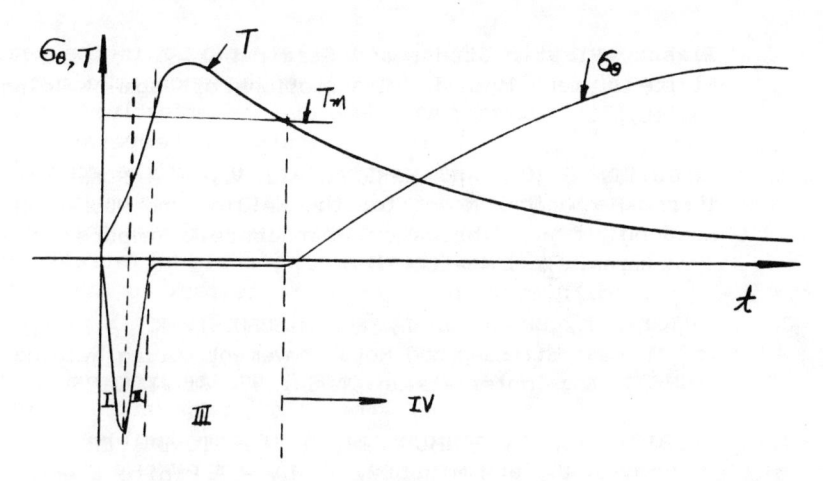

Fig. 3 Qualitative Description of stress $\tilde{\sigma}_\theta$ with time and Temperature

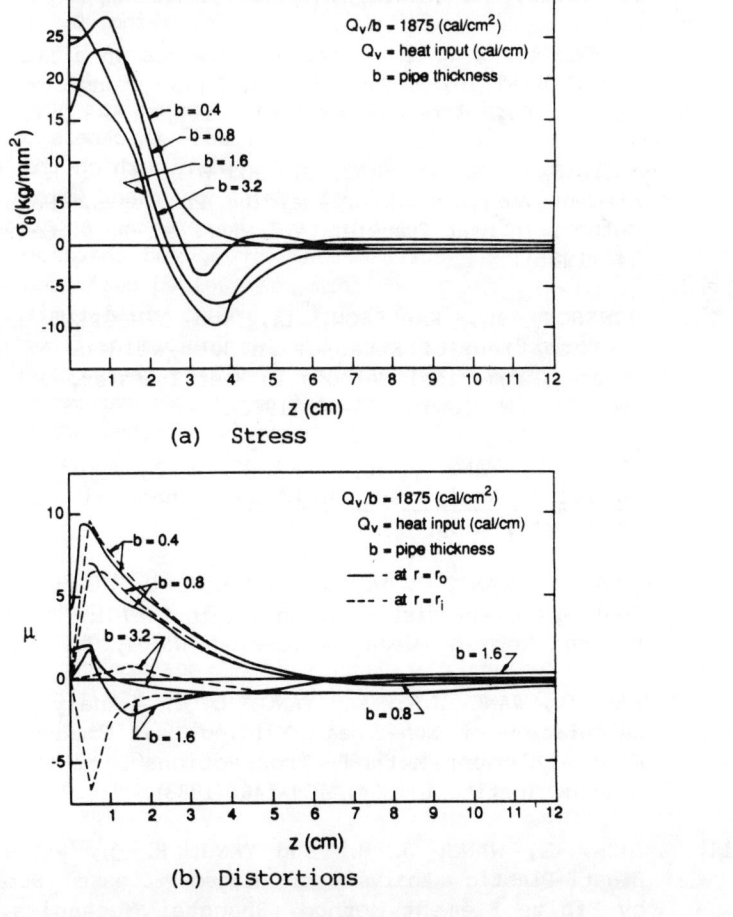

(a) Stress

(b) Distortions

Fig. 3 Dependence of Stress $\tilde{\sigma}_\theta$ and Distortions on b at fixed Q_V/b.

Elastic-Plastic Stress and Strain During Welding by Finite Element Method, Transactions of Japan Welding Society, 2, 90-100 (1971).

2. HIBBITT, H. D., and MARRCAL, P. V., - A Numerical Thermo-Mechanical Model for the Welding and Subsequent Loading of a Fabricated Structure, Computer and Structures, 3, 1145-1174 (1973).

3. MURAKI, T., BRYAN, J. J., and MASUBUCHI, K., - Analysis of Thermal Stresses and Metal Movement During Welding, ASME J. Eng. Materials and Tech., 97, 81-91 (1975).

4. RYBICKI, E. F., SCHMUESSER, D. W., STONESIFER, R. B., Groom, J. J., and MISHLER, H. W., - A Finite Element Model of Residual Stresses in Girth-Butt Welded Pipies, Pro. 1977 ASME WAM, Numerical Modeling of Manufacturing Processes, PVP-PB-025, 131-142 (1977).

5. ANDERSSON, B. A. B. - Thermal Stresses in a Submerged-Arc Welded Joint Considering Phase Transformations, ASME J. Eng. Material and Tech., 100, 356-362 (1977).

6. ARGYRIS, J. H., SZIMMAT, J., and WILLAM, K. J. - Finite Element Analysis of Arc-welding Processes, Numerical Methods in Heat Transfer, Vol. III, ed. R. W. Lewis, 1-34 (1985).

7. JONSSON, M., KARLSSON, L., and LINDGREN, L., - Deformations of Stresses in Butt-welding of Large Plates, Numerical Methods in Heat Transfer, Vol. III, ed. R. W. Lewis, 35-58 (1985).

8. CHEN, C., WANG, J. W., and YANG, H. Q., Applications of Numerical Analysis in Welding, Shanghai Jiao Tong University, 355p, 1985.

9. CHEN, C., WANG, J. W., and YANG, H. Q., -Analysis of Cooling Characteristics in Underwater Welding by Finite Element Method, Ocean Engineering, No. 4, 34-38 (1983).

10. CHEN, C., WANG, J.W., and YANG, H. Q., - Analysis and Calculation of Non-linear Welding Heat Conduction by Finite Element Method, Transactions . of the China Welding Institution, 4, 139-148 (1983).

11. CHEN, C., WANG, J. H., and YANG, H. Q., -Thermo Elasti-Plastic Analysis of Heated-Processed Stresses by Finite Element Method, Shanghai Mechanics, No. 4,(1984).

Continuous Composite Bridge Subjected to Negative Thermal Gradients
by
H.C.Fu[1], S.F.Ng[2] and M.S.Cheung[2]

[1] Civil Engineering, Royal Military College, Kingston, Canada
[2] Civil Engineering, University of Ottawa, Ottawa, Canada

ABSTRACT

This paper presents the analytical results obtained from a parametric study of a 0.354 scale bridge model under various "negative" thermal loads by means of finite element computer program ADINA-T (Automated Dynamic Incremental Nonlinear Analysis of Temperatures). Results reveal that (1) analytical solutions from ADINA-T, using two different mesh sizes, compare favourably with available experimental data; (2) temperature distribution in the bridge deck is very sensitive to the rate of ambient temperature rise or drop of up to 5 °F/min; (3) convection loss or gain from the air is small; (4) two-dimensional heat flow is limited within the boundaries of the steel stringers and that (5) the structural behaviour of a bridge under positive and negative thermal gradients are distinct.

1. INTRODUCTION

In a recent conference held in Ottawa, Canada in 1986 on "Bridge Structures", it was pointed out that about 130 or 1% of Ontario's bridges required rehabilitations, which will eventually cost hundreds of milions of dollars. The Federal Highway Administration, in an sbstract addressed to the June 1983 issue of "Public Roads" [1], emphasized and warned that "one out of every four bridge decks in the United States ,many of which are less than 20 years old, is badly deteriorated .... rehabilitation or replacement ... estimated as high as $25 billions ..." and that "major bridge deck deterioration problem was due to corrosion of reinforcing steel in chloride-contaminated concrete".

Highway bridges are exposed to daily and seasonal environmental temperature fluctuations throughout their service life. For continuous concrete-deck-steel-beam composite bridge, thermally induced stresses and temperature changes, are inherent because (1) differential thermal movements,due to the non-identical coefficients of thermal expansion of the two materials, induce stresses at the common boundaries of concrete and steel, (2) non-uniform temperature distribution due to the differential thermal conductivity properties of the two materials results in internal stresses and (3) continuity of spans provides external restraints to the thermal movements of the bridge. The first concern for thermal stresses that develop as a consequence of temperature differential in bridge structures is thermal cracks and deformations. The second concern for thermal loadings is the probability of the occurrence of the worst load combinations in bridge design.

Thermal loading depends upon local climate conditions and therefore is extremely variable. In order to understand thoroughly the thermal behaviour of concrete deck-steel beam composite bridges, temperature distributions in bridges under both short term temperature variations, daily or hourly cycles, and long term temperature variations, seasonally or annual cycles must first be determined.

A bridge deck is said to be subjected to positive thermal gradient when the temperature at the top surface of the bridge deck is greater than the temperature at any other points inside the bridge deck. For example, during a sunny and warm day, when the concrete deck is heated by solar radiation and the steel beams are relatively sheltered to maintain a lower and near uniform temperature as that of the surrounding air. As a result, the concrete deck is at a higher temperature compared to the almost uniform temperature in the steel girders. Service load stresses may tend to cancel some of the resulting thermal stresses.

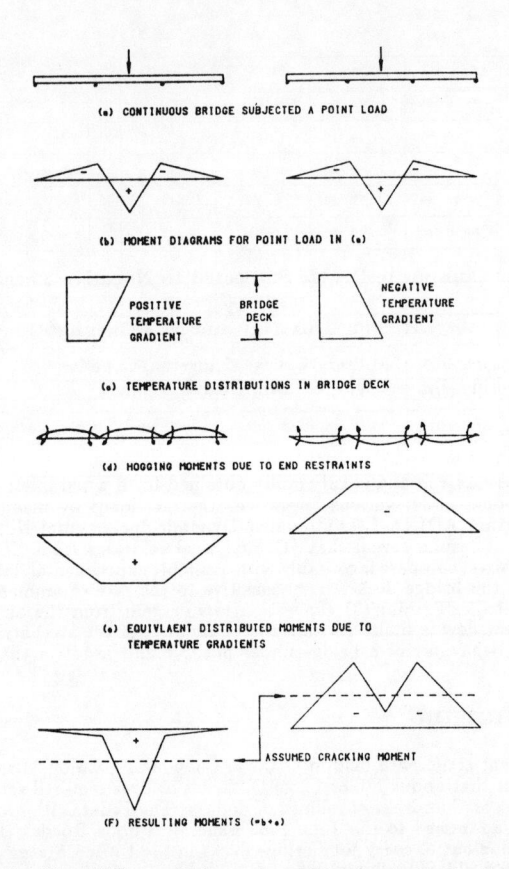

Figure 1 Combined Effects of Thermal and Live Loads in Continuous Bridge

A negative thermal gradient simply means the temperature at the bottom surface of the bridge deck is the maximum. In a sunny but cool day when the exterior beams are exposed to solar radiation or when the exterior and interior beams are subjected to a rapid increase in temperature due to hot air waves, steel with a higher thermal conductivity adjusts its temperature very quickly to the change of air temperature and the concrete deck, because of its poor thermal conductivity, maintains a much lower temperature, leading to a high-gradient temperature distribution through the bridge deck. Since it is the thermal gradient and not the thermal differential that contributes to most of the internal thermal stresses, hot-steel-beam and cool-concrete-deck could result in a more severe loading combination to the bridge than the cool-beam and hot-deck situation (Fig.1).

The objective of this investigation was to determine analytically the transient temperature distributions of a 0.354 scale three-span continuous concrete-deck-steel-girder composite bridge model under various negative thermal loads. In assessing the factors that are influential to the temperature distributions and thus the structural response in bridge, the following variables were relevant and considered : rate of air temperature changes, convection flow, finite element mesh size and two-dimensional nature of heat flow. A total of 17 different thermal loads were considered (Table 1).

2. PREVIOUS RESERACH

Over the past three decades, most investigations have been confined to either concrete bridges or thermal loadings in which the concrete deck is heated with the steel beams underneath maintaining a uniform temperature as that of the air, paucity of experimental and analytical studies directed to concrete-steel composite bridges subjected to negative thermal loading is noticeable. Zuk in 1961[2] conducted a theoretical study, based on one-dimensional thermoelasticity analysis, for determining the strains and stresses caused by various linear thermal gradients in a simple span concrete-and-steel composite

CASE	BEAM A	BEAM B	BEAM C	BEAM D	DUR. MIN.	MAXIMUM TEMPERATURE GRADIENT IN °F A	B	C	D
1	70 TO 120	70	70	70	25	38.0	0.0	0.0	0.0
2	70	70 TO 120	70	70	25	0.0	38.0	0.0	0.0
3	70 TO 120	70 TO 120	70	70	25	38.0	38.0	0.0	0.0
4	70 TO 90	70 TO 80	70	70	10	19.4	9.2	0.0	0.0
5	70 TO 100	70 TO 90	70	70	15	25.8	17.2	0.0	0.0
6	70 TO 110	70 TO 100	70	70	20	32.2	24.2	0.0	0.0
7	70 TO 120	70 TO 110	70	70	25	38.0	31.4	0.0	0.0
8	70 TO 120	70 TO 110	70 TO 100	70 TO 90	25	38.0	31.4	22.8	15.2
9	70 TO 120	70 TO 100	70 TO 90	70 TO 80	25	38.0	22.8	15.2	7.6
10	70 TO 120	70 TO 90	70 TO 80	70	25	38.0	15.2	7.6	0.0
11	70 TO 120	70 TO 120	70 TO 120	70 TO 120	25	38.0	38.0	38.0	38.0
12	70 TO 120	70 TO 120	70 TO 120	70 TO 120	12.5	44.7	44.7	44.7	44.7
13	70 TO 120	70 TO 120	70 TO 120	70 TO 120	6.25	48.5	48.5	48.5	48.5
14	70 TO 120	70 TO 120	70 TO 120	70 TO 120	5	48.6	48.6	48.6	48.6
15	70 TO 120	70 TO 120	70 TO 120	70 TO 120	50	30.9	30.9	30.9	30.9
16	70 TO 110	70 TO 110	70 TO 110	70 TO 110	20	32.2	32.2	32.2	32.2
17	70 TO 30	70 TO 30	70 TO 30	70 TO 30	20	32.2	32.2	32.2	32.2

THERMAL LOADS:
- TEMPERATURES IN °F
- ROOM TEMP. = 70 °F

Table 1 Thermal Loads Applied To Bridge Model

beam. Menzies in 1968[3] measured the "mean" temperatures of the Moat Street flyover, a concrete-steel composite bridge, by lowering thermometers into tubes which were embedded vertically to a depth of 127 mm in the concrete slab and partially filled with mercury. Based on differential equations for linear heat flow, Emerson[4] derived an iterative method for calculating the temperature distribution of bridges. Emanuel and Hulsey[5] investigated the effects of solar load and ambient air temperature extremes on the daily and yearly temperature distributions through an interior deck-stringer section of a concrete-steel composite highway bridge located in Missouri. Symko in 1979 [6] carried out a study to determine experimentally and analytically the thermal response of a 0.354 scale three-span concrete-steel composite highway bridge (the same bridge model that this paper was based upon) with emphasis on the transient thermal behaviour. Temperature distributions of a five-span continuous reinforced concrete-two steel-box composite bridge were monitored[7] and showed that a temperature differential between the concrete deck and the steel boxes reached 45 °F and that a maximum temperature differential of 73°F was noted due to solar radiation alone.

3. TRANSIENT HEAT CONDUCTION ANALYSIS

The analysis and design of thermal response of continuous composite bridges due to environmental changes is a complicated iterative process. Determination of temperature distribution in a bridge subjected to temperature changes is dependent upon a prediction of the boundary temperatures of the bridge, which is in turn affected by localities.

The flow of heat in a solid is linear and is governed by the three-dimensional partial differential equation,

$$\frac{\delta}{\delta X}\left(k_x \frac{\delta T}{\delta X}\right) + \frac{\delta}{\delta Y}\left(k_y \frac{\delta T}{\delta Y}\right) + \frac{\delta}{\delta Z}\left(k_z \frac{\delta T}{\delta Z}\right) = c\rho \frac{\delta T}{\delta t} \tag{1}$$

x,y,z = cartesian coordinates
t = time
c = specific heat of medium
ρ = density of medium
T = temperature of medium

and k_x, k_y, k_z are thermal conductivities corresponding to the x,y, and z cartesian axes.

For a bridge deck subjected to thermal changes, it can be assumed that there is no thermal variations along the longitudinal direction z. For temperature-independent material properties and replacing $k/c\rho$ by K, diffusivity of the medium, the flow of heat can be modelled with the following two-dimensional heat flow equation;

$$\frac{\delta}{\delta X}\frac{\delta T}{\delta X} + \frac{\delta}{\delta Y}\frac{\delta T}{\delta Y} = \frac{1}{K}\frac{\delta T}{\delta t} \qquad (2)$$

In order to solve this partial differential equation and to calculate the temperature fields in the solid, material properties such as conductivity, density, specific heat etc are requried and that the following initial and boundary conditions need to be specified : (a) initial temperatures; (b) specified temperature conditions, i.e. temperatures at specific points or surfaces of the body can be specified and (c) boundary conditions which can be defined as follows.

(1) heat flow boundary conditions - heat flow input at selected points or surfaces of the body, i.e. boundary heat flow input, q_s,

$$q_s = k_n \frac{\delta T}{\delta n} \qquad (3)$$

where k_n is the body thermal conductivity in a direction normal to the surface n.

(2) convection boundary conditions - due to convection flow to and from the surrounding as a result of temperature differences between the bridge surface and the air.

$$q_s = h(T(t) - T) \qquad (4)$$

where h is the convection coefficient or heat transfer coefficient, T is the surface temperature of the body in contact with the air and T(t) is the environmental temperature.

(3) radiation boundary conditions - due to heat gained or lost by solar radiation.

$$q_s = h_r(T_r^4 - T^4) \qquad (5)$$

where h_r is the coefficient determined from Stefan-Boltzmann constant (σ), the emissivity (ε) of the radiant and absorbing materials and the shape factor f, T_r is the temperature of the external radiation source and T is the surface temperature of the body.

According to Threlkeld [8], equation (5) can be transformed to,

$$q_s = \alpha I - \varepsilon\sigma(T_s^4 - T_*^4) \qquad (6)$$

α = bridge surface absorbivity
I = solar radiation incident on bridge surface
ε = bridge surface emissivity
T_s = shape temperature of the air in absolute scale (Rankin or Kelvin)
T_* = sky temperature in absolute scale (Rankin or Kelvin)

In general, heat transfer and stress analysis can be assumed to be uncoupled. That is, the nature of heat flow is independent upon the thermal stresses and vice versa. Therefore, heat transfer analysis can be carried out prior to stress and displacement analysis. When the temperatures of a body are knwon, the thermal stresses can be calculated using simple mechanics concepts or more sophisticated finite element displacement models.

4. 0.354 SCALE BRIDGE MODEL

The parametric study was conducted on a model bridge existing in the Structures Laboratory at the University of Ottawa. It is a 0.354 scale model of a three-span reinforced concrete-steel composite highway bridge that spans the Red River near Ste-Agathe, Manitoba, Canada. The model is approximately 53.1 ft.(16.20 m) long and 10.57 ft.(3.22 m) wide, with a reinforced concrete bridge deck of 2.43 in.(6.2 cm) deep. The model slab is continuous in the transverse direction over four equally placed longitudinal steel stringers. The geometry and the structural details of the bridge model are shown in Fig.2.

A normal weight concrete with specified 28-day compressive strength of 4000 psi(28 MPa), water/cement ratio of 0.54, slump of 3 in.(75 mm), 5% entrained air, a retarder (DERATARD HC type) for 60 minutes and maximum aggregate size of 0.4 in(10 mm) was used for the concrete slab. The average compressive strength and modulus of elasticity for

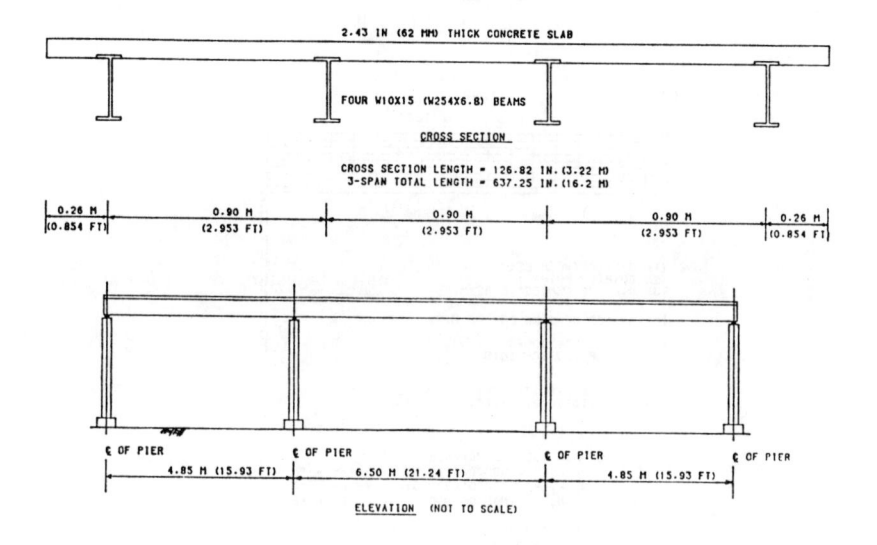

Figure 2 Three-Span Continuous Highway Bridge Model

the concrete were found to be 6900 psi(48 MPa) and 6300000 psi(43800 MPa) respectively. W10X15 structural steel sections were used with a specified yield strength of 43200 psi(300 Mpa) and a tensile strength of 64800 psi(450 MPa).

Material properties of most structural materials, concrete and steel for example, are temperature dependent. More often than not, practicing engineers and researchers believe it is reasonable to assume constant properties for both the heat transfer and thermoelastic analysis. In spite of the fact that ADINA-T[9] is capable of handling temperature dependent material properties, constant material properties were assumed in this study for concrete and steel.

5. ANALYTICAL RESULTS BY ADINA-T

ADINA-T and ADINA[10] are two commericially recognized and widely used compatible finite element computer programs developed jointly in Sweden and the United States. ADINA is a displacement and stress analysis program and ADINA-T is capable of linear and nonlinear steady-state and transient heat transfer analysis. To investigate the thermal behaviour of concrete-steel composite bridge subjected to temperature variations, ADINA-T was used to predict the transient temperature fields in the bridge model subjected to negative thermal loads in which the exterior and/or interior steel beams were subjected to temperature changes of various rates.

5.1 Comparison of ADINA-T Solutions and Experimental Data

In 1979, Symko conducted a cooling test on the same 0.354 scale bridge model at the University of Ottawa, Canada. The bridge, initially at a uniform room temperature of 78.3 °F, was rapidly cooled by covering the upper concrete surface with melting ice cubes within two minutes. Surface and room temperatures were monitored over a span of 77 minutes. The measured temperatures at the top surface of concrete slab and of the surrounding air were used as input into computer program ADINA-T to compute the temperature distribution through the depth of the bridge. Comparison between experimental results and analytical solutions at midspan of the bridge model, based on finite element idealizations 1 for transverse cross section shown in Fig.3 indicated excellent agreement as shown in Fig.4. Also shown in Fig.4 (bottom right) are the average measured boundary temperatures at the top surfaces of concrete slab and the air temperatures used in the analysis.

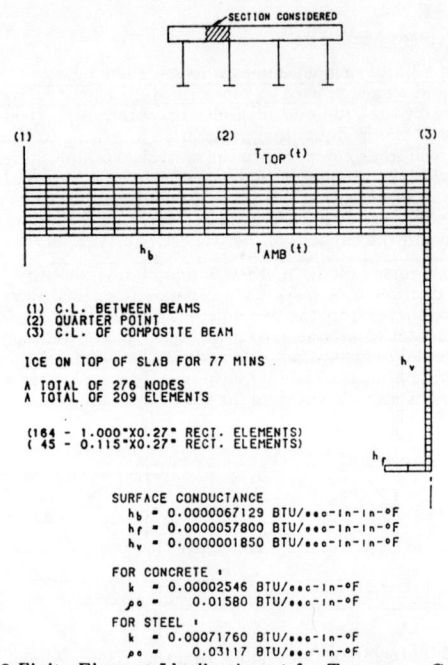

Figure 3 Finite Element Idealizations 1 for Transverse Cross Section

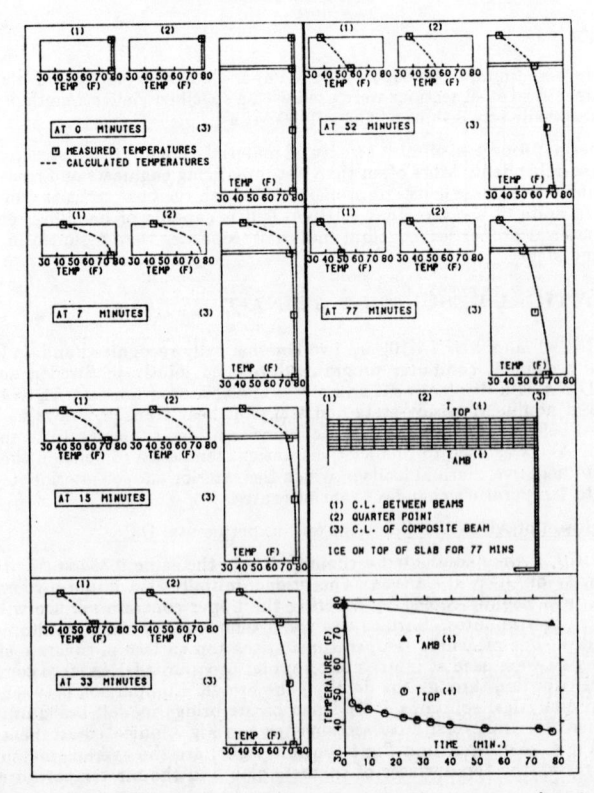

Figure 4 Instantaneous Temperature Fields in Cross section (Mesh 1)

5.2 Effect of Finite Element Mesh Size

The accuracy of a finite element analysis depends among other factors mainly upon the number of nodes and elements used in representing the geometry of the actual structure. ADINA-T results for the 276-node finite element mesh (Fig.3) which fully defined the geometry of about 1/8 of the bridge transverse section appeared to simulate the temperature distribution very well when compared to the measured temperatures. To represent the whole structure and with the number of nodes per transverse section exceeding 2000, the total number of nodes in representing this three-dimensional bridge structure divided into 16 sections would be well over 30000. Solving a 30000-plus node finite element problem would certainly be beyond the capacity of most electronic computers.

A second finite element model mesh was adopted as shown in Fig.5 with a total number of nodes per section of a mere 248, compared to 2000 for the previous mesh. Analytical temperature curves from the two different meshes (Figs.3 and 5) were compared with Sykmo's experimental results as shown in Fig.6. It can be seen that both models predicted the experimental data satisfactorily and that results from these two meshes were practically identical with a maximum difference of only 2.1 °F over a depth of 10.23 in. (260 mm). Finite element mesh 2 was used for subsequent analyses.

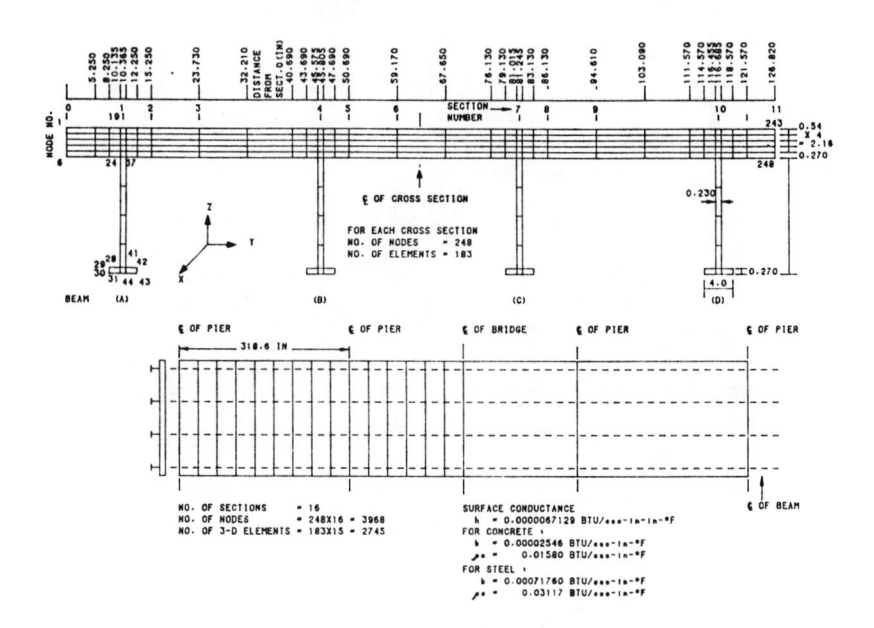

Figure 5 Finite Element Idealizations 2 for Transverse Cross Section

5.3 Convection Flow

The convection flow or the heat transfer between the surrounding air and the bridge surfaces can significantly affect the thermal response of the bridge subjected to environmental changes. Convection coefficient depends primarily upon the wind speed and according to Emanual and Hulsey, can range from 1.929×10^{-6} at still air condition to 9.654×10^{-6} BTU/sec-in^2-°F for very high wind speed condition. Four distinct convection coefficients were used in this study in predicting the temperature differentials of the bridge model when the steel beams were heated at a rate of 2°F per minute for a period of 25 minutes. Fig.7 presents the relationship between convection coefficient and the temperature differential when the steel beams were subjected to heat flow. The results of this series of study appear to confirm that the effects of convection flow on the temperature differentials exist but are insignificant (10%) when considering a high magnitude of temperature differential, say in the neighborhood of 40 °F.

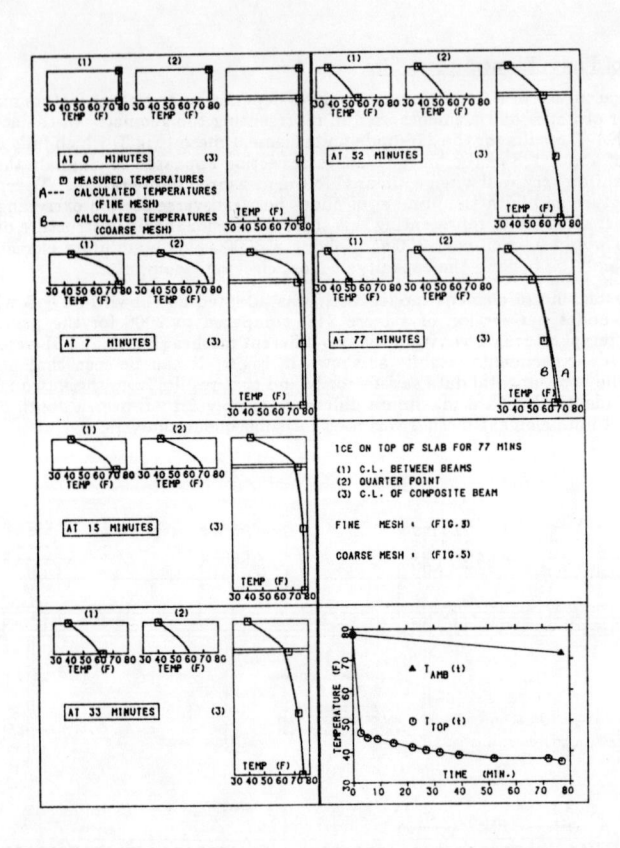

Figure 6 Instantaneous Temperature Fields in Cross section (Mesh 2)

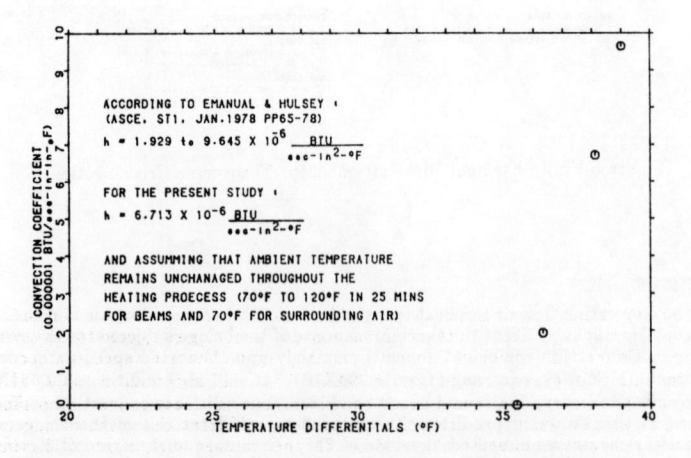

Fig.7 Temperature Differentials vs Convection Coefficients
(Constant Ambient Temperature)

5.4 Heat Flow in Two Dimensions

To understand the two-dimensional nature of heat flow in brdige deck when the steel beams are being heated, instantaneous temperature fields at various section around the heated steel beam or heat source was evaluated and depicted in Fig.8 for differing thermal load rates and constant or variable air temperature conditions. The comparative study indicate that the two-dimensional heat transfer field is limited to within a region of less than 3 in.(75 mm) from the heat source (edge of heated steel beam) for a deck thickness of 2.43 in.(62 mm). However, for deeper concrete decks and for lower rate of heating to the steel beams, larger affected area for two-dimensional heat flow is expected.

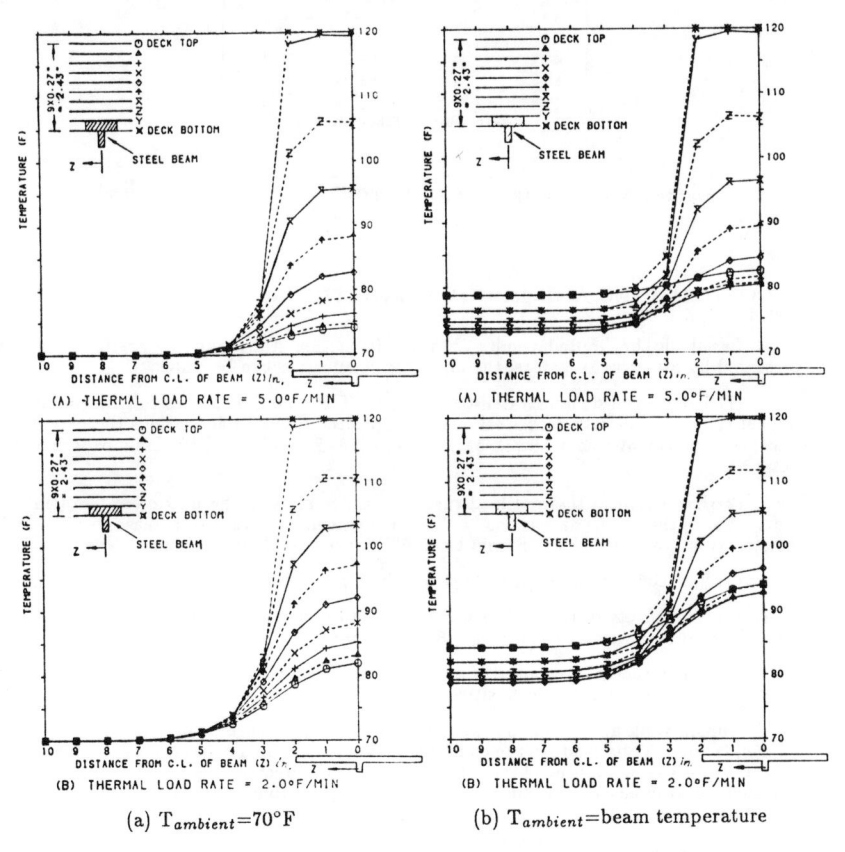

(a) $T_{ambient}=70°F$ (b) $T_{ambient}=$beam temperature

Figure 8 Two Dimensional Heat Flow in Concrete Deck
(duration = 25 minutes)

5.5 Rate of Temperature Change

The rate of thermal loading or the intensity of radiation for bridge decks being heated by solar radiation can substantially affect the thermal response of a bridge. A study was therefore carried out to evaluate the effects of thermal load rate on the temperature differential by subjecting one of the steel girders of the bridge model to seven different load rates, ranging from 0.1°F to 10.0°F per minute over a span of 25 minutes. The observed results are shown in Fig.9. Based on these results, it is apparent that thermal load rate does affect the temperature fields through the depth of a bridge deck. Fig.9 also indicates that the temperature differential is very sensitive to a thermal load rate of greater than 0.5 °F/min and that for a thermal load rate of greater than about 5.0°F the effects of the temperature differential diminish sharply.

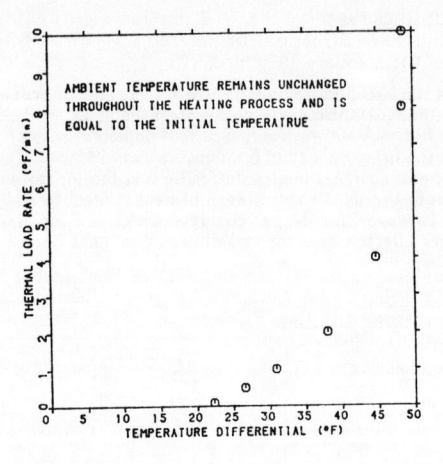

Figure 9 Maximum Temperature Differentials vs Thermal Load Rates

6. CONCLUSIONS AND RECOMMENDATIONS

Based on the limited results obtained from this study and the scattered results presented by other investigators, the following conclusions can be drawn :

(1) It has been demonstrated that ADINA-T can be used successfully in predicting the behaviour of concrete deck-steel beam composite bridges under various thermal loads. Excellent agreement was achieved between ADINA-T results and available experimental data.

(2) For a typical thermal load rate of 2°F per minute, the maximum temperature differential in the bridge deck is 38.0°F. The temperature differential increases from 30.9°F for a thermal load rate of 1.0°F/min to 48.6°F for a rate of 10.0°F/min.

(3) The convection loss/gain to/from the air is small.

(4) The effects of thermal load rate on the temperature differentials are negligible for a thermal load rate of greater than 5 °F per minute.

(5) The structural response of a bridge under negative thermal loads can be drastically different from that under positive thermal loads.

Considerable analytical and experimental work are needed before definite design recommendations can be made to the complex problem of thermal behaviour of indeterminate composite bridge structures. More studies should be directed to give full understanding of structural response of a continuous composite bridge subjected to simultaneous thermal and service loads.

REFERENCES

[1] U.S. Department of Transportation, *Public Roads*, Federal Highway of Administration, Vol.57, No.1, June 1983

[2] Zuk W - *Thermal and Shrinkage Stresses in Composite Beams*, Journal of American Concrete Institute, Detroit, Vol.58, No.3, September 1961, pp.327-340

[3] Menzies J.B., - *Structural Behaviour of the Moat Street Flyover*, Civil Engineering and Public Works Review, London, England, Vol.63, No.746, September 1968, pp.967-971

[4] Emerson M, - *The Calculation of the Distribution of Temperature in Bridges*, TRRL Report LR 561, Transport and Road Research Laboratory, Crowthorne, Berkshire, 1973

[5] Emanual J.H. and Hulsey J.L. - *Temperature Distributions in Composite Bridges*, Journal of Structural Engineering, American Society of Civil Engineering, Vol.104, January 1978, pp.65-78

[6] Symko Y., Transient Heat Conduction and Thermoelastic Analysis of Three Span Composite Highway Birdge, Ph.D. thesis, Department of Civil Engineering, University of Ottawa, Ottawa, Canada, 1979

[7] Dilger W., Beauchamp J.C., M.S. Cheung and Ghali A., - *Field Measurements of Muskwa River Bridge*, Journal of Structural Engineering, American Society of Civil Engineering, Vol.107, November 1981, pp.2147-2161

[8] Threlkeld J.L., Thermal Environmental Engineering, Prentice Hall Inc., 1970, pp.269-303

[9] ADINA Engineering Inc. (U.S.) and ADINA Engineering AB.(Sweden), A Finite Element Program for Automated Dynamic Incremental Nonlinear Analysis of Temperatures (ADINA-T), Report AE 81-2, Watertown, Mass., U.S. and Munkgatan 20D, Sweden, September 1981

[10] ADINA Engineering Inc. (U.S.) and ADINA Engineering AB.(Sweden), A Finite Element Program for Automated Dynamic Incremental Nonlinear Analysis (ADINA), Report AE 81-1, Watertown, Mass., U.S. and Munkgatan 20D, Sweden, September 1981

EFFECT OF ELEMENT SIZE ON THE SOLUTION ACCURACIES OF FINITE-ELEMENT HEAT TRANSFER AND THERMAL STRESS ANALYSES OF SPACE SHUTTLE ORBITER

William L. Ko and Timothy Olona

NASA Ames Research Center, Dryden Flight Research Facility, Edwards, California 93523-5000 U.S.A.

SUMMARY

The effect of element size on the solution accuracies of finite-element heat transfer and thermal stress analyses of space shuttle orbiter was investigated. Several structural performance and resizing (SPAR) thermal models and NASA structural analysis (NASTRAN) structural models were set up for the orbiter wing midspan bay 3. The thermal model was found to be the one that determines the limit of finite-element fineness because of the limitation of computational core space required for the radiation view factor calculations. The thermal stresses were found to be extremely sensitive to a slight variation of structural temperature distributions. The minimum degree of element fineness required for the thermal model to yield reasonably accurate solutions was established.

NOMENCLATURE

C	capacitance matrix
CQUAD2	quadrilateral membrane and bending element
CROD	two-node tension-compression-torsion element
C41	four-node forced convection element
E23	bar element for axial stiffness only
E25	zero length element used to elastically connect geometrically coincident joints
E31	triangular membrane element
E41	quadrilateral membrane element
E44	quadrilateral shear panel element
F_{ij}	view factor from element i to element j
FRSI	felt reusable surface insulation
H	convection load vector
HRSI	high-temperature reusable surface insulation
JLOC	joint location
K_h	convection matrix
K_k	conduction matrix
K_r	radiation matrix

This paper is declared a work of the U.S. Government and therefore is in the public domain.

K21	two-node line conduction element
K31	three-node area conduction element
K41	four-node area conduction element
K81	eight-node volume conduction element
NASTRAN	NASA structural analysis
Q	source load vector
R	radiation load vector
R31	three-node area radiation element
R41	four-node area radiation element
SIP	strain isolation pad
SPAR	structural performance and resizing
STS	space transportation system
T	absolute temperature, °R
TPS	thermal protection system
t	time, sec
x, y, z	rectangular Cartesian coordinates
X_0	station on x axis, in
Y_0	station on y axis, in
σ_x	normal stress in x direction (chordwise stress), ksi
σ_y	normal stress in y direction (spanwise stress), ksi
τ_{xy}, τ_{yz}	shear stresses, ksi

1 INTRODUCTION

In finite-element heat transfer analysis or finite-element stress analysis, it is well known that reduction of element sizes (or increase in element number) will improve the solution accuracy. For simple structures, the element sizes may be reduced sufficiently to obtain highly accurate solutions. However, for large complex structures, such as the space shuttle orbiter, the use of excessively fine elements in the finite-element models may result in unmanageable computations that exceed the memory capability of existing computers. This computational limitation is frequently encountered during radiation view factor computations in the three-dimensional finite-element heat transfer analysis of complex structures. Because of computational limitations in the past heat transfer analysis of the space shuttle orbiter, only small local regions of the orbiter structure were modeled. Several regions of the space shuttle were modeled by Ko, Quinn, and Gong. For the past several years, these finite-element models were used to calculate orbiter structural temperatures, which were correlated with the actual flight data during the initial orbit tests of the space shuttle Columbia.[1-7] Recently, Gong, Ko, and Quinn [4] conducted a finite-element heat transfer analysis of the orbiter whole wing (Fig.1) using a thermal model with relatively coarse elements. A similar whole wing finite-element structural model was used by Ko and Fields [8] in the thermal stress analysis of the orbiter whole wing. Both the thermal model and the corresponding structural model set up for the orbiter whole wing were too coarse to give sufficiently accurate structural temperature and thermal stress distributions. Before modifying the existing wing models by increasing the number of joint locations to improve the solutions, it is necessary to determine the minimum number of joint locations required for the modified wing thermal model (the corresponding wing structural model requires far fewer joint locations) to give reasonably accurate structural temperature distributions without causing the radiation view factor computations to become unmanageable. This report describes (1) heat transfer and thermal stress analyses of a single bay at the orbiter wing midspan using several different thermal and structural models having different numbers of joint locations (or different element sizes), (2) the effect of element sizes on the accuracies of solutions, and (3) the minimum number of joint locations required for the single-bay

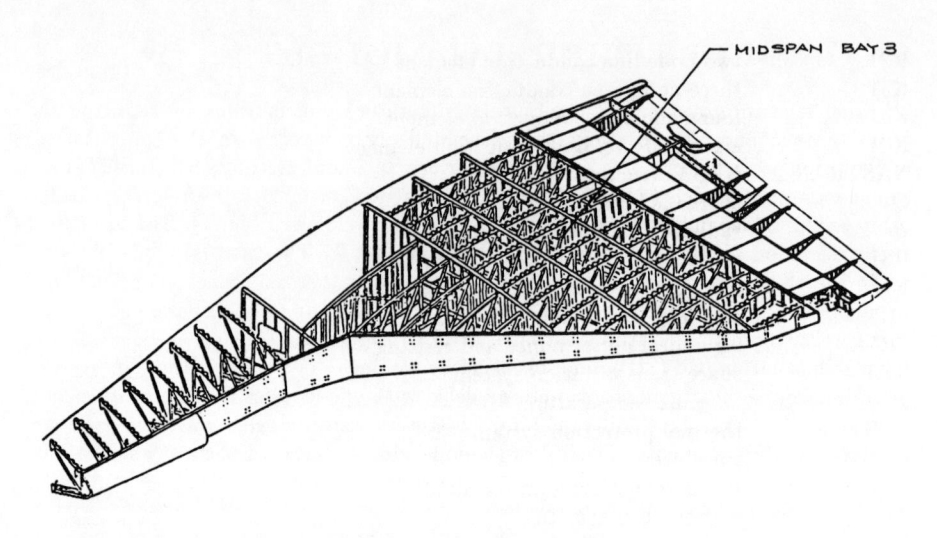

Fig. 1 Space shuttle wing.

basic criteria in remodeling the whole orbiter wing or modeling other types of hypersonic aircraft wings (hot structures).

2 WHOLE WING THERMAL AND STRUCTURAL MODELS

In finite-element thermal stress analysis of the space shuttle orbiter, the temperature input to the structural model for the calculation of thermal stresses is usually obtained from the results of finite-element (or finite-difference) heat transfer analysis using the corresponding thermal model. Since the thermal protection system (TPS) is not a major load-carrying structure, it is neglected in the structural model. Thus, the structural model has far fewer joint locations (JLOCs) than the corresponding thermal model. For the wing models, the thermal model contains 2289 JLOCs, while the structural model has only 232 JLOCs (see Table 1). Even though the thermal model has only one degree of freedom (temperature), because of the radiation view factor computations and the transient nature of heat transfer, the computer core space required by the thermal model is always many times more than that required for the structural model, which has six degrees of freedom. Thus, the thermal model is the one that limits how fine the element size can be reduced for improving the solutions.

Thermal model		Structural model	
Feature	Number	Feature	Number
JLOCs	2289	JLOCs	232
K21 elements	1696	E23 elements	498
K31 elements	84	E25 elements	10
K41 elements	485	E31 elements	19
R31 elements	84	E41 elements	181
R41 elements	568	E44 elements	67

Table 1 Comparison of finite-element thermal and structural models for space shuttle orbiter wing

3 ONE-CELL THERMAL MODELS

To study the improvement of structural temperature distributions by reducing the element sizes, and also to study the associated effort involved in the computations of radiation view factors, five structural performance and resizing (SPAR) [9] finite-element thermal models (with different degrees of element fineness) were set up for the orbiter wing midspan bay 3 bounded by Y_0-226 and Y_0-254 (see Fig. 1). The five SPAR thermal models A, B, C, D, and E are shown in Fig. 2. The thermal model A is set up to match the coarseness of the existing whole wing thermal model. The four-node area conduction (K41) elements were used to model the wing skins, spar webs, rib cap shear webs, room temperature vulcanized (RTV) rubber layers lying on both sides of the strain isolation pad (SIP), and TPS surface coatings. The aerodynamic surfaces for providing source heat generation were modeled with one layer of K41 elements of unit thickness. The spar caps, rib caps, and rib trusses were modeled with two-node line conduction (K21) elements. The TPS was modeled in 10 layers on the lower surface and 3 layers on the upper surface using eight-node volume conduction (K81) elements. The SIP layer was modeled with only one layer of K81 elements. The external and internal radiations were modeled by attaching a layer of four-node area radiation (R41) elements to the active radiation surfaces. The radiation into space was modeled with one R41 element of unit area. No radiation elements were attached to the surfaces of spar caps, rib caps, rib cap shear webs, and rib trusses because of small exposed areas. A layer of four-node forced convection (C41) elements were attached to the internal surfaces of the bay to model the internal convection of air resulting from the entrance of external cool air into the interior of the orbiter wing at 1400 sec after reentry (or at 100,000 ft altitude). The front and rear ends of the thermal models were insulated. Table 2 summarizes the sizes (joint location number, number of different types of elements) of the five SPAR thermal models A, B, C, D, and E.

SPAR thermal model	JLOC	K21	K41	K81	R41	C41
A	112	34	28	28	15	10
B	436	54	168	224	89	56
C	636	82	232	336	137	88
D	972	98	360	560	201	120
E	2076	146	848	1344	513	320

Table 2 Sizes of SPAR thermal models

3.1 Heat Input

The external heat inputs to the SPAR thermal models are shown in Fig. 3. These aerodynamic heating curves are associated with STS-5 flight trajectories and are taken from Ref. 4, which describes in detail the method of calculations of aerodynamic heating.

3.2 View Factors

The view factors used in the radiation to space were calculated by hand. However, for the internal radiation exchanges, the view factors were calculated by using a VIEW computer program, which is incorporated into the SPAR thermal analysis computer program.[9]

(a) Model A. *(b) Model B.*

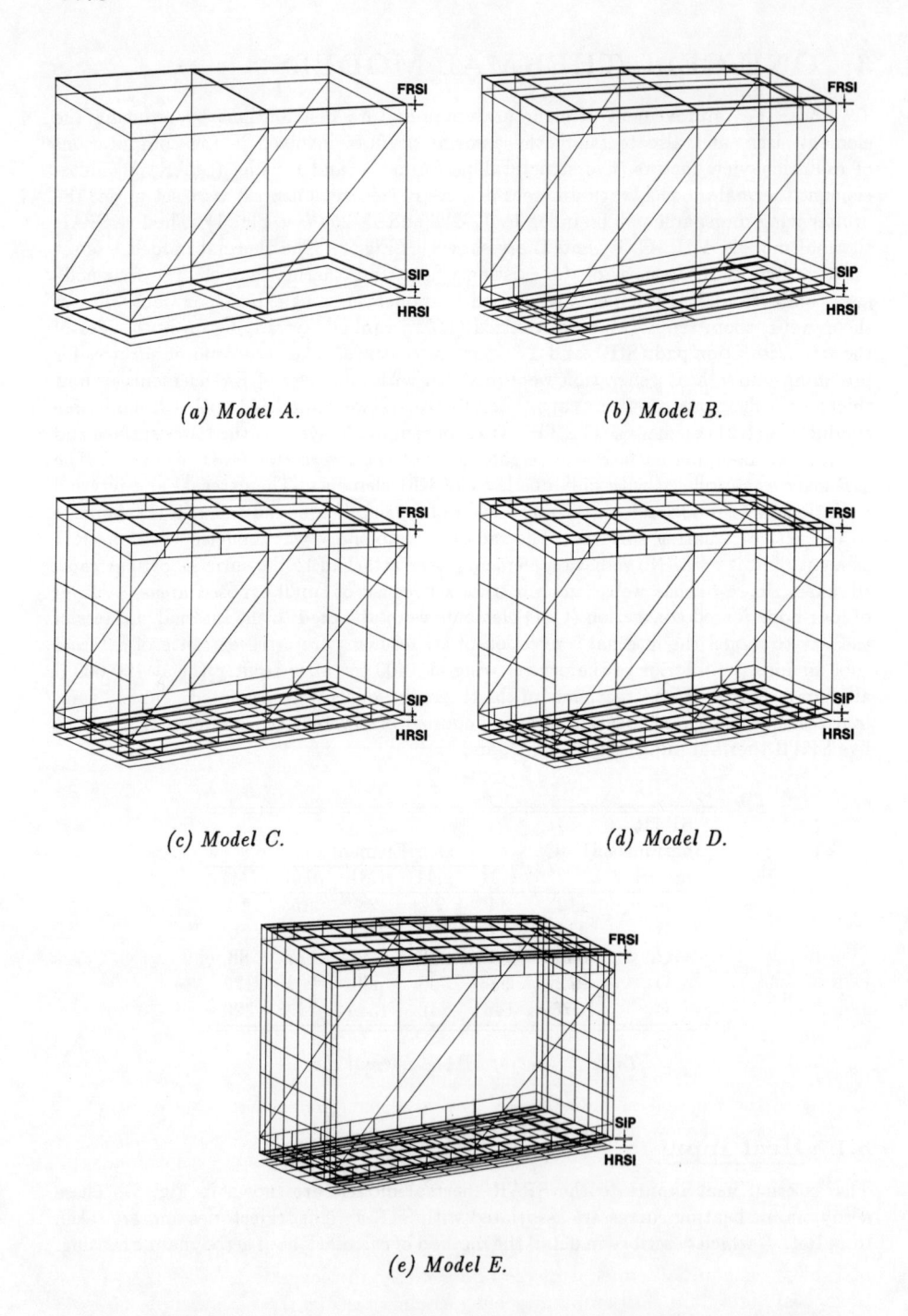

(c) Model C. *(d) Model D.*

(e) Model E.

Fig. 2 SPAR thermal models for bay 3 of orbiter wing bounded by Y_0-226 and Y_0-254. K81 elements for TPS and SIP not shown. TPS and SIP removed to convert to NASTRAN structural model.

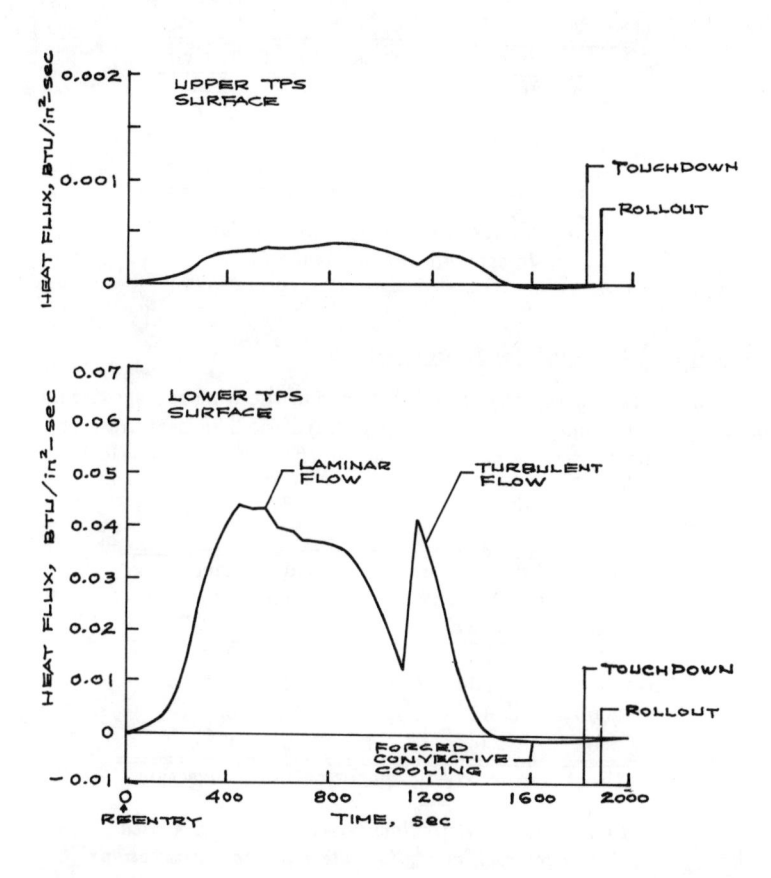

Fig. 3 Surface heating rates at midspan bay 3 of orbiter wing; STS-5 flight.

For both the external and the internal thermal radiation exchanges, all the view factors were calculated from the equation [9]

$$A_i F_{ij} = A_j F_{ji} \qquad (1)$$

where A_i is the surface area of radiation exchange element i and F_{ij} is the view factor, defined as the fraction of radiant heat leaving element i incident on element j. In the calculation of view factors for the external radiation exchanges (considering that element i represents the space element and element j any radiation exchange element on the wing surface), F_{ji} was taken to be unity; therefore, $F_{ij} = A_j/A_i$ according to Eq. (1).

Values of emissivity and reflectivity used to compute radiant heat fluxes are given in Table 3. The initial temperature distribution used in the analysis was obtained from the actual flight data. In thermal modeling, the majority of the time was consumed in the computations of view factors.

Surface	Emissivity	Reflectivity
Windward	0.85	0.15
Leeward	0.80	0.20
Internal structure	0.667	0.333
Space	1.0	0

Table 3 Emissivity and reflectivity values used to compute radiant heat fluxes

3.3 Internal Forced Convection

After opening the landing gear door and the vents at the wing roots, external air enters the shuttle wing and induces convective heat transfer. The heat transfer coefficients used for C41 elements were calculated using the effective air flow velocities inside the wing, listed in Table 4.[6]

Time (sec)	Effective air flow velocity (ft/sec)	Heat transfer coefficient (Btu/sec-in^2-°F)
1750	25	3.30×10^{-6}
1850	25	4.00×10^{-6}
2000	15	2.73×10^{-6}
3000	0	0.35×10^{-6}[a]

[a]Heat transfer coefficient for natural convection.

Table 4 Effective air flow velocities and associated heat transfer coefficients for internal forced convection

3.4 Transient Thermal Solutions

The SPAR thermal analysis finite-element computer program was used in the calculation of temperature time histories at all joint locations of the thermal models. The SPAR program used the following approach to obtain transient thermal solutions.

The transient heat transfer matrix equation

$$(K_k + K_r + K_h)T + C\dot{T} = Q + R + H \qquad (2)$$

where

K_k	is the conduction matrix,
K_r	the radiation matrix,
K_h	the convection matrix,
T	the absolute temperature,
C	the capacitance matrix,
Q	the source load vector,
R	the radiation load vector,
H	the convection load vector, and
$[\dot{\ }]$	denotes time derivative,

was integrated by assuming that the temperature vector T_{i+1} at time step t_{i+1} can be expressed as

$$T_{i+1} = T_i + \dot{T}_i \Delta t + \frac{1}{2!}\ddot{T}_i \Delta t^2 + \frac{1}{3!}\dddot{T}_i \Delta t^3 + \cdots \tag{3}$$

where T_i is the temperature vector at time step t_i and Δt is the time increment. The vector $\dot{T}_i$ is determined directly from Eq. (2) as

$$\dot{T}_i = -C^{-1}(K_k + K_r + K_h)T_i + C^{-1}(Q + R + H) \tag{4}$$

Higher order derivatives are obtained by differentiating Eq. (2) according to the assumptions that (1) material properties are constant over Δt, (2) Q and H vary linearly with time, and (3) R is constant over Δt:

$$\ddot{T}_i = -C^{-1}(K_k + 4K_r + K_h)\dot{T}_i + C^{-1}(\dot{Q} + \dot{H}) \tag{5}$$

$$\dddot{T}_i = -C^{-1}(K_k + 4K_r + K_h)\ddot{T}_i - 4C^{-1}\dot{K}_r\dot{T}_i \tag{6}$$

In the present computations, the Taylor series expansion (Eq. (3)) was cut off after the third term. The pressure dependency of the TPS and SIP thermal properties was converted into time dependency based on the trajectory of the STS-5 flight.

Time-dependent properties were averaged over time intervals (RESET TIME), which were taken to be 25 sec. Temperature-dependent properties were evaluated at the temperatures computed at the beginning of each time interval. The values Q, $\dot{Q}$, and R were computed every 2 sec.

4 ONE-CELL STRUCTURAL MODELS

For the thermal stress analysis, the NASA structural analysis (NASTRAN) [10] computer program was used because it can handle temperature-dependent material properties. The SPAR structural computer program lacks this capability. The five NASTRAN structural models (not shown) corresponding to the five SPAR thermal models A, B, C, D, and E (Fig. 2) are essentially the same except that the TPS layers are removed in the NASTRAN structural models. Thus, each set of thermal and corresponding structural models have identical joint locations so that the temperature distribution obtained from the thermal model can be input directly to the corresponding structural model for the calculations of thermal stresses. The wing skins, spar webs, and rib cap shear webs were modeled with quadrilateral membrane and bending (CQUAD2) elements. The spar caps, rib caps, and rib trusses were represented with two-node tension-compression-torsion (CROD) elements. To approximate the deformation field of the midspan bay 3 when it is not detached from the whole wing, the following boundary conditions were imposed on the NASTRAN structural models.

1. Y_0-226 plane fixed—The grid points lying in the Y_0-226 plane have no displacements in the y direction but are free to move in the x and z directions. The rotations with respect to the x, y, and z axes are constrained.

2. Y_0-254 plane free—The grid points lying in the Y_0-254 plane are free to move in the x, y, and z directions. The rotations with respect to the x, y, and z axes are constrained.

The thermal loadings to the NASTRAN structural models were generated by using the structural temperature distributions calculated from the corresponding SPAR thermal models. Table 5 summarizes the sizes of the five NASTRAN structural models. Because the TPS is removed, the structural models have far fewer joint locations as compared with corresponding SPAR thermal models (see Table 2).

NASTRAN structural model	Grid	CQUAD2	CROD
A	24	18	54
B	82	72	54
C	140	112	74
D	196	160	90
E	429	368	132

Table 5 Sizes of NASTRAN structural models

5 RESULTS

5.1 Structural Temperatures

Figure 4 shows the time histories of the midbay TPS surface temperatures calculated by using different SPAR thermal models. The five temperature curves respectively associated with the thermal models A, B, C, D, and E are so close as to be pictorially undiscernable. This implies that the element sizes in the substructure have negligible effect on the TPS surface temperatures. Figure 5 shows the time histories of the structural temperatures in the midbay regions of the lower and upper wing skins calculated from different thermal models. The thermal models B, C, D, and E yielded almost identical skin temperatures in the midbay regions. However, the thermal model A gave slightly lower wing skin temperatures because of coarseness of the model.

Figure 6 shows the calculated structural temperature distributions in the plane Y_0-240 of bay 3 at $t = 1700$ sec from reentry. The thermal model A definitely yielded inaccurate solutions. The structural temperature distributions given by thermal models B, C, D, and E are quite close. Especially, the thermal models D and E yielded very close structural temperature distributions. As shown in Section 5.2, a slight difference in the structural temperature distributions obtained from different thermal models could cause a "marked" difference in the induced thermal stress distributions. The structural temperature gradients are steepest near the lower spar caps because the spar webs function as heat sinks. Figure 7 shows the spanwise distributions of the wing skin temperatures at cross section X_0-1270 based on different thermal models. The thermal models B, C, and D yield almost identical structural temperature distributions because they have the same number of elements in the spanwise direction. The shapes of the skin temperature distributions given by model E approach circular arcs. The solutions given by the thermal model A are rather poor because of an insufficient number of elements.

5.2 Thermal Stresses

Figure 8 shows distributions of the spanwise stress σ_y calculated by using different NASTRAN structural models. Notice that the thermal stresses are very sensitive to the finite-element sizes (or structural temperature distributions). The coarser models A and B yielded peak compression in the midbay regions of both lower and upper skins. However, as the number of elements increased (models C, D, and E), the shallow U-shaped distributions of σ_y in the lower skin shifted to shallow W-shaped distributions, and the peak compression regions moved near the spar webs. The slight stress release in the midbay region of the lower skin, based on the structural models C, D, and E, is due to the bulging of the wing skin (described later in this section). For the upper skin, the zone of slight stress release showed up only for the stress distributions calculated from models D and E. These stress releases in the midbay regions of the wing skins were never

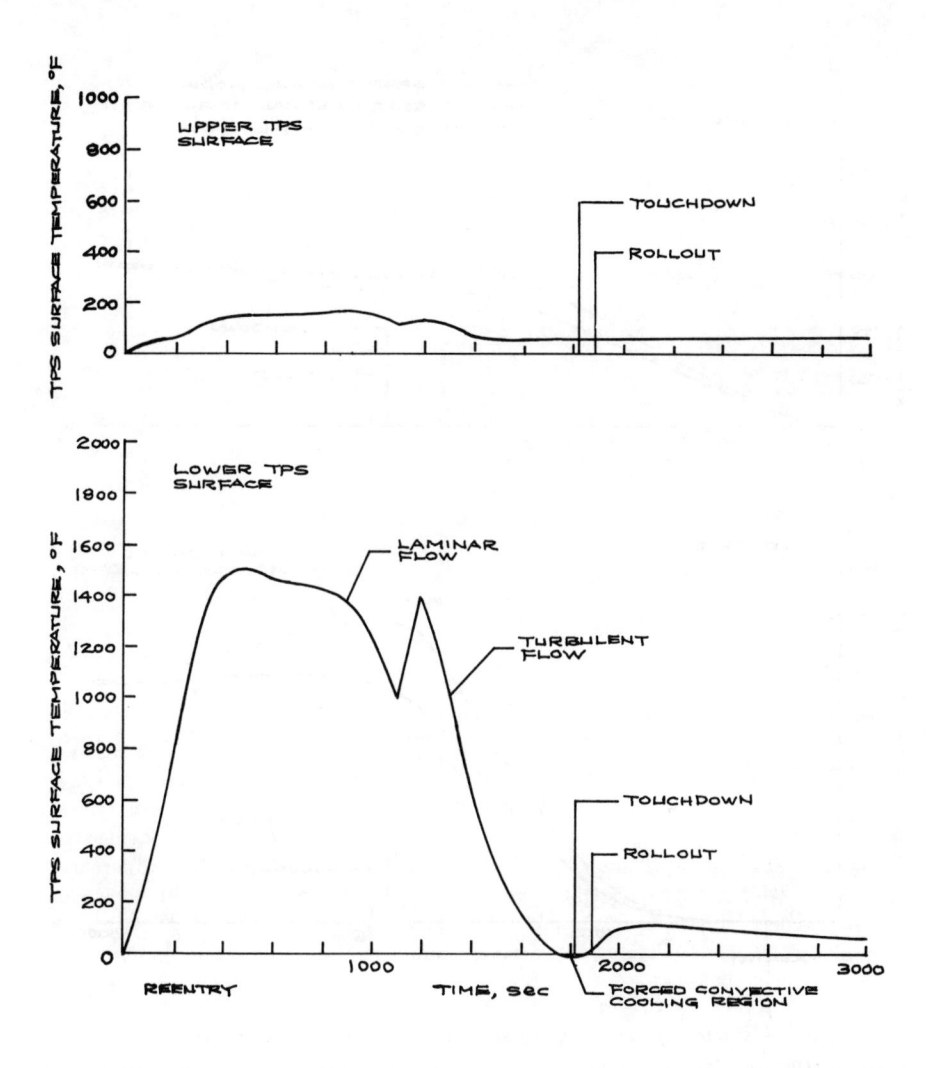

Fig. 4 Time histories of TPS surface temperatures calculated using different SPAR thermal models; STS-5 flight.

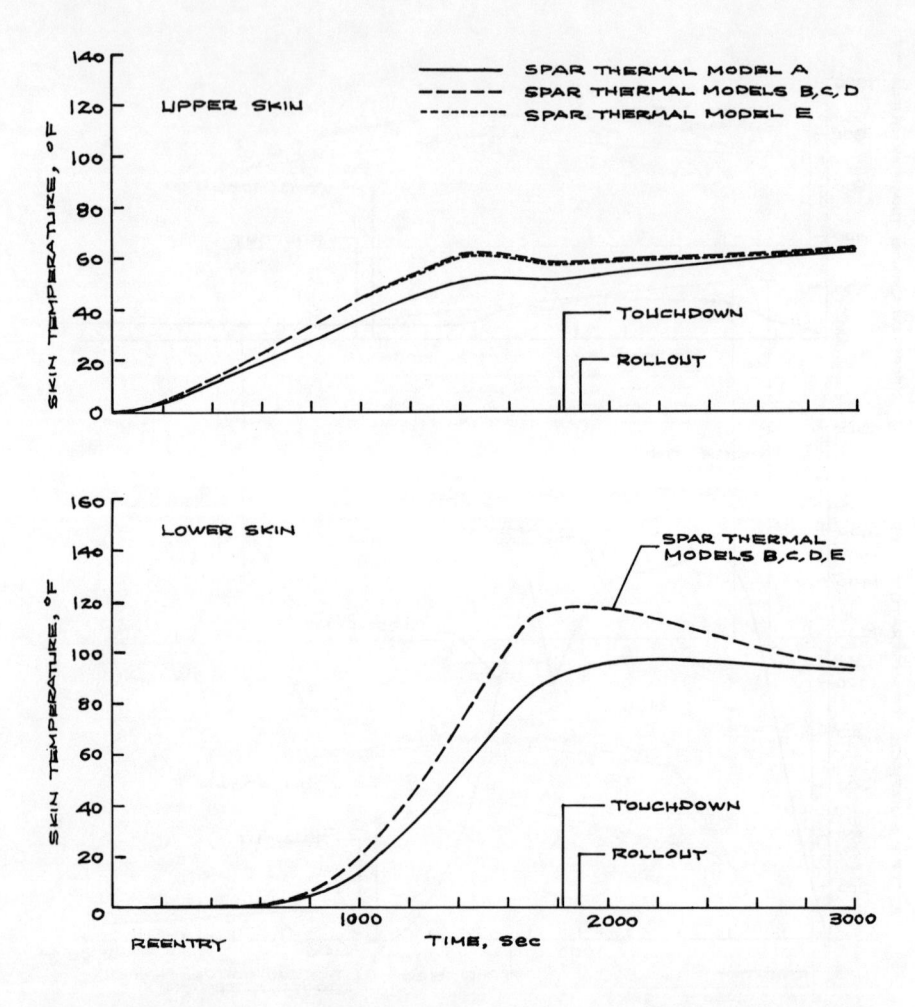

Fig. 5 Time histories of orbiter wing skin temperatures calculated using different SPAR thermal models; STS-5 flight.

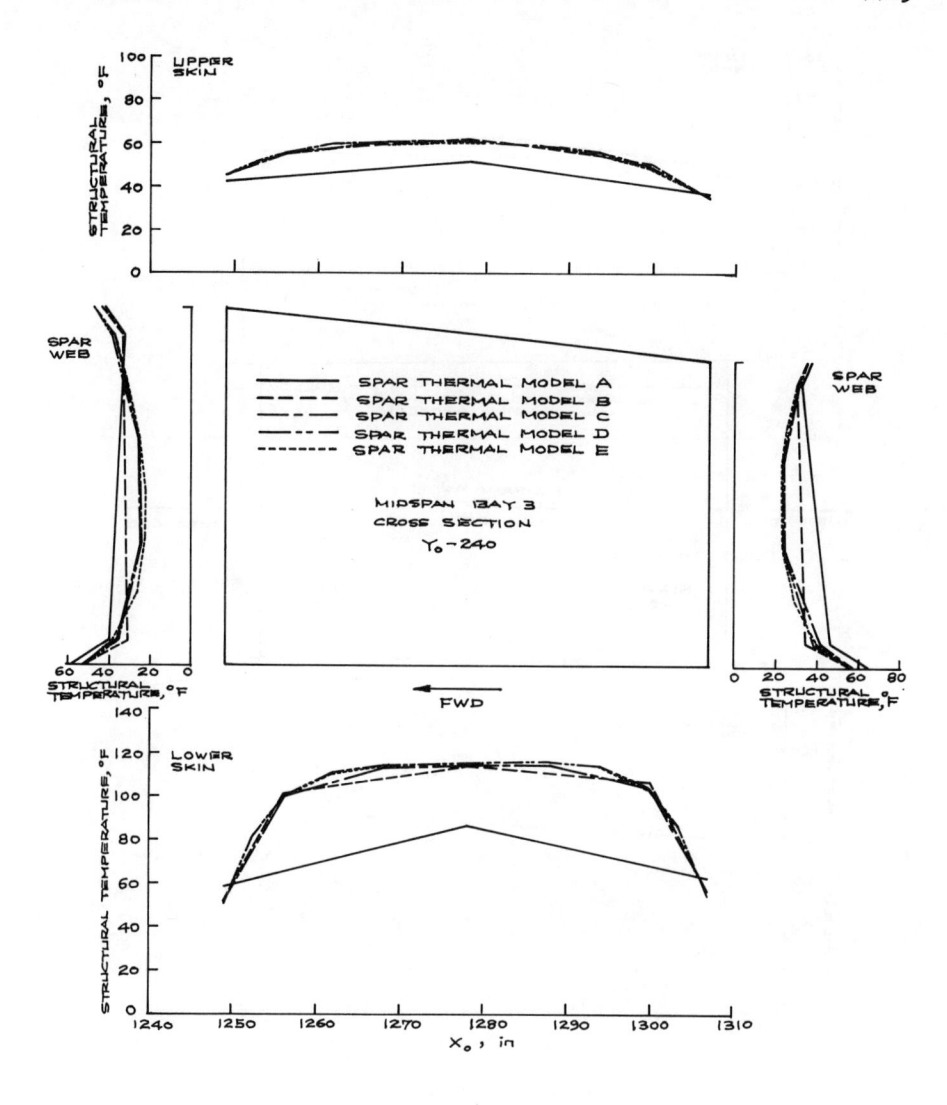

Fig. 6 *Structural temperature distributions in orbiter wing midspan bay 3 calculated using different SPAR thermal models; time = 1700 sec, STS-5 flight.*

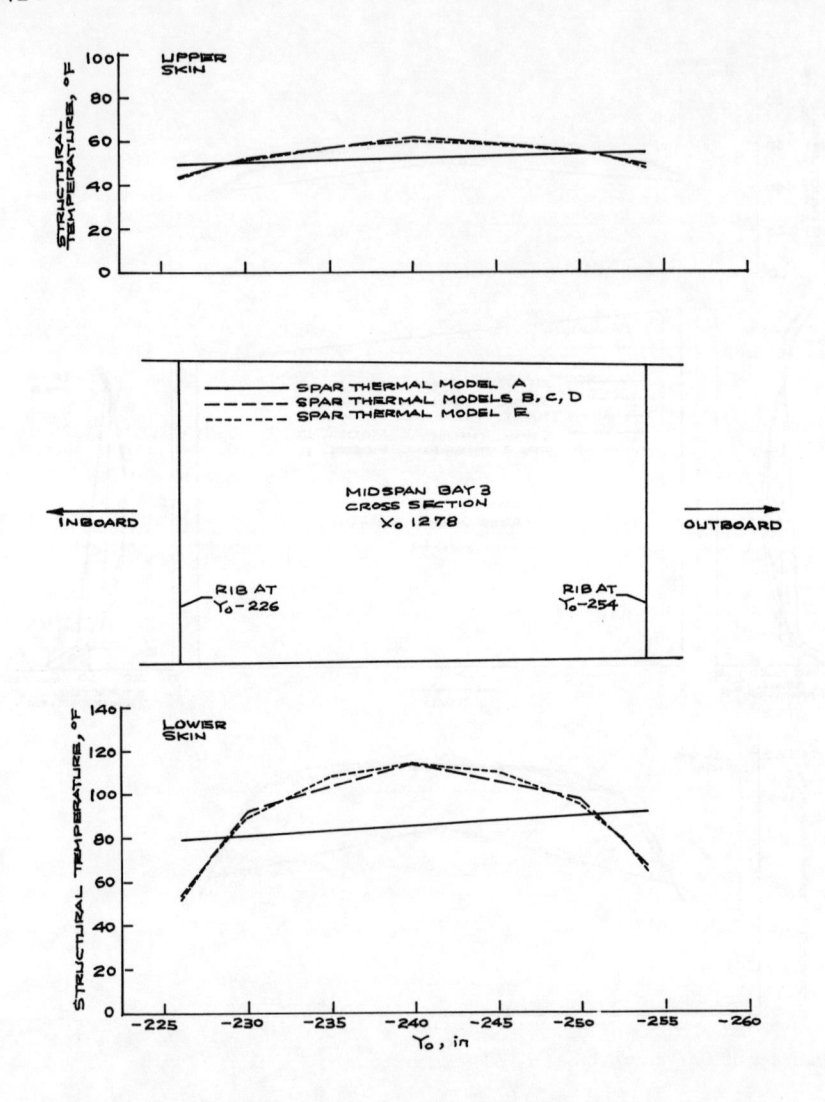

Fig. 7 Spanwise distributions of structural temperatures in orbiter wing midspan bay 3 calculated using different SPAR thermal models; time = 1700 sec, STS-5 flight.

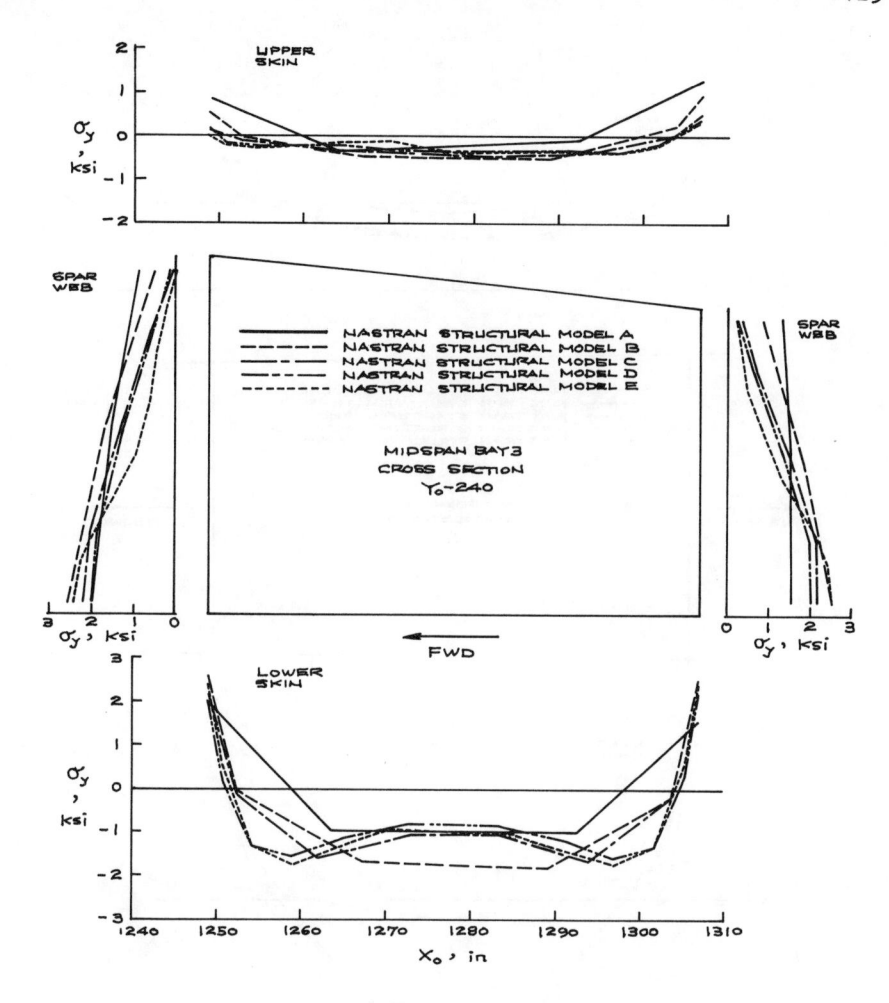

Fig. 8 Distributions of spanwise stress σ_y in orbiter wing midspan bay 3 calculated using different NASTRAN structural models; time = 1700 sec, STS-5 flight.

observed in the earlier thermal stress analysis, which ignored the three-dimensional deformations of the orbiter skins (that is, skin-bulging effect). Figure 9 shows the distributions of chordwise stresses σ_x calculated from the five structural models. Again, the solution given by the model A is quite poor. The distributions of σ_x given by the structural models B, C, and D (all of which have four elements in the spanwise direction) are quite close. The structural model E, which has eight elements in the spanwise direction, gave a magnitude of peak compressional stress about 1.2 ksi above those predicted from the structural models B, C, and D. The marked difference in the σ_x distribution given by model E and those given by models B, C, and D is due to the existence of a stress-increase zone, which appeared only in model E. Unlike the distribution of σ_y (Fig. 8), the distributions of σ_x calculated from all structural models did not exhibit stress release effects in the midbay regions of the wing skins. The magnitude of thermal stress σ_x (either in tension or compression) is higher than that of thermal stress σ_y shown in Fig. 8. Thus, σ_x is more critical than σ_y because the buckling strength of the wing skin in the x direction (normal to the hat stringers) is

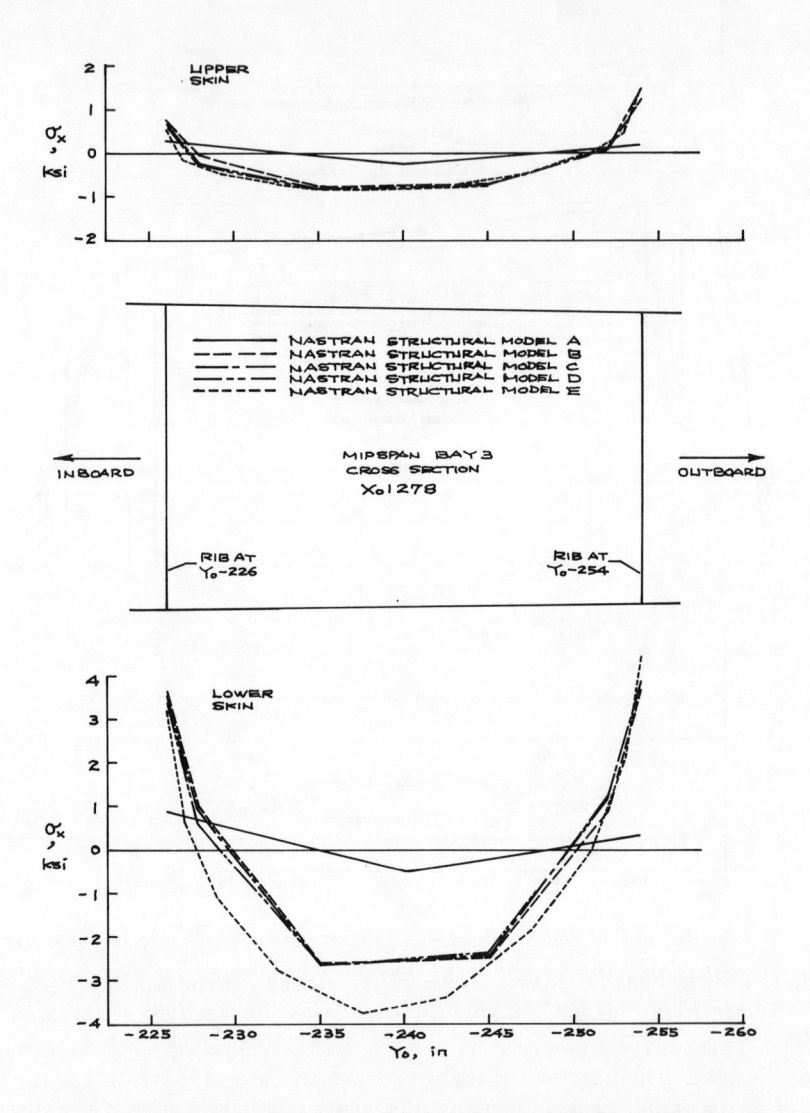

Fig. 9 Distributions of chordwise stress σ_x in orbiter wing midspan bay 3 calculated using different NASTRAN structural models; time = 1700 sec, STS-5 flight.

lower than that in the y direction (parallel to the hat stringers). The orbiter wing skin buckling stresses are in the neighborhood of $\sigma_x = -12$ ksi (normal to hat stringers) and $\sigma_y = -25$ ksi (parallel to hat stringers).

Figure 10 shows the distributions of shear stresses τ_{xy} and τ_{yz} in the cross section Y_0-252 (plane of highest shear) predicted from different NASTRAN structural models. The high shear-stress regions are near the lower spar caps.

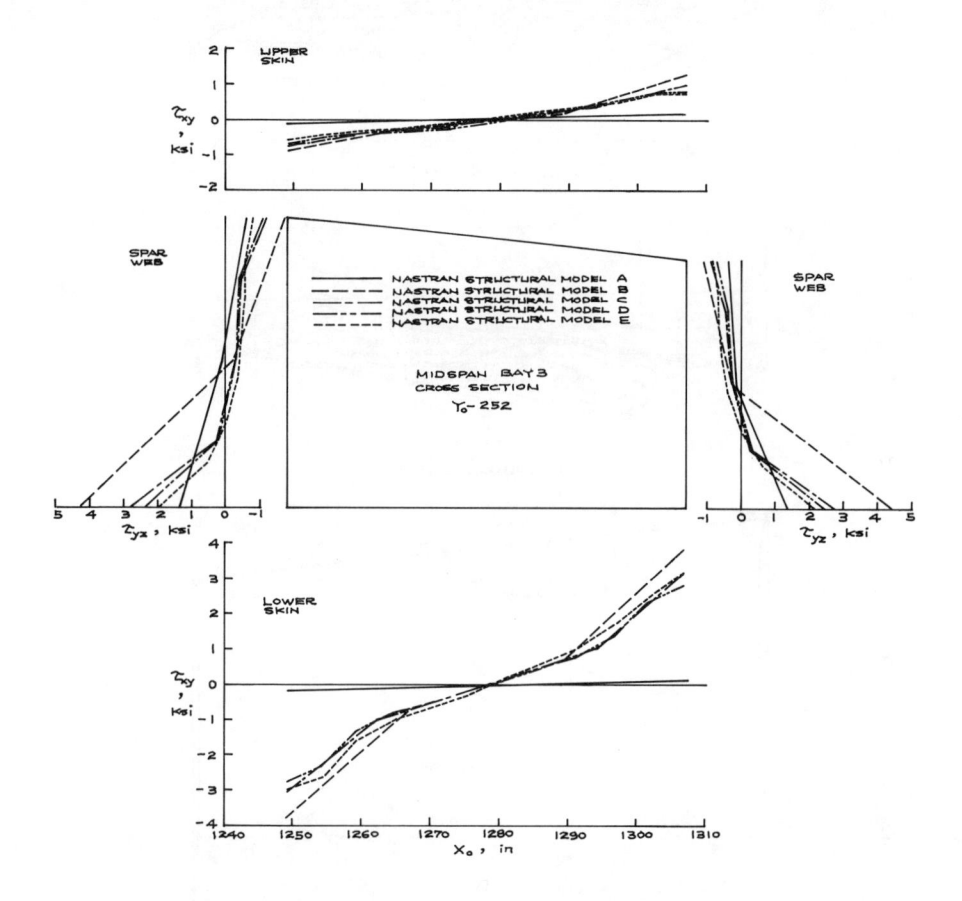

Fig. 10 Distributions of shear stresses τ_{xy} and τ_{yz} in orbiter wing midspan bay 3 calculated using different NASTRAN structural models; time = 1700 sec, STS-5 flight.

Figure 11 shows the deformed shape of the orbiter wing midspan bay 3 due to STS-5 thermal loading. The front half of the wing lower skin bulged inwardly, but the rear half bulged outwardly; almost the entire wing upper skin bulged outwardly with more severe deformations in the front half region.

5.3 Computation Time

Table 6 summarizes the number of internal radiation view factors F_{ij} needed for different SPAR thermal models and the computation time used in the transient heat transfer analyses associated with each thermal model. The data shown in Table 6 are plotted in Fig. 12. Both the SPAR computation time and the number of internal radiation view factors appear to increase almost exponentially with the increase in the number of JLOCs. The curves in Fig. 12 show how fast the computational "barrier" will be reached by accelerating the increase in the number of JLOCs.

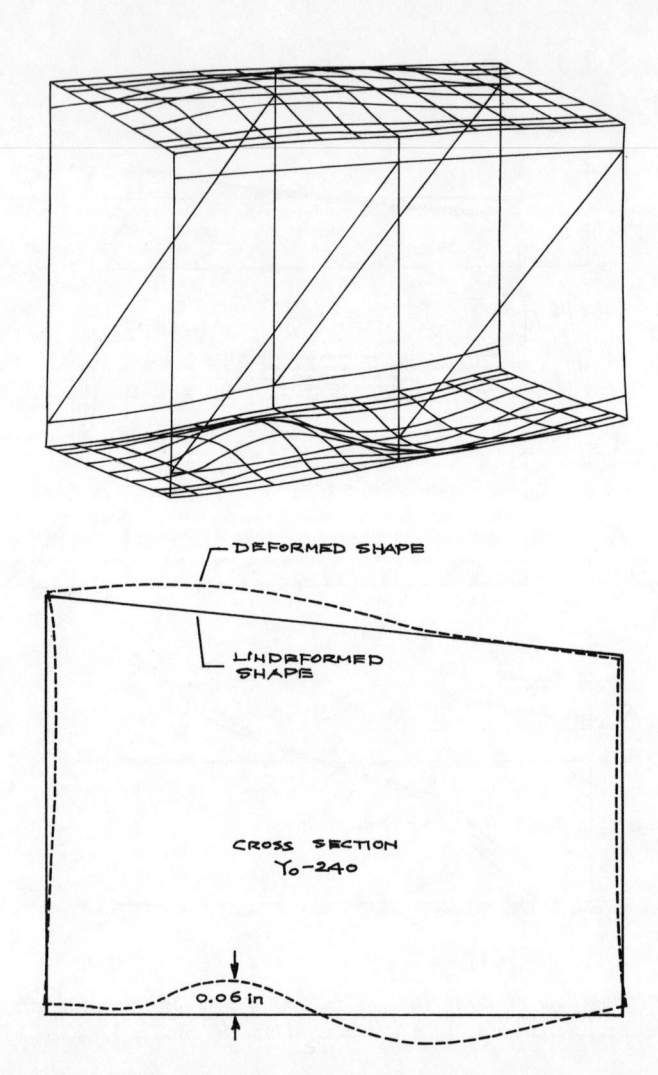

Fig. 11 Deformed shape of orbiter wing midspan bay 3 due to STS-5 thermal loading; time = 1700 sec.

SPAR thermal model	Number of joint locations	Number of internal radiation view factors F_{ij}	Thermal analysis computation time (hr)
A	112	78	0.25
B	436	2,816	1.25
C	636	6,894	3.5
D	972	13,500	9.0
E	2076	93,869	31.5

Table 6 Numbers of joint locations and internal radiation view factors and thermal analysis computation time associated with different SPAR thermal models

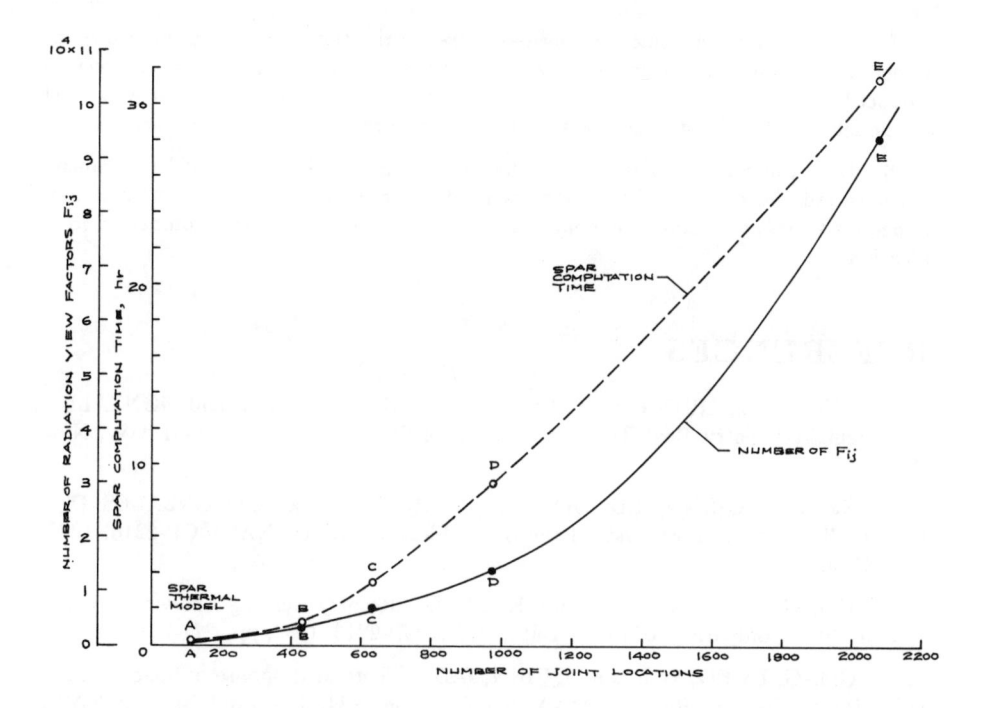

Fig. 12 Plots of number of radiation view factors F_{ij} and SPAR computation time as functions of number of joint locations.

6 CONCLUSIONS

Finite-element heat transfer and thermal stress analyses were performed on the space shuttle wing midspan bay 3 using several finite-element models of different degrees of element fineness. The effect of element sizes on the solution accuracy was investigated in great detail. The results of the analyses are summarized as follows:

    1. The finite-element model A (thermal or structural), which has the same coarseness as the earlier whole wing model, is too coarse to yield satisfactory solutions.

    2. The structural temperature distribution over the wing skin (lower or upper) surface of one bay was "dome" shaped and induced more severe thermal stresses in the chordwise direction than in the spanwise direction. The induced thermal stresses were very sensitive to slight variation of structural temperature distributions.

    3. The structural models with finer elements yielded spanwise stress distributions exhibiting a stress release zone (due to skin bulging) at the midbay region of the wing skin (lower or upper), and the peak wing skin compression occurred near the spar caps. However, the coarser models gave the peak skin compression in the midbay region.

    4. The front half of the wing lower skin bulged inwardly, but the rear half bulged outwardly. Almost the entire wing upper skin bulged outwardly with more severe deformations in the front half region.

5. For obtaining satisfactory thermal stress distributions, each wing skin (lower or upper) of one bay must be modeled with at least 8 elements in the spanwise direction (model E) and 10 elements in the chordwise direction (model D); each spar web must be modeled with at least 5 elements in the vertical directions (model D).

6. Both the computation time required for the SPAR finite-element heat transfer analysis and the number of view factors needed for internal radiation computations appeared to increase almost exponentially with the increase of the number of joint locations.

REFERENCES

1. KO, W.L., QUINN, R.D., GONG, L., SCHUSTER, L.S., and GONZALES, D.—Preflight Reentry Heat Transfer Analysis of Space Shuttle, AIAA-81-2382, Nov. 1981.

2. KO, W.L., QUINN, R.D., GONG, L., SCHUSTER, L.S., and GONZALES, D.—Reentry Heat Transfer Analysis of the Space Shuttle Orbiter, NASA CP-2216, 1982, pp. 295–325.

3. GONG, L., QUINN, R.D., and KO, W.L.—Reentry Heating Analysis of Space Shuttle With Comparison of Flight Data, NASA CP-2216, 1982, pp. 271-294.

4. GONG, L., KO, W.L., and QUINN, R.D.—Thermal Response of Space Shuttle Wing During Reentry Heating, AIAA-84-1761, June 1984. (Also published as NASA TM-85907, 1984.)

5. KO, W.L., QUINN, R.D., and GONG, L.—Finite-Element Reentry Heat Transfer Analysis of Space Shuttle Orbiter, NASA TP-2657, 1986.

6. KO, W.L., QUINN, R.D., and GONG, L.—Effect of Forced and Free Convections on Structural Temperatures of Space Shuttle Orbiter During Reentry Flight, AIAA-87-1600, June 1987.

7. GONG, L., KO, W.L., and QUINN, R.D.—Comparison of Flight-Measured and Calculated Temperatures on Space Shuttle Orbiter, NASA TM-88278, 1987.

8. KO, W.L. and FIELDS, R.A.—Thermal Stress Analysis of Space Shuttle Orbiter Subjected to Reentry Aerodynamic Heating, NASA TM-88286, 1987.

9. MARLOWE, M.B., MOORE, R.A., and WHETSTONE, W.D.—SPAR Thermal Analysis Processors Reference Manual, System Level 16, Volume 1: Program Execution, NASA CR-159162, 1979.

10. The NASTRAN User's Manual, Level 17.5, NASA SP-222(05), 1978.

PREDICTION OF DISTORTION IN OVERLAYED REPAIR WELDS

A.P. Chakravarti* and L.M. Malik*

A.S. Rao[+] and J.A. Goldak[o]

ABSTRACT

A finite element model is developed to predict the vertical distortions due to block welds in welding overlays on flat plate. The finite element predictions are compared to experimental measurements. The FEM analysis uses a three stage model to do the transient thermal analysis, transient thermo-elasto-plastic 3 dimensional analysis and elasto-plastic analysis of a plate containing six blocks using total strain from the blocks as initial strain in the plate. An approach is suggested to illustrate the possibility of integrating numerical and experimental results to construct an economical, but empirical model.

INTRODUCTION

The fusion welding process introduces distortion and residual stresses in welded structures, and due to their possible adverse effects on structural performance, codes and specifications may require control of these characteristics. Although various empirical or semi-empirical formulae are available to predict distortion associated with some typical welded joints, they are difficult to use for complex structures, and residual stress distributions are even harder to predict. The dependence of material properties (thermal conductivity, yield strength, etc.) on temperature is another complexity that reduces the usefulness of these empirical relationships.

---

*Materials Technology Centre, Arctec Canada Limited, Kanata
[+]B.C. Hydro Research and Development, Vancouver
[o]Professor, Carleton University, Ottawa

With continually declining computing costs, and the development of numerical techniques such as Finite Element Analysis, it is considered economical to apply this remarkable technique to welding research in general, and distortion and residual stress predictions in particular[1-4]. Recently, Ueda et al have summarized the application of computers in welding research[5]. The general approach followed is to develop a computational model, and subject it to experimental verifications at various stages[6,7]. Once this is done, the numerical model can be applied to any other situation as long as the temperature dependence of thermal and physical properties is known.

The objective of this study has been to model the distortion caused by overlay welding, a technique used to repair hydraulic turbines after damage due to cavitation erosion has been ground out. The Finite Element Model employed in this study for calculating heat flow and distortion, as well as the details of experiments for a single bead weld overlay situation, have been published before[8,9]. In the present paper, finite element methods have been applied to model:

(i)   three bead weld overlay;
(ii)   block weld comprising 18 beads; and
(iii)   series of six block welds (block welding*)

METHODOLOGY

Analysis Requirements

In order to predict distortion due to the welding process, one must simulate (a) the temperature history during and after welding, and (b) the resulting strain and stress history. Since plastic work due to welding is small, it is often convenient and sufficiently accurate to couple these simulations weakly by calculating the temperature history, and the stress history (distortion) consecutively[10].

Thermal Analysis

For analysis of transient heat flow, a cross-sectional model has been used[1,2,11]. This assumption implies that there is no heat flow in the longitudinal direction (i.e.

(*Block welding is a technique, wherein the surface of the whole plate is divided into a certain number of blocks, and each block is overlayed in a specific sequence. Done properly, block welding reduces distortion.)

in the direction of electrode movement). It is valid at a cross-section near the mid length of the plate, away from end effects and when the welding speed is high relative to the thermal diffusivity of the material.

The heat source model used is based on previous work done[12]. For thermal analysis, 2-dimensional, 8 node quadrilateral elements have been used for the finite element mesh. Since the subsequent distortion analysis uses these temperature histories to compute equivalent thermal nodal loads, it is desirable to have the same mesh for both thermal and distortion analysis. However, a 3-dimensional, 20 node brick element had to be incorporated in the finite element mesh for both thermal and distortion analysis, since with 2-dimensional elements, the subsequent distortion analysis gave unacceptable results.

Therefore, the 2-dimensional transient temperature field was mapped onto a 3-dimensional mesh[13]. For the three bead weld situation, individual passes were simulated consecutively. However, for block welds, two weld passes were combined and simulated as a single bead. This assumption is supported by studies in multi-pass welds[14].

For thermal analysis, material properties needed are thermal conductivity and volumetric specific heat, which are temperature dependent. For convection and radiation losses, an empirical formula was used[15] and no heat of fusion or solidification was considered.

Distortion Analysis

A time independent thermo-elasto-plastic model is adopted for the prediction of distortion. The thermal analysis done separately provides the transient temperature history and hence, the loads necessary for the stress analysis. The model is based on incremental plasticity. Beyond the elastic range, material is considered to yield, which is governed by the von Mises criterion and plastic flow by the Prandtl-Reuss flow law.

Work hardening has been modelled using a combination of isotropic and kinematic hardening rule on the assumption that most metal behaviour is closer to combined hardening. In the numerical solution, this is reflected by a parameter β and a value of 0.8 was adopted for the materials used.

The geometry was modelled by a slice of 3-dimensional, 20 noded brick elements parallel to the reference cross-section. This model allows the nodes on the midplane to have full 3-dimensional motion but prevents the nodes on the outer planes parallel to the slice surfaces from moving normal to those surfaces. If the thickness of the slice approached zero, this model would approach a plane strain model. Thus, the approach followed can be considered a step forward from plane strain model towards the desired full 3-dimensional analysis.

A typical finite element mesh is shown in Figure 1.

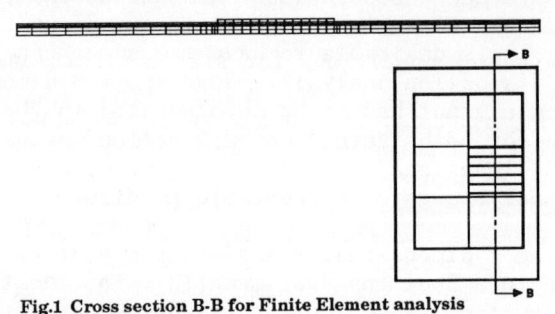

Fig.1 Cross section B-B for Finite Element analysis

Block Welding Sequence

The size of the plate being used for weld simulation was 457 X 305mm. A central area of 305 X 203mm was chosen for depositing the weld overlay. This area was divided into 6 square blocks of 101.5 X 101.5mm.

The simulation of a complete overlay sequence comprising several blocks, with each block containing 16 to 20 weld passes was an analyst's nightmare due to the magnitude of the problem. Therefore, further simplifications and assumptions were required.

The 6 square blocks are schematically shown in Figure 2.

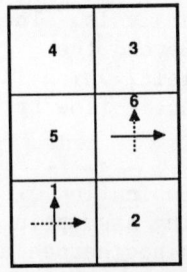

The two blocks (lower left corner, i.e. 1 and middle right, i.e. 6) were modelled for heat flow. They are the only two unique geometrical locations, for this six block division of the plate. The welding process considered is manual arc weld.

Fig.2 300mm x 200mm plate divided into 6 blocks of 100mm x 100mm square.

For the present analysis, only two welding sequences have been considered; they are shown with the solid arrows in Figure 2. Thus, it was assumed that the thermal history due to the other two welding directions (shown with dotted arrows in Figure 2) would be similar to that of the directions considered for analysis (shown with solid arrows).

The temperature fields obtained were used to calculate the
transient stresses during the thermal cycle and the
residual stresses after complete cool down in one 101.5 X
101.5mm block. This was done for two welding situations:
weld parallel and perpendicular to the long side (457mm).
Advantage was taken of the pseudo-symmetry of the block
locations (i.e. blocks 1, 2, 3 and 4 are symmetrically
placed - at the corners, and block nos. 5 and 6 are also
symmetrically placed - at the centre). Edges of the 101.5 X
101.5mm block were simply supported. This provided the
necessary boundary conditions for static stress analysis.

The residual stress and strain in the block computed for
welds parallel and perpendicular to the longitudinal
directions were then applied to the full plate (Figure 2).
This was done by mapping final strain in each block from
the above welds as the initial strain in the full plate.
With this initial strain, and the rigid body modes
constrained, the full plate was allowed to come to
equilibrium in a one step elasto-plastic analysis.

EXPERIMENTAL METHOD

The details of the experimental set-up, welding procedure
and measurements of distortion, have been described
elsewhere[9,18] and are not repeated here. The experimental
plate was 19mm thick and 305 X 457mm.

MATERIALS AND OTHER VARIABLES

For the three bead weld overlay system, two combinations of
base and filler materials have been studied in this
project. The distortion of carbon steel (ASTM A36)
overlayed with E-308L stainless steel and of martensitic
stainless steel (ASTM A743 Grade CA6NM) overlayed with
Stellite 21 have been measured. Manual arc welding process
has been used to deposit the overlays at two preheats, $20°C$
and $130°C$, respectively.

For the block weld simulations, the preheat was kept
constant ($20°C$) and the base material-filler material
combinations were A36/E308L and CA6NM/E308L. For each
combination, two blocks were analyzed for longitudinal and
transverse welding directions. The total strain from these
two blocks was used as initial strains in the three plate
patterns shown in Figure 3.

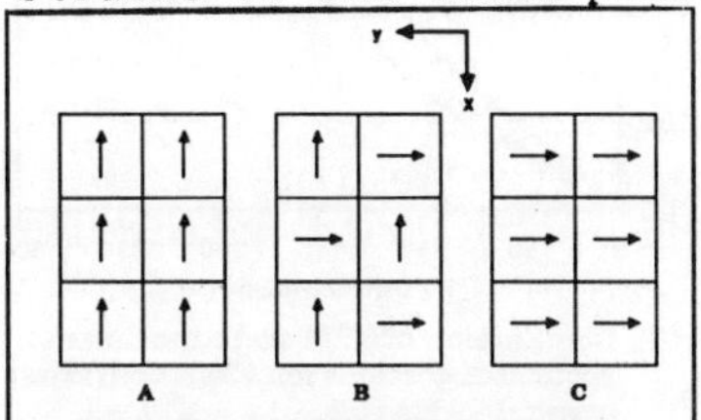

Fig. 3 Three Different Weld Patterns

RESULTS AND OBSERVATIONS

As mentioned before, the transient temperature fields were calculated for each welding condition. Figure 4 shows the typical peak temperature distribution for three bead overlay welds. Figure 5 shows the typical vertical distortion calculated by the finite element method and this is compared to the experimental measurements. Figure 6 shows the typical longitudinal residual stresses calculated by FEM.

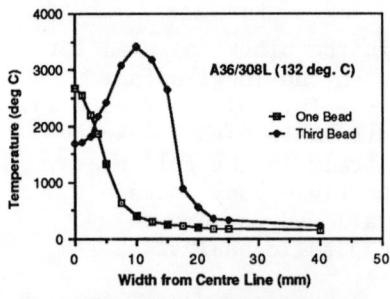

Fig.4 FEM prediction of Peak temperature distribution due to 1st and 3rd bead.

Fig.5 Comparison of FEM and vertical displacements for 3 bead welds.

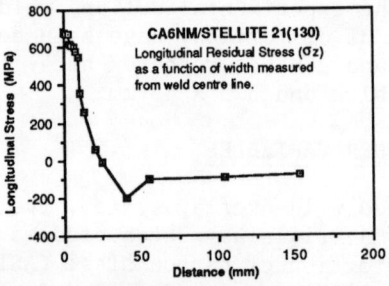

Fig.6 FEM prediction of longitudinal residual stress.

The vertical distortions calculated by FEM for welding the lower corner block and centre block (refer to Figure 2 - block number 1 and 6) are compared to experimental measurements for A36/E308L material combinations in Figures 7 and 8.

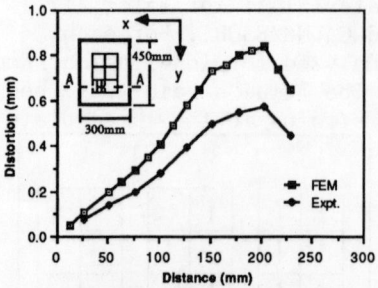

Fig.7 Comparison of FEM and experimental vertical distortions for A36/E308L lower corner block welds.

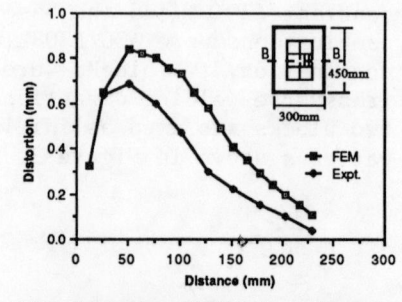

Fig.8 Comparison of FEM and experimental vertical distortion for A36/E308L centre block weld.

Figure 9 shows the FEM predicted final distortion due to all the six blocks overlayed, welds being parallel to the long side of the plate (longitudinal pattern).

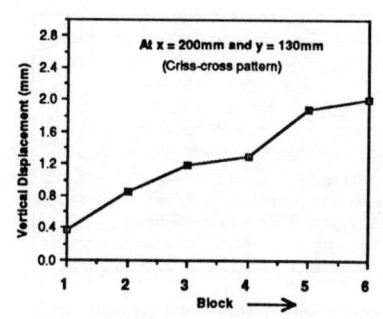

Fig. 9 Distortion due to different blocks on the whole plate.

A summary of all calculated distortions, together with the experimental measurements is given in Table 1.

TABLE 1

Maximum Distortion (mm)

Base Metal/ Filler Metal	Preheat(°C)	FEM	EXPT.	REMARKS
A36/E308L	20	1.75	0.94	3 beads
	130	1.67	1.05	3 beads
CA6NM/Stellite 21	20	0.65	0.60	3 beads
	130	0.43	0.49	3 beads
A36/E308L	20	2.44	1.63	Longitudinal Pattern
		2.00	1.29	Criss-Cross Pattern
		2.34	1.32	Transverse Pattern
CA6NM/E308L	20	1.42	--	Longitudinal Pattern
		1.13	--	Criss-Cross Pattern
		1.30	--	Transverse Pattern

Typical iso-distortion contours of the distorted CA6NM/E308L (3 seams) weld overlay test plate is shown in Figure 10. From Figure 10, it is seen that the iso-distortion contours are nearly elliptical in shape. The major axis of the ellipse is along the direction of welding. This implies that the cumulative strain due to lateral contraction of the welding passes is larger than that due to longitudinal shrinkage, by a factor of about 9.

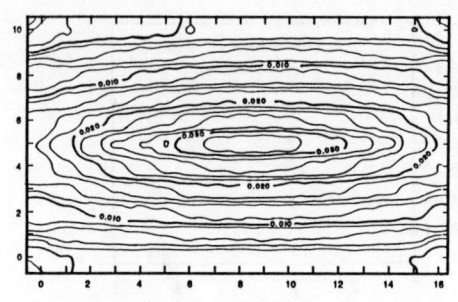

Fig. 10 Isodistortion contours of CA6NM/E308L (3 seams) weld overlay test plate.

DISCUSSION

The results presented in Table 1 show reasonable agreement between the finite element predictions and experimental measurements. This gives confidence in the F.E. model, and its potential to simulate more complex geometries where simple analytical expressions may be inadequate.

All the same, an attempt was made to develop an empirical relationship to estimate distortion for the simple case of an overlay weld on flat plate. Initially based on the finite element analysis, distortions were calculated for 1, 3, 10, 15 and 20 beads deposited side by side using A36/E308L and CA6NM/E308L material combinations. The FEM results, with experimental results for 1 and 3 bead welds, and for 20 bead welds (A36/E308L only) are shown in Table 1.

As mentioned earlier, in order to model the complete six block welding pattern in a plate, the thermal and stress analysis was performed on a single block. This was done for A36/E308L and CA6NM/E308L base metal/filler metal combinations. The strain resulting from a single block weld analysis was then assembled to build up the six block welding pattern in the full plate.

In both cases studied (A36/E308L and CA6NM/E308L), the criss-cross weld pattern (Figure 3b) produces the least distortion (2.0mm and 1.13mm respectively) and the longitudinal pattern (Figure 3a) produces the largest distortion (2.44 and 1.42mm respectively).

The FEM predictions for distortions are generally higher than the experimentally determined values, and there could be several reasons for this discrepancy. The distortions are primarily due to the plastic deformation (or thermal upsetting), which is flow path dependant. Due to the assumption that a single block can be modelled separately and the results used to predict distortions due to other combinations of blocks - might be the most serious one. When a block adjacent to a welded block is welded, there will be some release of plastic strain due to additional heat flow (annealing effect). This, in turn, will reduce distortion. Neglect of this effect will give higher predicted distortion.

Other possible explanations for higher predicted distortion values include cross-sectional heat transfer model that ignores the transient on starting the weld, and grouping thermal cycles of two passes for each thermo-elasto-plastic analysis leading to overestimating of temperature. The error in the temperature field due to assuming symmetry in each block, i.e. symmetrically placed welds are performed simultaneously, is expected to be small.

In contrast, one particular feature of the thermo-elasto-plastic analysis used, viz. only one layer of 20 node bricks, restricts deformation and thus is expected to underestimate the amount of distortion.

An exponential regression analysis was done between the distortion and number of beads deposited. The following two equations were obtained:

Distortion $Y = 0.932 \, Exp(0.083n)$, for A36/E308L   (1)

$Y = 0.594 \, Exp(0.78n)$, for CA6NM/E308L   (2)

where n = number of weld beads and Y (mm) is the vertical distortion.

The exponential co-efficients in Equations (1) and (2), i.e. 0.083 and 0.078, were averaged out and the equations below were obtained.

$Y = 0.932 \, Exp(0.08n)$ (3)

and $Y = 0.594 \, Exp(0.08n)$ (4)

Leggatt [16] has proposed an equation for calculating the maximum angular distortion for single pass welds as follows:

$$\beta_{max} = A \frac{(1+\mu)\alpha}{C} \cdot \frac{Q}{t^2} \qquad (5)$$

Where β_{max} = maximum free angular distortion (deg.)
  μ = Poisson's ratio
  σ = Co-efficient of thermal expansion (m/m°C)
  Q = Heat input (J/mm)
  t = plate thickness (mm)
  C = volumetric specific heat (J/mm^3) °C)
  A = a constant

Equation (5) has been derived for a full penetration butt weld with no external restraint. In order to apply the above equation to the present case, the co-efficient 'A' must be redetermined for relatively shallow, bead-on-restrained-plate overlay welds. For the present study, the value of A was computed by regression analysis to obtain the following equations:

$$\beta \text{ (deg)} = 36.1 \frac{(1+\mu)\alpha}{C} \frac{Q}{t^2} \quad \text{(for A36)} \qquad (6)$$

$$= 20.6 \frac{(1+\mu)\alpha}{C} \frac{Q}{t^2} \quad \text{(for CA6NM)} \qquad (7)$$

By converting angular distortions to vertical distortions, Equations (6) and (7), respectively become,

$$Y = 102.2 \frac{(1+\mu)\alpha}{C} \frac{Q}{t^2} \text{ Exp } (0.08n) \text{ for A36} \qquad (8)$$

$$Y = 53.5 \frac{(1+\mu)\alpha}{C} \frac{Q}{t^2} \text{ Exp } (0.08n) \text{ for CA6NM} \qquad (9)$$

Based on the observed inverse dependence of distortion on yield strength [18] and incorporating the known values of yield strength into Equations 8 and 9 (348 MPa for A36 and 750 MPa for CA6NM) and averaging out, we have

$$Y = 3.79 \times 10^4 \frac{(1+\mu)\alpha}{\sigma_y C} \frac{Q}{t^2} \text{ Exp } (0.08n) \qquad (10)$$

Table 2 shows the calculated distortions, using Equation 10.

TABLE 2

Distortion Analysis at 20°C

Weld/Filler Metal	No. of Beads	Expt.	Distortion (mm) FEM	From Eq.10
A36/E308L	1	0.80	0.71	0.99
	3	0.94	1.75	1.16
	10	--	2.15	2.04
	15	--	3.26	3.04
	20	2.54	4.75	4.53
CA6NM/E308L	1	0.49	0.49	0.56
	3	0.75	1.015	0.66
	10	--	1.29	1.15
	15	--	1.91	1.72
	20	--	2.75	2.57
A36/Stellite 21	1	0.74	0.79	0.95
CA6NM/Stellite 21	1	0.60	0.65	0.59

The above approaches illustrate a possibility of integrating numerical and experimental results to construct a simple economical empirical model.

SUMMARY

1.   A Finite Element Model is developed to predict the vertical distortions in welding overlays on flat plate. The FEM analysis used a three stage model: (a) Transient thermal analysis using 2D cross-sectional model in each block; (b) Transient thermo-elasto-plastic 3D analysis of each block in which 2 passes were grouped together to compute thermal loads; and (c) Elasto-plastic analysis of a plate containing six blocks using the total strain from the blocks as initial strain in the plate.

2.   An attempt has been made to use the FEM and experimental results to formulate an empirical relationship for prediction of vertical distortions due to overlay welds on a flat plate.

3.   An empirical relationship is presented for predicting angular distortions, which relates the angular distortion to heat input, plate thickness, Poisson's ratio, co-efficient of a thermal expansion, specific heat and yield strength of the material.

ACKNOWLEDGEMENT

The work described here was undertaken as part of a project funded by Canadian Electrical Association, and it is published with their approval.

REFERENCES

1. Friedman, E., "Thermo-mechanical Analysis of the Welding Process using the Finite Element Method". Transactions of the ASME, August 1975, pp.206-213.

2. Anderson, B.A.B., "Thermal Stresses in a Submerged Arc Welded Joint Considering Phase Transformations". Transactions of the ASME, October 1978, pp.356-362.

3. Argyris, J.H., Szimmat, J. and Willam, J., "Computational Aspects of Welding Stress Analysis'. Computer Methods in Applied Mechanics and Engineering, 33 (1982), pp.635-666.

4. Lindgren, L., "Deformation and Stresses in Butt Welding of Plates". Ph.D. Thesis, Lulea University of Technology, Sweden, 1985.

5. Ueda, Yukio, Murakawa, Hidekazu, "Applications of Computer and Numerical Analysis Techniques in Welding Research". Materials and Design, Vol. 6, No. 3, June/July 1985, pp.103-111.

6. Rybicki, E.F., Stonesifer, R.B., Merrick, E.A., and Haueter, J.R., "A Computational and Experimental Study of Deformations in a Draw Bead Welded Pipe". Welding Research Supplement, April 1986, pp.99s-105s.

7. Tokuzawa, Naoki and Horikwa, Kohsuke, "Mechanical Behaviours of Structural Members Under Loading". Trans. of JWRI, Vol. 10, No. 1, 1981, pp.95-101.

8. Chakravarti, A.P., Goldak, J.A. and Rao, A.S., "Thermal Analysis of Welds". Proceedings of the Fourth International Conference on Numerical Methods in Thermal Problems, July 15-18, 1985, Swansea, U.K.

9. Chakravarti, A.P., Goldak, J.A. and Rao, A.S., "Prediction of Thermal History of Repair Welds". Proceedings of the International Conference on Welding for Challenging Environment, October 14-18, 1985, Toronto, Canada.

10. Hibbitt, H.D. and Marcel, P.V., "A Numerical Thermo-Mechanical Model for the Welding and Subsequent Loading of a Fabricated Structure. Comput. Strut., 3 (1973), pp.1145-74.

11. Krutz, G.W. and Segerlind, L.J., "Finite Element Analysis of Welded Structures". Welding Research Supplement, July 1978, pp.211s-216s.

12. Goldak, J.A., Chakravarti, A.P. and Bibby, M., "A New Finite Element Model for Welding Heat Sources". Met. Trans. B, Volume 15B, June 1984, pp.299-305.

13. Patel, B., "Thermo-Elasto-Plastic Finite Element Analysis of Deformations and Residual Stresses Due to Welds". Ph.D. Thesis, Carleton University, Ottawa, 1985.

14. Ueda, Yukio and Nakacho, Keiji, "Simplifying Methods for Analysis of Transient and Residual Stresses and Deformations due to Multi-pass Welding". Trans. JWRI, Volume 11, No. 1, 1982.

15. Vinokurov, V.A., "Welding Stresses and Distortion". The British Library Lending Division, Translated from Russian into English by J.E. Baker.

16. Leggatt, R.H., "Distortion in Welded Plates". PhD. Thesis, Cambridge University, 1980.

17. Gray, T.G.F. and North T.H., "Rational Welding Design". Published by Newnes-Butterworths, London (U.K.)

18. Chakravarti, A.P., "Optimization of Overlay Welding Procedures for Hydraulic Turbines". AMCA International Report No. 7-184-016/6, July 1986.

FINITE ELEMENT SOLUTION FOR A ROUND CORNERED SQUARE PLATE
SUBJECTED TO THERMAL EFFECTS OVER A CENTRAL REGION

M.C. Singh, Professor and P.K. Chan, Research Assistant
Department of Mechanical Engineering
The University of Calgary
Calgary, Alberta, Canada

1. INTRODUCTION

Stress and displacement analysis of plates subjected to mechanical loading is extensively available in the literature. Thermoelastic analysis of plates however, is dealt with generally in thermoelastic stress analysis [1,2,3].

A clamped round cornered square plate subjected to a temperature distribution over a centrally located circular region is amenable to conformal transformations [4]. The plate deflection equation of equilibrium, the conditions of continuity at the juncture of loaded and unloaded regions and the conditions at the boundary of the plate form the basic equations of the plate for solution. Complex analysis of a plate thermally loaded over a centrally located circular region is available in the literature [5]. A finite element solution of this problem is developed and the results found to agree with those available by the complex variable approach. The Technique is then applied to temperature distribution over centrally located non circular regions.

2. BASIC EQUATIONS

The basic equations given herein are based on Kirchhoff's hypothesis that normal to the mid plane of the plate remains normal to the deflected mid plane with negligible change in its length. Further, normal stresses in the direction of plate thickness as well as membrane stresses are neglected. Equation of equilibrium in the presence of mechanical loading and thermal effects is expressible in the form

$$D\nabla^4 W = p - \frac{1}{1-\nu} \nabla^2 M_T ,\qquad(1a)$$

$$\text{where } D = \frac{Eh^3}{12(1-\nu^2)} ,\qquad(1b)$$

and $M_T = \alpha E \int_{-h/2}^{+h/2} T \, z \, dz$. (1c)

In equation (1) in usual notation W is the deflection of the plate, p is the mechanical loading normal to the plate, M_T is the thermal moment, $T(x,y,z,t)$ is the temperature as a function of position and time, E is the modulus of elasticity, ν is poisson's ratio and h is the thickness of the plate.

Assuming p = 0, and the temperature distribution to be antisymmetric with respect to mid plane of the plate

$$T(x,y,z,t) = - T(x,y, - z,t) ,$$ (2)

equation (1a) assumes the form

$$\nabla^4 W = - \frac{\nabla^2 M_T}{D(1-\nu)} .$$ (3)

Assuming the temperature distribution in harmony with equation (2) in the form

$$T = T_o + z [G(x,y) + H(t)] ,$$ (4)

Equation of heat conduction

$$\beta \nabla^2 T = \frac{\partial T}{\partial t}$$ (5)

is satisfied by (4) for

$$H(t) = t\beta \, \nabla^2 G(x,y) + C .$$ (6)

On the basis of (4) and (1c)

$$M_T = \frac{\alpha E h^3}{12} [G(x,y) + H(t)] .$$ (7)

and

$$\frac{\nabla^2 M_t}{(1-\nu)} = \frac{\alpha E h^3}{12(1-\nu)} \nabla^2 G(x,y) = B$$ (8)

where B is constant obtained on the basis of a choice of the function $G(x,y)$. On the basis of equations (8), equation (3), assumes the form

$$\nabla^4 W = - \frac{B}{D} ,$$ (9)

for the thermally loaded region.

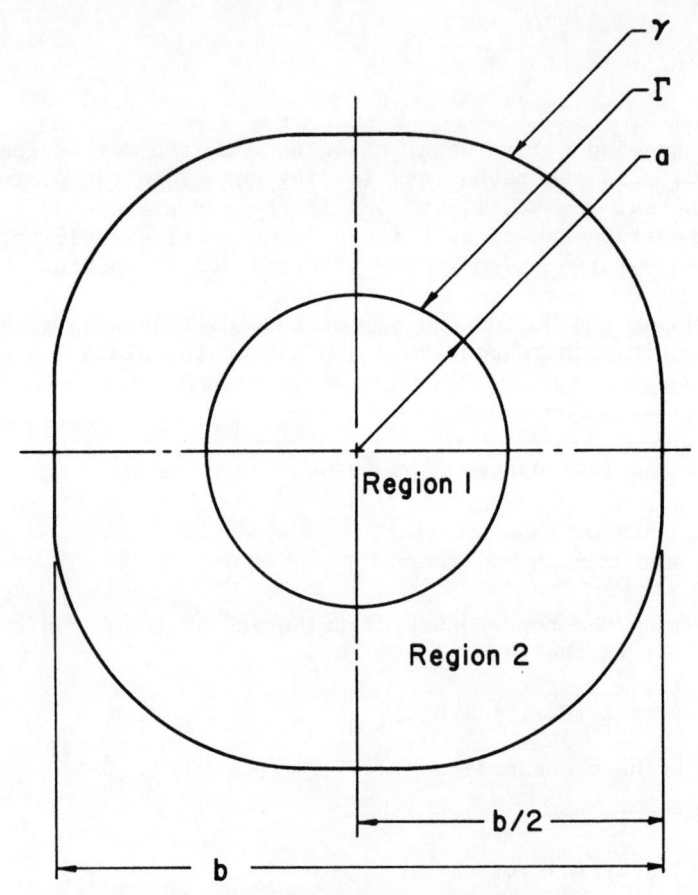

Fig. 1 Round Cornered Square Plate
Loaded Over a Circular Region

Basic equation for the round cornered square plate with fixed edges around the boundary and partially loaded as shown in Figure 1 can now be expressed in the form

$$\nabla^4 W = -\frac{B}{D}, \quad \text{in region 1, region of thermal loading} \quad (10a)$$

$$\nabla^4 W = 0, \quad \text{in region 2, region of no loading} \quad (10b)$$

$$W = \frac{\partial W}{\partial n} = 0, \quad (11a,b)$$

on the boundary γ.

The continuity conditions for deflection, slope, moment and shear are satisfied on the boundary Γ of thermally loaded region 1 and region 2 of no loading.

$$[W]_2^1 = \left[\frac{\partial W}{\partial n}\right]_2^1 = [M_n]_2^1 = [Q_n]_2^1 = 0, \qquad (12a,b,c,d)$$

where $[W]_1^2 = W_2 - W_1$, (13)

is the jump in displacement. Similar equations hold for other jump expressions in equations (12).

3. FINITE ELEMENT MODAL

An ANSYS finite element model of the plate with its boundary conditions is shown in Figure 2. The numbers at the center of the elements identify the elements. The four numbers around the elements identify the nodal point locations. The elements in the central region starts from the right toward the center of the region and the origin of the global Cartesian coordinate is located at nodal point 78. The central region is set up using the mesh generation procedure which requires the boundary and cornered keypoints of the plate to be defined, the line segments to be related between the keypoints, the region size to be specified in terms of number of nodal points in

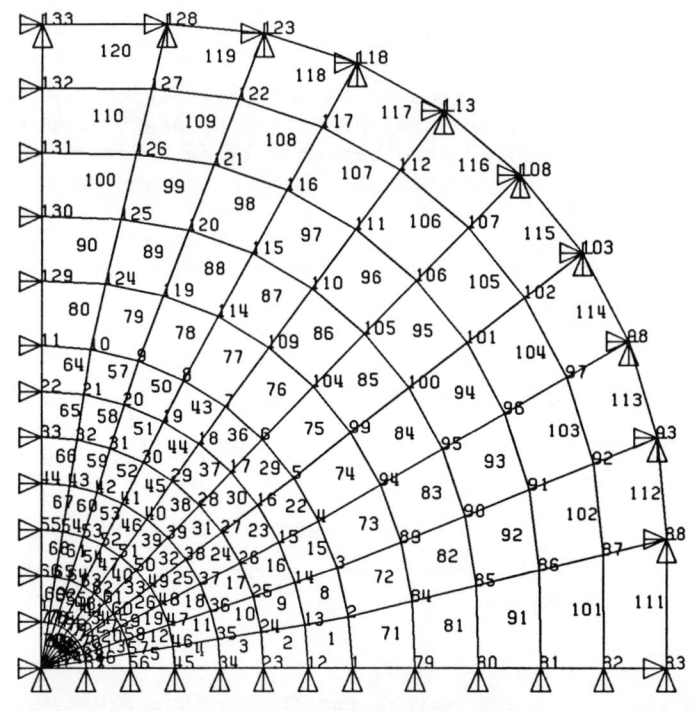

CSPL2A - ROUND CORNER SQ PLATE WITH FULL LOAD

Fig. 2

the mesh, and the area to be defined by the corner keypoints as desired. The mesh will be automatically generated if the above procedures are followed in proper order. The technique of mesh generation is shown in Figure 3. The keypoints 2 and 3 on a

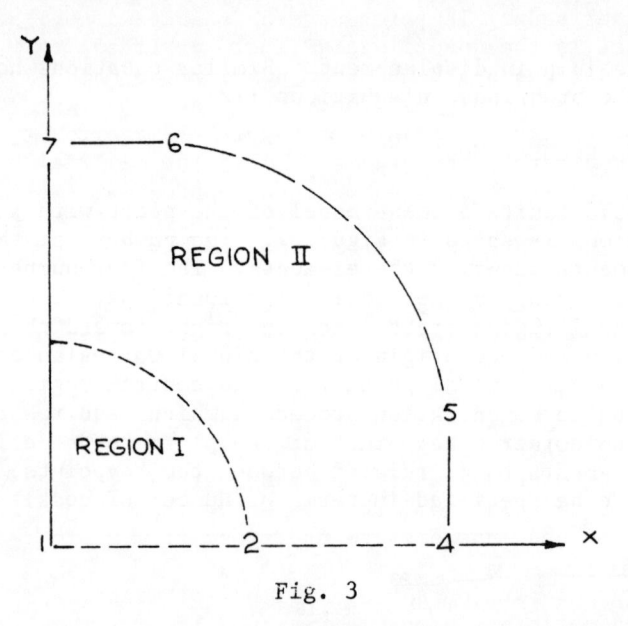

Fig. 3

surface define the locations of the Central region and the line segments are required to relate the boundary keypoints 1,2 and 2,3. A circular area region will be generated if the corner keypoint 1 is repeated in the AREA command as illustrated in the Manual [7].

A 'stretched out' rectangular view is shown in Figure 4 to identify the corner keypoints and the starting point of the region for the linear mesh interpolation.

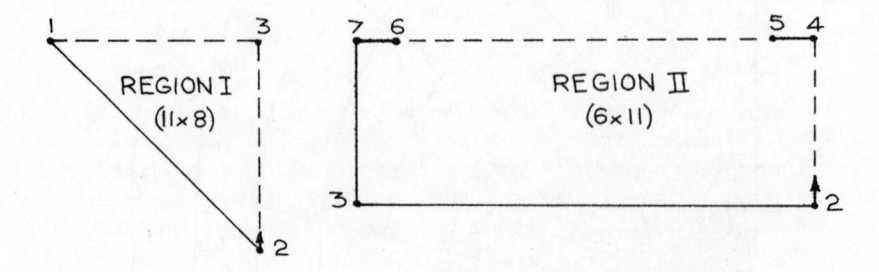

Fig. 4

The type of element used to describe the plate model is the quadrilateral flat shell element (STIF 63), which can be applied to plate structures.

4. INPUT DATA

The plate finite element model consists of 4-noded three dimensional quadrilateral flat shell elements with six degree of freedom per node. The element type selected from the ANSYS User's Manual is the quadrilateral shell stiffness type (STIF 63). It is defined by the ET command with the KEYOPT values specified for the bending stiffness only and for flat shell loading. The program starts with the /PREP7 preprocessing level to prepare for the data analysis. The material properties and thickness associated with the appropriate material number and element type are defined before the elements are generated. The plate elements are defined by four nodes excepted for Elements 7,14,21,28,35, 42,49,56,63, and 70 at the center which are described by three nodes. The fourth node is the same as the third node for generating the triangular element and the center nodal points are suppressed into a common nodal point 78. The element numbers in the central region are generated from 1 to 70 and the node numbers from 1 to 78. In the unloading region defined by keypoints 2, 4,7,3, as shown in Figure 4, the quadrilateral elements are adopted by the linear mesh interpolation. The element numbers begin from 71 to 120 and the node numbers are expressed from 79 to 133.

The overall data output control begins with the load step ITER command defining in one iteration. The reaction force is requested to be evaluated by defining the KRF command to 1. The clamped edge boundary conditions are specified by the D command to define the degrees of freedom or the specified displacements at the nodes. The D command is specified by the edge nodal points 83,88,93,98,103,108,113,118,123,128, and 133 to be completely constrained. Moreover, the displacements in y direction and rotations about x axis are restricted for the nodal points along the x axis. The displacements in x direction and rotations about y axis are also restricted for the nodal points along the y axis.

The thermal loading is evaluated to be equivalent to an externally applied load on the top surface of each element in the loaded region and is determined to be a constant of 827.4. The AFWRIT command is given before leaving the preprocessing level in order to prepare the current modal data and the load step data on a file to be analyzed. The /INPUT, 27 command will cause the data file 27 to be read and the evaluation will begin to be performed.

The post processing routine /POST1 is used to sort and to organize the output data of displacements, nodal stresses, and principal stresses on top and bottom surfaces of the plate.

5. NUMERICAL ANALYSIS AND SUMMARY OF RESULTS

The plate is made of steel having the material properties as follows:

$E = 30 \times 10^6$ psi
$\nu = 0.3$
$\rho = 7.24 \times 10^{-4}$ lb-sec^2/in^4
$G = 11.5 \times 10^6$ psi

The thickness of the plate is 0.25 in. The plate is rigidly clamped along the edge and is unstressed before applying load.

A 10" x 10" square plate with round corners of 4" radius is subjected to the thermal loading at the central region of 2.5" radius from the center of the plate as shown in Figure 1. When the thermal load is applied, the temperature of the central region is raised to a particular temperature. The thermal load is determined to be a constant which is equivalent to a mechanical load.

It is desired to analyze the displacements of the plate and the corresponding stresses after loading. The maximum displacements for various configurations of applied load on the round cornered square plate and on the sharp cornered square plate have been summarized in Table 3.1 and Table 3.2 respectively.

It is interesting to compare the results from the finite element method with those results from complex variable method and finite difference method. The maximum displacements at the center of the plate for all cases analyzed by the finite element method fall in between the results obtained by the other two methods. There are six different cases of various configurations of load applied on the round cornered plate. The maximum displacement evaluated in Case 1 of circular loaded plate is 0.140 in. For Case 2 of triangular loaded plate and Case 6 of round cornered plate having a loading region of 1.293" radius, the maximum displacements are 0.138 in. and 0.139 in. respectively. For Case 3 of elliptically loaded plate and Case 5 of round cornered plate having a loading region of 4" radius from x = -1.112" and y = -1.112", the maximum displacements are 0.141 in. and 0.142 in. respectively, which are larger than the result obtained in Case 1. The intensity of applied load is the same for all cases except in Case 4 for which the calculated area of the loading region is determined to be 4.9 in.^{2} and the thermal loading constant to be 827.4. For Case 4 of fully loaded plate, the maximum displacement is 0.23 in., which is 64% larger than that in Case 1.

Table 3.1 Maximum Displacements for Different Loading
Configurations on a Round Cornered Square Plate

Method	Description	Maximum Displacements(in.)
Complex Variable Method	Previous results[5]	0.151
Finite Difference Method	Previous results[5]	0.133
Finite Element Method (Case 1)	CSPL2A-Circular Load	0.140
(Case 2)	CSPL2B-Triangular Load	0.138
(Case 3)	CSPL2C-Elliptical Load	0.141
(Case 4)	CSPL2E-Full Load	0.230
(Case 5)	CSPL2F-Radius of Outer Corner(RAD=4")	0.142
(Case 6)	CSPL2F1-Round Corner (RAD=1.293")	0.139
(Case 6a)	CSPL2F1E-Full Load in Case 6	0.230

Table 3.2. Maximum Displacements for Different Loading
Configurations on a Sharp Cornered Square Plate

Method	Description	Maximum Displacements(in.)
Finite Element Method Case 7	SPL2A - Circular load	0.148
Case 8	SPL2E - Full load	0.254
Case 9	SPL2F	0.150
Case 10	SPL2F1	0.147
Case 11	SPL2FE - Full load in Case 9	0.254

The maximum displacements for various shape of applied loaded region or sharp cornered plate are presented in Table 3.2. For Case 8 and 11 of fully loaded plates where they have different element size and mesh, the maximum displacements are found to be 0.254 in. The maximum displacement for Case 7 of circular loaded sharp cornered plate is 0.148 in. For Case 9 and 10 of plates having loading regions of 4" and 1.293" radius, respectively, the maximum displacements are 0.150 in. and 0.147 in. respectively.

From the above results it may be concluded that refined mesh and element size used in the finite element method herein lead to reasonably accurate results for the round and sharp cornered square plates.

6. REFERENCES

1. Boley, B.A.; Weiner, J.H.; Theory of Thermal Stresses, John Wiley, 1960.

2. Parkus, H.; Thermoelasticity, Baisdell, 1968.

3. Nowinski, J.L.; Theory of Thermoelasticity With Applications. Sijthoff and Noordhof Alphen an Den Rijn, 1978.

4. Mansfield, E.H.; The Bending and Stretching of Plates, Pergamon Press, 1964.

5. Singh, M.C.; Maytum, J.N.; Deflection of a Round Cornered Square Plate Subjected to Thermal Effects Over a Central Circular Region. Proc. Canadian Congress of Applied Mechanics, Waterloo, 1969.

6. Weaver, W.; Johnston, P.R.; Finite Elements for Structural Analysis, Prentice Hall, 1984.

7. DeSalvo, G.J.; Swanson, J.A.; ANSYS User's Manual, Swanson Analysis Systems, Houston, Pa., 1982.

SECTION 8

Novel Computational Methods

ANALYTICAL AND COMPUTATIONAL TECHNIQUES IN THE TREATMENT OF
HEAT TRANSPORT PROBLEMS

Helmut F. Bauer

Head of the Institute of Spacetechnology,
University of the German Armed Forces Munich,
8014 Neubiberg, F.R.Germany

SUMMARY

Analytical and computational techniques for the determination of the local temperature distribution are presented for various engineering systems of different geometry. Solutions for steady heat conduction has been shown for conical, paraboloidal and spheroid geometries. In addition heat transport has been presented in flowing liquids confined to tubes of parallelogram cross-section, changing cross-sections as for flows in a conical or converging-diverging nozzle. In many cases, where no exact solution could be obtained, the procedure of solving the problem is based on the solution of an auxiliary differential equation followed by the Galerkin Method.

1. HEAT CONDUCTION PROBLEMS

In many engineering thermal problems, where heat passes through the substance of a body itself, the treatment for determining the local distribution of the temperature may be performed by considering only the conduction, thus neglecting convection and radiation. In quite a number of aeronautical- and aerospace engineeringproblems, as well as in thermal control studies for satellites, conical, paraboloidal and prolate spheroid geometries appear, such as the nose of a high-speed airplane or that of a reentry vehicle, missile or space capsule. The conduction of heat has been treated extensively for rectangular, cylindrical and spherical coordinates by Carslaw and Jaeger [1], but not much information is given for other geometries, where various bodies of annular-truncated-sector forms may appear.

1.1. CONICAL SYSTEMS [2]

In a conical system, in which the sector sides $\varphi=0$, and $2\pi\gamma$ ($\gamma \leq 1$) and the conical sides $\vartheta = \alpha$ and β exhibit constant temperature T_o, while at the spherical surfaces r=a and b the temperature distribution is given as $T(a,\vartheta,\varphi)=f_1(\vartheta,\varphi)$ and $f_2(\vartheta,\varphi)$ resp.. The solution of the heat conduction equation

$$\frac{\partial}{\partial r}(r^2 \frac{\partial T}{\partial r}) + \frac{1}{\sin\vartheta} \frac{\partial}{\partial \vartheta}(\sin\vartheta \frac{\partial T}{\partial \vartheta}) + \frac{1}{\sin^2\vartheta} \frac{\partial^2 T}{\partial \varphi^2} = 0 \qquad (1)$$

can be represented as an infinite double series with associated Legendre functions, i.e.

$$T(r,\vartheta,\varphi) = T_o + \sum_{m=1}^{\infty} \sum_{n=1}^{\infty} [P_{\lambda_{mn}}^{m/2\gamma}(\cos\vartheta) Q_{\lambda_{mn}}^{m/2\gamma}(\cos\alpha) - P_{\lambda_{mn}}^{m/2\gamma}(\cos\alpha)$$

$$Q_{\lambda_{mn}}^{m/2\gamma}(\cos\vartheta)] \cdot [A_{mn}(\frac{r}{a})^{\lambda_{mn}} + B_{mn}(\frac{r}{a})^{-(\lambda_{mn}+1)}] \sin(\frac{m}{2\gamma}\varphi) \qquad (2)$$

where the eigenvalues λ_{mn} are the roots of the determinant

$$\begin{vmatrix} P_\lambda^{m/2\gamma}(\cos\alpha) & Q_\lambda^{m/2\gamma}(\cos\alpha) \\ P_\lambda^{m/2\gamma}(\cos\beta) & Q_\lambda^{m/2\gamma}(\cos\beta) \end{vmatrix} = 0 \qquad (3)$$

with respect to the degree λ of the associated Legendre functions of the first and second kind [3,4]. By expanding the functions f_1 and f_2 in Legendre-Fourier-Sine-series the remaining constants A_{mn} and B_{mn} are obtained. This requires in addition the orthogonality relation of the cross-product Legendre-functions. The isothermal surfaces for a slit cone of apex angle 2α, subjected at r=a to the temperature distribution $T=T_o+T_1(\cos\vartheta-\cos\alpha)^\ell \cdot \sin\varphi/2$ and at $\vartheta=\alpha$ and $\varphi=0$ and 2π to T_o is shown for $\ell=1$ in Figure 1. For a given temperature distribution at the conical-surfaces, such as $T=T_o+T_1\overline{f}_j(r,\varphi)/\sqrt{r/a}$ for j=1,2 at $\vartheta=\alpha$ and β resp. ($\overline{f}_j=0$ for r=a,b) and constant temperature T_o at the other surfaces, the solution of equ.(1) is presented by

$$T=T_o + \sum_{m=1}^{\infty} \sum_{n=1}^{\infty} [A_{mn} P_{-1/2+i\nu}^{m/2\gamma}(\cos\vartheta) + B_{mn} Q_{-1/2+i\nu}^{m/2\gamma}(\cos\vartheta)] \cdot$$

$$\sin(\nu\ell n \frac{r}{a}) \sin(\frac{m}{2\gamma}\varphi) / \sqrt{\frac{r}{a}} \qquad (4)$$

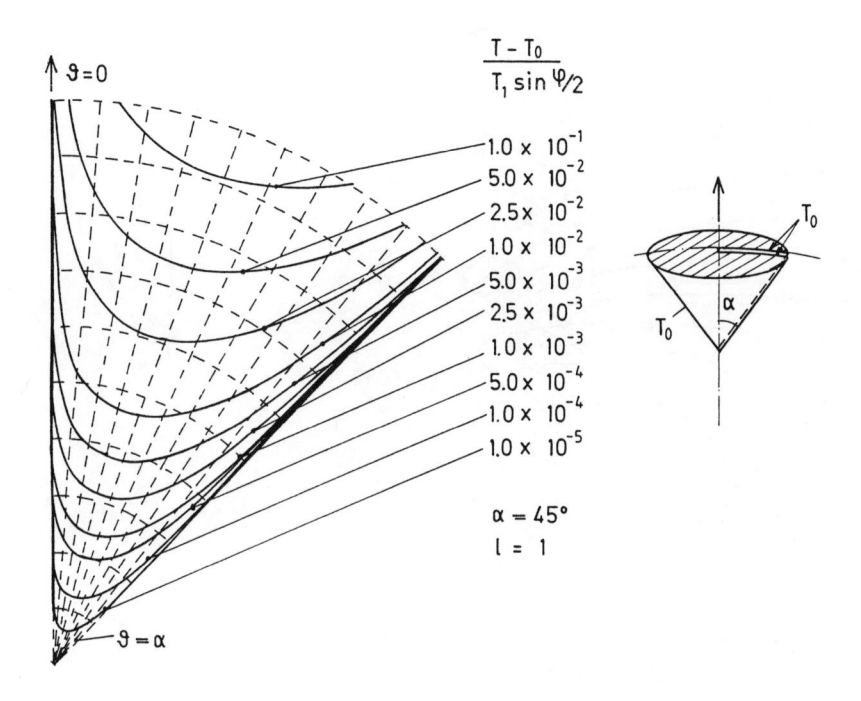

Figure 1: Isothermal surfaces for slit-cone with constant temperature T_0 at $\vartheta=\alpha$, $\varphi=0$, 2π and $T=T_0+T_1(\cos\vartheta-\cos\alpha)\cdot\sin\varphi/2$ at $r=a$

where $\nu=n\pi/\ln(b/a)$. A double Fourier-series expansion of $\bar{f}_j$ yields the remaining constants. For a cone-frustrum with constant temperature T_0 at $r=a,b$ ($b/a=0.2$) and a given temperature distribution of $T=T_0+T_1(r/a-1)^\ell[(r/a)^m-(b/a)^m]/\sqrt{r/a}$ at $\vartheta=\alpha$ the local temperature distribution is presented for $\ell=m=1$ in Figure 2.

1.2. PARABOLOIDAL SYSTEMS [5]

In a similar fashion paraboloidal systems may be treated, where for a truncated annular paraboloidal sector with constant temperature T_0 at $\varphi=0, 2\pi\alpha$ and $\xi=\xi_0,\xi_1$ and given temperature distribution $T=F_0(\xi,\varphi)$ and $F_1(\xi,\varphi)$ at $\eta=\eta_0$ and η_1 resp. the heat conduction equation

$$\frac{1}{\xi}\frac{\partial}{\partial\xi}(\xi\frac{\partial T}{\partial\xi})+\frac{1}{\eta}\frac{\partial}{\partial\eta}(\eta\frac{\partial T}{\partial\eta})+\frac{(\xi^2+\eta^2)}{\xi^2\eta^2}\frac{\partial^2 T}{\partial\varphi^2}=0 \qquad (5)$$

a solution

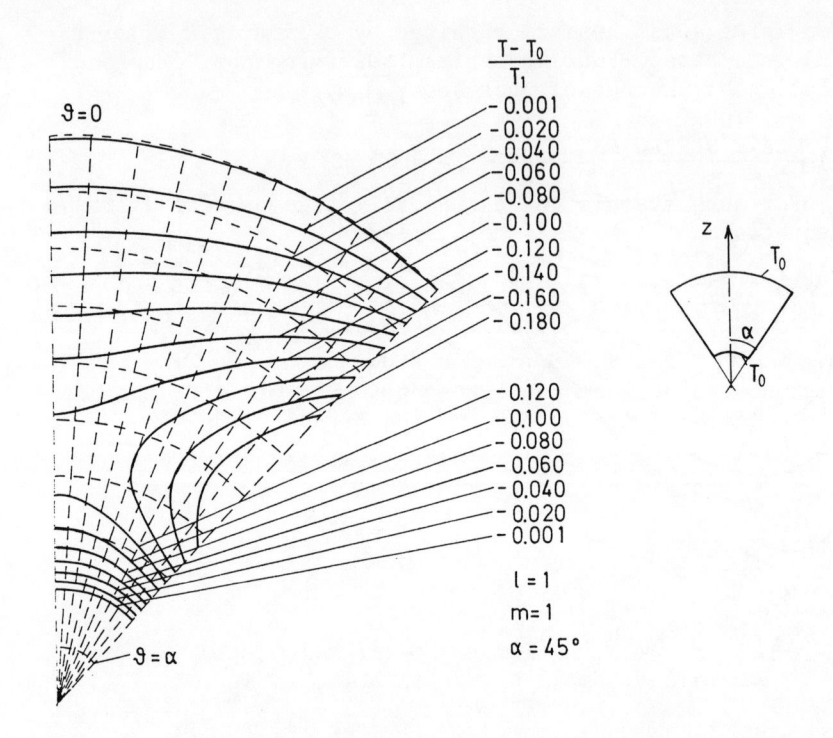

Figure 2: Isothermal surfaces for cone-frustrum with constant temperature T_o at $r=a,b$ and $T=T_o+T_1(r/a-1)^{\ell}[(r/a)^m-(b/a)^m]/\sqrt{r/a}$ at $\vartheta=\alpha$ ($m=\ell=1$; $b/a=0.2$)

$$T(\xi,\eta,\varphi)=T_o+\sum_{m=1}^{\infty}\sum_{n=1}^{\infty}[A_{mn}I_{m/2\alpha}(\varepsilon_{mn}\frac{\eta}{\xi_o})+B_{mn}K_{m/2\alpha}(\varepsilon_{mn}\frac{\eta}{\xi_o})]$$

$$C_{m/2\alpha}(\varepsilon_{mn}\frac{\xi}{\xi_o})\sin(\frac{m}{2\alpha}\varphi) \qquad (6)$$

with

$$C_{m/2\alpha}(\varepsilon_{mn}\frac{\xi}{\xi_o})\equiv J_{m/2\alpha}(\varepsilon_{mn}\frac{\xi}{\xi_o})Y_{m/2\alpha}(\varepsilon_{mn})-J_{m/2\alpha}(\varepsilon_{mn})Y_{m/2\alpha}(\varepsilon_{mn}\frac{\xi}{\xi_o})$$

may be obtained. The function $J_{m/2\alpha}$, $Y_{m/2\alpha}$ and $I_{m/2\alpha}$, $K_{m/2\alpha}$ are the Besselfunctions or modified Besselfunctions of $(m/2\alpha)$th order and first and second kind resp.. The eigenvalues ε_{mn} are the roots of the cross-product [6]

$$\begin{vmatrix} J_{m/2\alpha}(\varepsilon) & Y_{m/2\alpha}(\varepsilon) \\ J_{m/2\alpha}(\varepsilon\frac{\xi_1}{\xi_o}) & Y_{m/2\alpha}(\varepsilon\frac{\xi_1}{\xi_o}) \end{vmatrix} = 0 \qquad (7)$$

The remaining constants are obtained by expanding $F_i(\xi,\varphi)-T_0$ (i=1,2) into Bessel-Fourier-series, observing the orthogonality relation of the Besselfunctions.

1.3. PROLATE SPHEROIDAL SYSTEMS

For such systems one has to solve the partial differential equation

$$\frac{\partial}{\partial \xi}[(\xi^2-1)\frac{\partial T}{\partial \xi}] + \frac{\partial}{\partial \eta}[(1-\eta^2)\frac{\partial T}{\partial \eta}] + \frac{(\xi^2-\eta^2)}{(1-\eta^2)(\xi^2-1)}\frac{\partial^2 T}{\partial \varphi^2} = 0 \tag{8}$$

For a truncated prolate annular sector spheroid with the boundary conditions: $T=T_0$ for $\varphi=0, 2\pi\alpha$ and $\xi=\xi_0, \xi_1$ and given temperature distribution at the surfaces $\eta=\eta_0, \eta_1$, i.e. $T=g_i(\xi,\varphi)$, i=0,1, resp., the solution reads with

$$L_{\lambda_{mn}}^{m/2\alpha}(\xi) = P_{\lambda_{mn}}^{m/2\alpha}(\xi) Q_{\lambda_{mn}}^{m/2\alpha}(\xi_0) - P_{\lambda_{mn}}^{m/2\alpha}(\xi_0) Q_{\lambda_{mn}}^{m/2\alpha}(\xi) \quad (\xi_0 > \xi_1 \geq 1) \tag{9}$$

$$T(\xi,\eta,\varphi) = T_0 + \sum_{m=1}^{\infty}\sum_{n=1}^{\infty}[A_{mn}P_{\lambda_{mn}}^{m/2\alpha}(\eta) + B_{mn}Q_{\lambda_{mn}}^{m/2\alpha}(\eta)]L_{\lambda_{mn}}^{m/2\alpha}(\xi)\sin(\frac{m}{2\alpha}\varphi)$$

where λ_{mn} are the roots of

$$\begin{vmatrix} P_\lambda^{m/2\alpha}(\xi_0) & Q_\lambda^{m/2\alpha}(\xi_0) \\ P_\lambda^{m/2\alpha}(\xi_1) & Q_\lambda^{m/2\alpha}(\xi_1) \end{vmatrix} = 0 \tag{10}$$

Some roots of $P_\lambda^m(\xi_0)=0$ and $\partial P_\lambda^m/\partial \xi(\xi_0)=0$ have been determined to be $\lambda_{mn} = -\frac{1}{2} + i\alpha_{mn}$, where the α_{mn} have been tabulated. If the boundary conditions express constant temperature T_0 on $\varphi=0, 2\pi\alpha$ and $\eta=\eta_0, \eta_1$ and a given temperature distribution at $\xi=\xi_0$ and ξ_1, the above solution (9) must be changed by exchanging ξ with η. The determinant (10) also shows η instead of ξ ($-1\leq\eta\leq+1$). The procedure of solution is then similar as in the previous case. The eigenvalues λ_{mn} are obtained as from equ. (3).

2. HEAT TRANSPORT IN FLOWING LIQUIDS

For heat transport problems in flowing liquid the situation for solutions is quite different. The problem has been treated in channels and tubes for ideal and laminar flow. In all these investigations the appearing partial differential equation

$$\varkappa \Delta T - \vec{v} \cdot \text{grad } T = 0, \tag{11}$$

in which Δ represents the Laplacian operator and $\varkappa$ the diffusivity, could be separated, but the remaining ordinary differential equation had in most cases to be solved by some numerical method, which mainly was the Runge-Kutta procedure. In some cases this ordinary differential equation could by proper transformation be brought into the form of a confluent hypergeometric differential equation, for which solutions could be found in form of the confluent hypergeometric function $_1F_1$ (α,β;z). In other problems of flowing non-Newtonian liquids of the Ostwald-de-Waele type or for heat transport in conduits or tubes of varying cross-section, the remaining ordinary (or partial) differential equation cannot be solved in closed form. For problems of such a type the procedure for solving the remaining ordinary (or partial) differential equation for $\overline{T}$ exhibiting varying coefficients is as follows. One searches for a truncated auxiliary differential equation, which can be solved analytically with the given boundary conditions. This means, that the original ordinary differential equation

$$D[\overline{T};x]=0 \qquad \text{(or partial)} \qquad (12)$$

is truncated into an equation

$$D_o[\overline{T};x]=0 \qquad (13)$$

which renders an exact solution with the given boundary conditions, and which has its own eigenvalues. Assuming for the solution of the original equation a series expansion in the eigenfunctions of the auxiliary equation, which satisfies the boundary conditions renders with the Ritz-Galerkin condition finally an infinite system of homogeneous algebraic equations, which coefficient determinant set equal to zero yields the eigenvalue equation for the original differential equation. Truncating this determinant of infinite order to a finite order, renders the approximate values of the lower eigenvalues. If one does not request the local temperature distribution too close to the inlet, a relatively low order of the eigenvalue determinant shall suffice for good accuracy of the local temperature distribution in the conduit or tube.

2.1. HEAT TRANSPORT IN A TUBE OF PARALLELOGRAM-CROSS-SECTION [7]

For the determination of the local temperature distribution for ideal flow in a tube of parallelogram-cross-section the partial differential equation

$$\varkappa[\frac{\partial^2 T}{\partial \xi^2}+\frac{\partial^2 T}{\partial \eta^2} - 2\sin\alpha\frac{\partial^2 T}{\partial \xi \partial \eta}] - \omega_o \cos^2\alpha \frac{\partial T}{\partial z} = 0 \qquad (14)$$

with the boundary conditions $T=T_w$ at $\xi=0,a$ and $\eta=0,b$ and the initial condition at the inlet $T(\xi,\eta,0)=T_i$ at z=0. With $(T-T_w)/(T_i-T_w)=T^*e^{-\nu z}$ equation (14) may be written as

$$D[T^*;\xi,\eta] \equiv \frac{\partial^2 T^*}{\partial \xi^2} + \frac{\partial^2 T^*}{\partial \eta^2} - 2\sin\alpha \frac{\partial^2 T^*}{\partial \xi \partial \eta} + \lambda^2 T^* = 0 \tag{15}$$

where $\lambda^2 = \frac{w_o \nu}{\varkappa}\cos^2\alpha$ represents the eigenvalues (λ_n^2) of the problem. This differential equation cannot be separated. The auxiliary differential equation is chosen to be

$$D_o[T^*;\xi,\eta] \equiv \frac{\partial^2 T^*}{\partial \xi^2} + \frac{\partial^2 T^*}{\partial \eta^2} + \pi^2(\frac{m^2}{a^2}+\frac{n^2}{b^2})T^* = 0 \tag{16}$$

which yields a solution (satisfying the boundary conditions $\bar{T}=0$ for $\xi=0,a$ and $\eta=0,b$) of the form

$$T^*(\xi,\eta) = \sin(\frac{m\pi\xi}{a})\sin(\frac{n\pi\eta}{b}) \tag{17}$$

The solution of equ.(15) may therefore be a double-Fourier-series

$$T^*(\xi,\eta) = \sum_{m=1}^{\infty}\sum_{n=1}^{\infty} T_{mn}\sin(\frac{m\pi\xi}{a})\sin(\frac{n\pi\eta}{b}) \tag{18}$$

and satisfies the boundary conditions. The Ritz-Galerkin condition is then given by

$$\int_o^a \int_o^b D[T^*;\xi,\eta] \cdot \sin(\frac{k\pi\xi}{a})\sin(\frac{\ell\pi\eta}{b})d\xi d\eta = 0 \quad \text{for } \ell,k=1,2,\ldots \tag{19}$$

and yields a "double" determinant, where the diagonal elements are $[\lambda^2 - \pi^2((m^2/a^2)+(n^2/b^2))]$. This determinant is obtained from ∞^2 homogeneous algebraic equations for the ∞^2 unknown constants T_{mn}. The eigenvalues $a\lambda_{mn}$ are shown for skew angle $\alpha=40°$ as a function of the side ratio b/a for different modes (m,n) in Figure 3, while the mean temperature along the tube length z/a is presented for a/b =2 for various skew angles (Fig.4). In these cases the ratio of diffusivity to the plug flow velocity w_o was chosen to be $\varkappa/aw_o=10^{-3}$. With increasing α the mean temperature exhibits increased decay along the tube.

2.2. HEAT TRANSPORT OF NON-NEWTONIAN FLOW IN CIRCULAR CYLINDRICAL TUBE [8]

In this case the local temperature distribution is obtained from the solution of the differential equation

$$\varkappa[\frac{\partial^2 T}{\partial r^2}+\frac{1}{r}\frac{\partial T}{\partial r}] - \frac{\mu}{(\mu+1)}(\frac{|\partial p/\partial z|}{2K})^{1/\mu} a^{\mu+1/\mu}[1-(\frac{r}{a})^{\mu+1/\mu}]\frac{\partial T}{\partial z} = 0 \tag{20}$$

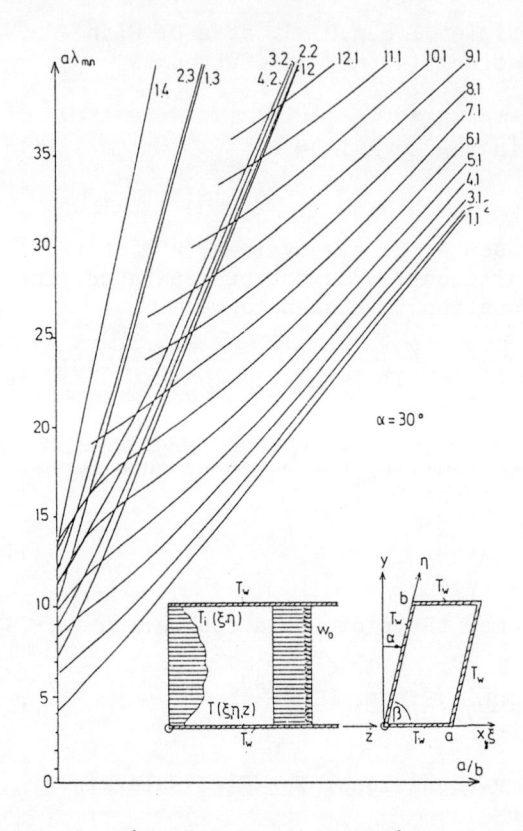

Figure 3: Eigenvalues as function of parallelogram side ratio a/b for skew angle α=30°

with the boundary condition $T=T_w$ at r=a and the initial condition $T=T_i$ at the inlet z=0. The Non-Newtonian flow is based on the flow law of Ostwald-de-Waele, where μ and K represent material constants. For μ<1 we deal with pseudoplastic and for μ>1 with dilatant liquid, while μ=1 renders K≡μ the laminar flow case. With y=r/a and $(T-T_w)/(T_i-T_w) = T^* e^{-\lambda z}$ we obtain

$$D[T^*;y] \equiv \frac{d^2 T^*}{dy^2} + \frac{1}{y}\frac{dT^*}{dy} + \beta^2(1-y^{\mu+1/\mu})T^* = 0 \tag{21}$$

for which the auxiliary differential equation is

$$D_o[T^*;y] \equiv \frac{d^2 T^*}{dy^2} + \frac{1}{y}\frac{dT^*}{dy} + \varepsilon_n^2 T^* = 0 \tag{22}$$

$\beta^2 \equiv \frac{\lambda a^{3\mu+1/\mu}}{\varkappa(\mu+1)} \left[\frac{|dp/dz|}{2K}\right]^{1/\mu} \equiv w_o \lambda a^2/\varkappa$ and ε_n are the roots of $J_o(\varepsilon)=0$. Equ.(22) exhibits the solution

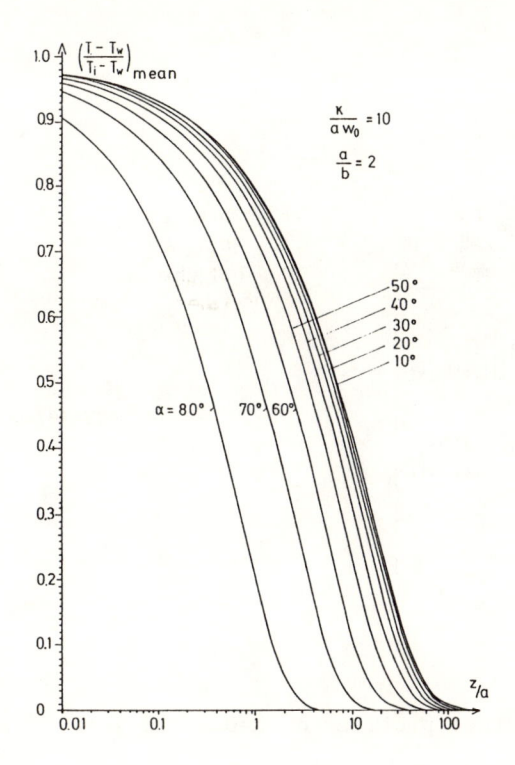

Figure 4: Mean temperature along tube of side ratio a/b=2 for various skew angles α

$$T^*(y) = J_o(\varepsilon_n y) \qquad (23)$$

which satisfies the boundary conditions $T^* = 0$ at the wall $y=1$. The solution of equ.(21) may therefore be of the form

$$T^*(y) = \sum_{n=1}^{\infty} T_n \cdot J_o(\varepsilon_n y) \qquad (24)$$

The Ritz-Galerkin condition

$$\int_0^1 D[T^*; y] y J_o(\varepsilon_m y) dy = 0 \qquad m=1,2,\ldots \qquad (25)$$

renders an infinite system of homogeneous algebraic equations of which the vanishing coefficient determinant yields the equation for the determination of the eigenvalues β_n^2. For $\mu = 0.1$ and $\varkappa/aw_o = 0.1$ the local temperature distribution is presented in Figure 5.

2.3. HEAT TRANSPORT IN A DIFFUSOR [9]

For a flow with changing cross-section we choose a cone

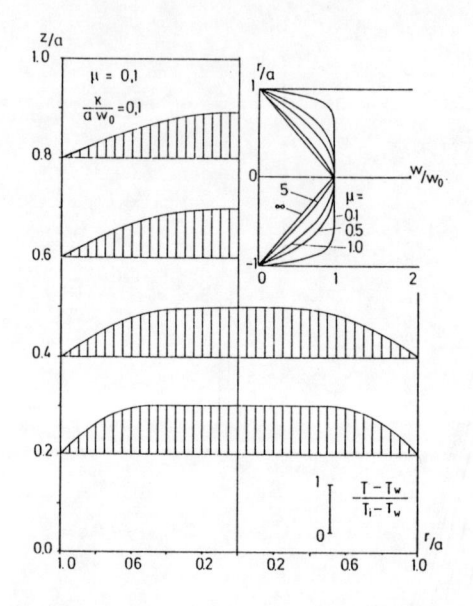

Figure 5: Local temperature profile for $\mu=0.1$ and $\varkappa/aw_o =0.1$.

with an inlet at r=a and an apex angle of 2α. For the determination of the local temperature, one has to solve

$$\varkappa[\frac{1}{r^2}\frac{\partial}{\partial r}(r^2\frac{\partial T}{\partial r}) + \frac{1}{r^2\sin\vartheta}\frac{\partial}{\partial\vartheta}(\sin\vartheta\frac{\partial T}{\partial\vartheta})] - u(r,\vartheta)\frac{\partial T}{\partial r} = 0 \qquad (26)$$

The flow velocity for creeping flow is

$$u(r,\vartheta) = \frac{3\dot{V}_o}{2\pi r^2}\frac{(\cos 2\vartheta - \cos\alpha)}{(3\cos\frac{\alpha}{2} - 3\cos\alpha + \cos\frac{3\alpha}{2} - 1)} \quad, \quad \dot{V}_o = \text{volumetric flow}$$

Neglecting radial conduction and introducing $(T-T_w)/(T_i-T_w) = T^* e^{-\lambda^2 r}$ the equation that has to be solved reads with $\xi \equiv \cos\vartheta$

$$D[T^*;\xi] \equiv \frac{d}{d\xi}[(1-\xi^2)\frac{dT^*}{d\xi}] + \gamma^2[\xi^2 - \cos^2\frac{\alpha}{2}]T^* = 0 \qquad (27)$$

where $\gamma^2 \equiv \dfrac{3\dot{V}_o \lambda^2}{\pi\varkappa[3\cos\frac{\alpha}{2} - 3\cos\alpha + \cos\frac{3\alpha}{2} - 1]}$

For the auxiliary differential equation we choose

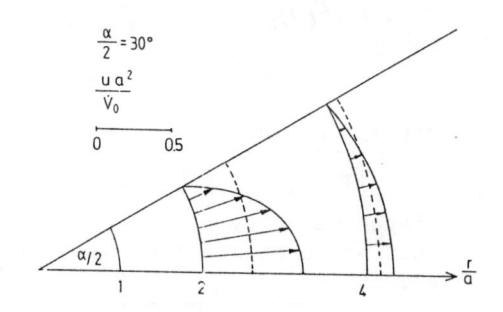

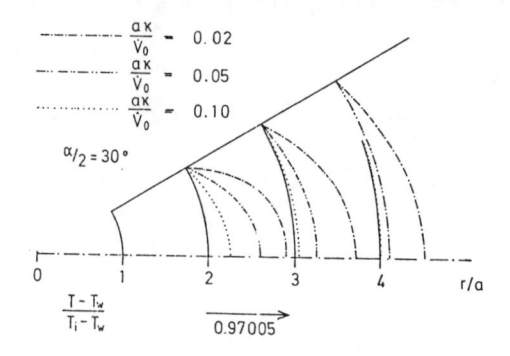

Figure 6: Local temperature profile for viscous liquid

$$D_o[T^*;\xi] \equiv (1-\xi^2)\frac{d^2T^*}{d\xi^2} - 2\xi\frac{dT^*}{d\xi} + \mu(\mu+1)T^* = 0 \tag{28}$$

which has with $T^*=0$ at $(\vartheta=\frac{\alpha}{2})$ $\xi=\cos\frac{\alpha}{2}$ the solution

$$T^* = P^o_{\mu_n}(\xi) \tag{29}$$

where μ_n are the zeros of the Legendre function of 0th order with respect to the degree μ, i.e. $P^o_\mu(\xi_o)=0$. The solution of the differential equation (27) is assumed to be

$$T^*(\xi) = \sum_{n=1}^{\infty} T_n P^o_{\mu_n}(\xi) \tag{30}$$

and the Ritz-Galerkin condition

$$\int_1^{\cos\alpha/2} D[T^*;\xi]P^o_{\mu_m}(\xi)d\xi = 0 \qquad m=1,2,\ldots \tag{31}$$

yields again an infinite algebraic system of equations for the determination of the eigenvalues γ_n. The local heat distribution is shown in Figure 6 for an vertex angle $\alpha=60°$ at various locations $r/a = 2,3$ and 4. It is obtained from

$$T(r,z)=T_w+(T_i-T_w)\sum_{n=1}^{\infty} T_n P_{\mu_n}^o (\cos\vartheta)e^{-\lambda_n^2(r/a)} \qquad (32)$$

2.4. HEAT TRANSPORT IN A CONVERGING-DIVERGING NOZZLE

For the transport of heat in a converging-diverging nozzle with a completely developed velocity profile the solution of the heat transport equation in oblate spheroid coordinates (ξ,η,φ) has to be solved, i.e.

$$\frac{\partial}{\partial\xi}[(1+\xi^2)\frac{\partial T}{\partial\xi}]+\frac{\partial}{\partial\eta}[(1-\eta^2)\frac{\partial T}{\partial\eta}]-\frac{au(\xi,\eta)}{\varkappa}\sqrt{(1+\xi^2)(\xi^2+\eta^2)}\frac{\partial T}{\partial\xi}=0 \qquad (33)$$

in which

$$u(\xi,\eta)=\frac{3\dot{V}_o}{2\pi a^2}\frac{(\eta^2-\eta_o^2)}{\sqrt{(1+\xi^2)(\xi^2+\eta^2)}[1-3\eta_o^2+2\eta_o^3]} \qquad (34)$$

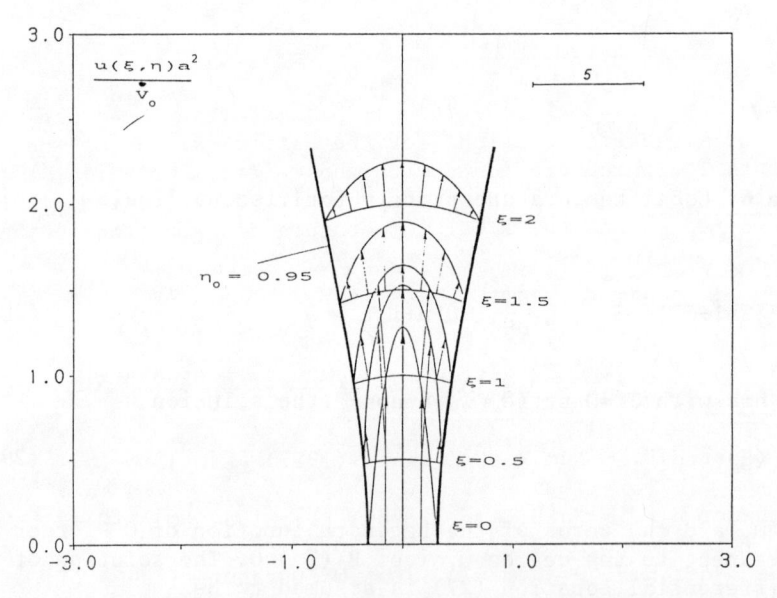

Figure 7: Velocity distribution in converging-diverging nozzle

$\dot{V}_o$ is the volumetric flow, $2a$ the throat diameter and $\eta=\eta_o$ is the wall of the nozzle. With $(T-T_w)/(T_i-T_w)=\bar{T}(\eta)e^{-\lambda^2\xi}$ and $\gamma^2=-3\lambda^2\dot{V}_o/2\pi\varkappa a[3\eta_o^2-2\eta_o^3-1]$ the partial differential equation (33) becomes

$$D[T^*;\eta]\equiv(1-\eta^2)\frac{d^2T^*}{d\eta^2}-2\eta\frac{dT^*}{d\eta}+\gamma^2(\eta_o^2-\eta^2)T^*=0 \qquad (35)$$

Representing the solution as an infinite series of the eigenfunctions $P_{\mu_n}^\theta(\eta)$ (Legendre functions), which satisfy the boun-

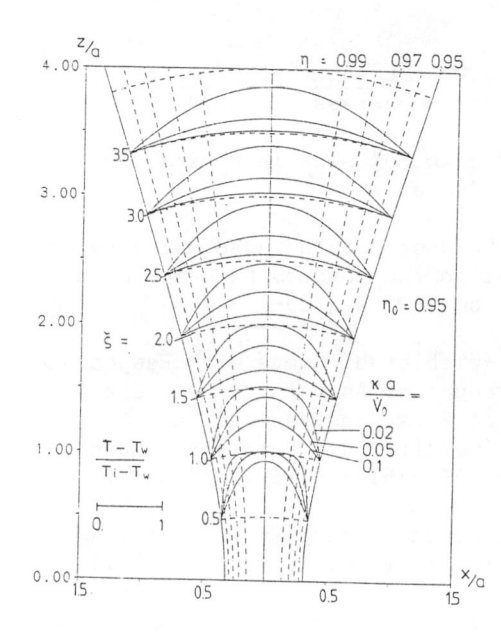

Figure 8: Local temperature distribution in converging-diverging nozzle

dary condition at the wall ($T=T_w$ at $\eta=\eta_o$), i.e. $T^*(\eta_o)=0$, yields with μ_n as the roots of $P^o_{\mu_n}(\eta_o)=0$

$$T^*(\eta) = \sum_{n=1}^{\infty} T_n P^o_{\mu_n}(\eta) \quad (36)$$

and the Ritz-Galerkin condition

$$\int_1^{\eta_o} D[T^*;\eta] P^o_{\mu_m}(\eta)d\eta = 0$$

$$m=1,2,\ldots \quad (37)$$

This presents an infinite system of algebraic equations for the determination of the constants T_n, which are found with the initial condition at the inlet $\xi=0$. Figure 7 shows the velocity distribution, while Figure 8 represents the local temperature distribution in the nozzle.

REFERENCES:

[1] Carslaw,A.S. and Jaeger,J.C. - Conduction of Heat in Solids. Oxford at the Clarendon Press, 2.Ed.1959

[2,5] Bauer,H.F. - Num.Heat Transf. (1987) [2], 1986 [5]

[3,6] Bauer,H.F. - Math.Comp. (1986) [3], 1964 [6]

[7,8,9] Bauer,H.F. - Wärme- u. Stoffübertragung (1980) [7], 1984 [8], 1986 [9]

[4] Bauer,H.F. and Eidel,W. - Tables of Roots with Respect to the Degree of a Cross-Product of Associated Legendre Functions (and derivatives) of First and Second kind. Forschungsbericht der Universität der Bundeswehr München LRT-WE-9-FB-9 (1986), FB-14 (1986) and FB-15 (1986)

REFERENCES:

[1] Carslaw,A.S. and Jaeger,J.C. - <u>Conduction of Heat in Solids</u>. Oxford at the Clarendon Press, 2.Ed.1959

[2] Bauer,H.F. - Steady Conduction of Heat in Conical Systems. Numerical Heat Transfer, (to appear)

[3] Bauer,H.F. - Tables of the Roots of the Associated Legendre Function with Respect to the Degree. Mathematics of Computation, 46, Nr.174, 601-602 and 29-41, 1986

[4] Bauer,H.F. and Eidel,W. - Tables of Roots with Respect to the Argument of the Cross-Product of Associated Legendre Functions (and derivatives) of First and Second kind. Forschungsbericht der Universität der Bundeswehr München LRT-WE-9-FB-9 (1986), FB-14(1986) and FB-15(1986)

[5] Bauer,H.F. - Steady Conduction of Heat in Paraboloidal Systems. Numerical Heat Transfer,10, No.4, 395-422 (1986)

[6] Bauer,H.F. - Tables of Zeros of Cross-Product Besselfunction. Mathematics of Computation, 18, 128-135, (1964)

[7] Bauer,H.F. - Diffusion und Konvektion im Rohr mit Parallelogramm-Querschnitt. Wärme- und Stoffübertragung 14, 49-58 (1980)

[8] Bauer,H.F. - Stofftransport im Kreisrohr bei Strömung einer nicht-Newtonschen Flüssigkeit. Wärme- und Stoffübertragung 18, 157-166 (1984)

[9] Bauer,H.F. - Masstransport in a three-dimensional diffusor or confusor. Wärme- und Stoffübertragung 21, 51-64 (1987)

THREE-DIMENSIONAL FINITE-DIFFERENCE CALCULATIONS OF
BUOYANT FLOW IN ARBITRARY PARALLELEPIPED ENCLOSURES
USING CURVILINEAR NON-ORTHOGONAL COORDINATES

H. Q. Yang[I], K. T. Yang[I], and J. R. Lloyd[II]

SUMMARY

A control volume-based finite-difference formulation is developed for heat transfer and fluid flow in arbitrary three-dimensional parallelepiped enclosures. The governing equations in Cartesian coordinates are first transformed to those in non-orthogonal curvilinear coordinates by using tensor transformation. After introducing the properties of the parallelepiped geometry, equations are obtained in the primitive variables for which all vectors and tensors are based on the curvilinear coordinates. With proper treatment of the heat flux and stress tensor terms, the finite difference equations analogous to those in the Cartesian coordinates are formed. Two examples are utilized to show the validity of the methodology.

1. INTRODUCTION

The control volume-based finite-difference formulation in Cartesian coordinates originally suggested by Patankar and Spalding [1] has now been used widely and successfully. The attractive features of the control-volume method are the general satisfaction of conservation properties over any group of cells and the whole calculation domain, and the clear physical interpretation of the discretized equations. As with other formulations, however, the application of the finite-difference method is not without its difficulties and uncertainties. One of the major shortcomings is that it requires the boundaries of the calculation domain to be regular and to coincide with the surface of the coordinate system, otherwise an elaborate treatment of the
----
I. Department of Aerospace and Mechanical Engineering
   University of Notre Dame, Notre Dame, IN 46556 USA
II. Department of Mechanical Engineering
    Michigan State University, East Lansing, MI 48824 USA

boundaries must be made. Many attempts have been made to extend the applicability of the finite difference formulation. In general, when flows in non-rectangular configurations have to be computed, two types of methods are possible according to the space in which the governing equations are approximated: one is in the physical space itself; and the other is in the transformed space. All the governing equations in the former method are written in Cartesian coordinates (x^1, x^2, x^3), and the discretizations are made directly in the physical space on uniform or nonuniform mesh adapted to the geometry, so that the irregular boundaries are approximated by a series of linear segments parallel to the axes. In the corresponding computations, therefore, some control volumes are rendered inactive, while the remaining active control volumes form the desired irregular domains [2].

In the latter method, however, the governing equations are first transformed into curvilinear coordinates (θ^1, θ^2, θ^3), and the resulting equations are then discretized by means of finite-difference control volumes. By doing so there are two choices for expressing the vectors and tensors. The first one is on a local basis, by which the vectors and tensors are based on the curvilinear coordinates. In other words, the components of a vector and a tensor are contravariant with respect to the curvilinear coordinates. Good examples are the control volumes for cylindrical-polar and spherical coordinates. This methodology has be sucessesfully applied to heat transfer and fluid flow problems in cylindrical and spherical geometries, such as, for instance, the horizontal concentric annulus [3]. In this regard, the recent methodology by Raithby et al. [4] by means of generalized orthogonal coordinates gives more flexibility in applications with rather general geometries. On the other hand, another approach expresses the vectors and tensors in a fixed Cartesian coordinate system while maintaining the governing equations in curvilinear coordinates. The body-fitted coordinates [5] are good examples.

The present paper describes a method of using control volume-based finite-difference formulation in the curvilinear coordinates (transformed ones) and on a local basis (vectors and tensors are all based on the curvilinear coordinates), for heat transfer and fluid flow calculations in parallelepipeds. The coordinates and control volumes are such that their boundaries coincide with the physical ones, which are basically skewed, thus simplifying the treatment of the boundary conditions. The objective of the present study is to combine the methodology for orthogonal coordinates in [4] and the one proposed here for skewed coordinates to develop an algorithm, so that the range of applicability of the control volume-based finite difference method can be greatly expanded.

2. MATHEMATICAL FORMULATIONS

2.1 Conservation Equations in General Coordinates

The equations governing three-dimensional laminar heat transfer and fluid flow are the conservation equations of mass, momentum and energy. As is well-known, they can be written in the Cartesian coordinate system in terms of tensors as follows,

$$\rho_t + (\rho u_i)_{,i} = 0 \tag{1}$$

$$(\rho u_i)_t + (\rho u_i u_j)_{,j} = -p_{,i} - \rho G_i + \sigma_{ij,j} \tag{2}$$

$$(\rho c_{pm} T)_t + (\rho u_i c_{pm} T)_{,i} = (kT_{,i})_{,i} + \mu \Phi + P u_{i,i} \tag{3}$$

where ρ is the fluid density, u_i the velocity vector, p the static pressure, G_i the gravity acceleration vector, σ_{ij} the stress tensor, c_{pm} the mean isobaric heat capacity, k the thermal conductivity, μ the dynamic viscosity, Φ the dissipation function and the subscript t denotes derivatives with respect to time. The shear stress tensor σ_{ij} is given by

$$\sigma_{ij} = \mu (u_{i,j} + u_{j,i} - 2/3 \delta_{ij} u_{k,k}) \tag{4}$$

and the dissipation function by

$$\Phi = 2(u^2_{i,j})\delta_{ij} + [u_{i,j}(1-\delta_{ij})]^2 - 2/3 (u_{i,i})^2 \tag{5}$$

To obtain the corresponding equations in the general curvilinear coordinates (θ^1, θ^2, θ^3), two rules in accordance with [6] are followed: (a) the partial differentiation symbol (,) is replaced by the covariant differentiation symbol (;); (b) the repeat indices are on the diagonal positions. Meanwhile, the physical components, $u_{(i)}$, $\sigma_{(i)}^{(j)}$ are utilized according to the relations,

$$u_i = u_{(i)} h_i \qquad u^j = u_{(j)} h_j \qquad \sigma_i^{\,j} = \sigma_{(i)}^{(j)} h_i / h_j \tag{6}$$

where h_i is the scale factor for the curvilinear coordinates in directions θ^i. It is not a component, therefore the summation rule does not apply to the index of h_i.

For simplification the parentheses are from now on omitted, and all the components are meant to be physical ones.

Equation (1)-(5) can now be recast into:

$$\rho_t + (\rho u^i/h_i)_{;i} = 0 \tag{7}$$

$$(\rho u^k g_{ki} h_i)_t + (\rho u^k g_{ki} u^j h_i/h_j)_{;j} = -P_{;i} - \rho G^j g_{ji}/h_j$$
$$+ (\sigma^{kj} g_{ki} h_i/h_j)_{;j} \tag{8}$$

$$(\rho c_{pm} T)_t + (\rho c_{pm} u^i/h_i T)_{;i} = (kT_{;j} g^{ji}/h_i)_{;i} + \mu \Phi$$

$$+ p\, u^i{}_{;i}/h_i \tag{9}$$

Shear stresses now become

$$\sigma^{kj} = \mu(u^j{}_{;m} g^{mk}/h_j + u^k{}_{;m} g^{mj}/h_k - 2/3 \delta_n{}^j g^{nk} u^m{}_{;m}/h_m) \tag{10}$$

and the dissipation function,

$$\Phi = 2(u^i{}_{;j}{}^2)\delta_i{}^j + [u^i{}_{;j}(1-\delta_i{}^j)]^2 - 2/3 (u^i{}_{;i})^2 \tag{11}$$

Here, g_{ij} and g^{ij} are the co- and contravariant metric tensors. All the components of vectors and tensors have been written in the contravariant form and are related to the covariant ones according to the rules,

$$u_i = u^k g_{ki} \quad \text{for vectors} \tag{12}$$

$$\sigma_i{}^j = \sigma^{kj} g_{ki} \quad \text{for tensors} \tag{13}$$

The covariant differentiation is the partial derivative of a scalar (say p), a vector (say u^i) or a tensor (say $\sigma_i{}^j$) with respect to the curvilinear system θ^i. For a scalar,

$$p_{;i} = p_{,i} \tag{14}$$

while for a vector,
$$u^i{}_{;j} = u^i{}_{,j} + \begin{Bmatrix} i \\ j\, m \end{Bmatrix} u^m \tag{15}$$

and for a tensor,
$$\sigma_i{}^j{}_{;j} = \sigma_i{}^j{}_{,j} - \begin{Bmatrix} m \\ i\, j \end{Bmatrix}\sigma_m{}^j + \begin{Bmatrix} j \\ j\, m \end{Bmatrix}\sigma_i{}^m \tag{16}$$

where $\begin{Bmatrix} i \\ j\, m \end{Bmatrix}$ is known as the Christoffel symbol of the second kind given by

$$\begin{Bmatrix} i \\ j\, m \end{Bmatrix} = \frac{\partial x^n}{\partial \theta^j \partial \theta^m} \frac{\partial \theta^i}{\partial x^n} = \frac{g^{in}}{2}\left[\frac{\partial g_{jn}}{\partial x^m} + \frac{\partial g_{mn}}{\partial x^i} - \frac{\partial g_{jm}}{\partial x^n}\right] \tag{17}$$

2.1 Conservation Equations in Skewed Coordinates

The general skewed coordinates together with the Cartesian coordinates are shown in Fig. 1. One of the special properties is that they are linearly related, which results in the independence of the base vector $\vec{g}_i$ in the skewed coordinates relative to the position. The general linear relation between the two coordinates can be obtained as;

$$x^i = b_{ij}\, \theta^j \tag{18}$$

here b_{ij} is constant coefficient. If $\vec{e}_i$ is the base vector in the Cartesian coordinates, then the covariant base vector $\vec{g}_i$ in the skewed system is,

$$\vec{g}_i = \frac{\partial x^j}{\partial \theta^i}\vec{e}_j = b_{ji}\vec{e}_j \tag{19}$$

Fig. 1 Parallelepiped Enclosure Geometry

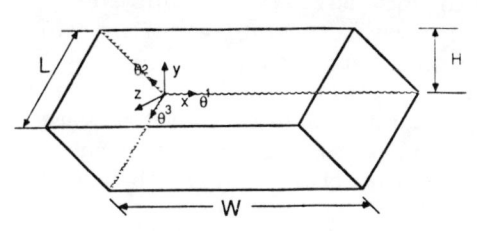

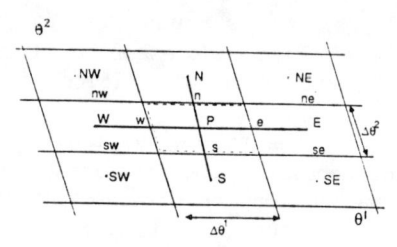

Fig.2 Finite-difference Control Volume

the metric tensor is thus:

$$g_{ij} = \vec{g_i} \vec{g_j} = b_{ki}\vec{e_k} \cdot b_{lj}\vec{e_l} = b_{ki}b_{lj}\delta^{kl} = b_{ki}b_{kj} \quad (20)$$

and $\quad h_i = (\vec{g_i} \cdot \vec{g_i})^{1/2} = (b_{ki} \cdot b_{ki})^{1/2} \quad (21)$

If the base vector $\vec{g_i}$ is chosen as unity, then we have

$$h_i = 1. \quad (i=1,2,3) \quad (22)$$

The contravariant metric tensor can be found by the properties of

$$g_{ik}g^{kj} = \delta^i_j \quad (23)$$

which leads to $\quad g^{ij} = \text{cof}(g_{ij})/g \quad (24)$

where g is the determinant of g_{ij} ; $\quad g = |g_{ij}| \quad (25)$

It can be seen that all the base vector and metric tensor are constant, such that the Christoffel symbol vanishes and the covariant partial differentiation is reduced to the usual partial differentiation, so that the momentum equation (8) can be simplified by multiplying both sides by g^{ik}. By properly choosing the indices, the set of conservation equations in the skewed coordinates can now be written as:

$$\rho_t + (\rho u^i)_{,i} = 0 \quad (26)$$

$$(\rho u^i)_t + (\rho u^i u^j)_{,j} = -p_{,j}g^{ji} + \rho G^i + (\sigma^{ij})_{,j} \quad (27)$$

$$(\rho c_{pm}T)_t + (\rho c_{pm}u^iT)_{,i} = (kT_{,j}g^{ji})_{,i} + \mu\Phi + p u^i_{,i} \quad (28)$$

$$\sigma^{ij} = u^j_{,m}g^{mi} + u^i_{,m}g^{mj} - 2/3\,\delta^{ij}u^m_{,m} \quad (29)$$

$$\Phi = 2(u^2_{i,j})\delta_{ij} + [u_{i,j}(1-\delta_{ij})]^2 - 2/3\,(u_{i,i})^2 \quad (30)$$

It is apparent that if g^{ij} is equal to δ^{ij}, a set of conservation equations in the Cartesian coordinates will be recovered.

In comparing equations in the non-orthogonal coordinates given by (26)-(30) with the ones in the Cartesian coordinates shown in

(1)-(5), it is found that they are very analogous except in the gradient terms (pressure gradient in (27), temperature gradient in (28)) and in the definition of the shear stresses (30).

In the following, the energy equation is used to demonstrate the necessary modifications, while the treatment on the momentum equations can be obtained by an analogous procedure so that only breif discussions are given.

3. CONTROL VOLUME FINITE-DIFFERENCE EQUATIONS

In the energy equation the appearance of a complicated temperature gradient term gives rise to difficulties. To eliminate them, the contravariant component of the conduction heat flux q^i is written as

$$q^i = -kT_{,j} g^{ji} \tag{31}$$

In the control volume method, the discretization equations are made by integrating the conservation equations over the non-orthogonal control volume $\sqrt{g}\, d\theta^1 d\theta^2 d\theta^3$, as shown in Fig.2, which shows a basic control volume and corresponding nodes. Here a staggered grid system is adopted, as explained in [2].

The energy equation, after integration over the basic control volume surrounding grid point P, becomes

$$(\rho c_{pm} T)_t \Delta\theta^1 \Delta\theta^2 \Delta\theta^3 + (J^1_e - J^1_w)\Delta\theta^2 \Delta\theta^3 + (J^2_n - J^2_s)\Delta\theta^3 \Delta\theta^1$$
$$+ (J^3_f - J^3_b)\Delta\theta^1 \Delta\theta^2 = S\Delta\theta^1 \Delta\theta^2 \Delta\theta^3 \tag{32}$$

In the equation, J^i is the total heat flux along the θ^i direction due to convection and diffusion. The subscript in J^i is the position where it is to be evaluated. Thus,

$$J^i = (\rho c_{pm} u^i T) + q^i \tag{33}$$

The source term S includes internal heat source and the energy generated by viscous dissipation and pressure work.

In the Cartesian coordinate system, q^i is in a simple Fourier form, while at the present time it is defined in such a way that multi-directional partial derivatives are involved. The idea of Raithby et al. [4] of using "stress-flux formulation" is introduced here, by which the conducting heat flux is given as:

$$q^i = -kT_{,i} g^{ii} + (q^i + kT_{,i} g^{ii}) \tag{34}$$

The contribution of the terms in the parenthsis is from the non-orthogonality of the coordinates utilized in the present study. They are evaluated from the information of the prior iteration, and are grouped into the source term, such that the equation can now be written as

$$(\rho c_{pm} T)_t \Delta\theta^1 \Delta\theta^2 \Delta\theta^3 + (J^1_e - J^1_w)\Delta\theta^2\Delta\theta^3 + (J^2_n - J^2_s)\Delta\theta^3\Delta\theta^1$$
$$+ (J^3_f - J^3_b)\Delta\theta^1\Delta\theta^2 = B \quad (35)$$

with $\quad J^i = (\rho c_{pm} u^i T) - kT_{,i} g^{ii} \quad (36)$

$$B = -((q^1 + kT_{,1} g^{11})_e + (q^1 + kT_{,1} g^{11})_w) \Delta\theta^2\Delta\theta^3$$
$$- ((q^2 + kT_{,2} g^{22})_n + (q^2 + kT_{,2} g^{22})_s) \Delta\theta^3\Delta\theta^1$$
$$- ((q^3 + kT_{,3} g^{33})_f + (q^3 + kT_{,3} g^{33})_b) \Delta\theta^1\Delta\theta^2$$
$$+ S_\Delta\theta^1\Delta\theta^2\Delta\theta^3 \quad (37)$$

Further approximation on the flux term J^i is the same as the one applied in the Cartesian coordinate system in our earlier paper on the buoyant flow in rectangular enclosures [7]. Basically, the convective terms are approximated by the QUICK scheme which is a higher order accuracy algorithm originally suggested by Leonard [8], and the diffusion terms are by central difference. The details will not be repeated here, and the reader is referred to reference [9].

The retaining of the conducting heat flux q^i in the equations as known quantity makes the treatment of boundaries easier and the discretized equations much simpler. The evaluation of q^i is based on the position where it is placed. From (37) q^i is found to be used at the center of the control volume for u^i which is staggered from the base cell. Therefore, interpolation is needed for the cross derivative terms. For example, to find q^1 at the west surface of the base cell, we have (Fig.2),

$$(q^1)_w = k(T_{,1} g^{11} + T_{,2} g^{21} + T_{,3} g^{31}) \quad (38)$$

$$(T_{,1})_w = (T_P - T_W)/(\theta^1_P - \theta^1_W) \quad (39)$$

Let $INTER_i(A,B)$ be the function of evaluating the interpolation of the values of A and B at point i, then

$$(T_{,2})_w = INTER_w ((T_{,2})_W, (T_{,2})_P) \quad (40)$$
where, $\quad (T_{,2})_W = INTER_W ((T_{,2})_{nw}, (T_{,2})_{sw}) \quad (41)$

$$(T_{,2})_P = INTER_P ((T_{,2})_n, (T_{,2})_s) \quad (42)$$

while $(T_{,2})_{nw}$, $(T_{,2})_{sw}$, $(T_{,2})_n$ and $(T_{,2})_s$ can be calculated directly from the central differences.

The treatment on the momentum equations is very similar. For example, in the u momentum equation, the pressure gradient term of $p_{,1} g^{11}$ is retained since it plays an important role in the velocity and pressure linkage. Because of the existence of g^{11}, the coefficients in the pressure correction equation are to be multiplied, while shear stresses $\bar{\sigma}^{11}$, $\bar{\sigma}^{12}$ and $\bar{\sigma}^{13}$ are also taken as known and evaluated from the previous iteration. For details the reader is referred to [4].

4. RESULTS AND DISCUSSIONS

Proper modifications have been made to facilitate three-dimensional natural-convection calculations in general parallelepiped enclosures by using the above mentioned methodology. Three major modifications are: 1. incorporating the new heat flux vector; 2. incorporating the new shear stress tensor, and 3. incorporating g^{ii} to the coefficients in the pressure correction equation. Several examples have been calculated for validating the calculations and also determine the physical differences in the resulting fluid flow and heat transfer in the parallelepiped enclosures from that of the rectangular enclosures.

4.1 Quasi-Two-Dimensional Parallelogrammic Enclosures

The quasi-two-dimensional enclosures here referred to are not real two dimensional enclosures in the computational model, but three-dimensional enclosures which are drag-free and adiabatic in the lateral walls of the depth, as shown in Fig. 3. There are several sets of experimental data available for comparison [10,11]. The two vertical walls are differentially heated, while the two inclined parallel walls are adiabatic.

The inclined angle ψ here is defined to be positive when the parallel adiabatic walls are inclined in the counter-clockwise direction. The perpendicular distance between the two isothermal walls is taken as the reference length H, which is independent of the inclined angle ψ. To compare with experimental data of Seki et al. [10], the present calculations are based on aspect ratio A_x, defined as W/H, of 1.44, and A_z, defined by L/H, of 1.00, a Rayleigh number based on H of 2.0×10^5, and a Prandtl number of 0.71. The unit aspect ratio A_z allows the possible reproduction of longitudinal rolls [12,13]. The whole calculation domain is divided into 36x45x10 (WxHxL) uniform cells.

The isotherms and streamlines at $\psi = 0$ for the case of the standard rectangular enclosure, are shown in Fig. 4. It is seen that boundary layers have developed near the vertical differentially-heated walls, and a stratified core region exists with almost stagnant flow. This is what can be expected in studies on rectangular enclosures when Rayleigh number is around 10^6 and Prandtl number around unity.

Shown in Fig. 5 are the isotherms and streamlines at several negative inclined angles of $\psi = -30°$, $-45°$ and $-60°$ where the hot surface is positioned vertically above the cold surface. As the

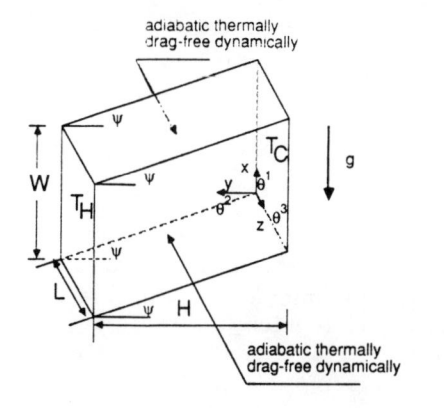

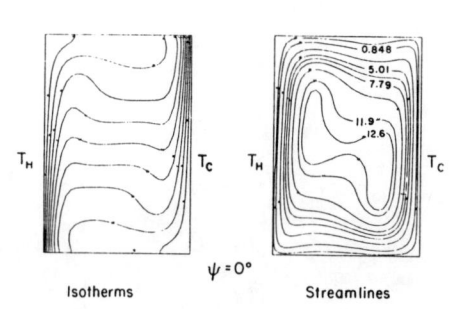

Fig. 3 Quasi-two-dimensional Paralleogrammic Enclosure

Fig. 4 Isotherms and Streamlines at Ra=2×10^5, ψ =0.

angle decreases from zero, stable stratification persists in the core region, and because of the geometry, the triangular corner regions becomes increasingly more isothermal, and the thermal boundary layers next to the hot and cold surfaces trend to disappear, as ψ becomes negatively large. The above behaviour is also reflected in the stream-lines patterns as shown in Fig. 5 (d), (e) and (f). In review of the indicated value of the stream functions, the circulation in the core region becomes increasingly weaker as ψ decreases, eventually resulting in the dual secondary cells because of the lack of fluid momentum to carry the fluid from one side of the enclosure to the opposite side.

For positive, as shown in Fig. 6, the physical behaviors are quite different and also appear to be more complex. From the isotherms, it is seen that thermal boundary layers persist as it increases. However, there is also a clear indication that these layers thicken for increasing ψ, especially in the lower and upper regions. This signifies reductions in heat transfer primarily due to the increased distance between the hot and cold walls. It is also of interest to note that the core region is not stratified, but becomes increasingly isothermal as ψ increases. Also in the region of positive ψ, there is another physical mechanism operating. This is related to the effect of thermal instability in view of the fact that the geometry dictates heavier fluid overlying warmer fluid in the region of circulation, thus leading to strong circulation and better heat transfer. However, it is to be noted that this instability effect is expected to be only significant for small positive ψ, and is overwhelmed by the reduction in heat transfer due to the increased distance between the isothermal walls as ψ becomes larger. Since the buoyant effects are concentrated in the isothermal wall regions, it is not difficult to observe the essentially uni-cell flow structure in Fig. 6 (e) and (f). The multi-cell structure for ψ =30° simply indicates a transition from the flow patterns in Fig. 4(b) to the uni-cell structure.

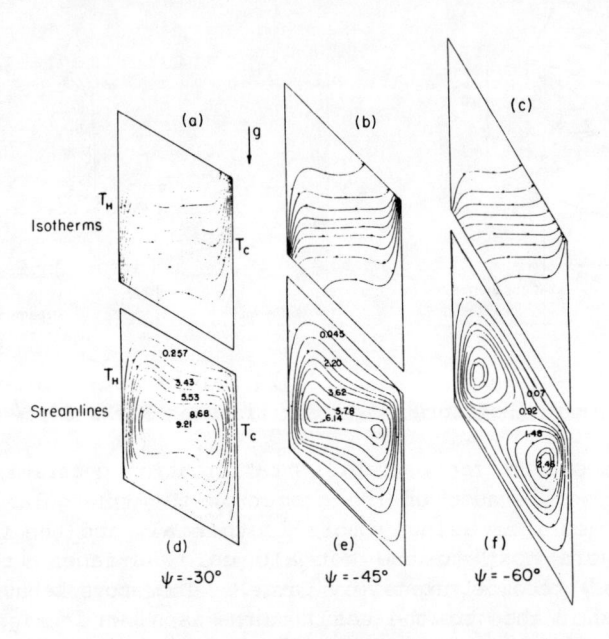

Fig. 5 Isotherms and Streamlines with A_x=W/H=1.44, Pr=0.71, at Inclined Angles of ψ=-30°, -45° and -60°

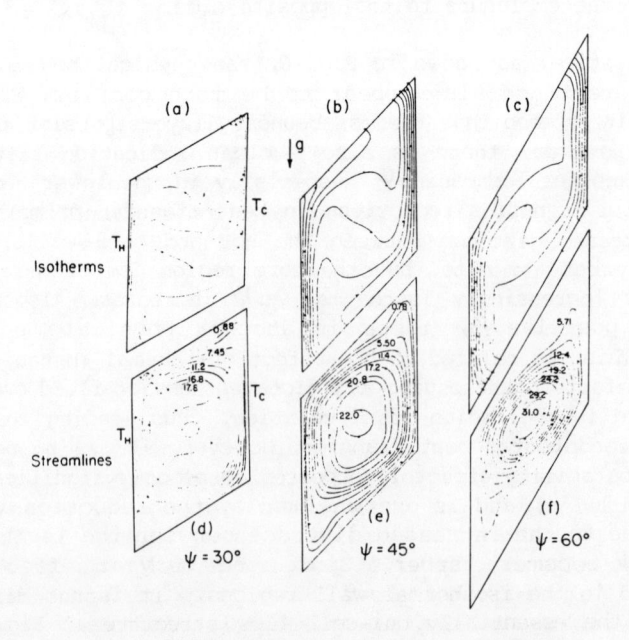

Fig. 6 Isotherms and Streamlines with A_x=W/H=1.44, Pr=0.71, at Inclined Angles of ψ=30°, 45° and 60°

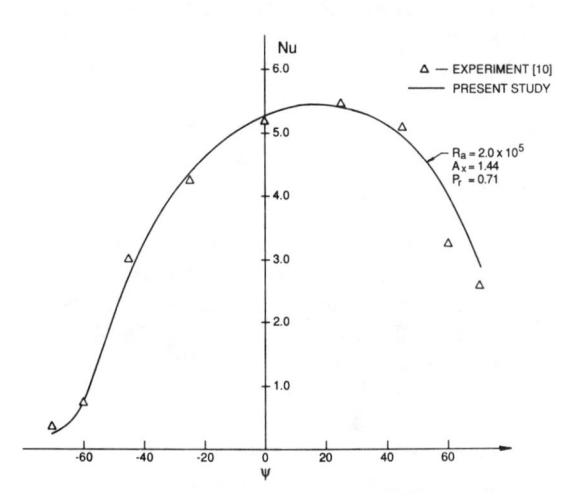

Fig. 7 Effect of Inclined Angle on Heat Transfer

The results of the present calculations can be directly compared to the experimental data of Ski, Fukusako, and Yamaguchi [10]. Flow visualization by means of particle injection has been made to delineate essential flow behaviors at different angles, both positive and negative. Both the unicell flow structure for positive and the dual-cell for negative from the calculation agree well with experimental observations. No temperature field data have been obtained in the experiments, except the overall Nusselt number, which are given in Fig.7 along with the calculated results. Here the Nusselt number is defined as the ratio of the total heat transferred across the enclosure to that by conduction along at ψ =0. The agreement between the two sets of results is seen to be very good. It is interesting to note that the Nusselt number has a maxima not at ψ = 0, but at an angle about 20°. As already discussed previously, this behavior has been noted in natural convection in tilted sqaure enclosures [14].

Finally it may be mentioned that throughout the angle range considered in the calculation ($-60° \leqslant \psi \leqslant 60°$) the flow fields are essentially two-dimensional, and three-dimensional transition to multi-cell structure is absent. In accordance with results of other similar three-dimensional natural convevction studies [12,13], this absence is primarily due to the small effects of thermal instability realized in the present problems. Neither has the transition been observed experimentally [10], even though the maximum angle considered is at 70°.

4. 2 Parallelepiped Enclosures Heated from Below

A second problem considered is that of confined Bénárd convection in a heated-from-below parallelepiped with two vertical parallel walls(z=0, L) and other two walls(x=0, W) inclined from the vertical position at an angle ψ as shown in Fig. 8. The fluid is air with a Prandtl number of 0.71, and Rayleigh number is taken to be 4×10^4 which is well above the critical Rayleigh number for rectangular enclosures. The aspect ratio is taken as A_x= W/H= 6.0 and A_z= L/H = 2.0. For the rectangular enclosure, this problem has been used in previous studies [13] to demonstrate the transient transition of multi-roll-cells with axis in the z-direction to a single cell with the axis in the x-direction. The isotherms and streamlines at $\psi=0$ in [13] are utilized here as the basis for comparison, and they are shown in Fig. 9(a). The primary motion as seen in Fig. 9 (a) consists of six near uniform rolls along the short dimension (z-axis). Results of several calculations with initial conditions given by no motion and linear conduction temperature profile have been depicted in a series of plots in Fig. 9. It is interesting to see that once the two parallel walls are inclined from the vertical in the angle range considered here (ψ from 0 to $60°$) the Bénard convection consists of multi-roll-cells with axis along the short dimensions like the rectangular ones, but the number of roll-cells is odd, different from those for the rectangular enclosure for which the number of roll-cells is even, because of the symmetry of the geometry and initial conditions. Basically, the appearance of the odd number roll-cells can be understood by the anti-symmetry of the geometry with respect to the central axis of the enclosure. As ψ increases, the roll-cells near the lateral walls change their shape while the cells in the central region remain undisturbed and in rectangular shapes, instead of parallelogrammic shapes. At $\psi=60°$, the effect of the end wall, which discourages flow development, become prominent so that the cells next to them are disappearing.

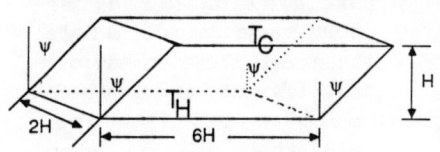

Fig. 8 Parallelepiped With Heated-from-below

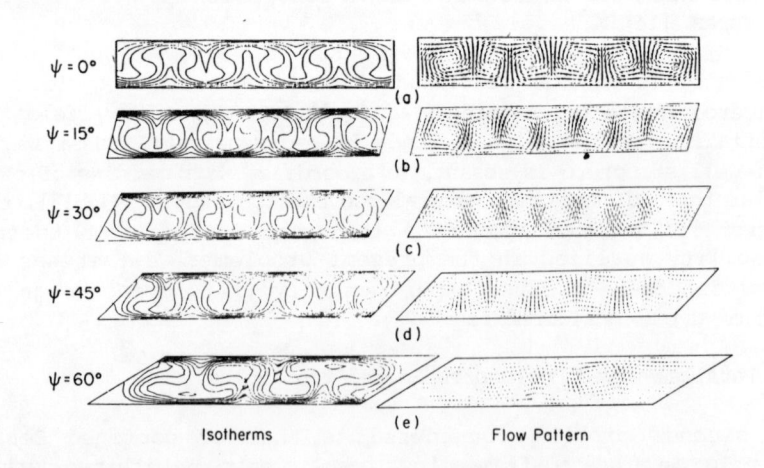

Fig. 9 Isotherms and Flow Patterns for Parallelepipeds with $A_x=6.0$, $A_z=2.0$ at $\psi=0°$, $15°$, $30°$, $45°$, $60°$

5. CONCLUDING REMARKS

Formulation of the control-volume finite-difference method applied to arbitrary three-dimensional parallelepipeds, is described and is demonstrated to be an efficient approach for skewed non-rectangular geometries. Two problems have been considered. One deals with a quasi-two-dimensional skewed enclosure with differentially-heated walls and skewed parallel adiabatic walls. The results show good agreement with existing experimental data in terms of flow structure and rate of heat transfer. A second problem treats the corresponding three-dimensional Benard-convection problem. The results show odd number multiple roll cells for angle inclination greater than zero, as compared to even number of cells for the zero-angle inclination case. Other applications of the present methodology are now being explored.

ACKNOWLEDGEMENT

The authors wish to acknowledge the support of the National Science Foundation under Grant CBT82-19158 to the Univerisity of Notre Dame.

REFERENCES

1. PATANKAR, S. V. and SPALDING, D. B., Int. J. Heat Mass Transfer, 15, 1787-1802 (1972).
2. PATANKAR, S. V. Numerical Heat Transfer and Fluid Flow, Hemisphere Publishing Corp., Washington, D. C., 1980.
3. NIECKELE, A. O. and PATANKAR, S. V. J., Heat Transfer, 107, 902-909 (1985).
4. RAITHBY, G. D., GALPIN, P. F. and VAN DOORMAAL J. P., Numerical Heat Transfer, 9, 125-142 (1986).
5. SHYY, W., TONG, S. S. and CORREA, S. M., Numerical Heat Transfer, 8, 99-113 (1985).
6. ERINGEN, A. C., Mechanics of Continua, John Wiley & Sons, Inc., 1967.
7. YANG, H. Q., YANG, K. T. and LLOYD, J. R., to appear in Int. J. Heat Mass Transfer, (1987).
8. LEONARD, B. P., Numerical Properties and Methodologies in Heat Transfer, ed. T. M. Shih, Hemisphere, Washington, D. C., 1983, p. 221.
9. YANG, H. Q., -Laminar Buoyany Flow Transitions in Three-Dimensional Tilted Rectangular Enclosures, Ph.D thesis, University of Notre Dame, 227 pp, 1987.
10. SEKI, N., FUKUSAKO, S. and YAMAGUCHI, A., J. Heat Transfer, 105, 433-439 (1983).
11. MAEKAWA, T. and TANASAWA, I., Pro. 7th Int. Heat Transfer Conf., Vol. 2, 1982, p. 227.
12. OZOE, H., FUJII, K., LIOR and CHURCHILL, S.W., J. Heat Mass Transfer, 26, p. 1427-1438 (1983).
13. YANG, H. Q., YANG, K. T. and LLOYD, J. R., Proc. 8th Int. Heat Transfer Conf., San Francisco, Vol. 4, 1986, p.1495.
14. ZHONG, Z. Y., YANG, K. T. and LLOYD, J. R., Numerical Method in Heat Transfer- Vol III, ed. R. W. Lewis and K. Morgan, John Wiley ltd, United Kingdom, 1985, p. 195.

AN UNCONDITIONALLY STABLE ONE-STEP FORMULA FOR THE DETERMINATION
OF TRANSIENT TEMPERATURES IN AERO/SPACE STRUCTURES *

C. Haberland and A. Lahrmann

Technical University of Berlin, Aero/Space Department

ABSTRACT

For the determination of transient temperatures in geometrically sophisticated structures such as aero/space structures a novel one-step finite difference method has been developed wich bases on a time dependent interpolation between two time steps. For this approach a linear factor was chosen wich can be split up into two parts. Each of them are evaluated through solving an ordinary differential equation to obtain a more stable integration scheme. Numerical accuracy and stability of this FDM with weighted time step (FDM-WT) are demonstrated for a one-dimensional slab heated with an constant heating rate and a convectively heated skin-web configuration.

NOTATION

$a_{0,1}$	Coefficient	q	Heat flux
A, B	Weight coefficient	t	Time
c	Specific heat	T	Temperature
d	Differential operator	x_i	Coordinates
Fo	Fourier number	α	Diffusivity, Interpolation factor
h	Convection coefficient	β	Angle
i	Mesh point	$\partial_{t,ii}$	Differential operator
k	Conductivity	Δ	Difference
m	Time step	ρ	Density
PF	Perturbation function	ξ	Amplification factor

1. INTRODUCTION

For the determination of temperature transients and the resulting nonlinear temperature-strain fields in complex structures such as stiffened plates, wing boxes or multilayer heat protection tiles finite methods (FDM, FEM, BEM) have to be applied. These, however, depend on recurrence formulae /1-6/ such as EULER, CRANK-NICOLSON, PURE IMPLICIT solution, wich only allow small time steps due to convergence and stability constraints.

On the other hand, the combination of FDM /7,8/ and the recurrence scheme of the semi-analytical method (SAM) has been shown to be a very

* Research was supported by the German Science Foundation

efficient numerical method for aero/space structures such as a skin-web configuration. Since the FDM is better suited for plate shaped structures, it could be expected that an effective calculation method for arbitrarily shaped structures can be obtained by inserting this recurrence formula into the FEM, as a consequence of the more convenient discretisation of these domains. That was performed in /10/ where a semianalytical recurrence approach has been presented wich allows the application of greater time steps even for fast transient conduction problems.

The resulting method has the advantage to show very good convergence and stability qualities but, unfortunately, the shortcoming of a high computational effort owing to the multistep integration scheme which requires the storage of the calculated temperature vectors. Hence, to cirumvent this numerical restriction and to yield a faster solution for the transients, an one-step solution on the basis of a modified difference-differential equation was formulated which, without perturbation function, can be exactly integrated. The solution schemes EULER, CRANK-NICOLSON, PURE IMPLICIT mentioned above are characterized by a distinct interpolation factor between two time steps. For example, the EULER yields $\alpha = 0$ and the PURE IMPLICIT method $\alpha = 1$.

To avoid a multistep recurrence formula, in our approach a time dependent α has been provided. This yields an approximate solution which allows the weighting of the time step. The weighting factor which in this first attempt was assumed to be linear can be split up into two parts which can be evaluated by means of an ordinary differential equation.

In this paper, the novel method will be formulated for the one-dimensional heat conduction problem, but it can be easily extended to 2D- or 3D- domains.

2. ONE-DIMENSIONAL INTEGRATION METHOD

In order to allow the comparision with the exact solution the method is derived for an interior element node of a slab and for boundary element nodes with convective heating, specified heating rate and perfect insulation. For a homogeneous material and temperature-independent properties the differential equation is

$$\partial_t T(x_i, t) = a\, \partial_i \partial_i T(x_i, t) \tag{1}$$

2.1 Interior element node

To develop a stable solution algorithm the temperature increase in Fig. 1

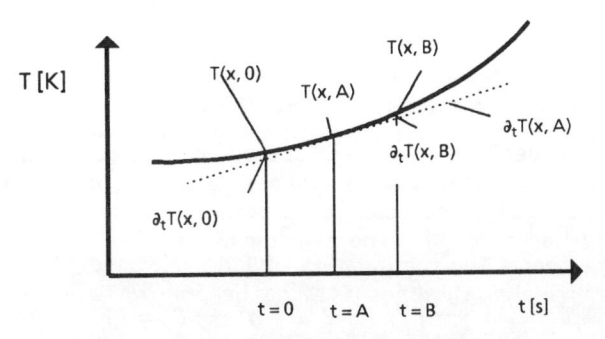

Fig.1: Linear time integration between two specific points

will be defined by the derivative with respect to time in two arbitrary points of the differential domain. Then the temperature within the interval can be be determined with the linear approach

$$\partial_t T = a_o + a_1 A$$

and the coefficients

$$a_o = \partial_t T(x_i, 0), \qquad a_1 = \frac{1}{B}\left[\partial_t T(x_i, B) - \partial_t T(x_i, 0)\right].$$

Then, the differential equation becomes

$$\partial_t T = \frac{A}{B}\partial_t T(x_i, B) + \left(1 - \frac{A}{B}\right)\partial_t T(x_i, 0).$$

Taking into account that this equation is valid for each time interval, the general equation is given by

$$B\partial_t T = A\partial_t T(x_i, t+\Delta t) + (B-A)\partial_t T(x_i, t) \qquad (2)$$

which by applying Eq.(1) yields

$$B\partial_t T = A\alpha\partial_i\partial_i T(x_i, t+\Delta t) + (B-A)\alpha\partial_i\partial_i T(x_i, t). \qquad (3)$$

The application of FDM requires the approximation of derivatives in terms of space and time differences which results in the well known difference operators

$$\partial_t T(x_i, t) \approx \frac{1}{\Delta t}\left[T_i^{m+1} - T_i^m\right], \qquad \partial_i\partial_i T(x_i, t) \approx \frac{1}{\Delta x^2}\left[T_{i\pm 1}^m - 2T_i^m\right].$$

Inserting these relations in Eq.(3) yields the difference-differential equation

$$B\partial_t T_i(t) + 2A\frac{\alpha}{\Delta x^2}T_i^{m+1} - A\frac{\alpha}{\Delta x^2}T_{i\pm 1}^{m+1} =$$

$$-2(B-A)\frac{\alpha}{\Delta x^2}T_i^m + (B-A)\frac{\alpha}{\Delta x^2}T_{i\pm 1}^m \qquad (4)$$

and the pure difference equation

$$(B + 2AFo)T_i^{m+1} - AFoT_{i\pm 1}^{m+1} =$$

$$(B - 2(B-A)Fo)T_i^m + (B-A)FoT_{i\pm 1}^m \qquad (5)$$

with the Fourier number $Fo = \alpha\Delta t/\Delta x^2$. Subsequently, the coefficients A and B which are still unknown have to be determined so that for the selected time step the error is as small as possible. This procedure can be started from the formulation in /9/. Hence, Eq. (4) can be rewritten as

$$B\partial_t T_i + 2A\frac{\alpha}{\Delta x^2}T_i^{m+1} + 2(B-A)\frac{\alpha}{\Delta x^2}T_i^m = PF$$

with the perturbation function

$$PF = A \frac{a}{\Delta x^2} T_{i\pm 1}^{m+1} + (B-A) \frac{a}{\Delta x^2} T_{i\pm 1}^{m}$$

For coinciding temperatures at the time steps $(m+1)\Delta t$ and $m\Delta t$, which means to introduce the differential quotient, the following ordinary nonhomogeneous differential equation will arise

$$d_t T_i + 2 \frac{a}{\Delta x^2} T_i = PF.$$

According to /9/ an exact solution of this differential equation is not possible since the time dependence of the adjacent nodal temperatures is yet not known. Thus, to obtain an approximate solution in this investigation a recurrence scheme has been applied which bases on a polygon approximation of the nodal temperatures and the boundary conditions by means of a triangle superposition (triangle method). However, the resulting multistep method has the disadvantage to require the storage of all temperatures of the previous iterations. For that reason, the method has been modified and the perturbation function assumed equal to zero. Thus, the homogeneous differential equation is given by

$$d_t T_i + 2 \frac{a}{\Delta x^2} T_i = 0$$

with the solution between the time steps $(m+1)\Delta t$ and $m\Delta t$

$$T_i^{m+1} = T_i^m \exp(-2 Fo). \qquad (6)$$

To evaluate the coefficients A and B the perturbation function in Eq.(5) is set equal to zero

$$T_i^{m+1}[B + 2 A Fo] = T_i^m [B - 2(B - A) Fo]. \qquad (7)$$

Comparing Eqs.(6) and (7) yields the coefficients

$$A = \frac{1}{4 Fo^2} \left[2 Fo - 1 + \exp(-2 Fo) \right] \qquad B = \frac{1}{2 Fo} \left[1 - \exp(-2 Fo) \right]. \qquad (8)$$

Rearranging Eq.(5) by dividing through B lead to a formulation of the interior node equation

$$\left[1 + 2 \frac{A}{B} Fo \right] T_i^{m+1} - \left[\frac{A}{B} Fo \right] T_{i\pm 1}^{m+1} =$$
$$\left[1 - 2 \left(1 - \frac{A}{B} \right) Fo \right] T_i^m + \left[\left(1 - \frac{A}{B} \right) Fo \right] T_{i\pm 1}^m \qquad (9)$$

which can generally be applied for all integration schemes considered here. The coefficients for the respective integration method can be written as

	A	B
EULER	0	1
CRANK-NICOLSON	1	2
PURE IMPLICIT	1	1
WEIGHTED TIME STEP	$[2 Fo - 1 + \exp(-2Fo)]/4Fo^2$	$[1 - \exp(-2Fo)]/2Fo$

3.2 Verification of Stability

A well known approach for the verification of the stability of a numerical method is the criterion of Neumann, which applies the Fourier expression in a nodal mesh point and requires that the propagation of the perturbation induced in this point does not increase with respect to time. Herewith, the influence of the boundary conditions is neglected. The Fourier transformation in the nodal point can be written as

$$T_i^m = \xi^m \exp(\sqrt{-1}\beta i \Delta x).$$

Substituting this relation into Eq.(8) renders

$$(1 + 2\frac{A}{B}Fo)\left[\xi^{m+1} \exp(\sqrt{-1}\beta i\Delta x)\right] - \frac{A}{B}Fo\left[\xi^{m+1} \exp(\sqrt{-1}\beta(i\pm 1)\Delta x)\right] =$$

$$(1 - 2(1-\frac{A}{B})Fo)\left[\xi^m \exp(\sqrt{-1}\beta i\Delta x)\right] + (1-\frac{A}{B})Fo\left[\xi^m \exp(\sqrt{-1}\beta(i\pm 1)\Delta x)\right]$$

which, after dividing with by $\xi^m \exp(\sqrt{-1}\beta i \Delta x)$ can be rearranged to

$$-1 \le \frac{1 - 4\left(1 - \frac{A}{B}\right)Fo}{1 + 4\frac{A}{B}Fo} = \xi \le 1$$

which reduces to the condition for the allowed Fourier-number range

$$Fo \le \frac{1}{2}\left(\frac{1}{1 - 2\frac{A}{B}}\right) \quad \text{if } \frac{A}{B} < 0.5 \quad (10), \qquad Fo \ge 0 \quad \text{if } \frac{A}{B} \ge 0.5. \quad (11)$$

The Fourier number is only dependent on the ratio A/B. Therefore, it has to be checked if the ratio

$$\frac{A}{B} = \frac{1}{1 - \exp(-2Fo)} - \frac{1}{2Fo} \tag{12}$$

satisfies the eqs. (10) or (11). The first term of the right-hand can be approximated by series expansion yielding

$$\frac{A}{B} = \frac{1}{2} + 2Fo\frac{B_1}{2!} + .. \quad + (-1)^{n+1}(2Fo)^{2n-1}\frac{B_n}{(2n)!} \tag{13}$$

with the Bernoulli number B_n. Thus, setting the Fourier-number equal to zero eq.(13) yields a minimum A/B = 0.5, consequently for small Δt the CRANK-NICOLSON-method. However, in case the Fourier-number tends towards infinity, eq.(12) leads to maximum A/B = 1. Hence from the Neumann-criterion follows that the present method is unconditionally stable.

3.3 Boundary conditions

For a one-dimensional boundary element which is heated by a heating rate which is arbitrary with respect to time the following differential equation is valid

$$\partial_t T(x_i, t) = a \partial_i \partial_i T(x_i, t) + \frac{1}{\rho c}\partial_i q(t) \tag{14}$$

which after conversion can be written as

$$\left[1 + 2\frac{A}{B}Fo\right]T_i^{m+1} - \frac{A}{B}Fo\,T_{i\pm1}^{m+1} =$$

$$\left[1 - 2\left(1 - \frac{A}{B}\right)Fo\right]T_i^m + \left(1 - \frac{A}{B}\right)Fo\,T_{i\pm1}^m \qquad (15)$$

$$\left[2\frac{A}{B}\frac{\Delta t}{\rho c \Delta x}\right]q_i^{m+1} + \left[2\left(1 - \frac{A}{B}\right)\frac{\Delta t}{\rho c \Delta x}\right]q_i^m.$$

The coefficients A and B follow from Eq.(8). For an element with perfect insulation, Eq.(15) without heat flux is valid. For a convectively heated element the governing equation is

$$\partial_t T - \alpha\,\partial_i\partial_i T(x_i, t) = \partial_i\left(\frac{h}{\rho c}(T_0 - T(x_i, t))\right)$$

which after approximation of derivatives in terms of differences yields

$$\left[1 + 2\frac{A}{B}(1 + Bi)Fo\right]T_i^{m+1} - \frac{A}{B}Fo\,T_{i\pm1}^{m+1} =$$

$$\left[1 - 2\left(1 - \frac{A}{B}\right)(1 + Bi)Fo\right]T_i^m + \left(1 - \frac{A}{B}\right)Fo\,T_{i\pm1}^m \qquad (16)$$

$$\left[2\frac{A}{B}Fo\,Bi\right]T_0^{m+1} + \left[2\left(1 - \frac{A}{B}\right)Fo\,Bi\right]T_0^m.$$

The coefficients can then be evaluated by comparing the difference-differential equation and the differential equation without perturbation function

$$A = \frac{1}{4[Fo(1+Bi)]^2}\left[2[Fo(1+Bi)] - 1 + \exp(-2[Fo(1+Bi)])\right] \qquad (17)$$

$$B = \frac{1}{2[Fo(1+Bi)]}\left[1 - \exp(-2[Fo(1+Bi)])\right].$$

With these boundary conditions a broad range of heat transfer problems can be dealt with.

4. NUMERICAL RESULTS

To verify the accuracy and stability of the novel formulation of the FDM the results obtained for some special structures (1D-web, 2D-skin-web-configuration as a typical aero/space structural element) and boundary conditions have been compared with those of well established solutions. The properties of stainless steel applied in the computational examples

Density $\rho = 7830$ [kg/m³]
Specific heat $c = 586$ [J/Kg K]
Conductivity $k = 19.3$ [W/ m K]

are assumed to be constant in the whole temperature range.

To illustrate the difference between the computational results of the present FDM-WT and the CRANK-NICOLSON as well as the PURE IMPLICIT method, a minimum Fourier-number of 5.77 was applied. First, for a web (thermally thick plate) which at one side is heated with a constant heating rate while the other side is perfectly insulated the transient temperatures calculated with the FDM-WT are compared with the results obtained with the FDM-CN and FDM-PI. The weighting factors applied for these calculations are

	A	B
EULER	0	1
CRANK-NICOLSON	1	2
PURE IMPLICIT	1	1
WEIGHTED TIME STEP	$[2Fo - 1 + \exp(-2Fo)]/4Fo^2$	$[1 - \exp(-2Fo)]/2Fo$

Due to the large oscillations resulting for Fourier-numbers larger then 0.5, the EULER method (FDM-EU) can not be applied. But, on the other hand, with one and the same program different solutions methods can be handled since in the recurrence formulae of the particular integration methods the coefficients are strictly denoted with A and B.

In the presented examples the time step was 120s which corresponds to a Fourier-number of 11.55. Fig.2 shows the transient temperatures of three characteristic points in the web calculated with the CRANK-NICOLSON solution. The parameter NX denotes the number of elements in the web, NT the number of time steps. At the surface, strong oscillations about the exact solution (solid line) occur whereas the temperature at the positions 2 and 3 show only small amplitudes and compare well with the exact solution. This behaviour is

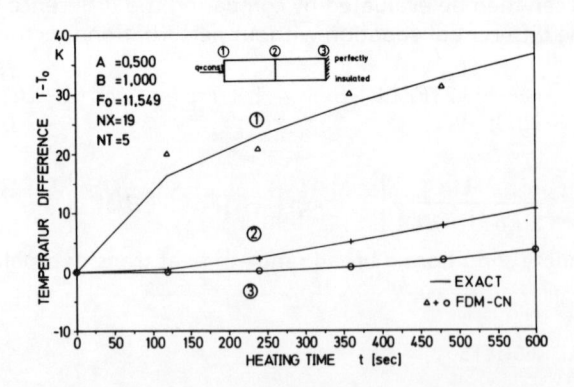

Fig. 2: Temperatures in a thick plate. Comparison of FDM-CN with exact solution.

corroberated in Fig.3 where the deviations to the exact solution are plotted. It indicates a maximum error of 3.7[K] at the surface which decays very slowly. In Fig.4 the temperatures in the slab calculated with the PURE IMPLICIT method of solution are presented. In contrast to the CRANK-NICOLSON implicit method which uses the arithmetic mean value of the derivatives at the beginning and the end of the time interval, they don't show numerical oscillations no matter how large a time step is taken, but, they are not as accurate as those obtained with the FDM-CN. On the other hand, the solution will become more accurate than the CRANK-NICOLSON method if very large time steps are being used. Fig.5 clearly indicates that at the begin of the iteration a strong deviation of the surface temperature occurs which is maintained during the whole integration range. The deviations in the other points are positiv. With increasing time the

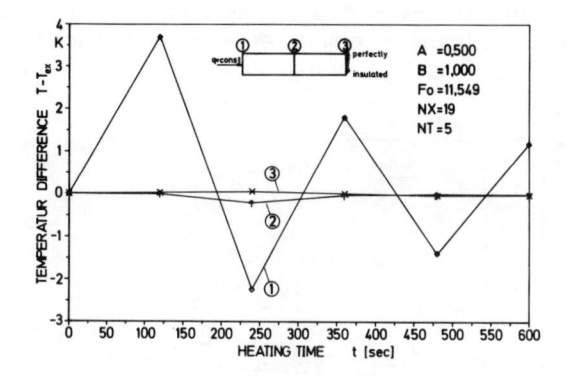

Fig. 3: Thick plate. Error of FDM-CN compared to exact solution.

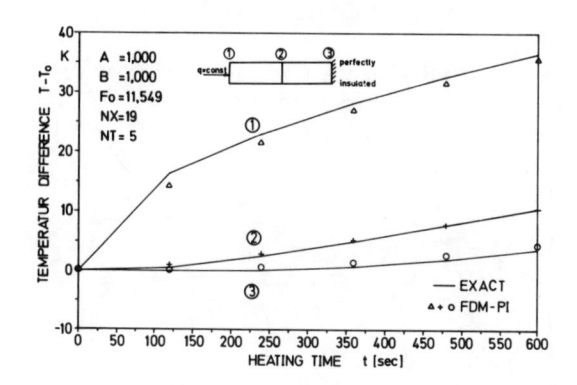

Fig. 4: Temperatures in a thick plate. Comparison of FDM-PI with exact solution.

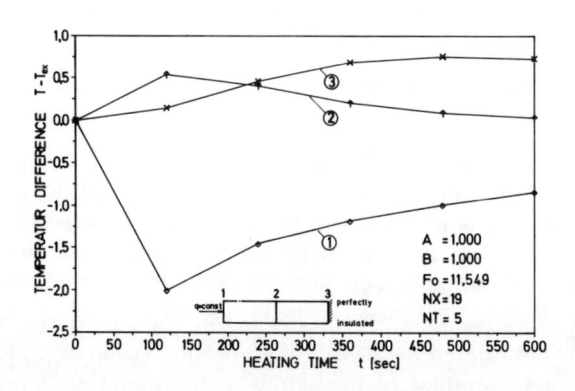

Fig. 5: Thick plate: Error of FDM-PI compared to exact solution.

error tends towards zero. The maximum error at the begin of the iteration is 2[K] that is 45% smaller than that of the FDM-CN.

The FDM with WEIGHTED TIME steps discussed in this paper has, of course, a behaviour which is very similiar to that of the PURE IMPLICIT method, Fig.6 , but it is more accurate than this method. That is illustrated in Fig.7 where again the devations are plotted. The maximum error of the surface temperature is 1.7[K], that is about 20% smaller than that obtained with the FDM-PI.

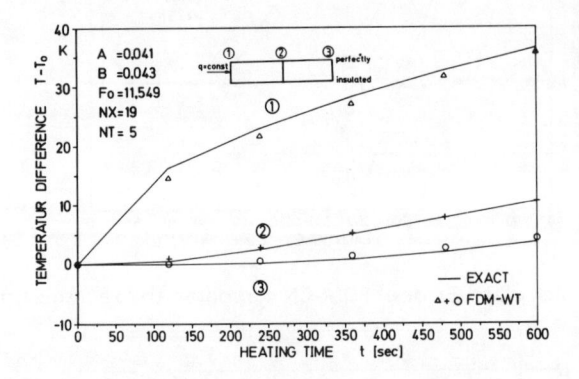

Fig. 6: Temperatures in a thick plate. Comparison of FDM-WT with exact solution.

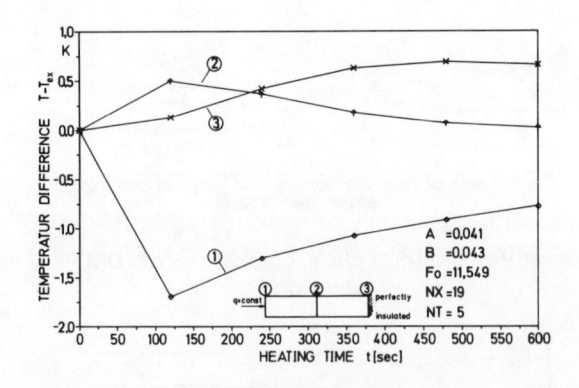

Fig. 7: Thick plate: Error of FDM-WT compared to exact solution

To demonstrate the influence of the time step, a Fourier-number of 5.77 was applied which corresponds to a time step of 60s. In Fig.8 the deviations of the surface temperature with respect to the exact solution are plotted for the three discussed integration schemes. It becomes obvious how the maximum error decreases and the damping qualities improve with reducing time step-

A thermally more complex structure is an integrally stiffened plate which has already been investigated for a M3.5-flight program used in /9/. Fig. 9 shows the characteristic temperatures of this skin-web configuration. Since no exact solution exists the results obtained with FDM-WT have been compared to the CN-solution showing very good agreement. The corresponding temperature distribution along skin and web is presented in Fig.10 again demonstrating the good agreement between the two methods.

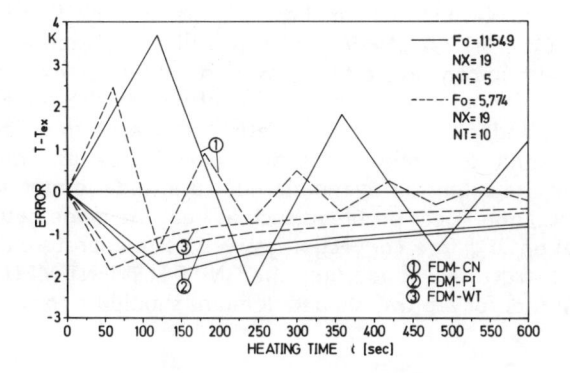

Fig. 8: Thick plate. Influence of Fourier-number on the error of the surface temperature.

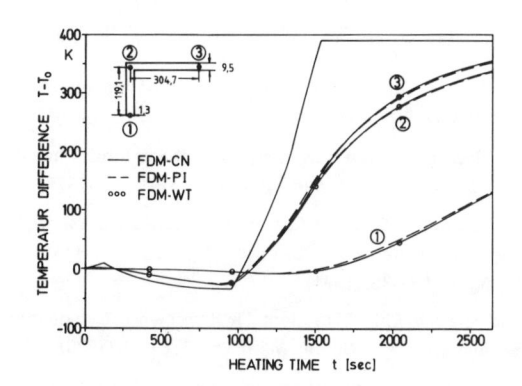

Fig.9: Temperatures in an integrally stiffened plate. Mach 3.5 flight program. Comparision of FDM-WT with FDM-CN and FDM-PI.

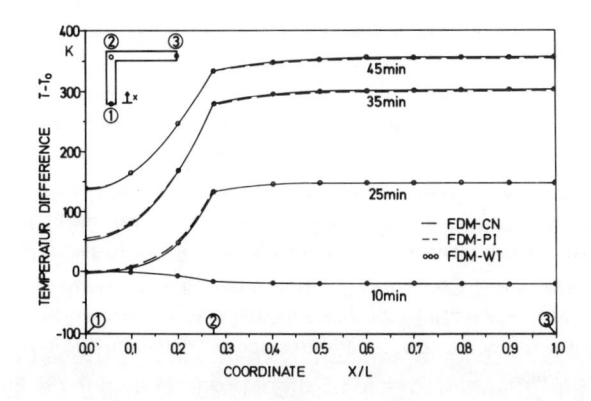

Fig. 10: Integrally stiffened plate. Temperature distribution of different flight times.

5. CONCLUDING REMARKS

Compared to the CRANK-NICOLSON implicit solution and the PURE IMPLICIT solution the present WEIGHTED ONE-STEP FDM provides very accurate and non-oscillating results. Also in contrast to solutions with multi-step time integration based on recurrence formulae with memory coefficients the novel method shows high accuracy and low computational effort. However, in order to make the method more useful for application to aero/space structures, future efforts will concentrate on the extension of the method to include jointed structures, composite structures, non-linear problems such as heat exchange between skin and web by radiation and free convection as well as temperature-dependent properties, and 3D-structures. In addition, the.FINITE ELEMENT METHOD /9/ as the more powerful tool for arbitray shaped domains should be formulated as a one-step solution.

REFERENCES

1. Nicolson, D.W. - 'On stable Num. Integration Methods of Maximum Time Step', Acta Mechanica 38, 191-8, (1981).
2. Zienkiewicz, O.C. -' A New Look at the Newmark, Houbolt and other Time Stepping Formulas. A Weighted Residual Approach', Earthq. Eng. Struct. Dyn. 5, 413-18, (1977).
3. Bettencourt, J.M., Zienkiewicz, O.C. and Cantin, G. - ' Consistent Use of Finite Elements in Time and the Performance of Various Recurrence Schemes of the Heat Diffusion Equation', Int. J. Numer. Meth. Eng. 17, 931-38, (1981).
4. Zienkiewicz, O.C. and Parakeh, C. J. - ' Transient Field Problems: Two-dimensional Analysis by Isoparametric Finite Elements', Int. J. Numer. Meth. Eng. 2, 61-71, (1970)
5. Myers, G.E. -Analytical Methods in Conduction Heat Transfer', McGraw-Hill, New York, (1972)
6. Zlatev, Z. and Thomsen, G.P.- ' Application of Backward Differentiation Methods to the Finite Element Solution of Time Dependent Problems', Ing. J. Numer. Meth. Eng. 14, 1051-61, (1979)
7. Czomber, L. - 'Ein hybrides Rechenverfahren zur Bestimmung der instationären Temperaturen in verzweigten Strukturen', Institut für Luft und Raumfahrt der TU-Berlin, ILR-Bericht 54, (1982).
8. Haberland, C. and Czomber. L. - ' A New Finite Difference and Hybrid Formulation for the Determination of Temperature Transients inBuilt-up Structures', Numerical Methods in Heat Transfer, Vol. III, Ed. Lewis, R.W., John Wiley, p.85, (1985).
9. Haberland C., Czomber L. - ' A Hybrid Calculation Method for the Determination of Transient Temperatures in Built-up Structures', Numerical Methods in Thermal Problems, Proceedings of the 3rd International Conference held in Seattle, U.S.A. on 2nd-5th August, (1983)
10. Czomber L., Haberland C. ,Lahrmann A.,-'A New Stable Finite Element Formulation for the Determination of Transient Temperatures in arbitrarily shaped structures', Numerical Methods in Thermal Problems, Proceeding of the 4th International Conference held in Swansea, U.K. on 15th-18th July, (1985)

AN ISOPARAMETRIC BOUNDARY SOLUTION FOR THERMAL RADIATION

M A Keavey

CEGB Berkeley Nuclear Laboratories, Berkeley, Gloucestershire

SUMMARY

Solutions to the problem of re-radiation and reflection in finite element thermal analysis usually assume a discretisation consisting of flat isothermal surfaces. Such an approximation is clearly inconsistent if an isoparametric model is used for any surrounding solid. In many cases geometric view-factors are required explicitly. Their calculation is generally considered to be tedious and error prone.

The representation of radiative exchange by a boundary integral equation is a standard analytical technique. The use of such a representation in conjunction with finite element analysis, however, appears to be quite recent. The present work describes such a scheme, whereby standard finite element - boundary element coupling theory may be applied to yield a method which is both simple and consistent. The use of isoparametric elements allows varying temperatures and curved surfaces. For an empty enclosure the kernel of the integral equation is simply the differential geometric view-factor. Re-radiation and reflection are implicit in the integral equation.

1. INTRODUCTION

The use of numerical methods to solve large scale practical problems in engineering has been standard practice for many years. Until recently, however, the various relevant analysis areas have tended to remain quite separate. Specifically, finite element conduction codes have supported only simple heat transfer boundary conditions based on prescribed ambient temperatures. Conversely, heat transfer and fluid flow codes, usually using finite differences, have tended not to consider solid regions as a participating component of the model. The larger conduction codes have often evolved as pre-processors attached to major finite element stress analysis packages.

In practical terms this uncoupling of the various analysis types was necessitated by limitations in available processing power. Often the different analyses would be carried out by different departments within an organisation, interfacing in only the most rudimentary manner. From a theoretical point of view this was acceptable as long as the problems were only loosely coupled. Even so, differences in the precise details of discretisation often led to inconsistencies which had to be 'smeared' when transferring results from one model to another.

Just as it is now becoming possible to model larger and larger geometries, however, it is also becoming possible to integrate the various analysis disciplines. For loosely coupled problems this gives consistency and ease of use. For closely coupled systems it is a necessity. When boundary element techniques began to emerge as a recognised numerical method in engineering it was natural to consider coupling with other methods from the outset. This paper describes the implementation of a coupled boundary element - finite element discretisation to solve the problem of radiating enclosures in a conducting solid.

2. THE RADIATION BOUNDARY CONDITION

Techniques for the analysis of thermal radiation have been developed, somewhat independently, in the separate disciplines of heat transfer and conduction. The most sophisticated modelling of the physics has generally been in the field of heat transfer. Conduction analysis has tended to seek only some sort of boundary condition. The earliest attempts to include surface to surface radiation in conduction programs employed the usual radiation to ambient term

$$q = h(\phi^4 - \phi_A^4) \qquad (1)$$

where q is the applied surface flux, h an effective heat transfer coefficient and ϕ the 'active' surface temperature. The ambient temperature ϕ_A was 'looked up' as the last known value for the appropriate surface. Such methods relied on complex house keeping and were, by their nature, explicit.

A slight improvement on the above is to discretise an n-sided enclosure with n^2 'radiation' elements. This method can be made implicit, but useful implementations rely heavily on mesh generation techniques. Furthermore, the computation requires not the geometric view factors but the modified, or Hottel [1], view factors to account for re-radiation and reflection. In practice such factors were often derived for a small number of gross areas and smeared over the many smaller element surfaces constituting the solid conduction model.

Far more sophisticated approaches have, of course, since been developed [2]. These generally involve taking the matrix equations for the Hottel view factors and incorporating them into the finite element derivation. Underlying all these methods, however, is the concept of the discrete isothermal surface. These methods take as their starting point a system of algebraic equations. Yet such systems are in fact only examples of the many possible discretisations of a more fundamental boundary integral equation. The use of such integral equations is well established in the field of heat transfer [3]. It appears, however, that they have only recently been incorporated directly into boundary element and finite element analysis programs [4,5].

3. BOUNDARY INTEGRAL REPRESENTATION

Consider the enclosure shown in Figure 1 and let J be the radiosity, that is, the total outgoing flux.

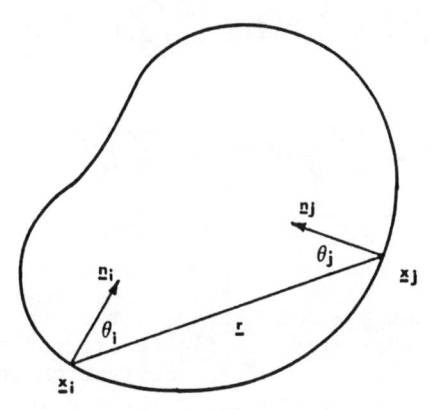

Figure 1 - Radiating Enclosure

Then at any point on the surface, J is the sum of the original emissive power plus the reflected part of the incident flux from all other points on the surface, i.e.

$$J = \sigma\epsilon\phi^4 + (1-\epsilon)\int_A FJdA \qquad (2)$$

where σ, ϵ and ϕ are Stefan's constant, emissivity and absolute temperature respectively, and F is a kernel which in three dimensions is given by

$$F = \frac{\cos\theta_i \cos\theta_j}{\pi|\underline{r}|^2} \qquad (3)$$

The net outgoing flux q, which is the quantity on which the heating or cooling of the surface actually depends, is the difference between the original emissive power and the part of the incident flux absorbed, i.e.

$$q = \sigma \epsilon \phi^4 - \epsilon \int_A FJ dA \qquad (4)$$

It is assumed that the absorptivity is equal to the emissivity. The net outgoing flux q should not be confused with net energy exchange, which is a concept meaningful only in the context of the discrete surface approach mentioned earlier.

The integral equation may now be discretised in any way which maintains consistency with other parts of the analysis. For incorporation into a general finite element system based on isoparametric elements the Galerkin method is used with shape function $\underline{N}$. Using Taylor's theorem to deal with the nonlinearity of the emissive power E, the final discretisation of equations (1) and (3) becomes

$$\sum_i \int_{A_i} \underline{N}_i^T \underline{N}_i \, dA_i \, \underline{J}_i - \sum_i \sum_j \int_{A_i} \underline{N}_i^T (1-\epsilon_i) \int_{A_j} F_{ij} \underline{N}_j \, dA_j \, dA_i \, \underline{J}_j \qquad (5)$$

$$= \sum_i \int_{A_i} \underline{N}_i^T \frac{\partial E_i}{\partial \phi_i} \underline{N}_i \, dA_i \, \underline{\phi}_i + \sum_i \int_{A_i} \underline{N}_i^T (E_i - \phi_i \frac{\partial E_i}{\partial \phi_i}) \, dA_i$$

$$\sum_i \int_{A_i} \underline{N}_i^T q_i \, dA_i = \sum_i \int_{A_i} \underline{N}_i^T \frac{\partial E_i}{\partial \phi_i} \underline{N}_i \, dA_i \, \underline{\phi}_i + \sum_i \int_{A_i} \underline{N}_i^T (E_i - \phi_i \frac{\partial E_i}{\partial \phi_i}) \, dA_i \qquad (6)$$

$$- \sum_i \sum_j \int_{A_i} \underline{N}_i^T \epsilon_i \int_{A_j} F_{ij} \underline{N}_j \, dA_j \, dA_i \, \underline{J}_j$$

The subscripts i and j simply refer to the ith and jth boundary elements and the usual rules of finite element assembly apply. A more detailed description of the derivation thus far is given in [5]. The way in which equations (5) and (6) may be utilised in practice is now considered in more detail.

4. ISOPARAMETRIC IMPLEMENTATION

The theory of isoparametric finite elements is fundamental [6], yet a completely general and consistent approach is not always employed. Consider the equation for global cartesian (x,y,z) derivatives in terms of local (ξ,η,ζ) curvilinear coordinates

$$\begin{Bmatrix} \frac{\partial}{\partial \xi} \\ \frac{\partial}{\partial \eta} \\ \frac{\partial}{\partial \zeta} \end{Bmatrix} = \begin{bmatrix} \frac{\partial x}{\partial \xi} & \frac{\partial y}{\partial \xi} & \frac{\partial z}{\partial \xi} \\ \frac{\partial x}{\partial \eta} & \frac{\partial y}{\partial \eta} & \frac{\partial z}{\partial \eta} \\ \frac{\partial x}{\partial \zeta} & \frac{\partial y}{\partial \zeta} & \frac{\partial z}{\partial \zeta} \end{bmatrix} \begin{Bmatrix} \frac{\partial}{\partial x} \\ \frac{\partial}{\partial y} \\ \frac{\partial}{\partial z} \end{Bmatrix} \qquad (7)$$

This is often solved using Cramer's rule with matrix multiplication, yet Gaussian elimination with multiple right hand sides is probably more efficient and may be applied to any number of dimensions and any number of nodes. The Jacobian determinant $|J|$ can be found from the product of the pivots. Surface integrals are often dealt with using cofactors. Again, a general n-dimensional alternative is available. Consider local cartesian coordinates (x',y',z'). Equation (7) may be rewritten

$$\begin{Bmatrix} \frac{\partial}{\partial \xi} \\ \frac{\partial}{\partial \eta} \\ \frac{\partial}{\partial \zeta} \end{Bmatrix} = \begin{bmatrix} \frac{\partial x}{\partial \xi} & \frac{\partial y}{\partial \xi} & \frac{\partial z}{\partial \xi} \\ \frac{\partial x}{\partial \eta} & \frac{\partial y}{\partial \eta} & \frac{\partial z}{\partial \eta} \\ \frac{\partial x}{\partial \zeta} & \frac{\partial y}{\partial \zeta} & \frac{\partial z}{\partial \zeta} \end{bmatrix} \begin{bmatrix} \frac{\partial x'}{\partial x} & \frac{\partial y'}{\partial x} & \frac{\partial z'}{\partial x} \\ \frac{\partial x'}{\partial y} & \frac{\partial y'}{\partial y} & \frac{\partial z'}{\partial y} \\ \frac{\partial x'}{\partial z} & \frac{\partial y'}{\partial z} & \frac{\partial z'}{\partial z} \end{bmatrix} \begin{Bmatrix} \frac{\partial}{\partial x'} \\ \frac{\partial}{\partial y'} \\ \frac{\partial}{\partial z'} \end{Bmatrix} \qquad (8)$$

where the columns of the transformation matrix $[\lambda]$ are the direction cosines of the local cartesian coordinate directions. Since the rows of $[J]$ constitute independent vectors in the local curvilinear coordinate directions, Gram-Schmidt orthonormalisation provides a general and convenient way of deriving the matrix $[\lambda]$. Let $\underline{\lambda}_m$ represent the columns of $[\lambda]$ and $\underline{J}_m$ the rows of $[J]$. Then

$$\underline{\lambda}_m = \frac{\underline{\lambda}'_m}{\|\underline{\lambda}'_m\|_2} \qquad \underline{\lambda}'_1 = \underline{J}_1 \qquad (9)$$

$$\underline{\lambda}'_m = \underline{J}_m - \frac{\underline{\lambda}'_{m-1} \cdot \underline{J}_m}{\underline{\lambda}'_{m-1} \cdot \underline{\lambda}'_{m-1}} \underline{\lambda}'_{m-1} - \ldots - \frac{\underline{\lambda}'_1 \cdot \underline{J}_m}{\underline{\lambda}'_1 \cdot \underline{\lambda}'_1} \underline{\lambda}'_1$$

For 3-D elements in 3-D space this technique may be used to deal with local anisotropy. For 2-D elements in 3-D space, $[J][\lambda]$ is 2x2 and Gaussian elimination may be applied as before, the resulting determinant being that required for surface integration. The local derivatives allow the modelling of conduction in the element surface yielding an Ahmad [7] type shell element. The same approach may be used for beams.

It can now be seen that only a trivial extension of the general isoparametric implementation is needed to evaluate the kernel (3) of the integral equation. The cosine terms are given by the dot products

$$\cos \theta_i = \frac{\underline{n}_i \cdot \underline{r}}{|\underline{r}|} \qquad \cos \theta_j = \frac{\underline{n}_j \cdot \underline{r}}{|\underline{r}|} \qquad (10)$$

The vector $\underline{r}$ is calculated directly from the integration point coordinates and the outward normal $\underline{n}$ is the vector product of the local in-plane direction cosines, i.e. the columns of $[\lambda]$. The kernel is, of course, evaluated for every combination of integration points in the enclosure surface. When the integration points coincide, the kernel F is zero.

For two-dimensional problems with infinite thickness, the equivalent kernel is given by

$$F = \frac{\cos\theta_i \cos\theta_j}{2|\underline{r}|} \quad (11)$$

The real world, however, is three-dimensional and problems with finite, and possibly different, thicknesses must be integrated correctly. The axisymmetric case raises questions of blocking. All these difficulties may be dealt with, but in a very much less general manner. More detail in this area is given in [8].

5. COUPLING

Equations (5) and (6) may be written

$$[B]\underline{J} = [A]\underline{\phi} + \underline{a} \quad (12)$$

$$\underline{q} = -[C]\underline{J} + [A]\underline{\phi} + \underline{a} \quad (13)$$

Coupling these with the usual finite element equations for heat conduction

$$[K]\underline{\phi} = -\underline{q} \quad (14)$$

gives the system

$$\begin{bmatrix} K+A & -C \\ -A & B \end{bmatrix} \begin{Bmatrix} \underline{\phi} \\ \underline{J} \end{Bmatrix} = \begin{Bmatrix} -\underline{a} \\ \underline{a} \end{Bmatrix} \quad (15)$$

If a fixed boundary temperature is assumed equations (12) and (13) may be solved on their own to give the radiative flux. If coupling is employed a number of alternative approaches may be taken.

The most straightforward approach is to solve (15) directly for $\underline{\phi}$ and $\underline{J}$, with the $\underline{J}$ regarded simply as additional nodal degrees of freedom. Alternatively the $\underline{J}$ may be eliminated to yield an equivalent conduction matrix $[K^*]$ given by

$$[K^*] = [CB^{-1}A] \quad (16)$$

This simply amounts to static condensation of the $\underline{J}$

$$[K + A - CB^{-1}A]\underline{\phi} = -\underline{a} + [CB^{-1}]\underline{a} \quad (17)$$

However, the separate inversion and multiplication and the additional complexity required to handle arbitrarily sized regions make this option less desirable.

Equations (5) and (6) represent a Newton-Raphson solution. Jacobian terms representing sources and boundary conditions are assumed to be already included in the matrix [K]. This arrangement is commonplace in most conduction finite element programs. It is significant, however, in that any approximation to the Jacobian becomes valid, provided convergence takes place. In particular, the matrix may be symmetrised, or alternatively, off-diagonal blocks can be ignored. The former perhaps explains the success of the commonly used boundary element symmetrisation

$$[K^*] = \tfrac{1}{2}\left\{[CB^{-1}A] + [CB^{-1}A]^T\right\} \qquad (18)$$

Off-diagonal terms can also be moved to derive explicit solutions. An example is given in the next section.

6. GAP ELEMENTS

An interesting by-product of the boundary integral theory is its reduction to a general theory for gap elements. Consider a region bounded by two coincident surfaces. The kernel F becomes the Dirac delta function, thus removing the double integral terms in (5) and (6)

$$\sum_i \sum_{j \neq i} \int_{A_i} \underline{N}_i^T \int_{A_j} \delta(\underline{x}_i - \underline{x}_j) \underline{N}_j \, dA_j \, dA_i = \sum_i \sum_{i'} \int_{A_i} \underline{N}_i^T \underline{N}_i \, dA_i \qquad (19)$$

where i and i' represent the opposing surfaces. Now let the net flux across a gap i-j be denoted by q_{ij}. On elimination of the J equations (5) and (6) become equivalent to

$$\begin{bmatrix} \dfrac{\partial q_{ij} G}{\partial \phi_i} & \dfrac{\partial q_{ij} G}{\partial \phi_j} \\[1em] \dfrac{\partial q_{ji} G}{\partial \phi_i} & \dfrac{\partial q_{ji} G}{\partial \phi_j} \end{bmatrix} \begin{Bmatrix} \underline{\phi}_i \\ \underline{\phi}_j \end{Bmatrix} = \begin{Bmatrix} -(q_{ij} - \phi_i \dfrac{\partial q_{ij}}{\partial \phi_i} - \phi_j \dfrac{\partial q_{ij}}{\partial \phi_j}) \underline{g} \\[1em] -(q_{ji} - \phi_i \dfrac{\partial q_{ji}}{\partial \phi_i} - \phi_j \dfrac{\partial q_{ji}}{\partial \phi_j}) \underline{g} \end{Bmatrix} \qquad (20)$$

where

$$[G] = \int_A \underline{N}^T \underline{N} \, dA \qquad \underline{g} = \int_A \underline{N}^T \, dA \qquad (21)$$

Various specific cases derived from this are described in [5]. In fact, it may now be seen that if $\hat{F}_{ij}$ is some appropriately defined net interchange factor, substituting

$$q_{ij} = \sigma \hat{F}_{ij}(\phi_i^4 - \phi_j^4) \qquad (22)$$

into (20) will give a conventional 'radiation' element. The form (20) is implicit in ϕ_i and ϕ_j. Transferring the off-diagonal terms in (20) to the right-hand side gives

$$\begin{bmatrix} \frac{\partial q_{ij}}{\partial \phi_i}G & 0 \\ 0 & \frac{\partial q_{ji}}{\partial \phi_j}G \end{bmatrix} \begin{Bmatrix} \underline{\phi}_i \\ \underline{\phi}_j \end{Bmatrix} = \begin{Bmatrix} -(q_{ij} - \phi_i \frac{\partial q_{ij}}{\partial \phi_i})\underline{g} \\ -(q_{ji} - \phi_j \frac{\partial q_{ji}}{\partial \phi_j})\underline{g} \end{Bmatrix} \qquad (23)$$

Assembly of all the possible combinations of this equation for an enclosure gives the explicit symmetric system used traditionally by older finite element codes.

7. FRONT SOLUTION

The choice of solution technique for the final system of linear equations is, in principle, unimportant. If the full implicit representation of equations (12) and (13) is employed, however, the method should ideally be capable of dealing with unsymmetric indefinite systems. The particular heat conduction code used by the author borrows its solution algorithm from a related linear stress analysis code. This is a frontal solver due to Hellen [9] and assumes a symmetric positive definite matrix. The necessary modifications to Hellen's algorithm are now described. Other suitable variations on the original front solution have been developed at Harwell [10].

7.1 Storage

The front solution described by Hellen [9] stores only the lower triangle. Extension to full matrix storage is trivial and actually simplifies the code. Furthermore, with modern computer architectures and particularly with vector processing, the time overhead is not great. The doubling of required storage, however, is inevitable.

7.2 Pivot Strategy

Of more consequence is the fact that the system of equations is not necessarily positive definite. The basic front technique relies on the fact that the ordering of the equations does not affect the solution numerically. It now becomes necessary to invoke a pivot strategy.

The Harwell routine [10] employs full pivoting. Only row interchange has been used here and is much simpler. Whenever a set of equations becomes ready for elimination the row with the maximum pivot is chosen first. For finite elements, freedoms become candidates for elimination with the assembly of each individual element. The scope for reordering is therefore small without the use of buffer space. For a boundary region, however, the off-diagonal terms mean that no freedoms are ready for elimination until the final boundary element is loaded. Full row pivoting within the subset representing the boundary integral equations is therefore automatic.

7.3 Conditioning

Two types of test may be employed to ensure accuracy. Each pivot may be tested for acceptability by comparing it with the other coefficients in its column

$$|a_{kk}| \geq \epsilon \max_i |a_{ik}| \qquad (24)$$

where the a_{ij} are the matrix coefficients and ϵ is some error criterion such that $0 < \epsilon < 1$. This test is taken from [10], and may be used either to abandon the analysis or to invoke the assembly of further elements into the front matrix. The latter, of course, places additional demands on storage.

Checks for ill conditioning, for example diagonal decay [11], pose particular problems when implemented with the front solution, since the coefficient matrix is never fully assembled. A test based on relative matrix norms [12] has therefore been tried. Using the maximum norm $\|A\|_\infty$ for an nxn matrix, the test becomes

$$|A| \geq \epsilon \|A\|_\infty \qquad \|A\|_\infty = \max_i \sum_{j=1}^{n} |a_{ij}| \qquad (25)$$

Only a vector of length the semi-frontwidth is required, the appropriate a_{ij} being incremented with every element assembly. A row is necessarily fully summed when its corresponding freedom is ready for elimination. The row-sum location is then released for the next new variable, just as rows are assembled and released in the front matrix itself. The norm is updated as the forward elimination proceeds, as is the product of pivots representing the determinant. Although the test is essentially a posteriori, in practice it is applied at each stage with current updated values and the elimination abandoned should it fail.

8. DISCUSSION

An alternative arrangement of equations (2) and (4) is given by

$$q - \epsilon \int_A \frac{(1-\epsilon)}{\epsilon} FqdA = \sigma\epsilon\phi^4 - \sigma\epsilon \int_A F\phi^4 dA \qquad (26)$$

This was used by Bialecki et al. [4] in combination with a boundary element discretisation of the conducting solid. This representation, however, can not be used unmodified for perfect reflectors. Other re-arrangements can similarly give rise to singularity for a perfect black body. For this reason the separate equations (2) and (4) were adopted. The method as stated is limited to grey, diffuse, convex and simply connected enclosures. The emissivities, however, may be both time and temperature dependent. More complex geometries can, in principle, be accommodated by special coding for blocking. To date no attempt has been made to involve a participating medium.

The modifications to the front algorithm were in fact made primarily to handle convective transport problems [13]. The matrices derived from the radiation problem tend to be diagonally dominant. Thus the modified front solution is probably not strictly necessary, as is born out by the fact that the older symmetric explicit approach, mentioned in section 6, is quite well behaved.

The actual implementation of the method as a computer code has presented some problems, particularly with regard to data structures. The sequential access of element data typical of many finite element codes is inappropriate to boundary element techniques. An internal database has therefore been constructed to deal with the complex data requirements of boundary elements, coupling and substructures. The prototype database manager (written in a form of Prolog [14]) is as yet incomplete. However, two very simple steady state examples are given in [5]. As yet no transient analyses have been attempted although the theory should be equally applicable. The program used by the author possesses a stiff solver based on Backward Euler with Newton-Raphson.

9. CONCLUSION

Thermally radiating enclosures can be incorporated into existing finite element codes using standard boundary element techniques. The differential geometric view factor appears as the kernel of the integral equation. The integral equation itself accounts for re-radiation and reflection and view factors do not need to be specified externally. Existing theory for isoparametric elements can easily be extended to allow for curved surfaces and varying temperatures. It is possible to derive an equivalent symmetric 'conduction' matrix for each enclosure. However, by considering both radiosity and temperature as variables, a radiating enclosure in a conducting solid can be regarded as a general nonlinear coupled problem.

ACKNOWLEDGEMENTS

This work was carried out at Berkeley Nuclear Laboratories and is published by permission of the Central Electricity Generating Board. The author is also grateful to A D Jackson for his contribution to the general implementation of isoparametric elements.

REFERENCES

1. MCADAMS, J. D. - *Heat Transmission*, McGraw-Hill, London, 1954.

2. OSNES, J. D. - 'A Method for Efficiently Incorporating Radiative Boundaries in Finite Element Programs', Numerical Methods in Thermal Problems Vol. III, Ed. Lewis, R. W., Johnson, J. A. and Smith, W. R., Pineridge Press, 1983, p.36.

3. VISKANTA, R. - 'Radiation Transfer and Interaction of Convection with Radiation Heat Transfer', Advances in Heat Transfer, Ed. Irvine, T. F. Jr. and Hartnett, J. P., Academic Press, 1966, p.175.

4. BIALECKI, R., NAHLIK, R. and NOWAK, A. J. - 'Temperature Field in Solid Forming an Enclosure where Heat Transmission by Convection and Radiation is Taking Place', Symposium Series No. 86, I. Chem. E., 1984, p.989.

5. KEAVEY, M.A. - 'A Boundary Integral Solution for Radiating Enclosures', Boundary Element Methods, IOP Short Meeting Series No. 1, Institute of Physics, London, 1986, p.111.

6. ZIENKIEWICZ, O. C. - The Finite Element Method, McGraw-Hill, London, 1977.

7. AHMAD, S. - 'Curved Finite Elements in the Analysis of Solid, Shell and Plate Structures', Ph.D. Thesis, University College of Swansea, 1969.

8. COLLIER, W. D. - 'Radiation of Heat in the Heat Transfer Program TAU', Report ND-R-555(R), United Kingdom Atomic Energy Authority, 1981.

9. HELLEN, T. K. - 'A Front Solution for Finite Element Techniques', Report RD/B/N1459, Central Electricity Generating Board, 1969.

10. DUFF, I. S. - 'MA32 - A Package for Solving Sparse Unsymmetric Systems Using the Frontal Method', Report AERE - R 10079, United Kingdom Atomic Energy Authority, 1981.

11. IRONS, B. M. - Roundoff Criteria in Direct Stiffness Solutions, AIAA Journal, $\underline{6}$, 1308-12 (1968).

12. PHILLIPS, G. M. and TAYLOR, P. J. - Theory and Applications of Numerical Analysis, Academic Press, London, 1973.

13. THORNTON, E. A. and WIETING, A. R. - 'Finite Element Methodology for Transient Conduction/Forced-Convection Thermal Analysis', Proc. 14th AIAA Thermophysics Conference, Orlando, 1979.

14. CLOCKSIN, W. F. and MELLISH, C. S. - Programming in Prolog, Springer-Verlag, New York, 1984.

A Multigrid Algorithm for the Investigation of Thermal, Recirculating Fluid Flow Problems

P.H. Gaskell[+]
Department of Mechanical Engineering,

N.G. Wright[++]
Department of Applied Mathematical Studies,

University of Leeds,
Leeds, U.K..

SUMMARY

A detailed description of a multigrid algorithm and an unsegregated solution technique, for the simulation of fluid flow problems, is given. This is then used in conjunction with a higher order discretisation scheme to find the solution of two test problems; namely a two-dimensional thermally driven cavity, and a three dimensional lid driven cavity, for various Rayleigh and Reynolds numbers, respectively.

1. INTRODUCTION

The investigation and accurate prediction of complex fluid flows is a topic of great importance in current engineering practice This paper summarises some recent work on the development of an efficient computer algorithm for the simulation of fluid flow phenomena occurring in different geometries, incorporating thermal effects, turbulence and chemical reaction. The analysis presented here builds on the knowledge gained in a recent investigation of two dimensional flows using a fully coupled solver and a multigrid algorithm[1]. The methodology described therein has been extended to handle thermal flows and flows in three- dimensions, the preliminary results of which are presented here.

2. CONSERVATION EQUATIONS

Although, a stream function - vorticity formulation is more desirable as a basis for simulating two-dimensional incompressible flows, it often the case in engineering situations that a "primitive variable (u,v,(w),p)" formulation is preferred. Consequently, the latter approach has been adopted here.

+ Lecturer, ++ Research Student

For steady state, incompressible laminar flows, the momentum conservation equation can be written (in Cartesian co-ordinates) as :

$$\frac{\partial}{\partial x_j}(\rho u_i u_j) = -\frac{\partial P}{\partial x_i} + \mu \nabla^2 u_i \quad i=1,2,3 \tag{1}$$

the mass conservation equation as:

$$\frac{\partial u_i}{\partial x_i} = 0 \tag{2}$$

and the scalar conservation equation as:

$$u_i \frac{\partial \phi}{\partial x_i} = \nabla^2 \phi \quad i=1,3 \tag{3}$$

3. NUMERICAL PROCEDURE

A normal staggered grid configuration is used and the conservation equations are integrated over a macro control volume. The QUICK discretisation has been chosen because of its high accuracy, and minimal numerical diffusion. Details of the QUICK scheme and associated results can be found elsewhere[2].

Here, a fully coupled solution procedure has been adopted. Unlike the popular "pressure-correction" technique[3] the fully coupled procedure solves the momentum, mass and energy conservation equations in their original form. This results in superior coupling between equations and makes for easier application of a multigrid technique. It has been observed that difficulty may be experienced with the "pressure-correction" approach when solving problems where velocity-pressure coupling is important, and it is believed that a fully coupled solver may alleviate this problem.

Many of the problems of interest in an engineering context exhibit flow patterns containing steep gradients. The flow characteristics in these regions are best resolved using a large number of mesh points. Obviously, the more points used the greater the time required to achieve a converged solution. In order to reduce the computational cost in these situations it was decided that a multigrid technique should be adopted. Such a procedure acts as a convergence accelerator and has the advantage that the cpu (central processor unit) time required is governed by the following relationship,

$$\text{cpu} \propto N^\alpha$$

where $\alpha \approx 1$. A distinct improvement over ordinary iterative methods for which $\alpha > 1.6$. The particular multigrid method used is a Full Approximation Storage method based on the ideas of Falle and Wilson[4]. This and the solution strategy involved are described below.

3.1. The Smoothing Technique - Symmetrical Coupled Gauss-Seidel (SCGS)

Various attempts have been made to adapt the "pressure-correction" technique for use with a multigrid procedure. However this approach has been seen to have problems associated with it[5,6]. The Symmetrical Coupled Gauss-Seidel method[7], on the other hand, is simple and offers the advantage of a low operation count and a minimal storage requirement. As a first step to assessing the use of a fully coupled solver with a multigrid technique, it is easily implemented and reasonably efficient.

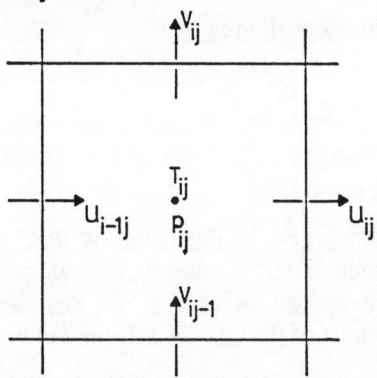

Fig. 1: The control volume with the four velocities and pressure shown.

For each control volume, four velocities (six in three dimensions) and one value of the pressure are updated simultaneously by inverting a 5x5 (or 7x7 in three dimensions) matrix, see Fig. 1. For a cell centred at i,j, this is represented by,

$$\begin{pmatrix} (A_p^u)_{i-1j} & & & & \frac{1}{h} \\ & (A_p^u)_{ij} & & & -\frac{1}{h} \\ & & (A_p^v)_{ij-1} & & \frac{1}{h} \\ & & & (A_p^v)_{ij} & -\frac{1}{h} \\ -\frac{1}{h} & \frac{1}{h} & -\frac{1}{h} & \frac{1}{h} & 0 \end{pmatrix} \begin{pmatrix} u'_{i-1j} \\ u'_{ij} \\ v'_{ij-1} \\ v'_{ij} \\ p'_{ij} \end{pmatrix} = \begin{pmatrix} r^u_{i-1j} \\ r^u_{ij} \\ r^v_{ij-1} \\ r^v_{ij} \\ r^c_{ij} \end{pmatrix} \qquad (4)$$

where dashed quantities (') represent updates, r^u_{ij} represents the residual in the equation for u at the point i,j and h is the mesh spacing. The matrix equation (3.1) is solved exactly using a form of LU-decomposition.

Relaxation of the updates is required, owing to the nonlinearity of the algebraic equations and the necessary use of old values for u,v and p when evaluating the matrix coefficients and the residuals. The velocity updates are multiplied by a factor α_1 and the pressure updates by α_2, once they have been calculated. This set of equations is solved for each control volume first in the direction of increasing i and then j, so that each velocity is updated twice. Vanka[7] observed that this ensured the stability which a single update method lacked.

At present the temperature field is updated by performing a seperate sweep over the solution domain after the u,v and p values have been updated. The plausibility and benefits of a simultaneous update for u,v,p and T are being investigated.

3.2. The Multigrid Scheme

The essence of the multigrid concept is that an error of wavelength λ is most easily eliminated on a mesh of size h, where $\lambda \approx h$. In view of this a hierarchy of grids of different mesh sizes is used to solve the fine grid problem. A representation of this problem is set up on the coarser grids and these are used to calculate corrections to the fine grid solution. Within this overall concept there are several different strategies that can be adopted. The one used here is based on the work of Falle and Wilson[4].

3.3. The Multigrid Strategy

If we call the current fine mesh size h_k, then the problem we wish to solve is,

$$L_k U^k = f^k \tag{5}$$

where, L_k is a finite difference operator and U_k is a solution vector. At any stage of the smoothing process we have a current approximation to the final solution U^k on grid k; let this approximation be u^k. Thus,

$$f^k - L_k u^k = r^k. \tag{6}$$

Put,

$$U^k = u^k + v^k,$$

such that,

$$L_k(u^k + v^k) = f^k, \tag{7}$$

and therefore,

$$L_k(u^k + v^k) - L_k u^k = r^k. \tag{8}$$

We wish to find v^k - the correction to u^k. Rather than linearise the operator L^k, we choose to solve for the quantity $u^k + v^k$ and then compute v^k. With a multigrid technique we attempt to solve this system on a coarser mesh of size h_{k-1} (usually $h_{k-1} = 2h_k$). i.e.,

$$L_{k-1}(I_k^{k-1}u^k + v^{k-1}) - L_{k-1}(I_k^{k-1}u^k) = I_k^{k-1}r^k \qquad (9)$$

I_k^{k-1} represents the restriction operator from grid k to grid k-1.

So, for a given set of meshes, k=1,....,M with $h_{k-1} = 2h_k$ we can set up a system of equations;

$$L_k u^k = F^k, \qquad (10)$$

where,

$$F^k = L_k(I_{k+1}^k u^{k+1}) + I_{k+1}^k (F^{k+1} - L_{k+1} u^{k+1})$$

and $u^k = I_M^k u^M$

In our case, $F^M = 0$.

In performing one multigrid cycle the following steps are taken:

(i) The equations are set up on all grids k = 1,....,M

(ii) The residual of the solution of the restricted problem on each mesh is calculated. The level with the highest residual is selected for smoothing.

(iii) The solution is smoothed on this mesh until the error has been reduced by a factor γ. The final and initial solutions are used to calculate a correction.

(iv) This correction is prolonged onto the next finer mesh and added to the current solution there. This solution is smoothed until the error has been reduced by a factor γ.

(v) Steps iii) & iv) are repeated until the finest mesh has been corrected and smoothed.

The above cycle is repeated until the modulus of the residual on the finest mesh is less than a prescribed value. For example, for a two-dimesional problem the modulus used is the $||.||_2$ norm, i.e.,

$$||r|| = \frac{(\sum_{i,j}(r_u^{ij^2} + r_v^{ij^2} + r_c^{ij^2}))^{1/2}}{3 \times in \times jn} \qquad (11)$$

where in and jn are the number of mesh points in the x and y direction respectively.

In our present strategy, the problem is solved on the coarsest grid (usually 5x5) and then an interpolation process (called prolongation) is used to establish an initial solution for the next mesh (with half the mesh size). This process is repeated until the solution is achieved on the required mesh.

3.4. Restriction and Prolongation

Linear interpolation is used for both restriction and prolongation, although it is hoped to implement quadratic interpolation with the QUICK scheme at a later stage. The restriction and prolongation operations are more difficult than they would be in the case of a stream function-vorticity formulation, because of the staggered grid arrangement, see Fig. 2. Here, the restriction is performed by averaging nearby values and prolongation by bilinear (or trilinear in the three dimensionl case) interpolation. When prolongating, the boundary values on the coarse mesh are not used. The near boundary values for u,v,(w) and p on the fine mesh are calculated using a zero derivative condition. This obviously affects the accuracy of the corrections in the near boundary region; it has been found that performing an update on all near boundary values after prolongation increases significantly the rate of convergence.

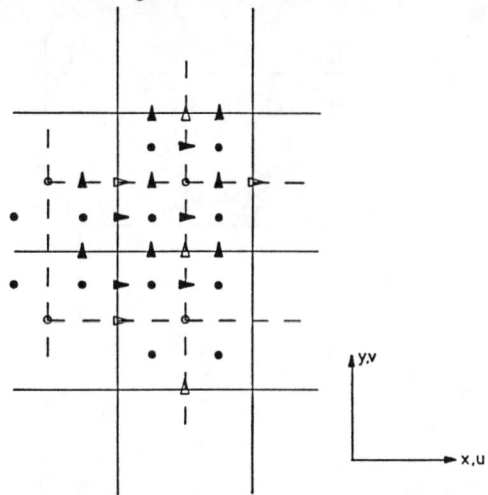

Fig. 2: The coarse and fine grid nodal configuration;
⊳ coarse grid velocity,▻ fine grid velocity,
o   coarse grid scalar,●fine grid scalar.

4. TEST PROBLEMS

The first results obtained using the above methods where for the problem of laminar flow in a two dimensional lid driven cavity[1]. The subsequent improvements to the original approach are associated mainly with the method of prolongation and the actual programming architecture. These improvements have been implemented in the present work, and will be reported in detail elsewhere. The following test problems are considered:

4.1. The Two Dimensional Thermally Driven Cavity

Fluid flows in the cavity defined by $x=0$, $y=1$, $x=1$ and $y=0$, see Fig. 3. The no slip condition is imposed on all boundaries for the velocity. i.e. $u=v=0$. For the temperature, the top and bottom walls are assumed to be adiabatic, imposing a zero temperature derivative condition there. The flow is driven by a temperature gradient across the cavity. For the non-dimensional equations used here, $T=1$ on $x=0$ and $T=0$ on $x=1$, see Fig. 3. The equations are :

$$\frac{\partial u^2}{\partial x} + \frac{\partial uv}{\partial y} = -\frac{\partial p}{\partial x} + \frac{1}{Gr^{1/2}}\left[\frac{\partial^2 u}{\partial x^2} + \frac{\partial^2 u}{\partial y^2}\right] \qquad (12)$$

$$\frac{\partial uv}{\partial x} + \frac{\partial v^2}{\partial y} = -\frac{\partial p}{\partial y} + \frac{1}{Gr^{1/2}}\left[\frac{\partial^2 v}{\partial x^2} + \frac{\partial^2 v}{\partial y^2}\right] + T \qquad (13)$$

$$\frac{\partial u}{\partial x} + \frac{\partial v}{\partial y} = 0 \qquad (14)$$

$$u\frac{\partial T}{\partial x} + v\frac{\partial T}{\partial y} = \frac{1}{Gr^{1/2}Pr}\left[\frac{\partial^2 T}{\partial x^2} + \frac{\partial^2 T}{\partial y^2}\right] \qquad (15)$$

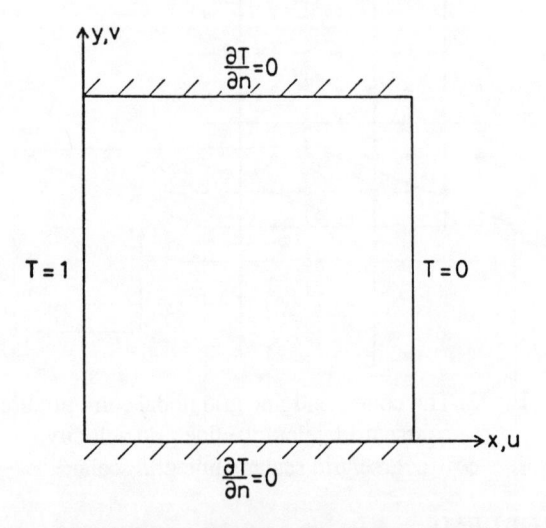

Fig. 3: The thermally driven cavity.

We take Pr=0.71 and solve the equations for Rayleigh Numbers of 10^3, 10^4, 10^5, and 10^6, in line with the bench mark solution provided by G. de Vahl Davis and I.P. Jones[8].

4.2. The Three Dimensional Lid Driven Cavity

This represents a natural extension of the well known two dimensional lid driven test problem considered elsewhere[1]. In the cubic cavity the velocities are zero on all faces except for the face y=0 where w=1, see Fig. 4. This test problem was also adopted by Vanka for studying the performance of a multigrid/hybrid solution strategy. However, Vanka solved for the whole cavity. Here we have made use of the symmetry of the problem; solving for one half of the cavity and applying a zero derivative boundary condition for v and w and a zero value condition for u at the symmetry plane. The equations solved are:

$$\frac{\partial u^2}{\partial x} + \frac{\partial uv}{\partial y} + \frac{\partial uw}{\partial z} = -\frac{\partial p}{\partial x} + \frac{1}{Re}\left[\frac{\partial^2 u}{\partial x^2} + \frac{\partial^2 u}{\partial y^2} + \frac{\partial^2 u}{\partial z^2}\right] \quad (16)$$

$$\frac{\partial uv}{\partial x} + \frac{\partial v^2}{\partial y} + \frac{\partial vw}{\partial z} = -\frac{\partial p}{\partial y} + \frac{1}{Re}\left[\frac{\partial^2 v}{\partial x^2} + \frac{\partial^2 v}{\partial y^2} + \frac{\partial^2 v}{\partial z^2}\right] \quad (17)$$

$$\frac{\partial uw}{\partial x} + \frac{\partial vw}{\partial y} + \frac{\partial w^2}{\partial z} = -\frac{\partial p}{\partial z} + \frac{1}{Re}\left[\frac{\partial^2 w}{\partial x^2} + \frac{\partial^2 w}{\partial y^2} + \frac{\partial^2 w}{\partial z^2}\right] \quad (18)$$

$$\frac{\partial u}{\partial x} + \frac{\partial v}{\partial y} + \frac{\partial w}{\partial z} = 0 \quad (19)$$

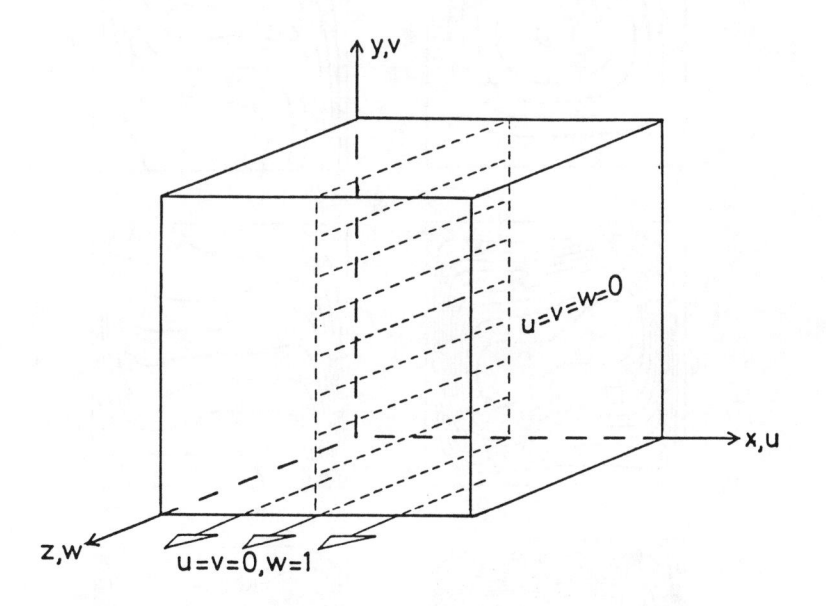

Fig. 4: The lid driven three-dimensional cavity.
Flow fields have been calculated for Reynolds Numbers of 100 and 1000.

5. RESULTS

5.1. Thermally Driven Cavity

Table 1 shows the maximum value of the stream function and its position for the four Rayleigh Numbers considered. These values by no means give a full account of all aspects of the problem, but it was decided that this was the best result to present as a basis for comparison within the space available. A full analysis of the flow field will be published elsewhere. The values in Table 1 are in good agreement with those of de Vahl Davis[8], and the percentage errors are as shown. Interestingly, these errors are among the smallest found for the finite

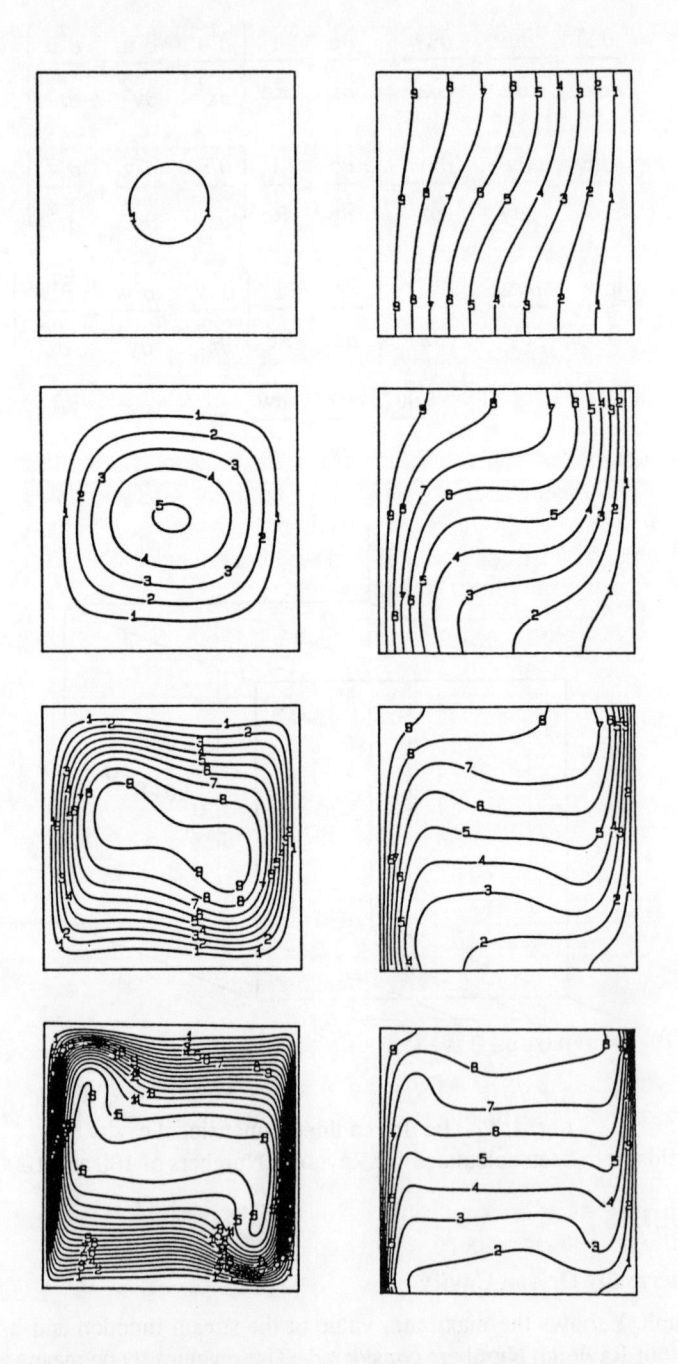

Fig. 5: Streamfunction(left) and temperature(right) contours for
Rayleigh numbers(from top to bottom) $10^3, 10^4, 10^5$ and 10^6.
Streamfunction contours from 0.01 in increments of 0.01, temperature
contours from 0.1 in increments of 0.1.

difference methods evaluated by de Vahl Davis and Jones.

Ra	ψ_{max}	x	y	Percentage error
10^3	1.178	0.500	0.500	+0.3
10^4	5.090	0.500	0.500	+0.4
10^5	9.603	0.283	0.600	-0.1
10^6	16.715	0.150	0.545	-0.2

Table 1 : Value and position of maximum streamfunction and error with respect to Ref.[8], for each Rayleigh number.

From Table 2 it can be seen that solutions are obtained for a large number of points in reasonably fast times. Even though cpu times become larger as the Rayleigh number is increased, they are still considerably less than those that would be found using ordinary methods. The linear relation between cpu time and the number of points, expected with a multigrid method, has been achieved in most cases. Deviations from this expected behaviour are thought to be due to deficiencies in the smoothing technique used.

Grid	Ra			
	10^3	10^4	10^5	10^6
4^2	0.25	0.44	-	-
8^2	0.80	1.47	3.65	-
16^2	3.15	7.00	14.44	39.00
32^2	11.43	45.93	77.39	96.39
64^2	43.38	143.23	275.86	319.73
128^2	170.87	619.40	859.54	1122.01
256^2	895.23	2897.92	3899.55	3985.47

Table 2 : CPU times for different Rayleigh numbers and different grids.

5.2. Three Dimensional Lid Driven Cavity

The results for flow at Reynolds number 1000 (See Fig. 6) are in good agreement with those of other authors[7]. The main features of the flow are clearly resolved on 32x32x16 mesh. Results have been obtained for Reynolds number 100 for a 64x64x32 mesh, but computing resources (in terms of a large storage requirement) are not readily availible. The results themselves differ only marginally from those on obtained on a coarser mesh.

Grid	Re	
	100	1000
4^3	0.27	-
8^3	2.93	8.79
16^3	17.23	60.86
32^3	94.58	496.93
64^3	610.28	-

Table 3 : CPU times for different Reynolds numbers. and different grids.

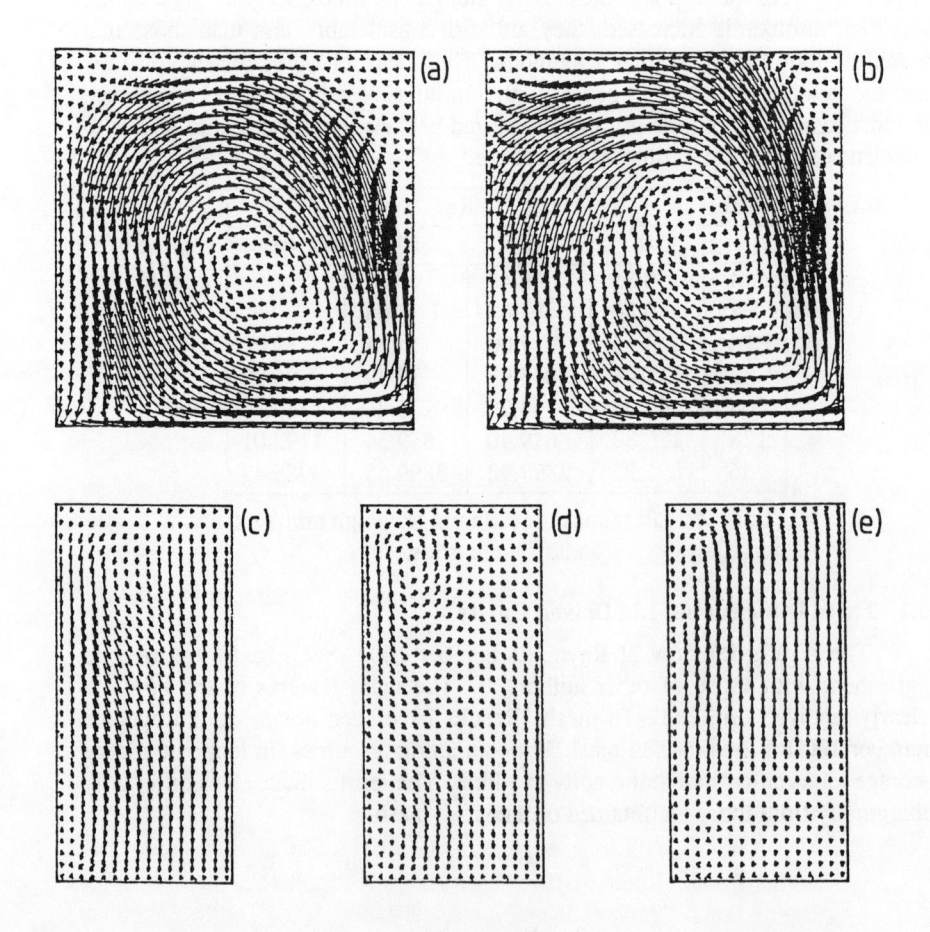

Fig. 6: Velocity vector plots for Re=1000; a) x=0.125, b) x=0.5, c) z=0.375, d) z=0.5, e) z=0.75.

The computing times are very low in view of the number of points used and compare well with those of Vanka[7] even though a higher order discretisation scheme has been employed. Also, optimal multigrid behaviour has been achieved.

6. CONCLUDING REMARKS

A multigrid, unsegregated solution technique has been developed for thermally driven and three-dimensional lid driven flows. This technique gives very accurate solutions with the aid of a higher order discretisation scheme and large numbers of points, in acceptable cpu times. These results are promising in relation to extending the technique to more complex geometries, with inclusion of effects such as turbulence. Work is presently underway on the use of quadratic interpolation and on the development of a more robust solution technique. The latter should be more suitable for use with higher order discretisation schemes and may incorporate full coupling between u,v,p and T. A more robust technique of this sort would reduce the convergence rate dependence on the relaxation factors employed.

7. REFERENCES

(1) GASKELL, P.H. and WRIGHT, N.G., "Multigrids Applied to an Efficient Fully Coupled Solution Technique for Recirculating Fluid Flow Problems", To appear in: The Proceedings of the IMA Conference on Simulation and Optimisation of Large Systems,(23-25 September,1986).

(2) LEONARD, B.P. "The QUICK Algorithm: A Uniformly Third Order Method for Highly Convective Flows", in Computer Methods in Fluids,ed. Morgan, Taylor & Brebbia, Pentech Press(1980).

(3) PATANKAR, S.P., Numerical Heat Transfer, McGraw-Hill, New York,(1980).

(4) FALLE, S.A.E.G. and WILSON, M.J., Private Communication, 1986

(5) BRANDT, A. and DINAR, N., "Multigrid Solution to Elliptic Flow Problems", pp. Academic Press in Numerical Methods in Partial Differential Equations,ed. S.V. Patankar,(1977).

(6) FUCHS, L. and ZHAO, H.-S., "Solution of Three Dimensional Viscous Incompressible Flow by a Multigrid Method", International Journal for Numerical Methods in Fluids 4 pp.539-555 (1984).

(7) VANKA,S.P., "Block-Implicit Multigrid Calculation of Three-Dimensional Recirculating Flows", in Proceedings of the Fourth International Conference on Numerical Methods for Thermal Problems. Pentech Press, Swansea(1985).

(8) DAVIS, G. de VAHL and JONES, I.P., "Natural Convection in a Square Cavity: A Comparision Exercise", International Journal for Numerical Methods in Fluids 3 pp. 227-248(1983).

The Application of Shape Preserving Splines
for the Numerical Solution of Differential
Equations.

M. Shoucri, P. Bertrand*, M. Feix**
IREQ, Groupe Tokamak de Varennes,
Varennes JOL 2P0, P.Q., Canada

* Physique Théorique, Université de Nancy I (France)
** PMMS/CNRS, Université d'Orléans, France

An important equation used as a test for many fluid and magnetofluid algorithms is the Burgers' equation:

$$\frac{\partial \upsilon}{\partial t} + \upsilon\frac{\partial \upsilon}{\partial x} = \nu\frac{\partial^2 \upsilon}{\partial x^2} \tag{1}$$

This equation contains a convection and a diffusion term and offers a simplified form of many much more complex and nonlinear equations such as the Navier-Stokes equations. An important problem in the numerical solution of Eq. (1) is the appearance of wiggles when $\nu \to 0$, especially at discontinuities, which are known as the Gibbs phenomenon. The addition of extra dissipation is usually the solution to keep these wiggles or oscillations under control, even in some of the most recently developed and advanced finite element algorithms used to solve Eq. (1), such as the characteristic finite element method[1], or the Ephemeral Particle-In-Cell method (EPIC)[2]. The purpose of the present note is to apply a simple Eulerian algorithm using a fractional step method which solves the convection term by interpolation along the characteristic with a second degree shape preserving splines. The scheme is stable, and the solution of Eq. (1) is free from wiggles and oscillations and compares favourably with what has been recently presented in the literature [1-2].

Shape preserving splines based on Bernstein polynomials have been successfully applied for the numerical solution of partial differential equations describing the propagation of shocks and discontinuities [3-4-5]. In the present algorithm, the second degree polynomial previously described is used to solve the left hand side of Eq. (1) by interpolating the value of υ along the characteristic as described in the following algorithm.

The numerical algorithm developed uses a splitting scheme (i.e. a fractional time-step method). To advance the system from t to $t + \Delta t$, we consider the following sequence:

a) We integrate the inviscid Burgers equation

$$\frac{\partial \upsilon}{\partial t} + \upsilon \frac{\partial \upsilon}{\partial x} = 0 \qquad (2)$$

from t to $t + \frac{\Delta t}{2}$, performing the shift

$$\upsilon\left(x, t + \frac{\Delta t}{2}\right) = \upsilon\left(x - \upsilon\frac{\Delta t}{2}, t\right) \qquad (3)$$

Equation (3) is solved numerically using shape-preserving splines to interpolate the value of υ at $x - \upsilon \Delta t/2$.

b) Then we integrate

$$\frac{\partial \upsilon}{\partial t} = \frac{\nu \partial^2 \upsilon}{\partial x^2} \qquad (4)$$

for a time Δt. Two methods have been used for this numerical integration:

i) In the first method Eq. (4) is integrated in the Fourier space (with the corresponding operators $\frac{\partial}{\partial x} \leftrightarrow -ik$) giving:

$$\upsilon_k(t + \Delta t) = \upsilon_k(t) \exp\left(-\int_t^{t+\Delta t} \nu k^2 \upsilon_k \, dt\right) \qquad (5)$$

where υ_k is the Fourier transform of υ. We then use inverse Fourier transform to calculate $\upsilon(t + \Delta t)$.

ii) In the second method, Eq. (4) is discretized using a time-centered space-centered Crank-Nicolson scheme:

$$\frac{\upsilon_j^{n+1} - \upsilon_j^n}{\Delta t} = \frac{\nu}{2}\left(\frac{\upsilon_{j+1}^{n+1} - 2\upsilon_j^{n+1} + \upsilon_{j-1}^{n+1}}{\Delta x^2} + \frac{\upsilon_{j+1}^n - 2\upsilon_j^n + \upsilon_{j-1}^n}{\Delta x^2}\right) \qquad (6)$$

where the superscript n denotes the time $t = n\Delta t$ and the subscript j denotes the grid point. The scheme in Eq. (6) gives a simple tridiagonal matrix which can be inverted to calculate υ_j^{n+1}

c) Finally, we repeat step a) from $t + \frac{\Delta t}{2}$ to $t + \Delta t$. So the resulting scheme is second order in Δt. Figure (1a, b) shows

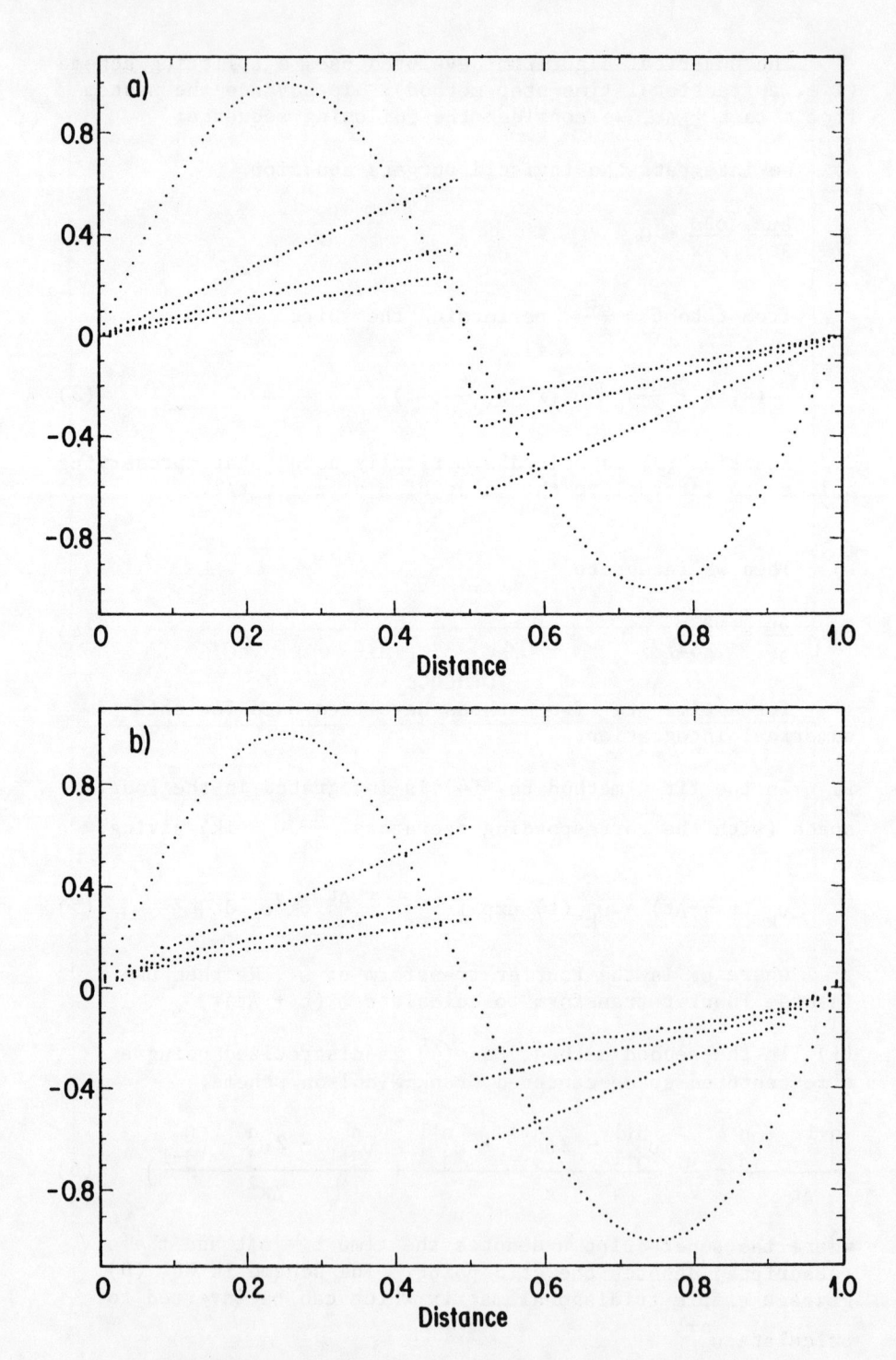

Figure 1 Sequences showing the evolution of Burgers' equation with method b) i). The curves are drawn at intervals of 40Δt, with Δt = 0.015. a) $\nu = 10^{-3}$; b) $\nu = 10^{-5}$.

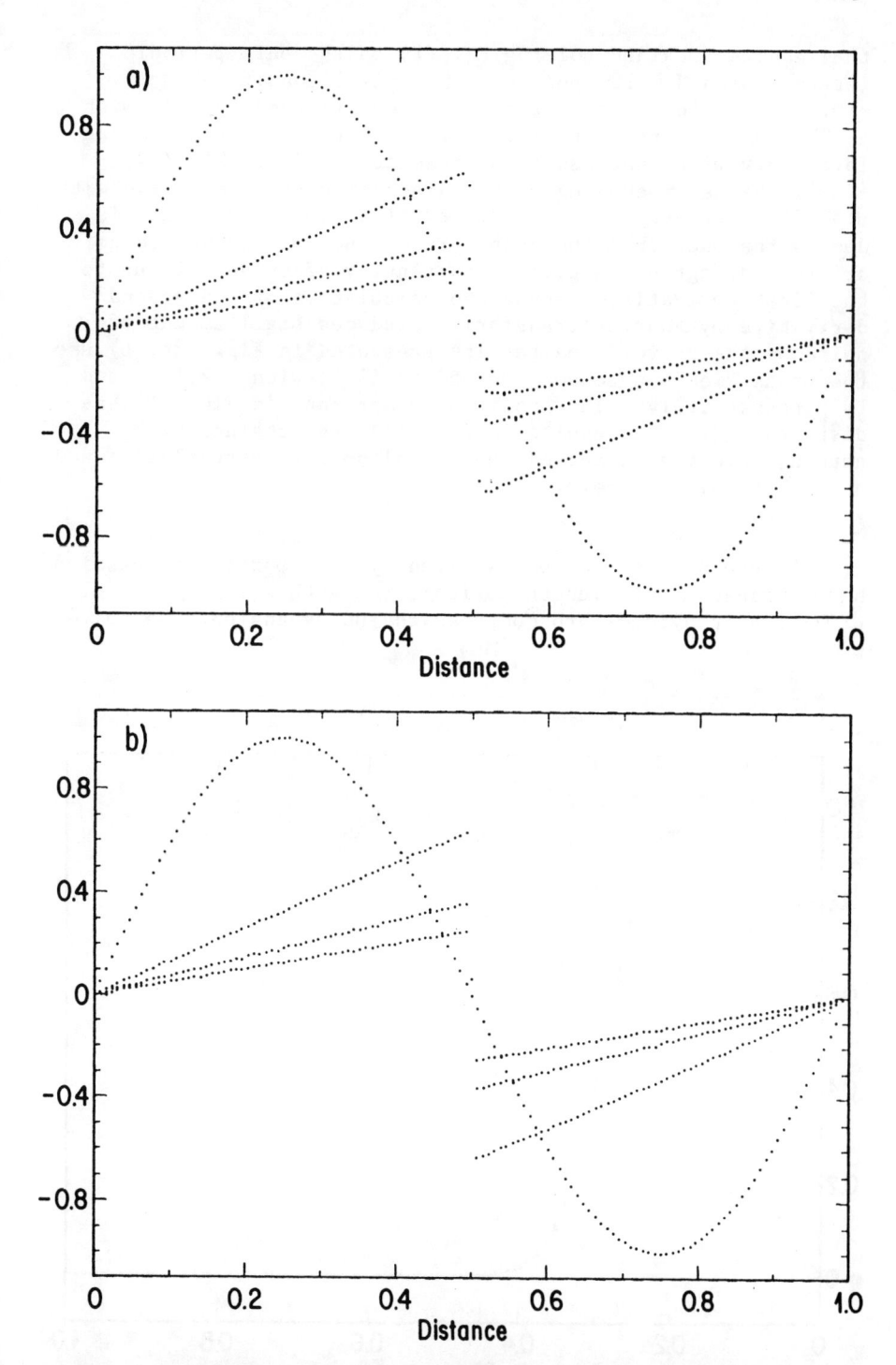

Figure 2 Sequences showing the evolution of Burgers' equation with method b) ii). The curves are drawn at intervals of $40\Delta t$, with $\Delta t = 0.015$. a) $\nu = 10^{-3}$; b) $\nu = 10^{-5}$.

the results obtained solving Eq. (1) on the unit periodic
interval with N = 100 points with initial condition $\upsilon =$
$-\sin 2\pi x$, using fast Fourier transform (method b) i)), with ν
$= 10^{-3}$ and 10^{-5} respectively. Although these curves compare
favourably with what has been presented in Ref. (2), (especially the well-behaving of the discontinuity), the curve with
$\nu = 10^{-5}$ show very small irregularities appearing. This is
due to the fact that the shape preserving splines we are using
are second degree and gives a continuity of curve only up to
the first derivative. Hence the calculation of the second
derivative by Fourier transform introduces small amount of
noise on the curve. The results presented in Fig. (2a, b) use
the Crank-Nicolson scheme (method b) ii)), with $\nu = 10^{-3}$ and
10^{-5} respectively. The small noise apparent in Fig. (1) has
disappeared. The results in Fig. (2) are obtained with
nothing more than a second degree polynomial interpolation and
a Crank-Nicolson scheme.

A second test problem is given by transporting a Gaussian
hump through a unit length periodic box with a velocity field
$\upsilon(x)$ that provides both compression and expansion. We solve

$$\frac{\partial f}{\partial t} + \upsilon \frac{\partial f}{\partial x} = 0 \qquad (7)$$

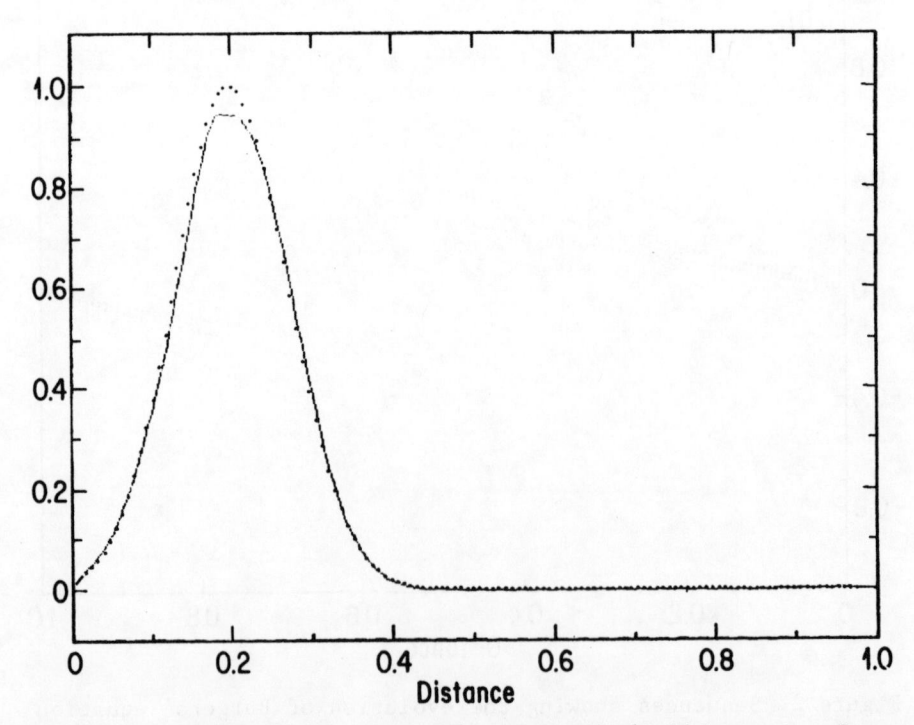

Figure 3 Initial (dashed curve) and final (solid curve, after
3 periods) profiles of a Gaussian convected by a
sinusoïdally varying velocity field.

with $v = 2 - \sin 2\pi x$, and a Gaussian density hump placed initially with its peak at $x = -0.2$ $\left(f(t = 0) = \exp(-(x + 0.2)^2/\tau^2),\ \tau = 0.1\right)$. $N = 100$ points are used. Fig. (3) show the results obtained by advancing Eq. (7) along the characteristic using the shape preserving splines previously described. The broken curve is the initial condition and the full curve is the result after transport through a distance of 3 periods, with $\Delta t = 0.004$ (around 434 time-steps). The initial area of the hump is 0.1769 and after transport through a distance of 3 periods the area was 0.1754.

Hence using a simple second degree polynomial interpolation and a Crank-Nicolson scheme, results obtained for the Burgers' equation and the transport equation compares favourably to what has been recently reported for similar problems[2].

REFERENCES

1. O. Pironneau, Numer. Math. 38, 309 (1982)

2. J. W. Eastwood, Computer Phys. Commun. 43, 89 (1986)

3. M. M. Shoucri, J. Comput. Physics 49, 334 (1983)

4. M. M. Shoucri and R. M. Shoucri, Int. J. Numerical Methods Engin. 20, 689 (1984)

5. M. Shoucri, P. Bertrand, M. R. Feix and B. Izrar, Computer Phys. Commun., to be published.

SAMPLED DATA MODEL FOR A TRANSIENT SOLUTION TO A NONLINEAR
HEAT TRANSFER SYSTEM

J. W. Graham and S. N. Sarwal
Technical University of Nova Scotia
Halifax, Nova Scotia Canada   B3J 2X4

SUMMARY

The problem of transient heat conduction in a solid subject to nonlinear boundary conditions is considered. An example would be when the heat source or the surface heat flux is a nonlinear function of the temperature as in the radiation boundary condition. A previous method of solution to this type of problem involved solving two simultaneous, singular nonlinear Volterra integral equations while another approach considered an iterative solution to the nonlinear partial differential equation.

The technique presented here applies generalized eigenfunction expansion which reduces the original partial differential equation into a set of ordinary, linear simultaneous equations having nonlinear input functions. The nonlinear input functions represent the nonlinear boundary conditions present on the original system. Each homogeneous equation (which is linear) represents the dynamics associated with each eigenvalue.

To develop the discrete-time state equations, the concept of a sampled-data system was employed. Using this method, the nonlinear boundary input function is sampled and held at discrete sampling times. The resulting difference equations for the dynamic system are then obtained. This set of equations is in normal or canonical form where each term in the system matrix represents the dynamics of each eigenvalue.

Having obtained the dynamic solution, the temperature at any location in the solid is given by a linear combination of these dynamic solutions and the corresponding eigenfunction for that location.

To illustrate the process, a one-dimensional heat transfer system was employed subject to the thermal radiation boundary condition at one of its boundaries. The results obtained were compared with previously published literature with excellent results.

INTRODUCTION AND MATHEMATICAL FORMULATION

This investigation is concerned with the transient temperature distribution in a solid, initially at any given prescribed temperature, and then suddenly subjected to thermal radiation and forced convection heat transfer at the surface. The following assumptions are made in the analysis:
1. The solid is isotropic, homogeneous and opaque to thermal radiation;
2. The diffusivity of the system is assumed to be constant;
3. The radiation exchange factor, F, the convection heat-transfer coefficient, h, the environment temperature, T_e, and the ambient fluid temperature, T_f, may be known functions of temperature and hence time;
4. The fluid is transparent to thermal radiation;
5. The number of boundary conditions must equal 2n where n is the spatial dimension of the problem.

Given a general linear parabolic diffential equation

$$\frac{\partial u}{\partial t}(\bar{x},t) = L[u(\bar{x}t)] \tag{1}$$

where L is any linear spatial operator in a domain D with the boundary conditions of the general type on the given boundary Γ of D.

$$M[u(x,t)] = B(\bar{x}/_{x=\Gamma}, t) = B(\Gamma,t) \tag{2}$$

where M is any general linear or nonlinear operator with at least one linear term in it. $B(\Gamma,t)$ represents the boundary input temperatures which can be either linear or nonlinear functions of time. For example, for the random boundary conditions given by

$$a_1 \frac{\partial u}{\partial x}(\bar{x},t)\big|_{x=\Gamma} + a_2 u(\bar{x},t)\big|_{x=\Gamma} + a_3 u^4(\bar{x},t)\big|_{x=\Gamma}$$
$$= a_4 T_e^{4}(t) + a_5 T_f(t) \tag{3}$$

In this example

$$M[u(\bar{x},t)] = a_1 \frac{\partial u}{\partial x}(\bar{x},t) + a_2 u(\bar{x},t) + a_3 u^4(\bar{x},t) \tag{4}$$

and $B(\Gamma,t) = a_4 T_e^{4}(t) + a_5 T_f(t) \tag{5}$

$a_1 \frac{\partial u}{\partial x}(\bar{x},t) + a_2 u(\bar{x},t)$ are the linear forms and $a_3 u^4(\bar{x},t)$ is the nonlinear term and $B(\Gamma,t)$ consists of given external boundary temperatures which can be time varying.

EIGENFUNCTION EXPANSION THEORY

The system is solved by generalized eigenfunction expansion for domain D which has boundaries coinciding with components of $\bar{x}$ = Constant. Since eigenfunction expansion theorem needs homogeneous linear boundary conditions, the eigenfunctions are found corresponding to boundary conditions

$$a_1 \frac{\partial u}{\partial x}(x,t) + a_2 u(x,t) = 0 \qquad (6)$$

If $\bar{x} = (x,y,z)$ and by separation of variable techniques the boundary value can be broken into Sturm-Liouville type equations in each of x, y, and z alone of the form

$$\frac{d}{dx}\left[r(x)\frac{d\psi}{dx}\right] + [q(x) + \lambda p(x)]\psi = 0 \quad a < x < b \qquad (7)$$

with boundary conditions

$$\begin{aligned} a_1 \psi(a) - a_2 \psi'(a) &= 0 \\ b_1 \psi(b) - b_2 \psi'(b) &= 0 \end{aligned} \qquad (8)$$

It has infinity of solutions $\psi_1(x), \psi_2(x) \ldots \psi_n(x) \ldots$ for values of $\lambda = \lambda_1, \lambda_2 \ldots \lambda_n$, respectively. $\psi_n(x)$ are called the eigenfunctions and λ_n are called the eigenvalues of the problem. Any arbitrary function f(x) can be expanded in terms of eigenfunctions $\psi_n(x)$ over the interval (a,b) i.e.

$$f(x) = \sum_{n=1}^{\infty} C_n \psi_n(x) \qquad (9)$$

where

$$C_n = \frac{\int_a^b p(x)\psi_n(x)f(x)dx}{\int_a^b p(x)\psi_n^2(x)dx} \qquad (10)$$

i.e. $\quad f(x) = \sum_{n=1}^{\infty} \psi_n(x) \left\{ \int_a^b p(x)\psi_n(x)f(x)dx / \int_a^b p(x)\psi_n^2(x)dx \right\}$ (11)

If we define $u_e(n,t) = \int_D u(x,t)\psi_n(x)dx$ (12)

then

$$u(x,t) = 2^\ell \sum_{m=1}^{\infty} u_e(\bar{m},t)\psi_{\bar{m}}(x) \qquad (13)$$

where ℓ is equal to the dimension of the vector $\bar{x}$. Applying this to the boundary value problem

$$\frac{\partial u}{\partial t}(\bar{x},t) = L[u(\bar{x},t)] \qquad (14)$$

the problem reduces to

$$\frac{d u_e(\bar{m},t)}{dt} + C(\bar{m}) u_e(\bar{m},t) = \sum_{\bar{m}=1}^{\infty} \chi(\bar{m}) \chi_\ell(t) \qquad (15)$$

where $C(\bar{m})$ are coefficients and $\chi_\ell(t)$ depends upon the non-homogeneous boundary conditions. $\bar{m}$ is a vector m,n,p where the dimension of each m,n,p is the set of non-negative integers.

Without any loss of generality, it is assumed that the non-homogeneous, nonlinear boundary condition exists on one boundary. Since the governing equation is linear, repeated application of the procedure to be given may be readily applied for other non-homogeneous boundary inputs. Equation (15) may now be written as

$$\frac{d u_e}{dt}(\bar{m},t) + C(\bar{m}) u_e(\bar{m},t) = d(\bar{m}) f(t) \qquad (16)$$

where $f(t)$ is a function of the non-homogeneous boundary condition existing.

SAMPLED-DATA MODEL

The discrete-time state equations are developed by employing the concept of a sampled-data system. The number of these discrete-time state equations required to adequately describe the system is readily determined by inspection of the eigenvalues of the physical system under consideration. Thus, the number is given by the product M_1, N_1 and P_1, denoted by E, where M_1, N_1 and P_1 are the upper limits of m, n and p respectively.

Using this method, the boundary input function $f(t)$, is assumed to be sampled at discrete times and then held at this constant value for the duration of the sampling period T. The original continuous signal $f(t)$ is now represented by the pricewise continuous function $f'(t)$. If the sampling period T is sufficiently small, then $f'(t)$ is a "good" approximation to $f(t)$.

Fig. 1 illustrates this process and the resulting signals

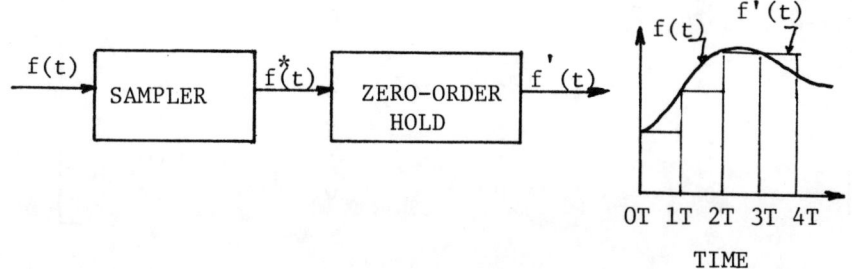

Fig. 1, SAMPLE & HOLD PROCESS

In the discrete-time model to the developed $f'(t)$ is the input function employed. Hence equation (16) is now written as

$$\frac{d\tilde{\phi}}{dt}(\bar{m},t) + C(\bar{m}) \phi(\bar{m},t) = d(\bar{m}) f'(t) \tag{17}$$

In vector notation the solution may be written as

$$u_e(\bar{m},t) = \phi(\bar{m},t-t_0) u_e(\bar{m},t_0) + \int_{t_0}^{t} \phi(\bar{m},t-\tau) d(\bar{m}) f'(\tau) d\tau \tag{18}$$

where

$$\underset{\sim}{u_e}(\bar{m},t) = \begin{bmatrix} u_e(\bar{m}_1 t) \\ u_e(m_2 t) \\ \cdot \\ \cdot \\ \cdot \\ u_e(E,t) \end{bmatrix} \qquad \underset{\sim}{d} = \begin{bmatrix} d(\bar{m}_1) \\ d(m_2) \\ \cdot \\ \cdot \\ \cdot \\ d(E) \end{bmatrix} \tag{19}$$

$$\phi(\bar{m},t-t_0) = \begin{bmatrix} e^{-C(\bar{m}_1)t} & 0 & 0 & \cdots & 0 \\ \cdot & e^{-C(\bar{m}_2)t} & 0 & \cdots & 0 \\ \cdot & \cdot & \cdot & & \vdots \\ \cdot & \cdot & \cdot & & -C(E)t \\ 0 & 0 & 0 & \cdots & e^{-C(E)t} \end{bmatrix} \tag{20}$$

It is noted that for the class of distributed parameter systems considered the state transition matrix is always in the normal or diagonal form.

Recalling that the function $f'(t)$ is a sampled and held function, it follows
$$f'(t) = f(KT) \text{ for } KT \leq t < (K+1) T \; ; \quad K = 0, 1, 2 \ldots \tag{21}$$
In systems employing A/D conversion a zero-order hold is usually incorporated in the device so the value of the signal is held constant until the next sample occurs. With this assumption the discrete-time state equations may be written as

$$u_e(K+1)T = \phi(\bar{m},T) u_e(\bar{m},KT) + D(\bar{m},T) f(KT) \tag{22}$$

where

$$D(T) = \int_0^T \phi(\bar{m},\tau) d(\bar{m}) d\tau \tag{23}$$

$$= \begin{bmatrix} [1-e^{-C(\bar{m}_1)T}]/C(\bar{m}_1) & 0 & 0 & 0 \\ & [1-e^{-C(\bar{m}_2)T}]/C(\bar{m}_2) & 0 & 0 \\ \vdots & \vdots & \vdots & \\ 0 & 0 & 0 & [1-E^{-C(E)T}]/C(E) \end{bmatrix} d(\bar{m}) \tag{24}$$

Using equation (13) $u(\bar{x},t)$ is given by

$$u(\bar{x}, KT) = 2^{\ell} \sum_{m=1}^{\infty} u_e(m, KT) \, \chi_{\underline{m}}(\bar{x}) \tag{25}$$

It should be noted that the functions $\phi(\bar{m},T)$ and $D(\bar{m},T)$ are exact as the exponential solution is in closed form. Hence this model is more accurate than those using forward or backward difference methods and Tustin's Method, as these are only polynomial approximations to the infinite series. The only approximations employed here are related to the sampling of the input function $f(t)$ and the number of terms kept in the infinite series solution. If the sampling period is small enough then the error resulting from this is negligible. Also it can be readily ascertained as to how many terms to keep in the infinite series by an inspection of the negative exponential terms. An example will now be given to illustrate the accuracy of this method.

EXAMPLE

The system as considered in [1] will be studied. The governing equation is given by

$$\frac{\partial T(x_1,t_1)}{\partial t_1} = \alpha \frac{\partial^2 T(x_1,t_1)}{\partial x_1^2} \; ; \quad t > 0, \; 0 < x_1 < L \tag{26}$$

The initial condition is given by $T(x_1,0) = T_i$ (27)

with the boundary conditions $\frac{\partial T}{\partial x_1}(0,t) = 0$ (28)

$$-k \frac{\partial T}{\partial x_1}(L,t_1) = F[T(L,t)^4 - T_e^4] + h[T(L,t) - T_f] \tag{29}$$

Using the dimensionless variables given by

$$t = \frac{\alpha t_1}{L^2}, \quad x = \frac{x_1}{L}, \tag{30}$$

the above can be written in terms of $T(x,t)$. The dimensionless temperature $u(x,t)$ is given by $T(x,t)/T_a$ where T_a is defined by

$$F \sigma T_a^4 + h T_a = F \sigma T_e^4 + h T_f \tag{31}$$

The problem may now be stated in terms of these variables by

$$\frac{\partial u(x,t)}{\partial t} = \frac{\partial^2 u(x,t)}{\partial x^2} \quad t > 0, \; 0 < x < 1 \tag{32}$$

with $u(x,0) = u_i$ $\quad 0 < x < 1$ (33)

and $\frac{\partial u}{\partial x}(0,t) = 0$ $\quad t > 0$ (34)

$$\frac{\partial u}{\partial x}(1,t) = -f(t) \tag{35}$$

where
$$f(t) = N_{Rh}(u(1,t)^4 - 1) + N_{Bi}(u(1,t) - 1) \tag{36}$$
where $N_{Rh} = F \sigma T_a^3 h/k$ and $N_{Bi} = hL/k$ (BIOT NUMBER) (37)

For this problem only the surface temperature at $x = 1$ need be sampled since both the environment and fluid temperatures and hence T_a are constant. However, if these temperatures are time varying then they would be sampled along with $u(1,t)$ and the function $f(t)$ would then be computed according to equation (29).

The eigenfunctions are given by
$$\chi_m(x) = \cos(m \pi x) \qquad m = 0, 1, 2 \ldots \tag{39}$$
Defining the function
$$u_e(m,t) = \int_0^1 u(x,t) \cos(m \pi x) \, dx \tag{40}$$
it follows that
$$\frac{du_e}{dt}(m,t) + m^2 \pi^2 u_e(m,t) = (-1)^{m+1} f(t) \tag{41}$$
Since we have a uniform (steady-state) initial condition, it follows that
$$u_e(m,0) = \begin{cases} u_i & m = 0 \\ 0 & m \neq 0 \end{cases} \tag{42}$$
for a non-uniform initial condition, say $f(x)$,
$$u_e(m,0) = \begin{cases} \int_0^1 f(x) \, dx; & m = 0 \\ \int_0^1 f(x) \cos m\pi x \, dx; & m = 1, 2, \ldots \end{cases} \tag{43}$$
The solution is given by ($t_0 = 0$)
$$u_e(m,t) = \Phi(m,t) u_e(m,0) + \int_0^t \Phi(m,t-\tau) d(m) f(\tau) \, d\tau \tag{44}$$
where
$$\Phi(m,t) = \begin{bmatrix} 0 & 0 & 0 & \cdots & 0 \\ 0 & e^{-\pi^2 t} & 0 & \cdots & 0 \\ \vdots & \vdots & \vdots & & \\ 0 & 0 & 0 & \cdots & e^{-m^2 \pi^2 t} \end{bmatrix} \tag{45}$$
In discrete form we have
$$u_e(m, (K+1)T) = \phi(m,T) u_e(m,KT) + D(m,T) d(m) f(KT) \tag{46}$$
where $\phi(m,T) = \phi(m,t)_{t=T}$ (47)
$$D(m,T) = \int_0^T \Phi(m,\tau) \, d\tau$$

$$D(m,T) = \begin{bmatrix} T & 0 & \cdots & 0 \\ 0 & \dfrac{1-e^{-\pi^2 T}}{\pi^2} & 0 & \cdots & 0 \\ \vdots & \vdots & \vdots & & \\ 0 & 0 & 0 & \cdots & \dfrac{1-e^{-m^2\pi^2 T}}{m^2\pi^2} \end{bmatrix} \quad (48)$$

The equation for the temperature $u(x,KT)$ at any location x and at a discrete time $t = KT$ is then given by

$$u(x,KT) = u_e(0,KT) + 2 \sum_{m=1}^{E} u_e(m,KT) \cos(m\pi x) \quad (49)$$

It is seen that the equation for the nth state variable is given by

$$u_e(n,(K+1)T) = e^{-n^2\pi^2 T} u_e(n,KT) + (-1)^{n+1} \dfrac{1-e^{-n^2\pi^2 T}}{n^2\pi^2} f(KT) \quad (50)$$

where $f(KT) = N_{Rh}(u(1,KT)^4 - 1) + N_{Bi}(u(1,KT) - 1)$ (51)

Due to the rapid exponential decay given by

$e^{-n^2\pi^2 T} \approx e^{-10n^2 T}$ it is not difficult to determine the number of terms $m = M$ to retain in the infinite series. For this example, 7 terms ($T = 0.0010$) were kept for comparison purposes. The equations for 7 state variables are given below

$u_e(0, (K+1)T) = 1.000000\ u_e(0,KT) - 0.001\ f(KT)$

$u_e(1, (K+1)T) = 0.990179\ u_e(1,KT) + 0.000995\ f(KT)$

$u_e(2, (K+1)T) = 0.961291\ u_e(2,KT) - 0.000981\ f(KT)$

$u_e(3, (K+1)T) = 0.915004\ u_e(3,KT) + 0.000957\ f(KT)$ (52)

$u_e(4, (K+1)T) = 0.853923\ u_e(4,KT) - 0.000925\ f(KT)$

$u_e(5, (K+1)T) = 0.781344\ u_e(5,KT) + 0.000886\ f(KT)$

$u_e(6, (K+1)T) = 0.700959\ u_e(6,KT) - 0.000615\ f(KT)$

where $f(KT) = 0.4\,(u(1,KT)^4 - 1) + 0.5\,(u(1,KT) - 1)$ and the surface temperature at $x = 1$ is given by

$$u(1,KT) = u_e(0,KT) + 2(u_e(1,KT) + u_e(2,KT) + u_e(3,KT) + u_e(4,KT) + u_e(5,KT) + u_e(6,KT)) \quad (53)$$

The results are presented in Table I along with those obtained by Crosbie & Viskanta [1].

TABLE I

Comparison of Solution with CROSBIE & VISKANTA [1]

Solution for u(1,t)

Time t	Crosbie & Viskanta	Sampled–Data Model
0.00	0.2000	0.2000
0.25	0.5563	0.5440
0.50	0.6670	0.6596
0.75	0.7488	0.7449
1.00	0.8115	0.8101
1.25	0.8592	0.8594
1.50	0.8951	0.8962
1.75	0.9219	0.9235
2.00	0.9419	0.9437

The results are all in excellent agreement. Due to the diagonal form of the state equations it is relatively straight-forward to program for any finite number of terms, M and to determine the effect of the sampling time T.

CONCLUSIONS

Using a generalized expansion function and the concept of a sampled-data system the solution to a class of distributed parameter systems subject to linear or nonlinear boundary conditions can be readily solved. The solution to the original partial differential equations yields a set of uncoupled ordinary differential equations. Using the state transition method (STM) the basic solution matrix $\phi(m,t)$ and the convolution matrix $D(m,T)$ were both in diagonal form, hence lending to a straight-forward computer solution.

REFERENCES

1. A. L. Crosbie and R. Viskanta: Transient heating or cooling of one dimensional solids by thermal radiation. Proceedings of the Third International Heat Transfer Conference. Vol. V, pp 146-153 A.I. ChE New York (1966)

2. R. Courant and D. Hilbert: Method of Mathematical Physics Vol. I, Interscience Publishers, New York (1953)

RECENT ADVANCES IN THE DEVELOPMENT OF TRANSFINITE ELEMENT FORMULATIONS FOR NONLINEAR/LINEAR TRANSIENT THERMAL PROBLEMS

Kumar K. Tamma* and Sudhir B. Railkar**
West Virginia University
Morgantown, West Virginia, 26506, USA

SUMMARY

Progress and recent advances in the development of a new transfinite element computational methodology is described with emphasis on applicability to nonlinear/linear transient thermal problems. The proposed methodology and concepts are new and unique and demonstrate the applicability to general transient nonlinear/linear thermal analysis by combining classical Galerkin schemes and transform approaches with contemporary finite element methods to preserve the modeling versatility — thereby, a hybrid computational methodology is proposed. Numerical test cases are presented to demonstrate the applicability of the proposed formulations. The proposed transfinite element methodology and concepts offer significant potential for extension to several areas of mathematical physics and engineering.

1. INTRODUCTION

Ever since the finite element method was first proposed for field problems [1], the applicability of the method to thermal problems has attracted considerable attention. To date, considerable progress has been made in the application of the finite element method to both linear and nonlinear thermal analysis situations involving complex geometries and boundary conditions.

---

*Associate Professor and Director-ICAD, Department of Mechanical and Aerospace Engineering.

**Graduate Research Assistant, Department of Mechanical and Aerospace Engineering.

In this paper, we present new and recent advances in the development and applicability of a novel 'transfinite element' computational methodology [2-3] for general nonlinear/linear thermal problems. The methodology and concepts are new and unique and demonstrate the applicability to general transient nonlinear/linear thermal analysis as it combines the classical Galerkin schemes and transform approaches with contemporary finite element methods to preserve the modeling versatility—thereby, a hybrid computational methodology is proposed. Infact, the methodology and concepts can be extended to finite differences, boundary elements, etc. A brief description of a generalized thermal problem and the associated boundary conditions is first presented. A general overview of the proposed transfinite element methodology and pertinent details of the approach and computational schemes for linear/nonlinear transient thermal problems is then described, wherein, for nonlinear thermal analysis, the nonlinearities may be due to temperature dependence of thermophysical properties and/or general nonlinear boundary conditions. Nonlinear effects due to radiation heat transfer are also considered. Finally, applicability of the proposed transfinite element methodology for nonlinear/linear transient thermal problems is demonstrated via comparative numerical test cases.

2. GENERALIZED TWO-DIMENSIONAL THERMAL PROBLEM

Consider the phenomenon of transient nonlinear heat conduction in a two-dimensional region Ω (Fig. 1), governed

$$(k_x \frac{\partial T}{\partial x} \ell_x + k_y \frac{\partial T}{\partial y} \ell_y) - q_s + q_h + q_r = 0$$

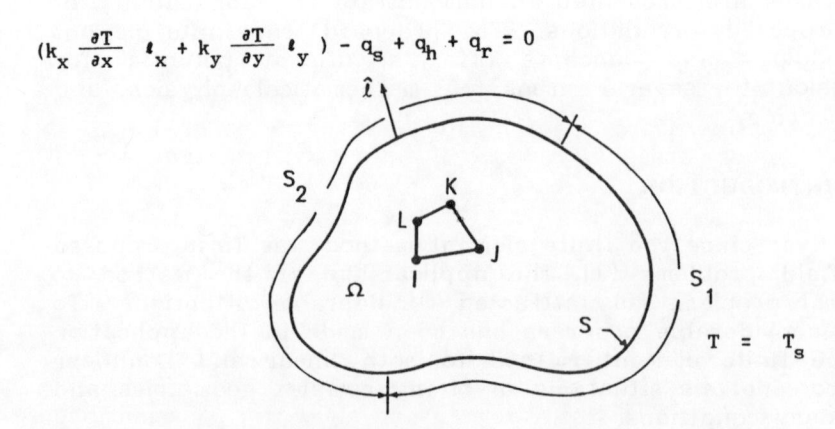

Fig. 1 General two-dimensional thermal problem and associated boundary conditions.

by the equation of the form

$$\rho c \frac{\partial T}{\partial t} + \nabla \cdot \mathbf{q} = Q \tag{1}$$

where

$$\nabla \cdot \mathbf{q} = \frac{\partial q_x}{\partial x} + \frac{\partial q_y}{\partial y} \qquad (2)$$

subject to boundary conditions:

$$T = T_s \qquad \text{on } S_1 \qquad (3)$$

$$(k_x \frac{\partial T}{\partial x} \ell_x + k_y \frac{\partial T}{\partial y} \ell_y) - q_s + q_h + q_r = 0 \quad \text{on } S_2 \qquad (4)$$

where Eqns. (1-4) refer to unsteady nonlinear thermal fields in materials with thermophysical properties dependent on temperature. In the above, ρ is the density; c is the specific heat; q is the heat flux vector; k_x and k_y are the thermal conductivities in the x and y directions respectively; and Q is the internal heat generation rate. The terms ℓ_x and ℓ_y are the direction cosines of the outward normal to the boundary surface. The terms q_s, q_h, and q_r represent the surface heating rate per unit area, the rate of heat flow per unit area due to convection, and the rate of heat flow per unit area due to radiation respectively. The terms q_h and q_r are given by the relations

$$q_h = h (T - T_h) \qquad (5)$$

and

$$q_r = \sigma \varepsilon (T^4 - T_r^4) \qquad (6)$$

where h is the heat transfer coefficient; T_h is the convective medium temperature; σ is the Stefan-Boltzmann constatn; ε is the surface emmisivity; and T_r is the radiation medium temperature.

3. PROPOSED COMPUTATIONAL METHODOLOGY AND SCHEME: TRANSFINITE ELEMENT APPROACH

General field problems (such as Eqns. 1-4) can be solved via the proposed transfinite element approach by discretizing a given region into finite domains or elements. As mentioned previously, the proposed methodology is a hybrid scheme which combines the modeling versatility of contemporary finite elements in conjunction with transform approaches and classical Galerkin schemes. The formulations within the transform domain are referred to as 'transfinite formulations' and the corresponding elements formulated in the transform domain are referred to as 'transfinite elements.' Briefly described, after discretizing the physical domain of interest, and selecting an appropriate transform and applying this to the governing heat conduction equations (1-4), the discretized equations on an element level are first obtained envoking the classical Galerkin schemes. Therein, numerical computations for each individual element yield element matrices; the total system contributions of which are obtained following the conventional procedures of finite element assembly. The actual solution is then obtained employing an inverse numerical transform [4-6]. Throughout this paper, the selected transform will be the Laplace transform method and the numerical inversion scheme of Durbin[6] is employed because of the high accuracy of the method.

For general nonlinear/linear transient thermal analysis, the dependent variable $\bar{T}$ in the transform domain for the physical region of interest Ω can be approximated throughout the solution domain by the relationship

$$\bar{T} = \sum_{i=1}^{n} N_i(x, y) \bar{T}_i = N T \tag{7}$$

where N_i are the usual interpolation functions defined piecewise on an element basis, and $\bar{T}_i$ are the values of the unknown nodal temperatures in the transform domain. In the above, the summation refers to contributions of each element Ω^e of the region Ω. The temperature gradients are expressed as

$$\bar{T} = B \bar{T} \tag{8}$$

For linear transient thermal problems, after discretizing the physical domain of interest Ω, and selecting the Laplace transform as the appropriate transform and applying this to the general linear heat conduction equations, the discretized system thermal-equilibrium equations following the method of weighted residuals or a variational approach lead to equations which can be cast in matrix form as

$$\bar{K} \bar{T} + \bar{F} = 0 \tag{9}$$

where the system matrices $\bar{K}$ and $\bar{F}$ have contributions from each element Ω^e of the region Ω. Specific details of the matrices in Eq. (9) are described in the next section. After developing the system equations (Eqns. 9) in the transform domain, these equations are solved directly in the transform domain itself. Therein, following Tamma and Railkar [2] the linear transient response is obtained employing an inverse numerical transform.

For general nonlinear transient thermal problems, we propose an iterative predict-correct scheme to converge the thermal response to its true solution. The nonlinearity due to radiation is approximated by expanding this term via Taylor series and using the concept of average element temperature. Hence, expanding the nonlinear term due to radiation via Taylor series we have

$$T^4 \approx T_{avg}^4 + 4 T_{avg}^3 (T - T_{avg}) \qquad (10)$$

To illustrate the basic concepts for applicability to nonlinear transient thermal problems, the approach proposed by Tamma and Railkar [3] is adapted in this paper. Following the discretization process by envoking the classical Galerkin formulations, the nonlinear transient system thermal-equilibrium equations in the transform domain reduce to the form represented in matrix notation as

$$\bar{K}(\bar{T}) \bar{T} + \bar{F}(\bar{T}) = 0 \qquad (11)$$

where now the system equations (Eqns. 11) are nonlinear because of the dependence of the system matrices $\bar{K}$ and $\bar{F}$, on the dependent variable $\bar{T}$. The specific details of the matrices in Eq. (11) are described in the next section. The approach proposed for the computational process is based on the application of a generalized Newton-Raphson technique in conjunction with an iterative predict-correct scheme as follows:

The general forms for the Jacobian and unbalanced heat load vector in the transform domain are obtained via setting

$$\bar{R} = \bar{K}(\bar{T}) \bar{T} + \bar{F}(\bar{T}) \qquad (12)$$

Thus, a typical unbalanced equation can be represented as

$$\bar{R}_i = \bar{K}_{ij} \bar{T}_j + \bar{F}_i \qquad \text{sum on } j \qquad (13)$$

The Jacobian matrix in the transform domain can be obtained from

$$\bar{J}_{ij} = \frac{\partial \bar{R}_i}{\partial \bar{T}_j} \qquad (14)$$

The computational algorithm proposed for obtaining the nonlinear transient thermal response is thus represented as

$$\bar{J}^m \Delta \bar{T}^{m+1} = \bar{R}^m \qquad (15)$$

and

$$T^{m+1} = T^m + \Delta T^{m+1} \qquad (16)$$

After obtaining the system equations for the n values of $\Delta \bar{T}$ in the transform domain, these equations are solved in the transform domain itself. Therein, by employing an inverse numerical transform in conjunction with an iterative application of a generalized Newton-Raphson scheme, the nonlinear transient thermal response is obtained.

Applications using transform methods in conjunction with numerical schemes has been in existence although much of the research has focused on linear problems in certain areas of mathematical and physical sciences and engineering [7-10]. However, research relevant to applicability to nonlinear problems is relatively scarce and/or non-existent. The transfinite element methodology advocated and proposed herein, successfully provides a viable computational methodology for general transient nonlinear/linear thermal problems via transform methods in conjunction with the classical Galerkin schemes and includes the modeling versatility and features of contemporary finite element formulations.

4. FORMULATIONS: TRANSFINITE ELEMENT APPROACH

The discretized equations are derived by applying the Laplace transform with respect to time to the governing heat conduction Eqns. (1-4) and therein envoking the classical Bubnov-Galerkin procedures. Thus, the resulting equations for the n values of $\bar{T}$ in the domain Ω are obtained typically for each node i by equating to zero the integral over the domain Ω of the product of the weighting function ($W_i = N_i$) and the residual in the transform domain.

Following the procedure mentioned above, for the general heat conduction problem described by Eqns. (1-4), this results in the following discretized equations written in matrix form as

$$\bar{K} \bar{T} + \bar{F} = 0 \qquad (17)$$

where $\bar{K}$ has contributions from conduction, a convective-like matrix matrix due to the transformed capacitance, convection, and radiation respectively. The heat load vector $\bar{F}$ has contributions from internal heat generation, initial conditions, specified nodal temperatures, surface heating, surface convection, surface radiation, etc. Typical transfinite element generalized formulations are thus obtained as

$$\bar{K} = \sum \int_{\Omega^e} [B]^T[k][B]d\Omega + \sum \int_{\Omega^e} \bar{\rho}\,\bar{c}\,[N]^T[N]d\Omega +$$

$$\sum \int_{S_2^e} (h + R_k)[N]^T[N]dS \qquad (18)$$

$$\bar{F} = -\sum \int_{\Omega^e} (\bar{Q} + \rho c T_i)\,[N]^T d\Omega -$$

$$\sum \int_{S_2^e} (\bar{q}\cdot\hat{n} + \bar{q}_s + h\bar{T}_h + R_r)\,[N]^T dS \qquad (19)$$

where the summations are taken over the contributions of each element Ω^e of the domain Ω, Ω^e refers to the element region, and S_2^e refers to those elements with external boundaries on which condition (Eq. 4) is specified. The terms R_k and R_r in Eqns. (18-19) are given by

$$R_k = 4\,\sigma\,\varepsilon\,T_{avg}^3 \qquad (20)$$

$$R_r = 3\,\sigma\,\varepsilon\,\bar{T}_{avg}^4 + \sigma\,\varepsilon\,\bar{T}_r^4 \qquad (21)$$

The set of Eqns. (17-21) which define the general discretized problem **for** thermal analysis can now be solved following the computational methodology outlined in the previous section.

5. EXAMPLES

To demonstrate the applicability of the transfinite element methodology for general nonlinear/linear transient thermal analysis, several comparative test problems are presented. Where possible, the results are compared with conventional finite element schemes and/or with those published in literature.

Transient linear thermal response in a square plate with hole – This problem analyses the transient linear response in a square plate with a hole subjected to conduction-convection-surface heating and internal heat generation. The plate is initially at zero degrees and the convective medium temperature is assumed to be 70. The problem is analyzed by both conventional finite element formulations using an implicit Crank-Nicolson algorithm and the proposed transfinite element formulations. In Fig. 2 are shown the comparative temperature distributions along the plate diagonal A-B. As shown in Fig. 2, the comparative results are in excellent agreement (within 1 percent accuracy) thereby demonstrating the applicability of the proposed formulations and methodology. In Fig. 3 is shown the temperature distributions over the entire plate obtained via the transfinite element formulations at t=4s. This test problem demonstrates the basic capabilities of the proposed formulations.

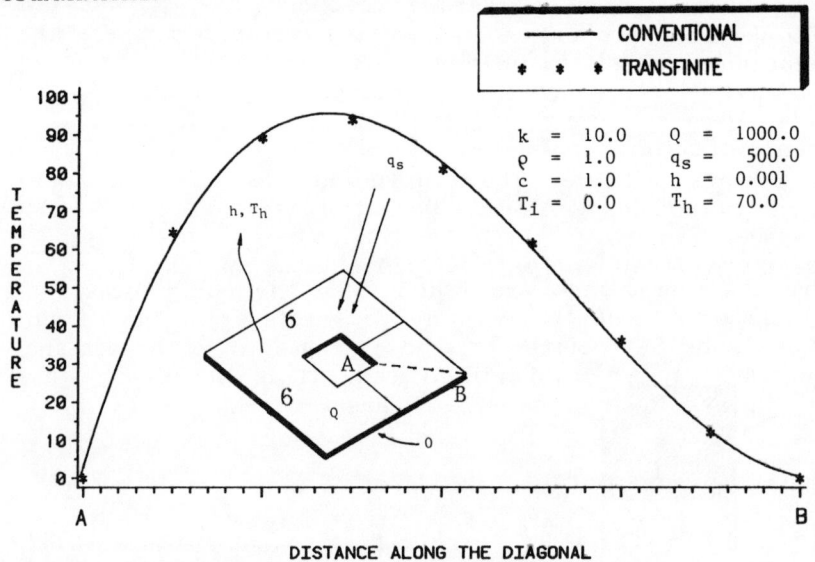

Fig. 2 Comparative temperature distributions along plate diagonal (A-B).

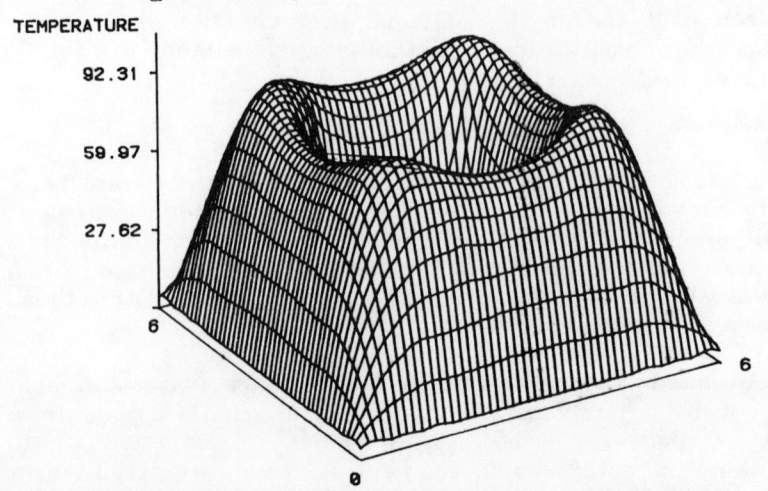

Fig. 3 Temperature distribution over plate obtained via transfinite element formulations (t=4s).

Nonlinear transient response in rectangular plate with natural and essential boundary conditions and varying thermal conductivity — This problem has been studied by Skerget and Brebbia [11] using the boundary element method. Because of symmetry considerations, only half of the plate is modeled. The thermal conductivity for the plate varies nonlinearly as shown in Fig. 4. The boundary conditions consist of prescribed heat fluxes on two boundaries and prescribed temperature on one of the plate's boundary as shown in Fig. 4. The problem was analyzed using the proposed transfinite element methodology to demonstrate the applicability of the approach for evaluating the nonlinear transient thermal response due to varying thermophysical properties. The comparative temperature distributions along the top edge A-B are shown in Fig. 4 at t=4s and t=8s respectively. As demonstrated in Fig. 4, the results agree extremely well and the accuracy is within 1-2 percent with those published by Skerget and Brebbia [11]. This example validates the applicability of the transfinite element methodology for evaluating the nonlinear thermal response where the nonlinearities are due to temperature-dependent material properties.

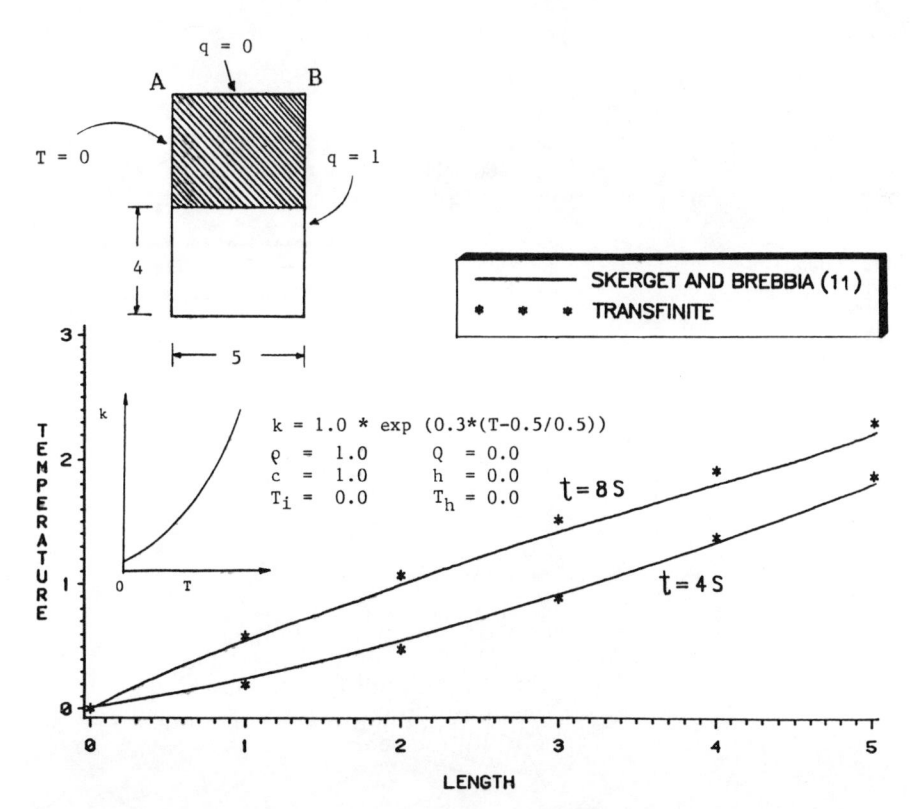

Fig. 4 Comparative nonlinear transient temperature distributions along the top edge A-B (Fig. 6) at

Conducting-Radiating structural member – This test problem analyses the nonlinear transient thermal response in a conducting-radiating structural member. The structural member has one end maintained at T=100 while the other end is maintained at T=0. The member is assumed radiating to space and the radiation medium temperature is $T_r=0$. The problem was modeled using two-noded elements and the description of the problem and numerical data assumed is shown in Fig. 5. For comparative purposes, the problem was also analyzed by conventional finite element formulations using an implicit Crank-Nicolson algorithm. The comparative nonlinear transient response of a point at the middle of the member (x = L/2) are shown in Fig. 5. The results for this problem are also in excellent agreement and within an accuracy of 1 percent, thereby validating the applicability of the proposed formulations.

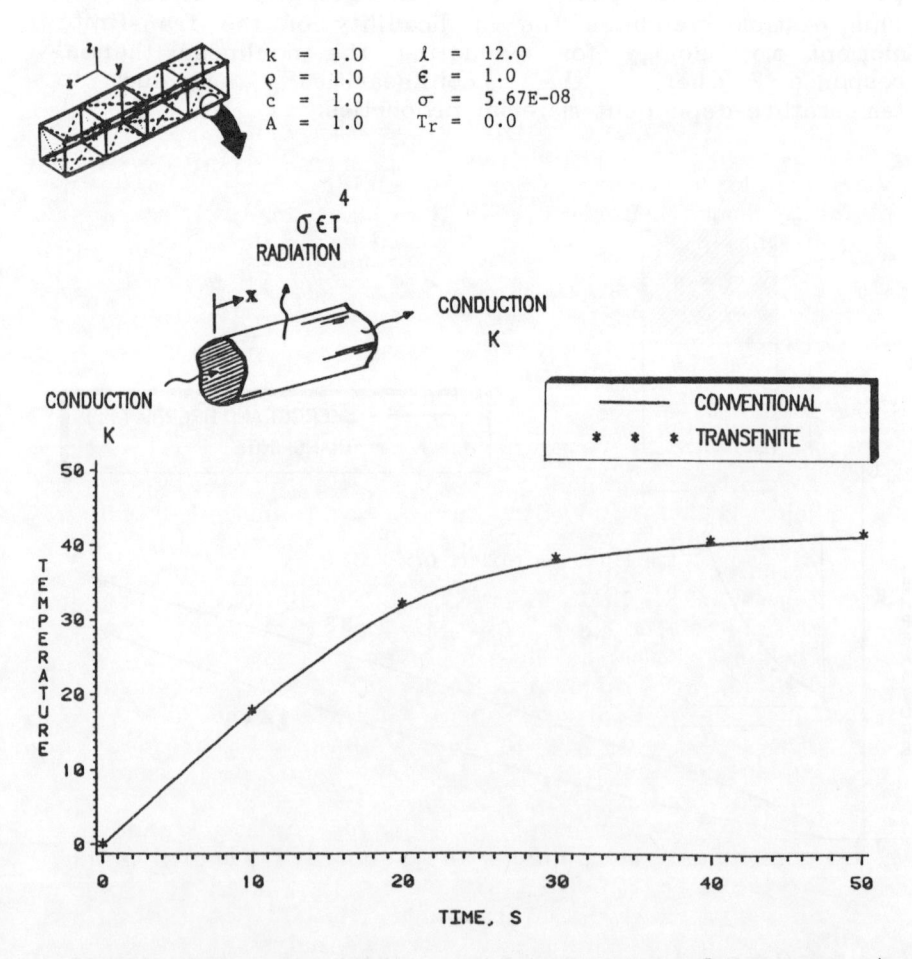

Fig. 5   Comparative nonlinear transient thermal response at x = L/2.

6. CONCLUSIONS

Recent progress and developments in the applicability of a new transfinite element computational methodology is described with emphasis on applications to nonlinear/linear transient thermal problems. The proposed methodology and concepts and computational schemes are described from basic concepts via generalized formulations for nonlinear and linear transient thermal analysis. The applicability of the methodology for general nonlinear/linear transient thermal problems is demonstrated via comparisons with several different test problems. The results obtained are in excellent agreement and validate the proposed developments. Currently research is underway for applicability of the transfinite element approach to interdisciplinary research areas of engineering and computational mechanics.

ACKNOWLEDGEMENTS

The authors greatly appreciate the continued encouragement and support of this research in part by NASA-Langley Research Center, Hampton, Virginia and the Flight Dynamics Laboratory, Wright Patterson Air Force Base, Dayton, Ohio.

REFERENCES

1. Zienkiewicz, O.C. and Cheung, Y.K. - Finite Elements in the Solution of Field Problems, The Engineer, 507-510 (1965).
2. Tamma, K.K. and Railkar, S.B. - Computational Aspects of Heat Transfer in Structures Via Transfinite Element Formulations, AIAA J. of Thermophysics and Heat Transfer (to appear).
3. Tamma, K.K. and Railkar, S.B. - Hybrid Transfinite Element Methodology for Nonlinear Transient Thermal Problems, Numerical Heat Transfer (to appear).
4. Bellman, R.E., Kalaba, R.E. and Lockett, J. - <u>Numerical Inversion of Laplace Transfer</u>, Elsevier, 1966.
5. Davis, B. and Martin, B. - Numerical Inversion of Laplace Transform: A Survey and Comparison of Methods, J. Comp. Physics, $\underline{33}$, 371-376 (1979).
6. Durbin, F. - Numerical Inversion of Laplace Transforms: An Efficient Improvement to Dubner and Abate's Method, Comp. J., $\underline{17}$, 371-376 (1979).
7. Shapery, R.A. - Approximate Method of Transform Inversion for Viscoelastic Stress Analysis, Proceedings of the Fourth U.S. National Congress of Applied Mechanics, 1961, pp. 1075-1085.

8. Beskos, D.E. and Boley, B.A. - Use of Dynamic Influence Coefficients in Forced Vibration Problems with the Aid of Laplace Transform, Computers and Structures, 5, 263-269 (1975).
9. Aral, M.M. and Gulcat, U. - A Finite Element Laplace Transform Solution Technique for the Wave Equation, Int. Journal of Num. Meth. in Eng., 11, 1719-1732 (1977).
10. Huntley, E., Pickering, W. and Zinober, S. - The Numerical Solution of Linear Time-Dependent Partial Differential Equations by the Laplace and Fast Fourier Transforms, J. Comp. Physics, 27, 256-271 (1978).
11. Skerget, P. and Brebbia, C.A. - 'Topics in Boundary Element Research', Time Dependent and Vibration Problems, Ed. Brebbia, C.A., Springer-Verlag Publ., 1985, p. 63.

TCR Method For Determinating The Thermal Transmittance Of 2-D Body

Ganquan Xie

Hunan Computer Technology Reseonch Institute; P.R.O.C.

Summary

In science and engineering, the problem for determinating the thermal transmittance of 2-D body is important. However the problem will be very difficult because it is disastrous ill posed and strong nonlinear problem. In 1-D case; the several research results and numerical methods have been obtained.

In this paper, the time convolution regularizing method (TCR) for determinating the thermal transmittance of 2-D body has been presented. In each iteration of TCR. 2-D heat equation with variable coefficient and a new time convolution regularizing integral equation will be solved.

TCR method is very effective accuracy and stable, we have got excellent computational results by this method.

1. Introduction

In scientific research, physical and chemical experiment and engineering, the problem for determinating the thermal transmittance is often metted. For example, thermal radiation pheonomenon in thermonuclear reaction, the thermal properties in atmosphere and cloud layer, the thermal transmittance control in burn of rocket; the control of heat conduction parameter in industral stove etc. The above problem can all be sum up to determinate the thermal transmittance of body. The mathematical physical mode of the above problem will be reduced to the coefficient inverse problem of multidimensional heat equation.

Then the research of mathematical and physical mode and computational method will be of important significant and width applications. However, the problem for determinating the thermal ill posed and strong nonlinear problem, so, it is very difficult in multidimensional case. In 1-D case, several results have been obtained [2] [3]. In this paper, a time convolution regularizing iteration (TCR) determinating the thermal transmittance of 2-D body has been presented. This is devolopment of idea of [1], in each iteration of TCR method, 2-D heat equation with variable coefficient and a new time convolution integral equaton will be solved. But, it is different from inverse problem of wave equation, even through under the condition with point thermal source, the above time convolution integral equation for determinating the increament of thermal transmittance will be still high ill posed. We analyized the ill posed degree by pseudo convolutional operation theory. The time convolution regularizing radon integral geometry equation was obtained. Then we constructed effective TCR method which is of high accuracy and fast convergent rate. The iteration is stable. Now, we have got the excellent simulation results of pratical thermal inverse problem by this method.

The organization of this paper is as follows. In section 1, we mentioned the inverse problem of thermal transmittance. The perturbation method and the linearzed equation of problem for determinating thermal transmittance was presented in section 3. The numerical method of TCR was in the section 4, approximation of function δ was in the section 5, finally we made the discussion about numerical remarks and numerical result.

2. Inverse problem for determinating the thermal transmittance.

2.1 2-D heat equation

(2.1) $\frac{\partial}{\partial x}(K(x,y)\frac{\partial u}{\partial x}) + \frac{\partial}{\partial y}(K(x,y)\frac{\partial u}{\partial y}) - \frac{\partial u}{\partial t} = 0$, $(x,y,t) \in R^+ \times R \times R^+$,

(2.2) $u(x,y,0) = 0$, $(x,y) \in R^+ \times R$,

(2.3) $\frac{\partial u}{\partial x}(0,y,t) = g(y,t)$, $(y,t) \in R \times R^+$.

where, $K(x,y) \geqslant K_0 > 0$, $K(x,y) \in L_\infty(R^+ \times R)$, and there exist a boundary domain $\Omega \subset R^+ \times R$, $K(x,y)$ is constant on $R \times R^+ - \Omega$, $g(y,t) \in C(R \times R^+)$.

2.2 Measure temperature on the surface

(2.4) $u(0,y,t) = f(y,t)$, $(y,t) \in R \times R^+$.

2.3 Inverse problem of thermal transmittance

The thermal transmittance inverse problem of heat equation is to recover the coefficient $K(x,y)$ from the measure temperature on the surface (2.4), and equaations (2.1)-(2.3). Suppose that functions f and g satisfy certain compatible conditions such that the solution of inverse problem to be exist.

3. The Linearization Equation of thermal transmittance Inverse problem

3.1 Linearization of Heat Equation

Suppose that there exist a small increament $k(x,y)$ for $K(x,y)$. So, The small increament $w(x,y,t)$ of solution of (2.1)-(2.3) will be arisen. From

(3.1) $\frac{\partial}{\partial x}((K+k)\frac{\partial}{\partial x}(u+w))$
$+\frac{\partial}{\partial y}((K+k)\frac{\partial}{\partial y}(u+w))$
$=\frac{\partial}{\partial t}(u+w)=0,$
$(x,y,t) \in R^+ \times R \times R^+,$

(3.2) $(u+w)(x,y,0)=0,$
$(x,y) \in R^+ \times R,$

(3.3) $\frac{\partial}{\partial x}(u+w)(0,y,t)=g(y,t),$
$(y,t) \in R \times R^+,$

we have

(3.4) $\frac{\partial}{\partial x}(K(x,y)\frac{\partial w}{\partial x})$
$+\frac{\partial}{\partial y}(K(x,y)\frac{\partial w}{\partial y})-\frac{\partial w}{\partial t}$
$=-[\frac{\partial}{\partial x}(k\frac{\partial u}{\partial x})+\frac{\partial}{\partial y}(k\frac{\partial u}{\partial y})],$
$(x,y,t) \in R^+ \times R \times R^+,$

(3.5) $w(x,y,0)=0, \quad (x,y) \in R^+ \times R,$

(3.6) $\frac{\partial w}{\partial x}(0,y,t)=0, \quad (y,t) \in R \times R^+.$

3.2 Auxiliary temperature under point source on the boundary

Let $v(x, y, t; \xi)$ be the solution of following heat equation and the domain is as before.

(3.7) $\frac{\partial}{\partial x}(K(x,y) \frac{\partial v}{\partial x}) + \frac{\partial}{\partial y}(K(x,y) \frac{\partial v}{\partial y}) - \frac{\partial v}{\partial t} = 0$,

(3.8) $v(x, y, 0) = 0$,

(3.9) $\frac{\partial v}{\partial x}(0, y, t) = \delta(y - \xi, t)$.

where $v(x, y, t; \xi)$ is auxiliary temperture under point source on the surface.

3.3 The Linearization Equation of thermal transmittance Inverse Problem

Theorem: let $u(x, y, t)$ be a solution of (2.1)-(2.3). $w(x, y, t)$ be solution of (3.4)-(3.6), $v(x, y, t; \xi)$ be solution of (3.7)-(3.9), then the linearization equation of thermal transmittance inversion is

(3.10) $K(0, y) w(0, y, t)$
$= \int_0^\infty \int_{-\infty}^\infty k(\xi, \eta) [\frac{\partial u}{\partial \xi} *_t \frac{\partial v}{\partial \xi} + \frac{\partial u}{\partial \eta} *_t \frac{\partial v}{\partial \eta}] d\xi d\eta$.

proof: Take convolution about ξ, η on the $R^+ \times R$, by integrating by part, we have

(3.11) $\int_0^\infty \int_{-\infty}^\infty [\frac{\partial}{\partial \xi}(K(\xi, \eta) \frac{\partial w}{\partial \xi})$
$+ \frac{\partial}{\partial \eta}(K(\xi, \eta) \frac{\partial w}{\partial \eta}) - \frac{\partial w}{\partial t}] *_t v d\xi d\eta$
$= -\int_0^\infty \int_{-\infty}^\infty [\frac{\partial}{\partial \xi}(k \frac{\partial u}{\partial \xi})$
$+ \frac{\partial}{\partial \eta}(k \frac{\partial u}{\partial \eta})] *_t v d\xi d\eta$.

(3.12) $\int_{-\infty}^\infty [K(0, \eta) \frac{\partial w}{\partial \xi} *_t v$
$- K(0, \eta) w *_t \frac{\partial v}{\partial \xi}] d\eta$
$= \int_0^\infty \int_{-\infty}^\infty k[\frac{\partial u}{\partial \xi} *_t \frac{\partial v}{\partial \xi} + \frac{\partial u}{\partial \eta} *_t \frac{\partial v}{\partial \eta}] d\xi d\eta$.

Finally, we have

(3.10) $K(0,y)w(0,y,t)$
$=\int_0^\infty \int_{-\infty}^\infty k(\xi,\eta)[\frac{\partial u}{\partial \xi}*_t \frac{\partial v}{\partial \xi}$
$+\frac{\partial u}{\partial \eta}*_t \frac{\partial v}{\partial \eta}]d\xi d\eta.$

(3.10) is the linearization equation of the thermal transmittance inversion.

4. Time Convolution regularizing Iteration.

4.1 Guess coefficient $K_0(x,y) \geqslant K_0 > 0$.

4.2 By induction, suppose that $K_n(x,y)$ has been obtained from $(n-1)$'th step, to replaced $K(x,y)$ by $K_n(x,y)$ and to solve (2.1)-(2.3) and (3.7)-(3.9), we got $u_n(x,y,t)$ and $v_n(x,y,t;\xi)$.

4.3 To solve the following convolution integral equation of first kind by regularizing method [7]

(4.1) $\|K(0,y)[u(0,y,t)-f(y,t)]$
$=\int_0^\infty \int_{-\infty}^\infty k(\xi,\eta)[\frac{\partial u}{\partial \xi}*_t \frac{\partial v}{\partial \xi}$
$+\frac{\partial u}{\partial \eta}*_t \frac{\partial v}{\partial \eta}]d\xi d\eta\|_H$
$+\alpha \|k(t,y)\|_{L2} = \min$
$0 < \alpha < \alpha_0.$

(4.2) $K_{n+1}(x,y) = K_n(x,y) + \beta k(x,y),$
when $K_{n+1} \geqslant K_0$, $0 < \beta < \beta_0$.
$K_{n+1}(x,y) = K_0$, when $K_{n+1} < K_0$.

5. Numerical Method

5.1 The Numerical Method for Solving heat Equation (2.1)-(2.3) and (3.7)-(3.9). For solving heat equation, in iteraction of the inversion, a new explicit implicit hybrid scheme combining FEM and FDM is suitable. In each iteration, once of discomposition LL^T will be only needed. Such that computer process is fast.

5.2 The Approximation of $\delta(y-\xi,t)$

In numerical method for solving (3.7)-(3.9),

$\delta(y-\xi, t)$ will be approximated by the following function, $D(y-\xi_i, t)$; where ξ_i is the node point in y direction.

(5.1) $D(y-\xi_i, t) = \sum D_j(y-\xi_i, t)$,

(5.2) $D_1(y-\xi_i, t)$
$= -(3/4)[(y-\xi_i)dt + tdy - dydt]/(dydt)$.
when $y > \xi_i$, $t > 0$ and
$(y-\xi_i, t)dt + tdy - dydt \leqslant 0$,
$D_1(y-\xi_i, t) = 0$, elsewhere.

Similary, D_j, j=2,3,4, can be given.

5.3 Radiational condition on the artifical boundary.

In fourth section, the time covolution regularizing iteration has been defined. Its domain is infinite. But numerical computation can not be performed in the infinite domain. Thus for solving (2.1)-(2.3) numerically, it has to do artifical boundary which contain Ω in it. The radiation condition on the artifical boundary are in [5],[6]. Here we used new radiational condition [4]. It is very effective for inversion.

6. Discusion And Numerical Results

6.1 Regularizing method

In the time convolution regularizing iterative method, the convolution integral equation is ill posed even through $\delta(y, t)$ is impulse temperature function. Thus, it is necessary to use regularizing method. But entire regularizing is too expensive and the integral operator is finite ill posed, we have analysized the ill degree by the pesudo convolution operator theory, then we use a regularizing method partly and trikly. By practical numerical tests, the regularizing convolution integral equation is effective, where $0<\alpha<\alpha_0$, which is sensitivity. In begining of the iteration, α can be larger than 1; after a few iteration it became small by experiences. We used automatical choice of finding α and orthgonal discompisition in solving (4.1) [8].

6.2 Simulation Test

The several simulation tests have been computed by this method. Details will be given in [8]. For compact sources on the boundary, the time convolution iteration of thermal transmittance inversion is very effective. The method can be used to atmosphere and control of the thermal transmittance in industral stove. We have got excellent numerical results by this method. The method is nice to make applied softwave and extension.

Acknowledgments. The authors would like to thank Professor P.D.Lax and Professor Feng Kang and Y.M.Chen for their encouragement and guidance. And we gratefully acknowledge the support of the National Science Foundation, Joint Seismological Science Foundation and Hunan Province Scientific and Technological Committee. This work was partly supported by Hunan Province Scientific and Technological Committee.

References

[1] Ganquan Xie, A New Iterative Method for Solving the Coefficient Inverse Problem of the Wave Equation. Communications on Pure and Applied Math. V.XXXIX. (1986)

[2] Ganquan Xie, Jianhua Li, Lunji Tang and Yingning Sun, TCR Iteration For Solving Coefficient Inverse Problem of 1-D heat Equation, Submit to J.Computational Methematics, China.

[3] Gauss-Newton-Regularizing Method for Solving Coefficient Inverse Problem of Partial Differential Equation and Its Convergence Jounal of Compayational Mathematics Vol 5. No 1 Jan 1987.

[4] Ganquan Xie, A New Radiational Contion On Artifical Boundary for heat Equation. In Preparation.

[5] Alvin bayliss and Eli Tarkel, Rediational Boundary Conditions for Wave Like Equation. Communications on Pure Math. and Applied Math. V. XXXIII,(1980).

[6] Kriegsman G. and Morawatz C.S., Numerical Solution of Exterior Problems With the Reduced Wave Equation J. Comp.Phys.28.1987.

[7] Tikhonov A.N. and Arsein V.Y., on the Solution of III-posed Problem. John Wiley and sons. New York (1977).

[8] Ganquan Xie, Jianhua Li, The Numerical Method of Time-Convolution Iteration for solving 2-D Inverse Problem of heat Equation.

A DIFFUSION OPERATOR FOR LAGRANGIAN MESHES

L. G. Margolin and A. Ellen Tarwater

Lawrence Livermore National Laboratory
University of California
Livermore, CA 94550

1.   INTRODUCTION

We describe a new diffusion operator for a Lagrangian mesh of irregular quadrilaterals. The operator diffuses cell-centered quantities, and in this sense is similar to the operators of Kershaw [1] and Pert [2]. However our operator differs from their previous works both in construction and in form.

We construct our diffusion operator by requiring it to exactly reproduce certain properties of the physical operator. These properties are conservation, self-adjointness and the ability to exactly capture quadratic fields. We express these properties as a system of linear equations whose solution yields the coefficients of the operator. In general, the system has more equations than unknowns - i.e., it is overdetermined and there is no solution that possesses all the properties we prescribe. Further, the condition that a solution does exist puts constraints on the geometry of the mesh.

It turns out that a sufficient condition for a solution to exist is the absence of hourglassing deformation in the mesh. Hourglassing is an unphysical distortion of the mesh, which we have discussed elsewhere [3]. Our construction allows us to understand the limits of accuracy of a numerical spatial operator (in this case, the gradient operator) on a distorted mesh. In fact, our construction introduces a natural measure of this distortion.

The calculation of the diffusion operator breaks naturally into two steps. We begin with a scalar field, such as temperature, stored at cell centers. In the first step, we calculate the vector flux at the cell vertices by defining the gradient. Second, we calculate the divergence of the flux,

which is again located at cell centers. These two steps are
easily combined to produce a nine point stencil for the
diffusion operator.

When we apply our new operator to the Kershaw challenge,
we find considerable improvement over the published results
of Kershaw [1] and Pert [2]. The Kershaw mesh does have
hourglassing deformation. A first order solution neglecting
this deformation exhibits small spatial inaccuracies -
wiggles - where the hourglassing is largest. A second order
correction further improves our results, supporting the
validity of the analysis.

Kershaw [1] describes five properties that he uses as
guidelines in constructing his operator. One of these is the
reduction of his operator to the standard five point scheme
on a regular mesh. This is not a property of the physical
operator, but more a matter of taste. Our operator does
not have this reduction. This is a direct consequence of
calculating the flux at the cell vertices rather than on the
edges. Calculating the flux at the vertices is more natural on
a staggered mesh, but may lead to checkerboarding. This is a
noise pattern superposed on the solution with a characteristic
wavelength of one cell dimension. We will show how a simple
filtering procedure can eliminate the noise without altering
the physical aspects of the solution.

2. THE DIVERGENCE OPERATOR

Here we summarize previously published work on spatial
differencing of the divergence operator [4]. We are given a
vector field F_i, stored at the vertices of a cell (see Fig. 1).
We equate the divergence $\partial_i F_i$ at the cell center to its average

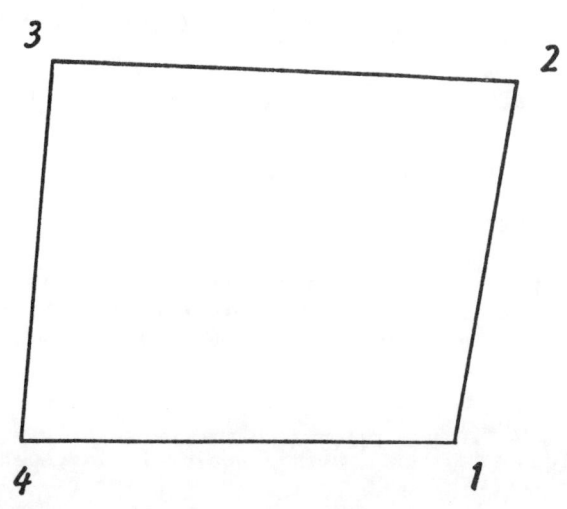

Fig. 1. A typical Lagrangian quadrilateral cell, showing the
vertices numbered in a counterclockwise pattern.

value over the entire cell. Then, using the divergence theorem, we convert to a surface integral

$$\overline{\partial_i F_i} = \frac{1}{vol} \int d^3x \, \partial_i F_i = \frac{1}{vol} \oint d^2x \, n_i F_i \qquad (1)$$

where n_i is the unit normal to the cell edges. Now the divergence of a flux has the meaning of a time rate of change of some conserved quantity. In the form of Eq. (1), exact conservation is obvious. Each cell edge is shared by two cells, and appears in two integrals, with opposite signs. Physically, this means the flux out of any cell through an edge is exactly the flux into the neighboring cell that shares that edge.

To perform the integral in Eq. (1), we need to interpolate the flux, which is known at the vertices, along the edge that joins them. A linear interpolation guarantees that the divergence operator exactly captures linear fields. In cylindrical coordinates (r,z), with counterclockwise numbering and flux components $F_i = (f,g)$ we derive

$$\frac{1}{r}\frac{\partial(rf)}{\partial r} = \frac{1}{6(vol)} [f_1(r_4 z_{14} + 2r_1 z_{24} + r_2 z_{21})$$

$$+ f_2(r_1 z_{21} + 2r_2 z_{31} + r_3 z_{32})$$

$$+ f_3(r_2 z_{32} + 2r_3 z_{42} + r_4 z_{43})$$

$$+ f_4(r_3 z_{41} + 2r_4 z_{13} + r_1 z_{14})] \qquad (2)$$

$$\frac{\partial g}{\partial z} = \frac{1}{6(vol)} [g_1(r_1 + r_2 + r_4)r_{42} + g_2(r_1 + r_2 + r_3)r_{13}$$

$$+ g_3(r_2 + r_3 + r_4)r_{24} + g_4(r_1 + r_3 + r_4)r_{31}] \qquad (3)$$

where the notation $z_{14} \equiv z_1 - z_4$.

3. THE GRADIENT OPERATOR

We are given a scalar field, such as temperature, which is stored at cell centers. We want to construct the gradient at the central node (see Fig. 2). We define coefficients a_i and b_i such that

$$\frac{dI}{dr} = a_i I_i \qquad i = 1,2,3,4 \qquad (4)$$

$$\frac{dI}{dz} = b_i I_i \qquad (5)$$

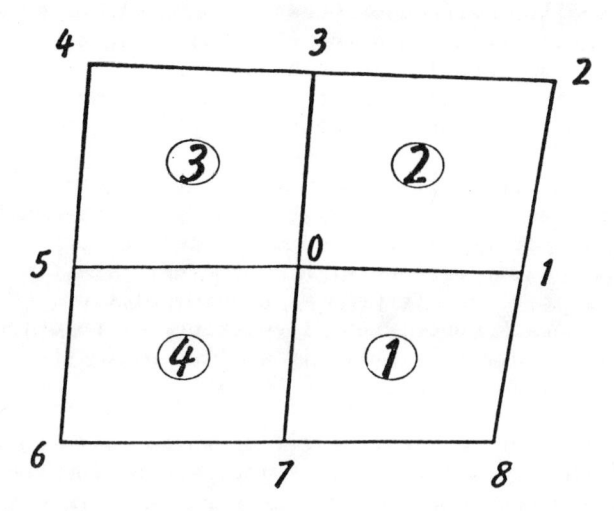

Fig. 2. The four cells surrounding a vertex (0). The cells are numbered counterclockwise 1 to 4 and the vertices 1 to 8.

Here we sum over repeated indices. The gradient is defined by the eight (as yet unknown) coefficients, by Eqs. (4) and (5).

Now there are many properties that the gradient has as a mathematical object. Any reasonable approximation to the gradient will approximately mimic these properties. However, it is possible to require the numerical operator to exactly reproduce some subset of these properties. This is the basis of our construction.

As an example, the gradient of a spatially constant field should vanish. This translates into conditions on a_i and b_i

$$a_1 + a_2 + a_3 + a_4 = 0 \tag{6}$$

$$b_1 + b_2 + b_3 + b_4 = 0 \tag{7}$$

Now if we enforce a sufficient number of conditions, we will create a system of equations whose solution uniquely determines the coefficients a_i and b_i, and thus the gradient.

In analogy to the previous section, we require the gradient to exactly capture linear fields. This leads to four additional equations, two of which are inhomogeneous. Also, we choose to require exact self-adjointness of the diffusion operator. This is, strictly speaking, not a condition on the gradient, but rather on the relation between the gradient and the divergence.

Numerically, self-adjointness corresponds to symmetry of the coupling matrix. Since the final solution of the diffusion equation for most applications involves inverting the coupling matrix (see Kershaw [1]), exact symmetry is important. Very efficient methods exist for inverting symmetric matrices [5].

There are six additional equations that guarantee self-adjointness, making our system total twelve equations. However, three of these are redundant, leaving us with a final tally of nine equations and eight unknowns. Thus the system is overdetermined, and in general has no solution. It is interesting to ask under what circumstances a solution does exist. Before answering this, it will be useful to introduce some additional notation.

For every vertex, there are four surrounding cells, which we label 1 through 4 counterclockwise. We define two vectors A_i and B_i, whose components are the r and z coordinates of the cell centers. Thus

$$A_1 = .25 \ (r_0 + r_1 + r_7 + r_8) \text{ and } B_1 = .25 \ (z_0 + z_1 + z_7 + z_8)$$

We define two additional vectors α_i and β_i. α_i is the r-component of the coefficient of the divergence operator of the vector stored at the vertex 0 with respect to the cell and β_i is the z-component. These are equivalent to the coefficients in Eqs. (2) and (3). However, we emphasize that Eqs. (2) and (3) are referenced with respect to the cell, whereas α_i and β_i are listed with reference to the vertex. For example (see Fig. 2 for numbering convention),

$$\alpha_1 = (r_1 z_{01} + 2r_0 z_{71} + r_7 z_{70}) \tag{8}$$

Note that we have deliberately omitted the volume factor from this definition.

With this notation, it is possible to show that our system of equations has a solution if and only if

$$\alpha_i B_i = \beta_i A_i \tag{9}$$

Summation is still implied over repeated indices. Now the corner coordinate, r_8 appears only in A_1 and hence only on the right side of Eq. (9). It is easy to see that the consistency condition of Eq. (9) will not be satisfied in general.

Suppose now that r_8 were expressible in terms of the other three vertex coordinates — for example $r_8 = r_1 + r_7 - r_0$

and similar relations for the other corner vertices 2, 4, and 6. This is sufficient to guarantee that Eq. (9) holds. The quantity $r_1 + r_7 - r_0 - r_8$ is called the hourglassing deformation of the cell [3]. It is basically a measure of the deformation of the mesh, and now we show that it is related to the accuracy of spatial approximation on the mesh.

When the consistency condition of Eq. (9) is satisfied, a general solution of our system of equations is

$$a_i = \left[\alpha_i [\beta_j B_j] - \beta_i [\alpha_j B_j]\right] / N \qquad (10)$$

$$b_i = \left[\beta_i [\alpha_j A_j] - \alpha_i [\beta_j A_j]\right] / N \qquad (11)$$

where the normalization

$$N = [\alpha_i A_i][\beta_j B_j] - [\alpha_j B_j][\beta_i A_i] \qquad (12)$$

Note that a_i and b_i are determined for any mesh in terms of the divergence operator.

Even when Eq. (9) does not hold, the dimensionless quantities

$$\varepsilon_1 = [\alpha_i B_i]/[\alpha_j A_j] \text{ and } \varepsilon_2 = [\beta_i A_i]/[\beta_j B_j] \ll 1 \qquad (13)$$

are usually small. Equations (10)-(12) can be viewed as the first term of a perturbation solution with respect to the small parameter. A natural question is to ask whether one can find the next order correction to Eqs. (10)-(12). The surprising answer is, in general, no. We have formulated an alternate approach to guarantee a unique solution that contains all the properties of consistency, capturing of linear gradients and self-adjointness on an arbitrary mesh. The idea is that the four equations that guarantee the capture of linear gradients are not all necessary. If one knew <u>a priori</u> the direction of the gradient, then only two equations are necessary. Of course this direction is itself dependent on a_i and b_i. Locally at any node, the field appears to have direction

$$\hat{n} = \cos\theta \, \hat{r} + \sin\theta \, \hat{z} \qquad (14)$$

where

$$\tan\theta = \frac{a_i I_i}{b_i I_i} \qquad (15)$$

In addition to the two equations that ensure capturing the linear gradient in the $\hat{n}$ direction, Eq. (15) must be enforced to guarantee self-consistency.

Thus, the net result is that we replace four equations with three equations. Our system then totals eight equations and eight unknowns, which guarantees a solution always exists. However, the addition of Eq. (15) makes the system nonlinear. This means that the coefficients a_i and b_i now depend weakly on the temperature field as well as the geometry of the mesh.

We have solved the new set of eight equations, generated by the self-consistent approach, by a perturbation technique using the small parameters of Eq. (13). To first order, we recover the solution of Eqs. (10)-(12). To second order, we introduce dependence (of order ε) on the temperature field. In the next section, we show that the self-consistent corrections can improve our results. It remains to be seen whether this correction is stable under all circumstances.

4. RESULTS

The Kershaw challenge [1] is shown in Fig. 3. The physical problem is a slab of material heated uniformly along the left (T = 1) and the right (T = 0). The other sides are insulating and the slab is initially uniformly cool (at T = 0). Thus, the

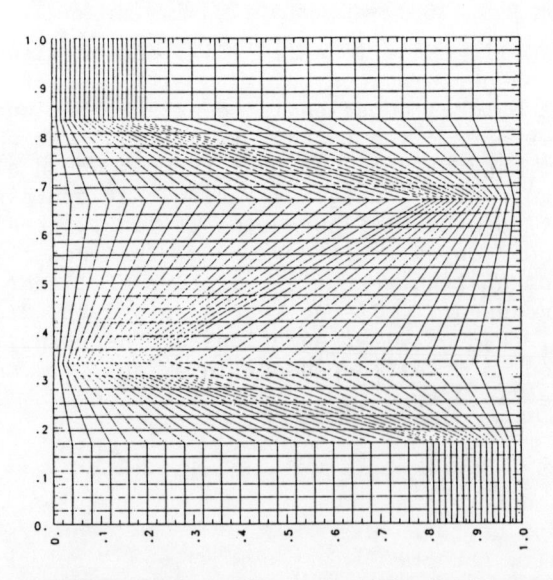

Fig. 3. The Kershaw mesh. The left side is heated (T = 1), the right side is cool (T = 0) and the top and bottom are insulating.

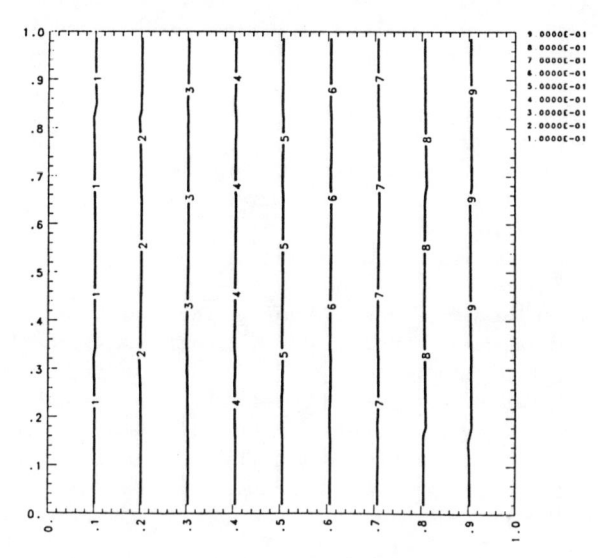

Fig. 4. A first order solution in steady state shows small wiggles in the isothermal contours.

physical problem is one-dimensional and the isothermal contours of the time-dependent solution should be vertical lines. In addition, in steady state, the contours should be equally spaced. The mesh is deliberately distorted to include large chevrons and also has significant hourglassing. The small ε parameters of Eq. (13) are about .05 in some cells. The challenge is to obtain vertical contours in this very distorted mesh.

We ran the problem in the 2-D code SHALE [6] using this new diffusion operator. We used a cylindrical coordinate system with the left vertical side being the axis of symmetry. The results in steady state are shown in Fig. 4. These are first order results using Eqs. (10)-(12). The results compare favorably with Kershaw's [1] and Pert's [2] results, but still have small wiggles. It is possible to show that these errors occur precisely where the hourglassing is largest. In Fig. 5 we show the results using the second order operator with self-consistent corrections. These contours are straight to the resolution of the plotter.

The steady state temperature is exactly linear in space, so it is not surprising that our operator does well. It was constructed to capture this solution. The dynamic problem is more difficult, for the field is not linear, but exponential. In Fig. 6 we show the contours at a short time after the problem begins. The quality of the solution is already apparent.

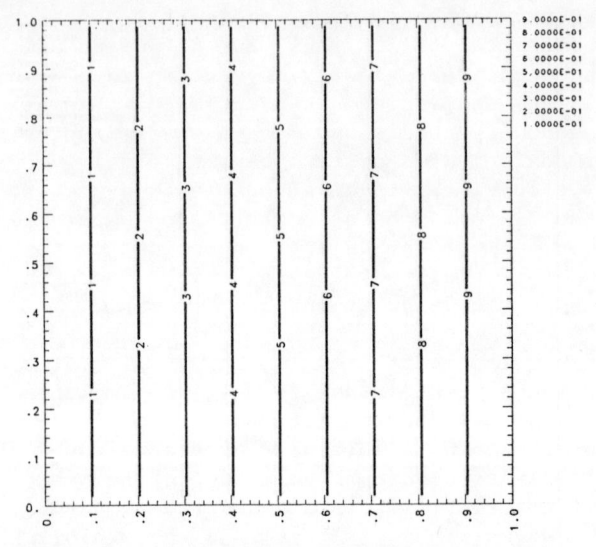

Fig. 5. A second order solution in steady state has straight isotherms.

5. FILTERING

On a regular mesh of rectangles, the diffusion operator stencil reduces to a five point scheme coupling each cell to its four diagonal neighbors. Each cell appears uncoupled to

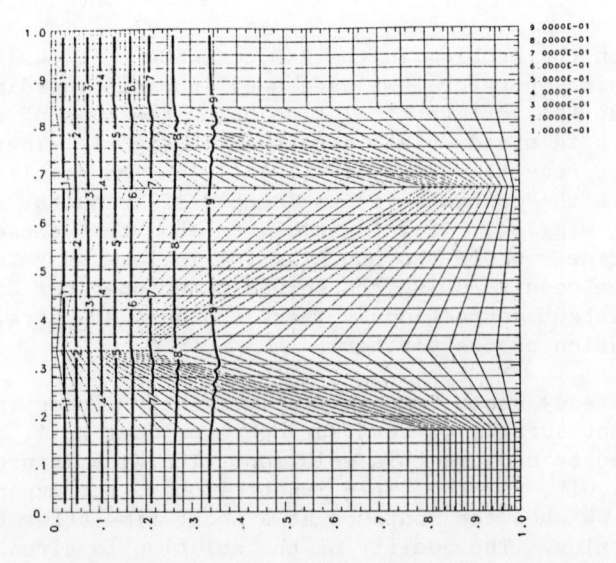

Fig. 6. The dynamic solution shows that the wiggles are largest where the hourglassing is largest.

its four nearest neighbors, i.e., those that share a common edge with the central cell. Thus, the diffusion process appears to proceed independently on two separate grids, like the red and black squares of a checkerboard. Variations in initial and boundary conditions with a wavelength of the order of a cell dimension, will diffuse on one grid, never seen by the other. Of course it is unreasonable to expect to resolve details of the diffusion on a length scale comparable to cell size. However, almost any initial conditions can have components of this small wavelength, which then propagate as noise through the calculation.

This effect is illustrated in Fig. 7. We calculate a problem in which a point source of heat is simulated by setting one cell - the bottom left corner cell - to a high temperature. The mesh is a regular orthogonal grid of rectangles. The isotherms of the expanding temperature field should become spherical at distances large compared to a cell dimension. However, these isotherms do not exhibit the expected symmetry.

Our analysis closely follows our previous work on hourglassing [3]. Mathematically, the noise corresponds to the existence of a null vector - a pattern of temperatures that is not uniform in space, and yet is calculated to have zero gradients. Such a situation is unphysical. Our strategy will be to identify this null vector in a local and approximate fashion and to filter it from the solution.

Fig. 7. Isothermal contours for a point expansion of a single heated cell in the lower left corner.

Consider a group of four cells as shown in Fig. 2. The gradient of temperature is given by

$$\frac{dI}{dr} = a_i I_i \qquad \frac{dI}{dz} = b_i I_i$$

Now we view these as inner products in an artificial vector space of four dimensions. We want to find a vector that is somehow orthogonal to a_i and b_i, and also to the uniform temperature distribution represented by a vector $\delta_i = (1,1,1,1)$.

Since we are in a four-dimensional space, a fourth linearly independent vector, h_i, always exists, no matter how the vectors a_i and b_i are chosen. Also, the set $(a_i, b_i, \delta_i, h_i)$ forms a basis set for the space with which we can always expand the temperature. The subspace generated by h_i is the null space of the gradient operator.

Now if we do expand the temperature, then the part of the field proportional to this vector h_i is unphysical. It does not lead to any gradient, since $a_i h_i = 0$ and $b_i h_i = 0$ by construction. Also, it does not affect the total energy since $\delta_i h_i = 0$. From our previous discussion, the part of the temperature field proportional to h_i is the noise and should be filtered from the solution.

The filtering is accomplished locally at each node by defining a filtered field

$$\tilde{I}_i = I_i - h_i (h_j I_j) \omega \qquad (16)$$

Here ω is a small number. In principal, we could choose $\omega = .25$ to completely filter the noise in one cycle. However, each cell is part of 4 different "sets". The filter we have constructed is approximate in the sense that we have assumed each "set" of cells is isolated. It is better to choose $\omega = .01$, representing relaxation in twenty-five cycles.

When the filter of Eq. (16) is introduced into the previous calculation of the hot spot, the isotherms of Fig. 8 result. These contours are now nicely spherical, demonstrating that our filter is a simple and effective method for eliminating checkerboarding.

Fig. 8. Use of a filter produces the point symmetric contours of the physical solution.

This work was performed under the auspices of the U.S. Department of Energy by Lawrence Livermore National Laboratory under Contract #W-7405-Eng-48.

REFERENCES

[1]   KERSHAW, D. S. - Differencing of the Diffusion Equation, J. Comp. Phys. 39, 375-395 (1981).

[2]   PERT, G. J. - Constraints in the Numerical Calculation of Diffusion, J. Comp. Phys. 42, 20-25 (1981).

[3]   MARGOLIN, L. G. and PYUN, J. J. - A Method for Treating Hourglass Patterns. Proceedings, 5th International Conference on Numerical Methods in Laminar and Turbulent Flow, July, 1987 (to be published).

[4]   MARGOLIN, L. G. and ADAMS, T. F. - Spatial Differencing for Finite Difference Codes. Los Alamos National Laboratory Report LA-10249, 1985.

[5]   KERSHAW, D. S. - The Incomplete Cholesby - Conjugate Gradient Method, J. Comp. Phys. 26, 43-65 (1978).

[6]   DEMUTH, R. B. et al. - SHALE: A Computer Program for Solid Dynamics. Los Alamos National Laboratory Report LA-10236, 1985.

MONTE CARLO METHOD FOR RADIATIVE HEAT TRANSFER ANALYSIS OF GENERAL GAS-PARTICLE ENCLOSURES

K. Kudo[1], H. Taniguchi[1], W.-J. Yang[2]
H. Hayasaka[1], T. Fukuchi[1] and I. Nakamachi[3]
1) Hokkaido University,   Sapporo 060, JAPAN
2) University of Michigan, Ann Arbor, Michigan 48109, U.S.A.
3) Tokyo Gas Co.,Ltd.,   Minato-ku, Tokyo 105, JAPAN

ABSTRACT

A new Monte Carlo method is developed to analyze multi-dimensional radiative heat transfer in an enclosure containing gray gas with anisotropically scattering gray particles. One set of variables, called READ, are introduced to represent radiation transfer, thus appreciably reducing repeated computations required by the conventional Monte Carlo technique. An 1m x 1m square duct is used as the enclosure with the upper and lower walls at different temperatures in the one-dimensional case and with adiabatic specular or diffuse side walls in the two-dimensional case. Results agree very well with the existing analytical solutions for one-dimensional, non-scattering cases. It is concluded that with anisotropic scattering, an increase in the absorption coefficient and/or single scattering albedo produces adverse effects on radiative heat transfer. Anisotropic scattering effects cannot be simulated by the use of the effective absorption coefficient.

1. INTRODUCTION

Radiative heat transfer in dispersed solid systems has important applications in many technological areas such as boilers and MHD generators using pulverized coal, insulations and packed and fluidized beds [1][2]. The heat transfer analysis in an absorbing, emitting and scattering media has an inherent difficulty in treating the governing integro-differential equations, which are derived from the remote effects of radiative heat transfer. Consequently, numerous assumptions are imposed to simplify the equations. Most of the existing analyses are limited to the one-dimensional system, taking into account only absorption or isotropic scattering of solid particles [2][3]. At present, no method other than the Monte Carlo technique can treat freely the multi-dimensional radiative transfer with anisotropic scattering, non-uniformity in properties and irregularity in system geometry and yet is compatible with finite-difference algorithms for solving fluid dynamics [2]. However, the Monte Carlo method is known to be time-consuming in numerical computations and its results periled by some statistical errors. In order to overcome these shortcomings, a set of constants called READ (Radiative Energy Absorption Distribution) is developed to treat combined radiative and convective heat transfer phenomena in absorbing-emitting media in a boiler furnace [4]. The READ represents radiative heat transfer between gas-gas, gas-surface and surface-surface elements. Its values can be calculated through a Monte Carlo technique without the iterational procedure which is generally required by the conventional Monte Carlo method [5].

In the present work, the READ is applied to treat radiative heat transfer in absorbing-emitting and scattering media. It is well known that the effects of isotropic scattering can be completely realized through the use of an equivalent absorption when heat is transferred by radiation alone (radiative equilibrium). Most of the scattering characteristics of real particles, however, are not isotropic. In the present study, anisotropic scattering effects are demonstrated in one- and two-dimensional systems. The validity of the new Monte Carlo method is borne out by its agreement with the existing analytical solutions in one-dimensional pure absorbing and emitting radiative transfer.

2. SYSTEM DESCRIPTION

Radiative heat transfer is analyzed in a two-dimensional rectangular (1m x 1m) duct shown in Fig.1. The upper wall is at a higher temperature than the lower wall. Both walls are black. The side walls are adiabatic, gray and diffuse or specular. Both one- and two-dimensional analyses are performed. In the one-dimensional case, both the upper and lower walls are assumed isothemal and the side walls are specular. In the two-dimensional case, the upper wall has a stepwisely higher uniform temperature region near the center, while the side walls are diffuse.

This rectangular enclosure contains a gray gas with uniformly dispersed gray spherical particles which absorb, emit and anisotropically scatter the radiative energy.

3. RADIATIVE CHARACTERISTICS OF WALL AND DISPERSED GAS
3.1 surrounding walls

The total radiative energy emitted per unit time from a

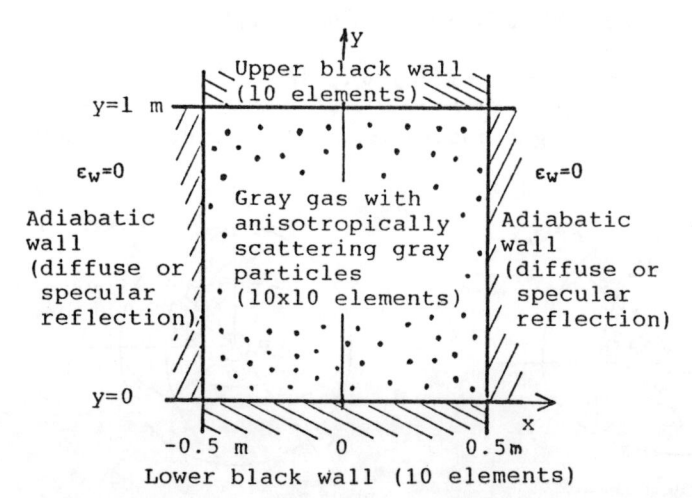

Fig. 1 Two-dimensional, rectangular duct for radiative transfer analysis

small gray wall element dA with an emissivity ε_w and a temperature T_w is

$$dQ_{we} = \varepsilon_w \sigma T_w^4 dA \tag{1}$$

where σ is the Stefan-Boltzmann constant. The corresponding radiative intensity i_{we}, being independent of direction, is

$$i_{we} = \frac{1}{\pi} \frac{dQ_{we}}{dA} \tag{2}$$

According to Lambert's cosine law, the energy emitted from dA in the direction (θ, η) within a solid angle $d\Omega$, Fig.2, is

$$d^2Q_{we} = i_{we} dA \cos\eta \, d\Omega \tag{3}$$

Out of the radiative energy rate dQ_{wi} incident upon dA, only $\varepsilon_w dQ_{wi}$ is absorbed by the wall, while the remaining $dQ_{wr}=(1-\varepsilon_w)Q_{wi}$ is reflected. The energy dQ_{wr} reflected in the direction (θ, η) within a solid angle $d\Omega$ can be expressed as

$$d^2Q_{wr} = \frac{1}{\pi} dQ_{wr} \cos\eta \, d\Omega \tag{4}$$

3.2 particles and gas

The temperature of the particles is generally assumed equal to that of the surrounding gas T_g. It implies an infinite heat transfer coefficient between the gas and the particle. This assumption is valid for small particles i.e. a large surface area-to-volume ratio and smaller radiative transfer to the particles. However, the case of a finite heat transfer coefficient between the gas and the particles can easily be treated using the present Monte Carlo technique. The radiative energy emitted per unit time from a small volume element dV containing gas and particles is

$$dQ_{ge} = 4a_g (dV - \frac{4}{3}\pi R^3 N_s \, dV) \sigma T_g^4 + \varepsilon_s 4\pi R^2 N_s \, dV \sigma T_g^4$$

$$= 4 [a_g (1 - \frac{4}{3}\pi R^3 N_s) + \varepsilon_s \pi R^2 N_s] \sigma T_g^4 dV \tag{5}$$

where a_g denotes the gas absorption coefficient; ε_s, particle

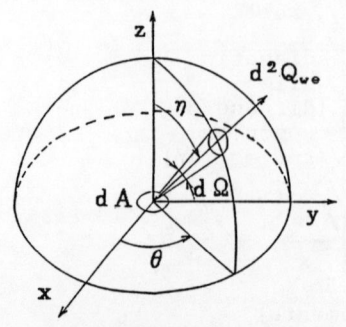

Fig. 2 Radiative energy emitted from dA on a solid wall

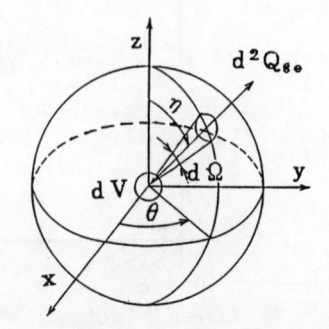

Fig. 3 Radiative energy emitted from gas volume dV

emissivity; R, particle radius; and N_s, particle number density. The energy emitted in the direction (θ,η) (shown in Fig.3) within a solid angle $d\Omega$ is given by

$$d^2Q_{s\,e} = dQ_{s\,e} \cdot \frac{d\Omega}{4\pi} \tag{6}$$

The attenuation of the intencity i of an incident radiation in the s direction (Fig.4) through a gas volume which contains dispersed particles due to the absorption of the gas and the absorption and scattering of the particles can be expressed by

$$di = -\beta i \, ds \tag{7}$$

where

$$\beta = a + \sigma_s \tag{8}$$
$$\omega = \sigma_s / \beta \tag{9}$$

represents the extinction coefficient; a, total absorption coefficient of gas and particles; σ_s, scattering coefficient; and ω, single scattering albedo. Out of the intensity attenuation di, ωdi is caused by scattering, while the remaining $(1-\omega)$di is absorbed. The total absorption coefficient a in equation(8) can be expressed as

$$a = a_g (1 - \frac{4}{3}\pi R^3 N_s) + \varepsilon_s \pi R^2 N_s \tag{10}$$

It includes the radiative equilibrium of emission (Eq.(5)) and the absorption of dispersed gas volume. The scattering coefficient σ_s reads

$$\sigma_s = (1 - \varepsilon_s) \pi R^2 N_s \tag{11}$$

when the radius of the particles is much larger than the wave length of the radiative energy. Equation(7) is integrated to yieled,

$$i(s) = i(0) e^{-\beta s} \tag{12}$$

which is Beer's law. Figure 5 depicts the composition of

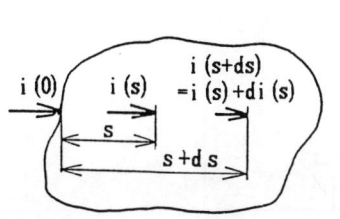

Fig. 4 Attenuation of intensity i

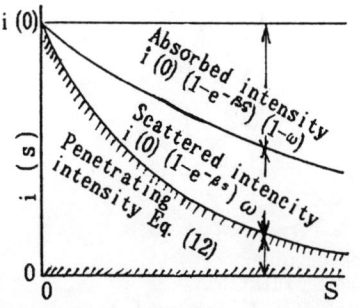

Fig. 5 Composition of intensity i

intencity. An anisotropic phase function given by the following equation is used in the present analysis to determine the angle of the scattered energy (Fig.6):

$$\Phi(\eta) = \frac{8}{3\pi}(\sin\eta - \eta\cos\eta) \qquad (13)$$

This expression is graphically illustrated in Fig.7. It is valid for a sphere with diffuse surface [6] and has strong backward scattering characteristics. Equation (1) through (13) can be used to determine the optical characteristics of dispersed gas as functions of three parameters, β, ω and Φ or a, σ_s and Φ.

4. ANALYSIS BY MONTE CARLO METHOD
4.1 heat balance

In Fig.1, the gas volume was divided into 100(10x10) square elements while both the upper and lower walls are divided into 10 elements. By prescribing the temperature profiles of the upper and the lower walls as the boundary conditions, the temperature profile in the gas region was obtained by heat balance for each gas element

$$Q_{out} = Q_{in} \qquad (14)$$

where Q_{in} and Q_{out} are the radiative heat input and output respectively. The heat input into each wall element Q_a was obtained as

$$Q_{out} = Q_{in} - Q_a \qquad (15)$$

It should be noted that

$$Q_{out} = (1-\alpha) \, 4a \, \sigma T_g^4 \, dV \qquad (16)$$

for the gas elements and

$$Q_{out} = (1-\alpha) \, \varepsilon_w \, \sigma T_w^4 \, dA \qquad (17)$$

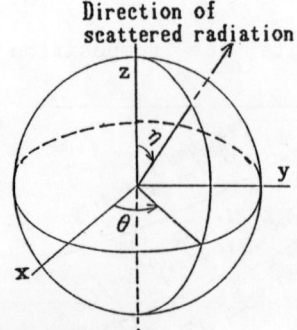

Fig. 6 Angle of scattered radiation η

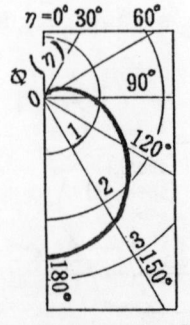

Fig. 7 Anisotropic phase function of a gray diffuse sphere as defined by Eq.(13)

for the wall elements. Here, α is the self absorption ratio, which represents the ratio of the energy absorbed by the element itself to the total energy emitted from the element. Hence, Q_{out} in both equations (16) and (17) represents the total energy emitted from an element and absorbed by other elements. The amount of the energy emitted from an element I and absorbed by another element J can be expressed by $Rd(I,J) \cdot Q_{out}(I)$, where $Rd(I,J)$ is defined as the ratio of the energy emitted from the element I and absorbed by the element J to the total emitted energy $Q_{out}(I)$. The magnitude $Rd(I,J)$, is called READ (Radiative Energy Absorption Distribution) [4]. It is independent of the temperature of the elements but depend only on the profile of the extinction coefficient in the gas region and the bounding surface characteristics. READ can be determined in a single computation by means of the Monte Carlo technique. It can be considered as a set of constants during the iterational procedure for temperature evaluation. Figure 8 is a flow diagram for the computation of temperature distribution in the gas region. The conventional Monte Carlo method for the iterational calculation of temperature is very time-consuming [5]. The radiative energy absorbed by an element Q_{in} can be obtained by adding all energy components transferred from all other elements, as

$$Q_{in}(J) = \sum_I Rd(I, J) \cdot Q_{out}(I) \tag{18}$$

The wall heat flux reads

$$q = Q_a / \Delta A \tag{19}$$

Figure 8 shows that, during the iterational calculation of temperature, the new values of temperature are obtained from equations (14)(16)(17) and (18) by the Newton-Raphson method.

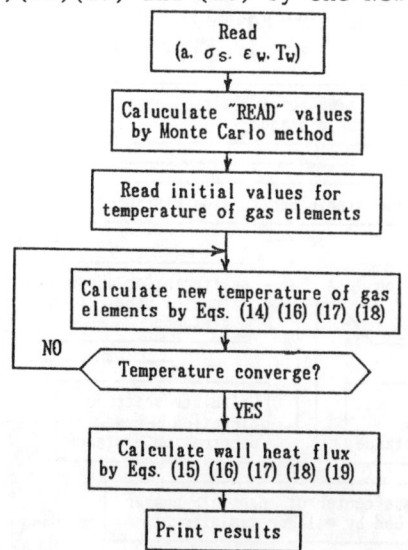

Fig. 8 Flow diagram for computation of radiation heat transfer

4.2 calculation of READ by Monte Carlo technique

For the calculation of Rd(I,J) and , many energy particles are emitted from an element I. The absorbed point of each particle is calculated following the computational procedure shown by the flow diagram in Fig.9. If n(J) particles are absorbed by an element J out of N particles emitted from an element I, α and Rd(I,J) can be written as

$$\alpha = n(I)/N \qquad (I = J) \qquad (20)$$

$$Rd(I.J) = n(J) / [N-n(I)] \qquad (I \neq J) \qquad (21)$$

$$Rd(I.I) = 0 \qquad (I = J) \qquad (22)$$

The procedure in Fig.9 is derived using random numbers in the Monte Carlo method [5][7]. However, the emitting point of the particles in an element should be distributed uniformly. Consequently, the coordinate of the emitting point in wall elements x can be expressed as

$$x = x_1 + (x_2 - x_1) RND_x \qquad (23)$$

where, x_1 and x_2 denote the two ends of the wall element and RND_x is a random number distributed uniformly between 0 and 1. For gas elements, equation(23) is used twice to determine the (x,y) coordinates of the emitting point in the element. The direction (θ, η) of emission of a particle from the wall elements shown in Fig.2 is evaluated using two random numbers as

$$\theta = 2\pi RND_\theta \qquad (24)$$

$$\eta = \cos^{-1} \sqrt{1-RND_\eta} \qquad (25)$$

With many particles, equations (24) and (25) are combined to give the Lambert's cosine law Eq.(3). For gas elements, the emitting direction (θ, η) shown in Fig.3 reads

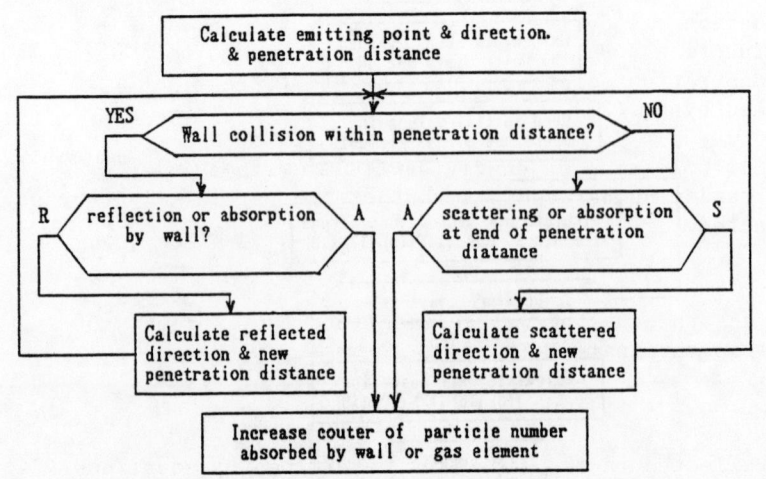

Fig. 9 Flow diagram for pursuit of an energy particle

$$\theta = 2\pi \text{RND}_\theta \tag{26}$$

$$\eta = \cos^{-1}(1-2\text{RND}_\eta) \tag{27}$$

which represent an isotropic emission. The penetration distance s is the length along which an energy particle travels before its extinction (absorption by gas or solid particle surface, or scattering by particles). It is determined by the equation

$$s = -\ln(1-\text{RND}_s)/\beta \tag{28}$$

which satisfies Beer's law Eq.(12) for energy particles. Within the penetration distance, whether the particles collide with a bounding wall or not is determined by simple geometrical consideration.

case (i): No collision of energy particles with a wall.
When the energy particle does not collide with a wall, a random number is used to determine whether it is scattered or absorbed. If

$$\text{RND}_{s,c} > \omega \tag{29}$$

the energy particle is absorbed by the gas or a particle at the end of its penetration distance and the pursuit of the energy particle is terminated. However, if

$$\text{RND}_{s,c} \leq \omega \tag{30}$$

the energy particle is scattered by an solid particle. The direction of scattering (θ, η) in Fig.6 is determined by

$$\frac{\eta}{2} - \frac{3}{8}\sin 2\eta + \frac{\eta\cos 2\eta}{4} - \frac{3\pi}{4}\text{RND}_\eta = 0 \tag{31}$$

The equation is derived by integrating the product of the phase function $\Phi(\eta)$ (Eq.(13)) and $\sin\eta$ from 0 to η [5]. The new penetration distance of the scattered particle is determined by Eq.(28).

case (ii): Collision of energy particles with a wall.
A random number decides whether the particle is absorbed or reflected. If

$$\text{RND}_r \leq \varepsilon_w \tag{32}$$

it is absorbed by the wall element and the pursuit of the particle is terminated. If

$$\text{RND}_r > \varepsilon_w \tag{33}$$

it is reflected in the direction determined by equations (24) and (25). The new penetration distance from the reflected point is obtained by subtracting the length along which the particle has already traveled from the original penetration distance. By nature, radiative heat transfer is three-dimensional. Hence, the pursuit of energy particles must be carried out in three dimensions even under one- or two-dimensional boundary conditions. The present method can easily meet the requirement of three-dimensional treatment.

5. RESULTS
5.1 one-dimensional radiative transfer

In order to ascertain the validity of the present method, some results of one-dimensional analyses are compared with the corresponding analytical results for the non-scattering case [8]. The upper-wall temperature of the rectangular duct shown in Fig.1 is 2000K, while the lower wall is 1000K. The adiabatic side walls are specular. The radiative properties of the dispersed gas confined within the walls are listed in Table 1, for case(a) through (e). The cases(a)(b) and (c) determine both the effects of the absorption coefficient under non-scattering condition. They are used to check the accuracy of the present method. The cases(b)(d) and (e) evaluate the effects of the scattering albedo under constant extinction coefficient.

Figure 10 and 11 depict the distributions of the temperature in the gas region and the wall heat flux, respectively. It is seen that the present results agree very well with the analytical ones. The validity of the present method is thus

Table 1 Radiative properties used in present study

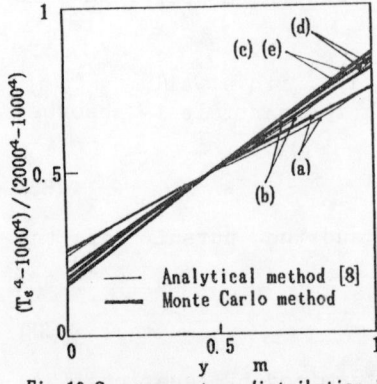

CASE	Without scattering (a)	(b)	(c)	With anisotropic scattering (b)	(d)	(e)
a (m^{-1})	1	2	3	2	1.5	1
σ_s (m^{-1})	0	0	0	2	0.5	1
β (m^{-1})	1	2	3	2	2	2
ω	0	0	0	0	0.25	0.5

Fig. 10 Gas temperature distribution in one-dimensional radiative transfer

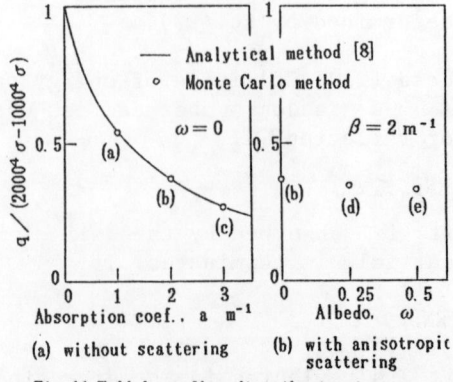

Fig. 11 Wall heat-flux distribution in one-dimensional radiative transfer

confirmed. The figures also show that increases in both the absorption coefficient and the scattering albedo result in steeper temperature gradients and reduce the wall heat flux. As in the case of radiative equilibrium, the isotropic-scattering albedo exerts little effect on radiative heat transfer when the extinction coefficient is kept constant. Therefore, the effects of the albedo shown in Fig.10 and 11 resulting from the differences among cases (b)(d) and (e) can be attributed to the anisotropic scattering phase function shown in Fig.7. In other words, the anisotropic scattering which is characterized by a strong backward scatter contributes to resisting radiative heat transfer.

5.2 two-dimensional radiative transfer

In order to extend a two-dimensionality into the rectangular duct system, a 0.2m-width higher temperature region at 2500K is provided at the center of the upper wall at 2000K. The adiabatic side walls are diffuse surfaces. Both the temperature profile and the wall heat flux are calculated under the same gas conditions shown in Table 1. Each case requires approximately 40seconds of computation time using a HITAC M-680H computer.

Figure 12 shows the effects of the absorption coefficient for cases(a) and (c), while Fig. 13 depicts the effects of the scattering albedo for cases(b) and (e). These figures reveal that the gas temperature profiles near the higher temperature wall region are very similar, within the ranges of the absorption coefficient and the albedo covered in the present study. Figure 14 illustrates the wall heat fluxes for all cases. All conclusions obtained for the one-dimensional case are also applicable to the two-dimensional one. Figure 14 shows that the upper wall heat fluxes fall near the ends of the higher temperature region, x=±0.1m, which is a typical two-dimensional effect. This is caused by the increase in the gas temperature near the higher temperature wall region in Figs. 12 and 13. Another two-dimensional effect shown in the figure

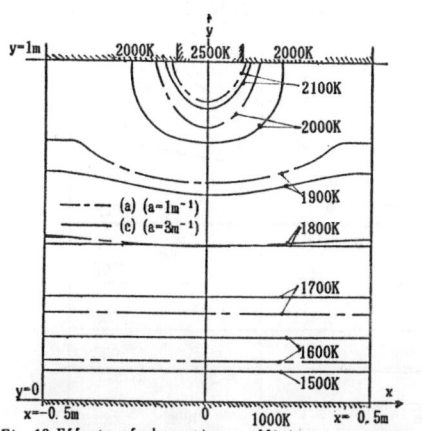

Fig. 12 Effects of absorption coefficient on temperature distribution in two-dimensional radiative transfer for no scattering (σ_s=0) case

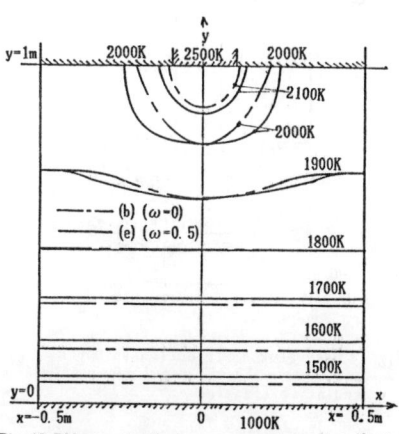

Fig. 13 Effects of albedo on temperature distribution in two-dimensional radiative transfer with anisotropic scattering (β=2m^{-1})

is a rise in the lower wall heat fluxes near the side walls, $x=\pm 0.5$m. This tendency is also detected in the case of uniform upper wall temperature (2000K) and diffuse adiabatic side walls, which is not presented here. Therefore, the latter effect is not caused by the higher temperature region of the upper wall, but rather is due to the diffuse side walls. Figure 15 compares qualitatively the effects of surface characteristics on radiative transfer between the specular and diffuse reflections. The diffuse wall tends to resist radiative heat transfer in the direction parallel to the emitting surface causing a reduction in lower-wall heat fluxes near the side walls as seen in Fig.14.

Figure 16 compares the temperature distributions in two-dimensional radiative heat transfer without scattering, case (c), and with anisotropic scattering, case (e). Like the one-dimensional system in Fig.10, the thermal profiles of both cases nearly coincide. This is not true for wall heat fluxes. It is seen in Fig.11 and 14 for one- and two-dimensional radiative transfer, respectively that the wall heat flux distribution for no scattering case (c) deviates appreciably from that for anisotropic scattering case (e). It is thus concluded that radiative heat transfer with anisotropic scattering cannot be simulated by non-scattering radiative transfer with an effective absorption coefficient.

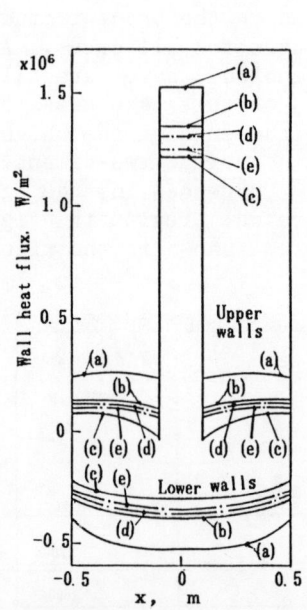

Fig. 14 Heat-flux distribution in upper and lower walls in two-dimensional radiative transfer

Fig. 15 Effects of surface characteristics on radiative energy transport

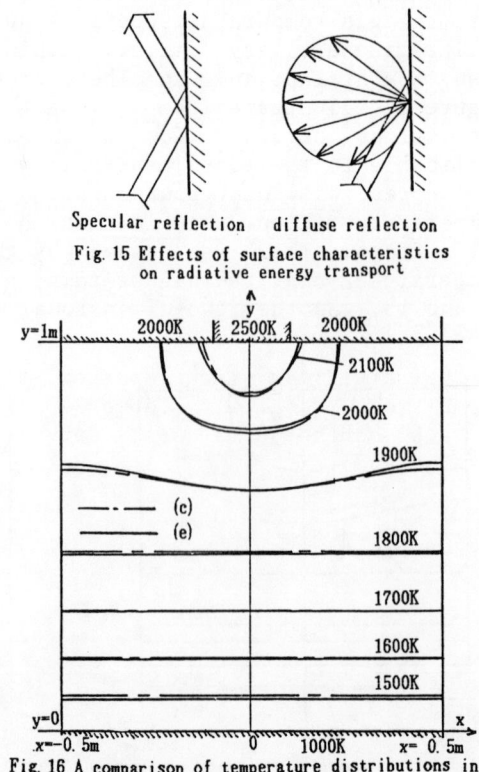

Fig. 16 A comparison of temperature distributions in two-dimensional radiative transfer without scattering, case (c), and with anisotropic scattering, case (e)

6. CONCLUSIONS
(1) An analytical method is developed to treat three-dimensional radiative heat transfer in absorbing-emitting and anisotropically scattering media by a new computation-time-saving, Monte Carlo technique.
(2) An increase in the absorption coefficient and/or the single scattering albedo has adverse effects on radiative heat transfer under anisotropic scattering condition. It results in an increase in temperature gradient in the gas region and a decrease in wall heat flux.
(3) Anisotropic scattering effects cannot be simulated through the use of an effective absorption coefficient.

The authors wish to acknowledge financial support by the Tanigawa Foundation. Numerical computations were performed at the Hokkaido University Computer Center.

REFERENCES
1. Vortmeyer, D. - Radiation in Packed Solids, Heat Transfer 1978(Proc.6th.Int.Heat Transfer Conf.),$\underline{8}$,525-39(1978).
2. Viskanta, R. - Radiation Heat Transfer: Interaction with Conduction and Convection and Approximate Methods in Radiation, Proc.7th Int. Heat Transfer Conf., $\underline{1}$,103-21 (1982).
3. Mengüç, M.P. and Viskanta, R. - Comparison of Radiative Transfer Approximations for a Highly Forward Scattering Planer Medium, J. Quant. Spectrosc. Radiat. Transfer, $\underline{29}$, 5,381-94(1983).
4. Taniguchi, H., Yang, W.-J., Kudo, K., Hayasaka, H., Oguma, M., Kusama, A., Nakamachi, I. and Okigami, N. - Radiant Transfer in Gas Filled Enclosures by Radiant Energy Absorption Distribution Method, Heat Transfer 1986 (Proc. 8th Int. Heat Transfer Conf.), $\underline{2}$,757-62(1986).
5. Siegel, R. and Howell, J.R. - Thermal Radiation Heat Transfer, 2nd ed., 751-62,McGraw-Hill, New York, 1981.
6. ibid., 582.
7. ibid., 371.
8. Usiskin, C.M. and Sparrow, E.M. - Thermal Radiation Between Parallel Plates Separated by an Absorbing-Emitting Nonisothermal Gas, Int. J. Heat Mass Transfer, 1,28-36(1960).

MODELING OF SEVERE THERMAL GRADIENTS NEAR FLUID PASSAGES USING SPECIAL FINITE ELEMENTS

J. H. Chin and D. R. Frank

Lockheed Missiles and Space Co., Inc.
Sunnyvale, California, U. S. A.

SUMMARY

Special Galerkin finite elements using exponential shape functions for the fluid and linear shape functions for the containing wall are shown to be effective in modeling of severe thermal gradients near fluid passages. The application of these special elements for engineering analysis of transient three- dimensional heat transfer problems is demonstrated.

1. INTRODUCTION

Many engineering heat transfer applications involve fluid passages in solid configurations of various complexity. It is well known that oscillations of fluid and solid temperatures often result in a finite difference or finite element analysis. These temperature oscillations may be reduced by using fine meshes in the region of severe temperature gradients. The use of upwind weighting functions [e.g., 1, 2] helps to reduce the temperature oscillations. In a previous paper [3], a special flowing fluid element was formulated, by using exponential shape functions for the fluid and linear shape functions for the solid wall of the fluid passage. By comparing the steady results of a conducting pipe flow problem, it was demonstrated that for $\beta = 1000$ (a large value of the ratio of element convective heating to mass transport heating), the grid must be refined by a factor of $1000/3$ if the basic upwind elements were used. For large values of β, the major fluid temperature change occurs near the fluid entrance region. Thus improved results may be obtained by using only one or two special elements with exponential shape functions at the entrance.

The present paper presents further experience of using this class of special flowing fluid elements in a transient heat transfer analysis. In

Section 2, the fluid rod and fluid quadrilateral elements are reviewed. The formulas for the special Galerkin elements are given, using exponential shape functions for the fluid and linear shape functions for the solid. The iterative scheme for numerical solution is presented. Then, an alternate approach for transient analysis using these special elements is discussed. The numerical results and applications are given in Section 3.

2. ANALYSIS

The mathematical model of the engineering fluid elements is described in this section.

2.1 Engineering fluid elements

Figures 1a and 1b show, respectively, the engineering Fluid Rod and Fluid Quadrilateral elements with the local node numbering. The fluid rod element models the longitudinal bulk temperature change of the fluid and the longitudinal conduction in the solid. The fluid quadrilateral element allows two-directional conduction in the solid as the heat transfer surface is one of the faces for a regular, eight-cornered, isoparametric brick element.

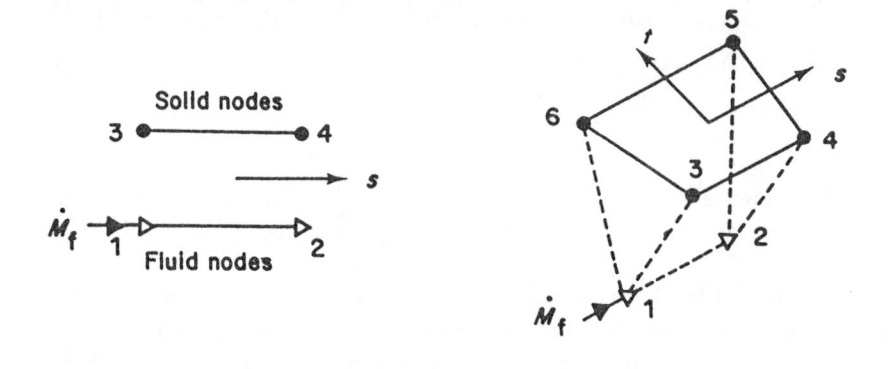

(a) Fluid Rod (b) Fluid Quadrileral

Fig. 1 Flowing Fluid Elements

For most engineering applications other than those for heat transfer to liquid metals, the thermal conductivity of the fluid is neglected.

The energy balance for the fluid in the fluid rod element is written as

$$\int_0^{\Delta x} \rho_f c_{vf} A_f \frac{\partial T_f}{\partial t} dx + \int_0^{\Delta x} \dot{m}_f A_f c_{pf} \frac{\partial T_f}{\partial x} dx \\ + \int_0^{\Delta x} hp(T_f - T_s) dx = 0 \,. \qquad (1)$$

where T_f and T_s are the fluid and solid temperatures, respectively; $\dot{m}_f A_f$ is the fluid mass flow rate which is constant over the element length Δx; h is the heat transfer coefficient and p is the heating perimeter.

For the steady state problem and when the solid temperature is linear within an element, the solution of Eq. (1) may be expressed as:

$$T_f = N_{f1} T_{f1} + N_{f2} T_{f2} \,, \qquad (2)$$

where N_{f1} and N_{f2} are the element shape functions for the fluid. The expressions for these shape functions are given below [3]:

$$N_{f1} = 1 - N_{f2} \,, \qquad (3)$$

$$N_{f2} = a \left(\frac{1 - e^{-\beta \xi}}{1 - e^{-\beta}} \right) + (1 - a) \xi \,, \qquad (4)$$

$$a = \frac{[\beta(T_{s1} - T_{f1}) - (T_{s2} - T_{s1})]}{[\beta(T_{s1} - T_{f1}) - (T_{s2} - T_{s1})] + \beta(T_{s2} - T_{s1})/(1 - e^{-\beta})} \,, \qquad (5)$$

$$\beta = hp \Delta x / \dot{m}_f c_{pf} A_f \,, \qquad (6)$$

$$\xi = (x - x_1)/\Delta x \,. \qquad (7)$$

The element shape functions for the solid rod are linear in ξ; they are given by $N_{s1} = 1 - \xi$ and $N_{s2} = \xi$, respectively.

In the finite element Galerkin formulation, the exponential shape functions for the fluid defined above are applied as weighting functions to Eq. (1). The linear shape functions are applied as weighting functions to the energy balance equations for the rod. The element matrix equations for the fluid rod element in terms of the local node numbering are given below [3]:

$$(\rho_f c_{vf} A_f \Delta x)^e \begin{pmatrix} C_{11} & C_{12} \\ C_{21} & C_{22} \end{pmatrix}^e \begin{pmatrix} \dot{T}_1 \\ \dot{T}_2 \end{pmatrix}^e + \left[(\dot{m}_f A_f c_{pf})^e \right. \\ \left. \begin{pmatrix} K_{11} & K_{12} \\ K_{21} & K_{22} \end{pmatrix}^e + (hp\Delta x)^e \begin{pmatrix} C_{11} & C_{12} \\ C_{21} & C_{22} \end{pmatrix}^e \right] \begin{pmatrix} T_1 \\ T_2 \end{pmatrix}^e$$

$$- (hp\Delta x)^e \begin{pmatrix} C_{13} & C_{14} \\ C_{23} & C_{24} \end{pmatrix}^e \begin{pmatrix} T_3 \\ T_4 \end{pmatrix}^e = 0, \tag{8}$$

$$Other\ Terms + (hp\Delta x)^e \begin{pmatrix} C_{33} & C_{34} \\ C_{43} & C_{44} \end{pmatrix}^e \begin{pmatrix} T_3 \\ T_4 \end{pmatrix}^e$$

$$- (hp\Delta x)^e \begin{pmatrix} C_{31} & C_{32} \\ C_{41} & C_{42} \end{pmatrix}^e \begin{pmatrix} T_1 \\ T_2 \end{pmatrix}^e = 0, \tag{9}$$

where *Other Terms* account for solid conduction and other modes of heat transfer for the rod. These *Other Terms* will be included during the assemblage.

The analytical expressions for the symmetric matrix C_{ij} and the asymmetric matrix K_{ij} are given below [3]:

$$C_{ij} = \int_0^1 N_i N_j d\xi, \qquad K_{ij} = \int_0^1 N_i \frac{dN_j}{d\xi} d\xi$$

$$C = \begin{pmatrix} 1 - 2C_2 + C_{22} & C_2 - C_{22} & \frac{1}{2} - C_2 + C_{24} & \frac{1}{2} - C_{24} \\ & C_{22} & C_2 - C_{24} & C_{24} \\ & & \frac{1}{3} & \frac{1}{6} \\ & & & \frac{1}{3} \end{pmatrix}$$

$$K = \begin{pmatrix} -(1 - K_{22}) & (1 - K_{22}) \\ -K_{22} & K_{22} \end{pmatrix}$$

$$C_2 = \int_0^1 N_2 d\xi = a \left(\frac{1}{b_1} - \frac{1}{\beta} \right) + \frac{c_1}{2}$$

$$C_{22} = \frac{c_1^2}{3} + c_0 \left(c_0 \left(1 + \frac{b_1(b_0 - 3)}{2\beta} \right) + c_1 (1 + 2c_2) \right)$$

$$C_{24} = \frac{c_1}{3} + c_0 \left(\frac{1}{2} + c_2 \right)$$

$$K_{22} = \frac{c_1^2}{2} + c_0 \left(c_1 + b_1 \left(\frac{b_1 c_0}{2} - \frac{c_1}{\beta} \right) - \beta c_1 c_2 \right)$$

$$b_0 = e^{-\beta}, \quad b_1 = 1 - b_0, \quad c_0 = \frac{a}{b_1}$$

$$c_1 = 1 - a, \quad c_2 = \frac{\beta b_0 - b_1}{\beta^2}$$

2.2 Solution scheme

According to Eqs. (3) to (7), except for the special case $T_{s1} = T_{s2}$ (hence $a = 1$), the fluid shape functions N_{f1} and N_{f2} depend nonlinearly on T_{s1} and T_{s2}. To provide an iterative solution procedure for nonlinear problems, Eqs. (8) and (9) are first reduced to the following

form:
$$C_{ij}^e \dot{T}_j + K_{ij}^e T_j + F_i^e = 0 \quad , (i = 1, N_p), \tag{10}$$

where N_p is the number of nodal points of element e. The global assemblage of Eq. (10) is integrated with respect to time according to a generalized mid-point scheme [4]:

$$\left(\frac{C_\theta^m}{\Delta t} + \theta K_\theta^m\right) \Delta T_{n+1}^m = -K_\theta^m T_\theta^m - \frac{C_\theta^m}{\Delta t}\left(T_{n+1}^m - T_n\right) - F_\theta^m. \tag{11}$$

The subscript θ denotes quantities evaluated at the weighted time and temperature (t_θ, T_θ) between times t_n and t_{n+1} as follows:

$$t_\theta = (1-\theta)t_n + \theta t_{n+1}, \tag{12}$$
$$T_\theta = (1-\theta)T_n + \theta T_{n+1}, \tag{13}$$

where θ ($0 \leq \theta \leq 1$) is a weighting factor. The correction between the iteration steps m and $m+1$ is given by:

$$\Delta T_{n+1}^m = T_{n+1}^{m+1} - T_{n+1}^m. \tag{14}$$

The steady state equation may be obtained by setting $\Delta t \to \infty$, $\theta = 1$, and dropping the subscripts θ and $n+1$ in Eq. (11).

In Eq. (11), C_θ^m and K_θ^m depend upon the element shape functions and are thus nonlinear functions of T_θ^m according to Eqs. (3) to (7). A modified Newton-Raphson scheme is used for the successive iteration of Eq. (11) until the corrections defined by Eq. (14) are below a specified small absolute value. With this iterative procedure, many calculations of the right-hand-side (RHS) of Eq. (11) may be performed per calculation of the left-hand-side (LHS) matrix coefficients for the corrections ΔT_{n+1}^m.

For transient analysis, in order to reduce the computer time, it is expedient to freeze the fluid shape functions during the RHS iterations, at their values computed using the known T_n instead of the iterative T_θ^m. With this approach, the number of RHS iterations required may be reduced. In fact, when the thermal and transport properties are temperature independent, K_θ^m, C_θ^m, and F_θ^m are constant so that only one RHS calculation is required to compute ΔT_{n+1} for advancement from $T_n (= T_{n+1}^0)$ to T_{n+1}. If Δt is constant for several time steps and the boundary conditions and temperature distribution do not change appreciably, a new calculation of the LHS matrix coefficients is only required when the iteration count for the RHS calculations exceeds a specified value.

3. NUMERICAL RESULTS AND APPLICATIONS

The algorithm described in Section 2 was incorporated into the CAFE computer program [5]. Section 3.1 presents the results of a simple pipe flow problem. A more complex model is then used in Section 3.2 to demonstrate the effectiveness of the special flowing fluid elements.

3.1 Conduction pipe flow problem

In a previous paper [3], results were presented for a steady state, coupled conduction-convection pipe flow problem, which was simulated by means of 5 fluid rod and 5 rod elements. The transient results for the same model problem are obtained here using the CAFE code. In order to verify the CAFE algorithm, fine-grid results with 200 grid intervals are calculated using an independent finite-difference code.

Figures 2a and 2b show the the time history results for the fluid and solid nodes, respectively, for $\beta = 1$. The node numbers for the finite element model and the values of the parameters used are given in the insert to Fig. 2a. The two ends of the pipe are maintained at 100 and 250 deg, respectively. The temperature is initially linear along the pipe. The fluid enters the pipe at 50 deg and instantaneously fills the pipe at time zero. The heat transfer coefficient has a value 0.1 to yield a unity value of β. The finite element results using all exponential elements show a slightly better agreement with the finite difference solution than the results using all linear Petrov-Galerkin elements.

To show the usefulness of the exponential elements to model problems with severe thermal gradients, the heat transfer coefficient is increased by a factor of 50 so that $h = 5$ and $\beta = 50$. The results for the fluid and solid temperatures are shown in Figs. 3a and 3b, respectively. The Petrov-Galerkin finite element solution shows an unacceptable over- or under-shooting behavior for fluid nodes 2, 3 and solid nodes 8, 9. In contrast, the exponential element algorithm yields a smooth solution.

3.2 Application to typical three-dimensional problem

The steady state results for a three-dimensional heating panel were presented in a previous paper [3] to demonstrate the effectiveness of the Galerkin elements utilizing exponential fluid shape functions. The transient results are demonstrated here using the same model. Figure 4 shows the geometry of the heating panel constructed with 840 linear solid elements and 60 flowing fluid quadrilateral elements. The upper and lower boundaries of the panel are maintained at 100 and 0 deg, respectively. The temperature is initially linear along the height of

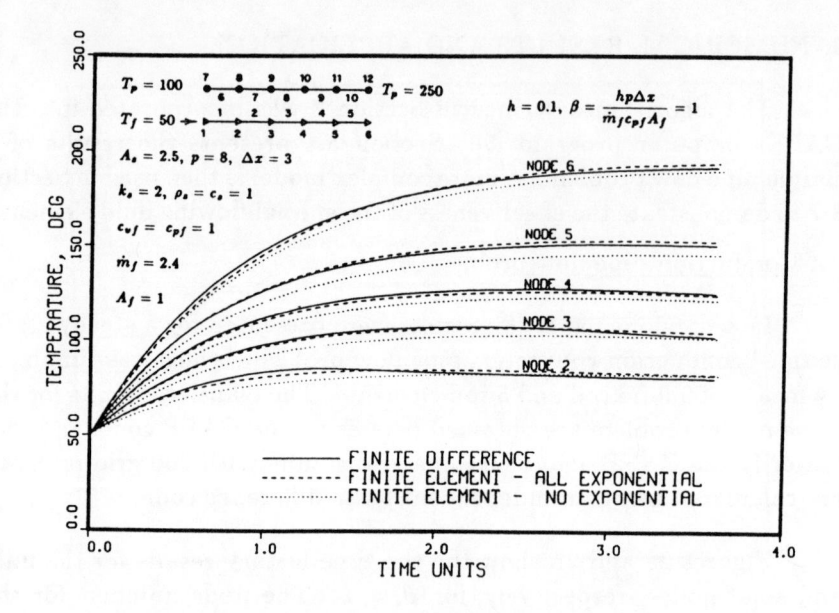

Fig. 2a Temperatures of Fluid Nodes along Pipe, $\beta = 1$

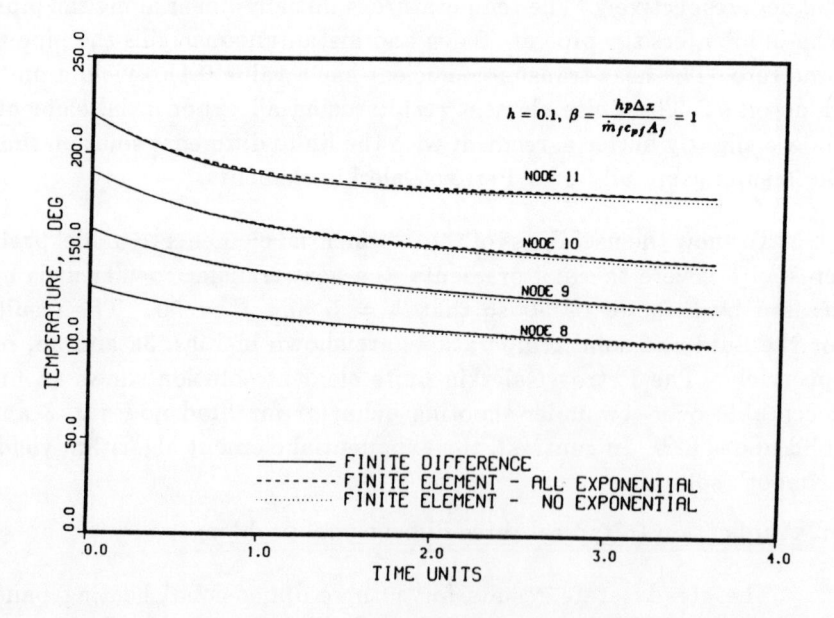

Fig. 2b Temperatures of Solid Nodes on Pipe, $\beta = 1$

the panel. The fluid is routed first through the upper channel, then the middle, and finally the bottom channel as indicated in Fig. 4. The fluid enters at 50 deg and instantaneously fills the channels at time zero. The flow conditions are selected to produce a *beta* value of 50.

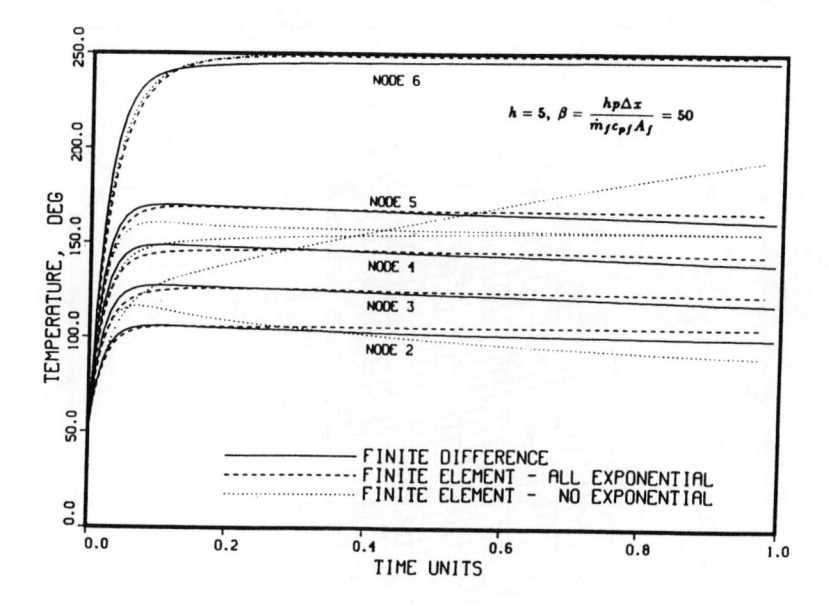

Fig. 3a Temperatures of Fluid Nodes along Pipe, $\beta = 50$

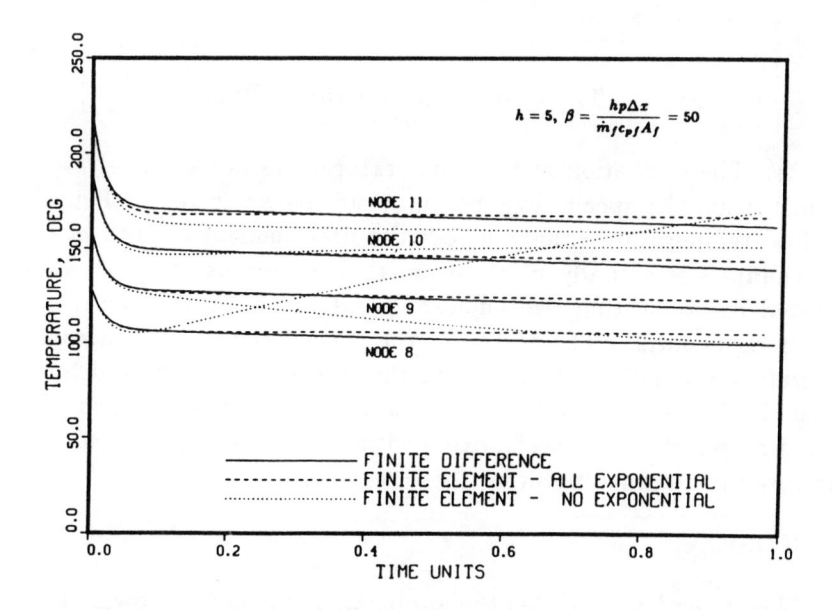

Fig. 3b Temperatures of Solid Nodes on Pipe, $\beta = 50$

Figures 5a and 5b show the fluid and solid temperature distributions along the channels at two different times: time 1 is early in the transient and time 2 is at a later time near the steady state. The results of Fig. 5a are obtained using the basic, linear upwind (Petrov-Galerkin)

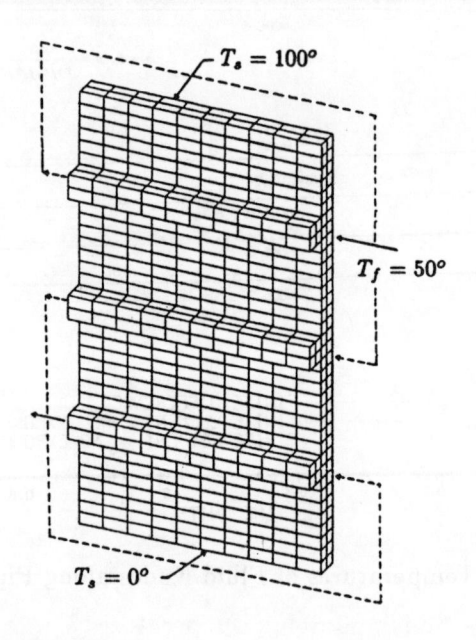

Fig. 4 Three-Dimensional Heating Panel

elements. The oscillation of the temperatures are noted. The results obtained using the special exponential elements are given in Fig. 5b. The fluid temperature change across the first element of each of the three channels is actually more severe than shown as the curves are generated by connecting two adjacent points by a straightline rather than an exponential. The ability of the special elements to suppress temperature oscillations near the channel entrances is clearly demonstrated. If the locations of large thermal gradients are known, as with the heating panel, these special exponential elements may be used as a substitute for a refined mesh for these regions.

CONCLUSION

This paper has extended the application of a special flowing fluid element, formulated by means of a Galerkin procedure with exponential shape functions for the fluid and linear shape functions for the solid, to the analysis of coupled transient conduction-convection problems. The effectiveness of this element to model the severe thermal gradients near fluid passages without temperature oscillations has been demonstrated. For large three-dimensional problems, a strategic application of this

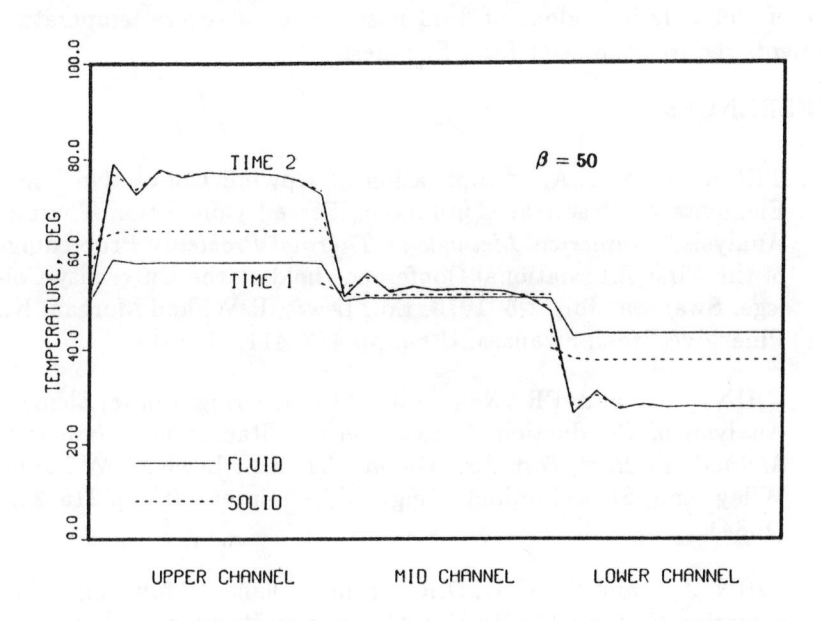

Fig. 5a Fluid and Solid Temperatures along Channels,
Using Basic Linear Elements

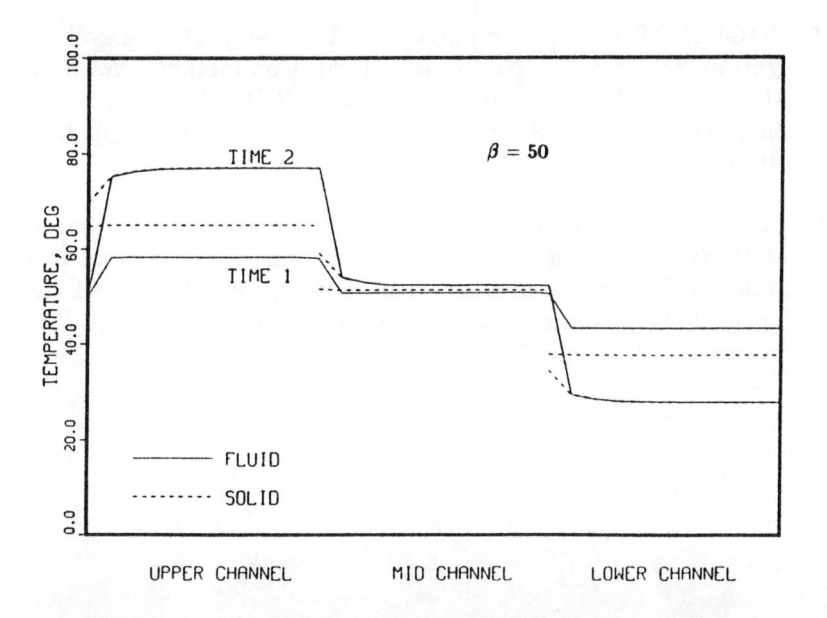

Fig. 5b Fluid and Solid Temperatures along Channels,
Using Special Exponential Elements

type of elements in regions of fluid passages with severe temperature gradients reduces the need for a fine mesh.

REFERENCES

1. THORNTON, E.A., "Application of Upwind Convective Finite Elements to Practical Conduction/Forced Convection Thermal Analysis," *Numerical Methods in Thermal Problems*, Proceedings of the First International Conference held at the University College, Swansea, July 2-6, 1979, Ed., Lewis, R.W. and Morgan, K., Pineridge Press, Swansea, U.K., pp 402-411, (1979).

2. CHIN, J.H. and FRANK, D.R., "Engineering Finite Element Analysis of Conduction, Convection and Radiation," *Numerical Methods in Heat Transfer, Volume III*, Ed. Lewis, L.W., John Wiley and Sons Limited, England, Chapter 10, pp 215-229, (1984).

3. CHIN, J.H. and FRANK, D.R., "Finite Element Modeling of Engineering Coupled Conduction-Convection Problems," *Numerical Methods in Thermal Problems*, Proceedings of the Fourth International Conference held at the University College, Swansea, July 15-18, 1985, Ed., Lewis, R.W. and Morgan, K., Pineridge Press, Swansea, U.K., pp 436-445, (1985).

4. HOGGE, M.A., "A Comparison of Two- and Three-Level Integration Schemes for Non-Linear Heat Conduction," *Numerical Methods in Heat Transfer*, Ed., Lewis, R.W., Morgan, K., and Zienkiewcz, D.C., John Wiley and Sons Limited, pp 75-90, (1981).

5. CHIN, J.H., "Charring Ablation by Finite Element," *Numerical Methods in Thermal Problems*, Proceedings of the Second International Conference held at Venice, Italy, July 7-10, 1981, Ed., Lewis, R.W., Morgan, K., and Schrefler, B.A., Pineridge Press, Swansea, U.K., pp 672-682, (1981).

ELEMENT-BY-ELEMENT SOLUTION OF STEADY STATE
HEAT CONDUCTION BY LANCZOS ALGORITHM

Alvaro L.G.A. Coutinho, José L.D. Alves,
Luiz Landau and Luiz C. Wrobel
COPPE/UFRJ - Civil Engineering Dept.
Federal University of Rio de Janeiro
Caixa Postal 68506
21945 Rio de Janeiro, Brazil

SUMMARY

Iterative methods for solving large sets of linear equations have been used as an alternative to direct methods of solution since the early beginning of numerical analysis. The conjugate gradient method (CGM), one of the most widely used, seeks a solution that minimizes the potential energy of the Finite Element assemblage. Recently, the use of Lanczos algorithm for the solution of large sets of linear equations has been examined. It has been shown that Lanczos and CGM share several properties but the former has the advantage of not needing to compute the approximate solution at each iteration. Jacobi preconditioning can also be employed in order to accelerate convergence. Following these considerations this paper presents an element-by-element (EBE) Lanczos procedure for the solution of the steady-state heat conduction equation discretized by finite elements. For all the addressed analyses the element-by-element Lanczos procedure presented outstanding computational efficiency.

1. INTRODUCTION

Iterative methods for solving large systems of algebraic equations arising from finite element and finite difference discretizations of partial differential equations are nowadays subject to a critical review facing the new generation of processors, the main feature of which is the possibility of parallelization. Generally these systems of equations can be expressed as,

$$\underline{A}\,\underline{u} = \underline{b} \tag{1}$$

where $\underline{A}$ is a large, sparse, positive definite and, restricting our discussion, symmetric matrix of order n, and u, b n-dimensional vectors, being b given. The solution sought, u, can be obtained by stationary iterative methods, such as SOR or SSOR (Hageman and Young [1]),parabolic regularization methods (Hughes [2],Zienkiewicz [3]) or variational methods as the steepest descent and conjugate gradients, introduced by Hestenes and Stiefel [4]. In the latter case it can be shown that solving (1) is equivalent to the minimization of the corresponding quadratic functional,

$$f(u) = \frac{1}{2} u^T A u - u^T b \qquad (2)$$

Another alternative is the Lanczos tridiagonalization procedure, introduced by Lanczos [5] in the same year of conjugate gradients. Lanczos method is an oblique projection method that provides a solution approximation which residual is orthogonal to a Krylov subspace. Contrary to CGM, Lanczos has the property of not needing to compute the approximate solution at each iteration. This method is also a powerful tool to solve large symmetric eigenvalue problems (Parlett[6], Cullum and Willoughby [7]). Furthermore, it is well known that Lanczos and conjugate gradients share the same properties in infinite precision arithmetic,the derivation of each method being possible from the other (see for a good presentation,Cullum and Willoughby [7]). Therefore, Lanczos procedures may be used as an alternative to conjugate gradients as suggested by Parllet [8] and even for nonsymmetric systems of equations (Hageman and Young [1], Saad [9] Paige [12]). Moreover, successful applications of Lanczos procedures have been reported, coupled with Newton-type iterations, for nonlinear optimization problems (Nash [10]) as well as nonlinear partial differential equations discretized by finite elements (Nour-Omid [11]).

Focusing on finite element systems of equations, recent works of Hughes [2] and Carey [13] suggest element-by-element (EBE) schemes. The EBE approach seems to be extremely suitable for vectorized and parallel machines. Following these considerations, this paper presents an element-by-element Lanczos procedure (EBELAN, for short) for the solution of large systems of finite element equations. The EBELAN and the EBELAN with Jacobi preconditioning (EBEJLAN) were applied to the solution of several two-dimensional steady-state heat conduction problems involving regular and irregular meshes. The results were compared with those obtained by EBE conjugate gradients procedures and, when possible, with the Accelerated Viscous Relaxation method.

2. LANCZOS ALGORITHM

The Lanczos algorithm constructs weak forms of equation (1) in the Krylov subspace,

$$K(\underset{\sim}{b}) = \text{span}\,(\underset{\sim}{b}, \underset{\sim}{A}\underset{\sim}{b}, \underset{\sim}{A}^2\,\underset{\sim}{b}, \ldots, \underset{\sim}{A}^{j-1}\,\underset{\sim}{b}) \qquad (3)$$

where j refers to j-th iteration, improving the Krylov sequence by Gram-Schmidt orthogonalization, leading to the set of vectors $\underset{\sim}{Q}_j = [\underset{\sim}{q}_1, \underset{\sim}{q}_2 \ldots, \underset{\sim}{q}_j]$ where $\underset{\sim}{Q}_j^t \underset{\sim}{Q}_j = \underset{\sim}{I}_j$. It can be shown that there is no need to proceed a complete Gram-Schmidt orthogonalization during each iteration (Parlett [8], Nour-Omid [11], Cullum [7]) only the orthogonalization against the two previous vectors being necessary. Therefore, in the j-th iteration the weak form of Eq. (1) can be written as,

$$\underset{\sim}{Q}_j^t\,\underset{\sim}{A}\,\underset{\sim}{u} = \underset{\sim}{Q}_j^t\,\underset{\sim}{b} \qquad (4)$$

and applying the coordinate transformation,

$$\underset{\sim}{u} = \underset{\sim}{Q}_j^t\,\underset{\sim}{x}_j \qquad (5)$$

the weak form is given by,

$$\underset{\sim}{T}_j\,\underset{\sim}{x}_j = \beta_1\,\underset{\sim}{e}_1^j \qquad (6)$$

where $\underset{\sim}{T}_j \equiv \text{tridiag}\,(\beta_i, \alpha_i, \beta_{i+1})$ is a matrix which entries α, β are the Gram-Schmidt orthogonalization parameters, $\underset{\sim}{e}_1^j$ is the first column of the identity matrix $\underset{\sim}{I}_j$, and each $\beta_i \geq 0$ is chosen so that $\|\underset{\sim}{q}_i\| = 1\,(i > 0)$. The sequence of operations needed to generate the vector $\underset{\sim}{q}_{i+1}$ can be summarized by,

$$\left.\begin{array}{l} \underset{\sim}{r}_i = \underset{\sim}{A}\,\underset{\sim}{q}_i - \beta_i\,\underset{\sim}{q}_{i-1} \\[4pt] \alpha_i = \underset{\sim}{q}_i^t\,\underset{\sim}{r}_i \\[4pt] \beta_{i+1}\,\underset{\sim}{q}_{i+1} = \underset{\sim}{r}_i - \alpha_i\,\underset{\sim}{q}_i \end{array}\right\} \quad i = 1, 2, \ldots \qquad (7)$$

where $\underset{\sim}{q}_o = 0$ and $\beta_1 \underset{\sim}{q}_1 = \underset{\sim}{b}$. If there were no rounding error the process would therefore terminate with $\beta_{m+1} = 0$ for some $m \leq n$. In practice, the solution sought will be generally obtained for $m \ll n$. Now it is necessary to introduce a termination criterium. Taking the residual vector at the m-th iteration,

$$\underset{\sim}{r}_m = \underset{\sim}{A}\,\underset{\sim}{u}_m - \underset{\sim}{b} \qquad (8)$$

as a convergence measure, and noting that,

$$\underset{\sim}{\hat{r}}_m = \underset{\sim}{A}\,\underset{\sim}{Q}_m\,\underset{\sim}{x}_m - \underset{\sim}{b} = \underset{\sim}{A}\,\underset{\sim}{Q}_m\,\underset{\sim}{x}_m - \underset{\sim}{Q}_m(\beta_1\,\underset{\sim}{e}_1^{(m)}) \qquad (9)$$

it can be shown that the residual vector $\underset{\sim}{\hat{r}}_m$ is a multiple of $\underset{\sim}{r}_m$,

$$\tilde{\underset{\sim}{r}}_m = \underset{\sim}{r}_m \phi_m \tag{10}$$

where ϕ_m is the m-th component of vector $\underset{\sim}{x}_m$. Further, the residual norm ρ_m is expressed by

$$\rho_m = (\tilde{\underset{\sim}{r}}_m^t \, \tilde{\underset{\sim}{r}}_m)^{1/2} = \beta_{m+1}|\phi_m| \tag{11}$$

Thus, the residual norm can be evaluated computing only the last component of vector x_m by noting that the scalar quantity β_{m+1} is determined as an entry of T_m during the Lanczos step. The determination of ϕ_m (Nour-Omid [11], Paige [12]) can be performed, without solving the tridiagonal system for x_m, by a scalar recursion similar to the one used in the SYMMLQ algorithm [14], and based on a QR factorization of T_m, at a negligible cost. Some remarks concerning the computer implementation should be made. With Lanczos, unlike conjugate gradients, there is no need to compute the approximate solution at each iteration. The required storage area, except for $\underset{\sim}{A}$ is only two n-dimensional vectors. The CG method needs, at least, four n-dimensional vectors. Finally, the solution of the weak form can be performed efficiently exploiting symmetry and the special structure of Eq. (6). The convergence rate of this procedure depends on the distribution of the eigenvalues of $\underset{\sim}{A}$. However, this rate can be improved by using a preconditioning technique. In this work, the Jacobi preconditioning matrix is employed,

$$\underset{\sim}{C} = \text{diag}(\underset{\sim}{A}) \tag{12}$$

Thus, the equation to be solved by the preconditioned Lanczos algorithm is,

$$\bar{\underset{\sim}{A}} \, \bar{\underset{\sim}{u}} = \bar{\underset{\sim}{b}} \tag{13}$$

where
$$\bar{\underset{\sim}{A}} = (\underset{\sim}{C}^{-1/2})^t \, \underset{\sim}{A} \, \underset{\sim}{C}^{(-1/2)} \tag{14a}$$

$$\bar{\underset{\sim}{b}} = (\underset{\sim}{C}^{-1/2})^t \, \underset{\sim}{b} \tag{14b}$$

$$\bar{\underset{\sim}{u}} = \underset{\sim}{C}^{1/2} \, \underset{\sim}{u} \tag{14c}$$

Applying Lanczos algorithm to the preconditioned system (13) no additional storage area is required and the final solution is easily obtained by Eq. (14c).

3. ELEMENT-BY-ELEMENT SCHEME

An important feature which is to be observed in the several iterative algorithms to solve Eq. (1) is the use of the coefficient matrix $\underset{\sim}{A}$ just as a linear operator. A typical iterative step performs vector-matrix products such as,

$$\underline{v} = \underline{A}\,\underline{r} \tag{15}$$

Having in mind the nature of a FEM approximation, the matrix A is computed as the sum of each finite element contribution over the problem domain, according to,

$$\underline{A} = \sum_{i=1}^{NE} \underline{A}_e^i \tag{16}$$

where NE is the total number of finite elements and $\underline{A}_e^i$ is a n x n matrix correspondent to the expansion of i-th finite element matrix to the global dimensions. Therefore, the vector-matrix product given in Eq.(7) can be evaluated exploiting the structure of matrix $\underline{A}$, element-by-element leading to,

$$\underline{v} = (\sum_{i=1}^{NE} \underline{A}_e^i)\,\underline{r} = \sum_{i=1}^{NE} (\underline{A}_e^i\,\underline{\bar{r}}_i) \tag{17}$$

where $\bar{r}_i$ is an expanded vector with element related nodal quantities. The relation (17) is the essence of the simple element-by-element scheme, as introduced by Hughes [2] and Carey [13]. The outstanding characteristics of such scheme can be readily seen, once nodal points ordering as well as elements ordering does not affect the number of required arithmetic operations while evaluating (17). Such scheme does not require the global coefficient matrix assemblage thus reducing the required storage area and further, proving a good mean for the treatment of global matrix sparsity. For the Lanczos algorithm with Jacobi preconditioning, the preconditioning matrix is assembled once and each element matrix is evaluated in this transformed form as,

$$\underline{\hat{A}}_e^i = (C^{-1/2})^t\,\underline{A}_e^i\,(C^{-1/2}) \tag{18}$$

$\underline{\hat{A}}_e^i$ being the preconditioned expanded matrix for the i-th finite element. However, it should be noted that the computations involved are performed at element level, the element matrices being sequentially stored.

4. NUMERICAL APPLICATIONS

This section presents numerical results obtained from the application of the standard Lanczos procedure (EBELAN) and the Jacobi preconditioned procedure (EBEJLAN) to typical steady-state heat conduction problems discretized by 4-noded isoparametric elements. Results are compared with EBE conjugate gradient procedures, the standard CG identified as EBECG, the Jacobi CG method (EBEJCG) and the modified CG, introduced by Gambolatti [18], also implemented using Jacobi preconditioning

(EBEMCG). Whenever possible, results are also compared with the EBE Accelerated Viscous Relaxation method [3]. All these procedures were implemented by the authors [15,16,17] using FORTRAN IV and the examples processed in the Burroughs B6800 machine from the Computer Centre of the Federal University of Rio de Janeiro.

4.1 - Heat Generation in a Square Domain

The heat generation in a 5 x 5 square domain considering Dirichlet and Neumann boundary conditions was analysed. This is a very well-conditioned problem. The ratio between the maximum and minimum values of the main diagonal of the conductivity matrix that bounds the condition number is just 4.0. Several analyses were performed for regular meshes with increasing number of degrees of freedom (NDOF). For a residual norm tolerance of 10^{-3}, Table 1 shows the obtained results.

MESH	NDOF	MAXIMUM NUMBER OF ITERATIONS					
		EBECG	EBEJCG	EBEMCG	EBELAN	EBEJLAN	EBEAVR
5x5	25	8	7	7	9	6	57
10x10	100	16	14	11	17	16	178
20x20	400	28	27	25	31	23	418
40x40	1600	55	51	47	57	40	*
60x60	3600	80	76	68	82	58	*
80x80	6400	103	99	88	106	76	*
100x100	10000	125	121	109	131	94	*

(*) No convergence after 1000 iterations.

Table 1 - Maximum Number of Iterations for the Heat Generation in a Square Domain

As it can be seen in Table 1 EBEJLAN requires less iterations to reach convergence than any other method. It also should be noted that EBELAN requires almost the same number of iterations than EBECG. The results presented for the EBEAVR employ the parameters recommended by Zienkiewicz [3].

4.2 - NASA Insulation Test Problem

This example, taken from Hughes [2], is posed on a

L-shaped domain with prescribed nodal temperatures in its ends, all the other edges being insulated. The main characteristics of this problem and the finite element mesh employed are given in Figure 1.

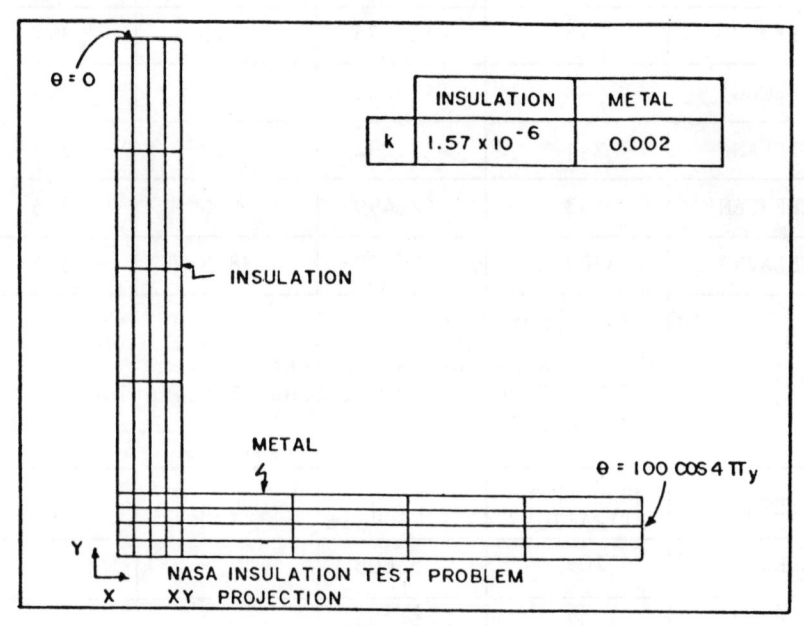

Fig. 1 - NASA Insulation Test Problem

Although this discretization involves a small number of degrees of freedom, this problem is very ill-conditioned since the finite element mesh is irregular and there are different materials. Moreover, the nonzero nodal prescribed temperatures were imposed by assigning a very large value in the corresponding degrees of freedom of the coefficient matrix. Two analyses were performed considering exact (2x2) and reduced (1x1) Gauss numerical integration for the evaluation of the element matrices. For the first analysis (2x2), the ratio between the maximum and minimum values of the conductivity matrix main diagonal is 1.3375×10^9. Table 2 presents, for this analysis, the required number of iterations for a preset residual tolerance of 10^{-12}, the decimal logarithm of the residual norm in the last iteration as well as the CPU and I/O times for each method. For the reduced integration solution, the computed bound of the conductivity matrix condition number is 1.78357×10^9. The results for this analysis are shown in Table 3.

From the results presented in Tables 2 and 3, it should be noted that, although EBEJLAN requires more iterations to reach convergence than the preconditioned CG methods (EBEJCG and EBEMCG) it is as fast as those methods. These computer performances are directly related to the fact that the main

METHOD	NUMBER OF ITERATIONS	$\log \|\underset{\sim}{r}\|_{min}$	CPU(s)	I/O(s)
EBECG	202	-13.661	18.9	1.9
EBEJCG	45	-12.325	9.2	1.7
EBEMCG	48	-13.652	9.1	1.9
EBELAN	241	-12.232	23.9	3.3
EBEJLAN	53	-12.499	9.0	2.5
EBEAVR*	3209	-3.009	238.3	8.4

(*) No convergence

Table 2 - NASA Insulation Test Problem
Results for 2x2 Gauss Integration

METHOD	NUMBER OF ITERATIONS	$\log \|\underset{\sim}{r}\|_{min}$	CPU(s)	I/O(s)
EBECG	209	-13.605	17.5	1.4
EBEJCG	57	-14.737	8.3	1.4
EBEMCG	57	-14.352	8.4	1.6
EBELAN	266	-13.540	23.7	3.3
EBEJLAN	62	-13.996	8.3	2.3
EBEAVR *	1542	- 6.981	114.1	4.6

(*) No convergence

Table 3 - NASA Insulation Test Problem
Results for 1x1 Gauss Integration

loop of the preconditioned Lanczos algorithm needs only two n-dimensional vectors. However, the task of assembling the final solution after convergence by Eq.(5) is the main responsible for the high I/O times of the Lanczos procedure. It can also be seen that the exact integration solutions required less iterations than the reduced integration solutions, once the condition number bound for the former is smaller than for the latter. The bad performance of the EBEAVR solution can be credited to the pseudo-time evaluation needed in this method, which is a function of the eigenvalues of the of the conductivity matrix. In our implementation the Gershgorin bound is used and it is well known that for very ill-conditioned matrices this bound is poor. Therefore, the computer performance of EBEAVR can be improved and further studies should be performed on this subject.

4.3 - Space Shuttle Orbiter Thermal Protection System

The last example is the thermal model of the Space Shuttle Orbiter Thermal Protection System given by Williams and Curry [19]. A sketch of this thermal model, the geometrical characteristics and material properties are shown in Figure 2.

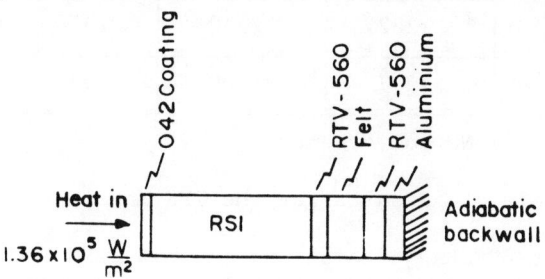

MATERIAL	THICKNESS (cm)	CONDUCTIVITY (W/mK)
042 coating	0.0381	0.8599
RSI	7.366	0.0495
RTV 560	0.01965	0.3113
Felt	0.4191	0.0294
Aluminium	0.1524	1.207906

Fig.2 - The TPS Model

The TPS model was discretized with 1262 equally spaced 4-noded isoparametric finite elements. The resulting mesh comprises 2524 degrees of freedom and halfbandwith 4. Using 2x2 Gauss integration for evaluation of the element matrices, the ratio between the maximum and minimum values of the conductivity matrix main diagonal is 4.1085×10^3. Therefore, this problem can be considered ill-conditioned. For a residual norm tolerance of 10^{-12}, Table 4 presents the number of iterations needed to reach convergence, the CPU and I/O times for each method.

As can be seen in Table 4, EBEJCG, EBEMCG and EBEJLAN required the same number of iterations to reach convergence within the preset tolerance. However, EBEJLAN presents a

METHOD	NUMBER OF ITERATIONS	CPU(s)	I/O(s)
EBECG	4620	4697	74
EBEJCG	1262	2365	74
EBEMCG	1262	2239	76
EBELAN	4567	4897	684
EBEJLAN	1262	2212	256
EBEAVR*	11917	21600	92

(*) No convergence

Table 4 - Results for the TPS Model

speed-up against EBEJLAN of 6.5% and against EBEMCG of 1.2%. The number of iterations needed for all preconditioned procedures is equal to the number of finite elements used in the discrete model. This is not a coincidence and it is related to the particular geometry and loading of the TPS model and the Jacobi preconditioning technique. The Jacobi technique spreads the residual in each iteration on one element after the other. This behaviour can be well understood by the examination of Figure 3. This Figure shows the plot of the residual norm evolution for each preconditioned method.

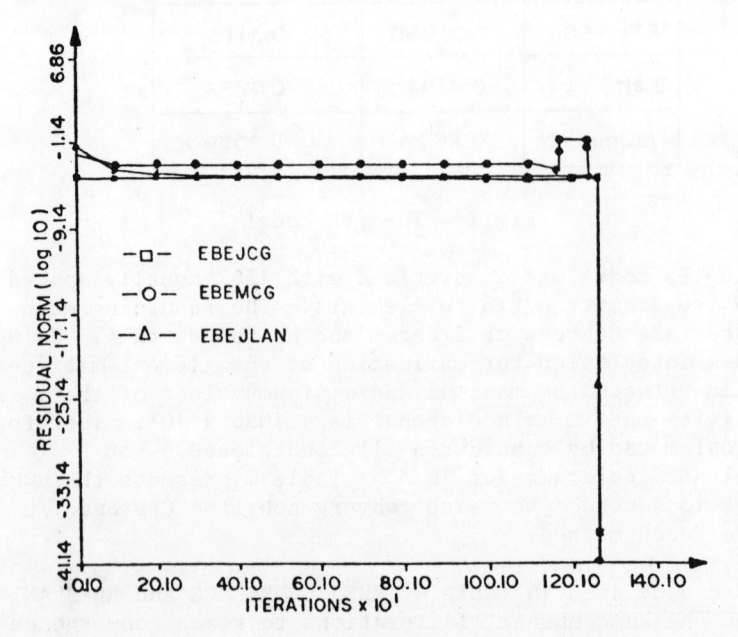

Figure 3 - Residual Norm Evolution

As can be seen in Figure 3, the residual norm is pratically constant and falls abruptly when the number of iterations reachs the number of elements.

5. DISCUSSION AND CONCLUSIONS

The EBE Lanczos procedures presented herein notably the Jacobi preconditioned procedure (EBEJLAN), achieved better, or at least equivalent computer performances than the EBE procedures based upon conjugate gradients (EBECG, EBEJCG and EBEMCG). The addressed numerical applications range from well to very ill-conditioned problems encompassing typical thermal engineering problems. Regarding the AVR solutions, it is our opinion that more detailed studies concerning the estimate of pseudo-time intervals and line searches shall be carried out trying to achieve computer performances similar to those reported in the related reference. Finally, it has been reported by several authors for instance, Kincaid et al [20] Seager [21], successful applications of conjugate gradients in high-speed vectorized and parallel environments. We do believe that Lanczos based procedures offer an interesting alternative which behaviour in such environments should be carefully examined.

6. REFERENCES

1. HAGEMAN, L.A., YOUNG, D.M., Applied Iterative Methods, Academic Press, New York, 1981.

2. HUGHES, T.J.R., WINGET, J.M., Solution Algorithms for Nonlinear Transient Heat Conduction Analysis Employing Element-by-Element Iterative Strategies, Comp. Meth. Appl. Mech. Engng., 52, 711-815, 1985.

3. ZIENKIEWICZ, O.C., LÖHNER, R., Accelerated Relaxation or Direct Solution? Future Prospects for FEM, Int. J. Num. Meth. Engng. 21, 1-11, 1985.

4. HESTENES, M.R. STIEFEL, E., Method of Conjugate Gradients for Solving Linear Systems, J. Res. Nat. Bur. Standards, Sect. B., 47, 471-478, 1952.

5. LANCZOS, C., Solution of Systems of Linear Equations by Minimized Iterations, J. Res. Nat. Bur. Standards, Sect. B. 49, 33-53, 1952.

6. PARLETT, B.N., The Symmetric Eigenvalue Problem, Prentice Hall, Englewood Cliffs, 1980.

7. CULLUM, J.K., WILLOUGHBY, R.A., Lanczos Algorithm for Large Symmetric Eigenvalue Computations, Vol. I Theory, Vol. II Programs, Birkhäuser, Boston, 1985.

8. PARLETT, B.N., A New Look at the Lanczos Algorithm for Solving Symmetric Systems of Linear Equations, Lin. Alg. Appl., 29, 323-346, 1980.

9. SAAD,Y., The Lanczos Biorthogonalization Algorithm and other Oblique Projection Methods for Solving Large Unsymmetric Systems, SIAM J.Num. Anal., 19, 485-506, 1982.

10. NASH,S.G., Newton-type Minimization via the Lanczos Method, SIAM J. Num. Anal., 21, 770-788, 1984.

11. NOUR-OMID, B., A Newton-Lanzos Method for Solution of Nonlinear Finite Element Equations, Comp.& Struct., 16 242-252, 1983.

12. PAIGE, C.C., Saunders, M.A., LSQR: An Algorithm for Sparse Linear Equations and Sparse Least Square, ACM Trans. Math. Soft., 8, 43-71, 1982.

13. CAREY, G.F., JIANG, B.-N., Element-by-Element Linear and Nonlinear Solution Schemes, Comm. Appl. Num. Meth., 2, 145-154, 1986.

14. PAIGE, C.C., SAUNDERS, M.A., Solution of Sparse Indefinite Systems of Linear Equations, SIAM, J. Num. Anal., 12, 617-629, 1983.

15. COUTINHO, A.L.G.A., The Lanczos Algorithm in Static and Dynamic Structural Analysis, D.Sc. Thesis, COPPE/UFRJ, to be submmited.

16. COUTINHO, A.L.G.A., ALVES, J.L.D., LANDAU, L., CONGR Program: An EBE Conjugate Gradient Solver for FEM Equations, Res. Report, COPPE/UFRJ, 1986 (in Portuguese).

17. COUTINHO, A.L.G.A., ALVES, J.L.D., LANDAU,L., AVR Program: An EBE Accelerated Viscous Relaxation Solver for FEM Equations, Res. Report, COPPE/UFRJ, 1986 (in Portuguese).

18. GAMBOLATTI, G., Fast Solution to Finite Element Flow Equations by Newton Iteration and Modified Conjugate Gradient Method, Int. Num. Meth. Engng.,15 , 661-675,1980.

19. WILLIAMS, S.D., CURRY, D.M., An Implicit-Iterative Solution of the Heat Conduction Equation with a Radiation Boundary Condition, Int. J. Num. Meth. Engng., 11,1605-1619, 1977.

20. KINCAID, D.R., OPPE, T.C., YOUNG, D.M., Vectorized Iterative Methods for Partial Differential Equations,Comm. Appl. Num. Meth., 2, 289-296, 1986.

21. SEAGER, M.K., Parallelizing Conjugate Gradient for Cray X-MP,Parallel Computing, 3, 35-47, 1986.

PREDICTION OF THE HEAT TRANSFER BETWEEN FLUID AND STRUCTURE IN TRANSIENT FLOWS USING LOCAL WALL MODELS

C. Lacroix, P.L. Viollet

Laboratoire National d'Hydraulique
Electricité de France
6 Quai Watier - 78401 CHATOU - FRANCE

Abstract : In many industrial flows the walls are usually considered in the computations as adiabatics or without heat capacity. In some cases, during transient flows the heat capacity of the boundaries can modify the temperature and the velocity fields of the flow.

To get the temperature and the heat transfer through the walls in fluid dynamics calculations a new local wall model has been developed.

In the flow calculation this method allows to get free from the logarithmic layer assumption and to reach better accuracy in the wall friction and the heat transfer prediction than the usual models.

In the wall temperature calculation, this method allows to take in account the heat capacity of the steel.

This method can be applied to two or three dimensional fluid calculations without increasing the size of the grid for the elliptic computation.

1. INTRODUCTION

The modification of the temperature and the velocity fields of the flow by the heat capacity of the boundaries are not usually considereted in fluid computation during transient.

To get the temperature and the heat transfer through the walls in fluid dynamics calculations a new local wall model has been developed. It consists of a one-dimensional calculation between each internal point near the boundary of the elliptic calculation and the external face of the wall.

The model has been tested in the two-dimensional numerical model SBIRE which solves the Navier-Stokes equations using a two equations turbulence model ($k - \varepsilon$).

A comparison has been performed with experimental results on a thermal transient in a vertical pipe.

The computation gives good results on the temperature prediction near the wall and on the decrease of the temperature gradient in the pipe due to the heat capacity of the walls.

2. DESCRIPTION OF THE NEW TECHNIQUE FOR NEAR-WALL AND WALL MODELLING

2.1 Near wall modelling

In order to obtain a better wall treatment for the wall-bounded flows, a new method is investigated. The purpose of the method is :

a) to use a local low-Reynolds number treatment leading to better accuracy than the usual wall functions (no logarithmic velocity profile to be assumed) ;

b) not to increase the size of the grid for the elliptic computation of the recirculating flows.

The basic assumption is that, inside the computing mesh close to the wall, advection terms as well as diffusion terms in the direction of the wall are negligible with respect to diffusion terms in the normal direction of the wall, x_n. Thus, the Navier-Stokes equations as well as the equations of the $k - \varepsilon$ turbulence model reduce to one-dimensional equations, along the normal direction x_n, for the tangential velocity u, k and ε,

$$0 = \frac{\partial}{\partial x_n}(\frac{\nu_T}{\sigma_k} + \nu)\frac{\partial k}{\partial x_n} + P - \varepsilon \qquad (1)$$

$$0 = \frac{\partial}{\partial x_n}(\frac{\nu_T}{\sigma_\varepsilon} + \nu)\frac{\partial \varepsilon}{\partial x_n} + \frac{\varepsilon}{k}(c_{\varepsilon 1} f_1 P - c_{\varepsilon 2} f_2 \varepsilon) \qquad (2)$$

$$0 = \frac{\partial}{\partial x_n}(\nu + \nu_T)\frac{\partial u}{\partial x_n} - \frac{1}{\rho}\frac{\partial p}{\partial x_\tau} \qquad (3)$$

The turbulence model equations (1) and (2) must include the low Reynolds number effects which appear inside the near-wall region. A review of the models avaible for that purpose may be found in Launder [1] for instance.

Here, following Lam & Bremhorst [2], the $c_{\varepsilon 1}$, $c_{\varepsilon 2}$ and c_μ constants are multiplied respectively by the functions f_1, f_2, f_μ, defined as :

$$\begin{aligned} f_\mu &= 0.5 \, (1 - e^{-AR_k})^2 \cdot \left[1 + \sqrt{1 + 50/R_t} \, (1 - e^{-AR_k})^2 \right] \\ f_1 &= 1 + (A'/f_\mu)^3 \\ f_2 &= 1 - e^{-R_t^2} \end{aligned} \qquad (4)$$

The local Reynolds numbers R_t and R_k are expressed as :

$$R_t = k^2/\nu\varepsilon \quad ; \quad R_k = \sqrt{k}\, x_n/\nu$$

and the constants chosen as : $A = 0.0119$; $A' = 0.05$.

The wall boundray conditions are :

$$u = 0 \; ; \; k = 0 \; ; \; \frac{\partial \varepsilon}{\partial x_n} = 0$$

The latter condition results from an assumption on the behaviour of velocity fluctuations inside the viscous sublayer : in thus sublayer, viscous effects are dominant and the fluctuation u' is generated by a fluctuation of the velocity at the outer boundary of the sublayer u' may thus be assumed to show a linear depedance versus x_n, from which the wall boundary condition for ε results.

The above equations define a one-dimensional wall model, which is coupled to the general elliptic computation of the recirculating flow in the following way (figure 1), at each time-step of the computation :

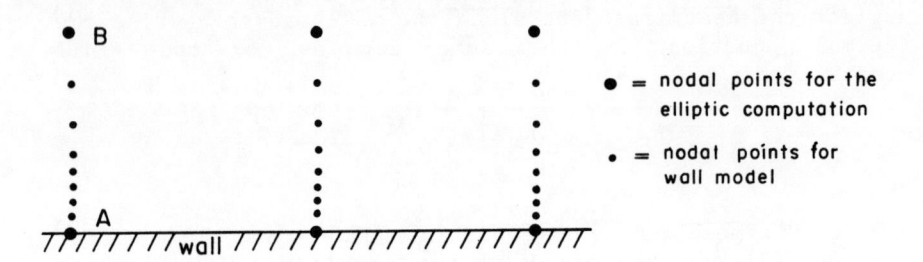

Figure 1 : Definition sketch for the wall treatment

- from the elliptic computation at step n, values of u (B), k (B), $\mathcal{E}$ (B) are taken as outer boundary conditions for the wall model ;

- from the wall model results, values of $\partial k/\partial x_n$ (B), $\partial u/\partial x_n$ (B), $\partial \mathcal{E}/\partial x_n$ (B) are taken as boundary conditions for the step (n + 1) of the elliptic computation.

In the results presented in the following section 3, as an additional simplification to the one-dimensional wall model, the pressure gradient term has been omitted in equation (3).

A hydraulic validation of this method can be find in [3].

2.2 <u>Modelling of the heat transfer</u>

In the fluid calculation we have :

$$\frac{\partial T}{\partial t} + u_j \frac{\partial T}{\partial x_j} = \frac{\partial}{\partial x_j}(Ke \frac{\partial T}{\partial x_j}) \qquad (5)$$

with Ke = diffusion therm including turbulent diffusion (calculated with the turbulent model k - $\mathcal{E}$).

Near the wall the advection term can be omitted :

$$\frac{\partial T}{\partial t} = \frac{\partial}{\partial x_j}(Ke \frac{\partial T}{\partial x_j}) \qquad (6)$$

This equation can be solved with a finite difference method using the local mesh, two boundary conditions can be used :

- temperature onto the internal face of the wall ;
- heat transfert with the wall.

To study the temperature inside the wall we can neglect the heat transfert along the wall and then write the thermal equation in the same way as in the fluid computation :

$$\text{div } \lambda \overrightarrow{\text{grad }} T - \rho C_p \frac{\partial T}{\partial t} = 0 \tag{7}$$

The mesh can be extended in the wall and the equations (6) and (7) can be solved at the same time with boundary conditions :

- temperature at the internal point of the fluid mesh coming form the elliptic computation ;

- heat flux at the external point of the wall calculated by a heat transfert coefficient.

3. VALIDATION OF THE MODEL

To get a validation on transient problem a two-dimensional case have been studied [4] . It consist in a thermal transient in a vertical pipe. At the beginning the flow and the walls are isothermal at a hot temperature and is going up in a vertical pipe. Then the temperature decreases in a few seconds. Temperature measurements in the fluid and on the walls are aviable along the pipe during the transient.

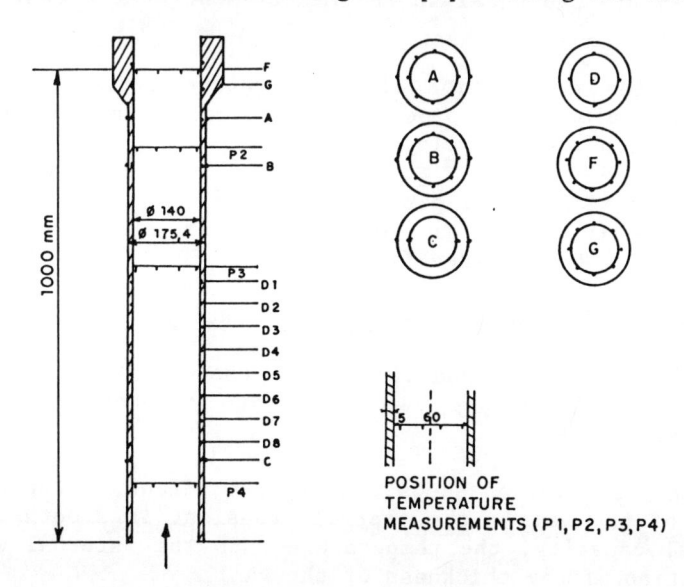

Figure 2 : Geometry of the model (DUPLEX) - (C.E.A. D.E.M.T.)

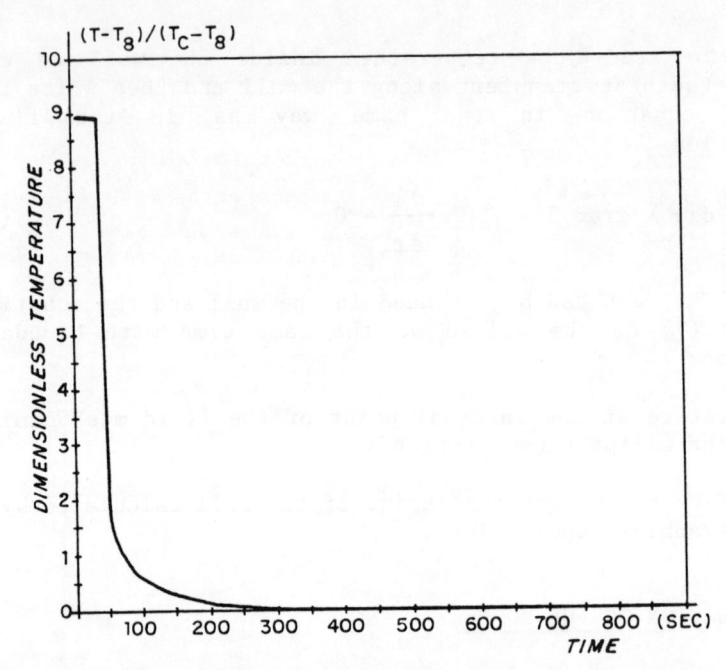

Figure 3 : Stepwise decreasing of the inlet temperature (P4)

3.1 Results whithout the heat capacity of the wall

The computation is done with a two-dimensional model with axisymetrics hypothesis. The boundary conditions at the bottom of the pipe are a turbulent profile of velocity and a uniform temperature equal to the measurement at the entrance of the pipe.

In the computation the temperature decreases very quickly all along the pipe, all temperature measurements decreasing with the same slope.

At the top of the pipe the thermal transient is quite the same as it is at the entrance while the experiment smoothes the transient (see figure 4).

3.2 Computation with the wall heat capacity

The boundary conditions are the same as in the previous calculation but is taken into account the heat capacity of the walls.

The calculation shows clearly the effect of the heat capacity of the wall : the thermal transient is smoothed by the termal capacity, the temperature onto the internal wall is a function of the thickness of the wall.

The comparison with the measurement is quite good for all measuring points.

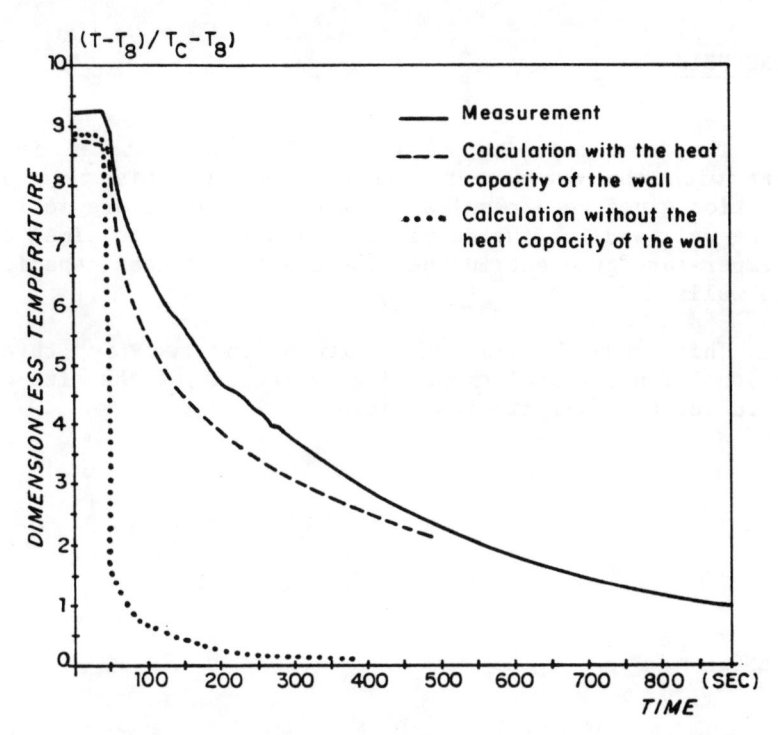

Figure 4 : Temperature at the wall near the outlet (capter G)

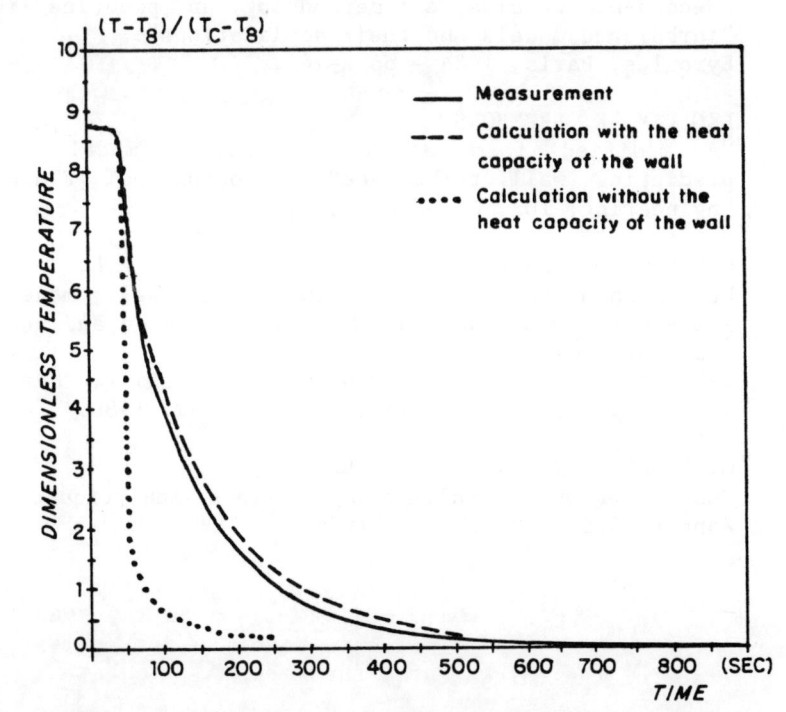

Figure 5 : Temperature in the fluid in the middle of the pipe (capter C)

4. CONCLUSION

The heat capacity of the walls can be taken into account with the one-dimensional wall model. This type of computation gives good results for the temperature prediction near the wall and for the decrease (whith respect to time) of the temperature gradient in the pipe due to the heat capacity of the walls.

This method can be applied in two or three dimensional fluid calculations without increasing the size of the grid for the elliptic computation.

REFERENCES

[1] B.E. LAUNDER
"Second-moment closure : methodology and practice" in "Turbulence models and their applications".
Eyrolles, Paris, 1984 - pp 2-147.

[2] LKG LAM, K. BREMHORST
"A modified form of the k-epsilon model for predicting wall turbulence" - Journal of Fluids Engineering, 103, 456-460, Spet. 1981.

[3] GARRETON D., LACROIX C, MAGUET Y., VIOLLET P.L.
Hot plenum computations using local wall models instead of wall functions for wall friction and heat transfer modelling.
IAHR 5th International meeting on Liquid Metal Thermalhydraulics - Grenoble - June 23-27 1986.

[4] G. CABARET, J. LAURENT, A. LE SECH
Essais Duplex. Résultats des essais Analytiques - Rapport C.E.A. D.E.M.T. 87/039.

SIMULATION OF GAPS AND CHANNELS IN CONDUCTIVE SOLID HEAT TRANSFER

Michael B. Hsu

MARC Analysis Research Corporation
Palo Alto, California, U.S.A.

ABSTRACT

The paper deals with two finite element modeling features (thermal contact gap and fluid channel flow) often encountered in conductive solid heat transfer analyses. Both the thermal gap and the fluid channel are simulated in the form of special elements and implemented into a general purpose finite element program. Two- and three-dimensional solids with gaps and channels in the model have been successfully analyzed. Numerical evaluations of the effects of gaps and/or channels on solid temperatures are presented.

1. INTRODUCTION

The demand of accurate modeling of heat transfer problems in the finite element analysis has been steadily increased in recent years. Various approximations to rather complex heat transfer problems have been widely adopted. An early study of coupled convective and conductive problems [1] and the recent work on coupled electrical-thermal analysis [2], as well as the continuous development of a general purpose nonlinear finite element program [3], make it possible for the current investigation on thermal contact gaps and fluid channels in conductive solid heat transfer analyses.

The occurrence of thermal contact gaps is typical in the casting process. The casting material is initially in a good contact with the mold, and gradually separated from the mold during cooling. A gap occurs between the solidified casting material and the mold at certain (high) temperature level, and both the convection and radiation are expected to exist in the gap. The thermal contact elements provide perfect conduction or radiation/convection between surfaces of the thermal contact gap, depending on the temperatures on the surfaces. The perfect conduction capability in the elements allows for the enforcement of equal temperatures at nodal pairs and the radiation/convection capability allows for nonlinear heat conduction between surfaces, depending on film coefficient and emissivity. The perfect conduction is simulated by applying tying constraint on temperatures of the corresponding nodal points and an automatic tying procedure has been developed for these elements. The radiation/convection capabilities in the elements are modeled by one-dimensional heat transfer in the thickness direction of the elements with variable thermal conductivity.

For the purpose of cooling, channels allowing coolant to flowing through often appear in the solid (e.g. cooling channels of the turbine blade). The fluid channel elements are designed for the simulation of one-dimensional fluid/solid convection conditions based on the assumptions which are reasonable for high velocity air flow: (i) heat conduction in the flow direction can be neglected compared to heat convection; and (ii) the heat flux associated with transient effects in the fluid (changes in fluid temperature at a fixed point in space) can be neglected. In the finite element analysis, each cooling channel is modeled using channel elements. On the sides of the channel elements, convection are applied automatically. The film coefficient is equal to the film coefficient between fluid and solid, whereas the sink temperature represents the temperature of the fluid. A two-step staggered solution procedure is used to solve the weakly coupled fluid and solid temperatures.

Both the thermal contact gap and the channel elements have the same topology of the regular heat transfer elements. Additional input is not needed for the mesh definition of these elements. However, parameters such as gap closure temperature, fluid mass flow rate, etc., must be provided for the characterizations of gaps and channels.

2. FINITE ELEMENT MODELING

As previously mentioned, both the thermal contact gap and channel conditions are simulated in the form of special elements. The modeling of these elements is described in this section.

2.1 Thermal Contact Gap Element

In problems involving thermal contact gaps, a major consideration is the open/close condition of the gaps. Although the heat transfer in the gap is rather complex, a simple gap open/close criterion has been chosen for this study. The criterion, based on the comparison of gap temperature with a prescribed gap closure temperature, appears to be satisfactory. As shown in Figure 1, the thermal contact element has two functions: (i) to serve as a radiation/convection link in gap open condition where the gap temperature is equal to or lower than gap closure temperature;

and (ii) to establish tying constraints between nodal points on element surfaces in gap close condition where the gap temperature is higher than gap closure temperature.

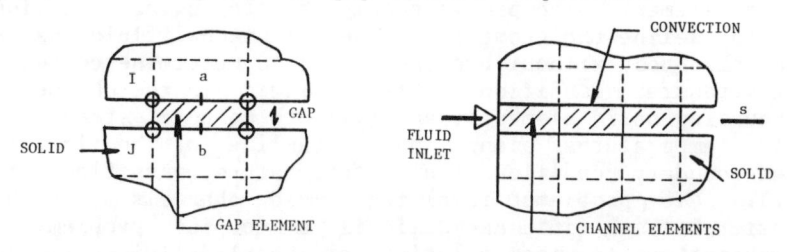

Figure 1. Thermal Contact and Channel Elements

In the gap open condition, two surface temperatures T_a and T_b at the centroid of the surfaces of a thermal contact element are interpolated from nodal temperatures. These two surface temperatures are used for the computation of an equivalent conductivity for the radiation/convection link. The expression of the equivalent thermal conductivity k_1 is:

$$k_1 = \varepsilon \cdot \sigma \cdot L \cdot (T_{ka} + T_{kb})(T_{ka}^2 + T_{kb}^2) + H \tag{1}$$

where ε is the emissivity, σ is the Stenfan-Boltzmann constant, L is the length of the element (distance between a and b), T_{ka}, T_{kb} are absolute temperatures at a and b converted from T_a and T_b; and H is the constant film coefficient.

The equivalent thermal conductivity k_1 for thermal contact element is assumed to be in the gap direction and the thermal conductivities in other two local directions are all set to zero. A coordinate transformation from local to global coordinate system allows the generation of the thermal conductivity matrix of the thermal contact element in the global system for assembly.

Similarly, in the gap close condition, tying constraints are automatically generated by the program for thermal contact elements. The constraint equation for each pair of nodes can be expressed as:

$$T_I = T_J \text{ if } T_{gap} > T_{close} \tag{2}$$

where $T_{gap} = \frac{1}{2}(T_I + T_J)$
T_I, T_J = nodal temperatures at nodes I and J
T_{close} = gap closure temperature.

2.2 Fluid Channel Element

The fluid channel element discussed in the paper is a special element which provides only the following functions: (i) to define the geometry of one-dimensional fluid channels in a solid body without branches, (ii) to determine convective heat transfer conditions between fluid in the channel and surrounding solid elements; and (iii) to compute steady-state fluid temperatures along the streamline of the flow, from given boundary conditions (inlet temperature, mass flow rate, etc.). In problems involving fluid channels, the heat transfer between fluid and solid is a coupled problem. An approximation to the solution of coupled fluid-solid heat transfer equations is discussed below.

The one-dimensional, steady-state, convective heat transfer in the fluid channel can be expressed as:

$$\dot{m}c \frac{\partial T_f}{\partial s} + \Gamma h (T_f - T_s) = 0 \tag{3}$$

$$T_f(0) = T_{inlet}$$

where $\dot{m}$ is mass flow rate, c is specific heat, T_f, is fluid temperature, T_s is solid temperature, s is streamline coordinate, Γ is circumference of channel, h is film coefficient, and T_{inlet} is inlet temperature.

Similarly, the conductive heat transfer in the solid region is governed by the following equation:

$$[c] \dot{T}_s + [k] T_s = Q \tag{4}$$

subjected to given initial condition and fixed temperature and/or flux boundary conditions. At the interface between the fluid and solid, the heat flux estimated from convective heat transfer is:

$$q = h (T_s - T_f) \tag{5}$$

In Equation (4), $[c]$ is the heat capacity matrix, $[k]$ is the conductivity matrix and Q is the heat flux vector. Equations (3) and (4) are coupled equations. The coupling is due to the unknown solid temperature T_s appearing in Equation (3) and unknown fluid temperature T_f in Equation (5) for the solution of Equation (4).

A two-step solution scheme is used to solve these coupled equations. At each time-step, the following two calculations are carried out: (i) from the known solid temperatures and the fluid inlet temperature, the fluid temperatures are calculated by integration from the inlet of the channel along the streamlines according to Equation (3); and (ii) based on the fluid temperatures computed from (i), perform a transient time-step for the solid temperatures. These two-step calculations are repeated for the next time-step until the final transient time is reached.

The solutions for Equations (3) and (4) are obtained from the introduction of a backward difference for the discretization of time variable in Equation (4) and of streamline distance in Equation (3). Let

$$\dot{T}_s = [T_s^i - T_s^{i-1}]/\Delta t \tag{6}$$

we obtain

$$(\frac{1}{\Delta t}[c] + [k]) T_s^i = Q^i + \frac{1}{\Delta t}[c] T_s^{i-1} \tag{7}$$

where Δt = time-step in transient analysis. Similarly, let

$$\frac{dT_f}{ds} = [T_f^j - T_f^{j-1}]/\Delta s \tag{8}$$

we obtain

$$T_f^j = [\Delta s \cdot \beta + T_f^{j-1}]/(1 + \Delta s \cdot \alpha] \tag{9}$$

where Δs is the streamline increment, $\alpha = \Gamma \cdot h/(\dot{m}c)$; and $\beta = \Gamma \cdot h \cdot T_s^{j-1}/(\dot{m}c)$.

3. CASE STUDIES

During this investigation, two studies were carried out for illustrating the use of thermal contact and channel elements, described in previous section, for heat transfer problems. In the first study, a simplified model of a casting problem is analyzed. The effect of thermal gaps on both the metal and mold temperatures is demonstrated. In the second study, temperatures of a generic fuel nozzle containing both radiation gaps and cooling channel have been reviewed. The finite element results are compared with that of finite difference.

3.1 Casting Problem with Thermal Gaps

A simplified casting problem during cooling is depicted in Figure 2a. The molten steel at 2845°F is assumed to be contained by a channel shaped mold made of sand. The mold is preheated to 400°F. The cooling of the steel-mold system is primarily due to the free convective heat transfer and radiative heat transfer on the inside and outside of the channel legs, respectively. The ambient temperatures are assumed to be room temperature at 70°F. In addition, a uniform temperature of 70°F is maintained at the base of the channel.

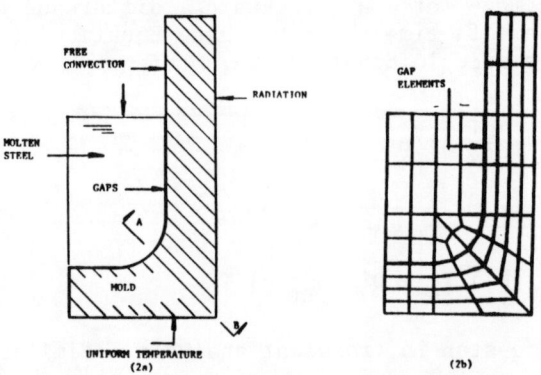

Figure 2. Simplified Casting Model

Thermal properties of the mold are assumed to be constant and those of steel are dependent on temperatures. The steel is assumed to have a solidus temperature at 2600°F; liquidus temperature at 2700°F; and a latent heat of 33.56 Btu/lb.

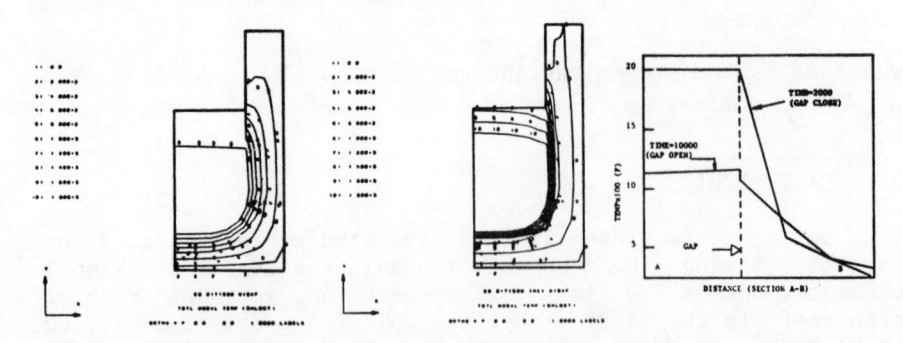

Figure 3. Temperatures in Steel and Mold

As shown in Figure 2b, a layer of thermal gap elements are placed in the mesh along the interface between steel and mold. Finite element results shown in Figure 3 clearly demonstrate the effect of thermal gaps on steel and mold temperatures.

3.2 Temperatures of a Generic Fuel Nozzle

Figure 4 represents a simplified two-dimensional model of a generic fuel nozzle for a gas turbine engine. The model made of steel containing two radiation gaps and a cooling channel, is assumed to be heated up from room temperature by a 200°F fluid flowing in the channel, as well as convective heat transfers from the exterior boundaries with ambient temperatures at 400 and 1600 degrees, respectively. Thermal properties of both the steel and fluid are dependent on temperatures and a steady-state heat transfer analysis is performed on the model.

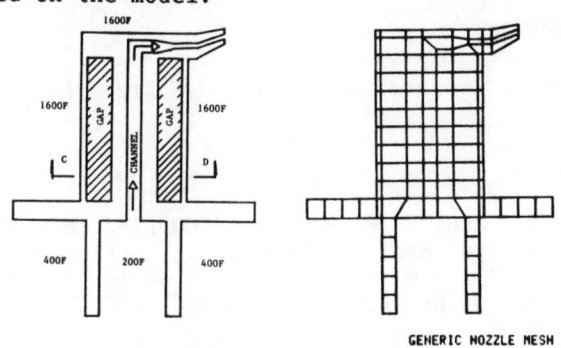

Figure 4. Simplified Nozzle Model

Steady state temperatures in the model with temperature dependent thermal properties can be obtained from a heat transfer analysis using: (i) several transient time-steps with large time increments or, (ii) one time-step with a number of iterations within the time-step. Both approaches converge to the same steady state solution.

As shown in Figure 5, comparisons of temperatures between finite element and finite difference analyses are favorable.

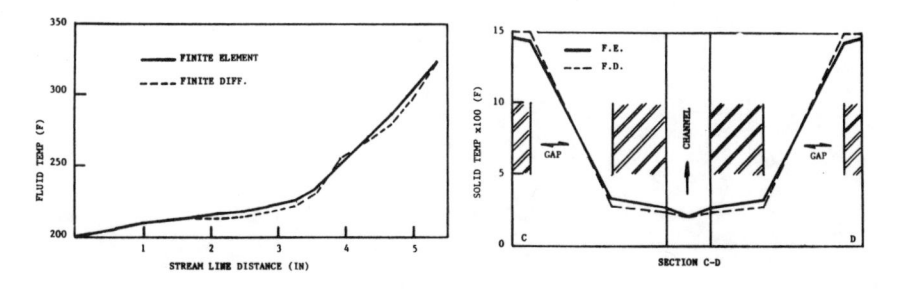

Figure 5. Solid and Channel Temperatures

4. CONCLUSIONS

Both the thermal contact gap and the channel elements described in the paper can be conveniently used in heat transfer problems involving contact gap and fluid channel conditions. Numerical results of various analyses performed to date using these elements are found to be satisfactory. During this study, simplified assumptions have been made for the heat transfers both in the gap and the channel. Better representations of gaps and channels in heat transfer analysis can be expected with improved models to be developed in future studies.

5. ACKNOWLEDGEMENT

The author is indebted to Pratt and Whitney Aircraft for the publication of this study.

6. REFERENCES

[1] M. B. Hsu, R. E. Nickell: "Coupled Convective and Conductive Heat Transfer by Finite Element Methods", <u>Finite Element Methods in Flow Problems</u>, UAH Press, Huntsville, Alabama, 1974.

[2] M. B. Hsu, "Modeling of Coupled Thermo-Electrical Problems by the Finite Element Method", Third International Symposium on Numerical Methods for Engineering, Paris, March, 1983.

[3] MARC General Purpose Finite Element Program", MARC Analysis Research Corporation, Palo Alto, California.

EVALUATION OF THE THERMAL RESPONSE FACTORS AND TRANSIENT
CONDUCTION HEAT FLOW OF MULTI-LAYERED SLABS BY HOPSCOTCH
FINITE DIFFERENCE ALGORITHM.

M. Y. Numan, B.Sc.(Arch), M.Bldg.Sc., Ph.D., M.Inst.E.

Department of Building Science and Technology
King Faisal University
Dammam, Saudi Arabia

SUMMARY

The paper presents a brief definition of the response factor
and its application in the simulation of transient conduction
heat flow. The Hopscotch procedure for the solution of one-
dimension conduction heat flow is outlined and an alternative
integrated simple and fast algorithm for the evaluation of the
response factors of multi-layered slabs by the Hopscotch
finite difference procedure is presented. The computational
efficiency and accuracy of the algorithm is illustrated by
worked examples of the response factors obtained for a variety
of composite constructions. These are compared with data obtained by alternative procedures. The general applicability
of the Hopscotch algorithm for building simulation is also
explored.

1.   INTRODUCTION:

   Complex and dynamic simulation computer models, increasingly used for the accurate and detailed analysis of buildings' thermal performances and energy demands, impose heavy
demands on machine resources, core storage and computational
time. Furthermore, the growing use of micro-computer and
micro-based computer models creates a need for efficient and
simple, but accurate, computational procedures for the evaluation of relevant heat flow modes.

   Dynamic simulation of transient conduction heat flow poses
particular difficulties. However, greater simplification
and practical computational advantages are to be gained
by the use of the response factor method [1, 2]. This method
limits the formulation of energy balance equations to surface

nodes. The procedure employed for the determination of the response factors of composite slabs involves matrix formulation of Laplace transforms and numerical inversion [3, 4]. This forms the basis of most programs [4, 5]. However, the complexity of the mathematics and numerical techniques involved is inhibiting. The quest has been therefore for an efficient numerical root-finding procedure [6, 7].

This paper describes the application of the Hopscotch finite difference algorithm for the evaluation of the response factors of composite slabs. Appropriate formulations and computational procedures for an integrated algorithm for the determination of the response factors are here outlined. The efficiency and accuracy of the algorithm is illustrated by worked examples of response factors for a variety of composite slabs. These are compared with data obtained by the NBSLD program [5]. The general applicability of the algorithm for building thermal simulation is also explored.

2. THE RESPONSE FACTORS, DEFINITION AND APPLICATION:

The response factor method is based on the super position principle and introduces time series expressions for the calculation of the one dimensional transient conduction heat flow. This expresses the overall heat flow at the surface of a slab, at a defined time level, as the sum of flux responses caused by thermal excitation pulses during preceding time intervals [1, 2]. Essential to this, is the definition of the fundamental unit response functions generated by a unit triangular surface temperature excitation pulse, in time series (1,0,0,0...). The resolution of the fundamental response functions in time series gives the response factor series. The response factors are thus unique for each constructional component and describe its inherent thermal conduction characteristics. Three basic response factor series are identified, illustrated by Fig. 1, as follows:

i) X series, which express the outside surface heat flow response due to an outside surface temperature excitation.

ii) Z series, which express the internal surface heat flow due to an internal surface temperature excitation pulse.

iii) Y series, express inside surface responses to outside excitation or vice versa.

For one-dimensional linear heat conduction, surfaces' heat fluxes are given by the fundamental convolution expressions for the response factors and the history of the surface temperature excitation, resolved in time series, as

$$Q_{1,t} = \sum_{j=0}^{\infty} X_j \cdot T1_{t-j} - \sum_{j=0}^{\infty} Y_j \, T2_{t-j} \qquad (1.1)$$

$$Q_{2,t} = \sum_{j=0}^{\infty} Z_j \cdot T2_{t-j} - \sum_{j=0}^{\infty} Y_j \, T1_{t-j} \qquad (1.2)$$

Where Q are the surfaces heat fluxes at time level 't';
X, Y and Z the response factor series; and
T1 and T2 surfaces temperature history.

In practice, the computation is limited to few finite response factor terms. This advantageously utilizes the common ratio 'CR' of successive terms of the response factor series exponential tail and past history of surface heat fluxes. The use of the response factors, thus, greatly simplifies the formulation of transient conduction heat flow of composite slab. With dynamic thermal modelling, the amount of computation involved is distinctly reduced as multi-layered constructional components are simulated by surface nodes only.

3. EVALUATION OF THE RESPONSE FACTORS BY FINITE DIFFERENCE TECHNIQUES:

The procedure commonly used for the evaluation of the response factors of multi-layered constructions employs matrix equation of Laplace transforms, for the solution of the conduction heat flow differential equation. The inversion of the Laplace transform matrix is obtained numerically for a boundary temperature excitation of a unit triangular pulse [4, 5]. However, the numerical search for the definition of the roots required by the procedure is complex and involved. This has promoted a quest for an improved root-finding procedure [6, 7].

It should be noted here that finite difference techniques are commonly used for thermal simulation model for the solution of transient conduction heat flow of composite slabs [11]. Finite difference techniques, however, offer an alternative simplified procedure for the evaluation of the response factors of composite slabs [8]. This essentially equates the heat fluxes at the surfaces, the response factors, by the heat fluxes through boundary elemental segments of the composite slab, as illustrated by Fig. 2. The heat fluxes through the boundary elemental segments are established from the knowledge of the temperature distribution across the slab, generated by the boundary excitation pulses, sampled at regular intervals. The temperature distribution across the slab is readily obtained by expressing the fundamental one-dimensional heat flow equation

$$\rho c \frac{\partial T}{\partial t} = K \frac{\partial^2 T}{\partial x^2} \qquad (2)$$

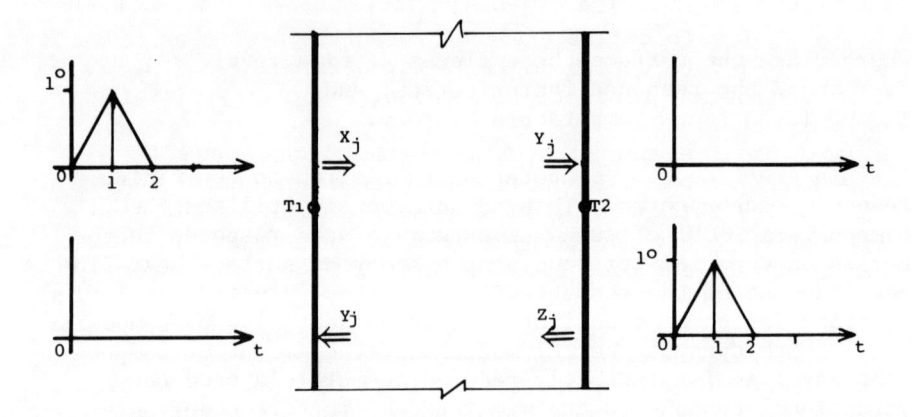

Figure 1   The Response Factors

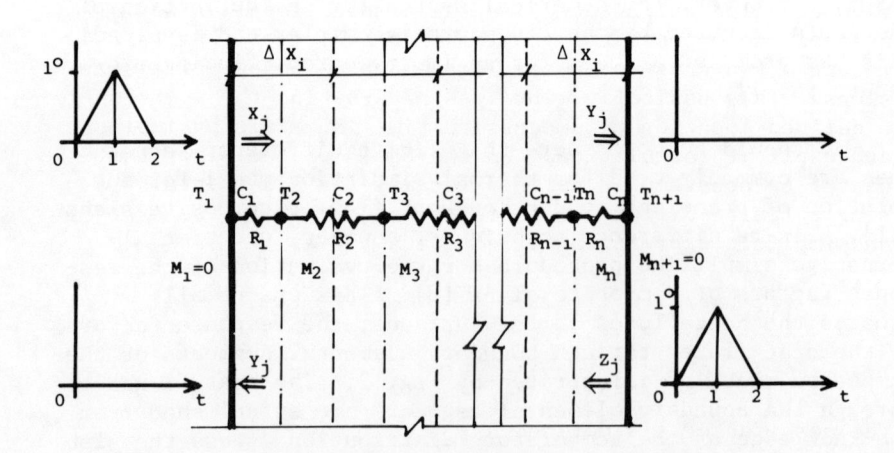

Figure 2.   Elemental Segments Representation of Multi-Layered Slab, for The Evaluation of The Response Factors by Finite Difference.

in finite difference form as

$$T_i^{j+1} = T_i^j + \Delta t/M_i [C_i(T_{i-1}^{j+m} - T_i^{j+m}) + C_{i+1}(T_{i+1}^{j+m} - T_i^{j+m})] \quad (3)$$

where $m = \begin{cases} 0 : \text{to define explicit formulation} \\ 1 : \text{to define implicit formulation.} \end{cases}$

K is the thermal conductivity, ρ density and c specific heat of slab material.
$M_i = \rho_i c_i \Delta X_i$: elemental segment thermal capacitance
$C_i = K_i/\Delta X_i$: thermal conductance between elemental segments.
'ΔX' is the thickness of an elemental segment of the slab and T_i^j and T_i^{j+1} are the temperatures of the segment at the beginning and end of a time interval 'Δt' and the superscript 'j' defines the time level.

Applying appropriate boundary conditions, the temperature distributions across the slab are obtained at regular intervals. The response factors are then defined on the basis of the temperature differences at boundary elemental segments. Computational stability, speed, storage economy and accuracy govern the choice of an appropriate finite difference technique if there is to be any practical advantage gained.

4. THE HOPSCOTCH ALGORITHM:

The Hopscotch algorithm is a fast second-order partial differential equation solver (Gourlay, [9, 10]). It provides a potent technique for the solution of the conduction heat flow equation which satisfies the computational criteria. It uniquely decomposes the problem into two simpler parts leading to a two stage process, independent of space-time dimensions. This basically, involves alternating the solution for a defined location at successive time intervals by implicit and explicit formulation. On the basis of the algorithm, the heat flow equation is defined locally for all points by a single finite difference formulation, combining the explicit and implicit expressions of equation (3) as

$$T_i^{j+1} - \lambda_i^{j+1}\{\Delta t/M_i[C_i(T_{i-1}^{j+1} - T_i^{j+1}) + C_{i+1}(T_{i+1}^{j+1} - T_i^{j+1})]\} =$$

$$T^j + \lambda_i^j\{\Delta t/M_i[C_i(T_{i-1}^j - T_i^j) + C_{i+1}(T_{i+1}^j - T_i^j)]\} \quad (4)$$

The mesh factor λ is defined as

$$\lambda = \begin{cases} 1 \text{ where } i+j \text{ is odd} \\ 0 \text{ where } i+j \text{ is even.} \end{cases}$$

This formulation enables projection of nodal temperature estimate into succeeding interval 'j+2', where i+j+2: even integer, by a simplified explicit form [9, 10]

$$T_i^{j+2} = 2T_i^{j+1} - T_i^j \quad (5)$$

This forms the basis of the fast Hopscotch procedure requiring only half the number of nodes, at each time level, for nodal temperature estimates by implicit formulation. These are readily solved as neighbouring nodes estimates are previously established. At the same time, a direct projection is obtained for nodal temperature estimates at the succeeding time interval. Throughout the calculation the procedure interchanges nodal temperature estimates at succeeding intervals by implicit and explicit formulation.

5. APPLICATION OF THE HOPSCOTCH PROCEDURE FOR THE EVALUATION OF RESPONSE FACTORS:

The Hopscotch procedure provides the basis for an integrated finite difference algorithm for the evaluation of the response factors. The algorithm, which is easy to encode and computationally efficient and economic, involves the following main main tasks:

5.1 Nodal Representation of Multi-Layered Slab:

Essential to the formulation of conduction heat flow in finite difference, is the consideration of the composite slab as discrete elemental segments which are conveniently represented by temperature nodes, as illustrated in Fig. 2. Nodal conductances and capacities are then defined according to the properties of the layers forming the slab. Two types of thermal capacitances are involved:

$M_i = 0.0$: for boundary massless nodes with no thermal capacity (6.1)

$M_i = \Delta X_i \rho_i . C_i$: for segmental node (6.2)

Five possible types of nodal conductances are identified:

i. For conductances between two mass nodes on same layer
$$C_i = K_i / \Delta X_i \quad (7.1)$$

ii. For conductance of end segment to boundary node
$$C_i = 2K_i / \Delta X_i \quad (7.2)$$

iii. For conductance between end segment and ambient boundary node with surface resistance 'R_o'
$$C_i = 2K_i / (\Delta X_i + 2K_i . R_o) \quad (7.3)$$

iv. For conductance across interface of adjacent layers
$$C_i = 2K_i K_{i+1} / (K_i \Delta X_{i+1} + K_{i+1} \Delta X_i) \quad (7.4)$$

v. For conductance across air cavity of resistance 'R_c'

$$C_i = 2K_i K_{i+1}/(K_i \cdot \Delta X_{i+1} + K_{i+1} \Delta X_i + 2K_i K_{i+1} \cdot R_c) \quad (7.5)$$

To insure computational stability, 'Δt' and 'ΔX' are defined by the relationship

$$\Delta X_i^2 = 2K_i \Delta t/\rho_i c_i \quad (8)$$

A predefined time interval, i.e. $\Delta t=120$ seconds, is initially used to establish the thickness of elemental segments of layers. The time interval and segment thickness are then re-adjusted to ensure a minimum of two segments per layer. Organization of the slab into elemental segments generates a total number of 'n' nodes. The first node 'n=1' assigned for the boundary excitation pulse. An additional node 'n+1' is required to define the second boundary condition.

5.2 Boundary Conditions:

A unit triangular temperature pulse is employed for the surface excitation, with the other boundary surface temperature maintained at constant zero. A one hour time interval is generally employed for the sampling of the response factors. This gives a base of two hours for the triangular pulse. However, a finer time mesh, of about two to three minutes, is required for the calculation of temperature distribution by the Hopscotch procedure [8, 12]. This necessitates resolution of the unit triangular pulse into smaller triangular components accordance to the time interval required, as illustrated by Fig. 3.

5.3 Temperature Distribution Computational Procedure:

Two sets of nodal temperature profiles, T_1 and T_2, are generated by the Hopscotch procedure, illustrated by the scheme shown in Fig. 4.

$T1_i$: $i=1,n$; for temperature excitation applied to the external surface.

$T2_i$: $i=1,n$; for temperature excitation applied to internal surface.

The following sequence is followed by the calculation:

i) Initial setting for nodal temperature profile, at time level, 'j=0', includes:

$T1_i^j$, $T2_i^j = 0$; for $i = 1,n$; $j = 0$

$T1_{n+1}^j$, $T2_{n+1}^j = 0$; constant for $j > 1$

Projecting estimate into succeeding interval 'j+1'

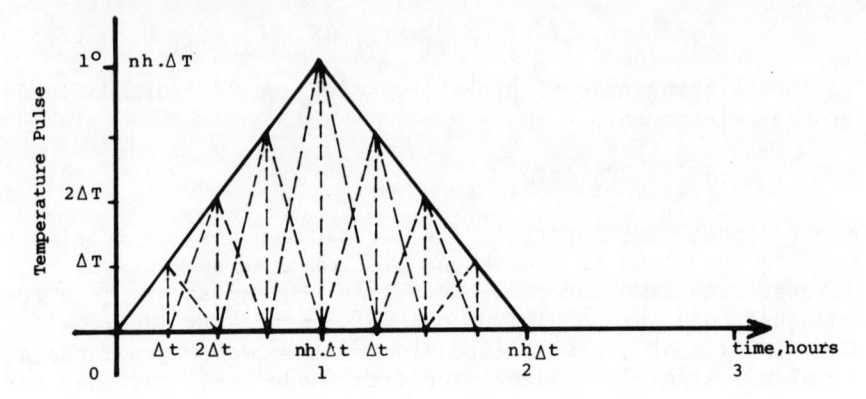

Figure 3: Resolution of Unit Triangular Temperature Pulse.

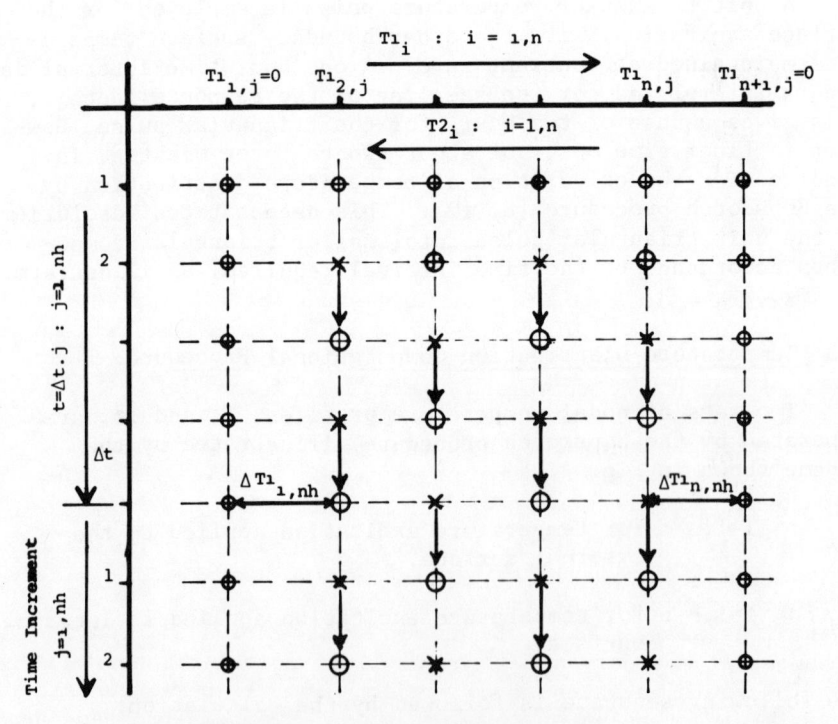

Figure 4: Hopscotch Scheme For the Evaluation of Temperature Distribution Profile of Slab.

$T1_i^{j+1}$, $T2_i^{j+1} = 0$: for 'i+j+1' even integer

ii) At succeeding intervals 'j≥1' internal nodes are sampled identifying every other node
$T1_i^j$, $T2_i^j$: i = 2,n,2 , for j≥1
Nodal temperatures are defined implicitly for 'i+j' odd and directly evaluated by equation (3) using previously determined values for neighbouring nodes (i±1) and defined excitation at boundary nodes '1' and 'n+1'.

iii) Nodal estimates are then projected into succeeding intervals 'j+1' by the explicit formulation, equation (5) for 'i+j+1' even. Thus nodal estimates are interchanged implicitly and explicitly at successive intervals.

5.4 Determination of The Response Factors

At the end of each hour, of 'nh' intervals, and before proceeding to step three the temperature differences at boundary elemental segments are sampled to determine the response factors by the expressions

$$X_m = C_1 (T1_1^{j+1} - T1_2^{j+1}) \qquad 9.1$$

$$Y_m = C_n \; T1_n^{j+1} = C_1 \; T2_n^{j+1} \qquad 9.2$$

$$Z_m = C_n (T2_1^{j+1} - T2_2^{j+1}) \qquad 9.3$$

The calculation is terminated when the common ratio 'CR' of successive terms (i.e. X_m, X_{m-1} and Z_m, Z_{m-1}) is established.

Conclusions

The Hopscotch Finite Difference procedure for the evaluation of response factors of multi-layered slabs achieves the twin objectives of accuracy and computational efficiency. Its accuracy compares favourably with results obtained using the NBSLD routine, illustrated by the examples Tables 1 and 2. The mean relative error is generally less than 1%. The algorithm used requires less storage and attains results 5 to 10 times faster (dependent upon the number of layers) than other techniques.

General application of the Hopscotch procedure for building thermal simulation would eliminate matrix inversion; facilitate thermostat switch modelling of short time step; cope with non-linearities [12]. To handle numerous convective and radiant exchanges at boundary nodes the explicit formulation of equation (5) would be extended to project nodal temperature estimates for interacting surrounding nodes. Temperature dependent coefficients can be expressed linearily with acceptable accuracy [11]. The direct solution of implicit formulation is readily performed and the interchanging Hopscotch routine maintained.

LAYER	W	K	ρ	c	R		
1	.0150	.6536	1830.0	619.0	.0000	Tile
2	.0150	1.1180	2100.0	619.2	.0000	Mortar
3	.1500	1.2040	2200.0	650.2	.0000	Concrete
4	.0000	.0000	.0	.0	.2442	Air Space
5	.0100	.0946	300.0	1548.0	.0000	Wood Board
6	.0130	.0748	200.0	619.2	.0000	Tex

W : Layer Thickness ,m.
K : Thermal Conductivity,W/m °C
ρ : Density ,Kg/m³
c : Specific Heat ,J/Kg °C
R : Resistance ,°C m²/W

HOPSCOTCH METHOD

NO. OF TERMS   NT : 9
COMMON RATIO   CR : .78199
U-VALUE   U : 1.46059

NBSLD PROGRAM

NO. OF TERMS   NT : 9
COMMON RATIO   CR : .78202
U-VALUE   U : 1.46059

RESPONSE FACTORS

J	X	Y	Z
0	19.72528	.00334	2.58317
1	-10.12894	.11981	-.85247
2	-2.16254	.24902	-.06407
3	-1.35790	.23094	-.04561
4	-1.01390	.18596	-.03507
5	-.78591	.14620	-.02734
6	-.61357	.11444	-.02137
7	-.47966	.08951	-.01671
8	-.37508	.07000	-.01306

RESPONSE FACTORS

J	X	Y	Z
0	19.67253	.00325	2.58473
1	-10.04748	.12646	-.85403
2	-2.18657	.24399	-.06391
3	-1.36320	.22969	-.04566
4	-1.01392	.18556	-.03507
5	-.78526	.14605	-.02733
6	-.61312	.11439	-.02137
7	-.47947	.08951	-.01671
8	-.37506	.07003	-.01308

TABLE 1: THE RESPONSE FACTORS OF MULTI-LAYERED SLAB BY THE HOPSCOTCH FINITE DIFFERENCE ALGORITHM AND NBSLD TOURINE.

LAYER	W	K	ρ	c	R	
1	.0300	1.1180	2100.0	619.2	.0000	Mortar
2	.0500	.6278	2000.0	619.2	.0000	Concrete
3	.0100	.5418	2120.0	681.0	.0000	Asphalt
4	.0300	1.1180	2100.0	619.2	.0000	Mortar
5	.1200	1.2040	2200.0	619.2	.0000	Concrete
6	.0000	.0000	.0	.0	.1634	Air Space
7	.0500	.0327	200.0	619.2	.0000	Glass Wool
8	.0100	.1548	1250.0	774.0	.0000	Gypsum Board

HOPSCOTCH METHOD

NO. OF TERMS   NT : 11
COMMON RATIO   CR : .89946
U-VALUE   U : .49789

NBSLD PROGRAM

NO. OF TERMS   NT : 12
COMMON RATIO   CR : .89912
U-VALUE   U : .49789

RESPONSE FACTORS

J	X	Y	Z
0	20.48045	.00001	3.70683
1	-13.06341	.00151	-3.10998
2	-1.56734	.01587	-.04287
3	-.77023	.03350	-.00857
4	-.53086	.03980	-.00525
5	-.42868	.03924	-.00436
6	-.37038	.03642	-.00383
7	-.32841	.03312	-.00342
8	-.29393	.02990	-.00307
9	-.26395	.02693	-.00276
10	-.23730	.02424	-.00248

RESPONSE FACTORS

J	X	Y	Z
0	20.49811	.00002	3.70702
1	-13.09078	.00184	-3.10806
2	-1.56104	.01630	-.04488
3	-.77007	.03343	-.00882
4	-.53134	.03960	-.00527
5	-.42873	.03909	-.00436
6	-.37011	.03632	-.00382
7	-.32802	.03304	-.00341
8	-.29353	.02985	-.00306
9	-.26359	.02689	-.00275
10	-.23700	.02421	-.00248
11	-.21319	.02179	-.00223

TABLE 2: THE RESPONSE FACTORS OF MULTI-LAYERED SLAB BY THE HOPSCOTCH FINITE DIFFERENCE ALGORITHM AND NBSLD ROUTINE.

REFERENCES

1. Mitalas, G.P. and Stephenson, D.G.   Room Thermal Response Factors. ASHRAE Trans., 73, Part I (1967).

2. Stephenson, D.G. and Mitalas, G.P.   Calculation of Heat Conduction Transfer Functions for Multi-Layer Slabs, ASHRAE Trans., 77, Part II (1971).

3. Kusuda, T.   Thermal Response Factors for Multi-Layer Structures of Various Heat Conduction Systems. ASHRAE Trans., 75, Part I, 246-271, (1969).

4. Mitalas, G.P. and Arseneault, J.G.   FORTRAN IV Program to Calculate Z-transfer functions for the Calculation of Transient Heat Transfer Through Walls and Roofs. The First Symposium on "Use of Computer for Environmental Engineering related to Building". NBRS Washington USA., Dec. 1970.

5. Kusuda, T.   NBSLD, The Computer Program for Heating and Cooling Loads in Buildings. NBS, Building Science Series 69, Washington, 1976.

6. Kimura, K.   Scientific Basis of Air Conditioning. Applied Science, London, 1977.

7. Hittle, D.   An Improved Root-Finding Procedure For Use in Calculating Transient Heat Flow Through Multi-Layered Slabs. Preprint, U.S. Army Construction Eng. Research Lab CERL, Champaign III, 1981.

8. Numan, M. Y.   Thermal Response Factors For Multi-Layer Slab By Finite Difference Techniques. The Martin Centre for Architectural and Urban Studies, University of Cambridge, working paper Oct. 1982.

9. Gourlay, A.R.   Hopscotch; a Fast Second-Order Partial Differential Equation Solver. J. Ins. Maths. Applics., 6, pp. 375-390, (1970).

10. Gourlay, A.R. and McGuire, G.R.   General Hopscotch Algorith for the Numerical Solution of Partial Differential Equations. J. Inst. Maths. Applics., 7, pp. 216-227, (1971).

11. Clarke, J.A.   Energy Simulation in Building Design. Adam Hilger Ltd., Bristol, 1985.

12. Basnett, P.   A Comparison of Numerical Method for Simulating the Thermal Response of Building Structures. IMA Symposium 'The Environment Inside Buildings', London, May 1979.

A FINITE DIFFERENCE CONVECTIVE ALGORITHM WITH REDUCED DISPERSION

V. P. Manno and J. A. Haubold
Department of Mechanical Engineering
Tufts University
Medford, MA 02155, U.S.A.

SUMMARY

The accurate simulation of convective and diffusive transport is important in many thermal problems. The problem addressed in this paper is the undesirable characteristics of currently used numerical analogs to the first order convective terms to be either excessively dispersive or produce non-physical results. The upwind differencing scheme is an example of the former while skew upwind differencing and the QUICK algorithm are illustrative of the latter. This contribution describes the more novel approach of maintaining the upwind logic but correcting for its 'numerical diffusion' through compensating reductions in the physical diffusional transport. This approach is predicated upon the accurate quantification of this effect. Further the implementation strategy derived and demonstrated through analytical stability analysis and validation simulations allows significant improvements even in problems of infinite Peclet number (no physical diffusion). This methods called Corrective Anti-Diffusion Scheme or CADS is compared to the major alternative schemes and is shown to be superior.

NOMENCLATURE

Symbols
D-diffusivity
I-root of -1
p-cell Peclet number
t-time
u-horizontal velocity
v-vertical velocity
x-horizontal direction
y-vertical direction
z-directional index
$\alpha = (1-u^+/abs(u^+))$
$b = (u^+/abs(u^+)+u^-/abs(u^-))$
$\epsilon = (1+u^-/abs(u^-))$
φ-transported scalar

Sub/Superscripts
CF-crossflow
e-effective
i-horizontal index
j-vertical index
n-time index
x-direction correction
y-direction correction
+-top or right face
--bottom or left face

1.0 INTRODUCTION

The accurate and efficient solution of physical problems which involve both convective and diffusive transport is an important part of thermal science problems. The improvement of current state-of-the-art algorithms is required in order to answer outstanding questions in such fields as metereology, combustion science and energy storage technology. This paper addresses the formulation and implementation of finite difference representations of first order convective terms which are stable and accurate. The specific concern at hand is the characteristic of many convective treatments such as pure upwind differencing to smear spatial gradients artificially especially in situations where the velocity vector is not aligned with the calculational mesh. This parasitic dispersion or 'numerical diffusion' remains an outstanding problem. It is also important to note that although this paper is focussed upon finite difference techniques, excessive dispersion is also a characteristic of many finite element treatments in computational thermal-fluid dynamics.

The following equation describes transient convection/diffusion in a two-dimensional Cartesian coordinate system:

$$\frac{\delta\varphi}{\delta t} + u\frac{\delta\varphi}{\delta x} + v\frac{\delta\varphi}{\delta y} = \frac{\delta}{\delta x}D\frac{\delta\varphi}{\delta x} + \frac{\delta}{\delta y}D\frac{\delta\varphi}{\delta y} \quad (1).$$

This is the model equation which is utilized as the focus of this paper. It is instructive to note that the terms which are the most problematic in the numerical solution of such equations are left hand side advective terms. The second order diffusive terms are less troublesome since they emboby a physical process with no directional ambiguity. The current work addresses transport of a scalar entity.

Historically, central differencing was applied first to the convective terms due to the second order truncation error associated with that discretization. In turn, it was discovered that numerical instabilities in the form of non-physical oscillations resulted when cell Peclet numbers exceeded 2 [1]. Thus, such an approach was limited to diffusion-dominated situations. Next, the conditional 'upwind' or 'donor cell' differencing was introduced. This physically-based logic exhibited excellent stability and no non-physical oscillations. However, this stability was achieved at the cost of excessive dispersion in that spatial gradients decayed faster than any physical mechanism could justify [2].

These limitations have spawned a significant research effort to find alternative treatments. One approach involves formulating a replacement for upwind differencing. Illustrative examples are the power law differencing scheme

[3], local analytical differencing scheme (LOADS) [4], influence coefficients [5], skewed-upwind differencing (SUDS) [6] and quadartic interpolation for convective kinematics (QUICK) [7]. All these methods have produced certain improved results but all suffer from one or more of the follwing limitations - spatial oscillations (wiggles), non-conservation, computational inefficiency and implementation complexity. Nevertheless, the QUICK technique does seem to be a very useful method and should be used as the standard against which alternatives are measured.

The approach described is representative of a second general approach in which upwind differencing is maintained but its excessive dispersion is 'corrected'. This 'anti-diffusion' [8] approach has received less attention than replacement schemes. The purpose of this contribution is to explore it more fully so that a better informed decision is possible. The approach is predicated on the accurate quantification of this error which is documented in [9] and reviewed briefly in the next section. An implicit formulation of Eq. (1) is then documented. The numerical stability implications of this formulation coupled with the correction is reviewed in section 4. Sample simulations are then presented and compared to some of the aforementioned techniques. A critique and conclusions complete the presentation.

2.0 CORRECTIVE SCHEME

Numerical diffusion can be divided into two types of inaccuracy -one arising from the truncation associated with the finite differencing and the other due to a crossflow dispersion typical of multidimensional problems. This has been explained elsewhere [2] and approximately quantified [10]. The diffusive nature of the crossflow error is illustrated in the thought experiment of Fig. 1 which shows a two-dimensional array of square cells in a purely convective flow. The scalar entity is transported along the streamline which intersects the origin point, A. In pure convection this entity would also appear at A_1, A_2, etc. However, with an upwind treatment, the scalar is dispersed as illustrated. The conservation property of upwind differencing is also illustrated in that the contributions of each plane intersecting the mesh points diagonally sum to φ.

During a transport time, Δt, which is required for the advection from A to A_1 there occurs an x-directed transport of $p\varphi$ to position $(\Delta x, 0)$ and a y-directed transport of $(1-p)\varphi$ to $(0, \Delta y)$. The x-directed transport occurs at a total speed $\Delta x/\Delta t$ which is the sum of both convective and diffusive (albeit non-physical) contributions. Therefore, an effective x-directed diffusional velocity is

$$u_{cf} = (\Delta x/\Delta t) - u \qquad (2).$$

An analogous y-directed diffusive velocity exists. An effective cross-flow diffusion constant can be defined on the basis of the diffusive velocity as is illustrated below for the x-component

$$u_{cf} p\varphi = -D_{cf}(\delta\varphi/\delta x) \qquad (3).$$

The gradient in Eq. (3) may be approximated by a finite difference as follows

$$(\delta\varphi/\delta x) = (-\varphi/\Delta x) \qquad (4).$$

Hence, the respective diffusivities are

$$D_{cfx} = u\Delta x(1-p) \quad \text{and} \qquad (5a)$$

$$D_{cfy} = vp\Delta y \qquad (5b).$$

From geometrical considerations the transport time can be estimated

$$\Delta t = ((u/\Delta x) + (v/\Delta y))^{-1} \qquad (6).$$

Note that the streamline is not required to be orthogonal to the cross-grid diagonal to obtain this result. This quantification can be extended easily to three dimensions. These numerical diffusivities are used to reduce the physical diffusive contribution in the model eqaution.

3.0 IMPLICIT FORMULATION

Eq. (1) is finite differenced using an Alternating Direction Implicit (ADI) formulation [11]. The cell-wise geometry and discretization nomenclature is summarized in Figure 2. A staggered mesh is utilized in which the passive scalar as well as material properties are defined at the cell centers while velocity components and numerical diffusivities are defined at the appropriate cell faces. The streamlines are assumed known a priori and are time invariant. Physical diffusivity is held constant spatially in the interest of presentation clarity.

The time derivative is forward differenced. The convective terms are upwind differenced. The x-component convective prescription is

$$u(\delta\varphi/\delta x) = (u_{ij}/2\delta x)[a_x \varphi_{i+1j} + b_x \varphi_{ij} - e_x \varphi_{i-1j}] \qquad (7).$$

An analogous vertical term is used. The diffusive terms are second central differenced with the effective diffusivity (ie. physical diffusivity less correction) defined uniquely at each face. The precise implementation of the corrective logic has important stability implications and is discussed in the next section.

A full time advancement of the ADI procedure involves advancing the time superscript, n, by 2 since δt is half the total time advancement per calculational cycle. The x-implicit and y-implicit steps for an internal node are

$$(\varphi_{ij}^{n+1}-\varphi_{ij}^{n})/\delta t + CON_x^{n+1} + CON_y^n = DIF_x^{n+1} + DIF_y^n \quad (8a)$$

and

$$(\varphi_{ij}^{n+2}-\varphi_{ij}^{n+1})/\delta t + CON_x^{n+1} + CON_y^{n+2} = DIF_x^{n+1} + DIF_y^{n+2} \quad (8b)$$

where $CON_z = u_z(\delta\varphi/\delta z)$ $DIF = \frac{\delta}{\delta z}(D-D_{cfz})\frac{\delta\varphi}{\delta z}$.

Eq. (8) is modified slightly for boundary cells. The introduction of the aforementioned discretization into Eq. (8) yields

$$a_1\varphi_{i-1j}^{n+1} + a_2\varphi_{ij}^{n+1} + a_3\varphi_{i+1j}^{n+1}$$
$$= b_1\varphi_{ij-1}^n + b_2\varphi_{ij}^n + b_3\varphi_{ij+1}^n \quad (9a)$$

$$c_1\varphi_{ij-1}^{n+2} + c_2\varphi_{ij}^{n+2} + c_3\varphi_{ij+1}^{n+2}$$
$$= d_1\varphi_{i-1j}^{n+1} + d_2\varphi_{ij}^{n+1} + d_3\varphi_{i+1j}^{n+1} \quad (9b)$$

where $a_1 = [(-u_{ij}\epsilon_x/2\delta x) - ((D-D_x^-)/\delta x^2)]$
$a_2 = [(1/\delta t) + (u_{ij}b_x/2\delta x) - ((D_x^+ + D_x^- - 2D)/\delta x^2)]$
$a_3 = [(u_{ij}\alpha_x/2\delta x) - ((D-D_x^+)/\delta x^2)]$

$c_i = a_i$ except replace x with y and u with v

$b_1 = -c_1$ $d_1 = -a_1$ $b_3 = -c_3$ $d_3 = -a_3$
$b_2 = ((2/\delta t) - c_2)$ $d_2 = ((2/\delta t) - a_2)$.

The expected two-step tridiagonal matrix problem results.

4.0 STABILITY IMPLICATIONS

The formulation derived above was analyzed for its stability characteristics using a von Neumann analysis and a confirmatory numerical study. A detailed discussion of this analysis as well as consideration of the general topic of negative diffusion and stability can be found in [12]. The von Neumann analysis proceeds by assuming that the finite difference solution can be described as the product of a time dependent amplitude and two spatial harmonics of arbitrary wave number such as

$$\varphi_{ij}^n = \rho^n e^{Ikx} e^{Iky} \quad (10).$$

Stability is determined by the parametric limits that yield amplification factors bounded by an absolute value of 1.

Substitution of Eq. (10) into Eq. (9) yields

$$\rho[a_1 e^{-kxI} + a_2 + a_3 e^{kxI}] = (2/\delta t) - [c_1 e^{-kyI} + c_2 + c_3 e^{kyI}] \quad (11a)$$

and

$$\rho[a_1 e^{-kxI} + a_2 + a_3 e^{kxI}] = (2\rho/\delta t) - \rho^2 [c_1 e^{-kyI} + c_2 + c_3 e^{kyI}] \quad (11b).$$

After some algebra, the stability criteria may be stated as

$$(\theta-1)(\theta-1)* < 1 \quad (12)$$

where * = complex conjugate and
$\theta = 2/(\delta t(c_1 e^{-kyI} + c_2 + c_3 e^{kyI}))$.

An alernative form of Eq. (12) that is useful for analysis is

$$\delta t > (1/A) \quad \text{where } A = (c_1 + c_3)\cos(ky) + c_2 \quad (13).$$

The criterion embodied in Eq. (13) can be studied by considering the nine bounding cases summarized in Table 1. The stabilty criteria are stated in terms of the upper limit of $(D_y^+ + D_y^- - 2D)$.

Table 1: Stability Analysis Cases and Resulting Criteria

CASE	cos(ky)	v	Stabilty Criterion
1	0	>0	<vδy
2	0	<0	<-vδy
3	0	=0	<0
4	1	>0	unconditional
5	1	<0	unconditional
6	1	=0	unconditional
7	-1	>0	<vδy
8	-1	<0	<-vδy
9	-1	=0	<0

The analytical results are remarkable for a number of reasons. First, the time step, which is the traditional variational parameter of stability, does not play a central role. Second, the six non-trivial stability test cases can be expressed in a single two part condition

$$D_y^+ + D_y^- - 2D \leq \text{abs}(v_{ij})\delta y \quad (14a)$$

and

$$D_x^+ + D_x^- - 2D \leq \text{abs}(u_{ij})\delta x \quad (14b).$$

This implies that numerical diffusion corrections can be imposed which exceed the magnitude of the physical diffusivity as long as the effective diffusion constant is less than the arithmetic sum of the two cell-faced horizontal or vertical components. These theoretical results were verified in extensive numerical experiments and found to be perfectly accurate. It is this positive finding that allows the overall corrective scheme described in this work to be useful.

5.0 SAMPLE ANALYSES AND COMPARISON WITH OTHER SCHEMES

Numerical experiments are required in order to address three issues. First, is the derived stability criteria correct? This question is answered affirmatively in [12]. Second, is the scheme proposed herein an improvement over the original methodology described in [9]? Third, how does the scheme compare to established techniques?

The pure convective transport of a scalar step at an angle not orthogonal to a Cartesian uniform finite difference mesh is utilized for the first test problem. An 11x11 grid is used in order to compare the results to [14]. An inflow at 45 degrees is used as it is the most challenging from a numerical diffusion standpoint. The flow enters from the lower left corner with boundary values for the scalar entity specified as 0 on the bottom and 1 on the left side. The physical diffusivity is 0 thus representing an infinite Pe flow.

The correction quantification and solution algorithm described above are employed and the composite method is called Corrective Anti-Diffusion Scheme or CADS. In the most general case, an independent correction factor must be computed for each cell face but usually fewer are required since a correction is only required if a node face shows an outflow. This computation is done at the same time as the ADI coefficient matrix assembly thereby minimizing computational effort. The additional calculations are usually a small fraction of the total matrix assembly time. In situations where a full correction is impossible due to the violation of the stability constraint (Eq.14), a partial correction strategy is required. The strategy which yields the best results is to reduce the maximum allowable overcorrection in the x or y direction (uδx/2 or vδy/2) by a fractional multiplier. The resulting correction factors yield:

$$D_x^+ = D_x^- = .5 \ c \ (u^+ + u^-) \ \delta x/2, \quad (15a)$$

$$D_y^+ = D_y^- = .5 \ c \ (v^+ + v^-) \ \delta y/2, \quad (15b)$$

where (0<c<1). This approach is different than that suggested in [9] but the two are coincidentally identical for a 45 degree flow. Also, while the approach explored in [9] does not violate the stability conditions for a 45 degree flow, it is unstable for other angles for corrections substantially below 100%.

Several numerical experiments were performed at various fractional corrections. As expected the c=101% case leads to instabilities and eventual solution divergence. The pure upwind (c=0%) solution illustrated in contour form in Fig. 3 and three-dimensionally in Fig.4 show severe gradient erosion. Analogous plots for a 100% correction are presented in Figs. 5 and 6. The dispersive effect is eliminated and an impressively steep gradient is maintained throughout the domain. Some 'wiggles' are observed but it is important to note that they are of small amplitude and stable. Further, a partial correction of c=90% still yields acceptable results but shows negligible spatial oscillation. Fig. 7 compares CADS to other methods using the predicted profile at the x=5 plane. CADS outperforms QUICK, PLDS and LOADS. SUDS yields the exact soltion for this problem but leads to overshoots which are considerably worse than the other methods at alternative flow angles. The accuracy of CADS at nodes 5 and 8 is noteworthy in that it is 3 to 4 times more accurate than the others.

A recent IAHR standard problem [15] is utilized as the next case. It consists of a solenoidal velocity field which transports a steep gradient scalar from the inlet left bottom boundary to the symmetric outlet plane on the lower right. Both convective and physically diffusive transprt is prescribed but only the infinite Pe case is addressed since it is the most restrictive for CADS and thus represents the greatest challenge. Several levels of correction were tested. The 101% correction is unstable which provides another validation of the stability analysis. Fig. 8 illustrates the dispersive upwind solution. The outlet state is remarkable when compared to the inlet profile. Figs. 9 and 10 show the scalar profiles at the symmetry plane and the outlet, respectively. The three curves in each figure are the analytical solution, upwind result and CADS using c=90%. The CADS result shows considerable improvement in maintaining the gradient. Figs. 11 can be compared to Fig. 8. Some small amplitude oscillations are observed but again they are stable and acceptable in most applications. The higher correction results do show greater oscillation but even these simulations are not unacceptable. It is also important to note that the wiggles are more pronounced at the matchline location than at the outlet. This has been apparently observed using other methods [15]. A comparison of CADS with other schemes confirms the findings of the first sample problem in that its predictions are more accurate and the degree of spatial oscillation usually less. Finally, it is important to note that if a problem involves even a small degree of physical diffusion, a 100% CADS solution is usually possible. Thus, the oscillations seen here represent the worst prediction possible using the technique.

6.0 CRITIQUE

The experience with CADS as illustrated in this presentation and demonstrated more extensively in [13] has shown it to be a competitive alternative to established finite difference convective algorithms. The results are significant improvements of the first method formulation reported in [9]. The implicit formulation as well as the new partial correction approach are the primary reasons for the improvement. One advantage of the method is the predictability of its stability limit, which allows utilization with good confidence and little need for numerical experimentation. The maintenance of the upwind logic is attractive for many reasons. First, its intrinsic stability and physical reasonableness is preserved nearly completely. Second, the computaional effort involved in the CADS implementation coupled with the simplicity of the upwind logic yields a compuationally efficient algorithm. This is in contrast with replacement convective schemes including the QUICK algorithm. Finally, the CADS approach is more easily 'backfit' to existing software which utilizes upwind differencing. Typical coding modifications consist of a new module to compute the corrections and some changes to the matrix assembly routines. This effort is considerably less than that required if a new convective algorithm is desired. This is not a trivial advantage given the investment involved in developing and utilizing a large software system such as metereological simulation tools.

This scheme is not perfect. It exhibits the tell-tale wiggles of non-upwind schemes. The magnitude of the oscillations seen to date have been small even at high levels of correction and they are universally stable. Further, realistic problems with finite Pe show no nonphysical results. Nevertheless, more testing is required to determine if this is a severe limitation in applications of interest. On a more fundamental level, improvements in the scheme to eliminate this disadvantage need to be investigated. The utilization of this approach in the solution of non-linear problems and the transport of vector quantitites such as is required in the solution of the Navier-Stokes equations is yet to be demonstrated although a scoping analysis of its implemention has shown no apparent restrictions to its utilization. Nevertheless, practical implementation is the more relevant and as yet unanswered question.

7.0 CONCLUSION

This paper documents the formulation and successful demonstration of an improved finite difference convective algorithm. The theoretical stabilty analysis of the formulation has been shown to be an accurate measure of stability. Numerous numerical experiments have demonstrated not only the viability of the technique but also its superioirity to established methodologies and thus further

investigation is justified. On a more general note, the
approach of correction of upwind dispersion versus complete
replacement of the convective logic is seen to be an avenue
of research which deserves more attention that it has
received previously.

8.0 REFERENCES
1. CHOW, L.C. and TIEN, C.L. - An Examination Of Four
   Differencing Schemes for Some Elliptic-type Convective
   Equations, Num Heat Trans, 1, 1978, p.87.
2. PATANKAR, S.V. - NUMERICAL HEAT TRANSFER AND FLUID FLOW,
   McGraw-Hill, 1980.
3. PATANKAR, S.V. - A Calculational Procedure for Two-
   Dimensional Elliptic Situations, Num Heat Trans, 2, 1979.
4. WONG, H.H. and RAITHBY, G.D. - Improved Finite
   Difference Methods Based On A Critical Evaluation Of The
   Approximating Errors, Num Heat Trans, 2, 1979, p.131.
5. STUBLEY, G.D., RAITHBY, G.D., STRONG, A.B. and WOOLNER,
   K.A. -Simulation Of Convection And Diffusion Processes By
   Standard Finite Difference Schemes And By Influence
   Schemes, Comp Meth in Appl Mech and Eng, 35, 1982, p. 153.
6. RAITHBY, G.D. - Skew Upstream Differencing Schemes For
   Problems Involving Fluid Flow, Comp Meth in Appl Mech and
   Eng, 9, 1976, p.153.
7. LEONARD, B.P. - A Stable And Accurate Convective
   Modelling Procedure Based On Quadratic Upstream
   Interpolation, Comp Meth in Appl Mech and Eng, 19, 1979,
   p.59.
8. HRS, A Numerical Model For Background Temperature
   Fields, Hydraulic Research Station Report No. EX806,
   Wallinford, England, 1978.
9. HUH, K.Y., GOLAY, M.W. and MANNO, V.P. - A Method For
   Reduction Of Numerical Diffusion In The Donor Cell
   Treatment Of Convection, J Comp Phys, 63-1, 1986, p.201.
10. DE VAHL DAVIS, G. and MALLINSON, G.D. - An Evaluation
    of Upwind and Central Difference Approximations by a Study
    of Recirculating Flow, Comp and Fluids, 4, 1976, p.29.
11. DOUGLAS, J. and RACHFORD, H.H. - On The Numerical
    Solution Of Heat Conduction Problems In Two and Three Space
    Variables, Trans Amer Math, 82, 1956, p.421.
12. HAUBOLD, J.A. and MANNO, V.P. - Stability Implications
    Of Negative Diffusivities In Convection-Diffusion Problems,
    submitted to Comp Meth in Appl Mech and Eng.
13. HAUBOLD, J.A. - Master of Science thesis, Tufts
    University, Medford, MA, 1987.
14. HUANG, P.G., LAUNDER, B.E. and LESCHZINER, M.A. -
    Discretization of Nonlinear Convective Processes: A Broad
    Range Comparison Of Four Schemes, Comp Meth in Appl Mech
    and Eng, 48, 1985, p.1.
15. SMITH, R.M. and HUTTON, A.G. - The Numerical Treatment
    Of Advection: A Performance Comparison Of Current Methods,
    Num Heat Trans, 5, 1982, p.439.

Fig. 1: COMPUTATIONAL CELL FOR AN INTERNAL GRID POINT (L)
Fig. 2: CROSSFLOW DIFFUSION ILLUSTRATION (R)

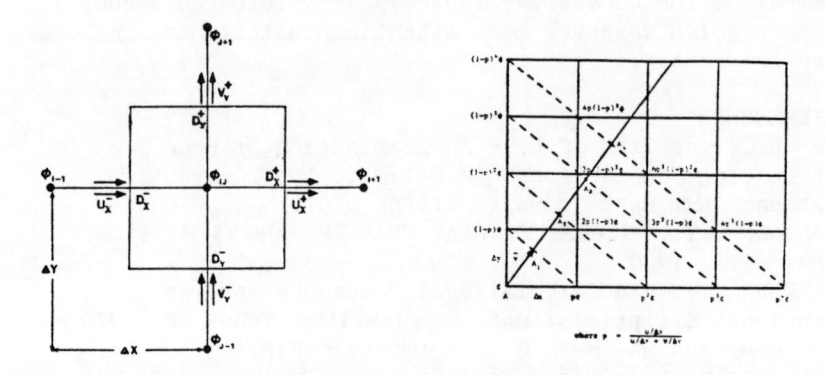

Fig. 3: UPWIND SOLUTION OF 45 DEGREE SCALAR STEP TRANSPORT (L)
Fig. 4: SCALAR TRANSPORT SURFACE OF UPWIND SOLUTION (R)

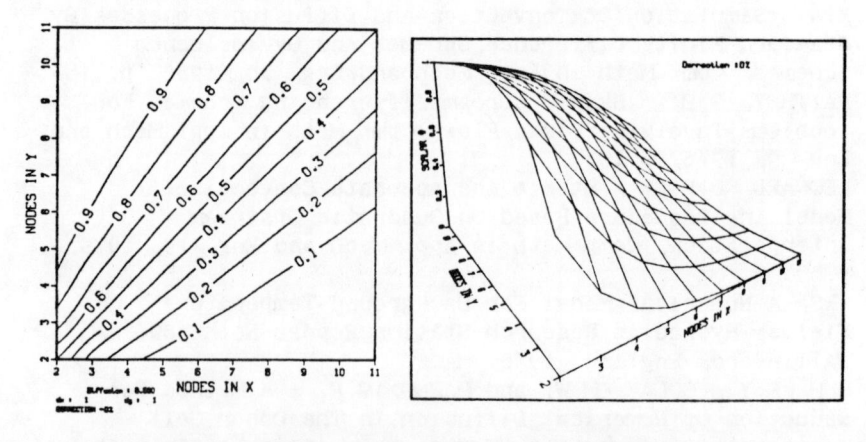

Fig. 5: 100% CADS SOLUTION OF SCALAR STEP TRANSPORT (L)
Fig. 6: SCALAR STEP TRANSPORT SURFACE OF 100% CADS SOLUTION (R)

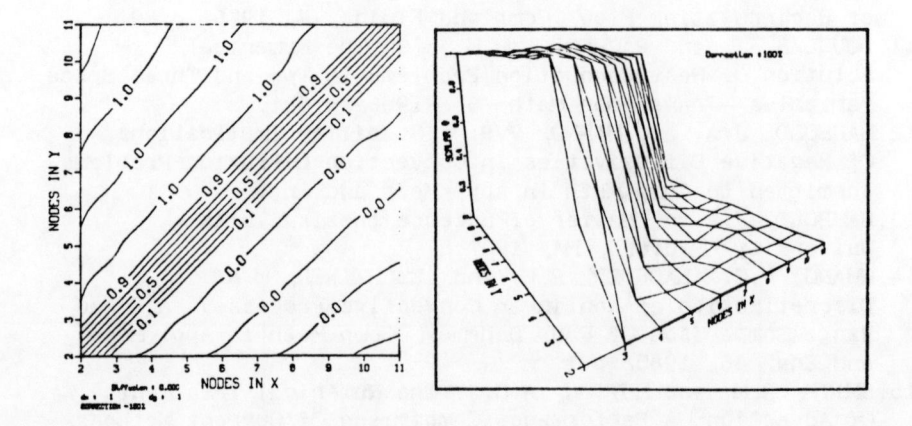

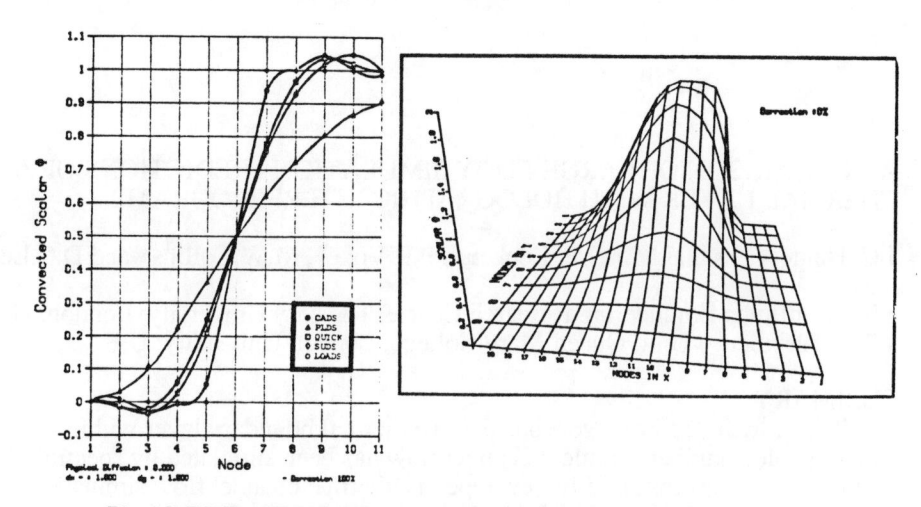

Fig. 7: COMPARISON OF VARIOUS ALGORITHM PREDICTIONS (L)
Fig. 8: UPWIND SOLUTION OF IAHR PROBLEM (R)

Fig. 9: EXACT, UPWIND AND 90% CADS SOLUTIONS AT SYMMETRY LINE (L)
Fig.10: EXACT, UPWIND AND 90% CADS SOLUTIONS AT OUTLET (R)

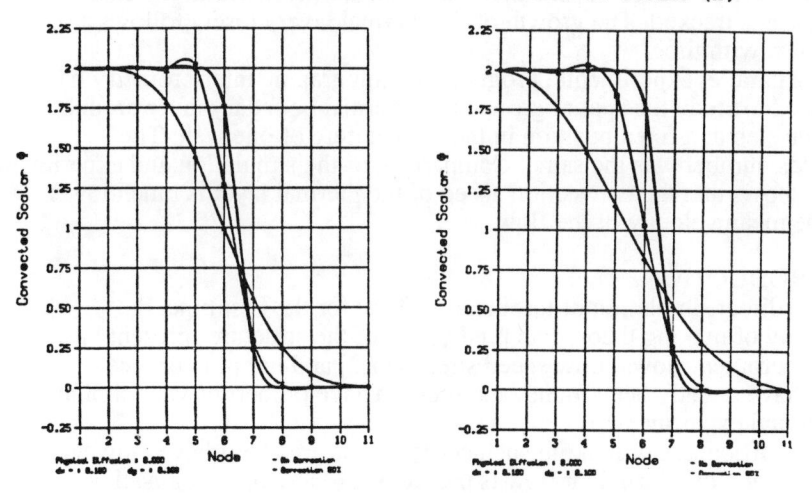

Fig.11: 90% CADS SOLUTION OF IAHR PROBLEM

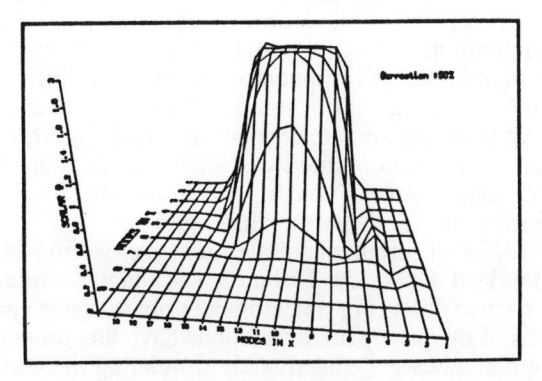

A COMPARISON OF LARGE-EDDY SIMULATION PREDICTIONS OF A THERMAL LAYER WITH HOLOGRAPHIC INTERFEROGRAMS.

J.C. Hunter*, Tsai H.M.†, J.F. Lockett*, P.R.Voke†, M.W.Collins* and D.C.Leslie†

* Thermofluids Engineering Research Centre, The City University, London, U.K.
† The Turbulence Unit, Queen Mary College, London University, U.K.

SUMMARY

The growth of a homogeneous passive thermal boundary layer within a low Reynolds number turbulent channel flow has been simulated by spectral methods. Starting from a fully-developed isothermal channel flow simulation, the temperature at one wall is suddenly increased to a non-zero value, and the subsequent evolution of the thermal boundary layer within the channel turbulence is tracked. The growth of the thermal layer closely follows a 0.8 power law with time.

In a parallel experimental programme, holographic interferometry has been used to study the spatial growth of a thermal boundary layer produced by a sudden step at a fixed position in the temperature of one wall. The Reynolds number was the same. Comparison of the simulation and experimental results shows that the convection speed of the thermal layer is much lower than the mean velocity of the flow.

1. INTRODUCTION

The direct simulation of turbulence offers, for the first time, the possibility of making theoretical predictions of the structures present in various turbulent flows. Low speed streaks [1,2] and hairpin vortices [3,4,5] have already been found, and seem to have properties very similar to those found experimentally.

The difficulties of making such comparisons are formidable, however. No realisation of a turbulent flow is precisely like any other realisation. In the same way (and for the same reasons) no simulation of turbulence can precisely match any experimental flow, or for that matter any other simulation. Two simulations will only produce identical results if the initial conditions and methods of computation are precisely identical. We have found that quite subtle changes in solution algorithms that affect nothing but rounding errors, or minute changes in initial conditions or in the systems software of a host computer, will lead to simulations in which the detailed fields diverge from each other — even though the time averages of all quantites are identical for the two realisations.

This unpredictability of turbulent simulations is reassuring (though inconvenient), and indeed necessary. Turbulence is chaotic. Minute changes in initial conditions produce diverging trajectories in phase space and hence different realisations of the flow. Our simulations have this property as a consequence of the fact that we are adequately simulating the nonlinear equations that govern real turbulence. In one sense it is never, and will never be, possible to compare directly the flow fields of a turbulent simulation with flow fields found experimentally. In the main, time averaged quantities are extracted from the simulations and compared with similar averages measured experimentally.

Nevertheless, it is important to see whether instantaneous simulated distributions of velocity, pressure and temperature can be shown to correspond to experimentally measured fields. By doing so, we can validate the simulation in a new way, and extend the range of available data. Experiments cannot produce the great wealth of data present in a simulation, where velocity, pressure, temperature and derived quantities such as vorticity are potentially measurable at hundreds of thousands of mesh points simultaneously over many thousands of time steps.

We have attempted to match instantaneous temperature fields predicted by spectral large-eddy simulation with holographic interferograms of the same flow. We are not yet able to identify common structures in the holograms and the simulated holograms: as noted above, the match can never be exact. However, we have been able to match the temporal growth of the simulation with the spatial growth in the experiment, and hence to identify the stages in the growth of the simulated layer corresponding to the holograms.

2. THE SIMULATIONS

Our spectral simulation code has been described extensively elsewhere in its application to isothermal channel flow [1,6,7]. The extension to the simulation of temperature is straightforward, since the temperature is treated as a passive scalar contaminant. The subgrid model used is identical to that of Moin and Kim [2] and the model for the temperature field is based on the eddy Prandtl number idea. For the homogeneous term in the subgrid thermal flux, the eddy Prandtl number was taken as 0.25, while for the inhomogeneous term it was taken as 1.0. Further details of our methods for simulating passive thermal and buoyant flows will be reported elsewhere.

The temperature field may safely be considered a passive scalar in this context. Buoyancy forces may be taken into account in the simulation, and we have repeated the simulations reported here with buoyancy forces active and temperature differences identical to those in the experiments. There is almost no change in the growth rate of the thermal layer, or of anything else, when buoyancy is introduced. The Richardson number in this flow is 0.013.

Because of the periodic boundary conditions used in the spectral code, it is not possible to simulate a spacially growing thermal layer. We take a fully developed turbulent channel flow at a Reynolds number of approximately 6200, and suddenly fix the temperature of one wall to a value of unity. At the same moment, the code starts to track the history of the temperature, which is initially set to be uniformly zero. The upper wall is also at zero temperature. The thermal boundary layer subsequently grows in time, but not in space. We are simulating a thermal layer produced by instantaneously raising the temperature of one wall of an infinite plane channel: the result is a streamwise-homogeneous time-dependent thermal layer.

3. THE EXPERIMENTS

The experimental part of this work, conducted at The City University, uses a wind tunnel in which the Reynolds number can be varied up to 60,000, and an image-plane holographic interferometer. The tunnel has a rectangular cross-section of aspect ratio 4, and an electrically heated surface in the working section. The entry length is 20 times the hydraulic diameter — sufficient to to eliminate flow entry effects. Both sides of the working section have float glass windows through which the interferometer object beam passes.

A continuous wave 3 Watt argon-ion laser acts as a coherent light source for the double exposure holograms. The image-plane interferometer is illustrated in figure 1. The holographic recording is made by exposing a light sensitive material to two coherent beams, one object and one reference. After photographic development, the holographic plate is replaced precisely in its original position.

A change in the object produces a change of phase in the object beam. Holographic interferometry involves the superposition of the phase-shifted object beam and the holographic image. The two wavefronts differ in phase owing to deformation or thermal changes in the object, and hence interfere to produce light and dark fringes. The technique gives a non-invasive means of measuring thermal changes, since the energy absorbed from the light beam is negligible. The temperature difference between the heated plate and the upstream flow was also sufficiently low (20K) that neither refraction of the light beam nor buoyancy in the fluid were important [8].

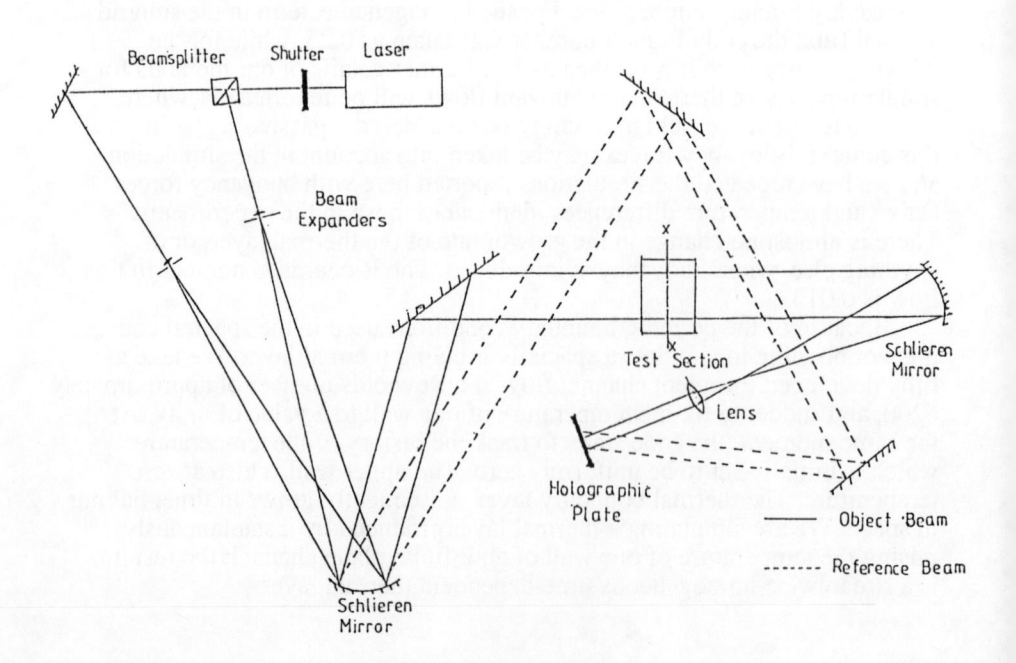

Figure 1. Holographic Interferometer — Layout

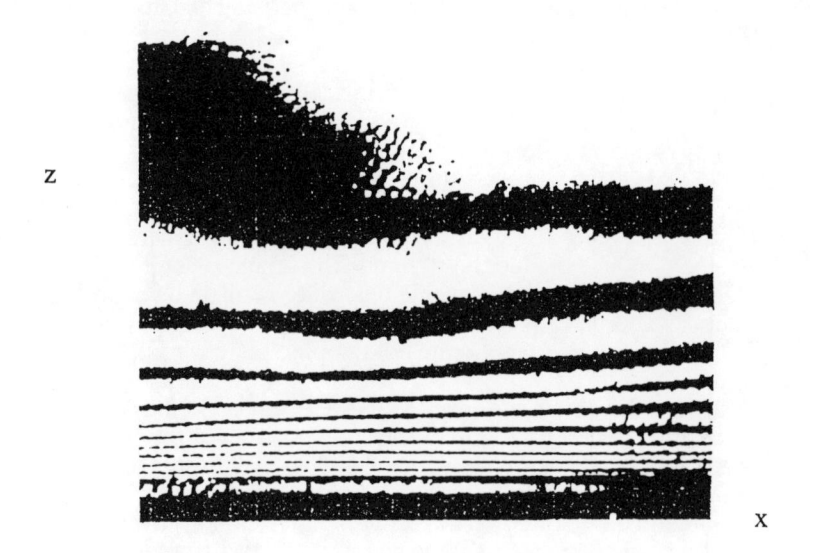

Figure 2. Digitised Image of Fringes (Downstream Station, x/h = 11.15)

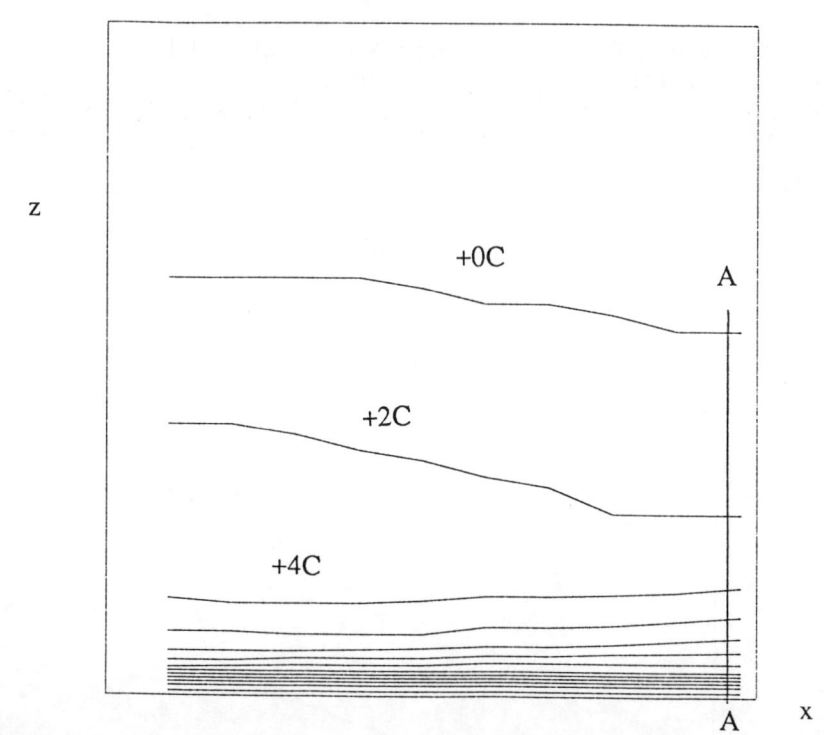

Figure 3. Derived Temperature Contours (Downstream Station, x/h = 11.15)

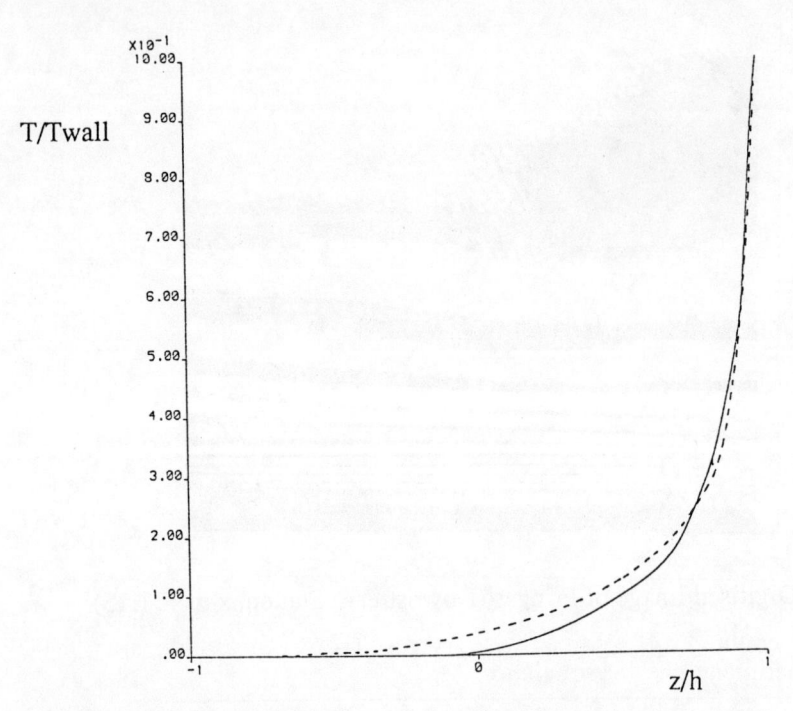

Figure 4. Temperature Profile (Downstream Section, x/h = 11.15)
────── experiment, section AA; - - - - - simulation

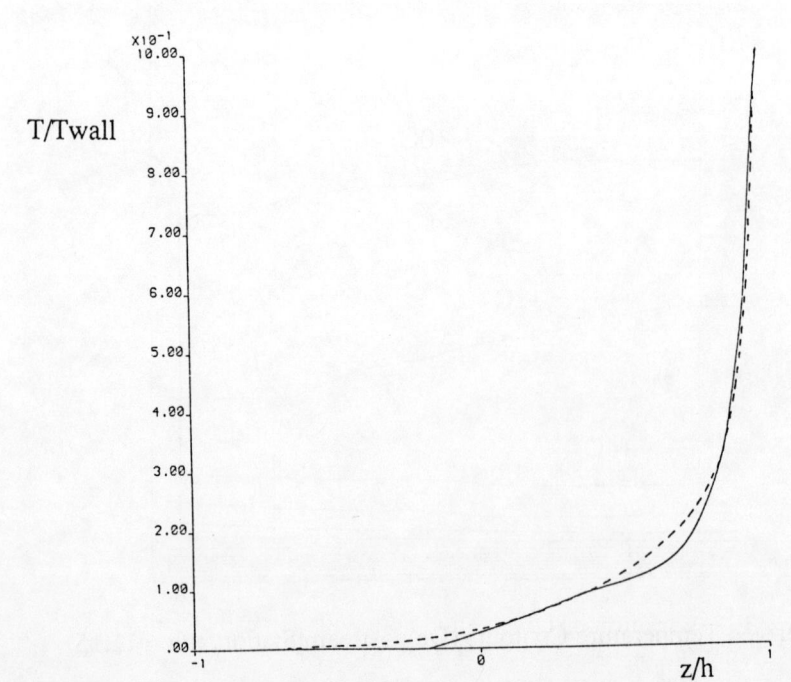

Figure 5. Temperature Profile, Streamwise Average (Downstream Station, x/h = 11.15)

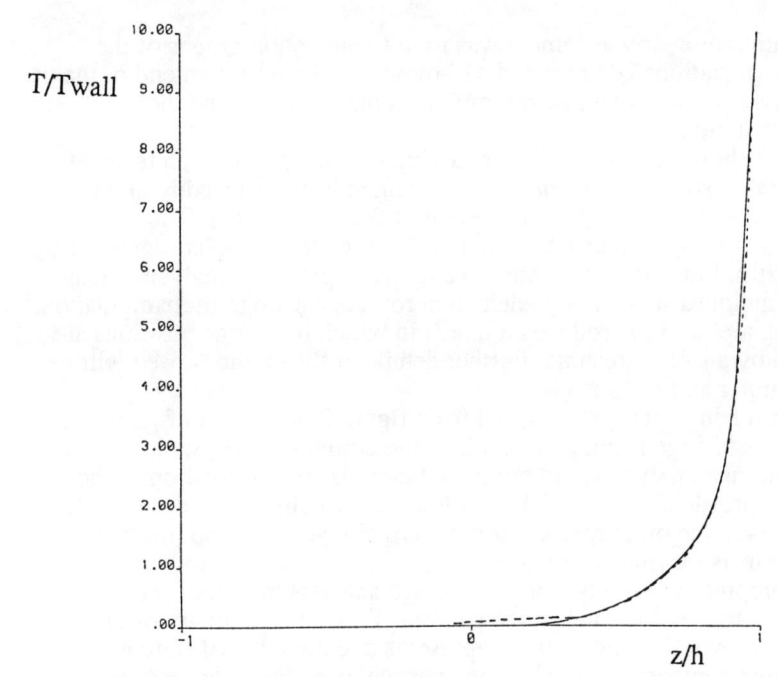

Figure 6. Temperature Profile, Streamwise Average (Upstream Station, x/h = 5.44)
———— experiment; - - - - - simulation

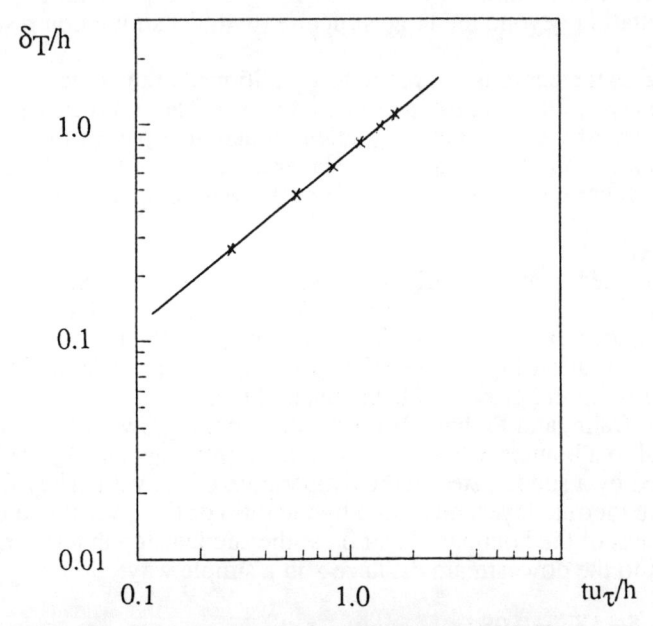

Figure 7. Variation of Layer Thickness with Time
× simulation; ———— $\delta_T / h = 0.74 (tu_\tau / h)^{0.8}$

A 35mm motor-driven camera was used to take photographs of the fringes, at two stations 204 mm and 418 mm from the upstream end of the heated plate. The two positions will be referred to as the upstream and the downstream stations.

Figure 2 shows a digitised image, derived from a photograph taken at the downstream station. (This particular example has a large eddy quite clearly positioned in the upper left portion of the photograph.) The photographs are digitised using a 1-dimensional charge-coupled-device array, and the digitised image is transfered to a minicomputer for analysis. Image analysis techniques such as edge detection routines, histogram manipulation, and filtering are used to produce an image in which the fringe positions are determined by a local threshold. Further details of the methods used will be found in Hunter and Collins [9].

The final fringe pattern extracted from figure 2 is shown in figure 3. The uppermost fringe is not positioned by the computer analysis in the same place as a human analyst would put it, and there is some distortion of the temperature profile as a result. This error arises principally from the width of the fringes in the outer regions of the thermal layer: the resolution of the holograms is inevitably low there.

The computer used to perform the image analysis may also extract temperature profiles from the fringe patterns. From the geometry of the tunnel, it is known that each fringe represents a temperature difference of 2K. From this, temperature profiles at a particular section (figure 4) or averaged over the image in the streamwise direction (figure 5) may be constructed. Figures 5 and 6 respectively give the profiles for the downstream and upstream stations, at particular instants. The downstream curve is clearly distorted by the presence of the large structure mentioned, and we therefore also compare the profile of figure 4, representing a streamwise station beyond the large structure, with the simulation results (dotted curves).

In all cases the curve is not accurately positioned at the outer edges of the layer, for the reasons discussed above. These profiles lead to an overestimate of the temperature gradient in the outer part of the layer, and a systematic underestimation of the layer thickness, if this is based on the position where the temperature is 1% of the wall temperature.

4. COMPARISONS

The simulated layer is clearly different from the experimental layer — indeed it is difficult to see how a streamwise-homogeneous thermal layer could be produced experimentally. The comparison of the two rests on the discovery of a relationship between the spacial growth of the experimental layer and the temporal growth of the simulated layer.

Antonia, Dahn, and Prabhu [10] have studied the growth of a thermal layer internal to a boundary layer that was itself growing. The thermal layer was produced by a sudden step in the temperature of the wall. They found the growth of the thermal layer, measured by the ratio of the layer thickness δ_T to the thickness of the boundary layer δ_o at the sudden step in temperature, to be related to the downstream distance x in a simple way:

(1) $$\delta_T / \delta_o = 0.08 \, (x / \delta_o)^{0.8}$$

Remarkably, our simulation shows a temporal growth rate obeying a similar law. We define δ_T as the distance from the lower wall at which the planar average temperature is 1% of the wall temperature step, and non-dimensionalise with respect to h, the channel half height. The behaviour of this quantity is shown in figure 7 as a function of tu_τ/h, the solid line representing the relationship

(2) $$\delta_T/h = 0.74\,(tu_\tau/h)^{0.8}.$$

This striking result leads us to assert that there is a close relationship between the temporal growth rate in the simulation and the spatial growth rate of experimental layers. Although not strictly entailed by the correspondence between the power laws (1) and (2), the natural inference is that there is a constant convective velocity which relates the simulation time to the distance the layer is convected downstream in the experiment. Unfortunately our channel experiment does not yield a relationship of the form (1): the power is too low. This is chiefly due to the underestimation of the layer thickness δ_T in the downstream profile. (It could partly arise from three-dimensional effects such as corner vortices which would enhance the heat transfer in the early stages of the layer growth, but there is no experimental evidence of such vortices being present at the later stages.)

To overcome this experimental bias, we took a new thickness measure, defined as the distance from the wall δ_{20} at which the planar average temperature is 20% (rather than 1%) of the wall temperature step. This brings us away from the region where the experimental curves are most likely to be in error. It is found that both simulation and experiment give power laws, the powers both being quite close to 0.7:

(3) $\delta_{80}/h = 0.053\,(x/h)^{0.7}$ (experimental)

(4) $\delta_{80}/h = 0.17\,(tu_\tau/h)^{0.7}$ (simulation)

The convective velocity implied by these power laws is $5.3u_\tau$. This velocity is very low compared to the mean velocity of the flow, which is $15.4u_\tau$. Nevertheless we find good matches between the experimental and theoretical temperature profiles (figures 4 and 6) at the corresponding times.

Having established this correspondence, we can construct simulated holograms (figure 8) by plotting contours of temperature averaged in the spanwise direction. These are broadly similar to the temperature contours extracted from the real holograms, though not, of course, in the detailed stucture. Of greater interest is the possibility of looking at the temperature contours (figure 9) at particular spanwise sections, revealing the details of the thermal eddies that are smeared out in the holograms. In a short article of this nature, we do not have space for further contour plots demonstrating the wealth of detail the simulation yields: a fuller analysis will be given elsewhere.

5. CONCLUSIONS

A correspondence has been demonstrated between a simulated internal thermal layer growing in time within a low Reynolds number turbulent channel flow, and a spatially developing layer in a duct at the same Reynolds number. The convective velocity relating simulation and experiment is found to be about one third of the mean velocity of the fluid. Such a low convection velocity could result from fast growth of the experimental thermal layer in its early stages, perhaps arising from three-dimensional effects such as

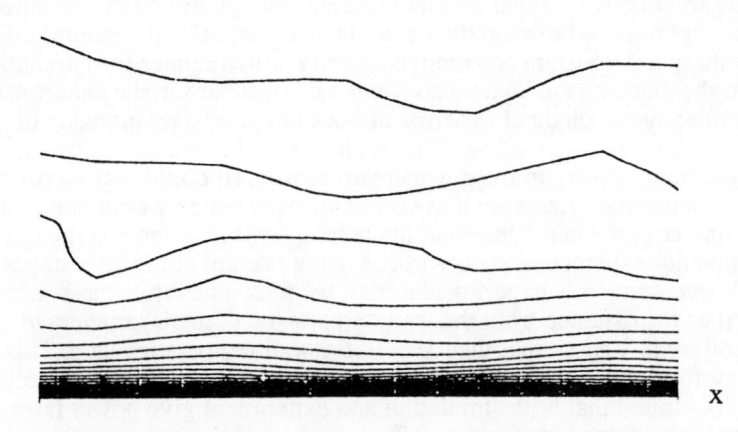

Figure 8. Simulated Hologram: Spanwise Averaged Temperature Contours (Downstream Station, $tu_\tau / h = 2.1$)

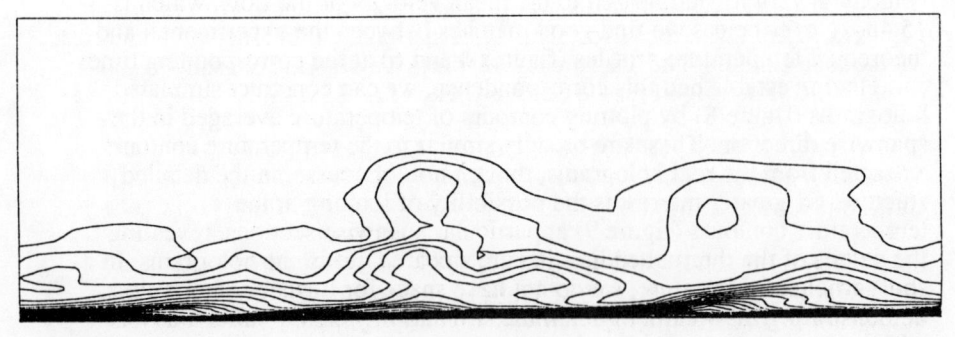

Figure 9. Instantaneous Temperature Contours at one Spanwise Position (simulation, $tu_\tau / h = 2.1$)

longitudinal corner vortices. Such effects would increase the heat transfer rate into the fluid.

Nevertheless, the agreement between simulated and experimental temperature profiles is excellent at the corresponding stations. The simulation can produce "holograms" looking much like the originals, and give a deeper insight into the underlying eddy structures due to the much greater detail in which the velocity fields are known.

REFERENCES

1. GAVRILAKIS, S., TSAI, H.M., VOKE, P.R. and LESLIE, D.C. — *Large-Eddy Simulation of Low Reynolds Number Channel Flow by Spectral and Finite Difference Methods,* Direct and Large Eddy Simulation of Turbulence, ed. Schumann, U. and Friedrich, R., Notes on Fluid Mechanics, **15**, Vieweg, Braunschweig, FRG, 105-118, (1986).

2. MOIN, P. and KIM, J. — Numerical Investigation of Turbulent Channel Flow, J. Fluid Mech., **118**, 341, (1982).

3. MOIN, P. and KIM, J. — The Structure of the Vorticity Field in Turbulent Channel Flow. Part 1: Analysis of Instantaneous Fields and Statistical Correlations, J.Fluid Mech., **155**, 441, (1985).

4. KIM, P. and MOIN, J. — The Structure of the Vorticity Field in Turbulent Channel Flow. Part 2: Study of Ensemble-Averaged Fields. J. Fluid Mech., **162**, 339, (1986).

5. ROGERS, M.M. and MOIN, P. — The Structure of the Vorticity Field in Homogeneous Turbulent Flows, J. Fluid Mech., **176**, 33-66, (1987).

6. ANTONOPOULOS-DOMIS, M. — *Numerical Simulation of Turbulent Flows in Plane Channels*, Proc. Third International Conference on Numerical Methods in Laminar and Turbulent Flows, ed. Taylor C., Johnson J.A., and Smith W.R., Pineridge Press, Swansea U.K., 113-123, (1983).

7. GAVRILAKIS, S., TSAI, H.M., VOKE, P.R. and LESLIE, D.C. — *Simulation Methods for Low Reynolds Number Channel Flow*, Proc. Fifth Int. Conf. on Numerical Methods in Laminar and Turbulent Flow, (1987).

8. LOCKET J.F., HUNTER J.C., VOKE, P.R. and COLLINS M.W. — Investigation of Convective Heat Transfer Enhancement using Real Time Holographic Interferometry, *Optical Measurents in Fluid Mechanics*, (Proceedings of the Sixth International Conference on Photon Correlation and Other Optical Techniques in Fluid Mechanics), Institute of Physics Conference Series, **77**, Adam Hilger, Bristol U.K., 129-134 (1985).

9. HUNTER J.C. and COLLINS M.W. — Holographic Interferometry and Digital Fringe Processing, J. Physics D, in press, (1987).

10. ANTONIA, R.A., DANH, H.Q. AND PRABHU, A. — Response of a Turbulent Boundary Layer to a Step Change in Surface Heat Flux, J. Fluid Mech., **80**, 153-177 (1977).

A NOVEL BOUNDARY ELEMENT FORMULATION FOR
NONLINEAR TRANSIENT HEAT CONDUCTION

L.C. Wrobel
COPPE/Federal University of Rio de Janeiro
Civil Engineering Department
Cx. P. 68506, 21945, Rio de Janeiro, Brazil

ABSTRACT

This work presents a formulation of the Dual Reciprocity Boundary Element Method (DRBEM) to nonlinear transient heat conduction problems. Two kinds of nonlinearities are considered: temperature - dependent material parameters and boundary conditions of the radiative type. The formulation makes use of Kirchhoff's transformation which introduces the integral of conductivity as a new variable to obtain a linear diffusion equation in the transform space. This equation involves a modified time variable which is itself a function of position. The problem is solved in an iterative way by using an efficient Newton-Raphson technique which is shown to be rapidly convergent even for strong nonlinearities of the radiative type.

1. INTRODUCTION

Many engineering problems are governed by a nonlinear diffusion equation containing coefficients which are dependent on the unknown variable, according to some prescribed law. This results in a nonlinear problem for which an iterative scheme of solution is essential. The most common problem of this type occurs in heat conduction when the material parameters are themselves function of temperature.

The classical way of numerically solving these problems consists of first writing the discretized equation assuming the thermal properties are fixed, and later adjusting them such that the appropriate temperature dependence is satisfied [1],[2].

The formulation developed herein employs Kirchhoff's transform [3] in an attempt to replace the original nonlinear differential equation by a linear diffusion equation in the transform space. However, contrary to steady-state problems

[4], this transformation is not sufficient in the sense that the equation in the transform variable still contains a temperature-dependent diffusivity coefficient. Some authors [5], [6] have employed a mean value of the diffusivity coefficient to linearize the resulting equation. In this work, a more general approach is developed which utilizes a further change of variables, resulting in a linear equation for the transformed temperature involving a modified time variable.

Nonlinearities will arise in the transform space due to the fact that the modified time is itself a function of position. A further nonlinearity appears if convective or radiative boundary conditions are prescribed at any point on the boundary. These are treated by using an efficient Newton-Raphson algorithm which is rapidly convergent even for strong nonlinearities [4].

Results of analyses are included to demonstrate the accuracy of the boundary element solutions obtained with the present formulation.

2. FORMULATION OF THE PROBLEM

The equation that describes heat conduction in a region Ω bounded by a surface Γ, with no internal heat generation, is of the form

$$\frac{\partial}{\partial x_1}(K \frac{\partial u}{\partial x_1}) + \frac{\partial}{\partial x_2}(K \frac{\partial u}{\partial x_2}) + \frac{\partial}{\partial x_3}(K \frac{\partial u}{\partial x_3}) = \rho c \frac{\partial u}{\partial t} \quad (1)$$

where u is the temperature, K the thermal conductivity of the medium, ρ its density and c the specific heat. It is assumed that K, ρ and c are all function of u.

The boundary conditions of the problem can be of the following types:

Prescribed temperature:	$u = \bar{u}$	(2a)
Prescribed flux	: $q = \bar{q}$	(2b)
Convection	: $q = h(u - u_c)$	(2c)
Radiation	: $q = \sigma\varepsilon(u^4 - u_r^4)$	(2d)

in which $q = -K \, \partial u/\partial n$, n is the outward unit normal vector, h the heat transfer coefficient, u_c the temperature of the medium surrounding the convective boundaries, σ the Stefan-Boltzmann constant and ε is the emissivity between the surface and the exterior ambient at temperature u_r.

The present formulation employs Kirchhoff's transformation [3], the basic idea of which is to construct a new dependent variable $U = T(u)$ so that Eq.(1) becomes linear in the new variable. Kirchhoff's transformation can be expressed, in integral form, as

$$U = T(u) = \int_{u_o}^{u} K(u)\,du \qquad (3)$$

where u_o is an arbitrary reference value. Thus, Eq.(1) in the new variable can be written as

$$\nabla^2 U = \frac{1}{k}\frac{\partial U}{\partial t} \qquad (4)$$

in which $k = k(u) = K/\rho c$.

It can be noticed that Equation (4) still contains a temperature-dependent diffusivity coefficient. Previous works on the subject [5], [6] assumed that the variation of k with u is usually not strong, and employed some kind of mean value of k to linearize Equation (4). Herein, a more general approach is pursued.

Following [7], one can write $k = k(x_1, x_2, x_3, t)$ since u is a continuous function of the coordinates as well as time. Thus, we can define a new variable τ by the relation:

$$\tau = \int_0^t k(x_1, x_2, x_3, t)\,dt \qquad (5)$$

The partial differentiation of τ with respect to t gives

$$\frac{\partial \tau}{\partial t} = k \qquad (6)$$

Substituting the above into Equation (4), we finally obtain

$$\nabla^2 U = \frac{\partial U}{\partial \tau} \qquad (7)$$

Equation (7) is to be solved by the Dual Reciprocity BEM, as described in [8], [9], [10]. However, since the modified time variable τ is now a function of position, an iterative solution process has to be employed. Before describing the Newton-Raphson algorithm developed in the present work, a comment is in order on the transformation of boundary conditions.

The boundary conditions in the transform space corresponding to (2a)-(2d) are:

Prescribed temperature: $U = \bar{U} = T(\bar{u})$ (8a)

Prescribed flux : $Q = \frac{\partial U}{\partial n} = \bar{Q} = K\frac{\partial u}{\partial n} = -\bar{q}$ (8b)

Convection: $Q = -h(u - u_c)$ (8c)

Radiation: $Q = -\sigma\varepsilon(u^4 - u_r^4)$ (8d)

One can notice that the convection condition also becomes nonlinear in the transform space, since $u = T^{-1}(U)$ is no longer the actual unknown. The initial conditions of the problem are also transformed in a similar way.

3. SOLUTION SCHEME

The application of the Dual Reciprocity BEM to Eq.(7) produces a system of equations which can be written in the form:

$$(\bar{C} + \theta H) U^{m+1} - \theta G Q^{m+1} =$$

$$= [\bar{C} - (1-\theta) H] U^m + (1-\theta) G Q^m \qquad (9)$$

In the above equation, U and Q represent values in the transform space and matrix $\bar{C}$ ($\bar{C}_{ij} = \tilde{C}_{ij}/\Delta\tau_j$) contains step values of the modified time variable at each node, i.e.

$$\Delta\tau_j = k_j \Delta t \qquad (10)$$

at node j. Matrices H, G and C all depend on geometrical data, thus are time invariant and need be formed only once. The parameter θ ($0 \leq \theta \leq 1$) positions the values of U and Q between time levels m and m+1.

The algorithm employed in the solution of the nonlinear system (9) is of the Newton-Raphson type, following the idea of Wrobel and Azevedo [4] for steady-state problems. The expression which has to be solved iteratively is of the form

$$J_{n-1} \Delta X_n = -\psi(X_{n-1}) \qquad (11)$$

in which $J = \partial\psi/\partial X$ is the tangent or Jacobian matrix, ΔX is the vector of increments and the subscript n means the approximate solution at the nth iteration.

Starting from the linear solution we determine, for each iteration, the updated solution through the incremental expression

$$X_n = X_{n-1} + \Delta X_n \qquad (12)$$

and the iteration proceeds until the residual vector

$$\psi(X^{m+1}) = (\bar{C} + \theta H)U^{m+1} - \theta G Q^{m+1} -$$

$$- [\bar{\underset{\sim}{C}} - (1-\theta)\underset{\sim}{H}]U^m - (1-\theta)\underset{\sim}{G}\,\underset{\sim}{Q}^m \qquad (13)$$

is sufficiently small.

The coefficients of the tangent matrix are computed by using the expression

$$J_{ij}^{m+1} = \frac{\partial \psi_i^{m+1}}{\partial x_j^{m+1}} \qquad (14)$$

From Equation (13), we have:

(i) When $X_j^{m+1} = Q_j^{m+1}$: $J_{ij}^{m+1} = -\theta G_{ij}$ (15)

(ii) When $X_j^{m+1} = U_j^{m+1}$: $J_{ij}^{m+1} = \frac{1}{\Delta \tau_j} C_{ij} + \theta H_{ij} -$

$$-\theta\, G_{ij}\, \frac{\partial Q_j^{m+1}}{\partial U_j^{m+1}} + C_{ij}(U_j^{m+1} - U_j^m)\, \frac{\partial}{\partial U_j^{m+1}}\, (\Delta \tau_j)^{-1} \qquad (16)$$

and the derivative of $(\Delta \tau_j)^{-1}$ with respect to U_j^{m+1} can be computed as follows:

$$\frac{\partial}{\partial U_j^{m+1}} (\Delta \tau_j)^{-1} = -\frac{\rho_j c_j}{K_j^3 \Delta t}\left(\frac{dK_j}{du_j^{m+1}} - \frac{K_j}{c_j}\frac{dc_j}{du_j^{m+1}} - \frac{K_j}{\rho_j}\frac{d\rho_j}{du_j^{m+1}}\right) \qquad (17)$$

In the present work the conductivity, density and specific heat dependence on temperature are all modelled by a piecewise linear representation, according to [11]. This means that the derivatives of K, ρ and c with respect to u, at any point j, are given by the slope of the segment of the curves to which u_j belongs.

The derivative of Q_j with respect to U_j in Equation (16) depends on the type of boundary condition at node j. For prescribed flux, relation (8b), the derivative is obviously zero since there is no dependence of Q on U. Boundary conditions of the convective and radiative types, however, produce the expressions:

Convection: $\dfrac{\partial Q_j^{m+1}}{\partial U_j^{m+1}} = -h\,\dfrac{\partial u}{\partial U} = -\dfrac{h}{K_j}$ (18a)

Radiation: $$\frac{\partial Q_j^{m+1}}{\partial U_j^{m+1}} = -\frac{4\sigma\varepsilon u_j^3}{K_j} \qquad (18b)$$

Another type of nonlinearity may appear if the region under consideration is made up of piecewise homogeneous subregions of different materials. Enforcement of the continuity condition along the interface between subregions produces discontinuity in the integral of conductivity U, the values of which have to be adjusted so that the physical requirement of temperature continuity is satisfied at the final stage. This feature can also be treated by the present Newton-Raphson scheme, as discussed in [11].

The iteration cycle will consist of the following steps:

(i)   Solve Equation (9) for a constant $\Delta\tau = \Delta t$, finding the linear solution. If a convection or radiation boundary condition exists at any point, the linear solution is carried out assuming $u = U$ in (8).

(ii) With the value of U at all points, an inverse transformation is performed to find values of u.

(iii) The conductivity, density and specific heat curves provide, for each value of u, values of K, ρ, c, dK/du, dρ/du, and dc/du.

(iv) Values of $\Delta\tau$ can now be calculated at each point through expression (10).

(v)   The tangent matrix and residual vector are updated, and a new solution carried out.

The iteration algorithm proceeds until a certain specified tolerance for the relative norm of increments $\Delta \underset{\sim}{X}$ is reached. With the converged solution for $\underset{\sim}{U}$ and $\underset{\sim}{Q}$, the process can be advanced to the next time step.

4. APPLICATIONS

The first example of application studies the temperature distribution along the thickness of a nonlinear wall. The wall is 20cm long, 1cm high, and is initially at 100°C. The temperature of the left end surface is suddenly raised to 200°C and kept at this value for 10s, after which it is decreased to 100°C again; temperature of the right end surface is kept at 100°C, while the other surfaces are assumed insulated. The thermal conductivity is $K = 2 + 0.01u$ W/(cm °C) and heat capacity $\rho c = 8$ J/(cm³ °C).

Boundary element results for a discretization of 22 equal constant boundary elements, taking into account symmetry with respect to the x_1 axis, are compared in Tables 1 and 2 with a finite element solution obtained with 20 equal linear finite elements [12]. The agreement of results is very good. The average number of iterations of the boundary element solution was 4, compared to a finite element number of 3. The time step value was 1s in both cases.

x	FEM	BEM
0	200.00	200.00
1	176.16	175.29
2	153.21	151.48
3	133.47	131.74
4	118.60	117.63
5	108.98	108.94
6	103.72	104.27
7	101.29	102.01
8	100.37	100.97
9	100.08	100.50
10	100.01	100.27

Table 1 — Temperature Variation (°C) of the Left End Surface of Nonlinear Wall, at Time t=10s.

x	FEM	BEM
0	100.00	100.00
1	128.53	130.15
2	139.97	138.95
3	136.95	132.47
4	124.72	121.71
5	114.40	112.64
6	107.18	106.71
7	103.24	103.36
8	101.29	101.62
9	100.45	100.77
10	100.13	100.36

Table 2 — Temperature Variation (°C) of the Left End Surface of Nonlinear Wall, at Time t=13s.

The second example deals with a slab of width 2L, initially at a uniform temperature u_i, the surfaces of which are suddenly exposed to simultaneous convection and radiation into a medium at zero temperature. The slab has linear thermophysical properties, permiting the problem solution be carried out directly in the real space.

Results for the surface temperature u_s and the

temperature u_o at the center of the slab are presented in Figure 1, in terms of nondimensional values. Parameters employed in the analysis are $k=1.0$ in^2/h, Biot number $Bi=hL/K=0.2$ and $R = \sigma \varepsilon u_i^3 L/K$ equals to 0, 1 and 4. The case $R=0$ corresponds to pure convection, representing a linear problem.

The results obtained by the present approach, employing 22 equal constant boundary elements, compare well with those of Haji-Sheik and Sparrow [13], who employed a Monte Carlo method. Results of the same order of accuracy have also been obtained with the ADINAT finite element code [2].

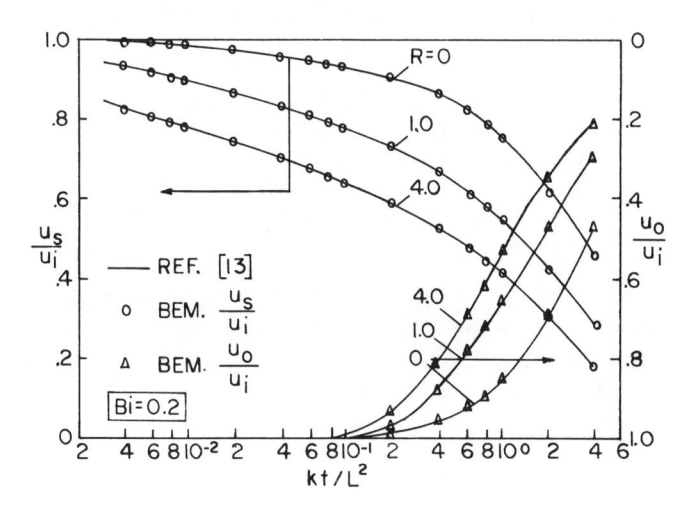

Figure 1 - Transient Temperature Results for a Slab with Convection - Radiation at Surfaces.

5. CONCLUSIONS

An effective boundary element solution scheme for the analysis of nonlinear transient heat conduction problems has been presented. The scheme treats temperature-dependent material parameters and boundary conditions of the radiative type in a uniform way, through a Newton-Raphson technique.

The results of the problems studied revealed accuracy comparable to well-known finite element codes, even though the time-stepping technique employed was a simple two-level scheme and the boundary element the crudest possible, namely the constant element.

6. REFERENCES

1. COMINI,G., DEL GUIDICE,S., LEWIS,R.W. and ZIENKIEWICZ,O.C., Finite Element Solution of Non-Linear Heat Conduction Problems with Special Reference to Phase Change, Int. J. Num. Meth. Engng, $\underline{8}$, 613-624, 1974.

2. BATHE, K.J. and KHOSHGOFTAAR, M.R., Finite Element Formulation and Solution of Nonlinear Heat Transfer, Nucl. Eng. Des., $\underline{51}$, 389-401, 1979.

3. CARSLAW,H.S. and JAEGER, J.C., Conduction of Heat in Solids, 2nd edn ,Clarendon Press, Oxford, 1959.

4. WROBEL, L.C., and AZEVEDO,J.P.S., A Boundary Element Analysis of Nonlinear Heat Conduction, in Numerical Methods in Thermal Problems, Vol.4 (R.W. Lewis and K. Morgan, Eds.), Pineridge Press, Swansea, 1985.

5. SKERGET,P., and BREBBIA,C.A., Time Dependent Nonlinear Potential Problems, in Topics in Boundary Element Research, Vol.2 (C.A. Brebbia, Ed.), Springer-Verlag, Berlin, 1985.

6. SKERGET,P. and ALUJEVIC,A., Boundary Element Method in Non-linear Transient Heat Transfer of Reactor Solids with Convection and Radiation on Surfaces, Nuclear Eng. Des.,$\underline{76}$, 47-54, 1983.

7. KADAMBI, V. and DORRI, B., Solution of Thermal Problems with Nonlinear Material Properties by the Boundary Integral Method, in BETECH 85 (C.A. Brebbia, Ed.), Springer-Verlag, Berlin, 1985.

8. WROBEL,L.C., BREBBIA, C.A. and NARDINI, D., The Dual Reciprocity Boundary Element Formulation for Transient Heat Conduction, in Finite Elements in Water Resources VI (A. Sá da Costa et al., Eds.), Springer-Verlag, Berlin, 1986.

9. WROBEL, L.C., BREBBIA,C.A. and NARDINI, D., Analysis of Transient Thermal Problems in the BEASY System, in BETECH 86 (J.J. Connor and C.A. Brebbia, Eds.) Computational Mechanics Public., Southampton 1986.

10. WROBEL, L.C., TELLES J.C.F. and BREBBIA,C.A., A Dual Reciprocity Boundary Element Formulation for Axisymmetric Diffusion Problems, in Boundary Elements VIII (C.A. Brebbia and M. Tanaka, Eds.) Springer-Verlag, Berlin, 1986.

11. AZEVEDO,J.P.S. and WROBEL,L.C., Nonlinear Heat Conduction in Composite Bodies: A Boundary Element Formulation, submitted for publication in Int. J. Num. Meth. Eng.

12. ORIVUORI,S., Efficient Method for Solution of Nonlinear Heat Conduction Problems, Int. J. Num. Meth. Engng, $\underline{14}$, 1461-1476, 1979.

13. HAJI-SHEIK, A. and SPARROW, E.M., The Solution of Heat Conduction Problems by Probability Methods, J. Heat Transfer, $\underline{89}$, 121-131, 1967.

A SOLUTION TO THE IHCP APPLIED TO CERAMIC PRODUCT FIRING.*

E. WEILAND and J.P. BABARY

Laboratoire d'Automatique et d'Analyse des Systèmes du C.N.R.S
7, Avenue du Colonel Roche
31077 TOULOUSE Cedex
FRANCE

Abstract.

We propose a solution to the Inverse Heat Conduction Problem by a new finite difference scheme using two steps in both time and space. The purpose of this paper is to study the stability and performance of this scheme applied to a real firing curve for an intermittent kiln and to a test problem.

INTRODUCTION.

In the control of processes of the ceramic industry, one must master the firing curve for product quality and implement an optimization scheme to reduce the expenditure of energy.

In intermittent kilns, where the product doesn't move, the firing curve depends on time and must respect two kinds of constraints: first, a constraint about the maximum thermal gradient where the physical and chemical product properties are changing (quartz points); second, a constraint about the minimum duration of the maximum temperature landing, so that the product is baked up to the heart.

Classical finite difference schemes solve the direct problem from successive estimations of the surface temperature versus time profile, thus determining the desired profile at heart. This way seems us tedious.

We pose the inverse problem which determines the surface versus time profile directly from the firing curve at heart. This is an ill-posed problem in Hadamard sense; we propose a solution by a new finite difference scheme and will study its properties and performances.

* Contract 'CNRS-Région Midi-Pyrénées'.

SETTING OF THE INVERSE PROBLEM.

We solve the diffusion equation:

$C(u) \cdot u_t = k \cdot u_{xx}$

Direct problem.
I.C. $u(x,0)=f(x)$
B.C. $u(L,t)=u(-L,t)=g(t)$
$\Rightarrow u(0,t)$

Inverse problem in x.
I.C. $u(x,0)=f(x)$
B.C. $u_x(0,t)=0$ and $u(0,t)=g(t)$
$\Rightarrow u(L,t)$

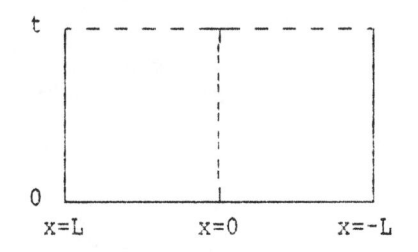

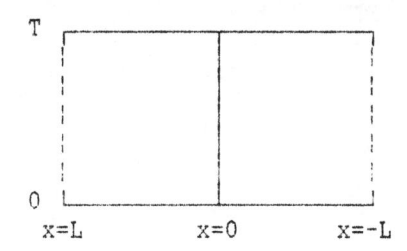

We prescribe to the inverse problem in space a final condition in time, equivalent to a condition at x=L [4], thus: $u(x,T)=h(x)$. We assign $h(x)=f(x)=u_0$.

The dimensionless form is:
$x^*=x/L$; $t^*=a_0 t/L^2$; $T^*=a_0 T/L^2$; $a_0=k/C(u_0)$; $C^*(u^*)=C(u)/C(u_0)$;
$k^*=1$; $u^*=(u-u_0)/(u_{max}-u_0)$.
L in m; T and t in s; C in J/m³°C; k in W/m°C; u in °C.

With the superscript * omitted:
$C(u) \cdot u_t = u_{xx}$ with the conditions:
$u(x,0)=0$ and $u(x,T)=0$
$u_x(0,t)=0$
$u(0,t)=g(t)$ and $u(1,t)$ is to be calculated.

STUDY OF THE IHCP.

One can find an analytical solution of the IHCP in the form of an integral equation [8], but the using of this solution poses some numerical difficulties [7].

We propose a solution by a weighted, time implicit, finite difference scheme, taking the form:

$$C(u_{m+1}^p) \left[\frac{u_{m+1}^{p+1} - u_{m+1}^{p-1}}{2Dt} \right] = \left[\frac{\theta(\delta^2 u)_m^{p+1} + (1-2\theta)(\delta^2 u)_m^p + \theta(\delta^2 u)_m^{p-1}}{Dx^2} \right]$$

and $(\delta^2 u)_m^p = u_{m+1}^p - 2u_m^p + u_{m-1}^p$ $0 \leq m \leq M$ with $M=1/Dx$
                                        $0 \leq p \leq P$ $P=T/Dt$

with
$u_0^p = g(p)$ and $\frac{u_{-1}^p - u_1^p}{2Dx} = 0$
$u_m^0 = 0$ et $u_m^P = 0$ and u_{m+1}^p is to be calculated.

We solve for the nonlinear term, u_{m+1} by successive approximations.
The scheme is consistent with the inverse problem at order 1 in t and order 2 in x, if the gradients at orders higher than 2 are bounded; consistency requires that $Dx \to 0$ as $Dt \to 0$.
The properties of the scheme depend on the value of the weighting parameter θ. The Well Conditioning Criteria [6], gives $0 \le \theta \le \frac{1}{2}$. One takes $\theta=0$ to simplify the scheme; the justification will be presented in a later section.

STABILITY STUDY.

There are different ways to caracterize the stability of the scheme. We choose to bind the power of the transit operator from the initial data to the solution [5].
The unweighted scheme is:

$$u_{m+1}^{p+1} - r u_{m+1}^{p} - u_{m+1}^{p-1} = r[-2u_m^p + u_{m-1}^p] \quad \text{with} \quad r = \frac{2Dt}{C(u_{m+1}^p)Dx^2}$$

Thus $u_{m+1} = (u_{m+1}^0, \ldots, u_{m+1}^p, \ldots, u_{m+1}^P)$ is the vector of temperature versus time at the point $m+1$. To simplify the writing, we set r constant.
The matrix form is:

$$T u_{m+1} = rId(-2u_m + u_{m-1}) \quad m=1 \text{ to } M-1 \quad \text{and} \quad T_0 u_1 = rId(-2u_0)$$

where T and T_0 are tridiagonals matrices of the form:

$T = (-1 \ -r \ 1)$ et $T_0 = (-1 \ -2r \ 1)$ of dimension $(P+1)*(P+1)$.

So $u_{m+1} = T^{-1}rId(-2u_m + u_{m-1})$ taking $w_{m+1} = \begin{bmatrix} u_{m+1} \\ u_m \end{bmatrix}$ and $w_0 = \begin{bmatrix} u_0 \\ 0 \end{bmatrix}$

One finds $\quad w_{m+1} = R w_m \quad$ with $\quad R = \begin{bmatrix} -2rT^{-1} & rT^{-1} \\ Id & 0 \end{bmatrix}$

Likewise $\quad w_1 = R_0 w_0 \quad$ with $\quad R_0 = \begin{bmatrix} -2rT_0^{-1} & 0 \\ Id & 0 \end{bmatrix}$

At the end $\quad w_M = R^{M-1} R_0 w_0$

We make clear R and R_0; matrix T is tridiagonal so that we write T^{-1} by its diagonals.

Upper Diagonals:
$a_{i,i-d} = (-1)^d D_{P-i} D_{i-d-1}/D_P$
$1+d \le i \le P$

Lower Diagonals:
$a_{i,i+d} = D_{P-d-i} D_{i-1}/D_P$
$1 \le i \le P-d$

d is the subscript of the diagonal (d=0,1,..,P-1 where d=0 is the main diagonal) and D_P is the determinant of order p of T constructed from the recursive algorithm:

$D_p = -rD_{p-1} + D_{p-2}$ with $D_0 = 1$ and $D_1 = -r$ then $D_p = \frac{1}{2\mu}\left(\frac{r}{2}\right)^p \phi(p)$

where $\phi(p) = \{-(-1+\mu)^{p+1} + (-1)^p(1+\mu)^{p+1}\}$ is a p+1 order polynom
and $\mu = \left(1 + \frac{4}{r^2}\right)^{1/2}$

Hence $a_{i,i-d} = \frac{(-1)^d}{2\mu}\left(\frac{2}{r}\right)^{d+1} \cdot \frac{\phi(P-i)\phi(i-1-1)}{\phi(P)}$

and $a_{i,i+d} = \frac{1}{2\mu}\left(\frac{2}{r}\right)^{d+1} \cdot \frac{\phi(P-d-i)\phi(i-1)}{\phi(P)}$

We suppose r>1 , $\phi(p) \# (-1)^p(1+\mu)^{p+1}$

hence $a_{i,i-d} = \frac{-1}{2\mu}\left(\frac{2}{r}\right)^{d+1} \cdot \frac{1}{(1+\mu)^d} \approx -\frac{1}{r^{d+1}}$

and $a_{i,i+d} = (-1)^d \cdot a_{i,i-d}$

by multiplying by -r, one finds the $-rT^{-1}$ elements:
$b_{i,i-d} = \frac{1}{r^d}$ and $b_{i,i+d} = (-1)^d \cdot b_{i,i-d}$

By the Taylor's series expansion at first order:
$-rT^{-1} \# Id + 1/r \cdot J + 1/r \cdot O(1/r)$ where J is the tridiagonal matrix (-1 0 1) .

Then one finds: $R \# \begin{bmatrix} 2Id & -Id \\ Id & 0 \end{bmatrix} + \frac{1}{r}\begin{bmatrix} 2J & -J \\ 0 & 0 \end{bmatrix} + \frac{1}{r} O(1/r)$

Likewise: $R_0 \# \begin{bmatrix} Id & 0 \\ Id & 0 \end{bmatrix} + \frac{1}{2r}\begin{bmatrix} J & 0 \\ 0 & 0 \end{bmatrix} + \frac{1}{r} O(1/r)$

So: $R^{M-1} \# \begin{bmatrix} MId & -(M-1)Id \\ (M-1)Id & -(M-2)Id \end{bmatrix} + \frac{1}{r}\begin{bmatrix} AJ & BJ \\ CJ & DJ \end{bmatrix} + \frac{1}{r} O(1/r)$

A,B,C,D are calculated by the recursive equations:
$A_{j-1} = 2A_{j-2} - C_{j-2} + j$ $C_{j-1} = A_{j-2}$
$B_{j-1} = 2B_{j-2} - D_{j-2} - (j-1)$ $D_{j-1} = B_{j-2}$ and $A_0 = B_0 = C_0 = D_0 = 0$

$R^{M-1} R_0 = \begin{bmatrix} Id & 0 \\ Id & 0 \end{bmatrix} + \frac{1}{r}\begin{bmatrix} [M/2 + A_{M-1} + B_{M-1}]J & 0 \\ [(M-1)/2 + C_{M-1} + D_{M-1}]J & 0 \end{bmatrix} + \frac{1}{r} O(1/r)$

Finally : $u_{M+1} = u_0 + \frac{1}{r}(M/2 + A_{M-1} + B_{M-1})Ju_0 + \frac{1}{r}O(1/r)u_0$

so $u_{M+1} = T_P u_0$ where T_P is the transit matrix from the initial data to the solution and its L^2 norm is:

$\| T_P \| = \max_i |\Sigma_j a_{i,j}| = 1 + \frac{2}{r}[M/2 + (A+B)_{M-1}]$

with $(A+B)_{M-1} = \frac{M(M-1)}{2}$

So $\|T_P\| = 1+\dfrac{M^2}{r}$ with $\dfrac{1}{r} = \Gamma.\dfrac{P}{M^2}$ and $\Gamma = \dfrac{C(u)}{2T}$

One bounds $\|T_P\| = 1+\Gamma.P \leq Cst.P^\alpha$

where α is real, ≥ 0 and finite. One takes $\alpha=1$, to assure the stability of the scheme, which is consistent and consequently convergent. The stability depends only on the choice of Dt and Dx to assure the condition $r \gg 1$; this condition results of the consistency.

COMPATIBILITY NOTION.

In several cases, the initial data $g(t)$ is a discrete function (resulting from measurements) or continuous by parts (prescribed temperature profile); then, there is incompatibility between $g(t)$ at $t=0$ and the initial condition $u(x,0)$; $g_t(0)=u_t(0,0)=0$ and $u_{xx}(0,0)=0$. One expects a particular behavior of the solution next to $t=0$.

NUMERICAL SIMULATIONS.

We restore a surface temperature versus time profile, from a heart temperature versus time profile calculated by a direct classical method (Crank-Nicholson). The Crank-Nicholson scheme is used with a constant step Dx=0.05 when Dt changes; the accuracy obtained is sufficient and have a little weight on the final result; the Dt step changes in the same manner as the inverse scheme.

The performances of the solution to the IHCP are judged by the parameter Σ and S:
_ Σ is the quadratic divergence between the two surface profiles, divided the calculated points number;
_ S is the divergence between the two surface profiles, divided the maximum value of the surface profile to be restored.
_ N is the calculated points number.
The better performance implies the less Σ and S. The value of S next to $t=0$ shows the incompatibility previously defined.

WEIGHT CHOICE.

We justify the choice of the weighting parameter $\theta=0$, through simulations on a real firing ceramic curve.
The results are:

θ	0.2	0.01	0.001	0
Σ	10.498	3.592	2.131	1.903

Dt=0.332
Dx=0.05

The later figures show the divergence between the two surface profiles divided the profile to be restored, versus time. One gives the divergence in %.

FIG 1.a : θ = 0.2   FIG 1.b : θ = 0

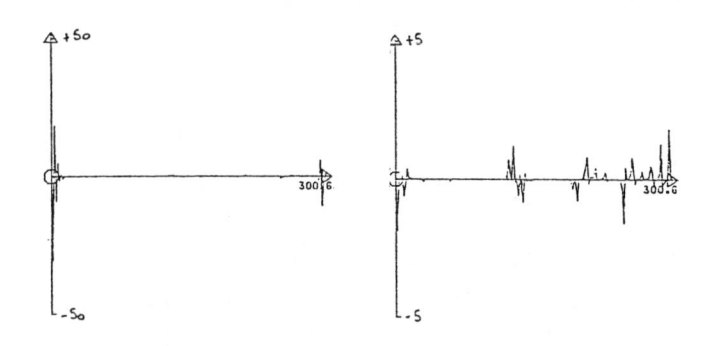

We now present investigations as to the choice of discretization steps in the case of a firing ceramic curve and in a test case. In twice, the final condition is h(x)=u_{max}, so: u(x,T)=1. One gives the parameter r and f resulting from the stability study.

CERAMIC FIRING CURVE.

The curve is bounded to a rise and to the landing of temperature. 1/C(u) takes the form $C_0+C_1u-C_2/u^2$.
Physical parameters in dimensionless form except L and T:
L=0.02m ; T=11700s ; $r^*=1/r_{min}$; $a_0=1.2296.10^{-6}$;
$C_0=1.1034$; $C_1=4.4842.10^{-4}$; $C_2=2.6105.10^4$.

Dt	0.332	0.0332
Σ	24.400	1.250

Dx=0.05
$0.01 \leq$ Γ ≤ 0.025

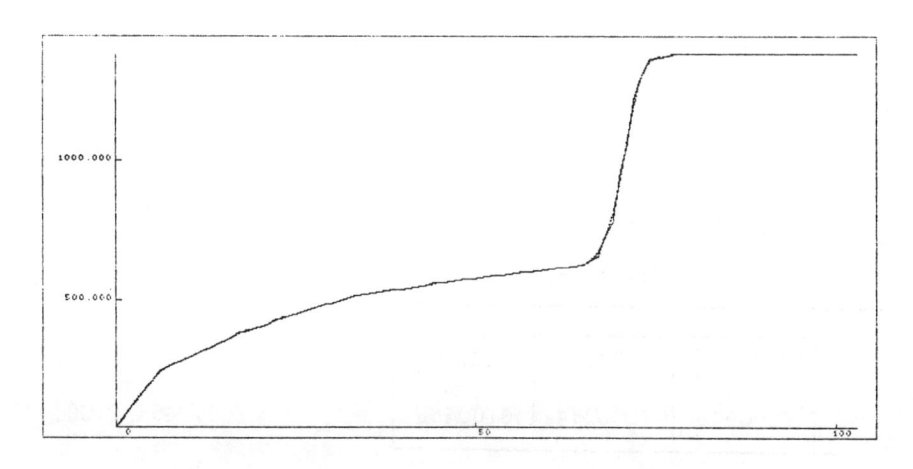

FIG 2.a : Surface profiles vs N : Dt = 0.332 .

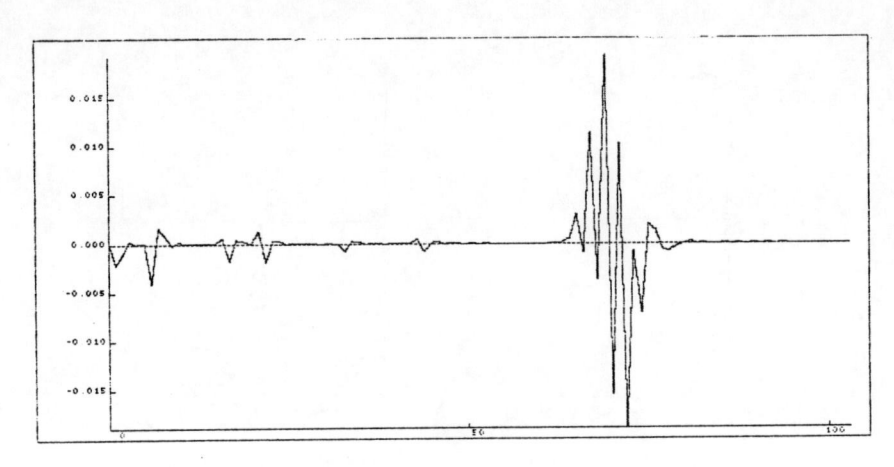

FIG 2.b : S vs N : Dt = 0.332

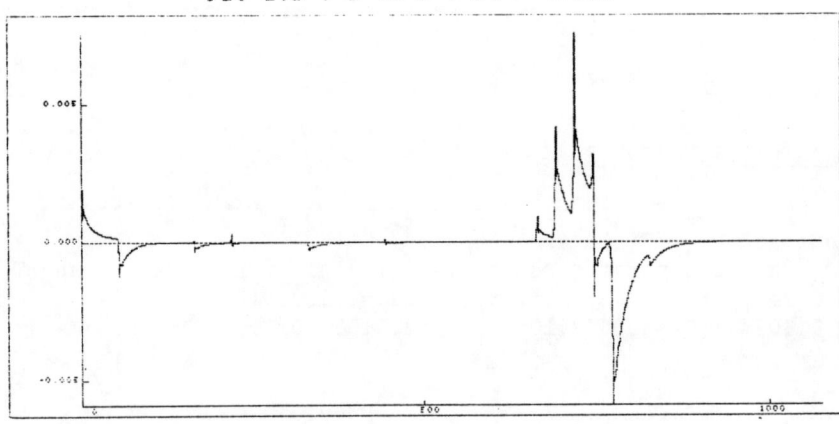

FIG 2.c : S vs N : Dt = 0.0332 .

TEST PROBLEM.

The surface profile to be restored is a temperature rise of 6000 °C/h lasting 10min and a landing lasting 50min. The parameters are identical with T=3600s.

Dt	0.1844	0.03074	0.01844	0.00922	
Σ	18.41	11.86	12.34	12.64	Dx=0.05
r^*	0.011	0.068	0.114	0.228	

17.66	0.54	0.46		Dx=0.01
0.00046	0.00274	0.00456		$0.045 \leq r \leq 0.075$

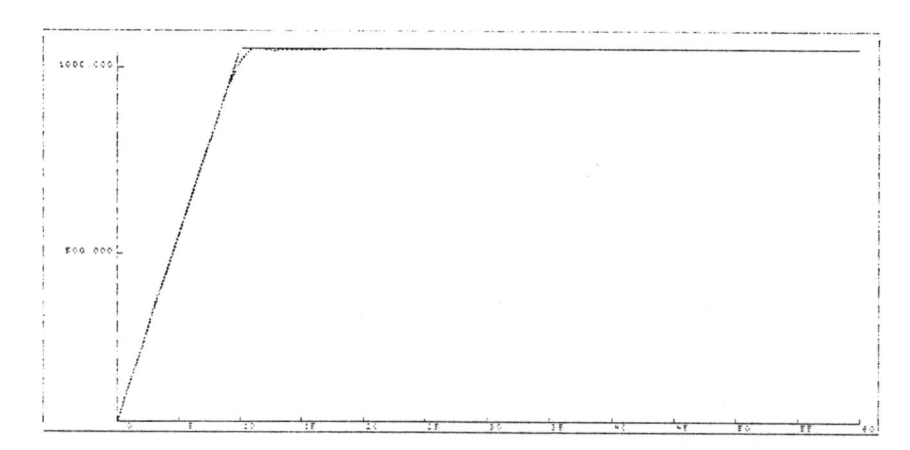

FIG 3.a : Surface profiles vs N : Dt = 0.1844 , Dx = 0.05 .

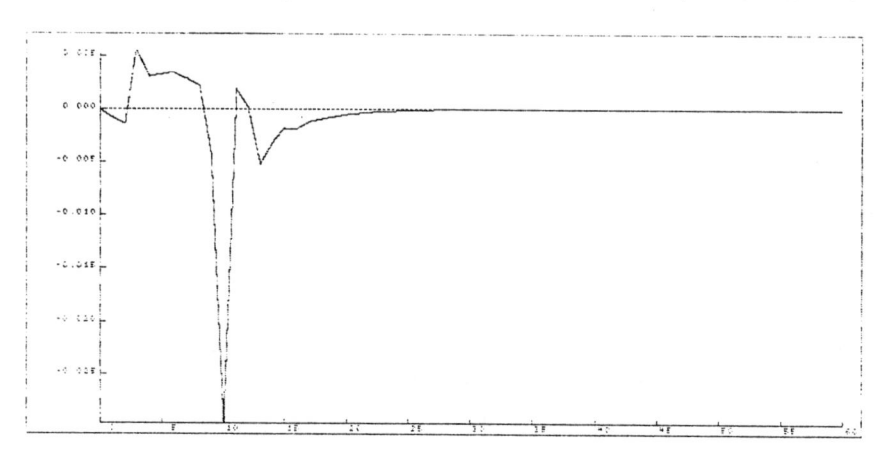

FIG 3.b : S vs N : Dt = 0.1844 , Dx = 0.05 .

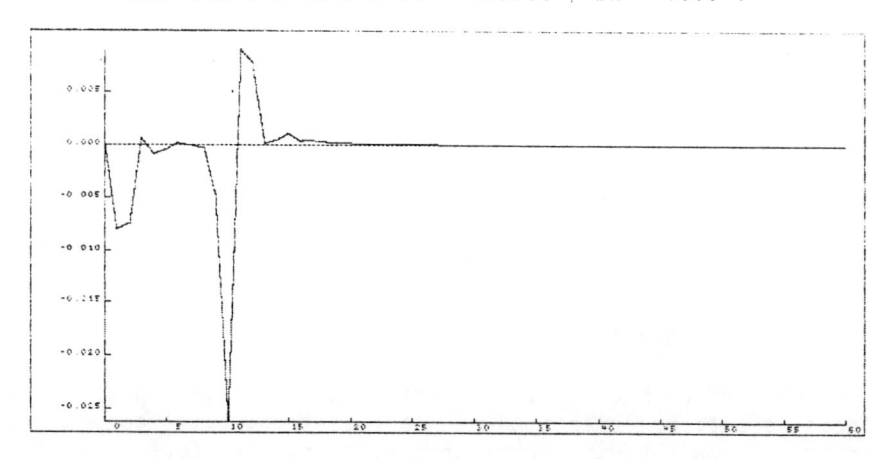

FIG 3.c : S vs N : Dt = 0.1844 , Dx = 0.01 .

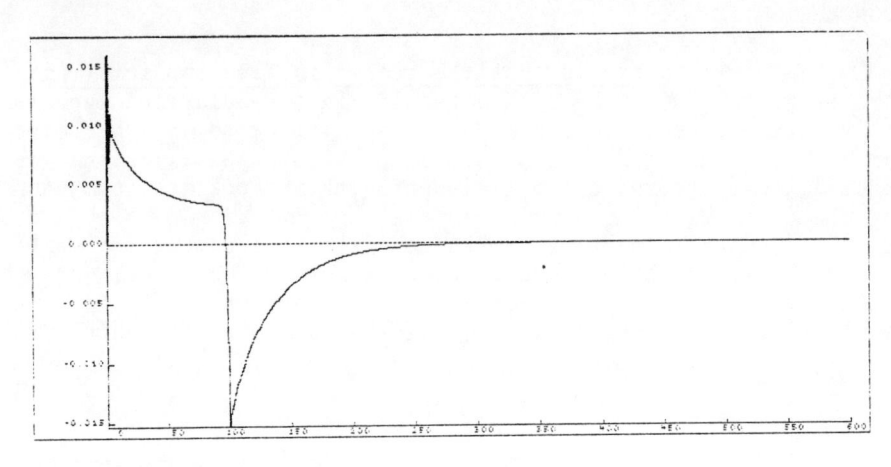

FIG 3.d : S vs N : Dt = 0.01844 , Dx = 0.05 .

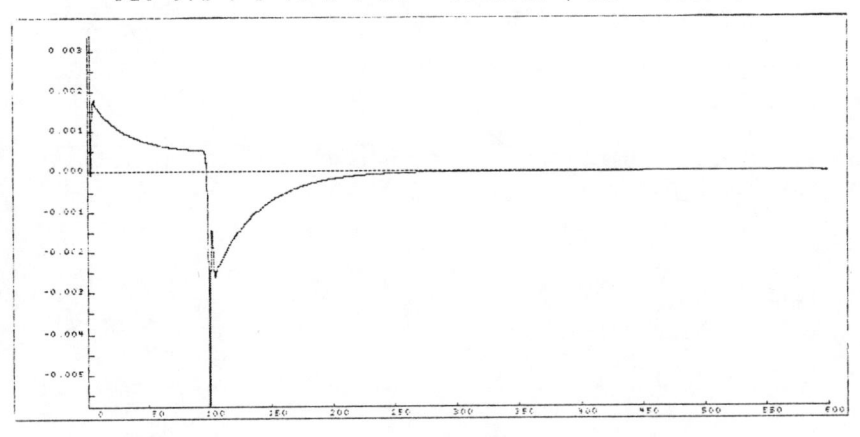

FIG 3.e : S vs N : Dt = 0.01844 , Dx = 0.01 .

ANALYSIS RESULTS.

A decrease of Dt implies a decrease of Σ inside the stability limit previously defined. On this side of Dt, it is convenient to decrease Dx so as to keep the condition r >> 1 . If Dt is not small enough in the direct resolution, then the accuracy may not be sufficient and the inverse resolution cannot restore the surface profile. Anyway, it stands a residue resulting from the incompatibility at t=0 and from the discontinuity points.

PERFORMANCES COMPARISON.

The performance results are effective in relation with the others solutions presented in the litterature. We show a comparison on the minimum dimensionless parameter Dt. In our simulation, we find Dt=0.009 .
[3] Dt=0.08; [1] Dt=0.06; [2] Dt=0.05; [7] Dt=0.013 .

CONCLUSION.

The originality is that the method is time implicit with two steps in time. The method results in a regularized type as if a finite difference scheme had been applied to an hyperbolic problem approximating our parabolic problem. The algorithm was found straightforward to implement with a moderate computer overhead.

The next step of our works consists in a comparison with the others inverse methods based on a methodology defined by BECK and in the inverse determination of the control law of the whole firing process, including all the parameters of the kiln.

REFERENCES.

[1] Alifanov: 'Numerical Solution of a Non Linear Reverse Problem of Heat Conduction'; J.Eng.Phys.(USSR) vol.25 1975.

[2] Beck: 'Criteria for Comparison of Methods of Solution of the Inverse Heat Conduction Problem'; Nucl.Eng.and Design vol.53 1979.

[3] Blackwell: 'Efficient Technique for the Numerical Solution of the One Dimensional Inverse Problem '; Num.Heat Transfer vol.4 1981.

[4] El Badia: 'Identifiabilité et Identification Numérique de Paramètres Répartis pour un Modèle Monodimensionnel'; Thèse 3e cycle UPS Toulouse 1985.

[5] Godounov-Riabenki: 'Schémas aux Différences'; Editions de Moscou 1977.

[6] Richtmayer-Morton: 'Difference Methods for Initial-Value Problems'; Interscience 1967.

[7] Stolz: 'Numerical Solutions to an Inverse Problem of Heat Conduction'; Trans.ASME.J.Heat Transfer vol.82C 1960.

[8] Tykhonov-Arsenin: 'Solutions of Ill-posed problems'; Winston & Wiley 1977.

[Beck-Litkouhi-S^{t}Clair]: 'Efficient Sequential Solution of the Non Linear Inverse Heat Conduction Problem'; Num.Heat Transfer vol.5 1982.

[Burgraff]: 'An Exact Solution of the Inverse Problem in Heat Conduction Theory'; J.Heat Transfer vol.86 1964.

[Garifo-Schrock-Spedicato]: 'On the Solution of the Inverse Heat Conduction Problem by Finite Differences'; Energ.Nucl. vol.22 1975.

SEMI-ANALYTIC SOLUTION OF THE HEAT CONVECTION EQUATION

W. S. Dunbar
Department of Metals and Materials Engineering
University of British Columbia
Vancouver, B. C., Canada V6T 1W5

and

A. D. Woodbury
Department of Geological Sciences
McGill University
Montreal, P. Q., Canada H3A 2A7

SUMMARY

Using a Lanczos algorithm developed for unsymmetric matrices and an eigenvalue eigenvector decomposition, the system of differential equations arising from a finite element or finite difference approximation to the heat convection equation is reduced to a small uncoupled system of first order differential equations. The method appears to be an efficient method of solving large problems, particularly when time intervals are very long.

INTRODUCTION

Convective heat (or mass) transfer in fluids is described by the partial differential equation

$$\nabla \cdot (\mathbf{D}\nabla u) - \mathbf{v} \cdot \nabla u = \frac{\partial u}{\partial t} \tag{1}$$

where the scalar u is some potential quantity, the vector $\mathbf{v}$ is the velocity of the fluid (assumed to be known) and $\mathbf{D}$ is a tensor of diffusion (or conduction) coefficients. Many processes, such as heat and mass transfer in molten metal or a plasma jet can be modeled by Equation 1. The motivation for the development of the method described in this paper arose as a result of some difficulties encountered in modeling contaminant transport in groundwater, where u would be contaminant concentration.

Solution of Equation 1 by finite element or finite difference methods leads to the system of first order differential equations

$$\mathbf{C}\dot{\mathbf{u}} + \mathbf{K}\mathbf{u} = \mathbf{f} \tag{2}$$

where $\mathbf{u}$ is a vector of temperatures (or mass) at the nodes of the mesh used to describe the problem and $\dot{\mathbf{u}} = \partial \mathbf{u}/\partial t$. $\mathbf{K}$ is the 'diffusion' matrix and $\mathbf{C}$ is the 'capacity' matrix.

In a source free domain, the vector **f** describes the boundary conditions of the problem (which may be time dependent). **C** is symmetric and positive definite; if $\mathbf{v} \neq 0$, **K** is unsymmetric. When the convective term $\mathbf{v} \cdot \nabla u$ dominates over the diffusion term $\nabla \cdot (\mathbf{D} \nabla u)$, care must be taken in the approximation used to compute **K** [1,2,3].

The solution of Equation 2 may be obtained by direct numerical time integration algorithms such as the Crank-Nicholson method:

$$\left(\mathbf{C} + \frac{\Delta t}{2}\mathbf{K}\right)\mathbf{u}^{s+1} = \left(\mathbf{C} - \frac{\Delta t}{2}\mathbf{K}\right)\mathbf{u}^s + \Delta t \mathbf{f}^{s+\frac{1}{2}} \tag{3}$$

where the superscript s denotes the quantity evaluated at time $t = (s-1)\Delta t$. When $\mathbf{v} \neq 0$, the time step must be small enough to track the solution between mesh points. Note that the solution of Equation 3 at each time step involves the solution of a system of equations. The computational effort is therefore directly related to the number of unknowns and the number of time steps required to obtain an accurate and stable solution.

One may reduce the size of the problem by introducing 'thermal modes'. This is done in the following way. Suppose the original size of the system of equations 2 is n. Given a n by m matrix **Q** (where $m \ll n$), whose columns are linearly independent, the solution **u** may be written $\mathbf{u} = \mathbf{Qw}$. Note that when convective terms dominate, **Q** may be complex-valued. Substituting into Equation 2 and multiplying by $\mathbf{Q}^h$, the Hermitian transpose of **Q**, gives

$$\mathbf{Q}^h \mathbf{C} \mathbf{Q} \dot{\mathbf{w}} + \mathbf{Q}^h \mathbf{K} \mathbf{Q} \mathbf{w} = \mathbf{Q}^h \mathbf{f} = \mathbf{g} \tag{4}$$

If **Q** is the matrix of eigenvectors of **K** which satisfy $\mathbf{Q}^h \mathbf{C} \mathbf{Q} = \mathbf{I}$ and $\mathbf{Q}^h \mathbf{K} \mathbf{Q} = \mathbf{D}$, where **D** is a diagonal matrix composed of the m largest eigenvalues of the generalized eigenproblem $\mathbf{KQ} = \mathbf{CQD}$, then Equation 4 reduces to the uncoupled system of order m:

$$\dot{\mathbf{w}} + \mathbf{Dw} = \mathbf{g} \tag{5}$$

for which an exact analytical solution is possible. Solution methods of this type were apparently first proposed for the diffusion equation ($\mathbf{v} = 0$) by Gallagher and Mallet [4]. Nickell et al [5] applied the method to the convection-diffusion problem where, in general, complex-valued eigenvalues and eigenvectors may arise. The main problem with this approach is that a large generalized eigenvalue problem must be solved.

More recently, Nour-Omid [6] used a recursive Lanczos algorithm to reduce the system of equations 2 arising from a heat conduction or diffusion problem to a small symmetric tridiagonal system. The advantage of this method is that one Lanczos recursion involves only the re-solution of a system of equations, a matrix vector multiplication, plus two inner products and is therefore amenable to vector processing. A further advantage is that the small matrix may be diagonalized thus uncoupling the system of differential equations and permitting an exact analytical solution in time. This is important in applications where time intervals are very long.

In this paper a Lanczos algorithm is developed for the convection-diffusion problem. This requires the development of a Lanczos algorithm to reduce a general unsymmetric matrices to tridiagonal form. It will be seen that under certain conditions, the resulting small tridiagonal matrix may be diagonalized to produce an uncoupled system of first order differential equations.

SOLUTION ALGORITHM

Lanczos algorithms for unsymmetric matrices have been proposed by Wilkinson [7] and Golub and Van Loan [8]. Recently, Cullum and Willoughby [9] developed a practical algorithm for large matrices which reduces an unsymmetric matrix to a (possibly) complex-valued tridiagonal matrix. It is important to note that the Lanczos algorithm is entirely analogous to an eigenvalue problem and that it is possible to define 'forward' iteration and 'inverse' iteration types of Lanczos algorithms. Each of the algorithms mentionned above is of the forward variety. The eigenvalues of the tridiagonal matrices resulting from such algorithms will therefore approximate the large eigenvalue or 'high frequency' thermal modes. However, the smaller eigenvalues or 'low frequency' modes are important in the convection-diffusion problem. Therefore, inverse type Lanczos procedures must be devised.

Continuing the analogy with the eigenvalue problem further, it is possible to define a generalized Lanczos algorithm. This was done by Nour-Omid [6]. Such an algorithm produces a set of Lanczos vectors $Q = (q_1, q_2, \ldots, q_m)$ which are C orthogonal (i.e., $Q^t C Q = I$). These are analogous to eigenvectors of the generalized eigenvalue problem $Kq = \lambda Cq$. In the case where K is unsymmetric, a Lanczos algorithm will produce two sets of Lanczos vectors $X = (x_1, x_2, \ldots, x_m)$ and $Y = (y_1, y_2, \ldots, y_m)$ which are C biorthogonal (i.e., $Y^t CX = I$). These vectors are analogous to right and left eigenvectors of an unsymmetric matrix.

Using the above analogies, it is relatively easy to derive an inverse type generalized Lanczos algorithm. Let $u = Xw$, where X is a n by m matrix of right Lanczos vectors $(m < n)$. Substitution into Equation 2 and multiplication by CK^{-1} gives

$$CK^{-1}CX\dot{w} + CXw = CK^{-1}f \tag{6}$$

Multiplication of Equation 6 by the transpose of the left Lanczos vectors, Y^t, and using the biorthogonality relationship gives the tridiagonal system of differential equations

$$T\dot{w} + w = g \tag{7}$$

where $g = Y^t CK^{-1} f$ and $T = Y^t CK^{-1} CX$ is the tridiagonal matrix

$$T = \begin{pmatrix} \alpha_1 & \gamma_1 & & \\ \beta_2 & \alpha_2 & \gamma_2 & \\ & \ddots & \ddots & \ddots \\ & & & \gamma_{m-1} \\ & & \beta_m & \alpha_m \end{pmatrix} \tag{8}$$

Note that all matrices are real-valued.

Using the definition of T and the biorthogonality relationship, the equations governing the reduction to tridiagonal form are

$$CX = KXT \tag{9a}$$
$$CY = K^t Y T^t \tag{9b}$$

Equating columns in Equations 9a and 9b results in

$$\beta_{j+1} K x_{j+1} = (C - \alpha_j K) x_j - \gamma_{j-1} K x_{j-1} = r_j \tag{10a}$$
$$\gamma_j K^t y_{j+1} = (C - \alpha_j K)^t y_j - \beta_j K^t y_{j-1} = p_j \tag{10b}$$

where $\mathbf{x}_j$ and $\mathbf{y}_j$ denote the jth columns of $\mathbf{X}$ and $\mathbf{Y}$ respectively. Multiplication of Equation 10a by $\mathbf{y}_j^t \mathbf{C} \mathbf{K}^{-1}$ and Equation 10b by $\mathbf{x}_j^t \mathbf{C} \mathbf{K}^{-t}$ results in two equations for α_j

$$\alpha_j = \mathbf{y}_j^t \mathbf{C} \mathbf{K}^{-1} \mathbf{C} \mathbf{x}_j \qquad (11a)$$

$$\alpha_j = \mathbf{x}_j^t \mathbf{C} \mathbf{K}^{-t} \mathbf{C} \mathbf{y}_j \qquad (11b)$$

These two values should be averaged as suggested by Cullum and Willoughby [9].

In the case of unsymmetric matrices, there is some freedom in choosing equations for β_j and γ_j. Since $\beta_{j+1} \mathbf{x}_{j+1} = \mathbf{K}^{-1} \mathbf{r}_j$ and $\gamma_j \mathbf{y}_{j+1} = \mathbf{K}^{-t} \mathbf{p}_j$, the biorthogonality relationship gives the following constraint on the values of γ_j and β_{j+1}:

$$\gamma_j \beta_{j+1} = \mathbf{p}_j^t \mathbf{K}^{-1} \mathbf{C} \mathbf{K}^{-1} \mathbf{r}_j \qquad (12)$$

Following Cullum and Willoughby [9], the choice

$$\gamma_j = \beta_{j+1} = (\mathbf{p}_j^t \mathbf{K}^{-1} \mathbf{C} \mathbf{K}^{-1} \mathbf{r}_j)^{\frac{1}{2}} \qquad (13)$$

is made resulting in a *symmetric* tridiagonal matrix. If the problem is not highly convection dominated, values of β_{j+1} will be real. A more precise condition for real values of β_{j+1} will be given in the next section.

If $\mathbf{T}$ is symmetric, then it may be decomposed by a standard eigenvalue eigenvector decomposition using an orthogonal matrix $\mathbf{Q}$

$$\mathbf{Q}^t \mathbf{T} \mathbf{Q} = \mathbf{D} \qquad (14)$$

where $\mathbf{D}$ is a diagonal matrix composed of the eigenvalues of $\mathbf{T}$. Substitution into Equation 7 gives

$$\mathbf{D}\dot{\mathbf{z}} + \mathbf{z} = \mathbf{h} \qquad (15)$$

where $\mathbf{z} = \mathbf{Q}^t \mathbf{w}$ and $\mathbf{h} = \mathbf{Q}^t \mathbf{g}$. Given initial conditions $\mathbf{z}^0$, and assuming $\mathbf{h}$ to be independent of time, the solution to the ith equation of Equation 15 is

$$z_i(t) = z_i^0 e^{-t/d_i} + h_i(1 - e^{-t/d_i}) \quad 1 \leq i \leq m \qquad (16)$$

where $\mathbf{z}^0 = \mathbf{Q}^t \mathbf{w}^0 = \mathbf{Q}^t \mathbf{X}^{-1} \mathbf{u}^0 = \mathbf{Q}^t \mathbf{Y}^t \mathbf{C} \mathbf{u}^0$, $\mathbf{u}^0$ being the initial conditions of the original problem. Once $\mathbf{z}$ is found, the original solution is given by

$$\mathbf{u} = \mathbf{X} \mathbf{Q} \mathbf{z} \qquad (17)$$

In terms of computer time, the semi-analytic algorithm is likely to be efficient. However, storage requirements are increased, the most significant being the storage for the arrays $\mathbf{X}$ and $\mathbf{Y}$.

EXAMPLE APPLICATION

The following one dimensional problem is used to illustrate some aspects of the proposed method:

$$D\frac{\partial^2 u}{\partial x^2} - v\frac{\partial u}{\partial x} = \frac{\partial u}{\partial t}$$
$$u(0,t) = 1 \quad u(1,t) = 0 \tag{18}$$
$$u(x,0) = 0$$

A consistent finite element scheme employing linear elements is used to reduce Equation 18 to the form of Equation 2. If the interval [0,1] is discretized into n_e elements of size h, the element capacity and diffusion matrices are

$$\mathbf{C}_e = \begin{pmatrix} h/3 & h/6 \\ h/6 & h/3 \end{pmatrix}$$

$$\mathbf{K}_e = \begin{pmatrix} D/h - v/2 & -D/h + v/2 \\ -D/h - v/2 & D/h + v/2 \end{pmatrix}$$

When assembled, the global capacity and diffusion matrices are tridiagonal. Elimination of the rows and columns corresponding to the boundary conditions produces a system of size $n = n_e - 1$.

Of particular interest is the performance of the method as the flow becomes more convection dominated. This is measured by the grid Peclet number $Pe_g = vh/D$. As Pe_g increases, the flow becomes oscillatory which is spurious and considered an artefact of the discretization. As velocity increases, the solution at a particular time becomes steeper and more difficult to model, even with linear elements. The steepness is related to the higher frequency modes of the system of equations 2. Consequently, it is expected that as Pe_g increases, more modes will be necessary to obtain an accurate solution.

A simple method of suppressing oscillations in the solution is to decrease h, i.e., use more nodes. This increases the size of the system to be solved. However, by means of the Lanczos procedure, the size of the system can be reduced. It is therefore of interest to know the amount of reduction possible.

An excellent description of the stability of solutions of Equation 2 is given by Carey and Sepehrnoori [10]. By means of an eigenvalue analysis of the system of equations 2, they show that for $Pe_g > 2$, oscillatory solutions will occur. This is manifested by complex eigenvalues of the matrix $\mathbf{K}_e$ shown above. Consequently, it is expected that for $Pe_g > 2$, the Lanczos algorithm described in the last section will break down, i.e., negative arguments of the square root in Equation 13 will occur.

To test these conjectures, a simple 21 node (i.e., $n_e = 20$) discretization of the interval [0,1] was used. This gives $h = 0.05$ and results in a system of equations of order $n = 19$. Values of v and D were used to yield Peclet numbers of 0.25 and 1.0. The results are shown in Figures 1 and 2 where the semi-analytic solution is compared with a Crank-Nicholson solution of the original system of equations. For the case of $Pe_g = 0.25$, excellent agreement to three digits was found using 5 modes. To obtain

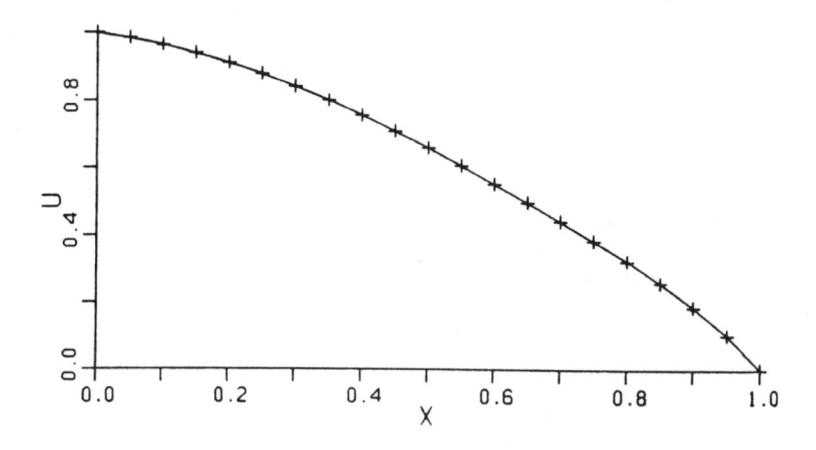

Figure 1 Comparison of Crank-Nicholson solution (solid line, $\Delta t = 0.05$) with the semi-analytic solution (crosses, number of modes $m = 5$) at time $t = 1.0$. $D = 0.1$, $v = 0.5$, $h = 0.05$, $Pe_g = 0.25$.

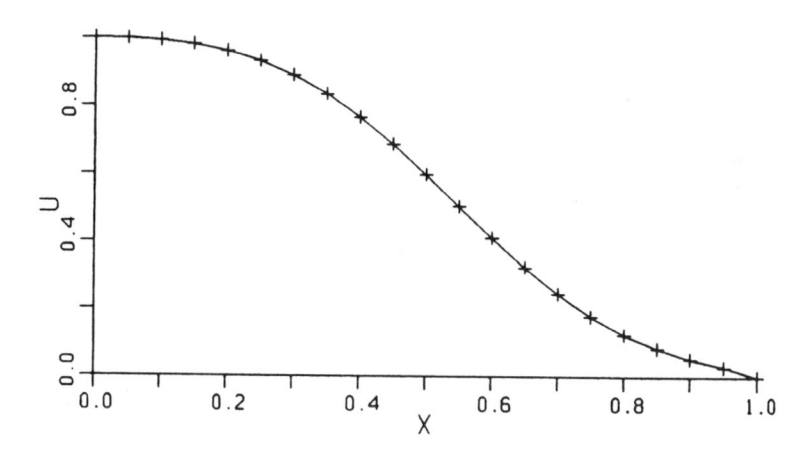

Figure 2 Comparison of Crank-Nicholson solution (solid line, $\Delta t = 0.01$) with the semi-analytic solution (crosses, number of modes $m = 10$) at time $t = 0.25$. $D = 0.1$, $v = 2.0$, $h = 0.05$, $Pe_g = 1.0$.

similar agreement for the case $Pe_g = 1.0$, ten modes were required. This confirms the first conjecture.

The behaviour of the algorithm for $Pe_g > 2$ was investigated by choosing values of D and v to yield Peclet numbers of 2.0, 2.5 and 3.0. Thus for $D = 0.1$ and $h = 0.05$, velocities of 4.0, 5.0 and 6.0 were used. It was found that, for the cases $Pe_g = 2.0$ and $Pe_g = 2.5$, the Lanczos algorithm did not break down, but the solution was extremely unstable no matter how many modes were included. For the case $Pe_g = 3.0$, the Lanczos algorithm did break down for all numbers of modes included.

The use of mesh refinement to decrease the Peclet number was investigated for the case $v = 5.0$ and $D = 0.1$. The number of elements was doubled to give $h = 0.025$ and $Pe_g = 1.25$. Twenty modes were required to obtain an acceptably accurate solution. The results are shown in Figure 3. In this example the velocity is very large and the solution is steep. More elements, higher order elements or upwinding could be used to improve accuracy.

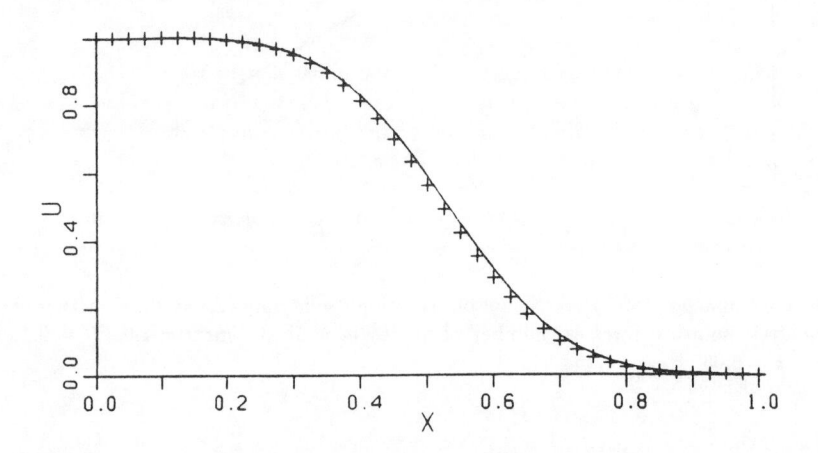

Figure 3 Use of mesh refinement. Comparison of Crank-Nicholson solution (solid line, $\Delta t = 0.002$) with the semi-analytic solution (crosses, number of modes $m = 20$) at time $t = 0.1$. $D = 0.1$, $v = 5.0$, $h = 0.025$, $Pe_g = 1.25$.

Based on these examples, it appears that the semi-analytic solution algorithm is working as expected. An important question is whether it is possible to predict the number of modes required to obtain a desired accuracy. In the case of a pure diffusion problem, Nour-Omid [6] terminates the Lanczos algorithm when the element β_{m+1} multiplied by a participation factor of the mth mode falls below a given value. However, such an approach may not be able to take the magnitude of the Peclet number into account. Consequently, a different termination criterion must be found.

CONCLUSIONS

A generalized Lanczos algorithm for unsymmetric diffusion matrices and symmetric and positive definite capacity matrices has been developed and used to reduce a system of n first order differential equations arising from a finite element approximation to the heat convection equation to a small symmetric tridiagonal system of size $m < n$. An eigenvalue eigenvector decomposition is then used to transform the tridiagonal system to an uncoupled system of first order differential equations for which, in the case of time independent boundary conditions, an exact analytical solution in time is possible. A series of matrix vector multiplications is then used to obtain the original solution.

It was found that

- The method is capable of very accurate solutions with a small number of modes, m in cases where the grid Peclet number is less than 1.0. More modes are required to obtain similar accuracy for large Peclet numbers. The method breaks down for Peclet numbers greater than 2.0, as do other methods. This is not a restrictive condition since non-oscillatory solutions are desired in any case.

- Using mesh refinement, the grid Peclet number can be reduced. Although this results in a larger system of equations, the Lanczos algorithm can be used to reduce the size of the system significantly. Thus the method appears to be an efficient means of solving large problems.

- The decoupled system of first order differential equations is particularly useful in cases where time intervals are long.

Further investigation of the advantages and limitations of the method is necessary. In addition, a method of terminating the Lanczos reduction must be devised.

ACKNOWLEDGEMENTS

The authors would like to express their gratitude to several people who (possibly unknowingly) greatly helped the progress of this work. Dr. Beresford N. Parlett of the Department of Computer Science at the University of California, Berkeley provided advice and programs during the original efforts. Dr. Jim Varah of the Department of Computer Science at the University of British Columbia suggested that the Lanczos algorithm be used and got us in touch with Dr. Jane Cullum of the IBM Thomas J. Watson Research Center whose original work on an unsymmetric Lanczos algorithm had much influence on the algorithm developed herein. Dr. Les Smith of the Department of Geological Sciences at the University of British Columbia provided examples with which to test the program. Computer funds were provided by the Centre for Metallurgical Process Engineering at the University of British Columbia.

REFERENCES

[1] Price, H. S., Varga, R. S. and Warren, J. E. - Application of oscillation matrices to diffusion-convection equations, *Journal of Mathematical Physics*, **45**, 301-311 (1966).

[2] Christie, I., Griffiths, D. F., Mitchell, A. R. and Zienkiewicz, O. C. - Finite element methods for second order differential equations with significant first derivatives, *International Journal for Numerical Methods in Engineering*, **10**, 1389-1396 (1976).

[3] Jensen, O. K. and Finlayson, B. A. - Oscillation limits for weighted residual methods applied to convective diffusion equations, *International Journal for Numerical Methods in Engineering*, **15**, 1681-1689 (1980).

[4] Gallagher, R. H. and Mallett, R. H. - Efficient solution processes for finite element analysis of transient heat conduction, *Transactions ASME, Journal of Heat Transfer*, **93**, 257-263 (1971).

[5] Nickell, R. E., Gartling, D. K. and Strang, G. - Spectral decomposition in advection-diffusion analysis by finite element methods, *Computer Methods in Applied Mechanics and Engineering*, **17/18**, 561-580 (1979).

[6] Nour-Omid, B. - Lanczos method for heat conduction analysis, *International Journal for Numerical Methods in Engineering*, **24**, 251-262 (1987).

[7] Wilkinson, J. - *The Algebraic Eigenvalue Problem*. Oxford University Press (1965).

[8] Golub, G. H. and Van Loan, C. F. - *Matrix Computations*. John Hopkins University Press (1983).

[9] Cullum, J. and Willoughby, R. A. - A practical procedure for computing eigenvalues of large sparse nonsymmetric matrices. In *Large Scale Eigenvalue Problems*, (J. Cullum and R. A. Willoughby, eds.). Elsevier (1986).

[10] Carey, G. F. and Sepehrnoori, K. - Gershgorin theory for stiffness and stability of evolution systems and convection-diffusion, *Computer Methods in Applied Mechanics and Engineering*, **22**, 23-48 (1980).

BOUNDARY INTEGRAL ANALYSIS WITH ORTHOGONAL BASIS FUNCTIONS

B. Dorri and V. Kadambi

General Electric Company
Corporate Research and Development
PO Box 8, Schenectady, NY 12301, USA

1. SUMMARY

Source strengths in an indirect single-layer source formulation of the Boundary Integral Method are described in terms of orthogonal functions rather than in terms of the conventional first order or second order polynomials. The least squares method is used to calculate the coefficients of the function. Basic formulation with Fourier series basis functions is presented for the two dimensional Laplace equation along with some representative results. Formulation is also presented for the two dimensional Diffusion equation. Application of the method to solution of the inverse heat conduction problem is discussed.

2. INTRODUCTION

The Boundary Integral Method has been used in the past to solve field problems described by partial differential equations. However, the inabilities of the method to handle complicated and general purpose problems has made it less attractive than Finite Difference and Finite Element Methods. In recent years, research has been extended in this area and techniques have been developed to further the applications of the method. Kadambi and Dorri [1,2] have demonstrated the use of Boundary Integral Method for the solution of nonlinear thermal problems. The technique has also been extended to inelastic stress analysis, Telles [3]. Other extensions of the method are in areas in which Finite Difference or Finite Element Methods have inherent difficulties. These areas include moving boundaries, thermodynamic phase change, and unbounded field problems. Hume et. al. [4] compared the Boundary and Finite Element Methods for solving moving boundary problems. O'Neill [5] describes the use of Boundary Integral Method for solving the phase change problem. Vallabhan et. al. [6] have used a combination of Finite Element and Boundary Integral formulations to handle soil-structure interaction. They have expressed the nonlinear structural behavior by the finite element and the unbounded soil by Boundary Integral.

In this paper, the aim is to expand on the idea of using trigonometric series, suggested by Liu [7], in place of polynomial basis functions. The source strength on the boundary is described in terms of one or more orthogonal functions and the method of least squares is used to determine the unknown coefficients in the function. The resulting coefficient matrix is symmetric and smaller than that for direct evaluation. The solution is also more smooth and unlike regular Boundary Integral models, it gives good results near boundary points. While the formulation is derived for both the two dimensional Laplace and Diffusion equations, representative results are only shown for the Laplace equation to demonstrate the capabilities and strength of the method. It is further shown that the method is applicable to the solution of inverse heat conduction problems where a few measured temperatures are used to calculate the missing boundary conditions.

3. METHOD

Mathematical derivation of the method is described for the two dimensional Laplace and Diffusion equations. These equations represent steady state and transient heat conduction problems respectively. While Fourier series representation of the source strength is used to illustrate the formulation for the Laplace equation, a general form of orthogonal function is used in the Diffusion equation application.

3.1. LAPLACE EQUATION

The indirect single-layer source formulation (Brebbia et. al. [8]) for two dimensional steady state heat conduction problem is used to demonstrate the mathematical development of the method. The temperature at any point i, T_i, inside or on the boundary of a two dimensional region, is represented by the following equation:

$$T_i = \int_S A(s) \ln(R) ds \tag{1}$$

Here $A(s)$ is the source strength which is a function of variable s (the boundary position variable), R is the distance between point i (collocation point) and integration point on the boundary, and $\ln(R)$ is the kernel which satisfies the two-dimensional Laplace equation. Normally, $A(s)$ is described by polynomial basis functions over boundary elements. But here, we assume one or more Fourier series to describe $A(s)$ over the boundary.

Let us assume that the boundary of a two-dimensional solid is represented by PN patches, Figure 1., where every patch p is divided into NE(p) boundary elements. The source strength on every patch p is represented by a Fourier series with M(p) alternating terms (2M(p)+1 total terms) as shown below:

$$A_p(s) = \frac{a_0(p)}{2} + \sum_{n=1}^{M(p)} \left[a_n(p) \sin \frac{2n\pi s(p)}{l(p)} + b_n(p) \cos \frac{2n\pi s(p)}{l(p)} \right] \tag{2}$$

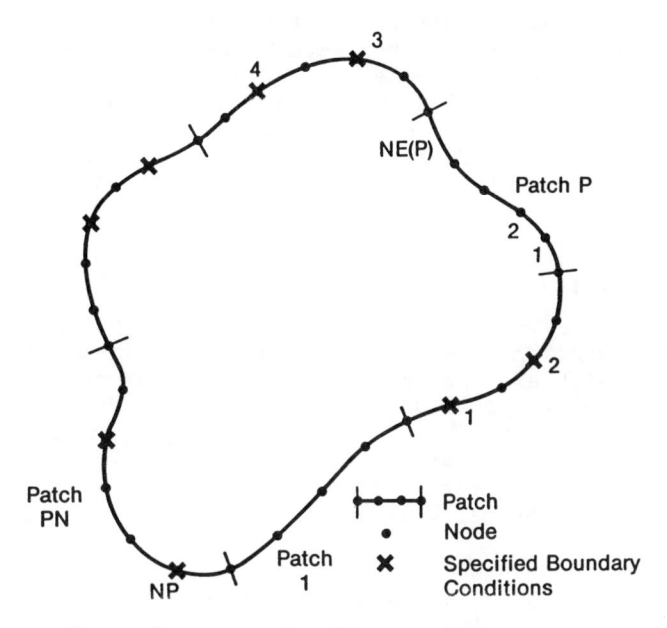

Figure 1. Schematic of a 2D Boundary Integral Mesh

Here s(p) is the boundary position variable for patch p and l(p) is the length of patch p, while $a_0(p)$, $a_n(p)$, and $b_n(p)$ are the coefficients of Fourier series describing the source strength for patch p. The integral in Equation (1) can be represented, in a numerical scheme, as a summation of integrals over all the elements on the boundary:

$$T_i = \sum_{p=1}^{PN} \sum_{e=1}^{NNE(p)} \int_e A_p(s) \ln(R) ds(p) \tag{3}$$

Here e is the index for element number. If the boundary temperatures are specified for all the boundary nodes, the problem reduces to the regular direct conduction analysis. However, it will be shown later on that they do not have to be specified at every point and therefore, the solution scheme is applicable to inverse conduction problems. To illustrate a general approach, let us assume that the temperatures at NP boundary nodes are given. Fourier series, describing the source strength, which best match the specified boundary conditions can be found by minimizing the following function.

$$\text{Minimize} \quad : \quad F = \sum_{i=1}^{NP} (T_i^* - T_i)^2 \tag{4}$$

The superscript * denotes the specified or measured value. Function F is the sum of squares of errors between the specified and calculated values of temperatures. The minimization of this function provides a least squares form which has also been used in many inverse conduction schemes, e.g. Beck et. al. [9]. Upon substitution of T_i from Equation 3 into Equation 4 and analytical differentiation with respect to the unknown Fourier series coefficients, sets of relations among the coefficients are obtained. These are set equal to 0 for minimization purposes. For illustration, a differentiation of function F (Equation 4) with respect to $a_m(q)$, which is the m th coefficient of sin in the Fourier series describing the source function over patch q of the boundary, is shown below:

$$\frac{d(F)}{d(a_m(q))} = \sum_{i=1}^{NP}\left[-(T_i^*-T_i)\sum_{e=1}^{NE(q)}\int_e \sin\frac{2n\pi s(q)}{l(q)}\ln(R)ds\right] = 0 \qquad (5)$$

T_i is the function given in Equation 3. Equation 5 and similar equations which are the results of differentiations of function F with respect to other unknown coefficients of the Fourier series, form a set of linear equations. Upon solving this system, the coefficients of the Fourier series are found. Therefore, the resulting series describe the source strengths on the boundary which optimally match the prescribed temperatures.

A computer program has been developed to study the above method. A six point Gaussian quadrature integration scheme is used to carry the integrals over the elements in Equation 5. The resulting coefficient matrix is symmetric, due to the least squares scheme, and has a size equal to:

$$\sum_{p=1}^{PN}(2M(p)+1) \quad X \quad \sum_{p=1}^{PN}(2M(p)+1) \qquad (6)$$

Representative results will be presented in section 4.

3.2. DIFFUSION EQUATION

For the Diffusion equation, as applied to transient heat conduction, both Kernel and source strength are time dependent as well as space dependent. A relationship for temperatures at any point (x,y) and time t is represented, similar to equation 1, in the following form.

$$T(x,y,t) = \int_s \int_0^t A(s,\tau).G(R,t,\tau)d\tau ds \qquad (7)$$

A is the source strength at the boundary location s and time τ and G is the time dependent fundamental solution which can be represented by the following relationship, Carslaw and Jaeger [10].

$$G = \frac{1}{4\pi\alpha(t-\tau)} e^{\frac{-R^2}{4\alpha(t-\tau)}} \qquad (8)$$

Here, R is the distance between the point source and the point for which the temperature is to be computed and τ is the time at which the point source has emitted its energy ($\tau<t$). Similar to the steady state formulation, the source strength (A) is represented by orthogonal basis functions. However, time dependency of these quantities is handled by introducing

polynomials of time as the coefficients of the orthogonal functions. In a general form, this is illustrated as

$$A(s_p, \tau) = \sum_{np} \left[a_{np}(t-\tau)^2 + b_{np}(t-\tau) + c_{np} \right] F_{np}(s_p) \qquad (9)$$

where, a_{np}, b_{np}, and c_{np} are the unknown parameters, p is patch number, s_p is the portion of the boundary representing the patch p, and np is the number of terms in the orthogonal function F for patch p. Function F can be Legendre polynomials, Fourier series or any orthogonal function. A few of the above functions are used to describe the sources over the entire domain. No attempt is made to ensure continuity of these functions across the boundary patches. Combining the equations 7 through 9, the relationship for temperature at any location within the solid or on the boundary is obtained.

Similar to the steady state formulation, a least squares method is used to minimize the error between the calculated and input values of temperature. However, the summation is carried over space and time.

Derivatives from the sum of the square of the errors with respect to the unknown coefficients a_{np}, b_{np}, and c_{np} are set equal to zero and the final system of equation is solved to compute these unknown coefficients. Mathematical derivation is similar to the steady state and, therefore, will not be shown here.

4. RESULTS

In this paper, results from the steady state formulation is shown to demonstrate the capabilities of the method as well as its sensitivity to some of the parameters. Transient results will be published separately.

A rectangular geometry with dimensions of 1X1 meter is assumed to have a fixed temperature of 10 °C on one side and 0 °C on the other 3 sides. To validate the method, two cases have been studied and the results have been compared with the analytical solution, which exists for this simple case. Results from a Boundary Integral program with linear elements are also presented.

First, the geometry is meshed (Figure 2) by placing 20 elements of equal size along the boundary. Temperature distributions over the diagonal line have been evaluated by: i) Exact solution, ii) A Boundary Integral program with linear basis functions, and iii) The current code where one Fourier series is assumed to represent the source strength over the whole boundary and temperatures are assumed to have been specified for all the boundary nodes. Figure 3, which exhibits these results, indicates that a 5 term Fourier series has some errors at the two ends, but a 13 term Fourier series gives a very good result. Further the Boundary Integral code with linear basis functions, as expected, gives unrealistic results very close to the boundaries.

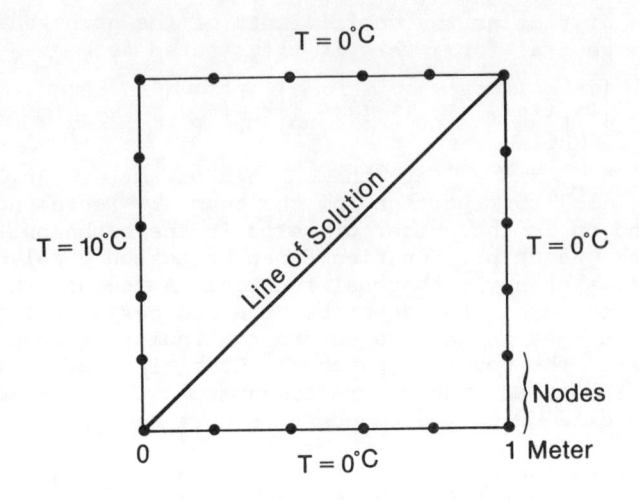

Figure 2. Test Case No. 1

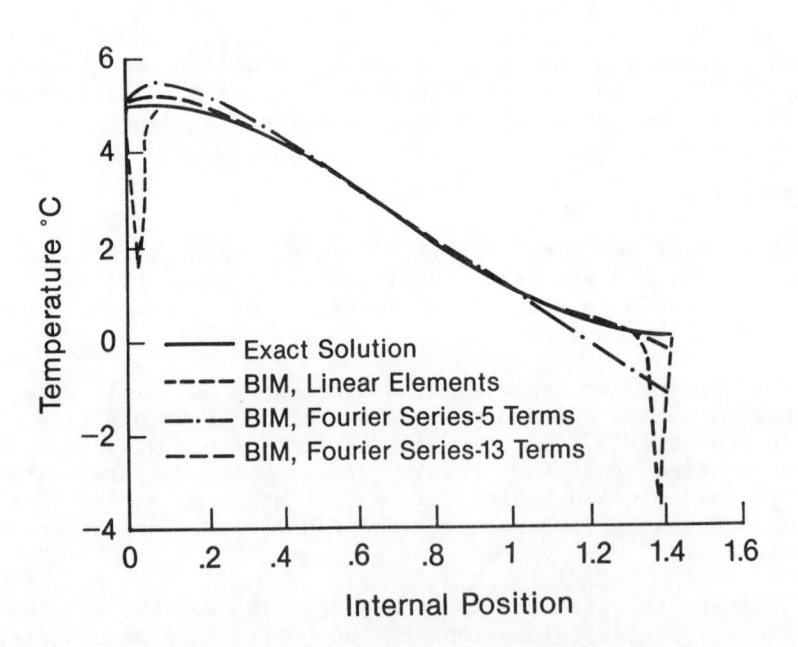

Figure 3. Results of Test Case No. 1

A closer look at the source strength distribution over the boundary can help to identify how the Fourier series should be applied. Figure 4 exhibits the source strength A(s) as computed from a Boundary Integral program which assumes constant values of source strength over each element. It is noted that over the three sides with 0 °C temperature the source strength behaves differently from that over the fourth side where the temperature is 10 °C. This, of coarse, is expected, since the source strength is a measure of heat flux, and the heat flux has different directions on side 1 and sides 2,3,4. A similar plot using Fourier series solutions is shown in Figure 5. It is noted that the series are trying to fit the discontinuous function. This figure suggests that if side 1 is described as one patch, represented by one Fourier series, and sides 2, 3, and 4 as a second patch with a different Fourier series, the discontinuity can be represented better. In general, it is suggested that patches with different signs for heat flux be represented by different Fourier series.

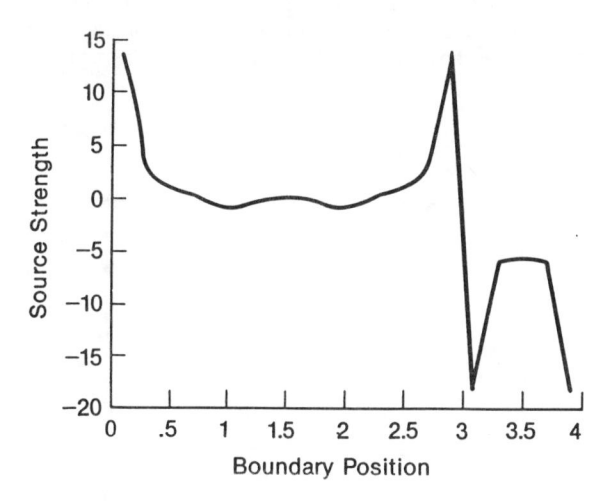

Figure 4. Source Strength Computed by Constant Element BIM

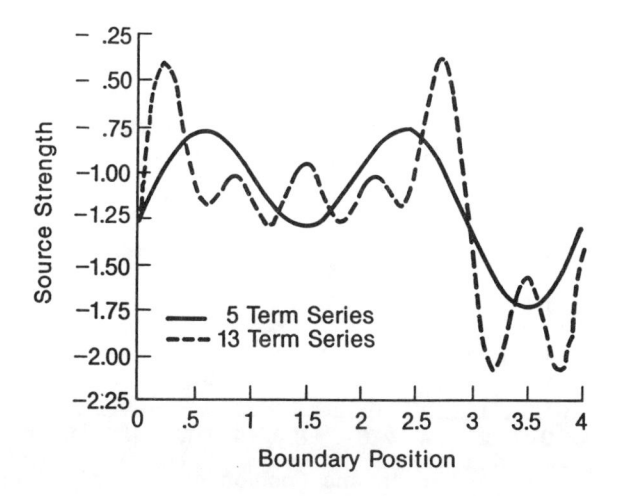

Figure 5. Source Strengths Computed by Fourier Series BIM

The second case study is for the same geometry and boundary conditions, but a different nodal structure. Here, side 1 is divided into 8 elements instead of 5. This has been done to allow a dedicated Fourier series with 5 terms for this side and another series with 5 terms for patch 2 which consists of sides 2,3, and 4. Another difference is that it is assumed that temperatures at only 13 out of 23 boundary nodes have been specified. Figure 6 shows the detail of nodal structure along with patch definitions. Figure 7 exhibits the temperature distribution over the diagonal line and compares it with the exact solution. It should be noted that a 10 by 10 symmetric matrix had to be solved for this analysis as compared to a 23 by 23 unsymmetric matrix for a regular Boundary Integral program.

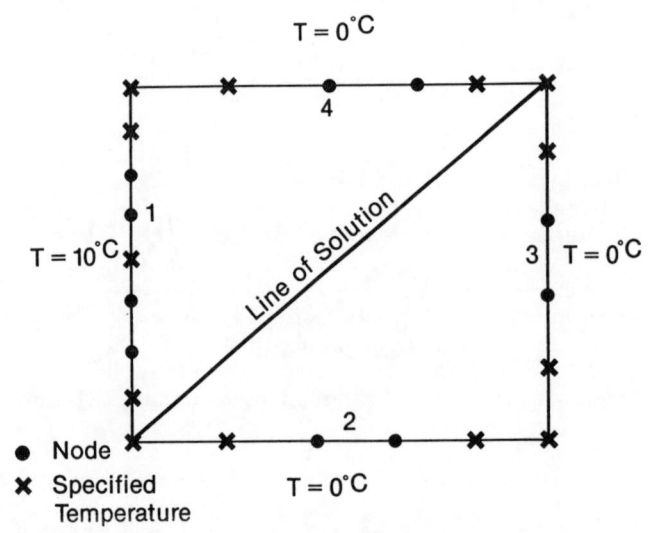

Figure 6. Test Case No. 2

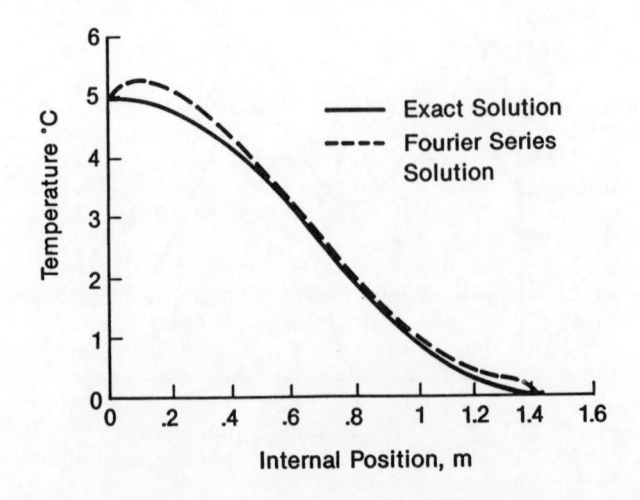

Figure 7. Results of Test Case No. 2

5. CONCLUSIONS

The capabilities of orthogonal functional representation of source strength have been demonstrated. A least squares method ensures the best fit for the function and also makes the coefficient matrix both symmetric and small. Formulations have been shown for both steady state and transient two dimensional heat conduction. Sample runs were made for the steady state case with the Fourier series basis functions. It has been shown that more than one Fourier series might be more appropriate to describe source strengths in cases where the flux exhibits discontinuities. The method is general and can be applied to other equations.

Another application of this method is in solving inverse heat conduction problems. Equation 3 and 7, which specify the temperature, holds for every boundary or internal point. Therefore, if instead of all boundary nodes, the temperatures are specified (i.e., measured) at a few internal points, the source strength which optimally fits these values will be computed. The advantages of this method over the Finite Element or Finite Difference schemes for solving the inverse conduction problems are: i) The solution is direct and unlike other methods, no iteration is needed. ii) The coefficient matrix is small and symmetric, and iii) The solution is optimum and measurement errors may be reduced by the least squares scheme. However, it should be noted that enough measured points are needed to restrain the orthogonal function from oscillating and giving unrealistic results.

The extension of this method to three dimensions might result in a great reduction in the matrix size and, therefore an increase in the computational efficiency. Also, other functions such as Legendre polynomials may be used in place of a Fourier series.

6. REFERENCES

1. Kadambi, V. and Dorri, B. - 'Solution of Thermal Problems with Nonlinear Material Properties by the Boundary Integral Method', Proceedings of the 1st Boundary Element Technology Conference, Eds. C.A. Brebbia and B.J. Noye, Springer-Verlag, 1985, p. 151.

2. Kadambi, V. and Dorri, B., - 'Solution of the Nonlinear Helmholtz Equation by the Boundary Integral Method', Innovative Num. Meth. in Engr., Eds. R.P. Shaw, J. Periaux, A. Chaudouet, J. Wu, C. Marino, and C.A. Brebbia, Springer-Verlag, 1986, p. 115.

3. Telles, J.C.F., - 'On Inelastic Analysis Algorithms for Boundary Elements', Advanced Topics in Boundary Element Analysis, Eds. T.A. Cruse, A.B. Pifko, and H. Armen, ASME, New York, 1985, p. 35.

4   Hume III, E.C., Brown, R.A. and Deen, W.M. - Comparison of Boundary and Finite Element Methods for Moving-Boundary Problems Governed by a Potential, Int. J. for Num. Meth. in Engr., 21, 1295, (1985).

5   O'Neill K. - Boundary Integral Equation Solution on Moving Boundary Phase Change Problems, Int. J. for Num. Meth. in Engr., 19, 1825, (1983).

6   Vallabhan, C.V.G., Sivakumar, J., and Radhakrishnan, N., - 'Application of Boundary Element Methods for Soil-Structure Interaction Problems', Proceedings of the 6th Int. Conf. on Boundary Elements, Southampton, 1984, P. 6.

7   Liu, H.K. - 'The Solution of Laplace's Equation by Integral Equation Technique', General Electric TIS Report No. 70-C-234, Class 1, 1970.

8   Brebbia, C.A., Telles, J.C.F., and Wrobel, L.C. - Boundary Element Techniques, Springer-Verlag, Berlin, P. 58, 1984.

9   Beck, J.V., Litkouhi, B. and St. Clair, C.R. - Efficient Sequential Solution of the Nonlinear Inverse Heat Conduction Problem, Num. Heat Transfer, 5, 275, 1982.

10 Carslaw, H.S. and Jaeger, J.C., - Conduction of Heat in Solids, Oxford Univ. Press, p. 361, 1959.

Chebychev Polynomial Solution of the Heat Equation

Patrick Hanley[1] and Wesley L. Harris[2]

[1]Graduate Research Assistant, Massachusetts Institute of Technology.
[2]Dean of Engineering, University of Connecticut.

1 Introduction

Solutions of the heat equation are obtained by utilizing a Chebychev polynomial series to represent the spatial dependence and an implicit finite difference method to approximate the temporal derivative. The steady state solutions represented by the Chebychev polynomial series are highly accurate and convergence is exponential for smooth solutions. The implicit temporal implimentation is very efficient since the dependent variable can be updated to the advance time level in $O(N)$ and $O(MN)$ operation for problems in one and two dimensions respectively; N and M is the maximum number of modes of the Chebychev series.

The one dimension heat equation is discretized in time using the trapezoid rule, however any accurate implicit scheme can be used. The resulting semi-discrete equation is an ordinary differential equation at the advance time level in the spatial variable, with a nonhomogenous term consisting of the quantities from the previous time levels. This equation is solved using a novel implimentation of Lanczos' Tau method in reference [1] and [2] to obtain the dependent variable at the advance time level within $O(N)$ operations.

The two dimension heat equation is discretized in time using an alternating direction implicit (ADI) method. The scheme split the problem into two independent semi-discrete equations in each spatial variables. Each of these equations is a separate ordinary differential equation in its respective spatial variable at the advance time level with the nonhomogenous term consisting of known quantities from previous time levels. The Tau method is again used to solve these equations, however the operational count is now within $O(N^2)$ to advance to the next time level since each equation must be solved along constant coordinate lines.

Finally, a technique is derived to handle problems in complex geometries using a multidomain method.

2 Temporal Discretization

Consider the one dimensional heat equation

$$\frac{\partial U}{\partial t} = \kappa \frac{\partial^2 U}{\partial x^2} \qquad (1)$$

with boundary conditions

$$U(-1) = f_0(t)$$
$$U(1) = f_N(t)$$

Integrating the left handside of equation [1] over a time interval of Δt and using the trapezoidal rule to approximate the right hand side gives

$$U^{i+1} - U^i = \frac{\kappa \Delta t}{2}(U_{xx}^{i+1} + U_{xx}^i) \qquad (2)$$

Rearanging equation [2] gives

$$\frac{\kappa \Delta t}{2} U_{xx}^{i+1} - U^{i+1} = -(U^i + \frac{\kappa \Delta t}{2} U_{xx}^i) \qquad (3)$$

Similarily the two dimensional classical heat equation

$$\frac{\partial U}{\partial t} = \kappa(\frac{\partial^2 U}{\partial x^2} + \frac{\partial^2 U}{\partial y^2}) \qquad (4)$$

subjected to

$$U(-1, y, t) = f_0(y, t)$$
$$U(1, y, t) = f_N(y, t)$$
$$U(x, -1, t) = g_0(x, t)$$
$$U(x, 1, t) = g_N(x, t)$$

can be discretized using the Peaceman-Rachford ADI scheme

$$U^{i+\frac{1}{2}} - U^i = \frac{\kappa \Delta t}{2}(U_{xx}^{i+\frac{1}{2}} + U_{yy}^i) \qquad (5)$$

$$U^{i+1} - U^{i+\frac{1}{2}} = \frac{\kappa \Delta t}{2}(U_{xx}^{i+\frac{1}{2}} + U_{yy}^{i+1}) \qquad (6)$$

The above equations can be rewritten as

$$\frac{\kappa \Delta t}{2} U^{i+\frac{1}{2}}_{zz} - U^{i+\frac{1}{2}} = -(U^i + \frac{\kappa \Delta t}{2} U^i_{yy}) \tag{7}$$

and

$$\frac{\kappa \Delta t}{2} U^{i+1}_{yy} - U^{i+1} = -(U^{i+\frac{1}{2}} + \frac{\kappa \Delta t}{2} U^{i+\frac{1}{2}}_{zz}) \tag{8}$$

The semi-discrete equations [3], [7], and [8] are ordinary differential equations of the form

$$A \frac{d^2 F}{d\xi^2} + CF = G(\xi) \tag{9}$$

where $F(\xi)$ represents the function at the advanced time level and $G(\xi)$ represents the terms from the previous time level. In subsequent sections the Chebychev polynomial method will be developed to solve equation [9]. The technique will applied to equations [3], [7] and [8] so the solution of the heat equations can be advanced one time step.

3 Chebychev Polynomials

Before going to the solution of the ordinary differntial equation some pertinent properites of Chebychev polynomials and series will be reviewed.
The Chebychev polynomial $T_n(x)$ is defined as

$$T_n(x) = cos(n cos^{-1}(x)) \tag{10}$$

where $-1 \leq x \geq 1$. $T_n(x)$ is implemented efficiently in numerical applications by using the recursion relation

$$T_{n+1}(x) - 2x T_n(x) + T_{n-1}(x) = 0 \tag{11}$$

The following orthoganality condition holds for Chebychev polynomial,

$$\int_{-1}^{1} \frac{T_m(x) T_n(x)}{\sqrt{1-x^2}} dx = 0 \tag{12}$$

for $n \neq m$. For $m = n$ the integral will be π if $n = 0$ and $\frac{\pi}{2}$ if $n \geq 0$. A function $f(x)$ can be approximated by a Chebychev polynomial series of the form

$$f(x) = \sum_{n=0}^{N} \frac{1}{c_n} a_n T_n(x) \tag{13}$$

where $c_n = 2$ if $n = 0$ and $c_n = 1$ otherwise, and N is some large integer. The coefficient of equation [13] are determined from

$$a_n = \frac{2}{c_n N} \sum_{i=0}^{N} f(x_i) T_n(x_i)$$

where $x_i = cos(\frac{\pi i}{N})$. A two dimensional function $F(x, y)$ can similarily be represented as

$$F(x,y) = \sum_{n=0}^{N}\sum_{m=0}^{M} \frac{1}{c_n c_m} a_{nm} T_n(x) T_m(y) \tag{14}$$

where the coefficients are determined from

$$a_{nm} = \frac{4}{c_n c_m} \sum_{i=0}^{N}\sum_{j=0}^{M} F(x_i, y_j) T_n(x_i) T_m(y_j)$$

The derivatives of the above functions can also be represented in terms of Chebychev polynomials. The derivative of $f(x)$ with respect to x is of the form

$$\frac{df}{dx} = \sum_{n=0}^{N} b_n T_n(x)$$

where

$$2n a_n = b_{n-1} - b_{n+1} \tag{15}$$

relates b_n to a_n. The partial derivative of $F(x,y)$ with respect to x is of the form

$$\frac{\partial F}{\partial x} = \sum_{n}\sum_{m} \frac{1}{c_n c_m} a^{(10)} T_n(x) T_m(y) \tag{16}$$

and the partial derivative with respect to y of the form

$$\frac{\partial F}{\partial y} = \sum_{n}\sum_{m} \frac{1}{c_n c_m} a^{(01)} T_n(x) T_m(y) \tag{17}$$

where

$$2n a_{nm} = a^{(10)}_{n-1,m} - a^{(10)}_{n+1,m}$$

and

$$2m a_{nm} = a^{(01)}_{n,m-1} - a^{(01)}_{n,m+1}$$

4 Solution of Ordinary Differential Equation

In this section a novel version of the tau method for the solution of ordinary differential equation will be developed. Since the only differential equations of interest is of the form of equation [9], a solution technique that is specific to second order constant coefficient equations will be presented.

Consider once again the ordinary differential equation

$$A \frac{d^2 F}{d\xi^2} + CF = G(\xi) \tag{18}$$

subjected to

$$F(-1) = K1$$
$$F(1) = K2$$

where A, C, $K1$ and $K2$ are constant. The function F, its second derivative, and G can be represented by Chebychev polynomial series and substitued into the above differential equation. Upon using the orthogonal relationship,

$$AF_n^{(2)} + CF_n = G_n \tag{19}$$

is obtained, where F_n, $F^{(2)}$ and G_n are the coefficients of the Chebychev expansion of F, the second derivative of F with respect to x, and G respectively. The coefficient of the Chebychev series expansion for the first derivative of F with respect to x, $F_n^{(1)}$, is related to F_n by

$$2nF_n = F_{n-1}^{(1)} - F_{n+1}^{(1)}. \tag{20}$$

Also $F_n^{(2)}$ is related to $F_n^{(1)}$ by

$$2nF_n^{(1)} = F_{n-1}^{(2)} - F_{n+1}^{(2)}. \tag{21}$$

Solving equation [19] for $F_n^{(2)}$ and substituting into equation [21] then substituting the result into equation [20] and rearranging terms give the following backwards recursion relations for F_n

$$F_{n-2} = n[\frac{2}{n+1} - 4(n-1)\frac{A}{C}]F_n - (\frac{n-1}{n+1})F_{n+2}$$
$$-(G_{n-2} - G_n) - \frac{1}{C}(\frac{n-1}{n+1})(G_n - G_{n+2}). \tag{22}$$

In order to start the backwards recursion, F_N and F_{N-1} are needed. To temporarily solve the problem, these two quantities are set identically equal to τ_0 and τ_1 respectively. Equation [22] is then solved for F_n in terms of τ_0 and τ_1.

The unknowns τ's are then subsequently determined for the boundary conditions on equation [9]. The way this is done numerically, is to let

$$F_n = \tau_0 \alpha_n + \tau_1 \beta_n + \gamma_n. \tag{23}$$

Substituting into the boundary conditions

$$F(-1) = \sum_n \frac{1}{c_n} F_n T_n(-1)$$

$$F(1) = \sum_n \frac{1}{c_n} F_n T_n(1)$$

gives to simultaneous equations for τ_0 and τ_1.

The quantities α_n, β_n and γ_n are determined by substituting equation [23] into equation [22] and setting the coefficients of τ_0, τ_1 and unity to zero which yields

$$\alpha_{n-2} = n[\frac{2}{n+1} - 4(n-1)\frac{A}{C}]\alpha_n - (\frac{n-1}{n+1})\alpha_{n+2}$$

$$\beta_{n-2} = n[\frac{2}{n+1} - 4(n-1)\frac{A}{C}]\beta_n - (\frac{n-1}{n+1})\beta_{n+2}$$

$$\gamma_{n-2} = n[\frac{2}{n+1} - 4(n-1)\frac{A}{C}]\gamma_n - (\frac{n-1}{n+1})\gamma_{n+2}$$
$$-(G_{n-2} - G_n) - \frac{1}{C}(\frac{n-1}{n+1})(G_n - G_{n+2})$$

where
$$\alpha_N = \beta_{N-1} = 1$$
$$\alpha_{N-1} = \beta_N = \gamma_N = \gamma_{N-1} = 0.$$

5 One Dimensional Heat Equation

To solve the semidiscrete equation[3] at the advance time level, let

$$U^{i+1} = \sum_{n=0}^{N} U_n^{i+1} T_n(x)$$

and

$$g^i(x) = U^i + \frac{\kappa \Delta t}{2} U_{xx}^i = \sum_{n=0}^{N} g_n^{i+1} T_n(x).$$

Using the analogy of equation[22] the coefficients of the Chebychev series expansion for the one dimensional heat equation at the advanced time level is

$$U_{n-2}^{i+1} = n[\frac{2}{n+1} - 4(n-1)\frac{A}{C}]U_n^{i+1} - (\frac{n-1}{n+1})U_{n+2}^{i+1}$$
$$-(g_{n-2}^i - g_n^i) - \frac{1}{C}(\frac{n-1}{n+1})(g_n^i - g_{n+2}^i) \qquad (24)$$

where
$$A = \frac{\kappa \Delta t}{2}$$

and $C = -1$.

The coefficients U_n^{i+1} are determined by letting

$$U_N^{i+1} = \tau_0^{i+1}$$

$$U_{N-1}^{i+1} = \tau_1^{i+1}.$$

The values of the τs are determined numerically by letting

$$U_n^{i+1} = \alpha_n^{i+1} \tau_0^{i+1} + \beta_n^{i+1} \tau_1^{i+1} + \gamma_n^{i+1}.$$

Upon substitution into equation [24], the values of α_n^{i+1}, β_n^{i+1} and γ_n^{i+1} are determined and τ_0^{i+1} and τ_1^{i+1} can be found from the boundary conditions

$$U^{i+1}(-1) = \sum_n \frac{1}{c_n} U_n^{i+1} T_n(-1) = u_0^{i+1}$$

$$U^{i+1}(1) = \sum_n \frac{1}{c_n} U_n^{i+1} T_n(1) = u_N^{i+1}.$$

6 The Two Dimensional Heat Equation

The dependent variable $U^{i+\frac{1}{2}}$ can be represented by

$$U^{i+\frac{1}{2}} = \sum_n \sum_m \frac{1}{c_n c_m} U_{nm}^{i+\frac{1}{2}} T_n(x) T_m(y) \qquad (25)$$

and the quantities at the previous time levels by

$$X^i(x,y) = -(U^i + \frac{\kappa \Delta t}{2} U_{yy}^i) = \sum_n \sum_m \frac{1}{c_n c_m} T_n(x) T_m(y). \qquad (26)$$

In light of equation [22] and taking into account that differentiation is respect to x, the coefficients for $U_{nm}^{i+\frac{1}{2}}$ are

$$U_{n-2,m}^{i+\frac{1}{2}} = n[\frac{2}{n+1} - 4(n-1)\frac{A}{C}]U_{n,m}^{i+\frac{1}{2}} - (\frac{n-1}{n+1})U_{n+2,m}^{i+\frac{1}{2}}$$
$$-(X_{n-2,m}^i - X_{n,m}^i) - \frac{1}{C}(\frac{n-1}{n+1})(X_{n,m}^i - X_{n+2,m}^i). \qquad (27)$$

There are now $(M+1)$ backwards recursion relations to solve and hence $M+1$ τ_0's and $(M+1)$ τ_1's to be determined. $\tau_{0m}^{i+\frac{1}{2}}$ and $\tau_{1m}^{i+\frac{1}{2}}$ are found from consideration of the boundary conditions at the faces $x = -1$ and $x = 1$. These conditions are

$$U^{i+\frac{1}{2}}(-1,y) = \sum_n \sum_m \frac{1}{c_n c_m} U_{nm}^{i+\frac{1}{2}} T_n(-1) T_m(y) = f_0^{i+\frac{1}{2}}(y)$$

$$U^{i+\frac{1}{2}}(1,y) = \sum_n \sum_m \frac{1}{c_n c_m} U_{nm}^{i+\frac{1}{2}} T_n(1) T_m(y) = f_N^{i+\frac{1}{2}}(y).$$

Using the orthogonality relation on the above equations yields

$$\sum_n \frac{1}{c_n} U_{nm}^{i+\frac{1}{2}} T_n(-1) = f_{0m}$$

$$\sum_n \frac{1}{c_n} U_{nm}^{i+\frac{1}{2}} T_n(1) = f_{Nm}$$

which are the additional $2(M+1)$ equations needed to solve for τ_{0m} and τ_{1m}. The coefficients of U^{i+1} are found in a similar manner. Let

$$U^{i+1} = \sum_n \sum_m \frac{1}{c_n c_m} U_{nm}^{i+1} T_n(x) T_m(y) \qquad (28)$$

and the quantities at the previous time level be

$$-(U^{i+\frac{1}{2}} + \frac{\kappa \Delta t}{2} U_{xx}^{i+\frac{1}{2}}) = Y^{i+\frac{1}{2}}(x,y) = \sum_n \sum_m \frac{1}{c_n c_m} Y_{nm}^{i+\frac{1}{2}} T_n(x) T_m(y).$$

In light of equation [22] and noting that differentiation is respect to y, the coefficients of U^{i+1} can be determined from

$$U_{n,m-2}^{i+1} = m[\frac{2}{m+1} - 4(m-1)\frac{A}{C}]U_{n,m}^{i+1} - (\frac{m-1}{m+1})U_{n,m+2}^{i+1}$$
$$-(Y_{n,m-2}^{i+\frac{1}{2}} - Y_{n,m}^{i+\frac{1}{2}}) - \frac{1}{C}(\frac{m-1}{m+1})(Y_{n,m}^{i+\frac{1}{2}} - Y_{n,m+2}^{i+\frac{1}{2}}). \qquad (29)$$

The $(N+1)$ τ_{0n} and τ_{1n} necessary to solve for U_{nm}^{i+1} can be found from the boundary conditions at $y = -1$ and $y = 1$

$$U^{i+1}(x,-1) = \sum_n \sum_m \frac{1}{c_n c_m} U_{nm}^{i+1} T_n(x) T_m(-1) = g_0^{i+1}(x)$$

and

$$U^{i+1}(x,1) = \sum_n \sum_m \frac{1}{c_n c_m} U_{nm}^{i+1} T_n(x) T_m(1) = g_N^{i+1}(x).$$

Using the orthogonality relation give

$$\sum_m \frac{1}{c_m} U_{nm}^{i+1} T_m(-1) = g_{0m}^{i+1}$$

$$\sum_m \frac{1}{c_m} U_{nm}^{i+1} T_m(1) = g_{Nm}^{i+1}$$

which are the additional $2(N+1)$ equations needed to solve for τ_{0n} and τ_{1n}. g_{0m}^{i+1} and g_{Nm}^{i+1} are the coefficients of the Chebychev polynomial expansion for $g_0^{i+1}(x)$ and $g_N^{i+1}(x)$.

7 Arbitrary Geometries

To apply the solution technique to a rectangular domain with arbitrary boundary in (x,y) space, a linear transformation is used to map this region to a computational domain in (ξ, η) space bounded by $(-1 \leq \xi \leq 1)$ and $(-1 \leq \eta \leq 1)$. This transformation is simply

$$x = \frac{(x_a - x_b)}{2}\xi + \frac{(x_a + x_b)}{2} \qquad (30)$$

$$y = \frac{(y_a - y_b)}{2}\eta + \frac{(y_a + y_b)}{2} \qquad (31)$$

where
$$x_b \leq x \leq x_a$$
and
$$y_b \leq y \leq y_a.$$

For nonrectangular geometries or problems with discontinuous boundary conditions, it is advisable to divide the domain of interest into a number of rectangular subdomains. The semidiscrete equations are solved independently in each subdomain up to the unknown τ_ns. The τ_ns for each subdomain are then subsequently determined from the boundary conditions and matching conditions across the interfac of each subdomain in a manner similar to the foregoing method. The conditions that are applied for rectangular subdomains in an (x, y) coordinate system are

$$[U] = 0$$

$$[\frac{\partial U}{\partial x}] = 0$$

across $x = constant$ faces and

$$[U] = 0$$

$$[\frac{\partial U}{\partial y}] = 0$$

across $y = constant$ faces. $[()]$ denotes the jump in $()$.

8 Results and Conclusion

The foregoing analysis shows that an operational count within $O(N)$ opertions per iteration is necessary to advance the coefficients $U(x,t)$ one time step. This operational count is comparable to finite difference solutions but with a vast improvement in accuracy and convergence.

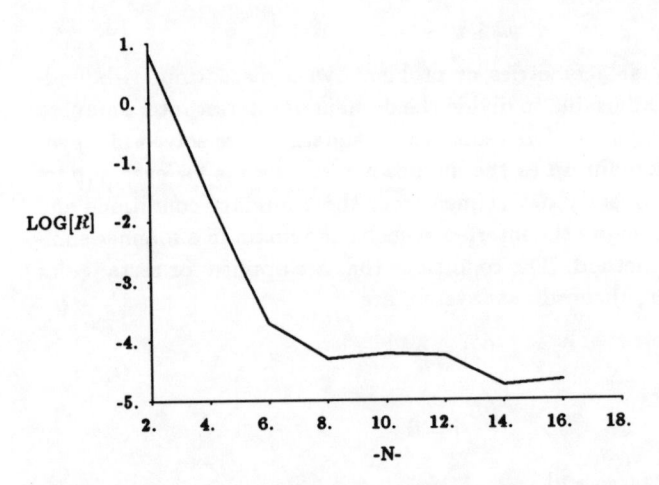

Figure 1: Convergence for 1-D heat equation where $R = U_{exact}(.5,.2) - U^{\frac{N}{2}}(.5)$

Figure [1] shows the log of the difference of the exact solution and the solution obtained with the above method at $t = .2$ for the problem

$$U_t = U_{xx}$$

with

$$U(x,0) = 100sin(\pi x)$$
$$U(0,t) = U(1,t) = 0$$

for $0 \leq x \leq 1$ and $t \geq 0$. The result shows that the convergence is superior to the algebraic convergence of finite difference. The convergence drops off rapidily at $N = 8$ because single precision is used for the calculations.

The ADI algorithm allow the coefficients of $U(x,y,t)$ to be advanced in time at an operation count within $O(MN)$ operations per temporal iteration.

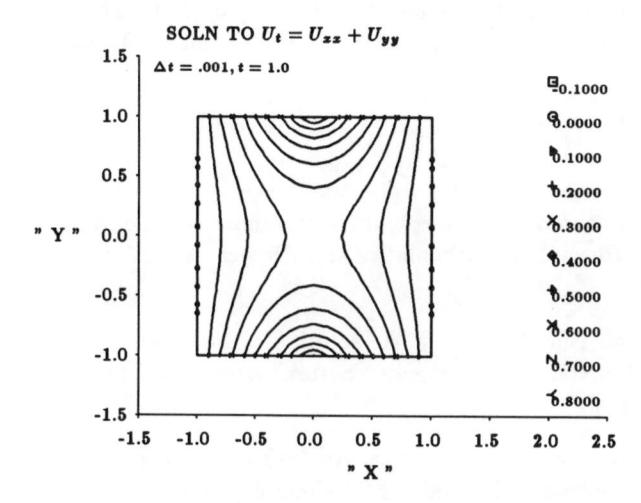

Figure 2: Chebychev ADI Solution of Classical Heat Equation

Figure [2] shows the solution of the problem

$$U_t = U_{xx} + U_{yy}$$

with

$$U(x,y,0) = 0$$

$$U(\pm 1, y, t) = 0$$

$$U(x, \pm 1, t) = \begin{cases} 1+x & \text{if } x \leq 0 \\ 1-x & \text{if } x \geq 0 \end{cases}$$

The solution shown is obtained for $N = M = 16$, $\Delta t = .001$ and $t = 1.0$.

The solution technique described above provides a superior alternative to finite difference and the classical separation of variables techniques. The advantage over the finite difference is obvious, accuracy up to four decimal places was attained for the one dimensional heat equation with $N = 6$ for the problem presented. This accuracy does not have a substantial cost over a finite difference scheme. The present method is superior to the classical separation of variable because of its flexibility in applying initial and boundary conditions. The separation of variable proceedure works well for problems with periodic boundary conditions but convergence is slow when non-periodic conditions are applied. The present method based on Chebychev polynomials converge rapidily for periodic and non-periodic boundary conditions.

The above technique can also be applied to the classical wave equation and solutions which exhibit the propper propagation speed and boundary behavior have been obtained for the one dimension wave eqution. Nonlinear problems can be addressed by collacating the nonlinear terms to obtain the coefficients at each time step then including these terms in the nonhomogenous part of the ordinary differntial equation. The technique has also been successfull applied to Burgers' equation and subsonic aerodynamic problems.

9 References

1. Lanczos, C.L., "Trigonometric Interpolation of Emperical and Analytical Functons," Journal of Mathematics and Physics. Vol.17, Sept. 1938 pp. 123-199

2. Clenshaw, C.W. and Norton, H.J., "The solution of Nonlinear Ordinary Differential Equations in Chebyshev Series," computer Journal, Vol. 6, 1963, pp. 88-92

3. Gottlieb, D., and Orszag, S.A., Numerical Analysis of Spectral Methods, Society for Industrial and Applied Mathematics, 1977

4. Gottlieb. D., Lustman, L. and Street, C.L., "Spectral Methods for Two-Dimensional Shocks," Icase Report No. 82-38 1982

5. Hanley, P. and Harris, W.L., Solutions of Fluid Flow Equations by Spectral Methods, FDRL Report No. 85-3 June 1985

SOLUTION OF NONLINEAR THERMAL PROBLEMS BY THE p-VERSION OF THE FINITE ELEMENT METHOD

N. Devanathan, [†] Y. Yang [†] and B. Szabó [‡]
Washington University,
St. Louis, Missouri 63130, U.S.A.

1. ABSTRACT

This paper marks the first application of the *p-version* of the finite element method to nonlinear problems. The application concerns mathematical modeling of the process of silicon crystal production by the float zone technique. Solution-dependent boundaries and radiative heat transfer are considered.

2. INTRODUCTION

The theoretical basis and performance characteristics of the p-version of the finite element method with respect to the solution of elliptic boundary value problems are well established [1]. When the exact solution is smooth then by using the coarsest finite element mesh permitted by the geometric configuration of the domain, and by increasing the polynomial degree of elements, exponential rates of convergence are realized. When the solution is not smooth then exponential convergence rates can be realized by combination of proper mesh design and increase of the polynomial degree of elements. It has been shown theoretically and demonstrated by a number of examples that very accurate solutions can be obtained with the p-version at very low computational cost (see [1] and the references listed therein). So far all applications of p-version finite element techniques have been concerned with linear problems. The substantial improvement in computational efficiency and reliability realized in the case of linear problems is magnified in the case of nonlinear problems, due to the iterative nature of solution procedures. In this paper we report our experience with p-version finite element technology in solving a nonlinear problem.

[†] *Department of Chemical Engineering*
[‡] *Center for Computational Mechanics*

The specific application considered herein concerns the mathematical modeling of the process of silicon crystal production by the float zone technique. This technique is used commercially for the production of high purity silicon used for integrated circuit fabrication. Zone refined crystals have a much higher purity than the ones produced by the conventional Czochralski (Cz) process in which the crystal is grown from the melt [2]. This is primarily due to the absence of the crucible in the zone refined process which in the Cz process is responsible for the impurities.

Figure 1 shows a schematic of the equipment. The polycrystalline feed rod to be purified is held in a vertical position and rotated. A small zone of the crystal (typically 1.5 cm long) is kept molten by means of a radio frequency heater which is moved so that this floating zone traverses the length of the bar. The molten zone is held in place by the combined action of the surface tension, gravity and electrodynamic forces. It resolidifies into a single crystal rod at the bottom section. The size of this zone is dependent on the heat transfer between the melt and the solid phases with the ambient.

In small floating zones the bulk flow in the melt is insignificant and does not influence heat transfer [3]. A conduction dominated model which is expected to be quantitatively accurate for this situation is discussed in the next section.

Two types of nonlinearity are considered: radiation heat transfer and moving boundaries due to phase change. The computer program *PROBE* [4] was used to solve Laplace's equations for the feed, melt and crystal. Since the interface shapes along the melt/feed and melt/crystal are not known a priori, the free boundary problem was solved iteratively by updating the position of the interface. The gas-liquid meniscus shape was obtained at every iteration by solving the Young-Laplace equation of capillary statics.

The high efficiency of p-extension procedures makes it feasible to perform quality control procedures at each iterative step in the nonlinear solution process.

3. MODEL DESCRIPTION

The following assumptions are incorporated in our model:

a) Steady growth;

b) Axisymmetric heating;

c) Constant thermophysical properties;

d) No convection due to heater movement/translation of the crystal and feed rods.

The domain consists of the feed crystal, E_f, melt, E_m, and the single crystal E_c. ∂E_f, ∂E_m and ∂E_c are the boundaries of interest and these are shown in Figure 2. The equations to be solved are:

$$k_i \nabla^2 T = 0 \quad \text{in } E_i \tag{1}$$

in which $i = f$ (feed crystal), m (melt), c (single crystal) and k_i are the respective thermal conductivities. The boundary conditions are as

Figure 1: Schematic of a float zone equipment

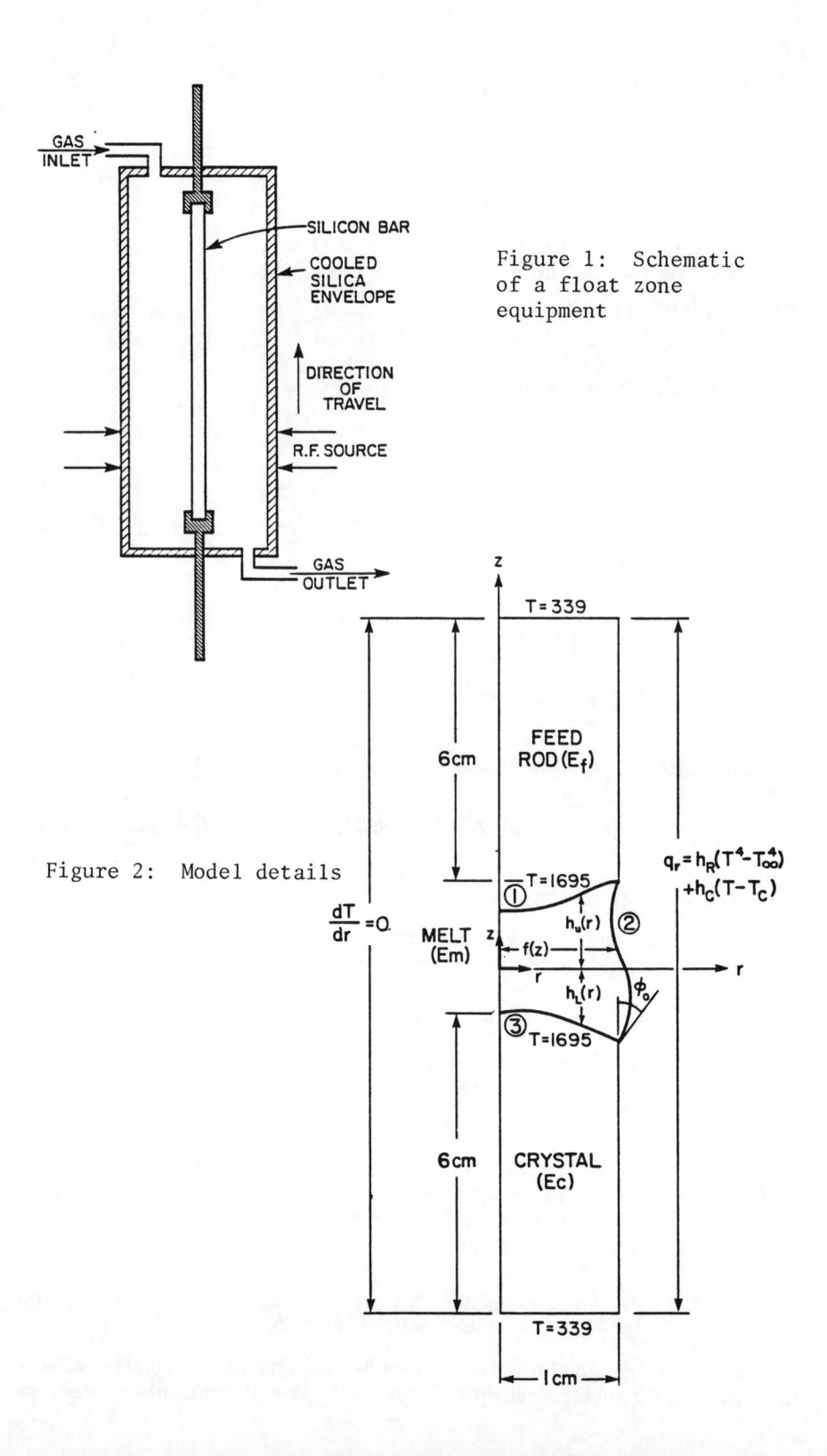

Figure 2: Model details

follows:

a) Along the feed-melt interface: (∂E_f)

(i) Energy balance:

$$(-k_m \frac{\partial T}{\partial n}) - (-k_f \frac{\partial T}{\partial n}) = (v \rho \Delta H) \cdot \vec{n} \qquad (2)$$

where v is the linear velocity of the interface, ΔH is the latent heat and $\vec{n}$ represents the normal to the interface.

(ii) Melting point:

$$T_f = T_m = T_{mp} \qquad (3)$$

b) Along the crystal-melt interface: (∂E_c)

(i) Energy balance:

$$(-k_f \frac{\partial T}{\partial n}) - (-k_m \frac{\partial T}{\partial n}) = (v \rho \Delta H) \cdot \vec{n} \qquad (4)$$

(ii) Freezing point:

$$T_c = T_m = T_{mp} \qquad (5)$$

c) Along the centerline $r = 0$:

$$\frac{\partial T_i}{\partial r} = 0 \quad i = f, \, m, \, c \qquad (6)$$

d) Along the outer boundaries:

$$-k_i \frac{\partial T}{\partial n} = h_R(T_i^4 - T_\infty^4) + h_c(T_i - T_\infty) \quad i = f, \, m, \, c \qquad (7)$$

Ambient temperature profile

Duranceau and Brown [3] recommend that the ambient temperature be modeled using a Gaussian distribution, about $z = 0$:

$$T_\infty = (T_{max} - T_\infty^\circ) \exp(-z^2/\sigma^2) + T_\infty^\circ. \qquad (8)$$

With this approach one can easily correlate experimental data with the model in terms of the two parameters T_{max} and σ.

Meniscus shape

As shown in Figure 2, the melt/feed, melt/gas and melt/crystal interfaces are represented by the functions $h_u(r), f(z)$ and $h_L(r)$ respectively. The shape of the melt/gas meniscus is governed by the Young-Laplace equation of capillary statics [3].

$$2\lambda + Gz = \frac{f_{zz}}{(1+f_z^2)^{3/2}} - \frac{1}{f(1+f_z^2)^{1/2}} \qquad (9)$$

where G is the gravitational Bond number measuring the relative effects of gravity and surface tension σ, and λ is the dimensionless pressure difference necessary to set the volume of the molten zone. The solution of the boundary value problem (9) requires two conditions plus an additional condition to determine λ. These are:

$$f[h_L(R_c)] = R_c \tag{10}$$

$$f[h_u(R_f)] = R_f \tag{11}$$

$$\frac{df}{dz}\bigg|_{z=h_L(R_f)} = \tan\phi_0 \tag{12}$$

Equations (10) and (11) require that the meniscus join the melt/solid interface at the trijunctions; ϕ_0 is the contact angle which is a material property, and equation (12) has to be satisfied. Solution of equation (9) with conditions (10) through (12) defines the menicus shape $r = f(z)$.

Radiation heat transfer

The heat transfer from crystal/melt to surroundings is linearized as follows:

$$\begin{aligned}h &= h_R(T^4 - T_\infty^4) + h_c(T - T_\infty) \\ &= \left[\sigma\epsilon(T_s^2 + T_\infty^2)(T_s + T_\infty) + h_c\right](T - T_\infty)\end{aligned} \tag{13}$$

Since the surface temperature, T_s, is not known a priori, a parabolic form is assumed:

$$T_s = a + bz + cz^2 \tag{14}$$

The constants a, b, c are chosen such that for the melt:

$$\begin{aligned}T_s &= T_{mp} & z &= -h_u(1) \\ &= T_{mp} & z &= h_L(1)\end{aligned}$$

and for the feed rod and the crystal:

$$\begin{aligned}T_s &= T_{mp} & z &= h_u(1) \text{ or } h_L(1) \\ &= T_\infty & z &= L\end{aligned}$$

Hence the temperature at one other point needs to be assumed to solve for constants a, b, c in equation (14). These constants are updated every iteration.

4. SOLUTION TECHNIQUE

The procedure adopted to update the interface shape is as follows:

1. Assume reasonable shapes ∂E_f and ∂E_c.
2. Solve for the meniscus profile, ∂E_m, governed by the boundary value problem [equation (9)] subject to conditions (10) through (12).
3. Solve the Laplace equation for heat conduction [equation (1)] for the three regions using the conditions (3), (5), (6) and (7).
4. Calculate the normal fluxes at the interface. Check if the energy balance [equations (2) and (4)] are satisfied.
5. Update the position of the interface and repeat steps 1 through 4.

6. Continue until convergence is realized.

The procedure for updating the interface shape at every iteration is discussed in the Appendix.

5. FINITE ELEMENT ANALYSIS

The parameters used in the model are summarized in Table I. Details of the finite element space are given in Table II. All computations were performed on a *VAX 11/750* computer (*VMS* operating system). Although *PROBE* allows computation with polynomial degree ranging from 1 to 8, a polynomial degree of 5 was used in order to balance accuracy and computation time/storage requirements.

Several quality control tests were performed. Table III shows a typical strain energy convergence data for the melt for the final iteration. The relative error in energy norm decreases uniformly from 3.83% (p=1) to 0.1% (p=8) as the polynomial degree is increased.

Table IV shows the heat flowing in or out of the molten zone at the boundary. Under steady state conditions the fluxes at the boundaries should satisfy:

$$\oint q \cdot \vec{n}\, ds = 0.$$

The heat flow sums to 0.127 watts.

Interelement continuity of temperature and fluxes can be checked using *PROBE*. In fact, all of the above quality control tests can be performed at each iterative step in the nonlinear solution process.

Table V shows that by systematic refinement of polynomial degree, the true interface shapes can be obtained.

6. RESULTS AND DISCUSSION

At each iteration the meniscus profile was obtained by solving equations (9) to (12) by Gear's backward differentiation. It is seen that the meniscus bulges outward near the bottom interface and is sucked inward near the top interface.

The finite element solution is not very sensitive to mesh design as long as there are a sufficient number of elements close to the interfaces and the meniscus in the melt. Figure 3 shows the mesh design in the melt and in the solids close to the interfaces. It is clear that most elements are located in the high flux regions. Steep gradients and singularity points warrant the use of more elements.

Figure 4 compares the normal heat flux at the interface on the melt side, Q_{mn}, and on the crystal side, Q_{fn}, for the final iteration. The difference, $Q_{mn} - Q_{fn}$ is almost a constant and equal to $Q_{\ell n}$, the latent heat flux, except at $r = 0$ and $r = 1$ which are the singular points. Solution at singular points could be improved by choosing a higher polynomial degree.

Figure 5 shows the assumed surface temperature, T_s, the predicted surface temperature and the ambient temperature profile. The assumed

Table I. Parameter values.

PARAMETER	VALUE
Radius of crystal and feed rods	1 cm
Growth velocity	0.25 cm/min
Growth angle, ϕ_0	11°
Heat of fusion	1882.57 J/g
Ambient temperature at $z = \pm\infty$	339 °K
Maximum heater temperature, T_{max}	2542.5 °K
Width of heater profile, σ	1.0
Thermal conductivity of melt, k_m	0.64 W/K·cm
Thermal conductivity of solid, k_c, k_f	0.22 W/K·cm
Melting temperature of silicon,	1695 °K
Emissivity of solid	0.7
Emissivity of melt	0.3
Density of melt	2.55 g/cm^3
Density of solid	2.33 g/cm^3
Surface tension	720 dynes/cm^2
Convective heat transfer coefficient for melt	0.06 W/Kcm^2
Convective heat transfer coefficient for solid	0.0011 W/Kcm^2
Boltzmann constant, σ	5.67 × 10^{-12} W/K^4 cm^2
Constants in equation (14):	
melt a	1740.0
melt b	-38.24
melt c	-134.17
crystal a	1789.53
crystal b	483.57
crystal c	40.27
feed rod a	1674.24
feed rod b	-445.08
feed rod c	37.09

Table II. Details of the finite element solution.

	N_{EL}	N_{NODE}	p	N	CPU sec / Iteration
Feed rod	48	63	5	647	226
Melt	36	49	5	493	179
Crystal	48	63	5	647	226

Table III. Estimated relative error in energy norm, e_r (percent).

p	N	Energy	e_r
1	35	0.201469×10^6	3.83
2	107	0.201758×10^6	0.55
3	179	0.201760×10^6	0.47
4	287	0.201763×10^6	0.28
5	431	0.201763×10^6	0.19
6	611	0.201764×10^6	0.15
7	827	0.201764×10^6	0.12
8	1079	0.201764×10^6	0.10

Table IV. Computation of flow rate at the boundary.

Element	Node 1	Node 2	Flow	Total flow
1	1	2	0.6370×10^0	0.6370×10^0
2	2	3	0.2364×10^1	0.3001×10^1
3	3	4	0.4992×10^1	0.7993×10^1
4	4	5	0.8852×10^1	0.1685×10^2
5	5	6	0.2818×10^1	0.1966×10^2
6	6	7	0.1223×10^2	0.3189×10^2
6	7	8	-0.9887×10^0	0.3090×10^2
7	8	21	-0.1196×10^2	0.1894×10^2
18	21	22	-0.2733×10^2	-0.8390×10^1
19	22	35	-0.1762×10^2	-0.2601×10^2
30	35	36	-0.9793×10^1	-0.3581×10^2
31	36	49	-0.8474×10^1	-0.4428×10^2
31	49	48	0.1446×10^2	-0.2982×10^2
32	48	47	0.4305×10^1	-0.2551×10^2
33	47	46	0.1406×10^2	-0.1145×10^2
34	46	45	0.7446×10^1	-0.4002×10^1
35	45	44	0.3290×10^1	-0.7120×10^0
36	44	43	0.8391×10^0	0.1271×10^0

Table V. Interface shapes at various polynomial degree levels.

Characteristic Points	$p=6$	$p=7$	$p=8$
$h_L(0)$	-0.6236	-0.6269	-0.6297
$h_L(1)$	-0.6967	-0.6948	-0.6928
$h_u(0)$	0.2122	0.2164	0.2199
$h_u(1)$	0.4707	0.4764	0.4831

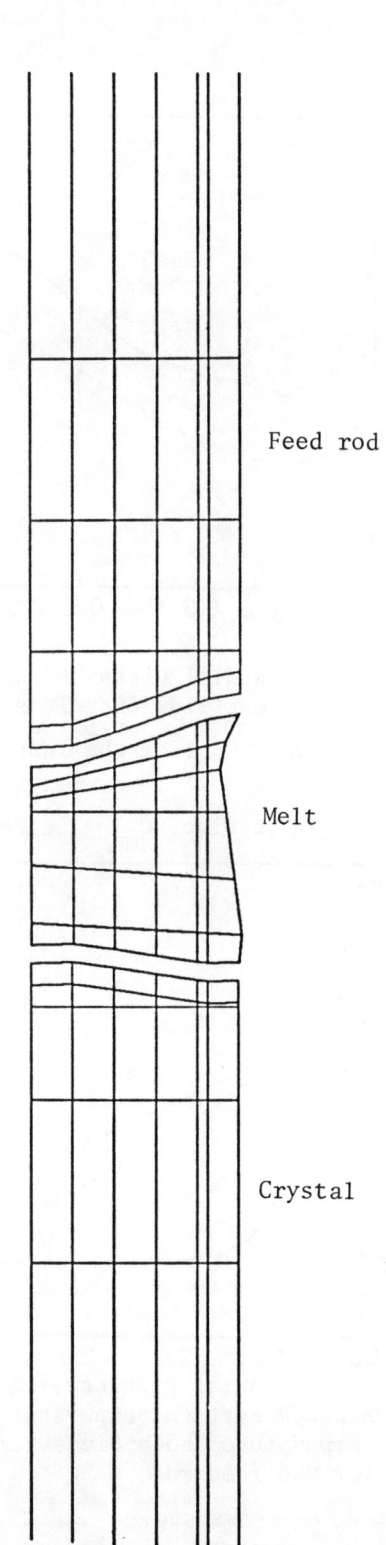

Figure 3: Mesh design close to the interface

Feed rod

Melt

Crystal

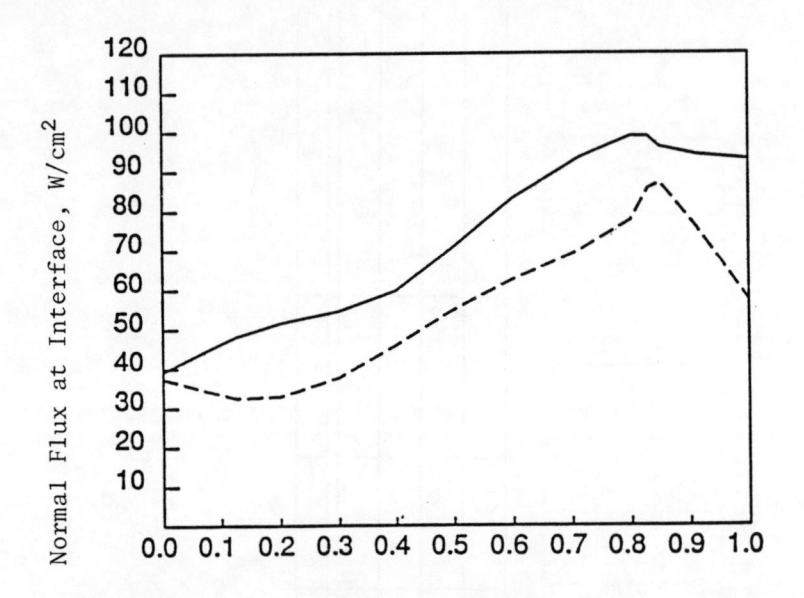

Figure 4: Normal heat flux on melt side (Q_{mn}) and feed rod side (Q_{fn}) of the interface for final solution

——— Melt side, Q_{mn} - - - - Feed side, Q_{fn}

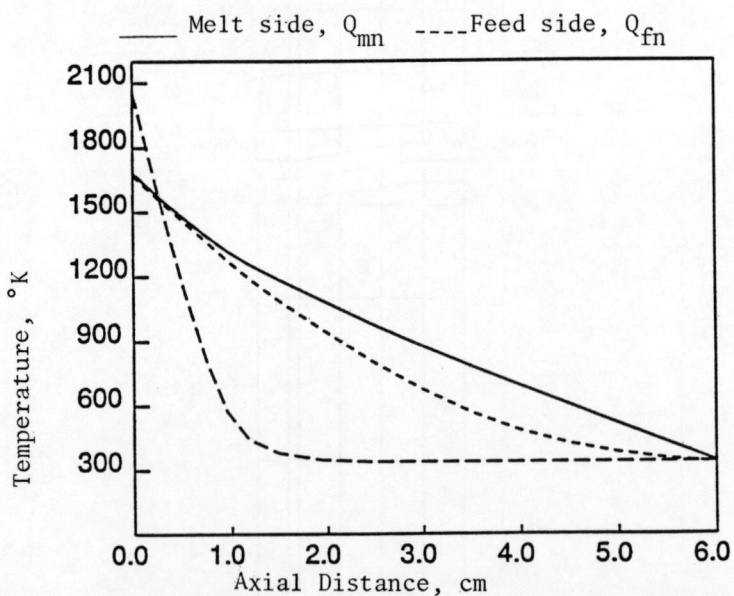

Figure 5: Assumed surface temperature, actual surface temperature and ambient temperature profiles for the feed rod

——— Predicted   - - - - - Assumed   — — — Ambient

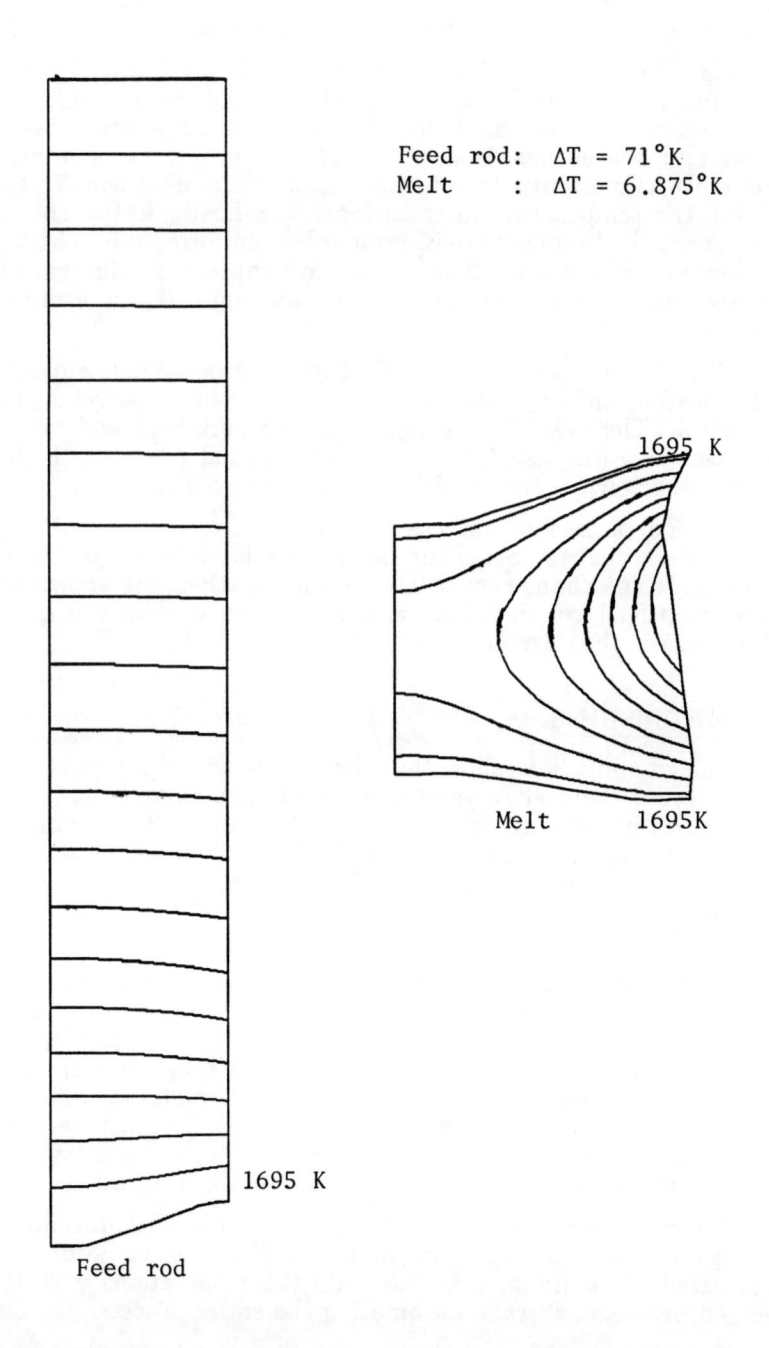

Figure 6: Temperature contours for feed rod and melt

and predicted surface temperatures agree well.

The solution is moderately sensitive to the polynomial degree. The difference between the fluxes for $p = 4$ and $p = 8$ is about 5% or less. The error is particularly noticeable at the singular points.

Figure 6 shows the temperature controls for the feed rod and the melt. In the feed rod significant radial temperature gradients exist close to the interface. The maximum difference of temperature between the center and the surface is about $31°K$. Away from the interface, the temperature is essentially constant in the radial direction. In the single crystal, the temperature distribution is similar to that of the feed rod. In the melt the temperature is more or less uniform, due to high thermal conductivity of the melt. The maximum temperature difference between any two points is around $65°K$. The maximum temperature is about $1760°K$.

Close to the interface in the feed rod, the radial flux is negative (the rod is heated) and approximately at $z = 0.3$ the flux changes sign (the rod is cooled). The axial flux is highest at the periphery and progressively decreases as z increases. In the melt the radial fluxes are high only at the periphery while the axial fluxes are much higher.

Some experimentation was required for a reasonable initial guess of the interface shapes. Since the updating scheme is linear, the interface shape does not change much after each iteration. As shown in Figure 3 the melt-feed rod interface is more distorted than the melt-crystal interface, i.e. they are not symmetric about $z = 0$.

7. CONCLUSIONS

The computations show that the melt/feed rod interface is strongly curved while the melt/crystal interface is less curved. It is clear that the profiles are not symmetric about the r-axis. The liquid meniscus bulges outward near the bottom (melt/crystal) interface and is sucked inward near the top (melt/feed rod) interface. The melt is more or less at a uniform temperature while significant radial temperature gradients exist in the crystal and feed rod close to the interface.

The present work illustrates that solution of moving bundary and radiative heat transfer problems can be performed very efficiently by the p-version of the finite element method. In the case of moving boundary problems subdomains with different material properties are delineated by contours which lie along finite element boundaries. In general, a coarse mesh is sufficient but points of singularity must be isolated by strongly graded local meshes. The high efficiency of p-extension permits one to perform quality control procedures at each iteration.

Because this was a preliminary investigation, the procedures of updating the interface shape and solving for the meniscus profile were not automated. It is possible to integrate these operations with the finite element program, thereby automating the entire process.

Our investigation has shown that the model described herein can be used to carry out parametric investigations and to determine optimal conditions for maximizing crystal productivity.

8. REFERENCES

[1] Babuška, I., "The p- and hp-versions of the finite element method; The state of the art", Technical note BN-1156, Institute for Physical Science and Technology University of Maryland, College Park MD, Sept. 1986.

[2] Keller, W. and Muhlbauer, A., "Floating zone silicon", Marcel Dekker, New york, NY, 1981.

[3] Duranceau, J. L., and Brown, R. A., "Thermal capillary analysis of small- scale floating zones: Steady state calculations", J. Crystal Growth, 75, 367 (1986).

[4] Szabó, B. A., PROBE, Theoretical manual, Noetic Technologies Corp., St. Louis, MO, 1986.

9. APPENDIX

In the following we outline the procedure used for updating the interface contour.

Let z_i $(i = 1, 2, \cdots n_r)$, where n_r is the number of nodes in the radial direction, denote the z-coordinates of nodes on the interface. z_1 refers to the center of the crystal $(r = 0)$ and z_{n_r} refers to the edge. Consider a fixed radial position corresponding to node i. Let dc_i = element size in the z-direction into the crystal at the node i. Assume that the interface needs to be moved a distance $f_i * dc_i$ in the positive z-direction to staisfy the energy balance. For the top interface in which melting occurs f_i can be obtained as follows:

At node i

$$Q_{fi} = -\frac{k_c(T_2 - T_m)}{dc_i} = \text{flux leaving interface into feed rod.} \qquad (15)$$

where T_2 is the temperature at $z_i + dc_i$

$$Q_{mi} = -\frac{k_m(T_m - T_1)}{dm_i} = \text{flux entering interface from melt.} \qquad (16)$$

where dm_i = element size in the z-direction into the melt and T_1 is the temperature at $z_i - dm_i$. After relocation of the interface:

$$Q_{fi}^{new} = \frac{-k_s(T_2 - T_m)}{(z_i + dc_i) - (z_i + f_i\ dc_i)} = \frac{-k_c(T_2 - T_m)/dc_i}{1 - f_i} = \frac{Q_{fi}^{old}}{1 - f_i}. \qquad (17)$$

Similarly:

$$Q_{mi}^{new} = \frac{-k_m(T_m - T_1)}{(z_i + f_i dc_i) - (z_i - dm_i)} = \frac{-k_m(T_m - T_1)/dm_i}{1 + f_i d_i} = \frac{Q_{mi}^{old}}{1 + f_i d_i} \qquad (18a)$$

where:

$$d_i = dc_i/dm_i. \qquad (18b)$$

Since the fluxes satisfy the energy balance (2) after relocation, we have

$$\frac{Q_{fi}}{1 - f_i} + Q_{li} = \frac{Q_{mi}}{1 + f_i d_i} \qquad (19)$$

where Q_{li} is the normal component of the latent heat term at node i. A semi-implicit rearranged form of equation (19) is:

$$f_i^{new} = \frac{Q_{li} - (Q_{mi} - Q_{fi})}{f_i^{old} d_i \, Q_{ti} - Q_{fi} d_i - Q_{mi} + Q_{li}(1 - d_i)}. \tag{20}$$

The new interface (top) is given by

$$z_i^{new} = z_i^{old} + f_i dc_i \qquad i = 1, 2 \cdots n_r \tag{21}$$

For the melt/crystal (bottom) interface the equation analogous to (20) is:

$$f_i^{new} = \frac{Q_{li} - (Q_{ci} - Q_{mi})}{f_i^{old} d_i Q_{li} + Q_{ci} d_i + Q_{mi} - Q_{li}(d_i - 1)}. \tag{22}$$

10. NOTATION

a	Constant defined by eq. (14)
b	Constant defined by eq. (14)
c	Constant defined by eq. (14)
dc_i	Element size in z direction into the crystal at the interface and at node i, cm
dm_i	Element size in z direction into the melt at the interface and at node i, cm
f	Function describing the meniscus shape
f_i	Factor for movement of interface
g	Acceleration due to gravity, cm/sec²
G	Gravitational Bond Number. [gR²Δ P/γ]
h_c	Convective heat transfer coefficient, W/cm² °K
h_R	Radiative heat transfer coefficient, W/cm² °K
$h_{L(r)}$	Function describing melt-crystal interface
$h_{u(r)}$	Function describing melt-feed rod interface
ΔH	Latent heat of fusion, J/g
k_i	Thermal conductivity of phase i, W/cm°K
N_{EL}	Number of elements
N_{NODE}	Number of nodes
N	Number of unknowns
p	Polynomial degree
p_0	Reference pressure, N/cm²
Q_{ci}	Flux leaving interface into crystal at node i, W/cm²
Q_{fi}	Flux leaving interface into feed rod at node i, W/cm²
Q_{li}	Latent heat flux, W/cm²
Q_{mi}	Flux entering interface from melt at node i, W/cm²
r	Radial coordinate, cm
R	Radius ratio, [R_c/R_f]
R_i	Radius of phase i, cm
T	Temperature, °K
T_{max}	Maximum temperatures of heater profile, °K
T_∞°	Ambient temperature at $z = \infty$
v	Growth rate, cm/min
z	Axial coordinate, cm

Greek Letters

γ	Surface tension, dynes/cm^2
ϵ	Emissivity
λ	Dimensionless reference pressure difference, $[\Delta p^\circ \cdot R_f/\gamma]$
ϕ_0	Contact angle
ρ	Density, g/cm^3
σ	Stefan-Boltzmann constant, W/K$^4 \cdot$ cm^2

Subscripts:

c	Crystal
f	Feed rod
m	Melt
mp	Melting point
s	Surface
∞	Ambient

USING THE APPROXIMATE OPTIMIZATION ALGORITHM IN THE ANSYS® PROGRAM FOR THE SOLUTION OF OPTIMUM CONVECTIVE SURFACES AND OTHER THERMAL DESIGN PROBLEMS

Mark C. Imgrund
Manager, Quality Assurance
Swanson Analysis Systems, Inc.
Houston, PA 15342

An optimization algorithm available within the finite element program ANSYS is applied to the problem of determining optimum shapes of extended surfaces. The optimization algorithm is briefly described and then used to solve several classical optimum fin problems. The method is shown to compare well with the traditional closed-form solutions. The technique is then applied to a more complicated fin array optimization design problem that would defy traditional optimization techniques. In addition, two other thermal design problems are posed and solved with the algorithm, including an example where temperature-dependent stress constraints are considered. The method presented is shown to be a viable tool for various types of heat transfer design.

1. INTRODUCTION

Solutions to the problem of optimizing convective fins for use in heat exchanging devices have been in existence since at least 1926, when Schmidt [1] first proposed a solution based on an intuitive approach. Much subsequent work has appeared since that time [2-7] addressing other geometries, various boundary conditions, and different assumptions regarding material properties and heat transfer coefficients. A great deal of work has also been undertaken to determine heat transfer coefficients analytically or experimentally for various finned design concepts.

The optimization problem of interest in the literature which is initially solved in this paper can be expressed as either: 1.) maximize the heat transfer subject to fin volume (geometric) constraints; or 2.) minimize the volume of a fin subject to a specified amount of heat transferred. These expressions of the problem are equivalent and will yield the same solution. The latter definition of the problem is considered in the present work. The algorithm investigated in this paper is applied to three classical problems from the literature including uniform profile fins of rectangular and circular shape, and circular fins with tapered cross-sections. The algorithm is subsequently used in solving more "realistic" heat transfer design problems to demonstrate capabilities beyond the solution of the elementary fin problem and beyond the present capabilities of sensitivity-based optimization algorithms.

2. COUPLING APPROXIMATION TECHNIQUES WITH OPTIMIZATION

The combined finite element – optimization technique used to solve the optimization problems considered in the present work is a significant departure from traditional optimization methods. Briefly, the problem is approached by creating a quadratic approximation to the problem at each design loop; this approximate subproblem is then minimized using sequential unconstrained minimization techniques (SUMT). The solution to the subproblem is used to predict the next trial design and the process continues until convergence. This is illustrated schematically in Figure 1. Parametric model representation is used to link the optimization module with the finite element analysis. Details of the algorithm and explanation of the theory can be found in Imgrund [8], Kohnke [9] and Beazley [13].

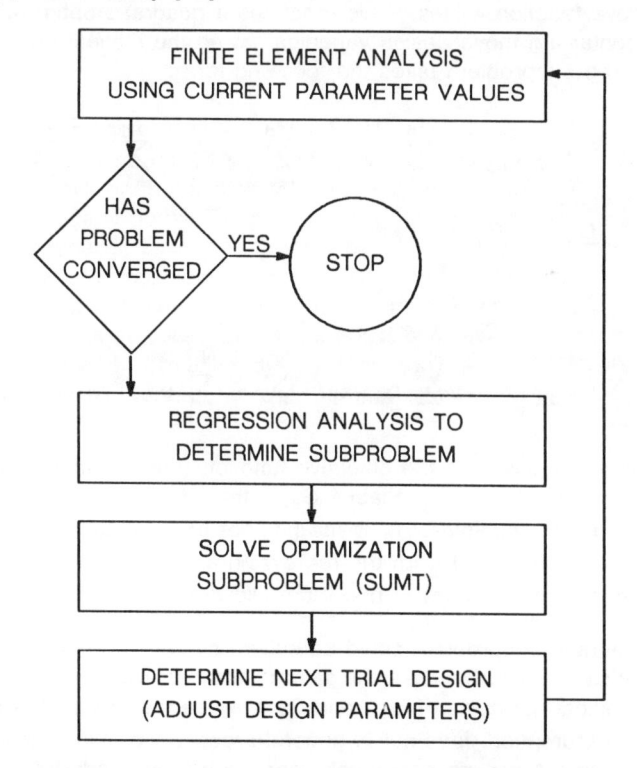

Figure 1 Simplified Flowchart of Solution Sequence

The wide range of optimization techniques available today can be divided into two major categories for the purpose of comparison. In the first category, the form of the expressions for the objective function and constraints are known, and the optimization algorithm makes use of explicit gradients with each evaluation of the design. The second category encompasses methods that don't require gradient information for the objective or constraints. These methods either calculate gradients using finite differences or optimize by making many function evaluations. Both of these strategies can have significant disadvantages when used in conjunction with a finite element solution. The first category limits, to a small subset, the items that may be chosen as design variables, constraints, and the function to be optimized. In the case of the second category, numerous finite element solutions may be required, and for many real problems, the number of function evaluations required can be prohibitively expensive and time-consuming.

The present method, closest in description to the second method above, is an attempt to glean the maximum amount of information for the optimizer from each finite element solution while preserving generality in the choices of design objective, design variables, and constraints. This is done by making successive approximations to the objective function and constraint functions as optimization progresses. Each approximation utilizes the total design history, and is based on a multiple regression analysis with a quadratic equation form assumed. The regression analysis makes use of various weighting factors to reduce the influence of designs remote from the current best strategy, to account for feasibility of designs, and to give more weight to designs with smaller objective function values. This produces a quadratic optimization subproblem containing those design variables having the highest correlation coefficients. The subproblem takes the following form:

$$\text{Minimize:} \quad F = a_{oo} + \sum_{i=1}^{N_d} a_{oi} x_i + \sum_{i=1}^{N_d} a_{ii} x_i^2 + \sum_{i=1}^{N_d-1} \sum_{j=1}^{N_d} a_{ij} x_j \quad (1)$$

Subject to: $\underline{G}_j < G_j < \overline{G}_j$

$\underline{x}_i < x_i < \overline{x}_i$

where:

$$G_k = b_{ok} + \sum_{i=1}^{N_d} b_{oik} x_i + \sum_{i=1}^{N_d} b_{iik} x_i + \sum_{i=1}^{N_d-1} \sum_{j=i+1}^{N_d} b_{ijk} x_i x_j \quad (k = 1, N_g) \quad (2)$$

In these equations, F is the objective function (the quantity to be minimized), x_i is the ith design variable, G_k is the kth constraint, N_d is the number of design variables, N_g is the number of constraints, a_{ij} and b_{ijk} are coefficients obtained from the regression analysis, and underbars or overbars represent minimum or maximum allowable values, respectively.

This quadratic subproblem could be solved by any number of methods in a relatively efficient way. Because the form of the subproblem is quite general, any logical choices of F, x_i, and G_k may be made with this method. The quadratic assumption does not in general present a problem, since most complicated functions can be adequately represented by quadratics over a small section of design space. The user interface with the algorithm in ANSYS is structured such that the form of the approximations may be chosen as linear, quadratic, or quadratic with cross-terms, as represented by the three summations in equations (1) and (2).

3. COMPARISON WITH CLOSED-FORM SOLUTIONS

To begin evaluating the performance of this algorithm on extended surface heat transfer problems, nine separate problems were optimized and their results compared to those of classical (1-D conduction) theory. The nine problems represent three different geometries, each solved for three different material properties. Each of the problems was defined using the ANSYS STIF55 element[10], a four node isoparametric thermal element with 2-D and axisymmetrical capabilities. In each case, symmetry along the fin centerline was utilized to reduce the finite element model.

The first three problems determine the optimum length and thickness of a rectangular convecting fin that must dissipate 100W per unit length when the base fin temperature is 100°C. The heat transfer coefficient is 200 W/m^2-°C and the fluid temperature is 0°C. The design variables are the fin length and half-thickness. The optimization results are shown in Table 1.

The second three problems seek to minimize the volume of a circular fin of rectangular cross-section. The fin's base radius is 0.05m, the required heat transfer is 500W per fin, and the design variables are the fin half-thickness and the tip radius. The last three problems are the same as the second three except that the fin thickness at the tip is added as a third design variable (the fin may taper). The base fin temperature is 100°C, the fluid temperature is 0°C and the heat transfer coefficient is 200 W/m^2-°C in all six of these problems. The theoretical solution to these problems is given in reference 4; these are compared with the ANSYS solutions in Table 1.

Case Description	Parameter	Copper (k = 382 W/m-°C) ANSYS	Copper Theory	Aluminum (k = 204 W/m-°C) ANSYS	Aluminum Theory	Carbon Steel (k = 54 W/m-°C) ANSYS	Carbon Steel Theory
Straight Rectangular Fin	thickness(m)	1.55×10^{-5}	1.66×10^{-5}	3.67×10^{-5}	3.10×10^{-5}	1.11×10^{-4}	1.17×10^{-4}
	length(m)	.0087	.0079	.0070	.0079	.0084	.0079
	volume(m^3)	2.69×10^{-7}	2.64×10^{-7}	5.14×10^{-7}	4.94×10^{-7}	1.86×10^{-6}	1.87×10^{-6}
Straight Circular Fin, Inner Radius = 0.5	thickness(m)	.00139	.00140	.00234	.00236	.0138	.0128
	tip radius(m)	.0956	.0946	.0936	.0938	.0843	.0860
	volume(m^3)	2.82×10^{-5}	2.83×10^{-5}	4.62×10^{-5}	4.68×10^{-5}	2.01×10^{-4}	1.96×10^{-4}
Tapered Circular Fin, Inner Radius = .05	base thickness(m)	2.42×10^{-3}	2.90×10^{-3}	2.84×10^{-3}	3.04×10^{-3}	1.55×10^{-2}	1.47×10^{-2}
	tip thickness(m)	1.41×10^{-5}	0	7.02×10^{-7}	0	3.24×10^{-5}	0
	tip radius(m)	.0901	.0979	.100	.0979	.0949	.0979
	volume(m^3)	1.99×10^{-5}	1.80×10^{-5}	3.01×10^{-5}	3.01×10^{-5}	1.43×10^{-4}	1.46×10^{-4}

Table 1: Results Comparison Table - Present Method vs. Theory

As can be readily observed from the table, ANSYS results agree quite well over the range of the tests. Results cannot be expected to agree perfectly for a number of reasons. Some of these reasons would be: 1) the ANSYS solution is discretized, and is based on 2-D conduction while the theoretical solution is continuous and assumes 1-D conduction; 2) the present method can converge due to small step size before reaching "true" optimum; and 3) it is possible for the present method to reach a local minimum in design space (or an apparent local minimum in approximate design space). In spite of these possibilities, all ANSYS results are within what could be characterized as "engineering accuracy" of the optimum. It is worthwhile to note that the initial designs provided to the ANSYS optimizer differed substantially from the optimum designs by at least a factor of two, and in some cases, by more than an order of magnitude. Tighter convergence tolerances and additional iterations were used in some of these cases to achieve the reported designs.

4. MINIMIZING COST FOR A MULTI-FIN ARRAY

In actual design situations, the conditions assumed for the previous nine cases do not exist. To assume that they do exist could lead to unacceptable designs in many instances. In order to show how the present method could be applied to problems encountered in engineering practice, a design problem concerning an array of circular fins is posed, similar to what one might encounter in a type of low speed finned-tube heat exchanger. In this example, it is assumed that the designer may specify not only the number and shape of the fins, but the flowrate of the cooling fluid (air in this case) as well. The problem and its associated design variables are shown in Figure 2.

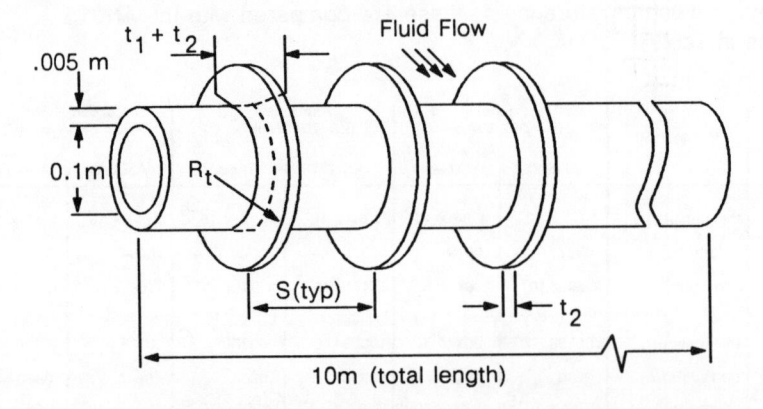

Figure 2   Design Variables for the Fin Array Design Problem

To create a problem conforming more to an actual design situation, a cost function is minimized rather than total fin volume. This cost function includes contributions from raw material cost per unit volume, fabrication costs per fin, and pumping costs in terms of the fluid velocity. The fabrication cost is assumed to be a fixed amount per fin, the pumping cost is assumed to increase linearly with flowrate, and the unit cost of the material is assumed to be constant. The cost function is then:

$$\text{Cost} = A(N_f)(\text{Vol}) + B(N_f) + C(V_f) \qquad (3)$$

where: $A \equiv$ cost per unit volume
$B \equiv$ fabrication cost per fin
$C \equiv$ pumping cost per unit velocity
$N_f =$ number of fins (10 meters $\div$ spacing distances)
$\text{Vol} =$ fin volume
$V_f =$ fluid velocity (free stream)

This cost function is admittedly arbitrary, and is not purported to represent any true costs of manufacture or operation. The equation form chosen is not essential to the optimization method being presented; any cost relationship known to the designer could have been input. The important concept is that optimization may be based on virtually any objective function that could be expressed parametrically in ANSYS input.

Other nonlinearities were also considered in this analysis: Thermal conductivity, for example is treated as variable with temperature. The heat transfer coefficient is evaluated as a function of temperature and fluid velocity by assuming that the following non-dimensional relationship [11] is sufficiently accurate.

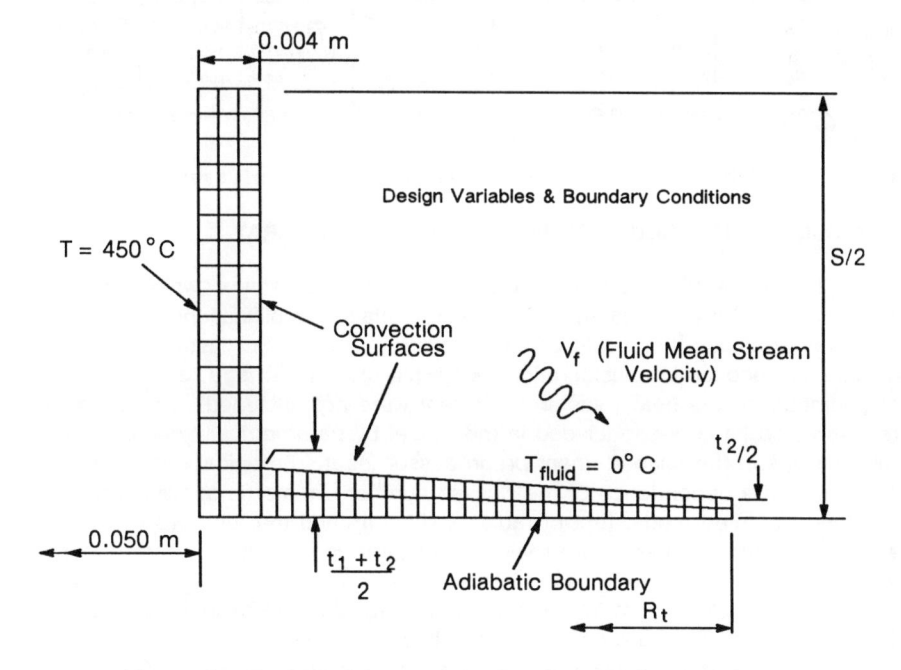

$$Nu_L = 0.66 \, Pr^{1/3} \, Re_L^{1/2} \tag{4}$$

Figure 3   Optimum Dimensions for a Radial Cooling Fin Array

The subscript on the Nusselt and Reynolds numbers in this expression indicates that it represents the laminar case. Design variable limits on the cooling fluid velocity were imposed to ensure that this assumption was not violated. Temperature dependence was considered in the determination of the Prandtl number, so the heat transfer coefficient is a function of both temperature and fluid velocity. Other empirical relationships that were not included in this problem's definition of the heat transfer coefficient are discussed in the following section.

Figure 3 shows a representative finite element mesh and boundary conditions used for the ANSYS solution. The total tube length for the fin array is defined to be 10 meters and is required to transfer 25000 W overall. The minimum spacing between fins is specified to be eight times the fin base thickness for fabrication purposes. Results of solving this model for several different starting designs and fin materials are shown in Table 2.

Case	Material	r_t $(0.05 < r_t < 0.20)$ Initial	Final	t_1 $(5 \times 10^{-6} < t_1 < 5 \times 10^{-3})$ Initial	Final	t_2 $(1 \times 10^{-7} < t_2 < t_1)$ Initial	Final	s $(8t_1 \le s < 0.25)$ Initial	Final	v_f $(0.1 < v_f < 2.0)$ Initial	Final	Cost Initial	Final
1	Cu	.100	.173	5×10^{-3}	1.33×10^{-4}	5×10^{-4}	7.78×10^{-6}	.0400	.0266	1.90	.163	(F) 1755	670
2	Cu	.195	.142	8×10^{-3}	3.22×10^{-4}	1×10^{-3}	1.27×10^{-5}	.1500	.0219	.150	.128	(I) 1171	746
3		.015	.183	2×10^{-3}	9.16×10^{-5}	4×10^{-4}	8.91×10^{-7}	.060	.0235	.800	.173	(I) 747	704
4		.100	.130	5×10^{-3}	6.76×10^{-4}	5×10^{-4}	1.84×10^{-5}	.0400	.0291	1.90	.320	(F) 1314	881
5	AL	.195	.129	8×10^{-3}	9.68×10^{-5}	1×10^{-3}	1.45×10^{-4}	.1500	.0310	.150	.415	(I) 424	901
6		.015	.170	2×10^{-3}	3.15×10^{-4}	4×10^{-4}	1.50×10^{-5}	.060	.0287	.800*	.186	(I) 588	733
7		.100	.132	5×10^{-3}	5.71×10^{-4}	5×10^{-4}	3.93×10^{-4}	.0400	.0315	1.90	.392	(F) 1358	1063
8	FE	.195	.129	8×10^{-3}	1.63×10^{-3}	1×10^{-3}	2.16×10^{-5}	.1500	.0320	.150	.388	(I) 499	1065
9		.015	.128	2×10^{-3}	2.96×10^{-4}	4×10^{-4}	2.93×10^{-4}	.060	.0276	.800	.512	(I) 604	1139

*Solutions were not all converged to chosen tolerances at termination. (F) and (I) labels indicate whether initial design is feasible.

Table 2: Summary of Fin Array Optimization Cases

5. RESULTS AND DISCUSSION FOR THE MULTI-FIN ARRAY

The model described and solved in the previous section, while certainly going beyond the basic case, nevertheless contains simplifications and ignores certain interactions that would occur in an array of fins. For instance, engineering economic considerations, possible fouling with age, and the effect of geometry on the heat transfer coefficient were not included. Any of these factors could have been included in the model by parametrically defining the relationships in the finite element preprocessor input. Empirical relationships expressing the dependence of the Nusselt number on interfin spacing and including the integrated effect of radial position around the tube are available and would have provided more realistic boundary conditions. A more general problem would also have included inter-tube spacing and stagger as design variables, and taken them into account in determining the overall heat transfer. These complications were not introduced into the example to maintain simplicity, since their inclusion is not necessary for demonstrating the practical utility of the method.

Studying the results in Table 2 underscores the approximate nature of this algorithm. Different starting designs yield significantly different solutions in several cases. This algorithm is clearly not intended for mathematicians interested in finding global optimums. It does show the ability to improve upon a given starting design, and that can be of much value to a designer. Comparisons of final solutions show that different design strategies can have virtually the same cost. These symptoms indicate that the objective function is fairly flat and/or has many local minima, or that the design space chosen for this solution is too large.

No conclusions are meant to be drawn about the relative merit of copper, aluminum, or carbon steel for this contrived design problem. Neither should any inferences be made regarding the relative desirability of adding fins, adding fin material, or increasing fluid flow in order to increase heat transferred. The only observations of any merit to be made deal with the performance of the algorithm given the problem and the starting designs. The starting designs were chosen arbitrarily with the only objective being to make the 3 initial cases substantially different from one another.

The most meaningful measuring stick that can be applied to an algorithm of this nature is its ability to improve upon the original design. Against this criterion, the algorithm gets acceptable marks. In addition, the first nine design optimization solutions presented in this paper show an ability to approach the global optimum, since in these cases it was known.

Other effects could be included in the finite element analysis in cases where they become important. Internal heat generation, contact resistance, radiation and material anisotropy can all be included in the numerical solution of the problem. Provided that these effects are correctly incorporated into the finite element model definition, there is no inherent reason for their inclusion to prevent a solution using the approximation technique described in this paper. Two other examples that follow show further practical applications of this algorithm to heat transfer design problems.

6. MINIMIZING THE COST OF AN LNG TANK INSULATION SYSTEM

The second example illustrates the application of this algorithm to the design of an insulation system for a low pressure liquified natural gas (LNG) storage tank. The tank consists of an inner liquid container and an outer tank serving as the pressure containment (see Figure 4) and insulation jacket. It is presumed that the inner tank has been sized as required to meet structural design specifications and will not be changed. LNG is stored at a temperature of −260°F. The insulation system must be designed to allow a maximum heat gain of 150,000 BTU/hr for a full tank with 100°F ambient air and 50°F base temperature. Heat gain in excess of this rate will exceed re-liquification capacity and cause pressure buildup in the tank.

The starting design variable values are those of an existing LNG tank that was built several years ago (to the nearest inch). The three primary insulating materials used are mineral wool, perlite, and load-bearing expanded glass foam for the roof, annular space, and base support respectively. The thicknesses of these materials are chosen as the three design variables (t_w, t_p, and t_g in Figure 4). The objective function is defined to be the sum of those costs influenced by the three chosen design variables; this includes insulation material, installation costs and the erected cost of the steel jacket (outer tank) that encloses the insulation. The cost is assumed to be:

$$\text{Cost} = C_p V_p + C_g V_g + C_w V_w + C_s V_s$$

The factors used to calculate the cost (C_p, C_g, C_w, C_s) were chosen to reflect approximate unit costs actually incurred during construction. V's represent material volumes; the subscripts indicate perlite, foam glass, mineral wool, and steel, respectively. The cost function for the original design had a value of $1,750,000. After ten optimization loops, the solution

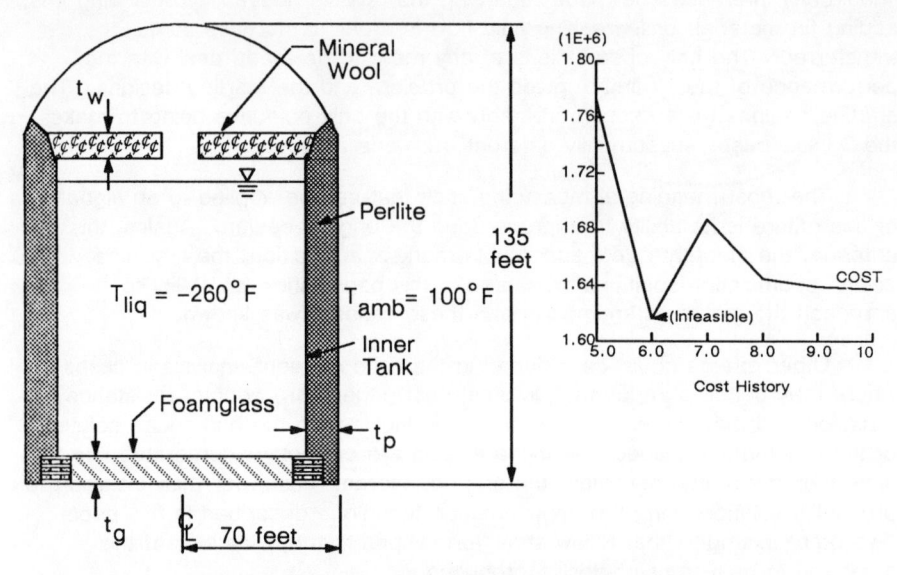

Figure 4 Sectional View of LNG Tank Showing Design Variables and Cost History

converged to a design having a cost function value of $1,640,000 a cost reduction of $110,000. The cost history is shown in the insert on Figure 4.

7. INTERDISCIPLINARY STRESS/HEAT TRANSFER DESIGN OPTIMIZATION

A problem is now considered which demonstrates the ability of the algorithm to handle a combined stress analysis and thermal design problem when the stress analysis includes temperature dependent plasticity, and the optimization includes temperature dependent stress constraints. The hypothetical design being considered is shown in Figure 5. A long straight section of insulated high pressure pipe is subjected to internal pressure of 3 KPa from a fluid at 500°C. As part of a design accident condition, the pipe is subjected to an ambient temperature of 1000°C for a short period of time, conservatively assumed to be long enough for steady-state to occur. Convection boundary condition at the interior and exterior walls of the pipe are imposed (h_f = 100 W/m^2-°C inside, h_f = 20W/m^2-°C outside). The problem is axisymmetric and is modeled with 20 axisymmetric elements as suggested by Figure 5. Conductivity of both materials was assumed to be variable with temperature.

The steady-state temperature solution was applied to the equivalent structural model along with internal pressure and equivalent mechanical loads. The insulation is assumed to carry no load, and the metal material properties are evaluated based on the temperature distribution; assumed stress-strain behavior is shown in Figure 5. The problem is to minimize the combined unit cost of the pipe-insulation assembly subject to the requirement that equivalent stresses may not exceed 85% of the (temperature dependent) yield stress within the pipe. The design variables are t_1 and t_2 as shown in Figure 5.

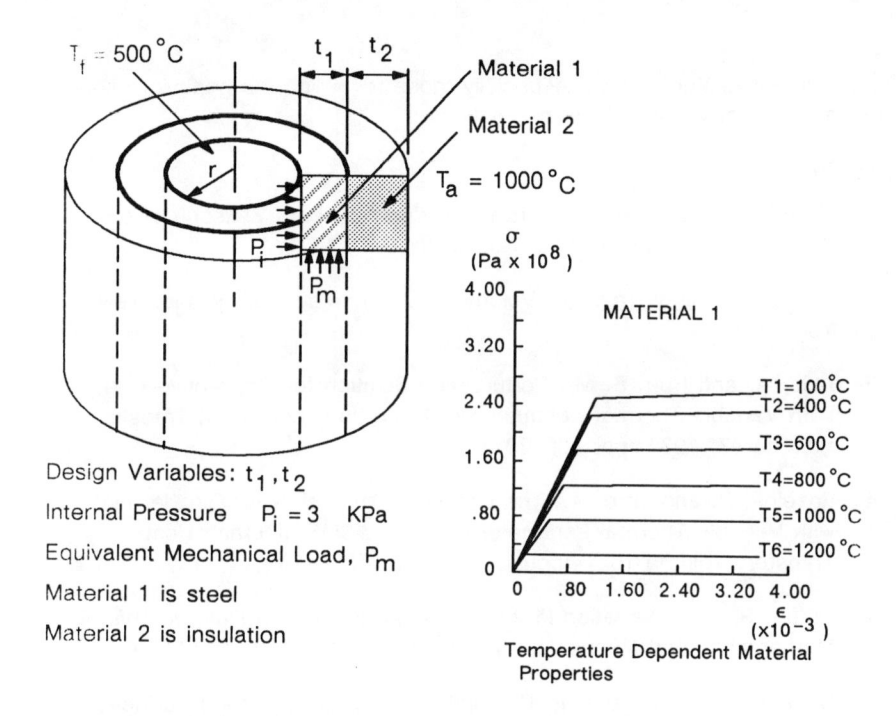

Figure 5 Example 3 - Thermal/Structural Design of an Insulated Plastic Pipe

The initial design was grossly overdesigned, and had a unit cost of 20.3 (for t_1 = .01, t_2 = .02). After 31 loops, the final design unit cost was 0.59 (for t_1 = .00024, t_2 = .0014). This problem is bounded by two cases: The non-insulated case, where no feasible solution exists; and the "fully" insulated case, when the unit cost is 18.3, both cases still being subject to the stress constraints mentioned earlier. This example shows that the lower cost lies between these extremes. While this is not a réalistic problem, it indicates a capability of this algorithm to approximate nonlinear thermal/structural behavior that can be applied to more complex analyses.

8. CONCLUDING REMARKS

An algorithm has been presented which incorporates two powerful numerical techniques, the finite element method and an approximation approach to design optimization. The combination of these methods for heat transfer problems provides a way to include complexities into the design model, solve the problem numerically using finite elements, and then simplify the optimization solution by using quadratic approximations of the functions. Results presented showed that theoretical solutions for several simple fin optimization problems were achieved to within small tolerances. When applied to a more complex fin problem with variable material parameters, significant reductions in the cost function were realized. Two additional examples provided evidence that the optimization technique can be applied to other thermal design problems and can successfully incorporate significant nonlinearities in the analysis. This algorithm was thus shown to be a workable means of helping to reach an improved or optimum design for various heat transfer problems. The method shows promise for applications in other areas of heat transfer design as well, especially those for which the behavior cannot be expressed easily in closed-form solutions.

9. REFERENCES

1. Schmidt, E., Die Warmeubertragung durch Rippen, Zeitscrift des, Volume 70, No. 26, p. 885, No. 25, p. 947, 1926.

2. Jakob, M., Heat Transfer, Volume 1, John Wiley & Sons, Inc. New York, 1956.

3. Aziz, A. and Hug, S. M., Perturbation Solution for Convective Fins With Variable Thermal Conductivity, ASME Journal of Heat Transfer, Volume 97, 1975, pp. 300-301.

4. Razelos, P. and Imre, K., The Optimum Dimensions of Circular Fins with Variable Thermal Parameters, Trans. ASME, Journal. Heat Transfer, Volume 102, August, 1980, pp. 420-425.

5. Duffin, R. J., A Variation Problem Related to Cooling Fins, Journal of Mathematics and Mechanics, Volume 8, No. 1, 1959, pp. 47-56.

6. Dhar, L. P., and Arora, P. C., Optimum Design of Finned Surfaces, Journal of Franklin Institute, Volume 301, No. 4, 1976, pp. 379-392.

7. Yudin, F. V., and Tokhtanova, S. L., Investigation of the Convection Factor for the Theoretical Effectiveness of a Round Fin, Thermal Engineering, Volume 21, No. 8, 1974, pp. 58-62.

8. Imgrund, M. C., Design Improvements Using Finite Elements and Optimizing with Approximating Functions, Proc. ASCE 9th Conference on Electronic Computation, Birmingham, Alabama, Feb. 1986, pp 507-518.

9. Kohnke, P. C., ANSYS Theoretical Manual, Swanson Analysis Systems, Inc., Houston, PA, 1986.

10. DeSalvo, G. J., and Gorman, R.W., ANSYS Revision 4.3 User's Manual, Volumes I and II, Swanson Analysis Systems, Inc., Houston, PA, 1987.

11. Chapman, A. J., Heat Transfer, Fourth Edition, Macmillan Publishing Co., New York, NY, 1984, pp. 217-226.

12. Imgrund, M. C. and Wheeler, M. J., Reducing Design Costs by Integrating Finite Element and Optimization Techniques, Proc. 1986 ASME Spring National Design Show, March, 1986, pp 79-92.

13. Beazley, P. K., Design Optimization, ANSYS Revision 4.3 Tutorials, Swanson Analysis Systems, Inc., March 30, 1987.

A Domain Partitioned Multigrid Method For The Solution Of The
Passive Scalar Advection-Diffusion Equation[i]

Arthur S. L. Shieh

Idaho National Engineering Laboratory
P.O.Box 1625, Idaho Falls, Idaho 83415, USA

   Many turbulent flow and convective heat transfer problems
require the solution of elliptic passive scalar
advection-diffusion equations. The presence of regions
containing large gradients of the solution variables, however,
necessitates the use of a fine mesh at least locally to
attain sufficient accuracy. One class of methods that has
been used successfully by several authors is multigrid methods
with local mesh refinement. The multigrid methods considered
in previous works, however, are general purpose solvers that
do not take full advantage of the special structure of the
problems. In this work we propose solving this class of
problems using a domain-partitioned multigrid method. It is a
special purpose solver that exploits the structure of the
problems. It eliminates the need of complicated up and down
cycling in a conventional multigrid method and thus makes the
algorithm particularly attractive for problems, the solution
variables of which have large gradients in certain regions of
interest. Various numerical examples and comparisons to other
multigrid methods are given.

(i) Work supported by the U.S. Nuclear Regulatory Commission,
    Office of Research, under DOE Contract No.
    DE-AC07-76ID01570.

1. INTRODUCTION

Many turbulent flow and convective heat transfer problems require the solution of elliptic passive scalar advection-diffusion equations. The presence of regions containing large gradients of the solution variables, however, necessitates the use of a fine mesh at least locally to attain sufficient accuracy. One class of methods that has been used successfully by several authors is multigrid methods with local mesh refinement. See e.g. [1] and [2]. Multigrid methods were studied by Brandt and others as an alternative to fast Fourier transform methods and cyclic odd even reduction methods for solving elliptic partial differential equations. Its main attractiveness lies in its versatility to handle almost any kind of general elliptic problem. See e.g. [7] for a good account of the method. Multigrid methods, however, typically use many up and down cycles between coarse and fine grid levels that make the methods less attractive especially for the supercomputers. In this work, we propose a domain partitioned multigrid method that minimizes the use of multigrid cycles without affecting the accuracy. Roughly speaking, the method virtually eliminates the need for multigrid cycling for subregions where the solution variables are smooth and it requires only a minimum number of multigrid cycles for subregions containing large gradients of the solution variables. The method therefore can be quite useful for the type of problems considered here.

2. THE DOMAIN-PARTITIONED MULTIGRID METHOD.

The method was previously used in [5] to solve coupled system of PDE's arising from VLSI simulation. It employs the use of local truncation error extrapolation and residual minimizing prolongation techniques. For completeness, these techniques are briefly described below.

2.1 Local Truncation Error (LTE) Extrapolation Technique.

Suppose

$$L^j U^j = F^j, \quad j = M-1, M$$

are discrete approximations of second-order accuracy to the continuous problem on the second finest mesh and the finest mesh respectively. Let U^{M*} be a computed approximation to U^M. It is known (see e.g. [3] or [4]) that

$$L^{M-1} U = F^{M-1} + (4/3) LTE \tag{1}$$

where

$$\text{LTE} = L^{M-1} I^{M-1}_U M^* - I^{M-1}_L M_U M^* \qquad (2)$$

can yield solutions of fourth-order accuracy to the continuous problem. Here I^{M-1} is the restriction operator from level M to level M-1 and it is assumed that both the finest and the second finest meshes are uniform with the mesh size of the second finest mesh twice that of the finest mesh. Because LTE is in general larger at regions containing large gradients of solution variables, the estimated values for LTE provide a natural basis for local mesh refinement or domain partitioning.

2.2 Residual Minimizing Prolongation (RMP)

A special prolongation technique is used to extend the fourth-order solution on the second finest mesh obtained by the LTE-extrapolation method to the finest mesh. The prolongation method is designed to preserve the order of accuracy on the finest mesh. In addition, the method also serves as a preconditioner in a two-level multi-grid setting. It preconditions the low- and medium-frequency components of the error on the coarse grid so that after prolongation they become mostly high-frequency components with smaller magnitudes on the fine grid. These high-frequency components of the error can then be removed efficiently by relaxation sweeps on the fine grid. See [6] for more detail when the prolongation technique is used in the solution of the Poisson equation.

Suppose we are solving the Dirichlet problem of the advection-diffusion equation

$$-\frac{1}{Pe} Lu + D_x(Pu) + D_y(Qu) = S(x,y) \qquad (3)$$

Here L denotes the Laplacian and Pe is the Peclet number. D_x and D_y denote partial differentiation with respect to x and y respectively and P and Q are functions of x and y. Let U be the discrete variable for u in Eq. (3). Let

$$-\frac{1}{Pe} L_h U + D_h(PU) + E_h(QU) = S$$

be the discrete analog of Eq. (3). Here L_h denotes the five-point discrete Laplacian and D_h and E_h are difference operators corresponding to the hybrid difference scheme for the advection terms. The hybrid scheme combines the central and upwind differencing schemes by switching from central to upwind differencing when the magnitude of the cell Peclet number

$$Pe_h = PhPe \text{ or } QhPe$$

is greater than two, where h is the mesh size. This is accomplished by defining $D_h(PU)$ at a point (x,y) to be

$$D_h(PU) = [P(x+h/2,y)U_e - P(x-h/2,y)U_w]/h \qquad (4)$$

where U_e is given by

$$U_e = \begin{cases} U(x,y) & PhPe \geq 2 \\ [U(x,y)+U(x+h,y)]/2 & -2 \leq PhPe \leq 2 \\ U(x+h,y) & PhPe < -2 \end{cases} \qquad (5)$$

Similar expressions are used for U_w and for $E_h(QU)$.

Suppose there are only two grids with uniform meshes where the coarse grid is obtained from the fine grid by deleting every other grid line of the fine grid. Assume that the solution is already known on the coarse grid. Below we describe how the solution is prolongated to the fine grid.

Let h(2h) be the fine (coarse) grid mesh size. Let A_h be the set of mesh points whose mesh neighbors in the y-axis directions are coarse grid mesh points. Let B_h be the set of remaining mesh points in the complement of the coarse grid. At each point P(x,y) on A_h, the following equation is formed

$$[-U(x,y-h)+2U(x,y)-U(x,y+h)]/h^2 + \\ [-U(x-2h,y)+2U(x,y)-U(x+2h,y)]/4h^2 + D_{2h}(PU) + \\ E_h(QU) = S(x,y)/2 + [S(x-h,y)+S(x+h,y)]/4 \qquad (6)$$

Here $D_{2h}(PU)$ is defined in the same way as $D_h(PU)$ with h in Eq. (4) and (5), replaced by 2h. The cell Peclet number is still taken to be PhPe. Once U is known on the coarse grid, its values on A_h are obtained by solving linear tridiagonal equations of the form.

$$[-1-2PehP \quad 10+6PehP \quad -1] U = G$$

Here for simplicity we've assumed that 2PhPe is greater than 2 throughout A_h. Once U is determined on A_h, its values on

B_h can easily be determined by operator based prolongation by solving tridiagonal equations of the form

$$[-1-PePh \quad 4+2PePh \quad -1] U = G.$$

In the case when U is the fourth-order solution on the second finest grid, the S in Eq. (6) should be replaced by $S+(4/3)LTE$, where LTE is defined by (2). Note that LTE is only defined on the second finest grid. We extend the definition of LTE to the finest grid by defining LTE to be

$$LTE := (I_{M-1}LTE)/4$$

on the complement of the second finest grid.

Here I_{M-1} denotes linear interpolation from level M-1 to M. This definition is motivated by the fact that the fine grid local truncation error is approximately one-quarter of that on the coarse grid. We shall refer to this prolongation technique as the quarter-LTE RMP technique.

2.3 Domain Partitioning

The approach used here is the same as the domain partitioning techniques described in [5]. It makes use of tensor product grids. The coarsest grid covers the entire domain. The problem is solved to second-order accuracy, and then the accuracy is improved using the LTE-extrapolation technique. At this point we assume that the solution is sufficiently accurate in large portions of the domain where the solution is smooth. We then consider one or more small subregions of the original domain where the solution is still deemed inaccurate. On these subregions, we solve reduced problems, using cubic interpolation methods to set the boundary conditions and provide the initial iterate.

2.4 Multi-grid Method

Suppose we are solving Eq. (3) on the unit square. Let $h=1/32$. Assume that four grid levels ($h = 1/4$, $1/8$, $1/16$, and $1/32$) are used. Level 1 is the coarsest grid level. Then the basic scheme can be described as follows:

Basic scheme: 5 SOR*1 -> 4 SOR*2 -> 3 SOR*3
                2 G.S.*4 - LTE-extrapolation -
                1 V-cycle - quarter-LTE RMP -> level 4   (7)

Below we give the definition for the terms used in (7).

m SOR(G.S.)*k: m SOR (Gauss-Seidel) relaxation sweeps on level k.
->: RMP prolongation
One V-cycle: 3 G.S.*3 - weighted transfer of residuals to levels 1 and 2 - 5 SOR*1 -> 4 SOR*2 -> update.

The basic scheme is used to obtain a solution of fourth-order accuracy on the entire region. This is accomplished by getting an accurate estimation of the local truncation error LTE using a very simple multi-grid process. Note that there is no up and down cycling before LTE-extrapolation. Hence this simple multi-grid process is not a true multi-grid method and there is some question on the robustness of this method. In our numerical experiments, we've found that for h=1/32 and four grid levels, this simple pseudo multi-grid process gives an accurate determination of LTE. Roughly speaking, the use of RMP makes up for the lack of up and down cycling as long as h is not too small and the number of grid levels is not too large. See [6] for a detailed account of this. Because domain partitioning is used, the basic scheme is repeated for the one or more subregions where the solution is still deemed inaccurate until the norms of the residuals are within a certain tolerance on the entire region. So in essence true multi-grid methods with a minimal amount of up and down cycling are used for regions where large gradients of solution variables are present. The method is therefore robust and efficient for the type of problems considered here.

3. NUMERICAL EXPERIMENTS

Numerical results for problems similar to those considered in [2] are given here. The test runs are made on Cyber 176 with NOS operating system at the Idaho National Engineering Laboratory. The test problems involve Eq.(3) with various source terms and velocity fields. The domain of interest is always the unit square. For problem 2 below, Gauss-Seidel relaxation sweeps are used on all grid levels.

Problem 1.

$S(x,y) = (-4Pe+Pe^2(4x^2+4y^2))\exp(-Pe(x^2+y^2))$
$P(x,y) = Q(x,y) = 0$
$U(x,y) = 1-\exp(-Pe(x^2+y^2))$

Problem 2.

$$S(x,y) = 4Pe \cdot \exp(-Pe(x+y))$$
$$P(x,y) = Q(x,y) = 1$$
$$U(x,y) = 1 - \exp(-Pe(x+y))$$

The following numerical results are obtained with the basic scheme.

Table 1

Problem No.	Pe	e_{max}	e_2	e_{2D}	e_D
1	200	5.7E-3	3.3E-4	6.5E-4	6.4E-5
2	10	7.7E-4	1.0E-4	1.5E-4	1.5E-4
2	20	6.7E-3	1.1E-3	1.4E-3	1.4E-3
2	50	3.9E-2	4.6E-3	7.5E-3	7.5E-3

Here e_{max} and e_2 denote the maximum norm and the Euclidean norm of the errors. The Euclidean norm used here is taken to be the square root of the sum of the squares divided by the number of the mesh points in the region. The e_{max} and e_2 are computed over the entire region while e_{2D} is the Euclidean norm computer over the subregion D: D:0<x<0.5, 0<y<0.5. The e_D is the same as e_{2D}, except that the difference equations are solved to machine accuracy using many multi-grid cycles. The subregion D is chosen because of the large gradients of solution variables there. In the domain partitioned multi-grid method, we repeat the basic scheme once on the region D. The numerical results are given in the following table.

Table 2

Problem No.	Pe	e_{max}	e_2	e_{2D}	e_D
1	200	1.7E-3	1.4E-4	2.7E-4	2.7E-4
2	10	2.4E-4	7.5E-5	4.5E-5	4.5E-5
2	20	2.6E-3	8.1E-4	1.7E-4	1.7E-4
2	50	1.9E-2	2.8E-3	2.0E-3	2.0E-3

We note the significant improvements in the numerical results on the region D after domain partitioning. This process may be repeated until e_{max} is about the same as the maximum error in the subregions where the solution variables are smooth.

4. CONCLUSIONS

We have found that the basic scheme described in this work is useful in providing an overall accurate solution for problems that have subregions containing large gradients of the solution variables when it is used in conjunction with a domain partitioning scheme.

5. REFERENCES

1. R. E. Philips and F. W. Schmidt, Multi-Grid Techniques for the Numerical Solution of the Diffusion Equation, Numer. Heat Transfer, $\underline{7}$, 251-268 (1984).

2. R. E. Philips and F. W. Schmidt, Multigrid Techniques for the Solution of the Passive Scalar Advection-Diffusion Equation, Numer. heat Transfer, $\underline{8}$, 25-43 (1985).

3. N. Dinar, Fast Methods for the Numerical Solution of Boundary Value Problems, Ph.D. Dissertation, Weizmann Inst. of Science, Rehovat, Israel, 1979.

4. J. Gary, On Higher Order Multigrid Methods with Application to a Geothermal Reservoir Model, Int. J. Numer. Meth. in Fluids, $\underline{2}$, 43-60 (1982).

5. A. Shieh, On the Solution of Coupled System of PDE by a Multigrid Method, IEEE Transactions of Computer-Aided Design, Vol. CAD-4, $\underline{4}$, Oct., 1985.

6. A. Shieh, On the Theoretical and Experimental Aspects of a Highly Compact High Order Multi-Grid Algorithm, INEL Report EGG-RST-6832, March 1985.

7. A. Brandt, Multi-level Adaptive Solutions to Boundary Problems, Math. Comput., $\underline{31}$, 333-390 (1977).

NOTICE

This report was prepared as an account of work sponsored by an agency of the United States Government. Neither the United States Government nor any agency thereof, or any of their employees, makes any warranty, expressed or implied, or assumes any legal liability or responsibility for any third party's use, or the results of such use, of any information, apparatus, product or process disclosed in this report, or represents that its use by such third party would not infringe privately owned rights. The views expressed in this report are not necessarily those of the U.S. Nuclear Regulatory Commission.

A 3-D Finite Element Investigation of Non-Newtonian Thermal Fluid Flow

M. Gadbois and P.A. Tanguy
Département de génie chimique
Université Laval, Québec, G1K 7P4
Canada

The numerical simulation of molten polymer flows requires the simultaneous solution of the momentum, mass and energy equations which are associated with more or less complex rheological equations of state to describe the fancy behavior of the fluid. Due to the very poor thermal conductivity of the polymer, the energy equation is advection-dominated with yields essentially an hyperbolic problem. In a finite element context, the solution of such a problem is generally obtained using some kind of upwinding to kill the wiggles generated during the solution process. Indeed, the standard solution procedures in finite element are well-suited for elliptic problems (diffusional problems) but are quite sensitive to ill-conditioned matrix systems hence the numerical oscillations mentioned before, when convective transport dominates diffusion. Methods adapted to hyperbolic problems should be preferred for the solution of advection problem but this approach has been surprizingly much less used than the standard upwinding method.

The aim of the present paper is to show that the method of characteristics can be used as a smart alternative for the solution of the 3-D energy equation at high Peclet number in the presence of viscous dissipation. Such a problem is typical of polymer processing applications. The method of characteristics is coupled to the solution of the diffusion problem by the standard Galerkin procedure. Iterative solvers (conjugate gradient type method) are used instead of a classical factorization to reduce computational cost. The present approach is compared to the streamline-upwind Pedrov-Galerkin in terms of accuracy for the solution of polymeric fluid flow in complex geometry.

SECTION 9

Industrial and Scientific Applications

SPORTS-M: A THERMALHYDRAULICS COMPUTER MODEL FOR POOL-TYPE REACTORS

S.Y. Shim, P.J. Mills and S. Meyer

Atomic Energy of Canada Limited
Whiteshell Nuclear Research Establishment
Pinawa, Manitoba, Canada ROE 1L0

SUMMARY

SPORTS-M is a thermalhydraulic computer code that performs steady-state and transient analysis in a piping network. The code solves the hydrodynamic equations for a one-dimensional transient flow of a homogeneous two-phase mixture, and is capable of modelling nonequilibrium effects such as subcooled boiling. The hydrodynamic equations used in SPORTS-M are solved iteratively in their primitive form by a fully implicit, forward-marching, finite-difference method. The heat transfer and heat conduction models are also treated in a fully implicit manner for wall temperatures, consistent with the hydrodynamic model. The code performance has been demonstrated using sample cases of steady-state/transient, natural/forced convection conditions with/without the presence of subcooled boiling in a typical MAPLE research reactor.

NOMENCLATURE

A = area, m^2
a = speed of sound, $m \cdot s^{-1}$
CHF = critical heat flux
c_p = specific heat at constant pressure, $J \cdot kg^{-1} \cdot K^{-1}$
D = diameter, m
f = friction factor
g = gravitational acceleration, $m \cdot s^{-2}$
h = enthalpy, $J \cdot kg^{-1}$
h_{ht} = heat transfer coefficient, $W \cdot m^{-2} \cdot K^{-1}$
H_p = pump head, m
K = flow loss coefficient
k = thermal conductivity, $W \cdot m^{-1} \cdot K^{-1}$

u = flow velocity, $m \cdot s^{-1}$
W = pump work, $m \cdot s^{-2}$
X = flow quality
x = space coordinate, m
z = elevation above an arbitrary datum, m
α = void fraction
β = thermal expansion coefficient, K^{-1}
Δt = time step, s
Δx = spatial increment, m
Δz = height increment, m
Φ^2 = two-phase multiplier
ρ = density, $kg \cdot m^{-3}$
τ = shear stress per unit length, $N \cdot m^{-3}$

ONB = onset of nucleate boiling
OSV = onset of significant void
p = pressure, Pa
Q = heat generation rate, W
q" = heat transfer/generation rate, $W \cdot m^{-3}$
Re = Reynolds number
r = radius, m
T = temperature, K
t = time, s

Subscripts

c = core
f = fluid
ht = heat transfer
i = space index
sat = saturation
w = wall

Superscript

n = time level index

1. INTRODUCTION

Analysis of thermalhydraulic performance for a pool-type reactor is expensive using most existing semi-implicit computer codes, especially for single-phase/two-phase natural circulation calculations in the presence of a large volume of fluid. This is mainly due to a time-step limitation from numerical stability considerations. Also, it is difficult to incorporate a nonequilibrium effect, such as subcooled boiling, into a homogeneous-equilibrium type model that uses fluid property derivatives. To overcome these concerns, the SPORTS-M code was developed.

The hydrodynamic equations used in SPORTS-M are the conservation equations of continuity, momentum, and energy, together with the equation of state for a one-dimensional transient flow of a homogeneous two-phase mixture. The basic computational method of solving the hydrodynamic equations follows the scheme of [1]. It is a fully implicit, forward-marching, finite-difference scheme, which uses an iterative procedure from node to node throughout an open or closed system.

A heat transfer package covering a full boiling curve and a radial heat conduction model were developed and coupled with the hydrodynamic model. The heat transfer package describes the subcooled boiling process in detail. The heat transfer coefficient and heat conduction equation are solved in a fully implicit manner for wall temperatures, consistent with the fully implicit hydrodynamic variables. The two-phase density during subcooled boiling is described through the homogeneity definition using a true void.

This paper presents the overall methodology used in the SPORTS-M code including the numerical scheme, heat transfer model, heat conduction model, constitutive relations, and convergence scheme. Code accuracy and computational efficiency are demonstrated with sample cases for steady-state and transient natural circulation predictions for MAPLE (Multipurpose Applied Physics Lattice Experimental) [2], a pool-type research reactor.

2. MATHEMATICAL MODELLING

The mathematical model for homogeneous two-phase flow in a piping network consists of the hydrodynamic equations, constitutive relations and boundary conditions.

2.1 Hydrodynamic Equations

SPORTS-M uses a simple one-dimensional two-phase flow model. The flow in a pipe is represented by variables averaged over the cross section. The model assumes that vapor and liquid velocities and temperatures are the same at any cross section. The governing equations for a one-dimensional homogeneous two-phase mixture are:

$$\frac{\partial \rho}{\partial t} + \frac{\partial \rho u}{\partial x} = 0 \tag{1}$$

$$\frac{\partial \rho u}{\partial t} + \frac{\partial \rho u^2}{\partial x} = -\frac{\partial p}{\partial x} - \rho g \frac{\partial z}{\partial x} - \tau_w + \rho W \tag{2}$$

$$\frac{\partial}{\partial t}(\rho H - p) + \frac{\partial}{\partial x}(\rho u H) = q'' + \rho u W \tag{3}$$

$$\rho = \rho(p, h, \alpha) \tag{4}$$

where

$$H = h + \frac{u^2}{2} + gz \tag{5}$$

$$W = \frac{gH_p}{\Delta x} \tag{6}$$

$$\tau_w = \rho \Phi^2 \left(\frac{K}{\Delta x} + \frac{f}{D}\right) \frac{|u|u}{2} \tag{7}$$

$$q'' = \frac{h_{ht} A_{ht} (T_w - T_f)}{A_f \Delta x} \tag{8}$$

2.2 Constitutive Equations

Wall Momentum Transfer

To evaluate the wall shear force in Equation (7), the following relations are used in SPORTS-M:

Friction Factor: $f = 64/Re$ for laminar flow
                   Colebrook Equation for turbulent flow [3]
Two-phase Multiplier: Homogeneous model or
                      Jones [4]

The local pressure drop due to an abrupt area change is also taken into account using the loss coefficients given by [5].

Wall-Fluid Heat Transfer

The heat transferred from the heated wall to the fluid is obtained from Equation (8) with the appropriate heat transfer correlations for h_{ht}. The heat transfer package in SPORTS-M covers a full boiling curve consisting of the heat transfer criteria and correlations as summarized in Table 1.

HEAT TRANSFER MODE		CRITERIA	REFERENCE
Single Phase Flow	Laminar	Re < 2000	Collier [6]
	Transition	2000 < Re < 2500	Interpolation
	Turbulent	2500 < Re	Dittus-Boelter [7]
ONB Temperature			Davis-Anderson [8]
Partially Subcooled Boiling		$T_{ONB} < T_w < T_{OSV}$	Interpolation
OSV Temperature			Saha-Zuber [9]
Fully-Developed Subcooled Boiling		X < 0.005 X > 0.005	Thom [10] Chen [11]
Saturated Boiling		$T_f = T_{sat}$	Chen
CHF Temperature			Groeneveld-Rousseau [12]
Post CHF Heat Transfer	Transition Boiling	$T_{CHF} < T_w < T_{rewet}$	Bjornard-Griffith [13]
	Rewet		Groeneveld-Rousseau
	Film	$T_w < T_{rewet}$	Groeneveld-Rousseau

Table 1: Heat Transfer Correlations and Criteria

Particular emphasis was also given to the detailed modelling of subcooled boiling. It is important to model the subcooled boiling process, which is the dominant heat transfer and hydrodynamic mechanism in the MAPLE core. The subcooled boiling heat transfer regime is divided into partially and fully developed subcooled boiling. A partially subcooled boiling regime, defined to be between ONB and OSV, is important thermally because very effective heat transfer from the heated wall is achieved without the presence of significant void in the system. A fully developed subcooled boiling regime, defined to be between OSV and CHF, is both hydrodynamically and thermally important, since the generated void influences pressure losses in the system and the resulting nucleate boiling is an effective heat transfer mode.

Void Model for Subcooled Boiling

To model nonequilibrium effects such as subcooled boiling, the two-phase density was described through the homogeneity definition using a true void. The void generation during subcooled boiling is predicted using a drift-flux model [14]. Void along a heated channel is generated when the local heat flux exceeds the OSV criterion. The void during subcooled boiling is assumed to collapse when entering a non-heated channel.

Heat Conduction Equation

To calculate radial temperature distributions in a heating element such as a nuclear fuel, SPORTS-M solves the one-dimensional, transient heat conduction equation:

$$\rho c_p \frac{\partial T}{\partial t} = \frac{1}{r} \frac{\partial}{\partial r} (kr \frac{\partial T}{\partial r}) + q'' \qquad (9)$$

The boundary condition at the fluid-wall interface is given by:

$$k \frac{\partial T}{\partial r} = h_{ht}(T_w - T_f) \qquad (10)$$

The heat conduction equation is solved by a fully implicit scheme for wall temperature. This scheme is consistent with fully implicit hydrodynamic variables such as density, velocity, pressure and enthalpy. Since boiling heat transfer is usually described as a function of a wall temperature, this implicit treatment is important when an abrupt boiling regime transition takes place.

To facilitate modelling of the fuel pin, the fuel pin may be divided into five radial regions of different materials. Each region can be discretized into a user-specified number of segments. The material properties such as density, specific heat and thermal conductivity may be input as a function of nodal temperature. A heat generation rate for each region may also be input as a function of time. An axial heat generation distribution may be specified as a function of axial distance along the fuel channel.

2.3 Initial and Boundary Conditions

To complete the mathematical formulation of Equations (1) to (4), boundary conditions must be specified. For a steady-state solution, SPORTS-M allows three types of boundary conditions: (a) inlet enthalpy, inlet pressure and outlet pressure for an open system, (b) inlet mass flow rate, inlet enthalpy and outlet pressure for an open system, and (c) pressure and enthalpy anywhere in a closed system.

Because of the node-to-node forward-marching advantage, a set of initial conditions is not needed. Instead, it requires only a velocity estimate at the boundary node.

3. NUMERICAL MODELLING

3.1 Finite-Difference Formulation

The hydrodynamic equations (1) to (3) are discretized by taking the forward difference in space. The variables in the transient terms of these equations are evaluated to be the average between the two adjacent spatial nodes. All the variables are evaluated at the current time step. A detailed derivation of the discretized equations is given in [1].

As in [15], the heat conduction equation is discretized by taking forward difference in time and space.

3.2 Solution Methods

Overall Solution Method

The discretized equations of the hydrodynamics and heat conduction are solved iteratively in their primitive form. No matrix algorithm is needed to solve the hydrodynamic equations, while a simple matrix algorithm TDMA (Tri-Diagonal Matrix Algorithm) is used to solve the heat conduction equation.

The overall solution method of SPORTS-M resembles that of [1]. It is a fully implicit, forward-marching, finite-difference scheme. Figure 1 illustrates the solution scheme used in SPORTS-M.

ρ_{i+1}^{n+1} → u_{i+1}^{n+1} → p_{i+1}^{n+1} → h_{i+1}^{n+1} → ρ_{i+1}^{n+1}
Estimate — Continuity equation — Momentum equation — Energy equation — Equation of state

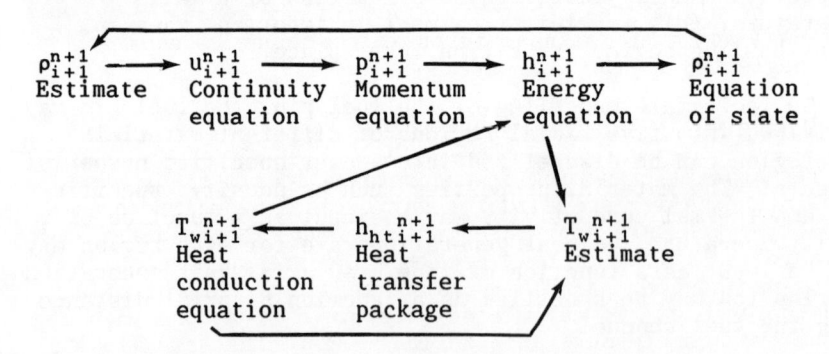

T_{wi+1}^{n+1} ← h_{hti+1}^{n+1} ← T_{wi+1}^{n+1}
Heat conduction equation — Heat transfer package — Estimate

Figure 1:   Solution Scheme Used in SPORTS-M

Let the solution procedure begin at the current time step, n+1, and the spatial step, i+1, following the boundary node, i. The conservation equations are solved sequentially for ρ, u, p, h, and T_w where necessary, for node i+1 at time n+1. The continuity equation is solved for u_{i+1}^{n+1} using an estimate of ρ_{i+1}^{n+1} and the user-supplied initial estimate of u_i^{n+1}. The momentum and energy equations yield p_{i+1}^{n+1} and h_{i+1}^{n+1}, respectively, and an improved ρ_{i+1}^{n+1} is obtained from the equation of state.

Calculation at a node is done by iteration until convergence is obtained for density. Following the density convergence at a node, the solution proceeds from node to node until it reaches the exit pressure boundary node. The estimate of velocity at the entrance pressure boundary node is corrected, based on the difference between the calculated and boundary condition pressures. Once the density and velocity converge to a unique solution, the iteration process stops.

When a heated channel is encountered in a system, the fluid energy equation, wall-fluid heat transfer, and heat conduction equation are solved iteratively in series until convergence is obtained. As illustrated in Figure 1, the hydrodynamic variables are passed to the heat transfer package where a heat transfer coefficient is determined. The heat conduction equation is then solved for the wall temperature using the heat transfer coefficient and fluid temperature. The resulting wall-fluid heat flux is returned to the energy equation. The energy equation is then solved and an updated fluid temperature becomes available. This process continues until the fuel-wall temperature used in the heat transfer correlation agrees with the fuel wall temperature obtained from the conduction equations, and the nodal fluid temperature used in the heat transfer package agrees with the temperature resulting from the energy equation.

The numerical stability analysis for the discretized hydrodynamic equations is given in [1] as $\Delta t \geq \Delta x/2a$. However, the time step required for numerical stability was found to be higher than that given. This may be due to frictional and gravitational terms in the momentum equation being neglected in the analysis done in [1].

Solution Method for Parallel Branching

If parallel branching is encountered in a system, mass flows into each leg are adjusted until the flow split results in the same pressure drop through each leg. The solution procedure is similar to that used for the pressure-velocity convergence scheme. SPORTS-M calculates a pressure error as the difference of the calculated pressure drop across each leg and the average pressure drop for all legs. The mass flows into each leg are adjusted based on the pressure error.

3.3 Convergence and Accuracy

With appropriate boundary conditions in a system, an initial steady-state condition is obtained prior to a transient analysis. The solution at each time step is then obtained based on the previous solution. For transient solutions, the velocity at the beginning node is based on the previous value and the calculations proceed through the remainder of the system. If the calculated pressure at the boundary agrees within a specified tolerance with the boundary pressure, the calculations proceed to the next time step.

SPORTS-M uses four iterative steps: (1) the density convergence at every node, (2) the pressure boundary convergence for the overall system, (3) the pressure drop convergence in a parallel branching, and (4) the wall and fluid temperature convergence in a heated channel. Item (1) is attained when the equation of state is satisfied at every node. To speed up convergence for Items (2) and (3), SPORTS-M uses a second-order least-squares fit. It has been found that a linear fit using two data pairs is more effective when a solution is close to the final solution, but otherwise the least-squares fit using four data pairs is more effective.

The solution accuracy can be controlled by user-specified tolerance values for the above four items. For instance, the tolerance value for the pressure convergence has to be smaller for a thermosiphoning flow than that for a pumped flow.

4. SAMPLE CASES

The following predictions were made on the basis of a simplified representation of a MAPLE research reactor. The case was chosen to demonstrate the SPORTS-M performance and to validate the predictions. A schematic flow diagram of a typical MAPLE reactor is shown in Figure 2.

Name	Length (m)	Height (m)	Diameter (m)	Flow area (m^2)	K
Inlet piping	10.1	6.7	0.57	0.259	2.0
Heat exchanger	2.5	0.0	0.0064	0.0072	1250.0
Outlet piping	10.0	-9.0	0.48	0.1793	2.0
Inlet plenum	0.85	0.85	1.4	1.5394	0.0
Core	0.7	0.7	0.0062	0.0027	14.0
Chimney	0.75	0.75	0.66	0.35	0.0

Table 2: Hydraulic Information of Sample Case

The hydraulic information used for the cases is summarized in Table 2. The core and the heat exchanger carry 20 and 146 equivalent parallel channels, respectively. Each core channel contains a 36-element fuel bundle as in [2]. The initial power used was 30 MW and the initial pump head was 60 m.

Steady-State Thermosiphoning

An analytical solution for a steady-state natural circulation can easily be derived from Equations (1) to (4) by removing the transient and pump terms. The resulting equation becomes:

$$u_c = \left[\frac{2\beta Q g \Delta z}{\rho_c c_p} \frac{1}{\Sigma(\frac{f_i}{D_i} + \frac{K_i}{\Delta x_i}) \frac{A^3}{A_i^2} \Delta x_i}\right]^{1/3} \quad (11)$$

The SPORTS-M predictions for the sample case of a single-phase steady-state condition agreed within two significant digits of the core velocity calculated by Equation (11).

It is of practical interest to represent natural circulation mechanisms such that the reactor can be designed to possess a passive means of cooling. Furthermore, when a thermal limit is imposed to maintain desirable thermal margins at the hottest point in the reactor core, the maximum natural circulation cooling will be limited by the heat transfer criteria such as ONB, OSV and CHF.

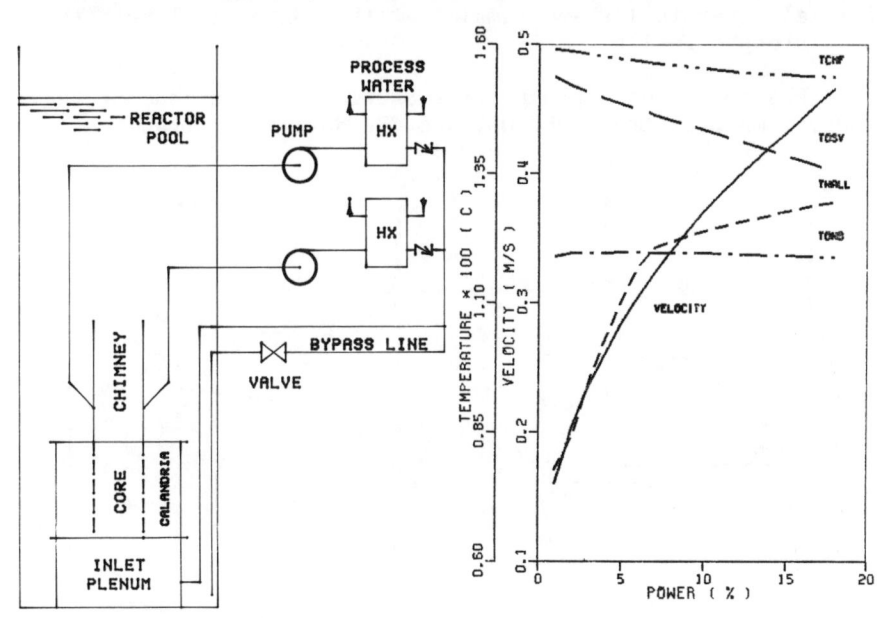

Figure 2: Schematic Flow Diagram of a MAPLE Research Reactor

Figure 3: Steady-State Flow and Temperature Predictions of SPORTS M

Figure 3 shows the thermalhydraulic performance of a simplified MAPLE reactor based on predictions made by the SPORTS-M code. As shown in the figure, the natural circulation flow rate increased as the driving force due to density difference between the hot and cold columns increased. Figure 3 shows that a switch-over from a single-phase to a partially subcooled boiling heat-transfer regime takes place with increasing power. The switch-over is indicated by the sharp slope change in the the locus of wall temperature. The natural circulation cooling capability without subcooled boiling is small, but the cooling capability is much enhanced by subcooled boiling heat transfer. During this partially subcooled boiling regime, vapor generated in the core is attached to the wall and, thus, does not influence the system hydraulics.

Each simulation of the above cases, that has 54 calculation nodes, took less than 3 s on a VAX 785 computer.

Transient Simulation of a MAPLE Reactor

A pool-type reactor like MAPLE contains a large fluid volume, which has a large thermal inertia during a long-term transient. During a long-term transient such as a loss of pump scenario, it must be ensured that a transition from forced to natural convection flow would not lead to adverse thermal stratification.

This section demonstrates the performance of the SPORTS-M code for the case where the core power decays to 10% of the initial power in 1 s and remains at the 10% power level; the pump stopped in 1 s.

Figures 4 and 5 present the SPORTS-M predictions of the thermal margins from ONB, OSV and CHF during the transient.

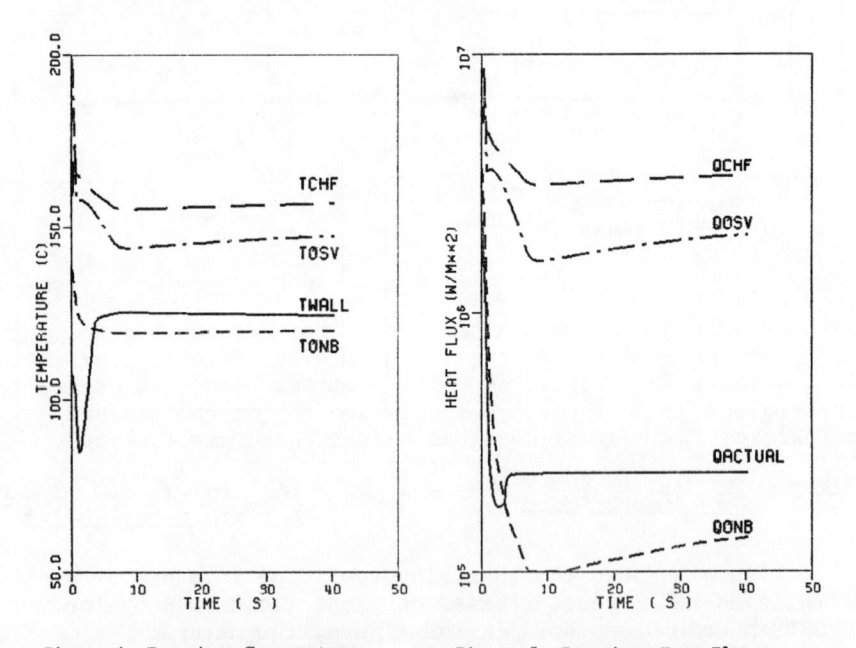

Figure 4: Transient Temperature Predictions of SPORTS-M

Figure 5: Transient Heat Flux Predictions of SPORTS-M

Figures 6 and 7 show the comparison of the predictions of SPORTS-M with those of CATHENA [16], an inhomogeneous nonequilibrium two-fluid code. The fuel temperature predictions between the two codes differ slightly due to the differences in initial coolant temperatures and nodal representations.

As shown in Figure 6, a smooth transition from pumped to natural circulation flow was predicted. Figure 7 shows the transient temperatures of fuel centerline, fuel sheath and coolant at the core exit. During the transient, the fuel is overcooled at the early stage and then is cooled by natural

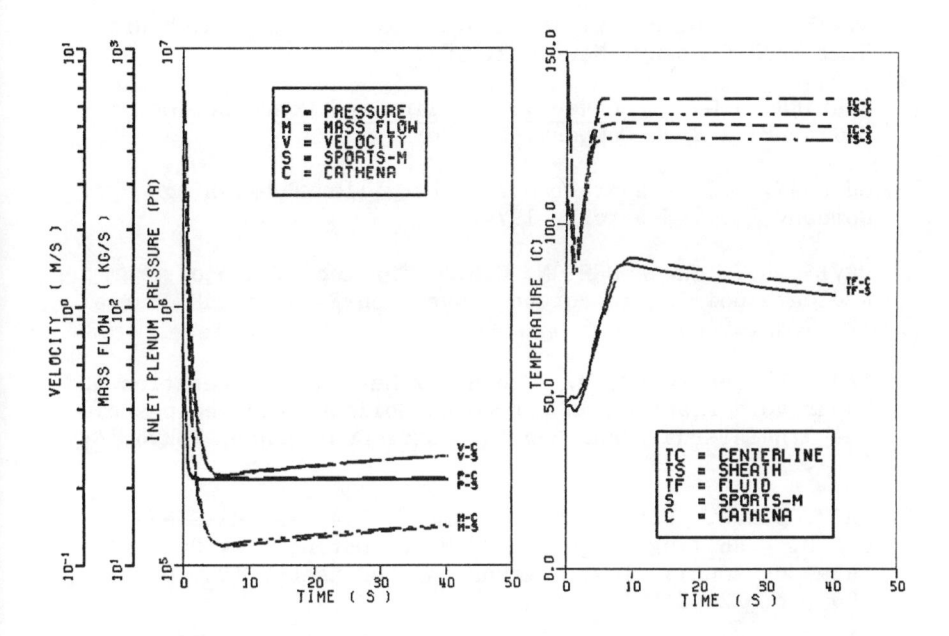

Figure 6: Comparison of Transient Pressure and Flow Predictions

Figure 7: Comparison of Transient Temperature Predictions

circulation flow in the partially subcooled boiling regime. The core-exit temperature increases sharply at the beginning, and then decreases slowly due to the slowly increasing natural circulation flow. This trend will continue with time as the core cooling will improve gradually due to the continuous increase of the buoyant head (as hotter core-exit coolant fills the hot column of the system).

REFERENCES

1. CHATOORGOON, V. and THIBEAULT, P.R. - SPORTS - An Advanced Thermalhydraulic Stability Code", presented at the 23rd ASME/AIChE/ANS National Heat Transfer Conference, Denver (1985).

2. HEEDS, W. - The MAPLE-X Concept Dedicated to the Production of Radio-Isotopes, presented at the 6th Canadian Nuclear Society Annual Conference, Ottawa, 1985 June.

3. COLEBROOK, C.F. - Turbulent Flow in Pipes with Particular Reference to the Transition Region Between the Smooth and Rough Pipe Laws, J. Inst. Civ. Eng. Lond., 11, 133-156 (1938).

4. JONES, A.B. - Hydrodynamic Stability of a Boiling Channel, KAPL-2170, Knolls Atomic Power Laboratory, 1961.

5. STREETER, V.L. and WYLIE, E.B. - *Fluid Mechanics*, McGraw-Hill Book Company, New York, 1975.

6. COLLIER, J.G. - *Convective Boiling and Condensation*, McGraw-Hill Book Company, New York, 1972.

7. CHAPMAN, A.J. - *Heat Transfer*, Macmillan Publishing Company Inc., New York, 1974.

8. DAVIS, E.J. and ANDERSON, G.H. - The Incipience of Nucleate Boiling in Forced Convection Flow, AIChE Journal, $\underline{12}$, 774 (1966).

9. SAHA, P. and ZUBER, N. - Point of Net Vapor Generation and Vapor Void Fraction in Subcooled Boiling, Proceedings of the 5th International Heat Transfer Conference, Tokyo, $\underline{4}$, 1974, p. 175-179.

10. THOM, J.R.S., WALKER, W.M., FALLON, T.A. and REISING, G.F.S. - Boiling in Subcooled Water During Flow Up Heated Tubes or Annuli, Inst. Mechanical Engineers, $\underline{180}$, part 3c, 1966, p. 226-246.

11. CHEN, J.C. - Correlation for Boiling Heat Transfer to Saturated Liquids in Convection Flow, Ind. Eng. Chem. Process Design and Development, $\underline{5}$, 322 (1966).

12. GROENEVELD, D.C. and ROUSSEAU, J.C. - CHF and Post/CHF Heat Transfer: An Assessment of Prediction Methods and Recommendations for Reactor Safety Codes, NATO Meeting on Advances in Two Phase Flow and Heat Transfer, $\underline{1}$ (NATO ASI Series, Series E, No. 63), Nijhoff Publishers, 1983, p. 203-239.

13. BJORNARD, T.A. and GRIFFITH, P. - PWR Blowdown Heat Transfer, ASME Topical Meeting on Thermal and Hydraulic Aspects of Nuclear Reactor Safety, Atlanta, $\underline{1}$, 1977, p.17-41.

14. HETSRONI, G. - *Handbook of Multiphase Systems*, McGraw-Hill Book Company, New York, 1982.

15. PATANKAR, S.V. - *Numerical Heat Transfer and Fluid Flow*, McGraw-Hill Book Company, New York, 1980.

16. RICHARDS, D.J., HANNA, B.N., HOBSON, N. and ARDRON, K.H. - ATHENA: A Two-Fluid Code for CANDU LOCA Analysis, presented at the 3rd International Topical Meeting on Reactor Thermalhydraulics, Newport, R.I., USA (1985).

SOME EXCEPTIONAL NONLINEAR TEMPERATURE FIELD PROBLEMS AND THEIR SOLUTION BY THE FINITE ELEMENT METHOD

J.F.Stelzer and R.Welzel

Research Centre (Kernforschungsanlage, KFA), D-5170 Juelich, F.R.Germany

SUMMARY

This paper shows after a short mathematical survey the treatment of non-everyday temperature field problems which are characterized by heat radiation and temperature dependent heat sources. Furthermore, a phase change propagation case is calculated with a classical method and with temperature dependent heat sources to compare both schemes.

INTRODUCTION

About the solution of nonlinear temperature fields many publications exist. Very detailed Huebner and Thornton [1] deal with this matter. From the amount of contributions especially those of Hogge [2, 3, 4, 5, 7] deserve to be mentioned, and a paper of Bathe and Khoshgoftaar [6]. With this paper the authors continue their paper given in [8]. The presented examples could not be solved with the direct iteration but demanded the Newton Raphson iteration.

MATHEMATICAL ASPECTS

The presentation of formulas may remain restricted to the principles. For the calculation of a transient nonlinear temperature field for every time step from t_n to t_{n+1} the following recursion eq. must be solved for $\{T_{n+1}\}$

$$(\Theta[k(T_\Theta)] + \frac{1}{\Delta t}[C(T_\Theta)])\{T_{n+1}\} = (-(1-\Theta)[k(T_\Theta)] + \frac{1}{\Delta t}[C(T_\Theta)])\{T_n\} + \{F(T_\Theta)\} \quad . \tag{1}$$

Θ is a constant between 0.5 and 1, most frequent equal 2/3 (Galerkin scheme), [k] and [C] the conductivity matrix and the capacity matrix, $\{F\}$ a vector regarding heat sources, heat fluxes and emitted or absorbed heat radiation which all are temperature dependent. Besides $\{T_{n+1}\}$ also $\{T_\Theta\}$ is unknown which is the temperature vector at an intermediate time when 2/3 of the time interval are elapsed. It is calculated by the N.R. iteration (m is the iteration counter)

$$[J(T_\Theta^m)]\{\Delta T_\Theta^m\} = -\{r(T_\Theta^m)\} \tag{2}$$

$$\{T_\theta^m\} = \{T_\theta^{m-1}\} + \{\Delta T_\theta^m\} \tag{3}$$

with $\{r\}$ being the residuum of an eq. similar to eq.(1), however marching only from T_n to T_θ

$$\{r(T_\theta^m)\} = [\tilde{k}(T_\theta^m)]\{T_\theta^m\} - \{f(T_\theta^m)\}. \tag{4}$$

The matrix and the vector of the right hand side look like

$$[\tilde{k}(T_\theta)] = (\frac{1}{\theta \Delta t} [C(T_\theta) + k(T_\theta)])\{T_\theta\} \tag{5}$$

$$\{f(T_\theta)\} = \{F(T_\theta)\} + \frac{1}{\theta \Delta t} [C(T_\theta)]\{T_n\} . \tag{6}$$

The matrix $[J]$ in eq.(2) is called Jacobian or tangent matrix and contains the derived terms

$$[J(T_\theta^m)] = [k(T_\theta^m)] + [\Delta k(T_\theta^m)] - [\Delta f(T_\theta^m)] . \tag{7}$$

The matrix $[k]$ is added from the conductivity matrix $[k_\lambda]$, the heat transfer matrix $[k_\alpha]$ and the capacity matrix $[k_c]$

$$[k_\lambda] = \iiint_V [B]^T[\lambda][B] \, dV \tag{8}$$

$$[k_\alpha] = \iint_S \alpha\{N\}\{N\}^T \, dS \tag{9}$$

$$[k_c] = \frac{\rho c}{\theta \Delta t} \iiint_V \{N\}\{N\}^T \, dV \tag{10}$$

$[\lambda]$ being the matrix of the direction dependent conductivities, $[B]$ the temperature gradient interpolation matrix, $\{N\}$ the vector of the shape functions, α the heat transfer coefficient, V the volume of an element and S one of its surfaces.

The vector $\{F\}$ comprises the heat source vector $\{F_Q\}$, the heat flux vector $\{F_q\}$, the heat transfer vector $\{F_\alpha\}$, the heat radiation vector $\{F_r\}$

$$\{F_Q\} = \iiint_V \dot{Q}\{N\} \, dV \tag{11}$$

$$\{F_q\} = \iint_S \dot{q}_s\{N\} \, dS \tag{12}$$

$$\{F_\alpha\} = \iint_S \alpha T_a\{N\} dS \tag{13}$$

$$\{F_r\} = -\iint_S \sigma\epsilon T^4\{N\} \, dS \quad , \tag{14}$$

with $\dot{Q}$ being the heat source rate per volume, $\dot{q}$ the heat flux entering an element surface, σ the Stefan-Boltzmann radiation constant and ϵ the emission coefficient.

The *derived terms* of eq. (7) are as following

$$[\Delta k_{\lambda x}] = \iiint_V \frac{\partial\{N\}}{\partial x} (\frac{\partial\{N\}}{\partial x}^T \{T\})\{N \frac{d\lambda_x}{dT}\}^T \, dV \tag{15}$$

$$[\Delta F_Q] = \iiint_V \{N\}\{N\frac{d\dot{Q}}{dT}\}^T \, dV \tag{16}$$

$$[\Delta F_r] = -4\sigma\varepsilon \iint_S (\{N\}^T \{T\})^3 \{N\}\{N\}^T \, dS \tag{17}$$

$$[\Delta k_c] = \frac{1}{\theta \, \Delta t} \frac{1}{i} \frac{dc}{dT} \iiint_V \{N\}\{N\}^T \, dV \, [T_\theta] \tag{18}$$

$$[\Delta f_c] = \frac{1}{\theta \, \Delta t} \frac{1}{i} \frac{dc}{dT} \iiint_V \{N\}\{N\}^T \, dV \, [T_n] \,, \tag{19}$$

i being the number of nodes per element, $[T_\theta]$ and $[T_n]$ matrices with i rows and columns having in every column the node temperatures at the time t_θ or t_n, respectively.

The *Newton-Raphson iteration* is accomplished by calculating the expressions (4) and (7), inserting them into eq.(2) and solving eq.(2) for $\{\Delta T_\theta^m\}$. This temperature vector is added to the previous one according to eq.(3). The iteration finishes if $\{\Delta T_\theta^m\}$ is sufficient small.

TEMPERATURE DEPENDENT PROPERTIES

The physical properties in the above equations like thermal conductivity, λ, specific heat capacity, c, are to be expressed by functions, e.g. polynomials, see the book of Stelzer [9]. These and also the derivations hereof are to be written into the special subroutines of the software to which the program jumps with the just existing temperature vector. If there are temperature dependent heat sources the appropriate mathematical expressions must be put in a special subprogram, too, together with its derivation to the temperature. We will describe a typical case in the following.

TEMPERATURES IN AN ELECTRICAL RESISTANCE

A wire of tungsten is twisted to a coil as shown in Fig. 1. The ends are constrained to a constant temperature of 500 centigrade. The environment is vacuum. The coil surfaces send energy by heat radiation to corresponding surfaces of 20 centigrade. The radius of the coil is 3 cm and the wire radius is one eighth of that.

The temperature dependent thermal conductivity is put in by the algorithm

$$\lambda(T) = D(AT + B)/(T+C) \quad \text{in W/(mK)} \tag{20}$$

with A=0.207, B=124.14, C=161.511, D=418.68, T temperature in Kelvin, and the derivation by

$$\frac{d\lambda}{dT} = \frac{D(AC-B)}{(T+C)^2} \quad \text{in W/(m K}^2\text{)} \,. \tag{21}$$

The temperature dependence of the specific heat capacity of tungsten is linear

$$c(T) = E + FT \quad \text{in J/(kg K)} \tag{22}$$

with E=127.96, F=0.018561, T here in centigrade, and the derivation is simply

$$\frac{dc}{dT} = F \quad \text{in J/(kg K}^2\text{)} \,. \tag{23}$$

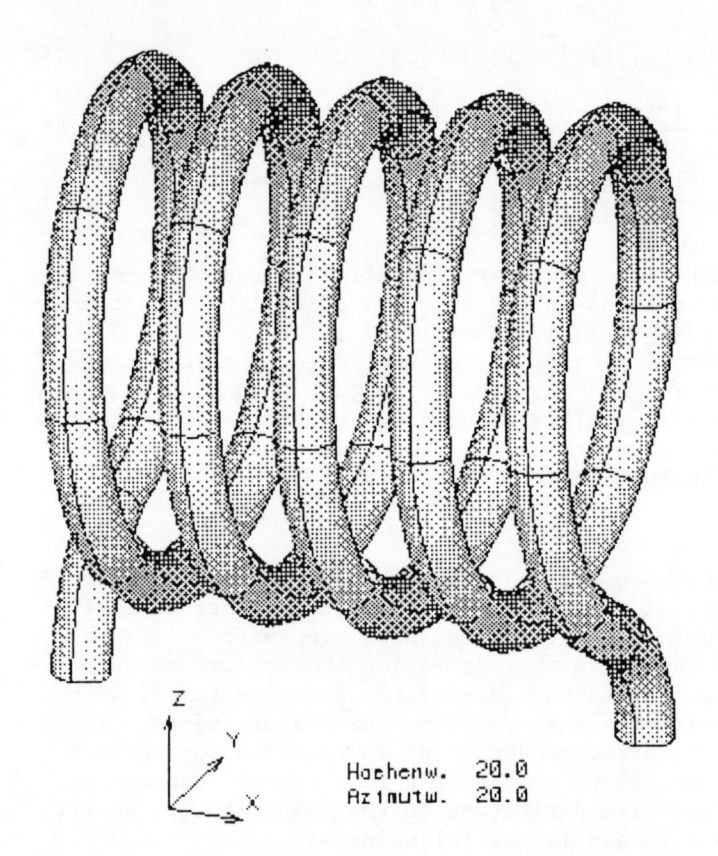

Fig.1
Model of an electrical resistance in coil form. The model consists of 42 spatial 20-node elements (Serendipity)

The electrical resistivity of tungsten shows also an approximately linear function of the temperature. The internal heat sources are coupled with the resistivity and the conducted current. For the temperature field calculation it is advantageous to assign the temperature dependence of the heat sources to the thermal conductivity, λ, which of course is related to the electrical resistivity by the law of Wiedemann, Franz and Lorenz. In our case we arrived at a relationship for the heat sources of

$$\dot{Q}(T) = \lambda(2 + 0.05T), \qquad (24)$$

T in centigrade. For the derived heat source term according to eq.(16) also the following expression is to be put in into the program

$$\frac{d\dot{Q}}{dT} = 0.05\lambda \quad . \qquad (25)$$

The density ρ = 19.3 g/cm^3 remains temperature invariant, as well as the emissivities of the emitting and absorbing heat radiators (ε=0.8). If the temperature dependence of the density should be respected then a coupling with a deformation calculation would be adequate since only then the volume changements due to the thermal expansion could be calculated correctly.

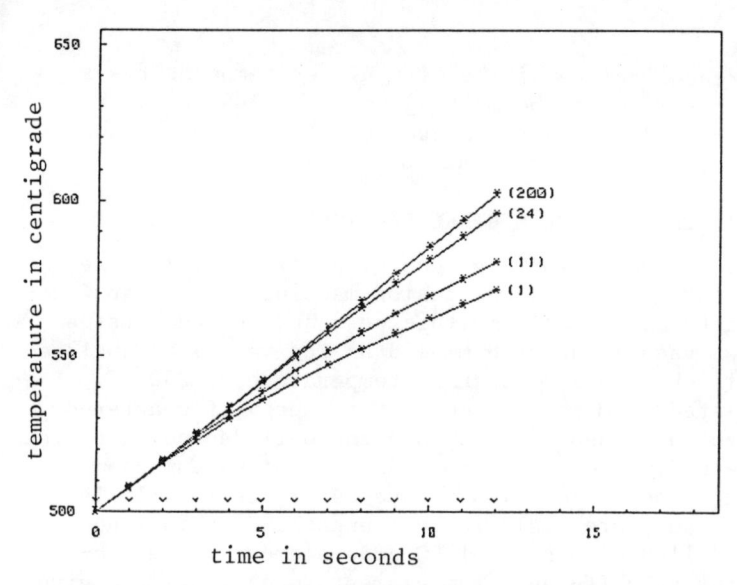

Fig.2
Time dependent temperature development at selected nodes during the first twelve seconds

Results. It is advisable first to execute a steady-state run to see where the final temperature is. Unfortunately, the problem did not converge indicating that no steady-state solution exists. Indeed, the transient solution shows, see Figure 2, that the temperature increases with about 8 K/s from the beginning. An intermediate temperature field is shown in Figure 3.

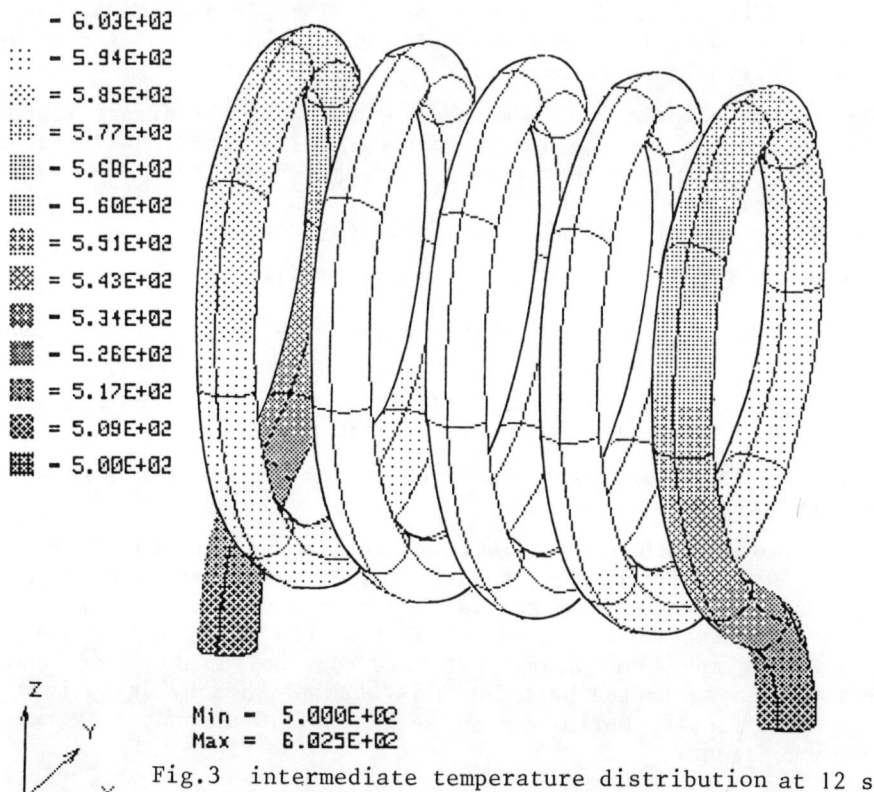

Fig.3 intermediate temperature distribution at 12 s

The slope decreases only a little with higher temperatures so that this coil will reach the melting point of 3650 K after about 10 minutes and be destroyed at the hottest spot in the middle region. The applied current is too high.

TEMPERATURES IN THE DUCT OF NEUTRAL PARTICLES INJECTORS

This is a conductor for high energy particles which are shot into the torus of a tokamak fusion machine. It is manufactured from tungsten sheet of 2 mm thickness and on the outside surrounded by a vacuum gap of 8 mm width. The vaccum is enclosed by a steel wall which has a constant temperature of 250 ^{o}C. The duct is heated from the inside by the energy of shattered particles which carry heat fluxes into the wall depending on the geometry between 0.01 to 0.43 W/mm^2. Across the vacuum gap heat is transferred by heat radiation to the outer steel wall. The steady state temperature field in the tungsten duct is sought. The duct is modelled with spatial 20-node elements. The prescibed heat fluxes, different from element to element, are given on the inside surfaces using the vector (12). The tungsten wall is modelled by one element layer, however, on the outside another element layer of 8 mm thickness is put on to simulate the vacuum gap. The outside nodes of this layer obtain prescribed temperatures (Dirichlet boundary conditions) of 250 ^{o}C to simulate the steel wall heat sink. Of course, the thermal conductivity of these vacuum elements is zero. The radiation across them is managed in form of a Stefan-Boltzmann transport term as was reported in reference [8].

During the calculation it turned up that with the direct iteration no convergence could be achieved. The Newton-Raphson iteration procedure was necessary. For accomplishing that, the temperature derivations of the transport term which substitutes the thermal conductivity had to be included, too:

transport term: $\lambda_r = C_r(T_1^3 + T_1^2 T_2 + T_1 T_2^2 + T_2^3)$ (26)

derivation to T_1: $\frac{d\lambda_r}{dT_1} = C_r(3T_1^2 + 2T_1 T_2 + T_2^2)$, (27)

with $C_r = \sigma/[(1/\varepsilon_1) + (1/\varepsilon_2) - 1]$, T_1 and T_2 being the hot and cool radiation partner surface temperatures (in Kelvin).

Now, the convergence proceeded very quick. *Results* are shown in Figure 4.

Another duct with narrowing cross section and fabricated from copper was analysed, too, see the result presentation of Figure 5. Here, the neutral particles occur in a pulsing way. The transient temperature field calculation was executed for one pulse. During the short heat input time the cooling does not become effective. Again the heat input is accomplished by using invading heat fluxes. During the pulse time of 10 seconds they remain constant.

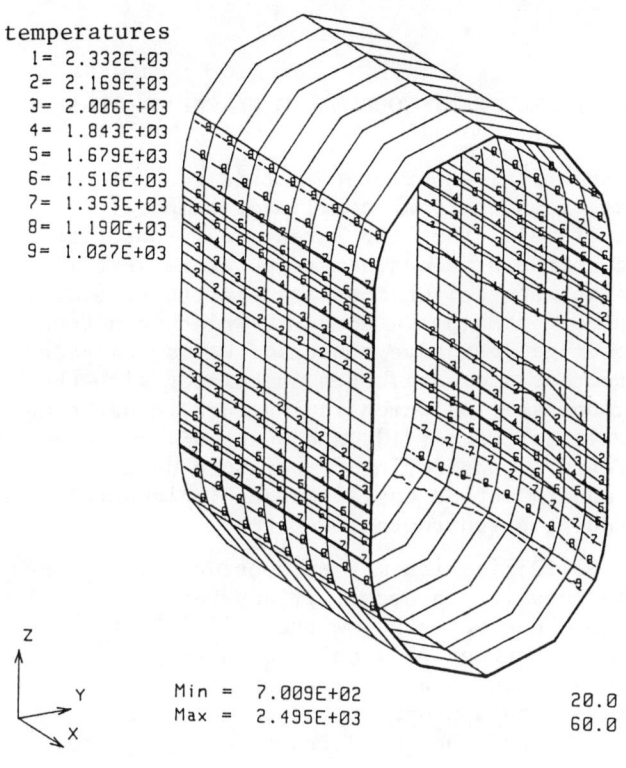

temperatures
1= 2.332E+03
2= 2.169E+03
3= 2.006E+03
4= 1.843E+03
5= 1.679E+03
6= 1.516E+03
7= 1.353E+03
8= 1.190E+03
9= 1.027E+03

Min = 7.009E+02
Max = 2.495E+03

20.0
60.0

DUCT

Fig.4 (left)
Duct from tungsten conducting high energy neutral particles, results of the temperature field calculation (outside elements representing the surrounding vacuum are not shown)

Fig.5 (below
A duct from copper exposed to a pulsing energy input; temperature field

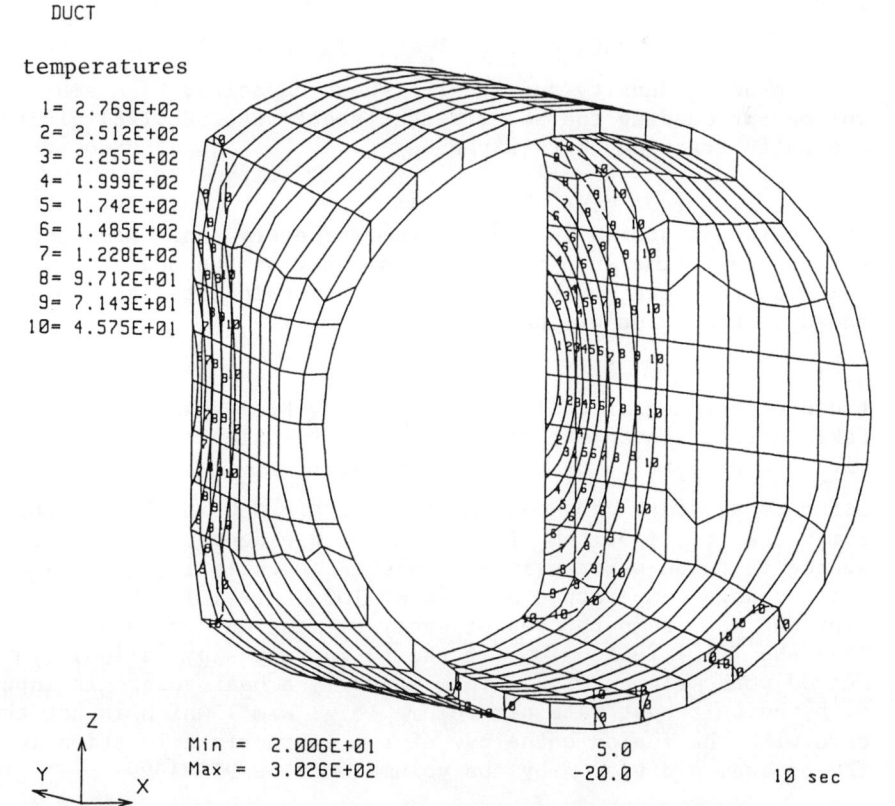

temperatures
1= 2.769E+02
2= 2.512E+02
3= 2.255E+02
4= 1.999E+02
5= 1.742E+02
6= 1.485E+02
7= 1.228E+02
8= 9.712E+01
9= 7.143E+01
10= 4.575E+01

Min = 2.006E+01
Max = 3.026E+02

5.0
-20.0

10 sec

USING TEMPERATURE DEPENDENT HEAT SOURCES TO CALCULATE PHASE CHANGE PROPAGATION

We executed a numerical experiment in this respect which suits to the context of exceptional non-linear temperature field problems. When a stuff melts then the latent heat is set free which causes a heat source at a certain temperature just like a temperature dependent source. On contrary, in the case of freezing the same amount of enthalpy is consumed which is a temperature dependent heat sink (negative source). We may compare the results of an appropriate case, once calculated with the method described by Ohnaka [10] which is included in our temperature field analysis software [11], and a second time computed with our newer feature considering temperature dependent heat sources. The Ohnaka method easily can be implemented into a program just by adding a subprogram.

The Ohnaka method. Solidification may be considered. The method observes the relationships in nodes. If anywhere in a cooling melt a node temperature decreases below the solidification temperature the difference between this calculated temperature and the phase change temperature is a measure for a fraction of frozen stuff, Δf_s. Between the appropriate heat energy, ΔQ_i, the temperature span $T_s - T_i$ and the solid fraction Δf_s the relationship exists

$$\Delta Q_i = \rho c V (T_s - T_i) = \rho r V \Delta f_s \ , \qquad (28)$$

ρ and c being density and specific heat capacity, V an ambient volume surrounding the node, r the latent heat. Eq.(28) gives the solid fraction in this region

$$\Delta f_s = c(T_s - T_i)/r \ . \qquad (29)$$

This value is stored in the program for every concerned node. For the next time step this node is reset on the phase change temperature. With every time step Δf_s is calculated anew and added to the value of the previous turn. If

$$f_s = \sum \Delta f_s = 1 \qquad (30)$$

the book keeping of that node is closed. From now on the node takes part again on the usual temperature calculating procedure, with the properties of the solid phase.

Using temperature dependent heat sources. Now in the subprogram where the algorithm of the temperature depence of the heat source is to be written in a simple IF check is inserted. The phase change temperature may lie at 100 degree. In the loop over the node temperatures of every element we implement: IF T>99 AND T<101 THEN Q=100 (if the fusion enthalpy is 100). This method works elementwise. Since normally a heat source is input with the unit heat rate per volume (e.g. W/m^3) which is not the case with the fusion enthalpy, at the appropriate location in the program a division by the volume must be provided.

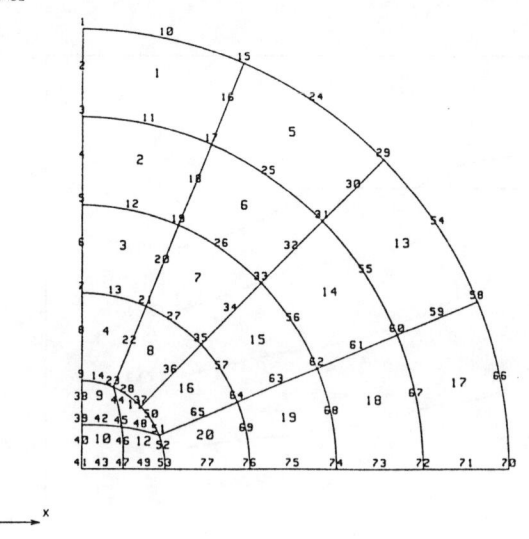

Fig.6 The mesh for the phase change calculation

Figure 7 shows a plane mesh which was used for the experiment. An overall start temperature of 120 degree is prescribed. Along the outer arc convection boundary conditions exist characterized by an ambient temperature of 5 degree and a heat transfer coefficient of 1 from where the structure is cooled. There is a invariant thermal conductivity of 1.5, the density is 2 and the specific heat capacity 4. At 100 deg the phase change takes place with a fusion enthalpy of 100. An intermediate temperature distribution is presented in Figure 7 where over every node of the projected mesh in the third dimension the temperature is drawn. The plateau of the region just in phase change can be seen. However, the better interpretation give pictures with the time dependent pattern of local temperatures. In Figure 8 the relationships of the Ohnaka algorithm are displayed and in Figure 9 those of the algorithm using the heat source method. The node point numbers are assigned (70 - 73, compare Fig.6). With the Ohnaka method the plateaus are more distinctly marked, and also the influence of a halted temperature development to the neighboured regions.

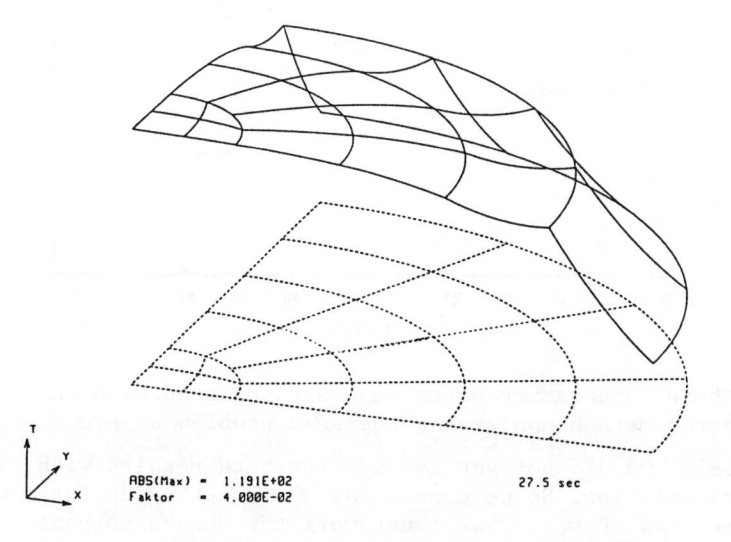

Fig.7 Intermediate temperatures in the mountain aspect

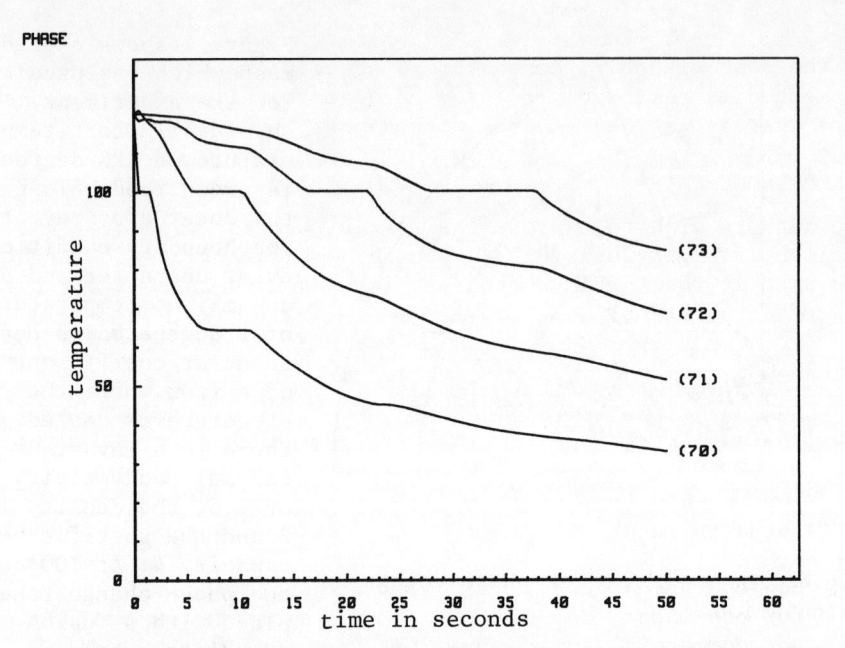

Fig.8 Temperatures vs. time for four nodes. Calculation took place with the Ohnaka method

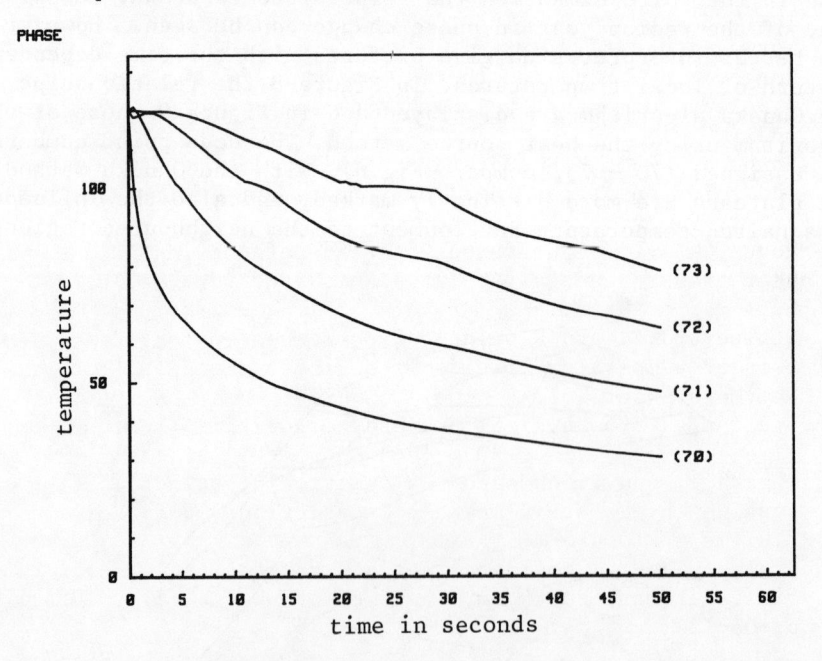

Fig.9 These temperature patterns are calculated with the heat source method concerning the same problem as Fig.8

Unfortunately, we did not yet compare the both results with an analytic calculation. So we cannot say which method delivers more exact results. However, we trust more the Ohnaka method.

In the case with no temperature dependent properties the Ohnaka method is remarkably quicker because iterations are unnecessary whereas the heat source method must iterate.

ACKNOWLEDGEMENTS

The authors wish to express their gratitude to A.Sievers, KFA Juelich, for very helpful assistance. The described features are part of the FEMFAM software which was used for solving the described problems.

REFERENCES

[1] Huebner,K.H. and E.A.Thornton, The finite element method for engineers, 2nd ed., New York, John Wiley & Sons (1982)
 2   Hogge, M.A., A comparison of two- and three-level integration schemes for nonlinear heat conduction, in: Num.Methods in heat transfer, edited by R.W.Lewis, New York, J.Wiley & Sons (1981)
[3] Hogge,M.A., Accuracy and cost of integration techniques for nonlinear heat transfer, in: Finite element methods in the commercial environment, ed.J.Robinson, Dorset (1978)
[4] Hogge, M.A. and C.Nyssen, Nonlinear heat transfer by finite element tangent conductivity methods, in Proceedings Int.Conf.FEM in nonlinear solid and struct.mech., Geilo, Norway, August 1977
[5] Hogge, M.A., Modelled thermal response of structures in fire environments, Appl.Math.Modelling Vol.1 (1977), p.319 -324
[6] Bathe,K.J. and M.Koshgoftaar, Finite element formulation and solution of nonlinear heat transfer, Nucl.Eng.& Design 51 (1979), p.385-401
[7] Hogge, M.A., Secant versus tangent methods in nonlinear heat transfer analysis, Int.J.Num.Meth.Engn.Vol.16(1980), p.51-64
[8] Stelzer, J.F. and R.Welzel, Experiences in nonlinear analysis of temperature fields with finite elements, Int.J. Num.Meth.Engn. Vol.24 (1987), 59-73
[9] Stelzer, J.F., Physical property algorithms, Munich, Hanser (1984)
[10] Ohnaka,I. and T.Fukusako, Calculation of solidification of castings by a matrix method, Transactions ISIJ, Vol.17 (1977), p.410-418
[11] Stelzer, J.F., Considerations and strategies in developing finite element software, Engn.Comput. Vol.1 (2), (1984), p.106-124

THE REDUCTION OF ROOM RADIANT AND CONVECTIVE EXCHANGE TO STAR BASED SYSTEMS

M.G.Davies, School of Architecture and Building Engineering, The University, Liverpool L69 3BX, United Kingdom

The traditional method of handling heat exchange in a room for design purposes was to suppose that all heat from the plant, radiative as well as convective, was input at a so called 'air temperature'. This is an evident misnomer since air temperature as such cannot drive longwave radiation as the model actually assumed. The 'environmental temperature' (t_{ei}) concept has been introduced in the UK to get round the difficulty. This paper presents an analysis to examine when an approach along these lines may be logically acceptable. The surface-to-surface system of radiant exchange is first reduced to a surface-to-star point (T_r) exchange by a least squares fit. It is then shown that the space averaged observable radiant temperature can be approximated by the value of the temperature generated at T_r when the radiant output from heating appliances and casual gains is taken to act at T_r. It is normally assumed that convective gains can be treated as though input at the space averaged air temperature, T_a. An equivalence theorem is then demonstrated so show that the two star system, based on T_r and T_a, can in certain well defined conditions be replaced by a single star system, centred on an index temperature, T_{ra}, ('rad-air' temperature). T_{ra} in fact serves as 'air temperature' in the old fashioned sense. t_{ei} is a form of T_{ra}. The model based on T_{ra} is workable, but is physically unattractive and a model which handles convection and radiation separately may provide a better design procedure.

1 INTRODUCTION

The four walls of a room together with its floor and ceiling can be regarded as a control volume. Outside it lies the fabric of the building which provides resistance to and storage for the flow of heat, the ventilation process and the external environment. Within it convective and longwave heat transfer mechanisms serve to move heat between the various sources - radiators, lighting, equipment and occupants - and the air and walls of the room. This article is concerned with the interplay of mechanisms within the control volume. Further, the treatment is presented at a level appropriate to design methods for sizing heating and cooling equipment in a room; these processes can be modelled with any degree of detail using a computer model, but such treatment is unnecessarily complicated in a design context.

The traditional standpoint is set out in for example the 1963 ASHRAE Guide and Data Handbook, Fundamentals and Equipment, and also the 1965 IHVE Guide. Calculations centred around a global room temperature T_i, say, termed 'air temperature'. T_i was taken to drive a flow of heat through a transmittance Eh_r+h_c (of about 8 W/m²K) to the inner surface of an outer wall at T_s, (Fig. 1) so T_i had radiative properties as well as convective properties, despite its name. T_i served too to drive the ventilation loss of heat and its value was also taken to estimate the comfort temperature in the room. Both radiative and convective components of all internal sources of heat, were taken to be input at T_i. The model provides a very coarse account of room heat transfer, but it remains the basis of most plant design calculations today.

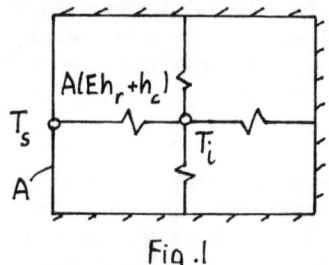

Fig. 1

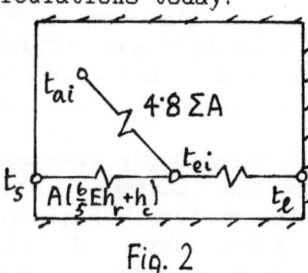

Fig. 2

In the UK in the 1960's a move was made to provide a better index temperature than T_i, which would take more explicit account of the longwave radiant exchange in a room. The concept of 'environmental temperature' t_{ei} was developed. It was a linear mix of mean air temperature t_{ai} and surface temperature t_m:

$$t_{ei} = (1/3).t_{ai} + (2/3).t_m$$

This model will be termed the 'environmental temperature model', (ETM, Fig. 2). (In discussing the ETM, temperature will be denoted by t, following the IHVE Guide notation. The author's thesis will be presented using T.) According to the model, t_{ei} drives a heat flow from the room as a whole to a surface, (internal or external), through a transmittance of $(6/5)Eh_r+h_c$. The ventilation loss is driven by the air temperature, t_{ai} and there is a conductance of $4.8\Sigma A$ (W/K) between t_{ei} and t_{ai}. ΣA is the total internal area of the room. (The conductance was later notated as $h_a\Sigma A$).

In the ETM, convectively input heat was taken to act at t_{ai}, as expected, but the longwave radiation Q_r from all sources was taken to act as the value 1½.Q_r at t_{ei}, and at the same time a quantity ½Q_r was taken to be *extracted* from t_{ai}. The comfort temperature – dry resultant temperature, t_c – was taken to be given by a node on the $h_a\Sigma A$ conductance with a value

$$t_c = (1/4).t_{ai} + (3/4).t_{ei}$$

This system was adopted by the UK Institution of Heating and Ventilating Engineers, (now the Chartered Institution of

Building Services Engineers, CIBSE), as its recommended procedure for conducting heating design calculations in the 1970 revision of its Guide, Section A5 [1]. It was revised in 1979 [2]. (The factor of (6/5) in the quantity $(6/5)Eh_r+h_c$ was given in the papers supporting [1] but not in [1] itself.)

The present author suspected from the outset that the reasoning used to arrive at t_{ei} was flawed and that its handling of radiant exchange was oversimplified. It was not immediately apparent however exactly where the reasoning broke down, nor what degree of simplification of radiant exchange was acceptable in a design context. These problems have now been solved. The present paper is intended to give an overview of how radiant exchange can be so simplified, how a model based on a global room temperature such as T_i can be arrived at logically, and where the shortcomings in the ETM lie.

2 THE BINARY STAR MODEL

In this section it is to shown how for design purposes, the convective and radiant exchange mechanisms can be modelled with good accuracy using two superposed star based thermal networks, one for convective and one for radiant exchange.

2.1 The convective network

The convective network is based on a central mean air temperature, T_a, whose value is arrived at by making measurements of air temperature at uniformly separated locations in the room, and averaging. There may be steadily sustained temperature differences between say floor and ceiling, but these are disregarded in the design model as far as global behaviour is concerned. We assume that we know the h_c values at the mean surface temperatures T_1, T_2, .. of the room, though we recognise that there is uncertainty in what values to select. The conductance linking T_a to the jth surface is then $A_j h_{cj}$. Heat input convectively to the room is taken to be input at the T_a node. A measuring device such as a thermometer is linked to T_a through a conductance $A_p h_{cp}$, where the subscript p denotes the thermometer or probe properties. The $A_p h_{cp}$ conductance is of course very, very much smaller that of a typical room conductance $A_j h_{cj}$.

This is the conventional model for convective exchange in a room. It is to be argued that radiation can be so manipulated that it can be handled in virtually the same way. This is not obvious, nor is it exact, but it is adequate for design purposes.

2.2 Radiation as a surface-surface exchange

Consider an empty room with six surfaces at mean temperatures T_1 to T_6. If they are blackbody, the direct radiant exchange between surfaces j and k is proportional to $\sigma(T_j^4-T_k^4)$ and on linearisation, this can be written as $4\sigma T_m^3 \cdot (T_j-T_k)$ or $h_r \cdot (T_j-T_k)$. h_r is about 5.7 W/m²K. The radiation conductance between these surfaces is given as

$A_j F_{jk} h_r$, where F_{jk} is the viewfactor between them and is related to room geometry. There are 15 such conductances if the six surface temperatures are specified. Fig. 3 shows the pattern for a four sided enclosure.

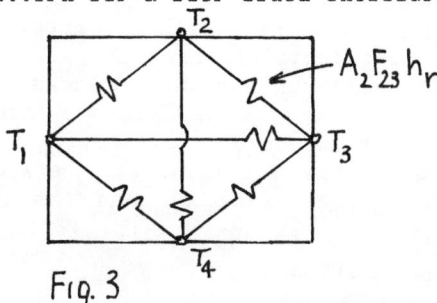

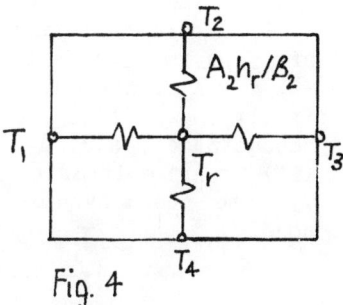

Fig. 3

Fig. 4

2.3 Radiation as a surface-star point exchange

The above system is too complicated for design use and the traditional and environmental temperature models both tacitly assume that radiation can be exchanged via a star point of some sort. The present author has attempted to give a logical foundation for this assumption [3].

If the surface j radiates to an enclosing blackbody surface at a uniform temperature, the radiant conductance is simply $A_j h_r$. If the surface 'radiates' to an intermediate node, the radiant star node T_r, (Fig. 4) its conductance will be greater than $A_j h_r$ and will be written as $A_j h_r / \beta_j$, where $\beta_j < 1$. We have to have some sort of logical procedure to find values of β such that, seen from the outside, the behaviour of the equivalent star circuit (Fig 4) is as like that of the parent (or delta) circuit (Fig. 3) as is possible.

To do this, we write down an expression for the net resistance R_{jk}^Δ between nodes j and k in the parent network when the other four nodes are taken to be adiabatic. (The direct resistance is $1/A_j F_{jk} h_r$. The net resistance can be found as the ratio of two determinants.) The resistance of the star network is simply

$$R_{jk}^* = \beta_j/A_j h_r + \beta_k/A_k h_r$$

If the difference $R_{jk}^\Delta - R_{jk}^*$ were zero the two circuits would be identical in their external effects. This cannot be achieved simultaneously for all 15 pairs of nodes, so we form the sum of the squares of the non dimensionalised differences,

$S = \Sigma\Sigma (1 - R_{jk}^*/R_{jk}^\Delta)^2$,   j = 1... 5,   k = j+1.... 6,

and by simultaneous adjustment of β_j, S can be minimised.

By examining a range of rectangular enclosures of all shapes, we find that β_j is largely determined by the ratio

$$f_j = A_j/(\text{total surface area})$$

Then $\beta_j \simeq 1-f_j-3.53(f_j^2-\tfrac{1}{2}f_j)+5.04(f_j^3-\tfrac{1}{4}f_j)$

with a standard deviation of 0.0070. The root mean square difference between the delta and star circuits is given as $(S/15)^{1/2}$ and this is for the most part less than 0.02. (Details are given in [3]).

Thus we conclude that the surface to surface or delta pattern which provides an exact description of the geometrical aspects of radiant exchange in a rectangular enclosure can, as far as its external effects are concerned, be replaced with good accuracy by a suitably designed surface to star pattern.

2.4 The emittance conductance

If a surface is not blackbody, but is grey with an emittance of ϵ_j we have to include two further features (Fig. 5):

(i) A blackbody equivalent node T_j' which replaces T_j itself as the termination of the geometrically based conductances, $A_j F_{jk} h_r$, or the single conductance $A_j h_r/\beta_j$.

(ii) An emittance based conductance $A_j \epsilon_j h_r/(1-\epsilon_j)$ between T_j, the thermodynamic temperature of surface A_j, and its T_j' node.

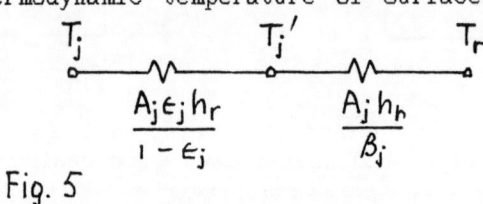

$$T_j \;\; \xrightarrow{\;\; \dfrac{A_j \epsilon_j h_r}{1-\epsilon_j} \;\;} \;\; T_j' \;\; \xrightarrow{\;\; \dfrac{A_j h_r}{\beta_j} \;\;} \;\; T_r$$

Fig. 5

T_j' is the linearised equivalent of radiosity in conventional radiant exchange theory. It has the property that all longwave radiation falling on the surface A_j is to be completely absorbed at T_j', not partly absorbed and partly reflected at T_j itself.

In the star based system, the emittance conductance is in series with the geometrical conductance and they can be combined as

$$[(A_j \epsilon_j h_r/(1-\epsilon_j))^{-1} + (A_j h_r/\beta_j)^{-1}]^{-1} \quad (= A_j E_j^* h_r, \text{ say})$$

2.5 The average radiant temperature

The temperature T_r is a fictitious temperature with no physically significance. It simply serves as a convenient device to model the *external* effect of the real radiant exchange. The radiant inputs should be taken to act at the several T_j' nodes but suppose for the moment that their total, Q_r, were to act at T_r. This is physically meaningless, but it is very easy to calculate T_r as an increment above wall temperature, which can be conveniently taken to be zero.

Let us set aside such fictions for the moment and consider what radiant effect can actually be observed in an enclosure, (taken as air free so as to avoid consideration of convection). Suppose we have an enclosure with blackbody surfaces all at a reference temperature of zero, (e.g $0°C$). A pure radiant source of strength Q_r is placed at its centre and the temperature is sensed by a probe of some kind at points over a uniform array of points within the room, (exactly as air temperature was supposed to be sensed). The average radiant temperature T_{avr} can be found from the local values. In relation to the labour of performing design calulations, evaluation of T_{avr} for a room of given dimensions is a laborious task

2.6 The two global radiant temperatures

We thus have two global measures of the radiant temperature in the room relative to its walls, due to the presence of an

internal radiant source:
(i) The radiant star temperature T_r, network based, fictitious but very easily evaluated,
(ii) The average radiant temperature, T_{avr}, physically based and on the same footing as T_a, but laborious to evaluate (and indeed, specific to situation).

The computations based on a wide variety of enclosure shapes show that T_{avr} tends to be a little larger - some 14% on average - than T_r though it varies less with shape. If the radiant source is placed at the wall of the enclosure - a more realistic position for a radiator - T_{avr} is reduced somewhat and it turns out that for practical design purposes, T_r provides a satisfactory estimate of the physical parameter, T_{avr}.

2.7 The radiant star model

The star model for radiant exchange thus consists in a series of conductances of type $A_j E_j * h_r$ linking the surface nodes T_j to the fictitious radiant star node T_r. The longwave radiant heat input can be treated as though input at T_r.

2.8 The binary star model

It was noted at the outset that the model for convective exchange consists in a series of conductances of type $A_j h_{cj}$ linking the surface nodes T_j to the physically based mean air temperature node T_a. Since the radiant and convective processes procede quite independently of each other within the enclosure and only interact at solid surfaces - those of the room itself, or of furnishings or sensors - a physically based model of the enclosure must consist of the two networks superposed so as to form a binary star pattern.

T_a and T_r denote respectively the average perceptible air and radiant temperatures in the enclosure. The resultant perceived temperature, whether by a sensor or an occupant must be a linear mix of the two. Dry resultant temperature will be

$$T_c = \tfrac{1}{2} T_a + \tfrac{1}{2} T_r$$

The T_c node is of course linked to these nodes by very small conductances, because of the very small dimensions of say a thermometer bulb.

3 THE RAD-AIR MODEL

It is the purpose of this section to show that the binary star model of the last section, in which convective and radiative processes are handled separately, can be transformed in restricted circumstances to a single star model centred on a node to be called the 'rad-air' node, T_{ra}, since it will be found to be a linear combination of T_a and T_r. In Section 4, it will be shown that the environmental temperature model is in most respects the same as the rad-air model.

It is convenient first to state an equivalence theorem.

3.1 The equivalence theorem

Consider a very simple thermal (or electrical) circuit (circuit A) comprised of three nodes, T_a, T_1 and T_r. A conductance C links T_a with T_1 and a conductance R links T_1 with T_r. (T_a and T_r will be given the meanings they had in Section 2, and C and R are to denote convective and radiant conductances, but this interpretation is not needed for the moment.) A heat flow Q_r is supposed input at T_r and there are heat losses of Q_f from T_1 and Q_v from T_a. If T_a is fixed in some way, it is an elementary calculation to find T_1 and T_r.

Fig. 6 - circuit A   **Fig. 7 - circuit B**

Consider now another circuit - circuit B - which consists of T_a, T_1 and C as in circuit A, but which lacks T_r, R and the heat input. Suppose that a node T_{ra} is located on C, so as to define conductances $(C+R).C/R$ to T_a and $(C+R)$ to T_1. Suppose that a heat input of $Q_r.(1+C/R)$ is input at T_{ra} and at the same time a flow of $Q_r.C/R$ is *extracted* from T_a.

It is easily shown that the temperature established at T_1 and the heat flows from T_1 and T_a are identical with those in circuit A. T_{ra} can of course be found. T_r does not appear explicitly in circuit B but its value can be constructed as

$$T_r = T_{ra}.(1+C/R) - T_a.C/R$$

Thus circuit B provides exactly the same information - values of T_1, T_r, Q_f and Q_v - as did the parent circuit A.

If there are further heat inputs in circuit A at T_a and T_1, they are simply included in circuit B without change.

This theorem can be used in connection with an elementary building model.

3.2 The model for a basic enclosure

We consider the most elementary building enclosure possible. See Fig. 8. It consists of an internal surface, all of whose area A is at a single uniform temperature T_1; the fabric provides a conductance F_1 between T_1 and the ambient

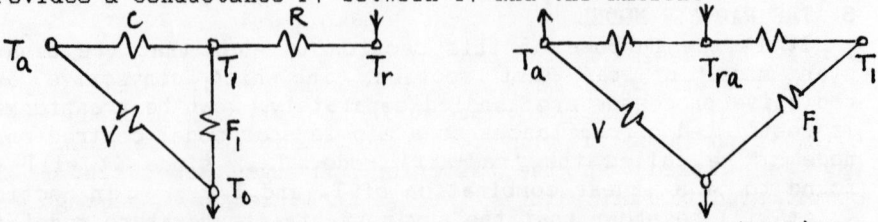

Fig. 8 - circuit A'   **Fig. 9 - circuit B'**

temperature of T_o. The air temperature T_a is linked to ambient by the ventilation conductance V and to T_1 by the convective conductance C_1, equal to Ah_c. A pure radiant source Q_r is present within the enclosure and its output is taken to act at T_r, linked to T_1 through the radiant conductance R_1. The thermal circuit for the enclosure (circuit A') is that of circuit A, together with the two loss mechanisms F_1 and V.

If the equivalence theorem is applied to circuit A', it transforms to circuit B' (Fig.) which consists of the following sequence of nodes and conductances:

T_a, $(C_1+R_1).C_1/R_1$, T_{ra}, (C_1+R_1), T_1, F_1, T_o, V, back to T_a.

In this configuration, we input $Q_r.(1+C_1/R_1)$ at T_{ra} and withdraw $Q_r.C_1/R_1$ from T_a. According to the theorem, the real temperatures T_a and T_1 have the same values as they did in circuit A', and of course T_r can be constructed from information provided by circuit B'.

3.3 Restrictions to the rad-air model

An enclosure consisting of a single isothermal surface is too idealised to be of any value. Suppose instead that we have an enclosure consisting of an outer wall of inside temperature T_1 and five internal surfaces at another uniform temperature T_2. T_a and T_r exist as before but we now have additional convective and radiant links C_2 and R_2. There will be a heat loss mechanism F_1 from T_1. There may or may not be a loss mechanism from T_2.

It can be shown - after some algebra - that if
$$C_2/R_2 = C_1/R_1 \quad (= \alpha, \text{ say})$$
the equivalence idea holds: We can replace T_r by a node T_{ra} which has links

C_1+R_1 to T_1,
C_2+R_2 to T_2 and
$(C_1+C_2+R_1+R_2).(C_1+C_2)/(R_1+R_2)$ or $\Sigma(C+R).\alpha$ to T_a.

An input of Q_r at T_r in the physical circuit can be replaced by an input of $Q_r.(1+\alpha)$ at T_{ra} together with an extract of $Q_r.\alpha$ from T_a in the equivalent circuit. T_a, T_1 and T_2 in circuit B' then have their circuit A' values and the value of T_r in circuit A' can be constructed from the values of T_a and T_{ra}.

We assume that the same result holds when the enclosure surface consists of three or more portions at different temperatures. If $C_2/R_2 \neq C_1/R_1$, the equivalence does not hold exactly.

3.4 Discussion of the rad-air model

The rad-air model is a single star model: it is centred on T_{ra}, which is linked to the surface j through a conductance C_j+R_j which lumps the convective and radiative mechanisms. T_j may be linked to the exterior through some simple or complicated thermal path. T_{ra} is linked to T_a through the conductance $(C+R).\alpha$, (C denotes ΣC_j and R denotes ΣR_j), and T_a is linked to the exterior by the ventilation conductance. As a working tool, the model has a number of points both to commend and to deprecate it.

In its favour we may note:
(i) The model retains the familiar U value concept for conduction losses. If wall j consists of a uniform slab of material, thickness d and conductivity λ, and an outside film coefficient of h_∞, its U value is given simply as

$$\frac{1}{A_j U_j} = \frac{1}{C_j + R_j} + \frac{d}{A_j \lambda} + \frac{1}{A_j h_\infty}$$

(ii) The conductance $(C+R).\alpha$ prevents a radiant input being too readily 'lost' by the ventilation process. (This will be explained more fully later.) This feature is useful in performing calculations on overheating due to solar gains. These are very quick and rough estimates, and it is sufficient to assume that all the solar gain is input at T_{ra}.

The model however contains a number of unattractive features:
(i) The conductances of type $C_j + R_j$ lump together totally unlike physical processes. In no real sense can convective and radiative energy fluxes be said to 'flow together' to a surface.
(ii) The $(C+R).\alpha$ conductance defies any kind of physical interpretation.
(iii) It is physically meaningless to input an augmented energy flux into some node and to extract part of it from a neighboring node.
(iv) The rad-air temperature can be given an interpretation of a kind: it a weighted mean of the average air and radiant star temperatures.

$$T_{ra} = \frac{C.T_a}{C+R} + \frac{R.T_r}{C+R}$$

but this is of formal rather than of substantive value.
(v) T_{ra} is not a generally accepted measure of comfort temperature. Comfort or dry resultant temperature is usually taken as

$$T_c = \tfrac{1}{2}.T_a + \tfrac{1}{2}.T_r$$

As such, it can be represented by a node on the $(C+R).\alpha$ conductance. This however is conceptually wrong. T_c is a measured or perceived temperature at a thermometer bulb or the human body, it is a local value, and the conductances linking T_c to the room are orders of magnitude smaller than the room conductances themselves. Thus T_c should be linked to T_a and T_r by very small conductances. The rad-air model does not include T_r explicitly and so is unable to provide a low conductance link to T_c. If T_c is placed on $(C+R).\alpha$, as it must be in the rad-air model, it is linked to the room by high conductances.

If a thermometer bulb is placed in a narrow beam of strong sunshine, it will register a high value, although the heating effect in the enclosure as a whole may be negligible. The binary star model can handle this situation but the rad-air model cannot.
(vi) It is easy in a design context to calculate the effect of a heat input at T_{ra} since all the paths from it are in parallel. If heat is input elsewhere, at T_a or T_1 say, some thermal circuit analysis is needed. To the research worker

this complication is trivial but to a thermal services engineer, for whom the intricacies of enclosure heat transfer are quite peripheral, the complication is a hindrance. A procedure has been developed in connection with the environmental temperature model - which is a form of rad-air model - in which the real heat input Q_1 at T_1 say can be scaled down somewhat to the value $F'.Q_1$; if the surface factor F' is chosen appropriately the reduced input $F'.Q_1$, if applied at T_{rm}, (ie, not at T_1 itself), has the same effect everywhere in the thermal circuit as does the application of Q_1 at T_1, except in the part of the circuit that contains T_1; the temperature at T_1 is underestimated and a post-scaling adjustment has to be applied to restore T_1 to its proper value.

This procedure works, but it is a messy operation, and it is unattractive since the procedure breaks the basic assumption of conservation of heat flow.

(vii) Finally, it may be remarked that in stating the equivalence theorem, T_r was replaced by T_{rm} and T_m was left intact. This represents an unsymmetrical handling of the elementary circuit. (We could have left T_r and replaced T_m.) This choice of transformation has the effect that when it is applied to the elementary enclosure, we can accommodate an external loss of heat from T_m - the ventilation loss in fact - but no longer an external loss from T_r, the radiation loss that would occur from an open window, for example. This does not matter in a cold climate where ventilation losses are important but where the effect of open windows can be ignored, but it limits the use of a rad-air like model to situations where the enclosure concerned has no open apertures.

4 EXISTING SINGLE STAR MODELS

Section 2 showed how enclosure heat transfer could be handled in a design context by keeping separate the convective and radiative heat transfer mechanisms and expressing them in the form of a binary star model. Section 3 showed how these mechanisms could be combined in a formal manner to form a one star model. In the present section, the status of the existing one star models, mentioned in Section 1, is to be examined.

4.1 The traditional model

The traditional model, it will be recalled, simply spoke of a 'room temperature' T_1, at which all heat was input and from which all heat was lost by conduction through the fabric and by ventilation. This situation can be derived from the binary model, simply by superposing the T_m and T_r nodes. It is obvious that this is only possible if T_m and T_r happened to be equal, and in general this is not so. The physical inappropriateness may be seen by noting that longwave radiation is handled as though input at T_r. The ventilation loss is driven from T_m. If T_m and T_r are superposed, the circuit would allow radiantly input heat to be lost directly by ventilation, without the necessary intervention of a solid surface. This is absurd. Thus the traditional model can only be a rather crude method of handling room heat exchange. Whether it is adequate in a design context is another matter. The author believes that it is adequate for many design purposes.

The radiant conductances of the traditional model were not set up correctly but that is better discussed in connection with the environmental temperature model.

4.2 The environmental temperature model

The environmental temperature model (ETM) is closely similar to the rad-air model. Unfortunately the ETM was set up on the basis of an oversimplification of radiant exchange, illegal reasoning in arriving at the concept of environmental temperature, and a misconception of what constituted 'mean radiant temperature'. The defects are fully discussed in [4]. The model consisted of a cubic enclosure with an outer surface, area A, emittance E, at a temperature t_s, five internal surfaces (5A) at t_l, (subscript l, not 1), and we are forced to assume that they have an emittance of unity - they are blackbody surfaces - although this was probably not intended. The radiant conductance between t_s and t_l was given as AEh_r, without distinguishing between the emittance and geometrical components of this quantity.

Environmental temperature t_{ei} itself was arrived at by finding that temperature which would drive the same heat flux to t_s as was physically driven by t_l and the air temperature t_{ai}. The argument included mention however of *mean surface temperature* which is irrelevant as far as the radiant exchange is concerned. Constancy of heat flux proves to be an insufficient principle with which to establish t_{ei}. It turns out that t_{ei} as it is defined is an absurd quantity, and the ETM, if one adheres rigidly to the logic by which it was set up, leads to some ridiculous conclusions. It can be shown that t_{ei} may 'depend' on the value of the emittance of a non-existant surface; furthmore, the model asserts that there will be no radiant exchange between floor and ceiling if the outer wall has zero emittance.

The radiant conductance between the surface at t_s and the t_{ei} node as expressed in the ETM is

$$R_{ETM} = (6/5)AEh_r$$

The correct value as given in the binary star model of Section 2 is

$$R_{BSM} = [(A\epsilon h_r/(1-\epsilon))^{-1} + (Ah_r/\beta)^{-1}]^{-1} = AE^*h_r$$

In the ETM, E is taken as 0.9. In the binary star model, we can chose ϵ arbitrarily, and we will take the same value, 0.9. For a cube, $\beta = 5/6$. So

$$R_{ETM} = 1.08\ Ah_r$$
$$R_{BSM} = 1.06\ Ah_r$$

and so the ETM value is near enough correct numerically, if not in principle. ($E^* = 1.06$ for one surface of a cube.)

With values of $h_c = 3$ W/m^{2}K and h_r of 5.7 W/m^{2}K we should have for a cubic enclosure,
$C/R = \Sigma C_j/\Sigma R_j = 6Ah_c/(6AE^*h_r) = 3/(1.06 \times 5.7) = 0.497 \simeq \frac{1}{2}$

Thus α or C/R is here equal to about ½ and a radiant input at T_r in the binary star model has to be replaced by an input of $Q_r \cdot (1+\alpha) = 1\frac{1}{2}.Q_r$ at T_{ra} and a withdrawal of $\frac{1}{2}.Q_r$ at T_a.
But these are precisely the values acting at the t_{ei} and t_{ai} nodes of the ETM. Thus *operationally* the ETM is a rad-air model. The value of t_{ei} as arrived at by performing the operations recommended in the CIBSE 1979 Guide Section A (p A5-8) is to be identified with T_{ra}. The value of t_{ei}(operational) is greater than, and so conflicts with, the value of t_{ei}(defined).
Thus the environmental temperature model is fundamentally a rad-air model, whose derivation has been marred by a number of logical flaws. It will have the strengths and weaknesses of the rad-air model listed in Section 3.4.

5 DISCUSSION

The traditional model of enclosure heat exchange can be understood at one level with little effort and it provides a procedure of sufficient accuracy to cover most heating design exercises. If the $4.8\Sigma A$ conductance of the ETM is incorporated empirically into the traditional model it makes it well suited to checking the likelihood of overheating due to solar gains.

If however the traditional model is felt to be dubious for some application, the binary star model provides a more flexible and accurate model, with very little extra computational cost. At one level, it too can be understood with little effort by a design engineer.

For those who wish, the logic of the binary star model can be followed in detail. If the designer wishes to conduct calculations using a single star model but with better accuracy than the traditional model can give, the rad-air model - or the ETM - is available. But to understand this, the designer must first follow through the theory of the binary model, and then come to terms with the equivalence theorem and its limitations, only to arrive at conceptually recondite model.

REFERENCES
1 IHVE Guide Book A, (mainly Section A5), Institution of Heating and Ventilating Engineers, London, 1971.
2 CIBS Guide, Section A5, 1979 Revision, Chartered Institution of Building Services Engineers, London, 1979.
3 M.G.Davies, Optimal designs for star circuits for radiant exchange in an enclosure, Building and Environment, **18**, 135-150, 1983.
4 M.G.Davies, A critique of the environmental temperature model, Building and Environment, **21**, 155-170, 1986.

HEAT TRANSFER CHARACTERISTICS OF SPLIT-FILM SENSOR

H. Yano; Dept. of Mechanical Engineering, Doshisha University, Kyoto 602, Japan

Y. Tanaka; Sanyo Electric Co., Ltd., Osaka 570, Japan

A. Kieda; Dept. of Mechanical Engineering, Doshisha University, Kyoto 602, Japan

SUMMARY

The heat transfer characteristics of a two-dimensional (dual-sensing) split-film sensor near a wall were investigated by means of a finite difference scheme with a numerical mapping based on Laplace equations. The model for computation is a constant-temperature circular cylinder placed in a laminar uniform shear flow at local Reynolds numbers Re of 0.5 to 65, where Re is based on the cylinder diameter and on-coming flow velocity to the cylinder. The near-wall characteristics were clarified in terms of Nusselt numbers and near-wall measuring errors involved in split-film anemometry in laminar boundary layers. Additionally, a comparatively good agreement was observed between computational and experimental results.

1. INTRODUCTION

Recently, split-film anemometry has attracted strong attentions in the field of two or three-dimensional flow measurements [1]. It is well known that this technique has many advantages over the conventional cross-wire probe anemometry; a wider angle of approach, better spatial resolution, less sensitivity to the flow along the sensor axis, and so on. However, the near-wall characteristics of split-film sensors have not yet been so understood as in the case of a single sensor hot wires [2], [3]. For example, Fig. 1 illustrates apparent velocity distribution u/u_∞ (the flow direction component) and v/u_∞ (the perpendicular component) measured by a split-film sensor in laminar boundary layer flow on an aluminum plate, in comparison with hot-wire data. As is clear, the measuring errors in u/u_∞ and v/u_∞ with the split-film

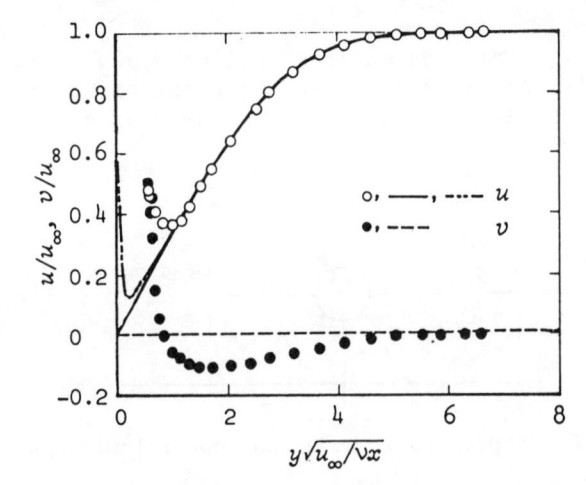

Fig. 1. Apparent velocity distributions measured by a split-film sensor and hot-wire in a laminar boundary layer flow of air on an aluminum plate; o, ● present experiment (split-film sensor, u_∞ = 1.16 m/s); ——, ---- Blasius solution; —·— Okui's Experiment [7] (hot-wire, u_∞ = 6.35 m/s).

sensor are not negligible, behaving differently from hot-wire data. Especially, v/u_∞ indicates a very interesting variation.

In this study, from the foregoing point of view, the wall effects on the heat transfer characteristics of a dual-sensing split-film sensor (composed of constant-temperature circular cylinder) placed in a laminar uniform shear flow on a conducting or non-conducting wall, were investigated by means of a finite difference scheme for the equations of vorticity stream-function and temperature at various shear flow gradients. From the view-point of numerical technique, SI scheme or strongly implicit scheme [4] was used for the vorticity transfer equation and energy equation, and SOR scheme for the Poisson equation.

In addition, an experiment was conducted in a laminar boundary layer flow developed on an aluminum wall to obtain the near-wall errors of the velocity components measured with a split-film sensor. And then, they were compared with the predicted values.

2. METHOD OF COMPUTATION

2.1 Computational model and governing equations

Fig. 2 shows the model for computation and co-ordinate system, where a constant-temperature circular cylinder of diameter d is placed in a uniform shear flow of velocity distribution $u = u_o y/y_o$, at a height of y_o from the wall. Here u_o denotes the approaching velocity to the center of the cylinder. Additionally, the temperature boundary conditions are such that $T = T_\infty$ at infinity, $T = T_o$ at the cylinder surface and $T = T_\infty$ at a conducting wall or $\partial T/\partial y = 0$ at a non-conducting wall. Additionally, the the plane of splits is supposed to be parallel to the wall or the x-axis.

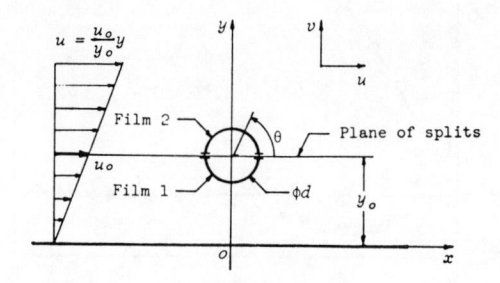

Fig. 2. Computational model and co-ordinate system.

The governing equations for the present problem can be expressed in terms of vorticity ζ, stream function ψ and temperature T, provided that the fluid is incompressible without any dissipation. Namely,

$$\frac{\partial \zeta}{\partial t} + \frac{\partial \psi}{\partial y}\frac{\partial \zeta}{\partial x} - \frac{\partial \psi}{\partial x}\frac{\partial \zeta}{\partial y} = \frac{1}{Re}\left(\frac{\partial^2 \zeta}{\partial x^2} + \frac{\partial^2 \zeta}{\partial y^2}\right) \qquad (1\text{-a})$$

$$\frac{\partial T}{\partial t} + \frac{\partial \psi}{\partial y}\frac{\partial T}{\partial x} - \frac{\partial \psi}{\partial x}\frac{\partial T}{\partial y} = \frac{1}{RePr}\left(\frac{\partial^2 T}{\partial x^2} + \frac{\partial^2 T}{\partial y^2}\right) \qquad (1\text{-b})$$

$$\frac{\partial^2 \psi}{\partial x^2} + \frac{\partial^2 \psi}{\partial y^2} = -\zeta \qquad (1\text{-c})$$

with $Re = u_o d/\nu$, $Pr = \nu/\kappa$,
where ν denotes the kinematic viscosity, and κ the thermal diffusivity of the fluid. Here, all variables in Eqs. (1-a) to (1-c) are non-dimensionalized as follows, and then the superscripts * are omitted hereafter. Namely,

$$x^* = \frac{x}{d}, \quad y^* = \frac{y}{d}, \quad t^* = \frac{t u_o}{d}, \quad \zeta^* = \frac{\zeta d}{u_o}, \quad \psi^* = \frac{\psi}{d u_o}, \quad T^* = \frac{T-T_\infty}{T_o-T_\infty}$$

In addition, the boundary conditions are as follows:
On the cylinder surface: $\psi = $ const., $\partial \psi/\partial n = 0$, $T = 1$.
On the wall: $\psi = 0$, $\partial \psi/\partial n = 0$, $T = 0$ (for conducting wall) or $\partial T/\partial n = 0$ (for non-conducting wall).
At infinity: $\psi = y^2 d/2y_o$, $\zeta = -d/y_o$, $T = 0$.
Here, n means the external normal to the boundary.

2.2 Grid generation

In the present finite difference technique, a boundary-fitted grid is generated numerically by solving an elliptic boundary-value problem, according to Thompson et al. [5]. More specifically, the numerical mapping of this type between the physical plane (x, y) and the computational plane (ξ, η) in Fig. 3, can be governed by

$$\frac{\partial^2 \xi}{\partial x^2} + \frac{\partial^2 \xi}{\partial y^2} = P(x, y), \quad \frac{\partial^2 \eta}{\partial x^2} + \frac{\partial^2 \eta}{\partial y^2} = Q(x, y) \quad (2\text{-}a), (2\text{-}b)$$

subject to proper Dirichlet boundary conditions. According to the usual grid generating technique, Eqs. (2-a) and (2-b) are transformed into

$$\alpha \frac{\partial^2 x}{\partial \xi^2} - 2\beta \frac{\partial^2 x}{\partial \xi \partial \eta} + \gamma \frac{\partial^2 x}{\partial \eta^2} = -J^2 \left(P \frac{\partial x}{\partial \xi} + Q \frac{\partial x}{\partial \eta} \right) \quad (3\text{-}a)$$

$$\alpha \frac{\partial^2 y}{\partial \xi^2} - 2\beta \frac{\partial^2 y}{\partial \xi \partial \eta} + \gamma \frac{\partial^2 y}{\partial \eta^2} = -J^2 \left(P \frac{\partial y}{\partial \xi} + Q \frac{\partial y}{\partial \eta} \right) \quad (3\text{-}b)$$

$$\alpha = \left(\frac{\partial x}{\partial \eta}\right)^2 + \left(\frac{\partial y}{\partial \eta}\right)^2, \quad \beta = \frac{\partial x}{\partial \xi}\frac{\partial x}{\partial \eta} + \frac{\partial y}{\partial \xi}\frac{\partial y}{\partial \eta},$$

$$\gamma = \left(\frac{\partial x}{\partial \xi}\right)^2 + \left(\frac{\partial y}{\partial \xi}\right)^2, \quad J = \frac{\partial x}{\partial \xi}\frac{\partial y}{\partial \eta} + \frac{\partial x}{\partial \eta}\frac{\partial y}{\partial \xi}$$

In the present method, $P = Q = 0$ is assumed, and boundary collocation points are chosen properly in the physical plane regularly in the computational plane; M points on Γ_1 (the boundary ABC) and Γ_1' (the boundary A'B'C'), M points on Γ_3 (the boundary EFG) and Γ_3' (the boundary E'F'G'), N points on Γ_2 (the boundary CDE) and Γ_2' (the boundary C'D'E'), and N points on Γ_4 (the boundary GHA) and Γ_4' (the boundary G'H'A').

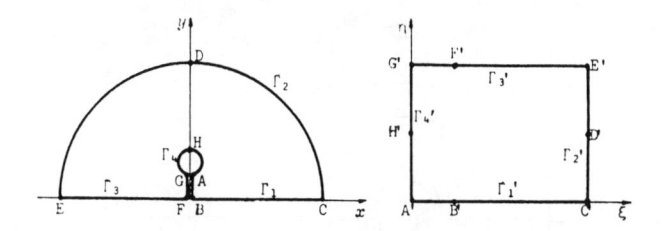

Fig. 3. Numerical mapping.

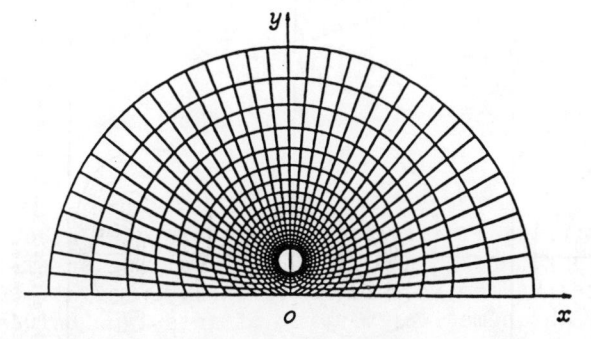

Fig. 4. Grid generation by numerical mapping.

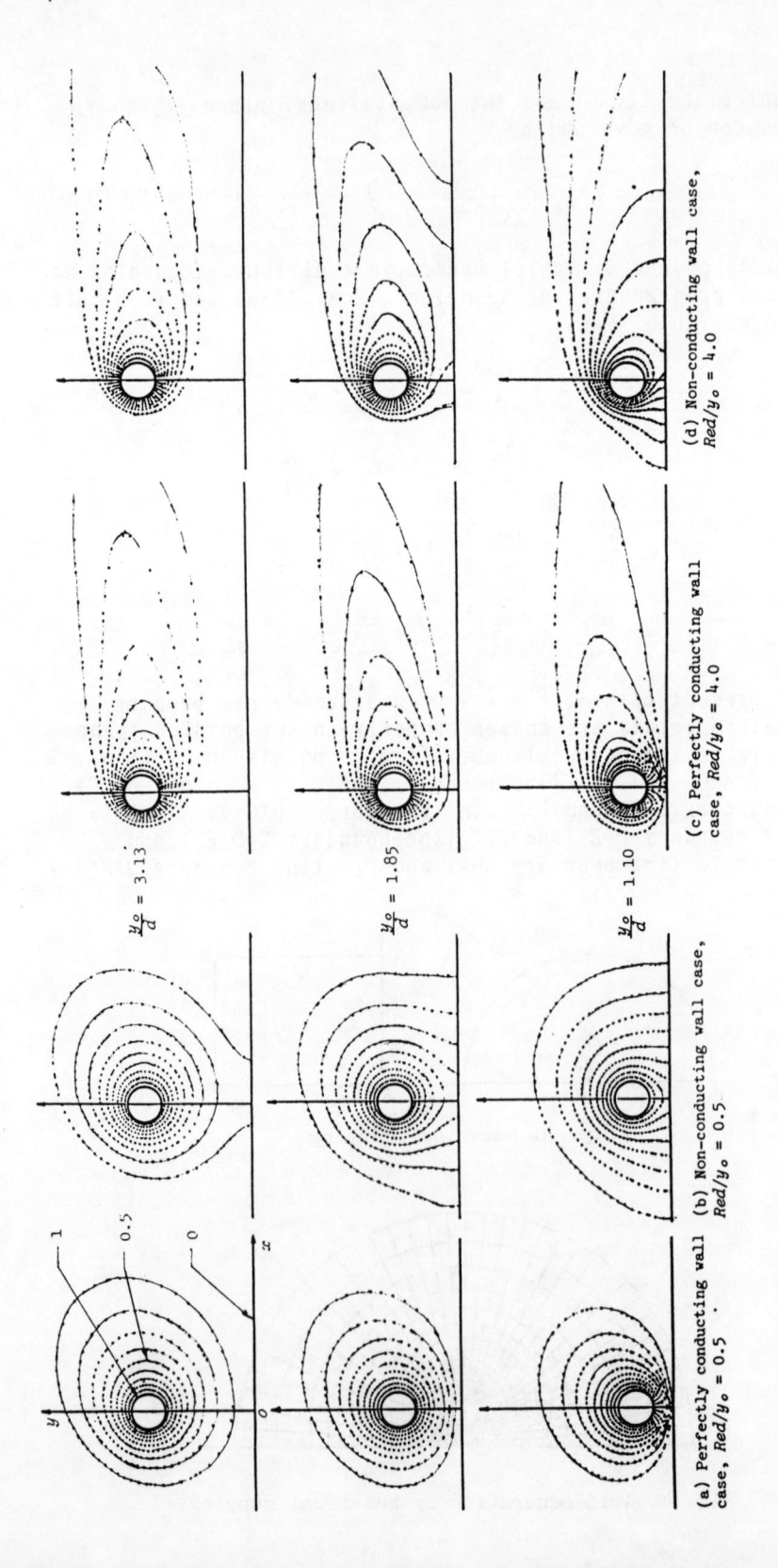

Fig. 5. Temperature distributions in isotherms.

With the virtual outer boundary CDE (Γ_2) in the physical plane being a semi-circle of a radius of more than $50d$ with $M = 36$ and $N = 48$, Eqs. (3-a) and (3-b) are solved by SOR scheme, with an example shown in Fig. 4 with M and N much less than those used in practical computations. As is obvious, this type of boundary-fitted grid is not an orthogonal one.

With this mapping, Eq. (1) can be rewritten as

$$\frac{\partial \zeta}{\partial t} + \frac{1}{J}\frac{\partial \psi}{\partial \xi}\frac{\partial \zeta}{\partial \eta} - \frac{1}{J}\frac{\partial \psi}{\partial \xi}\frac{\partial \zeta}{\partial \eta} = \frac{1}{ReJ^2}\left(\alpha\frac{\partial^2 \zeta}{\partial \xi^2} - 2\beta\frac{\partial^2 \zeta}{\partial \xi \partial \eta} + \gamma\frac{\partial^2 \zeta}{\partial \eta^2}\right) \quad (4\text{-}a)$$

$$\frac{\partial T}{\partial t} + \frac{1}{J}\frac{\partial \psi}{\partial \xi}\frac{\partial T}{\partial \eta} - \frac{1}{J}\frac{\partial \psi}{\partial \xi}\frac{\partial T}{\partial \eta} = \frac{1}{RePrJ^2}\left(\alpha\frac{\partial^2 T}{\partial \xi^2} - 2\beta\frac{\partial^2 T}{\partial \xi \partial \eta} + \gamma\frac{\partial^2 T}{\partial \eta^2}\right) \quad (4\text{-}b)$$

$$\alpha\frac{\partial^2 \psi}{\partial \xi^2} - 2\beta\frac{\partial^2 \psi}{\partial \xi \partial \eta} + \gamma\frac{\partial^2 \psi}{\partial \eta^2} = -J^2 \zeta \quad (4\text{-}c)$$

3. NUMERICAL DISCUSSIONS AND COMPARISON WITH EXPERIMENTAL DATA

Eqs. (4-a) to (4-c) were solved in time-marching process to obtain a steady-state solution; Eqs. (4-a) and (4-b) by SI scheme or strongly implicit scheme [4], and Eq. (4-c) by SOR scheme with a relaxation factor of 1.65, using the following parameters:

Prandtl number $Pr = 0.72$.
Dimensionless shear flow gradient $Red/y_o = 0.5$ to 9.5.
Dimensionless distance from wall $y_o/d = 1.10$ to 6.85.
Time division $\Delta t = 0.001$ to 0.08.
Grid parameter $M = 36$ and $N = 48$.

And the tolerances for iterations at each time step were taken as $\Delta \zeta < 10^{-3}$, $\Delta T < 10^{-3}$ and $\Delta \psi < 10^{-4}$, and that for time steps for steady-state solution as $\Delta T < 10^{-5}$. Various computations were made on a HITAC M-280H computer in single precision, taking a time of nearly 200 sec. (corresponding to nearly 1800 iterations) for the case of $Red/y_o = 1.0$, $y_o/d = 3.13$ and $\Delta t = 0.004$ (for example).

Fig. 5 shows the computed temperature distributions in isotherms around the heated cylinder at various heights with different shear flows of $Red/y_o = 0.5$ and 4.0 on perfectly conducting and non-conducting walls. It is clear that the thermal conductivity of the wall considerably affects the temperature distribution especially when the cylinder is closer to the wall in weaker shear flows; the conducting wall gives a greater wall effect on the heat transfer from the cylinder than the non-conducting wall does. Furthermore, at $y_o/d = 3.13$ with $Red/y_o = 4.0$, the conducting wall data and non-conducting wall data are almost the same, though having a slight asymmetry with respect to the horizontal diameter of the cylinder, occurring due to the shear flow itself.

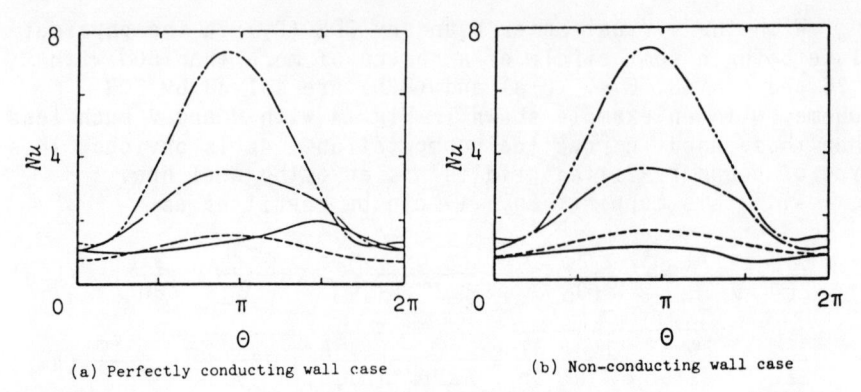

Fig. 6. Distributions of local Nusselt number Nu; —— Red/y_o = 0.5, y_o/d = 1.10; --- 0.5, 5.28; —·— 9.5, 1.10; —··— 9.5, 5.28.

(a) Perfectly conducting wall case   (b) Non-conducting wall case

Fig. 6 indicates the results of local Nusselt number distributions Nu on the sensor surface at dimensionless shear flow gradients Red/y_o = 0.5 and 9.5 with dimensionless distances of sensor y_o/d = 1.10 and 5.28; Fig. 6-(a) for the case of perfectly conducting wall, and Fig. 6-(b) for the case of non-conducting wall. Here the local Nusselt number is defined as

$$Nu = d\frac{\partial T}{\partial n}\bigg|_s \qquad (5)$$

with n indicating the external normal to the sensor surface, and "s" the sensor surface itself. There are not so much differences between the perfectly conducting wall data non-conducting wall data, except for the case of Red/y_o = 0.5 with y_o/d = 1.10 as is expected from the plots of isotherms in Fig. 5.

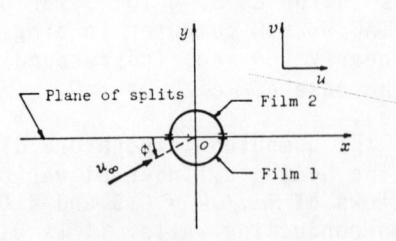

Fig. 7. Computational model for the split-film sensor placed in uniform flow.

Before discussing the near-wall heat transfer characteristics of the split-film sensor, it is necessary to obtain its characteristics in a uniform flow. A finite-difference computation for the model in Fig. 7 was conducted in a similar way as before, on an orthogonal grid in the physical plane shown in Fig. 8.

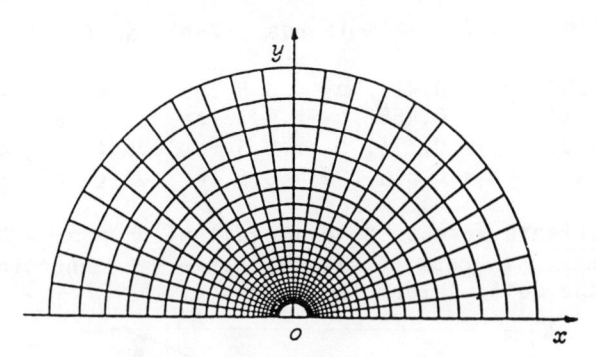

Fig. 8. Conformal mapping.

The results can be expressed in terms of mean Nusselt numbers in the form:

$$\overline{Nu}_m = \frac{\overline{Nu}_1 - \overline{Nu}_2}{2}, \quad \Delta \overline{Nu}_m = \overline{Nu}_1 - \overline{Nu}_2 \qquad (6\text{-a}), (6\text{-b})$$

where

$$\overline{Nu}_1 = \frac{1}{\pi}\int_{\pi}^{2\pi} Nu d\theta, \quad \overline{Nu}_2 = \frac{1}{\pi}\int_{0}^{\pi} Nu d\theta$$

$\overline{Nu}_1$ and $\overline{Nu}_2$ can be considered as the mean Nusselt numbers for the film 1 and film 2 respectively. It is well known that the variables $\overline{Nu}_m$ and $\Delta \overline{Nu}_m$ depend on the Reynolds number Re and the angle of approach ϕ of the uniform flow, as follows:

$$\overline{Nu}_m = a_1 + b_1 Re^{\alpha_1} \qquad (7\text{-a})$$

$$\Delta \overline{Nu}_m = f(Re)\sin\phi, \quad |\phi| \leq \pi/3 \qquad (7\text{-b})$$

with $f(Re) = a_2 + b_2 Re^{\alpha_2}$
where $Re = u_\infty d/\nu$

The results of computation of $\overline{Nu}_m$ and $f(Re)$ with $Pr = 0.72$ (for the assumed film temperature $T_f = 360$ K and uniform flow temperature $T_\infty = 290$ K), are shown in Fig. 9 in comparison with the Collis' experimental formula [6]

$$\overline{Nu} = (0.24 + 0.56 Re^{0.45})\left(\frac{T_f}{T_\infty}\right)^{0.17}, \quad 0.02 \leq Re \leq 44 \qquad (8\text{-a})$$

$$\overline{Nu} = 0.48 Re^{0.51}\left(\frac{T_f}{T_\infty}\right)^{0.17}, \quad 44 \leq Re \leq 140 \qquad (8\text{-b})$$

As is obvious from Fig. 9, the present results of $\overline{Nu}_m$ agree very well with the Collis' experimental formula over the entire computational range of $Re = 1$ to 60, thus insuring the effectiveness of the present finite difference scheme. Then, we obtained the coefficients in Eqs. (7-a) and (7-b) as

$$\begin{aligned} a_1 &= 0.500, & b_1 &= 0.417, & \alpha_1 &= 0.5 \text{ for } 0.5 < Re \le 4 \\ a_1 &= 0.424, & b_1 &= 0.455, & \alpha_1 &= 0.5 \text{ for } \phantom{0.5<} 4 \le Re < 65 \\ a_2 &= -0.291, & b_2 &= 0.471, & \alpha_2 &= 0.5 \text{ for } 0.5 < Re \le 4 \\ a_2 &= -0.455, & b_2 &= 0.553, & \alpha_2 &= 0.5 \text{ for } \phantom{0.5<} 4 \le Re < 65 \end{aligned} \quad (9)$$

These coefficients were so determined that $\alpha_1 = \alpha_2 = 0.5$ in a practical sense, thereby causing a change of connecting point $Re = 44$ in the Collis' formula to $Re = 4$.

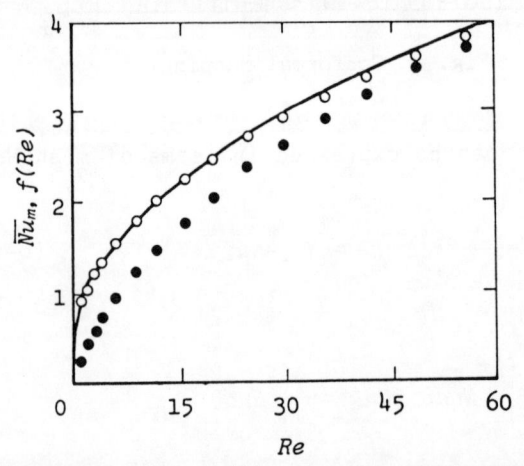

Fig. 9. Re dependencies of $\overline{Nu}_m$ and $f(Re)$ in uniform flow case; $Pr = 0.72$; o present results of $\overline{Nu}_m$, ● present results of $f(Re)$, —— Collis's formula [6] for $\overline{Nu}_m$ with $T_\infty = 290$ and $T_f = 360$ K.

(a) Perfectly conducting wall case

(b) Non-conducting wall case

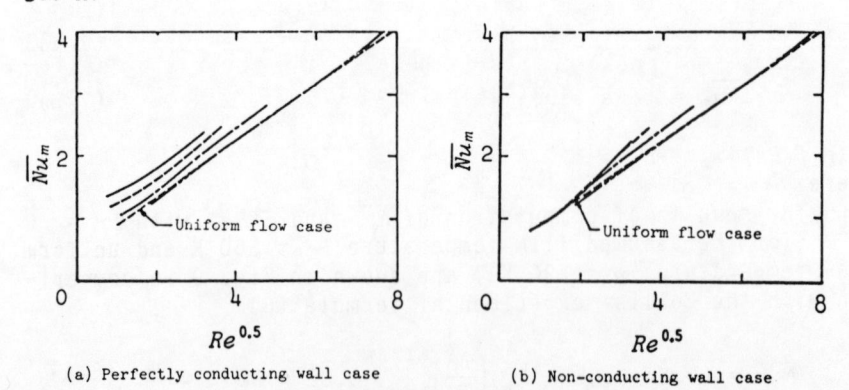

Fig. 10. Relation between mean Nusselt number $\overline{Nu}_m$ and Reynolds number Re; $Pr = 0.72$; —— $y_o/d = 1.10$, --- 1.42, -·- 2.42, -··- 6.85.

Fig. 10 illustrates near-wall characteristics of the sensor, the relationships between the mean Nusselt number $\overline{Nu}_m$ and Re with a parameter of y_o/d; Fig. 10-(a) for the case of a perfectly conducting wall, and Fig. 10-(b) for the case of a non-conducting wall. The general tendency in Fig. 10-(a) is the same as that of Wills [2] who conducted an experiment to study the metallic wall effect on hot wires for $Re \leq 1$, and also the same as that of Okui et al. [7] who performed a computation for the same purpose. Whereas, Fig. 10-(b) indicates an interesting feature that the less the local Reynolds number Re, the less becomes the wall effect on $\overline{Nu}_m$, or that the thermal conductivity effect of the wall appears at lower Reynolds numbers.

(a) Perfectly conducting wall case

(b) Non-conducting wall case

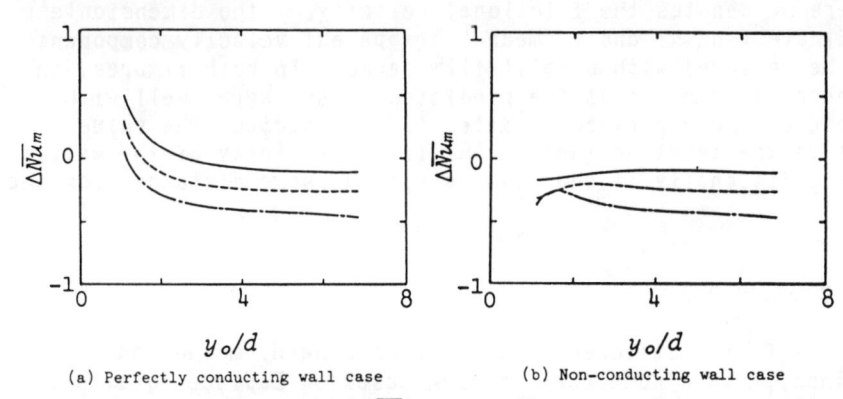

Fig. 11. Relation between $\Delta\overline{Nu}_m$ and y_o/d; $Pr = 0.72$; —— $Re d/y_o$ = 0.5, --- 2.0, —·— 5.0.

Fig. 11 shows the dependencies of $\Delta\overline{Nu}_m$ on the dimensionless distance y_o/d with a parameter of dimensionless shear flow gradient $Re d/y_o$; Fig. 11-(a) for the perfectly conducting wall case, and Fig. 11-(b) for the non-conducting wall case. The wall effect is remarkable in the case of the conducting wall. And in both plots, it seems that $\Delta\overline{Nu}_m$ dose not approach zero as y_o/d increases; this is because the shear flow still exists at infinity in the present model.

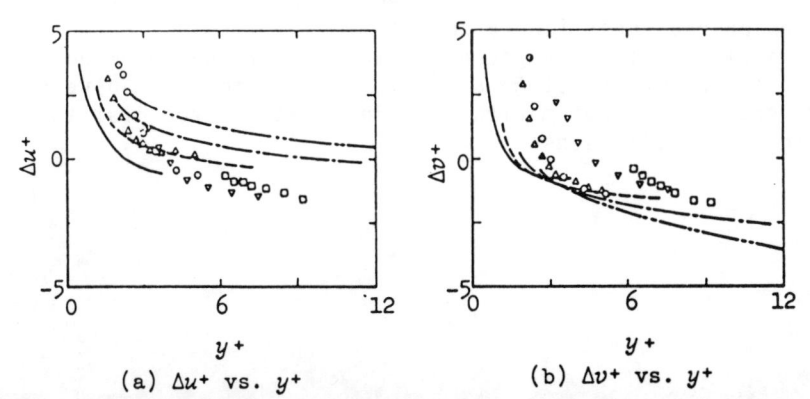

(a) Δu^+ vs. y^+

(b) Δv^+ vs. y^+

Fig. 12. Comparison between predicted and experimental values of Δu^+ and Δv^+ in the vicinity of conducting wall; —— $Re d/y_o$ = 0.5 (calculated), --- 2.0, —·— 5.0, —··— 9.5; O $Re d/y_o$ = 0.68 (experimental), △ 1.85, ▽ 4.59, □ 12.56.

Lastly, the predicted measuring errors Δu^+ and Δv^+ for the perfectly conducting wall case are presented against y^+ in Fig. 12, in comparison with experimental data obtained in laminar boundary layers on an aluminum wall. Here,

$$\Delta u^+ = \frac{u_a - u}{u_\tau}, \quad \Delta v^+ = \frac{v_a - v}{u_\tau}$$

with $y^+ = y d u_\tau / \nu$
where u_τ denotes the frictional velocity, y the dimensionless ordinate, and u_a and v_a mean the apparent velocity components to be measured with a split-film sensor. In both figures, the general tendencies of the predicted values agree well with those of the experimental date. As is expected, the value of Δv^+ is greater than that of Δu^+ in the vicinity of the wall, and additionally it changes the sign at some distance from the wall.

4. CONCLUSIONS

A finite difference scheme combining SI method and SOR method, with a numerical mapping based on Laplace equations, was applied to investigate the wall effect on the heat-transfer characteristics of a split-film sensor at Re = 0.5 to 65. As a result, it was clarified that the thermal conductivity of the wall affects the heat-transfer coefficient of the sensor more and more as the sensor comes closer to the wall, especially at lower Reynolds numbers. And we found that large measuring errors Δv^+ occur in the vicinity of a conducting wall, with positive errors very close to the wall ($y^+ < 2.5$), and negative error relatively apart from the wall ($2.5 < y^+ < 16$) within the range of the presented shear flow gradients. There are also some measuring errors Δu^+ in a close vicinity of a conducting wall ($y^+ < 10$), which depend somewhat on the dimensionless shear flow gradient.

Additionally, the present scheme should be improved to cover a wider range of Re and y_o/d, and further to solve a more practical problem with a uniform heat source distribution in the split-film sensor.

REFERENCES

1 STOCK, D.E., MICHENER, T.E., SEETHARAM, H.C. and MACK, M.D. - 'Measurement in Three-Dimensional Flows with a Dual-Dual Split-Film Probe', Symposium on Turbulence, 8th, Ed. Reed, X B, Patterson, G.K. and Zakin, J.L., Dept. of Chemical Eng., Univ. of Missouri-Rolla, 1984, pp. 197-205.

2 WILLS, J.A.B. - The Correction of Hot-Wire Readings for Proximity to a Solid Boundary, J. Fluid Mech., 12, 388-396 (1962).

3 BHATIA, J.C., DURST, F. and JOVANOVIC, J. - Corrections of Hot-Wire Anemometer Measurements near Walls, J. Fluid Mech., 122, 411-431 (1982).

4 LIN, C.L., PEPPER, D.W. and LEE, S.C. - Numerical Methods for Separated Flow Solutions around a Circular Cylinder, AIAA Journal, 14-7, 900-907 (1976).

5 THOMPSON, J.F., THAMES, F.C. and MASTIN, C.W. - Automatic Numerical Generation of Body-Fitted Curvilinear Co-ordinate System for Field Containing Any Number of Arbitary Two-Dimensional Bodies, J. Comp. Phys., 15, 299-319 (1974).

6 COLLIS, D.C. and WILLIAMS, M,J. - Two-Dimensional Convection from Heated Wires at Low Reynolds Numbers, J. Fluid Mech., 6, 357-384 (1959).

7 OKUI, K. and MIKAMI, F. - On the Reading Errors of Hot-Wire Close to the Wall, Trans. JSME, B-50, 2738-2743 (1984), in Japanese.

NUMERICAL PREDICTION OF TEMPERATURE DISTRIBUTIONS IN A STEEL
BAR DURING HOT ROLLING

J.A. Visser and E.H. Mathews

Department of Mechanical Engineering
University of Pretoria, Pretoria, 0002, R.S.A.

SUMMARY

Knowledge of the temperature distribution in a steel bar at any position in the hot rolling process is useful when the process is to be optimised. This paper presents a finite difference simulation of the temperature distribution. The numerical procedure accounts for thermal radiation from the bar, convection heat transfer to the atmosphere and cooling water, as well as conduction between the moving bar and mill rolls and manipulators. Measured values for convection heat transfer coefficients are used at the boundaries. The following effects are also described in the procedure: deformation of the bar during rolling, tilting the bar between passes through a mill, cropping the bar and contraction due to heat losses. Predicted and measured temperatures correlated favourably.

NOTATIONS (SI-Units)

a	Coefficient in the discretised equation
a_p	Σa_{nb}
A	Area
b	Constant term in the discretisation equation
C	Specific heat of the material
F_{WB}	Geometric view factor for radiation from bar surface to building surface
h_c	Convection heat transfer coefficient to surrounding air
h_r	Radiation heat transfer coefficient
h_w	Convection heat transfer coefficient for water boiling on a surface
k	Thermal conductivity of the material

k_{int}	Thermal conductivity at the interface between two materials
$\Delta \ell$	Contraction due to heat losses
Pr	Prandtl number
$\dot{q}$	Production rate of heat in the material
Re	Reynolds number
t	Time
T	Temperature
ΔT	Temperature differential
T_B	Temperature of the surrounding building
T_w	Temperature of the steel bar surface
T_∞	Temperature of the environment
x,y,z	Space coordinates in cartesian system
α	Contraction coefficient
ϵ_w	Emisssivity of the bar surface
ρ	Material density
σ	Stefan Boltzmann constant

Subscript

nb	General neighbouring grid point
p	Central grid point under consideration

1. INTRODUCTION

The temperature distribution in a steel bar during the hot rolling process has an important influence on the quality of the final product [1,2,3,4]. Effective control of the temperature during the process will also optimise the reheating, and therefore production rate and energy costs. A method which accurately predicts the temperature distribution in steel bars during the hot rolling process is therefore of considerable importance to both the operating personnel and designers of the process and equipment.

A literature survey produced only a limited number of publications on the numerical prediction of temperature distributions in steel. In these papers the heat transfer was assumed unsteady and one-dimensional. Lu and Williams [5], Cook [6], and Wick [7], for example numerically studied the ingot handling process by assuming the ingot to be cylindrical with one-dimensional heat transfer. This approximation is acceptable for modelling the reheating process. However, when hot rolling is investigated a three-dimensional approach is needed to model the effect of steel bar deformation on the temperature distribution. A one-dimensional approach also cannot model edge temperatures. Edge temperatures are the lowest temperatures in the steel bar and therefore the limiting temperatures during hot rolling. If the edge temperatures drop below a certain minimum temperature, cracks will occur. For the abovementioned reasons a three-dimensional approach was followed.

This paper outlines a method to predict three-dimensional temperature distributions in steel bars during the hot rolling process. Predicted and measured data are also discussed.

2. OUTLINE OF THE MODELLING PROCEDURE

2.1 Governing and finite difference equations

The general conduction equation for three-dimensional unsteady heat transfer in Cartesian co-ordinates is given by the following:

$$\frac{\partial}{\partial x}\left(k\frac{\partial T}{\partial x}\right) + \frac{\partial}{\partial y}\left(k\frac{\partial T}{\partial y}\right) + \frac{\partial}{\partial z}\left(k\frac{\partial T}{\partial z}\right) + \dot{q} = \rho C \frac{\partial T}{\partial t} \qquad (1)$$

The finite difference equations for the numerical procedure are derived by integrating the partial differential equation (1) over control volumes surrounding a grid point. The general finite difference equation for equation (1) can be written as follows [8]:

$$a_p T_p = \sum a_{np} T_{np} + b \qquad (2)$$

The finite difference equations at the boundaries, however, differ slightly from equation (2) [9].

2.2 Boundary conditions

(a) Conduction boundaries

The general discretised conduction equation (2) can also be used when different materials are in contact. However, the interface conductivity must be calculated by the harmonic mean [8] of the conductivity values of the two materials in contact. The equation for the interface conductivity between materials (1) and (2) is given by the following [8]:

$$k_{int} = \frac{2k_1 k_2}{k_1 + k_2} \qquad (3)$$

Significant contact between the steel bar and other materials occurs at the manipulators and mill rolls. At the manipulators however, the contact is not perfect. To account for this imperfect contact, equation (3) is multiplied by a constant. Experiments showed that a value of 0,64 is a good approximation for the constant [9]. In this study heat conduction between the steel bar and roller table was neglected as suggested by Khloponin et al [10].

(b) Convective boundaries

The convective heat transfer coefficient to the surrounding air was calculated by assuming that the steel bar surfaces can be approximated as flat plates. For a flat plate the mean convection coefficient can be calculated by the following equation [11]:

$$h_c = \text{constant} \cdot (Pr)^{1/3} (Re)^{1/2} \qquad .(4)$$

The value of the constant in equation (4) was derived from measurements [9] as 0,362 which compares favourably with the theoretical value of 0,332 [11].

Convective heat transfer from the bar to the cooling water of the mills must also be taken into account. The following equation for the convective heat transfer coefficient for this case was suggested by Hatta et al [12] and was used in this study:

$$h_w = 0{,}362 \, (Pr)^{1/3} (Re)^{1/2} (T_w - T_\infty)^{-1/2} \qquad .(5)$$

(c) Radiation boundaries

For the analysis of radiation heat transfer from the boundaries, a radiation heat transfer coefficient approach was used. The radiation heat transfer coefficient is given by the following equation [11]:

$$h_r = \frac{(T_w^2 + T_B^2)(T_w + T_B)}{(1-\epsilon_w)/\epsilon_w + 1/F_{WB}} \qquad .(6)$$

The radiation heat flow from the boundaries can now be written as follows:

$$q = h_r \, A \, (T_w - T_B) \qquad .(7)$$

2.3 Solution of grid

(a) Solution procedure

The simulation started with the steel bar leaving the furnace. It was therefore assumed that initially, the temperature throughout the bar was constant. From this initial condition the simulation was progressed in time, with one second time intervals. The Gauss Seidel iteration scheme was used to solve the finite difference equations (2) at each time level. The solution was fully converged at each time level before moving on to the next.

(b) Stability, convergence and computing effort

The solution procedure is unconditionally stable because the finite difference equations are implicit [8]. The solution is assumed to have converged if the largest difference between the temperature at the previous and present iteration for any grid point is smaller than a prescribed value.

To reduce the number of iterations to convergence the grid was solved from the eight numbered corners towards the centre of the grid (see Figure 1). In effect the steel bar was divided into eight segments which were solved individually. In this way the boundary conditions are carried into the grid from all directions at the same time. With this procedure convergence at each time level was usually attained within two or three iterations.

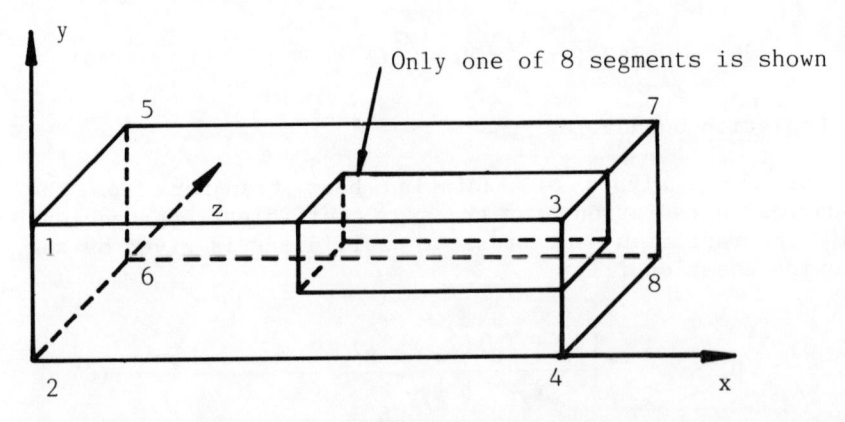

FIGURE 1 - METHOD OF SOLVING THE GRID FROM THE EIGHT CORNERS

A typical grid independent solution of the complete hot rolling process for a steel bar with a 13 x 13 x 13 grid (2197 grid points) was achieved within 42 hours on a HP 200 personal computer. This running time could drastically be reduced by using PASCAL instead of BASIC as the programming language.

(c) Position of grid points

A major problem when simulating the rolling of steel is to establish the position of grid points during deformation in the rolls, at tilting, at cropping and after contraction of the bar.

The prediction of the grid points in the steel bar at the rolls will be discussed with reference to Figures 2 and 3. According to Dieter [13] it is a fair assumption to assume that cross sectional vertical planes stay vertical during the rolling process (see Figure 2). At each cross sectional area the grid points in the y- and z- directions are evenly distributed as shown in Figures 2 and 3. The distance between these vertical planes can be determined from the known, constant mass flow rate. At the contact area in the rolls the assumption is made that deformation in the z-direction occurs in a straight line (see line AB in Figure 3).

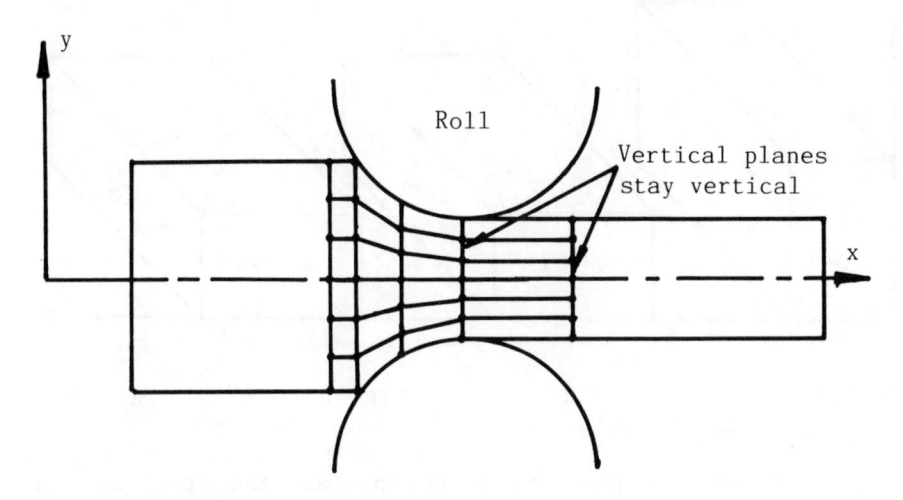

FIGURE 2 - SIDE VIEW OF DEFORMED BAR IN A MILL

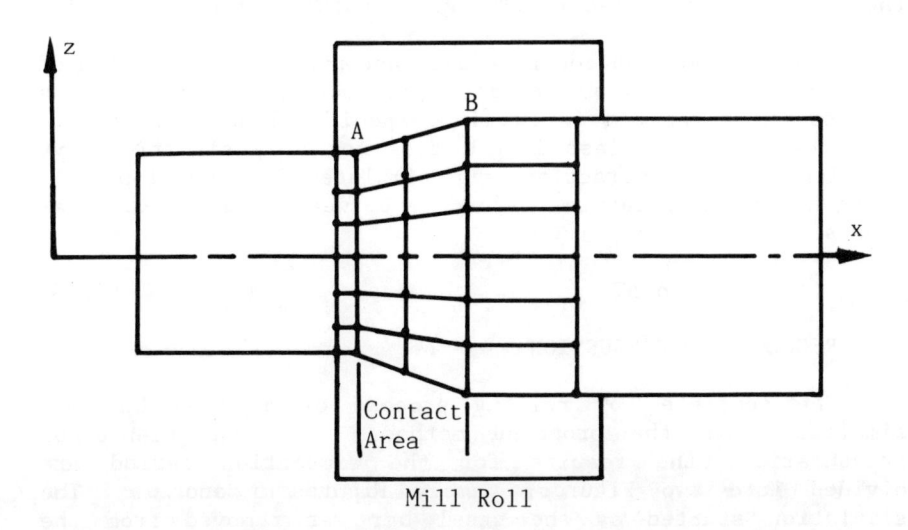

FIGURE 3 - TOP VIEW OF DEFORMED BAR IN A MILL

At the mills, rolling pressure is only exerted in the vertical direction. The bar must therefore be repeatedly tilted and passed through the mill to ensure a square cross sectional shape. When the bar is tilted through 90° between passes through the mill, the grid must be renumbered in such a way that the numbering sequence is similarly orientated towards the environment as before tilting (see Figure 4). The reason is that the solution procedure requires that the boundaries, being related to the environment, must be solved in a specific order.

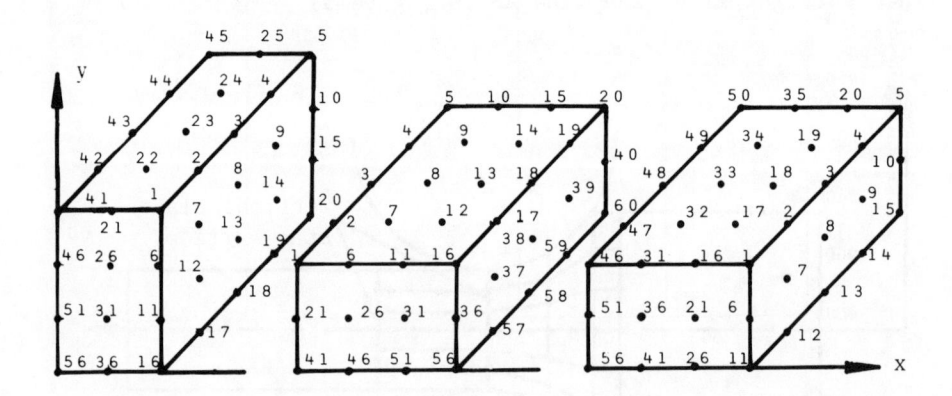

FIGURE 4 - RENUMBERING OF GRID AFTER TILTING A STEEL BAR THROUGH 90°

At a specific position in the process the steel bar is cropped to remove misformed ends. Provision is therefore made in the modelling procedure to remove gridpoints belonging to the piece that is cut-off. The grid is then renumbered again.

The grid must be contracted to accommodate contraction of the bar. This is of practical importance at the end of the process where the bar is cut into specific lengths to ensure that the first and last length does not vary significantly. The lengthwise contraction was simulated by equation (8), using a mean temperature difference between adjacent vertical planes.

$$\Delta \ell = \alpha \, \Delta T \qquad \qquad (8)$$

3. RESULTS AND DISCUSSION

The complete hot rolling process of a steel bar was simulated with the proposed method. For convenience of presentation, the results for the simulation period are divided into two Figures, namely Figures 5 and 6. The simulation started as the steel bar was removed from the furnace. This time was taken as time zero (see Figure 5).

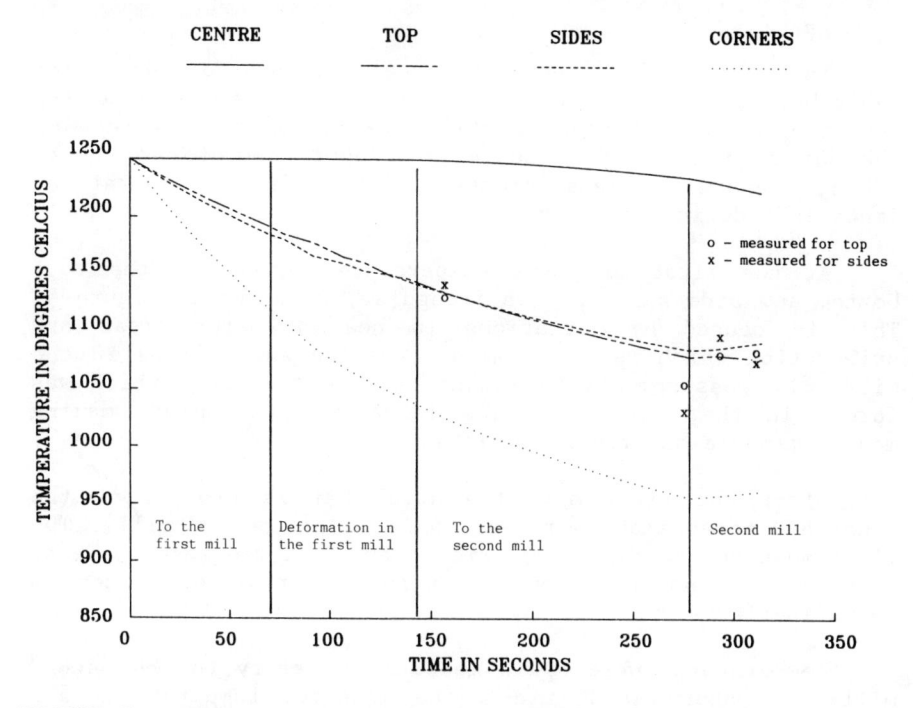

FIGURE 5 - TEMPERATURE DISTRIBUTION IN THE STEEL BAR

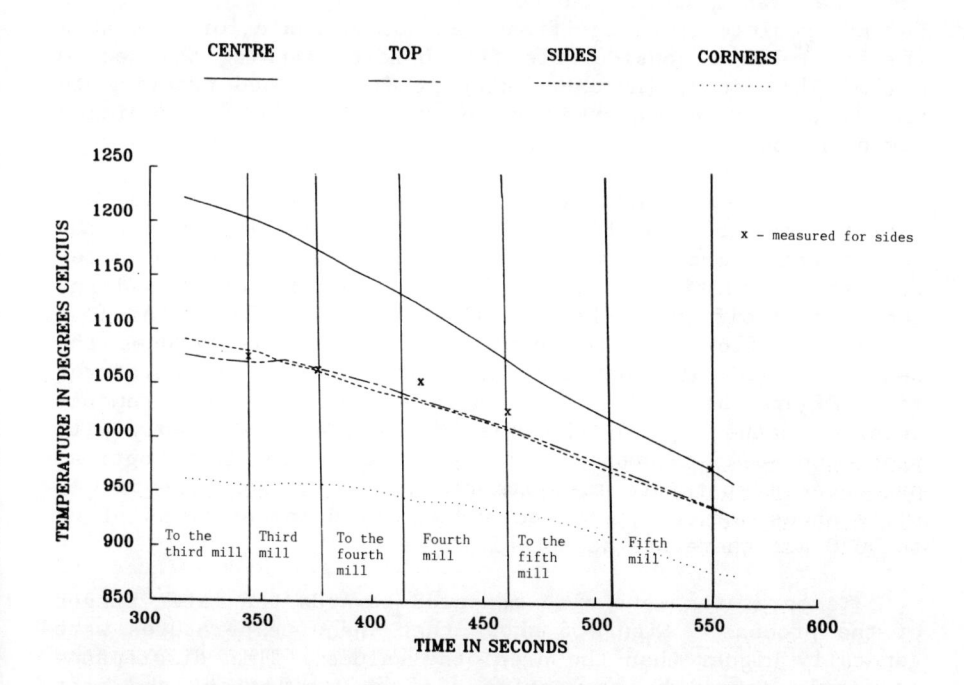

FIGURE 6 - TEMPERATURE DISTRUBUTION IN THE STEEL BAR

In the first 70 seconds the steel bar moves on the roller table before entering the first mill. The scale formed during reheating in the furnace is still present on the surfaces of the bar at this stage. Scale is a poor conductor of heat, making accurate measurements of surface temperatures impossible during this time.

At the first mill the temperature curves for the top, bottom and side surfaces are irregular, as shown in Figure 5. This is caused by variations in heat transfer areas and orientation which result from deformation and tilting in the mill. No measurements were done because the scale which was formed in the furnace becomes loose at this point, making measurements a hazardous operation.

After the first mill the steel bar is cropped at the shear and moved along a roller table to the second mill. The first measured temperatures were recorded at the shear. These temperatures compare favourably with the predicted values as shown in Figure 5.

Temperatures were again measured on entry to the second mill. As shown in Figure 5 the measured temperatures are considerably lower than the predicted ones. The reason for this discrepancy is that the temperature of the scale on the steel bar was measured and not the surface temperature of the bar as predicted. A significant amount of scale forms because the bar takes a considerable time before reaching the second mill. This study indicates that it is not good practice to use temperatures measured on entry to a mill for decision making purposes.

At the second mill the centre temperature starts to fall rapidly as shown in Figure 5. This is partly due to the centre being much closer to the surfaces of the bar after further deformation in this mill. Together with the large temperature difference between the centre and the surfaces, a large heat flow from the centre results. This reduces the centre temperature and stabilises the surface temperatures (see Figure 5). Figure 5 also shows that measurements recorded between passes in this mill compare favourably with predicted temperatures. It must be noted that optical pyrometer temperature measurements are only accurate to 1%. Differences between predicted and measured temperatures of up to 10°C are therefore acceptable.

Temperatures were also measured towards the later stages of the process. Figure 6 shows that these temperatures were generally higher than the predicted values. This discrepancy is partly caused by neglecting, in the simulation, the heat produced as a result or work done on the bar during deformation. As the cross sectional areas get smaller, this effect becomes more dominant [10].

4. CONCLUSIONS

The numerical simulation of the hot rolling process provides complete temperature information of the steel bar throughout the process. Edge temperatures, which are difficult to measure in practice, can therefore be predicted. By ensuring correct edge temperatures, cracks can be prevented.

With the proposed procedure it is now possible to simulate the process for other materials than steel as well as for different environmental conditions. This information is useful when designing a new mill.

Up to the present only measured surface temperatures could be used to calculate the contraction where the bars are cut to specific lengths. This method presents the possibility to predict the contraction more accurately by using the predicted temperature distribution in the bar.

Several optimisation studies can be carried out in future using the developed program. One such a study is to investigate the effect of non-uniform initial temperature profiles. It may be unnecessary to initially heat the bar centre to the same temperature as the surfaces. A lower acceptable centre temperature will reduce the reheating time and energy input.

It is concluded that numerical simulations could, in many cases, offer an attractive alternative to empirical approaches in the steel industry.

REFERENCES

1. ZHELEZNOV, YU.D., TSIFRINOVICH, B.A., LYAMBAKH, R.V., ROMASHKEVICH, L.F., and SAVICHEV, G.T. - Variation in temperature over the length of a strip as it passes through a continuous hot rolling mill, STAL USSR, 10, 854-858 (1968).

2. VOROBEI, S.L. and MAZUR, V.L. - Stability of temperature conditions in hot rolling of wide strip, STEEL IN THE USSR, 12, 375-377 (1982).

3. SUZUKI, H. and OHNO, J. - Recent trends in temperature measurement technology in iron and steel industry, Transactions of the Iron Steel Institute of Japan, 19 440-447 (1978).

4. CHELYUSTKIN, A.B., TSIFRINOVICH, B.A., ROMASHKEVICH, L.F., DORMAN, A.I. and GALIEV, F.M. -Rise in temperature at end of rolling strip in wide strip mill, STEEL IN THE USSR, 8, 37-39 (1978).

5. LU, Y. and WILLIAMS, T.J. - Energy savings and productivity increases with computers - A case study of the steel ingot handling process, Computers in Industry, 4, 1-19 (1983).

6. COOK, J.R., LONGWELL, E.J. and NACHTIGAL, C.L. - Algorithms for on-line computer control of steel ingot processing, IFAC 7th Triannual World Congress, Helsinki, Finland, 1, 131-137 (1978).

7. WICK, H.J. - On-line sequential estimation of ingot center temperatures in a soaking pit, Iron and Steel Engineer, 59, 56-59 (1982).

8. PATANKAR, S.V. - *Numerical Heat Transfer and Fluid Flow*, Hemisphere Publication Corp., New York, 1980.

9. VISSER, J.A. - *The numerical prediction of temperature distribution in steel billets during hot rolling*, M.Eng thesis, University of Pretoria, Pretoria, 1986.

10. KHLOPONIN, V.N., SAVCHENKO, V.S., HENZEL, A. AND WEBER, K.H. - Investigation of the development of the stock temperature during hot rolling, STEEL IN THE USSR, 3, 146-148 (1976).

11. HOLMAN, J.P. - *Heat Transfer*, McGraw-Hill International book Company, Tokyo, 1981.

12. HATTA, N., KOKADO, J. AND HAWASAKI, K. - Numerical analysis of cooling characteristics for water bar, Transactions of the Iron and Steel Institute of Japan, 23, 555-564 (1983).

13. DIETER, G.E. - *Mechanical Metallurgy*, McGraw-Hill International Book Company, Singapore, 1981.

APPLICATION OF THE GALERKIN METHOD TO ANALYSIS OF HEAT TRANSFER IN A LOW-HEAT-REJECTION DIESEL ENGINE

John Dewey Jones

Engine Research Department
General Motors Research Laboratories
Warren, MI 48090, U.S.A.

Summary

Determination of the temperature fluctuations in the combustion-chamber walls of an engine is one of many engineering applications requiring calculation of the temperature distribution in a body exposed to periodically varying boundary conditions. In this paper it is shown that, under suitable circumstances, the temperature distribution within such a body can be resolved into spatially varying and temporally varying terms. Repeated application of the Galerkin method allows this temperature distribution to be approximated by the solution of a complex matrix equation. The solution method is rapid, stable, and can be taken to any desired accuracy. A sample calculation is shown for the cylinder liner of an insulated diesel engine, taking into account the effects of time-varying convective and radiative heat transfer and frictional heating.

GLOSSARY

Scalars

A_w Area of a patch of the combustion chamber inner surface.
M There are $2M + 1$ terms in the complex Fourier series expression for the temporal variation in temperature.
N Number of terms in polynomial expression for the spatial variation in temperature.
T_g Cyclically varying gas temperature.
T Temperature of wall.

c Specific heat of the wall material.
d Distance over which wall temperature is considered to vary.
h Cyclically varying film heat transfer coefficient.
k Conductivity of wall material
x Distance below surface of combustion chamber wall.

α One-half the crank angle.
β Time-varying function of the boundary conditions.
δ $\dfrac{k}{\rho c d^2 \omega}$
γ Dummy variable of integration.
η Time-varying function of the film heat transfer coefficient.
ϱ Test function for spatial variation in temperature.
ς Test function for temporal variation in temperature.
λ Dimensionless distance into the wall.
τ Time.
ρ Density of wall material.
θ Time-varying coefficient in the polynomial expression for temperature.
$\dot{\theta}$ Derivative of θ with respect to time.
θ' Derivative of θ with respect to ω.
ω One-half the angular crank velocity (rad/s).

Vectors

t   Vector of the N coefficients θ_n.
u   Vector of length N all of whose elements are unity.
Θ Vector of length $2M + 1$ whose elements are the coefficients in the Fourier expansion of **t**.
Υ Vector of length $2M + 1$ whose elements are the products β_m**v**.

1. INTRODUCTION

This paper develops a technique for rapid solution of temperature variations in solids exposed to periodically varying boundary conditions. An important application of this technique is in calculating the time-varying temperatures on the inner surfaces of the combustion chamber of a reciprocating engine. Previous research has shown [1,2] that the ceramic surface temperatures in an insulated engine may fluctuate over a range of several hundred degrees over an engine cycle. Prediction of the amplitude of this fluctuation and its propagation into the ceramic is essential for the thermal and structural design of low-heat-rejection engines. While it is possible to predict surface temperatures by iterative use of a simple finite-difference model, as reported in [2], this approach demands an excessive quantity of computer time for solution. By observing that the periodicity of the boundary conditions enforces the same periodicity on the solution, the solution may be restricted to functions expressible as a sum of harmonics.

2. MATHEMATICAL ANALYSIS

The equation governing temperature distribution in the engine structure is

$$\nabla(k\nabla T) = \rho c \frac{\partial T}{\partial \tau} \tag{1}$$

together with the boundary conditions on the engine surfaces exposed to combustion gas or coolant. If it is assumed that conditions on the coolant side do not vary with time, it may be shown [1] that the temporal variation in temperature is restricted to the few millimeters of the engine structure closest to the combustion gas. In this paper a method for calculating the temperature distribution within those few millimeters is developed. This solution may then be linked with a time-independent finite-element model of the temperature distribution in the remainder of the engine structure. The formulation of such a finite-element model is straightforward and will not be discussed here. The finite-element model and skin model are linked by the iterative technique described in [2]: the skin model calculates the mean heat flow through a particular area of the engine inner surface, and the finite-element model finds the sub-surface temperature corresponding to that flow.

The inner surfaces of the engine may now be divided into patches, each patch a few millimeters deep and covering one element of the sub-surface finite-element model. It will be assumed that the heat transfer parallel to the surface within each patch is negligible in comparison with the heat transfer orthogonal to the surface. Thus, denoting the direction orthogonal to the surface by x, Eq. 1 may be re-written for a patch of area A_w as

$$\rho c A_w \frac{\partial T}{\partial \tau} - k A_w \frac{\partial^2 T}{\partial x^2} = 0 \tag{2}$$

assuming that the variation in conductivity with depth is negligible. Since this analysis will deal with the operation of a four-stroke reciprocating engine, it will be found convenient to replace time, τ, by the variable α, defined by $\alpha = \omega\tau$, where ω is one-half the angular velocity of the crank. Derivatives with respect to α will be denoted by a prime, as T'. Eq. 2 then becomes

$$\rho c \omega T' - k \frac{\partial^2 T}{\partial x^2} = 0 \tag{3}$$

1.1 Decomposition of the Spatial Variation

Let the thickness of each patch be d. Following the approach set forth by Caulk in [3], assume that the variation in temperature with depth can be represented by a polynomial of degree N having time-dependent coefficients.

$$T(x,a) = \theta_0 + \sum_{n=1}^{N} \theta_n(a)\chi^n \tag{4}$$

where $\chi = (1 - x/d)$. Note that $T(d,a) = \theta_0$, which by hypothesis is constant with time. The differential equation (3) may be rewritten in variational form as [4]

$$\int_0^1 \left(\rho c \omega dT' - \frac{k}{d}\frac{\partial^2 T}{\partial \chi^2}\right)\phi \, d\chi = 0 \tag{5}$$

where ϕ denotes any function from some test space $\{\phi_n\}$.

Choose this test space to be $\{1, \ldots, \chi^N\}$. For $\phi = 1$.

$$\int_0^1 \left(\rho c \omega d \sum_{m=1}^{N} \theta'_m \chi^m - \frac{k}{d} \sum_{m=1}^{N} \theta_m \frac{\partial^2 \chi^m}{\partial \chi^2}\right) d\chi = 0$$

$$\Rightarrow \rho c \omega d \sum_{m=1}^{N} \frac{\theta'_m}{m+1} + \frac{k}{d}\theta_1 = \frac{k}{d} \sum_{m=1}^{N} \theta_m \frac{\partial \chi^m}{\partial \chi}\bigg|_{\chi=1}$$

The second and third terms in this equation are recognized as the heat flux at $x = d$ and $x = 0$ respectively, from the equation

$$q = -k\frac{\partial T}{\partial x} = \frac{k}{d}\frac{\partial T}{\partial \chi}$$

The second term will be used at the end of the analysis, after θ_1 has been found, to calculate the heat flow into the sub-surface engine structure for the finite-element model. Write the third term as q_1. Then

$$\rho c \omega d \sum_{m=1}^{N} \frac{\theta'_m}{m+1} + q_2 = q_1 \tag{6}$$

Now, choosing test functions $\phi = \chi^n, n = 1, N$

$$\int_0^1 \left(\rho c \omega d \sum_{m=1}^{N} \theta'_m \chi^m - \frac{k}{d} \sum_{m=1}^{N} \theta_m \frac{\partial^2 \chi^m}{\partial \chi^2}\right)\chi^n \, d\chi = 0$$

Integrating the second term by parts leads to:

$$\sum_{m=1}^{N} \frac{\theta'_m}{m+n+1} + \frac{k}{\rho c \omega d^2} \sum_{m=1}^{N} \frac{mn}{m+n-1}\theta_m = \frac{q_1}{\rho c \omega d} \tag{7}$$

Eqs. 7 give N expressions for the N unknown coefficients $\{\theta_m\}$ in terms of the heat flux at the boundary, q_1. The remaining coefficient, θ_0, is constant and may be determined from static conditions immediately below the surface layer. Appropriate boundary conditions for the inner surfaces of a reciprocating engine may now be specified.

The heat flux at x_0, q_1, is given by

$$q_1(\alpha) = h(\alpha)\left(T_g(\alpha) - \sum_{m=0}^{N} \theta_m(\alpha)\right) + q_r(\alpha) \qquad (8)$$

where h is the time-varying film heat transfer coefficient, T_g is the gas temperature, and q_r is the total additional incident heat flux due to radiation and frictional heating. The choice of ω as one-half the angular velocity of the crank ensures that the boundary conditions will vary with period 2π.

Substituting Eq. 8 in the set of Eqs. 7 gives

$$\sum_{m=1}^{N} \frac{\theta'_m}{m+n+1} + \frac{k}{\rho c \omega d^2} \sum_{m=1}^{N}\left(\frac{mn}{m+n-1} + \frac{d}{k}h(\alpha)\right)\theta_m$$
$$= \frac{1}{\rho c \omega d}\left(h(\alpha)\left(T_g(\alpha) - \theta_0\right) + q_r(\alpha)\right)$$

This may be written as a single vector equation:

$$K\mathbf{t}' + \delta(L + \eta(\alpha)Q)\mathbf{t} - \beta(\alpha)\mathbf{u} = \mathbf{0} \qquad (9)$$

where K, L and Q are N by N matrices, defined as follows:

$$K_{j,k} = \frac{1}{j+k+1} \quad . \quad L_{j,k} = \frac{jk}{j+k-1} \quad . \quad Q_{j,k} = 1$$

for all j, k. (It will be noted that all of these matrices are symmetric.) $\mathbf{t}$ is the vector made up of the N time-varying quantities $\{\theta_j\}$ and $\mathbf{u} = (1, 1, \ldots, 1)^T$. δ, $\eta(\alpha)$ and $\beta(\alpha)$ are scalar quantities defined by

$$\delta = \frac{k}{\rho c \omega d^2} \quad . \quad \eta(\alpha) = \frac{d}{k}h(\alpha)$$

$$\beta(\alpha) = \frac{1}{\rho c \omega d}\left(h(\alpha)(T_g(\alpha) - \theta_0) + q_r(\alpha)\right)$$

1.2 Decomposition of the Temporal Variation

The only unknowns in Eq. 9 are the N components of the vector **t**. Each of these components will be a function of α.

The definition of α has been chosen such that for steady-state operation of the engine, the boundary conditions will vary with period 2π. Complex Fourier series for the quantities $\eta(\alpha)$ and $\beta(\alpha)$ may be derived by standard techniques from the boundary conditions. The unknown vector **t** may similarily be represented by a Fourier series with complex vector coefficients:

$$\eta(\alpha) = \sum_{m=-M}^{M} \eta_m e^{im\alpha} \quad , \quad \beta(\alpha) = \sum_{m=-M}^{M} \beta_m e^{im\alpha}$$

$$\mathbf{t}(\alpha) = \sum_{m=-M}^{M} \mathbf{t}_m e^{im\alpha}$$

Now recast Eq. 9 in the weak variational form [4]

$$\int_0^{2\pi} \left(K\mathbf{t}' + (L + \eta(\alpha)Q)\mathbf{t} - \beta(\alpha)\mathbf{u} \right) \varsigma(\alpha)\, d\alpha = 0 \qquad (10)$$

where the test functions ς are drawn from the space $\{e^{im\alpha}\}_{m=-M}^{M}$. It is now possible to obtain $2M+1$ equations determining the $\{\mathbf{t}_m\}$ by choosing $\varsigma = e^{im\alpha}$, $m = -M, \ldots, M$. In the case $m = 0$

$$\int_0^{2\pi} K\mathbf{t}' + (L + \eta(\alpha)Q)\mathbf{t} - \beta(\alpha)\mathbf{u}\, d\alpha = 0$$

$$\Rightarrow L\mathbf{t}_0 + Q \sum_{m=-M}^{M} \eta_{-m}\mathbf{t}_m - \beta_0 \mathbf{u} = 0 \qquad (11)$$

The remaining $2M$ equations are derived as follows for the test function $e^{-il\alpha}$ with $l \in [-M, M] \setminus \{0\}$.

$$\int_0^{2\pi} e^{-il\alpha} \left(K\mathbf{t}' + (L + Q\eta)\mathbf{t} - \beta\mathbf{u} \right) d\alpha = 0$$

$$\Rightarrow ilK\mathbf{t}_l + L\mathbf{t}_l + Q \sum_{m=M_1}^{M_2} \eta_{(l-m)}\mathbf{t}_m = \beta_l \mathbf{u} \qquad (12)$$

where

$$M_1 = \begin{cases} l - M & \text{if } l > 0 \\ -M & \text{if } l < 0 \end{cases} \quad M_2 = \begin{cases} M & \text{if } l > 0 \\ l + M & \text{if } l < 0 \end{cases}$$

These $2M+1$ equations can be summarized in a single matrix equation.

$$D\Theta = \Upsilon \tag{13}$$

where Θ and Υ are vectors of length $2M+1$ defined by

$$\Theta = \begin{pmatrix} t_{-M} \\ \vdots \\ t_M \end{pmatrix} \quad , \quad \Upsilon = \begin{pmatrix} \beta_{-M}\mathbf{v} \\ \vdots \\ \beta_M\mathbf{v} \end{pmatrix}$$

where the t_m are themselves complex vectors of length N, the β_m are complex scalars and $\mathbf{v}$ is a real vector of length N.

The $(2M+1)$ by $(2M+1)$ matrix D can be written as

$$D_{m,n} = \begin{cases} L + iK + Q\eta_0 & \text{if } m = n \\ 0 & \text{if } m \neq n \text{ and } |n - m| > M \\ Q\eta_{(n-m)} & \text{otherwise} \end{cases}$$

where K, L and Q are real, symmetric N by N matrices, and the $\{\eta_m\}$ are complex numbers.

3. SOLUTION AND VALIDATION

A FORTRAN program was written to assemble the matrix D and the vector Υ from given boundary conditions and material properties. In this study, the solution algorithms CSIFA and CSISL from the 'LIN-PACK' software package were used [5] for its solution. This minimized the coding effort involved and led to a concise, efficient program.

As a first test of the method described above, the model was supplied with artificial boundary conditions, consisting of a constant film heat transfer coefficient, h, and a gas temperature varying sinusoidally between 100 K and -100 K. Under these conditions, an exact analytic solution is available, namely [6]

$$T(x,\tau) = 100 \times \frac{1}{\sqrt{1 + \sqrt{2}/N + 1/N^2}} e^{-ax} \cos(\omega\tau - ax - \lambda) \tag{14}$$

where $\quad a = \sqrt{\dfrac{\omega\rho c}{2k}} \quad , \quad N = \dfrac{h}{\rho c k \omega} \quad , \quad \lambda = \tan^{-1}\dfrac{1}{1+N}$

Equation 14 predicts that the temperature fluctuation in the solid dies away to 20 % of its surface value within a distance $d = \ln(5)/a$ of the surface. For an assumed engine speed of 2 400 r/min and solid properties corresponding to cast iron, d is about 1 mm. Using a skin depth much below this figure, it was impossible to obtain a good match with the analytic solution. However, for skin depths greater than 1 mm, excellent agreement was obtained on the surface temperature fluctuation.

Comparisons between the analytic and computed variations in temperature with depth at one point in the cycle are shown in Figure 1. It is seen that reasonably good agreement is obtained using a quadratic expression for x. Varying the number of terms in the Fourier analyses of the polynomial coefficients had no effect on the accuracy of the solution; all harmonic terms but the first were zero.

As a more stringent test of the model's usefulness, a comparison was made with the predictions of a finite-difference model [2]. The boundary conditions supplied to both models were calculated from a heat release model reported previously [7]. Again, an engine speed of 2 400 r/min and material properties of cast iron were used.

The effect of the number of terms used in the harmonic analysis of the polynomial coefficients is shown in Figure 2. It is seen that, for the boundary conditions used, good agreement is obtained using the first nine harmonic terms; even using only six harmonics, a reasonably good fit can be obtained. More terms might be necessary if the engine being modeled imposes boundary conditions with a steeper gradient, as, for example, a sharp peak in radiative emissions.

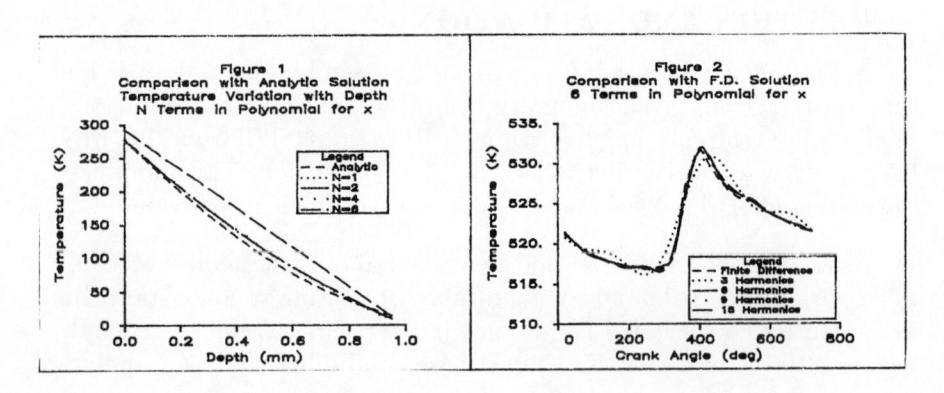

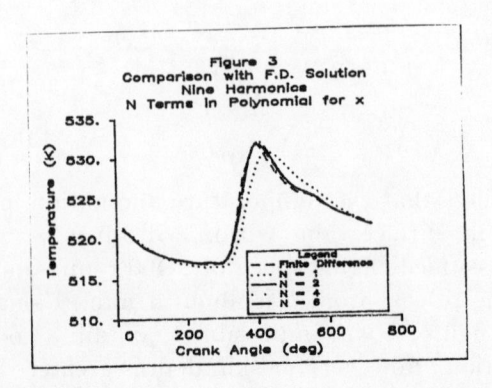

Figure 1
Comparison with Analytic Solution
Temperature Variation with Depth
N Terms in Polynomial for x

Figure 2
Comparison with F.D. Solution
6 Terms in Polynomial for x

Figure 3
Comparison with F.D. Solution
Nine Harmonics
N Terms in Polynomial for x

The effect of varying N, the number of terms in the polynomial expansion in x, is shown in Figure 3, and may also be observed in the contrast between Figures 4 and 5. Figure 3 shows the variation in surface temperature for different N, taking a skin depth of 1.0 mm. From this figure, it appears that good results can be obtained by taking $N = 2$. However, as the latter two figures show, the accuracy of the results is sensitive to the assumed skin depth, and this sensitivity is greater for lower values of N.

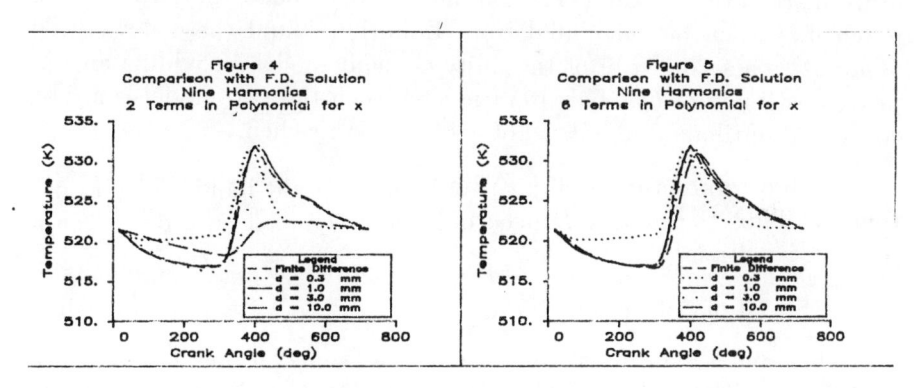

Figure 4
Comparison with F.D. Solution
Nine Harmonics
2 Terms in Polynomial for x

Figure 5
Comparison with F.D. Solution
Nine Harmonics
6 Terms in Polynomial for x

Permissible values for skin depth are constrained in two ways. Firstly, too shallow a skin depth forces too rapid extinction of the sub-surface temperature oscillation. This makes a physically realistic solution impossible. The second limitation is numerical in origin. As x grows without limit, e^{-x} tends to zero, while any polynomial in x will tend to plus or minus infinity, depending on the sign of the coefficient of the highest power. Thus any approximation of e^{-x} by a polynomial must break down at some point as x increases. This sets an upper limit on the skin depth.

In the present case, there is a narrow range between these two limits where $N = 2$ gives acceptable results. For $N = 6$, the acceptable range is much broader. The appropriate value for any given application must be decided by the boundary conditions and the user's requirements.

4. LINKING TO FINITE-ELEMENT MODEL

This analysis has represented temperature by Equation 4.

$$T(x, \alpha) = \theta_0 + \sum_{n=1}^{N} \theta_n(\alpha) \chi^n$$

Solution of the matrix equation 13 supplies the $\{\theta_n\}$, but not θ_0: indeed, θ_0 must be known in order to define the variable β in Equation 9. This is not unreasonable; θ_0 is the time-constant temperature below the skin, and must depend on conditions deeper in the solid.

While the top 1 mm of the combustion chamber surfaces may be modeled as one-dimensional, the complex geometries of the piston and cylinder head require a two- or three-dimensional model. Assume that such a model is available. Using time-averaged values of the boundary conditions, this model can supply a first approximation to θ_0 for each of the patches making up the inner surfaces of the combustion chamber. Substituting this value in Equation 9 permits solution for the temperature distribution in the skin. The heat flow through the base of each patch may then be calculated from Equation 6 and these flows used as input to a second run of the finite-element model, providing an improved estimate of θ_0. This iteration between the two models may be continued until a suitable level of accuracy is reached.

It will be noted that θ_0 appears only on the right-hand side of Equation 13. Thus the matrix D need only be assembled once for use in a series of iterations.

4.1 Speed of Solution

When one-dimensional analysis of the time-varying temperature fluctuations in patches of the skin is coupled with two- or three-dimensional analysis of the underlying structure, the speed of solution becomes important. By taking advantage of the periodic nature of the boundary conditions, the method presented here permits solution in a fraction of the time required for convergence of a finite-difference model. It is also somewhat faster than the Updike method [8], while maintaining an acceptable level of accuracy.

5. CONCLUSIONS

An analytical technique has been developed and shown to permit rapid approximate solution of heat transfer problems involving periodic boundary conditions. An example of its application to the low-heat-rejection engine is given. The technique may also be applied to a variety of other practical problems.

ACKNOWLEDGEMENTS

This work benefited greatly from several discussions with my colleagues at G.M. Research Labs, in particular David Caulk and Cleve Ashcraft.

REFERENCES

1. MOREL, T.M., BLUMBERG, P.N., FORT, E.F. and KERIBAR, R. - Methods for Heat Analysis and Temperature Field Analysis of the Insulated Diesel, DOE/NASA/0342-1, NASA CR-174783, (1984).
2. JONES, J.D. - 'Shuttle Heat Transfer in the Insulated Diesel,' Eighth International Heat Transfer Conference, San Francisco, 1986.
3. CAULK, D.A. - An Approximate Method for Analyzing Transient Heat Conduction in Permanent Moulds for Metal Casting, Winter Annual Meeting of the ASME, Anaheim, CA. Paper 86-WA/HT-92, (1986).
4. BAKER, E.B., CAREY, G.F., and ODEN, J.T. - Finite Elements An Introduction, Volume 1, Prentice-Hall, New Jersey, 1981.
5. DONGARRA, J.J., BUNCH, J.R., MOLER, C.B., and STEWART, G.W. - LINPACK Users' Guide, SIAM, Philadelphia, 1979.
6. CARSLAW, H.S. and JAEGER, J. - Conduction of Heat in Solids, 2nd Ed., Oxford University Press, Oxford, 1959.
7. AHMAD, T. and ALKIDAS, A.C. - A Simple Correlation for Mixing g and Combustion Rates in Diesel Engines, International Symposium on Diagnostics and Modeling of Combustion in Reciprocating Engines, COMODIA, Tokyo, (1985).
8. UPDIKE, W.A. - Computation of Surface Temperatures for Time-Dependent Periodic Convection Boundary Conditions, 84-WA/HT-40, Winter ASME Meeting, New Orleans, (1984).

AN ALGORITHM FOR SOLVING THE EIGENVALUE PROBLEMS
ASSOCIATED WITH HEAT DIFFUSION IN COMPOSITE SLABS

M. BOUZIDI and P. DUHAMEL
Laboratoire de Thermique du CNAM
292, rue Saint-Martin
75141 PARIS Cédex 03, FRANCE

ABSTRACT

The analogy between the problems connected with one-dimensional heat diffusion in composite layers and those dealing with vibrations of skeletal structures is shown : both cases lead to a singular Sturm-Liouville problem.

A known property of the natural frequencies of vibrations of linearly elastic structures is then used for calculating the number of eigenvalues which lie below a chosen positive bound in any singular Sturm-Liouville system. This counting technique permits us to build a reliable algorithm for determining the eigenvalues in the case of heat conduction through multilayered composite walls. We consider the simple coupling case (one composite wall) and the multiple coupling case (enclosures including multilayered walls). One example in each configuration is presented.

1. INTRODUCTION

The need of controlling the temperatures inside enclosures such as industrial furnaces, buildings etc..., and the concern for energy saving induce the designers to build enclosures made of multilayered composite walls. A good knowledge of the thermal losses through this kind of wall requires then to study heat diffusion in composite media.

Numerous method for solving the thermal conduction problems exist presently. Among them, the integral transform method is suited to linear or linearized problems. Its application to unidirectional laminated media leads to a singular Sturm-Liouville system, the solution of which goes through the computation of a set of eigenvalues.

The objective of this paper is in proposing a reliable

method to find theese eigenvalues by adding a safe counting technique to the numerical calculation.

2. ANALOGY BETWEEN DIFFUSION AND VIBRATION PROBLEMS

The well-known analogy in the case of continuous media is based on the spectral properties of differential operators [1]. We consider here piecewise continuous media.

Let a mechanical structure be an assembly of linearly elastic continuous components. We suppose that the governing displacement equation of a given point inside a component involves only one space variable (it is a wave equation). At each joint of the assembly appears a discontinuity, the origin of which is either in the change of physical (or geometrical) properties or in the properties of the joints. The search of natural frequencies of the structure (sinusoïdally varying solutions of the equations), leads to a singular Sturm-Liouville problem where the variable is the space variable (cf refs [2], [3], [4]).

Let us consider now a multilayered wall with n layers. In the ith. layer, the temperature, T_i, obeys the unidirectional equation of conduction :

$$\frac{\partial T_i}{\partial t} = \frac{\partial^2 T_i}{\partial x_i^2} \qquad x_i \in\ ]0, e_i[\ ; i = 1,n ; t > 0.$$

We make use of the dimensionless variables proposed in ref. [5] for obtaining equations independent of the diffusivity coefficients of the layers. The method of separation of variables gives :

(1) $$\frac{d^2 X_i}{dx_i^2} + \mu^2 X_i = 0 \qquad x_i \in\ ]0, e_i[\ ; i = 1,n.$$

This equation has the same form as the vibration equation of a beam [1] or a compressed composite plate [3]. When each layer is submitted to homogeneous linear boundary conditions, the equation (1), associated with the whole conditions leads to a singular Sturm-Liouville problem.

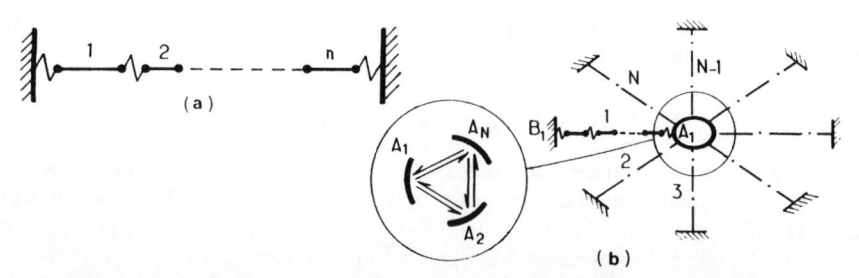

Figure 1 - Mechanical analogous to thermal problem (a) Multilayered wall, (b) Multicoupled composite walls

More precisely, the mechanical analogous to a n-layer composite wall [6], [7], [8] may be considered as an assembly of beams in which one member is linked to its only two neighbours (fig. 1a) through appropriate joints such as springs (wether there are contact resistances at the interfaces or not). The mechanical analogous to the N walls with coupling at the internal faces [9] is an assembly of rods making a N-branch star (fig.1b). The analogy of fig.1a has been exploited by Mikhailov and al. [8] who applied the methology of research of natural frequencies proposed by Wittrick and al. [2] to diffusion problems in layered composites. A direct application might be too considered in the scope of the analogy 1b. However, the method which we use for obtaining the eigenvalue equation (refs. [5], [6], [7] and [9]) is very different of that presented in refs. [2] and [8]. It is then necessary to re-examine the problem long before the findings of ref. [2].

3. COUNTING OF EIGENVALUES

Let an assembly be made of n components in which the physical function, X_i (x_i), obeys an equation of type (1). Besides, this system is subject to homogeneous linear boundary conditions so that the resulting problem be a singular Sturm-Liouville problem. The eigenvalues, μ_k, are the roots of the transcendental equation :

$$K_{NC}(\mu) = 0.$$

Let us impose upon this system the constraint :

$$X_q(y_q) = 0 \; ; \; 1 < q < n \text{ with } y_q = 0 \text{ or } e_q,$$

without changing the other boundary conditions. The new problem admits the eigenvalues λ_k which are the roots of the equation :

$$K_C(\mu) = 0.$$

The Rayleigh's theorem [10], which apply to vibrations of piecewise continuous systems [2] and thus, holds in any singular Sturm-Liouville problem permits us to write the serie of inequalities :

$$(2) \quad 0 < \ldots \lambda_{k-1} \leq \mu_k \leq \lambda_k \leq \mu_{k+1} \leq \lambda_{k+1} \ldots,$$

where for convenience a multiple eigenvalue is repeated a number of times equal to its order of multiplicity (equality case in (2)).

Let μ^* be any positive bound of μ and let $J_{NC}(\mu^*)$ and $J_C(\mu^*)$ be the numbers of eigenvalues which are less than μ^*, respectively in the unconstrained and constrained cases. We

set :
$$T(\mu^*) = K_{NC}(\mu^*)/K_C(\mu^*)$$

Following ref. [2], let us consider fig. 2. By comparing the signs of $K_C(\mu^*)/K_C(0)$, fig. 2a, and of $K_{NC}(\mu^*)/K_{NC}(0)$, fig. 2b, one sees that,

Property :

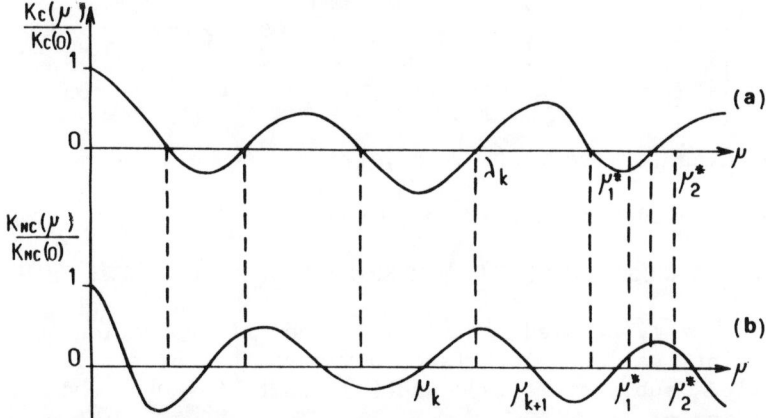

(3)
If $T(\mu^*)/T(0) < 0$,
then $J_{NC}(\mu^*) = J_C(\mu^*) + 1$ ($\mu^* \equiv \mu_1^*$ on fig. 2)
or else $J_{NC}(\mu^*) = J_C(\mu^*)$ ($\mu^* \equiv \mu_2^*$ on fig. 2)

Figure 2 - Sketch of the variations of the eigenvalue equations (a) With constraint, (b) Without constraint

Some remarks may be noted :

rem. 1 - An important particular case is when the insertion of the constraint, $X_q(y_q) = 0$, splits up the original system into two uncoupled (independent) sub-systems (α) and (β). In that case :
$$K_C(\mu) = K_\alpha(\mu) \, K_\beta(\mu).$$

rem. 2 - The property (3) stands if the unconstrained system possesses only one nember. By adding the constraint one simply changes one of the boundary condition of the system.

rem. 3 - The case with multiple eigenvalues is included.

4. APPLICATION TO THE EIGENVALUE CALCULATION

4.1. MULTILAYERED COMPOSITE WALL

a. STURM-LIOUVILLE PROBLEM - EIGENVALUE EQUATION

We consider the case where the n layers are in perfect

contact. The problem is defined by eq.(1). The associated boundary conditions are :

(4a) $F_i(e_i) = F_{i+1}(0)$ (4b) $X_i(e_i) = X_{i+1}(0)$, $i = 1, n-1$

(4c) $F_1(0) + h_1 X_1(0) = 0$ (4d) $F_n(e_n) - h_n X_n(e_n) = 0$,

we took $F_i(x_i) = -\beta_i dX_i/dx_i$; where β_i represents the adimensional effusivity of the ith. layer.

We use a method [5], [6], [7] permetting us to deduce the eigenvalue equation from the product of the transfer matrices of each layer. With :

$$\begin{bmatrix} \xi_n(\mu) & \eta_n(\mu) \\ \zeta_n(\mu) & \chi_n(\mu) \end{bmatrix} \equiv \prod_{i=n}^{1} \begin{bmatrix} \cos \mu e_i & -\dfrac{\sin \mu e_i}{\mu \beta_i} \\ \mu \beta_i \sin \mu e_i & \cos \mu e_i \end{bmatrix}$$

one obtains (cf. refs. [5], [6], [7]) :

(5) $K_n(\mu) = h_n(\xi_n(\mu) - h_1 \eta_n(\mu)) - (\zeta_n(\mu) - h_1 \chi_n(\mu))$

b. CALCULATION OF THE NUMBER OF EIGENVALUES LESS THAN μ^*

In the studied assembly, the introduction of a constraint at an interface divides the system into two uncoupled sub-systems (α) and (β). Let (j) be the system consisting of :

- The sub-system (α) where j layers obeys eqs. (1), (4,a,b,c) and :

$$X_j(e_j) = 0$$

- The sub-system (β) where n-j layers obeys eq. (1) and :

$$X_i(0) = X_i(e_i) = 0 \; ; \; i = j+1, n.$$

The eigenvalue equation of the system (j) writes (by using rem. 1) :

$$K_C(\mu) \equiv K_j(\mu) \, \varepsilon_{j+1}(\mu) \ldots \varepsilon_n(\mu) = 0.$$

$K_j(\mu)$ is easily deduced from eq. (5) by setting $n = j$ and $h_n^j \to \infty$. The $\varepsilon_i(\mu)$ term writes :

(6) $\varepsilon_i(\mu) = \sin \mu e_i / \mu \beta_i$ (the ith layer is independent)

Let (j+1) be the system obtained by suppressing the constraint $X_j(e_j) = 0$ in the (j) system. The condition at the interface j/j+1 is then :

$$X_j(e_j) = X_{j+1}(0) \neq 0$$

In that new system, j+1 layers are coupled to eachother and n-j-1 layers are uncoupled. The characteristic equation writes

$$K_{NC}(\mu) \equiv K_{j+1}(\mu) \ \varepsilon_{j+2}(\mu) \ldots \varepsilon_n(\mu) = 0$$

The number, $J_{j+1}(\mu^*)$, of eigenvalues which are less than μ^* in the (j+1) system is deduced from the property (3) by setting :

$$T(\mu^*) \equiv T_{j+1}(\mu^*) = K_{j+1}(\mu^*)/K_j(\mu^*) \ \varepsilon_{j+1}(\mu^*)$$

The algorithm is initialized with the number J_0 corresponding to the case where all the layers are independent. Then by applying the remark 2 to the first layer and next by using the algorithm one finds $J_n(\mu^*)$. At this stage the wall is still constrained by $X_n(e_n) \equiv 0$, the final searched number, $J'_n(\mu^*)$, results from the removal of the last constraint by prescribing the condition (4d).

c. CONTACT RESISTANCES AT THE INTERFACES

The counting method fits without any difficulty, but it must be noticed that each interface is treated through two steps :

- First step

By suppressing the constraint at $x_j = e_j$ in the (j) system (as previously defined), one determines with property (3) a number, $J'_j(\mu^*)$, of eigenvalues which is not equivalent to $J_{j+1}(\mu^*)$ for $X_j^j(e_j) \neq X_{j+1}(0)$ in that case. Fig. (1a) shows that this is due to the spring (or contact resistance) which is inserted between both faces.

- Second step

By suppressing the constraint $X_j(e_j) = 0$, one determines the number, $J_{j+1}(\mu^*)$ corresponding to the (j+1) system.

d. COMPUTATION OF THE EIGENVALUES

The process is akin to that describes in ref. [8]. We content ourselves with showing how the calculation of $J'_n(\mu^*)$ supplements the classical dichotomy method (binary successive division of the interval $]\mu_m, \mu_M[$ containing one root).

Let $k \geq 0$ be the number of roots which have been already computed.

We set : $\mu_o = 0$, $\mu_m = \mu_k$ and $\mu = \mu_m$

If $k \neq 0$, the increment of μ is defined by $\mu = \mu_k - \mu_{k-1}$ or else by $\mu = \delta\mu_o$ (arbitrary positive value).

The application of the algorithm 2.1.b may give three possible results :

i. $J'_n(\mu + \delta\mu) = k$

One sets $\mu_m := \mu + \delta\mu$ and $\mu := \mu_m$, a new value of $J'_n(\mu + \delta\mu)$ is then computed.

ii. $J'_n(\mu + \delta\mu) = k + 1$

One sets $\mu_M := \mu + \delta\mu$, the dichotomy method is started on the interval $]\mu_m, \mu_M[$.

iii. $J'_n(\mu + \delta\mu) = k + p$ with $p > 1$

One sets $\delta\mu := \delta\mu/2$ and $\mu_M := \mu + \delta\mu$, a new value of $J'_n(\mu + \delta\mu)$ is then computed.

A comparison of $|\mu_M - \mu_m|/\mu_M$ to a given accuracy number permits us to stop the computation.

It clearly appears that the case of multiple eigenvalues (and notably the even multiplicity case) is treated (iii). The classical method is in fact used when the interval $]\mu_m, \mu_M[$ contains only one root. The known inconvenience of classical methods, namely the missing of eigenvalues[I] is suppressed.

c. EXAMPLE

The preceding technique is applied to the example of ref. [11] already used in refs. [8], [12]. Some results are reported in table 1. We use the International System, the unit of eigenvalues is the $s^{-1/2}$ (the results of refs. [11], [12] use the (hour)$^{-1/2}$). The accuracy is 10^{-6}.

n	1	2	3	4	5
μ (s$^{-1/2}$)	.277606	.736687	.972431	1.39473	1.66626
n	10	20	30	40	50
μ (s$^{-1/2}$)	3.32821	6.83918	10.0797	13.6094	16.9186

Table 1. Multilayered wall of ref. [11] . Eigenvalues.

[I] Such a missing may be corrected after the computation of a great number of eigenvalues and the computation of the development of the initial condition of the non-stationary problem on the eigenbasis which permits to compare the true initial condition to its development (cf. [6] [7]).

One knows that the accuracy is finally limited by the characteristics of the computer and is entirely external to the counting process which only give, with certainty, the suffix of the calculated eigenvalue.

4.2. MULTICOUPLED WALLS

a. STURM-LIOUVILLE PROBLEM - EIGENVALUE EQUATION

The problem appears in each wall constituting the enclosure, as described in § 4.1.a. It is simply necesserary to add a supplementary suffix, referring to the wall, to the various functions, variable and physical parameters. The first suffix (generally i) stands for the wall, the second one (generally j) stands for the layer. The last layer possesses the n_i layer suffix. The interface conditions are similar to (4a), (4b). The external boundary ($x_{ij} = 0$) conditions write :

(5c) $\qquad F_{i1}(0) + h_{i1} X_{i1}(0) = 0 \quad i = 1, N.$

The internal boundary ($x_{ij} = e_{i,n_i}$) conditions may be written in matrix form (cf. [9]) :

(5d) $\qquad [F_N] = [H_N] [X_N],$

where the elements of the column matrices $[F_N]$ and $[X_N]$ are the N values, respectively of the flux $F_{i,n_i}(e_{i,n_i})$ and of the temperature $X_{i,n_i}(e_{i,n_i})$ on the internal faces of the enclosure, $[H_N]$ is a symmetrical square matrix, the off-diagonal elements of which are the coupling coefficients H_{ij}.

We showed [9] that the eigenvalue equation writes :

$$\text{Det } ([H_N][\gamma_N(\mu)] - [\delta_N(\mu)]) = 0,$$

where $[\gamma_N(\mu)]$ and $[\delta_N(\mu)]$ are diagonal matrices with elements $\gamma_i(\mu)$ and $\delta_i(\mu)$. They connects the properties of external faces with those of internal faces :

$$[F_N] = [\delta_N(\mu)] [X_o] \quad \text{and} \quad [X_N] = [\gamma_N(\mu)] [X_o],$$

$[X_o]$ stands for the column matrix with elements $X_{i1}(0)$.

b. CALCULATION OF THE NUMBER OF EIGENVALUES LESS THAN μ^*

Let (L) be the system consisting of :

- the sub-system (α) containing the first L walls which are coupled to eachother through their internal faces (eq. 5d in which L takes the place of N) and subject to condition (5c) (with i = 1,L).

- the sub-system (ß) containing the N-L other walls subject to condition (5c) (with i = L+1, N) and constrained by :

$$X_{i,n_i}(e_{i,n_i}) = 0 \quad i = L + 1, N.$$

The sub-systems are uncoupled and, provided that $\gamma_i(\mu) \neq 0$ when $i \leq L$, the eigenvalue equation writes :

$$\text{Det}([H_L][\gamma_L(\mu)] - [\delta_L(\mu)]) \cdot \prod_{i=L+2}^{N} \gamma_i(\mu) = 0.$$

When the constraint on the internal face of the L + 1 th. wall is removed, the eigenvalue equation of the (L+1) system writes :

$$\text{Det}([H_{L+1}][\gamma_{L+1}(\mu)] - [\delta_{L+1}(\mu)]) \cdot \prod_{i=L+2}^{N} \gamma_i(\mu) = 0,$$

thus, by setting :

$$T_{L+1}(\mu^*) = \frac{\text{Det}([H_{L+1}][\gamma_{L+1}(\mu^*)] - [\delta_{L+1}(\mu^*)])}{\text{Det}([H_L][\gamma_L(\mu^*)] - [\delta_L(\mu^*)]) \cdot \gamma_{L+1}(\mu^*)}$$

and by using the property (3), one deduces an algorithm suitable for the calculation of the number $J_N(\mu^*)$ of eigenvalues not exceeding μ^* :

If $T_{L+1}(\mu^*)/T_{L+1}(0) < 0$, then $J_{L+1}(\mu^*) = J_L(\mu^*) + 1$ or else $J_{L+1}(\mu^*) = J_L(\mu^*)$.

The process is initialized with the number J_0 which is obtained as described in § 4.1.b. since all the walls are independent.

c. EXAMPLE

We consider the two multicoupled walls of ref. [9]. The physical characteristics of the assembly are given in table 2.

	Wall 1			Wall 2			Steel
Layer	1	2	3	1	2	3	4
ρ_{ij}	144.	48.	96.	200.	700.	1000.	7800.
C_{ij}	961.	1045.	1045.	1045.	1045.	1045.	460.
λ_{ij}	.1084	.2419	.2468	.0896	.2703	.4981	40.
e_{ij}	.075	.05	.05	.075	0.1	0.1	.05

Table 2. Multicoupled walls of ref. [9] : Physical characteristics. External exchange coefficients : $h_{11} = h_{12} = 10$. Coefficients of the coupling matrix : $H_{11} = 125.19$, $H_{22} = 133.65$, $H_{12} = H_{21} = -29.55$ (S.I. units)

The counting technique fits without any trouble into a calculation program of the eigenvalues in accordance with the process which is described in § 4.1.d. Some results appear in table 3.

n	1	2	3	4	5
μ (s$^{-1/2}$)	.00655869	.0138624	.0213469	.0220677	.0252254
n	10	20	30	40	50
μ (s$^{-1/2}$)	.0506328	.106247	.159405	.214028	.263761

Table 3. Multicoupled walls of ref. [9] : Eigenvalues

Figs. 3 and 4 illustrate the whole method. We simulate an elementary heat treatment of steel operation in a batch-working electrical furnace.

The furnace being at ambient temperature, a steel load which is itself at the ambient temperature is put down on the furnace sole (floor). Then, the temperature of resistances is abruptly increased.

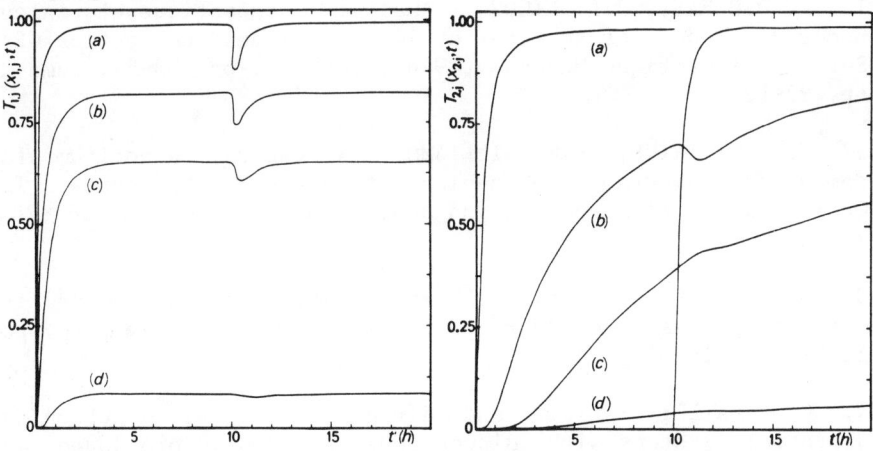

Fig. 3. Temperature variations in the ceiling :(a) internal face, (b), (c) interfaces, (d) external face

Fig. 4. Temperature variations in the floor (a) internal face, (b), (c) interfaces (d) external face

One follows the temperature rising of the ceiling (fig. 3) during 10 hours. At this time a "cold" load is introduced into the "hot" furnace. The new load induces temperature drops of the layers with more or less pronounced amplitudes and phase-lags.

In the same way one follows the temperature rising of the sole (fig. 4). The steel temperature which is homogeneous within 0,2 % cannot be distinguished from the internal face temperature (curve (a)). The double-mode heating of the second load (conduction in the sole and radiation from the resistances and from the ceiling) may be observed.

5. CONCLUSION

The counting of the number of eigenvalues smaller than a given bound makes the integral transform technique completely reliable even when applied to piecewise continuous media. Only one arbitrary datum, the initial increment of the eigenvalue parameter, permits one to initialize the eigenvalue computation. This parameter fits then itself the local properties of the eigenvalue equation depending of the multiplicity order of the roots. That property truly differentiates the used method from classical one.

One may think that, thanks to its safety the proposed method will soon open onto true thermal application packages.

REFERENCES

1. R. DAUTRAY, J.L. LIONS - Analyse mathématique et calcul numérique pour les sciences et les techniques. Collection CEA, Série scientifique Masson (1984), tome 1, pp. 33-59, tome 2, pp. 22-120

2. W.H. WITTRICK, F.W. WILLIAMS - A general algorithm for computing natural frequencies of elastic structures, The Quart. Jour. of Mechanics and applied maths., 24, pp. 263-284 (1971)

3. M.H. COBBLE - Dynamic vibrations and stresses in composite elastic plates, The Jour. of the Acoust. Soc. of Am., 46, pp. 1175-1179 (1969)

4. C.D. BAILEY - Exact and direct analytical solution to vibrating systems with discontinuities, Jour. of Sound and Vib., 44, (1), pp. 15-25 (1976)

5. J. GOSSE - Sur la conduction thermique variable dans un mur composite à propriétés constantes, C.R. Acad. Sc. Paris, série B, 286, pp. 303-306 (1978)

6. P. DUHAMEL, J. GOSSE - Analyse thermique d'un mur composite soumis sur une face à une variation en créneaux de la température, Int. J. Heat Mass Transfer, 23, pp. 1663-1671 (1980)

7. M. BOUZIDI, P. DUHAMEL - Study of the transient heat conduction regime in the composite wall of an anode firing furnace, Int. Heat Transf. Conf., Munich (1982)

8. M.D. MIKHAILOV, M.N. OZISIK, N.L. VULCHANOV - Diffusion in composite layers with automatic solution of the eigenvalue problem, Int. J. Heat Mass Transfer, 26, pp. 1131-1141 (1983)

9. M. BOUZIDI, P. DUHAMEL - Non-stationary Heat conduction in composite slabs with coupling. Application to enclosures : numerical computation of the analytical solution. Comp. and Maths with appls, 11, (10), pp. 1043-1055 (1985)

10. J.W.S. RAYLEIGH - The theory of Sound, Dover publ. N.Y. (1945) vol. I, pp. 119-126

11. G.P. MULHOLLAND, M.H. COBBLE - Diffusion through composite media, Int. J. Heat Mass Transfer, 15, pp. 147-160 (1972)

12. YI-HSU JU, WEN-CHIEN LEE - A method for solving transient diffusion problems, Num. Heat Transfer, 9, pp. 663-676 (1986)

DETERMINATION OF TEMPERATURE GRADIENTS IN A CIRCULAR CONCRETE SHAFT

L. Claes

Civil Engineering Departement
Katholieke Universiteit Leuven, Belgium

SUMMARY

Many circular concrete shafts have vertical cracks on their inner side due to solar radiation heat. In order to perform crack risk analysis for given environmental conditions, a finite difference model is proposed, calculating the internal temperature distributions on the condition that surface temperatures are known. These surface temperatures can be measured on an existing tower, using thermocouple wires. In that way actual temperature distributions can be computed which are the necessary data to check the crack risk of a shaft with its particular characteristics.

1. INTRODUCTION

In literature several papers can be found dealing with the problem of considerable cracking of constructions due to internal non-homogeneous temperature distributions. These temperature fields can have different origins. In thick concrete structures (e.g. foundations, dikes, dams) important temperature differences have to be dealt with due to the released hydration heat during concrete hardening[4]. Another kind of thermal load exists of temperature differences between the inner and outer environment of the construction, inducing an internal temperature gradient, e.g. off-shore reservoirs filled with hot oil and surrounded with cold water, silos filled with hot cement, chimneys conducting hot gas flows [3]. But even solar radiation heat has damaged several structures, e.g. bridges[2] and chimneys[3]. Many towers with a circular concrete shaft have two or three vertical cracks on the inside orientated to the south-west and west due to solar radiation heat[5].

In the past, the effect of temperature fields was very often treated in a simplified way or was not taken into account at all[1]. Reasons for doing so are propably the fact that little temperature measurement data were available and the lack of computer aid. The aim of our research is the determination of the actual temperature fields in circular concrete shafts exposed to solar radiation heat. Starting from these fields one can perform internal stress calculations and crack risk analysis, taking into account the characteristics of the reinforced concrete.

2. TACKLING THE PROBLEM

We will assume that the actual temperature fields do not depend on the longitudinal direction of the shaft. As we deal with axial symmetric structures, it is appropriate to write the differential equation, governing the heat conduction in a solid body, in cylindrical polar co-ordinates [9,10]:

$$\frac{\partial^2 T}{\partial r^2} + \frac{1}{r}\frac{\partial T}{\partial r} + \frac{1}{r^2}\frac{\partial^2 T}{\partial \theta^2} = \frac{1}{\beta}\frac{\partial T}{\partial t} \qquad (1)$$

in which : T is the temperature in point (r,θ) (°C);
r is the radial co-ordinate (m);
θ is the tangential co-ordinate (rad);
t is the time since the beginning of the calculation (s);
β is the thermal diffusion coefficient (m^2/s).

Equation (1) can be solved when the initial conditions and the boundary conditions are known. A study of literature concerning the environmental parameters which influence the thermal behaviour of a structure, like solar radiation and convection (depending on wind velocity) [e.g. 6,7,8], demonstrates that these phenomena constitute rather complex boundary conditions. A first consequence of this fact is that one has to use numerical methods when solving eq.(1). Thurston, Priestly and Cooke[11] indicated that finite element temperature analyses are in many cases unnecessarily sophisticated. For this reason we decided to use a finite difference model.

In order to determine the boundary conditions we decided to measure surface temperatures at one level on an existing tower. In this way we should be able to determine actual temperature fields, something which avoids the use of complex boundary conditions considering radiation and convection phenomena. Besides, those sophisticated formulas can deliver anything but approximations.

In the present paper we intend to describe the way we chose an appropriate finite difference model, taking into account the purpose of this model and its convergence and stability behabiour.

3. CONSTRUCTION OF FINITE DIFFERENCE MODELS

We consider a ring with inner radius a and outer radius b (fig. 1.). This ring is provided with a finite difference mesh which has P intervals Δr in the radial direction and Q intervals $\Delta\theta$ in the tangential direction. For the time we choose a time step Δt.

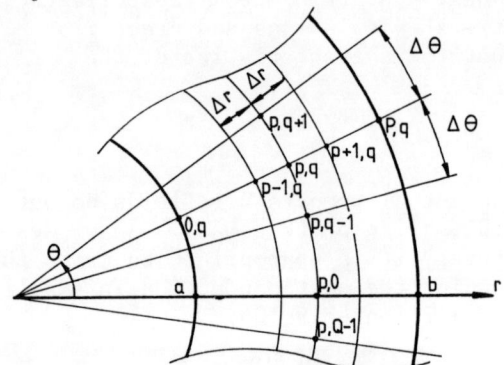

Fig. 1 : Finite difference mesh.

The partial derivatives in eq.(1) can be approximated by the following finite difference formulas :

$$\frac{\partial^2 T}{\partial r^2} \simeq \varphi \frac{T^{n+1}_{p-1,q} - 2T^{n+1}_{p,q} + T^{n+1}_{p+1,q}}{(\Delta r)^2} + (1-\varphi) \frac{T^n_{p-1,q} - 2T^n_{p,q} + T^n_{p+1,q}}{(\Delta r)^2} \quad (2a)$$

$$\frac{1}{r}\frac{\partial T}{\partial r} \simeq \varphi \frac{1}{r} \frac{T^{n+1}_{p+1,q} - T^{n+1}_{p-1,q}}{2\Delta r} + (1-\varphi) \frac{1}{r} \frac{T^n_{p+1,q} - T^n_{p-1,q}}{2\Delta r} \quad (2b)$$

$$\frac{1}{r^2}\frac{\partial^2 T}{\partial \theta^2} \simeq \psi \frac{1}{r^2} \frac{T^{n+1}_{p,q-1} - 2T^{n+1}_{p,q} + T^{n+1}_{p,q+1}}{(\Delta\theta)^2} + (1-\psi) \frac{1}{r^2} \frac{T^n_{p,q-1} - 2T^n_{p,q} + T^n_{p,q+1}}{(\Delta\theta)^2} \quad (2c)$$

$$\frac{1}{\beta}\frac{\partial T}{\partial t} \simeq \frac{1}{\beta} \frac{T^{n+1}_{p,q} - T^n_{p,q}}{\Delta t} \quad (2d)$$

In these formulas :

$T^n_{p,q}$: is the temperature in point $(r = a + p.\Delta r, \theta = q.\Delta\theta)$ after a time $t = n.\Delta t$;

φ and ψ : are two parameters of which the values still have to be chosen. Normally they vary between 0 and 1.

The approximations (2) deliver the following model :

$$\frac{1}{\beta}\frac{T^{n+1}_{p,q} - T^n_{p,q}}{\Delta t} = \varphi R^{n+1}_{p,q}(T) + (1-\varphi) R^n_{p,q}(T) + \psi \theta^{n+1}_{p,q}(T) + (1-\psi)\theta^n_{p,q}(T) \quad (3a)$$

in which :

$$R^n_{p,q}(T) = \frac{T^n_{p-1,q} - 2T^n_{p,q} + T^n_{p+1,q}}{(\Delta r)^2} + \frac{1}{r}\frac{T^n_{p+1,q} - T^n_{p-1,q}}{2\Delta r} \quad (3b)$$

$$\theta^n_{p,q}(T) = \frac{1}{r^2} \frac{T^n_{p,q-1} - 2T^n_{p,q} + T^n_{p,q+1}}{(\Delta\theta)^2} \quad (3c)$$

4. SOME DEFINITIONS ABOUT FINITE DIFFERENCE MODELS

Concerning the definitions of stability, convergence and compatibility of finite difference models, we rely on [12,13]. These definitions are shortly mentioned here and they are adapted to the difference model (3).

4.1. Stability

A model is called stable if for fixes values of Δr, $\Delta \theta$ and Δt the deviation between the exact solution of (1) and the approximating solution of (3) stays small when the number of time steps (n) tends to infinity.

4.2. Convergence

If for a fixed time $t = n \cdot \Delta t$ the approximating solution of (3) tends to the exact solution of (1) while Δr, $\Delta \theta$ and Δt tend to zero (which means that n tends to infinity), the model is called convergent. In other words, if our model is convergent we can reach any accuracy when choosing the values of Δr, $\Delta \theta$ and Δt sufficient small.

4.3. Compatibility

The truncation error TE for a given point and time (r,θ,t) or (p,q,n) is defined as follows :

$$TE = \left(\frac{1}{\beta} \frac{v_{p,q}^{n+1} - v_{p,q}^{n}}{\Delta t} - \varphi R_{p,q}^{n+1}(v) - (1-\varphi) R_{p,q}^{n}(v) - \psi \theta_{p,q}^{n+1}(v) - (1-\psi) \theta_{p,q}^{n}(v) \right) - \left(\frac{1}{\beta} \frac{\partial v}{\partial t} - \frac{\partial^2 v}{\partial r^2} - \frac{1}{r} \frac{\partial v}{\partial r} - \frac{1}{r^2} \frac{\partial^2 v}{\partial \theta^2} \right)_{p,q}^{n} \quad (4)$$

In (4) $v(r,\theta,t)$ is any function containing continuous partial derivatives. The model (3) is called compatible with eq.(1) if the truncation error TE tends to zero while Δr, $\Delta \theta$ and Δt tend to zero. A very interesting example when studying the compatibility of difference models is given by Du Fort and Frankel [12].

4.4. Lax's theorem

In practice it is much easier to prove the compatibility of a model than its convergence. Lax's equivalence theorem [13] states that under certain conditions for a compatible model stability is the necessary and sufficient condition for convergence.

We will examine the compatibility and the stability of (3) and assume that the model (3) fulfils the conditions to verify Lax's theorem. Afterwards the convergence itself is checked by means of a practical example.

5. COMPATIBILITY OF THE FINITE DIFFERENCE MODEL (3)

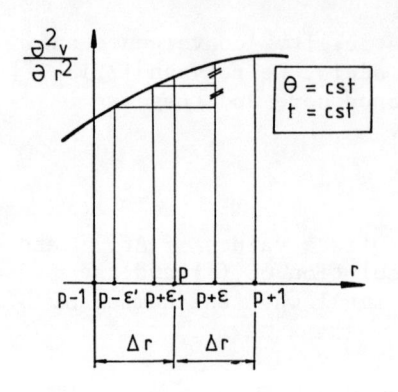

Fig. 2 : Illustration of eq. (7).

By Taylor's theorem and using Lagrange's rest term we can write

$$v^n_{p+1,q} = v^n_{p,q} + \Delta r (\frac{\partial v}{\partial r})^n_{p,q} + \frac{(\Delta r)^2}{2!} (\frac{\partial^2 v}{\partial r^2})^n_{p+\varepsilon,q} \qquad (6a)$$

$$v^n_{p-1,q} = v^n_{p,q} - \Delta r (\frac{\partial v}{\partial r})^n_{p,q} + \frac{(\Delta r)^2}{2!} (\frac{\partial^2 v}{\partial r^2})^n_{p-\varepsilon',q} \qquad (6b)$$

in which : $0 \leq \varepsilon \leq 1$
$0 \leq \varepsilon' \leq 1$

As $\frac{\partial^2 v}{\partial r^2}$ is supposed to be continuous, we can derive following equation (see fig. 2):

$$\frac{v^n_{p-1,q} - 2 v^n_{p,q} + v^n_{p+1,q}}{(\Delta r)^2} = (\frac{\partial^2 v}{\partial r^2})^n_{p+\varepsilon_1,q} \qquad ; -1 \leq \varepsilon_1 \leq 1 \qquad (7a)$$

In the same way one can derive :

$$\frac{1}{r} \frac{v^n_{p+1,q} - v^n_{p-1,q}}{2 \Delta r} = \frac{1}{r} (\frac{\partial v}{\partial r})^n_{p+\varepsilon_3,q} \qquad ; -1 \leq \varepsilon_3 \leq 1 \qquad (7b)$$

$$\frac{1}{r^2} \frac{v^n_{p,q-1} - 2 v^n_{p,q} + v^n_{p,q+1}}{(\Delta\theta)^2} = \frac{1}{r^2} (\frac{\partial^2 v}{\partial \theta^2})^n_{p,q+\varepsilon_5} ; -1 \leq \varepsilon_5 \leq 1 \qquad (7c)$$

$$\frac{1}{\beta} \frac{v^{n+1}_{p,q} - v^n_{p,q}}{\Delta t} = \frac{1}{\beta} (\frac{\partial v}{\partial t})^{n+\varepsilon_7}_{p,q} \qquad ; 0 \leq \varepsilon_7 \leq 1 \qquad (7d)$$

Using (7) the truncation error, defined in (4), can be written as :

$$TE = \frac{1}{\beta} \left((\frac{\partial v}{\partial t})^{n+\varepsilon_7}_{p,q} - (\frac{\partial v}{\partial t})^n_{p,q} \right) - \varphi \left((\frac{\partial^2 v}{\partial r^2})^{n+1}_{p+\varepsilon_2,q} + \frac{1}{r} (\frac{\partial v}{\partial r})^{n+1}_{p+\varepsilon_4,q} \right)$$

$$- (1-\varphi) \left((\frac{\partial^2 v}{\partial r^2})^n_{p+\varepsilon_1,q} + \frac{1}{r} (\frac{\partial v}{\partial r})^n_{p+\varepsilon_3,q} \right) + \left((\frac{\partial^2 v}{\partial r^2})^n_{p,q} + \frac{1}{r} (\frac{\partial v}{\partial r})^n_{p,q} \right)$$

$$-\psi \frac{1}{r^2} \left(\frac{\partial^2 v}{\partial \theta^2}\right)^{n+1}_{p,q+\varepsilon_6} - (1-\psi) \frac{1}{r^2} \left(\frac{\partial^2 v}{\partial \theta^2}\right)^{n}_{p,q+\varepsilon_5} + \frac{1}{r^2} \left(\frac{\partial^2 v}{\partial \theta^2}\right)^{n}_{p,q} \qquad (8)$$

in which : $-1 \leq \varepsilon_i \leq 1$ for $i = 1, 2, \ldots 6$
$0 \leq \varepsilon_7^i \leq 1$

Obviously

$$\lim_{\substack{\Delta r \to 0 \\ \Delta \theta \to 0 \\ \Delta t \to 0}} TE = 0 \qquad (9)$$

which proves the compatibility of (3) with (1).

6. STABILITY OF THE FINITE DIFFERENCE MODEL (3)

It is assumed that the temperature field for $t = 0$, that is $T°_{p,q}$, is known. $T°_{p,q}$ can be written as follows (see fig. 3)

$$T°_{p,q} = \sum_{\ell=0}^{Q} \sum_{k=0}^{P} a_{k,\ell} \cos k\pi \frac{r-a}{b-a} \cos \ell \frac{\theta}{2} \qquad (10a)$$

which assumes that $T°_{p,q}$ is an even function in the area ABCD.

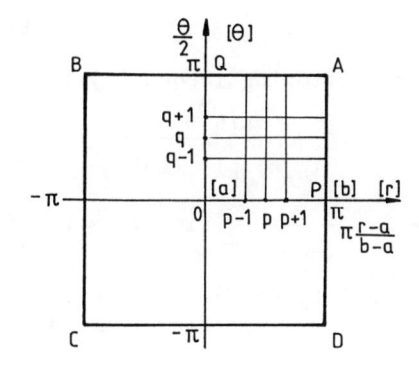

Fig. 3 : Illustration of eq. (10).

Eq. (10a) delivers a set of $(Q+1).(P+1)$ independent simultaneous equations by which the coefficients $a_{k,\ell}$ can be determined. Instead of its real form we prefer to use the complex notation of (10a) :

$$T°_{p,q} = \sum_{\ell=-Q}^{Q} \sum_{k=-P}^{P} \alpha_{k,\ell} e^{ik\pi \frac{r-a}{b-a}} e^{i\ell \frac{\theta}{2}} \qquad (10b)$$

in which $i = \sqrt{-1}$

We now investigate the possibility to write the temperature field for $t = n.\Delta t$ as :

$$T^n_{p,q} = \sum_{\ell=-Q}^{Q} \sum_{k=-P}^{P} \alpha_{k,\ell} e^{ik\pi \frac{r-a}{b-a}} e^{i\ell \frac{\theta}{2}} \xi(k,\ell)^n \qquad (11)$$

As $T_{p,q}^n$ is the solution of the finite difference model (3), substituting (11) into (3) delivers the condition which has to be fulfilled to verify (11). One can check that this condition is the following :

$$\xi(k,\ell) = \frac{1 + (1-\varphi)\varepsilon_1 + (1-\psi)\varepsilon_2}{1 - \varphi\varepsilon_1 - \psi\varepsilon_2} \qquad (12)$$

in which :

$$\varepsilon_1 = -4\tau \sin^2\mu + i \frac{\tau}{p_a + p} \sin 2\mu \qquad (13a)$$

$$\varepsilon_2 = -\frac{4\tau}{(p_a + p)^2 (\Delta\theta)^2} \sin^2\nu \qquad (13b)$$

and

$$\tau = \frac{\beta\Delta t}{(\Delta r)^2} \ ; \ \mu = \frac{1}{2} k \frac{\pi\Delta r}{b-a} \ , \ \nu = \frac{\ell}{4} \Delta\theta, \ p_a = \frac{a}{\Delta r} \qquad (14)$$

If for fixed values of Δr, $\Delta\theta$ and Δt the number of the time steps n tends to infinity, the temperature fields $T_{p,q}^n$ do not tend to infinity (which means that the deviations between approximating and exact solutions remain small) provided that the modulus of $\xi(k,\ell)$ is not greater than 1 :

$$|\xi(k,\ell)| \leq 1 \qquad (15)$$

Working out this condition delivers following inequality :

$$-8\left(1 - 2\tau(1-2\varphi)\sin^2\mu - 2\tau(1-2\psi)\frac{\sin^2\nu}{(p_a+p)^2(\Delta\theta)^2}\right)\left(\sin^2\mu + \frac{\sin^2\nu}{(p_a+p)^2(\Delta\theta)^2}\right) + (1-2\varphi)\frac{\tau}{(p_a+p)^2}\sin^2 2\mu \leq 0 \quad (16)$$

For given values of φ and ψ the stability condition (16) can be investigated more deeply, which delivers simplified conditions. Some cases are tabulated in table 1.

Group	φ, ψ	condition	example
1	$0,5 \leq \varphi \leq 1$ $0,5 \leq \psi \leq 1$	none	$\varphi = \psi = 0,5$ like Crank-Nicolson implicit model
2	$\varphi = \psi$ $0 \leq \varphi \leq 0,5$	$\beta\Delta t \left(\frac{1}{(\Delta r)^2} + \frac{1}{(a\Delta\theta)^2}\right) \leq \frac{1}{2(1-2\varphi)}$	$\varphi = \psi = 0$: full explicit model
3	$0,5 \leq \varphi \leq 1$ $0 \leq \psi \leq 0,5$	$\frac{\beta\Delta t}{(a\Delta\theta)^2} \leq \frac{1}{2(1-2\psi)}$	$\varphi = 0,5$ $\psi = 0$
4	$0 \leq \varphi \leq 0,5$ $0,5 \leq \psi \leq 1$	$\frac{\beta\Delta t}{(\Delta r)^2} \leq \frac{1}{2(1-2\varphi)}$	$\varphi = 0$ $\psi = 0,5$

Table 1 : Stability conditions

7. CHOICE OF APPROPRIATE VALUES OF φ AND ψ

The stability conditions impose restraints to the ratio of Δt to Δr and/or $\Delta \theta$. We intend to determine the temperature fields in a ring with :
- inner radius a = 2,90 m;
- outer radius b = 3,30 m.

In order to achieve a good knowledge concerning the actual temperature gradients we desire a finite difference mesh with a rather small value of Δr, e.g.

Δr = 2 cm, which means P = 20.

For $\Delta \theta$ we choose :

$\Delta \theta$ = 10°, which means Q = 36.

The greater the thermal diffusity β, the more severe the stability condition. As Neville [14] indicates that for normal weight concrete β varies between

$0,002 < \beta < 0,006$ m^2/h

we choose : β = 0,005 m^2/h

For Δt we choose :

Δt = 0,5 h.

The greater the time step Δt is, the less measurements data have te be recorded and worked up afterwards.
With these values we obtain :

$$\frac{\beta \Delta t}{(\Delta r)^2} = \frac{0,005 \cdot 0,5}{(0.02)^2} = 6,25 \gg 0,5$$

$$\frac{\beta \Delta t}{(a \Delta \theta)^2} = \frac{0,005 \cdot 0,5}{(2,9 \cdot \frac{\pi}{18})^2} = 0,0098 \ll 0,5$$

The great advantage of models belonging to group 2 and 4 (see table 1) is that one can choose φ = 0. But, in these cases the stability condition becomes too severe. To fulfil that condition Δt should drop to about 2 minutes or Δr should raise to more than 7 cm. Neither of these situations seems to be suitable for the object in view. It is true that for models of group 1 no stability condition has to be fulfilled, but it must be mentioned that for each time step (P-1).Q= 19.36 = 684 simultaneous equations have to be solved (for the case that the surface temperatures are known).
The stability condition of models of group 3 can very easily be fulfilled. Besides, when ψ = 0 is chosen, for each time step 36 sets of 19 simultaneous equations have to be solved, something which can be computed many times faster and much more elegantly than solving 684 equations. The matrix of each set of simultaneous equation appears to have only three diagonals different from zero and appears to be independent on the value of q and n. In [12] a very interesting algorithm is given to solve that kind of simultaneous equations. It is concluded that the Gauss-elemination have to be executed only one time. Afterwards, for each new q and n, the coefficients of the right hand terms have to be simply adapted.

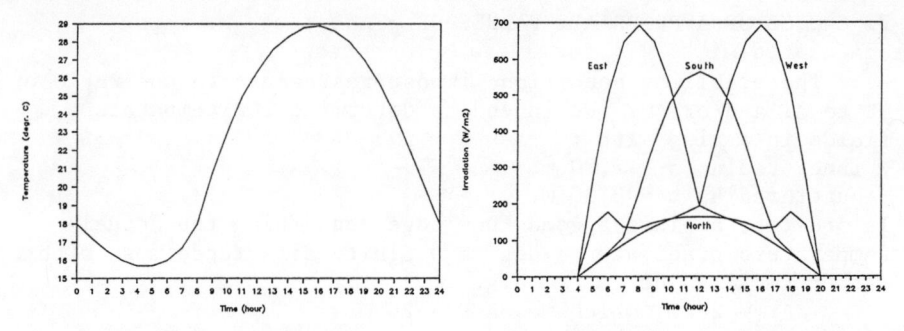

Fig. 4 : Typical course of air temperature

Fig. 5 : Irradiation on vertical planes (July 15th, Brussels)

8. PRACTICAL EXAMPLE

On the basis of a typical course of the air temperature for a summer day [8] (fig. 4) and calculated irradiations on vertical planes for clear days by means of a computer program from the Belgian Royal Meteorological Institute [15] (fig. 5) we made an estimation of what the surface temperatures for the outside of the given ring could be. These surface temperatures were developed in a double Fourier series of the following kind:

$$T(b,\theta,t) = \alpha_{o,o} + \sum_{n=1}^{2} [\alpha_{o,n} \cos \frac{n\pi}{\ell} t + \beta_{o,n} \sin \frac{n\pi}{\ell} t]$$

$$+ \sum_{m=1}^{2} [\alpha_{m,o} \cos m\theta + \gamma_{m,o} \sin m\theta]$$

$$+ \sum_{m=1}^{2} \sum_{n=1}^{2} [\alpha_{m,n} \cos m\theta \cos \frac{n\pi}{\ell} t + \beta_{m,n} \cos m\theta \sin \frac{n\pi}{\ell} t$$

$$+ \gamma_{m,n} \sin m\theta \cos \frac{n\pi}{\ell} t + \delta_{m,n} \sin m\theta \sin \frac{n\pi}{\ell} t] \qquad (17)$$

in which $2\ell = 24$ hours.
We determined the coefficients of table 2.

$\alpha_{m,n}$	n = 0	n = 1	n = 2	$\beta_{m,n}$	n = 1	n = 2
m = 0	26,910	-3,261	-0,241	m = 0	-4,771	-0,153
m = 1	1,114	-1,427	0,867	m = 1	-0,780	0,549
m = 2	-0,795	0,157	0,656	m = 2	0,088	0,416
$\gamma_{m,n}$	n = 0	n = 1	n = 2	$\delta_{m,n}$	n = 1	n = 2
m = 1	-0,002	1,701	-1,003	m = 1	-3,100	1,585
m = 2	0,001	0,079	-0,293	m = 2	-0,143	0,461

Table 2 : Fourier coefficients (°C)

Figures 6 and 7 show the outside surface temperatures
calculated with (17) for a few orientations.

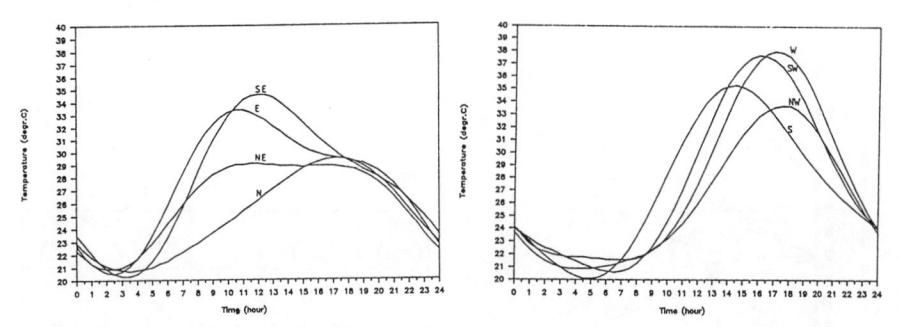

Fig. 6 : Outside surface
temperature

Fig. 7. : Outside surface
temperature

Fot the inside surface temperature we chose for simplicity :
$$T(a,\theta,t) = \alpha_{o,o} = 26,91°C. \tag{18}$$
As the surface temperatures given by (17) and (18) are periodic
functions of θ and t, one can calculate the analytical solution
of the differential equation (1), provided that one assumes that
the concrete shaft is in the steady state. This means that
after a period of 24 hours one obtains identical temperature
fields :
$$T(r,\theta,t) = T(r,\theta,t + n.24.3600); \quad n = 0,1,2,... \tag{19}$$
Figures 8, 9 and 10 show some of the calculated temperature
gradients.

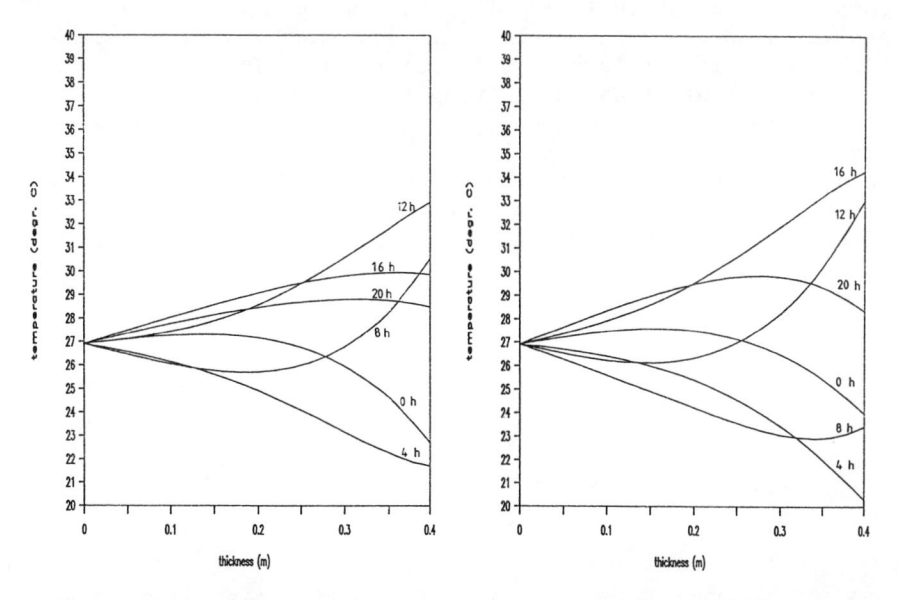

Fig. 8 : Temperature fields for
$\theta = -90°$ (East)

Fig. 9 : Temperature fields for
$\theta = 0°$ (South)

Fig. 10 : Temperature fields for θ = 90° (West)

Table 3 : Deviations ΔT between analytical and finite difference solutions

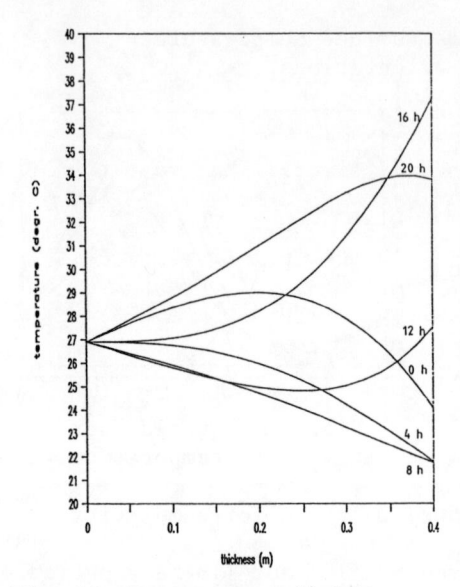

Δr (cm)	Δθ (°)	P	Q	ΔT (°C)
8	30	5	12	0,10
4	15	10	24	0,03
2	10	20	36	0,01

If the boundary conditions (17) and (18) are used for the chosen finite difference model ($\varphi = 0,5$ and $\psi = 0$), after a certain time t (in practice t is less than 10 days) the computed temperature fields $T_{p,q}^n$ have to be equal to the analytical steady state solution. We compared that solution with the finite difference solution for the 10th day for a few combinations of Δr and Δθ, giving the deviations of table 3. We find that the difference model is convergent for the given example.

We also calculated the finite difference solution for the 15th day and compared it with the finite difference solution of the 10th day. We concluded that there was no deviation at all which proves the stability of the model for the given example.

REFERENCES

1. LEONHARDT, F., LIPPOTH, W. - Folgerungen aus Schäden an Spannbetonbrücken,Beton- und Stahlbetonbau, 65.Jahrgang, Heft 10, October 1970, pp.231-244.
2. ELBADRY, M., GHALI, A. - Thermal stresses and cracking of concrete bridges, ACI-Journal, November-December 1986, pp. 1001-1009.
3. KUPFER, H., NOAKOWSKI, P. - Untersuchung der vertikalen Temperaturrisse in Schäften von Industrieschornsteinen aus Stahlbeton, Technische Mitteilungen - Organ des Hauses der Technik e.V. Essen, 72. Jahrgang, Heft 2/3/4, February, March, April, pp. 155-159
4. KURZAWA, J., KIERNOZYCKI, W. - Thermische Spanningen, hervorgerufen durch den Hydratationsprozess des Zements in massigen Fundamenten von Ingenieurskonstructionen, Bauingenieur 57, pp. 357-361 (1982).
5. LEONHARDT, F. - The art of building reinforced concrete towers, International Civil Engineering, July 1969, pp. 289-308.
6. DOGNIAUX, R. - Les rayonnements solaire, atmosphérique et terrestre, Institut Royal Météorologique de Belgique, Brussels, 1982.
7. GARG, H.P. - Treatix on solar energy - Vol. 1 : Fundamentals of solar energy, J. Wiley & Sons, Chichester/New York/Toronto, 1982.
8. KEHLBECK, F. - Einfluss der Sonnenstrahlung bei Brückenbauwerken, Werner Verlag, Düsseldorf, 1975.
9. GRÖBER, ERK, GRIGULL - Die Grundgesetze der Wärmeübertragung, Springer-Verlag, Berlin/Göttingen/Heidelberg, 1963.
10. CHAPMAN, A.J. - Heat transfer, the Macmillan Company, New York, 1967.
11. THURSTON, S.J., PRIESTLY, M.J.N., COOKE, N. - Thermal analysis of thick concrete sections, ACI-Journal, September October 1980, pp. 347-357.
12. SMITH, G.D. - Numerical solution of partial differential equations, Oxford University Press, London, 1969.
13. RICHTMEYER, R.D., MORTON, K.W. - Difference methods for initual - value problems, Interscience publishers, John Wiley & Sons, New York/London/Sidney, 1967.
14. NEVILLE, A.M. - Properties of concrete, Pitman publishing, London/Massachusetts, 1981.
15. DOGNIAUX, R. - Programme général de calcul des éclairements solaires énergétiques et lumineux des surfaces orientées et inclinées-ciels clairs, couverts et variables (Misc. série C - n°21), Institut Royal Météorologique de Belgique, Brussels, 1985.

FINITE ELEMENT ANALYSIS OF THE THERMAL FIELD INDUCED BY A BURIED POWER CABLE

Albino BUIZZA, Michele CARNIMEO and Marcello SYLOS LABINI(*)

ABSTRACT

A computer program for the evaluation of the thermal field around a buried power cable is described. The laying procedures are taken into account for the current-carrying capacity calculation in steady state condition. An algorithm for automatic mesh generation is reported on.

1. INTRODUCTION

The goal of this paper is to report on the use of a numerical technique - the Finite Element Method (FEM) - to analyze the stationary thermal field induced by a buried single-pole electric cable. Among the complex aspects related to the usage of electric cables the most complex one is its current-carrying capacity evaluation under specified conditions.
The allowable load depends on the air temperature and the laying and operating conditions considering the highest temperature admitted by the Standards [1] for the influence of temperature on the mechanical properties decay of the insulation material.
From the above ensues the need of a calculation method to determine with adequate precision the thermal field generated by a cable under given loading conditions.
This is important for the designer as well as for the user. For the designer, because it becomes a guide to the

---

(*) Facolta' di Ingegneria, Universita' di Bari, Italy.

selection of materials and building-layouts which are the best ones for a given application. For the user, because it helps him to choose the kind of cable to use in a given installation.

2. CALCULATION METHOD

The heat transfer under steady state conditions is described by the differential Poisson's equation. For isotropic media it assumes the following form:

$$\lambda \nabla^2 T(x, y, z) + Q(x, y, z) = 0 \qquad (1)$$

where:
λ = thermal conductivity (W/(m.K)),
T = temperature (K),
Q = heat generation rate (W/m^3).

A close solution to the above differential equation is unattainable when complex geometries and boundary conditions are involved as in most actual applications.

The diffusion of computers makes it possible to use numerical methods like the FEM which is the most versatile one. Fundamental aspects and details of the FEM theory can be found in existing literature [2, 3].

Essentially the method consists in subdividing the region of interest into finite elements in order to achieve an approximate solution of the above differential equation. A bidimensional region is considered here and linear triangular elements are selected.

To compute the temperature distribution in the domain the following assumptions are introduced.
a) An infinitely long and straight cable is considered;
b) a two-dimensional analysis is carried out;
c) the thermal conductivity of cable strand, insulation and ground is different;
d) the heat rate generation in the strands of the cable is different from that in the layer of insulation;
e) all the materials show thermal properties independent of temperature;
f) a steady state is considered.

From assumptions e) and f), the differential equation of thermal conduction under steady state conditions becomes the Poisson's equation within each homogeneous part of the ana-

lyzed region.
From b) the equation in cartesian coordinates becomes:

$$\frac{\partial^2 T}{\partial x^2} + \frac{\partial^2 T}{\partial y^2} = \frac{-Q}{\lambda} \qquad (2)$$

3. PROBLEM STATEMENTS

A power carrying cable shows resistance and dielectric losses leading to a non-negligible operating temperature. The computer code setted up in this research computes the thermal field generated when a steady state is reached by a single-pole electric cable laid at a finite depth.
The main aspects of the program are the following:
a) It computes the distribution of temperature in conductor, insulation and ground;
b) A boundary condition of convective exchange on the air-to-ground surface is assumed.

4. BOUNDARY CONDITIONS

Only a few researchers have dealt with the problem of determining the temperatures caused by buried power cables with the FEM.
It is known that the distribution of temperature depends on the following factors:
- heat generation rates,
- physical properties of the materials,
- boundary conditions.

By dividing the space into finite elements the FEM makes it possible to study in a simple manner a structure consisting of different materials. The heat generation rate per unit volume Q and the thermal conduction coefficient depend on the type of material. Then, if the structure is an inhomogeneous one it is necessary to subdivide into finite elements each homogeneous part.
About the boundary conditions it is necessary to remember that our problem concerns with a two-dimensional field with open boundary. In fact, the temperature field covers a whole half-plane (Fig. 1).
What kind of boundary conditions are to be imposed is a problem to be solved.
There are three methods to solve the problem of field

clousure. The first two of them are in fact mathematical techniques and we will make only a list of their merits and limitations. The most physical is the third one and we will dwell longer upon it.

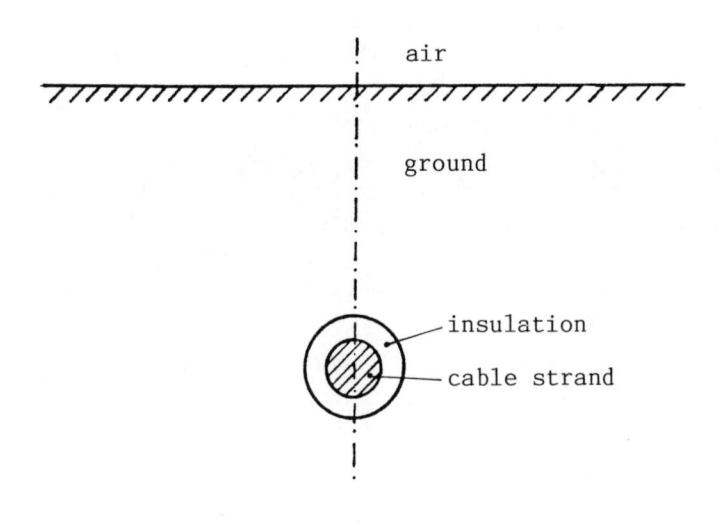

Fig. 1

1) Trefftz's method

In this method [4] the exact solution at some distance from the source in the form of series has to be known.
Thus the application range of the method is limited to simple shaped sources with a uniform environment around them. This makes the method inapplicable for real fields.

2) Second Green's formula

This method was used by Sroka [5] in developing a computer program for a cable buried under a homogeneous medium in steady state condition. The cable was treated as a line source and ground to air convective exchanges was not taken into account. The author observed after several numerical experiments that Dirichelet's and Newmann's condition at the boundary in the earth converge to the exact one. He points out however that the extension of the method to more complex configurations is a difficult task.

3) Dirichelet's condition

Several authors use the FEM in unbounded fields assuming a bounded region to study. They assume Dirichelet's condition on the boundary.

In [6] for example the author considers a square domain 6.1 m wide to study a buried cable. In so doing the computed temperature distribution would not be significantly different from that obtained for an unbounded field.

Of the three above methods we preferred the last one considering a finite domain with Dirichelet's conditions. The domain under consideration is illustrated in Fig. 2 where the system simmetry across the vertical plane through the cable axis is taken into account.

On the side AB the nodes reach a temperature different from that of the air and the air-to-ground boundary condition may be assumed as a heat convection exchange according to the following equation:

$$\lambda \frac{\partial T}{\partial n} = - h (T - Ta) \quad (3)$$

where:

λ = thermal conductivity of the ground (W/(m.K)),
h = ground to air convection coefficient (W/(m^2.K)),
Ta = air temperature (K),
T = surface temperature (K).

The side AD lies on the line of simmetry of the physical configuration. Owing to the simmetry no heat losses along the normal to the side can be assumed. Then,

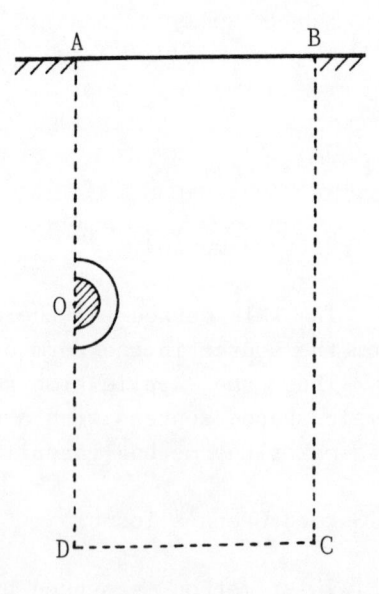

Fig. 2

$$\lambda \frac{\partial T}{\partial n} = 0 \quad (4)$$

The points lying on the side DC show the same underground level. If we are far enough from the cable we can assume approximately that all points are at the same temperature T.

A constant underground temperature (15 °C), of the same value as the mean annual temperature of the place where the cable is buried, was assumed.

That simplification was stated after performing a simulation of the daily temperature oscillation [7, 8] and observing that it reduces to about 2% of the initial value at a depth of 0.8 m and its mean value is that of the above temperature as the geothermal gradient is negligible at short distances from the surface.

The side BC lies so far from the buried cable that an adiabatic condition can be assumed.

5. AUTOMATIC MESH GENERATION

There are several advantages in generating automatically the mesh:
a) a reduced geometrical input data;
b) less possibilities for input mistakes;
c) a drastic reduction of the data input time;
d) an easy mesh refinement to improve the accuracy of the results.

To solve the problem of the buried cable a fine algorithm for the automatic mesh generation was set up (Fig. 3).

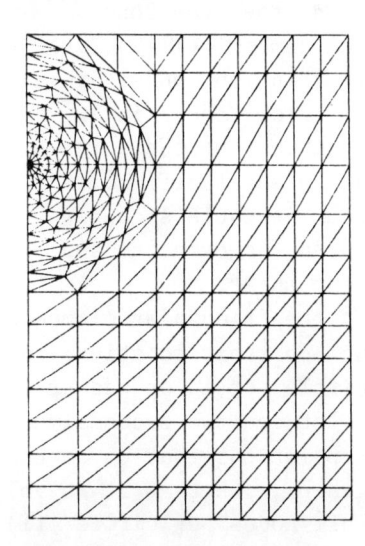

Fig. 3

The following input data are necessary:
- the cable's laying depth,

- both the conductor and the insulation radius (R1 and R2),
- the number of concentric circles to be assumed either in the conductor (N1) as in the insulation (N2),
- the total number of circles (SE),
- the number of nodes lying on the odd circles (TT),
- the domain width (LL),
- the number of nodes on the x-axis (NX),
- the number of nodes on the y-axis (NY),
- the domain height (PP).

A schematic mesh subdivision is presented in Fig. 4. It will be useful to describe the mesh generation.

At the end of the input phase the computer code carries out the generation of the nodes (numeration and coordinates) and the triangular elements (numeration and vertices' definition) for the zone A (Fig. 5).

To get the triangles of the zone A the same number of nodes is assumed on each circle. Even and odd circles are differentiated: on the odd one a given number of nodes (TT) is assumed, whereas on the even one a different number (TT+2) is assumed. Then the nodes are obtained from the intersection between the odd circles and a sheaf of half-lines from the cable's center and between the even one and a different sheaf.

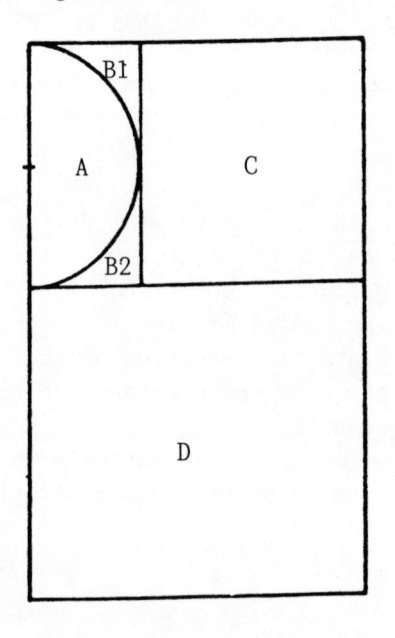

Fig. 4

For the triangular elements the generation is different depending on their location, namely if they are inside the first circle or in an odd or even ring. A ring is the mesh enclosed between two successive circles.

We point out that our method for generaring the mesh is quite general whereas it is able to produce a mesh with any number of nodes per circle (TT) and any number of circles (SE).

As the zones B1 and B2 are symmetric we may describe only one of them (Fig. 6). The figure shows how the closure problem in the corners of the above zones can be solved.

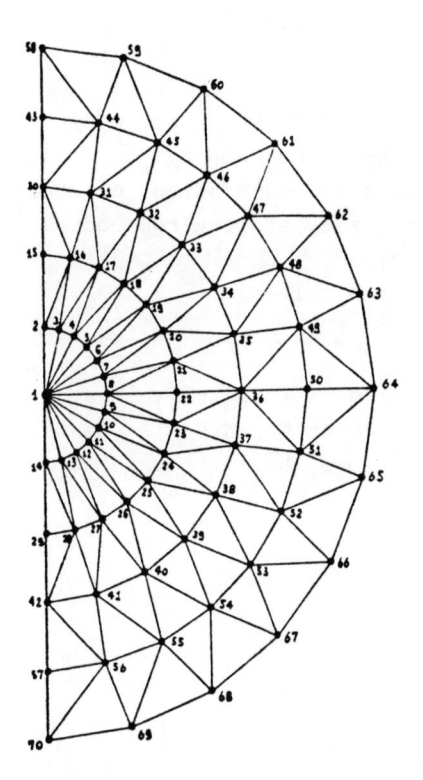

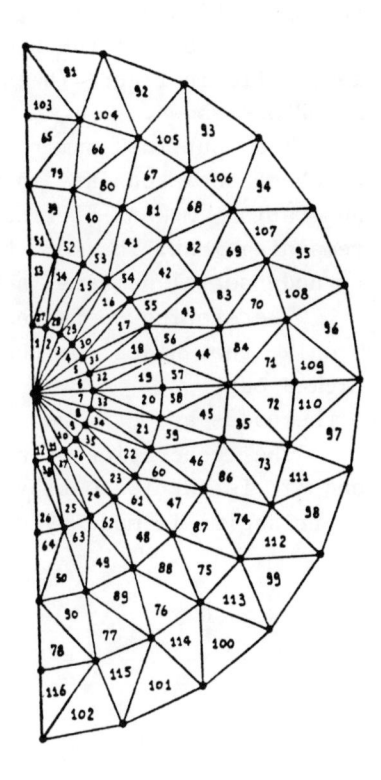

Fig. 5

   In order to get these nodal points it is necessary to make the following distinctions.
1) Forced nodes (59-63-65-69) - They lie on the last circle. Their coordinates are modified from the previously obtained values (as intersection of one sheaf of half-lines and the last circle). The change of coordinates shifts the nodal points where the segments bounding the zones B1 and B2 are tangent to the circle. That shifting avoids the generation of too many thin triangles.
2) Additional nodes (71...74-76...79) - They are generated by making them to lie on the boundary of the zones B1 and B2. The nodes on the horizontal side show the same y-coordinate of the corresponding nodes on the last circle. The ones on the vertical boundary, instead, show the same x-coordinate.

3) Vertex nodes (75-80) - They are located on the vertices of the zones B1 and B2.

In figure 7a the numeration of the triangles is presented, while in figure 7b the nodes and triangles of the zone A-B1-B2 are shown.

The outlined numeration procedure for the zone A-B1-B2 reaches the boundary in a random way. The keeping of that numeration makes difficult the automatic generation of the remaining mesh. The work becomes very easy storing the boundary node serial numbers in two arrays.

The storing operation followes two criteria:
- only the nodes on the boundary of the zone A-B1-B2 are picked out,
- the nodes are arranged for decreasing coordinates values.

The generation of nodes in the zones C and D is easier and it is carried out by horizontal rows while the generation of triangles is made by submeshes.

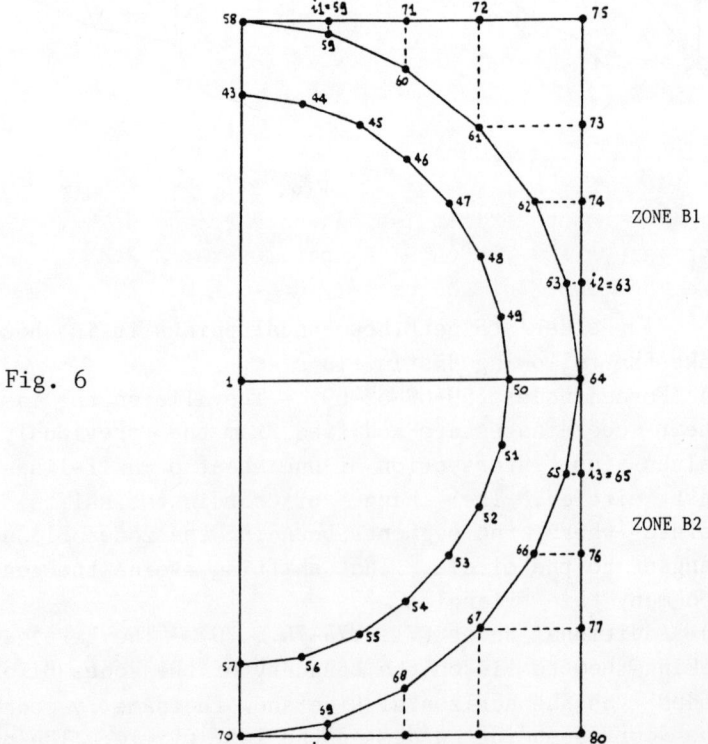

Fig. 6

a)

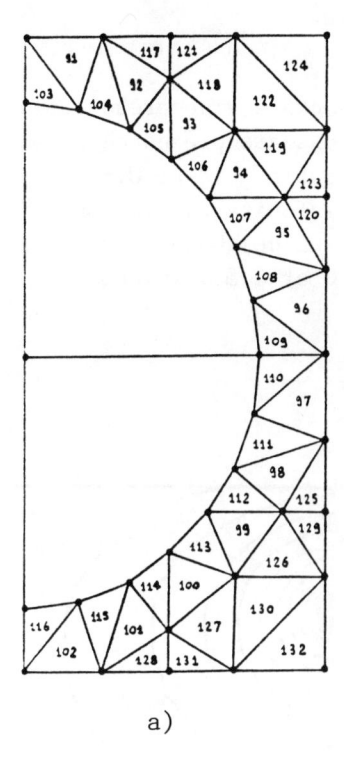

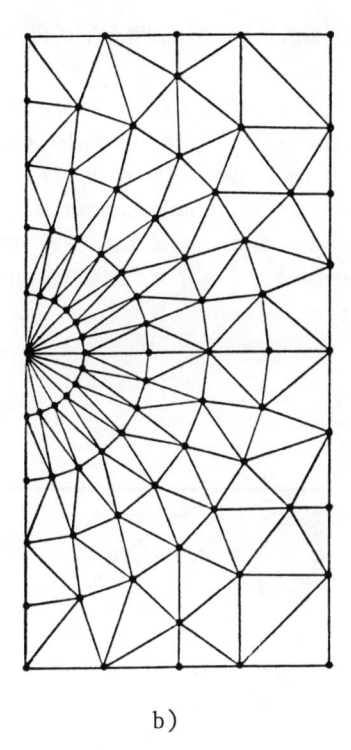

b)

Fig. 7

6. THE COMPUTER SIMULATION

The computer simulation was carried out with the following values for the physical parameters:
- Laying depth of the cable, H = 0.8 m. It is the distance between the cable centre and the ground surface: Its value is suggested by CEI Standards [1].
- Radius of electrical conductor, r1 = 0.0126 m. This value corresponds to the standard section S = 500 mm .
- Insulation radius, r2 = 0.015 m.
- Thermal conductivity of the conductor, λ_1 = 380 W/(m.K).
- Thermal conductivity of the insulation, λ_2 = 0.2 W/(m.K).
- Thermal conductivity of the ground, λ_3 = 0.7 W/(m.K).
- Heat generation rate in the conductor, Q1 = 38100 W/m^3. This value comes from the equation $Q = \rho G^2$, where ρ is the electric resistivity of copper and G = 1.33*10^6 A/m^2 is the current density in the conductor.
- Heat generation rate in the insulation, Q2 = 381 W/m^3. It was assumed equal to 1% of that in the conductor.

7. RESULTS AND CONCLUSIONS

The described computer code presents a widespread application field. As example two applications are presented here.
In the first one (Fig. 8) the cable laid at 0.8 m in depth in an homogeneous field of half-infinity exstension. In the second one (Fig. 9) the same laying depth is assumed in presence of a phreatic stratum of 15 °C at 2.5 m in depth.

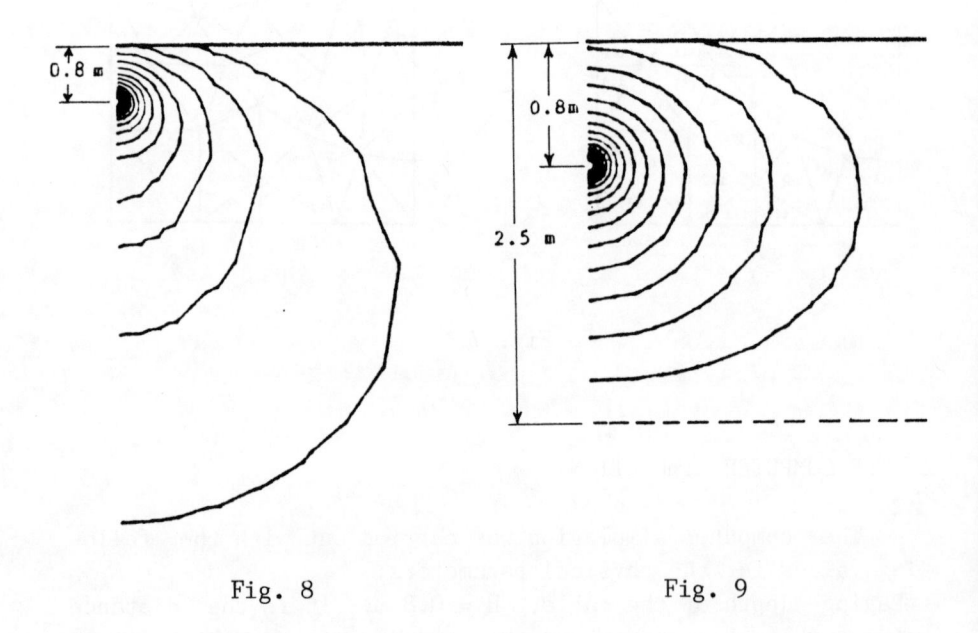

Fig. 8   Fig. 9

The different trend of the isotherms for the two cases can be seen in the above drawings.
In brief the main feature of the described computer code are the following:
- it is suitable to be implemented on personal computers;
- the precision we can acheive even with coarse meshes is widely acceptable for the final utilization of the results even accounting for the limited precision in the values of the physical properties of materials;

- complex problems of computing can be solved improving the quantity and quality of the data at users disposal in the current-carrying cables design;
- it is possible to analyze the variation of the thermal field around the cable as a function of several parameters (i.e. conductivity, convection coefficient, current rate, etc.) by the drawing of the isotherm lines;
- the mesh is automatically generated;
- the interaction of several cables can be analyzed.

ACKNOWLEDGEMENTS

This research has been funded by the CNR (Italian National Research Council).

REFERENCES

1. Italian Standards, NORME CEI 20-21 - Norme per la portata dei cavi elettrici in regime permanente. (Guidelines on the current carrying capacity of electrical cables in steady state conditions).
2. ZIENKIEWICZ, O. C. - The Finite Element Method, Third edition, Mc Graw-Hill, London, 1971.
3. DESAI, C. S. and ABEL, J. F. - Introduction to the Finite Element Method, Van Nostrand Reynold Company, New York, 1972.
4. KRZEMINSKI, S., SROKA, J., SIKORA, J. and WINCENCIAK, S. - Application of boundary solution procedure in thermal field analysis, Arch. Elektrotechnik, 62, 195 (1980).
5. SROKA, J. - The finite elements methods in unbounded temperature field of power cables, Arch. Elektrotechnik, 67, 1 (1984).
6. MITCHELL, J. K. and ABDEL HADI, O. N. - Temperature distributions around buried cables, IEEE Transactions on Power Apparatus and Systems, 98, 4 (1979).
7. BONDI, P. and SACCHI, A. - Experimental of thermal oscillations on large walls, Istituto di Fisica Tecnica del Politecnico di Torino, Internal Report n. 366/b, 1968.
8. BONDI, P., BUIZZA, A. and CIRILLO, E. - Sulla misura in opera dei parametri rappresentativi delle pareti in regime di temperatura costante e variabile periodicamente, Atti e Rass. Tecnica Soc. Ingg. ed Arch. in Torino, 32, 119 (1978).

NUMERICAL SIMULATION OF THE OPERATION OF NATURAL DRAFT COOLING TOWERS

Y. CAYTAN

Département I.T.A.
Direction des Etudes et Recherches
Electricité de France
6, Quai Watier - 78400 CHATOU (FRANCE)

ABSTRACT :

The operation of natural draft cooling towers is analysed by use of a numerical model which solves the Navier-Stokes bidimensional and "heat" equations in axi-symetrical geometry, as applied to the calculation of aerodynamics and to the thermodynamic exchanges inside a natural draft cooling tower.

The equation are solved by applying a method of finite differences to a non-regular rectilinear grid, using a fractionnal step algorithm ; the computation domain includes the near field of the tower and the inside of the tower, so that the boundary conditions are easy to determine ; the modelling of turbulence using the k-epsilon model allows for good modelling of hot and ambient air mixing.

The STAR model has been recently used to solve several problems raised by the assistance to the design and the operation of large cooling towers : the influence of the partial sealing of the air inlet and the influence of non-homogeneous water distribution on the minimum cold water temperature has been determined.

1. INTRODUCTION

The two-dimensional numerical model STAR (Thermodynamic and Aerodynamic Simulation of Cooling Towers) has been designed for the study of the operation of natural draft cooling tower without wind.

The first version of STAR developed solved the Navier-Stokes and heat equations using a finite difference method as applied to a regular grid, the simulation of turbulence being of the "mixing length" type. It is described in reference [1].

This version of the STAR model in the case of natural draft and cross-flow cooling towers was validated, in 1982 [1] by comparing it with on-site measurements taken by EDF on the cooling towers of BUGEY and DAMPIERRE.

However, the need to simulate increasingly complex cases of operating conditions (eg. anti-freeze operation) required the improvement of the STAR model. These developments secured interesting results relating to the influence of the shape of the shell on flow at tower outlet, described in reference [2]. The STAR model is presently used for analysis of two main ways for anti-freeze operation :

- the partial sealing of the air inlet by rolling shutters or icing of a wire netting,
- the sprayed surface reduction system.

These two devices are used separately or together to prevent icing of exchange surfaces by decreasing the efficiency of the tower, that means increasing the cold water temperature.

2. MAIN CHARACTERISTICS OF THE STAR MODEL

The STAR model involves the use of methods designed by EDF to simulate two-dimensional non-isothermal unsteady flows [3] [4].

2.1 Equations solved

The equations describing the phenomena entering the case of a cooling tower without wind are described in reference [1]. They are :

- the two-dimensional Navier-Stokes equations in an axi-symmetrical geometry ;
- the mass preservation equation of wet air ;
- the absolute humidity preservation equation of air ;
- the simplified energy balance equation for wet air ;
- the equation of state of wet air ;

- the circulation water preservation equation ;
- the simplified energy balance equation for circulation water ;
- the two equations of the k-ε turbulence model described in paragraphe 2.4. and
- the equation of water drops velocity in the spray below the packing, if there is one.

These equations are solved by a finite difference method applied to an orthogonal grid using a fractional step method.

2.2 Variable orthogonal grid and consideration of curved walls

The specific sizes of the "passageways" of the air circuit of a cooling tower vary substantially from one point of the installation to the next : the air inlet of a 2600 "thermal" MW natural draft tower is approximately 12 metres high, whereas the radius of the outlet of the tower is approximately 40 metres and the height of the shell exceeds 150 metres.

For the computation of flow to be accurate, each "passageway" should comprise at least ten computation points. Obviously, the use of a variable grid allows, for identical accuracy of description, a marked reduction in the number of computation points, or preferably, a better description for an identical number of computation points.

In addition to a better distribution of computation points, the variable orthogonal grid also takes account of curved walls by giving an approximate representation in a series of segments formed of the diagonal lines of the half-meshes of the wall. The computation of boundary conditions relating to normal and tangential speed at the wall automatically takes account of this curvature effect.

It is therefore possible to take account of curved walls, but this method is less flexible than finite element or finite difference methods using a curvilinear grid ; nevertheless, the advantage of the orghogonal rectilinear grid resides in the simplicity of discretization of equations.

2.3 Computational domain and boundary conditions

As indicated in figure 1, the computational domain comprises the inside of the tower and the near field.

The boundary conditions for enthalpy, humidity of the air and pressure are :

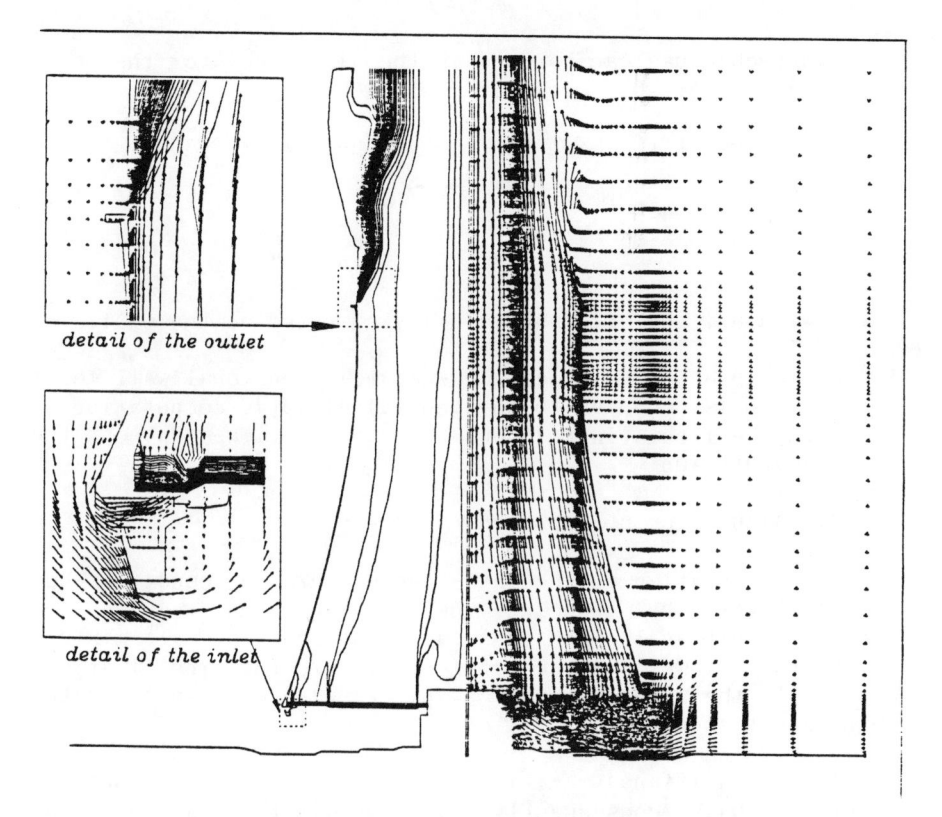

Figure 1 : Belleville cooling tower, without sealing of the air inlet ; air velocity and air temperature fields computed by the STAR model.

$$\frac{\partial h}{\partial n} = 0, \quad \frac{\partial x}{\partial n} = 0, \quad \frac{\partial p}{\partial n} = 0,$$ everywhere except for the "right-hand"

boundary of the domain. At this point, it is assumed that it is not disturbed by the cooling tower, pressure is constant and equal to the reference pressure value, and enthalpy and humidity of the air are equal to the values corresponding to the selected meteorological conditions.

There are two types of boundary conditions for velocity :

- non-disturbance over a boundary located in the fluid :

$$\frac{\partial \vec{v}}{\partial n} = \vec{0}$$

- friction on the wall.

Two techniques can be used in the STAR model for the near-wall modelling :

- the use of the wall function technique :

$$\vec{V}.\vec{n} = 0 \text{ and } \frac{\partial (\vec{V}.\vec{\tau})}{\partial n} = \frac{u^2_*}{\nu + \nu_T}$$

with u_*, the friction velocity computed at the wall.

- the use of a local low-Re one-dimensional wall model which is coupled to the general elliptic computation of the recirculating flow, at each time-step of the computation.

2.4 Simulation of turbulence in k-ε

The simulation of the anti-freeze operating conditions of cooling towers by reducing the sprayed area requires an accurate "portrayal" of the turbulent hot-cold air mixture inside the tower ; the efficiency of the installation, especially the minimum cold water temperature, which is the main result sought.

The traditional mixing length schematic turbulence simulation was consequently abandoned for a k-ε type simulation. This method, described in reference [4], is a good compromise between mixing length models and the sophisticated models developed today which solve a transport equation for each turbulent stress.

The values of the "universal" constants of the k-ε model used in STAR are standard values ; they are indicated in reference [4] and considered as "physical constants" in the STAR model, not as parameters for fitting the model.

2.5 Data and results

The data of the STAR model are limited to :

- vertical profile of temperature and moisture of ambient air,
- horizontal profil of hot water flow and temperature at the inlet of the fill,
- local coefficients of drag for the fill, drift eliminators and struts,
- local exchange coefficient for the fill,
- droplet's diameter.

The result are :

- velocity, temperature, water content and pressure fields of the moist air,
- water flow and temperature fields which give the mean cold water temperature.

3. STUDY OF THE PARTIAL SEALING OF THE AIR INLET

To decrease the efficiency of the tower, the simplest idea is to decrease the air flow crossing the tower by partial sealing of the air inlet.

For that purpose, two devices can be used :

- icing of a wire netting,
- mechanical rolling shutters.

The last system allows maintaining the temperature of the cooled water in a range where the efficiency of the installation is at its maximum, whatever the outside temperature.

For the design of such a system, several questions have to be answered :

- For given meteorological conditions and cooling range, what is the sealed height required to avoid icing of the fill ?
- What are the operating penalties with such a method ?
- Is the sealing of the upper part more efficient than the sealing of the lower part of the air inlet ?
- What about the effect of the channel located in the air inlet of some cooling towers equipped with a water recovery system ?

The use of the STAR model, combined with the use of more simple models, provides answers to those questions. The present case of study is the cooling tower of Belleville ; it is a counter-flow natural draft wet cooling tower, coupled to a 1300 MWe nuclear unit. The water is recuperated directly beneath the exchange fill so as to dispense with the pumping height corresponding to the rain which play a small thermal role.

The figure 1 provides the results of the computation (air velocity and temperature fields) without sealing of the air inlet, for given operating conditions. The computation time is about 2 hour of CRAY XMP with a grid of 7000 points.

sealing 'up'
⊙———⊙ mean temperature
△-----△ min. local temperature

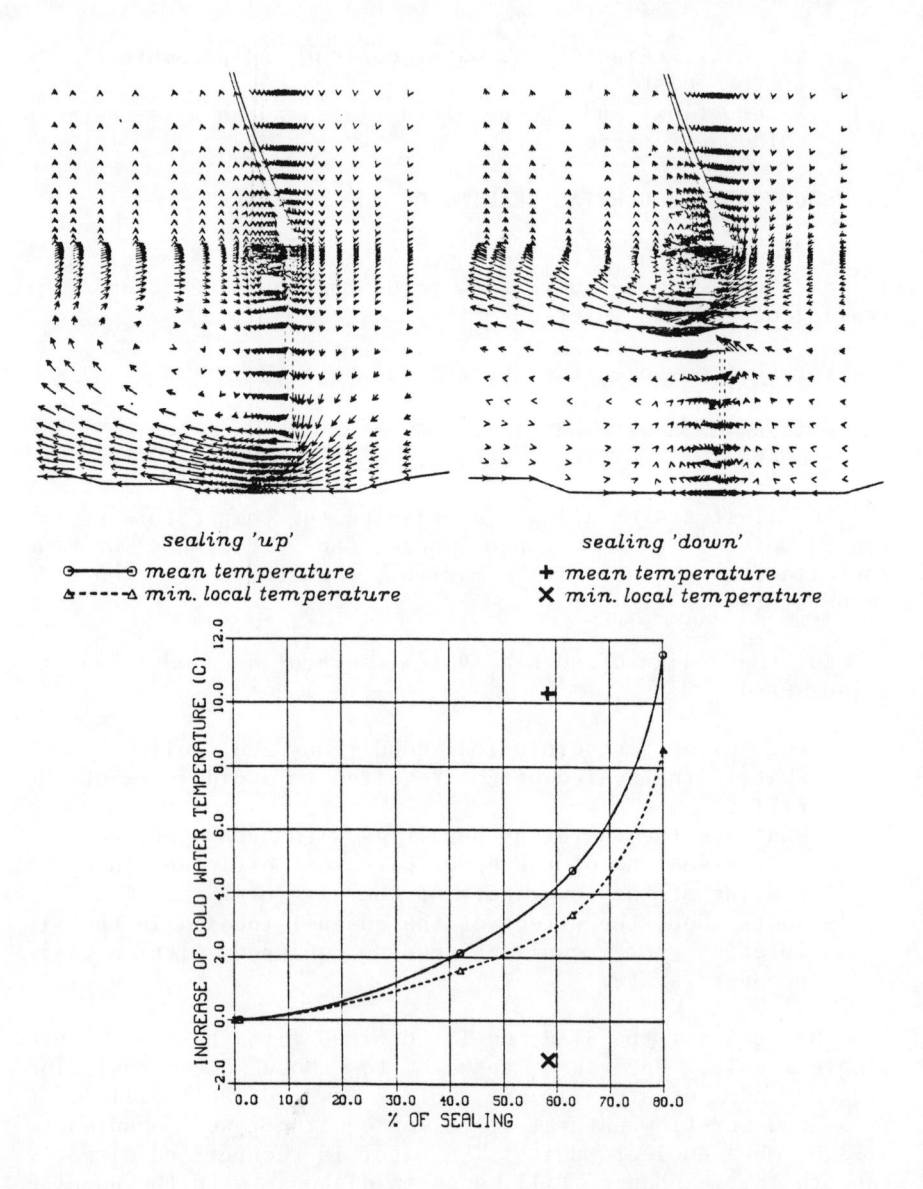

sealing 'down'
+ mean temperature
× min. local temperature

Figure 2 : Belleville cooling tower ; effect of the sealing of the air inlet computed by the STAR model.

The figure 2 provides the results concerning the influence of the air inlet sealing on the cold water mean temperature and on the cold water minimum local temperature. It can be seen that the sealing of the lower part of the tower is not efficient : the minimum local water temperature decreases, due to the presence of the channel located in the upper part of the air inlet.

About the sealing of the upper part of the air inlet, the difference between the cold water mean temperature and the cold water minimum temperature allowed to compute the operating penalties with this system.

4. STUDY OF THE SPRAYED SURFACE REDUCTION SYSTEM

In cold weather, the central area of the exchange fill is isolated (approximately 45 % of the total fill) by transferring all the hot water flow to the periphery. The entire zone or merely half can be isolated.

In this configuration, the air flow in the tower comprises streams of hot air which have crossed the sprayed surface and streams of cold air which have crossed the unsprayed surface. Frontal velocity is always higher in the cold zone than in the sprayed zone. This is due to the decline in the load loss coefficient of the unsprayed zone. Velocity and temperature gradients are therefore very high and flow is clearly heterogeneous.

The efficiency of the system is due to two main factors :

- the exchange surface is reduced,
- the effect of natural draft in the tower is also reduced by the cold air flow.

The present case of study is the Dampierre cooling tower. It is a counter-flow natural draft wet cooling tower, coupled to a 900 MWe nuclear unit. In this tower, there is a rain zone under the packing, due to the cold water falling down from the packing to the cold water basin. Thus, the "cold" air which crosses the unsprayed surface, has already crossed the rain zone under the sprayed surface. Due to the high water loading on the sprayed surface, the heat and mass exchange and the momentum exchange in the rain zone are higher than in normal cases.

The figure 3 provides the results of the computation in the case of operating conditions measured on the seal tower. The ambient temperature was higher than zero Celcius.

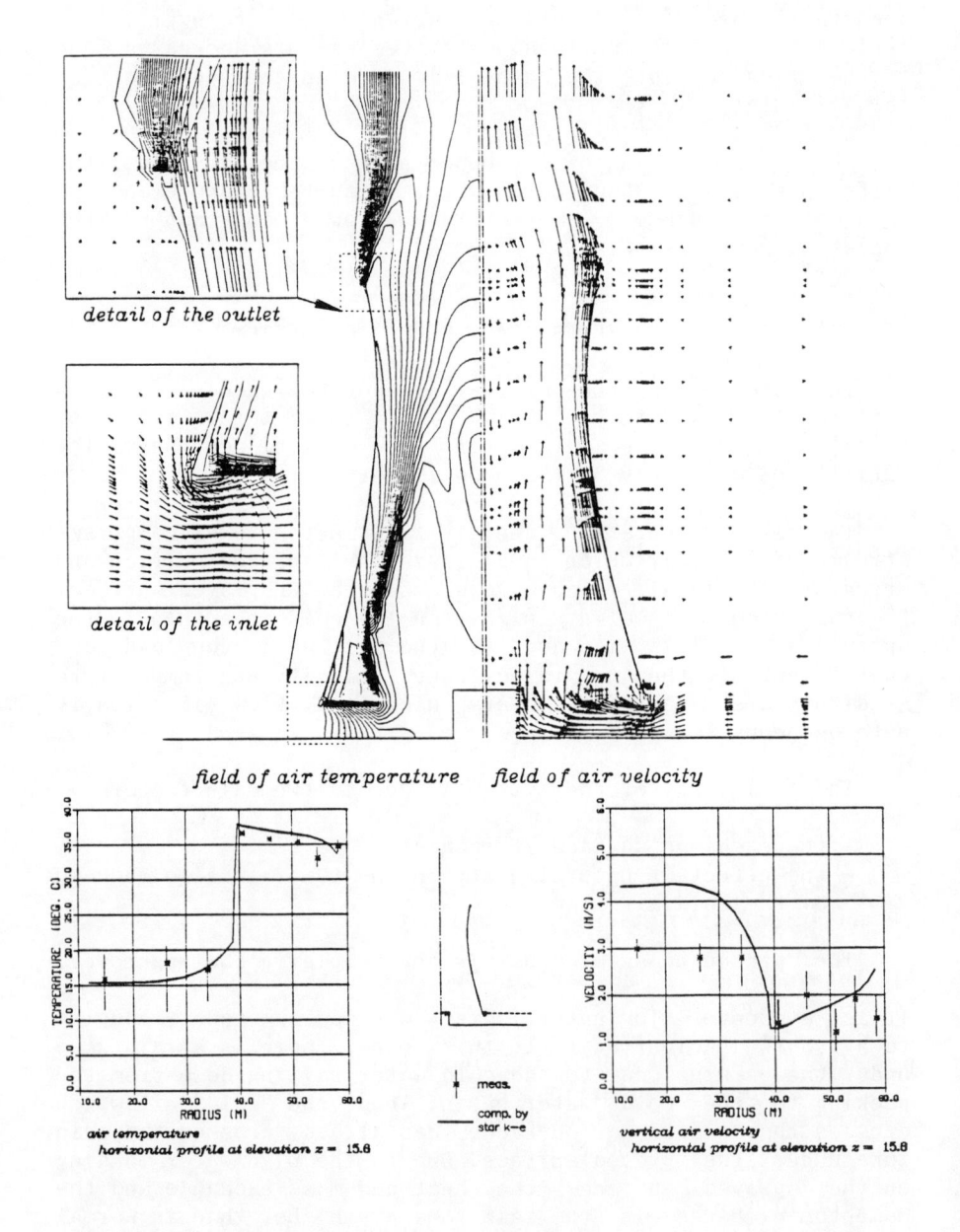

Figure 3 : Dampierre cooling tower ; anti-freeze operation computed by the STAR model

The air velocity and temperature profiles above the exchange surface are compared to the on-site measurements. The temperature profile is well represented but the velocity above the unsprayed zone is overestimated. Despite this fact, the mean cold water temperature is predicted with reasonable accuracy (less than 0.3°C). The computation time is about 1.2 hour of CRAY XMP with a grid of 4000 points.

5. CONCLUSION

Since its validation through its comparison with on-site measurements taken on the cooling towers of BUGEY and DAMPIERRE, the STAR numerical model has been used to verify the sizes proposed by manufacturers and to examine some two-dimensional effects.

Requests for the study of increasingly complex phenomena (taking account of the exact geometry of the project, high heterogeneities in air temperature) have required that the model be improved.

The use of a variable grid results in the specific consideration of the geometry of the shell, while k-ε modelling of turbulence and the near-wall modelling yields and accurate simulation of mixing in the tower and the interaction between the fluid and the wall.

BIBLIOGRAPHICAL REFERENCES

1. CAYTAN Y.
   Validation of the two-dimensional numerical model "STAR" developed for cooling tower design.
   IAHAR Cooling Tower Workshop, Budapest, october 1982

2. CAYTAN Y.
   Numerical simulation of the natural draft of a cooling tower.
   International Symposium on Buoyant Flows, Athens, Greece, September 1986

3. VIOLLET P.L., BENQUE J.P., GOUSSEBAILE J.
   2D numerical modelling of non isothermal flows for unsteady thermal - hydraulic analysis.
   Nuclear Science and Engineering 84, 350-372 - 1983

4. VIOLLET P.L.
   The modelling of turbulent recirculating flows for the purpose of reactor thermal-hydraulic analysis.
   Nuclear Engineering and Design 99, 1987

MICROCOMPUTER SOFTWARE FOR PRELIMINARY ANALYSIS AND
DESIGN OF COOLING OF ELECTRONIC EQUIPMENTS.

P.Ciresa
SMA,Firenze,Italia

SUMMARY

This paper deals with the problem of the cooling
of the electronic packing.
It is a major challenge for the designers,due to its
difficulty and impact on the reliability and
performance of the equipment.
Sophisticated computational techniques can
be available,but,as will be shown,sometimes they
could be not cost-effective.In these cases,generally
preliminary stages where rather accurate but quick
calculations are needed,traditional methods of
calculations can be enhanced by means of the
microcomputer.
In this paper a microcomputer software package is
shortly described,based on the classical
methods,very easy-to-use,very interactive in order
to be effectively used by not specialists also.
Some preliminary test results show an acceptable
agreement with the predictions.

NOTATIONS

T   :   temperature
Q   :   heat power
h   :   convection heat transfer coefficient
S   :   surface area
μ   :   efficiency of a fin
Cp   :   specific heat
m   :   mass air flow
x   :   fin thickness
y   :   fin spacing
p   :   pressure
V   :   air speed
L   :   length of the duct
k   :   thermal conductivity of the air

De   : equivalent diameter
U   : overall heat transfer coefficient
ΔTm   : log mean temperature difference between
          hot & cold fluids
ΔTio: inlet-outlet temperature rise/drop in the
          cold/hot fluid.

1. INTRODUCTION

The cooling requirements of electronic packing are often a major problem for the designers.
The decreasing size of the hardware,furthermore,implies additional efforts in order to achieve an effective heat removal with acceptable temperatures rises and pressure drops.
The task of the designers can be considerably supported by the computer aided design methods and facilities,that have reached a very high level, like the related numerical computation and/or simulation.It appears quite obvious that the importance of the cooling problems be faced by means of advanced computer methodologies.In the reality,sometimes the things go differently. Often, pragmatically experienced designers or draftmen need to perform quick and reasonably accurate calculations in order to check very preliminary design hypotheses.
In these cases it's not possible or effective to use large,sophisticated computer programs and facilities,due to:
-cost
-required time for the calculations
-required skilled and specifically experienced operators
-uneffectiveness of detailed calculations at the very early stage of the project.
The use of the microcomputer,a powerful and inexpensive tool,can be the evolutional link between the old style cumbersome hand calculations or manipulations of empirical nomographes and the advanced methods,e.g. FEM.
In this paper a practical application of this concept is shown,consisting in an easy-to-use software package for the analysis/design of devices for forced air cooling of electronic equipments,with an emphasis on finned cold plates and heat exchangers.The traditional methods for forced convection,[1],[2],[3],[4],[5],have been translated into microcomputer programs.

2.GENERAL CHARACTERISTICS OF THE PROGRAMS

The software for microcomputers,in the author's opinion,needs to have some peculiarities,in order to encourage also the users at first less experienced;it should be:
-easy to use
-highly interactive (self explaining)
-involving as low memory occupation as possible.
The programs have been written in IBM BASIC interpreter under MS/DOS operating system.
The BASIC language is the optimum,in the author's opinion,for this kind of application,because of:
-full capability to program engineering problems at any level of complexity,
-high degree of interactivity
-advanced graphics features.
As said before,the package is intended for designers and draftsmen,in order to enable them to quickly analyse their preliminary layouts,without any special skill or knowledge in computer sciences and/or thermophysics.
The following tasks are performed by the modules of the package:
1) optimization,with respect to the max allowable temperature rise of the hardware and/or the cooling air,of the thickness and spacing of the rectangular fins,given:power,overall size of the cold plate,acceptable pressure drop,air flow;
2) analysis of the finned cold plate,calculating the max temperature rise of the cooling air and hardware,pressure drop, given: geometry of fins and plate,power,air flow;
both aforesaid modules are based on the evaluation of the coefficient of convective heat transfer for turbulent air flow,and the efficiency of the fin,taking into account the optimum balance between conductive (along fin) and convective heat transfer mechanisms,[C];
3) analysis of the ducted fin cold plate,i.e. a finned plate with a cover plate,resulting in side-by-side ducts,with calculation of the temperature rises and pressure drop in both cases of laminar and turbulent air flow.The heat conduction along the fin is taken into account.

3 THEORETICAL BACKGROUND

The problems of the forced convection cold plate and air-air heat exchanger have been faced by means of traditional methods.The cold plate can be simple fin or ducted fin type.The heat exchanger is made by cells in which alternately cold and hot air flow.

3.1 Optimization/analysis of the finned plate

Given: width, height, length of the cold plate, dissipated heat power, air flow, air pressure drop, air properties (temperature, thermal conductivity, density, kinematic viscosity, specific heat), hardware material conductivity.

is calculated: the optimum fin thickness and spacing related to a certain allowable max. temperature difference between the cold plate hardware and inlet air. The fin efficiency is taken into account. The algorithm for optimization,[6], is shortly described here. Starting from the basic equation:

$$\Delta T = Q/(hS\mu) + (Q/Cp*\dot{m}) \qquad (1)$$

where $\Delta T = (Tmax\ hardware - Toutlet\ air) - (Toutlet\ air - Tinlet\ air) = Tmax\ hardware - Tinlet\ air$,
after algebraic manipulations, Q can be expressed in terms of $\Delta T, x, y$ and a parameter E. Differentiating Q with respect to x and equating to zero, the condition is found for which Q is maximum as a function of x;

$$(\sinh \alpha)/\alpha = (5y+x)/(5y-x) \qquad (2)$$

where $\alpha = 2*E(x+y)^{1.4}/(x^{0.5}xy^{1.5})$.

Equation (2) can be plotted, obtaining the locus of max values of Q as a function of x,y and parameter E. In order to have the design compatible with the air pressure drop, one can start from an expression of Δp for a turbulent flow:

$$\Delta p = constant*\rho*V^{1.2}*L/(y*Re^{0.2}) \qquad (3)$$

obtaining, after algebraic manipulations, a relation between x, y and the parameter P, function of Δp. This can be plotted, giving curves whose intersections with the E curves yield values for x and y which satisfy both pressure drop compatibility and heat transfer optimization.

The process can be reversed:
given: overall geometry of the cold plate, dissipated power, air and hardware properties, fin thickness and spacing;
E, P are evaluated from the curves described above and air flow, air pressure drop, temperature rise are calculated.

3.2 Cold plate with ducted fins

A greater efficiency is obtained by means of "ducted fins", i.e. putting a closing plate on the top of the fins. Uncertainties whether the flow is laminar or turbulent can rise, due to the variability of the thickness of the boundary layer of the flowing fluid, in a certain range of values of Re.

The following variables are taken into account, [5]:
-coefficient of convection heat transfer
 for laminar flow:

$$h = 1.86 * Pr^{.33} * k/De * (De/L)^{.33} * Re^{.33} \qquad (4)$$

for turbulent flow:

$$h = .023 * Pr^{.33} * k/De * Re^{.8} \qquad (5)$$

- pressure drop
 The Darcy-Weisbach equation expresses the pressure drop in terms of height of a column of the flowing fluid:

$$\Delta p = f * L/De * V^2/2/g \qquad (6)$$

The coefficient f is evaluated according to the Hagen-Poiseuille formula: $f = 64/Re$ for laminar flow, $Re < 2000$, and the Prandtl formula $f = .184/Re^{.2}$ for turbulent flow, $10,000 < Re < 3,000,000$;

-temperature rises
 ΔT between outlet and inlet air $= Q/\dot{m}/Cp$
 ΔT " wall and fluid $= Q/h/S$
The temperature drop due to heat conduction along the base plate and fin height is also taken into account.

3.3 Heat exchanger

In the considered heat exchanger the cold and hot air flow in parallel and alternate ducts.
The rate and energy equations simultaneously apply to the heat exchanger design:

$$Q = U * S * \Delta Tm \qquad (7)$$
$$Q = m * Cp * \Delta Tio \qquad (8)$$

In this kind of application, generally are:
known: inlet and outlet temperatures of cold air
unknown: "   "   "   "   " hot ".

The hot air temperatures are calculated by means of the Mikheyev method,[4].

4. SOME TEST RESULTS

Some tests have been carried out in order to check the accuracy of the predictions.

Only a few test results are available because the activity is in the initial stage. They are synthesized here.

4.1 Finned cold plate

A typical cold plate with ducted fins, shown in fig.1, has been tested.

G= 631 Watt

	measured		calculated
T air out - T air in	14.5	degree C	14.4
Tmax wall - T hot air	15	"	16.4

4.2 Air-air heat exchanger

The heat exchanger of the type described above is shown in fig.2.

	meas.		calc.
T cold air out - T cold air in	16	deg.C	14.8
T hot air in - T hot air out	17	"	13.8
(T hot air - T cold air) inlet	48.1	"	41.8
(T " " - T " ") outlet	15.1	"	13.2

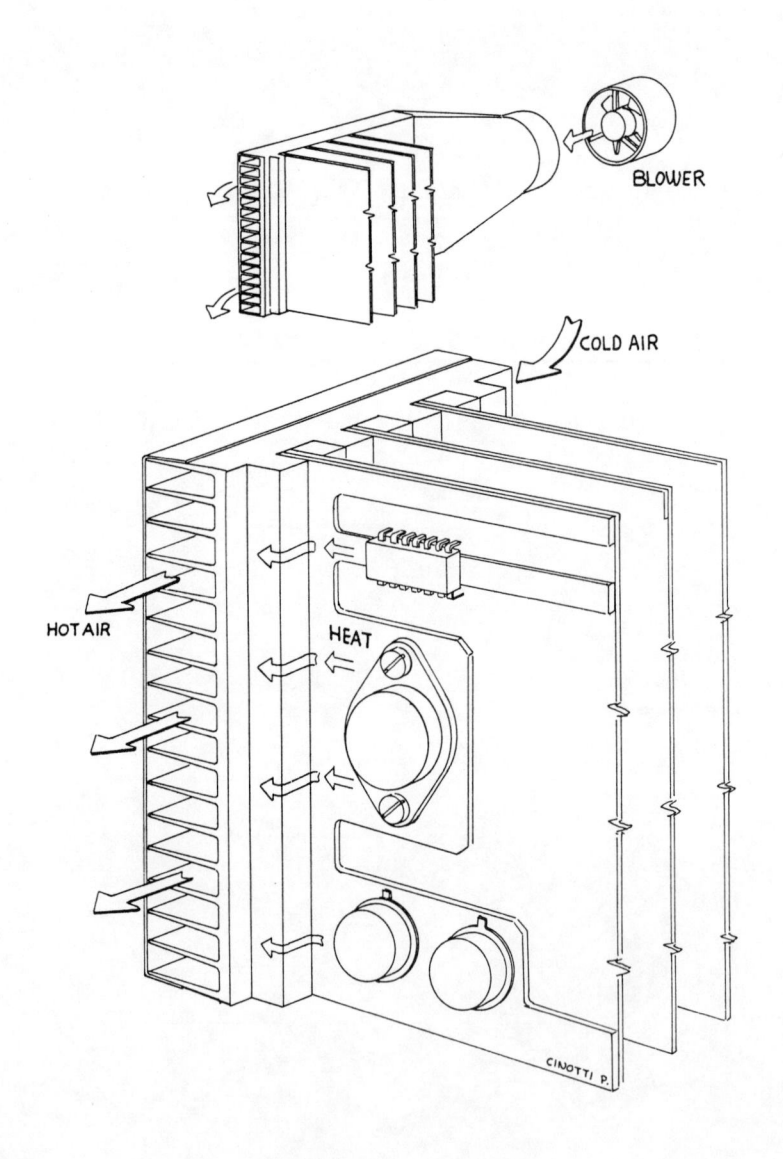

FIG. 1

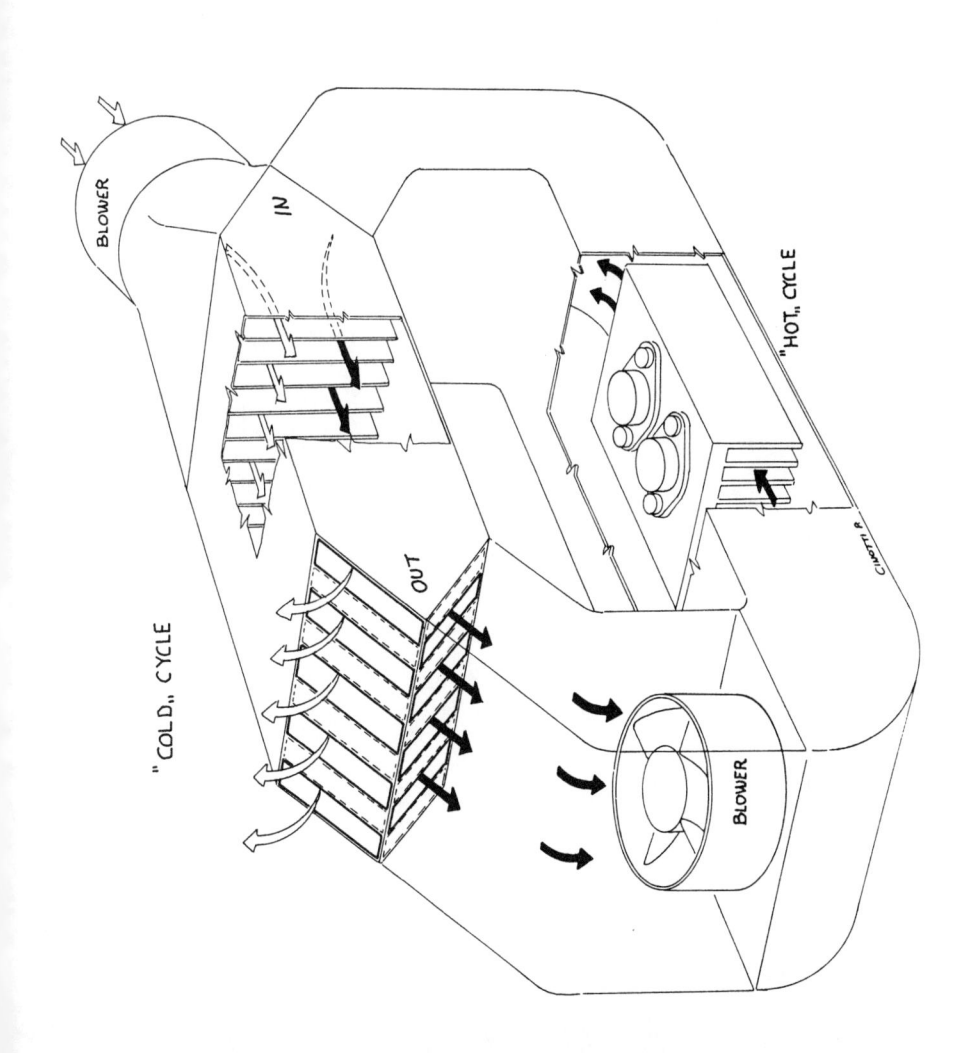

FIG. 2

REFERENCES

1) MC ADAMS,W.H.- <u>Heat transmission</u>,McGraw Hill Kogakusha Ltd.,Tokyo,1969

2) SCOTT,A.W.- <u>Cooling of electronic equipment</u>, J. Wiley & Sons,New York,1974

3) HARPER, C.A.-<u>Handbook of electronic packaging</u>,Mc Graw Hill,New York,1969

4) MIKHEYEV,M.-<u>Fundamentals of heat transfer</u>,MIR Publishers,Moskow,1968

5) MIL-HDBK-251 Reliability/Design Thermal Applications,1978

6) TADMOR,M. and FRENKEL H.-Optimum performance heat sink procedure, IEEE Mechanical Engineering in Radar

A MATHEMATICAL MODEL OF THERMAL FIELD IN SEMICONTINUOUS COPPER CASTING

A. Del Puglia, F. Fontanella, O. Quilghini, B. Tesi and G. Zonfrillo
Engineering Faculty, University of Florence, Italy
P. Lombardi
La Metalli Industriale, Florence, Italy

SUMMARY

To simulate a semicontinuous casting process, a mathematical model has been studied and solved numerically by a finite differences technique and by the Strongly Implicite Procedure; non-convergence problems of the numerical solution are examined.
With this model it is possible to determine - in function of the selected casting conditions and alloy chemical composition temperature distribution in the forming ingot, consequent thermal flows and liquid-solid interface configuration. It is also possible, with this method, to evaluate the projected working capacity of different size moulds.

1. INTRODUCTION

This work deals with the numerically solved mathematical model of the industrial semicontinuous casting process. In this casting process the moulds consist of a hollow copper, graphite lined, ring into which cooling water flows (fig. 1). When the process begins, the moving block closes the bottom port of the mould. After the liquid metal has filled the mould and a sufficiently thick solid case is formed, the moving block is lowered at constant speed V (casting rate); at the same time new liquid metal is fed into the mould. As the ingot forms, secondary cooling by direct water jet is carried on outside the mould; cooling is then completed in the water tank. A subsequent stationary condition is reached as shown in fig. 1. The developed model refers to circular section ingot of Desoxidised High Phosphor (DHP) Copper generally designed for subsequent hot plastic working.

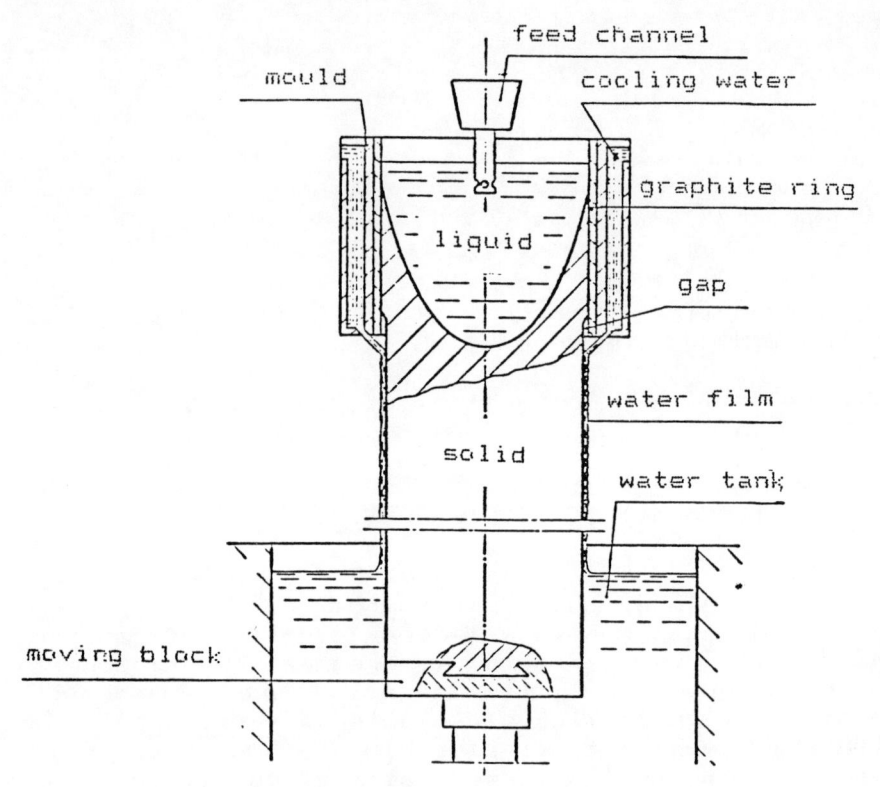

Fig. 1. Semicontinuous casting apparatus scheme

A detailed description of the process and of the results, obtained by means of the numerical simulation, are reported in [1,2].

2.MATHEMATICAL MODEL
In the process the main physical aspect concerns heat exchange - in the forming ingot as well as between the ingot and the cooling system - in the presence of the liquid-solid transformation and with material thermal characteristics variable with temperature.
Since the casting rate is constant, the solidification process, after the starting period, is independent from time; the temperature field T is stationary with regard to a reference system fixed at the mould.
Furthermore, the process is axial-symmetric; so a bidimensional cylindrical coordinate system - r, x -, with the origin in the centre of the liquid metal surface, has been used as reference.
Thus the problem can be studied in the rectangular field: $0 \leq r \leq R$, $0 \leq x \leq L$, where R is the internal mould radius and L is a reference height assumed equal to three times the mould height. In this field the

adopted scheme permits evaluations of heat exchange in both radial and axial directions, taking into account the various heat exchange conditions on the side ingot surface.
Though the liquid-solid transformation occurs at a supercooling value, in function of the thermal gradient and hence of the radius, it has been assumed constant to overtake analytical difficulties in the mathematical model.
Mathematically, this is a typical Stefan's problem: heat conduction in a material that presents a phase transformation with a heat emission or absorption, and with a previously undefined boundary between the two different phases.
Both analytical [3,4,5] and numerical [6,7,8] research has been carried out to solve technical problems of this type.
For stationary conditions, the whole heat exchange process including the state transformation can be represented [9] by:

$$(1) \quad C(T) V \frac{\partial T}{\partial x} = Qi + \frac{1}{r}\frac{\partial}{\partial r}[\, rK(T)\frac{\partial T}{\partial r}\,] + \frac{\partial}{\partial x}[\, K(T)\frac{T}{x}\,]$$

where C and K are the thermal capacity and the thermal conductivity respectively, and Qi is the internal heat generation. To solve this nonlinear parabolic equation, a new function — integral conductivity — has been defined

$$(2) \quad \Phi(r,x) = \int_{Tr}^{T} K(n)\,dn$$

where Tr is a reference temperature.
Thus, the equation (1) can be re-written as

$$(3) \quad \frac{\partial^2 \Phi}{\partial r^2} + \frac{1}{r}\frac{\partial \Phi}{\partial r} + \frac{\partial^2 \Phi}{\partial x^2} + Qi - \frac{C(T)}{K(T)} V \frac{\partial \Phi}{\partial x} = 0$$

2.1. Boundary conditions

Boundary conditions of a Neumann type have been fixed, but a selected constant temperature value has been applied only for the zone affected by the feed channel.
In particular, in the side ingot surface, the general boundary condition can be expressed by

$$(4) \quad K(T) \left(\frac{\partial T}{\partial r}\right)_{r=R} = U(T - Tw)$$

where the total heat exchange coefficient U and the cooling water temperature Tw are functions of the height x.
Thus the different modes of heat flow, as shown in fig. 1, depend on the x coordinate. On the basis of

the process scheme and of the hypothesis that in the mould wall only radial thermal flow is to be considered, six different zones in the heat exchange process can be identified:
- zone of contact between liquid metal and mould wall;
- zone of contact between solid metal and mould wall;
- zone of detachment between ingot and mould wall, consequent on volume contraction effect;
- zone between the mould and the water jet;
- zone where the water film flows along the external ingot surface;
- zone where the ingot is dipped into the water tank.

For each of these zones the process is realized and clearly set out by using a definible value of the coefficient U, determined on the basis of the thermal resistances to the heat flow from the forming ingot to the cooling system water.

3. NUMERICAL FORMULATION

The physical problem is reduced to the integration of (3) with suitable boundary conditions. Such a mathematical problem must be solved by numerical techniques: a finite differences method has been chosen.

Grid size selection was necessary in order to define with sufficient accuracy the position of the liquid-solid interface and, at the same time, to avoid a too large computer time. Further, a constant radial step δr and an axial step δx piece-wise constant, were established using δx minimum, where the thermal gradient is forecast maximum, in order to determine the isotherm configuration in a satisfactory way.

Thus, each node is marked with two indices n and m, n=1,2,...N, m=1,2,...M; these indices define the node position along the x and r directions respectively.

The last radial grid points (m = M) have been placed half a step outside the actual ingot section. The value assumed by the Φ function in these nodes, to which no actual material volume is associated, is defined on the basis of the boundary conditions expressed by the equation (4).

The finite difference equation for the node (n,m), n=1,2...N-1, m=1,2...M-1, is given by

(5) $D1\ \Phi_{n,m+1} + D2\ \Phi_{n,m-1} + D3\ \Phi_{n+1,m} +$
       $+ (D4 + D5)\ \Phi_{n-1,m} - (D5 + D6)\ \Phi_{n,m} = -D7$

where the coefficients Di, i=1,2...7 depend on the

node position, but D5 and D7 depend also on the casting rate and on the thermal characteristics of the material.
In this way the physical problem is reduced to the solution of the non linear algebraic system

(6) $$A(\Phi) \cdot \Phi = B(\Phi)$$

where A is a non-symmetric dominant diagonal matrix of order gxg, with $g=(N-1) \cdot (M-1)$ and with only five diagonal elements different from zero. In the examined cases $g \approx 1600$.
For solving (6), an iterative method has been used: at step "i" a system like (6) has been solved, but in this system A is replaced by a constant matrix W:

(7) $$W \cdot \Phi^{(i)} = B(\Phi^{(i)})$$

$$W = A(\Phi^{(i-1)})$$

The vector B must be considered as a function of $\Phi^{(i)}$, taking into account the internal heat generation and the boundary heat exchange.
Therefore the B components are different from zero only for the nodes close to the liquid-solid interface and for the nodes closest to the boundary. Because most of the input data are only in these non zero terms, the problem would be quite different if B were taken as a constant vector.
After several attempts, a solution of the system (7) was reched using the Strongly Implicit Procedure (SIP) method [10]. This was then specifically developed to solve large systems of algebraic equations derived from the approximation of partial differential equations with finite differences methods.
However, in the SIP technique, which involves the solution of an equation set that is a modification of the original system, the new coefficient matrix (indicated as W+S) must be factorizable in the product of a lower triangular matrix E for an upper triangular matrix F; E and F having at most only three non zero elements in each row. Moreover, the modified matrix must guarantee a quick convergence if it is used in an iterative method. While the equation of the original system, relating to the node (n,m), involves only the Φ value of the examined point and those of the four adjacent points, the modified equation, considering the form of the matrix $E \cdot F$, also involves the Φ value of two further diagonally adjacent points. The two new terms are approximated by means of Taylor expansion, and their influence is minimized through

an appropriate definition of the terms of the matrixes E and F in relation to the terms of the original matrix W, using a numerical corrective factor σ $(0 \leq \sigma < 1)$.
The generic iteration is given by

(8) $\quad (W + S) \bar{\Phi}^{(k+1)} = (W + S) \bar{\Phi}^{(k)} - (W \bar{\Phi}^{(k)} - B)$

Further, in order to reduce the round-off errors and to increase the convergence rate the algorithms suggested in [10] have been used.
Still, in the solution of (7) some problems arose concerning the convergence of $\{\bar{\Phi}^{(i)}\}$. In fact, though limited to a few nodes, those with non zero B components, there were non-convergence problems.
It has been noted, however, that there were fewer or no non-convergence problems if $B(\bar{\Phi})$ inputs corresponded to either a stronger heat exchange, which pre-supposes a more efficient mould than the actual one, or a hypothetical less latent heat value. For these reasons, the liquid-solid interface changes, during subsequent iterations, can be said to over-influence the temperature field.
To overcome this difficulty the latent heat was considered as generated in more times rather than in only one, and a similar procedure was adopted regarding the boundary changes.
From a mathematical point of view, the B variations between two subsequent iterations have been limited through two suitable coefficients p_i, $i=1,2$ $(0<p_i \leq 1)$, where $i=1,2$ refers respectively to latent heat and boundary changes.
For $p_i=0$ B is a constant vector; for $p_i=1$ the latent heat generation and the boundary changes are considered just in one time.
An example of the effect induced by these coefficients is shown in fig. 2. Here, for a grid node involved in the state transformation, both the internal heat generation, actually used in the calculus ($p_1=0.1$), and the non-corrected one ($p_1=1$) are reported at each iteration. It can be noted that the internal heat generation is equal to zero at the start iterations, because the liquid-solid interface is far from the considered node; at the subsequent iterations, owing to the interface crossing the heat generation is reduced and delayed by the coefficient p_1.
Moreover, the coefficient p_1 influence on the residual vector, defined by

(9) $\quad\quad\quad\quad H = B - W \cdot \bar{\Phi}$

has been analysed. The fig. 3 shows, for different

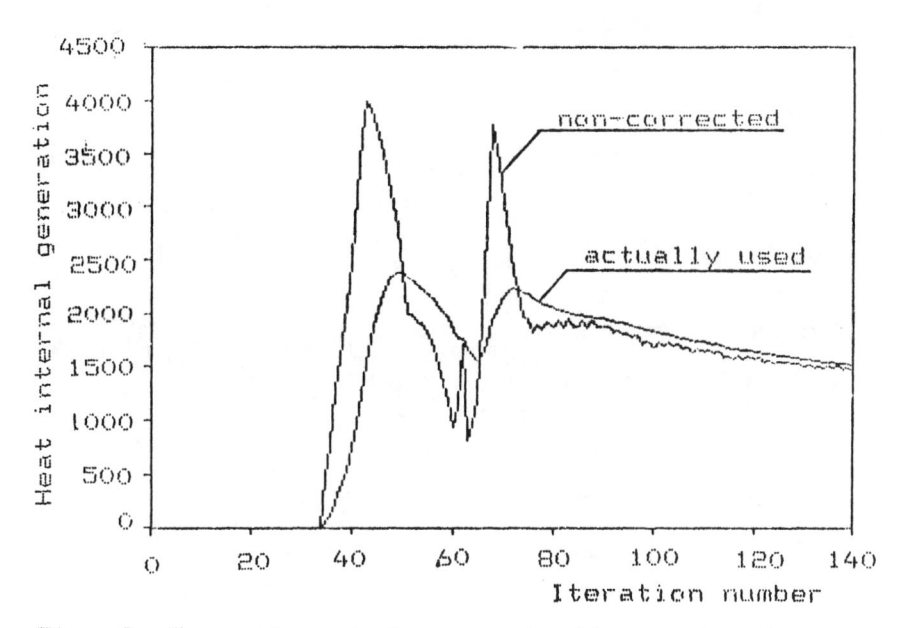

Fig. 2. Comparison between actually used and non-corrected internal heat generation.

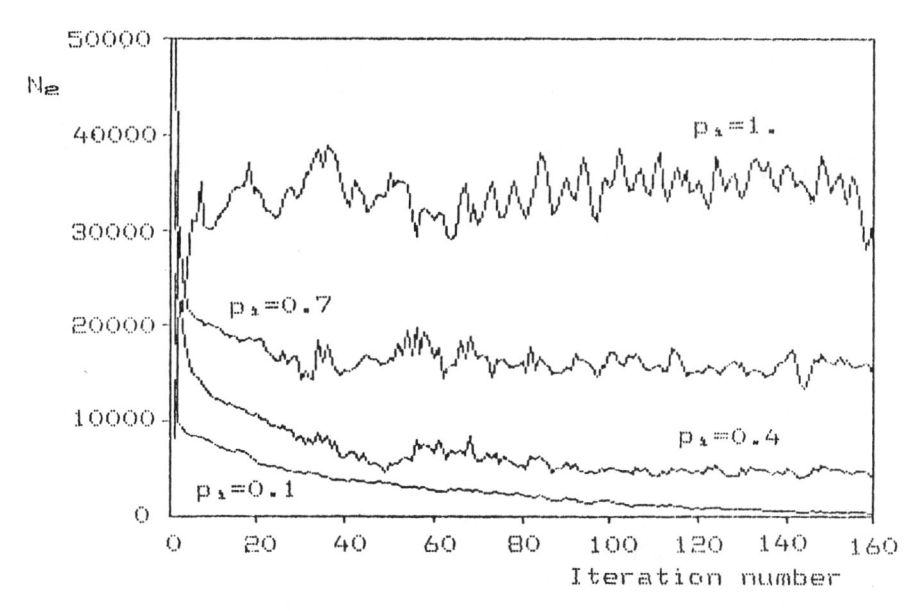

Fig. 3. Euclidean norm of the residual vector vs. iteration number.

p_1 values, the Euclidean norm N_2 of the residual vector as a function of the iteration. The fig. 4 shows a similar diagram for the magnitude N_∞ of the largest element of the residual vector.
As it can be seen from these figures, to obtain a

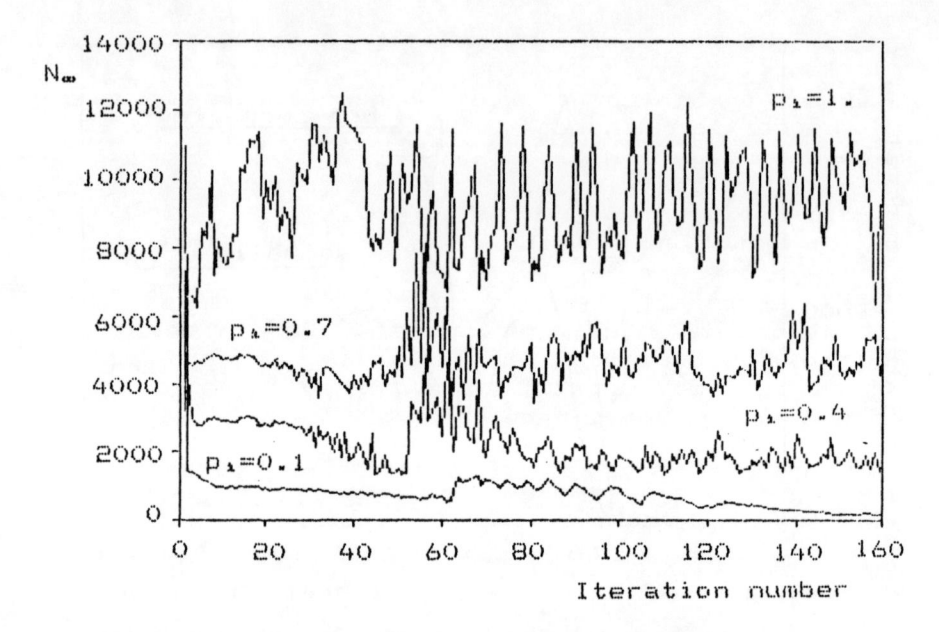

Fig. 4. Magnitude of the largest element of the
residual vector vs. iteration number.

small enough error, p_1 values which are not too
large must be chosen. In all the numerical texts no
convergence for $p_1 > 0.5$ was ever obtained.
To complete the study of the influence of p_1 on the
vector H, the values of appropriated parameters
after 160 iterations have been reported in table 1.
These parameters are exactly N_2, N_∞ and the number
of H vector elements greater, in absolute value,
than $N_\infty/2$ (d_1) and $N_\infty/20$ (d_2).

p_1	N_2	N_∞	d_1	d_2
0.1	343	166	4	66
0.2	757	475	3	45
0.3	2218	938	5	77
0.4	4420	1637	9	83
0.5	7649	2732	10	104
0.6	11402	3645	13	117
0.7	15627	4654	12	110
0.8	19911	5255	18	127
0.9	26617	6551	23	125
1.	31743	9649	15	124

Tab. 1. Influence of the coefficient pi on the
residual vector.

4. EXPERIMENTAL TESTS

The results of the simulation were verified by experimental tests, carried out at the foundry of L.M.I factory in Fornaci di Barga, in cooperation with the Centro Ricerche S.M.I.-L.M.I.. These tests regard an evaluation of:
- temperature distribution in the ingot;
- liquid-solid interface configuration;
- mould behaviour with reference to the heat exchange process.

The fig. 5 reports the experimental temperatures of a point inside the ingot and the equivalent ones calculated by numerical simulation vs. ingot height: the numerical model results show excellent agreement with the experimental ones.

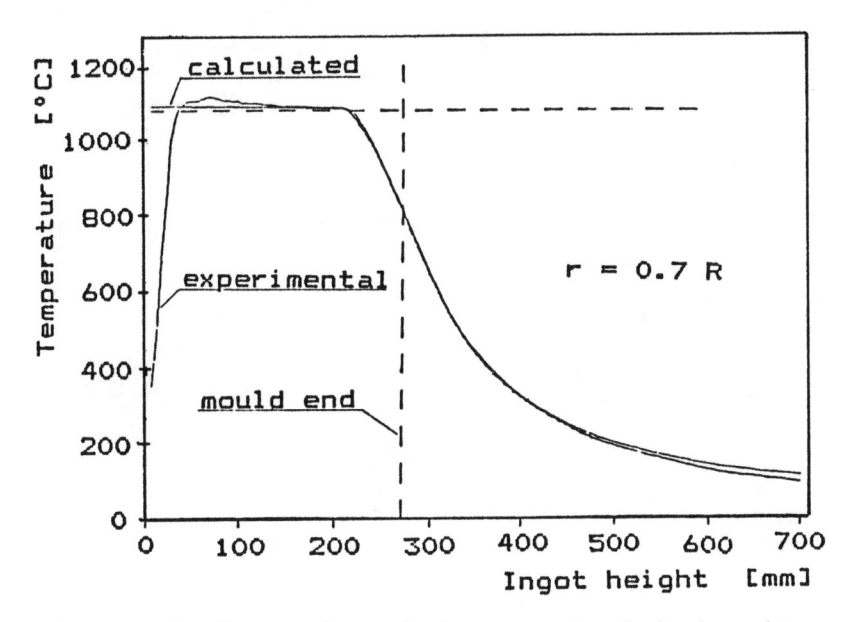

Fig. 5. Comparison between calculated and experimental temperatures.

6. RESULT DISCUSSION AND CONCLUSIONS

With the mathematical model, defined above, it is possible to evaluate the temperature distribution in an ingot as a function of the thermal characteristics of the cast material, the shape and sizes of the casting apparatus and the working conditions.

The results of the numerical simulation, which are reported here, refer to the mould apparatus and the casting parameters used for the experimental tests.

Fig. 6 shows a diagram of calculated thermal flow vs. ingot height.

Fig. 7 shows a diagram of the temperatures of

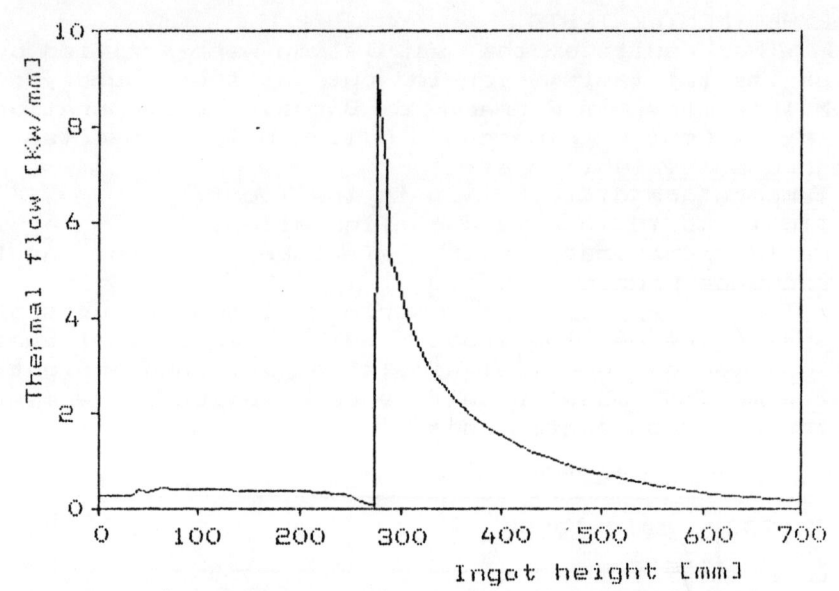

Fig. 6. Diagram of calculated thermal flow vs. ingot height.

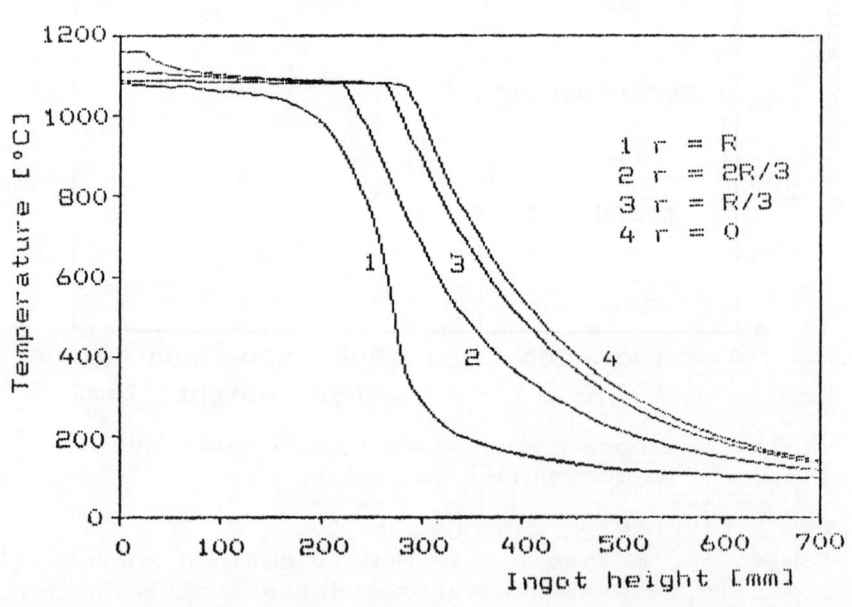

Fig. 7. Temperatures of different radial position inside the ingot

different radial positions inside the ingot vs. ingot height.
Fig. 8 represents a calculated temperature map in a longitudinal ingot section: the section is subdivided into zones where the temperature is comprised within well defined and pre-fixed ranges.
In conclusion, it can be stated that the

mathematical scheme, calculus procedure, heat exchange coefficients and boundary condition spectrum for the numerical simulation when correlated with the specific examined process and with the data obtained by direct temperature

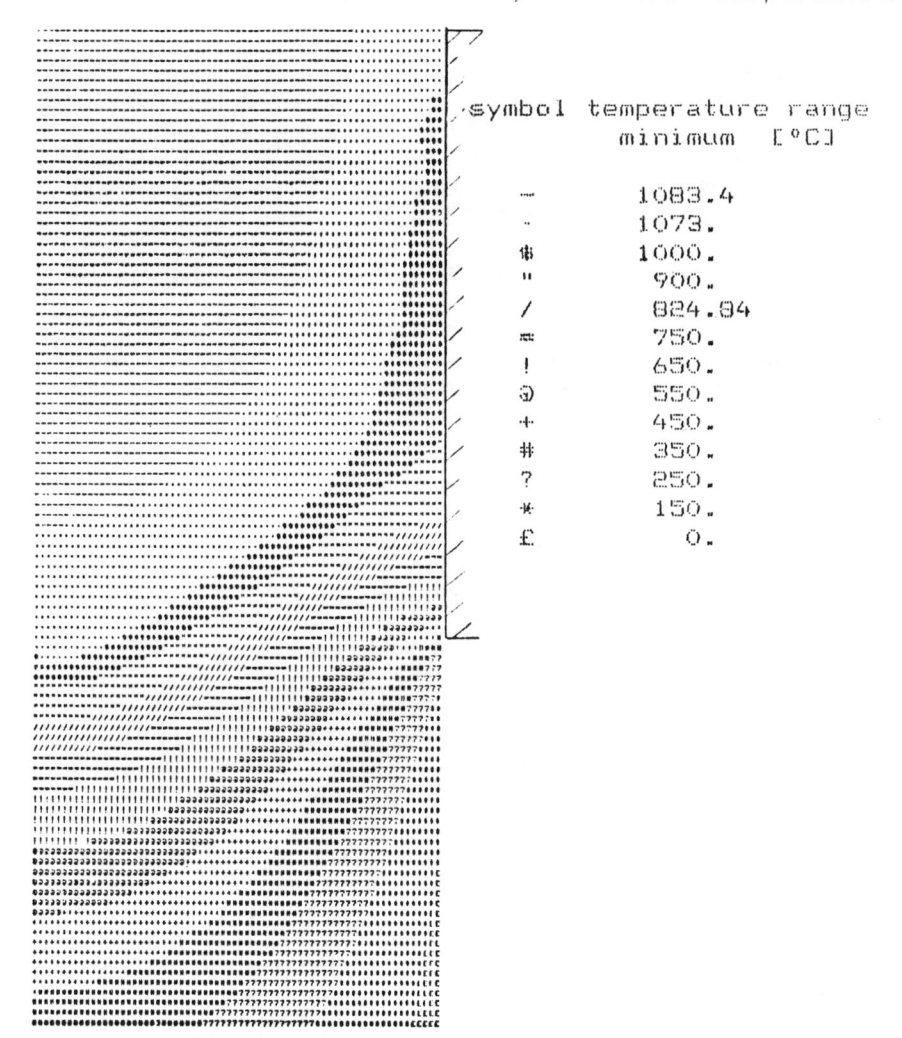

symbol	temperature range minimum [°C]
—	1083.4
.	1073.
₤	1000.
"	900.
/	824.84
=	750.
!	650.
@	550.
+	450.
#	350.
?	250.
*	150.
£	0.

Fig. 8. Temperature distribution in a longitudinal ingot section.

measurements in the tested equipment, give results that are in agreement.
Further an efficient mathematical model definition for semicontinuous casting process permits an analysis of technological parameters which influence the casting process development and, in consequence, a method for optimizing the use of the mould apparatus. Moreover, this model serves, during the

design of a new mould, to predict — at least as a first approximation — ingot temperature fields (on which the final ingot quality depends) under various, possible working conditions.

6. ACKNOWLEDGMENTS

This research has been partially supported by the "Fondi di Ricerca" of the "Ministero della Pubblica Istruzione" and of the Florence University.

7. REFERENCES
1. DEL PUGLIA, A., QUILGHINI, D., TESI, B., ZONFRILLO, G., LOMBARDI, P. and BILLI, A. - Numerical Simulation of Semicontinuous Copper Casting Process, Conf. "Cu 86, Copper Tomorrow", Barga, sept. 1986.
2. DEL PUGLIA, A., QUILGHINI, D., TESI, B., ZONFRILLO, G. and LOMBARDI, P. - Heat Transfer and Ingot Temperature Distribution Analysis in Semicontinuous Copper Casting Process, submitted to Mater. Sc. Techn. (1987).
3. PRIMICERIO, M., - Problemi di Diffusione a Frontiera Libera, Bollettino U.M.I., 18A, (5), 11-68 (1981).
4. TARZIA, D.A., - Una Revision sovre Problemas de Frontera Movil y Libre para la Equacion del Calor. El Problema de Stefan, Separata de Mathematicae Notae, Univ. Nac. de Rosario, 1981/1982.
5. FASANO, A. and PRIMICERIO, M. - Free Boundary Problems: Theory and Applications, vol. I and II, Pitman Adv.ed Publ. Program, Boston, 1983.
6. MARCHENKO, I.K., GUMENNYY, N.V., DENISOV, V.A. and SUROVOV, V.V. - Computer Simulation of the Solidification of a Semicontinuously Cast Ingot of Large Cross-Section, Russian Metall., 3, 55-58 (1981).
7. WECKMAN, D.C. and NIESSEN, P. - A Numerical Simulation of the D.C. Continuous Casting Process Including Nucleate Boiling Heat Transfer, Metall. Trans. B, 13B, (12), 593-602 (1982).
8. SZARGUT, J. and SKOREK, J. - Analysis of Ingot Temperature Field in Continuous Casting of Copper, Met. Technol., 7, 36-40 (1980).
9. BONACINA, C., COMINI, G., FASANO, A. and PRIMICERIO, M. - Numerical Solution of Phase-Change Problems, Int. J. Heat Mass Tranfer, 16, 1825-32 (1973).
10. STONE, H.L. - Iterative Solution of Implicit Approximation of Multidimensional Partial Differential Equations, SIAM J. Numer. Analysis, 5, 530-58 (1968).

A SOLUTION METHOD FOR MODELS OF THERMAL PROCESSES
IN PACKED BEDS OF SOLIDS

J.A. Thurlby, E.A. Kowalczyk, M.J. Cumming and G.J. Thornton

Commonwealth Scientific and Industrial Research Organisation
Division of Mineral Engineering, Melbourne, Australia

SUMMARY

The contacting of gases with solids in packed beds usually involves heat transfer and often rapid chemical and physical changes. Mathematical models of such processes may be quite complex, requiring efficient techniques for their solution.

In this paper, a solution method developed for an iron ore sintering model is presented. The method is applicable to systems of partial differential equations which can be reduced to ordinary differential equations -- albeit coupled and highly non-linear. A key feature is an adaptive method of grid splitting, by which the integration step sizes are adjusted to satisfy accuracy requirements with minimum computation. Other techniques used to reduce unnecessary computation are also discussed.

INTRODUCTION

Industrial processes in which packed beds of solids are contacted with gases are common. The bed may be stationary, on a moving grate (where it is subjected to cross-flowing gases), or in a shaft where it moves counter-currently or co-currently with respect to the contacting gases. In some cases heat transfer between the gas and solid may be the only objective, as in the recovery of heat from waste gases in pebble recuperators. Other applications may concern drying of either the solid or gas. Chemical reactions are frequently involved, e.g. in gas phase cracking, hydrogenation and reforming, where the solid is a catalyst. Reactions are also important in processes such as induration (hardening) of iron ore pellets and sintering of minerals, where physical and chemical transformation of the solid material is the objective.

THERMAL PROCESSES IN PACKED BEDS

Such processes, particularly where physical and chemical changes are involved, are usually quite complex and mathematical simulation can frequently lead to improved understanding and assist in their optimization and control (e.g. 1,2). Simulation provides estimates for chosen state variables, many of which may be difficult to measure, at points throughout the bed. These state variables include gas and solid temperatures, the concentrations of components in each of these phases and other characteristics such as gas pressure and solid particle size.

This paper discusses techniques chosen and developed for the solution of a model (3,4) of a particular process, namely the sintering of iron ore fines to produce an agglomerate with the chemical and physical characteristics required of blast furnace feed. As heat transfer, drying, condensation, and a number of chemical reactions and physical changes are featured in this model, the solution techniques are expected to be generally applicable to gas/solid systems.

IRON ORE SINTERING

In iron-ore sintering, ore fines, coke breeze or fine coal, and fluxes such as limestone and dolomite are mixed and then granulated by the addition of water. The granulated feed is deposited on a moving grate and gases are drawn through the bed. Initially, under a short ignition hood, combustion gases at about 1300°C are employed to heat the top of the bed sufficiently for it to sinter and to ensure that combustion of the solid fuel will continue when ambient air is introduced. During this latter stage the heat front continues through the bed, the solids above it exchanging heat with the incoming air and those below being progressively heated, dried and calcined. Within the high temperature zone, melting of the fluxes occurs, with subsequent dissolution of some of the iron oxide and gangue constituents. The state of the bed can be represented by a number of zones, shown in Figure 1.

MODELLING OF PACKED BED PROCESSES

A single model can be derived which represents the transient behaviour of either a stationary packed bed or a thin slice of a bed on a moving grate. In the latter case the equations are written with respect to a frame of reference which moves with the grate.

Assumptions commonly made in modelling packed bed processes, and which are made in the above sintering model, include:

1. There is no lateral dispersion of gas in the bed.
2. Direct heat transfer between particles comprising the bed is negligible, i.e. radiation and conduction between particles can be ignored.

THERMAL PROCESSES IN PACKED BEDS

3. The properties of the initial bed material and the properties and velocity of the inlet gas do not vary across the bed.

4. The residence time of the gas is negligible compared with that of the solid.

5. There is no heat loss from the sides of the bed.

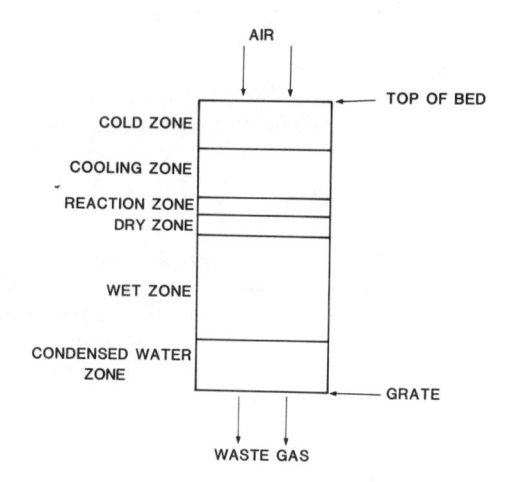

Figure 1. Zones within a typical column of sinter plant burden or within a laboratory pot charge during a sintering test.

Where these assumptions are made, it follows that there is no heat or mass transfer from the sides of the bed, or in a lateral direction within the bed. Hence, all state variables are invariant with respect to the lateral coordinate throughout the bed, i.e. the problem is 2-dimensional. The dimensions are time and distance (measured parallel to the gas flow) from one of the bed boundaries. In the case of a moving grate, the position of a chosen column of bed material (i.e. the distance it has moved) is directly related to time.

Assumptions 1, 2 and 4 allow second order derivatives to be ignored, and hence the process can be represented by two inter-related sets of partial differential equations for the state variables. These are of the form:

$$\partial X_\theta / \partial \theta = f(X_\theta, X_z) \qquad (1)$$

and

$$\partial X_z / \partial z = f(X_\theta, X_z) \qquad (2)$$

where θ is time and z is distance into the bed, usually measured from where the gas enters the bed; X_θ is the set of state variables associated with the solid, including solid temperature and component concentrations; X_z is the set of state variables associated with the gas, including gas temperature, flowrate, concentrations and pressure losses.

They will be referred to as 'solid' and 'gas' state variables, respectively. The initial conditions for the solid phase, i.e. the feed temperature and composition, are assumed to be known. Initial conditions for the gas might vary with time but would also normally be known, except perhaps the gas velocity.

Of all state variables, the solid and gas temperatures are frequently the most important. The two temperature derivatives are typically expressed as:

and
$$\partial T_s/\partial \theta = \left(hA(T_g - T_s) - \Sigma R_i \Delta H_i\right)/S_s \qquad (3)$$

$$\partial T_g/\partial z = -hA(T_g - T_s)/(VS_g) \qquad (4)$$

where h and A are the coefficient and area, respectively, for convective heat transfer; R_i is the rate of formation of species i (usually dependent on solid temperature and the concentration of one or more gas species) and ΔH_i the associated heat release; V is the gas velocity and S_g and S_s are the heat capacities of gas and solid, respectively.

SOLUTION METHOD

Where few state variables are involved, expressing the D.E's in finite difference form may allow rearrangement to give each state variable explicitly. Successive solution of these new equations over bed elements of fixed size may be the simplest and most efficient method of solving the model. This approach has been used successfully for the iron-ore pellet induration process (1,5). Usually, however, a numerical integration technique is employed - such as the finite difference method used by Dash and Rose (6,7) or the method of characteristics (8) used by Young (9), Yoshinaga and Kubo (10), and Cumming et al (3,11) for iron ore and lead/zinc sintering.

In general, the responses of state variables are highly non-linear, particularly where strongly exo- or endo-thermic reactions are involved. Indeed, some variables may not even change monotonically, sometimes displaying pronounced maxima or minima, as exemplified by the sinter bed temperature profiles shown in Figure 2. In regions of rapid non-linear change, small integration steps are required. For large models, involving numerous physical property and thermochemical calculations, it is important that computation methods be efficient.

The solution method discussed in this paper is applicable for processes involving a fixed bed or a moving bed on a grate or in a shaft. The essential requirement is that second order

THERMAL PROCESSES IN PACKED BEDS

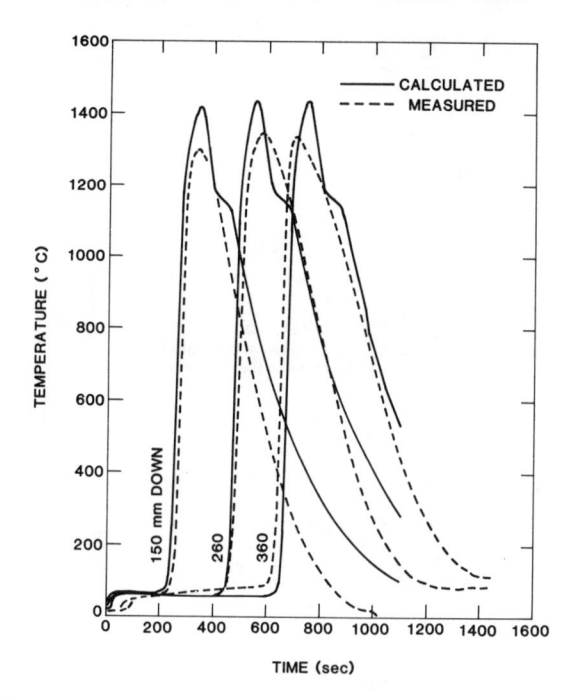

Figure 2. Calculated sinter bed temperature profiles and temperatures measured in a laboratory pot.

derivatives can be ignored, allowing the process to be represented by two sets of partial D.E.'s which can be reduced to ordinary D.E.'s.

The method is based on a coarse grid of integration points covering the required time and distance ranges. Each coarse mesh of the grid is sub-divided into fine meshes as necessary to achieve the required accuracy. To cater for shrinkage or slumping of the bed caused by e.g. condensation of water vapour and by melting, the model equations may be derived in terms of a 'transformed' frame of reference and include a 'shrinkage factor'(12). This does not affect the solution method.

The grid is traversed column by column - i.e. the equations are solved over the whole range of z for time increments up to the maximum corresponding to a coarse mesh, and this is repeated for successive columns until the required time range has been covered. Figure 3 illustrates the procedure. Where the gas flowrate is not known and it is necessary to satisfy a specified gas pressure drop, the flowrate is determined iteratively for each column before proceeding to the next. To do this the relationship between gas pressure and velocity developed by Ergun (13) can usually be applied. The two terms of the Ergun equation (the viscous and inertial pressure losses) are integrated as separate state

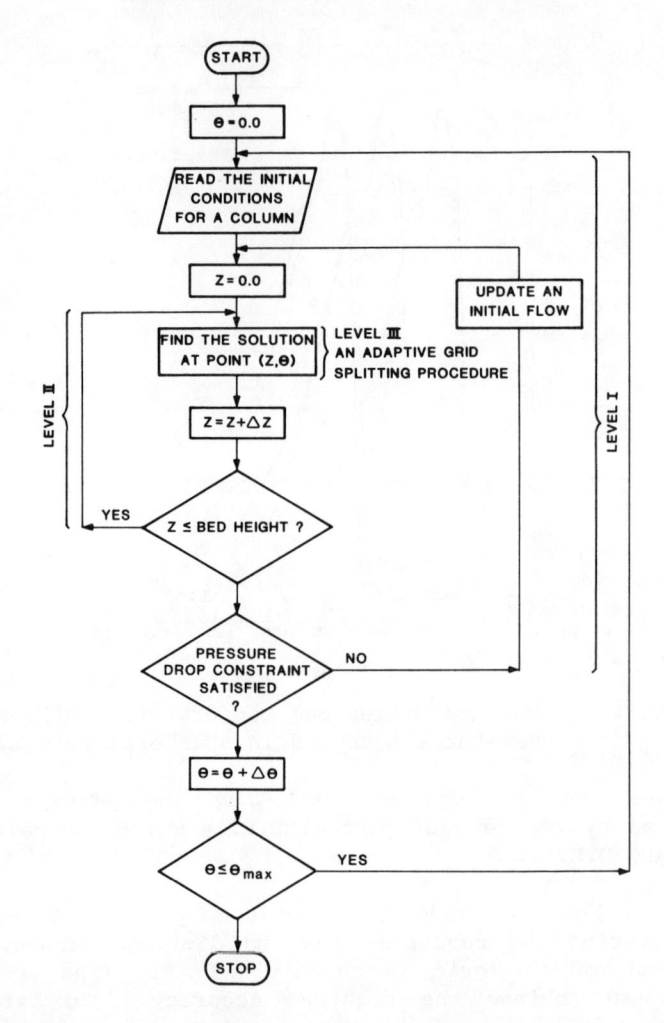

Figure 3. Procedure for traversing the coarse grid.
  Level I   — Meeting required pressure drop
  Level II — Traversing the column
  Level III — Solving coarse mesh, splitting
              as necessary

variables, giving the pressure losses through the bed. Factors A and B relating the current pressure losses to the current inlet gas velocity are determined by:

$$A = \Delta P_{visc}/V \quad (5a)$$

and

$$B = \Delta P_{inert}/V^2 \quad (5b)$$

The values of A and B for one iteration are used to estimate the inlet gas velocity for the next by solving equation (6), in which ΔP_{req} is the required pressure drop.

$$\Delta P_{req} = AV + BV^2 \quad (6)$$

THERMAL PROCESSES IN PACKED BEDS

INTEGRATION TECHNIQUE

The numerical procedure for solving the two sets of partial D.E's involves the method of characteristics (8) which employs 'characteristic curves' in the Θ-z plane, along which solutions can be obtained by the integration of ordinary D.E's. Thus the equation sets to be solved are reduced to:

and
$$dX_\theta/d\Theta = f(X_\theta, X_z) \qquad (7)$$

$$dX_z/dz = f(X_\theta, X_z) \qquad (8)$$

These sets of equations are solved, respectively, along paths of constant z and constant Θ, i.e. their characteristic curves.

When starting a new column, the solid state variables where gas enters the bed are first integrated across the column. This gives the complete set of solid state values for this level of the bed at the end of the current time step. These values are required for integration of the gas states (along the gas path) at that time. If the inlet gas velocity or any inlet gas conditions for the new column are significantly different to those of the previous column, the previously calculated values of the gas states and solid state derivatives at the end of the last column do not apply for the start of the new column. In such cases the gas states and solid state derivatives are re-evaluated.

A predictor-corrector integration method is used. The difference between the predicted and first-corrected value of each state variable provides a relative assessment of the accuracy of the procedure, and a basis for changing the step size(s). In general, derivatives of state variables in one set may be dependent on state variables contained in both sets. Hence, alternate integration of the equation sets at the predictor and subsequent corrector stages is employed when integrating over meshes. This greatly improves accuracy for given integration step sizes and allows the use of much larger step sizes than would otherwise be the case. Each equation set is first integrated along the appropriate path (constant z or constant Θ) in the predictive mode and then the corrective mode. If the difference between the two values calculated for any of the state variables exceeds the tolerance for that variable, the integration of all variables is repeated using the corrector results as a prediction for the new corrective phase. This action continues with alternate integration at constant z and constant Θ until either the convergence criterion is satisfied or the maximum allowed corrector steps have been taken. In the latter case one or both step sizes are reduced.

First-order (modified Euler) and third-order (Adams-Bashforth-Moulton) integration methods are used, which respectively require either one or three previous values of the state variable derivatives. For the single-step integration of coarse meshes where enough previously calculated relevant coarse mesh values are available, the choice is made by the user. With its higher accuracy, third-order integration allows the maximum step size (i.e. the coarse mesh size) to be increased, but it cannot be applied close to boundaries and is impractical for regions where variable step sizes are used. The latter limitation is due to the loss of accuracy involved in interpolating to obtain the three prior values for each derivative and the complexity of storing and retrieving them. First-order integration is automatically used in such regions.

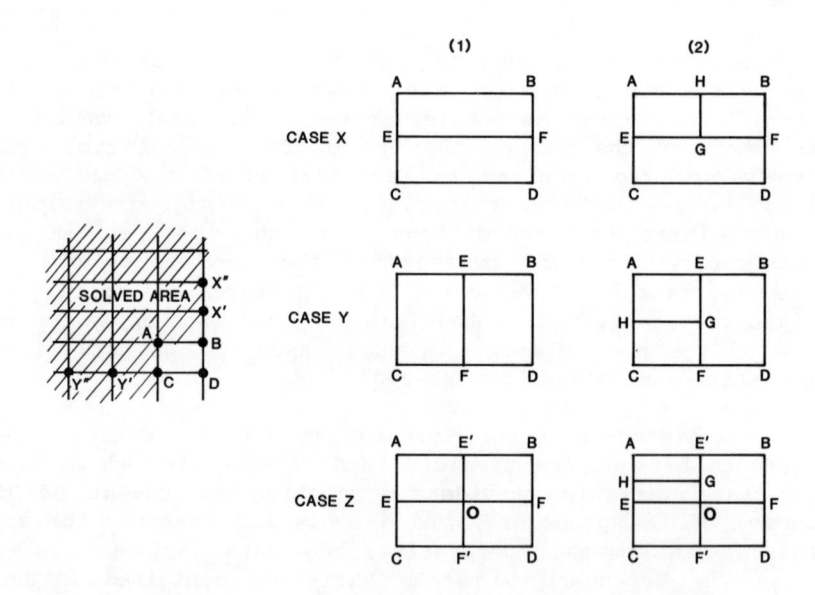

Figure 4. Some possible ways a coarse mesh can be sub-divided to meet required accuracy.

INTEGRATION STEP SIZE CONTROL

The procedure for changing the step sizes for alternating integration within the bed can be illustrated with reference to Figure 4. The solution at point D on the coarse grid is sought, all other labelled points having been solved. Solution is attempted in one step, alternately integrating along the paths CD and BD, as discussed above. If the specified error and convergence criteria are not satisfied for all state variables in each set, the relevant step size is halved. Thus the mesh is halved (either horizontally or vertically) or quartered. Figure 4 shows some of the possibilities for further splitting if this is required. Each time a mesh is split, the next fine mesh to be integrated is that closest to the boundary between solved and unsolved meshes. This ensures that the maximum possible information is available from previous integrations. However, interpolation is required for some of the points, i.e. those mesh corners which have not been solved previously. For the first stage of splitting in cases X, Y and Z in Figure 4, the next meshes to be attempted would be ABFE, AEFC and AE'OE, respectively. Interpolation might be necessary to obtain initial conditions at E and E'.

A special case of step size reduction should be mentioned. This occurs where a solution is acceptable mathematically but is physically infeasible. Negative concentrations, resulting from 'over-shooting', are common examples.

In general the halving of the integration steps may continue until either or both are, say, 1/256 of the coarse mesh values.

A binary tree was found to be particularly suitable as an information structure for representing the process of splitting. Each node in the tree represents a mesh and comprises co-ordinates and a variable describing the manner in which the mesh has been split. A stack information structure has been used to implement the binary tree structure in computer memory.

When integration is not alternated, i.e. when obtaining conditions at the column boundaries, the step size may be increased to improve efficiency. The compromise between accuracy and computation time requires

$$C_{min} \leqslant C_o \leqslant C_{max} \tag{9}$$

where C_o, C_{max} and C_{min} are arrays of the integration error and specified maximum and minimum error limits for each state variable, respectively.

If C_o for each solid or gas state variable is less than the appropriate C_{min}, the step size of the relevant independent variable (θ or z, respectively) may be doubled. To avoid instability, this is only done when the criterion has been satisfied for a number of successive steps.

SELECTIVE CALCULATION

Two other approaches are also used to reduce calculations. The main one is the avoidance of integration (with its associated derivative evaluation) of state variables which cannot change significantly across the mesh being solved. The identification of these states is based on temperature and concentrations at the two points from which integration would proceed.

THERMAL PROCESSES IN PACKED BEDS

Chemical reactions and processes such as drying can only occur at a significant rate when the solid or gas temperature is within a certain range. Thus the complete equation set can be sub-divided into sets appropriate to each temperature 'zone'. Within each set, some of the equations may not be required at a given time and position in the bed. Equations for irreversible reactions are one example, where integration is pointless if a reactant concentration is negligible.

Hence, use is made of a set of zone flags defining the possible phenomena and a set of integration flags which specify whether each individual state is to be integrated or held at its current value.

The second approach concerns calculation of temperature-dependent physical property and thermochemical data. Considerable time savings can be achieved by omitting re-evaluation until the relevant temperature has changed significantly. Different tolerances can be applied to different categories of data.

EXPERIENCE WITH SINTERING MODEL

The combination of techniques outlined has proved successful for the solution of the sintering model, which typically involves determining the values of 26 state variables at 14,000 coarse grid points and about the same number of fine grid points. It is planned to discuss performance aspects in a forthcoming paper, but some comments can be made on initial experiences.

The choice of coarse mesh dimensions is a compromise. If either dimension is too small, unnecessary calculations will be performed in regions where the state variables are changing slowly. On the other hand, if the coarse meshes are too large more splitting is required where appreciable changes occur. The effects can be seen from computing times for the sintering model, shown in Figure 5.

Third-order integration did not give the expected time savings for the sintering model. This is thought to be primarily due to the unsuitability of high order integration where sudden changes occur, offsetting its advantages in other regions.

One other technique which could reduce computing times is 'a priori' grid splitting, i.e. prior estimation of how each grid will need to be split, based on experience with neighbouring meshes. It has been attempted, but not found to be of any advantage in the case of the sintering model. This is due to the large range of step sizes required within a coarse mesh. For the method to be effective for this model it

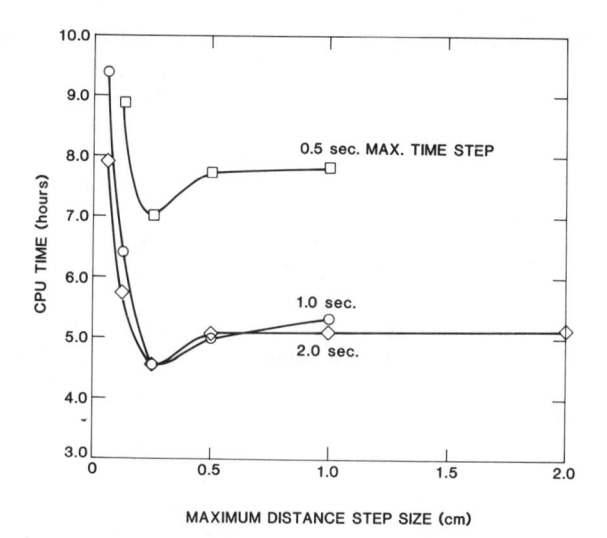

Figure 5. Effect of coarse mesh dimensions (maximum integration step sizes) on computing times for sintering model.

would be necessary not just to estimate the average degree of splitting required, but to estimate the extent of splitting in different regions of the coarse mesh.

ACKNOWLEDGEMENTS

This work was undertaken as part of a collaborative research project between CSIRO and the Broken Hill Proprietary Company Ltd. The company's support is gratefully acknowledged, as are the contributions made by CSIRO colleagues - R.J. Batterham, W.J. Rankin and J.R. Siemon.

REFERENCES

1. Turner, R.E., Philp, D.K., Batterham, R.J., Thornton, G.J. and Thurlby, J.A. Computer assisted control of pellet induration, AIME Annual Meeting, Las Vegas, 1980, Paper No.80-41.

2. Thurlby, J.A., Batterham, R.J., and Turner, R.E. Minimizing energy costs in straight-grate iron ore pelletization, Ironmaking and Steelmaking, 9, 1982, pp. 200-206.

3. Cumming, M.J., Rankin, W.J., Siemon, J.R., Thurlby, J.A., Thornton, G.J., Kowalczyk, E.A., and Batterham, R.J. Modelling and simulation of iron ore sintering, in Agglomeration 85, (editor: C.E.Capes), 4th International Symposium on Agglomeration, Toronto, 1985, pp. 763-776.

4. Rankin, W.J., Siemon, J.R., Cumming, M.J., and Batterham, R.J. The modelling of iron ore sintering, Proc. 11th Australian Conference on Chemical Engineering, Brisbane, 1983, pp. 209-216.

5. Thurlby, J.A., Batterham, R.J., and Turner, R.E. Development and validation of a mathematical model for the moving grate induration of iron ore pellets, Int. J. Mineral Processing, 6, 1979, pp. 43-64.

6. Dash, I.R., Carter, C.E. and Rose, E. Heat wave propagation through a sinter bed; a critical appraisal of mathematical representations, Proc. SIMAC Conference on Measurement and Control of Quality in the Steel Industry, Sheffield, 1974, pp. 8/1-8/7.

7. Dash, I.R. and Rose, E. Simulation of a sinter strand process, Ironmaking and Steelmaking, 5, 1978, pp. 25-31.

8. Lapidus, L. Digital computation for chemical engineers, McGraw-Hill, New York, 1962.

9. Young, R.W. Dynamic mathematical model of sintering process, Ironmaking and Steelmaking, 4, 1977, pp. 321-328.

10. Yoshinaga, M. and Kubo, T. Approximate simulation model for sintering process, Sumitomo Search, 20, 1978, pp. 1-14.

11. Cumming, M.J., Barton, G.W. and Batterham, R.J., Dynamic simulation of a moving-grate combustion process, Proc. 4th IMACS International Symposium on Computer Methods for Partial Differential Equations, Bethlehem, 1981, pp. 286-292.

12. Cumming, M.J., Simplification of a two-dimensional heat and mass transfer model by a transformation of co-ordinates, Mathematics and Computers in Simulation, 27, 1985, pp. 141-145.

13. Ergun, S., Fluid flow through packed columns, Chem. Eng. Prog., 48, 1952, pp. 89-94.

NUMERICAL SIMULATION OF THE TRANSPORT PROCESSES IN A HEAT TREATMENT FURNACE

Y. Jaluria
Mechanical Engineering Department
Rutgers University, New Brunswick, NJ 08903, USA

SUMMARY

The numerical simulation of the heat transfer and fluid flow processes in a furnace is carried out using finite difference techniques. A heat treatment furnace for the annealing of steel is considered as an example and a mathematical model of the system is obtained by simplifying the equations that govern the transport in the various elements that constitute the system. The resulting set of equations then governs the time-dependent temperature distribution in the furnace. The transport processes undergone by individual components of the furnace are simulated first because of the complexity of the full problem. These individual simulations are considered in terms of the underlying physical mechanisms for validation. Finally, these are combined to allow the simulation of the full, coupled problem. The temperature cycle for the material is computed, along with those for the other components. A comparison with experimental data indicates close agreement, lending strong support to the numerical model.

INTRODUCTION

An important problem of considerable practical and fundamental interest is the numerical simulation of a furnace, such as a heat treatment furnace, which is heated by means of hot combustion products and in which the material under consideration must undergo a specific thermal cycle. The problem is complicated by the presence of all three modes of heat transfer - convection, conduction and radiation - which are coupled through the boundary conditions. Material properties are functions of the temperature and the flow must be considered to obtain the convective transport due to the gases.

This paper presents a study on the numerical simulation of such a thermal system. An annealing furnace is considered. An inert environment is provided at the material being annealed by means of a closed system for inert gas flow. Also the walls, insulation and the hot flue gases form other components of the system. Interest lies in developing a numerical scheme for the simulation, so that the temperature cycle undergone by the material can be computed. This information will be useful in the design and optimization of such systems. The method adopted proceeds with the simulation of individual components, using finite difference methods and assuming simple boundary conditions so that the components are decoupled from the transport mechanisms in other elements of the system. Following this simulation of individual components, the actual boundary conditions that couple all the components are incorporated.

Not much detailed effort has been directed at this problem, despite its considerable importance in practical applications, such as those related to the processing of metals. Most of the earlier studies have been concerned with the operation of the furnace and the rate of cooling needed for the annealing process [1-3]. These studies provide detailed and valuable information on the geometry, dimensions, operating temperatures and gas flow rates in furnaces employed in practice. This information is important in the mathematical modeling of the system and also in the validation of the numerical scheme. A few studies have considered the mathematical and numerical simulation of some of the processes involved, particularly the cooling of the material, using important simplifications in the governing equations [4,5]. A detailed mathematical model of the system was developed by the present author and some of the relevant results have been reported earlier [6,7]. This paper continues the effort to obtain further insight into this complex problem.

Several very interesting results are obtained. First, the study presents a method for the numerical simulation of such complicated thermal systems. Secondly, the effect of the various input parameters, such as gas flow rates, inflow temperature of the flue gases, geometry, etc., is studied and related to the basic heat transfer mechanisms. The importance of these results in the design of such furnaces is outlined, considering the relevant parameters involved. The effect of various numerical parameters introduced into the solution in order to obtain the results, such as grid spacing, time step, etc., is also discussed. The study, thus, provides many valuable inputs for studies on the simulation of thermal systems. The transient response of the various elements that constitute the system is investigated in detail to determine the appropriate time step that may be employed in the numerical scheme. Also studied in detail is the temperature variation in the steel which is being annealed and in the gases. The temperature profile in the material being treated is extremely important from metallurgical considerations [8].

Finally, the results from the numerical simulation are compared with measurements. A very good agreement is observed, thus validating the numerical scheme and the mathematical model developed for this problem.

MATHEMATICAL MODEL

The industrial system considered for the numerical simulation in this study is a furnace for the annealing of steel sheets which are rolled into coils, with an inner diameter D_i and outer diameter D_o. Figure 1 shows a qualitative sketch of the system, indicating the various components that comprise the system. These include the coils, which are stacked vertically, the convector plate, the cover, the wall consisting of fire clay and insulation, the inert gases and the hot gases from the burners. Energy input into the system is provided by the combustion of the flue gases at the burners, where the temperature is known and denoted by T_o. The inert gases include a large amount of nitrogen and are driven by a fan. Burners are located circumferentially at several points and the flow enters tangentially. The furnace is like a vertical cylinder about 5 m in height and 2.5 m in outer diameter. Because of the cylindrical geometry of the furnace, coils and cover and because of the circumferential symmetry of the burners, the problem may be treated as axisymmetric, with the local temperature T dependent only on radial distance r, height z and time τ as $T(r,z,\tau)$. For further details on the system, the references mentioned above may be consulted.

The thermal process that the steel coil is subjected to involves heating it up to the recrystallization temperature, which is around 723°C for the material considered, maintaining the temperature at around this value so that the material is uniformly at this temperature level and a slow cooling process for the material microstructure to settle down gradually, relieving any residual stresses that existed in the material. Figure 2 shows the typical temperature variation undergone by the material at a given spatial location, along with an envelope representing the deviation from the ideal in practice. In many instances, rapid cooling is employed to accelerate the process once the material has cooled to a particular critical value by which the annealing process in the material has been completed [1,2].

The axial conductivity k_z of the steel coil is often much larger than the radial conductivity k_r, because of contact resistance resulting from gaps between the coils [8]. Also, both k_r and k_z are functions of the temperature T and are approximated in this study as $k_z = k_i(1+aT+bT^2)$, where a and b are constants and k_i is a reference value taken as the value at the initial temperature T_i. This expression was obtained by curve fitting of available data [4,8] and a similar equation was derived for k_r. Then, the governing energy

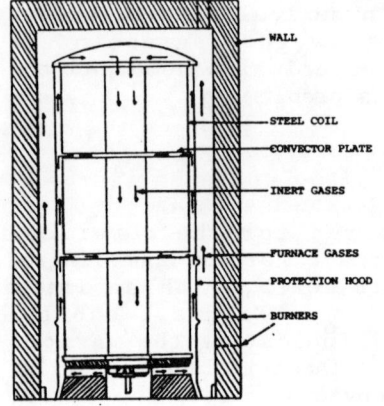

Fig. 1. Sketch of the heat treatment furnace considered.

Fig. 2. Typical temperature cycle of the material undergoing heat treatment.

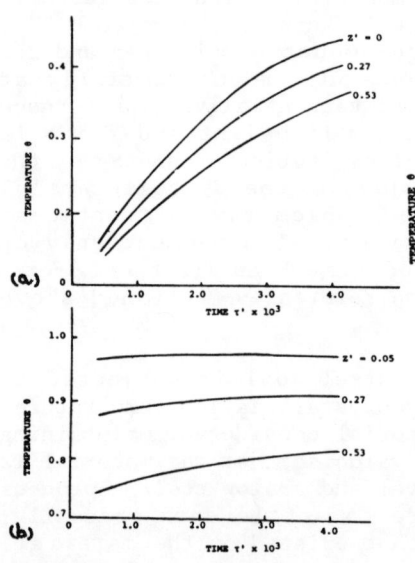

Fig. 3. Variation of the cover and gas temperatures with time, when these are treated as uncoupled problems in the numerical simulation.

Fig. 4. Convector plate and steel coil temperature response, with these individual, uncoupled, problems.

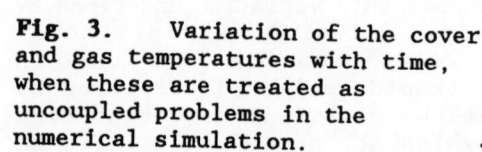

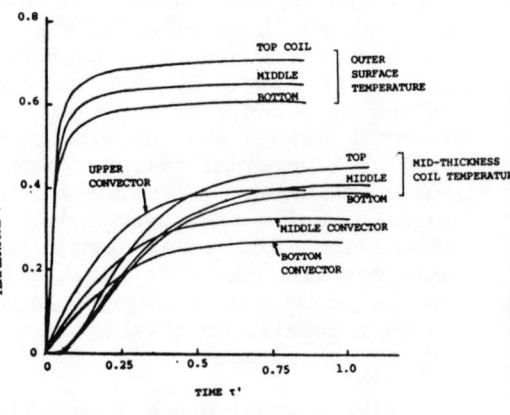

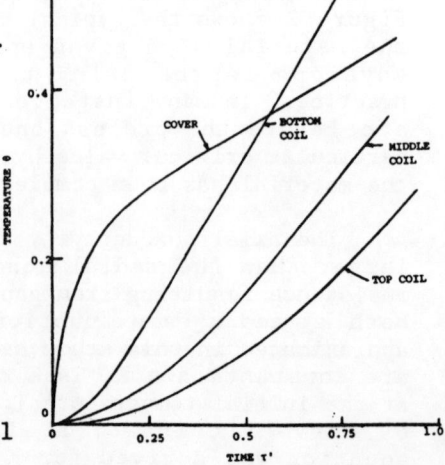

Fig. 5. Transient temperature variation of the cover and the steel coils at mid-height and mid-radial distance, for the full coupled problem, $k_r/k_z = 0.28$.

equation for the time-dependent temperature T_m in the material (steel coils) is:

$$\rho_m C_m \frac{\partial T_m}{\partial \tau} = \frac{\partial}{\partial z}\left(k_z \frac{\partial T_m}{\partial z}\right) + \frac{1}{r}\frac{\partial}{\partial r}\left(rk_r \frac{\partial T_m}{\partial r}\right) \qquad (1)$$

where ρ_m is the density, C_m the specific heat and $T_m(r,z,\tau)$ the local temperature in the coil. Both ρ_m and C_m are temperature dependent and curve fitting, similar to that for k_r and k_z, is employed to express these in terms of the temperature. The initial and boundary conditions for the above equations are

$$\tau \leq 0: \quad T_m(r,z) = T_i \qquad (2)$$

$$\tau > 0: \quad k_r \frac{\partial T_m}{\partial r} = h_o(T_g - T_m) + \epsilon_m F\sigma(T_c^4 - T_m^4) \quad \text{at } r = D_o/2$$

$$-k_r \frac{\partial T_m}{\partial r} = h_i(T_g - T_m) \quad \text{at } r = D_i/2 \qquad (3)$$

$$T_m = T_p \text{ at } z = 0, H$$

where the subscript g refers to the inert gases, c to the cover and p to the convector plate. H is the height of the coil, ϵ the emissivity of the coil surface, F a geometry factor for radiative transport, σ the Stefan-Boltzmann constant, h_o the convective heat transfer coefficient at the outer surface and h_i that at the inner surface.

The energy equation for the wall is similar to Eq. (1) and may be written as

$$\rho_w C_w \frac{\partial T_w}{\partial \tau} = \nabla \cdot (k_w \nabla T_w) \qquad (4)$$

where ρ_w, C_w, T_w and k_w refer to the wall. Boundary conditions, similar to Eq. (3) may also be derived for the wall. The cover is very thin and temperature may be taken as uniform across its thickness. Also, axial conduction is small and may be neglected, to yield the governing equation as [6,7]:

$$\rho_c C_c \ell \frac{dT_c}{d\tau} = \sigma(\epsilon_f T_f^4 - \epsilon_c T_m^4) - \epsilon_c F\sigma(T_c^4 - T_m^4)$$
$$+ h_f(T_f - T_c) + h_g(T_g - T_c) + \epsilon_c F_1 \sigma(T_w^4 - T_c^4), \qquad (5)$$

where the subscript f refers to the flue gases. The diffuse gray approximation is employed for radiative transport and F_1 is a geometry factor [9]. The gas emissivity ϵ_f is obtained for transport to the adjoining cover surface. The effect of the inclusion of axial conduction was found to be small.

The equation for the convector plate temperature T_p is obtained by assuming uniform temperature across its thickness, since the Biot number based on this thickness is less than 0.1. Thus,

$$\rho_p C_p \frac{\partial T_p}{\partial \tau} = \frac{1}{r}\frac{\partial}{\partial r}\left[r\, k_p \frac{\partial T_p}{\partial r}\right] \tag{6}$$

Again, the boundary conditions may be derived for this equation on the basis of energy transfer at the surfaces. The initial condition is taken as $T_p = T_i$ at $\tau = 0$. For the inert and flue gases, the temperature is assumed to be uniform in the radial direction because of mixing and to vary in the vertical direction only, i.e., $T(z,\tau)$. The governing equation for the inert gases, for instance, is

$$\rho_g (C_p)_g \, UA \, \frac{dT_g}{dz} = -Ph_i(T_m - T_g) \quad \text{at } r = D_i/2 \tag{7a}$$

$$= Ph_o[(T_m - T_g) + (T_c - T_g)] \quad \text{at } r = D_o/2 \tag{7b}$$

where A is the cross-sectional area, U the average flow velocity and P the perimeter for heat transfer. The negative sign in Eq. (7a) arises because of downward flow at the center. Similarly, the equation for the flue gases may be derived considering convective and radiative energy loss to the cover and the wall. Thus,

$$\rho_f (C_p)_f \frac{UA}{P} \frac{dT_f}{dz} = -\sigma(2\epsilon_f T_f^4 - \epsilon_c T_c^4 - \epsilon_w T_w^4) - h_f(T_f - T_c) - h_f(T_f - T_w) \tag{8}$$

where U, A and P now refer to the annular region over which the flue gases flow. The flue gases radiate to both the cover and the wall and the radiative loss is doubled to take into account the total loss, since ϵ_f was determined for radiation to a surface element of the cover.

NUMERICAL SCHEME

It is seen from the above set of governing equations that all the equations are coupled to each other through the boundary conditions. Thus, these have to be solved simultaneously. Equations (5), (7) and (8) are ordinary differential equations. Being initial-value problems, they are easily solved by a one-step method, such as Runge-Kutta [9]. Note that T_c is also a function of z as $T(z,\tau)$, the axial dependence being brought in by the axial variation of the temperatures. Similarly, T_f and T_g are also time dependent, $T_f(z,\tau)$ and $T_g(z,\tau)$. The time dependence is introduced into the calculations by the transient variation of the temperatures. Thus, these ordinary differential equations are solved using the fourth-order Runge-Kutta scheme, starting with the initial conditions and with the computed temperatures of the neighboring system components being employed at each axial location and time step τ. For instance, the solution of Eq. (8) requires the time-dependent temperatures at a given time τ to proceed to the next time interval $\tau+\Delta\tau$. These temperatures are obtained from the simultaneous solution of the coupled equations for T_c and T_w.

The remaining equations are partial differential equations. Equation (6) is a one-dimensional transient diffusion equation and is parabolic in nature. Equations (1) and (4) are parabolic in time and elliptic in the spatial coordinates r and z. Thus, the solution to these equations is obtained by marching in time. The Crank-Nicolson implicit finite difference method was employed [9]. It yields a tridiagonal system for the one-dimensional problem governed by Eq. (6). The tridiagonal matrix algorithm, generally known as Thomas algorithm, was employed to invert this matrix and advance the solution to the next time step. For the two dimensional problems governed by Eqs. (1) and (4), a tridiagonal system is not obtained and the Gauss-Seidel iterative method was employed to solve the resulting system of equations. The Crank-Nicolson method is second-order accurate in space as well as in time and is unconditionally stable for linear equations. It was preferred over other available methods because of this reason.

A 51x51 grid was employed for each coil and similar mesh sizes were employed for other components. The transient response for each component was considered initially in simple analytical terms [10]. It was found that, as expected, the coils were the slowest and the gases and the cover the fastest in response. Since the response times for the two were substantially different, a considerable reduction in the computational effort was obtained by employing much larger time steps for the coils, as compared to those for the cover and the gases. Intermediate values were used for the wall and the convector plate. The time steps chosen were varied, as were the mesh sizes, to ensure that the numerical results were essentially independent of the values employed.

It must be noted that the partial differential equations, Eqs. (1), (4) and (6), are all nonlinear because of the temperature dependence of the properties ρ, C and k. As mentioned earlier, curve fitting of available property data was employed to obtain $\rho(T)$, $C(T)$ and $k(T)$ as polynomial functions. These functions are substituted in the governing equations. To linearize the resulting nonlinear problems, the properties at the previous time step were employed in the calculation of the temperatures at the present time step. This approach limits the time step $\Delta\tau$ that may be employed, because of accuracy considerations. Iteration may also be used to obtain a better estimate of the properties in the computation. The values at the previous time step then provide the starting values and the computed temperatures are used to obtain a better approximation of the properties. The calculation for temperatures is repeated and so on. This approach may be employed if larger time steps are desired.

Because of the complexity of the system under consideration, individual components were first simulated using constant boundary conditions so that a given component becomes decoupled from the others. This individual simulation allows one to study the transient response of the component, investigate the resulting temperature distributions and confirm that the numerical scheme is satisfactory in terms of the physical nature of the observed trends. The individual simulation programs were also used to study simple problems, such as flow in a pipe and heat transfer in a rectangular bar, for which the analytical results are available. After all these individual programs were satisfactorily developed, they were assembled to obtain the numerical simulation code for the complete system. The physical inputs needed for the simulation of the system are the flue gas composition and flow rate, inert gas flow rate, dimensions and materials involved in the system, initial temperature T_i, which is generally higher than the room temperature because of previous operation of the furnace, and material property data. The heat transfer coefficients were obtained from the literature and k_r/k_z was taken as a parameter which varies from 0.1 to 0.3. Typical values were obtained from the literature [4]. For further details on these input values, see Refs. [6,7]. Once the simulation is completed satisfactorily, these parameters and input variables may be varied to optimize the design of the system and its operation.

NUMERICAL RESULTS AND DISCUSSION

The numerical results are presented in terms of the following dimensionless variables

$$\theta = \frac{T-T_i}{T_o-T_i}, \qquad z' = \frac{z}{H_T}, \qquad r' = \frac{r}{L}, \qquad \tau' = \frac{\alpha\tau}{L^2} \qquad (9)$$

$$\text{with } L = \frac{D_o - D_i}{2}$$

Thus, L is a characteristic length dimension, α is the thermal diffusivity of the material of the steel coil, H_T is the total height of the furnace and θ, z', r' and τ' are dimensionless. This nondimensionalization is frequently employed in unsteady-state conduction [10] and allows one to present results for a wide range of governing parameters in generalized terms.

Figure 3 shows the variation of the cover and gas temperatures with time when these two are simulated individually, as uncoupled from the other components. Constant boundary conditions are imposed and the transient behavior is studied. The observed trends are very much as expected. The temperature level increases with time as the furnace heats up. Also, the temperature decays with height since the gas loses energy to the wall and the cover. The gas temperature rises from the initial value almost instantaneously as the burners are turned on and then varies only slightly with increasing time. The cover temperature rises almost exponentially, which is expected for a lumped mass subjected to convective heating. The radiative input makes the temperature rise faster.

Figure 4 shows the time-dependent temperature distributions in the three steel coils, stacked vertically, and in the convector plates. First, the response is seen to be much slower than that of the cover, as expected. The outer surface rises to the steady-state value much more rapidly than the interior of the coils. As mentioned earlier, the boundary conditions are held constant for this uncoupled problem. The temperatures decrease from the outer surface toward the center. The inert gases heat up as they flow upward and their temperature increases with z' at the outer surface of the coil and decreases with z' on the inside.

Figure 5 shows the transient variation as computed from the full coupled problem. The individual simulations are assembled and the time-dependent distributions in the various components are obtained. Clearly, the bottom coil responds the fastest and rises to the highest temperature level at mid-height and mid-thickness. The response of the cover is faster than that of the coils and the results are shown for the temperature at mid-height. All the observed trends are physically reasonable and lend support to the validity of the numerical model. Several simple problems, particularly plates and cylinders, were also studied using the developed numerical scheme and the results compared with available analytical results. A good agreement was observed.

Figures 6-8 show several other numerical results obtained from the simulation of the furnace. The time-dependent temperature distributions in the bottom coil are shown in Fig. 6 for two values of the parameter k_r/k_z. A decrease in the value of this parameter results in a greater temperature variation across the thickness of the coil, which is expected for a lower thermal conductivity in the radial direction. With increasing time τ', the temperature level increases. The value of k_r/k_z for a given application can be estimated from the tightness of the wrapping, the thickness of the sheets and the material. The decay of the flue gas temperature with height z' is seen in Fig. 7. The temperature is uniform in the region between the burners due to the mixing generated by the swirling inflow. Very little temperature increase is observed with time in the flue gas, indicating the rapid approach to a quasi-steady circumstance. The axial temperature variation in the steel coils at mid-thickness is shown in Fig. 8. The temperature level increases with time and the bottom and middle coil distributions are similar in form. The temperature at the upper surface of the top coil is

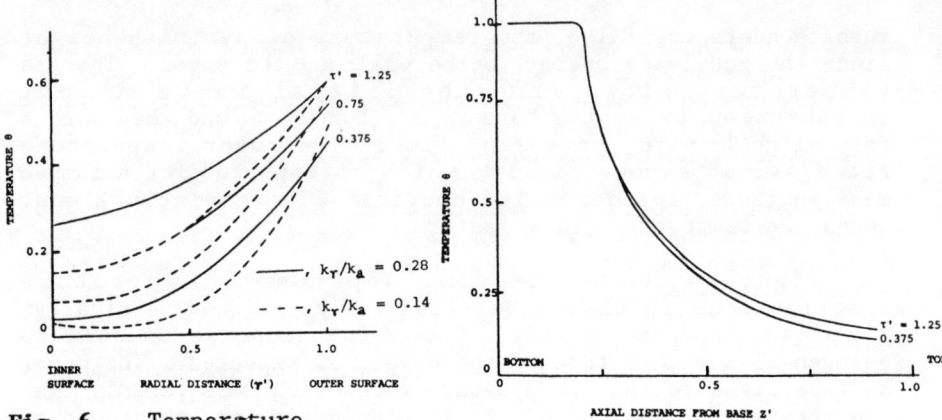

Fig. 6. Temperature distribution in the bottom coil at the ght for two values parameter k_r/k_z.

Fig. 7. Decay of the flue gas temperature with height.

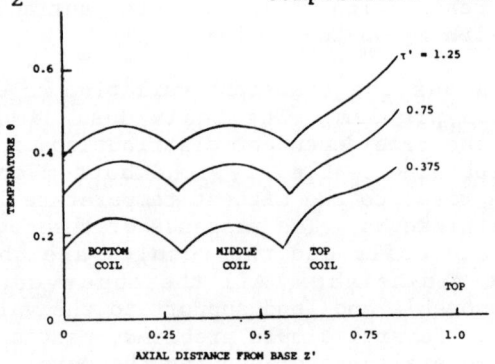

Fig. 8. Axial temperature variation in the steel coils at mid-radial distance, $r = (D_o+D_i)/4$, for $k_r/k_z = 0.28$.

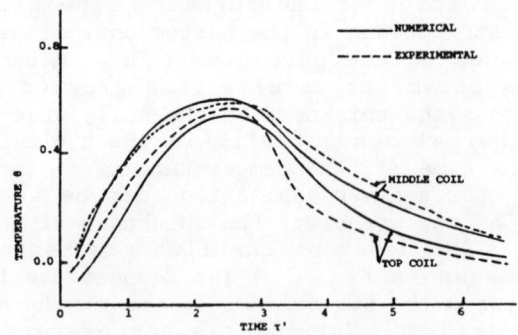

Fig. 9. Comparison of the results from the numerical simulation of the system with those from experiment on the furnace at two points in the steel coils.

relatively high because of the large amount of mixing that occurs there and because of the heating of the inert gases as they flow upward, resulting in the highest temperature near the top [7]. The lowest temperatures are at the convector plates which tend to slow down the response of the coils.

Several temperature measurements were also taken and compared with the numerical simulation results. Figures 9 and 10 show some of these comparisons. The spread in the experimental data is shown in Fig. 10. The exact location where the thermocouples were located were employed in the numerical code to generate the temperature variation with time. Two points within the coils were employed for Fig. 9 and a point on the bottom surface of the lower most coil was used in Fig. 10. Clearly, the comparison is very good, providing a strong support to the mathematical and numerical modeling of the system.

CONCLUDING REMARKS

A detailed numerical simulation of a heat treatment furnace, which represents a complex thermal system for manufacturing, is carried out. The governing equations are simplified and the individual components of the system are first simulated to check the validity of the model and to guide the choice of the relevant numerical parameters such as grid size and time step. These individual simulation programs are then assembled to obtain the numerical model for the complete system. The time-dependent temperature distributions in the system are computed and found to follow the expected trends. Comparison with experimental data on temperatures at several points indicates good agreement, lending support to the approach and the numerical model.

The numerical simulation of the system can be employed for the optimization of the thermal process and for the control of the system, say, by controlling the flow rate of the flue gas. The process is a fairly complicated one and knowledge-based expert systems may be employed to ensure that the relevant constraints, such as maximum allowable temperature for the coils and the wall, are not violated [11]. Computer aided design may be employed to adjust the input variables so that the best solution for design, within the given constraints, is obtained.

ACKNOWLEDGEMENTS

The author acknowledges the financial support provided by the Center for Computer Aids for Industrial Productivity and by the Mechanical Engineering Department, Rutgers University, for this work. The help of Dr. K. Kapoor and Dr. S.M. Aeron is also acknowledged.

REFERENCES

1. GORDON, G. - Rapid Cooling for Coil Annealing, Iron Steel Eng., 151-161 (1965).

2. WARGO, W. - Accelerated Cooling After Coil Annealing, Iron Steel Eng., 118-124 (1966).

3. KUNIOKA, K., KURIHARA, K. and TADA, T. - Heat Transmission Analysis of Tight Coils in Single Stack Bell Type Annealing Furnace, Proc. ICSTIS, Suppl. Trans. Iron Steel Inst., Japan, 11, 796-800 (1971).

4. STIKKER, U.O. - Numerical Simulation of the Coil Annealing Process, Math. Models Metal Process Dev., Iron and Steel Inst., Spec. Rept., 123, 104-113 (1970).

5. HARVEY, G.F. - Mathematical Simulation of Tight Coil Annealing, J. Aust. Inst. Met., 22, 28-37 (1977).

6. JALURIA, Y. - Heat Transfer Modeling of Batch Annealing Furnaces, Repts. C&T/ME/YJ/73/1 to 3, Mech. Engg. Dept., Indian Inst. of Tech., Kanpur, India (1979-1980).

7. JALURIA, Y. - Numerical Simulation of the Thermal Processes in a Furnace, Num. Heat Transfer, 7, 211-224 (1984).

8. LISOGOR, A.A. and MITKALINNYI, V.I. - Thermal and Physical Properties of Stacks of Cold Rolled Steel, STAL Met. Tech., U.S.S.R., 996-998 (1970).

9. JALURIA, Y. and TORRANCE, K.E. - *Computational Heat Transfer*, Hemisphere Pub. Corp., Harper and Row, NY, 1986.

10. INCROPERA, F.P. and DEWITT, D.P. - *Fundamentals of Heat Transfer*, Wiley, NY, 1981.

11. RUSSO, M.F., PESKIN, R.L. and KOWALSKI, A.D. - A Prolog Based Expert System for Partial Differential Equation Modeling, to be published in Special Issue of SIMULATION (1987).

Fig. 10. Comparison of the time-dependent variation of the temperature, at a point on the bottom surface of the lower most steel coil, obtained from the numerical simulation with the corresponding measurements.

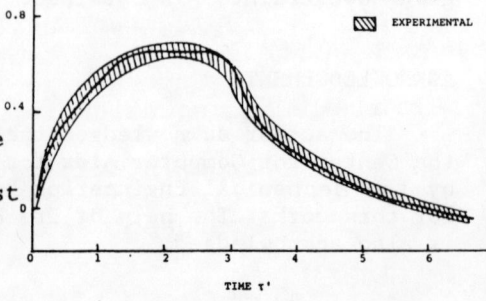

THERMAL MODELLING OF HIGH POWER SEMICONDUCTORS

By M.S. Khanniche[I] and A. Hooper
University College of Swansea.
Electrical Engineering Department,
Singleton Park, Swansea UK.

SUMMARY

A 3-D finite element thermal model is developed to deal with steady-state and transient analysis of high power semiconductor electronic modules. In the transient solution finite difference method is used to approximate the time dimension using θ-method with a diagonal thermal capacity matrix. The application of the model to the solution of a 4kW phase change cooling unit is presented and also the solution of the transient behaviour of a power gate turn-off thyristor device under a fault condition is given.

1. INTRODUCTION

The advent of high power semiconductor devices has led to simplified circuitry with high voltage and high current capability. However, the maximum permissible junction temperature of 125°C (typical) has become a limiting factor in using these devices at their rated values. Thus the need of developing efficient cooling techniques which require accurate modelling.

Because of the potential complex geometry of heatsink-device assemblies, basic thermal models based on the concept of thermal resistance and thermal capacitance are no longer valid thus a 3-D finite element solution is presented which deals

[I] Robert Gordon's Institute of Technology,
Electrical Engineering Department.
Schoolhill, Aberdeen, Scotland UK.

with steady-state and transient analysis. The latter uses finite differences to approximate the relatively simple time dimension. The thermal capacity matrix is made diagonal such that all the elemental specific heat, density and volume terms are preserved, this formulation has led to a better numerical stability and accuracy than the consistent finite element formulation. The solution of some typical practical problems are presented with a special reference to phase change cooling systems.

2. PROBLEM FORMULATION

The governing differential equation of the heat flow in a three dimensional medium can be derived as follows:

Consider a small control volume as shown in Figure 1. The energy balance equation can be stated during a time interval, dt, as follows:

Heat inflow + heat generated = heat outflow + change in internal energy   (1)

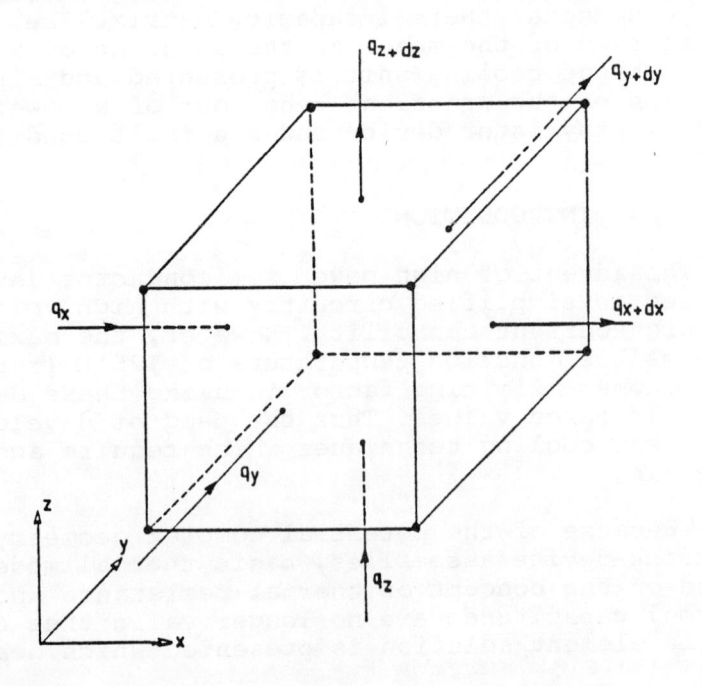

Figure 1 : 3-Dimensional Elemental Volume

The various terms in this equation have been evaluated [1] using the heat rate equations and the Fourier's law of conduction which leads to the following second order differential equation.

$$\frac{\partial}{\partial x}\left(K_x \frac{\partial T}{\partial x}\right) + \frac{\partial}{\partial y}\left(K_y \frac{\partial T}{\partial y}\right) + \frac{\partial}{\partial z}\left(K_z \frac{\partial T}{\partial z}\right) + Q = \rho C_p \frac{\partial T}{\partial t} \quad (2)$$

Where
K_x, K_y, K_z = thermal conductivities in the x-, y- and z-directions.
Q = heat generation per unit volume
ρ = density
C_p = specific heat

2.1 Boundary Conditions

Since equation (2) is a second order differential equation only two boundary conditions are required these are in the form of
(a) a derivative which can be:
  - a heat flux
  - convection
  - an insulated boundary (adiabatic).
(b) a prescribed nodal temperature on a specific boundary.
This can be written in a differential form as follows:

$$K_x \frac{\partial T}{\partial x} L_x + K_y \frac{\partial T}{\partial y} L_y + K_z \frac{\partial T}{\partial z} L_z + q + h(T_s - T_a) = 0 \quad (3)$$

Where
L_x, L_y, L_z = direction cosines of the outward normal to the boundary surface
q = heat flux per unit area.
h = heat transfer coefficient
T_s = heatsink surface temperature
T_a = ambient temperature

Where the heat transfer coefficient, h, can be approximated depending on the mode of the convective heat transfer. In the case of phase change cooling it approximated using Rohsenow correlation [2] in the nucleate boiling regime as shown in Figure 2.

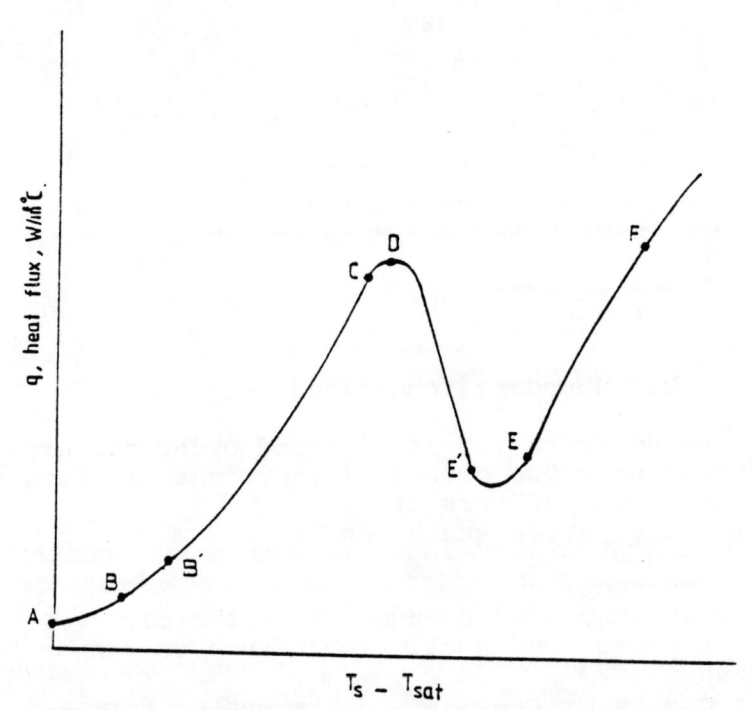

Figure 2: Regimes in boiling heat transfer

```
Where
A-B   = natural convection
B-B'  = subcooled pool boiling
B'-C  = nucleate boiling
C-D   = transition
D     = critical point
D-E'  = partial film boiling
E'-E  = saturated film boiling
E-F   = film boiling
```

Using Freon 113 as the cooling liquid, its heat transfer coefficient was found to be proportional to the square of the the temperature difference (Ts-Tb) where Ts and Tb correspond to the heatsink surface temperature and to the freon 113 boiling point respectively [3]. This can be approximated in transient problems at each time step thus avoiding the need for an iterative solution.

Note that the heat transfer by radiation has been omitted, because it is negligible in the temperature range considered.

2.2 Heat Source

The heat generation Q represents the inherent internal power loss of the power semiconductor device. The heat flux, q, represents the power loss per square meter of the heatsink-device contact area.

3. FINITE ELEMENT FORMULATION

In the present formulation, numerically integrated isoparametric elements are used as shown in Figure 3 : 20-noded elements are used for the heatsink/device structure where as the 8-noded 2-D elements are used for the convective and heat flux source boundary conditions. Gaussian integration scheme is used to evaluate the volume and area integrals. The details of these elements have been reported in the literature [4-6]. However, in the case of a boundary subjected to a heat flux or to convection, one has to apply differential geometry to compute the surface integrals in terms of the 3-dimensional structure coordinates.
Let the local 2-D coordinate system be r,s, then the differential area can be written as follows:

$$dA = \sqrt{a(r,s)} \, dr \, ds \qquad (4)$$

$$\text{Where } a(r,s) = \left[\frac{\partial(x,y)}{\partial(r,s)}\right]^2 + \left[\frac{\partial(y,z)}{\partial(r,s)}\right]^2 + \left[\frac{\partial(x,z)}{\partial(r,s)}\right]^2 \qquad (5)$$

Provided that the three functional determinents are not zero.

$$\frac{\partial(x,y)}{\partial(r,s)} = \frac{\partial x}{\partial r} \cdot \frac{\partial y}{\partial s} - \frac{\partial x}{\partial s} \cdot \frac{\partial y}{\partial r} \qquad (6)$$

$$\frac{\partial(y,z)}{\partial(r,s)} = \frac{\partial y}{\partial r} \cdot \frac{\partial z}{\partial s} - \frac{\partial y}{\partial s} \cdot \frac{\partial z}{\partial r} \qquad (7)$$

$$\frac{\partial(z,x)}{\partial(r,s)} = \frac{\partial z}{\partial r} \cdot \frac{\partial x}{\partial s} - \frac{\partial z}{\partial s} \cdot \frac{\partial x}{\partial r} \qquad (8)$$

Again an isoparametric formulation is adopted.

$$x(r,s,t) = N_i^{(e)} \cdot x_i \qquad (9)$$

$$y(r,s,t) = N_i^{(e)} \cdot y_i \qquad (10)$$

and $\quad z(r,s,t) = N_i^{(e)} \cdot z_i \qquad (11)$

Where N_i are the usual shape functions

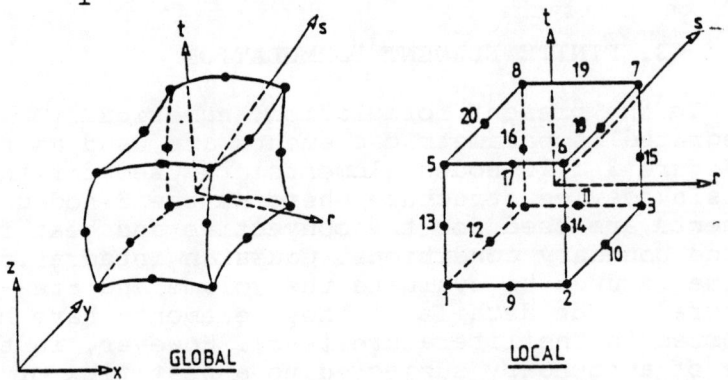

(a) 20-noded Element.

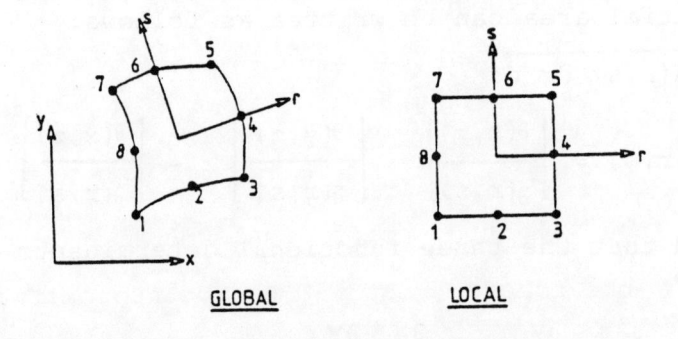

(b) 8-noded Element.

Figure 3: Isoparametric Finite Elements.

4. TRANSIENT SOLUTION

The heat flow equation can be written in a matrix form [7] as follows:

$$C\dot{T} + KT = F \tag{12}$$

Where
C = Thermal capacity matrix
T = Temperature unknown vector
K = Stiffness matrix which includes convective terms.
F = Total load vector including heat generation, convection and heat flux contributions.

4.1 Thermal Capacity Matrix

The thermal capacity matrix is made diagonal by lumping all the coefficients to the diagonal terms such that all the physical properties of the system are preserved. This is implemented as follows:

Let C_{ij} be the square thermal capacity matrix and the scaling coefficient XS be defined as

$$XS = \frac{\Sigma\Sigma C_{ij}}{\Sigma C_{ii}} \tag{13}$$

Then terms of Cij become

$$\begin{aligned} C_{ij} &= 0 \qquad \text{for } i \neq j \\ C_{ij} &= C_{ii} \cdot XS \qquad \text{for } i = j \end{aligned} \tag{14}$$

4.2 Theta-Method

Assuming a linear temperature variation during each time step, the temperature at time n+θ is given as follows:

$$T^{n+\theta} = T^n + \theta \Delta T = T^n(1-\theta) + \theta T^{n+1} \tag{15}$$

Where θ is a weighting coefficient $0 \leq \theta \leq 1$

Similarly the forcing vector F can be approximated at time n+θ as follows:

$$\begin{aligned} F^{n+\theta} &= F^n + \theta \Delta F \\ &= F^n + \theta [F^{n+1} - F^n] \end{aligned} \tag{16}$$

Also $\dfrac{dT}{dt} = \dfrac{T^{n+1} - T^n}{\Delta t}$ (17)

At time $t = t^{n+\theta}$

$C\dot{T} + KT = F$ (18)

Substituting expressions (15),(16) and (17) into (18) leads to :

$T^{n+1} = [\dfrac{C}{\Delta t} + K\theta]^{-1} \cdot \{ [\dfrac{C}{\Delta t} + K(\theta - 1)] T^n$
$+ (1 - \theta) F^n + \theta F^{n+1} \}$ (19)

This computes temperature at time t^{n+1} knowing the temperature at time t^n. Therefore the transient temperature distribution can easily be obtained if the initial nodal temperatures are known. Note that by setting the value of theta standard finite differences algorithms can be recovered. The merit of each scheme is assessed with respect to its accuracy and numerical stability. This is described in details by Zienkiewicz et al [8]. Note that with theta=0 and C diagonal the solution of Tn+1 will become fully explicit but conditionally stable, however in applications requiring very small time steps, this can lead to computer storage and time saving. Also in the case of implicit schemes of linear systems, C and K need to be formed and assembled only once. Thus using Gauss-Choleski method the system of equations can be factorized and only the forcing vector needs to be modified at each time step, leading to computer time saving [9].

5. APPLICATIONS

5.1. A 4-kW Phase Change Cooling Unit

The 3-D doubled-sided cooled heatsink is shown in photo 1. Assuming that the heatsink is symmetrical and the power dissipation at the anode and cathode of the power semiconductor are equal, then only 1/8th of one heatsink needs to be analysed. The power dissipation of the device is translated to an equivalent heat flux (W/m^2). The

Photo 1: Double-Sided Heatsink/Device Assembly

finite element mesh comprises of 48 20-noded isoparametric elements and 381 nodes. The isothermal plot of the device face side is given in Figure 4.

5.2. Transient Analysis

The transient temperature distribution of the example 5.1 is as shown in Figure 5, note that only the extreme critical points are plotted.

The transient temperature distribution of a 50mm gate turn off thyristor is analysed. Because of the axisymmetrical nature of the device slice, 8-noded isoparametric elements are used. Table 1 shows the device transient results with the consistent and the modified (lumped) thermal capacity matrix respectively.

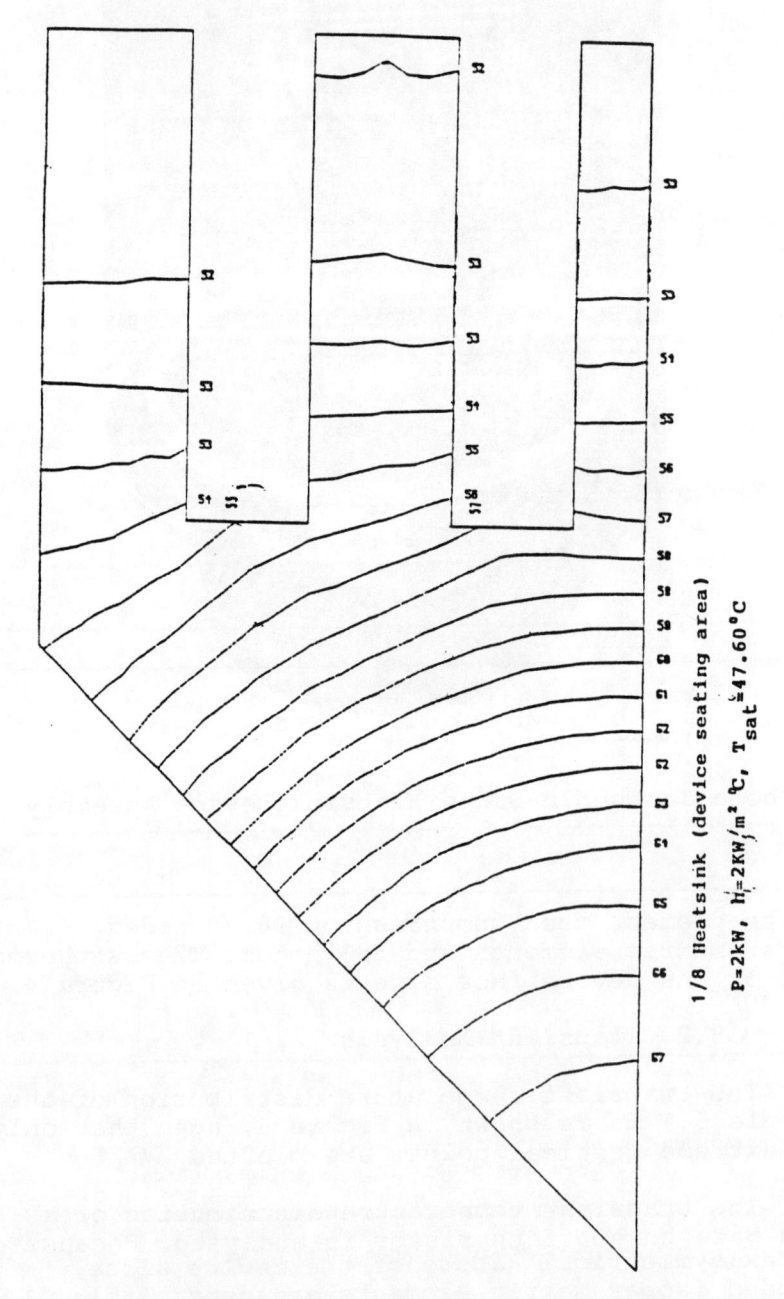

Figure 4: Heatsink Isothermals

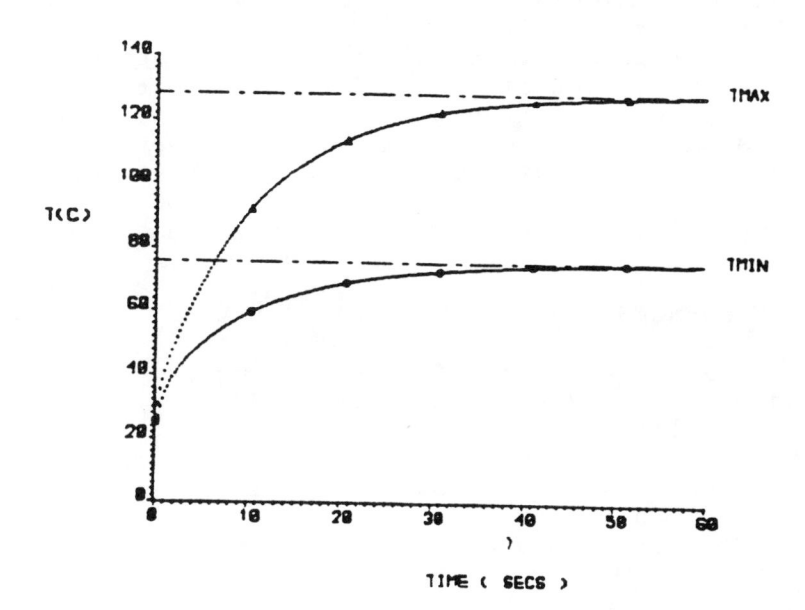

Figure 5: Heatsink Transient Temperatures

NODE	R	C.FEM	M.FEM	C.FEM	M.FEM	ANALYTICAL
		t=.25 Secs		t=.50 Secs		
1	0.	100.7	98.7	95.2	89.9	-
2	0.5	100.7	97.6	90.6	89.9	-
3	1.0	97.	92.7	79.	76.9	-
4	1.5	80.8	78.4	58.3	58.	-
5	2.	46.	46.8	30.1	30.7	-
6	2.5	0.	0.	0.	0.	-
		t=.75 Secs		t=1.0 Secs		
1	0.	80.2	76.2	64.9	62.3	62.64
2	0.5	75.0	72.6	60.3	59.0	59.25
3	1.	63.	61.9	50.	49.5	49.52
4	1.5	44.6	44.6	34.9	35.0	34.76
5	2.	22.4	22.6	17.3	17.4	17.28
6	2.5	0.	0.	0.	0.	0.

ABBREVIATIONS:

    C.FEM = Consistent finite element solution.

    M.FEM = Modified finite element solution.

    R = Radial distance, cm

Table 1: Device Transient Temperatures

6. CONCLUSIONS

The application of the finite element method to the thermal modelling of power electronic systems under steady-state or transient (or fault) operating conditions has proven to be valuable. Because of the ease and flexibility of the finite element method, the feasibility of complicated systems can be carried out without the need of expensive test rigs.

REFERENCES

[1] M.S. Khanniche. -Phase Change Cooling of Power Semiconductor Devices-, Ph.D. thesis 1985. University College, Swansea, U.K.
[2] W.M. Rohsenow. -A Method of Correlating Heat Transfer Data For Surface Boiling of Liquids-, ASME Trans. No 48, July 1952.
[3] J.G. Collier. -Convective Boiling and Condensation- McGraw-Hill, London, 1972.
[4] O.C. Zienkiewciz. -The Finite Element Method-, Third Edition, McGraw-Hill, 1977.
[5] S.S. Rao. -The Finite Element Method in Engineering-, Pergamon Press, 1982.
[6] E. Hinton and D.R.J. Owen. -Finite Element Computations-, Pineridge Press Ltd., Swansea, U.K.
[7] R.W. Lewis and O.C. Zienkiewicz. -Finite Element Solution of Non-linear Heat Conduction Problems with Special Reference to Phase Change-,Int.J. for Num. Methd Eng. Vol.8, 613-624, 1974.
[8] O.C. Zienkiewcz and K. Morgan. -Finite Elements and Approximation-, John Willey and Sons, 1983.
[9] A.Hooper and M.S. Khanniche. -Phase Change Cooling of Power Electronic Devices-, Int. 2nd Int. Conf. in Power Electronics and variable-Speed Drives, Nov., 1986.

FINITE ELEMENT / FINITE DIFFERENCE ANALYSIS OF A DEEP-WELL OXIDATION PROCESS

T.E. Tezduyar *, H.A. Deans ** and J. Marble **

* Department of Mechanical Engineering
** Department of Chemical Engineering

University of Houston
Houston, TX 77004

ABSTRACT

A deep-well oxidation technique has been proposed to process dilute aqueous waste streams. The well assembly, consisting of three concentric tubes, is used in order to generate high pressure by hydrostatic head. The heat of reaction of the waste with oxygen raises the temperature of the fluid above the critical temperature of water (704 °F). Under these conditions, the oxygen and waste are essentially miscible with transcritical water, and the oxidation reaction is very rapid. A residence time of a few minutes is sufficient for almost complete oxidation.

The time-dependent analysis of the well-earth system involves computation of temperature inside the reactor as well as in the earth surrounding it. The energy balance for the earth is described by the heat equation with temperature boundary conditions at the interface with the reactor; it is solved by a Galerkin finite element formulation. For the reactor down and up tubes, the energy balance is represented by convection-reaction equations with heat transfer between the tubes. The heat flux computed from the outside problem is used as a source term for the inside problem. The equations for the reactor tubes are discretized by a finite difference formulation derived from a Petrov-Galerkin approach.

NOMENCLATURE

$B(z)$ reaction rate, assumed independent of time
$T_E(z)$ initial temperature profile in the earth
H depth of the well
$Q(z)$ heat flux from the reactor to the earth
r radial spatial coordinate
s_d metal cross section area of the down tube
s_f flow cross section area, assumed to be equal for down and up flow
s_u metal cross section area of the up tube

$T_d(z,t)$ temperature of fluid in the down tube
$T_u(z,t)$ temperature of fluid in the up tube
t   time
U   overall heat transfer coefficient between the tubes
V   velocity of the fluid in the tubes
z   vertical spatial coordinate
$(\rho C_p)_d$ volumetric heat capacity of the down tube
$(\rho C_p)_E$ volumetric heat capacity of the earth
$(\rho C_p)_f$ volumetric heat capacity of the fluid
$(\rho C_p)_u$ volumetric heat capacity of the up tube
κ thermal conductivity of the earth, assumed to be isotropic
σ heat exchange area between the pipes per unit length

1. INTRODUCTION

A wide variety of hazardous wastes can potentially be processed by oxidation in a deep-well reactor [1]. To be complete in a reasonable residence time, the reactions need to be carried out at transcritical conditions. The residence time at high-rate conditions can be as much as several minutes with the configuration given below. Total throughput is then determined by the flow cross section.

A reliable computer model is needed to predict accurately the temperature profiles in the well and in the surrounding earth as functions of time. The model can be used to predict startup heat requirements and transient time; control strategies can also be tried. Shutdown for cleanout also needs to be simulated.

Well Configuration

The well may be several thousand feet deep and may consist of three concentric tubes. One possible configuration is given in Figure 1. The central tube delivers the oxygen down to the lower section of the reactor. The slurry of water and waste enters the well by the next tube (inner), turns the corner at the bottom and exits by the third tube (outer).

The three-tube reactor assembly is suspended in an outer casing which is cemented in the borehole. The annulus between reactor and outer casing may be filled with a high-temperature non-corrosive oil, the hydrostatic medium. The pressure difference between the earth outside and this outer annulus is minimized, as is the pressure difference between the outer annulus and the reactor fluids. This configuration minimizes the thickness of the reactor tubes, which must be highly corrosion resistant material.

Reactor Profile   Reactor Cross section

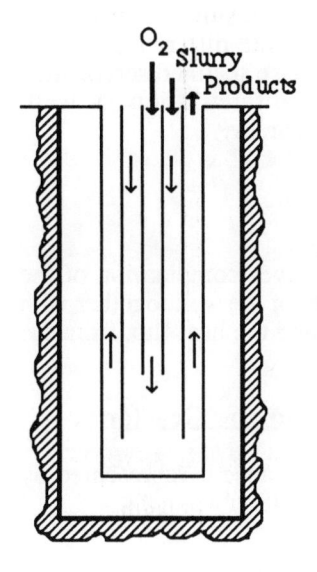

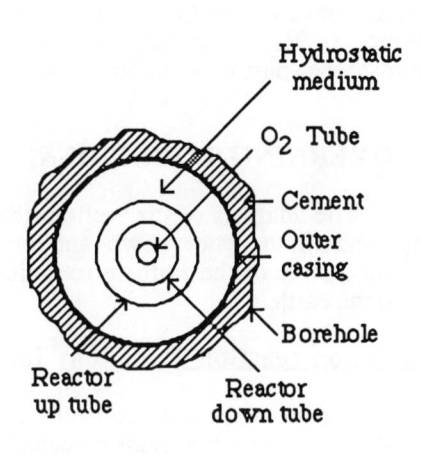

Figure 1. Well schematic.

Reactor Modeling

The details of the reaction kinetics are not considered here. For this example, we assume that the reaction occurs at constant rate over the lowest ten percent of the down tube and goes to completion. Once initiated, the exothermic reaction combined with countercurrent heat exchange fuels the process.

For the tubes, convection occurs due to the motion of the fluid. One source term for the outer tube is computed from the heat flux at the interface of the outer tube and the earth. Another source term, both for the inner and outer tubes is the heat transfer between the two tubes. We assume that the local heat flux between the up and down tubes is given by $q = U(T_u - T_d)$, where U is the overall heat transfer coefficient depending on the velocity in the tubes. At this time, we consider the physical properties of the fluid to be constant.

For the earth, the temperature is governed by the heat equation. The temperature at the outer casing/earth interface is computed from the equations of the tubes. The temperature and heat flux values are matched at the interface of the outer tube and the earth.

Numerical Treatment of the Governing Equations of the Well

The governing equations of the well involve convection and source terms. It is well known that for convection-dominated problems Galerkin (or central difference) formulations result in spurious node-to-node oscillations [2,3,4]. Because of the presence of time-dependent and source terms, use of backward difference or classical upwind methods result in substantial degradation in accuracy. Therefore, we employ a finite-difference scheme derived from a Petrov-Galerkin approach [5,6,7]. These spatial discretization schemes [8,9], which may depend on temporal discretization as well, minimize the spurious oscillations without loss of accuracy.

2. GOVERNING EQUATIONS

The analysis of the well-earth system involves computation of the temperature in the inner (down) and outer (up) tubes of the well together with the temperature of the earth surrounding the well and the heat flux from the well to the earth.

Convection-reaction Equations for the Fluid in the Reactor

For the inner and outer tubes, the energy balance is represented by the following convection-reaction equations:

$$[s_d (\rho C_p)_d + s_f (\rho C_p)_f] \frac{\partial T_d}{\partial t} + s_f (\rho C_p)_f V \frac{\partial T_d}{\partial z} + \sigma U (T_d - T_u) = B \quad (2.1)$$

$$[s_u (\rho C_p)_u + s_f (\rho C_p)_f] \frac{\partial T_u}{\partial t} + s_f (\rho C_p)_f V \frac{\partial T_u}{\partial z} + \sigma U (T_u - T_d) = Q \quad (2.2)$$

The boundary and initial conditions are given as:

$$T_d (0, t) = T_E (0) \quad (2.3)$$

$$T_d (z, 0) = T_u (z, 0) = T_E (z) \quad (2.4)$$

In dimensionless form these equations can be written as:

$$\gamma_d \frac{\partial T_d^*}{\partial \tau} + \frac{\partial T_d^*}{\partial x} + \beta (T_d^* - T_u^*) = B^* \quad (2.5)$$

$$\gamma_u \frac{\partial T_u^*}{\partial \tau} + \frac{\partial T_u^*}{\partial x} + \beta (T_u^* - T_d^*) = Q^* \quad (2.6)$$

The corresponding boundary and initial conditions are:

$$T_d (0, \tau) = 0 \quad (2.7)$$

$$T_d^*(x,0) = T_u^*(x,0) = T_E^*(x) \qquad (2.8)$$

The dimensionless time, spatial coordinate, and temperature are defined by

$$\tau = \frac{tV}{H} \qquad x = \frac{z}{H} \qquad T^* = \frac{T - T_E(0)}{\Delta T_0} \qquad (2.9)$$

Note that $x = 1$ represents the bottom of the reactor assembly. Furthermore, we define the following dimensionless groups:

$$\beta = \frac{H \sigma U}{V s_f (\rho C_p)_f} \qquad (2.10)$$

$$B^* = B \cdot \frac{H}{V s_f (\rho C_p)_f \Delta T_0} \qquad Q^* = Q \cdot \frac{H}{V s_f (\rho C_p)_f \Delta T_0} \qquad (2.11)$$

$$\gamma_d = 1 + \frac{s_d (\rho C_p)_d}{s_f (\rho C_p)_f} \qquad \gamma_u = 1 + \frac{s_u (\rho C_p)_u}{s_f (\rho C_p)_f} \qquad (2.12)$$

Heat Equation for the Surrounding Earth

The energy balance for the earth is the heat equation,

$$\frac{\partial T}{\partial t} = \alpha \left[\frac{1}{\rho} \frac{\partial}{\partial \rho} \left(\rho \frac{\partial T}{\partial \rho} \right) + \frac{\partial^2 T}{\partial z^2} \right] \quad , \quad \alpha = \frac{\kappa}{(\rho C_p)_E} \qquad (2.13)$$

where κ is the thermal conductivity of the earth, which is assumed to be isotropic; ρ and z are the radial and axial coordinates. In the finite element formulation, anisotropic thermal conductivity is readily introduced.

The boundary and initial conditions are

$$T(\rho_0, z, t) = T_u(z, t) \quad , \quad 0 < z < H \qquad (2.14)$$
$$T(\rho, 0, t) = T_E(0) \quad , \quad \rho_0 < \rho < R \qquad (2.15)$$
$$T(R, z, t) = T_E(z) \quad , \quad 0 < z < H+h \qquad (2.16)$$
$$T(\rho, H+h, t) = T_E(H+h) \quad , \quad 0 < \rho < R \qquad (2.17)$$

$$\left(\frac{dT}{d\rho}\right)_{\rho=0} = 0 \qquad , \quad H < z < H+h \qquad (2.18)$$

$$T(\rho,z,0) = T_E(z) \qquad (2.19)$$

where ρ_o is the radius of the outer casing; R is the radius of the computational domain; H is the reactor assembly length; and h is the vertical "heat penetration" depth below the reactor assembly.

In dimensionless form equations (2.13) and (2.14) become

$$\frac{\partial T^*}{\partial t^*} = \frac{1}{\rho^*}\frac{\partial}{\partial \rho^*}\left(\rho^* \frac{\partial T^*}{\partial \rho^*}\right) + \left(\frac{R}{H}\right)^2 \frac{\partial^2 T^*}{\partial X^{*2}} \quad 0 < X^* < 1 \qquad (2.20)$$

$$\frac{\partial T^*}{\partial t^*} = \frac{1}{\rho^*}\frac{\partial}{\partial \rho^*}\left(\rho^* \frac{\partial T^*}{\partial \rho^*}\right) + \left(\frac{R}{h}\right)^2 \frac{\partial^2 T^*}{\partial x^{*2}} \quad 1 < x^* < 2 \qquad (2.21)$$

where the dimensionless temperature, time, radial and axial coordinates are defined as follows :

$$T^* = \frac{T}{\Delta T_0} \qquad t^* = t \cdot \frac{\alpha}{R^2} \qquad \rho^* = \frac{\rho}{R} \qquad (2.22)$$

$$\begin{aligned} X^* &= \frac{z}{H} & 0 < z < H \\ x^* &= 1 + \frac{z}{h} & H < z < H+h \end{aligned} \qquad (2.23)$$

There are two different axial scalings for the earth. Adjacent to the well, z is scaled with the well depth H. Below the bottom of the well, z is scaled with the heat penetration depth h, which is much less than H.

3. FINITE ELEMENT/FINITE DIFFERENCE FORMULATION OF THE PROBLEM

Spatial Discretization of the Governing Equations of the Well

The weak form of (2.5)-(2.7) can be written as follows :

$$\gamma \int_\Omega \tilde{w}\left(\frac{\partial T^*}{\partial \tau}\right) dx + \int_\Omega \tilde{w}\left(\frac{\partial T^*}{\partial x}\right) dx + \beta \int_\Omega \tilde{w}(T^* - \overline{T}^*) dx =$$

$$\int_\Omega \tilde{w}(B+Q) dx \qquad (3.1)$$

The external heat flux Q is zero for the down tube and the reaction heat source B is assumed to be zero in the up tube. The Petrov-Galerkin weighting function $\tilde{w}$ is obtained by a perturbation of the Galerkin weighting function w, where w belongs to the Sobolev space H^1 and satisfies the boundary condition (2.7). The weighting function $\tilde{w}$ is interpolated by the basis functions given below [5,8]:

$$\tilde{N}_a = N_a + C_{2\tau} \frac{\Delta x}{2} \text{ sign } V \cdot \frac{\partial N_a}{\partial x} \qquad (3.2)$$

where N_a is the weighting function (associated with the element node a) leading to Galerkin formulation and $C_{2\tau}$ is the algorithmic Courant number. Various choices for $C_{2\tau}$ are possible, including $C_{2\tau} = 1$ and $C_{2\tau} = C_{\Delta t}$ [6,8], where $C_{\Delta t}$ is the element Courant number based on the convection speed and the temporal as well as spatial step sizes.

Spatial discretization of equation (3.1) leads to the following set of semi-discrete equations:

$$\mathbf{M}\dot{\mathbf{d}} + \mathbf{K}\mathbf{d} = \mathbf{F} \qquad (3.3)$$

$$\mathbf{d}(0) = \mathbf{0} \qquad (3.4)$$

where **M**, **K** and **F** are respectively the mass matrix, the stiffness matrix and the force vector. The matrix **K** is composed of two parts: **K'** and **K"**; where **K'** is due to the convection terms and **K"** is due to the heat transfer between the tubes. Nodal values of the temperature and their temporal derivatives are represented by vectors **d** and $\dot{\mathbf{d}}$.

The reactor assembly is modelled by a one-dimensional system along the x-axis, which starts from the surface, goes down to the bottom along the inner tube and comes back to the surface along the outer tube. For the purpose of implementation, we build 4 by 4 element level matrices. These matrices include the two one-dimensional elements which face each other in the up and down tubes as shown in Figure 2. Because of this, the matrix **K"** is a six-diagonal matrix. The element level constituents of the matrices **M**, **K'** and **K"** can be derived by exact integration:

$$\mathbf{m}^{(e)} = \gamma \Delta x \begin{bmatrix} 1/3 - C/4 & 0 & 0 & 1/6 - C/4 \\ 0 & 1/3 + C/4 & 1/6 + C/4 & 0 \\ 0 & 1/6 - C/4 & 1/3 - C/4 & 0 \\ 1/6 + C/4 & 0 & 0 & 1/3 + C/4 \end{bmatrix} \qquad (3.5)$$

$$k'(e) = \begin{bmatrix} -1/2 + C/2 & 0 & 0 & 1/2 - C/2 \\ 0 & 1/2 + C/2 & -1/2 - C/2 & 0 \\ 0 & 1/2 - C/2 & -1/2 + C/2 & 0 \\ -1/2 - C/2 & 0 & 0 & 1/2 + C/2 \end{bmatrix} \quad (3.6)$$

$$k''(e) = \beta \Delta x \begin{bmatrix} 1/3 - C/4 & -1/3 + C/4 & -1/6 + C/4 & 1/6 - C/4 \\ -1/3 - C/4 & 1/3 + C/4 & 1/6 + C/4 & -1/6 - C/4 \\ -1/6 + C/4 & 1/6 - C/4 & 1/3 - C/4 & -1/3 + C/4 \\ 1/6 + C/4 & -1/6 - C/4 & -1/3 - C/4 & 1/3 + C/4 \end{bmatrix} \quad (3.7)$$

where Δx is the element length and $C = C_{2\tau} \cdot \text{sign } V$.

For the value $C = 1$, from these matrices, we can derive the difference equations for nodes i and j (see Figure 2).

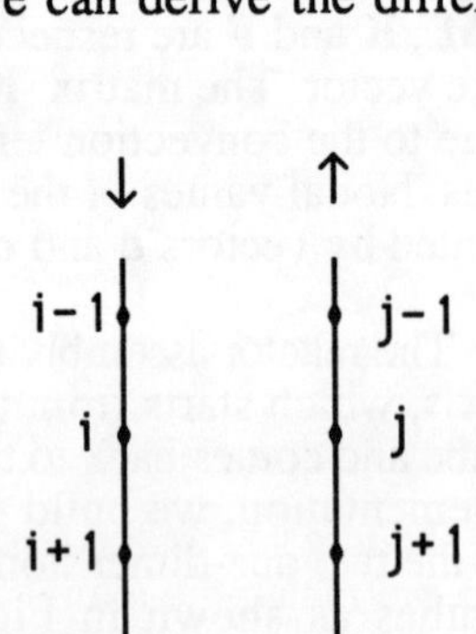

Figure 2. Node numbering convention for the down (left) and up (right) tubes

$$\gamma (5 d_{i-1} + 8 d_i - d_{i+1}) / (12\Delta\tau) + (-d_{i-1} + d_i) / \Delta x$$
$$+ \beta (5 d_{i-1} - 5 d_{j-1} + 8 d_i - 8 d_j - d_{i+1} + d_{j+1}) / 12 = B_i^* \quad (3.8)$$

$$\gamma (-d_{j-1} + 8 d_j + 5 d_{j+1}) / (12\Delta\tau) + (d_j - d_{j+1}) / \Delta x$$
$$+ \beta (-d_{j-1} + d_{i-1} + 8 d_j - 8 d_i + 5 d_{j+1} - 5 d_{i+1}) / 12 = Q_j^* \quad (3.9)$$

where B_i^* and Q_j^* are the dimensionless heat of reaction and heat flux contributions to nodes i and j.

Spatial Discretization of the Governing Equation for the Surrounding Earth

A regular Galerkin finite element formulation [10] is employed for this part of the computational domain. In the finite element method thermophysical properties varying as functions of temperature and other earth characteristics can easily be handled [11].

The weak form of equations (2.20) and (2.21) is written as :

$$\int_{\Omega'} w \left(\frac{\delta T^*}{\delta t^*} \right) d\Omega' + \alpha \int_{\Omega'} \nabla w . \nabla T^* d\Omega' = 0 \qquad \forall w \in H^1 \qquad (3.10)$$

Spatial discretization of (3.10) leads to a set of semi-discrete equations similar to equations (3.3)-(3.4).

Temporal Discretization

The entire semi-discrete equation system is time integrated by a Crank-Nicolson scheme which can be described as follows :

Given d_n and $\dot{d}_n$ at time $n\Delta t$, find d_{n+1} and $\dot{d}_{n+1}$ such that :

$$\dot{d}_{n+1} = (M + \frac{\Delta t}{2} K)^{-1} \{ F_{n+1} - K (d_n + \frac{\Delta t}{2} \dot{d}_n) \} \qquad (3.11)$$

$$d_{n+1} = d_n + \frac{\Delta t}{2} (\dot{d}_n + \dot{d}_{n+1}) \qquad (3.12)$$

4. RESULTS

The internal and external problems are solved independently. At each time step, we match the temperature and heat flux values at the interface of the outer pipe and the earth by iteration.

Our computations are performed in a dimensionless way for the first seven hours of the startup of the reactor. Time scaling is changed, and the computation is continued up to 700 hours of operation. The reaction is then turned off. The inputs variables of the program are the depth of the reactor H, the dimensionless rate of reaction B^* and the dimensionless heat transfer coefficient between the up and down tubes β.

Physically, we are interested in the temperature profile in the tubes and the heat loss to the earth. The maximum temperature reached at the bottom of the well and the exit temperature are two important values to control. The former determines the maximum reaction rate achieved. The latter is a measure of the heat removed in the exit stream. Both are plotted for four different time intervals.

With a 7000 feet deep well and with the values $\beta = 3.888$ and $B^* = 7.51$, we obtain the results given by Figures 3-8.

Figure 3. Finite element mesh for the earth problem

Figure 4. Temperature contours for the earth after seven hours of operation

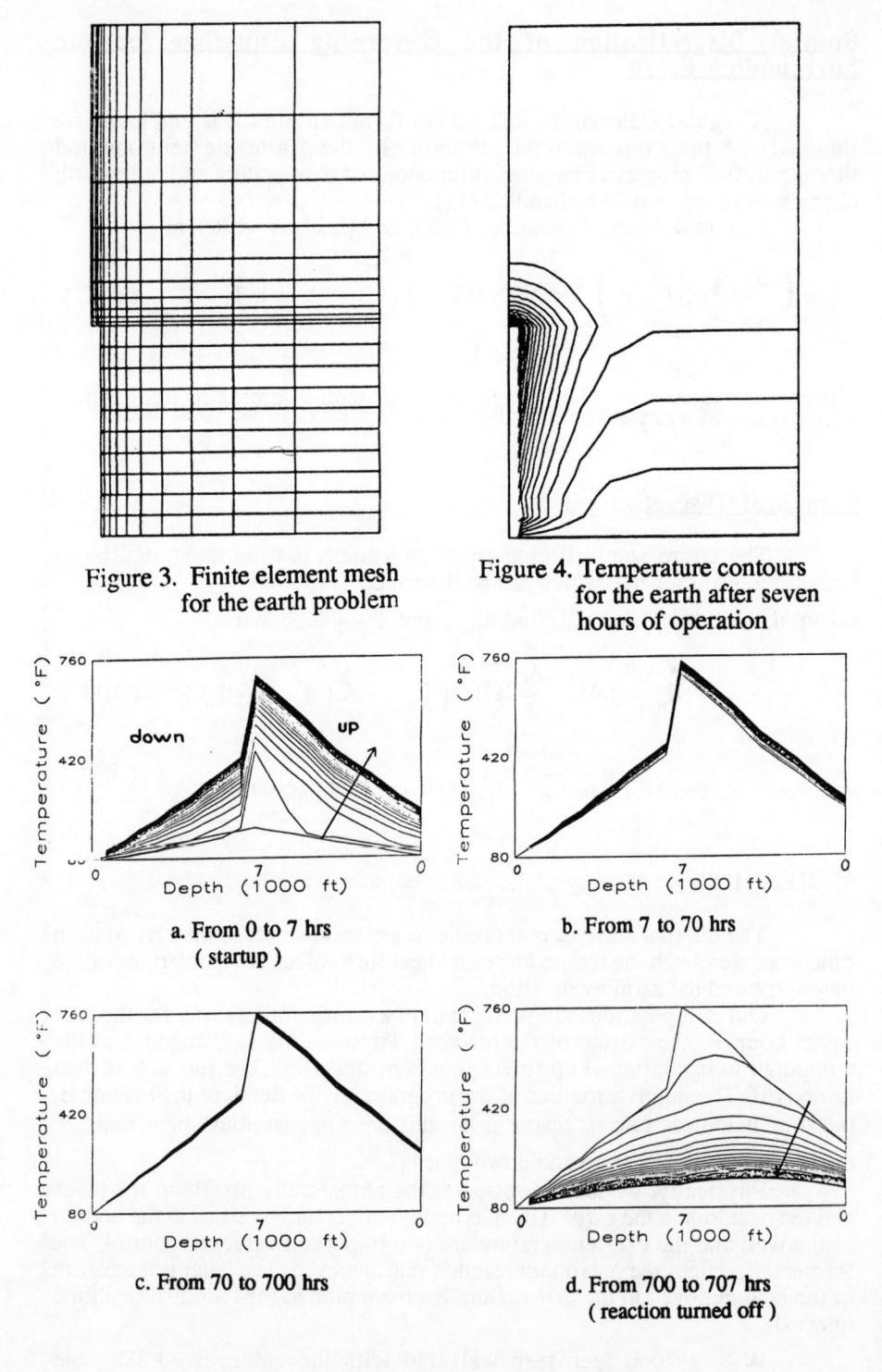

a. From 0 to 7 hrs (startup)

b. From 7 to 70 hrs

c. From 70 to 700 hrs

d. From 700 to 707 hrs (reaction turned off)

Figure 5. Temperature history in the well

1619

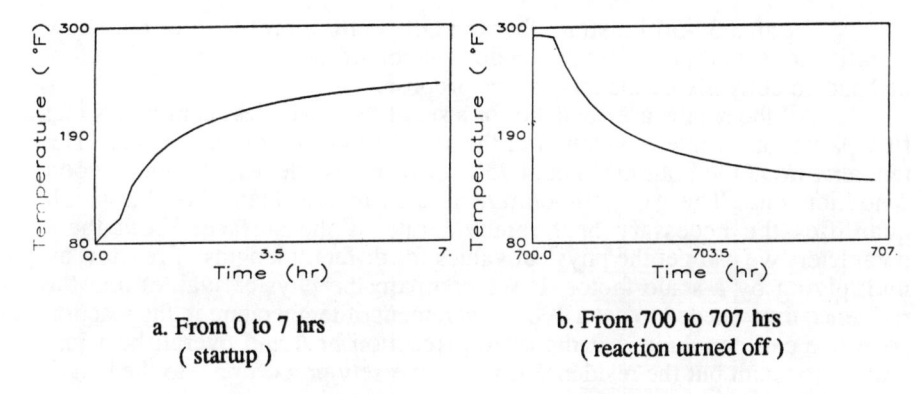

a. From 0 to 7 hrs
(startup)

b. From 700 to 707 hrs
(reaction turned off)

Figure 6. Temperature history at the exit of the well

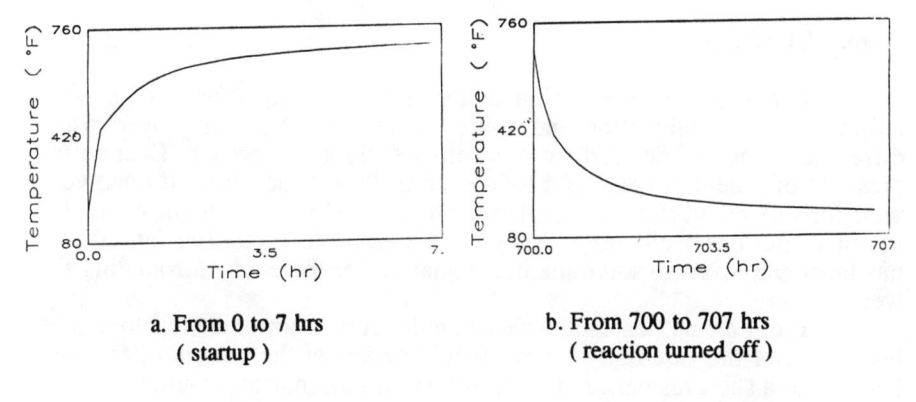

a. From 0 to 7 hrs
(startup)

b. From 700 to 707 hrs
(reaction turned off)

Figure 7. History of the maximum temperature in the well

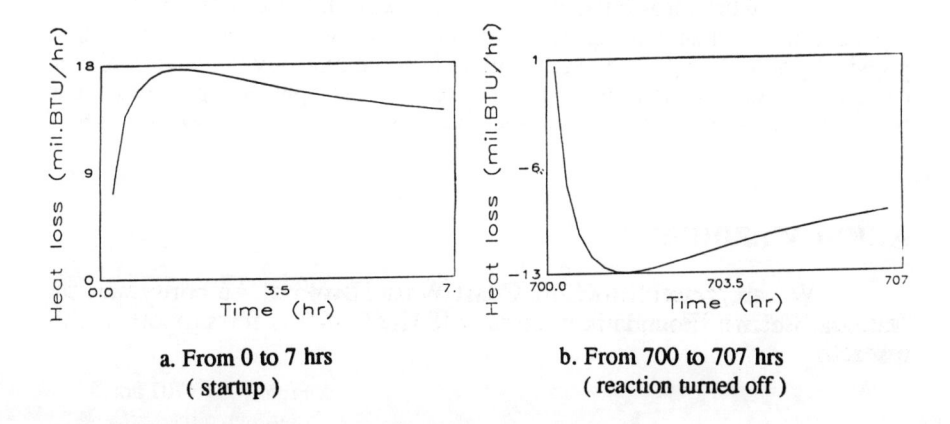

a. From 0 to 7 hrs
(startup)

b. From 700 to 707 hrs
(reaction turned off)

Figure 8. Heat loss history

Figures 5a-5d illustrate that the temperature profile does not show significant oscillations. Convection domination of the equation can therefore be handled correctly by the method we propose.

All the results are obtained for a set of dimensionless parameters that fit a particular problem which is in the set of probable reactor designs. The temperature at the bottom is about 750 °F, which is adequate to obtain good oxidation rate. The exit temperature is a little less than 300 °F, which quantifies the necessary heat removal rate at the surface. From these parameters we can get the physical values for different lengths of reactor by multiplying by a scale factor. If we compare the physical values and the residence time needed to get a 200 °F increment of temperature in the reaction zone, we can conclude that the overall reaction heat and overall heat loss remain constant but the residence time is inversely proportional to the length of the reactor.

These results can be used to design the reactor depending on the kinetics of the reaction and the residence time desired.

CONCLUSIONS

For spatial discretization of the governing equations of the well (which involve convection and source terms), we have employed finite difference schemes derived from a Petrov-Galerkin approach. Despite the presence of time-dependent and source terms, these schemes performed very well for this convection-dominated problem. The results obtained involve minimal spurious oscillations. A regular Galerkin finite element formulation has been employed to solve the heat equation for the earth surrounding the well.

From the model results, we can tell that the shorter the reactor is, the less effect of the side reactions we would expect at the bottom of the well because of a short residence time. However, a short reactor would mean that the reaction has to be very fast and the inlet pressure needs to be high in order to reach critical water temperature and pressure at the bottom.

The work presented in this paper was the first attempt to get a transient model of such a reactor. Several improvements can be made. From a physical point of view, we need to include some hydrodynamics in order to calculate pressure so that the assumption of constant physical properties can be relaxed. Another improvement would be to introduce some real kinetics of the reaction.

ACKNOWLEDGMENT

We are grateful to Gulf Coast Waste Disposal Authority and the National Science Foundation (grant MSM-8552479) for support of this research.

REFERENCES

1. SMITH, J. M., and RAPTIS, T. J., Supercritical deep well wet oxidation of liquid organic wastes, International Symposium on Subsurface Injection of Liquid Wastes, March, 1986.

2. HUGHES, T.J.R., A simple scheme for developing "upwind" finite elements, Intern. J. Num. Meth. Eng., 12 (1978) 1359-1365.

3. GRESHO, P.M. and LEE, R.L., Don't suppress the wiggles-they are telling you something!,in : T.J.R. Hughes,ed.,Finite Element Methods for Convection Dominated Flows, AMD-34 (ASME,New York,1979) 37-61.

4. KELLY, D.W., NAKAZAWA, S. and ZIENKIEWICZ, O.C., A note on upwinding and anisotropic balancing dissipation in finite element approximations to convective diffusion problems, Intern. J. Num. Meth. Eng., Vol.15, 1705-1711 (1980).

5. BROOKS, A.N. and HUGHES, T.J.R., Streamline upwind / Petrov-Galerkin formulations for convection-dominated flows with particular emphasis on the incompressible Navier-Stokes equations, Computer Methods in Applied Mechanics and Engineering, 32 (1982) 199-259.

6. TEZDUYAR, T.E. and HUGHES, T.J.R., Finite element formulations for convection dominated flows with particular emphasis on the compressible Euler equations in : Proceedings AIAA 21st Aerospace Sciences Meeting, AIAA Paper 83-0125, Reno, NV, 1983.

7. HUGHES, T.J.R., MALLET, M. and FRANCA, L., New finite element methods for the compressible Euler equations, in : R. Glowinski and J.L. Lions, eds., Computer Methods in Applied Sciences and Engineering VII (North-Holland, Amsterdam, 1986).

8. TEZDUYAR, T.E. and GANJOO, D.K., Petrov-Galerkin formulations with weigthing functions dependent upon spatial and temporal discretization: applications to transient convection-diffusion problems, Computer Methods in Applied Mechanics and Engineering 59 (1986) 49-71.

9. TEZDUYAR, T.E. and PARK,Y.J., Discontinuity-capturing finite element formulations for nonlinear convection-diffusion-reaction equations, Computer Methods in Applied Mechanics and Engineering 59 (1986) 307-325.

10. ZIENKIEWICZ, O.C., The Finite Element Method, 3rd edn., McGraw-Hill, New York, 1977.

THE EFFECT OF INLET VELOCITY ORIENTATION ON THE SCAVENGING EFFICIENCY IN A UNIFLOW SCAVENGED ENGINE

A. Carapanayotis
Dept. of Mech. Engineering,
University of Ottawa.

M. Salcudean
Dept. of Mech. Engineering,
University of British Columbia.

SUMMARY

The scavenging efficiency in a two-stroke uniflow scavenged engine is strongly affected by the velocity orientation of the gas entering the engine cylinder through the inlet ports. This paper presents the effect of the inlet velocity orientation on the scavenging efficiency for the General Motors GM6V53T six cylinder, two-stroke, uniflow scavenged, direct injection turbocharged diesel engine. The only way the scavenging process can be studied accurately is by using multi-dimensional modeling because the process is geometry dependant. A two-dimensional model with swirl is used with the turbulence effects simulated using the K-ϵ model. The momentum, continuity, energy, concentration, turbulence kinetic energy and energy dissipation rate conservation equations are written in finite difference form and solved explicitly in time. The scavenging efficiency is computed for four different swirl angles and four different inclination angles. The optimum swirl angle was found to be between 20 and 30 degrees. The scavenging efficiency increased with the flow inclination.

NOMENCLATURE

C concentration
K turbulence kinetic energy
P pressure
Pr Prandtl number
r radial coordinate
Sc Schmidt number
t time
T temperature
u radial velocity
v axial velocity
w circumferential velocity
y axial coordinate

Greek Symbols
- ϵ energy dissipation rate
- μ effective viscosity
- μ_l fluid dynamic viscosity
- ρ density

INTRODUCTION

The scavenging process in two-stroke uniflow scavenged engines takes place when the exhaust valves and inlet ports are open and the incoming fresh air replaces the combustion products. The process is very important because the available oxygen for combustion is strongly dependent on the scavenging efficiency, which is defined as the mass of air trapped in the cylinder divided by the mass of air required to fill the cylinder volume at air-box pressure and temperature. The scavenging efficiency is a function of the scavenging ratio, which is the mass of air supplied to the cylinder divided by the mass of air required to fill the cylinder volume at air-box pressure and temperature. The amount of air retained in the cylinder during scavenging is very difficult to calculate because it depends on the cylinder geometry (inlet ports and exhaust valves), inlet port and exhaust valve timing, pressure and temperature in the cylinder, the exhaust manifold and the air-box. Recently, a few studies have been reported based on multi-dimensional analysis studying the scavenging process in two-stroke engines [1-3]. Sher [1], using a code similar to the one used by Gosman and Watkins (TEACH), calculated the exhaust gas-air mixture in a loop-scavenged engine. He calculated the overall scavenging efficiency for the engine for certain flow conditions. Sung [2] studied the air motion in a uniflow-scavenged engine theoretically and experimentally. He measured experimentally the three velocity components across the cylinder radius at 2.5 cm from the Top Dead Center (TDC) for two different flow rates, using laser doppler anemometry. Theoretically, he computed the velocity field by solving the vorticity function for four different exhaust openings. The predicted results were not in good agreement with the experimental results. Recently, Diwakar [3] studied the scavenging process in a two-stroke uniflow-scavenged engine using the "CONHAS SPRAY" two dimensional code. The study concentrated on the effect of the swirl angle on the scavenging efficiency assuming the flow entering the cylinder horizontally. Carapanayotis and Salcudean [4] studied the scavenging process using a two dimensional code that they developed and they derived a scavenging efficiency-scavenging ratio equation which was used in a thermodynamic model for engine performance predictions. Later, a three-dimensional model was developed by Carapanayotis to study the process [5]. He compared the results of the two-dimensional with the three-dimensional model and concluded that the two dimensional model overestimates the recirculation zones in the cylinder close to the cylinder axis but the overal efficiency close to the end of the process was similar. The three dimensional effects were significant mainly close to the valves. Based on this observation it was decided to carry out the investigation of the effects of inflow parameters two dimensionally in order to avoid excessive computer cost.

The objective of the present work is to study the effect of the inlet flow orientation on the scavenging efficiency. The engine selected for the scavenging studies is the GM6V53T two-stroke, uniflow scavenged diesel engine. The engine specifications and operating conditions are given is Table 1. The GM6V53T engine has four overhead exhaust valves and 18 inlet ports. The inlet ports are inclined at 22 degrees to the radius and are located in the cylinder sleeve close to the BDC.

TABLE 1. ENGINE SPECIFICATIONS

Manufacturer	Detroit Diesel Allison
Model	6V53T, Turbocharged
Compression ratio	17.5:1
Number of cylinders	6
Bore	98.4 mm
Stroke	114.3 mm
Piston displacement	5.22 ℓ
Number of valves per cylinder	4
Valve face radius	13.89 mm
Valve to cylinder center dist.	33.65 mm
Number of ports	18
Height of ports	21.33 mm
Width of ports	12.7 mm

MULTIDIMENSIONAL MODEL

The transient two-dimensional code with swirl is used for studying the inflow orientation on the scavenging efficiency [4]. The code is based on the Implicit Continuous-fluid Eulerian (ICE) numerical technique for solving the continuity and momentum equations [6]. The energy equation is solved and the density is computed from the ideal gas equation. The scavenging efficiency-scavenging ratio relationship is derived from the solution of the species equation. These equations are written as follows for two-dimensional flows with swirl:

MASS CONSERVATION

$$\frac{\partial \rho}{\partial t} + \frac{1}{r}\frac{\partial}{\partial r}(r\rho u) + \frac{\partial}{\partial y}(\rho v) = 0 \tag{1}$$

MOMENTUM EQUATIONS
r - component

$$\frac{\partial}{\partial t}(\rho u) + \frac{1}{r}\frac{\partial}{\partial r}(\rho r u u) + \frac{\partial}{\partial y}(\rho v u) = -\frac{\partial P}{\partial r} + \frac{\partial}{\partial y}(\mu \frac{\partial u}{\partial y})$$

$$+ \frac{2}{r}\frac{\partial}{\partial r}(\mu r \frac{\partial u}{\partial r}) + \frac{\partial}{\partial y}(\mu \frac{\partial v}{\partial r}) + \frac{2\mu u}{r^2} - \frac{\rho w w}{r} \tag{2}$$

y - component

$$\frac{\partial}{\partial t}(\rho v) + \frac{1}{r}\frac{\partial}{\partial r}(\rho r u v) + \frac{\partial}{\partial y}(\rho v v) = -\frac{\partial P}{\partial y} + 2\frac{\partial}{\partial y}(\mu \frac{\partial v}{\partial y})$$

$$+ \frac{1}{r}\frac{\partial}{\partial r}(\mu r \frac{\partial v}{\partial r}) + \frac{1}{r}\frac{\partial}{\partial r}(\mu r \frac{\partial u}{\partial r}) \qquad (3)$$

θ - component

$$\frac{\partial}{\partial t}(\rho w) + \frac{1}{r}\frac{\partial}{\partial r}(\rho r u w) + \frac{\partial}{\partial y}(\rho v w) = \frac{1}{r}\frac{\partial}{\partial r}(\mu r \frac{\partial w}{\partial r})$$

$$+ \frac{\partial}{\partial y}(\mu \frac{\partial w}{\partial y}) + \frac{\rho w u}{r} + \frac{1}{r^2}\frac{\partial}{\partial r}(\rho \mu) \qquad (4)$$

Three velocity components are calculated but no angular variation is present.

ENERGY EQUATION

$$\frac{\partial}{\partial t}(\rho T) + \frac{1}{r}\frac{\partial}{\partial r}(\rho r u T) + \frac{\partial}{\partial y}(\rho v T) = \frac{1}{r}\frac{\partial}{\partial r}(r\frac{\mu}{Pr_t}\frac{\partial T}{\partial r}) +$$

$$\frac{\partial}{\partial y}(\frac{\mu}{Pr_t}\frac{\partial T}{\partial y}) \qquad (5)$$

CONCENTRATION EQUATION

$$\frac{\partial}{\partial t}(\rho C) + \frac{1}{r}\frac{\partial}{\partial r}(\rho r u C) + \frac{\partial}{\partial y}(\rho v C) = \frac{1}{r}\frac{\partial}{\partial r}(r\frac{\mu}{Sc_t}\frac{\partial C}{\partial r}) +$$

$$\frac{\partial}{\partial y}(\frac{\mu}{Sc_t}\frac{\partial C}{\partial y}) \qquad (6)$$

The conservation equation of the turbulent kinetic energy and turbulence dissipation rate are used for the computation of the effective viscosity. These equations are written as follows:

TURBULENCE KINETIC ENERGY EQUATION

$$\frac{\partial}{\partial t}(\rho K) + \frac{1}{r}\frac{\partial}{\partial r}(\rho r u K) + \frac{\partial}{\partial y}(\rho v K) = \frac{1}{r}\frac{\partial}{\partial r}(\frac{\mu}{S_k}r\frac{\partial K}{\partial r})$$

$$+ \frac{\partial}{\partial y}(\frac{\mu}{S_k}\frac{\partial K}{\partial y}) - \rho \epsilon C_d + G \qquad (7)$$

where G is the turbulence generation written as:

$$G = \mu[2((\frac{\partial v}{\partial y})^2 + (\frac{\partial u}{\partial r})^2 + (\frac{u}{r})^2) + (\frac{\partial v}{\partial r} + \frac{\partial u}{\partial y})^2 + (r\frac{\partial}{\partial r}(\frac{w}{r}))^2$$
$$+(\frac{\partial w}{\partial y})^2] - \frac{2}{3}\nabla \cdot \vec{u}(\mu\nabla \cdot \vec{u} + \rho K) \tag{8}$$

and

$$\nabla \cdot \vec{u} = \frac{1}{r}\frac{\partial}{\partial r}(ru) + \frac{\partial v}{\partial y} \tag{9}$$

ENERGY DISSIPATION RATE EQUATION

$$\frac{\partial}{\partial t}(\rho\epsilon) + \frac{1}{r}\frac{\partial}{\partial r}(\rho ru\epsilon) + \frac{\partial}{\partial y}(\rho v\epsilon) = \frac{1}{r}\frac{\partial}{\partial r}(\frac{\mu}{S_e}r\frac{\partial\epsilon}{\partial r})$$
$$+\frac{\partial}{\partial y}(\frac{\mu}{S_e}\frac{\partial\epsilon}{\partial y}) + \epsilon(C_1 G - C_2\rho\epsilon)/K \tag{10}$$

$$\mu = \mu_l + C_\mu \rho K^2/\epsilon \tag{11}$$

where:

S_k, S_e, C_d, C_1 and C_2 - are constants.

A numerical method based on the Implicit Continuous-fluid Eulerian (ICE) technique is used for solving the momentum and continuity equations (eqns (1) to (3)) [6]. Equations (4) to (7) and (10) are solved explicitly in time. The governing differential equations are approximated by finite difference equations and are solved algebraically. In the computations non-uniform grid spacing is used to allow detailed calculations in places where large gradients are expected. In the finite difference equations the hybrid differencing scheme is used for the convective and diffusive terms.

INITIAL AND BOUNDARY CONDITIONS

The equations presented are time-dependent, therefore, initial conditions must be specified in the computational domain for all variables. The fluid is assumed to be at rest at time zero (zero velocity). The turbulence kinetic energy and the energy dissipation rate are assumed to be equal to zero and the effective viscosity to be equal to the molecular viscosity. The temperature of the gas in the cylinder is initially set equal to 700 K, which is an average value obtained from the thermodynamic model. The concentration of the residual gases in the cylinder is set equal to zero. The air supplied to the engine during scavenging is assumed to have a concentration of 1.0. Therefore, when the gases in the cylinder have been fully replaced by the air, the concentration will be equal to 1.0.

The assumption of zero initial velocity will affect only a short period at the beginning of the scavenging process. After that, the strong inflow will be dominant, eliminating the effect of the initial zero velocity.

The finite difference form of the governing equations have to be solved together with the relevant boundary conditions. The boundary treatments are outlined in the following for the type of boundaries encountered in the present problem.

Axis: At the axis of the cylinder, no transport occurs and the scalar and velocity gradients are set equal to zero.

Inlet: The flow velocity components and the temperature and concentration are specified.

Outlet: The outlet area is an annulus opening at the cylinder head. The area changes because of the valve motion during scavenging. This motion is modeled by changing the annulus opening area in fine steps.

Wall: The solid walls are treated as impermeable. The tangential velocity at the wall is set equal to zero. The "wall function" method based on one-dimensional Couette flow yielding the universal logarithmic law is used to model the near-wall conditions. Expressions for the corresponding shear stresses have been implemented in the finite difference form of the momentum equations. Appropriate modifications based on the near-wall conditions are made to the turbulence kinetic energy equation and the turbulence generation term. The energy dissipation rate close to the wall is obtained from the assumption that the turbulence production rate is equal to the dissipation rate. The treatment of the energy equation is similar to the momentum equations.

MODEL ASSUMPTIONS

In the present model the piston is assumed to be stationary at the Bottom Dead Center (BDC) during scavenging and the flow area through the ports is assumed to change with time. This is carried out by utilizing a fine mesh at the inlet ports and assuming that some cells behave as wall or port openings depending on the crank angle. The actual inlet and outlet areas have been used in the computations. The assumption that the piston is stationary is acceptable as the piston displacement during scavenging is small. Diwakar [3] assumes a constant pressure in the air-box and exhaust manifold and calculates the inflow and outflow rates using the one dimensional equation for compressible flow through an orifice. In the present study the inlet and outlet flow conditions are established from a thermodynamic model developed by the authors [4].

Initially a zero velocity is assumed in the cylinder. The cylinder residual gases are at a temperature of 700 K, the air-box at 320 K and the pressure is 115.1 kPa [4]. A difficult task is the estimation of the inlet velocity orientation. Diwakar assumes the inlet velocity as being in the horizontal plane and inclined to the radius so as to impose the engine swirl. The engine sleeve has 18 inlet ports with a height of 2.13 cm and a width of 1.24 cm at a 22 degree inclination to the radial direction. From computations for laminar flow, modeling the engine cylinder and the air-box the authors have shown [4] that the flow enters the engine cylinder with a 20 degree inclination to the horizontal plane. Also, it is expected that the inclination of the flow entering the cylinder to the radius will not be 22 degrees because of the relatively large width of the inlet ports. It is understood, therefore, that the prediction of the inflow orientation is not easy. In this work the effect of the inflow orientation on the scavenging efficiency and on the engine performance is studied.

DISCUSSION

The objective of the scavenging process is to replace the residual gases in the cylinder with air at air-box pressure and temperature. By increasing the scavenging efficiency more oxygen becomes available for combustion, which means that more fuel can be burned, thus increasing the engine output power.

The scavenging efficiency depends on the engine operating conditions (air-box pressure, exhaust manifold pressure, etc.) and on the engine geometry (cylinder geometry, piston shape, inlet ports geometry, etc.) These parameters affect the process because they influence the generation of the fluid recirculation zones in the cylinder. Here, the effect of the inflow orientation on the scavenging efficiency is studied. Figure 1 shows the inflow velocity geometry used in boundary conditions. θ is the angle between the inlet velocity with the horizontal and ϕ is the angle between the inlet velocity projection to the horizontal and the radius. To study the effect of the inflow the following two cases were considered: a) variation of the ϕ angle with constant θ and, b) variation of the θ angle with constant ϕ.

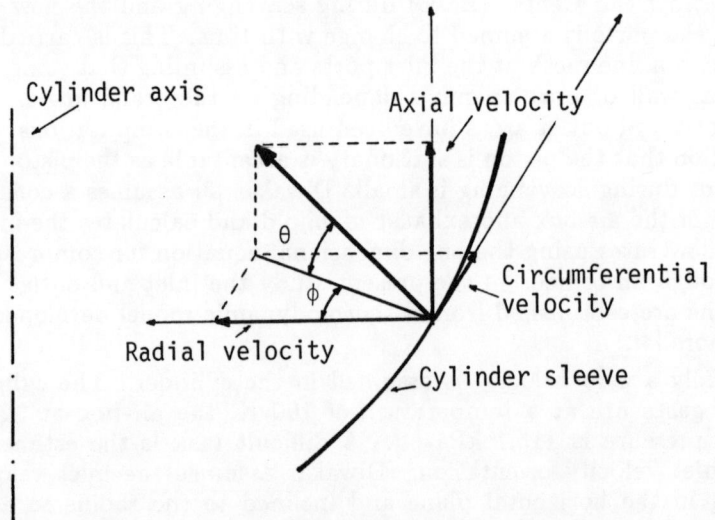

Figure 1. Inlet velocity geometry.

In the computations an unequal grid with 24 grid points in the radial direction and 28 grid points in the axial direction is used. Figure 2 shows the flow development during the scavenging process at 192 degree crank angle after top dead center (ATDC) for different ϕ angles

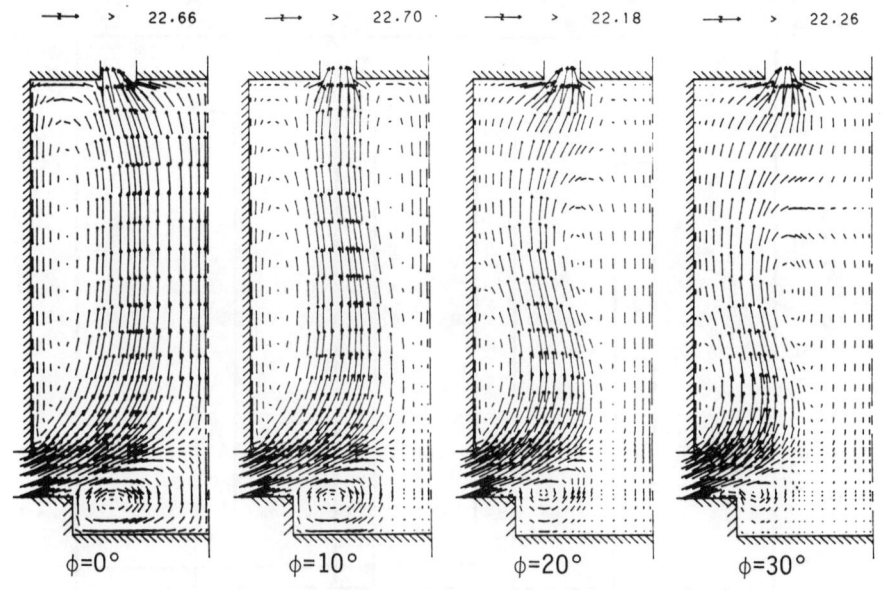

Figure 2. Velocity patterns during scavenging at 192 degrees ATDC for different φ angles (θ=20°).

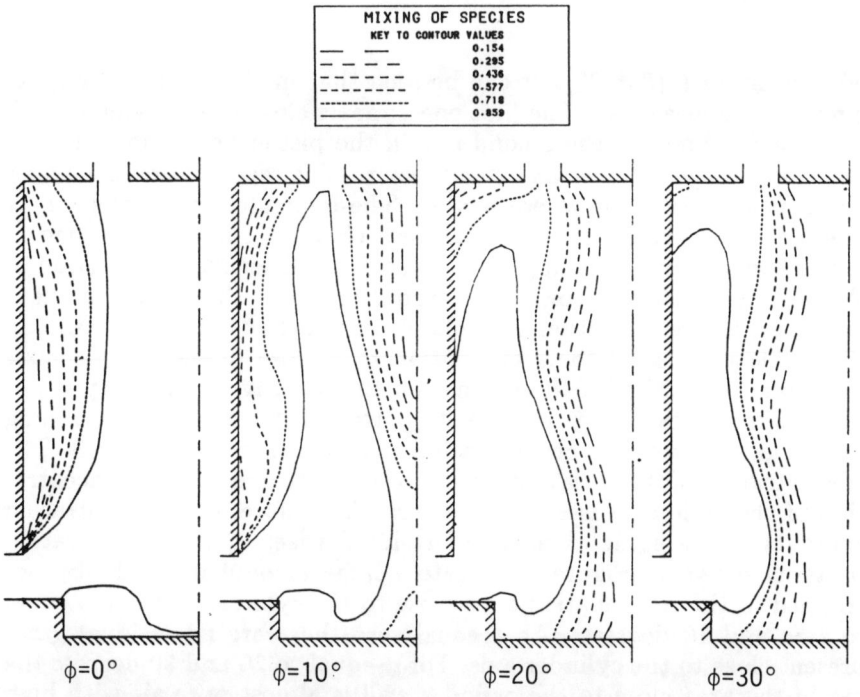

Figure 3. Air concentration contours during scavenging at 192 degrees ATDC for different φ angles (θ=20°).

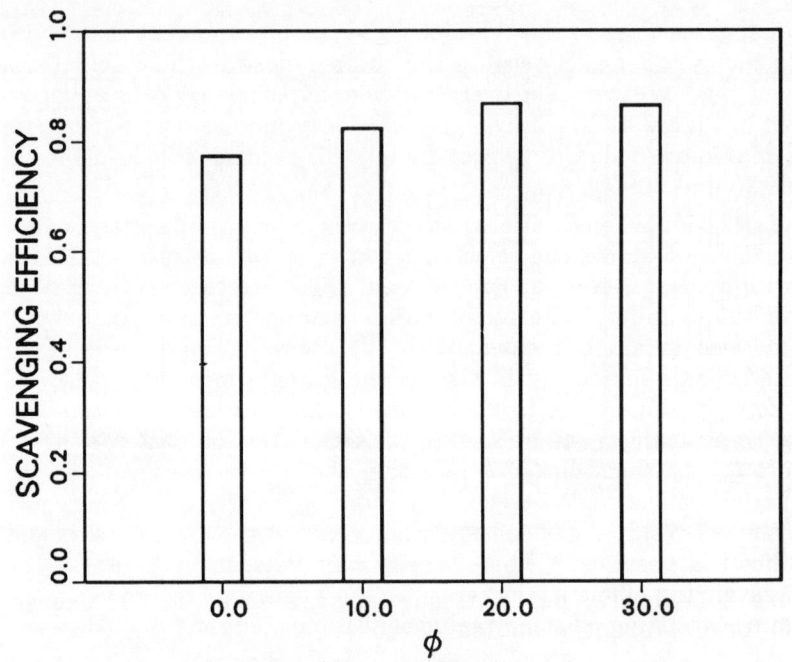

Figure 4. Scavenging efficiency as a function of angle ϕ.

with constant θ ($\theta = 20$). It can be seen that in the cylinder three recirculation zones exist. The first one appears close to the cylinder wall (flow reattachment), the second one in the piston bowl and the third one close to the cylinder axis due to swirl (this recirculation zone does not appear in the case where $\phi = 0$). The recirculation zones strongly affect the scavenging efficiency because residual gases are trapped in the cylinder, thus, reducing the scavenging efficiency. Comparison of the velocity patterns shown in Figure 2 indicates that as the angle ϕ increases, the recirculation zone close to the cylinder wall decreases in size while the size of the recirculation zone close to the cylinder axis increases. The effect of these zones on the scavenging efficiency is not obvious from examining the velocity patterns. A better understanding can be gained from the concentration contours shown in Figure 3. The concentration of the air is assumed to be equal to one and the concentration of the residual gases equal to zero. Therefore, low concentration contours indicate the presence of residual gases. The concentration patterns shown in Figure 3 indicate a large amount of residual gases trapped in the recirculation zone close to the cylinder wall for ϕ equal to zero and 10 degrees. At ϕ equal zero there are no residual gases present close to the cylinder axis. For ϕ equal to 20 and 30 degrees the gas in the area close to the cylinder wall is almost pure air with high residual gas concentration close to the cylinder axis. The concentration of residual gases close to the cylinder wall is more important than those close to the cylinder axis because the volume close to the cylinder periphery is much greater. The overall scavenging efficiency of the process is shown in Figure 4. The scavenging efficiency increases as ϕ increases with a maximum value of approximately 87 percent appearing for ϕ between 20 and 30 degrees.

The situation when θ varies with constant ϕ ($\phi = 15$ degrees) is very similar. Figure 5 shows the velocity patterns at 192 degrees ATDC for different θ angles. Again, as can be seen, there are three recirculation zones in the cylinder. The recirculation zone in the cylinder bowl is the same size in all four cases, with the recirculation zone close to the cylinder axis increasing in size as the θ angle increases. Also, as θ increases the zone close to the cylinder wall decreases in size. The effect of these changes can be seen in the concentration patterns shown in Figure 6. As the value of θ increases the gas in the area close to the cylinder wall is replaced with air more efficiently but an increase in the value of residual gas concentration close to the cylinder axis and piston bowl is observed. Therefore, as can be seen in Figure 7 the overal scavenging efficiency increases as the angle θ increases, with the maximum scavenging efficiency value of 88.15 percent at $\theta = 30$ degrees.

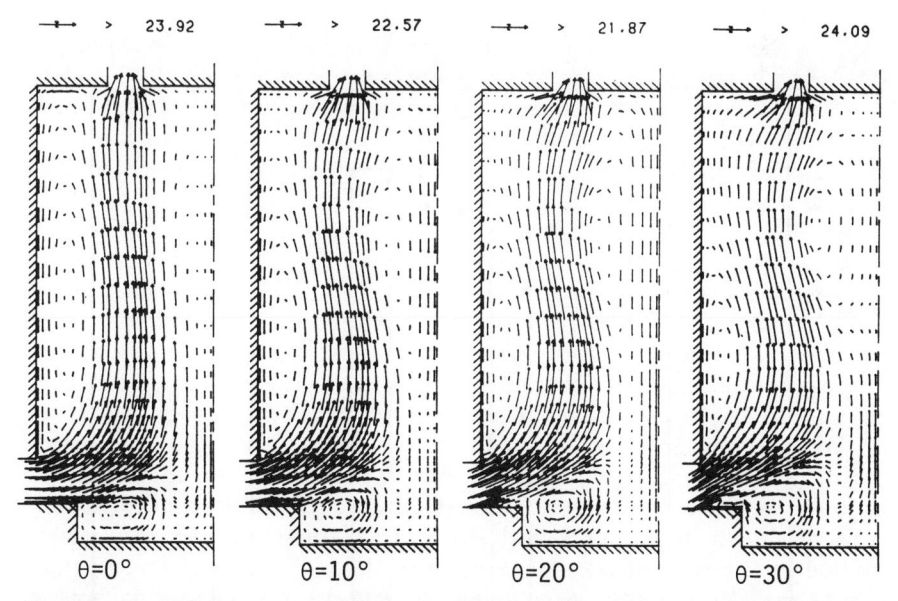

Figure 5. Velocity patterns during scavenging at 192 degrees ATDC for different θ angles (φ=15°).

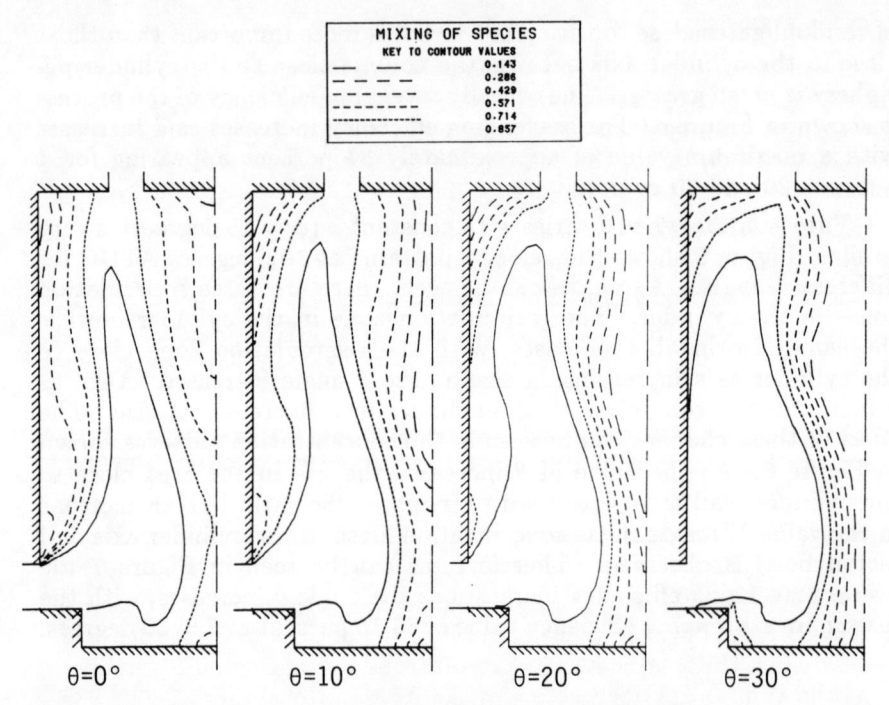

Figure 6. Air concentration contours during scavenging at 192 degrees ATDC for different θ angles. (ϕ=15°).

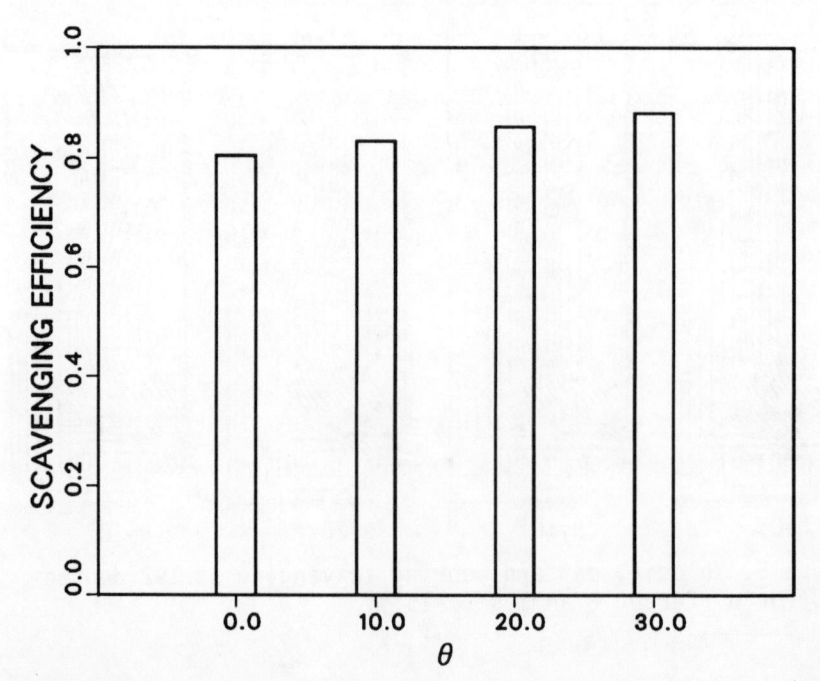

Figure 7. Scavenging efficiency as a function of angle θ.

CONCLUSIONS

The effect of the inflow orientation on the scavenging efficiency was studied using a two-dimensional code. It was found that the optimum scavenging efficiency occurs when the swirl angle ϕ has a value between 20 to 30 degrees. Also, the scavenging efficiency increases as the flow inclination angle θ to the horizontal increases.

ACKNOWLEDGEMENTS

The authors wish to acknowledge the financial support of the Department of National Defense of Canada.

REFERENCES

1 Sher, E., "An Improved Gas Dynamics Model Simulating the Scavenging Process in a two Stroke Cycle Engine", SAE Paper 800037 1980.

2 Sung, N. W., and Patterson, D. J., " Air Motion in a Two Stroke Engine Cylinder- The Effects of the Exhaust Geometry", SAE Paper 820751, 1982

3 Diwakar, R., "Multidimensional Modeling of the Gas Exchange Processes in a Uniflow-Scavenged Two-Stroke Diesel Engine", presented at the Winter annual meeting of the ASME, November 17-22, 1985.

4 Carapanayotis, A. and Salcudean, M., "Thermodynamic Simulation of a Two-Stroke Direct Injection Turbocharged Diesel Engine and Comparison with Experimental Measurements", ASME-JSME Thermal Engineering Join Conference, pp. 209-217, March 1987.

5 Carapanayotis, A., "Modeling of Fluid Flow and Heat Transfer Processes in an Engine", Ph.D Thesis, University of Ottawa, Canada, 1987.

6 Harlow, F. H., and Amsden, A. A., "A Numerical Fluid Dynamics Method for All Flow Speeds", Journal of Computational Physics 8, pp. 197-213, 1971.

DESIGN OF A LATENT HEAT STORAGE USING NUMERICAL SIMULATION AND
EXPERIMENTAL RESULTS, A INDUSTRIAL CASE STUDY

J. Persson and A. Solmar

Department of Mechanical Engineering
Division of Energy Systems
Linköping Institute of Technology, S-581 83 Linköping, SWEDEN

ABSTRACT

Substantial amounts of energy can be saved in industry by
using short term energy storage. Energy audits conducted in
Sweden show possibilities of saving 18 PJ annually in small
and middle sized industries. One way of taking care of surplus
heat from industrial processes is by using a latent heat
storage.

This paper describes design of a storage at a corrugated board
mill. On site measured parameters of a surplus energy flow is
used in numerical simulation and experimental tests. Economical as well as technical performance of the storage is studied.

The lumped heat capacity method is used in the simulation
model. A general purpose program for dynamic simulation ,
Simnon, is used to describe the latent heat storage.

The simulation model is shown to be a useful tool in designing
latent heat storages. Possible annual energy savings amounts
to 360 GJ. Pay back periods vary from 2.6 to 8.5 years,
depending on energy price and installation cost.

1. INTRODUCTION

Industrial and public buildings are characterised by
strongly varying power demand due to the activity in the
premises. Power demand may vary with production rate for
different processes and with climatic parameters, mainly the
outside temperature.

Knowledge of these characteristics make it possible to
draw conclusions as to what energy conservation measures to be

taken. The main reason for energy conservation measures is to reduce the need of primary energy. This can be done in various ways e.g. heat exchanging, heat pumping and by using a heat storage. Heat exchanging is applied when the supplied surplus energy from the source process is available simultaneous with the energy demand, the sink process. This requires an adequate temperature level of the source, if this is not fulfilled heat pumping is required. However simultaneousness is not always at hand, which implies using an energy storage in order to overcome the time lag between source and sink. The interaction between two processes can be shown as in figure 1.

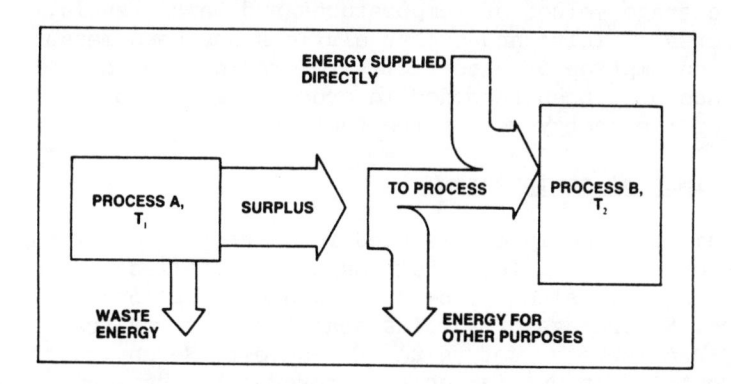

Figure 1. A source-sink pair and the attached nomenclature (Söderström, Johansson, [8]).

Provided that this energy storage can be achieved at reasonable cost the potential for short term energy storage is substantial (Söderström, Johansson, [8]). Energy audits conducted in Sweden, show a potential for short term energy storage of 18 PJ per annum. In a full scale experiment (Solmar, [5]) a latent heat storage was installed in a bakery. Experiences from this project shows that there is a need of tools for latent heat storage design.

1.1 Corrugated board mill

The corrugated board mill is one out of six plants in the Esselte Well Corporation. The factory employs 50 people and the production is about 8000 tons of corrugated board annually. Corrugated board produced is mainly used for wrapping.

The factory is in operation between 6 am and 10 pm, divided in two working shifts. Waste material is evacuated through a pneumatic tube system where material and air is separated by a cyclone. The evacuation system is in operation during both working shifts, that is, there is a need for heating inlet air replacing the evacuated air. The corrugated board machine, emitting airborne surplus energy, is in operation only in the morning shift, 6 am to 2 pm. This indicates that surplus energy can used by heat exchanging combined with heat storage during the morning shift (Persson, [3]). Discussed in this paper is installation of a storage only.

2. METHOD

2.1 Measurements in the corrugated board mill

Measurements has been made in 15 minute intervals, generating average values of temperatures and power levels. Temperatures of inlet and outlet airflows has been measured as well as consumption of electricity. Operation time of oil boilers has also been recorded in order to get an overall picture of the energy use of the factory.

2.2 Experimental equipment

An air flow equipment is used for performance testing of energy storage units. Test programs are controlled by a computer allowing set values to be defined according to a test procedure. Recordings of surplus heat temperatures from an industry can serve as set values for a test. Hereby possible installments of industrial energy storages can be studied in the laboratory. The main parts of the equipment are shown in figure 2.

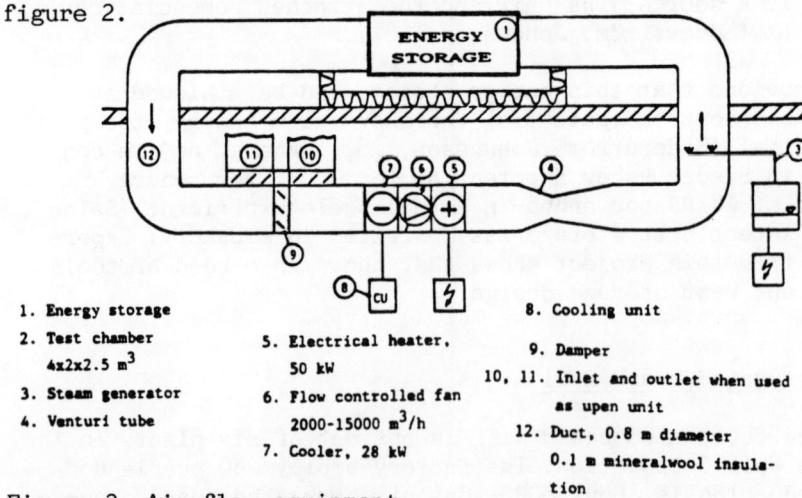

1. Energy storage
2. Test chamber 4x2x2.5 m^3
3. Steam generator
4. Venturi tube
5. Electrical heater, 50 kW
6. Flow controlled fan 2000-15000 m^3/h
7. Cooler, 28 kW
8. Cooling unit
9. Damper
10, 11. Inlet and outlet when used as upen unit
12. Duct, 0.8 m diameter 0.1 m mineralwool insulation

Figure 2. Air flow equipment.

Temperatures are measured with type T thermocouples. Mass flow is measured with a venturi tube. The datalogger scans the measurements continously, storing mean values over a specified time interval in data base.

2.3 Simulation model

Numerical simulation aims to calculate the the temperatures of the dynamic process of charging and discharging the storage. Knowledge of inlet and outlet temperature ables us to calculate power levels and amount of energy stored and discharged. A lumped heat capacity model is used. See e.g. (Holman, [2]). To large extents the used model agrees with (Solmar, Loyd, [6]), although enhancements are introduced.

PCM can be encapsulated in various ways, for example cylinders, spheres and flat rectangular capsules. For understanding the lumped heat capacity model applied to PCM materials encapsulation in flat rectangular capsules is discussed. The heating fluid surrounds both sides of the plate. A number of parallel plates make a module through which the heating fluid passes in a uniform way. Due to symmetry only a part of the module is studied.

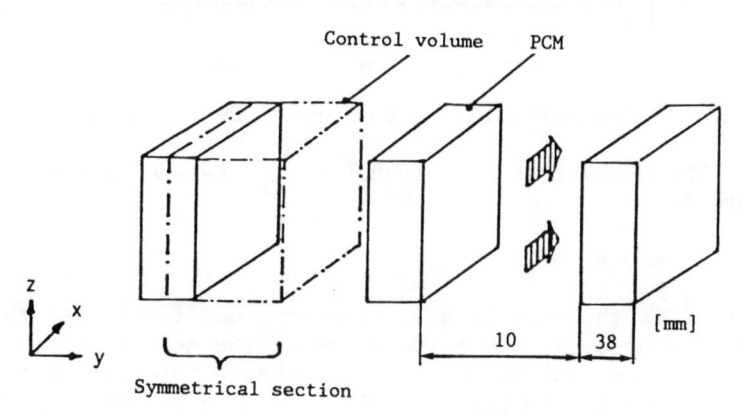

Figure 3. Geometry and control volume (Solmar, [7]).

The following assumptions are used.

1. The heat capacity and the density of the heating fluid are constant.
2. The temperature of the heating fluid is constant in each lump.
3. The heat capacity of the encapsulation material is neglected and the thermal conductivity is infinite.
4. The thermal conductivity of the PCM is infinite.
5. The heat transfer between two PCM lumps in the x-direction is neglected.
6. The phase transition occurs at a fixed temperature.
7. The heat loss from the storage is neglected.

Heat transfer between air and plates is convective, radiation is of less importance. Equations presented here discuss a charging situation. The balance of the energy reduction in the air and the increase of energy in the PCM is described. In figure 4 a cross section of a plate is shown, indicating energy transfer to the plate. The model used in this paper consists of five lumps. However the principle is the same as in figure 4.

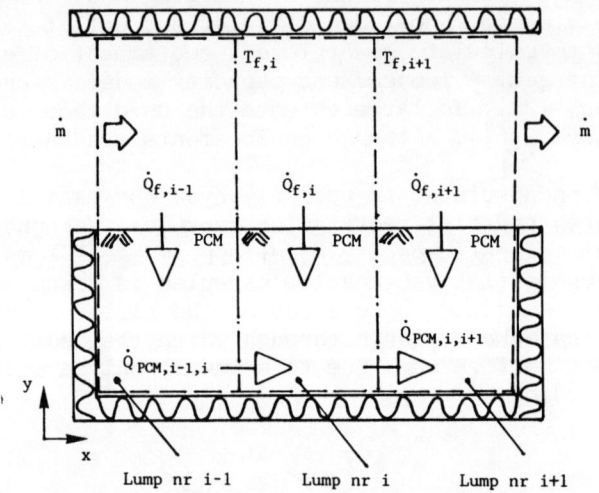

Figure 4. Control volume with three lumps, (Solmar, [7]).

The convective heat transfer, $\dot{Q}_f$, between the air and a plate is

$$\dot{Q}_f = U A_1 (T_f - T_{PCM}) \qquad (1)$$

where U is the overall heat transfer coefficient, A_1 the heat transfer area, T_f the temperature of the heating fluid and T_{PCM} the temperature of the phase change material. Due to the assumptions the temperature diffrences in the PCM and the capsule are neglected. The inverse of the overall heat transfer coefficient U describes the resistance to heat transfer from heating fluid to the <u>phase change front</u>. The distance is denominated d. The overall heat transfer coefficient is expressed as,

$$U = \frac{1}{(1/h + d/k_1)} \qquad (2)$$

where k_1 is thermal conductivity in liquid phase. At discharge k_1 is replaced by k_s, s meaning solid.

Included in the overall heat transfer coefficient is the convective heat transfer coefficient h. As h varies with velocity of the heating fluid the convective heat transfer coefficient is expressed as,

$$h = h_o (w_a/w_i)^n \qquad (3)$$

where h_o is the convective heat transfer coefficient observed at a known velocity, w_i, of the heating fluid, w_a is actual velocity and n is the factor governing the change in convective heat transfer coefficient due to different velocities of the heating fluid, n is identified for each specific test of PCM energy storage. When the PCM is in solely liquid or solid phase, U is calculated on basis of the distance d/2, using corresponding k_1 or k_s. The temperature of the heating fluid is calculated as a mean value,

$$T_f = \frac{1}{2}(T_{f,1} + T_{f,2}) \qquad (4)$$

where $T_{f,1}$ is the inlet temperature of the heating fluid at the lump and $T_{f,2}$ is the corresponding outlet temperature. The heat transfer is

$$\dot{Q}_f = U A_1 (T_f - T_{PCM}) = c_{pf} \dot{m} (T_{f,1} - T_{f,2}) \qquad (5)$$

where c_{pf} is the specific heat of the heating fluid and $\dot{m}$ the mass flow of the fluid. The equations (4) and (5) give

$$T_{f,2} = T_{f,1} \frac{2 c_{pf} \dot{m} - U A_1}{2 c_{pf} \dot{m} + U A_1} + \frac{2 U A_1 T_{PCM}}{2 c_{pf} \dot{m} + U A_1} \qquad (6)$$

$$\dot{Q}_f = \frac{2 U A_1 \dot{m} c_{pf}}{2 c_{pf} \dot{m} + U A_1} (T_{f,1} - T_{PCM}) \qquad (7)$$

During the phase transition the temperature of the PCM is constant until all material is melted. The share of the liquid phase of the PCM is β, $\beta = 1$ indicates solely liquid phase and $\beta = 0$ solely solid phase. The energy balance of PCM lump number i is

$$\dot{Q}_{f,i} = m_{PCM,i} F \qquad (8)$$

where F is

$$F = c_{p,PCM,solid} \frac{d T_{PCM,i}}{dt} \qquad \beta_i = 0 \qquad (9)$$

$$F = q_{tr} \frac{d \beta_i}{dt} \qquad 0 < \beta_i < 1 \qquad (10)$$

$$F = c_{p,PCM,liquid} \frac{d T_{PCM,i}}{dt} \qquad \beta_i = 1 \qquad (11)$$

where $m_{PCM,i}$ is the mass of the PCM of lump number i and q_{tr} the latent heat.

2.4 Simulation program

The program used for simulating the PCM storage is Simnon (Åström ,[10]). Simnon is a special programming language for simulating dynamical systems. Systems may be described as ordinary differential equations, as difference equations or as combinations of such equations. The program is a general purpose program, that is, there is no energy system component library. The language has an interactive implementation which makes it easy to work with. Simnon also allow optimization, introduction of experimental data and parameter fitting. The result of a simulation is displayed as a diagram on a terminal. Parameters of interest are stored in a data base.

Simnon is provided with different integration methods. Hammings predictor corrector with automatic step length adjustment is normally used. The initial value of the step length is chosen as one hundredth of the integration interval. In the algoritm the step length is reduced until the difference between the prediction and correction is sufficiently small. It is also possible to choose the Runge-Kutta algorithm of the fourth order with a fixed or variable time step.

3. RESULTS

3.1 Finite Element Method

A finite element program, THAFEM (Andersson et al, [1]), is used to analyse the flat rectangular encapsulation of phase change material. An example of these calculations is shown in figure 5.

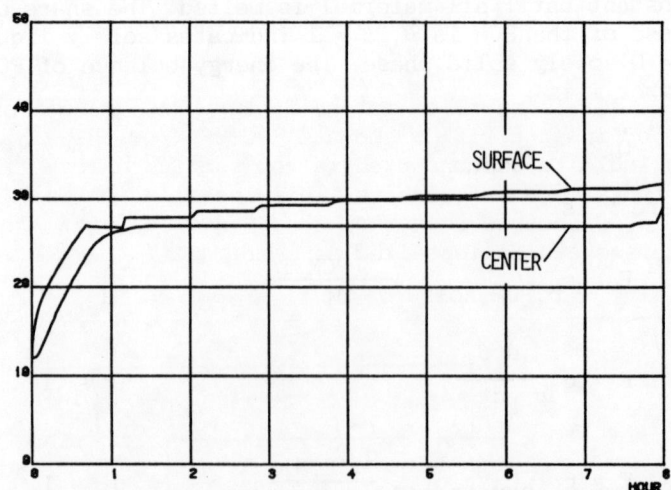

Figure 5. FEM calculation of surface and center temperature in flat rectangular capsule.

The initial temperature is 12°C, and the temperature of ambient air is 40°C. The convective heat transfer coefficient is 20 W/m^{2}K. The heat of fusion is 166 kJ/kg and the temperature of phase transition 29.5 °C, values given by a calorimetrical test. The curves show the temperature at the surface and at center of the capsul. The temperature gradient inside the capsul is less than 4°C.

3.2 Corrugated board machine

Measurements has been carried out in two periods during 1986 and 1987. Surplus energy air flow from the corrugated board machine amounts to 28000 m^3/h. Temperature level as of figure 6.

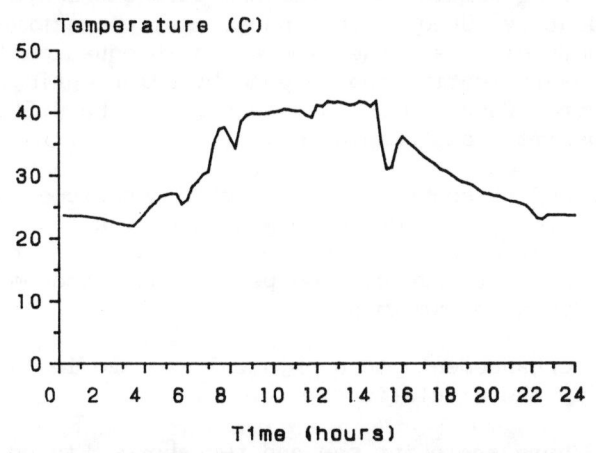

Figure 6. Temperature level of outlet airflow from corrugated board machine, measured on the 18th of February, 1987.

Air is supplied to the factory in a number of places. One major inlet air flow amounts 33000 m^3/h. This inlet flow is suitable to be connected to the surplus energy flow from the corrugated board machine. This installation is favourable due to a reasonable amount of tube installations. A latent heat storage can be installed as in figure 7.

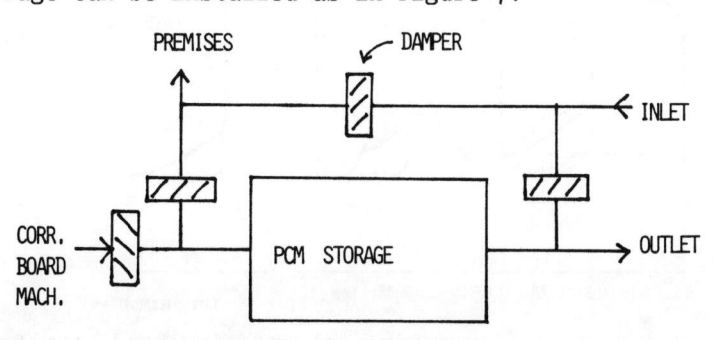

Figure 7. Installation of PCM storage.

Recreating the surplus energy flow combined with a energy storage is done in laboratory and by mathematical simulation. Discussed here are results from flat rectangular capsules. This is due to the fact that the simulation model is verified only for this geometry.

In finding out possible performance of a storage there are three stages of execution;

1. Tests are carried out in the laboratory with the PCM storage in question. This test is used in the simulations to identify important parameters. Parameter identification is done to achive good accordance between experimental results and mathematical simulations. Various parameters are possible to identify. Used in this paper are, convective heat transfer coefficient and the exponent n in equation (3). Parameter identification can be done by using optimization. Hereby an error function is used describing the difference between experiment and simulation.

2. Having achieved reasonable accordance as described above, the size of storage is scaled to make use of actual surplus energy flow. Variations in e.g. geometry and size are tested to find a storage of good performance. Economical performance is also examined.

3. Apply found enhancements to a physical storage in laboratory and verify the simulations.

Discussed here are point one and two above. Figure 8 and 9 shows results after identifying parameters on an storage with flat rectangular capsules. Variation in convective heat transfer coefficient according to equation (3) is shown in figure 8.

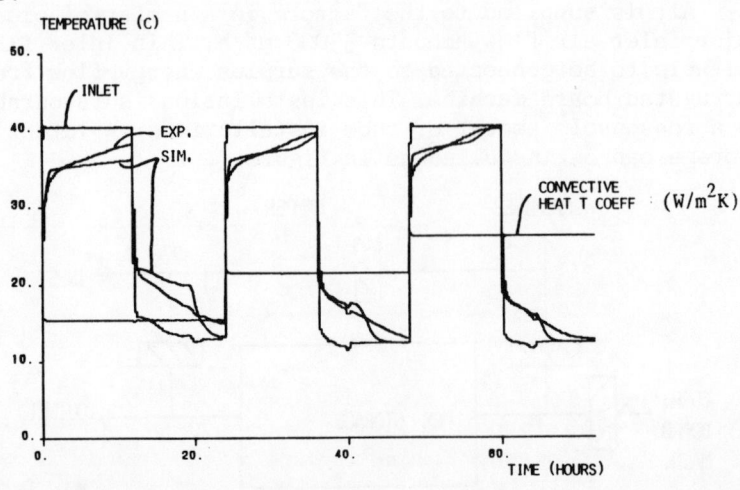

Figure 8. Comparision of experimental and simulated results of a PCM storage.

In figure 9 inlet temperature varies according to collected data from the currugated board machine. Parameters are found to be, h = 17 W/m^{2}K at a volume flow of 2000 m^3/h, and n = 0.5.

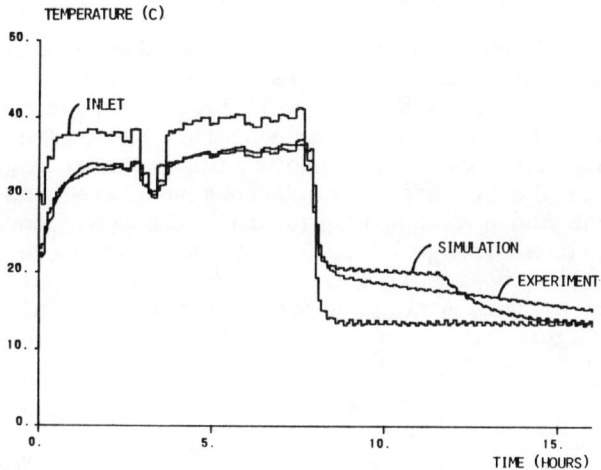

Figure 9. Comparision of experimental and simulated results of a PCM storage, using on site collected data.

Scaling to actual surplus energy flow is done combined with increased convective heat transfer coefficient by introducing enlargement of heat transfer surface. Here it is assumed that it is possible to introduce surface enlargement doubling the convective heat transfer coefficient compared to the results in figure 8 and 9.

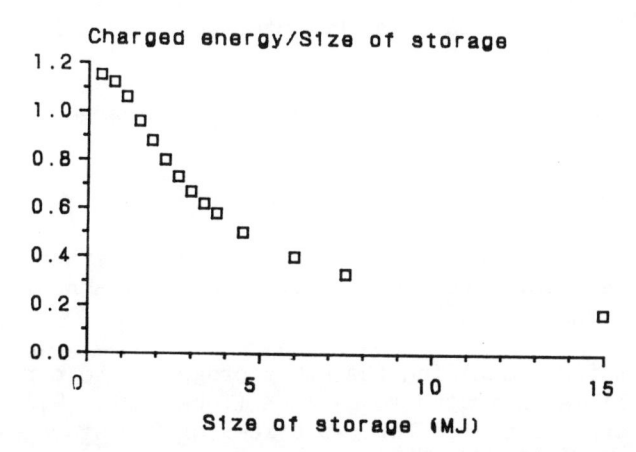

Figure 10. Performance of latent heat storage at the corrugated board mill as a function of size of storage.

Using identified value of parameter n, combined with increased convective heat transfer coefficient, numerous simulations are performed. In the simulations a temperature level from the corrugated board machine as of figure 6 is used. The result is presented as actual charged energy divided by size of storage (latent heat), figure 10.

From figure 10 it is understood that a small PMC storage will be fully charged and discharged. However, the amount of energy saved on each cycle will be correspondingly small. On the other hand, using a large storage will save more energy and thus more money. Therefore sizing the storage is done to maximize the profitability for a given life time of the investment and a given interest rate. Economic considerations show that a storage of 2.25 GJ is advisable (Persson, Karlsson, [4]). The result from simulating an installation of a storage of this size is shown is figure 11. Inlet temperature as of figure 6.

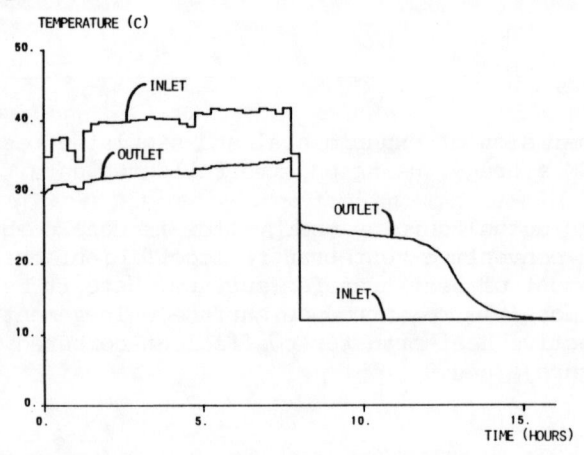

Figure 11. Simulation of an PCM storage, size 2.25 GJ latent heat.

The results show that charged and discharged energy amounts to about 1.8 GJ.

3.3 Economics of PCM storage installation

The outdoor temperature duration, on an annual basis, is used to calculate the annual energy savings. Annually energy savings amounts to 360 GJ.

Cost for installing the PCM storage at the corrugated board mill is estimated to 64 kSEK (Tube plåt, [9]). Pay off periods are calculated for two prices of PCM storages, 83 SEK/MJ and 28 SEK/MJ. The energy savings are worth either 0.08 or 0.14 SEK/MJ. This results in four different pay off periods showing various profitability of an installment.

Combination cost + saving	28 + 0.14 [SEK/MJ]	83 + 0.14 [SEK/MJ]	28 + 0.08 [SEK/MJ]	83 + 0.08 [SEK/MJ]
Pay off period [years]	2.6	5.1	4.3	8.5

DISCUSSION AND CONCLUSIONS

The lumped heat capacity model is shown to be a useful tool in designing a latent heat storage for an industrial process. Experimental equipment is needed to identify parameters of the PCM storage. Having identifying important parameters, simulations show the performance of various storage designs.

However scaling and altering parameters in simulations to improve performance of the storage should be done with concideration. A FEM model of the encapsulated PCM is useful for examining assumptions in the model.

Future work include improving the model for PCM encapsulated in flat rectangular plates. Developing and verifying a model for other geometries, spherical and cylindrical. Future work also include connecting the model to a economical analyses model for thermal energy storages in general.

REFERENCES

[1] Andersson, G., Fröier, M. and Loyd, D., - Heat Transfer Analysis, Finite Element Systems - A Handbook, Ed Brebbia C.D., 3rd rev ed, Springer Verlag, Berlin, 1985.

[2] Holman, J.P., - Heat Transfer, McGraw-Hill, Tokyo, 1981 (5th ed).

[3] Persson, L.J, (1986). Inledande mätningar vid Esselte Well AB, Vikingstad (Measurements at the Esselte Well Corp., Vikingstad), In Swedish, LiTH-IKP-R-449. Linköping Inst. of Tech., Dept. of Mech. Eng., Div. of Energy Systems, Linköping.

[4] Persson, L.J., and B.G. Karlsson, (1987b). Industrial Case Study of a Latent Heat Storage, Proc. of the Innovation for Energy Efficiency Conf., Newcastle (To be published)

[5] Solmar, A. (1983). Energy Storage in a Bakery - Full Scale Experience, Proc. of 2nd BHRA Fluid Engineering

International Conference on Energy storage: Energy Storage for Energy Management. Published by BHRA Fluid Engineering, Cranfield, Bedford.

[6] Solmar, A., Loyd, D., - Heat Transfer at a Latent Heat Storage, Proc. of the 3rd Int. Conf.; Computational Methods and Experimental Measurements, Eds Keramidas G.A. and Brebbia C.A., Computational Mechanics Publications, Springer Verlag, Berlin, 1986.

[7] Solmar, A., (1986), Smältvärmelager, simulering och experiment (Latent Heat Storage, Simulations and Experiments). In Swedish, LiU-Tek-Lic-1986:34. Linköping Inst. of Tech., Dept. of Mech. Eng., Div. of Energy Systems, Linköping.

[8] Söderström, M., and T. Johansson (1983). Use of Energy Storage in the Industry, Proc. of 2nd BHRA Fluid Engineering International Conference on Energy storage: Energy Storage for Energy Management. Published by BHRA Fluid Engineering, Cranfield, Bedford.

[9] Tube plåt, (1986), Installation were discussed with C. Andersson at Tube Plåt Corp., a heat recovery installation company, Linköping.

[10] Åström, K.J, (1982). A Simnon Tutorial, Dept. of Automatic Control, Lund Inst. of Tech., Lund.

THREE-DIMENSIONAL NATURAL CONVECTION IN A HIGH-PRESSURE MERCURY DISCHARGE LAMP

Wei Shyy
James T. Dakin

General Electric Corporate Research and Development
Schenectady, New York 12301, USA

Natural convection plays a significant role in the physics of the high-pressure Hg discharge. As shown in [1-3], convection influences the stability and location of the discharge, and transports energy away from the discharge. Here we report a study for a three-dimensional natural convection process which solves the discharge equations so as to predict the shape of a high-pressure mercury discharge, with arbitrary orientation, based on simple boundary conditions and first principles.

The model provides a solution to the coupled discharge equations—the energy and momentum equations for the discharge fluid, a simplified treatment of radiation, the current continuity equation, and Ohm's law. These equations are solved using finite volume techniques and the method of successive approximations. The discharge medium is assumed to exhibit local thermodynamic equilibrium, and to be steady in time. Inputs to the model include the arc tube geometry, the orientation of the arc tube relative to gravity, and the electrical power input. The self-consistent solution includes the fluid temperature and velocity, the electrostatic potential, and the electric current density, all of which are determined at each node in the staggered three-dimensional grid. The temperature-dependent fluid properties are taken from Zollweg [3] for the case of a 3 atm discharge in Hg.

A schematic view of the arc tube appears in Figure 1. The fused quartz arc tube is characterized by a cylindrical center section of constant bore, which is capped by end chambers. The end chambers are ellipses in the r-z plane. The electrodes protrude into the arc tube by a distance called the insertion length. The separation between the electrode tips is called the arc gap. The calculations are done for a 400 W discharge and a typical commercial arc tube shape. Specific arc tube parameters are summarized in Table I.

The fluid energy, momentum, and continuity equations are solved to determine the temperature and velocity fields using an algorithm developed for solving the Navier-Stokes type of flows using the general non-orthogonal curvilinear coordinates [4-6]. The volumetric source terms driving the flow are the ohmic heating and radiation cooling terms, which enter

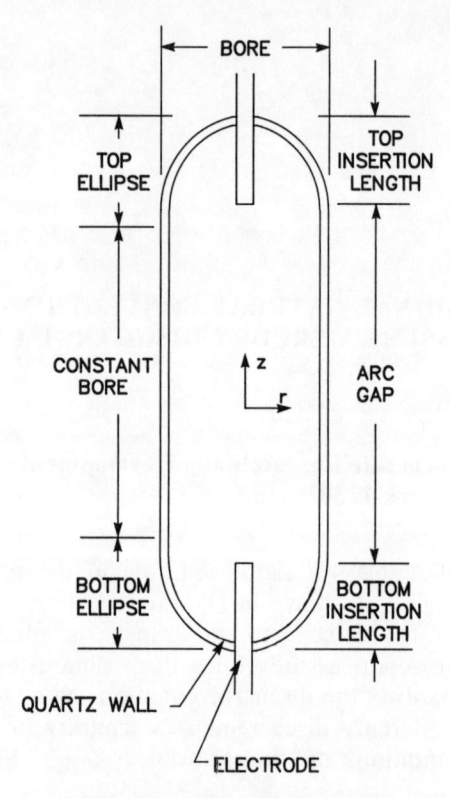

Figure 1. Arc tube schematic

Table I

ARC TUBE GEOMETRY

parameter	value (mm)
bottom insertion length	8.7
gap	42.6
top insertion length	8.7
bottom ellipse length	13.0
cylindrical section length	34.0
top ellipse length	13.0
bore	20.0
effective electrode diameter	3.0

the energy equation, and the gravitational term, which enters the momentum equation. Boundary conditions imposed at the arc tube and electrode surfaces include specified temperatures, and no velocity slip. The convective flow in the discharge is assumed to be purely laminar; this assumption is supported by both the experimental evidence and the calculated Reynolds number. For all the derivatives in the transport equations, convection terms included, the second-order central differencing scheme is employed for discretization. A multigrid method [7] is employed for solving the resulting difference equations.

The ohmic heating term in the fluid energy equation is driven by the electric field, which is the gradient of the electrostatic potential. Electric current continuity requires that this potential satisfy Laplace's equation with a source term to account for the non-uniform electrical conductivity of the discharge fluid. In the numerical algorithm, the treatment of the electrostatic potential is analogous to the treatment of other transport variables. The temperature dependence of the electrical conductivity produces a strong coupling between the temperature and the electrostatic potential.

To handle the imposed geometrical constraints and to maintain a flexible grid distribution, a three-dimensional grid generation code has been developed for the arc tube. The algorithms use a zonal approach to account for the various geometrical components in the arc tube, e.g., electrodes, top/bottom elliptic sections, and mid-section. These various components are then patched together by solving a system of elliptic partial differential equations with multiple outer and inner boundaries. Representative views of the grid for the 400 W arc tube are shown in Figure 2.

Three different arc tube orientations have been investigated — vertical ($\theta = 0°$), oblique ($\theta = 45°$), and horizontal ($\theta = 90°$), where θ is the angle between the axis of the arc tube and the direction of gravitational force. Side views of the solutions for these three orientations are shown in Figures 3 to 5 respectively. End views of the solutions are shown in Figure 6.

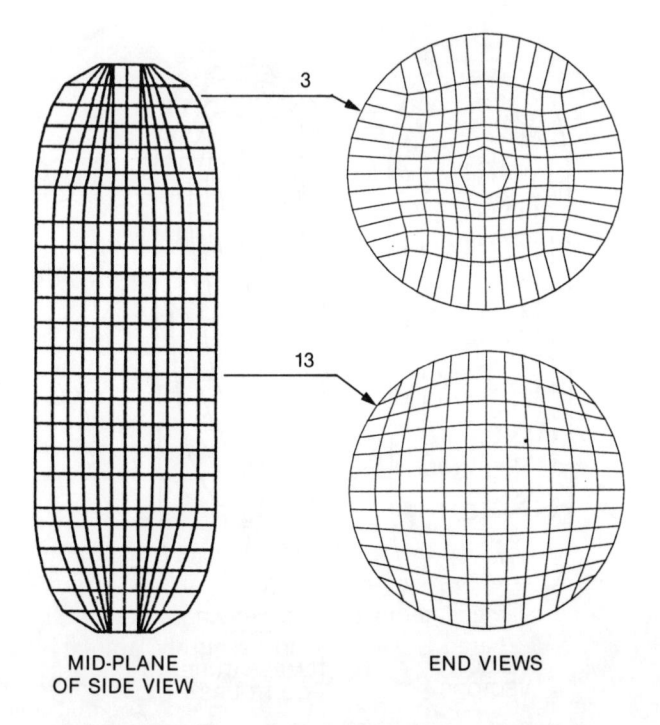

Figure 2. Cross sectional views of arc tube grid

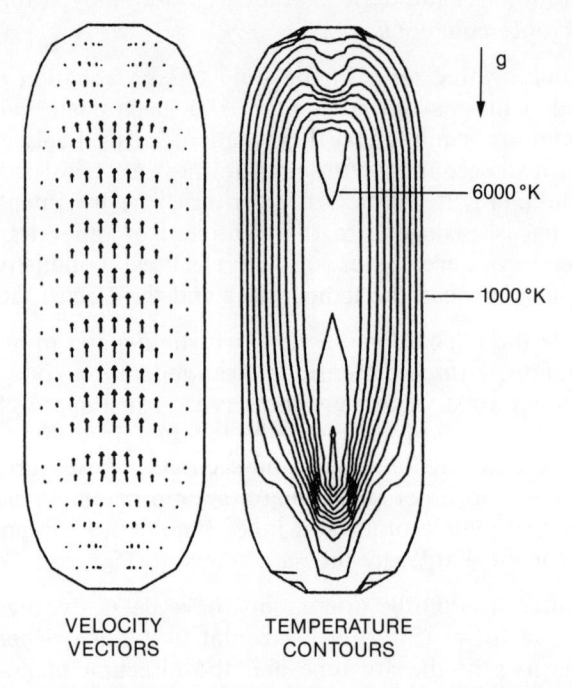

Figure 3. Vertical ($\theta = 0°$) arc tube solution as seen in mid-plane of side view

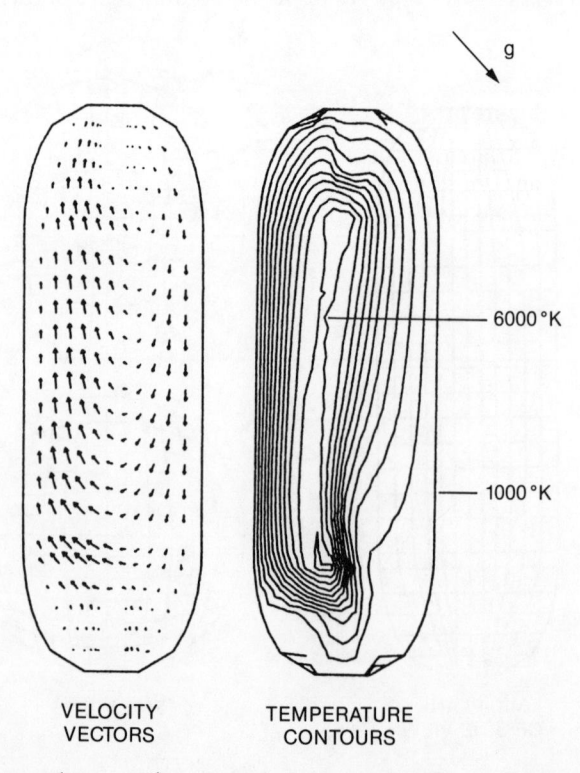

Figure 4. Oblique ($\theta = 45°$) arc tube solution as seen in mid-plane of side view

VELOCITY
VECTORS

TEMPERATURE
CONTOURS

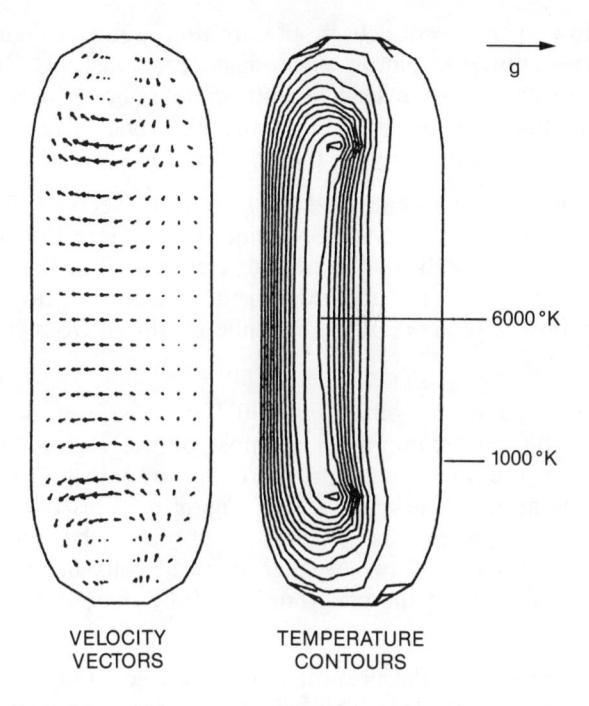

Figure 5. Horizontal ($\theta = 90°$) arc tube solution as seen in mid-plane of side view

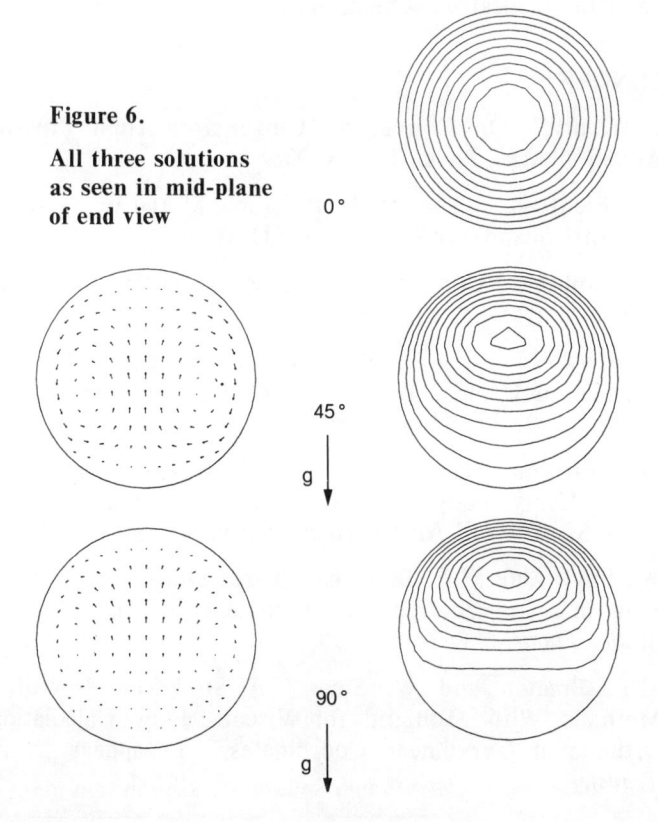

Figure 6.

All three solutions as seen in mid-plane of end view

0°

45°

90°

The flow in the vertical ($\theta = 0°$) arc tube, seen in Figures 3 and 6, is a vertical axi-symmetric plume as modeled previously by Zollweg [3]. The temperature profile is constricted immediately above the bottom electrode, where the fluid flowing up around the electrode is relatively cooler, and lower in electrical conductivity, than fluid in the shadow of the electrode.

The flow in the oblique ($\theta = 45°$) arc tube, seen in Figures 4 and 6, is more complicated and fully three-dimensional. Here the convective effects again cause a plume above the bottom electrode, but the plume is not axi-symmetric since the arc tube axis is not parallel to gravity. A transverse secondary flow is seen when the arc tube is viewed from the end, as in Figure 6.

The flow in the horizontal ($\theta = 90°$) arc tube, seen in Figures 5 and 6, has very little axial component in most of the region between the electrodes. Here the arc is buoyed toward the top wall by the transverse convective flow seen in the end view of Figure 6. This transverse convective flow has two strong cells which pass through the hottest part of the discharge, and signs of two weaker induced cells below them. The end flows in the vicinity of the electrodes are fully three-dimensional in character.

The calculated results presented here are generally consistent with the observations of Kenty [1]. This calculation technique provides a means for predicting the complex convective flows and variations in wall heat loading, which occur in practical arc tube designs.

REFERENCES

[1] C. Kenty, "On Convection Currents in High Pressure Mercury Arcs," *J. Appl. Phys.*, 9, 53 (1938).

[2] W. Elenbass, "The High Pressure Mercury Vapor Discharge," (North-Holland, Amsterdam, 1951).

[3] R.J. Zollweg, "Convection in Vertical High-Pressure Mercury Arcs," *J. Appl. Phys.*, 49, 1077-1091 (1978).

[4] W. Shyy, S.S. Tong and S.M. Correa, "Numerical Recirculating Flow Calculation Using a Body-Fitted Coordinate System," *Numer. Heat Transf.*, 8, 99-113 (1985).

[5] M.E. Braaten and W. Shyy, "A Study of Recirculating Flow Computation Using Body-Fitted Coordinates: Consistency Aspects and Mesh Skewness," *Numer. Heat Transf.*, 9, 559-574 (1986).

[6] W. Shyy and M.E. Braaten, "Three-Dimensional Analysis of the Flow in a Curved Hydraulic Turbine Draft Tube," *Inter. J. Numer. Meths. Fluids*, 6, 861-882 (1986).

[7] M.E. Braaten and W. Shyy, "A Study of Pressure Correction Methods With Multigrid for Viscous Flow Calculations in Non-Orthogonal Curvilinear Coordinates," to appear in *Numer. Heat Transf.*

TEMPERATURE FIELD OF VEHICLES' CLUTCHES AND BRAKES

Marko Tirović and Dragan Stevanović

Mechanical Engineering Faculty,
The University of Belgrade
27.Marta 80, 11000 Belgrade, Yugoslavia

SUMMARY

The paper deals with theoretical and experimental determination of the temperature field of vehicles' clutches and brakes. The theoretical calculations were performed by using semi-implicit Crank-Nicolson's method of finite differences. An appropriate program for microcomputers was developed enabling easy data input, short calculation time and graphical presentation of results. Certain real service conditions were simulated on the inertia dynamometers and temperatures were measured by using thermocouples. Although relatively simple models were used, good concordance between calculated and experimental results was achieved.

1. INTRODUCTION

The performances and reliability of the car clutches and brakes are directly influenced by their temperature fields. Temperatures of clutch and brake elements can be significantly changed in different service conditions, or in the case of material characteristics changes (especially friction material characteristics), elements dimension changes, as well as vehicle's maximum mass and speed changes. Achieved temperature fields can cause unexpected distortion of elements, noise, wear of friction material and friction coefficient reduction. These reasons ask for paying full attention to the determination of the temperature field for real service conditions.

Therefore we have developed a computer program and carried out appropriate temperature measurements to predict and check temperatures reached during car braking and clutch engaging. To make precise measurements and get repeatable results all measurements were taken on the test stands. Real service conditions were simulated on inertia dynamometers at The Motor Vehicle Department of Mechanical Engineering Faculty (The University of Belgrade).

2. MATHEMATICAL MODELING

Cranck-Nicolson's semi-implicit method of finite differences was used to calculate temperature fields. This method is well known [1] and often utilized in nuclear engineering [2]. But here this method is specifically applied to the observed assemblies of the vehicles.

Temperature changes in vehicles' clutches and brakes are far the greatest in the direction perpendicular to the friction surface. Since the heat fluxes in other two directions can be neglected, the one-dimensional models were used. The metal part and friction material of the clutch and the brakes were divided into elements parallel to the friction surface. For one dimensional models Fourier's number has to be less than or equal to 0.5. The characteristics of metal parts (made of cast iron) and friction materials are very different, so for the same Fourier's number the thickness of these elements have also to be different.

The heat is generated on the friction surface, and the developed power can be calculated by multiplying the torque and the sliding (rotation) speed on the friction surface. However, it is very important to divide properly the known developed heat between the metal part and the friction material. This was performed in the following manner. It is supposed that there is the common temperature on the friction surface (marked by To in Figure 1), while elements in contact (metal and friction) can have different temperatures (T1' and T1" respectively in Figure 1).

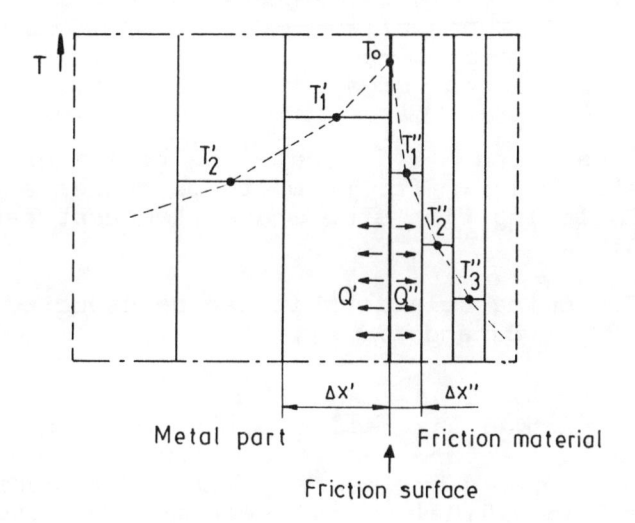

Figure 1. The friction surface model

If Q is the known generated heat, it is equal to the sum of the heats transferred to the metal element (Q') and the friction element (Q") in contact:

$$Q = Q' + Q'' \qquad (1)$$

Further, the following relations are well known:

$$Q' = \left(\frac{\partial T}{\partial x}\right)'_0 k' A'$$
$$Q'' = \left(\frac{\partial T}{\partial x}\right)''_0 k'' A'' \qquad (2)$$

where $\left(\frac{\partial T}{\partial x}\right)_0$ is temperature gradient on the friction surface, k [W/m2K], thermal conductivity, and A [m2] the element surface. Signs ' and " mark the metal and friction elements respectively.

Thermal gradients can be written in the form of the finite differences:

$$\left(\frac{\partial T}{\partial x}\right)'_0 = \frac{\frac{T_{0,n}+T_{0,n+1}}{2} - \frac{T'_{1,n}+T'_{1,n+1}}{2}}{\frac{\Delta x'}{2}}$$

$$\left(\frac{\partial T}{\partial x}\right)_o'' = \frac{\frac{T_{o,n}+T_{o,n+1}}{2} - \frac{T_{1,n}''+T_{1,n+1}''}{2}}{\frac{\Delta x''}{2}} \tag{3}$$

where indices n and n+1 indicate the time step n and the time step n+1 (n for known temperatures and n+1 for the following time step where temperatures are not known).

The following relationship can be deducted from equations (3), (2) and (1):

$$Q = \frac{k'A'}{\Delta x'}(T_{o,n} + T_{o,n+1} - T_{1,n}' - T_{1,n+1}') +$$
$$+ \frac{k''A''}{\Delta x''}(T_{o,n} + T_{o,n+1} - T_{1,n}'' - T_{1,n+1}'') \tag{4}$$

The above equation can be further developed and arranged in the same manner as the equations of heat balance for each element. In this case the coefficients of unknown temperatures do not disturb the three diagonal matrix (there is only one more equation and one more unknown temperature) and usual procedures can be used for calculating the temperature field. It is important to note that equation (4) allows different sizes of surfaces in contact to be considered, being very important for the vehicles brakes. Characteristics of the friction material are given as the functions of temperature.

3. CLUTCH

The section of a typical vehicle's single-plate dry friction clutch is shown in Figure 2. The way of temperature measuring by using a thermocouple placed in the pressure plate near to the friction surface can also be seen.

The temperature field was calculated and the temperature on the friction surface was measured for the clutch operated on the inertia test stand. An inertia test stand is usually used for clutch testing enabling accurate measurements of the clutch performances. The acceleration of a vehicle during clutch engagement is simulated by acceleration of a flywheel of an appropriate inertia moment. The heat is generated on the friction surfaces (there is two surfaces in the case of a single plate clutch) during the clutch engagement.

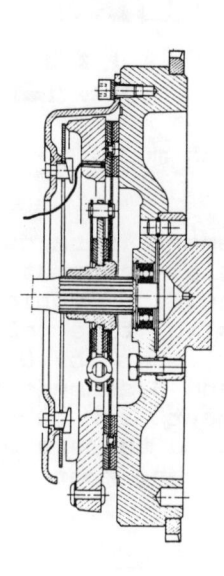

Figure 2. The clutch

The generated power can be calculated by multiplying the developed torque with the slip speed on the friction surface. For that purpose the torque and rotation speeds of the plate and the clutch assembly have to be measured. The typical power-time dependence is shown in Figure 3. A half of the developed power is generated on the each friction surface. So, it can be considered that there is no thermal flux throughout the central plane of the clutch's driven plate. Assuming that the pressure on the friction surface is uniformly distributed, the problem becomes axisymmetric.

Since the difference between sliding speeds on the inner and outer plate radii can be neglected, one-dimensional model can be formed, as shown in Figure 4. To get the same Fourier's numbers, pressure plate (metal disc) was divided in 15 elements of 1 mm thickness, while friction material (plate) was divided in 30 elements of 0.1 mm thickness each. The friction generates the heat on the friction surface (coordinate 0). On the left side of the metal disc the heat is transferred by convection, while there is no thermal flux through the central plane of the clutch's (driven) plate (Q=0).

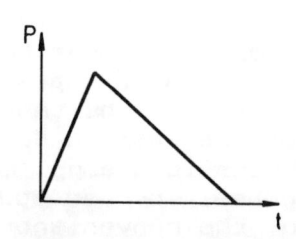

Figure 3. Typical power-time dependence during clutch engagement or during braking

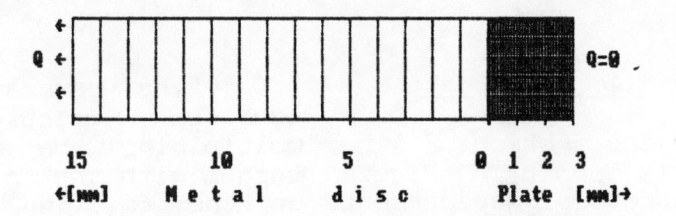

Figure 4. The clutch model

The calculation and measurement were carried out for the "cold" clutch (initial temperature 20 oC) simulating a start of a specified vehicle (a tractor with a trailer) for the slip time of 3 s. Calculation was done with the time step of 0.02 s. Temperatures obtained for 1, 2 and 3 seconds after the start of the clutch engagement are shown in Figure 5. The peak of the measured temperature is marked with the sign "*" and is very near to the calculated one at the same point.

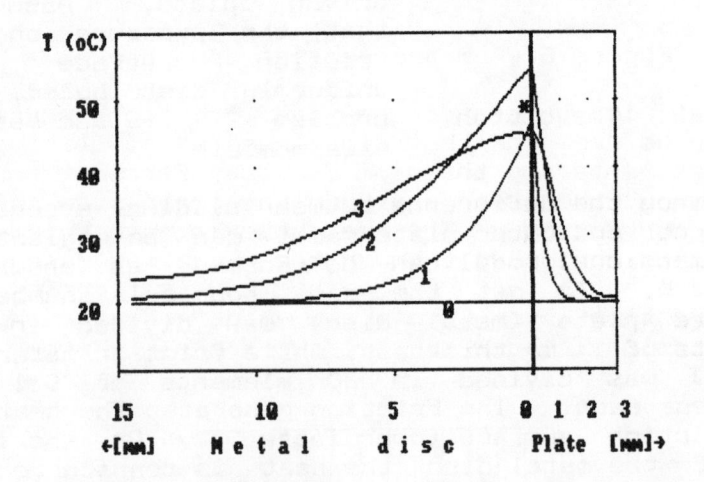

Figure 5. Temperature field of the clutch

4. BRAKES

The investigations were made for the front disc and the rear drum brakes of a passenger car. The temperatures were measured by thermocouples during the brake testing on the inertia dynamometers. On a such test stand, kinetic energy of a vehicle is simulated by a flywheel of an appropriate inertia moment, brought to the convenient rotation speed. The power developed during braking is time dependent as shown in Figure 3, and can be calculated by multiplying braking torque and the sliding speed (which were also measured).

4.1. Drum brake

The section of the drum brake of a passenger car is shown in Figure 6. The thermocouple for temperature measurements was placed in the drum, near to the friction surface.

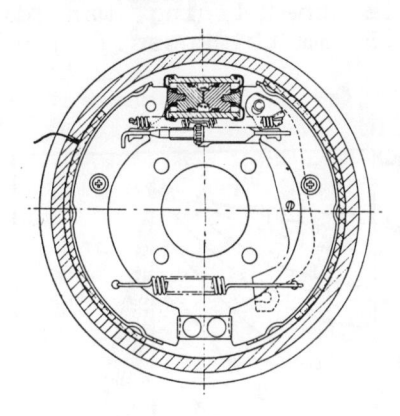

Figure 6. A passenger car drum brake

Heat is generated on the friction surface between the drum and the friction material (lining). The heat entering the drum is transferred through the drum by conduction and from the outer side of a drum by convection. The heat entering the friction material is conducted through the friction material lining and metal part of the shoe and is transferred by convection from the inner side of the shoe. One-dimensional model of the drum brake is shown in Figure 7.

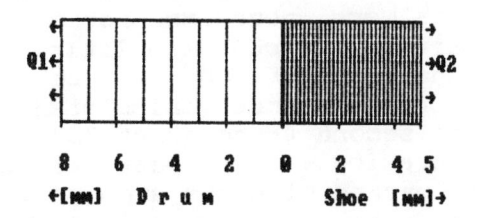

Figure 7. The drum brake model

The coordinate O marks the friction surface where the heat is generated, while Q1 and Q2 are heat fluxes on the outer side of the drum and on the inner side of the shoe lining respectively (these boundary conditions are taken from literature [2]). Assuming the uniform distribution of the interface pressure through the width and the circumference of the drum, one-dimensional model was formed. The real sizes of friction surfaces of the drum and the lining were simulated. Part of the heat transfered through the drum to the hub was taken into account with the bigger accumulation of the heat in the drum. The drum was divided in 8 elements of 1 mm thickness, while the lining was divided in 33 elements of .15 mm thickness. The time step was 0.04 second.

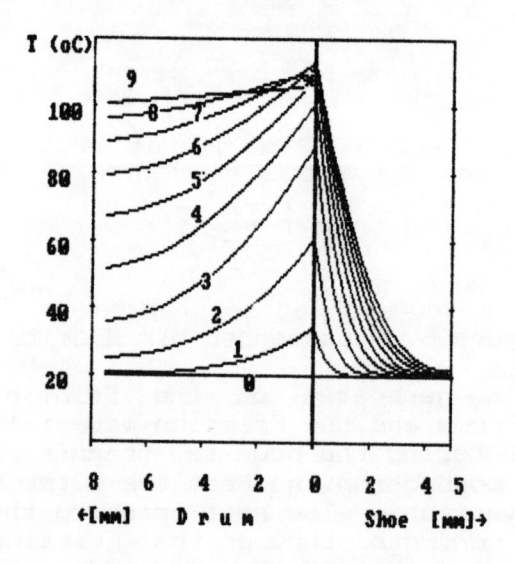

Figure 8. Temperature field of the drum brake

The calculation of the brake temperatures and the measurements were performed for a "cold" brake (initial temperature 20 oC), simulating braking of a vehicle from the speed of 100 km/h to the stop in 9 seconds. Figure 8 shows calculated temperature fields for each second (from start "0" to the stop "9") during braking. The peak of the measured temperature is marked with the sign "*".

4.2. Disc brake

A disc brake is shown in Figure 9. Heat is generated on the friction surface between the disc and the friction material (pad). The heat entering the disc is partially conducted to the hub and partially dissipated by convection (and by radiation on higher temperatures) from the part of the disc surface not covered by the pad. The heat entering the pad is conducted through the pad, then partially transferred by convection and partially conducted through the piston and the caliper.

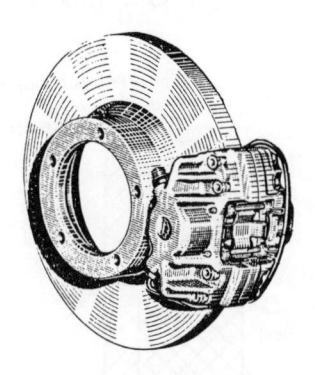

Figure 9. A passenger car disc brake

It can be considered that there is no thermal flux throughout the central plane of the disc (perpendicular to the disc axis). This makes the problem symmetrical. Assuming the uniform interface pressure between the disc and the pad, and neglecting the heat dissipated from the disc during braking, one-dimensional model can be formed, as shown in Figure 10. Part of the heat conducted to the hub was taken into account by increasing the amount of the heat accumulated in the disc. Although the model is one-dimensional, the real sizes of the friction surfaces of the disc and the pad were simulated. Many experiments were performed in the course of which the temperatures were carefully measured [4] proving that such a model can be accepted. Thickness of the elements and the time steps are the same as for the drum brake.

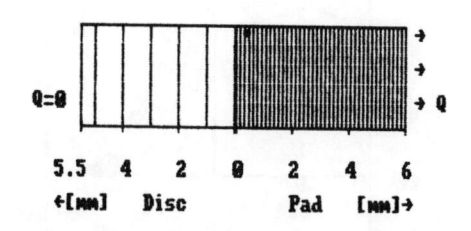

Figure 10. The disc model

Four thermocouples placed in the pad on the different distances from the friction surface were used for temperature measurements, as shown in

Figure 11. Full attention was paid to ensure good contact between the thermocouple and the measuring point. The special high thermal conductivity paste was used.

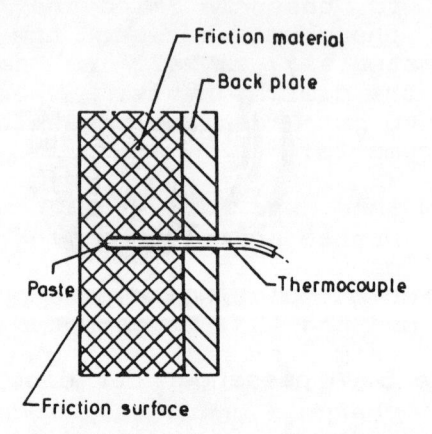

Figure 11. Measuring of the temperatures in the pad

The calculation of the temperature field and measurements were carried out for the same braking conditions as for the drum brake. Figure 12 shows temperature fields calculated for each second during the braking. Temperatures measured in the pad for the 9th second of the braking are marked with the sign "*".

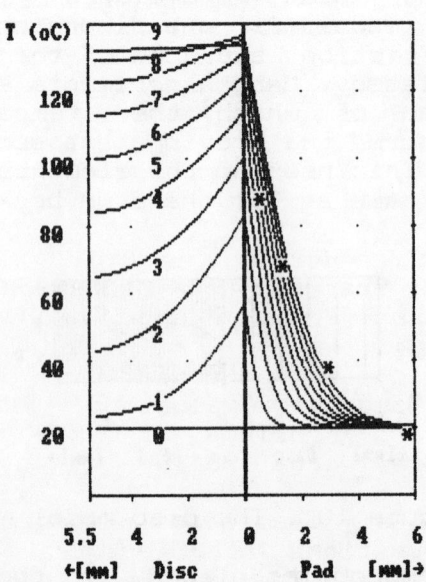

Figure 12. Temperature field of the disc brake

5. CONCLUSION

The comparison of the calculated and measured temperatures showed their good concordance. That means that very simple models with appropriate boundary conditions can give good results. It is necessary to bear in mind the complexity of the friction phenomena between the metal and the friction material, as well as deformations of the brake and the clutch parts. Therefore it is very difficult to get "identical" results by calculations and measurements.

The method used for calculating temperatures in vehicles' brakes and clutches makes it possible to get significant results with a microcomputer. The program for calculating temperatures was developed and used on the IBM PC computer. The calculation time is very short making the calculations attractive and economical. In this way it is very easy to change characteristics of materials, element-dimensions, boundary and working conditions (vehicle's mass, speed etc.). The graphical presentation of the results offers a convenient way of examining the influence of specified parameters.

Further improvements ought to include different boundary conditions as well as more complex model. Therefore it is necessary to take further measurements and program improvements enabling calculation of the temperature field for repeated braking and clutch engagement. But the idea is still to have "a small and practical program" for microcomputers.

REFERENCES

1. Dahlquist, G., Bjorck, A.- <u>Numerical Methods</u>, Prentice Hall, 1974.

2. Day, A.J., Harding, P.R.J., Newcomb, T.P.- Combined Thermal and Mechanical Analysis of Drum Brakes, Proc. Instn. Mech. Engrs.,1984.

3. Stevanović, D., Studović, M. - Method for Fuel Element Response Evaluation, Heat and Nuclear Engineering Problems of PRB, 5th National Conference, Varna 1981.

4. Tirović, M. - Theoretical-Experimental Analyses of Deformations of Disc Brake Pads Due to Mechanical and Thermal Loads, MSc Thesis, Mech. Engr.Faculty, The University of Belgrade, 1982.

NUMERICAL COMPUTATION OF CZOCHRALSKI BULK FLOW FOR LIQUID METALLIC SILICON

Hiroyuki Ozoe and Toshio Matsui
Institute of Advanced Material Study
Kyushu University, Kasuga 816, Japan

Abstract

In the Czochralski method, the silicon crystal is pulled up and rotated at the same time, to maintain uniformity of the crystal composition. The melted metal is subject to both buoyancy and rotational centrifugal forces. The resulting convection pattern depends on the combination of the Reynolds, Grashof and Prandtl numbers. A stable computational code, using primitive variables, was developed for a three-dimensional field in a cylindrical coordinates, with zero gradient in the azimuthal direction. One of the cases computed is characterized by Re=1000, Gr=1.85x10^7, Pr=0.054 (metallic silicon), and an angular velocity Ω=0.0216 (rad./s), with Archimedes number Gr/Re^2 of 18.5. This case was found to be buoyancy-force-dominated. Two other representative cases were computed with Archimedes numbers of 1 and 2.31 and the results were visually presented. It was found that the Archimedes number has important influence on the flow and temperature fields, and consequently on the shape of the crystal rod.

Nomenclature

a_i	coefficient, eq.(8)
D_i	coefficient, eq.(8)
F_i	flux
g	acceleration due to gravity, [m/s^2]
Gr	Grashof number = $g\beta(T_h{'}-T_c{'})h^3/\nu^2$
H	=h/r$_0$
h	height of the crucible, [m]
L	= $R_{out}-R_{in}$
ℓ_s	radius of a crystal rod [m]
L_s	=ℓ_s/r$_0$
P	=P'/P$_0$
P'	perturbed pressure due to convection [Pa]
Pr	Prandtl number = ν/α
R	=r/r$_0$

Ra   Rayleigh number = Gr Pr
Re   Reynolds number = $\rho_0 v_{max} r / \nu$
r   radius of the crucible   [m]
T   = $(T'-T_0')/(T_h'-T_c')$
T_0' = $(T_h'+T_c')/2$ [K]
T'   temperature   [K]
t   time   [s]
U   = u/u_0
u   velocity component in the radial direction [m/s]
V   = v/v_0
v   velocity component in the circumferential direction [m/s]
v_{max} maximum rotational velocity at the perimeter of the crucible [m/s]
V_{max} = v_{max}/v_0
W   = w/w_0
w   velocity component in the axial direction [m/s]

Greek letters
α thermal diffusivity   [m²/s]
β volumetric coefficient of expansion   [1/K]
Δ difference
θ circumferential coordinate   [radian]
ν kinematic viscosity   [m²/s]
ρ density   [kg/m³]
τ dimensionless time = t/t_0
Ω angular velocity of a crystal rod   [rad./s]

Subscripts
c   cold wall
e   half point from a point P to a point E in the east direction
E   point in the east direction
h   hot wall
in   inner radius
n   half point from a point P to a point N in the north direction
N   point in the north direction
nb   neighborhood points
out outer radius of a domain
p   central point to represent pressure
s   half point to S in the south direction
S   point in the south direction
w   half point from a point P to a point W in the west direction
W   point in the west direction
0   reference value for a dimensionless variable

1. Introduction

   Semiconducting materials such as metallic silicon and Gallium Arsenide are processed in a crucible at high temperatures. Due to inevitable temperature gradients, melted metal undergoes convection under the gravitational field. In the Czochralski method as shown schematically in Fig.1. the crystal silicon is pulled up with a simultaneous rotation, to keep the uniformity of the crystal composition. The metallic silicon in a crucible is heated from the crucible wall. The melted metal is thus subject to both the buoyancy and the rotational centrifugal forces. The resulting convection pattern depends on the combination of the primary parameters such as the Reynolds. Grashof and Prandtl numbers. The convection of the melted metal has been found to affect the shape of the crystal rod [1]. For example, depending on the radius of the crystal, bottom shape of the crystal becomes concave or convex downwards. When a rotational speed exceeds some value, the crystal rod can not be grown up stably any more. When the length of the crystal rod exceeds some value, the shape of the crystal becomes a spiral mode rather than a real circular-mode. These unpreferable effects have been found to become dominant for a large crucible. Kobayashi[1-6] has been quite active in the simulation of this Czochralski-process bulk flow. He employed a vorticity equation in the vertical cross section of the crucible, and the momentum equation in the azimuthal direction as well as an energy and a mass balance equation in a vertical cross section. His publications cover Grashof numbers up to 10^6. Reynolds numbers

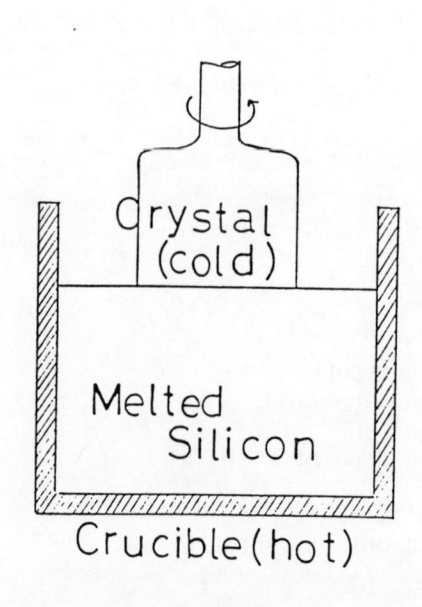

Fig.1 Schematics of the Czochralski method.

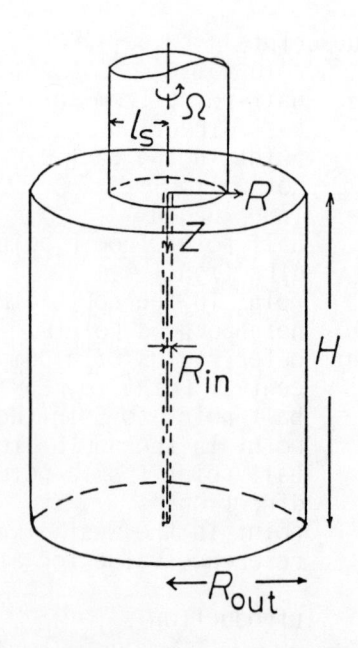

Fig.2 System coordinate and dimensions.

up to 1000, and aspect ratios of the crucible from 1 to 5. The combined effect of the Grashof number and the Reynolds number has, however, not been studied in detail at low values of the Prandtl number. The computational scheme he used was a classical relaxation method. Other related papers are Langlois[7-8], C.E.Chang[9], Van der Hart and Uelhoff[10], Nikitin et al.[11], Polezhayev et al.[12] and Tsukada et al.[13]. However, many more studies have been conducted on this subject.

In the present work, the primitive variables of u,v,w and p are employed. This is in contrast with most of the former studies which employ vorticity and a stream function, and hopefully provides physical understanding in a more direct way. The solution scheme developed by the authors of this paper employs the SIMPLER procedure proposed by Patankar[14] for a forced convection in a duct. The computational code was developed for Czochralski bulk flow for liquid metallic silicon, and sample computations were carried out for a number of extreme combinations of parameters. In this paper, three distinct flow modes were found to exist depending on the Archimedes number for liquid metallic silicon.

2. The mathematical model

The flow field is in a crucible of cylindrical shape. A part of the top of the liquid metal is the crystal which is rotated. This introduces an assumption that azimuthal derivatives of all physical properties are zero.
Following Ozoe et al.[15], the dimensionless variables are defined as follows.
$R = r/r_0$, $Z = z/z_0$, $U = u/u_0$, $V = v/v_0$, $W = w/w_0$, $\tau = t/t_0$.
$P = p'/P_0$, $T = (T' - T'_c)/(T'_h - T'_c)$.
$r_0 = z_0 = [g\beta(T'_h - T'_c)/(\alpha\nu)]^{-1/2} = h/Ra^{1/2}$
$u_0 = v_0 = w_0 = \alpha/r_0$, $P_0 = \rho_0 u_0^2$, $t_0 = r_0^2/\alpha$.
These variables are selected for the computation at high Rayleigh number because the height of the enclosure is scaled with the Rayleigh number so that a large dimensionless height of the enclosure can be studied in detail. The whole system of equations in dimensionless form is thus:

$$\frac{\partial U}{\partial \tau} + U\frac{\partial U}{\partial R} + W\frac{\partial U}{\partial Z} - \frac{V^2}{R} = -\frac{\partial P}{\partial R} + Pr[\frac{1}{R}\frac{\partial}{\partial R}(R\frac{\partial U}{\partial R}) + \frac{\partial^2 U}{\partial Z^2} - \frac{U}{R^2}] \quad (1)$$

$$\frac{\partial V}{\partial \tau} + U\frac{\partial V}{\partial R} + W\frac{\partial V}{\partial Z} + \frac{UV}{R} = Pr[\frac{1}{R}\frac{\partial}{\partial R}(R\frac{\partial V}{\partial R}) + \frac{\partial^2 V}{\partial Z^2} - \frac{V}{R^2}] \quad (2)$$

$$\frac{\partial W}{\partial \tau} + U\frac{\partial W}{\partial R} + W\frac{\partial W}{\partial Z} = -\frac{\partial P}{\partial Z} + Pr[\frac{1}{R}\frac{\partial}{\partial R}(R\frac{\partial W}{\partial R}) + \frac{\partial^2 W}{\partial Z^2}] - Pr \cdot T \quad (3)$$

$$\frac{\partial T}{\partial \tau} + U\frac{\partial T}{\partial R} + W\frac{\partial T}{\partial Z} = \frac{1}{R}\frac{\partial}{\partial R}(R\frac{\partial T}{\partial R}) + \frac{\partial^2 T}{\partial Z^2} \quad (4)$$

$$\frac{1}{R}\frac{\partial}{\partial R}(RU) + \frac{\partial W}{\partial Z} = 0 \quad (5)$$

The system configuration is shown in Fig.2. The crystal rod rotating at the top is cooled, and the outside and bottom walls are heated and kept isothermal. The crystal rod at the top is rotated at an angular velocity Ω. An inner radius R_{in} is shown, but its size is essentially zero. The inner rod is assumed to be a dragless fluid boundary. The annular fluid region at the top between (R_{in} + 0.5L) and R_{out}, is assumed to be a dragless boundary, thermally insulated. The boundary conditions are given as follows.

1) Temperature
   $T = 0.5$ at $R = R_{out}$, and at $Z = H$.
   $T = -0.5$ for $R_{in} < R < (R_{in} + 0.5L)$ at $Z = 0$
   $\partial T/\partial R = 0$ at $R = R_{in}$
   $\partial T/\partial Z = 0$ for $R > (R_{in} + 0.5L)$ at $Z = 0$

2) Velocity
   $V = R\Omega$ for $R_{in} < R < (R_{in} + 0.5L)$ at $Z = 0$
   $\partial U/\partial Z = \partial V/\partial Z = 0$ for $R > (R_{in} + 0.5L)$ at $Z = 0$
   $U = V = \partial W/\partial R = 0$ at $R = R_{in}$
   $W = 0$ at $Z = 0$
   $U = V = W = 0$ at $R = R_{out}$, and at $Z = H$

3. The numerical scheme

The numerical scheme is a SIMPLER(Semi-implicit Method for Pressure Linked Equations Revised) method proposed by Patankar. However, the computer code itself was developed by ourselves. Figure 3(a) shows the typical grid for computing a scalar variables, such as temperature and pressure. The domain defined by hatched lines is called a control volume. The temperature T, or pressure P, are defined at the points named as P, W, E, N & S by capital letters. The velocity components are defined at the intermediate points indicated by lower case letters, w,e,n and s. Equation (4) can be rewritten as follows with the use of Eq.(5):

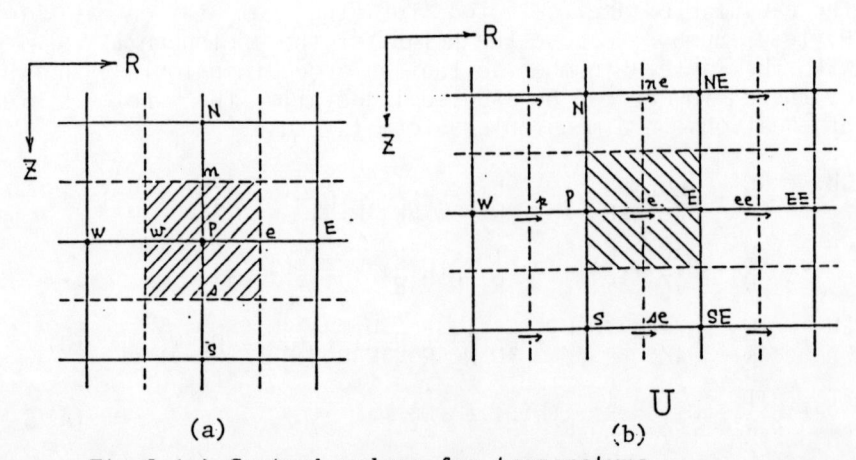

Fig.3 (a) Control volume for temperature.
(b) Control volume for velocity component U.

$$\frac{\partial T}{\partial \tau} + \frac{1}{R}\frac{\partial}{\partial R}(RUT) + \frac{\partial}{\partial Z}(WT) - \frac{1}{R}\frac{\partial}{\partial R}(R\frac{\partial T}{\partial R}) - \frac{\partial^2 T}{\partial Z^2} = 0 \qquad (6)$$

This equation is integrated over the hatched region of Fig.2 as follows.

$$\int_{\tau}^{\tau+\Delta\tau} \int_{n}^{s} \int_{w}^{e} \int_{\theta}^{\theta+\Delta\theta} [Eq.(6)] \ R \ d\theta \ dR \ dZ \ d\tau \qquad (7)$$

where n, s, w and e indicate the boundary locations, and $\Delta\theta$ is a unit azimuthal angle in the circumferential direction. Integrating term by term, the first term of Eq.(7), becomes, under an assumption that T is constant in a control volume.

$$[T_P]_{\tau}^{\tau+\Delta\tau}\Delta V = [T_P - T_P"]\Delta V \ .$$

where $\Delta V = R \cdot \Delta\theta \cdot \Delta Z \cdot \Delta R$ and $T_P"$ is the temperature at time τ. The second term of Eq.(7) becomes, with an assumption that RUT is constant in the Z- and the θ- directions and takes the value at $\tau+\Delta\tau$.

$$\int_{\tau}^{\tau+\Delta\tau}d\tau \int_{n}^{s} dZ \int_{\theta}^{\theta+\Delta\theta}d\theta \int_{w}^{e}\frac{1}{R}\frac{\partial}{\partial R}(RUT)RdR$$

$$= \Delta\tau \cdot \Delta Z \cdot \Delta\theta \ [RUT]_{w}^{e} = \Delta\tau\Delta Z\Delta\theta[\frac{(RU)_e(T_E+T_P)}{2} - \frac{(RU)_w(T_P+T_W)}{2}] \ .$$

The other terms were similarly approximated to give the following form finally from Eq.(7).

$$a_P T_P = a_E T_E + a_W T_W + a_S T_S + a_N T_N + a_P^0 T_P" \qquad (8)$$

where

$a_E = D_e - F_e/2$
$a_W = D_w + F_w/2$
$a_S = D_s - F_s/2$
$a_N = D_n + F_n/2$
$a_P = a_E + a_W + a_S + a_N + F_e + F_s - F_w - F_n + a_P^0$

$a_P^0 = \Delta V/\Delta\tau$
$F_e = \Delta Z(RU)_e, \ F_w = \Delta Z(RU)_w$.
$F_s = (R_e^2 - R_w^2)W_s/2, \ F_n = (R_e^2 - R_w^2)W_n/2$
$D_e = \Delta Z \cdot R_e/\Delta R_e, \ D_w = \Delta Z \cdot R_w/\Delta R_w$.
$D_s = (R_e^2 - R_w^2)/(2\Delta Z_s), \ D_n = (R_e^2 - R_w^2)/(2\Delta Z_n)$

$\Delta\theta = 1$ was used in these expressions because all physical properties are assumed to be constant in the θ-direction and a unit angle in the azimuthal direction was taken, while the velocity component in the azimuthal direction was considered. Equation (8) can be solved by an Alternating Direction Implicit scheme alternating between the R-direction and the Z-direction, and taking into consideration the temperature boundary conditions. In the above process of discretization, the central-difference and the upwind-difference approximations can be combined, depending on the magnitude of the velocity, in a "hybrid scheme." The discretization for momentum equations can be carried out similarly except that the control volume is different. Fig.3 (b) is applicable for velocity component U. More details on this numerical scheme can be found in [14].

4. The results

The values of the parameters used in the computation are listed in Table 1. The computation was carried out till the convergence of the unsteady implicit scheme. Convergence was successfully obtained. The steady state solutions are graphically shown in Fig.4. The velocity profile and streamlines in the vertical cross section show that the liquid descends along the central core regime and ascends along the lower half of the outer radius of the crucible. The liquid near the crystal rod (at the top) does not flow outwards despite the rotational centrifugal force. On the other hand, the liquid near the bottom of the crucible flows outwards. This means that buoyancy force governs in determining the main bulk flow as also confirmed by the fact that the Archimedes number Gr/Re^2=18.5. The isotherms shown in Fig.4(b) reflect this bulk flow. Most of the liquid at the outer perimeter is at a high temperature and a steep temperature gradient occurs near the outer perimeter of the rotating crystal. The isotherms are convex downwards. This suggests that the shape of the bottom part of the crystal will be convex downwards. The stream functions of Fig.4(c) show the steady state traces of the particles in this case. The core of the convection is located at the lower part of the crucible, and the bulk flow is asymmetric relative to the central height. This asymmetry is caused by the rotational boundary condition on the upper half of the region.

Table 1 Summary of the computed system.

system	1	2	3
R_{out}/H	0.5	0.5	0.5
R_{in}/R_{out}	10^{-10}	10^{-10}	10^{-10}
Ra	10^6	5.4×10^4	1.25×10^5
Pr	0.054	0.054	0.054
Ω	0.0432	0.3024	0.1728
Re	1000	1000	1000
Gr	1.85×10^7	10^6	2.31×10^6
Gr/Re^2	18.5	1	2.31
$2L_s$	50	18.9	25
V_{max}	1.08	2.857	2.16

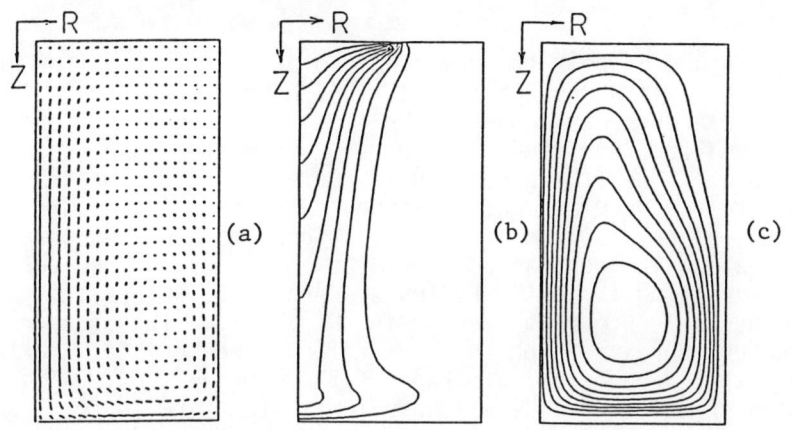

Fig.4 Computed results for system 1 at Ra=10^6, Gr/Re^2=18.5 and Pr=0.054. (a)velocity vectors (b)isotherms (c)streamlines

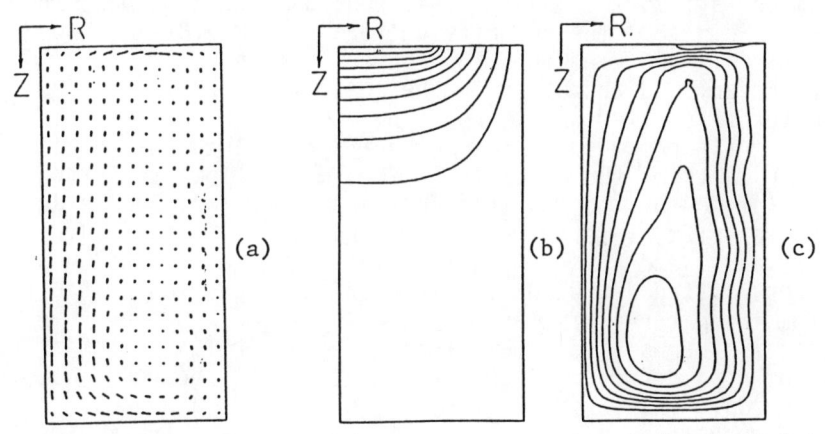

Fig.5 Computed results for system 2 at Ra=5.4x10^4, Gr/Re^2=1 and Pr=0.054. (a) to (c) are the same as Fig.4.

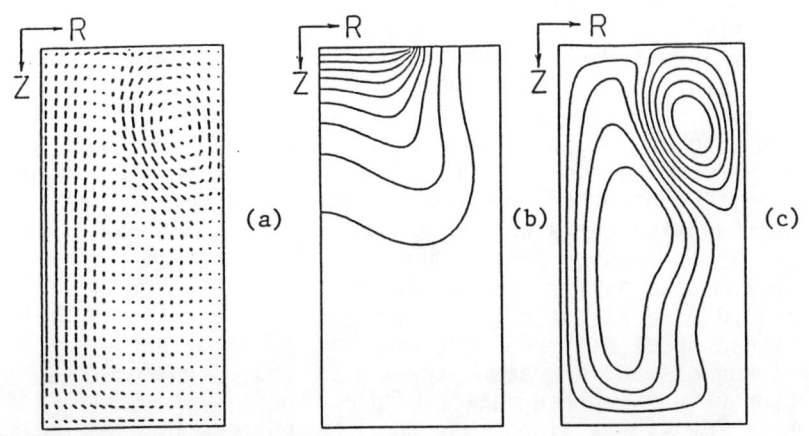

Fig.6 Computed results for system 3 at Ra=1.25x10^5 and Gr/Re^2=2.31 and Pr=0.054. (a) to (c) are the same as Fig.4.

Figure 5 shows the numerical solution obtained for case 2. The Rayleigh number is 5.4×10^4, Reynolds number is 1000, and the Archimedes number $Gr/Re^2=1$. The buoyancy force is much smaller than case 1. The velocity vectors are shown in Fig.5(a). The flow direction is reversed in comparison to that in the previous case. Strong outward radial flow is seen along the crystal rod at the top. There is a strong upward flow along the central core of the cylinder. The velocity vectors along the outer radius appears to be vibrating. The isothermal contours are shown in Fig.5(b). The temperature gradient appears to be uniform along the crystal surface. This will produce a flat shape of the crystal rod. Fig.5(c) shows stream lines. These wavy lines along the outer radius perimeter are eminent, reflecting the unstable downward flow along the hot vertical wall.

Figure 6 shows the contour maps for case 3. The Rayleigh number is 1.25×10^5, again Re = 1000, and the Archimedes number is $Gr/Re^2=2.3$. The velocity vectors in Fig.6(a) shows two vortices, one in the top upper corner near the outer perimeter, and the other in the rest of the region. The top vortex is smaller and is rotating counter-clockwise. It is apparently produced by the buoyant force along the hot upper outer walls of the crucible. The main flow is along the crystal rod at the top in an outward direction, and is due to the centrifugal force. This outward flow changes direction downwards, and strengthens the smaller vortex. Consequently a strong tilted downward flow is seen to have been established from the outer perimeter of the crystal rod. Observations of the isotherms indicates that the vertical temperature gradients decrease from the center of the crystal rod outwards. This suggests the development of a concave shape for the bottom of the crystal rod. The stream function reveals more clearly the flow mode as shown in Fig.6(c). The flow and thermal phenomena observed in this case 3 fall between those observed in cases 1 and 2, and so do their respective Archimedes numbers. This indicates the significance of this number in determining the flow field and the crystal shape. The lower vortex in Fig.6(c) shows wavy flow along the lower side wall of the crucible. This flow is against to the buoyancy force. It is resulted because of the subtle force balance between the centrifugal and the buoyancy forces. When the buoyancy force prevails over the centrifugally-generated one because of some reasons, then this flow easily changes to the buoyancy dominant mode shown in case 1. Kobayashi[3] reported various temperature profiles in a liquid of Pr=1 but not reported for silicon oil of Pr=0.054 at these three different extreme conditions of parameters.

For case 3, a tracer streak line was computed to give a better visual representation of the flow. The computational scheme for streak lines has been reported by Yamamoto et al.[16]. The tracer was placed at $(R,\theta,Z)=(18,0,4)$. With the dimensionless time step $\Delta\tau=0.5$, its path was computed for 1000

time steps. A perspective view is shown in Fig.7. The
starting point is indicated by a black circle. The tracer
rotates with the up and down movement along the axis. This
indicates a complicated flow in the Czochralski manufacturing
process. Table 2 lists up sample numerical values to indicate
a dimensional size of this system. The size of a crystal is
only 5cm in its diameter for this case, but computations are
expected to be carried out with finer grid sizes for higher
Reynolds and Rayleigh numbers.

5. Conclusions

A computational code employing primitive variables using
the ADI method for the balance equations for energy and
momentum and the SIMPLER scheme for the pressure derivation
was developed for a cylindrical quasi three-dimensional
coordinate system. It was applied to the Czochralski bulk
flow of liquid metallic silicon. Three radically-different
flow modes were found to exist depending on the Archimedes

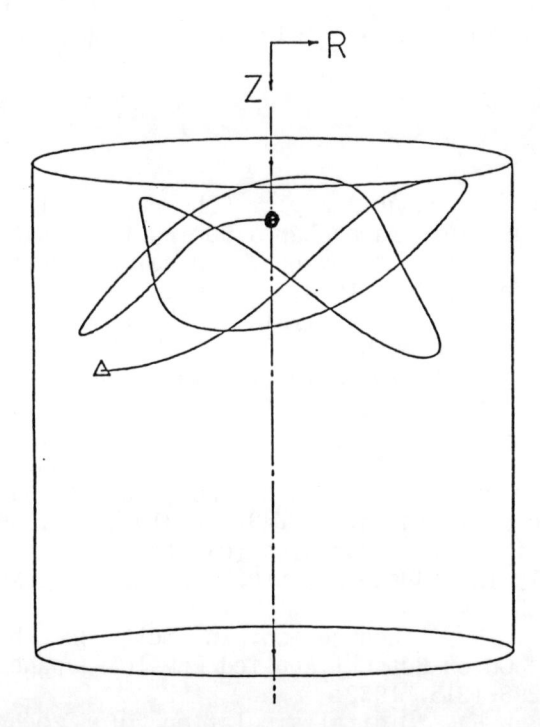

Fig.7 Computed streak lines for system 3 at a steady state.
The starting point of the tracer is indicated by a black
circle located at (R, θ, Z)=(18, 0, 4), and it was
followed for 1000 time-steps with a step size of $\Delta\tau=0.5$.

number at Pr=0.054. The temperature profile under the crystal rod was found to take a concave or convex shape, depending on the Archimedes number. This would also be the shape of the bottom of the crystal rod which is pulled up and rotated at the same time.

Acknowledgement
The authors are grateful for valuable advices provided by Prof.Noam Lior, University of Pennsylvania in preparing this paper.

Table 2. Dimensional sample numerical values for case 3.
Ra=1.25×10^5, Gr=2.315×10^6, Re=1000, Pr=0.054.

height of a crucible, h	0.1 (m)
radius of a crystal rod, ℓ_s	0.025 (m)
reference length, r_0	0.002 (m)
reference time, t_0	0.8 (s)
rotational speed, ω	0.216 (rad./s)
maximum velocity of a rod, $v_{max}=\ell_s\omega$	0.0054 (m/s)
temperature difference	$1.72 \times 10^{-5}/\varepsilon$ (K)
elapsed time of a tracer in Fig.7	400 (s)

References

1. Kobayashi,N. and Arizumi,T., "The numerical analyses of the solid-liquid interface shapes during the crystal growth by the Czochralski method." Japanese Journal of Applied Physics, vol.9, No.4, pp.361-367, 1970.
2. ibid. "Computational analysis of the flow in a crucible." Journal of Crystal Growth, vol.30, pp.177-184, 1975.
3. Kobayashi, N., "Computational simulation of the melting flow during Czochralski growth.", Jouranl of Crystal Growth, vol.43, pp.357-363, 1978.
4. Kobayashi, N. and Arizumi, T., "Computational studies on the convection caused by crystal rotation in a crucible." Journal of Crystal Growth, vol.49, pp.419-425, 1980.
5. Kobayashi, N., "Hydrodynamics in Czochralski growth-computer analysis and experiments." Journal of Crystal Growth, vol.52, pp.425-434, 1981.
6. Kobayashi, N., "Convection in melt growth-Theory and Experiments." Ouyou Butsuri(Applied Physics-Japanese article) vol.51, pp.1206-1215, 1982.
7. Langlois, W.E., "Digital simulation of Czochralski bulk flow in a parameter range appropriate for liquid semiconductors." Journal of Crystal Growth, vol.42, pp.386-399, 1977.

8. Langlois, W.E., "Digital simulation of Czochralski bulk flow in microgravity." Journal of Crystal Growth, vol.48, pp.25-28, 1980.
9. Chang, E. Chong, "Computer simulation of convection in floating zone melting. I. Pure rotation driven flows. II. Combined free and rotation driven flows." Journal of Crystal Growth, vol.44, pp.168-186, 1978.
10. Van der Hart, A. and Uelhoff, W., "Macroscopic Czochralski growth. I. Theoretical investigation, heat flow and growth model.", Journal of Crystal Growth, vol.51, pp.251-266, 1981.
11. Nikitin, S.A., Polezhayev, V.I. and Fedyushkin, A.I., "Mathematical simulation of impurity distribution in crystals prepared under microgravity conditions." Journal of Crystal Growth, vol.52, pp.471-477, 1981.
12. Polezhayev, V.I., Dubovik, K.G., Nikitin, S.A., Prostomolotov, A.I. and Fedyushkin A.I., "Convection during crystal growth on earth and in space." Journal of Crystal Growth, vol.52, pp.465-470, 1981.
13. Tsukada, T., Imaishi, N., Hozawa, M. and Fujinawa, K., "Numerical analysis of crystal growth in Czochralski method." Preprints of the 19th autumn annual meeting of the Society of Chemical Engineers, Japan, D104, p.154, 1985.
14. Patankar, S.V., "Numerical heat transfer and fluid flow." Hemisphere Pub. Corp., Washington, 1980.
15. Ozoe,H., Mouri, A., Ohmuro M., Churchill, S.W. and Lior, N., "Numerical calculations of laminar and turbulent natural convection in water in rectangular channels heated and cooled isothermally on the opposing vertical walls." Int. J. Heat Mass Transfer, vol.28, No.1, pp.125-138, 1985.
16. Yamamoto, K., Ozoe, H., Chao, P. and Churchill, S.W., "The computational and dynamic display of three-dimensional streaklines for natural convection in enclosures." Computers and Chemical Engineering, vol.6, no.2, pp.161-167, 1982.

NUMERICAL SIMULATION OF THE DC PLASMA JET TWO-PHASE REACTOR

Pierre Proulx, Javad Mostaghimi and Maher I. Boulos
Université de Sherbrooke, Sherbrooke
Québec, Canada, J1K-2R1

ABSTRACT

A numerical simulation of the two-phase plasma jet reactor is performed. The results of the model are compared with the available data for single and two-phase plasma jets. The model is in good agreement with the experimental data for the single phase measurements. The prediction of the efficiency of spheroidization for significant two-phase loadings are also in agreement with the experimental data.

INTRODUCTION

The plasma jet is an important tool in the thermal treatment of powdered refractories and high melting and boiling point materials. It is widely used for spray coating of protective layers on the heat, corrosion or abrasion sensitive surfaces. It is also used as a two-phase reactor both for thermal and/or reacting systems. The complex two-phase interactions in the plasma reactor has been studied theoretically by only a few authors [1,2,3,4]. In the present paper, a turbulent model of a plasma jet reactor has been developed and the results of the model for low and high temperature jets are compared with the available experimental measurements.

1-THE MODEL: TURBULENT GAS PHASE

The conservation equations for the plasma jet reactor are written in their boundary layer form, with the assumption of axial symmetry. In the time averaged conservation equations, the density fluctuations are neglected. This hypothesis has been used by Mckelliget et al[5] and in spite of the strong gradients in a plasma jet they had a good agreement with the experimental data. The plasma is optically thin and the radiation losses from the jet are small compared with the other mechanisms of energy transfer.

The time averaged conservation equations are:

Continuity
$$\frac{1}{r}\frac{\partial}{\partial r}(r\rho v) + \frac{\partial}{\partial z}(\rho u) = 0 \qquad (1)$$

Axial momentum
$$\frac{\partial}{\partial z}(\rho uu) + \frac{\partial}{\partial r}(\rho uv) = \frac{1}{r}\frac{\partial}{\partial r}\left(r(\mu+\mu_t)\frac{\partial u}{\partial r}\right) + S_p^{Mz} \qquad (2)$$

Energy
$$\frac{\partial}{\partial z}(\rho uh) + \frac{\partial}{\partial r}(\rho vh) = \frac{1}{r}\frac{\partial}{\partial r}\left(r\left(\frac{k}{C_p} + \frac{\mu_t}{\sigma_h}\right)\frac{\partial h}{\partial r}\right) \qquad (3)$$
$$+ \frac{1}{r}\frac{\partial}{\partial r}\left(r\rho D(1-\frac{1}{Le})(h_1 - h_2)\frac{\partial \omega_1}{\partial r}\right) + S_p^E$$

Conservation of a chemical species
$$\frac{\partial}{\partial z}(\rho u\omega_1) + \frac{\partial}{\partial r}(\rho v\omega_1) = \frac{1}{r}\left(\frac{\partial}{\partial r} r(\rho D + \frac{\mu_t}{\sigma_m})\frac{\partial \omega_1}{\partial r}\right) + S_p^m \qquad (4)$$

The S_p terms in the above equations represents the two-phase coupling term which are evaluated from the well-known PSI-Cell model[2]. The definition of the other symbols is given in the nomenclature. Equations 1-4 are subject to the following boundary conditions:

$$r=R_\infty \ldots \ldots \ldots \frac{\partial u}{\partial r} = \frac{\partial h}{\partial r} = \frac{\partial \omega_1}{\partial r} = \frac{\partial v}{\partial r} = 0$$

$$r=0 \ldots \ldots \ldots \frac{\partial u}{\partial r} = \frac{\partial h}{\partial r} = \frac{\partial \omega_1}{\partial r} = \frac{\partial v}{\partial r} = 0$$

$$z=0,\ r<R_o \ldots \ldots u_o = u_{max}\left(1 - \frac{r}{R_o}\right)^\alpha$$

$$h_o = h_{max}\left(1 - \frac{r}{R_o}\right)^\beta,\quad \omega_1 = 1$$

$$z=0,\ r>R_o \ldots \ldots u_o = h_o = \omega_1 = 0$$

Two models were used to evaluate the turbulent viscosity. The first is the mixing length hypothesis of Prandtl and the second is the k-ε model. In the case of a two-phase flow, the particulate phase also affects the mobility of the turbulent eddies. One of the first models proposed to take this effect into account was the Abramovitch[6] model:

$$\mu_t = \frac{\mu_t'}{(1 + \rho/\rho_d)} \qquad (5)$$

where μ_t is the effective two-phase turbulent viscosity, ρ_d is the apparent two-phase density and μ_t' is the single phase turbulent viscosity predicted by the turbulent model.

2-THE MODEL: PARTICULATE PHASE

Assuming that the only force affecting an individual particle trajectory is the drag force, the momentum equations for a single particle injected in a plasma can be written.

$$\frac{du_p}{dt} = -\frac{3}{4} C_d \left(\frac{\rho}{\rho_p d_p}\right)(u - u_p) U_r \qquad (6)$$

$$\frac{dv_p}{dt} = -\frac{3}{4} C_d \left(\frac{\rho}{\rho_p d_p}\right)(v - v_p) U_r \qquad (7)$$

Where U_r is the relative velocity between the plasma and the particles.

The particle temperature and thermodynamic state is determined by an energy balance. This approach is valid for a particle injected into a thermal plasma if the Biot number is less than 0.04, i.e. internal conduction can be neglected[4].

$$Q_p = h_c A_p (T - T_p) - \sigma \epsilon_p A_p (T_p^4 - T_a^4) \qquad (8)$$

$$\begin{aligned}
Q_p &= \frac{1}{6} \rho \pi d_p^3 C_{ps} \frac{dT_p}{dt} & T < T_m \\
Q_p &= \frac{1}{6} \rho \pi d_p^3 C_{pl} \frac{dT_p}{dt} & T_m < T < T_v \\
Q_p &= \frac{1}{6} \rho \pi d_p^3 H_m \frac{dX_p}{dt} & T = T_m \\
Q_p &= -\frac{1}{2} \rho \pi d_p^2 H_b \frac{dd_p}{dt} & T = T_b
\end{aligned} \qquad (9)$$

The drag correlations used are those proposed by Boulos[8,9], the Nusselt correlation was proposed by Vardelle et al[10], and the corrections for rarefaction effects and strongly varying properties are those proposed by Lee et al[11] and Bourdin et al[7]. The initial conditions of the particles are distributions of velocity, diameter and position of injection according to the measurements of Vardelle et al[10].

3-THE MODEL: COUPLING

The PSI-Cell model of Crowe et al[12] is used to couple the

gas phase and the particulate phase. This method uses an iteration loop between the solution of the gas phase equations and the particles individual trajectories to solve the two-phase system. The formulation of the algorithm was modified here to take advantage of the parabolic boundary layer equations so that in a single pass through the computational domain the two-phase solution could be obtained. The coupling was taken into account in the momentum and the energy equation.

To include the effect of turbulence on the particles, the model of Shuen et al[13] was used. This model simulates the turbulent dispersion of the particles due to the random nature of the movement of eddies in the gas.

4-SOLUTION TECHNIQUE

The solution of the equations 1 to 4 coupled with the equations describing the turbulence is obtained using an implicit marching technique. The gas phase equations were discretized using a control volume technique. The solution was obtained for a simple cylindrical non-uniform grid of 72 radial grids and 40 axial marching steps. The highest density of grid points was of course in the initial shear layer of the plasma jet, which is a zone extending to a few nozzle diameters from the jet exit. This grid was sufficiently accurate to study the initial zone of the plasma jet, where extreme temperature and velocity gradients exist. Since the gradients decrease very rapidly, adaptation of the grid was not necessary and the same grid was used for the plasma jet which was studied up to about 10 nozzle diameters. The computations were made on an IBM-4381 main frame computer and the CPU time was less than 5 minutes even for the two-phase cases. The model has also been implemented on an IBM-PC and for the same simulations the CPU time was less than 1 hour.

The marching technique incorporated the PSI-Cell iterations at every marching step, therefore the two-phase solution was also obtained after one 'march' through the calculation domain. The number of PSI-Cell iterations for each forward marching step was less than 5 for most of the cases studied here. This parabolic PSI-Cell was very efficient and the CPU time for the complete two-phase solution was roughly less than twice that of the single phase case.

The number of particle trajectories calculated for each case was approximately 600 and the solution of the trajectory equations was made through a second order Runge-Kutta integration algorithm. The basic idea of Boulos and Gauvin[6] to simulate the actual powder through discretized distributions of size, injection velocity and injection position was used. The influence of the number of trajectory calculations was not studied.

5-COMPARISON WITH EXPERIMENTAL DATA

5.1-COLD JETS

5.1.1-SINGLE PHASE

The axial decay of centerline velocity and the radial velocity profiles predicted by the mixing length model compare very well with the measurements of Boguslawski and Popiel[14] and of Lesinski et al[15] as shown in Figure 1.

The velocity profiles predicted by the k-ϵ model and the mixing length model are compared in Figure 2. Maximum differences of 5 % occur on the axis. The spread of the jet and its external boundaries are quite in good agreement.

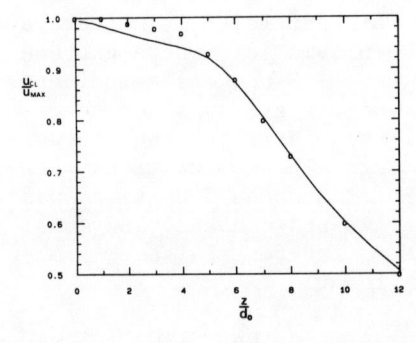

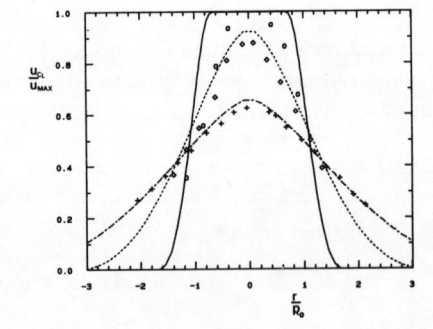

Figure 1 a
Axial decay of the axial velocity for the turbulent free jet. Model (_____) and measurements of Boguslawski-Popiel[14] (o).

Figure 1 b
Radial profiles of the axial velocity for the turbulent free jet. Model (_____) and measurements of Lesinski et al[15] (o , 2 d_o), (◊ , 5 d_o), (+ , 10 d_o)

The axial profile of kinetic energy is compared with the measurements of Boguslawski and Popiel[14] in Figure 3. The model slightly over-predicts the maximum but the position of the initial, the transition and the developed zones are well represented.

For the case of the argon plasma jet the dependance of the mixing length on the radial position is rather small, so that the Prandtl's uniform mixing length is a good approximation. The mixing length hypothesis is therefore used for the following calculations.

5.1.2-TWO-PHASE

The angle of dispersion of the two-phase cold jet have been measured by Wall et al.[16] for different particle loadings. Table 1 shows the comparison between the present model and the measurements. The agreement is excellent.

Two-phase load. kg(pa.)/kg(gas)	ref[16]	Model
0.22	8.6	8.7
0.34	8.2	8.3
0.50	7.8	7.8

Table 1: Angle of spread of the two-phase jet (degrees)

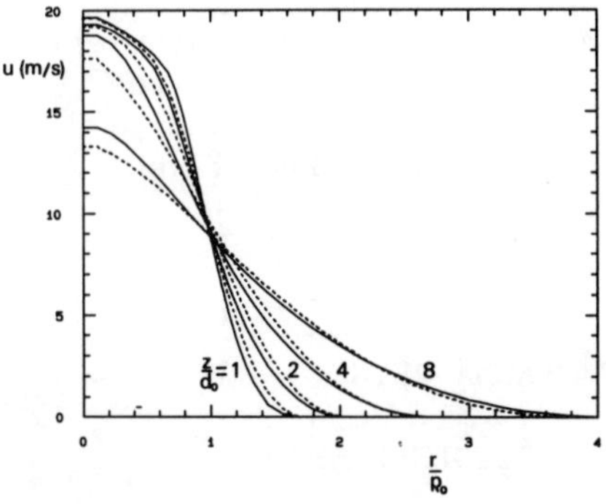

Figure 2. Comparison of the radial profiles of velocity calculated with the k-ϵ model (_____) and with the mixing length model (_ _ _)

The ratio of the velocity of the two phases of a jet to that of a clean jet 20 diameters from the jet exit is compared with the measurements of Modaress et al[17] (Figure 4). The difference of velocity between the two phases is almost constant for the range of the two-phase loadings studied. The model agrees with the experimental values practically within the scatter of the data.

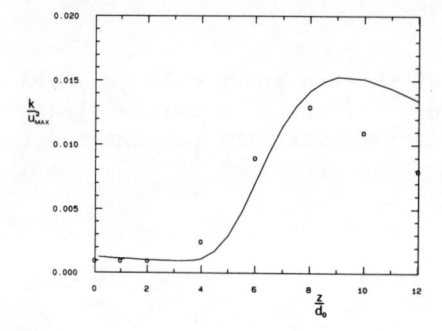

Figure 3.
Axial profile of the turbulent kinetic energy for the free jet. Model (___), and measurements of Boguslawski and Popiel (o)

Figure 4.
Ratio of the two-phase velocities to the single phase velocity 20 diameters from the jet exit. Model (___) and measurements of Modaress[17], (◆, Gas), (◇, Particles)

5.2-PLASMA JETS

5.2.1-SINGLE PHASE

The measurements of Vardelle[10] for the argon-hydrogen and the nitrogen-hydrogen plasma jets have been widely cited and used to validate plasma jet models. From these measurements one can determine the exponents of the temperature and velocity distributions at the tip of the plasma torch and use them for the model. Figure 5 shows a comparison between the measured[10] and the predicted temperature and velocity profiles. While the velocity profiles are in excellent agreement, for the temperature profile the model predicts a faster decay than the measurements. Overall the agreement is good between the model and the measurements.

Beaulieu et al[18] measured the temperature in the initial region of the discharge of an argon plasma. Since they did not measure the velocity we used the power and flow rate of argon from their experiments to determine the initial temperature and velocity profiles needed for our model. Figure 6 shows that while close to the axis the comparison between the model and [18] is excellent, however this is not the case for the plasma fringes. This disagreement is due to the fact that the plasma edges are in a state of non-LTE (Local Thermodynamic

Equilibrium), where the electron and atom/ion temperatures could be substantially different. Since spectroscopic measurements are close to the electron temperature and our model predicts the atom/ion temperatures, the comparison between the two is meaningful in regions where LTE prevails, i.e. close to the axis of symmetry.

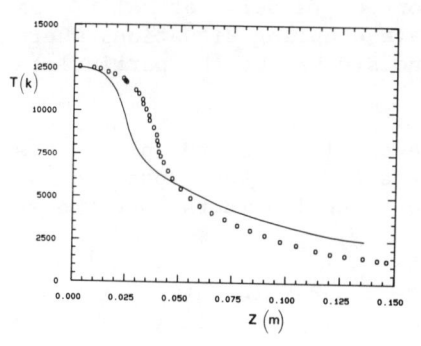

Figure 5a
Axial temperature profile of the plasma jet. Measurements of Vardelle[10] (o), and model (____).

Figure 5b
Axial velocity profile of the plasma jet. Measurements of Vardelle[10] (o), and model (____).

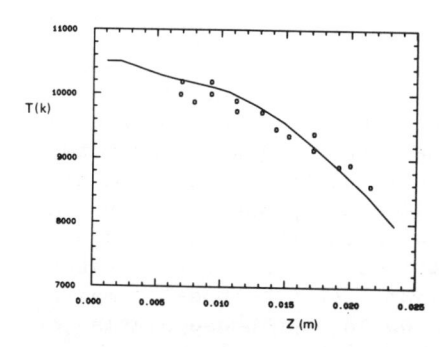

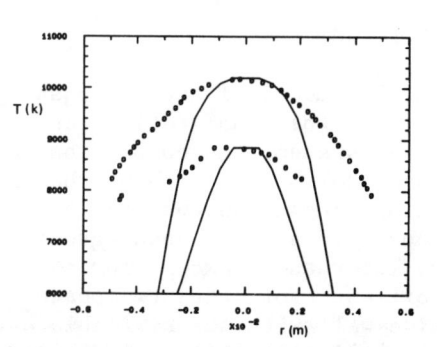

Figure 6a
Axial temperature profile of the plasma jet. Measurements of Beaulieu[18] (o), and model (____)

Figure 6b
Radial temperature profiles of the plasma jet. Measurements of Beaulieu[18] (o) at 7 and 21 mm, and model (____).

5.2.2-TWO-PHASE

The two-phase plasma jet has been under very little experimental investigation. The first experimental study on the parameters of a two-phase argon plasma jet was made by Surov[19] in 1969 but can not be used for other than qualitative comparison. Thorough studies of the temperature and velocity fields in a two-phase plasma jet have not been made to the knowledge of the authors. The effort of Vardelle et al[10] has been mostly devoted to low two-phase coupling situations where the plasma can be considered as undisturbed by the particulate phase.

The loading effect of the particles on the plasma has been studied experimentally by Bokhari[20] for the spheroidization efficiency of alumina and nickel powders in nitrogen plasmas.

The model was used to simulate a spheroidization reactor and to predict the spheroidization efficiency. It was assumed that a completely molten particle was completely spheroidized. Since it is difficult to know exactly the thermal efficiency of the plasma torch used for the experiments of [20], the two-phase behavior of the plasma-particle reactor was calculated for effective powers of 10 to 15 kW, this covers the actual powers used by Bokhari[20], the thermal efficiencies of a DC plasma torch being typically around 50 percent. The results of these simulations are compared with the measurements in Figure 7. The curves of 10 and 11 kW correspond approximately to the measurements at a nominal power of 20 kW. The agreement is quite good.

CONCLUSION

The model of the two-phase plasma jet reactor is able to predict the qualitative and the quantitative behavior of an actual system. The comparisons made with the single and two-phase colds jets show the validity of the simple Prandtl's mixing length hypothesis. The single phase plasma jet predictions are also in good agreement with the available data from the literature, except for some differences in the temperature profiles. For the two-phase plasma jet the model compares quite well with the available data on the two-phase plasma jet spheroidization of alumina powders in nitrogen plasmas.

Further developments of this model should include three dimensional modeling to remove the axial symmetry hypothesis. The turbulence modeling of the single and the two-phase plasma jet should also receive attention, since their treatment was simplified, e.g. density fluctuations of the single and two-phase plasma jet should be investigated.

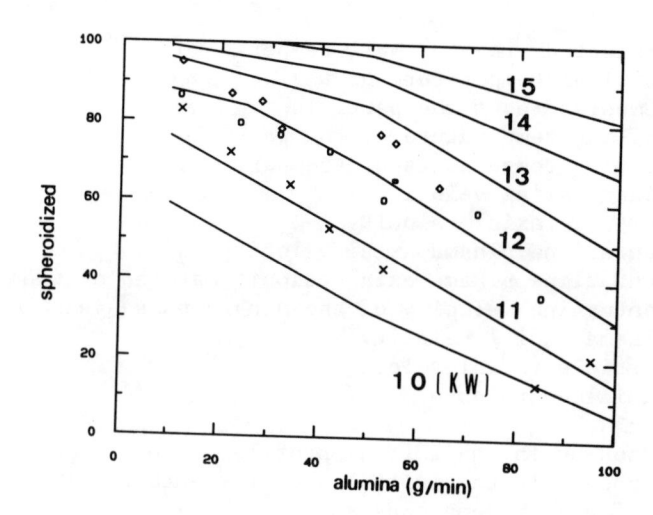

Figure 7.
Spheroidized fraction of alumina particles injected into a nitrogen plasma as a function of the particle loading. Measurements of Bokhari[20] (X , 20 kW) , (o , 25 kW), (◇ , 34 kW) and model (_____).

ACKNOWLEDGEMENTS
The support of the Conseil de Recherche en Sciences Naturelles et en Génie du Canada (CRSNG), the Ministère de l'éducation du Quebec, Echanges France-Québec and the Laboratoire de thermodynamique de l'Université de Limoges(France) is gratefully aknowledged.

NOMENCLATURE
C_p : heat capacity of the plasma gas
C_{ps} : heat capacity of the particle, solid
C_{pl} : heat capacity of the particle, liquid
C_d : drag coefficient
d_p : diameter of a particle
D : diffusion coefficient
D_o : diameter of the plasma torch or jet nozzle
H_b : latent enthalpy of boiling
H_m : latent enthalpy of melting
Q_p : net heat transfer rate to particle
r : radial coordinate
R_o : radius of the plasma torch or the jet nozzle
S_p^E : two-phase coupling term, energy
$S_{p_{Mz}}$: two-phase coupling term, momentum
S_p^m : two-phase coupling in the mass equation
t : time
T : plasma temperature

T_a : ambient or reactor wall temperature
T_{ci} : critical temperature of the i_{th} species
T_p : temperature of the particle
T_b : boiling temperature of the particle
T_m : melting temperature of the particle
u : plasma axial velocity
u_p : particle axial velocity
U_{cl} : centerline plasma axial velocity
U_{max}: centerline plasma axial velocity at the exit of the jet
U_s : centerline velocity of the single-phase jet
v : plasma radial velocity
v_p : particle radial velocity
z : radial coordinate

Greek
α : exponent in the initial profile, velocity
β : exponent in the initial profile, enthalpy
σ : Stefan-Boltzmann constant
σ_h : turbulent Prandtl number for enthalpy
σ_m : turbulent Schmidt number for mass
ρ : density of the plasma gas
ρ_p : density of the particles
ω_1 : mass fraction of specie 1
μ : molecular viscosity
μ_t' : turbulent viscosity of the single phase
μ_t : turbulent viscosity of the two-phase flow

BIBLIOGRAPHY
1- Lee Y.C., Hsu K.C., Pfender E., ISPC-5, p. 795, Edinburgh (1981)
2- Proulx P, Mostaghimi J.T., Boulos M.I., International Journal of Heat and Mass Transfer, Vol. 28, p. 1327, (1985).
3- Chyou Y.P., Young R.M., Pfender E., ISPC-7, Eindhoven, p.892, (1985)
4- Wei D., Apelian D., Farouk B., ISPC-7, p. 810, Eindhoven, (1985)
5- McKelliget J., Szekely J., Vardelle M., Fauchais P., Plasma Chemistry and Plasma Processing, Vol. 2, No. 3, p. 317, (1982)
6- Abramovitch G.N., International Journal of Heat and Mass Transfer, Vol. 14, p. 1039, (1971)
7- Bourdin E., Fauchais P., Boulos M.I., International Journal of Heat and Mass Transfer, Vol. 26, No. 4, p. 567 (1983)
8- Boulos M.I., IEEE Transactions on Plasma Science, Vol. PS-6, No. 2, June (1978)
9- Boulos M.I., Gauvin W.H, The Canadian Journal of Chemical Engineering, Vol. 52, June (1974)
10- Vardelle A., Vardelle M., Pateyron B, Fauchais P., ISPC-7, Eindhoven, p. 898, (1985)
11- Lee Y.C. Chyou Y.P., Pfender E., Plasma Chemistry and Plasma Processing, Vol. 5, No. 4, p. 391 (1985)
12- Crowe C.T., Sharma M.P., Stock D.E., Journal of Fluids Engineering, p.325, June (1977)

13- Shuen J-S., Chen L.-D., Faeth G.M., AIChE Journal, Vol. 29, No. 1, p. 167, (1983)
14- Boguslawski L., Popiel Cz. O., Journal of Fluid Mechanics, Vol. 90, p. 531, (1979)
15- Lesinski J., Lesinska B., Fanton J.C., Boulos M.I., AIChE Journal, Vol. 27, No. 3, p. 358, (1981)
16- Wall T.F., Subramanian V., Howley P., Trans. Inst. Ch. Eng., Vol. 60, p. 231, (1982)
17- Modaress D., Tan H., Elgobashi S., AIAA Journal, Vol. 22, No. 5, p. 624, (1984)
18- Beaulieu M., Gravelle D.V., Boulos M.I., Pfender E.,Report for the contract #00874 (PEC), Sherbrooke,(1986)
19- Surov N.S., High Temperature, p. 276, (1969)
20- Bokhari A., unpublished results, Université de Sherbrooke, Canada (1986).

SOLID PARTICLE INJECTION IN A LAMINAR BOUNDARY LAYER

P. Durbetaki[†]

The George W. Woodruff School of Mechanical Engineering
Georgia Institute of Technology
Atlanta, Georgia, 30332-0405, U.S.A.

SUMMARY

This work is concerned with the problem of laminar flow of a viscous gas over a semi-infinite flat plate with wall-slot injection of solid particles. The boundary layer effects are studied. The analysis treats the particles as a continuum. Thus, a set of conservation equations is used to represent each phase. The coupling of the momentum equations by the shear between the two phases results in the necessity of solving simultaneous partial differential equations. The finite difference technique is employed to obtain the velocities and densities necessary to describe the flow of the two phases. The injection velocity and angle are varied to study various flow situations. Flow conditions with the parallel component of the injection velocity both less and greater than the free-stream velocity are investigated. The temperature profiles are obtained by similarity considerations.

NOMENCLATURE

c_p	constant pressure specific heat
c_s	specific heat of the solid particles
E_t	truncation error
F	Frictional parameter between the phases, $9\mu/2\rho_s \sigma^2$
k	thermal conductivity
L	characteristic length
Nu	Nusselt number

[†] Professor

Pr	Prandtl number, $c_p \mu / k$
Re	Reynolds number, $\rho u_\infty L / \mu$
T	temperature of the gas
u,v	velocity components
x,y	Cartesian coordinates
α	injection angle
δ	boundary layer thickness
θ	dimensionless temperature, $(T-T_w)/(T_\infty - T_w)$
μ	viscosity of the mixture of gas and solid
μ_o	viscosity of the gas
ρ	density
σ	radius of the particle

Subscripts

iv	injection velocity
p	particle
s	pertaining to the solid phase
w	at the wall
∞	free stream

1. INTRODUCTION

Gas-solid particle flows have been the object of research for many years. Many publications have been written on the subject but very few deal with two-phase flow in boundary layers. The most important of these are papers by Soo [1-4], Marble [5] and Singleton [6].

Soo [1-2] develops the differential equations of conservation mass, momentum and energy for the two-phase boundary layer. He investigates flow over a flat plate using the integral form of the conservation equations. A major conclusion is that the solid particles could effect an increase or decrease in the boundary layer thickness, depending on the fluid Reynolds number and particle diameter. The other two papers of Soo deal with Brownian motion of small particles in suspension [3] and particle size distribution [4].

Like Soo, Marble [5] develops the boundary layer differential equations of conservation. In his analysis of laminar flow over a flat plate these equations are combined and rewritten in terms of the velocity differences between the two phases. With the assumption that the y-component of the gas velocity is equal to the y-component of the particle velocity, Marble expresses the equations in terms of a stream function. This enables a power series solution to be used.

Singleton [6] uses the conservation equations to solve the problem of flow over a flat plate and flow perpendicular to the axis of a cylinder. He reduces these partial differential equations to ordinary differential equations with use of the stream function. This enables him to solve the equations numerically by power series expansion.

The problem considered in this work follows logically from the gas-solid flow over a flat plate. It is the laminar flow of a viscous gas over a flat plate with wall-slot injection of solid particles. This situation is investigated to determine the boundary layer behavior of the two-phase system. The velocity and temperature distributions are of primary concern. Various flow situations are created by varying the slot injection velocity and angle.

The solid particles are injected into the gas at the plate's leading edge. Since the particles are not injected parallel to the plate and are not injected at the free-stream velocity, the particles must slip with respect to the gas. The magnitude of the particle slip velocity (velocity of the particles with respect to the gas) is dependent upon the region of the boundary layer under consideration. Near the plate, the slip velocity is large, but it decreases to zero in the free stream.

The apparent particle density, defined as the mass of the particles per unit volume of mixture of both phases, is assumed to be sufficiently low and the particles are assumed to move at so nearly the same speed that they do not collide with each other. The interaction of the flow fields around the individual particles is neglected. The particles are assumed to have no random motion and exert no pressure. Since their individual behavior is of no interest, the particles are considered as a continuum.

With no collision assumption, the behavior of the two-phase system is entirely dependent upon the interaction between fluid and particles. The particle Reynolds number and the molecular mean free path of the fluid are considered to be small enough that Stokes drag law for spheres is a reasonable approximation. From Stokes law the drag force of the fluid on the particle phase, and the force of the particles on the fluid phase, may be evaluated.

The x-injection ratio, the ratio of the x-component of the particle injection velocity to the free-stream velocity, is an important factor in this two-phase flow system. For small x-injection ratios the injection velocities exert relatively little influence downstream as their effects are damped out in short distances. However, at larger x-injection ratios the effect of the particle injection velocities is important farther downstream.

2. CONSERVATION EQUATIONS

The particle phase is treated as a continuum in the development of the equations for this two-phase flow system. This treatment enables the establishment for the particle phase equations of conservation of mass, momentum and energy. Conservation equations of mass, momentum and energy are also written for the gas phase.

These conservation equations are written for a two-dimensional, laminar flow, boundary layer system [1]. Using the boundary layer approximations, fluid-phase incompressibility, and a steady-state analysis [7], the non-dimensional forms of the equations are written as follows:

$$\frac{\partial u'}{\partial x'} + \frac{\partial v'}{\partial y'} = 0 \qquad (1)$$

$$u'\frac{\partial u'}{\partial x'} + v'\frac{\partial u'}{\partial y'} = \frac{1}{Re}\frac{\partial^2 u'}{\partial y'^2} + \frac{FL}{u_\infty}\frac{\rho_p}{\rho}\left[u'_p - u'\right] \qquad (2)$$

$$\frac{\partial}{\partial x'}(\rho_p u'_p) + \frac{\partial}{\partial y'}(\rho_p v'_p) = 0 \qquad (3)$$

$$u'_p\frac{\partial u'_p}{\partial x'} + v'_p\frac{\partial u'_p}{\partial y'} = -\frac{FL}{u_\infty}\left[u'_p - u'\right] \qquad (4)$$

$$u'_p\frac{\partial v'_p}{\partial x'} + v'_p\frac{\partial v'_p}{\partial y'} = -\frac{FL}{u_\infty}\left[v'_p - v'\right] \qquad (5)$$

where the primes for the position coordinates indicate normalized variables with respect to the characteristic length L and for the velocities with respect to the free stream velocity. Furthermore, it has been shown that the viscosity of the mixture of gas and particles is approximately equal to the pure gas viscosity. Thus, the viscosity is taken to be the gas phase viscosity.

The establishment of the appropriate boundary conditions along with equations (1) through (5) completes the mathematical description of the hydrodynamic problem. For the gas phase the boundary conditions are:

$$u'[0,y] = 1; \quad v'[0,y] = 0 \qquad (6)$$

$$u'[x,0] = 0; \quad v'[x,0] = 0 \tag{7}$$

$$\lim_{y\to\infty} u'[x,y] = 1 \tag{8}$$

For the particle phase the boundary conditions are:

$$u_p'[0,0] = u_{iv}'\cos(\alpha); \quad v_p'[0,0] = u_{iv}'\sin(\alpha);$$

$$\rho_p[0,0] = \rho_s(0.1) \tag{9}$$

$$v_p'[x,0] = 0 \tag{10}$$

$$\lim_{y\to\infty} u_p'[x,y] = 1 \tag{11}$$

$$\rho_p[x,0] = 0; \quad \lim_{y\to\infty} \rho_p[x,y] = 0 \tag{12}$$

The particle density condition in equation (9) is arbitrary and represents the particle loading. Any similar condition could be used provided that the no collision assumption is valid.

The two-dimensional conservation of energy equation for the gas and for the particle are also established and with the use of the assumptions employed for the momentum equation plus the assumption that k and μ are independent of temperature, they are written in the non-dimensional form as

$$u'\frac{\partial \theta}{\partial x'} + v'\frac{\partial \theta}{\partial y'} = \frac{1}{Pr}\frac{1}{Re}\frac{\partial^2 \theta}{\partial y'^2} + \frac{1}{3}\frac{\rho_p}{\rho}\frac{Nu}{Pr}\frac{FL}{u_\infty}\left[\theta_p - \theta\right] \tag{13}$$

$$u_p'\frac{\partial \theta_p}{\partial x'} + v_p'\frac{\partial \theta_p}{\partial y'} = -\frac{1}{3}\frac{Nu}{Pr}\frac{c_p}{c_s}\frac{FL}{u_\infty}\left[\theta_p - \theta\right] \tag{14}$$

The boundary conditions for these two equations are as follows:

$$\theta[0,y] = 1; \quad \theta[x,y] = 0; \quad \lim_{y\to\infty}[x,y] = 1 \tag{15}$$

$$\theta_p[0,0] = \text{constant}; \quad \lim_{y\to\infty} \theta_p[x,y] = 1 \tag{16}$$

3. SOLUTION

The finite difference technique is employed to solve numerically the differential equations (1) through (5). The criteria of convergence and stability must be satisfied if the solution of these five equations is to be a reasonable approximation to the solution of the partial differential equations describing the two-phase flow. The problem of convergence and stability of the numerical solution for non-linear partial differential equations with variable coefficients can be handled in only a few particular cases. However, if the coefficients of the derivative terms are always at least one order lower than the derivatives themselves, then the non-linear equations are considered quasi-linear. The variable coefficients are treated as constants throughout the analysis. They take on their most adverse values in order to determine the restriction on the increment size [8]. The von Neumann stability analysis uses the quasi-linear form of the partial differential equations [9]. This method expresses the solution in terms of finite Fourier series. Richtmyer [10] has extended the von Neumann stability analysis to include systems of equations.

Rewriting equations (1) through (5) in quasi-linear form [11] we obtain a system to be analyzed consisting of five variables and a set of five linear equations. The stability of the system difference equations is established if

$$0 \leq \frac{\Delta x'}{\Delta y'} \frac{v_p'}{u_p'} \leq 1 \qquad (17)$$

Since the x-component of the particle velocity is always greater than or equal to the y-component for injection angles of 45° or less, and the y-component of the particle velocity is always positive, equation (17) is satisfied if

$$\Delta x' = \Delta y' \qquad (18)$$

Lax [10] has shown the equivalence of convergence and stability for problems which satisfy the consistency conditions. For linear equations the consistency condition states that the truncation error must vanish as $\Delta x'$ and $\Delta y'$ approach zero. The truncation error in the continuity and momentum equations are

$$E_t = O[(\Delta x'),(\Delta y')] \qquad (19)$$

$$E_t = O[(\Delta x'),(\Delta y')] \qquad (20)$$

The truncation error vanishes in all cases as $\Delta x'$ and $\Delta y'$ approach zero, and the difference equations converge.

In conclusion, the finite difference equations are shown to converge and to be stable. The von Neumann method is used to establish stability, and the Lax consistency condition is used in the convergence analysis.

4. RESULTS

The digital computer is used to solve numerically the problem of wall-slot particle injection into a gas flowing over a flat plate. The solution is based on air as the viscous gas with

$u_\infty = 30.48$ m/s, $L = 0.3048$ m, $\sigma = 5$ μm, $\rho = 1.1358$ kg/m^3,

$\rho_s = 2,319.64$ kg/m^3, $\rho_p[0,0] = 231.964$ kg/m^3,

$\mu_0 = 1.9098 \times 10^{-5}$ kg/m·s, $Re = 5.5253 \times 10^5$, $FL/u_\infty = 14.8197$

The above conditions are not unique. Other sets will yield results as long as the assumptions and criteria of the problem are met.

The results of primary concern are the x-velocity profiles and the boundary layer thicknesses. In order to discuss these results it is necessary to define the fluid boundary layer thickness and the particle boundary layer thickness. In this work the boundary layer thicknesses are defined as the distances for which

$$|u - u_\infty| = (0.00001)u_\infty, \quad \text{for } \delta \tag{21}$$

$$|u_p - u_\infty| = (0.00001)u_\infty, \quad \text{for } \delta_p \tag{22}$$

This definition enables similar treatment of situations when the x-component of the injection velocity is less than or greater than the free-stream velocity.

Figure 1 shows a comparison of the boundary layer thickness at different injection conditions. From an observation of these graphs it is apparent that the boundary thickness increases considerably with increasing particle injection velocity and particle injection angle. An examination of Figures 1 and the y-component of the injection velocity in Table 1 reveals a definite dependency of the boundary layer thickness on the y-component of the injection velocity.

Schlichting [7] demonstrates that for flow over a flat plate with no particle injection the boundary layer thickness is

$$\delta_{\text{no injection}} = A(\mu/\rho u_\infty)^{0.5} x^{0.5} \tag{23}$$

where A is a constant. The fluid and particle boundary layer thicknesses are also proportional to some power of x. By analogy with the no-injection case

$$\delta = B(\mu/\rho u_\infty)^{0.5} x^a, \qquad \delta_p = C(\mu/\rho u_\infty)^{0.5} x^b \tag{24}$$

where a and b are fractional, positive constants, and B and C are proportional to the y-injection velocity.

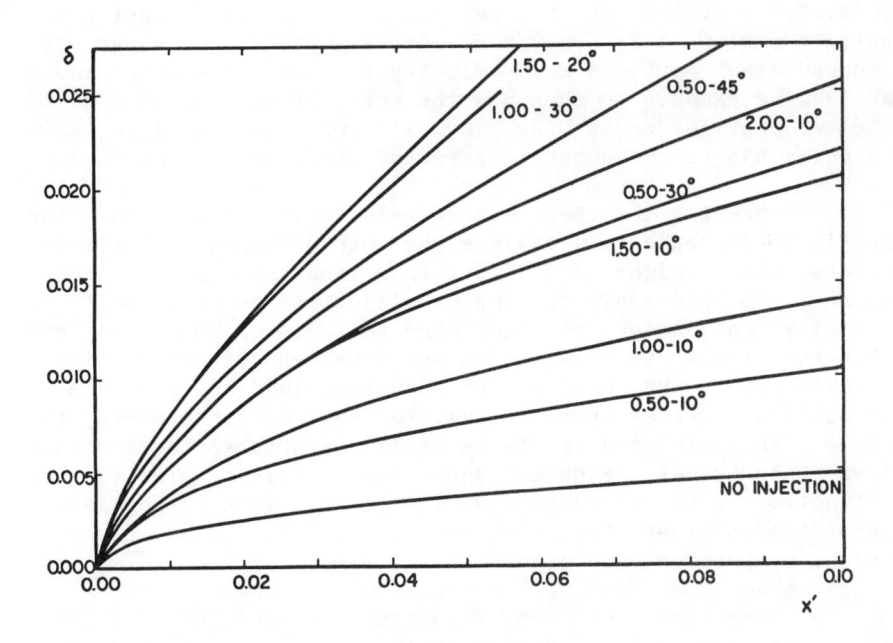

Figure 1. Effect of particle injection velocity and particle injection angle on the boundary layer thickness.

Table 1. Injection parameters investigated

Injection Velocity	Injection Angle	x-Injection Velocity	y-Injection Velocity
0.050	10	0.49240	0.08682
	30	0.43301	0.25000
	45	0.35355	0.35333
1.00	10	0.98481	0.17365
	30	0.86603	0.50000
1.50	10	1.47721	0.26047
	20	1.40954	0.51303
2.00	10	1.96962	0.34730

The x-velocity profiles with the x-injection velocities less than free stream are of the same form as the no-injection profiles. The x-velocities increase from their wall values to the free-stream velocity. Since the particles slip at the plate, their wall velocities near the leading edge are greater than the fluid's. In spite of this they are slower to accelerate to the free-stream velocity because the solid material is so much denser than the gas. Therefore, the x-velocity profiles of the particles and the fluid must cross until the flow is far enough downstream that the shear has reduced the particle wall velocity to zero. Figures 2 and 3 are a good example to show how the profiles cross. At $x'=0.02$ the two profiles cross; but by $x'=0.05$ the particle wall velocity has been reduced to zero and there is no crossing.

For greater than free-stream x-injection velocities the profiles increase to a maximum and then decrease to the free-stream value. Figure 4 displays this behavior. In this graph there is no crossing of the profiles since the particle velocity can never be decreased fast enough to become less than the fluid velocity. An examination of this figure reveals that the maximum of both the fluid and particle x-velocities occur at approximately the same distance above the plate. This distance is always small compared to the boundary layer thicknesses. Figure 5 shows these maximum velocities as a function of the distance down the plate. This graph clearly demonstrates that the particle x-velocity and the fluid x-velocity approach each other.

A comparison of the velocity profiles of different injection conditions, Figures 2 through 4, demonstrates that at constant injection angle and at a given distance downstream, the x-velocity profiles become flatter as the x-injection velocity approaches the free-stream velocity. This behavior results from the decrease in slip velocity as the free-stream velocity is approached.

Figures 2 through 4 demonstrate that the particle boundary layer thickness is always thicker than the fluid thickness for a given injection velocity and angle. Furthermore, an examination of the x-velocity profiles of Figures 2 though 4 reveals that the fluid x-velocity and the particle x-velocity approach each other as the flow proceeds downstream. This characteristic results from the momentum transfer between the two phases.

The particle density profiles are displayed for two x'-positions on Figures 2 through 4. Once the flow has moved downstream all profiles appear to achieve a similar form. The density is zero at the plate and at the edge of the particle boundary layer reaching a maximum at some midpoint. The particles disperse as the flow proceeds downstream reducing the particle density to values which are considerably less

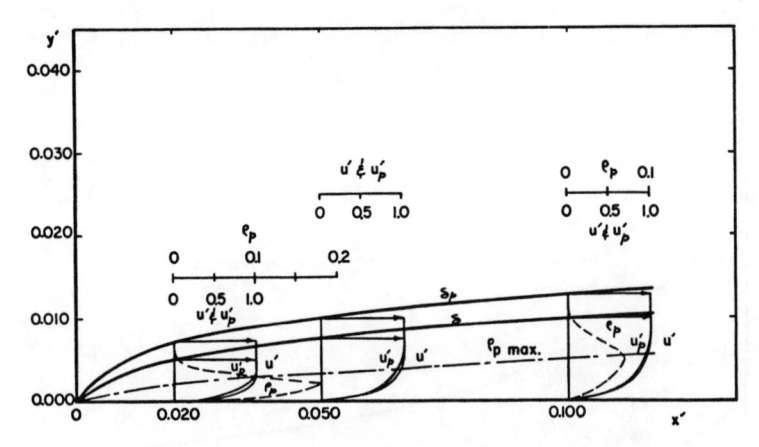

Figure 2. Velocity and density profiles at injection velocity of 0.50 and injection angle of 10°.

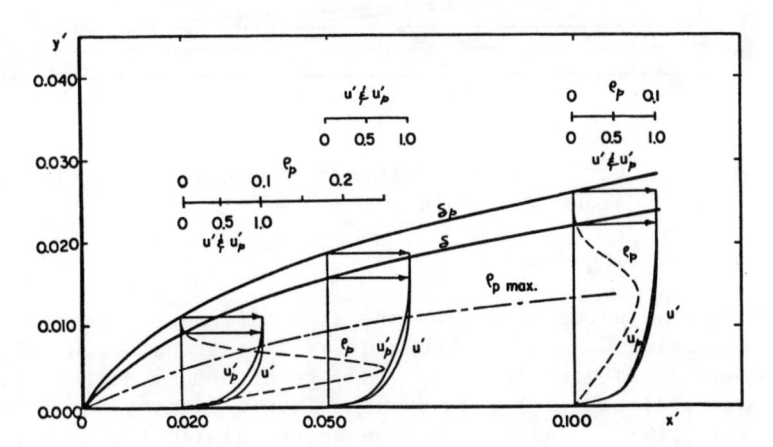

Figure 3. Velocity and density profiles at injection velocity of 0.50 and injection angle of 30°.

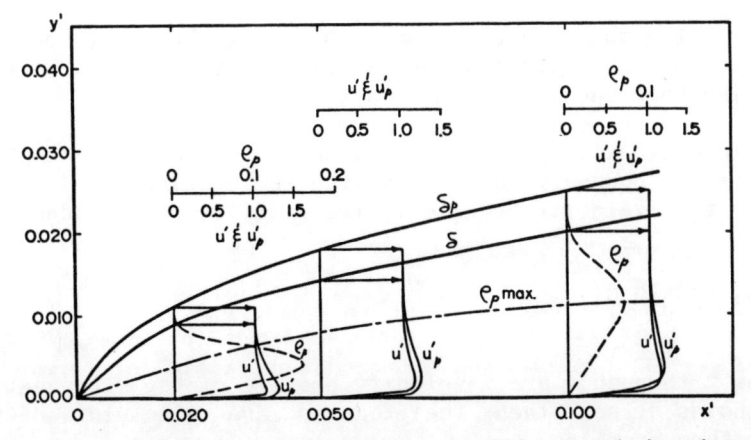

Figure 4. Velocity and density profiles at injection velocity of 1.50 and injection angle of 10°.

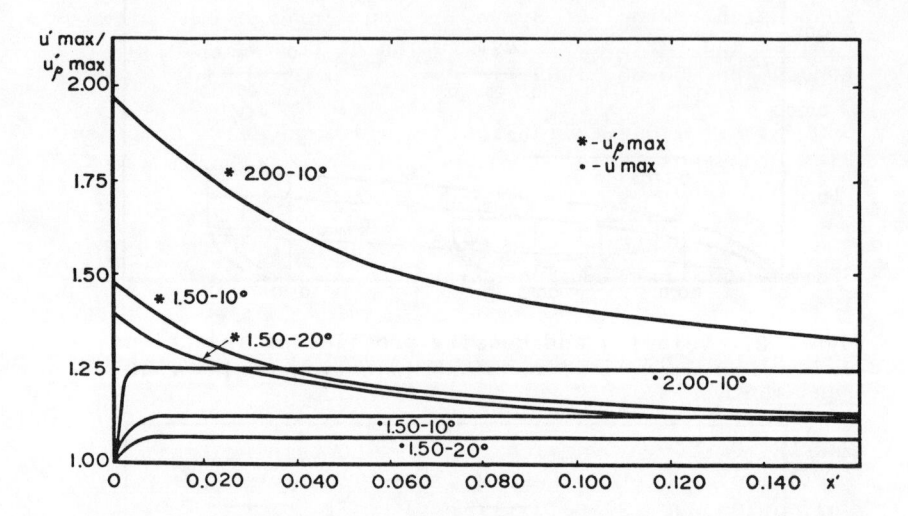

Figure 5. Variation of maximum x-component velocities along the plate.

than the injection density.

If the injected particles are at a different temperature than the viscous gas, then heat will be transferred. This heat transfer is described by the transformed differential equations (13) and (14) and the associated boundary conditions (15) and (16). These equations become similar to equations (2) and (4) if

$$Pr = 1, \quad Nu = 3, \quad c_s = c_p \qquad (25)$$

The Nusselt number can be shown to be about three by applying the various correlations for gas flow over a sphere. Kreith [12] gives one such a correlation that yields Nu=2.68.

Since equations (13) and (14) and equations (2) and (4) have similar boundary conditions, the temperature profiles are similar to x-velocity profiles. The profiles become identical if

$$u_p'[0,0] = \theta_p[0,0] \qquad (26)$$

Figures 2 through 4 are x-velocity profile plots. If equation (26) holds true, then the figures are also plots of the temperature profiles.

REFERENCES

1. SOO, S. L. - Boundary Layer Motion of a Gas-Solids Suspension. *Proc. of Symposium on Interaction between Fluids and Particles*, Institution of Chemical Engineers, London, pp. 50-63, 1962.

2. SOO, S. L. - Gas-Solid Flow. *Single and Multi-Component Flow Processes*, Rutgers University, New Brunswick, NJ, pp.1-52, 1965.

3. SOO, S. L. - Laminar and Separated Flow of a Particulate Suspension. *Astronautica Acta*, Vol. II, pp. 422-431, 1965.

4. SOO, S. L. - Dynamics of Multiphase Flow Systems. *I & EC Fundamentals*, Vol. 4, pp. 426-433, 1965.

5. MARBLE, F. E. - Dynamics of a Gas Containing Small Solid Particles. *Fifth AGARD Combustion and Propulsion Colloquium*, pp. 175-215, 1962.

6. SINGLETON, R. E. - *Fluid Mechanics of Gas-Solid Particle Flow in Boundary Layers*, Ph.D. thesis, California Institute of Technology, 1964.

7. SCHLICHTING, H. - *Boundary-Layer Theory*, McGraw-Hill, New York, 1979.

8. SMITH, G. D. - *Numerical Solution of Partial Differential Equations*, Oxford University Press, London, 1965.

9. O'BRIEN, G. G., HYMAN, M. A., and KAPLAN, S. - A Study of the Numerical Solution of Partial Differential Equations. *J. Mathematics and Physics*, Vol. 29, pp. 223-251, 1950.

10. RICHTMYER, R. D. - *Difference Methods for Initial Value Problems*, Interscience Publisher Inc., New York, 1964.

11. URQUHART, J. B., III - *Laminar Boundary Layer Motion of a Gas with Solid Particle Injection*, MS thesis, Georgia Institute of Technology, 1968.

12. KREITH, F. and Bohn, M. S. - *Principles of Heat Transfer*, 4th Edition, Harper & Row, New York, NY, 1986, pp. 360-363.

A NUMERICAL STUDY OF FREEZING IN A FINNED ENCLOSURE WITH A CONVECTIVE BOUNDARY CONDITION

Richard N. Smith
Department of Mechanical Engineering,
Aeronautical Engineering and Mechanics
Rensselaer Polytechnic Institute
Troy, New York 12180-3590 USA

James R. Warnot, Jr.
I.B.M. Corporation
Kingston, NY 12401
USA

SUMMARY

A finite difference numerical analysis of soliification heat transfer in a phase change material (PCM) within a finned enclosure and with a convective boundary condition at the cooled surface has been carried out. An enthalpy formulation has been applied to the PCM/fin/base wall system and a straightforward iteration scheme has been employed to transform the system of finite difference equations involving temperature and enthalpy into equations involving only enthalpy.

The effects of Biot number, fin thickness to spacing ratio, fin to PCM thermal conductivity ratio, geometrical aspect ratio, Stefan number, and the fin to PCM thermal capacity ratio have all been examined. The results for cases with varying ratios of fin and PCM thermal conductivities and fin thickness suggest a new dimensionless parameter which combines these effects. Further, the potential for a new similarity variable which allows the effect of Biot number to be examined easily for a wide range of problems through interpolation on a simple graph has been discussed.

1. INTRODUCTION

Heat transfer involving solid-liquid phase change have been studied with increasing regularity in recent years. Manufacturing processes such as welding and casting, nuclear reactor core meltdown simulation, cryogenic systems, and phase change energy storage are common engineering examples.

The present paper is motivated by an application in phase change thermal energy storage (TES) systems, in which it is desired to enhance the heat transfer rate occurring during the solidification (discharging) process by containing the phase change material (PCM) in an enclosure having highly conducting fins. The added surface area within the PCM domain may offset somewhat the thermal conduction resistance provided by the solid layer as it grows in thickness.

Previous numerical studies of the interactions of fins with solid/liquid phase change have been concerned only with isothermal or constant heat flux cooling conditions, so that conclusions may not be directly applicable to a practical energy storage system, in which a cooling fluid interacts with a base wall through convection.

The development of numerical methods for solving multidimensional phase change heat transfer problems has received a great deal of attention within the last decade. Of particular concern is utilizing a problem formulation which accounts for the conditions of the moving phase front, which in the multidimensional case cannot be described by a closed form analytical expression. Fairly complete reviews of the literature of conduction heat transfer with change of phase have recently appeared in [1-3]. Therein are presented detailed descriptions of analytical and numerical methods for solving such problems. Shamsundar [4] has pointed out many of the difficulties associated with numerical solutions to multidimensional problems.

The numerical approach selected in this study is the use of an enthalpy formulation of the governing equations as a basis for finite differencing [5, 6]. The enthalpy rather than the temperature of the phase change material is the primary dependent variable, so that the location and shape of the phase front are not needed in advance to perform computations. However, the phase front location, fraction frozen, and heat transfer quantities can be determined from the enthalpy distribution. The advantages of this method are that solidification over a range of temperatures is easily accommodated, a small time solution is not required for starting the computation, and thermally complex boundaries are easily incorporated into the domain. A disadvantage is that the solution must be obtained over the entire domain of interest, including those areas where no temperature gradients may exist. Smith et al. [7-9] used the enthalpy formulation for several problems of solidification without liquid phase convection, but including heat transfer in the container walls surrounding the phase change material.

Experimental studies of solidification or melting adjacent to finned surfaces have been carried out by several investigators [10-12]. Schneider [13] used a modified enthalpy method to examine melting in a geometry similar to that of the present paper, however, with only Neumann and Dirichlet boundary conditions. Griffin and Smith [14] performed a two-dimensional integral analysis of a rectangular fin/PCM system. Later, Smith and Koch [7] used an enthalpy method to prepare a finite difference solution for the same geometry with a Dirichlet boundary condition. The container wall was specified and its heat capacity effects were neglected. Dimensionless constants were established which considered both physical and geometric parameters. The present paper extends this work to consider the more realistic convective boundary condition as well as wall heat capacity.

2. ANALYSIS

2.1 Problem Statement and Assumptions

The geometry considered is shown in Figure 1. This represents a symmetry section of a phase change heat exchanger in which two walls extending infinitely in the x- and z-directions are connected by fins. The spatial variables used in the problem formulation, as well as the geometric constants for a given case, are indicated on the figure.

A PCM at the phase change temperature fills the enclosure between the walls and fins. Liquid superheat is not considered in this study since natural convection in the liquid phase, which likewise is not considered, becomes increasingly important as the amount of superheat increases. At some initial time the flow of a cooling fluid at a constant temperature $T_b < T_f$ commences external to the top and bottom walls, exchangeing energy with the wall via convection with a film coefficient h. Heat transfer takes place from the solidifying PCM to both the bottom wall and the fin.

A two dimensional analysis is considered since there are no z-axis effects due to the nature of the boundary conditions. The boundaries at x=0, x=s, and y=L are adiabatic because of the symmetry of the problem. A number of assumptions have been incorporated into the analysis. They include uniform properties within each phase and the conducting walls, phase change at a single temperature, constant external convection coefficient and external fluid temperature. None of these are strictly required by the numerical formulation to be described; however, interpretation of the results in terms of key parameters is greatly simplified.

2.2 Governing Equations

The equation for conservation of energy is identical in the fin, base wall, and the PCM. In non-dimensional terms, the

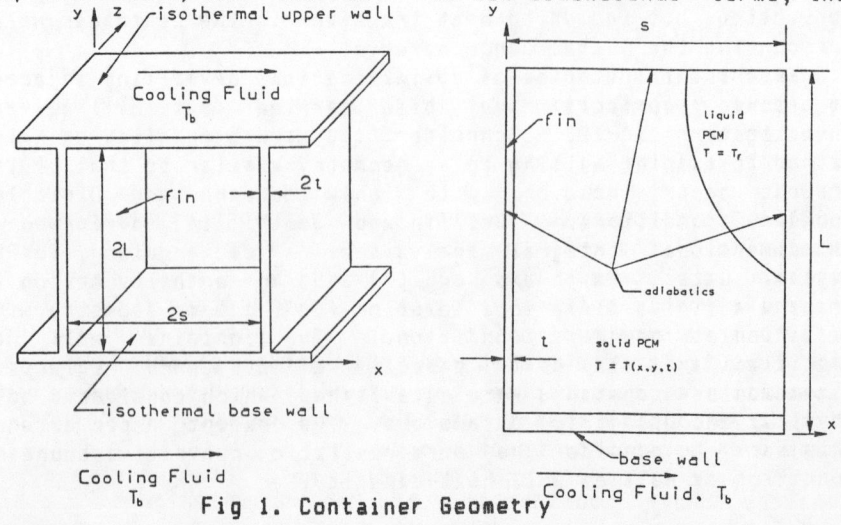

Fig 1. Container Geometry

equation may be written as

$$\overline{\rho c} \frac{\partial \theta}{\partial \tau} = \frac{\partial}{\partial X}\left(\overline{k}\frac{\partial \phi}{\partial X}\right) + \frac{\partial}{\partial Y}\left(\overline{k}\frac{\partial \phi}{\partial Y}\right) \qquad (1)$$

subject to the following boundary conditions:

$$X = 0 : \quad \frac{\partial \phi}{\partial X} = 0 \qquad (2a)$$

$$X = \frac{S}{L} : \quad \frac{\partial \phi}{\partial X} = 0 \qquad (2b)$$

$$Y = 1 : \quad \frac{\partial \phi}{\partial Y} = 0 \qquad (2c)$$

$$Y = 0 : \quad \frac{\partial \phi}{\partial Y} = Bi\,(\phi_w - \phi_b) \qquad (2d)$$

$$\tau = 0 : \quad \phi(X,Y,0) = 0 \qquad (2e)$$

where the dimensionless quantities appearing are defined as follows:

$$\phi = \frac{c(T - T_f)}{\lambda} \quad ; \quad \theta = \frac{h - h_f}{\lambda} \qquad (3a,b)$$

$$Bi = \frac{hL}{k} \quad ; \quad Ste = \frac{c(T_f - T_b)}{\lambda} \qquad (4,5)$$

$$\overline{(\rho c)}_{\ell,w} = \frac{(\rho c)_{\ell,w}}{\rho c} \quad ; \quad \overline{k}_{\ell,w} = \frac{k_{\ell,w}}{k} \qquad (6a,b)$$

$$\tau = \frac{\alpha t}{L^2} \qquad (7)$$

A relationship between the enthalpy and the temperature is required to complete the formulation. Within the PCM, this is provided by

	ϕ Range	$\phi(\theta)$	
Solid Phase	$\phi < 0$	$\phi = \theta$	(8a)
Phase Change	$\phi = 0$	$1 < \theta < 0$	(8b)
Liquid Phase	$\phi > 0$	$\phi = \frac{c}{c_\ell}(\theta-1)$	(8c)

Within the fin and the base wall, the enthalpy and temperature (θ and ϕ) are proportional to each other for all temperatures.

2.3 Numerical Formulation

The development of finite difference equations and resolution of the enthalpy field was done following procedures outlined in Patankar [15], with modifications to account for the relationship between enthalpy and temperature. Many of the details are contained in [16,17] and will not be repeated here. The computer program used for this study contains the full capability of calculating the natural convection flow field within the liquid phase, as well as accommodating change of phase over a range of temperatures.

The grid structure and nomenclature of the two-dimensional geometry under consideration are shown in Figure 2. The

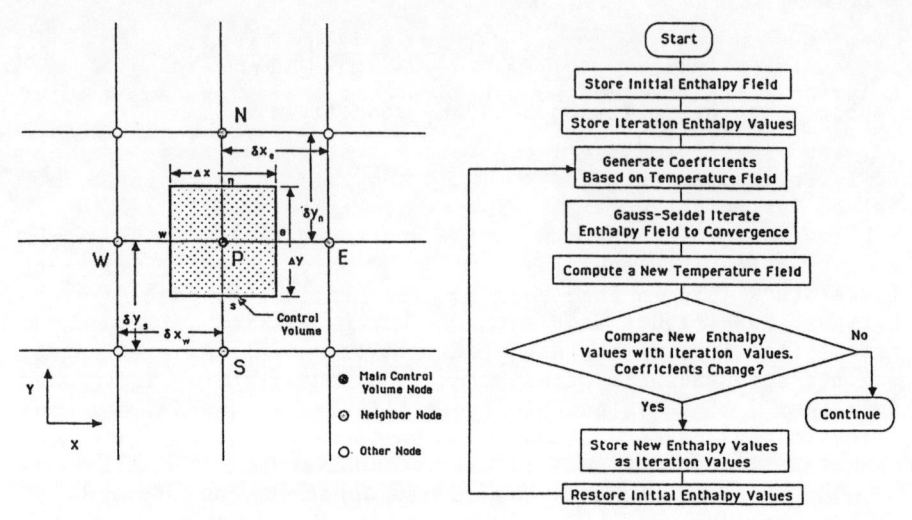

Fig 2. F-D Control Volume   Fig 3. Coefficient Iteration Flowchart

enthalpy is assumed to vary in a piecewise linear fashion between control volumes. The conductivity across control faces is calculated by a harmonic mean averaging method.

The non-dimensionalized energy equation (1) can be written in discretized form for an interior node as

$$a_p \theta_p = a_e \phi_e + a_w \phi_w + a_n \phi_n + a_s \phi_s + b \qquad (9)$$

where

$$a_e = \bar{k}_e \Delta Y / \delta X_e \quad ; \quad a_w = \bar{k}_w \Delta Y / \delta X_w \qquad (10a,b)$$

$$a_n = \bar{k}_n \Delta X / \delta Y_n \quad ; \quad a_s = \bar{k}_s \Delta X / \delta Y_s \qquad (10c,d)$$

$$a_{po} = \bar{\rho c} \Delta X \Delta Y / \Delta \tau \quad ; \quad b = a_{po} \theta_{po} \qquad (10e,f)$$

$$a_p = a_e + a_w + a_n + a_s + a_{po} \qquad (10g)$$

This general set of equations must be modified at the four boundaries:

Left boundary (fin symmetry line): $a_w = 0$ \qquad (11)

Top boundary (container symmetry line): $a_n = 0$ \qquad (12)

Right boundary (PCM symmetry line): $a_e = 0$ \qquad (13)

Bottom boundary:

$$b = a_{po} \theta_{po} + \frac{\Delta X \, Ste}{(1/Bi + \Delta Y_s)} \qquad (14a)$$

$$a_p = a_e + a_w + a_n + a_{p_0} + \frac{\Delta X}{(1/Bi + \delta Y_s)} \qquad (14b)$$

The modifications at the bottom boundary account for the heat transfer with the cooling fluid, through an energy balance across the bottom surface.

The coefficients of the discretization equation are now fully defined for all nodes in the domain. The enthalpy method solves for enthalpy as the dependent variable while Eq. (9) is written in terms of both enthalpy and tempeature. To develop a finite difference equation in terms only of enthalpy, each temperature (θ) in the equation is replaced by the relation between temperature and enthalpy which existed at the most recent iteration of the calculation (Eqs. (8)). The effect is an iteration on the coefficients of the finite difference equations. The values of the coefficients change from one iteration to another only if the phase of the node in question changes. The iteration sequence is shown in Figure 3.

The finite difference equations, now in terms of enthalpy alone, are solved by a line-by-line tri-diagonal matrix algorithm. An outer, Gauss-Seidel iteration is used to converge the solution. At each iteration, a new temperature field may be calculated and from this, a new set of coefficients.

The cases examined in this thesis were run on an IBM 3081 processor unit using the VM Operating System. Run times ranged from 20 seconds for a constant temperature boundary condition case, to 2 minutes for a typical boundary condition case. A notable exception was the case with Bi = 0.1, Ste = 0.1. To generate a smooth fraction frozen curve, an extremely small time step and fine grid were required. This case required 17 minutes of CPU time to reach a fraction frozen of 0.95.

3. RESULTS

All results presented in this section are derived from the basic calculation of enthalpy throughout the field. The phase front profiles are determined by calculating the points where enthalpy is equal to zero. The fraction frozen is calculated from the volume ratio of nodes where θ ≤ 0 to the overall volume. A linear interpolation is performed on nodes which are partially frozen. The wall heat flux is calculated by

$$Q = \sum_{base} \frac{(\phi_w - \phi_b)}{(\phi_f - \phi_b)} \qquad (15)$$

for the entire wall. These dimensionless temperatures are calculated from the node enthalpies.

Two additional geometric parameters naturally arise through the calculation. They are the dimensionless fin and base wall thickness defined as $T = t/L$ and the container aspect ratio $\beta_2 = L/s$. For all cases run, the dimensionless base wall thickness was taken to be 0.02, while the thickness of the fin itself was varied. The appropriate time scale for presentation of the results is the product of Ste and Fo. This tends to minimize the independent effects of the Stefan number, especially for the small values of this parameter which are of interest in this study.

Because of the large number of parameters, a baseline case was selected, with the following values for the governing parameters:

$\bar{k}_w = 1000$; $\overline{\rho c}_w = 1$; Ste = 0.1 ; T = 0.005
Bi = 1 ; $\beta_2 = 2$

These values were selected as typical of those which would exist in a phase change energy storage system.

To verify the accuracy of the algorithm in this application, runs were made to compare with the numerical results of Saitoh [18] for solidification in a square domain using a phase front immobilization technique, with excellent agreement. In addition, the previous results of Smith and Koch [7] in which an enthalpy method with a different numerical algorithm was used for the isothermal base case, were compared again with excellent agreement.

The parameter having the greatest effect on the extraction of heat from the PCM is the external convection coefficient, the Biot number. Regardless of the other parameters which influence the freezing process, this ultimately determines how much energy can be taken from the entire system. Figures 4 show the phase front location for three different values of Bi. Figure 4a is the baseline case. The shape of the phase front location remains similar but the rate of freezing increases by approximately five times for each order of magnitude increase in the Biot number. Figures 5 show fraction frozen and boundary heat flux versus time for the three values of Bi examined. In addition to showing the greatly increased time to

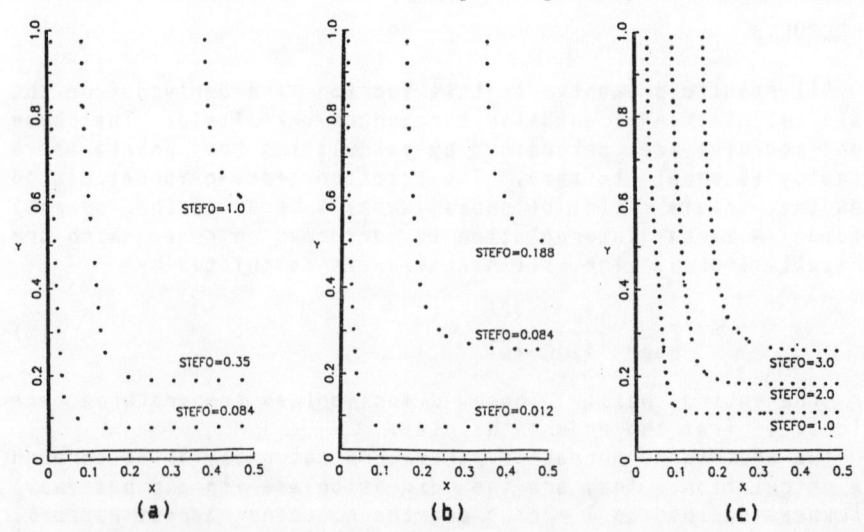

Figs 4. Effect of Convection on Phase Front Profiles; Ste=0.1, $\bar{k}_w$=1000, $\overline{\rho c}_w$=1.0, T=0.005, β_2=2.0
(a) Bi=1.0, (b) Bi=10.0, (c) Bi=0.1

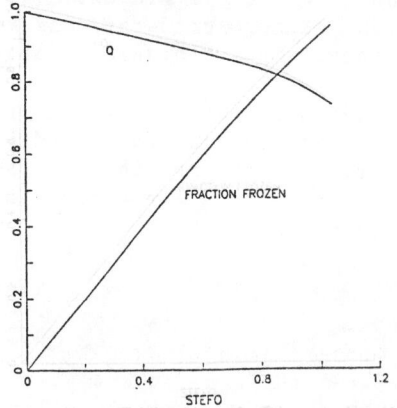

Fig 5a. Effect of Bi on Heat Transfer and Fraction Frozen, Base Case

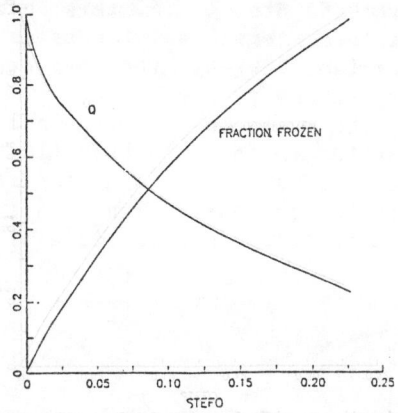

Fig 5b. Effect of Bi on Heat Transfer and Fraction Frozen, Bi = 10

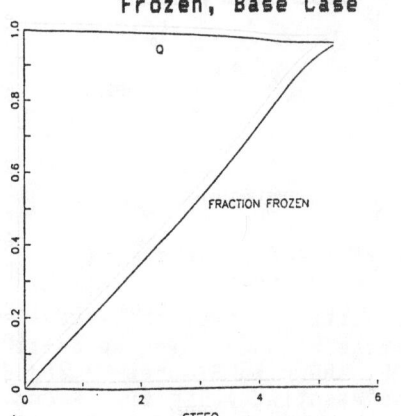

Fig 5c. Effect of Bi on Heat Transfer and Fraction Frozen, Bi = 0.1

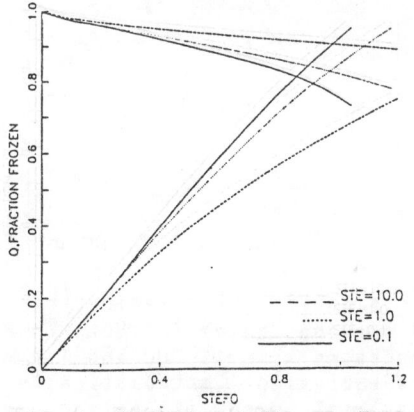

Fig 6. Effect of Ste on Fraction Frozen and Heat Transfer; Bi=1, $\overline{\rho c_w}$=1, $\overline{k_w}$=1000, T=.005, β_2=2

complete freezing as Bi decreases, these figures show a decrease in boundary heat flux versus time as Bi increases. This occurs because the wall temperature approaches the fluid temperature quickly in the case of high Bi. For Bi = 0.1, the wall temperature stays very close to the PCM phase change temperature. The slowness in freezing is further compounded by the low Ste value, which indicates both a high PCM latent heat capacity and a small temperature potential between the PCM and the cooling fluid.

Regarding the effect of the dimensionless heat capacity, $\overline{\rho c_w}$, on the fraction frozen and heat transfer, changes of several orders of magnitude are necessary to affect significantly the freezing process. The results for values of .001 and 1.0 fall over each other. Only for very large values of this parameter is any effect evident. Such a situation would not be typical of a TES system, however.

The independent effect of the Stefan number is shown in Figure 6. For applications anticipated by the present study, involving Ste < 1.0, this independent effect is minimal. In absolute terms an increased Stefan number produces faster freezing, but at the expense of requiring a higher system temperature gradient.

The effects of fin conductivity and fin thickness on fraction frozen and heat flux are shown in Figures 7 and 8. These parameters have an identical effect on the system freezing process. They can be combined into a fin conduction parameter, β_1 (nomenclature of [7]),

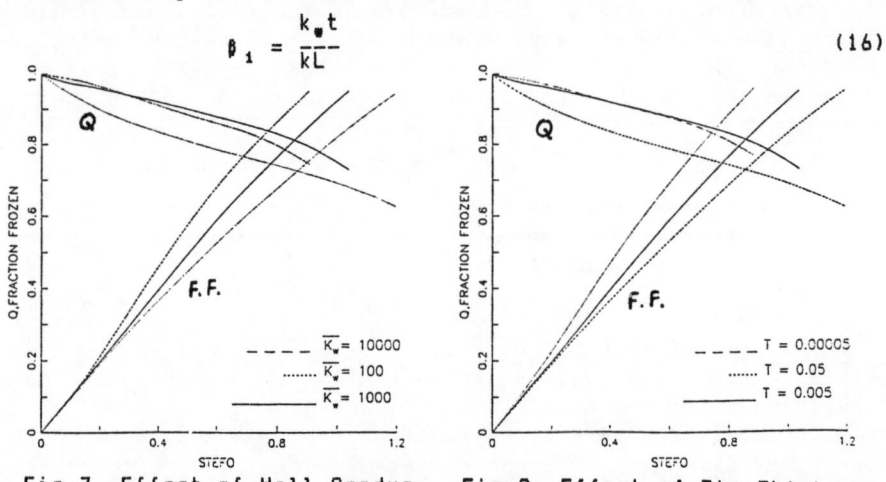

$$\beta_1 = \frac{k_w t}{kL} \qquad (16)$$

Fig 7. Effect of Wall Conductivity

Fig 8. Effect of Fin Thickness

It is possible to examine this parameter without the necessity of solving cases for both parameters k_w and T independently. The cases $\overline{k}_w$ = 10000 and T = 0.05, shown in Figures 7 and 8, are equivalent mathematically, representing β_1 = 50. From a practical sense it is more desirable to have a higher fin conductivity, since a thick fin is both difficult to manufacture and consumes volume in a system which could be better used to contain PCM.

The fraction frozen curves for the $\overline{k}_w$ = 10000, T = 0.05 cases show a rapidly increasing fraction frozen at the initiation of the freezing process which then has a decreasing slope when the fraction frozen reaches approximately 0.35. The base heat flux curves for these cases also cross the curves for $\overline{k}_w$ = 1000, T = 0.005 at the same time. These phenomena are caused by the increasing thermal resistance of the solid PCM as solidification occurs rapidly along the fin. Once this poiont is reached, the solidification rate is no longer enhanced by a highly conductive fin.

The final parameter of importance to this problem is the fin aspect ratio, β_2, which reflects fin spacing variation or fin height variation for a fixed spacing. Although no plots are presented in this paper, the normally expected behavior of an increased rate of freezing with higher values of β_2 is in fact observed. For cases of high aspect ratio, combined with low Ste values and low Bi values, some numerical difficulties were encountered. Specifically, oscillations in the computed enthalpy field which did not permit satisfactory convergence were observed. The physical explanation is that the combination of low temperature potential, low external heat transfer potential, and high aspect ratio leads to a necessity of an excessive number of nodes in the Y- direction to achieve satisfactory solutions. For the β_2 = 10 case, no satisfactory solution was obtained with Ste = 0.1 and Bi = 0.1. Changing any of these three facors eliminates the numerical difficulty.

It is useful to examine the effects of Biot number on fraction frozen and base heat flux on a single plot, rather than individually for each Biot number. This has the effect of removing time from the problem. Shamsundar and Srinivasan [19] proposed the use of a similarity variable which would produce a single curve for each geometric condition. From the curve heat transfer versus fraction frozen may be determined for any values of the boundary conditions. This variable is defined as

$$S = \frac{1}{Bi} \left(\frac{1}{Q} - 1 \right) \qquad (17)$$

The validity of this approach is limited to cases in which the Stefan number is small, and where temperature variations around the boundary of the domain are not large.

The variation of S with fraction frozen was calculated for the three different Biot numbers examined in the present study, with results shown in Figure 9. The curves do not correspond to one another as closely as those in [19] but are useful. By interpolating between these curves, any one of a variety of boundary condition values can be quickly examined without the necessity of re-running the computer model. A set of these curves can be generated for different geometries, providing a useful reference for the design of phase change energy storage systems. Additional work remains, however, to establish fully the limits on the applicability of this similarity analysis for solidification within a finned enclosure.

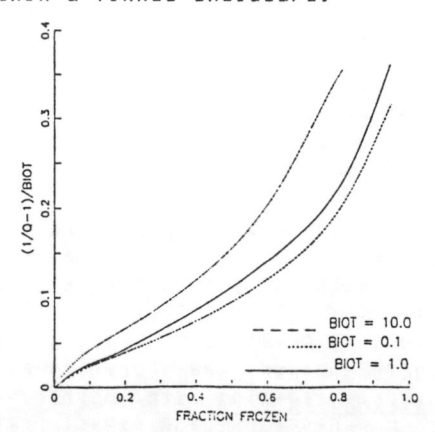

Fig 9. Similarity Variable Plot
Ste=0.1, $\overline{k}_w$=1000, $\overline{\rho c}_w$=1,
T=.005, β_2=2

4. CONCLUSIONS

A numerical analysis has been carried out for the problem of solidification within an enclosure containing highly conducting sidewalls. The analysis extends earlier work in that a convective external cooling boundary condition has been accommodated.

The additional parameter which results, namely the Biot number, plays a vital role in the solidification and heat transfer process. Two physical properties of the fin were found to combine usefully together into a single fin conduction parameter, consistent with an assumption used in earlier work.

A similarity variable, which examined multiple values of the boundary condition sumultaneously, was found to be great potential utility. By combining the use of this variable with the optimization of the material and geometric parameters, the design of a PCM-TES system can be greatly simplified.

The need for several other areas of investigation becomes evident. Experimental verification of freezing in finned rectangular enclosures with a variety of boundary conditions is needed. The effects of liquid superheat and the resulting natural convection in the liquid phase need to be explored, both analytically and experimentally. The analytical approach will be computationally intensive, but is within the capabilities of the existing program.

5. ACKNOWLEDGEMENTS

The use of computational facilities of the IBM Corporation by one of the authors (JRW) is gratefully acknowledged. This work was support in part by the National Science Foundation, through Grant No. MEA-8406042.

REFERENCES

1. Lane, G.A., ed., *Solar Heat Storage: Latent Heat Materials*, Vol. I, CRC Press, Inc., 1983.
2. Lunardini, V.J., *Heat Transfer in Cold Climates*, Van Nostrand & Reinhold Co., New York, 1981.
3. Crank, J., *Free and Moving Boundary Problems*, Clarendon Press, Oxford, 1984.
4. Shamsundar, N., "Comparison of Numerical Methods for Diffusion Problems with Moving Boundaries," in *Moving Boundary Problems*, Solomon, A.D. and Boggs, P.T., eds., Academic Press, 165-185 (1978).
5. Shamsundar, N. and Sparrow, E.M., "Analysis of Multidimensional Conduction Phase Change via the Enthalpy Model," *ASME, J. Heat Transfer*, 97, 333-340 (1975).
6. Schneider, G.E. and Raw, M.J., "An Implicit Solution Procedure for Finite Difference Modeling of the Stefan Problem," *AIAA J.*, 22, 1685-1690 (1984).

7. Smith, R.N. and Koch, J.D., "Numerical Solution for Freezing Adjacent to a Finned Surface," *Proc. 7th Int. Heat Transfer Conf.*, Hemisphere Press, Washington, D.C., 2, 69-84 (1982).
8. Smith, R.N., Pike, R.L., and Bergs, C.M., "Numerical Analysis of Solidification in a Thick-Walled Cylindrical Container," *AIAA J. Thermophysics and Heat Transfer*, 1, 90-96 (1987).
9. Boucheron, E.A., and Smith, R.N., "An Enthalpy Formulation of the SIMPLE Algorithm for Phase Change Heat Transfer Problem," ASME Paper No. 85-HT-7 (1985).
10. Bathelt, A.C. and Viskanta, R., "Heat Transfer and Interface Motion During IMelting and Solidification Around a Finned Heat Source Sink," ASME Paper No. 80-HT-10 (1980).
11. Sparrow, E.M., Larson, E.D. and Ramsey, J.W., "Freezing on a Finned Tube for Either Conduction-Controlled or Natural Convection Controlled Heat Transfer," *Int. J. Heat Mass Transfer*, 24, 273-278 (1981).
12. Smith, R.N. and Boucheron, E.A., "Analytical and Experimental Study of Freezing Adjacent to a Single Pin Fin," ASME Paper No. 83-HT-17 (1983).
13. Schneider, G.E., "A Numerical Study of Phase Change Energy Transport in Two-Dimensional Rectangular Enclosures," *AIAA J. Energy*, 7, 652-759 (1983).
14. Griffin, F.P. and Smith, R.N., "Approximate Solution for Freezing Adjacent to an Extended Surface," ASME Paper No. 80-HT-8 (1980).
15. Patankar, S.V., *Numerical Heat Transfer and Fluid Flow*, Hemisphere, Washington, D.C., 1980.
16. Boucheron, E.A. and Smith, R.N., "Application of the Enthalpy Method to Multidimensional Solidification Problems with Convection," ASME Paper No. 86-WA/HT-42 (1986).
17. Warnot, J.R., "A Numerical Study of Freezing in a Finned Enclosure with a Convective Boundary Condition," M.S. Thesis, Rensselaer Polytechnic Institute, Troy, NY, 1986.
18. Saitoh, T., "Numerical Method for Multidimensional Freezing Problems in Arbitrary Domains," *ASME, J. Heat Transfer*, 100, 294-299 (1978).
19. Shamsundar, N. and Srinivasan, R., "A New Similarity Method for Analysis of Multi-Dimensional Solidification," *ASME J. Heat Transfer*, 101, 585-591 (1979).

Heat Exchange in the Presence of Nonlinearities : Comparison between a One-Dimensional Model and a Two-Dimensional Model

G. Joly, J.P. Kernévez, R. Luce

Université de Compiègne, B.P. 233, 60206 Compiègne, FRANCE.

Abstract.

Multiple steady-state solutions have been found in the behavior of a heat exchanger, modelled by nonlinear parabolic and hyperbolic equations. However, the temperature of the wall, that separates the heating liquid from the cooling liquid, was assumed to depend only on the axial position in the tube, and a global heat exchanger coefficient took into account the wall thickness.

Therefore, it seemed to us very interesting to see whether the so obtained results depend only on this simplifying assumption. So a two-dimensional mathematical model is studied and we will show how this model gives qualitatively, and even quantitavely, the same results.

1. Introduction.

In a tube containing a boiling liquid and heated by another fluid, the exchanged heat flux is not the same throughout the tube and there is a thermal coupling between the two-phase mixture and the heating fluid. In a first study [1] the tube wall temperature was assumed to only depend on the axial coordinate. This paper considers that the wall temperature also depends on the radial coordinate and so the problem is treated as a two dimensional one.

The particularity of the heat exchanger considered is that the heat flux density q transferred from the heating surface to boiling water is a highly non linear function of the temperature difference between the internal tube and the liquid-vapor water mixture[2].

The study of the sodium heated evaporator consists in determining how, under steady state conditions, the distribution of temperature of the heating wall and the distribution of temperature of the sodium, as well as the quality of the water, change when the sodium inlet temperature is made to vary.

The one-dimensional model has showed the existence of very extreme temperature fronts along the wall and the existence of multiple stationary states for the same sodium inlet temperature. It would be interesting to investigate if a two-dimensional model confirms these phenomena.

The numerical method applied is the continuation method which has been developed since the seventies to study non-linear systems [3,4]. Thanks to an adaptive mesh, the dry-out front is correctly accounted for and observed in its progression when the sodium inlet temperature varies.

2. GOVERNING EQUATIONS

A schematic of the physical situation is shown in figure 1.

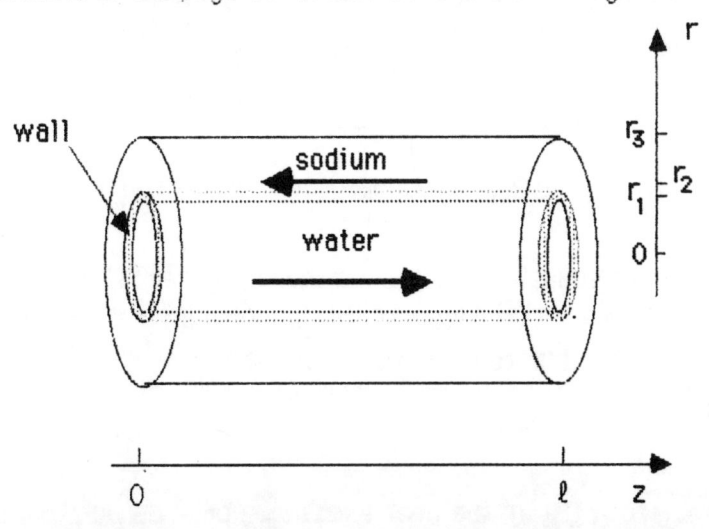

figure 1. Schematic of the Physical Situation

The sodium enters the exchanger at a temperature s_e, leaves it at a temperature s_s and presents a temperature stationary distribution $s(z)$. The thermal exchange, between the sodium and the internal tube wall, is represented by a linear relation

$$q_e = h_s (s(z) - w(r_2,z,\varphi))$$

where r_2 is the external radius of the internal tube, h_s is the thermal tranmission coefficient and $w(r,z,\varphi)$ is the temperature in the wall. The thermal exchange between the boiling liquid and the external tube wall is described by a non linear function

$$q_i(w(r_1,z,\varphi)-T_{sat}, X(z))$$

where r_1 is the internal radius of the internal tube, $X(z)$ is the mass quality of water and T_{sat} is the saturated vapor temperature. A closed-form analytical expression is given by

$$q_i(\theta,z) = 1.107\, \theta + 47.273\, (1 - X)\, \theta^3 / (1 + 1.7\, \theta).$$

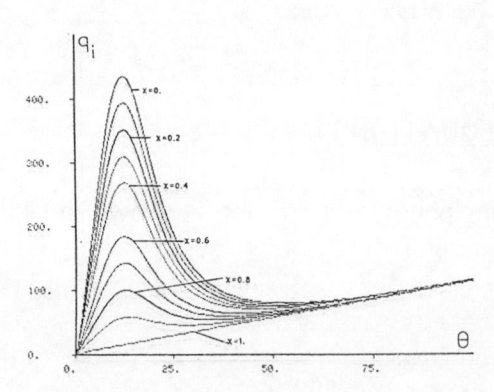

figure 2. The isoquality curves $q_i(\theta,X)$

The system of equations expresses energy conservation in the two fluids and the wall. Because of the axisymmetry of the problem, the various distributed variables do not depend on the center angle φ.

Wall.

(2.1) $\quad \lambda\, \Delta w(r,z) = 0$, $\qquad\qquad r_1 < r < r_2,\ 0 < z < \ell$,

(2.2) $\quad \lambda\, \dfrac{\partial w}{\partial r}(r_2,z) = h_s(s(z) - w(r_2,z))$, $\ 0 < z < \ell$,

(2.3) $\lambda \frac{\partial w}{\partial r}(r_1,z) = q_i(w(r_1,z)-T_{sat}, X(z))$, $0 < z < \ell$,

(2.4) $\frac{\partial w}{\partial z}(r,0) = 0$, $\frac{\partial w}{\partial z}(r,\ell) = 0$, $r_1 < r < r_2$

The parameter λ is the thermal conductivity of the wall and Δ is the Laplacian operator. The first two boundary conditions denote respectively the exchange with both sodium and water.

Sodium.

(2.5) $-G C_p s'(z) + 2 \frac{r_2}{r_3^2 - r_2^2} h_s(s(z) - w(r_2,z)) = 0$, $0 < z < \ell$,

(2.6) $s(\ell) = s_e$.

Water.

(2.7) $-G' C'_p X'(z) + \frac{2}{r_1} q_i(w(r_1,z)-T_{sat}, X(z)) = 0$, $0 < z < \ell$,

(2.8) $X(0) = 0$.

With the magnitude of the numerical values, water is always at the saturated state, so the mass quality is less or equal to one.

3. THE ONE-DIMENSIONAL MATHEMATICAL MODEL.

In a first study we have considered the thickness of the wall as infinitely thin and the temperature of the wall is reduced at one longitudinal temperature distribution. Hence, the system of equations was the following one

(3.1) $\lambda w''(z) + 2 \frac{r_2}{r_2^2 - r_1^2} [\tilde{h}_s(s(z) - w(z)) - q_i(w(z)-T_{sat}, X(z))] = 0$,

$w'(0) = w'(\ell) = 0$,

(3.2) $-G C_p s'(z) + 2 \frac{r_2}{r_3^2 - r_2^2} \tilde{h}_s (s(z) - w(z)) = 0$, $0 < z < \ell$,

$s(\ell) = s_e$,

(3.3) $-G' C'_p X'(z) + \frac{2}{r_1} q_i(w(z)-T_{sat}, X(z)) = 0$, $0 < z < \ell$,

$X(0) = 0$.

It is important to note that the coefficient $\tilde{h}_s$ is not equal to the coefficient h_s since $\tilde{h}_s$ integrates the thermal transmission coefficient of the sodium as well as this one of the wall. So the coefficient $\tilde{h}_s$ permits to take into account the thickness of the wall but it is calculated by the expression :

(3.4) $\quad (\tilde{h}_s)^{-1} = (h_s)^{-1} + \dfrac{r_2}{\lambda} \operatorname{Log} \dfrac{r_2}{r_1}$,

which is valid only when the axial conduction is unimportant. But when the front of temperature becomes important, the axial conduction is not negligible. This reason motivates the study of the two-dimensional representation.

4. Methodology

The principle, which is used to solve the two-dimensional model, is to operate a discretization according to the radial axis in the equation of the wall to get a system of mono-dimensional equations which could be solved by the same software that is used for the one-dimensional model.

Given the equation (2.1), with the boundary conditions (2.2) and (2.3), we multiply the equation (2.1) by a function $\varphi(r)$ of the set $H^1(]r_1, r_2[)$ and integrate it

$$\lambda \int_{r_1}^{r_2} \dfrac{\partial}{\partial r}(r \dfrac{\partial w}{\partial r}) \varphi(r) \, dr + \lambda \int_{r_1}^{r_2} \dfrac{\partial^2 w}{\partial z^2} \varphi(r) \, r \, dr = 0$$

Then the separation variable method gives the approximation $\tilde{w}(r,z)$

$$\tilde{w}(r,z) = \sum_{i=1}^{n} w_i(z) \, u_i(r)$$

where the functions $u_i(r)$ are the "roof functions". So we replace the function $\varphi(r)$ by the functions $u_i(r)$ and integrate by part, and we obtain the following matrix equation

$$A \begin{bmatrix} w_1'' \\ w_2'' \\ \\ w_n'' \end{bmatrix} = B \begin{bmatrix} w_1 \\ w_2 \\ \\ w_n \end{bmatrix} + \begin{bmatrix} -r_1 q_1 \\ 0 \\ 0 \\ r_2 q_e \end{bmatrix}$$

where A and B are two three-diagonal matrices. So we get a system of differential equations.

The method applied to solve this system is based on a continuation method implemented in a software package AUTO [5], which makes it possible to follow the changes of the solution of a distributed parameter system against a given parameter. A continuum of solutions is thus determined for the system when the parameter varies. All the possible states of the system (multiple states), their stability and, thus, the potential hysteresis phenomena can be found.

5. Comparison between the two models

To compare the results we will consider different figures : the power as a function of the sodium inlet temperature, the wall and sodium profiles and the mass quality profile of the liquid.

The results with the two-dimensional model give n axial profiles in the wall of the tube. The tests with different values of n have showed that the radial discretization is hardly sensitive to n; the radial profiles are almost linear. That is the reason why we chose n=3. So the first profile corresponds with the temperature of the wall in contact with the boiling water, the second one corres-

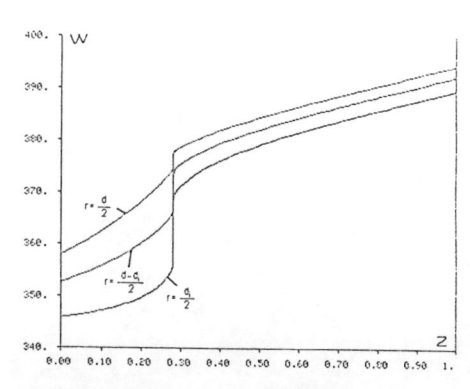

figure 3. Three longitudinal profiles of the wall.

ponds with the wall middle temperature and the last one with the temperature of the wall in contact with the sodium (fig.3).

If the two models are coherent, the longitudinal profiles of the first model must agree with the first profiles of the second one since the one-dimensonal model integrates the thermal transmission coefficient.

The thermal power P of the exchanger is a linear function of the difference between the sodium inlet and outlet temperatures.

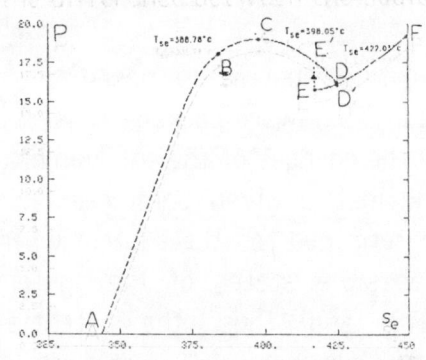

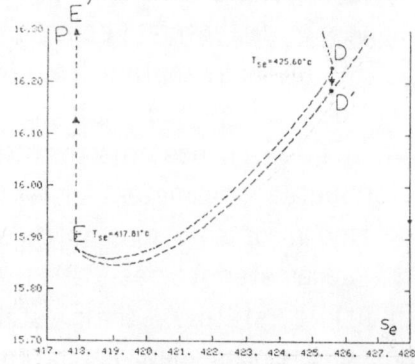

figure 4. Thermal power curve.

figure 5. Thermal power curve (zoom).

Both the thermal power curves present the same particularities for the same values of the parameter T_{se}, so we give only the resuls of the two-dimensional model (fig.4, fig.5). Between the points A and B, P is linear. In all the tube the water is in nucleate boiling. The wall and sodium temperature profiles are regular (fig.6, fig.7).

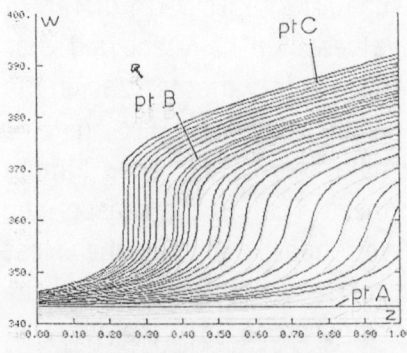

figure 6. Wall temperature profiles.

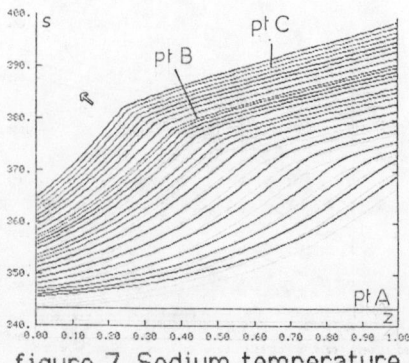

figure 7. Sodium temperature profiles.

Then the slope of the curve P starts decreasing from point B at a regular rhythm and becomes zero at point C where the maximum thermal power is exchanged. The part of the tube in film boiling increases but the transition between the nucleate boiling and the film boiling is smooth (fig.6, fig.7). There is no boiling crisis. The profiles of the two models are identical.

From C to D, the slope of the curve becomes negative. A dry-out front can be seen getting formed and moving towards the evaporator inlet.

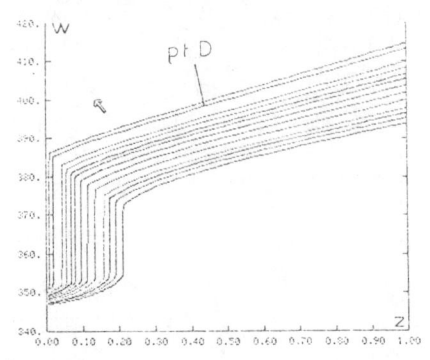

figure 8. Wall temperature profiles.

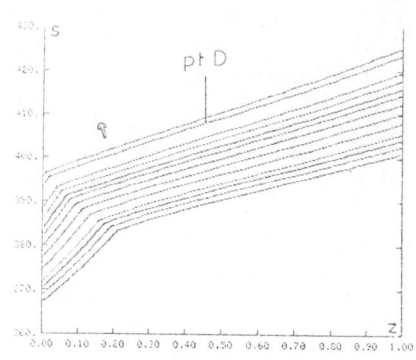

figure 9. Sodium temperature profiles.

The profiles seem equal but numerically the slope of the front is twice as important with the two-dimensional model. The front is very narrow and the differences can not be seen in the figures (fig.8, fig.9). It is the only difference between both models due to the approximation of the global exchange coefficient $\tilde{h}_s$ (3.4).

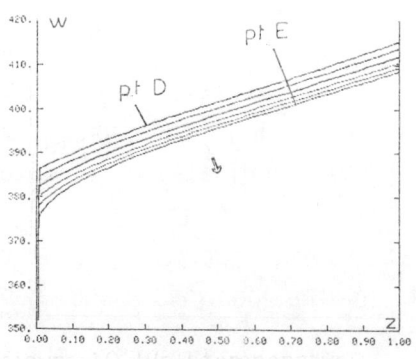

figure 10. Wall temperature profiles

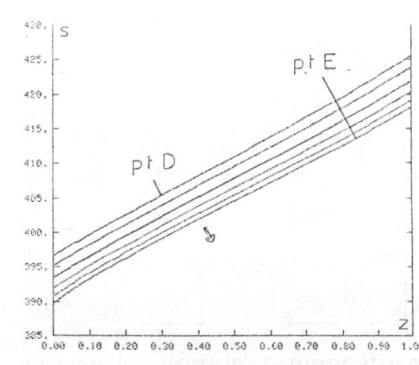

figure 11. Sodium temperature profiles

From D to E, the sodium inlet temperature decreases because the point D is the first turning point of the bifurcation diagram and the states of the branch DE are dynamically unstable. This corresponds to the disappearance of the wall temperature front at the inlet of the tube (fig. 10, fig. 11).

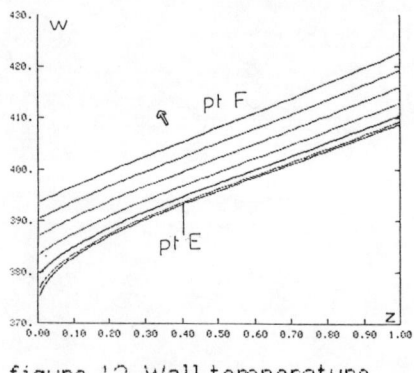

figure 12. Wall temperature profiles

figure 13. Sodium temperature profiles

The point E is the second turning point of the continuation curve. After this point, the sodium inlet temperature is increasing, the curve P is linear again, and the wall and sodium profiles are regular (fig. 12, fig. 13).

The hysteresis phenomenum is represented in the figures 4 and 5 between the points E', D, D', E.

We did not speak about the mass quality of water because it does not give additional information.

6. Conclusion

The results obtained with both models taking into account the axial conduction must still be definitely validated by precise experimental investigations.

The hysteresis effects observed are not due to the one-dimensional formulation but to the nonlinear and non monotonous feature of the boiling curves.

Aknowledgement

The authors wish to thank the EDF (Electricité de France) society, whose financial support allowed this study to be accomplished.

References

1. M. LLORY, G. JOLY, J.P. KERNEVEZ, "Theoretical Considerations on the Boiling Crisis in Boiling Systems under High Heat Flux Conditions" - the 8th International Heat Transfer Conference and Exhibition, Ed. C.L. Tien, V.P. Carey, J.K. Serrell, H.P.C., vol 5, pp. 2191-2195, San Francisco 1986.

2. D.C. GROENVELD, A.S. BORODIN, "Occurence of Slow Dryout in Forced Convective Flow", CERI Conf. Miami Beach (USA) 1979.

3. H.B. KELLER, "Numerical Solution of Bifurcation and Nonlinear Eigenvalue Problems" in Applications of Bifurcation Theory, P.H. Rabinowitz, ed. Academic Press, New-York 1977, pp. 359-384.

4. R. SEYDEL, "Numerical Computation of Branch Points in Ordinary Differential Equations", Numer. Math., 32(1979), pp. 51-68.

5. E.J. DOEDEL, J.P. KERNEVEZ, "AUTO : Software for Continuation and Bifurcation Problems in Ordinary Differential Equations", Applied Mathematics Technical Report, CALTEC, 1986, 226 pages.

NUMERICAL SOLUTION OF LOW REYNOLDS NUMBER FLOWS WITH HEAT ADDITION

Y.-H. Choi, S. Venkateswaran, C. L. Merkle
The Pennsylvania State University
Department of Mechanical Engineering
University Park, Pennsylvania, 16802, U.S.A. and

C. W. Larson
University of Dayton Research Institute
Edwards AFB, California, 93535, U.S.A.

1. ABSTRACT

The computation of very low speed flows made compressible by heat transfer is considered. The presence of strong diffusive effects require modification of a previously developed low Mach number formulation for inviscid flows. Applications to the flow in a plasma spectroscopy cell are considered.

2. INTRODUCTION

Computational algorithms for compressible flows have progressed rapidly during the past two decades. Most of this progress has been focused on transonic flows where the change of type of the equations and the presence of shock waves have generated unique challenges to the computational techniques. The presence of compressibility effects in these problems causes the energy equation to be coupled with the continuity and momentum equations so that all three conservation laws must be considered simultaneously. Consequently, viscous transonic flow solutions by necessity include the heat transfer characteristics along with the fluid dynamic effects. Nevertheless, heat transfer generally has a minor impact on both the flowfield and the computation in this type of problem. The solution of the energy equation is primarily concerned with assessing changes in the density that occur because of velocity differences. In the present paper we consider the application of this type of algorithm to flows that are compressible because of heat transfer effects, not velocity changes.

Our interest in these low speed compressible flow problems arises from two applications that are associated with laser [1,2] and solar [3] propulsion concepts for spacecraft. The first has to do with understanding the interaction between the incoming laser or solar radiation and the working fluid in the rocket engine. The second concerns an analysis of the flowfield in an experimental plasma spectroscopy cell (PSC) that is being designed [4] to measure physical properties of candidate working fluids for these propulsion applications. Both of these flowfields are characterized by low velocities and low Reynolds numbers but the presence of significant heat addition makes compressibility effects of dominant importance in the flowfield. The low velocities coupled with low Reynolds numbers make these problems particularly challenging for contemporary compressible flow algorithms.

The flowfield in either a laser or solar rocket is characterized by high volumetric heating rates in the center of the flowfield with cool external walls. The flowfield in the plasma spectroscopic cell is characterized by very hot walls that are used to heat the flowing gas to elevated temperatures for diagnostic purposes. Consequently, the heat transfer characteristics of the two problems are almost "mirror images" of each other. Because our interests here are primarily with the performance of compressible flow algorithms for these low speed problems, we limit our examples to calculations in the plasma spectroscopy cell.

The working fluid of choice for either laser or solar propulsion is hydrogen because of its low molecular weight. Direct absorption of laser energy in pure hydrogen can be accomplished easily [5] but the lower temperatures associated with solar radiation require the addition of trace amounts of other gases. Alkali metal vapors appear to be an excellent candidate for this purpose [6]. The purpose of the plasma spectroscopy cell is to measure the absorptivity and related spectroscopic properties of such mixtures. The flow path inside the cell is lined with rhenium as shown in Fig. 1. This refractory metal allows wall temperatures to 3000K which give a wide range of conditions that cannot be reached in present spectroscopy experiments. An understanding of the flowfield and particularly the rate of diffusion of heat and molecular species is an important design of the cell as well as later evaluations of the data. Because our emphasis in the present problem is on identifying numerical algorithms that are appropriate for these low speed, compressible flows, we consider the diffusion of only heat and momentum here. The addition of molecular diffusion is not expected to require changes in the algorithm.

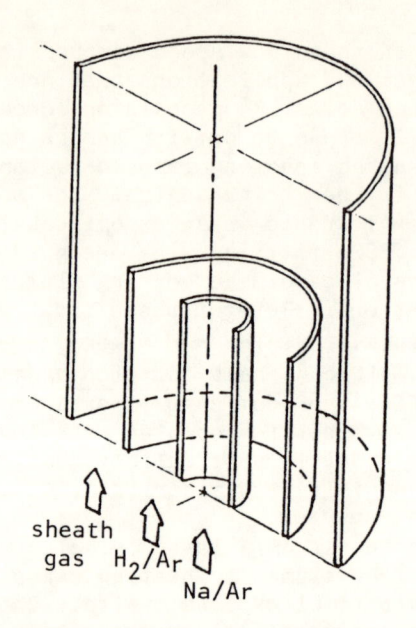

Fig. 1 Schematic diagram of plasma spectroscopy cell

3. PROBLEM FORMULATION

3.1 Review of Inviscid Procedure

The low Mach number perturbation expansion given in Ref. 7 for inviscid flows with heat addition results in the system of equations,

$$\frac{\partial Q}{\partial t} + \frac{\partial E}{\partial x} + \frac{\partial F}{\partial y} = H \qquad (1)$$

where,

$$Q = (\rho, \rho u, \rho v, p_1)^T$$
$$E = (\rho u, \rho u^2 + p_1, \rho u v, \gamma p_0 u)^T$$
$$F = (\rho v, \rho u v, \rho v^2 + p_1, \gamma p_0 v)^T \qquad (2)$$
$$H = (0, 0, 0, \rho q)^T$$

In these expressions, p_0 is the zeroeth order pressure and satisfies the relation, $\nabla p_0 = 0$, p_1 is the perturbation pressure, and q is a volumetric heat source. All other notation is standard. Inspection of the system shows that the continuity equation is unchanged from its traditional form, and the momentum equations are also unchanged except the pressure is replaced by p_1. The energy equation, however, undergoes a substantial change and because the energy equation is the most troublesome in viscous flow, we look at it in more detail.

The result of the low Mach number perturbation is to reduce the energy equation to the form,

$$\gamma p_o \left(\frac{\partial u}{\partial x} + \frac{\partial v}{\partial y} \right) = \rho q \qquad (3)$$

As can be seen, this equation has no time derivative and is identical to the incompressible continuity equation with a source term, $\rho q/\gamma p_o$. To enable the use of a time-marching procedure, an arbitrary time derivative, $\partial p_1/\partial t$, (as indicated in Eqns. 1 and 2) was added to this equation in Refs. 7 and 8. The results show that this artificial time derivative is effective for inviscid flow calculations. This, however, says nothing about its usefulness for viscous flow. The requirements placed on an artificial time derivative in inviscid flow are three. First, the resulting system must contain purely real eigenvalues. Second, the signs of the eigenvalues must be such that physically meaningful boundary conditions can be imposed. (For example, the eigenvalues may not transform a split boundary value problem in space to a one-sided boundary value problem.) Third, the eigenvalues must be well-conditioned to provide rapid convergence in the time iteration. These requirements remain true for the convective operators in a viscous flow, but the diffusion terms also add further requirements as discussed below.

3.2 Extension to Viscous Problems

A straightforward extension of the perturbation procedure to the viscous flow equations shows that the terms to be added to Eqn. 1 are the viscous diffusion terms in the momentum equations and the heat conduction terms in the energy equation. In the low Mach number limit, the viscous terms in the momentum equations are unchanged from their traditional form, whereas the viscous dissipation in the energy equation can be shown to be negligible so that only heat conduction is left. Temporarily retaining $\partial p_1/\partial t$ as an artificial time derivative in the energy equation, we write the low Mach number viscous equations as,

$$\frac{\partial Q}{\partial t} + \frac{\partial E}{\partial x} + \frac{\partial F}{\partial y} = H + L(Q_v) \qquad (4)$$

where Q, E, F and H are unchanged from the quantities given above, and L is the differential operator,

$$L = \frac{\partial}{\partial x} R_{xx} \frac{\partial}{\partial y} + \frac{\partial}{\partial x} R_{xy} \frac{\partial}{\partial y} + \frac{\partial}{\partial y} R_{yx} \frac{\partial}{\partial x} + \frac{\partial}{\partial y} R_{yy} \frac{\partial}{\partial y} \qquad (5)$$

The quantities R_{xx} and R_{yy} are diagonal matrices,

$$R_{xx} = \text{diag}(0, 4/3\mu, \mu, k) \qquad R_{yy} = \text{diag}(0, \mu, 4/3\mu, k) \qquad (6)$$

while R_{xy} and R_{yx} are sparse matrices with only two elements filled. For R_{xy}, the non-zero elements are $r_{23}=\mu$, $r_{32}=-2/3\mu$, while for R_{yx} they are $r_{23}=-2/3\mu$ and $r_{32}=\mu$. The vector Q_V is,

$$Q_V = (p_1, u, v, T)^T \tag{7}$$

As indicated before, the choice of $\partial p_1/\partial t$ as an artificial time derivative in the energy equation is adequate for the inviscid terms. Stability analyses of the full viscous equations, however, show it is unstable at low Reynolds numbers. Detailed study of this instability shows that it depends primarily upon the Prandtl number. In particular, stability analyses of approximate factorization procedures for Eqn. 4 showed them to be highly unstable for Peclet numbers below 500, and in the absence of approximate factorization, Euler implicit schemes retained weak instabilities. The obvious conclusion is that this artificial time derivative does not meet appropriate criteria for the diffusive terms.

The major requirement of artificial time derivatives in diffusive equations is that they be added in such a way that the equations are well posed. Most specifically, the time derivative must be chosen so that the diffusivity remains positive. Because the $\partial p_1/\partial t$ term and its sign were chosen without regard to the diffusion terms, there is no assurance that the resulting viscous system is acceptable. The coupled nature of the equations make the verification of this difficult, and it is useful to re-write them in nonconservative form for this purpose.

The conversion from conservative to nonconservative form is standard for the continuity and the momentum equations so we concentrate on the energy equation. Starting from the Eqn. 3 with diffusive terms added, we first place p_o (a constant) inside the derivatives and use the perfect gas law, $p_o = \rho R T$ to obtain,

$$\gamma R \left(\frac{\partial}{\partial x} \rho u T + \frac{\partial}{\partial y} \rho v T \right) = \rho q + \nabla \cdot (k \nabla T) \tag{8}$$

Expanding and subtracting out the continuity equation gives,

$$\gamma R \rho \left(\frac{\partial T}{\partial t} + u \frac{\partial T}{\partial x} + v \frac{\partial T}{\partial y} \right) = \rho q + \nabla \cdot (k \nabla T) \tag{9}$$

The time derivative in Eqn. 9 enters through the continuity equation where the perfect gas relation has been used to replace $\partial \rho/\partial t$ by $\partial T/\partial t$. This time derivative appears better suited to the heat conduction term than does the artificial $\partial p_1/\partial t$ term. Comparison of the complete system of equations, however, shows two problems exist. The term p_1 does not appear in the equations while $\partial \rho/\partial t$ and $\partial T/\partial t$ (which are uniquely related through the perfect gas law) both appear.

To circumvent this, we use the classical artificial compressibility procedure from incompressible flow, and place the pressure derivative in the continuity equation. The resulting non-conservative form of the equations can be written,

$$\tilde{\Gamma}\frac{\partial \tilde{Q}}{\partial t} + \tilde{A}\frac{\partial \tilde{Q}}{\partial x} + \tilde{B}\frac{\partial \tilde{Q}}{\partial y} = H + L(Q_v) \qquad (10)$$

where,

$$Q = (p_1, u, v, T)^T$$

$$\Gamma = \text{diag}(\rho_R/p_0, 1, 1, \gamma\rho R) \qquad (11)$$

and $\tilde{A}$ and $\tilde{B}$ follow directly from the above. Note $Q=Q_v$, suggesting the diffusion terms are properly treated.

The eigenvalues of the inviscid part of Eqn. 10 are $\lambda = u$, u, $u \pm [(u/2)^2 + p_0/\rho_R]^{1/2}$ (ρ_R is a reference density used to maintain proper units). These eigenvalues are well conditioned, real, and have three positive and one negative value as is fitting of subsonic flow. This suggests the system should be well-behaved for inviscid flows. The time derivatives in the energy equation and the momentum equations have their traditional form, so we anticipate this formulation should perform adequately in both viscous and inviscid limits.

Placing Eqn. 10 in conservation form results in a time derivative that contains p_1 in all four equations. Because $Q=Q_v$, we use the former notation throughout. The result is,

$$\Gamma\frac{\partial \tilde{Q}}{\partial t} + \frac{\partial E}{\partial x} + \frac{\partial F}{\partial y} = H + L\tilde{Q} \qquad (12)$$

where,

$$\Gamma = \begin{pmatrix} \rho_R/p_0 & 0 & 0 & 0 \\ u\rho_R/p_0 & \rho & 0 & 0 \\ v\rho_R/p_0 & 0 & \rho & 0 \\ T\rho_R/p_0 & 0 & 0 & \gamma\rho_R \end{pmatrix} \qquad (13)$$

Stability checks of this modified equation system reveals it is substantially better than the system given in Eqn. 4, although there still appears to be minor difficulties at a narrow range of intermediate Reynolds numbers (between 10 and 100). Experimental tests of the system also verified this difficulty, however, it was traced to the scaling of the pressure term. The low Mach number scaling in Eqn. 12 assumes the entire pressure gradient is determined by inertial forces. As the Reynolds number is reduced, the viscous forces become more dominant and also support a pressure gradient. Rather than rescaling the pressure for this effect, we rescaled the time by a constant K as was done in Ref. 8. The resulting equation system is identical to Eqn. 12 except that Γ now becomes,

$$\Gamma = \begin{pmatrix} \rho_R/p_o & 0 & 0 & 0 \\ u\rho_R/p_o & \rho & 0 & 0 \\ v\rho_R/p_o & 0 & \rho & 0 \\ T\rho_R/p_o & 0 & 0 & \gamma\rho_R \end{pmatrix} \qquad (14)$$

Appropriate values of K that were used to enhance convergence are shown in the Figures.

4. RESULTS

The modified artificial time formulation given in Eqn. 12 with K-scaling as noted in Eqn. 14 was used to compute low Reynolds number diffusion of heat and mass through a straight duct, simulating the plasma spectroscopy cell. The trapezoidal wall temperature distribution shown in Fig. 2 was used for all cases with the maximum level being used as a parameter. Peak temperatures of 300 (no wall heating) 1000, 2000 and 3000K were used. The inlet gas temperature was 300K. For this range of temperatures, results have been obtained over five decades of Reynolds numbers ranging from 0.16 to 1600. In all cases, the Prandtl number was taken as 0.35. This Reynolds number range spans the regime where the heat is transmitted "instantaneously" in y so that the flow remains fully developed, to the regime where the thermal energy is retained in a boundary layer near the wall and the velocity profiles are different at each axial location. All cases shown here are for maximum wall temperature, 3000K.

Figure 3 shows velocity profiles at several axial locations and a corresponding pseudo-three-dimensional plot of the temperature at a Reynolds number of 0.16. At all stations, the velocity profile is nearly "developed" although the centerline velocity first increases and then decreases in concert with the trapezoid in wall temperature. The temperature throughout the duct is almost constant at each axial station showing diffusion is faster than the convection. These calculations are for an inlet Mach number of about 7.5 x 10^{-6}.

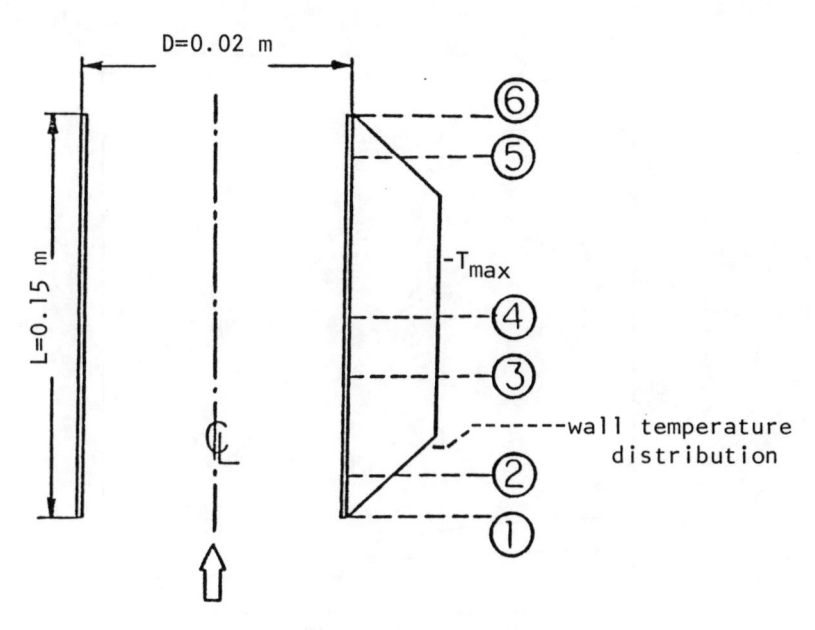

Fig. 2 Schematic diagram of constant area duct with wall temperature distribution.

The comparable set of velocity and temperature results on Fig. 4 are for a Reynolds number of 16. At this larger Reynolds number, the velocity profiles shift at each axial location, but there is considerable nonuniformity in the cross-stream temperature profiles.

The results at a Reynolds number of 1600 are far from fully developed as can be seen from Fig. 5. The wall heating causes a bulge near the wall in the velocity contours, while the temperature in the middle part of the duct is almost unaffected by the heat addition. This spectrum of results bound the flow regime of interest for the PSC experiments from above and below and demonstrate the capability of the time-dependent algorithm in this low Mach number, low Reynolds number regime.

Applications of this same algorithm to flows with species diffusion are presently being made, and results thus far verify that the species equations do not affect convergence rates or stability of the algorithm.

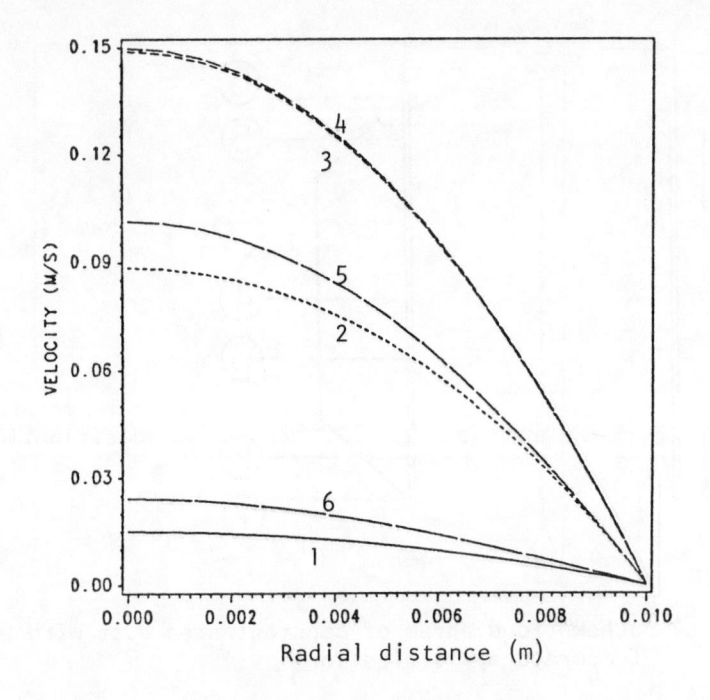

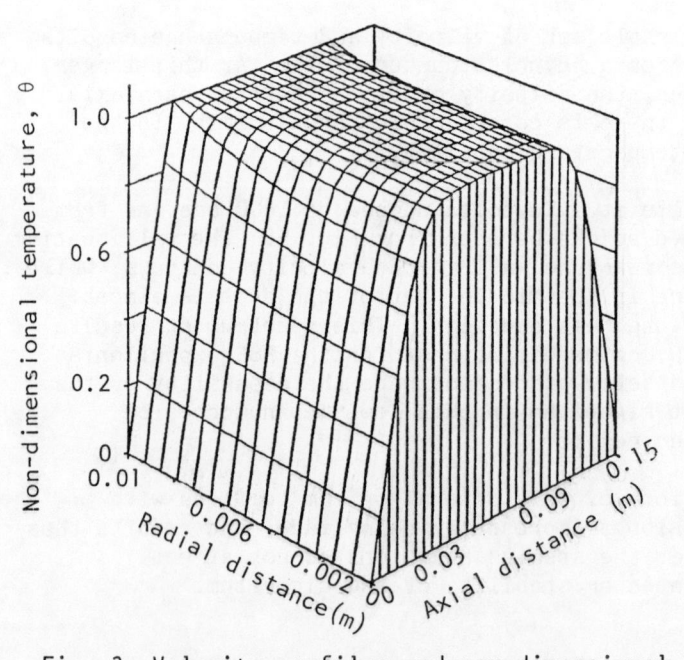

Fig. 3 Velocity profiles and non-dimensional temperature profile for R_e=0.16, P_r=0.35, $T_{wall_{max}}$=1000K. $\theta=(T_{max}-T)/(T_{max}-T_{in})$

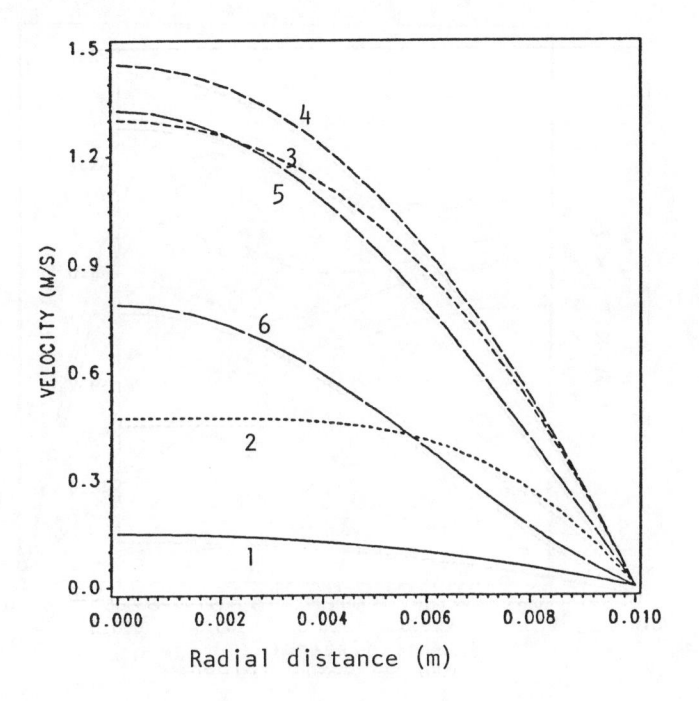

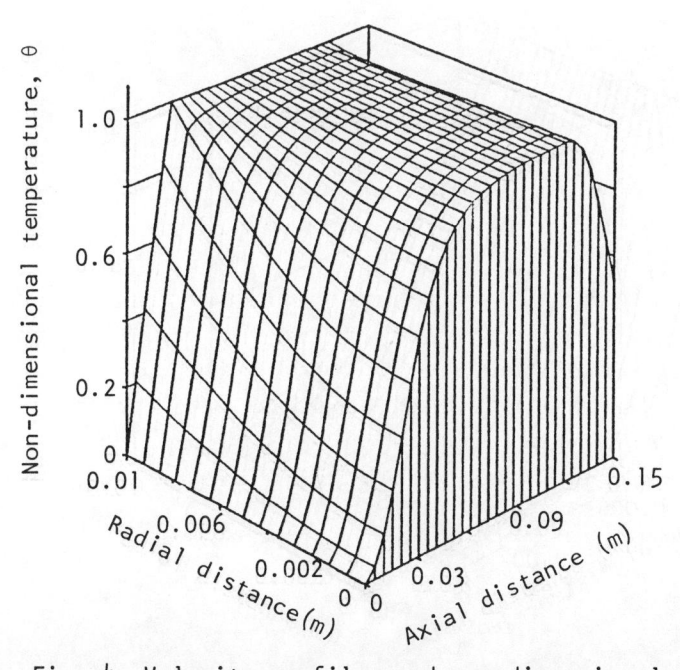

Fig. 4 Velocity profiles and non-dimensional temperature profile for $R_e=16$, $P_r=0.35$, $T_{wall_{max}}=3000K$. $\theta=(T_{max}-T)/(T_{max}-T_{in})$

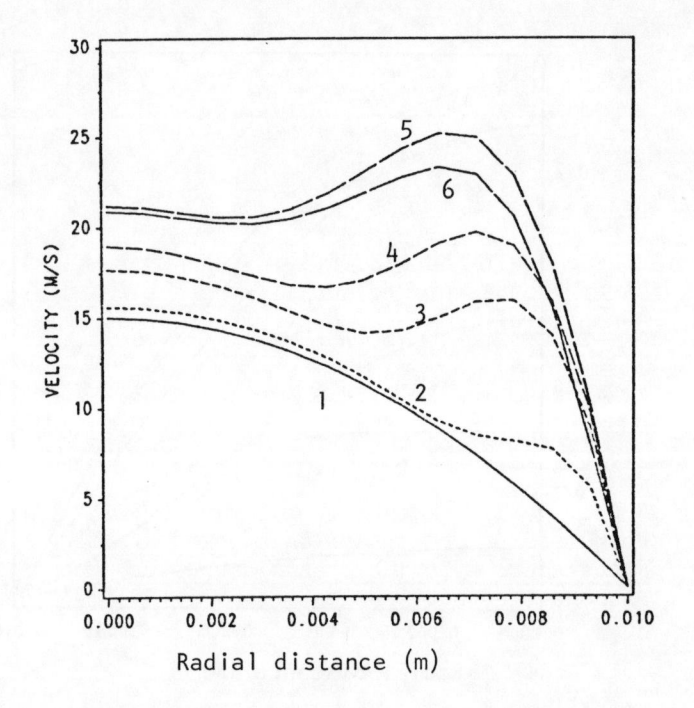

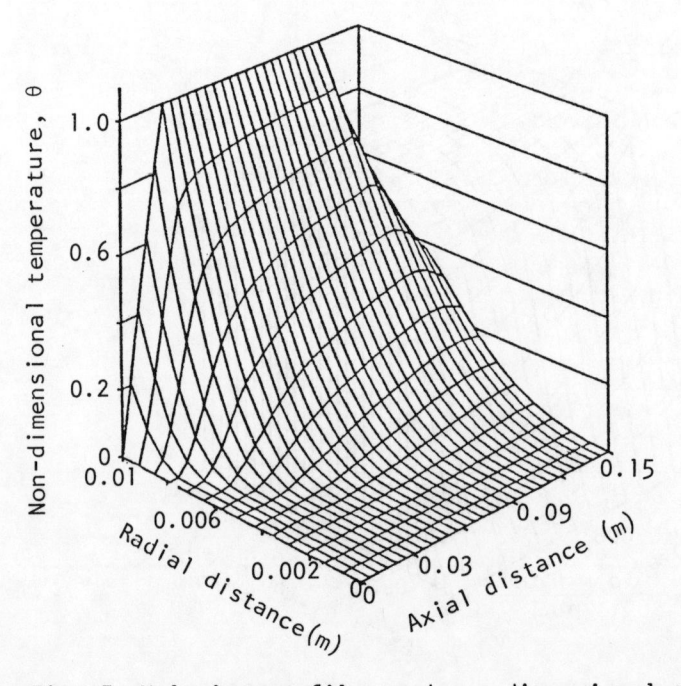

Fig. 5 Velocity profiles and non-dimensional temperature profile for R_e=1600, P_r=0.35, $T_{wall_{max}}$=3000K. $\theta=(T_{max}-T)/(T_{max}-T_{in})$

5. ACKNOWLEDGEMENT

This work was sponsored by the Air Force Rocket Propulsion Laboratory under the direction of Mr. Gerald Naujokas, Contract Monitor.

6. REFERENCES

    1. JONES, L.W., and KEEFER, D.R., "NASA's Laser-Propulsion Project", Astronautics & Aeronautics, Vol. 20, Sept. 1982, p. 66.

    2. KEMP, N.H. and ROOT, R.G., "Analytical Study of Laser-Supported Combustion Waves in Hydrogen", Journal of Energy, Vol. 3, Jan. 1979, pp. 40-49.

    3. SHOJI, J.M., "Potential of Advanced Solar Thermal Propulsion", Orbit-Raising and Maneuvering Propulsion: Research Status and Needs, Progress in Astronautics and Aeronautics, Vol. 89, 1984, pp. 30-47.

    4. LARSON, C.W., "The Plasma Spectroscopy Cell", presented at the Solar Plasma Propulsion Workshop, Bergamo Center, Dayton Ohio, Jan. 1986.

    5. WELLE, R., KEEFER, D., and PETERS, C., "Energy Conversion Efficiency in High-Flow Laser-Sustained Argon Plasmas", AIAA Paper No. 86-1077, AIAA/ASME 4th Fluid Mechanics, Plasma Dynamics and Lasers Conference, May 12-14, 1986, Atlanta, GA.

    6. MATTICK, A.T., "Absorption of Solar Radiation by Alkali Vapors", Radiation Energy Conversion in Space, Progress in Astronautics and Aeronautics, Vol. 61, 1978, pp. 159-171.

    7. MERKLE, C.L., and CHOI, Y.-H., "Computation of Compressible Flows at Very Low Mach Numbers", to appear in AIAA Journal, May 1987.

    8. MERKLE, C.L., and CHOI, Y.-H., "Computation of Low Mach Number Flows with Buoyancy", Proceedings of the 10th International Conference on Numerical Methods in Fluid Dynamics, Beijing, China, June 23-27, 1986, pp. 169-173.

HEAT TRANSFER IN FORCED CONVECTION BOUNDARY LAYER FLOWS OF VISCOINELASTIC FLUIDS

N. L. Kalthia,
Deptt.of Mathematics,
S.V.Regional College of
Engg.& Tech.,Surat-7 INDIA

M. G. Timol
Deptt.of Mathematics,
B.K.M. Science College
Bulsar, INDIA

ABSTRACT

Heat transfer in forced convection flow of non- Newtonian viscoinelastic fluids past a wedge is a case study in this present work. The velocity distribution and temperature variation for three different fluid models, namely, Powell-Eyring, Prandtl-Eyring and Williamson are studied through the extended form of numerical technique, satisfaction of asymptotic boundary condition method due to Nachsteims et al. The mutual comparison of these three different fluid models is also discussed with their graphical representation. Properties like skin friction, velocity and temperature distribution for all different fluids are also compared. It is hoped that this numerical study represents a wide-cross section of these fluids encountered in the chemical process industries.

1.INTRODUCTION

Fluids are encountered at each step of life, some fluids are Newtonian in behaviour other come under non-Newtonian type. In non-Newtonian fluids shearing stress bears non-linear relation to the rate of strain. These type of fluids found to be of great commercial importance. They are handled extensively by chemical industries, namely Plastics and Polymers, Biological and Biomedical devices like homodialyser make use of the rehological aspects of non-Newtonian fluids. The wide usage and application of these fluids in various industrial fields have prompted modern researchers to explore extensively in the field of non-Newtonian fluids.

The non-Newtonian fluids are generally subdivided into two categories : viscoelastic fluids and viscoinelastic fluids. Viscoelastic fluids are those in which stress tensor is related to both the instantaneous strain rate component and the past strain history, whereas in viscoinelastic fluids the stress tensor depends upon the current state of strain only.

There are several types of non-Newtonian fluid models proposed by the researchers $\underline{/}1,2,3\underline{/}$. Among them the non-Newtonian viscoinelastic fluids are most versatile and they are correctly reduced to Newtonian behaviour for low and high shear rates. The current research on the boundary layer flows of non-Newtonian fluids treat these type of models almost extensively.

In this paper similarity solutions for momentum and heat transfer of different non-Newtonian viscoinelastic fluids about a right angle wedge is obtained. A numerical solution is obtained for forced convection flow of Powell-Eyring, Prandtl-Eyring and Williamson fluids past a right angle wedge with an isothermal surface.

2. FORMULATION OF THE PROBLEM

The following normal assumptions are made in this paper :

(i) The fluid is assumed incompressible and viscoinelastic.
(ii) The flow is two-dimensional and steady.
(iii) The constant pressure specific heat C_p is assumed to be constant with respect to temperature or velocity changes.
(iv) The fluids under consideration are of the form in which shearing stress must be related to the rate of strain by an arbitrary continuous function of the type.

$$F(\tau_{ij}, e_{ij}) = 0 \qquad \ldots(2.1)$$

Under usual boundary layer assumptions, the dimensionless basic partial differential equations for low velocity forced convection wedge flow of non-Newtonian fluids with stream function ψ can be derived as follows $\underline{/}3\underline{/}$:

$$\frac{\partial \psi}{\partial y} \frac{\partial^2 \psi}{\partial x \partial y} - \frac{\partial \psi}{\partial x} \frac{\partial^2 \psi}{\partial y^2} = \frac{\partial}{\partial y}(\tau_{yx}) + U \frac{dU}{dx} \qquad \ldots(2.2)$$

$$\frac{\partial \psi}{\partial y} \frac{\partial \theta}{\partial x} - \frac{\partial \psi}{\partial x} \frac{\partial \theta}{\partial y} = \frac{1}{P_r} \frac{\partial^2 \theta}{\partial y^2} \qquad \ldots(2.3)$$

with the stress-strain relationship

$$F(\tau_{yx}, \frac{\partial^2 \psi}{\partial y^2}) \qquad \ldots(2.4)$$

Subject to the boundary conditions,

$$y = 0 \Rightarrow \frac{\partial \psi}{\partial y} = 0, \quad \frac{\partial \psi}{\partial x} = 0; \quad \theta = 0 \qquad ,\ldots(2.5a)$$

$$y \to \infty \Rightarrow \frac{\partial \psi}{\partial y} \to U(x) \; ; \; \text{and} \; \theta \to \theta_w(x) \quad \ldots (2.5b)$$

where non-dimensional quantities used are:

$$u = \frac{u'}{U_o} \; ; \; v = \frac{v'}{U_o} \sqrt{Re} \; ; \; x = \frac{x'}{L} \; ; \; y = \frac{y'}{L} \sqrt{Re}$$
$$\ldots (2.6)$$
$$\tau_{yx} = \frac{\tau'_{y'x'}}{\rho U_o^2} \sqrt{Re} \; ; \; \psi = \frac{\psi'}{U_o L} \; ; \; U = \frac{U'}{U_o}$$

with $P_r = \frac{\mu C_p}{K}$ (Prandtl number) and $Re = \frac{\rho U_o L}{\mu}$ (Reynolds number).

Here μ - viscosity of fluids, ρ = density of fluid, K - heat conductivity, U - free stream velocity, C_p - constant pressure specific heat, θ_w - temperature of fluid at the wall.

System (2.2)-(2.5a,b) can be reduced to similarity form by the transformation [4]:

$$\eta = x^{\frac{m+1}{2}} F(\eta) \; ; \; \theta = x^k G(\eta) \; , \; \tau_{yx} = x^{\frac{3m-1}{2}} H(\eta) \quad \ldots (2.7)$$
$$U = e x^m \; \text{and} \; \eta = y x^{\frac{m-1}{2}}$$

where m, k, e are arbitrary constants, η is similarity independent variable and $G(\eta)$, $H(\eta)$, $F(\eta)$ are similarity dependent variables. Using (2.7), equations (2.2)-(2.4) become

$$m F'^2 - \left(\frac{m-1}{2}\right) F F'' = H'(\eta) + m \quad \ldots (2.8)$$

$$k F'G - \left(\frac{1+m}{2}\right) FG' = \frac{1}{P_r} G'' \quad \ldots (2.9)$$

$$F\left[x^{\frac{3m-1}{2}} H(\eta) \; ; \; x^{\frac{3m-1}{2}} F''(\eta)\right] = 0 \quad \ldots (2.10)$$

with (2.5 a,b) giving the boundary conditions,

$$\eta = 0 \Rightarrow F = 0, \; F' = 0, G = 0$$
$$\eta \to \infty \Rightarrow F' = 1 \; \text{and} \; F = 1 \quad \ldots (2.11 \; a,b)$$

(here dashes denote differentiation with respect to η).

For present flow situation similarity solution exist it,

the free stream velocity outside the boundary layer is varying with the powerlaw coordinate of x i.e.

$$U(x) = x^m \quad \ldots(2.12)$$

and temperature at the surface of the wedge should be of the form :

$$\Theta_w = x^k \quad \ldots(2.13)$$

It is interesting to note that there exist certain physical situations for three different values of k, viz. k=0, 1/3 ; 2/3 respectively corresponds to the isothermal wedge flow; constant heat flux at the edge of boundary layer and heat generating due to high velocity flows. All these cases are treated here extensively.

The similarity equations (2.8)-(2.11 a,b) are more general than those obtained by Lee et al [3].

In case of m=1/3 i.e. for special type of free stream velocity, similarity transformations (2.7) together with similarity equations (2.8)-(2.10) will reduce to :

$$\psi = x^{2/3} F(\eta) \; ; \; \tau_{yx} = H(\eta) \; ; \; \Theta = G(\eta) x^k$$
$$U = e\, x^{1/3} \quad \eta = y/x^{1/3} \quad \ldots(2.14)$$

$$F'^2 - 2FF'' = 3H + 1 \quad \ldots(2.15)$$

$$G'' + 3P_r(FG' - 3kF'G) = 0 \quad \ldots(2.16)$$

$$f(H, F'') = 0 \quad \ldots(2.17)$$

with the same boundary conditions (2.11 a,b).

Similarity transformations (2.14) can be applied to all different non-Newtonian viscoinelastic fluids (given in Table-2.1).

3. ANALYSIS FOR POWELL-EYRING, PRANDTL-EYRING AND WILLIAMSON FLUIDS

Among the different models proposed so far in Table-2.1 it would be interesting to study theoretically the fluid properties of these three different models and to compare them.

3.1 Powell-Eyring Fluid :

The study of Powell-Eyring fluid is preferred because it can be deduced from the Kinetic theory of liquids rather than imparical relation. Moreover it correctly reduces to Newtonian behaviour for low and high shear rates though this fluid model

TABLE-2.1

No.	Model	Stress vs. rate of strain
1	Prandtl	$\tau'_{y'x'} = A\sin^{-1}\left(\dfrac{1}{C}\dfrac{\partial u'}{\partial y'}\right)$
2	Powell-Eyring	$\tau'_{y'x'} = \mu\dfrac{\partial u'}{\partial y'} + \dfrac{1}{B}\sinh^{-1}\left(\dfrac{1}{C}\dfrac{\partial u'}{\partial y'}\right)$
3	Prandtl-Eyring	$\tau'_{y'x'} = A\sinh^{-1}\left(\dfrac{1}{B}\dfrac{\partial u'}{\partial y'}\right)$
4	Power-law	$\tau'_{y'x'} = m\left|\dfrac{\partial u'}{\partial y'}\right|^{n-1}\dfrac{\partial u'}{\partial y'}$
5	Sisco	$\tau'_{y'x'} = A\dfrac{\partial u'}{\partial y'} + B\left|\dfrac{\partial u'}{\partial y'}\right|^{n}$
6	Williamson	$\tau'_{y'x'} = \left[\dfrac{A}{B+\dfrac{\partial u'}{\partial y'}} + \mu_{\infty}\right]\dfrac{\partial u'}{\partial y'}$
7	Ellis	$\tau'_{y'x'} = \dfrac{\partial u'/\partial y'}{A + B\left|\tau'_{y'x'}\right|^{\alpha-1}}$
8	Sutterby	$\tau'_{y'x'} = \mu_{0}\left[\dfrac{\sinh^{-1}(B\partial u'/\partial y')}{B\partial u'/\partial y'}\right]^{A}\dfrac{\partial u'}{\partial y'}$
9	Reiner-Philippoff	$\tau'_{y'x'} = \left[\mu_{\infty} + \dfrac{\mu_{0}-\mu_{\infty}}{1+(\tau'_{y'x'}/\tau'_{0})^{2}}\right]\dfrac{\partial u'}{\partial y'}$

is mathematically more complex.

Using stream function Ψ in Powell-Eyring model [Model no. 2 Table-2.1], and also introducing non-dimensional quantities (2.6), similarity variables (2.14) and substituting resulting form of $H'(\eta)$ in the right hand side of equation (2.15), yields

$$F''' = \frac{(1 + \beta' F''^2)^{1/2} (F'^2 - 2FF'' - 1)}{(1 + \beta' F''^2)^{1/2} + \frac{1}{\alpha'}} \quad \ldots(3.1)$$

($\alpha' = \mu BC$; $\beta' = \dfrac{\rho U_o^3}{3C^2 L \mu}$ are non-dimensional number)

together with same energy equation (2.16) and boundary conditions (2.11 a,b).

3.2 Prandtl-Eyring Fluid :

Again after proper simplification, substituting Prandtl-Eyring fluid model with two constant (Model no.3 Table-2.1) in right hand side of equation (2.15) yields :

$$F''' = \frac{1}{\alpha'}(F'^2 - 2FF'' - 1)(1 + \beta' F'''^2)^{\frac{1}{2}} \quad \ldots(3.2)$$

($\alpha' = A/\mu B$ and $\beta' = \dfrac{\rho U_o^3}{3B^2 L \mu}$ are non-dimensional numbers)

together with same energy equation (2.16) and boundary conditions (2.11 a,b).

3.3 Williamson Fluid :

The purely viscous non-Newtonian behaviour in shear rate range is not accurately described by Power law model which has two constants, but by Williamson model which has three constants. Also it is interesting to note that for A = 0 Williamson model (Model no.6, Table-2.1) will reduce to Newtonian fluid as a particular case. Following same procedure as described above we shall obtain here

$$F''' = \frac{(F'^2 - 2FF'' - 1)(1 + \sqrt{\beta'} F'')^2}{\alpha' + (1 + \sqrt{\beta'} F'')^2} \quad \ldots(3.3)$$

(wherein $\alpha' = \dfrac{A}{\mu}$ $\beta' = \dfrac{\rho U_o^3}{3B^2 L \mu}$ are non-dimensional numbers) together with same energy equation (2.16) and boundary conditions (2.11 a,b).

4. NUMERICAL SOLUTION :

Equations (3.1),(3.2) and (3.3) with common energy equation (2.16) are non-coupled ordinary non-linear differential equations. The numerical solutions obtained for these equations is based on the satisfaction of asymptotic boundary conditions at the edge of boundary layer, due to Nachtsheim et al [5]. In this method the initial conditions are so adjusted that the mean square error between the computed variables and the asymptotic values is minimum. This method is well applicable to study the relationship of properties of solution with the parameter-variation.

4.1 Solution of Powell-Eyring Fluid :

The equation (3.1) governing the motion of Powell-Eyring fluid flow is solved first. The overall procedure for solution is briefly described below.

The asymptotic boundary condition states that $F' \to 1$ as $\eta \to \infty$. For practical purposes this can be put as $F' = 1$ at $\eta = \eta_{edge}$ where η_{edge} is the value of η at the edge of the boundary layer. The boundary value problem is equivalent to the problem of finding

$$F'_{edge}(x) = 1 \quad \ldots\ldots(4.1)$$

where $x = F''(o)$.

Expanding left hand side to the first approximation gives

$$F' + \dfrac{\partial F'}{\partial x}\Delta x = 1 \quad \ldots\ldots(4.2)$$

Differentiating (3.1) with respect to x, yields :

$$F'''_x = \dfrac{(F'^2 - 2FF'' - 1 - F''')\left[2(1+\beta'F''^2)(F'F'_x - FF''_x - F_x F'') + (F'^2 - 2FF'' - f - F''')(\beta'F''F''_x)\right]}{(\dfrac{1}{\alpha'})^2 F''' + (1+\beta'F''^2)(F'^2 - 2FF'' - 1 - F''')}$$

$$\ldots\ldots(4.3)$$

where subscript x denotes differentiation with respect to x

The initial conditions become,

at $\eta = 0 \Rightarrow F_x = F'_x = 0$; $F'''_x = 1$ .....(4.4)

Assuming a value for $x = F'(o)$, equation (3.1) and (4.3) are integrated. The integration is stopped at a point wherein the asymptotic boundary conditions are satisfied i.e. both $F' = 1$ and $F'' = 0$ are to be satisfied for some value of η_{edge} correction x, to the initial value of x is so taken that the equation (4.2) and equation

$$F'' + F'''_x \Delta x = 0 \quad \ldots (4.5)$$

are satisfied at $\eta = \eta_{edge}$.

Since there are two equations with one unknown, the least square method is applied for finding Δx. Thus

$$\Delta x = \frac{F'_x(1-F') - F'''_x F''}{F'^2_x + F'''^2_x} \quad \ldots (4.6)$$

The error E, between asymptotic conditions and the computed values at $\eta = \eta_{edge}$ is given by

$$E = (1 - F')^2 + F''^2 \quad \ldots (4.7)$$

Finally, once minimum E is attained, no more change in the intial conditions is required and this ensures the satisfaction of asymptotic boundary condition at a finite value of η.

In order to find variation of $F''(o)$ with respect to $1/\alpha'$ and β' respectively, exactly similar procedure is followed.

Further using the values of F and F', equation (2.16) is solved for different values of $G(o)$ using following perturbation equation :

$$G''_x = \frac{P_r}{3} \left[K(F'G_x + F'_x G) - 2(FG'_x + F_x G') \right] \quad \ldots (4.8)$$

5. SOLUTION SCHEME :

A general programme in Fortran 77, for VAX.11/980 is developed for the solution of forced convection flow past $90°$ wedge in the case of any non-Newtonian fluid model. In the present algorithm, equation (2.15) is solved first for F, F', F_x and F'_x and stored for their respective values of η and these values are used in solution of equation (2.16), which remains same for all fluid models.

Three fluid models are considered here. To include any other model from table-2.1, the differential equation governing the flow of such fluid model can be introduced in separate sub-routines.

A modified version of subroutine is used here for integrating the functions, following Runga-Kutta scheme for the first three increments of step size 0.5 and for further increments using Adams-Moulton scheme, modification are introduced to suit the algorithm of the problem.

6. RESULTS AND DISCUSSIONS :

The dimensionless velocity profiles for forced convection flow of Powell-Eyring, Prandtl-Eyring and Williamson fluids is presented in Fig.(6.1) for one set of value $1/\alpha' = 0.1$ and $\beta' = 5 \times 10^3$ and corresponding numerical data is given in table 6.1. The numerical comparison between the results of these three different fluids shows that the Powell-Eyring fluid is faster than Prandtl-Eyring and Williamson fluids.

The slope of velocity profile $F''(o)$ on the surface of the forced convection flow past $90°$ wedge is calculated for $1/\alpha' =0.1$ and $\beta' = 5 \times 10^3$. It is interesting to notice that in case of Powell-Eyring and Williamspn fluids the value initial slope are 1.3102 and 1.2999 respectively. These values are slightly smaller than that of Newtonian fluids which has a value of 1.3120 [6] whereas in case of Prandtl-Eyring fluid this value is very large i.e. 5.6560. This shows that Prandtl-Eyring fluid reduces to Newtonian behaviour for low and high shear rates.

In figures (6.2)(6.3) and (6.4), temperature distribution G for Powell-Eyring, Prandtl-Eyring and Williamson fluids is given for one set of value $P_r=0.71$, $1/\alpha'=0.1$ and $\beta'=5 \times 10^3$ for K=0, 1/3, 3/3 respectively. It is observed that the temperature of all fluids decreases faster than that of (i) flow with constant heat flux (ii) high speed velocity flows.

In case of flow past an isothermal surface, the temperature of Prandtl-Eyring fluid increases comparatively faster than Powell-Eyring fluid and Williamson fluid. But there is somewhat reverse case when heat is generating due to high velocity flows i.e. for K=2/3. In this case temperature variations in Powell-Eyring fluid is considerably faster than those of Prandtl-Eyring fluid and Williamson fluid.

Finally in case of constant heat flux at the edge of boundary layer (i.e.for K=1/3 case) temperature of Prandtl-Eyring fluid is slower than the other two fluids.

TABLE-6.1

Values of F' for forced convection Powell-Eyring, Prandtl-Eyring and Williamson fluids (for $1/\alpha'=0.1$ $\beta'=5\times10$).

η	Powell-Eyring fluid	F' Prandtl-Eyring fluid	Williamson fluid
0.50	0.5308	0.8395	0.5257
1.00	0.8288	0.9286	0.8187
2.00	0.9912	0.9757	0.9768
4.00	0.9997	0.9966	0.9977
6.00	1.0000	0.9997	0.9997
8.00	-	0.9999	0.9999
10.00	-	1.0000	1.0000

TABLE-6.2

Values of G for Powell-Eyring, Prandtl-Eyring and Williamson fluids (for $P_r = 0.71$; $1/\alpha'=0.1$; $\beta'=5\times10$ and $K=0$).

η	Powell-Eyring fluid	G Prandtl-Eyring fluid	Williamson fluid
1.00	0.4261	0.4681	0.4241
2.00	0.7706	0.8081	0.7680
4.00	00.9876	0.9916	0.9871
6.00	0.9998	0.9999	0.9998
8.00	0.9999	1.0000	1.0000
10.00	1.0000	-	-

TABLE-6.3

Values of G for Powell-Eyring, Prandtl-Eyring and Williamson fluids (for $P_r=0.71$, $1/\alpha'=0.1$; $\beta'=5\times10^3$, $K=1/3$).

η	Powell-Eyring fluid	G Prandtl-Eyring fluid	Williamson fluid
1.00	0.1200	0.09671	0.0846
5.00	0.4701	0.3425	0.3331
10.00	0.6945	0.4923	0.4874
15.00	0.8603	0.6073	0.6044
20.00	0.9738	0.7034	0.7018
25.00	1.0000	0.7877	0.7671
30.00	-	0.8638	0.8639
35.00	-	0.9337	0.9345

TABLE - 6.4

Values of G for Powell-Eyring, Prandtl-Eyring and Williamson fluids (for P_r=0.71, $1/\alpha'$=0.1, β'=5x10^3 K=2/3).

η	Powell-Eyring fluid	Prandtl-Eyring fluid	Williamson fluid
1.00	0.0369	0.0216	0.0182
5.00	0.2275	0.12077	0.1135
10.00	0.4843	0.2443	0.2371
15.00	0.7473	0.3705	0.3634
20.00	0.9477	0.4965	0.4894
25.00	1.0000	0.6224	0.6153
30.00	-	0.7482	0.7411
35.00	-	0.8741	0.8670

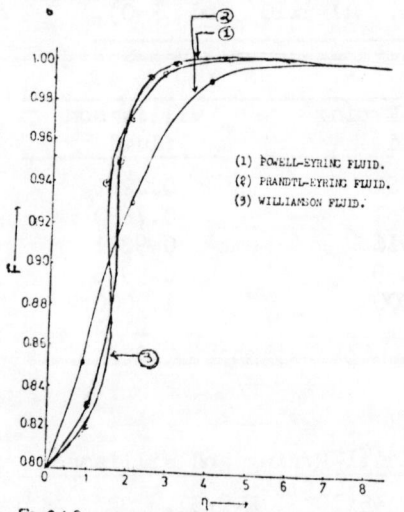

Fig. 6.1 Dimensionless velocity distribution for $1/\alpha$=0.1 and β=5×10^3

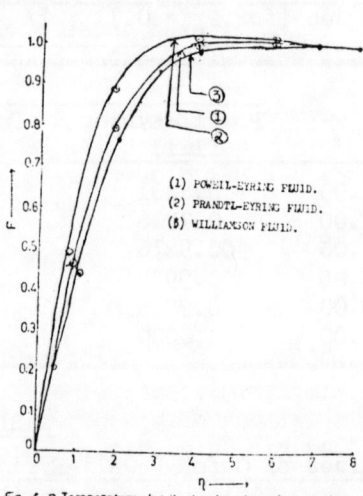

Fig. 6.2 Temperature distribution for forced covection wedge flow for Pr=0.71, $1/\alpha'$=0.1, β'=5×10^3, K=0

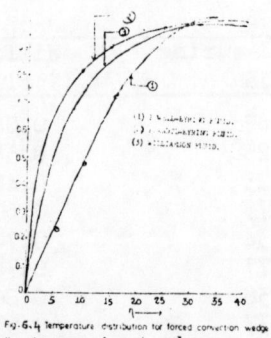

Fig. 6.4 Temperature distribution for forced convection wedge flow for Pr=0.71, $1/\alpha'$=0.1, β'=5×10^3, K=2/3

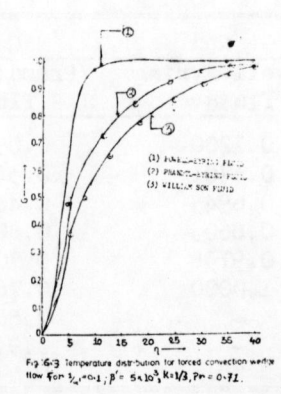

Fig. 6.3 Temperature distribution for forced convection wedge flow for $1/\alpha'$=0.1, β'=5×10^3, K=1/3, Pr=0.71

7. CONCLUSIONS :

We have developed a numerical method for the treatment of non-linear differential equation of higher order. Further appropriate general computer software is also developed for the study of the flow characteristics and heat transfer properties of different non-Newtonian viscoinelastic fluid models given in Table-2.1. The models considered are for Powell-Eyring, Prandtl-Eyring and Williamson fluids and it is hoped that these models represent a wide cross-section of fluids encountered in many process industries. Therefore this study provides valuable data alongwith the procedures developed for the numerical analysis of these fluids.

Acknowledgement :

We would like to express our thanks to Dr.Mrs. V. Sirohi for helping us in computer work. Constant encouragement from the Principal, S.V.Regional College of Engineering & Technology Surat-7 and financial support from University Grants Commission New Delhi granted under No.F.8-10/84(SR III) are also gratefully acknowledged.

References :

1. Bird,R.B., Stewart,W.E.,and Lightfoot, "Transport Phenomena" John Wiley and Sons Inc.,1960.
2. Skelland,A.H.P.,"Non-Newtonian Flow and Heat Transfer", John Wiley and Sons Inc.,P.8,1967.
3. Lee,S.Y. and Ames,W.P.,"Similarity Solutions for non-Newtonian Fluids",A.I.Ch.E.J 12, P.700,1966.
4. Timol,M.G.,"Group theoretic Approach to the Similarity Solutions of Viscoinelastic Flows", Doctoral Thesis, South Gujarat University,Surat,India,1986.
5. Nachtsheim,P.R. and Swigert,P.,"Satisfaction of Asymptotic Boundary Conditions in Numerical Solutions of Non-Linear equations of Boundary Layer Type",NASA TN D-3004.
6. Hansen,A.G. and Na,T.Y.,"Similarity Solutions of Laminar Incompressible Boundary Layer Equations of non-Newtonian Fluids", Trans. ASME, J.Basic Engn,90,P.71,1968.

NUMERICAL STUDY IN HEAT TRANSFER OF NON-NEWTONIAN FLOWS PAST A VERTICAL PLATE

M. G. TIMOL,
Department of Mathematics,
B.K.M. Science College,
Tithal Road,
VALSAD-396001   India

N. L. KALTHIA,
Department of Mathematics,
S.V. Regional College of
Engineering & Technology,
SURAT-395007   India

SUMMARY :

    The present work is expected as a contribution in obtaining numerical solutions for the problems of heat transfer. The importance of non-Newtonian fluids such as molton plastics (Williamson fluids), emulsions, slurries, polymer pulps (Power-law fluids), calcium acid, hydrophobic sillica, lithium hydroxy sterate, sodium tellow (Sisco fluids), Solutions of polymer oxide, carboxymethyl cellulose (Sutterby fluids, Powell-Eyring fluids) etc. in everyday chemical engineering practices has motivated many investigators to analyze the behaviour of these fluids in motion.

    The problem considered in the present paper is that of free convection boundary layer flows of general non-Newtonian fluids on a vertical infinite flat plate when a constant plate temperature is not attained directly at the leading edge but at finite distance along the plate. We have transformed the coupled non-linear partial differential equations governing the flow to the similarity equations. The novel spline collocation method combined with linearization techniques is developed to solve coupled non-linear similarity equations. This linearization in spline method is discussed giving the anology between finite difference method. The temperature and velocity distributions for two different fluids (Powell-Eyring and Power-law) for different values of parameters are presented in tabular as well as graphical form. Significant advantage is evident by the use of present numerical method, as the calculations do not consume much labour and computational time.

    Over and above industrial applications, the present problem has an interesting features like non-linearities in coupled partial differential equations. Moreover the interest in spline method is justified due to its simplicity though this method gives a bit slow convergence in iterative approximations.

1. INTRODUCTION :

Recently under the impact of both academic cariority and practical necessity, considerable effort has been extended to investigate the non-Newtonian effects. The class of solution known as similarity solution plays an important role to find out such effect especially in the boundary layer flows. The similarity method involves the determination of similarity parameters which reduce the system of partial differential equations to ordinary differential equations. In addition to the fact that it is only the class of exact solution for the governing differential equations. It serves also as a reference point to check the approximate solutions. However these solutions have its limitations that it exists for the limited cases of velocity at the edge of boundary layer only.

Soundalgekar et al[1] have studied the flow past an accelerated vertical infinite plate in the presence of free convection currents. They have considered the flow over uniformly accelerated vertical infinite plate in the presence of pure natural circulation. Also using similarity transformations they have solved their governing ordinary differential equations by variation of parameter technique. But their entire problem was limited to the Newtonian fluids with uniform velocity only.

In this paper we shall cover here non-Newtonian fluids of all possible models characterized by the property that tensor component of its shearing stress τ_{ij} can be related to the rate of strain component e_{ij} by an arbitrary continuous function of the type

$$F(\tau_{ij}, e_{ij}) = 0 \qquad \ldots\ldots(1.1)$$

The consideration of non-uniformly accelerated plate relaxes the restriction on mainstram velocity $U(t)$ which ultimately leads to the class of solution. In order to obtain the similarity solutions the new group theoretic generalized dimensional analysis is extended to applied.

Numerical solution is obtained for viscous Newtonian fluids, power-law non-Newtonian fluids and Powel-Eyring fluids for different flow indicies and of parameters and as well as for different Prandtl Greshor numbers by spline collocation method.

2. GROUP THEORETIC GENERALIZED DIMENSIONAL ANALYSIS :

The group theoretic method is wide applicable and well accepted technique to find the similarity solutions in many physical situations. The similarity method involves the determination of similarity variables which reduces the given system into ordinary differential equation. The foundation of group theoretic method is contained in general theories of continuous transformation group, that were introduced and treated extensively by Lie [2] in the latter part of the last century. However later on numerous contribution have been made in this area by Morgan [3]. Bluman et al [4] etc. Moran and his co-worker [5,6] have applied group theory to dimensional analysis and develop exclusively a generalization of dimensional analysis method. Recently Timol et al [7] have extended theory of Moran et al [6] for non-Newtonian fluids.

In the next section application of group theoretic generalized dimensional analysis has been presented followed by discussion.

3. MATHEMATICAL ANALYSIS :

Considering laminar, two-dimensional incompressible boundary layer equations with a cartesian coordinate system, we take x' axis is taken normal to it. If T_w and T be the temperature of the plate and the fluid far away from the plate, then the boundary layer equations governing the flow and heat transfer in dimensionless form are given by,

$$\frac{\partial u}{\partial t} = \frac{\partial}{\partial y}(\tau_{yx}) + G\theta \qquad \ldots\ldots(3.1)$$

$$\frac{\partial \theta}{\partial t} = \frac{1}{P_r}\frac{\partial^2 \theta}{\partial y^2} \qquad \ldots\ldots(3.2)$$

and under the boundary layer assumption the stress-strain relationship (1.1) will be

$$F(\tau_{yx} ; \frac{\partial u}{\partial y}) = 0 \qquad \ldots\ldots(3.3)$$

The boundary conditions are

$y = 0 \Rightarrow u = 0 ; T_w = T \to T$ for $t < 0$ $\ldots\ldots(3.4)$

$y = 0 \Rightarrow u = U(t) ; T = T_w$ $\ldots\ldots(3.5a)$

$y = \infty \Rightarrow u = 0 ; T \to T$ for $t > 0$ $\ldots\ldots(3.5b)$

wherein u - velocity component in x direction, y = rectangular coordinate, τ_{yx} = component of shearing stress. F = arbitrary function, t = dimensionless time G = Gresh of number, Pr = Prandtl number, θ = dimensionless temperature, U(t) = dimensionless free stream velocity.

With dimensionless quantity used are

$$u = \frac{u'}{U_o} \; ; \; y = \frac{y'}{L} \; ; \; \tau_{yx} = \frac{\tau'_{y'x'}}{U_o^2} \; ; \; t = \frac{t'U_o}{L} \; ;$$

$$\theta = \frac{\theta'}{T'_w - T'_\infty} = \frac{T' - T'_\infty}{T'_w - T'_\infty} \; , \; U = \frac{U'}{U_o} \; ; \; Pr \; \frac{\mu c p}{K}$$

$$Gr = \frac{L}{U_o^2} \, g\beta(T'_w - T'_\infty) \qquad \qquad \ldots\ldots(3.6)$$

where U_o = constant with the dimension of velocity; P = density of the fluid; K = thermalconductivity; Cp = specific heat at constant pressure; μ = viscosity; g = acceleration due to gravity.

The gole of reducing the system of equations (3.1)-(3.3) by group-theoretic generalized dimensional analysis will impose restrictions on the function $U(t)$ and $\theta(t)$ in such a way that boundary conditions can be transformed in to meaningful form.

This will lead to possible form of $U(t)$ and $\theta(t)$.

Introducing following group of transformation G_1 :

$$G_1 : \begin{cases} \bar{u} = A_1 u & \bar{\tau}_{yx} = A_4 \tau_{yx} & \bar{\theta} = A_5 \theta \quad \text{dependent variables} \\ \bar{y} = A_3 y & \bar{t} = A_2 t & \text{independent variables} \\ \bar{P} = A_6 p & \bar{U} = A_7 U & \text{physical variables} \end{cases}$$

where A_i's $(i = 1,\ldots,7)$ in G_1 are positive real parameters.

At a first step of similarity requirement A_i's of G_1 must be inter-related in order to systems (4.1)-(4.5 a,b) to be an invariant in form. In deed this requirement will be met under following three parameter group of transformations :

$$\Gamma_1 \begin{cases} \bar{y} = A_1^o A_2^o A_3^1 \, y \; ; & \bar{t} = A_1^o A_2^1 A_3^o \, t \\ \bar{P} = A_1^o A_2^1 A_3^{-2} P \; ; & \bar{U} = A_1^1 A_2^o A_3^o \, U \\ \bar{\tau}_{yx} = A_1^1 A_2^{-1} A_3^1 \, \tau_{yx} & \bar{u} = A_1^1 A_2^o A_3^o \, u \end{cases}$$

$$\bar{\theta} = A_1^1 A_2^{-1} A_3^o \, \theta$$

Now in order to apply theorem 3, a Pi-theorem stated by moran et al [6], the rank of dimensional matrix associated with independent and physical variables BC and physical variables C of Γ_1 is required to be determined.

The associated dimensional matrix will be

$$BC : \begin{bmatrix} 0 & 0 & 1 \\ 0 & 1 & 0 \\ \hdashline 0 & 1 & -2 \\ 1 & 0 & 0 \end{bmatrix} \quad \ldots\ldots(3.7)$$

By inspection, rank of $[BC]$ $r = 3$ and rank$[C] = s = 2$ and since $r \le S$ in the light of theorem – 3 [6] following set of pi's can be obtained :

$$\Pi_1 = u/U \; ; \; \Pi_2 = \frac{\theta t}{U} \; ; \; \Pi_3 = \frac{\tau_{yx}}{U}(t/P)^{\frac{1}{2}} \quad \text{and}$$
$$\hat{\pi} = y(P/t)^{\frac{1}{2}} \quad \ldots\ldots(3.8)$$

Clearly $\hat{\pi}$ is similarity independent variable whereas Π_1, Π_2 and Π_3 are similarity dependent variables in the usual notation.

$$\Pi_1 = f_1(\hat{\pi}) \; ; \; \Pi_2 = f_2(\hat{\pi}) \; ; \; \Pi_3 = f_3(\hat{\pi}) \quad \ldots\ldots(3.9)$$

Now for nonuniformly accelerated plate, if we consider $U(t) = t^k$ and put $\hat{\pi} = \eta$ then set of absolute invariants (3.8) will be

$$\eta = y(P/t)^{\frac{1}{2}} \; ; \; f_1(\eta) = u/t^k \; ; \; f_2(\eta) = \theta/t^{k-1} \; ;$$
$$f_3(\eta) = \frac{\tau_{yx}}{t^{k-1/2} p^{1/2}} \quad \ldots\ldots(3.10)$$

Under this set of transformations, system (3.1)–(3.5 a,b) will be reduced to the following set of ordinary differential equations :

$$K f_1 - 1/2 \, \eta \, f_1' = Pr \frac{d}{d\eta}(f_3) + G_{\varepsilon} f_2 \quad \ldots\ldots(3.11)$$
$$(K-1) f_2 - \tfrac{1}{2} \eta f_2' = f_2'' \quad \ldots\ldots(3.12)$$

where prime denotes differentiation with respect to η. The boundary conditions will cause the restriction on the wall temperature. It will be $\theta(t) = t^{k-1}$. Also

$$\eta = 0 \Rightarrow f_1 = 1 \; ; \; f_2 = 1 \quad \ldots\ldots(3.13a)$$
$$\eta = \infty \Rightarrow f_1 = 0 \; ; \; f_2 = 0 \quad \ldots\ldots(3.13b)$$

The stress-strain relationship reduces to

$$F(p^{\frac{1}{2}} f^{k-\frac{1}{2}} f_3 \; ; \; p^{\frac{1}{2}} t^{k-\frac{1}{2}} f_1') = 0 \quad \ldots(3.14)$$

Systems (3.11)-(3.13 a,b) suggest that similarity solutions of general Newtonian fluids exist for non-uniformly accelerated vertical plate with variable wall temperature. Clearly for K=1 i.e. $U(t) = t$ and $\theta = 1$ i.e. in case uniformly accelerated plate with constant wall temperature equations (3.11) and (3.12) with boundary conditions (3.13 a,b) will generate same equations obtained by Soundalgekar et al [1] for Newtonian fluids.

For $K=\frac{1}{2}$, the strain-stress relationship (3.14) will be free from independent variable. This will lead to an interesting case, of a right angle flow geometry. In such case, similarity solution for all non-Newtonian fluids will exist.

In this case Equations (3.11)-(3.12) will be

$$f_1 - \eta f_1' = 2F \frac{d}{d\eta}(f_3'') + 2G\, F_2 \quad \ldots(3.15)$$

$$f_2 + \eta f' + 2f_2'' = 0 \quad \ldots(3.16)$$

with the same boundary conditions as (3.13 a,b).

4. ANALYSIS FOR DIFFERENT MODELS :

4.1 Newtonian Case :

In case of general Newtonian fluids the stress-strain relationship (3.3) will come to

$$\tau_{y'x'}' = \mu \frac{\partial u'}{\partial y'} \quad \ldots(4.1.1)$$

Using non-dimensional quantities (3.6), transformations (3.10) in (4.1) substituting these in right hand side of equation (3.11), the systems (3.11)-(3.13 a,b) one gets

$$Kf_1 + \tfrac{1}{2}\eta f_1' = f_1'' + f_2 \quad \ldots(4.1.2)$$

$$(K-1)f_2 + \tfrac{1}{2}\eta f_2' + f_2'' = 0 \quad \ldots(4.1.3)$$

with the boundary conditions,

$$f(0) = 1 \; ; \; f(0) = 1 \quad \ldots(4.1.4a)$$
$$f(\infty) = 0 \; ; \; f(\infty) = 0 \quad \ldots(4.1.4b)$$

4.2 Non-Newtonian Power-Law Case :

Mathematically power-law model can be written as

$$\tau'_{y'x'} = m \left| \frac{\partial u'}{\partial y'} \right|^{n-1} \frac{\partial u'}{\partial y'} \quad \ldots(4.2.1)$$

where m and n are fluid consistancy indices. After non-dimensionalizing, this equation is substituted in equation (3.15) which brings basic equation as :

$$f_1 - \tfrac{1}{2}\eta f'_1 = nP_r^{\frac{n+1}{2}} (f'_1)^{n-1} f''_1 + Gr\, f_2 \quad \ldots(4.2.2)$$

$$f_2 + \tfrac{1}{2}\eta f'_2 + 2f''_2 = 0 \quad \ldots(4.2.3)$$

with the same boundary conditions (4.1.4 a,b).

4.3 Non-Newtonian Powell-Eyring Case :

Mathematically powell-Eyring model can be written as

$$\tau'_{y'x'} = \mu \frac{\partial u'}{\partial y'} + 1/B \sinh^{-1}\left(\tfrac{1}{c}\frac{\partial u'}{\partial y'}\right) \quad \ldots(4.3.1)$$

where μ, B, C are the parameter of fluids. Here we obtain

$$f_1 - \eta f'_1 = P_r f''_1 \left[1 + \frac{1}{\alpha'(1 + \beta' f'^2_1)^{\frac{1}{2}}} \right] + 2G_r f_2 \quad \ldots(4.3.2)$$

$$f_2 + \eta f'_2 + 2f''_2 = 0 \quad \ldots(4.3.3)$$

with the same boundary conditions (4.1.4 a,b) and

$$\alpha' = \mu BC \; ; \; \beta' = \frac{\rho U_o^3}{2C^2 L \mu} \quad \text{(non-dimensional quantities).}$$

5. NUMERICAL SOLUTION OF THE PROBLEM :

This method is based upon the theory of splines. The spline functions are piecewise polynominals and they were first used for interpolation in a data. A very smooth curve passing through the data points can be obtained and reliability of the approximation is observed very clearly. In 1969 Bickley [8] gave a series of expression of spline function which is useful in interpolating the solution of a linear differential equation

$$y''(x) + P(x)y'(x) + q(x)y(x) = t(x) \; ; \; a \leq x \leq b \quad \ldots(5.1)$$

with boundary conditions

$\alpha_0 y_0 + \beta_0 y_0' = \gamma_0$ at $x = a$

$\alpha_n y_n - \beta_n y_n' = \gamma_n$ at $x = b$

Ahlberg et al [9] were also involved in the study of spline functions. According to their definition, the spline functions are consisting their derivative. The linearity of the function is used to construct the spline function of desired degree. In 1961 Blue [10] suggested the use of spline functions as a solution of the boundary value problem (BVP)

$$y'' = f(x,y,y') \; ; \quad 0 \leq x \leq 1 \qquad \ldots(5.2)$$

with boundary conditions

$$G_1\bigl[y(0), y'(0)\bigr] = 0 \; ; \quad G_2\bigl[y(1), y'(1)\bigr] = 0$$

Equation (5.2) may be linear or non-linear. Here we shall confine our study regarding the cubic splines only. Let us divide the interval [0,1] into N equal subintervals. Now as $S(x)$ is cubic splines, $S''(x)$ is linear and for this reason it can be written as follows :

$$S''(x) = M_{i+1} \frac{x-x_i}{h} + M_i \frac{x-x_i}{h} \quad \text{for} \quad x_i \leq x \leq x_{i+1} \qquad \ldots(5.3)$$

where $M_i = S''(x_i)$, $M_{i+1} = S''(x_{i+1})$ and h is a length of the subinterval (x_i, x_{i+1}).

The cubic spline $S(x)$ is an interpolant to $y(x)$ of equation (5.2) and hence $S(x_i) = y(x_i)$. With these expressions and two successive integrations of $S''(x)$, the cubic $S(x)$ is obtained as :

$$S(x) = M_{i+1} \frac{(x-x_i)^3}{6h} + M_i \frac{(x_{i+1}-x)^3}{6h} + (y_i - \frac{h^2}{6} M_i) \frac{(x-x_i)}{h}$$

$$+ (y_{i+1} - \frac{h^2}{6} M_{i+1}) \frac{(x_{i+1} - x)}{h} \qquad \ldots(5.4)$$

where $i = 1,2,\ldots,N$.

The continuity of $S'(x)$ at $x=x_i$ yields following recurrence relation :

$$M_{i+1} + 4M_i + M_{i-1} = 6/h^2 (y_{i+1} - 2y_i + y_{i-1}) \qquad \ldots(5.5)$$

The equation (5.5) along with BCs yield a system of algebraic equations whose coefficient matrix is a tridiagonal matrix. Thus numerical procedure becomes easier. The convergence of the solutions depends upon the variations of BCs. Namely, if the functions are specified at end points then the convergence becomes faster than the other cases where the derivatives at either or both ends are given. However, the linearization of equation (5.5) gives faster convergence. This linearization of equation (5.5) is much similar to the finite differences expression.

The numerical solution of the system of equations (4.1.2)-(4.1.4 a,b), (4.2.2)-(4.2.3) and (4.3.2)-(4.3.3) is obtained through this cubic spline collocation method. The spline solutions converge to the numerical values of this closed form solution and fairly represent the actual flow behaviour curve (see Tables 1,2 and 3)

Table-1 : Values of f-Power-law case.

η/f	Pr=7 and Gr=10				
	n=0.2	n=0.4	n=0.6	n=0.8	n=1
0.00	1.0000	1.0000	1.0000	1.0000	1.0000
0.25	0.1829	0.2937	0.3879	0.4899	
0.50	0.1566	0.2589	0.3428	0.4333	
0.75	0.1263	0.2303	0.3278	0.4221	
1.00	0.0000	0.0000	0.0000	0.0000	

Table-2 : Values of f Powell-Eyring case for $\alpha'=0.1$ and $\beta'=10^4$

η/f	Gr=5		Gr=10	
	Pr=7	Pr=0.71	Pr=7	Pr=0.71
0.00	1.0000	1.0000	1.0000	1.0000
0.25	0.4236	0.6077	0.5249	0.6538
0.50	0.5576	0.7255	0.6440	0.8125
0.75	0.3140	0.5535	0.1545	0.3533
1.00	0.0000	0.0000	0.0000	0.0000

Table-3 : Solution of Equations (4.1.2)-(4.1.4 a,b) for $K=\frac{1}{2}$.

η/f	Pr=7	Pr=0.71
0.00	1.0000	1.0000
0.25	0.0208	0.1996
0.50	0.0274	0.2645
0.75	0.0202	0.1939
1.00	0.0000	0.0000

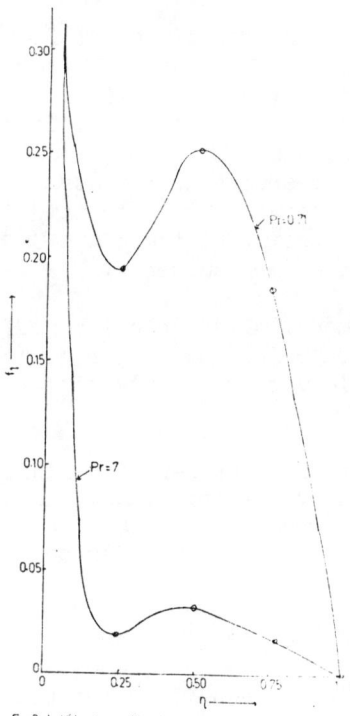

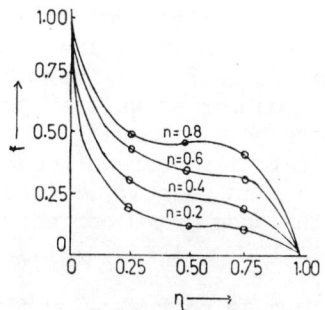

Fig. 2 Velocity distribution for power-law fluid flow past suddenly accelerated vertical plate. Pr=7, Gr=10

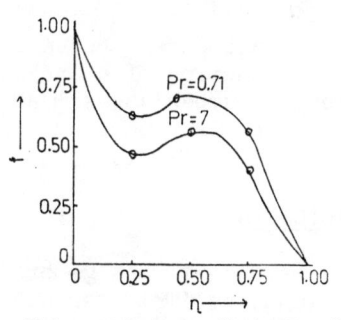

Fig 3 Velocity profiles for Newtonian fluid Gr=5

Fig. 3a Velocity profile for Powell-Eyring fluid. Gr=5

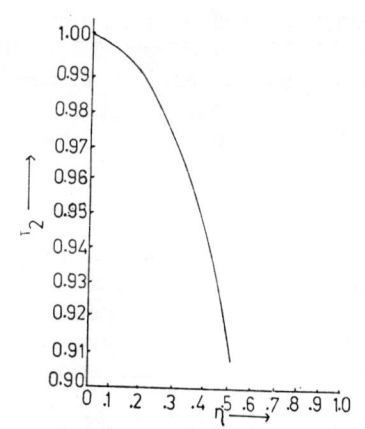

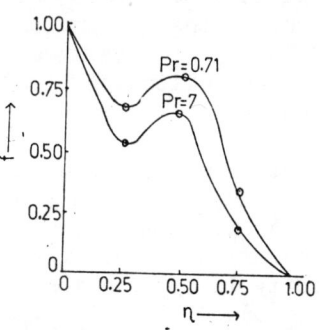

Fig. 4 Temperature distribution obtained by cubic Spline method

Fig. 3b Velocity profile for Powell Eyring fluid. Gr=10

6. DISCUSSION OF RESULTS :

Dimensionless velocity distributions for different fluids model is shown in Figures 1,2 and 3a,b. It is evident from the curves represent velocity profiles that the role of Prandtl number as well as Greshof number is quite significant. An increment in Prandtl number is responsible for the reduction in velocity whereas reduction in Greshof number causes the decreases in velocity. Here it will be worthwhile to note that the case $Gr > 0$ corresponds the cooling of the plate $Gr < 0$ the heating of the plate by free convection currents.

In case of Power-law fluids dimensionless velocity profiles are presented in Fig.2 for Pr=7, Gr=100 and 1 for several flow indices n. The velocity profile vary regularly with the parameter n.

Fig.3 a,b shows the velocity of the Powell-Eyring fluid flow as a function of n and different values of Prandtl number and Greshof number as well as $\alpha' = 0.1$ and $\beta' = 10^4$. In this case velocity profile resembles similarity to the Newtonian behaviour as shown in Fig.1.

7. CONCLUSION :

We have developed general similarity equations for present flow situations. This technique can be applied to non-Newtonian fluids of all those models in which shearing stress is directly related to the rate of strainy by any arbitrary continuous function. The extended form of most promising group-theoretic generalized dimensional analysis method has a main advantage over conventional dimensional analysis and other similarity method is that, it relaxes the restriction over main stream velocity and ultimately leads to the general form of similarity equations.

Further we conclude that spline collocation method is most effective and gives the good convergence for the solution of non-linear problem as it leads the solution to the linear algebraic equations. Here very compact calculations involved. This simplicity in application of splines justifies a wide scope of this method.

Acknowledgement :

The authors are greatful to University grant commission (UGC,India) for the provision of financial assistance to carry out this research under the grant no. F.8-10/84 (SR III).Thanks also due to the principal, S.V.R. College of Engn. & Tech., Surat, India for providing necessary research facilities.

References :

1. Soundalgekar V.M. and Pop I., 'Flow past an accelerated vertical infinite plate in the presence of free convection currents,'Reg. J. Energy Heat Mass Trans.,1980,2,2,127.

2. Lie S., 'Uber Differential Invarianten,'Math Ann, 1884, 24,52.

3. Morgan A.J.A., 'The reduction of one of the numbers of independent variables in the some systems of partial differential equations' Qut.J.Maths.,1952,2,250.

4. Blueman G.W. and Cole J.D.,'The general similarity solution of the heat equation', J.Math.and Mech.,1969,18,11, 1025.

5. Moran M.J.,'A generalization of dimensional analysis', J. Franklin Inst.,1971, 292,423.

6. Moran M.J. and Marshek K.M., 'Some matrix aspects of generalized dimensional analysis', J.Engg.Maths.,1972,6,291.

7. Timol M.G. and Kalthia N.L.,'A note on some matrix aspects of generalized dimensional analysis, 30th ISTAM Congress, IIT Delhi, 1986.

8. Bickley W.G.,'Piecewise cubic interpolation and two point boundary value problems', Comp.J.,1968,11,206.

9. Ahlberg , Nilson and Walsh , 'Theory of splines and their applications', Academic Press, Newyork,1969.

10. Blue J.L.,'Spline function methods in non-linear boundary value problems',CACM,1969,12,No.6,327.

ON THE NONLINEAR INSTABILITY OF THE BUOYANCY-LAYER FLOW IN A HEATED VERTICAL SLOT

M. Weinstein* and P.G. Daniels

Department of Mathematics, the City University, Northampton Square, London EC1V OHB, U.K.

SUMMARY

This work deals with the nonlinear stationary instability of the buoyancy-layer flow between heated vertical planes, that are subject to constant horizontal temperature and constant vertical temperature gradient. The analysis is concerned with the high Prandtl number limit when the critical disturbance occurs as stationary convection. The interaction between core and boundary layers leads to a reduced form of the boundary-layer stability equations, the solution of which enables the determination of the neutral curve. By expanding the functions involved in powers of $\varepsilon^{1/2}$, where ε is the elevation above the critical Rayleigh number and by adding long length and time scales $z=\varepsilon^{-1}Z$, $t=\varepsilon^{-1}\tau$, respectively, the $O(\varepsilon^{1/2})$, $O(\varepsilon)$ and $O(\varepsilon^{3/2})$ systems are obtained. The full problem involves the numerical solution of a nonlinear system of 60 ordinary differential equations integrated between $-\frac{1}{2}$ and 0 by a Runge-Kutta initial-value routine, where proper care has to be taken in order that the boundary conditions at x=0 are satisfied. The $O(\varepsilon^{1/2})$ system enables the numerical calculation of the critical wave number α_c and critical Rayleigh number A_c as a function of the vertical stratification parameter γ. A solvability condition of the $O(\varepsilon^{3/2})$ system involving the adjoint will determine actually the amplitude equation.

INTRODUCTION

The buoyancy-driven flow between vertical planes that are maintained at different temperatures has many applications in astrophysical, geophysical and industrial domains and has been studied intensively. More recent applications include the solar collectors and one of the most relevant problems in our times, the cooling of nuclear reactors.

----
* On leave from Raphael, P.O.Box 2250, Haifa 31021, ISRAEL.

In many cases the flow will occur in the presence of a vertical stratification which, in the case of convection in a vertical slot,as an example, is induced by the combination of the applied horizontal temperature differential and the presence of horizontal end-walls. Elder (1965) in an idealized model of the flow included the vertical temperature variation as an underlying stratification and succeeded to obtain an exact solution of the Boussinesq equations in an infinite vertical layer. The solution depends on a Rayleigh number, R, based on the width of the layer, ℓ^*, and the vertical temperature gradient, $\Delta T_v^*/\ell^*$. The flow is antisymmetric, unidirectional and independent of the vertical coordinate z^*, with upflow near the hot plane and downflow near the cold plane, the temperature field relative to the vertical stratification being also independent of z^*. In a narrow enclosure, three distinct regimes of flow can occur, each corresponding to a different range of values of the Rayleigh number. If R_a is small (conduction regime) there is little or no variation of the fluid temperature with height, and heat is transferred between the vertical walls primarily by conduction. In this case the temperature has a linear dependence on the horizontal coordinate x^*, and the velocity profile is cubic in x^*. Detailed study of this flow has been performed by Gershuni 1953, Rudakov 1967, Vest and Arpaci 1969 and Korpela et al 1973. Two-dimensional disturbances will induce an instability of the basic flow at a critical value of the horizontal Rayleigh number A based on ℓ^* and the horizontal temperature difference ΔT_h^*. The critical value of A depends on the Prandtl number, σ, and is connected with stationary convection if $\sigma<12.7$ and travelling waves if $\sigma>12.7$. However, as R_a is increased, a stable vertical temperature gradient develops in the core of the flow (transition regime), and vertical velocities are getting lower. Finally, if R_a is increased sufficiently (boundary-layer or convection regime), the flow is confined to boundary layers at the side walls, and the dominant mode of heat transfer is convection. The stability characteristics of this flow have studied by Vest & Arpaci (1969), Gill & Kirkham (1970), Barcilon & Pedlosky (1967), Gill (1966) and comprehensively by Bergholz (1978).

His results give the critical value of the Rayleigh number A as a function of σ and R, by using the notation

$$4\gamma^4 = R$$

The numerical results of Bergholz (1978) show that at high Prandtl numbers, increasing the vertical stratification (large values of γ) leads to a transition from travelling-wave to stationary instability, indicating that the travelling waves of Gill & Davey (1969) are not in fact the preferred mode of instability in the vertical layer at high Prandtl numbers. This conclusion was confirmed later by Daniels (1985).

In the case of oscillatory mode structure discussed by Gill & Davey (1969), the disturbance is completely contained within each buoyancy layer and has vertical wavelength $O(\gamma^{-1})$, comparable with the width of the layer. It has been found that the stationary modes have longer wavelengths $O(\gamma^{-\frac{1}{2}})$. This time the disturbance is not confined to the boundary layers, but presents an extension, in the form of elongated horizontal rolls, across the core. A reduced form of the boundary-layer stability equations is obtained by the interaction between core and boundary layers. The solution of these equations will enable the determination of the neutral curve and the completion of the core solution. Daniels (1985) solved the linear stability equations and the object of this work is to solve the nonlinear boundary-layer stability equations.

CHAPTER I - GOVERNING INSTABILITY EQUATIONS

This analysis deals with the nonlinear stationary instability of the buoyancy-layer flow between heated vertical planes, that are subject to constant horizontal temperature and constant vertical temperature gradient. The analysis is concerned with the high Prandtl number limit when the critical disturbance occurs as stationary convection.

If $\sigma=\nu/k$ is the Prandtl number, where ν denotes the kinematic viscosity and k the thermal diffusivity, α^* the coefficient of thermal expansion and

$$A = \alpha^* g \Delta T_h^* \ell^{*3} / \nu k \tag{1.1}$$

where g is the acceleration due to gravity. x,z, time t, velocity components u,w and the reduced pressure are non-dimensionalized by the quantities $\ell^*, \ell^{*2}/k$, k/ℓ^* and $\rho \times k/\ell^{*2}$ respectively (ℓ^* the width of the layer).

The stationary case for finite γ and $\sigma \to \infty$ is considered. Under the Boussinesq approximation the equations which describe the two-dimensional flow in the region between the planes can be written in non-dimensional form as follows:

$$\frac{\partial u}{\partial x} + \frac{\partial w}{\partial z} = 0$$

$$0 = -\frac{\partial p}{\partial x} + \nabla^2 u \tag{1.2}$$

$$0 = -\frac{\partial p}{\partial z} + \nabla^2 w + AT$$

$$\frac{\partial T}{\partial t} + u\frac{\partial T}{\partial x} + w\frac{\partial T}{\partial z} = \nabla^2 T$$

with boundary conditions

$$\psi = \frac{\partial \psi}{\partial x} = 0 \quad (x=\pm\tfrac{1}{2}) \tag{1.3}$$

$T = \beta z \pm \tfrac{1}{2}$ $(x = \pm \tfrac{1}{2})$

where ψ is the stream function

$$u = \frac{\partial \psi}{\partial z}$$
$$w = -\frac{\partial \psi}{\partial x}$$
(1.4)

and the parameter β denotes the temperature ratio.

$$\beta = \frac{\Delta T_v^*}{\Delta T_h^*} \tag{1.5}$$

The parameter of vertical stratification γ is defined by

$$\gamma^4 = \tfrac{1}{4} A \beta \tag{1.6}$$

It is convenient to replace the parameter β by the new parameter γ, which can be seen to be essentially a Rayleigh number based on the vertical temperature scale

$$R = 4\gamma^4 = \alpha^* g \Delta T_v^* \ell^{*3} / \nu k \tag{1.7}$$

CHAPTER II - STABILITY FORMULATION

The stream and temperature functions can be formulated as follows:

$$\begin{cases} \psi = A\tilde{\psi}(x,z,t) \\ T = \beta z + \tilde{T}(x,z,t) \end{cases} \tag{2.1}$$

It yields a system of differential equations

$$\begin{cases} \nabla^4 \tilde{\psi} = \dfrac{\partial \tilde{T}}{\partial x} \\ \dfrac{\partial \tilde{T}}{\partial t} + A \dfrac{\partial(\tilde{T},\tilde{\psi})}{\partial(x,z)} = \nabla^2 \tilde{T} + 4\gamma^4 \dfrac{\partial \tilde{\psi}}{\partial x} \end{cases} \tag{2.2}$$

Boundary conditions

$$\tilde{\psi} = \frac{\partial \tilde{\psi}}{\partial x} = 0, \quad \tilde{T} = \pm \tfrac{1}{2} \quad (\text{at } x = \pm \tfrac{1}{2}) \tag{2.3}$$

The following perturbation expansion is used for the nonlinear problem

$$\begin{cases} \tilde{\psi} = \tilde{\psi}_0 + \varepsilon^{1/2} \tilde{\psi}_1 + \varepsilon \tilde{\psi}_2 + \varepsilon^{3/2} \tilde{\psi}_3 + \ldots \\ \tilde{T} = \tilde{T}_0 + \varepsilon^{1/2} \tilde{T}_1 + \varepsilon \tilde{T}_2 + \varepsilon^{3/2} \tilde{T}_3 + \ldots \end{cases} \tag{2.4}$$

where

$$A = A_c + \varepsilon \tag{2.5}$$

so that ε is the elevation above the critical Rayleigh number.

$O(1)$ system

The base flow of interest is as follows

$$\begin{cases} \tilde{\psi}_0 = \Phi(x) \\ \tilde{T}_0 = \theta(x) \end{cases} \tag{2.6}$$

so that (2.2)-(2.3) system becomes

$$\Phi^{IV} = \theta'$$

$$\theta'' + 4\gamma^4 \Phi' = 0 \tag{2.7}$$

B.C.s.

$$\Phi = \Phi' = 0, \quad \theta = \pm\tfrac{1}{2} \quad (x = \pm\tfrac{1}{2})$$

Solution for Φ' is as follows:

$$\Phi' = \frac{\sinh[\gamma(x+\tfrac{1}{2})]\sin[\gamma(x-\tfrac{1}{2})] - \sinh[\gamma(x-\tfrac{1}{2})]\sin[\gamma(x+\tfrac{1}{2})]}{4\gamma^2(\cosh\gamma - \cos\gamma)} \tag{2.8}$$

Φ is even and θ is odd.

Additional effects

Long length-time scales

$$z = \varepsilon^{-1/2} Z; \quad t = \varepsilon^{-1}\tau$$

$$\frac{\partial}{\partial z} \to \frac{\partial}{\partial z} + \varepsilon^{1/2} \frac{\partial}{\partial Z}$$

$$\frac{\partial^2}{\partial z^2} \to \frac{\partial^2}{\partial z^2} + 2\varepsilon^{1/2}\frac{\partial^2}{\partial z \partial Z} + \varepsilon \frac{\partial^2}{\partial Z^2} \tag{2.9}$$

$$\frac{\partial}{\partial t} \to \varepsilon \frac{\partial}{\partial \tau}$$

$O(\varepsilon^{1/2})$ system

$$\nabla^4 \tilde{\psi}_1 - \frac{\partial \tilde{T}_1}{\partial x} = 0$$

$$\nabla^2 \tilde{T}_1 + 4\gamma^4 \frac{\partial \tilde{\psi}_1}{x} + A_c \left(\frac{\partial \tilde{T}_0}{\partial x} \frac{\partial \tilde{\psi}_1}{\partial z} - \frac{\partial \tilde{\psi}_0}{\partial x} \frac{\partial \tilde{T}_1}{\partial z} \right) \tag{2.10}$$

$$\tilde{\psi}_1 = \frac{\partial \tilde{\psi}_1}{\partial x} = \tilde{T}_1 = 0 \quad (x = \pm\tfrac{1}{2})$$

it yields the solution

$$\tilde{\psi}_1 = B(Z,\tau)e^{i\alpha_c z}\bar{\psi}(x) + B^*(Z,\tau)e^{-i\alpha_c z}\bar{\psi}^*(x)$$
$$\tilde{T}_1 = B(Z,\tau)e^{i\alpha_c z}\bar{\theta}(x) + B^*(Z,\tau)e^{-i\alpha_c z}\bar{\theta}^*(x) \tag{2.11}$$

The differential equations system for $\bar{\psi}(x)$ and $\bar{\theta}(x)$ is identical to the stationary linear stability solved[3] for infinite Prandtl number. The numerical solution of this system enables the calculation of the critical Rayleigh number A_c and the critical wavenumber α_c as functions of the vertical stratification parameter γ.

By following the method of Vest & Arpaci 1969, the solutions for $\bar{\psi}, \bar{\theta}$ can be conveniently expressed as

$$\bar{\psi} = (\psi_0 + i\psi_e)(x), \quad \bar{\theta} = (\theta_e + i\theta_0)(x) \tag{2.12}$$

where ψ_0, θ_0, ψ_e and θ_e are odd and even real functions of x, respectively.

O(ε) system

$$\nabla^4 \tilde{\psi}_2 - \frac{\partial \tilde{T}_2}{\partial x} = -4\frac{\partial^4 \tilde{\psi}_1}{\partial z^3 \partial Z} - 4\frac{\partial^4 \tilde{\psi}_1}{\partial x^2 \partial z \partial Z}$$

$$\nabla^2 \tilde{T}_2 + 4\gamma^4 \frac{\partial \tilde{\psi}_2}{\partial x} - A_c\left(\frac{\partial T_0}{\partial x}\frac{\partial \tilde{\psi}_2}{\partial z} - \frac{\partial \psi_0}{\partial x}\frac{\partial \tilde{T}_2}{\partial z}\right) = -2\frac{\partial^2 \tilde{T}_1}{\partial z \partial Z} + A_c \frac{\partial \tilde{T}_0}{\partial x}$$

$$+ A_c\left[\frac{\partial T_0}{\partial x}\frac{\partial \tilde{\psi}_1}{\partial Z} - \frac{\partial \psi_0}{\partial x}\frac{\partial \tilde{T}_1}{\partial Z} + \frac{\partial \tilde{T}_1}{\partial x}\frac{\partial \psi_1}{\partial z} - \frac{\partial \tilde{\psi}_1}{\partial x}\frac{\partial \tilde{T}_1}{\partial z}\right] \tag{2.13}$$

$$\tilde{\psi}_2 = \frac{\partial \tilde{\psi}_2}{\partial x} \quad \tilde{T}_2 = 0 \quad (x = \pm \tfrac{1}{2})$$

$$\tilde{\psi}_2 = \frac{\partial B}{\partial Z}e^{i\alpha_c z}\hat{\psi}(x) + \frac{\partial B^*}{\partial Z}e^{-i\alpha_c z}\hat{\psi}^*(x) + |B|^2\hat{\hat{\psi}}(x) + B^2 e^{2i\alpha_c z}\hat{\psi}(x) +$$
$$+ B^{*2}e^{-2i\alpha_c z}\hat{\psi}^*(x) \tag{2.14}$$

$$\tilde{T}_2 = \frac{\partial B}{\partial Z}e^{i\alpha_c z}\hat{T}(x) + \frac{\partial B^*}{\partial Z}e^{-i\alpha_c z}\hat{T}^*(x) + |B|^2\hat{\hat{T}}(x) + B^2 e^{2i\alpha_c z}\hat{T}(x) +$$
$$+ B^{*2}e^{-2i\alpha_c z}\hat{T}^*(x)$$

Real systems are obtained by using even real parts and odd imaginary parts for the various ψ functions, the vice versa being true for the corresponding θ functions.

At this stage it is still impossible to get the amplitude equation and in order to achieve this goal there is need to continue to the O($\varepsilon^{3/2}$) system.

$O(\varepsilon^{3/2})$ system

This system will lead to the amplitude equation only if an orthogonality condition involving the adjoint will be satisfied.

The adjoint condition arises from the fact that there is a non-zero solution

$$\begin{cases} \bar{\psi}^{IV} - 2\alpha^2 \bar{\psi}'' + \alpha^2 \bar{\psi} - \bar{\theta}' = 0 \\ \bar{\theta}'' - \alpha^2 \bar{\theta} + 4\gamma^4 \bar{\psi}' - i\alpha A(\theta' \bar{\psi} - \Phi' \bar{\theta}) = 0 \\ \bar{\psi} = \bar{\psi}' = \bar{\theta} = 0 \quad (x = \pm \tfrac{1}{2}) \end{cases} \quad (2.15)$$

so that

$$\bar{\psi}_3^{IV} - 2\alpha^2 \bar{\psi}_3'' + \alpha^3 \bar{\psi}_3 - \bar{\theta}_3' = \bar{\chi}_3$$
$$\bar{\theta}_3'' - \alpha^2 \bar{\theta}_3 + 4\gamma^4 \bar{\psi}_3' - i\alpha A(\theta' \bar{\psi}_3 - \Phi' \bar{\theta}_3) = \bar{\chi}_4 \quad (2.16)$$
$$\bar{\psi}_3 = \bar{\psi}_3' = \bar{\theta}_3 = 0$$

has a solution only if the orthogonality condition,

$$\int_{-\frac{1}{2}}^{\frac{1}{2}} (\bar{\chi}_3 g + \bar{\chi}_4 h) dx = 0 \quad (2.17)$$

involving the adjoint functions g and h is satisfied. χ_3 and χ_4 are expressions containing terms in $d^2B/dZ^2, B, dB/d\tau$ and $B|B|^2$.

The adjoint functions g and h satisfy the system:

$$h'' - \alpha^2 h + g' + i\alpha A \Phi' h = 0$$
$$g^{IV} - 2\alpha^2 g'' + \alpha^4 g - 4\gamma^4 h' - i\alpha A \theta' h = 0 \quad (2.18)$$
$$g = g' = h = 0 \quad (x = \pm \tfrac{1}{2})$$

The adjoint functions g and h can be expressed as

$$g = g_0 + ig_e; \quad h = h_e + ih_0 \quad (2.19)$$

where g_0, h_0, g_e and h_e are odd and even real functions of x, respectively. Their contribution to the condition (2.17) leads to the amplitude equation

$$\hat{a}_1 \bar{B}_\tau + \hat{a}_2 \bar{B}_{ZZ} + \hat{a}_3 \bar{B}|B|^2 + \hat{a}_4 \bar{B} = 0 \quad (2.20)$$

after some normalisation for convenience purposes.

Numerical Scheme

The full stability problem involves the numerical solution of a nonlinear system of 60 ordinary differential equations integrated between $-\frac{1}{2}$ and 0 by Runge-Kutta initial-value routine, by taking care that the boundary conditions at x=0 are satisfied. The $O(\varepsilon^{1/2})$ system of the main system enables the numerical calculation of the critical number α_c and the critical Rayleigh number A_c as a function of the vertical stratification parameter γ. A certain determinant has to vanish and this leads to the determination of A_c and α_c. The results are identical with those obtained by the consideration of the $O(\varepsilon^{1/2})$ adjoint system and Bergholz (1978) results. The $O(\varepsilon)$ inhomogeneous system consists of 36 equations and is solved by matrix calculus, involving the search for a particular solution. The $O(\varepsilon^{3/2})$ system involving the adjoint determines the amplitude equation. Due to the large variaion of the $\hat{a}_3$ ($|\bar{B}|\bar{B}|^2$ coefficient) mainly it was imperative to compute the coefficients appearing in the amplitude equation for many values of γ. The results for the nonlinear coefficient $\hat{a}_3$ present a small region of subcritical instability near $\gamma=10$.

REFERENCES

[1] Barcilon, V. & Pedlosky, J. 1967, J. Fluid Mech. 29, 609.

[2] Bergholz, R.F. 1978 J. Fluid Mech. 84, 743.

[3] Daniels, P.G. 1985 Proc. R. Soc. London A401, 145-161.

[4] Elder, J.W. 1965 J. Fluid Mech. 23, 77.

[5] Gershuni, G.Z. 1953 Zh. tekh. Fiz. 23, 1838.

[6] Gill, A.E. 1966 J. Fluid Mech. 26, 515.

[7] Gill, A.E. & Davey, A. 1969 J. Fluid Mech. 35, 775.

[8] Gill, A.E. & Kirkham, C.C. 1970 J. Fluid Mech. 42, 125.

[9] Korpela, S.A., Gozum, D. & Baxi, C.B. 1973 Int. J. Heat Mass Transfer 16, 1683.

[10] Rudakov, R.N. 1967 Prikl. Mat. Mekh. 31, 349.

[11] Veronis, G. 1967 Tellus 19, 326.

[12] Vest, C.M. & Arpaci, V.S. 1969 J. Fluid Mech. 36, 1.

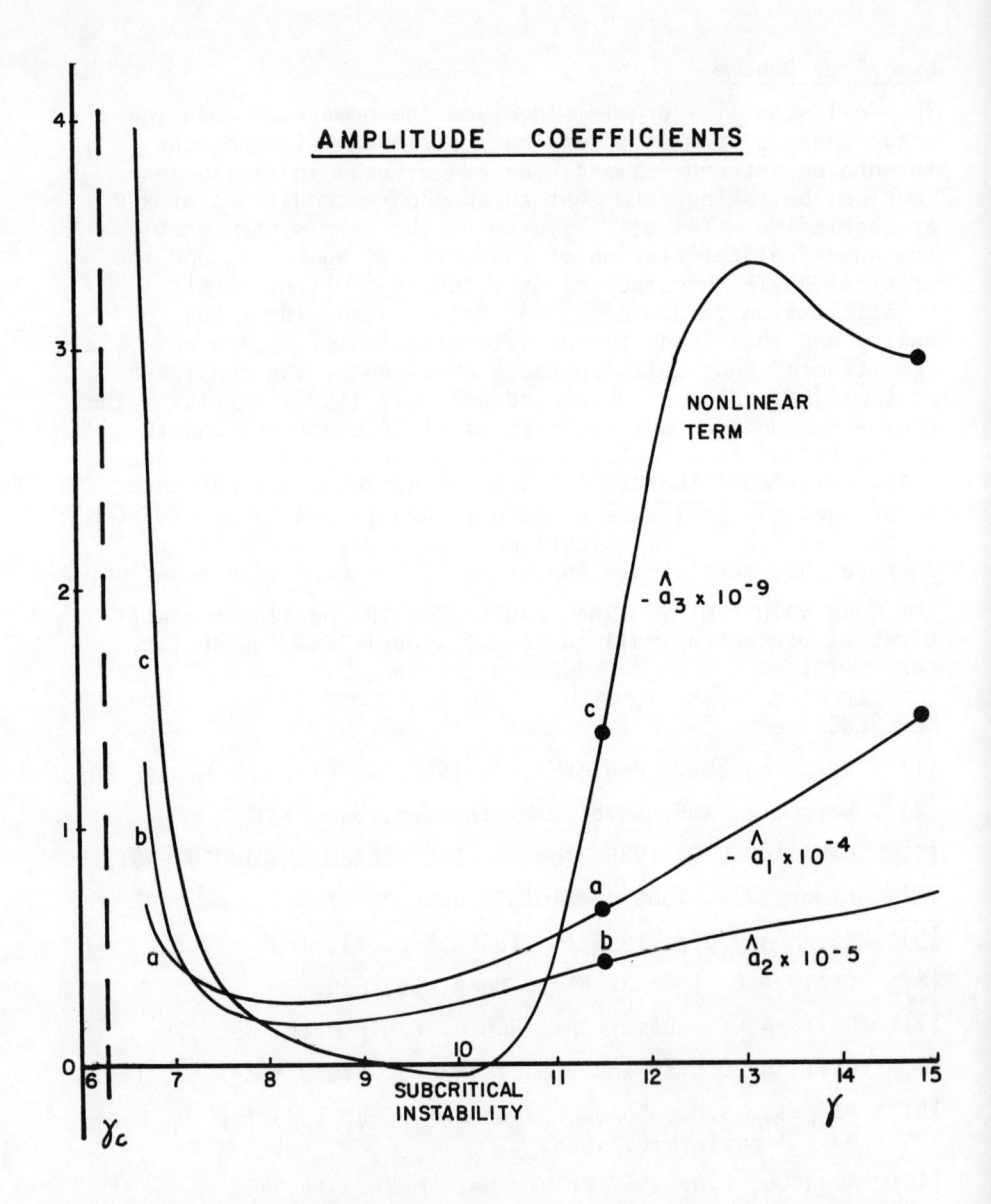

Fig. 1: The coefficients appearing in the nonlinear amplitude equation.

ANALYSIS OF F.E.M. AND PRACTICAL RESEARCH ON TALL COLD STORAGE BUILDING

Chen Xiangfu

Design Institute of Ministry of Commerce
Beijing, China

SUMMARY

Based on the characteristics of the tall cold storage buildings and its special conditions, this paper researches structural system by F.E.M. According to these useful results, some new reliable seismic resistant measures are adopted and some important problem are explored.

Particularly, the calculation and analysis of thermal stresses and deformations in tall and multi-storeys cold storage buildings by F.E.M. and also discussed first of all. Its results quite approach the data of practical measurement, provide the basis for the compile state code.

INTRODUCTION

The differential equation of an elastic thin plate is

$$D\left(\frac{\partial^4 w}{\partial x^4} + \frac{\partial^4 w}{\partial y^4} + \frac{\partial^4 w}{\partial z^4}\right) = g(x,y) \quad in \, \Omega \quad (1)$$

But, the distribution of bending moments and stresses in the structural system of the slab-column or flat slab beamless can not satisfactorily be determined by the elastic theory. The structural system of the slab-column has some clear advantages and be widely used in department buildings house buildings, and storage buildings, especially, cold storage buildings. This paper is based on the requirements of refrigeration and the special conditions of the practical engineering design as well as the structural programmes for the biggest tall cold storage building in China (Fig. 1), and researches the structural system of the slab column - shear wall on tall buildings. Concretely, the quasi-frame method of the finite element method has been adopted to calculate the seismic forces and periods, stresses and displacements, besides of these also considered the compositions of various loads(including thermal loads) and various interior forces. According to the results of calculations some new reliable earthquake proof construction measures are applied on tall cold storage building. For example, it is dealed with stub posts and used by rib slab of seismic resistance and so forth.

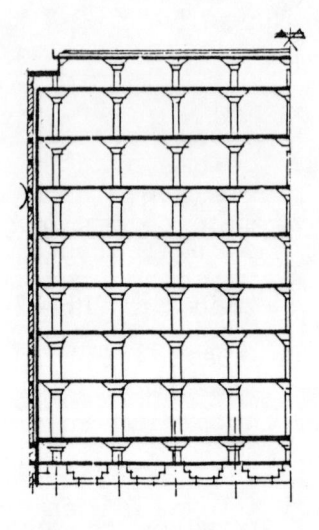

Fig.1 Section of the biggest tall cold storage building in China

Particularly, the thermal stresses and deformations are also calculated by the finite element method on the biggest tall cold storage building first of all. the thermal deformations by practical measurement are tallied with the results of this paper. It can guide the structural design of structural engineers and provide the basis data for the compile of state code.

THE MODEL AND RESULTS OF F.E.M. FOR THE AYSTEM OF SLAB COLUMN-SHEAR WALL

In general, the establishment of numerical calculation model is rather difficult for the system slab column-shear wall. When F.E.M. is applied, there are four ways of solution as follows:
(i)   The sub-structure method;
(ii)   The equivalent frame method;
(iii) The quasi-frame method (It is considered the work in coordinating space.);
(iv)   The 3D analysis method.

Here, the quasi-frame method is assumed to be plane frame, but it is coordinating space of the slab-column and shear wall. When the finite element method is applied, it is the model of the calculation as follows:
(i)   The stub posts are considered to be quasi-

shear members and increase its section.
(ii) The column heads are assumed to be regid members;
(iii) The rigid members are used to combine between the shear wall and the structure of slab - column. So, the displacements of shear wall should be tallied with the structure of slab-column, and the shear wall bears shear forces of 80 per cent;
(iv) The column bands are supposed to be equasi-equivalent beam and the boundary conditions are considered to be the effect of support due to column head;
(v) The pile foundation is assumed to be fixed end.

According to these conditions, the structural programme of the biggest cold storage building and the model of calculation have been established with the finite element method to calculate the bending moments and shear forces of quasi-frame on the horizontal loads and the vertical loads (The live load, q= 2000 kg/m^2), and the bending moments and shear forces of the shear wall (See Fig.2), and the periods and displacements of quasi-frame(See Fig.3)

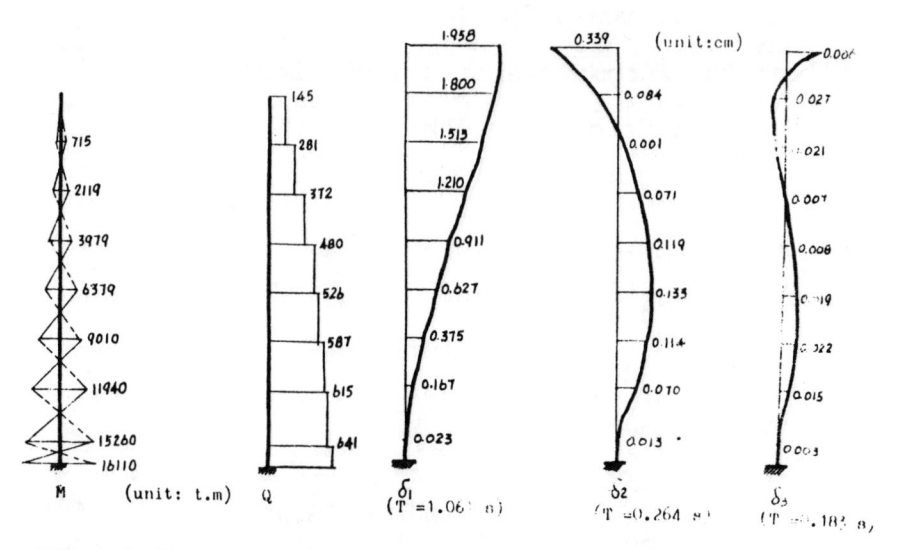

Fig. 2 The moments and forces of the shear wall

Fig. 3 the displacements of quasi-frame

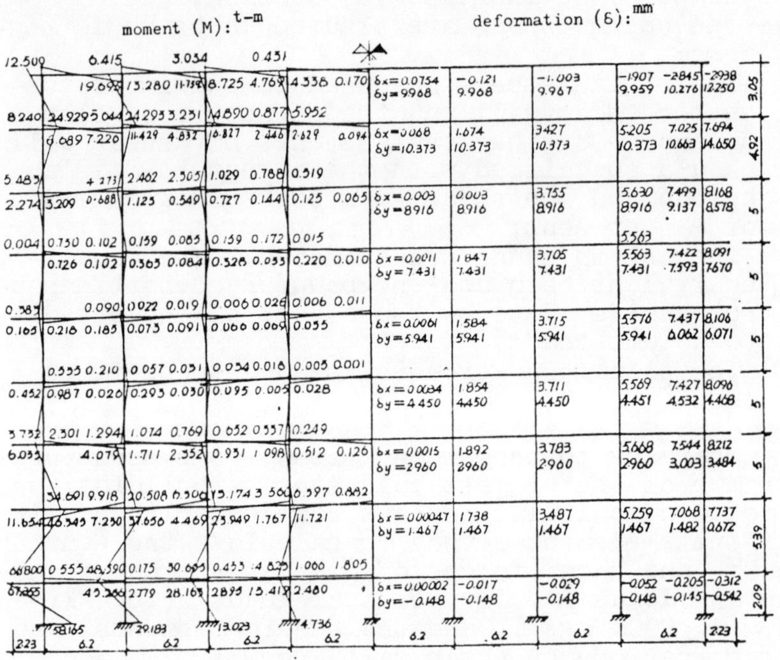

Fig.4 The moments and deformations of structure due to application "temperature load"

Notice: The parameters of calculation are as follows:
(I) Coefficient of thermal expansion,

$$\alpha = 1.2 \times 10^{-5} \frac{1}{°C}$$

(II) Yang's module, $E = 2 \times 10^5$ kg/cm^2
(III) Construction temperature of concrete,
Tc = 12 °C;
(IV) Temperature of roof slab,
T upper = 17°C
T lower = 13°C
(V) Temperature of atlic,
T a = 17°C
(VI) Temperature of floor built on stilts,
T upper = T lower = 18.3 °C
(VII) Temperature of stilts,
T s = 20.3°C
(VIII) Temperature of sold storage,
T cs = 20°C

THE CALCULATION OF THERMAL STRESSES AND DEFORMATIONS FOR TALL COLD STORAGE BUILDING

In the light of the differential equation of conductive heat transfer is:

$$\frac{\partial T}{\partial t} = \alpha \left(\frac{\partial^2 T}{\partial x^2} + \frac{\partial^2 T}{\partial y^2} + \frac{\partial^2 T}{\partial z^2} \right) + \frac{W}{c\rho} \quad (2)$$

and the equations of thermal elasticity are:

$$\left. \begin{array}{l} \dfrac{\partial^2 u}{\partial x^2} + \dfrac{1-\mu}{2} \cdot \dfrac{\partial^2 u}{\partial y^2} + \dfrac{1+\mu}{2} \cdot \dfrac{\partial^2 v}{\partial x \partial y} - (1+\mu) \cdot \alpha \cdot \dfrac{\partial T}{\partial x} = 0 \\[6pt] \dfrac{\partial^2 v}{\partial y^2} + \dfrac{1-\mu}{2} \cdot \dfrac{\partial^2 v}{\partial x^2} + \dfrac{1+\mu}{2} \cdot \dfrac{\partial^2 u}{\partial x \partial y} - (1+\mu) \cdot \alpha \cdot \dfrac{\partial T}{\partial y} = 0 \end{array} \right\} \quad (3)$$

as well as the boundary conditions of desplacement are:

$$\left. \begin{array}{l} \ell \left(\dfrac{\partial u}{\partial x} + \mu \dfrac{\partial v}{\partial y} \right) + m \dfrac{1-\mu}{2} \left(\dfrac{\partial u}{\partial y} + \dfrac{\partial v}{\partial x} \right) = \ell (1+\mu) \alpha \cdot T \\[6pt] m \left(\dfrac{\partial v}{\partial y} + \mu \dfrac{\partial u}{\partial x} \right) + \ell \dfrac{1-\mu}{2} \left(\dfrac{\partial v}{\partial x} + \dfrac{\partial u}{\partial y} \right) = m (1+\mu) \alpha \cdot T \end{array} \right\} \quad (4)$$

Hence, the thermal stresses canbe solved out. And then, the quasi-body forces are:

$$\left. \begin{array}{l} X = -\dfrac{E\alpha}{1-\mu} \cdot \dfrac{\partial T}{\partial x} \\[6pt] Y = -\dfrac{E\alpha}{1-\mu} \cdot \dfrac{\partial T}{\partial y} \end{array} \right\} \quad (5)$$

and quasi-plane stress is:

$$\sigma_n = \frac{E\alpha T}{1-\mu} \quad (6)$$

For the theory of thermal elasticity, the deformations and stresses of slab and column can be calculated by the finite element method. But it is very difficult to calculate thermal stresses for the total structural system of slab column-shear wall. In my another paper, it has been used by the method of moment distribution to calculate thermal stresses of slab-column for the tall cold storage building due to application of "temperature load".

This paper is applied by the finite element method to calculate the deformations and moments for the biggest tall cold storage building due to action of difference of temperature (See Fig. 4). However, the results are tallied with the method of

moment distribution and more than ever accurate.
The relations between the moment (M) and span number (Ne), the most moment (Mmax) the deformations (δ) at the attic and floor built on stilts are shown in Fig.5 and Fig.6

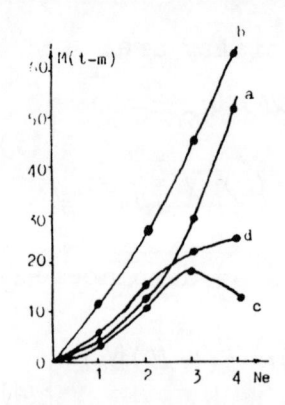

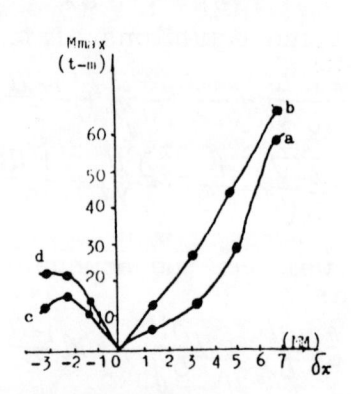

Fig.5 Curve of M-Ne Fig.6 Curve of Mmax-δ
Notice: (I)　　a-- Upper column end of floor
　　　　　　　　　built on stilts;
　　　　(II)　 b-- Lower column end of floor
　　　　　　　　　built on stilts;
　　　　(III)　c-- Upper column end of attic;
　　　　(IV)　 d-- Lower column end of attic.

THE ANALYSIS OF CALCULATION RESULTS AND THE RESEARCH OF STRUCTURAL DESIGN

This has been researched by F.E.M. in various boundary conditions and various structural programmes. According to comparision and analysis of various results, it is the basic conclusions as follows
(i) The tall cold storage buildings can be adopted the model of quasi-frame method by the finite element method. The results of calculation can completely satisfactory the requirements of engineering design. But, how can the results of F.E.M. be believed and applied darely by civel engineers?
(ii) The displacement of the roof top, Δ is 1.958 cm hence,

$$\frac{2\Delta}{H} = \frac{1}{1032.43} < \frac{1}{450}$$

(iii) From the Fig.9, it is discovered that the stub post of floor built on stilts is shear members and no opposite moment point. Therefore, it is especially paied attention to design of structural section and should be increased the reinforcement of resistant shear forces. The all con-

juctions lf main bars and stirrups should be welded.
(iv) The shear wall (N/B ≑ 2) is belong to the model of bending-shear and the system of slab-column is belong to shear model. The top of shear wall has negative action for bearing capacity and it is difficult to deal with "cold bridge". The height of shear wall did not build to the roof-slab. For preventing the stresses concentration of the resistant rib slab is applied (See Fig.7). Thus it solves above two difficult problems with success in structural design.

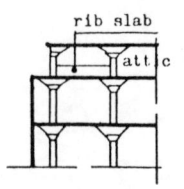

Fig.7 Resistant rib slab of attic

(v) The outside wall was designed to be the structure of quasi-weak frame. At the four angles of outside wall, the bolt beams were designed be live member (See Fig.8). Thus, it is not only satisfactory the stability of outside wall but also solves the problem of slab beamless deformation due to application of "temperature load".

(vi) The thermal stresses of tall cold storage building should not be neglected. The thermal building of the attic and floor built on stilts are rather high, and the adoptions of resistant method are not economic and scientific ways. It can only use some structural measures. For example, it is quite effective to reduce the demansions of structure and difference of temperature (See Fig.9).

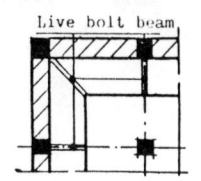

Fig.8 Live bolt beam of four angles of outside wall

If the relaxation of thermal stresses is considered (reduced coefficient, Ks = 0.5, you can see Fig.10) it can be easily discovered that the time of temperature drop for cold storage should be one month over. In addition, based on the research of calculation results, when the first floor isl natural ground of thermal insulation, the composition of thermal stresses and interior forces due to application of vertical load is very advantageous. Conversely, when the first floor is built on total stilts, their composition will not be advantageous.

(vii) The thermal deformations of calculation are tallied with results of practical measurement (See Fig.4 and 11).

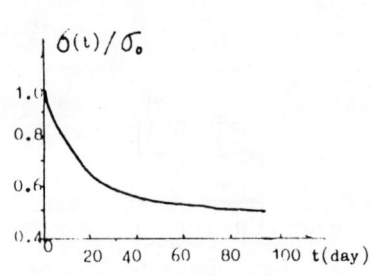

Fig.10 The curve of the relaxation of thermal stresses

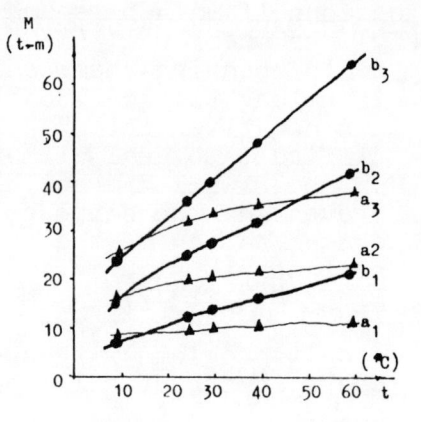

Fig.9 The relation between moments and defference of temperature drop
Notice;
(I) a_i -- lower column end of floor built on stilts;
(II) b_i -- upper column end of floor built on stilts;
(III) c_i -- span number

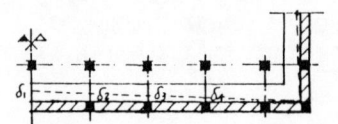

Fig.11 (a) The deformation of outside wall

Fig.11 The deformations of practical measurement

Notice:
(I) The horizontal deformation of slab beemless δ max = 9.925 mm
(II) The vertical deformation of column, δ max = 10.02 mm

(viii) The difference subsidence homogeneous and also tallied with results of practical measurement (Area of floor is 2750 m^2), $\Delta j \leq$ 1 cm

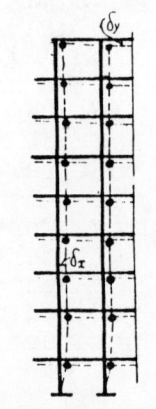

Fig,11 (b) The deformation of practical measurement

ACKNOWLEDGEMENT

The auther wishes to express his thanks to Tianjin 2th refrigeration factory, and to acknowledge particularly the help of professor Luo Songfa,

Lin Fenying and Ruan Zhimin and other fellows.

REFERENCES

1. Chen Xiangfu, The Calculation and Analysis by the Finite Element Method for Thermal Stresses of Multi-Storey Cold Storage, Journal of Refrigeration, Vol.3. 1981, pp 27-44.

2. Chen Xiangfu, The Structural Design of Cold Storage, Tianjin Commercial College Press, 1983.

3. Chien Weizang & Ye Kaiyuan, Elasticity, Press, 1956.

4. Zienkiewicz, O.C. & Cheng, Y.K., The Finite Element Method in Structural & Continum Mechanics, McGraw-HI11 Publishing Company Limited 1967.

5. Chen Anxi & Chen Xiangfu, Spline interpolation of the Stress-Strain Curves for Concrete in Compression, Second Int. Conference on Computational Methods and Experiemental Methods, 1984.

6. Chen Xiangfu, Boundary Element Method and its Application, 1980, Beijing.

7. Chen Xiangfu, Simplified Method of the Calculation and Analysis of Thermal Fields and Thermal Stresses on Cold Storages of Underground Rock Caverns, Large Rock Caverns, Proc. of the Int. Symposium, Helsinki, Finland, 25- 28 August, 1986.

THREE DIMENSIONAL THERMAL FIELD AND IT'S BOUNDARY CONDITIONS

Kuang Wenqi, Zhu Junjie, Gu Song and Di Yunzhen
Department of Civil Engineering, Tsinghua University
Beijing, China

Summary

When the temperatures of the interiol portion and outside environment of a three dimensional (3-D) body are different, thermal strain and thermal stress would be induced. Firstly the thermal field of this body must be Calculated. And the boundary conditions are shown very important for this kind of problem. In this paper, the 3-D stability thermal field was calculated by isoparametric elements.

1. Formula for Computation of 3-D Stability Thermal Field

According to the heat transfer theory, the temperatures $T(x,y,z)$ must Satisfy the following differentia equation:

$$\frac{\partial^2 T}{\partial x^2} + \frac{\partial^2 T}{\partial y^2} + \frac{\partial^2 T}{\partial z^2} = 0 \tag{1}$$

which is so-called heat transfer equation.
Equation (1) is equal to this limit condition:

$$\mathcal{L} = \iiint_V \frac{1}{2}\left[\left(\frac{\partial T}{\partial x}\right)^2 + \left(\frac{\partial T}{\partial y}\right)^2 + \left(\frac{\partial T}{\partial z}\right)^2\right] dxdydz = min \tag{2}$$

where $\mathcal{L}$ is the function of temperature $T(x,y,z)$
V is the whole region of the space elastic body
The 20-node isoparametric element is used, the whole region is divided into a number of elements and in every element we have

$$T = \sum_{i=1}^{20} N_i T_i \tag{3}$$

Formula of transformation of coordinate is:

$$X = \sum_{i=1}^{20} N_i X_i$$
$$y = \sum_{i=1}^{20} N_i y_i \qquad (4)$$
$$z = \sum_{i=1}^{20} N_i z_i$$

Ni is the shape function of element that found at the local coordinate system. Because the whole region of the elastic body is divided into element, formula (2) becomes as following:

$$J = \sum_e J^e = \min \qquad (5)$$

$\sum_e$ in formula (5) inplys add together for all elements. But

$$J^e = \iiint_{v^e} \frac{1}{2}\left[\left(\frac{\partial T}{\partial x}\right)^2 + \left(\frac{\partial T}{\partial y}\right)^2 + \left(\frac{\partial T}{\partial z}\right)^2\right]dxdydz \qquad (6)$$

We know that the function J is determined by the temperatures of nodes. So formula (2) can be written as

$$\frac{\partial J}{\partial T_i} = \sum_e \frac{\partial J^e}{\partial T_i} = 0 \qquad (7)$$

(number of nodes i=1,2,3,......,20)

Firstly, we calculated the $\frac{\partial J^e}{\partial T_i}$ of a typical element, and then for every element.

From formula (6) we have

$$\frac{\partial J^e}{\partial T_i} = \iiint_{v^e}\left[\frac{\partial T}{\partial x}\frac{\partial}{\partial T_i}\left(\frac{\partial T}{\partial x}\right) + \frac{\partial T}{\partial y}\frac{\partial}{\partial T_i}\left(\frac{\partial T}{\partial y}\right) + \frac{\partial T}{\partial z}\frac{\partial}{\partial T_i}\left(\frac{\partial T}{\partial z}\right)\right]dxdydz \qquad (7')$$
etc.

(i=1,2,3,......,20)

Substituting formula (3) into (7') we get

$$\frac{\partial J^e}{\partial T_i} = [h_{i1}, h_{i2}, \cdots, h_{i20}][T_1, T_2, \cdots, T_{20}]^T \qquad (8)$$

in which

$$h_{ij} = \iiint_{v^e}\left(\frac{\partial N_i}{\partial x}\frac{\partial N_j}{\partial x} + \frac{\partial N_i}{\partial y}\frac{\partial N_j}{\partial y} + \frac{\partial N_i}{\partial z}\frac{\partial N_j}{\partial z}\right)dxdydz \qquad (9)$$

From the character of isoparametric element we also have

$$h_{ij} = \int_{-1}^{1}\int_{-1}^{1}\int_{-1}^{1}\left(\frac{\partial N_i}{\partial x}\frac{\partial N_j}{\partial x} + \frac{\partial N_i}{\partial y}\frac{\partial N_j}{\partial y} + \frac{\partial N_i}{\partial z}\frac{\partial N_j}{\partial z}\right)|J|d\zeta d\eta d\xi \qquad (10)$$

The Gauss Integrat Method is utilized for the calculation for hij. For the element e there are

$$\left[\frac{\partial J^e}{\partial T_1} \quad \frac{\partial J^e}{\partial T_2} \quad \cdots \quad \frac{\partial J^e}{\partial T_{20}}\right]^T = [h]\{T\}^e \qquad (11)$$

Substituting formula (11) into (7), we get

$$\sum_e \sum_{i=1}^{n} h_{ij} T_j = 0 \qquad (12)$$

From formula (12) the linear equation group of unknown nodal temperature is obtained. Solution of this linear equation group is the temperature of different point in the 3-D thermal field.

2. The Boundary Conditions of 3-D Thermal Field

There are infinite solutions we can obtained from formula (1). When the boundary conditions are determined, the solution becomes unique. Correct determination of boundary conditions is shown very important for this problem.

One example is the moving platen of prestressed concrete hydraulic press (show Fig.1). There are a circular hole (ϕ =1.6m) in the center of moving platen, and a hot bloom is through the circular hole. The temperature of inner wall of moving platen is $80°C \sim 100°C$, and the temperature of environemnt is $0°C$. Now we calculate the thermal field of this 3-D body (moving platen).

For the calculation of 3-D thermal field the boundary condition is the temperature of outer surface of 3-D body, but the temperature of outer surface is 2-D thermal field problem.

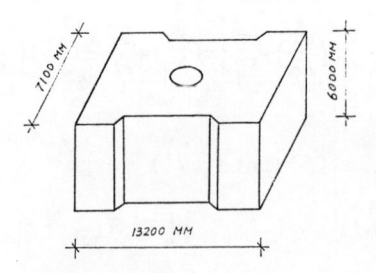

Fig.1 moving platen of Prestressed concrete hydraulic press

The heat transfer equation of 2-D thermal field problem is

$$\frac{\partial^2 T}{\partial x^2} + \frac{\partial^2 T}{\partial y^2} = 0 \qquad (13)$$

This kind of problem is easy to solve by difference method.

3. Calculation Program and Example

A 3-D thermal field calculation program has been developed

by FORTRAN-77.
The flow diagram is

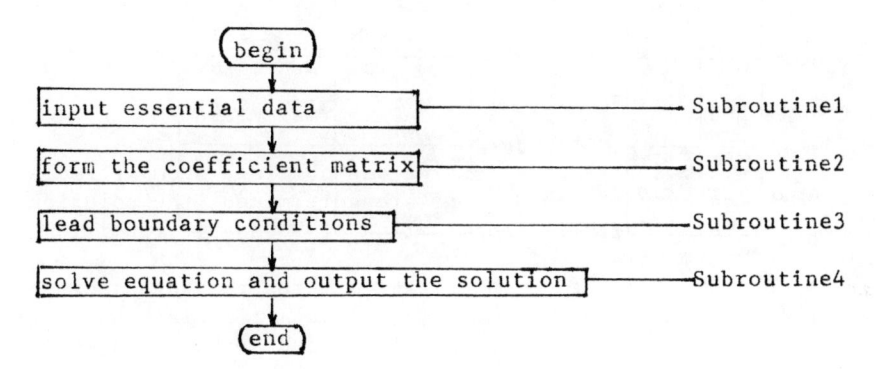

```
           (begin)
              ↓
   ┌──────────────────────┐
   │ input essential data │ ─────────── Subroutine1
   └──────────────────────┘
              ↓
   ┌──────────────────────────┐
   │ form the coefficient matrix │ ───── Subroutine2
   └──────────────────────────┘
              ↓
   ┌────────────────────────┐
   │ lead boundary conditions │ ───────── Subroutine3
   └────────────────────────┘
              ↓
   ┌─────────────────────────────────────┐
   │ solve equation and output the solution │ ── Subroutine4
   └─────────────────────────────────────┘
              ↓
            (end)
```

Program: (be omitted)
Example:
 The shape and size of the example is shown in Fig. 1, and the idealization of the body (becouse the platen structure is symmetry of X-Y axis, so we take 1/4 of platen structure) is in Fig. 2.

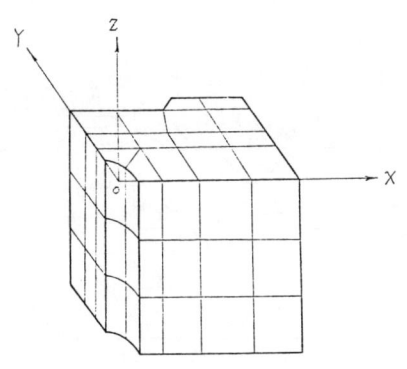

Fig.2 Idealization of the body.

 The temperature distribution is shown in Fig.3 a,b,c.

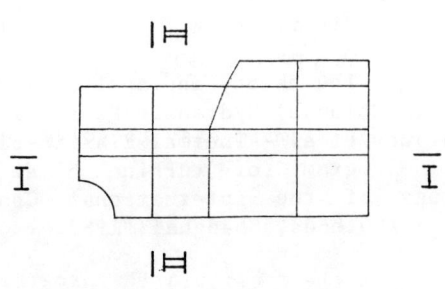

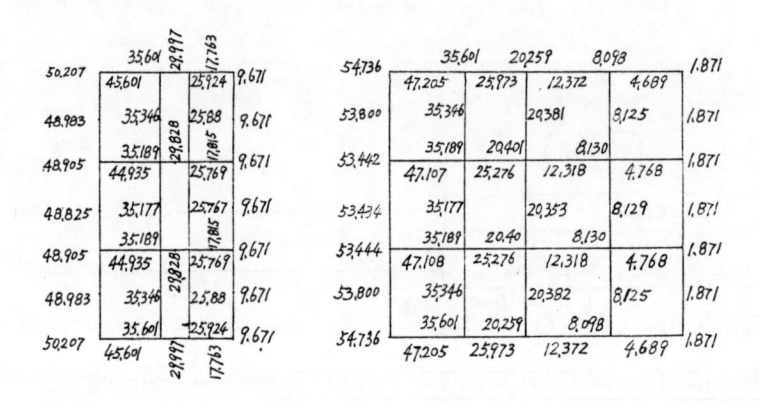

II—II Section   I—I section

(a)   (b)

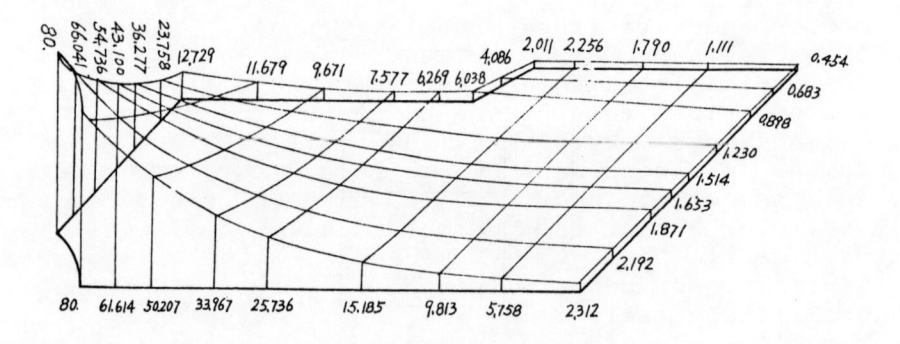

(c) temperature distribution of top surface

Fig.3 temperature distribution

References
1. Zhu Defang - The theory and application of Finite Element Method. (in Chinese) Hydraulic Press, 1979.
2. Sung Waijung et al - Thermal Elastic-plastic Aualysis and computer program for turbine blade root fastenings, Proceedings of the International Conference on Finite Element Methods, Shanghai, 1982.

A NUMERICAL INVESTIGATION OF DOUBLE DIFFUSIVE CONVECTION USING THE FINITE ELEMENT METHOD

V. D. Murty
School of Engineering, University of Portland
Portland, OR 97203

D. B. Paul and M. E. Pajak
Flight Dynamics Laboratory
Wright Patterson Airforce Base, Dayton, OH 45433

ABSTRACT

This investigation is concerned with the application of the finite element method to the stability problem associated with double diffusive convection. The flow of a Boussinesq fluid is studied in a rectangular box of aspect ratio 1.4142, which corresponds to linear stability theory. Three types of hydrodynamic boundary conditions are considered. The solutal Rayleigh number is held fixed at 1000 and numerical results are presented for thermal Rayleigh number varying from 2000 to 10000. The Prandtl number is varied from 0.1 to 10.0. The element used is the eight node type in which all the dependent variables except the pressure are interpolated quadratically and the pressure is interpolated linearly. For certain range of Rayleigh numbers, oscillatory solutions are obtained although the algorithm used is steady state. This unsteady nature of the solution is demonstrated by means of the numerical results which oscillate about a mean value.

INTRODUCTION

Double diffusive convection is the pheonomenon in which the driving density gradients are caused by temperature as well as concentration differences. The physical mechanisms and the governing equations of both of these variables are very similar and hence can be treated easily by methods used to analyze natural convection processes. Double diffusive convection is important in many engineering applications like energy storage in solar ponds, rollover in storage tanks containing liquid natural gas, casting of metal alloys etc. Chen and Johnson [1] provided an overview of the several disciplines in which double diffusive convection is important. This problem poses a challenge to both applied mathematians and physicists due to the stability phenomenon associated with it. There have been several theoretical and numerical investigations of this problem in the past. Zengrando et al [2] used the Fourier series method and obtained results for high solute Rayleigh numbers. Turner [3] and Veronis [4] studied finite amplitude motion for two dimensional flows. Huppert and his coworkers [5] obtained bounds for unsteady and steady motions on the thermal and solutal Rayleigh numbers. Baines and Gill [6], while investigating the stability problem showed that convection can occur at fairly low values of thermal Rayleigh number. This subject has also been investigated experimentally by several researchers in the past. Chen et al [7] have studied double diffusive convection by lateral heating. Gebhart and his coworkers [8] experimentally studied the melting of ice in saline water. There have been relatively few studies involving numerical methods to solve the complete continuum equations and obtain the cut off values for thermal and saline Rayleigh numbers for stability. This paper uses the finite element method to solve the steady state equations of motion and obtain the flow field and heat transfer rates.

GOVERNING EQUATIONS AND FORMULATION

The equations that govern the motion of steady, viscous flow are the conservations of mass, linear momentum, thermal energy, and concentration. They are given as follows (using nondimensional variables):

$$\frac{\partial u'}{\partial x'} + \frac{\partial v'}{\partial y'} = 0 \qquad (1)$$

$$\frac{1}{Pr}\left[u'\frac{\partial u'}{\partial x'} + v'\frac{\partial u'}{\partial y'}\right] = -\frac{\partial p'}{\partial x'} + \nabla^2 u' \qquad (2)$$

$$\frac{1}{Pr}\left[u'\frac{\partial v'}{\partial x'} + v'\frac{\partial v'}{\partial y'}\right] - Ra_T T' - Ra_S C'$$
$$= -\frac{\partial p'}{\partial y'} + \nabla^2 v' \qquad (3)$$

$$u'\frac{\partial T'}{\partial x'} + v'\frac{\partial T'}{\partial y'} = \nabla^2 T' \qquad (4)$$

$$u'\frac{\partial C'}{\partial x'} + v'\frac{\partial T'}{\partial y'} = \nabla^2 C' \qquad (5)$$

where u' and v' are the velocity components in directions x' and y'; T', C', and p' are temperature, concentration, and pressure; Pr, Ra_T, and Ra_S are the Prandtl, thermal and solutal Rayleigh numbers respectively. In the above equations, the effects of generation of heat due to viscous dissipation are neglected and all the properties are treated to be constant. The fluid is assumed to be incompressible to within the Boussinesq approximation that is, density is a constant everywhere except in the buoyancy force term. To complete the continuum description of the problem, a suitable set of boundary conditions are needed. These could be the specification of velocities, temperatures, concentrations, tractions, fluxes, or some combination thereof.

An eight node serendipity element has been used in this study. The element has a total of 36 degrees of freedom (see Fig. 1). The interpolation of the variables is as follows:

$$u' = \phi^t \underset{\sim}{U} \; ; \quad v' = \phi^t \underset{\sim}{V} \; ; \quad T' = \phi^t \underset{\sim}{T} \tag{6}$$

$$c' = \phi^t \underset{\sim}{C} \; ; \quad p' = \psi^t \underset{\sim}{P}$$

where ψ is the shape function for pressure, and ϕ is the shape function vector for velocities, temperature, and concentration; $U, V, T, P,$ and C are the vectors of nodal point values of the dependent variables. If the above equation is introduced into the Galerkin's formulation of Eqs. (1) – (5) the following set of matrix equations is obtained:

$$[\underset{\approx}{C}(\underset{\sim}{\theta}) + \underset{\approx}{K}]\{\underset{\sim}{\theta}\} = \{F\} \tag{7}$$

Due to the nonlinearity of Eq. (7) an iterative scheme is needed. In this study Picard's method of successive substitutions has been used with satisfactory results. The starting vectors are the creeping flow and diffusion solutions (with the convective terms dropped). The following convergence criterion was used for temperature in the j'th iteration:

$$\max\left\{\frac{|T_i^{(j)} - T_i^{(j-1)}|}{T_{max}^{(j-1)}}\right\} \leq 10^{-2} \tag{8}$$

$1 \leq i \leq$ No. of nodes

$$T_{max}^{(j-1)} = \max |T_i^{(j-1)}|$$

$1 \leq i \leq$ No. of nodes

with similar criteria used for all the other dependent variables.

NUMERICAL RESULTS

The problem considered is a rectangular box of height H, and length to match that of a unit cell, 1.4142 as determined by Veronis [4] and other investigators in the past from a linear stability analysis. The physical description of the problem and some boundary conditions is shown in Fig. 2a. The following three cases are considered:

Case	Sides 1,2	Sides 3,4
I	traction free ($v' = 0$)	traction free ($u' = 0$)
II	traction free ($v' = 0$)	no slip ($u' = v' = 0$)
III	noslip ($u' = v' = 0$)	traction free ($u' = 0$)

The thermal boundary conditions in all three cases are the same as shown in Fig. 2a. The vertical boundaries are insulated both thermally and solutally. The solutal Rayleigh number is fixed at 1000 throughout the investigation. The Prandtl number is varied from 0.1 to 10. Numerical results are presented for values of thermal Rayleigh number as high as 10000.

The finite element grid is shown in Fig. 2b. All the numerical solutions took about 16-18 iterations to converge. A plot of streamlines, isotherms, and isohals for Rayleigh number varying from 2000 to 10000 for case I is shown in Fig. 3. It appears that the grid is barely able to model the concentration field although the streamlines and isotherms are quite well represented. The general features of the flow field are that the cell tends to become rectangular with more flow near the walls than the center. Also, notice the thermal and solutal boundary layer formed near the horizontal walls at high Rayleigh numbers. Similar results were obtained for Prandtl numbers of 0.1 and 10 and also for the remaining two cases. Since the plots differed very little from those shown in Fig. 3, they are not reproduced here. A comparison of average Nusselt number with previous investigations is given in Table 1.

This investigation is concerned with steady state convection. Veronis [4], and Huppert [5] have observed in their studies that monotonic convection is not possible for Rayleigh numbers below 2000. They have also observed that for a certain range of Rayleigh numbers only oscillating solutions are possible. The results from this investigation seem to contradict this. This phenomenon was observed accidentally while studying the numerical results for thermal Rayleigh number of 1900. For this value, the convergence process (nonconvergence process in this case) is shown in Table 2 in which the error measured according to (8), the variable, and the change in the variable are shown. It can be easily seen that from iteration number 12, the error is almost constant while there is no change in the variable nor its increment. The numerical solution might be oscillating between successive iterations. With this in mind, solutions were subsequently obtained for iteration numbers 31,32, and 33. Plots of streamlines, isotherms, and isohals after 30,31,32, and 33 iterations are shown in Figure 4. Although the direction of flow is not obvious from the streamlines, the isotherms and isohals clearly indicate an oscillatory solution. That the solution oscillates about a mean value can be seen in Figure 5 in which the velocity, temperature and concentration profiles are drawn along lines of symmetry. This is consistent with the results of Veronis [4] and Huppert and Moore [5] who reported oscillatory solutions for thermal Rayleigh number of 1900. However, in constrast to their investigations in which they found that for low values of thermal Rayleigh number (below 1797) the solution is one of pure conduction, oscillatory solutions were obtained for Rayleigh numbers as low as 1300. It is not known if this an artifact of the numerical scheme since the results of one Rayleigh number are used as the starting vector for the next Rayleigh number. However, the oscillatory solutions do vary with the Rayleigh number.

CONCLUSIONS

The finite element method has been used to analyze steady double diffusive convection in a single cell. Three different cases, based on the boundary conditions have been studied and the results agree quite well with values available in literature. For Rayleigh numbers below 1900, oscillatory solutions were obtained. It is interesting to note that although the numerical algorithm is based on steady state equations, the solution oscillates from iteration to iteration about a mean value, thus confirming that steady solution does not exist. Extensions of the present work to include aspect ratio, and infinite width of channel are underway and will be reported in future.

REFERENCES

1. Chen, C. F., and Johnson, D. H., J. of Fluid Mechancics, Vol. 138, pp. 405-416.

2. Zengrando, F., Ph. D. Dissertation, University of New Mexico, 1979.

3. Turner, J. S., and Stommel, H., Proceedings of U.S. National Academy of Sciences, Vol. 52, pp. 49-53.

4. Veronis, G., J. of Fluid Mechanics, Vol. 34, pp. 315-336.

5. Huppert, E. H., and Moore, D. R., J. of Fluid Mechanics, Vol. 78, pp. 821-854.

6. Baines, P. G., and Gill, A. E., J. of Fluid Mechanics, Vol. 37, pp. 289-306

7. Chen, C. F., Briggs, D. G, and Wirtz, R. A., *Intl. J. of Heat and Mass Transfer*, Vol. 14, pp. 57-65.

8. Gebhart, B., and Pera, L., *Intl. J. of Heat and Mass Transfer*, Vol. 14, 25-2050.

Table I: Saline and Thermal Nusselt number vs. Rayleigh and Prandtl numbers

Ra_T		Pr = 1.0 Veronist [4]	Pr = 1.0 PRESENT Case I	Pr = 1.0 PRESENT Case II	Pr = 1.0 PRESENT Case III	Pr = 10.0 Veronist [4]	Pr = 10.0 PRESENT Case I	Pr = 10.0 PRESENT Case II	Pr = 10.0 PRESENT Case III	Pr = 0.1 Veronist [4]	Pr = 0.1 PRESENT Case I	Pr = 0.1 PRESENT Case II	Pr = 0.1 PRESENT Case III
1900	Nu_S	Osc	Osc	-	-	Osc	-	-	-	-	-	-	-
	Nu	Osc	Osc	-	-	Osc	-	-	-	-	-	-	-
2000	Nu_S	3.307	3.412	1.000*	1.0	3.279	3.421	1.109	1.0	3.393	3.464	1.0	Osc
	Nu	1.787	1.836	1.000*	1.0	1.769	1.840	1.011	1.0	1.844	1.872	1.0	Osc
2500	Nu_S	4.451	4.52	3.021	1.0	4.46	4.544	3.114	1.0	4.450	4.523	1.0	Osc
	Nu	2.505	2.513	1.664	1.0	2.506	2.523	1.714	1.0	2.507	2.515	1.0	Osc
5000	Nu_S	6.490	6.74	4.972	1.0	6.436	6.692	5.272	1.0	6.544	6.774	4.494	Osc
	Nu	3.709	3.743	2.832	1.0	3.681	3.722	2.959	1.0	3.736	3.753	2.615	Osc
10000	Nu_S	8.360	9.37	6.635	1.0	-	8.914	7.139	1.0	-	9.361	5.9	Osc
	Nu	4.965	5.098	3.717	1.0	-	4.881	3.950	1.0	-	5.087	3.40	Osc

+ Veronis [4] considered only case I.
* Ra_T for this case is 2100.

Table II: Convergence Process for Ra_T of 1900

Iteration	Error	Change in Variable	Variable
1	0.11007E+00	0.864600BE+00	0.7854791E+01
2	0.11227E+00	0.1887762E+02	0.1681446E+03
3	0.10850E+00	0.1824435E+02	0.1681446E+03
4	0.12579E+00	0.2115041E+02	0.1681446E+03
5	0.17161E+00	0.2885485E+02	0.1681446E+03
6	0.29452E+00	0.3790305E+01	0.1286926E+02
7	0.35207E+00	0.4530867E+01	0.1286926E+02
8	0.51153E+00	0.5115271E+00	0.1000000E+01
9	0.58372E+00	0.5837213E+00	0.1000000E+01
10	0.59020E+00	0.5901967E+00	0.1000000E+01
11	0.58970E+00	0.5896966E+00	0.1000000E+01
12	0.59000E+00	0.5899962E+00	0.1000000E+01
13	0.58987E+00	0.5898716E+00	0.1000000E+01
14	0.58992E+00	0.5899229E+00	0.1000000E+01
15	0.58990E+00	0.5899020E+00	0.1000000E+01
16	0.58991E+00	0.5899106E+00	0.1000000E+01
17	0.58991E+00	0.5899071E+00	0.1000000E+01
18	0.58991E+00	0.5899085E+00	0.1000000E+01
19	0.58991E+00	0.5899079E+00	0.1000000E+01
20	0.58991E+00	0.5899082E+00	0.1000000E+01
21	0.58991E+00	0.5899081E+00	0.1000000E+01
22	0.58991E+00	0.5899081E+00	0.1000000E+01
23	0.58991E+00	0.5899081E+00	0.1000000E+01
24	0.58991E+00	0.5899081E+00	0.1000000E+01
25	0.58991E+00	0.5899081E+00	0.1000000E+01
26	0.58991E+00	0.5899081E+00	0.1000000E+01
27	0.58991E+00	0.5899081E+00	0.1000000E+01
28	0.58991E+00	0.5899081E+00	0.1000000E+01
29	0.58991E+00	0.5899081E+00	0.1000000E+01
30	0.58991E+00	0.5899081E+00	0.1000000E+01

• U,V,T,C

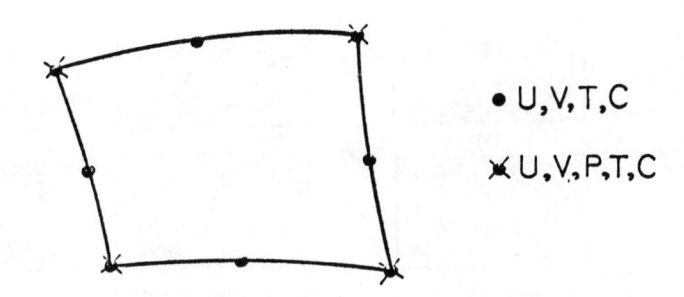

✕ U,V,P,T,C

Figure 1: Eight node element used in this study

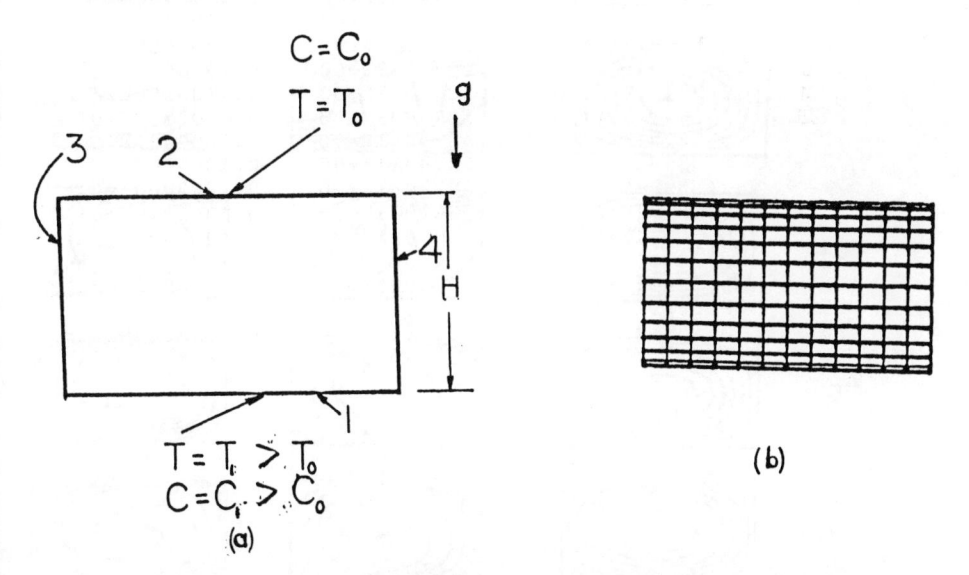

Figure 2: a) Description of the continuum problem with boundary conditions and b) Finite Element grid

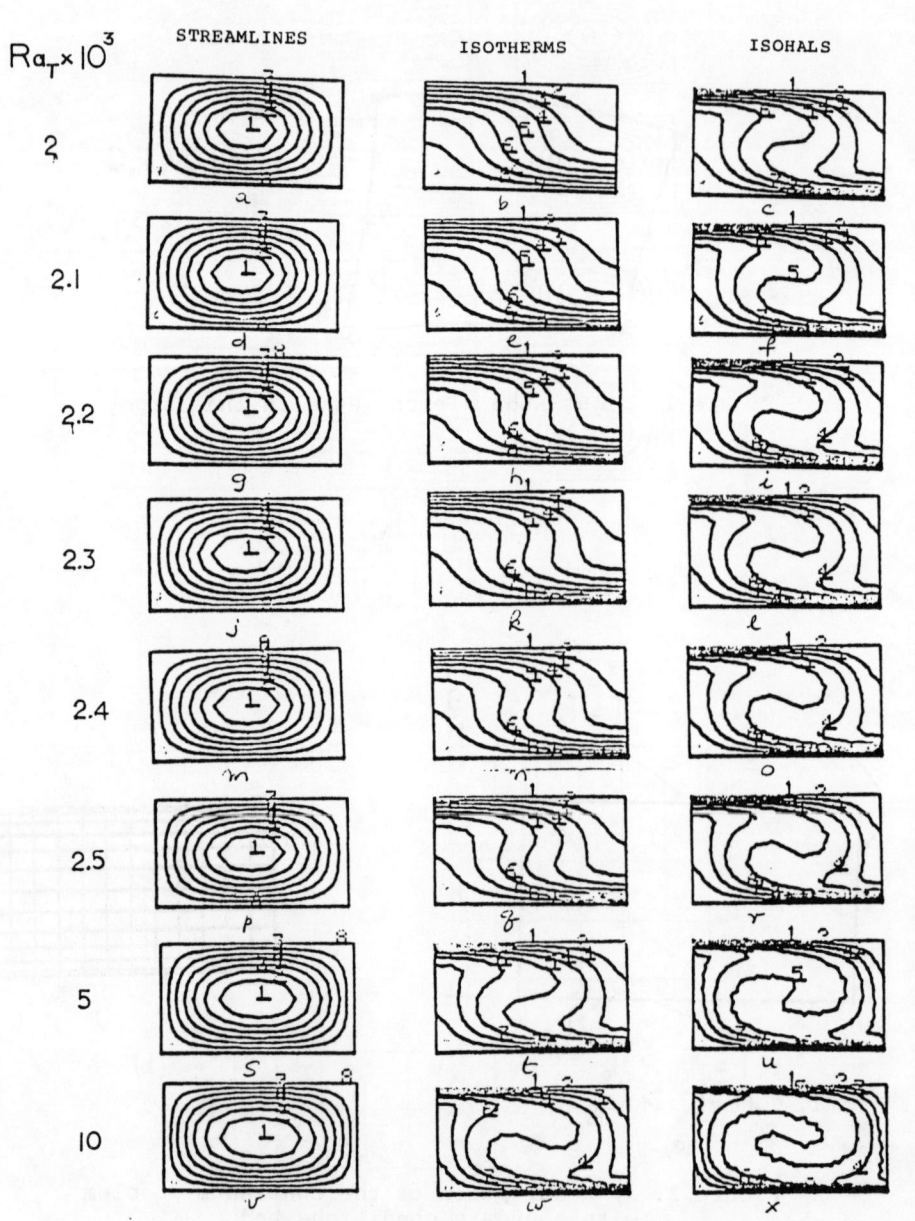

Figure 3: Streamlines, isotherms, and isohals for various values of Ra_T.

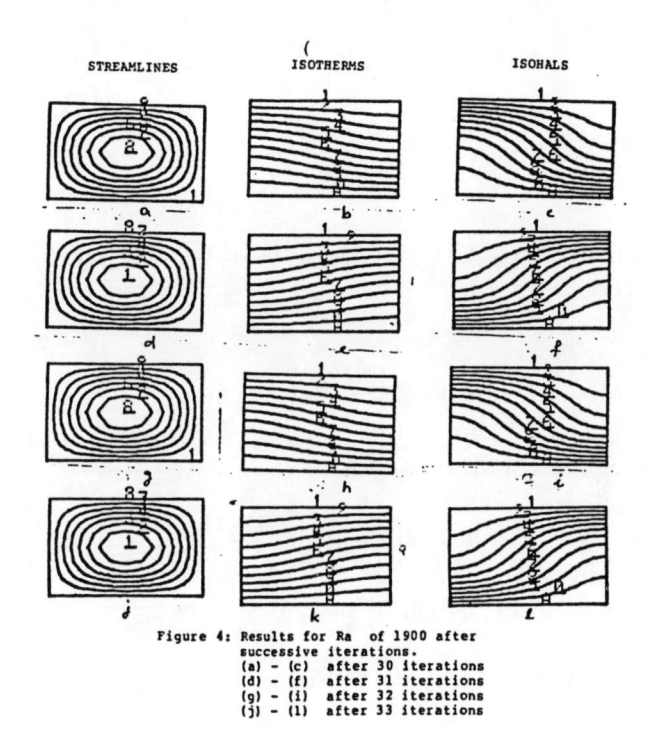

Figure 4: Results for Ra of 1900 after successive iterations.
(a) - (c) after 30 iterations
(d) - (f) after 31 iterations
(g) - (i) after 32 iterations
(j) - (l) after 33 iterations

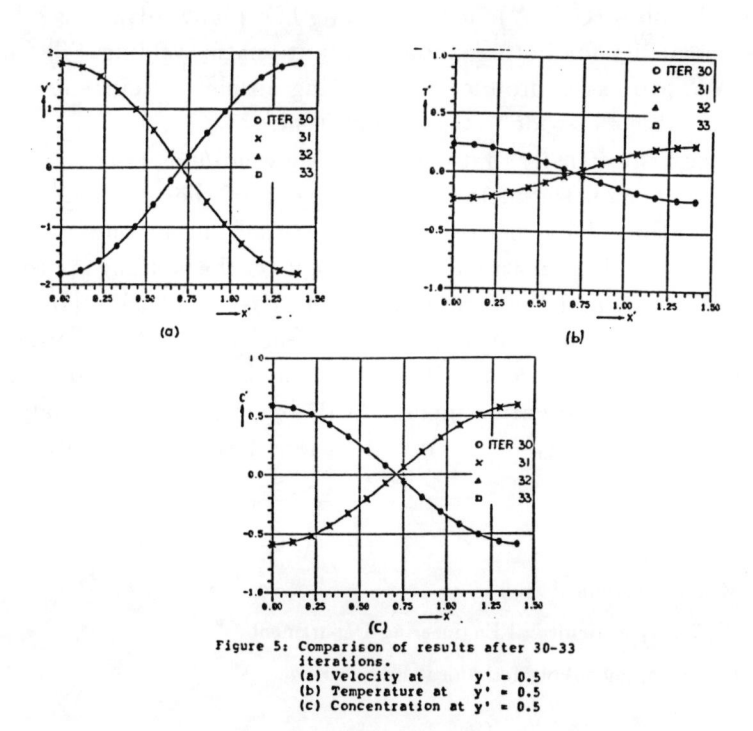

Figure 5: Comparison of results after 30-33 iterations.
(a) Velocity at $y' = 0.5$
(b) Temperature at $y' = 0.5$
(c) Concentration at $y' = 0.5$

EXPERIMENTAL AND ANALYTICAL STUDY OF AN ARC WELDING PROCESS

R. L. Akau [1]
Fluid Mechanics and Heat Transfer Division
Sandia National Laboratories
Albuquerque, NM, 87185, USA

G.W. Krutz [II] and R.J. Schoenhals [III]
Purdue University
West Lafayette, IN, 47906, USA

1. INTRODUCTION

Welding is an important process in industry for manufacturing and fabrication purposes. There are many types of welding. Gas Tungsten Arc Welding (GTAW), in particular, has many advantages because of its versatility for both manual and automatic welding. The arc of the GTAW process is produced by passing an electric current from a nonconsumable electrode to a base material.

The heat transfer that occurs during welding is important because of its influence on the metallurgical makeup of the base material and the modification of its mechanical properties. The amount of heat transferred to the base material is dependent on the welding speed, electrical power input, and the type of shielding gas. A considerable amount of research has been performed on heat transfer modeling in order to determine the temperature distributions in the base material during welding. Methods of investigation include both experimental and analytical techniques. A majority of the investigations have been purely theoretical,

---

[1] Member Technical Staff

[II] Professor, Agricultural Engineering Department

[III] Professor, School of Mechanical Engineering

which include exact and approximate solutions. Approximate methods, which include numerical schemes such as finite difference and finite element [1-5], are used extensively because of their ability to handling variable properties, phase change, and complex boundary conditions. The thermal and mechanical processes which occur during welding are very difficult to model because of melting and solidification, motion of the liquid metal in the weld pool, irregular weld bead surface, surface tension and magnetic forces.

This investigation involves both an experimental study of a GTAW process and the development of a numerical heat transfer model for the base material. Measured and calculated results will be compared to validate the thermal model. Actual welding (i.e. joining of metals) is not performed in this study. The welding process consists of just passing the arc over the base material. However, the results obtained can be readily applied to actual welding conditions.

2. EXPERIMENTAL APPARATUS

The experimental apparatus consisted of a welder power supply. Some of its features are: alternating current(AC) or direct current (DC), voltmeter and ammeter, current control dial, selector switch, straight polarity (SP) or reverse polarity (RP), high frequency start, and post flow timer. The high frequency start permits striking the arc by means of an impressed voltage that causes a spark to jump from the electrode to the base material. For AC welding the high frequency start is continuously used to restrike the arc when the voltage goes through the zero point of the AC cycle. On the other hand, for DC welding the high frequency control is terminated after arc initiation. The post flow timer allows the shielding gas to flow for a short period of time after the arc is turned off. This prevents the hot tip of the electrode from oxidizing when it is exposed to the atmosphere. For each experimental test, the welder power supply was set at DC and SP welding conditions (the welding electrode was negative polarity).

Argon gas was used as the shielding gas. The pressurized argon gas was stored in a steel cylindrical tank, and the flowrate was controlled with a regulator and measured with a flowmeter. The welding torch was a lightweight, water cooled, and general purpose torch. The torch holds a non-consumable electrode which has a pointed tip. The thoriated tungsten electrode has a diameter of 0.381 cm. The welding torch was held to a traversing mechanism which provided different constant velocity settings and maintained a constant arc gap between the electrode tip and the base material. The mechanism could move forward or backward along a horizontal aluminum rack. A two channel strip chart recorder was used to obtain transient recordings of welder supply voltage and current, and total welding time. A foot pedal provided arc initiation, and manual control of the magnitude of the current flow.

Temperatures in the base material were recorded with 30 gage (0.025 cm diameter) chromel-alumel thermocouples embedded in a plane normal to a line oriented along the path traced out by the moving welding torch. The thermocouples were capable of recording temperatures up to 1370°C. Each thermocouple had a beaded tip to improve temperature response, and a ceramic insulator (0.1588 cm diameter) covered with a metal sheath to provide structural support during installation. The thermocouples were inserted into 0.2 cm diameter holes drilled in a steel plate which was 1.27 cm thick, 10.16 cm wide, and 15.24 cm long. Before each test the surface of the plate was cleaned with acetone and the holes cleaned with a jet of pressurized air. The thermocouples were then inserted into the holes and attached to the plate with a ceramic cement capable of withstanding temperatures up to 1100°C. A mechanical stepping switch was used to scan each of the thermocouples continuously. The stepping was driven by a logic signal from a PDP 11/70 computer. A computer terminal accessed a data acquisition program which interactively controlled the experimental procedure.

The current setting dial on the welder power supply was calibrated along with the welder power supply ammeter and voltmeter. Five different current dial settings were calibrated and the results are shown in Table 1. The power input to the welder power supply ($\dot{Q}_{elec.}$) was calculated from the product of the voltage and current values. The traversing mechanism, which holds the welding torch, was also calibrated at four different velocity settings. The average velocities ranged from 0.27 cm/sec to 0.62 cm/sec, respectively. Water flows through the electrical cable and the welding torch to prevent the torch from overheating. The average mass flowrate of the water was 1.5 kg/min. The inlet and outlet water temperatures were also recorded with chromel-alumel thermocouples. The thermocouples were located before and after the water

Current Setting Dial (amps)	Observed Current (amps)	Welder Supply Voltage (volts)
125	149.5±5	13.5±1
150	182.0±5	14.3±1
170	209.4±5	15.9±1
200	241.3±5	17.8±1
220	262.8±5	19.3±1

Table 1. Current and Voltage Values for Power Supply.

left the welder power supply. The thermocouples were inserted in the mainstream so as to obtain a better average water temperature. The signals from these thermocouples were recorded simultaneously with the thermocouples imbedded in the base material.

The welding parameters used in this investigation are given in Table 2. For each test, the electrode extension, arc gap, and electrode tip included angle was 0.3175 cm, 0.03125 cm, and 30 degrees, respectively. These values were chosen to insure arc initiation, adequate gas shielding, and a clear observation of the arc. Glickstein et al [6], and Niles and Jackson [7] experimentally verified that a 30 degree electrode tip angle provides a higher arc voltage and a larger weld pool than other tip angles.

Torch Velocity (cm/sec):			
0.27	0.40	0.50	0.62
Current Dial Setting (amps):			
125	170	200	220
Argon Shielding Gas Flow (m^3 per hour):			
0.28	0.42	0.47	

Table 2. Welding Parameters.

3. EXPERIMENTAL RESULTS

Measurements were conducted with the previously described experimental apparatus and welding parameters shown in Table 2. Time-temperature data from the thermocouples located in the base material were needed as input to a heat transfer model to determine surface heat input. The heat transferred to the water coolant was calculated knowing the inlet and outlet water temperatures and mass flowrate of the water coolant. A photographic study of the arc provided information for determining the diameter of the arc at the plate surface.

3.1 Heat Transfer to Water Coolant

The primary mode of heat transfer to the water coolant was by heat conducted through the torch head and then convected to the water, and Joule heating (I^2R) from the power cable. The heat transfer to the water coolant assuming a constant flowrate, no work input, negligible boundary work, incompressible fluid, and no rate of change of kinetic and potential energies is

$$\dot{Q}_{water} = \dot{m} \, c_p \, (T_{out} - T_{in}) \qquad (1)$$

where $\dot{m}$ is the mass flowrate of the water, c_p is the specific heat of the water and T_{out} and T_{in} are the steady-state outlet and inlet water temperatures. Usually, the water temperature reached steady-state approximately 20 seconds after the arc was started. A plot of $\dot{Q}_{water}$ versus the electrical power input ($\dot{Q}_{elec.}$) showed a linear relationship. For shielding gas flowrates of 0.28 m^3 per hour and 0.42 m^3 per hour, there was not a very large deviation of heat transfer rate to the water for a particular power setting. However, the velocity did have a moderate effect on the heat transfer rate at a shielding gas flowrate of 0.57 m^3 per hour. Results also showed that the magnitude of heat transfer rate to the water ranged from 15 to 50 percent of $\dot{Q}_{elec.}$.

3.2 Temperature Measurements in Base Material

The thermocouple nearest to the heat source (TC2), located 0.635 cm below the path of the arc, experienced an increase in temperature first and also the highest maximum temperature. A plot of maximum temperatures versus $\dot{Q}_{elec.}$ showed a linear relationship. Increasing the torch velocity reduced the peak temperature because the energy per unit length is reduced. The effect of shielding gas flowrate on the peak temperature was very small as compared to torch velocity and $\dot{Q}_{elec.}$.

3.3 Welding Arc Observations

Photographs were taken of the arc in order to determine the width of the arc at the surface of the base material as a function of different welding parameters. Because the surface of the metal plate was fairly smooth, there was considerable reflection of light which made it difficult to distinguish the light emanating from the arc from that reflected from the surface. Nevertheless, the photographs illustrated that the overall size of the arc was not significantly affected by the shielding gas flowrate or torch velocity.

Since the shielding gas flowrate did not affect the size of the arc, photographs of the front view of the arc were taken for a shielding gas flowrate of 0.57 m^3 per hour. The arc was observed to be symmetric, and the torch velocity did not change the overall size of the arc. From the front view photographs, the arc diameter was measured with a micrometer and the results are shown in Table 3.

Glickstein [8] observed from arc photographs that the arc diameter varied linearly with current. From the values in Table 3, a 75 percent increase in average current generated an 80 percent increase in the average arc diameter.

Average Voltage (volts)	Average Current (amps)	Arc Diameter (cm)
13.5	149.5	0.5 ± 0.10
15.9	209.4	0.74 ± 0.15
17.8	249.3	0.85 ±0.17
19.3	262.8	0.90 ±0.18

Table 3. Arc Diameter at Plate Surface.

4. THERMAL MODEL

A heat transfer model was developed for the base material. The model assumes that the base material moves at a constant velocity, V, and is subjected with a stationary, two-dimensional Gaussian heat flux distribution as shown in Figure 1. The thermal model also incorporates variable properties (thermal conductivity and specific heat), and phase change. The phase change is implemented by using the enthalpy method. This approach alleviates the sudden change in specific heat with temperature when melting occurs. The motion in the weld pool is accounted for by specifying an effective thermal conductivity value (k_{eff}) which is one and a half times larger than the thermal conductivity at the melting point, 1755°K. The base material is AISI 1020 steel and the properties

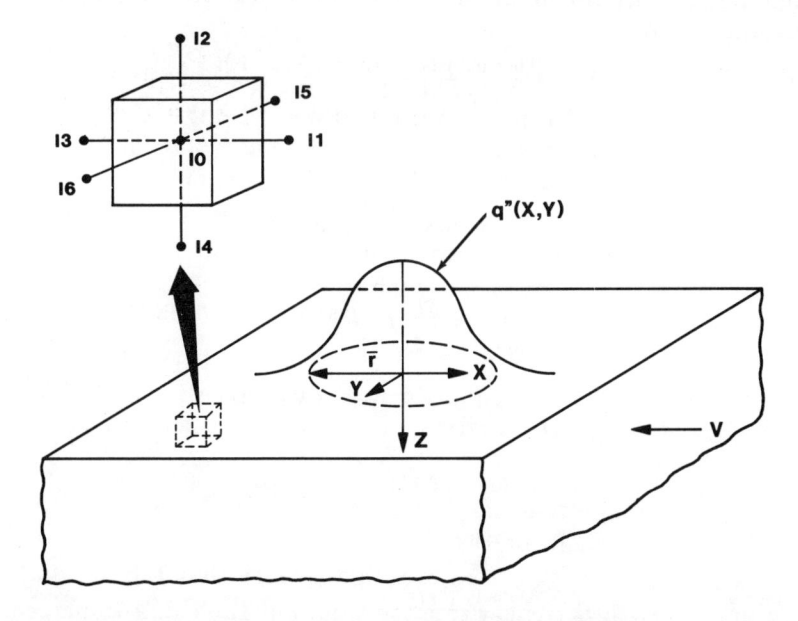

Figure 1: Medium imposed with stationary heat source.

are given in Table 4. Performing an energy balance on a control volume in Figure 1, the energy equation is expressed as

$$\frac{\partial}{\partial x}(k\frac{\partial T}{\partial x}) + \frac{\partial}{\partial y}(k\frac{\partial T}{\partial y}) + \frac{\partial}{\partial z}(k\frac{\partial T}{\partial z}) + V\frac{\partial}{\partial x}(\rho c T) = 0 \qquad (2)$$

where k, T, and ρ and c are the thermal conductivity, temperature, density and specific heat of the medium, respectively. The first three terms in Equation (2) define the heat conduction and the fourth term is the advection heat transfer due to the bulk motion of the medium.

The boundary conditions are

$$\frac{\partial T}{\partial y} = 0, y = 0$$

and

$$q''(x,y) = q''_{max} e^{-3(x^2 + y^2)/\bar{r}^2}, z = 0$$

where q''_{max} is the maximum heat flux at the center of the heat source defined as

$$q''_{max} = \frac{3\dot{Q}}{\pi \bar{r}^2} \qquad (3)$$

The parameter $\dot{Q}$ in Equation (3) is the total heat input from the source and $\bar{r}$ is the radial distance within which 95 percent of the heat input is contained. The first boundary condition is due to symmetry at the mid-plane (y=0).

Thermal Conductivity (W/m-°K):
k = 77.32 - 0.0451 T, T $\leq$ 1100°K
k = 16.05 + 0.0106 T, 1100°K < T $\leq$ 1755°K
k = k_{eff}, T > 1755°K

Specific Heat (W-sec/kg-°K):
c = 236.9 + 0.58 T, T $\leq$ 1000°K
c = 1000, 1000°K < T $\leq$ 1200°K
c = 660, T > 1200°K
c = 660 + $\frac{\Delta H_F}{T_L - T_S}$, 1755°K $\leq$ T $\leq$ 1796°K
c = 660, T > 1796°K

Latent Heat of Fusion (W-sec/kg):
ΔH_F = 271400

Solidus and Liquidus Temperatures (°K):
T_S = 1755
T_L = 1796

Density (kg/m^3):
ρ = 7850

Table 4. Property Data for AISI 1020 Base Material.

Equation (2) is an elliptic equation representing a boundary value problem. The medium is discretized into nodal points in the x, y, and z directions. The nodal arrangement is also shown in Figure 1. The nodal temperature at the center node, I0, is obtained from temperatures at six neighboring nodes located at I1, I2, I3, I4, I5, and I6. Equation (2) is solved numerically by finite difference. The second order derivatives in Equation (2) were estimated with centered difference equations and the advection term was estimated using an upwind scheme. This avoids the stability restriction on velocity that result from a central difference approach but does introduce some numerical diffusion into the solution. The temperatures in the medium were calculated by sweeping the nodes continuously using a Gauss-Seidel iteration until the difference between the present nodal temperature and the nodal temperature at the previous iteration was less than a convergence criteria $(0.1°K)$.

The heat transfer model also assumes there is no heat loss at the boundaries. Of course, during an actual welding process heat loss occurs by radiation and convection at the surface of the base material. However, this assumption implies that the heat loss is considered negligible as compared to the total heat input, $\dot{Q}$, and heating occurs over a short duration of time. An added feature of the model is the use of a variable grid generator for the finite difference grid. This provides a large number of nodal points in regions of high temperature gradients and where melting occurs. The finite difference grid used in this investigation is shown in Figure 2. Due to symmetry, only half of the medium is shown.

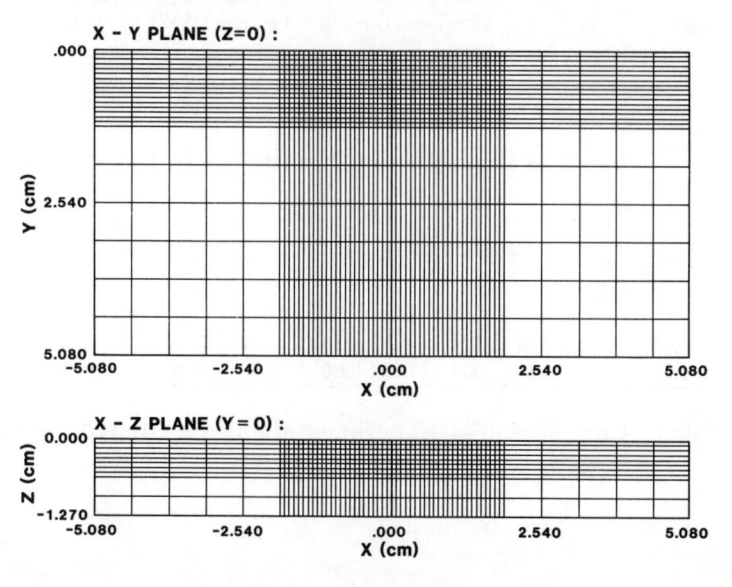

Figure 2: Finite difference grid.

The grid consisted of 59 nodes in the x-direction, 23 nodes in the y-direction, and 11 nodes in the z-direction. The values of $\dot{Q}$ and $\bar{r}$ were determined, for each test, by changing the value of $\dot{Q}$ until the calculated temperature and the measured maximum temperature at thermocouple TC2 were approximately the same, and changing $\bar{r}$ until the calculated width of the weld pool compared to the measured value. The calculated values for TC2 and $\bar{r}$ should match exactly with the measured results. However, this would be a tedious iterative process.

5. DISCUSSION OF RESULTS

The experimental results are shown in Table 5 and the comparison of measured and calculated results are given in Table 6 for two tests. A metallurgical investigation of the weld cross-sections normal to the direction of the heat source were conducted to determine the width (W) and depth (d) of the weld bead. The weld bead cross-sections were observed to be symmetric with respect to the mid-plane (y=0). The results in Table 6 show a good comparison between measured and calculated results. Test 1 shows an over-predicted weld bead width due to the additional heat input ($\dot{Q}$) needed to raise the temperature at the location of TC2 closer to the maximum experimental value. However, the ratio of W/d is in very good agreement. Also, the values of $\bar{r}$ in Table 6 are within the arc size values which were obtained photographically in Table 3.

Test	Torch Velocity (cm/sec)	Average Current (amps)	Average Voltage (volts)	$\dot{Q}_{elec.}$ (kW)	$\dot{Q}_{water}$ (kW)
1	0.27	149.5	13.5	2.02	0.39
2	0.27	262.8	19.3	5.07	1.56

Table 5. Experimental Results for Tests 1 and 2.

Test	$\bar{r}$ (cm)	$\dot{Q}$ (kW)	TC2 (°K) Exp.	TC2 (°K) Calc.	Weld Bead Width (cm) Exp.	Weld Bead Width (cm) Calc.	Weld Bead Depth (cm) Exp.	Weld Bead Depth (cm) Calc.	W/d Exp.	W/d Calc.	η
1	0.18	1.5	530	524	0.472	0.555	0.226	0.238	2.09	2.33	0.74
2	0.55	3.3	743	718	1.024	1.050	0.331	0.325	3.08	3.23	0.65

Table 6. Comparison of Experimental and Calculated Results.

The overall welding efficiency is defined as

$$\eta = \frac{\dot{Q}}{\dot{Q}_{elec.}} \qquad (4)$$

The values of η shown in Table 6 indicate a decrease in welding efficiency as the welding current increases. This trend is consistent with those obtained by Niles and Jackson [7]. However, the values of η were 10 percent higher than those obtained in reference [7].

A typical temperature contour plot calculated from the thermal model is shown in Figure 3 for Test 1 for the top surface (z=0) and at the midplane (y=0). The heat source center is located at x=y=z=0. The region enclosed by the 1755°K isotherm (cross-hatched area) defines the weld pool. The isotherms are tear dropped shaped with the maximum weld pool width and depth located behind the heat source center. The temperature gradients are highest in front of the heat source center.

The thermal model has several shortcomings in that it does not have a detailed model of the weld pool motion, assumes no heat loss at the base material surface, and neglects vaporiztion. However, the model does provide guidance in obtaining weld pool shapes and temperature distributions.

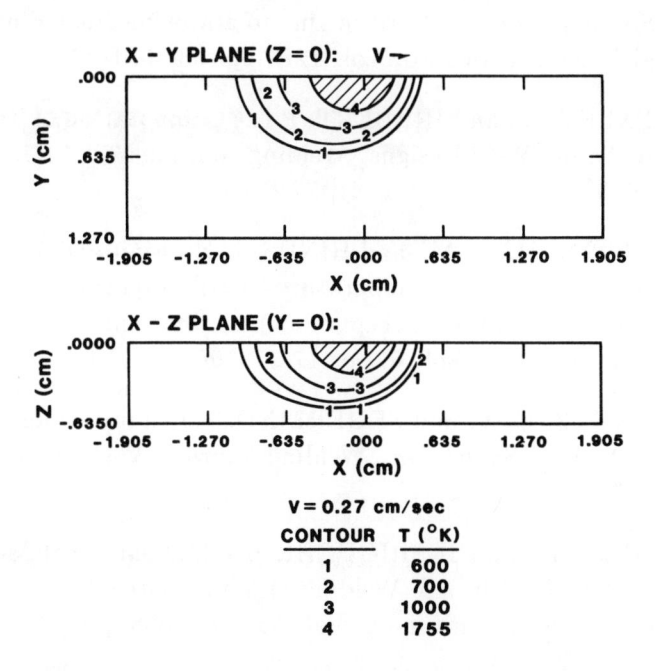

Figure 3: Temperature contour plots for Test 1.

6. CONCLUSIONS

An experimental apparatus was constructed for a gas tungsten arc welding process. Experimental results showed that the heat transfer to the water coolant is a significant fraction of the electrical power input to the welder power supply. Photographs were taken of the arc, and the measured arc diameter at the surface of the base material varied linearly with welder power supply current.

A three-dimensional thermal model was developed for the base material. A comparison of the calculated and measured weld pool widths and depths were in good agreement.

ACKNOWLEDGEMENT

This work was performed at Sandia National Laboratories, supported by the U.S. Department of Energy, under Contract No. DE-AC04-76DP00789.

REFERENCES

1. KOU, S., HSU, C., and MEHRABIAN, R.- Rapid Melting and Solidification of a Surface Due to a Moving Heat Flux. Metallurgical Transactions B, Vol. 12B, pp. 33-45, 1981.

2. PALEY, Z., and HIBBERT, P.D.- Computation of Temperatures in Actual Weld Designs. Welding Journal, Vol. 54, pp. 385s-392s, 1975.

3. ALAMEIDA, S.M., and HINDS, B.K.- Finite Difference Solution to the Problem of Temperature Distribution Under a Moving Heat Source Using the Concept of a Quasi-Stationary State. Numerical Heat Transfer, Vol. 6, pp. 17-27, 1983.

4. KRUTZ, G.W. and SEGERLIND, L.J.- Finite Element Analysis of Welded Structures. Welding Journal, Vol. 57, pp. 211s-216s, 1978.

5. OHJI, T., and HISHIGUCHI, K.- Mathematical Modeling of a Molten Pool in Arc Welding of Thin Plate. Technology Reports of the Osaka University, Vol. 33, No. 1688, pp. 35-43, 1983.

6. GLICKSTEIN, S.S., FRIEDMAN, E., and YENISCAVICH, W.- Investigation of Alloy 600 Welding Parameters. Welding Journal, Vol. 54, pp. 113s-122s, 1975.

7. NILES, R.W., and JACKSON, C.E.- Weld Thermal Efficiency of the GTAW Process. Welding Journal, Vol. 54, 25s-32s, 1975.

8. GLICKSTEIN, S.S.- Basic Studies of the Arc Welding Process. Trends in Welding Research, Ed. DAVID, S.A., proceedings of a conference sponsored by the Joining Division of the American Society for Metals, New Orleans, Louisiana, Nov. 16-18, 1981.

HEAT TRANSFER AND SOLIDIFICATION OF A STREAM OF MOLTEN METAL DURING ATOMISATION BY AN IMPINGING GAS JET

S. Rogers and L. Katgerman

Alcan International Limited,
Southam Road, Banbury, Oxon, OX16 7SP, England.

1. SUMMARY

To obtain new alloys for use in the metals industry atomisation processes have been developed instead of conventional casting techniques.

A mathematical model has been developed to investigate the heat and mass transport processes occurring when a stream of molten metal is atomised by an impinging high speed gas jet. The metal stream breaks up into droplets which are cooled by the gas and solidification of these droplets can take place.

The process is modelled as being two phase, steady state and the flow is turbulent and the effect of process variables on the history of the droplets and the overall flow pattern is calculated.

2. INTRODUCTION

Spray deposition is a relatively new manufacturing technology to produce solid billets, tubes, coatings, and strip directly from the melt. The major advantage of the process is that a rapidly solidified, near-net shape product can be fabricated in a single operation with significant energy earnings. Similar to powder metallurgy processing a liquid stream of alloy is atomised to form a spray of molten droplets of about 70-100 microns in diameter.

The molten droplets are splatted onto a substrate after a short flight distance. At the substrate they consolidate and build-up to form a preform. By manoeuvering the substrate in a number of different ways underneath the metal spray a variety of shapes can be produced. The process was pioneered by Singer [1] and is currently commercialised by Osprey Metals Limited in Great Britain, and a number of their licencees worldwide.

Injection of ceramic particulate into the metal spray allows fabrication of metal matrix composites with a variety of exciting properties [2].

In order to gain an understanding of the transfer mechanisms involved in the process a model has been developed to investigate the heat and mass transport between a stream of liquid metal and an impinging gas jet.

Figure 1 gives an outline description of the process and indicates the geometry used in the model.

Essentially the metal jet is atomised at region A and droplets that are formed are cooled by the gas and should arrive at the substrate B in a semi-liquid state and further freezing takes place here resulting in a compact final product.

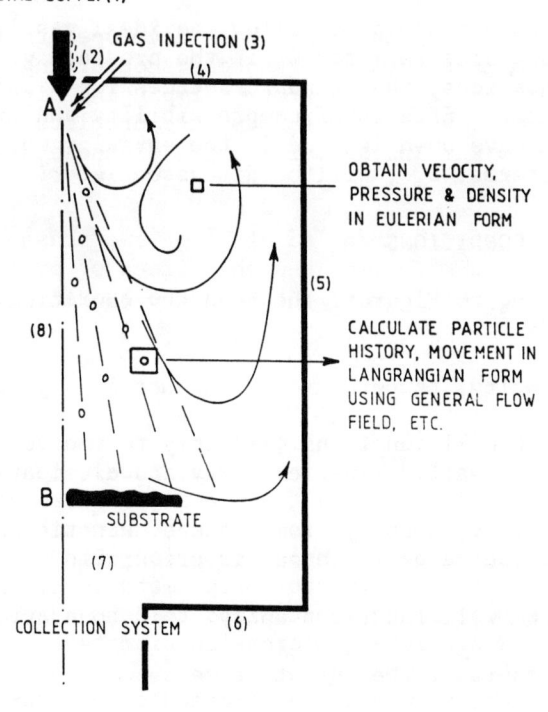

Figure 1. Schematic of Process

3. MATHEMATICAL FORMULATION

The flow is considered to be two dimensional, axisymmetric, steady-state and two phase with the gas considered to be the first phase and the metal the second phase. The flow in the gas is considered to be turbulent and the k-ε model of turbulence [3] is used for evaluating the turbulent viscosity of the gas. The equations are cast in an Eulerian frame and the continuity equations, the equations of motion, the enthalpy equations for the two phases as well as the equations for the kinetic energy of turbulence and the dissipation rate can all be written in the generalised form;

$$\text{div}\,(r_i\,\rho_i\,v_i\,\phi - r_i\,\Gamma_i\,\text{grad}\,\phi) = r_i\,S_\phi + r_i\,S_{\phi_T} \qquad (1)$$

Here r_i is the volume fraction of phase;
ρ_i is the density of phase;
v_i is the velocity of phase;
ϕ is the dependant variable in question
Γ_i is an exchange coefficient
S_ϕ is a source term for ϕ per unit phase volume
S_{ϕ_T} is the source term for ϕ due to interphase processes

The gas is assumed to follow the ideal gas law, PV = RT and S_ϕ, the source term for ϕ, is the pressure gradient in the momentum equations, the standard sources for k and ε, and is zero elsewhere. Effects of compressibility and viscous dissipation have been ignored in the enthalpy equations. The interphase terms, S_{ϕ_T}, will be discussed later.

4. BOUNDARY CONDITIONS

Referring to Figure 1 the boundary conditions are that at regions:

(1) the velocity, temperature of the metal is prescribed

(2) turbulent wall functions [3] apply to the velocities (i.e. no slip at wall), kinetic energy and dissipation rate

(3) the velocity, density, temperature, kinetic energy and the dissipation rate of the gas is prescribed

(4), (5), (6) wall functions apply as in boundary (2)

(7) the pressure at the outlet is zero

(8) is a symmetry boundary

Conditions at the substrate are that as any metal droplets that hit the substrate are likely to stick there it will be assumed that there is a sink of mass (of the second phase) which is equal to:

$$\dot{m} = -r_2 \, \rho_2 \, v_2 \tag{2}$$

where v_2 is the second phase velocity normal to the substrate.

The substrate will be considered to be impervious to the first phase and that wall functions apply there.

5. INTERPHASE TERMS

The process of droplet formation due to a liquid jet being atomised by a high speed gas jet is a complex one [4] and no attempt will be made to model this process here. Instead it will be assumed that the metal consists of droplets of given size and their spatial distribution is defined in terms of their volume fraction in any given computational cell.

The interphase transfer terms for momentum and energy can then be defined in terms of this droplet size.

The frictional force between the phases per unit volume is defined as:

$$F = 0.5 \, C_D \, A_D \, \rho_1 \, v_{SLIP}^2 \tag{3}$$

A_D is the projected area of the droplets per unit volume, v_{SLIP} is the slip velocity between the phases, ν the kinematic viscosity of the gas.

C_D is the drag coefficient and is the following function of the local Reynold's number [5]:

$$C_D = \frac{24}{Re} \times (1 + 0.15 \, Re^{0.687}) + \frac{0.42}{(1 + \frac{4.25E4}{Re^{1.16}})} \tag{4}$$

where $Re = \dfrac{v_{SLIP} \, d}{\nu}$

This formula is approximately valid for spheres up to a Reynold's number of 3×10^5.

The interphase heat transfer terms have to be considered carefully to allow for the effect of latent heat. The conservation equations are written in terms of the enthalpy of each phase and these are related to the temperature of each phase as follows;

$$H_1 = c_1 \, T_1 + \tfrac{1}{2} v^2 \tag{5}$$

$$H_2 = c_2 T_2 + (1-f)L \tag{6}$$

Here c_i is the specific heat of phase i, L the latent heat of the metal and f the fraction solid which is defined simply as the following function of temperature.

$$\begin{aligned} f &= 0 & T_2 &> T_L \\ f &= \left(\frac{T_L - T_2}{T_L - T_S}\right)^\alpha & T_S &< T_2 < T_L \\ f &= 1 & T &< T_S \end{aligned} \tag{7}$$

Here T_L and T_S are the liquidus and solidus temperatures for the metal and are typically 620°C and 520°C respectively. α is an experimentally determined constant.

The kinetic energy of the gas ($\frac{1}{2} v^2$) has been included in the expression for the enthalpy of gas as it is not negligible at the gas speeds considered.

The interphase heat flux, Q, between the two phase per unit area of interface is determined by:

$$Q = h (T_1 - T_2) \tag{8}$$

where h is a heat transfer coefficient which is given by correlations [5] in terms of the local Reynold's and Prandtl numbers.

Relating T_1 and T_2 to the enthalpy by means of equations (5) and (6) complete the set of equations.

6. SOLUTION PROCEDURE

The two continuity equations, four momentum equations, two enthalpy equations and the k and ϵ equations can all be written in the conservation form (1). The solution procedure employed is the IPSA algorithm [6] as employed in the PHOENICS software package.

In brief, the hydrodynamic solution cycle takes the form:

(a) solve for volume fractions from the continuity equations, ensuring they sume to one

(b) with guessed pressures, solve for the velocities

(c) insert into a joint continuity equation to establish a representation continuity error

(d) solve for the pressure corrections using the joint continuity equation

(e) repeat the process until convergence

Because of the large velocities encountered large under relaxation devices have to be employed to obtain a converged solution resulting in large computational times.

Although the formulation is an Eulerian one, it is useful to track variables such as temperature along the phase streamlines to attain their time history characteristics.

This is done by linearly interpolating the various velocity components to obtain the velocity at a particular point and then using a simple finite difference scheme to obtain the new position of the phase, a time δt later; that is:

$$x_n = x_0 + u \, \delta t$$
$$y_n = y_0 + v \, \delta t$$

where (x_0, y_0) is the initial position of the particle, (x_n, y_n) the new position and (u, v) the velocity.

Scalar variables (such as temperature) are linearly interpolated from their field values to give values at the points (x_n, y_n).

By chosing δt to be sufficiently small, the solution procedure can be iterated to give proper particle tracks.

7. RESULTS

a) Effect of droplet size

Figure 2 shows streamlines of the gas phase for a fixed gas velocity but for two different metal droplet sizes of 10 and 100 microns. Both plots show a large recirculation zone in the main chamber and are broadly similar in overall shape. However in the 100 micron case the core of the recirculation zone is closer to the gas inlet position showing the energy damping effect of the larger droplets.

Figure 3 shows the path of single 2, 10 and 100 micron droplets within the gas flow field. As might be expected the small droplets follow the gas flow lines and do not impact onto the substrate profile while the larger ones will. In practice droplet sizes are of the order of 50-100 microns and hence, on average, will impact onto the substrate.

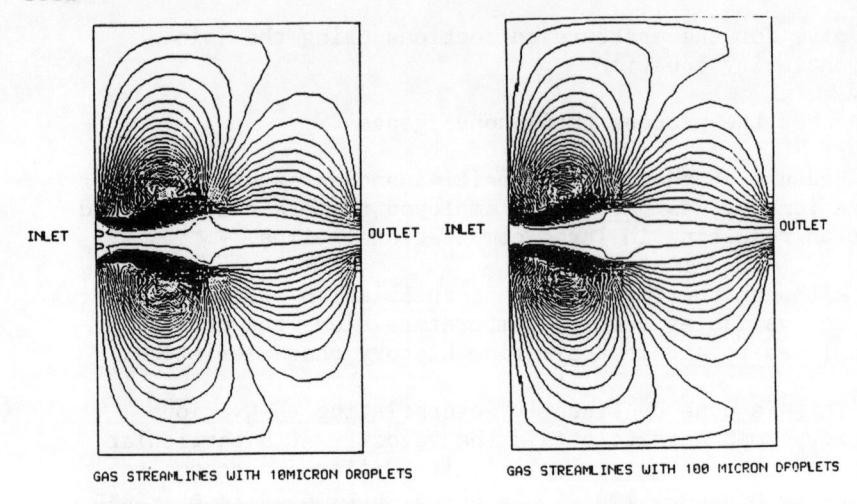

Figure 2. Gas Streamlines

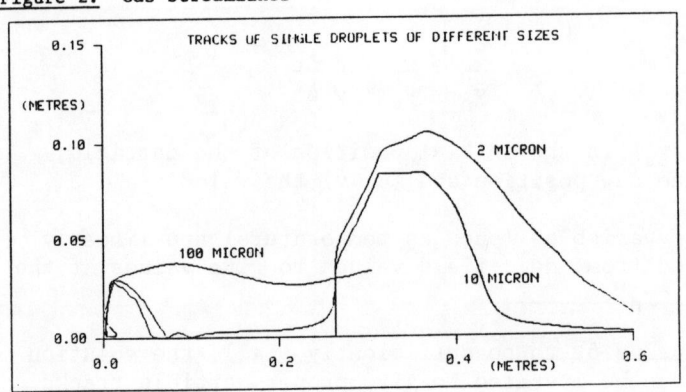

Figure 3. Tracks of single droplets

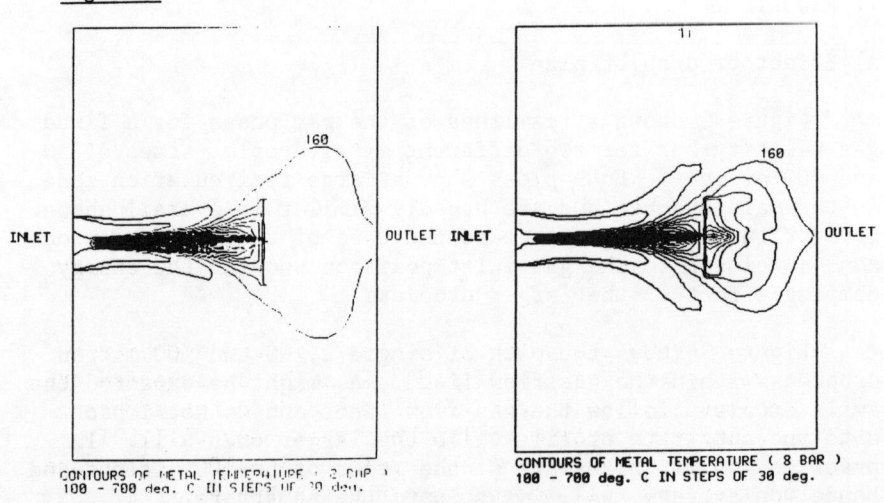

Figure 4. Metal Temperature Contours

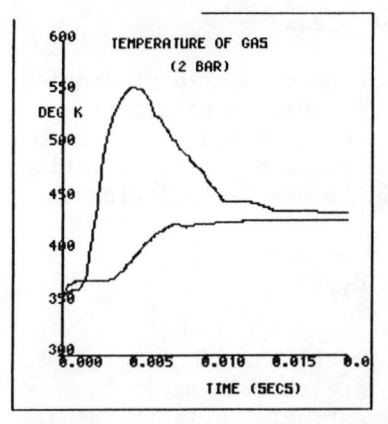

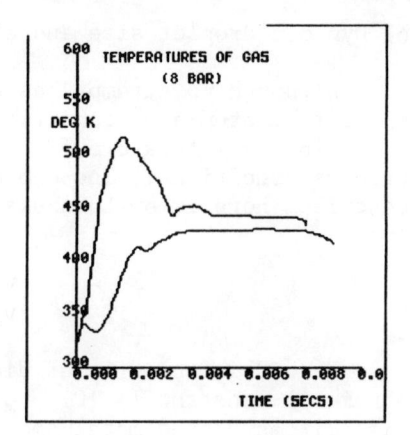

Figure 5. Temperature of Gas against time

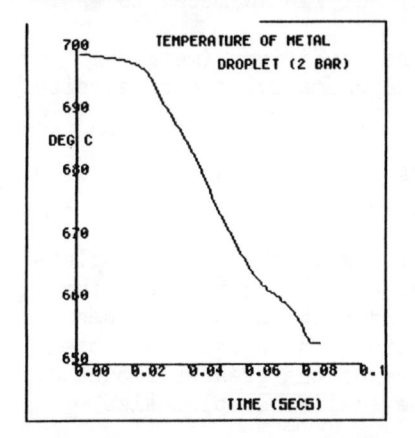

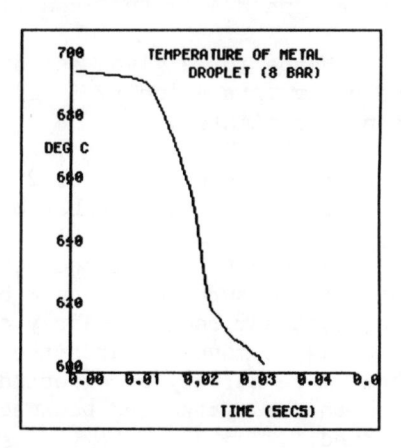

Figure 6. Temperature of Metal against time

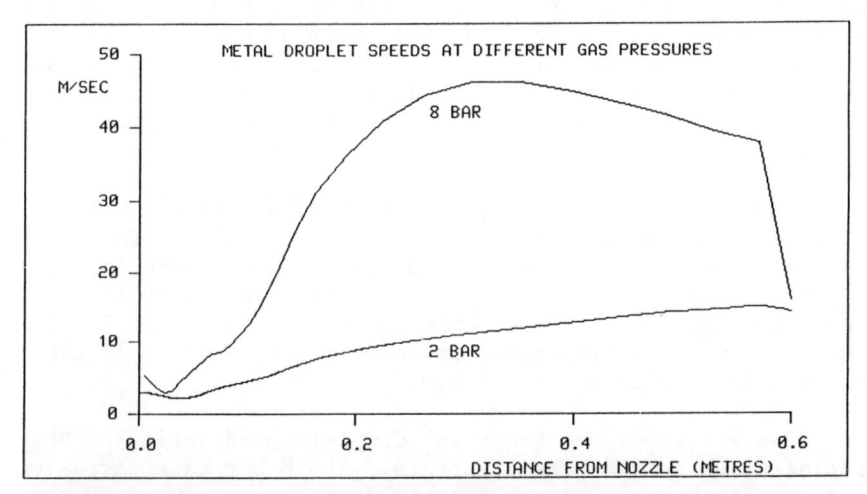

Figure 7. Metal droplet speed against distance

b) Dynamic droplet size and effect of pressure

Although no attempt has been made here to predict droplet sizes from atomising conditions experimental correlations exist in the literature [7] for estimating droplet sizes from process conditions. The one most appropriate to the situation described here is by Lubanska [8] and is of the following form:

$$\frac{d}{D} = k \left[\frac{\nu_M}{\nu_G W} \left(1 + \frac{M}{A}\right)\right]^{\frac{1}{2}}$$

Here d is the droplet diameter, D the nozzle diameter, k a numeric constant (∿ 10), ν_M and ν_G are the kinematic viscosities of the metal and gas respectively, W is the Weber number and M/A is the mass flow ratio of liquid metal to gas.

Hence for a given set of process conditions (metal velocity, gas velocity (pressure) and geometry) a droplet size can be estimated.

At a gas pressure of 2 bar this results in a droplet size of 110 micron and at 8 bar of 50 microns.

Figure 4 shows temperature contours of the metal droplets at gas pressures of 2 and 8 bar. As can be clearly seen the metal stream remains fairly central in both cases and loses heat as it moves towards the substrate. The 8 bar case loses heat more quickly as expected because of the greater amount of gas in the system and because of the smaller droplet size formed.

Figure 5 shows temperature-time tracks of the gas for the two cases of 2 and 8 bar. The two tracks in the plots are for points starting at the edges of the incoming gas jet. The increase in temperature is due to heat transfer between metal and gas, and to the decrease in gas velocity resulting from the expansion of the gas and the drag caused by the metal stream.

Figure 6 shows how the temperature of a typical metal droplet decays in time, this loss caused by mixing with the gas jet. The 8 bar case shows a quite rapid decrease in temperature (> 10^3 °C/s) and then a partial arrest as it passes through the liquidus temperature (620 °C). The 2 bar case has a much smaller temperature drop and is still liquid by the time it hits the substrate.

Figure 7 shows how the metal droplet speed varies as the droplets move towards the substrate. The 8 bar case increases more rapidly because of the greater drag force induced by a combination of higher speed gas and smaller metal droplets.

Figure 8 shows the actual gas particle tracks for the cases described for Figure 3. The substrate is located on the x-axis at x = 0.6. The tracks for the 8 bar case remain more central than those for the 2 bar case.

Figure 9 shows how the speed of the gas for the before mentioned tracks varies with distance from the nozzle. The two cases are comparable in the shape of the curves but not in the magnitude of the speed. The shape of the curves are due to a combination of factors; expansion of the gas as it leaves the nozzle; drag forces induced on the gas by the metal droplets; heating of the gas (reducing the density of the gas); acceleration around the substrate and showing down in the quiescent region behind the substrate; acceleration towards the outlet.

Figure 10 shows the temperature distribution of the metal droplets as a function of distance from the substrate centre. The 2 bar case is much more centralised as a result of less mixing with the gas stream.

Figure 11 shows the volume fraction of metal at the substrate for the two cases. While the 2 bar case is again much more centralised, the 8 bar case is more diffuse and shows an identation at the centre. This is because of the high velocities and lower droplet sizes involved, dragging droplets away from the centre.

8. ACKNOWLEDGEMENTS

The authors are grateful to Alcan International Limited for the permission to publish this work.

9. REFERENCES

1. SINGER, A.R.E. - The Principle of Spray Rolling of Metals, Metals and Materials, 4, 246 (1970)

2. FEEST, E.A. - U.K. Patent Application, 2 172 826 (1986)

3. LAUNDER, B.E. and SPALDING, D.B. - The Numerical Computation of Turbulent Flows. Comp-Meth Appl. Mech. Eng., 3 (1974) 269-289

4. SELLENS, R.W. and BRZUSTOWSKI, T.A. - A Prediction of the Drop Size Distribution in a Spray from First Principles. Atomisation and Spray Technology, 1, 89-103 (1985)

5. CLIFT, R., GRACE, J.R. and WEBER, M.E. - Bubbles, Drops and Particles. Academic Press, Page 111

6. SPALDING, D.B. - Numerical Computation of Multiphase Fluid Flow and Heat Transfer. Contribution to Recent Advances in Numerical Methods in Fluids. Ed. Taylor, C. and Morgan K., pp 139-167

7. JONES, H. - Production and Characterisation of Rapidly-Solidified Particulates, Science and Technology of the Undercooled Melt. Ed. Sahm, P.R., Jones, H., Adams, C.M., Martinus Nijhoff Publ., 1986, p. 156

8. LUBRANSKA, H. - Correlation of Spray Ring Data for Gas Atomisation of Liquid Metals, Journal of Metals, 22 (2), 45-49 (1970)

Figure 8. Tracks of gas particles

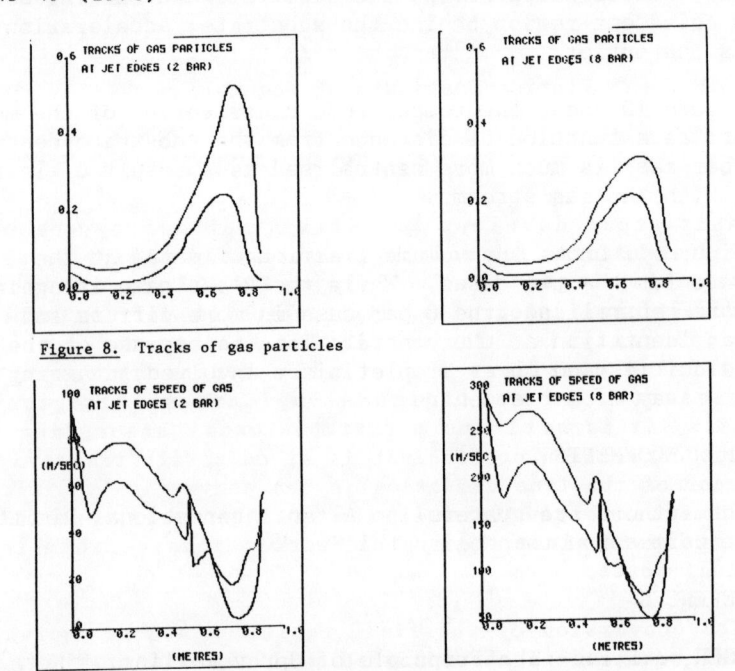

Figure 9. Tracks of speed of gas particles

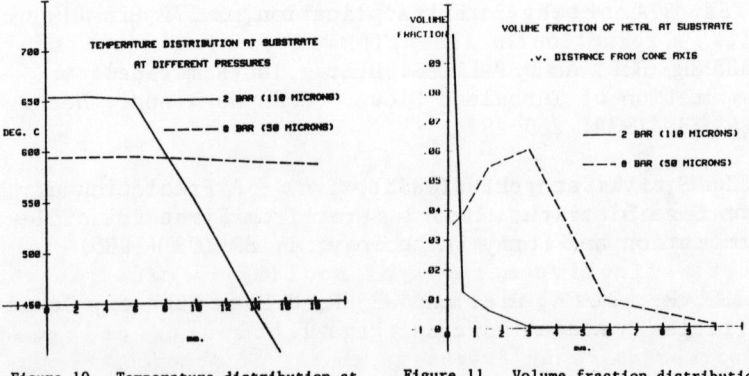

Figure 10. Temperature distribution at substrate

Figure 11. Volume fraction distribution at substrate

The Use of the Finite Integral Transform Technique
for Thermal Analysis in Microelectronic Chip Modules

Jay I. Frankel
Tai-Ping Wang
Gary W. Howell

Florida Institute of Technology
Melbourne, FL 32901

SUMMARY

This paper develops a solution methodology based on integral transforms for solving nonseparable elliptic partial differential equations. This methodology produces a multidimensional integral equation which is very amenable to numerical analysis. The particular heat transfer problem investigated considers a spatially dependent convection coefficient. In multidimensional and multiregion fin analysis, it is well-known that the usual assumption of a constant convection coefficient is at odds with reality. The inclusion of the true variation in the convection coefficient in the mathematical formulation will generally preclude an exact analysis since the system becomes nonseparable in the classical sense.

The conversion of the field equation into an equivalent integral equation is accomplished with the aid of the finite integral transform technique. The orthogonal basis functions are obtained by treating the undesirable terms as effective sources or nonhomogeneities. An integral equation approach permits a reduction in dimensionality. Finally, concurrent processing can easily be implemented in this approach.

1. INTRODUCTION

Heat transfer considerations are of great importance in the design of today's microelectronic chip modules. The high heat fluxes associated with these components require efficient and cost effective methods of cooling in order to remove excessive heat generated. In general, the complex configurations developed by many computer chip module designers require an extensive amount of computer (numerical) simulation to accurately determine the temperature and thermal stress distributions.

In hopes of identifying important design parameters and characteristics associated with these modules, a model is proposed which emulates the behavior of common modules used by Hitachi and Mitsubishi (see Fig. 1). The model and proposed solution method promotes physical understanding and provides an accurate solution which permits parallel processing to be incorporated. The solution method is based on a generalized finite integral transform technique which permits the conversion of a field equation into an equivalent integral equation. The resulting integral equation is shown to be advantageous for many reasons including the capability of handling nonseparable and discontinous boundary conditions while permitting the use of concurrent processing in determining the temperature distribution.

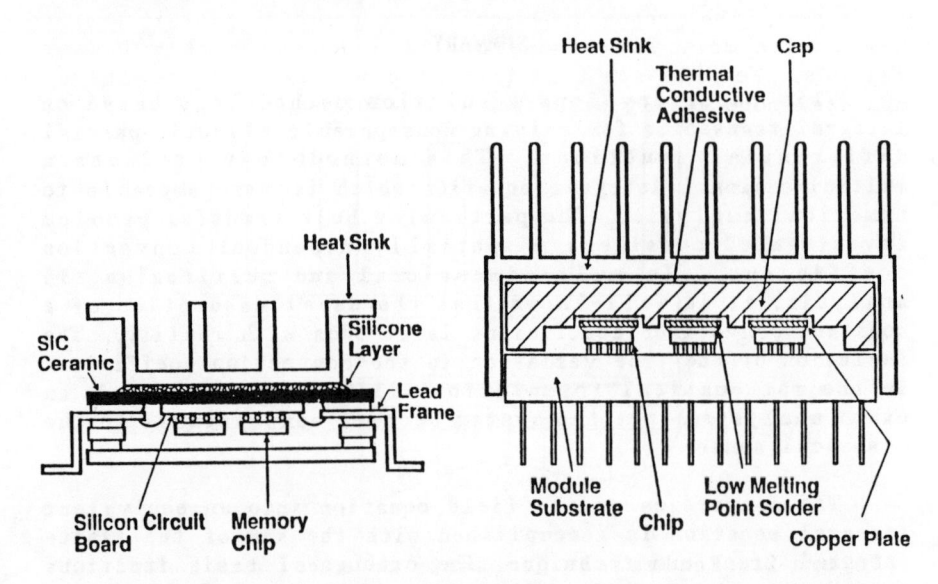

Figure 1. The Hitachi and Mitsubishi chip modules.

Generally, integral equation formulations and solution methods have been somewhat neglected compared to more popular finite difference and finite element methods. Nevertheless, integral solution methods present several advantages over these methods. In the first place, a properly designed integral solution should be less susceptible to the truncation errors inherent in numeric differencing. Secondly, the solution given by an integral method is evaluated at each point by numeric integration of a surface, that is, a reduction in dimensionality will occur. As the integral for each point may be evaluated separately, the algorithm lends itself quite naturally to parallel processing.

2. FORMULATION OF MATHEMATICAL MODEL

In hopes of identifying important design parameters, a simplified mathematical model is proposed which permits an analytic investigation. Both the Hitachi and Mitsubishi modules shown in Fig. 1 may be idealized as laminated composites with extended surfaces (see Fig. 2). Each extended surface can be regarded as an enhancement device which brings about a local (periodic) increase of the heat transfer coefficient for convection heat exchange from the fin base to the environment [1-3]. That is, one may model the equivalent system as a laminate composite subjected to a spatially dependent convection coefficient along the surface where the fins are placed. An idealized variation in the convection coefficient would be a sequence of high and low values of the convection coefficient correspoinding to the finned and unfinned regions (see Fig. 3). Even when a linear system is considered, the proposed discontinuous periodic pattern wreaks havoc with most finite difference [1] and finite element methods. We will see that the integral transform technique can deal more easily with such a discontinuity.

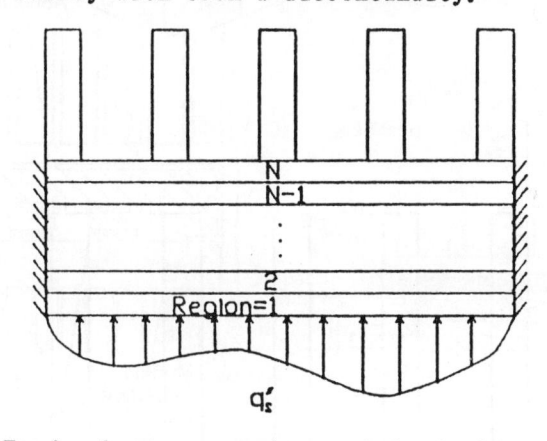

Figure 2. Equivalent composite model of chip modules.

The use of the integral transform technique in solving nonstandard, i.e., nonlinear and nonseparable, parabolic [2-5] and hyperbolic [6] partial differential equations have produced promising results. In addition to these studies, it appears possible to connect solutions in one region using the integral transform technique to a purely numerical solution in another region. This "hybriding" takes advantage of the best features inherent in both techniques in certain circumstances.

To illustrate the proposed analytical procedure, attention is given to the one region (N=1) problem corresponding to Fig. 3 at steady state conditions. A later study will consider the effects associated with the composite problem, both for the two and three dimensional cases. When the material properties are assumed constant, the governing dimensionless temperature field equation is

$$\nabla^2 \theta(\eta,\xi) = 0, \quad (1)$$

where ∇^2 is the two-dimensional Laplacian in the cartesian coordinate system.

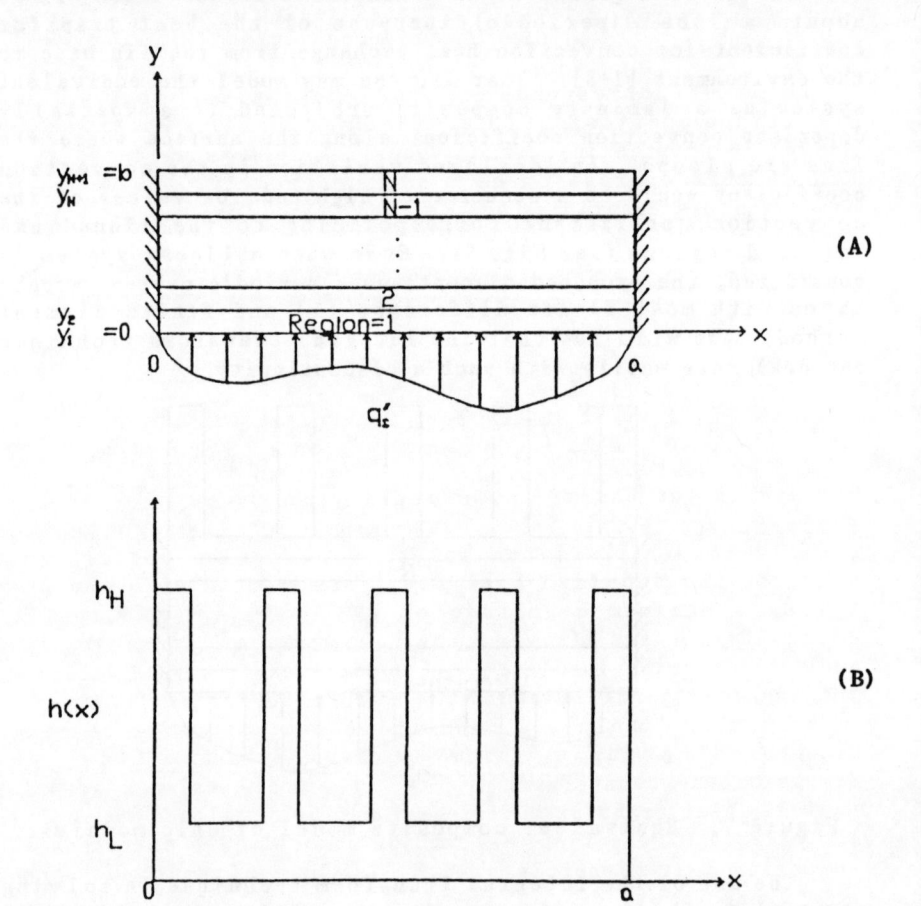

Figure 3. Idealized composite model (A) with variable convection coefficient (B).

The dimensionless boundary conditions are chosen as

$$\frac{\partial \theta}{\partial \eta}(0,\xi) = 0 \quad (2a)$$

$$\frac{\partial \theta}{\partial \eta}(1,\xi) = 0, \quad \xi \varepsilon (0,\gamma), \quad (2b)$$

$$\frac{\partial \Theta}{\partial \xi}(\eta,0) = -Q(\eta), \qquad (2c)$$

$$\frac{\partial \Theta}{\partial \xi}(\eta,\gamma)+Bi_L\Theta(\eta,\gamma)=[Bi_L-Bi(\eta)]\Theta(\eta,\gamma), \quad \eta\varepsilon(0,1), \qquad (2d)$$

where

$$Bi(\eta) = \begin{cases} Bi_H, & \text{finned region} \\ Bi_L, & \text{unfinned region} \end{cases} \qquad (2e)$$

where $Bi(\eta)$ corresponds to a dimensionless form of the convection coefficient, i.e., the Biot number. Observe that Eq.(2d) is written in a form suggesting that we will treat the RHS of this expression as a nonhomogeneity. The various dimensionless quantities are defined as follows

$$\Theta(\eta,\xi)=\frac{T-T_\infty}{T_{ref}}, \quad Q(\eta)=q_s''(x)/q_o'', \quad T_{ref}=q_o'' a/k$$

$$(3a-h)$$

$$\eta=\frac{x}{a}, \quad \xi=\frac{y}{a}, \quad \gamma=\frac{b}{a}, \quad Bi_L=\frac{h_L a}{k}, \quad Bi(\eta)=\frac{h(x)a}{k},$$

where k is the thermal conductivity, T_∞ is the environment temperature, h_L is the low value of the convection coefficient, h(x) is the general convection coefficient, q''(x) is the supplied heat flux associated with the chip having a maximum magnitude of q_o'', (x,y) are the spatial coordinates, and T(x,y) is the temperature. Equations (2a) and (2b) describe insulated end conditions, Eq. (2c) describes the induced thermal disturbance due to the chips themselves, and Eq. (2d) displays a convective heat transfer boundary condition having a variable Biot number. Using the dimensionless counterpart to Fig. 3, a model using a stepwise periodic heat transfer coefficient can be developed to emulate the effects of the fins. A low Biot number would correspond to an unfinned area whereas a high Biot number would correspond to a finned area.

3. SOLUTION METHOD

The integral transform technique shall now be used to resolve Eqs.(1) and (2). In a nutshell, this method removes one spatial variable in the original field equation as expressed in Eq. (1) and reduces it to a second order ordinary differential equation in the transform variable. The ordinary differential equation is then subject to the transformed boundary conditions. Once the transform has been resolved, the original dependent variable is reconstructed through the inversion formula.

Recalling that this technique finds its antecendents in separation of variables [7], the integral transform pair is developed by considering the representation of an arbitrary function in terms of the eigenfunctions corresponding to the eigenvalue problem. Therefore, the eigenfunctions, eigenvalues, and normalization integrals are developed from the solution of the associated homogeneous version of Eqs. (1) and (2). The integral transform pair may then be constructed.

In this one region problem, the orthogonal basis functions are the eigenfunctions required in a separation of variables type solution. The basis functions may be developed in the ξ or η directions. If we choose the η direction, we obtain the standard separation of variables solution. However, in composite analysis, it is more convenient to develop the eigenfunctions in the ξ-direction. Therefore, we choose to develop the eigenfunctions in the nonexact ξ-direction. This case will then act as a precursor to later analysis of composites. The availability of the standard (exact) solution is convenient as it offers a check on the accuracy and numeric efficiency of the proposed scheme.

The required transform pair is developed from considering the associated eigenvalue problem

$$\frac{d^2\psi_m(\xi)}{d\xi^2} + \lambda_m^2 \psi_m(\xi) = 0, \quad \xi\varepsilon(0,\gamma) \tag{4a}$$

subject to

$$\frac{d\psi_m(0)}{d\xi} = 0 \quad \text{and} \quad \frac{d\psi_m(\gamma)}{d\xi} + \text{Bi}_L \psi_m(\gamma) = 0, \tag{4b,c}$$

where the nonseparable term expressed in Eq. (2d) is treated as an effective nonhomogeneity. The orthogonality property may be readily established as

$$\int_{\xi=0}^{\gamma} \psi_m(\xi)\psi_n(\xi)d\xi = \begin{cases} 0 & m \neq n \\ N(\lambda_m) & m = n \end{cases} \tag{5}$$

where $N(\lambda_m)$ is the normalization integral. The eigenfunctions $\psi_m(\xi)$, are an infinite sequence (m=1,2,...) and form a complete orthogonal set of functions in the interval $\xi\varepsilon(0,\gamma)$.

The eigenfunctions, $\psi_m(\xi)$, can be readily established as

$$\psi_m(\xi) = \cos(\lambda_m \xi) \tag{6a}$$

which correspond to the discrete set of eigenvalues defined by the positive roots of

$$\lambda_m \tan(\lambda_m \gamma) = Bi_L . \tag{6b}$$

From the orthogonality relation, the normalization integral is found to be

$$N(\lambda_m) = \frac{\gamma}{2} + \frac{\sin(2\lambda_m \gamma)}{4\lambda_m} . \tag{6c}$$

Using the orthogonality property, we define the integral transform pair as

Inversion Formula:

$$\Theta(\eta,\xi) = \sum_{m=1}^{\infty} \frac{\psi_m(\xi)}{N(\lambda_m)} \cdot \bar{\Theta}_m(\eta), \tag{7a}$$

Integral Transform:

$$\bar{\Theta}_m(\eta) = \int_{\eta=0}^{\gamma} \Theta(\eta,\xi) \psi_m(\xi) d\xi \tag{7b}$$

where $\bar{\Theta}_m(\eta)$ is the integral transform of $\Theta(\eta,\xi)$ and $\psi_m(\xi)$ represents the kernel of the transformation. The general solution to Eq. (1) subject to the boundary conditions expressed in Eq. (2) is now available.

We now remove the ξ-direction by operating on Eq. (1) with

$$\int_{\xi=0}^{\gamma} \psi_m(\xi) d\xi , \tag{8}$$

to get

$$\int_{\xi=0}^{\gamma} \nabla^2 \Theta(\eta,\xi) \psi_m(\xi) d\xi = 0 . \tag{9}$$

The eigenfunctions, eigenvalues, and normalization integrals are developed from the associated linear operator [2-5]. Performing the indicated integration, making use of the definition for the integral transform, and recalling the eigenvalue problem, we find

$$\frac{d^2\bar{\theta}_m(\eta)}{d\eta^2} - \lambda_m^2 \bar{\theta}_m(\eta) = f_m(\eta,\theta(\eta,\gamma)), \qquad (10a)$$

where

$$f_m(\eta,\theta(\eta,\gamma)) = -\psi_m(\gamma)[Bi_L - Bi(\eta)]\theta(\eta,\gamma) - \psi_m(0)Q(\eta) \qquad (10b)$$

subject to the transformed boundary condition. Observe that $f_m(\eta,\theta(\eta,\gamma))$ contains the unknown temperature at $\xi = \gamma$ for $\eta\varepsilon(0,1)$. One could substitute the inversion formula, with a new dummy index, into Eq. (10a) and solve for $\bar{\theta}_m(\eta)$ numerically. An alternative procedure is to convert Eq. (10a) into an integral equation directly treating $\theta(\eta,\gamma)$ as a known function. The transform can be easily resolved with the aid of the Green's function method [8,9]. Omitting the details, we arrive at

$$\bar{\theta}_m(\eta) = \int_{\eta_o=0}^{1} G_m(\eta,\eta_o) f_m(\eta_o,\theta(\eta_o,\gamma)) d\eta_o \qquad (11a)$$

where $G_m(\eta,\eta_o)$ is the Green's function and is found to be

$$G_m(\eta,\eta_o) = \begin{cases} \dfrac{-\cosh[\lambda_m(1-\eta)]\cosh\lambda_m\eta_o}{\lambda_m \sinh\lambda_m}, & 0 \leq \eta_o \leq \eta \\[2ex] \dfrac{-\cosh[\lambda_m(1-\eta_o)]\cosh\lambda_m\eta}{\lambda_m \sinh\lambda_m}, & \eta \leq \eta_o \leq 1 \end{cases} \qquad (11b)$$

Here $G_m(\eta,\eta_o)$ is the Green's function where the arguments are written to represent the "effect/cause" relation for fixed eigenvalue λ_m.

Finally, we may reconstruct the temperature distribution through the inversion formula. Substituting Eq. (11a) into Eq. (7a) yields

$$\theta(\eta,\xi) = \sum_{m=1}^{\infty} \frac{\psi_m(\xi)}{N(\lambda_m)} \cdot \int_{\eta_o=0}^{1} G_m(\eta,\eta_o) f_m(\eta_o,\theta(\eta_o,\gamma)) d\eta_o, \qquad (12)$$

where $G_m(\eta,\eta_o)$ is defined in Eq. (11b) and $f_m(\eta_o,\theta(\eta_o,\gamma))$ is defined in Eq. (10b).

At this junction, we can make some observations concerning the formal solution, as displayed in Eq. (12) for $\theta(\eta,\xi)$. First up to this point, the formal solution is exact, i.e., no approximations have been made. Second, this approach permits us to determine the temperature along the region where the nonseparable boundary condition is present, namely $\xi=\gamma$, $\eta\varepsilon(0,1)$ without knowing the entire field. Thus, we may evaluate Eq. (12) at $\xi=\gamma$ and obtain a one-dimensional Fredholm integral equation of the second kind with a semi-degenerate kernel for $\Theta(\eta,\gamma)$. Many numerical solution methods are available [10] for this type of integral equation. Once the temperature has been established along this line, the entire temperature distribution can be established concurrently. In contrast, a finite difference or finite element method would require the entire field to be determined simultaneously. One advantage of this approach is now clear. In general, at steady state conditions, this approach would reduce the dimensionality by one variable, i.e., $O(n)$ points need to be evaluated in determining the temperature $\Theta(\eta,\xi)$ at $\xi=\gamma$, for all η. Again purely numerical methods would require $O(n^2)$ grid points to be evaluated simultaneously to find the temperature at $\xi=\gamma$, $\eta\varepsilon(0,1)$. In this method, only numerical integration will be required in obtaining the desired temperature distribution. Finally, the numerical integration for various points can proceed in parallel. These benefits are the result of using analytic developments prior to initiating any numerical procedure.

4. RESULTS AND DISCUSSION

In this section, we present some numerical results for various sources and source arrangements. The upper surface temperature is obtained by evaluating Eq. (12) at $\xi=\gamma$. As can be seen, a reduction in dimensionality requiring a numerical approximation occurs immediately. Once $\Theta(\eta,\gamma)$ has been resolved numerically, the temperature at any ξ for all η can be obtained concurrently. The resulting one variable Fredholm equation was solved using a simple trapezoidal method. This method has a global truncation of $O(\Delta\eta^2)$. Higher order block-by-block methods using Simpson's rule and internal quadratic interpolations could also be incorporated. The proposed solution method took a minimal amount of computer (CPU) time to generate the figures presented. In addition, one would expect the N-region composite problem (as shown in Fig. 3) to require roughly the same amount of computer time since an integral equation along the top surface can be developed in a similar manner as presented in this paper. The only differences in the composite formulation is that the eigenfunctions, eigenvalues, and normalization integrals are a bit more complicated. The results presented were compared to the available exact analytic solution where it was shown possible to obtain 3-4 significant figures quickly. A maximum of thirty eigenvalues were required to obtain the desired accuracy along with 81 points in the numerical (trapezoidal) integration. The high Biot number, Bi_H, was set to 5 while the low Biot number, Bi_L, was set to 1 for all the figures. A unfinned region was considered twice the width of a finned region. Five fins were considered throughout this analysis.

Figure 4 displays the temperature distributions at two locations for two dissimilar thermal disturbances, $Q(\eta)$ being applied at $\xi=0$. For this figure and the remaining, the height of the plate is 0.2, i.e., $\gamma = 0.2$. In this and the remaining figures, only two thermal sources shall be represented. The two models chosen to represent the chip are a delta function and square (step) source. The reason for chosing these simple sources is to demonstrate the numercial accuracy of the proposed scheme. That is, we choose

$$Q(\eta) = \begin{cases} Q_1 \delta(\eta-\eta_i), & \text{otherwise zero} \\ Q_2, & \eta_i < \eta < \eta_{i+1}, \text{otherwise zero} \end{cases}$$

where the total energy content of a source or sources are held constant, i.e., (the integral of the source(s) over space is unity). Figure 4 shows the effect of the two sources. The delta function (concentrated) source at $\eta=.5$ predicts much higher temperatures than the square source situated $0.25 < \eta < 0.75$. This of course is expected since a higher energy content is locally available. At $\xi=\gamma=.2$, the effects of the variable convection coefficient (fins) causes a waviness in the temperature distibution along the top surface corresponding to the high-low biot number variation.

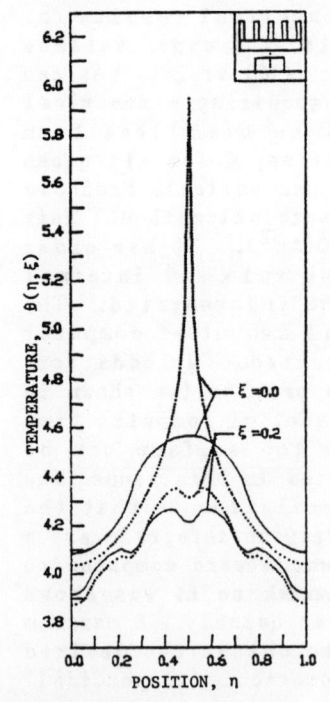

Figure 4.
Effect of different thermal disturbance.

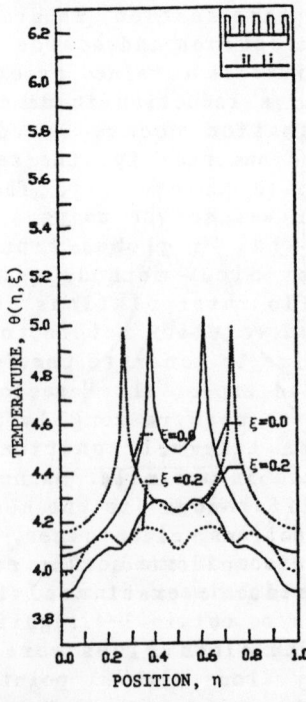

Figure 5.
Effect of concentrated source location.

Figure 5 considers the effects of source location with regard to the fins. In this figure, only a concentrated delta function is chosen in order to display the physical trends. As expected, by placing the sources directly beneath a fin (high Biot region), a lower temperature will occur at $\xi=.2$. Figure 6 compares two sources (concentrated-delta function sources and step sources) centered between the second and fourth fins. The square pulse width was chosen to be twice that of the fin thickness and again the total energy content is constant. It is interesting to note that a similar temperature distribution results at $\xi=.2$ for dissimlar sources. It appears that by modelling the chip with a concentrated source (when chip is two or even three times that of the fin thickness) yields indentical results. For purely numerical methods, a delta function is undesirable however for analytical methods a generalized function is quite useful since the integrations become trivial. Figure 7 displays similar trends but the sources are centered directly under the high Biot region.

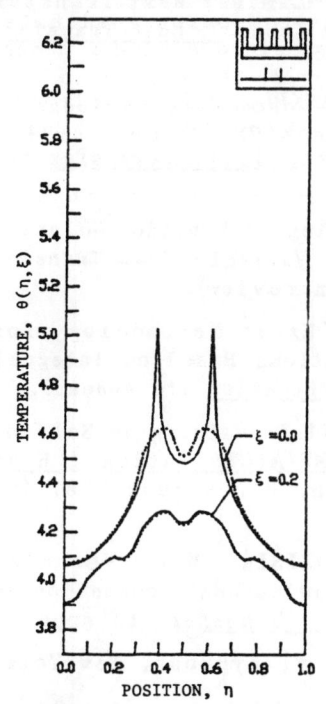

Figure 6.
Comparison of two sources centered between fins.

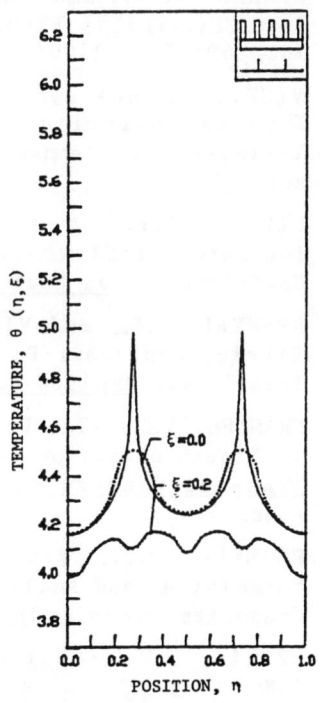

Figure 7.
Comparison of two sources centered below fins.

5. CONCLUSIONS

The use of integral transforms in solving nonseparable partial differential equations appears quite promising where high accuracy can be easily achieved. This initial study demonstrates that multidimensional composites with nonseparable boundary conditions may be investigated by the finite integral transform technique. The solution to the two-dimensional composite is a straightforward generalization of the one region problem presented here. Only the eigenfunctions, eigenvalues, and normalization integral would differ. The numerical procedure would be identical. The three-dimensional composite will also permit a reduction in dimensionality along $\xi=\gamma$. That is, a two-dimensional integral equation may be established at the top surface. More general boundary conditions at the chips locations can be simulated and connected to purely numerical solutions in other parts of the modules if necessary. Finally, the use of the integral transform technique to convert a partial differential equation into an equivalent integral equation is shown to be a viable alternative to purely numerical methods. It appears to be a modernization and generalization to a classical method.

6. REFERENCES

1. SPARROW, E.M., and CHARMCHI, M., -Laminar Heat Transfer in an Externally Finned Circular Tube, *J. Heat Transfer*, 102, 605-611, 1980.

2. VICK, B., BEALE, J.H. and FRANKEL, J.I., -Integral Equation Solution for Internal Flow Subjected to a Variable Heat Transfer Coefficient, *J. Heat Transfer* (to appear).

3. FRANKEL, J.I., -An Alternative Integral Equation Solution for Internal Flow Subjected to a Variable Heat Transfer Coefficient', *J. Heat Transfer* (in review).

4. FRANKEL, J.I., and VICK, B., -An Exact Methodology for Solving Nonlinear Diffusion Equations Based on Integral Transforms, *Applied Numerical Mathematics* (to appear).

5. FRANKEL, J.I., -The Use of Integral Transforms in Solving Nonlinear Diffusion Equations, *ASME/AIChE National Heat Transfer Conference*, Pittsburgh, PA, August 1987 (to appear).

6. FRANKEL, J.I., VICK, B., and OZISIK, M.N., -General Formulation and Analysis of Hyperbolic Heat Conduction in Composite Media, *Int. J. Heat Mass Transfer* (to appear).

7. OZISIK, M.N., -*Heat Conduction*, Wiley&Sons, New York, 1980.

8. ROACHE, G.F., -*Green's Functions* 2nd ed., Cambridge University Press, London, 1982.

9. FRANKEL, J.I., and WANG, T.P., -Radiative Exchange From an Array of Gray Fins Using a Dual Integral Equation Formulation, *J. Thermophysics and Heat Transfer* (in review).

10. BAKER, C.T.H., -*The Numerical Treatment of Integral Equations*, Clarendon Press, Oxford, 1978.

NUMERICAL MODELLING OF HEAT AND MOISTURE FLOW IN A POROUS SOLID

A.P.S. Selvadurai and M.C. Au
Department of Civil Engineering
Carleton University, Ottawa,
Ontario, Canada K1S 5B6

SUMMARY

A Galerkin finite element scheme is used to solve the partial differential equations describing the coupled processes associated with heat and moisture flow in a porous solid. Numerical results presented in the paper illustrate the transient temperature rise and moisture loss in an axisymmetric porous buffer region that isolates a cylindrical nuclear waste container which emitting a constant heat output.

1. INTRODUCTION

Porous buffer materials which are composed of mixtures of clay and crushed sand have been proposed for use in high level nuclear waste disposal endeavours. Especially in the Canadian proposals for the disposal of high level nuclear wastes, such buffer regions isolate the waste container from the disposal vault borehole (Figure 1). The functional requirements of the buffer region are to conduct the radiogenic heat from the container to the host rock; suppress the detrimental effects of corrosive water in the host rock and enhance the life of the waste container; serve as a geomechanical filter for the sorption of radionuclides released either by accidental damage or natural breakdown of the waste container, provide sufficient mechanical strength to support the waste containers; possess sufficient plasticity to isolate the containers from detrimental rockmass movements and be capable of swelling under fluid influx to seal gaps and cracks that may be developed by moisture depletion during the heat and moisture transfer process. The Canadian high level nuclear waste management program [1-4] has emphasized the importance of maintaining the integrity of the buffer region, which forms the focal point of the multiple engineered barrier scheme.

The slow influx of water into the repository area is considered a plausible situation in the initial stages of operation of the waste disposal vault. At these early stages of waste emplacement, the partially saturated (or unsaturated) buffer materials will be subjected to sustained temperature gradients. It is anticipated that among these initial stages of operation of the vault, the buffer material may tend to dry out due to the slow influx of water from the host rock. The drying out of the buffer can lead to cracking and dessication of the buffer region. Such shrinkage action can lead to heat conduction in the separation or gap regions. The explusion of moisture from the buffer region can also lead to the reduction in its thermal conductivity properties. These phenomena potentially have the adverse effect of thermally isolating the waste container from the host rock which is intended to act as the main heat sink in the disposal scheme. Such a situation could lead to the development of thermal runaway conditions in which the heat energy would be dissipated by premature degradation of the disposal vault and its components. The assessment of the thermal performance of the buffer material, particularly in the initial stages of the vault system, is therefore considered a factor of crucial importance to the waste isolation effort. In this paper we shall examine the thermal performance of the buffer material under coupled effects of the heat and moisture flow, by solving the governing partial differential equation using a finite element technique.

2. EQUATIONS GOVERNING HEAT AND MOISTURE FLOW IN POROUS MEDIA

The mechanistic approach proposed by Philip and deVries [5] is based on the microscopic interaction processes that exist between the liquid, vapour and porous structure and employs physical laws governing transport of matter, energy conservation and equilibrium thermodynamics. The differential equation governing moisture movement in the unsaturated porous solid under the combined action of thermal and moisture gradients takes the form

$$\frac{\partial \theta}{\partial t} = \nabla \cdot (D_T \nabla T) + \nabla \cdot (D_\theta \nabla \theta) + \frac{\partial K_\theta}{\partial Z} \qquad (1)$$

where ∇T denotes the gradient of the temperature field; $\nabla \theta$ denotes the gradient of the volumetric moisture content field; D_T is the thermal moisture diffusivity; D_θ is the isothermal diffusivity; K_θ is the unsaturated hydraulic conductivity and Z is the spatial co-ordinate. The terms D_T and D_θ are made up of two components each, one for the vapour pressure and one for liquid flow, i.e.,

$$D_T = D_{T\ell} + D_{Tv} \quad ; \quad D_\theta = D_{\theta\ell} + D_{\theta v} \tag{2}$$

in which the subscripts ℓ and v refer to the liquid and vapour phases respectively. The two liquid diffusivities tend to be the most important at high moisture contents and the two vapour diffusivities tend to dominate at low moisture contents [5]. The last term on the right-hand-side of (1) accounts for gravitational effects.

The differential equation governing the heat flow through the porous medium is given by,

$$\rho c \frac{\partial T}{\partial t} = \nabla \cdot (\lambda \nabla T) - \rho L \nabla \cdot (D_{\theta v} \nabla \theta) \tag{3}$$

where ρ is the bulk density, c is the specific heat of solid, λ is the thermal conductivity, L is the latent heat of vaporization and $D_{\theta v}$ is the isothermal vapour diffusivity.

The general boundary conditions are

$$T = \bar{T} \quad \text{on} \quad x_i \in S_1 \tag{4}$$

$$\lambda \frac{\partial T}{\partial n} - \rho L D_{\theta v} \frac{\partial \theta}{\partial n} + \bar{h} = 0 \quad \text{on} \quad x_i \in S_2 \tag{5}$$

$$\theta = \bar{\theta} \quad \text{on} \quad x_i \in S_3 \tag{6}$$

$$D_T \frac{\partial T}{\partial n} + D_\theta \frac{\partial \theta}{\partial n} + K_\theta n_z + \bar{q} = 0 \quad \text{on} \quad x_i \in S_4 \tag{7}$$

where $\bar{T}$, $\bar{h}$, $\bar{\theta}$ and $\bar{q}$ are respectively the prescribed temperature, heat flux, volumetric water content and moisture flux at a boundary indicated by S_1 to S_4. n is the unit vector in the radial (r) and vertical (z) directions and n_z is the unit vector in the z direction only. The initial conditions when $t = t_0$ are

$$T = \bar{T}_0 \quad \text{and} \quad \theta = \bar{\theta}_0 \quad \text{in} \quad \Omega \tag{8}$$

where t_0, $\bar{T}_0$ and $\bar{\theta}$ are the initial time, temperature and volumetric moisture content respectively in the region, Ω.

3. FINITE ELEMENT FORMULATIONS

There are a number of ways in which the finite element versions of differential equations can be formed: these are fully documented by Zienkiewicz [6], O'Connor and Brebbia [7] and Lewis et al. [8,9]. The integral or weak form of the governing equations (1) and (3) can be obtained via the Galerkin weighted residual technique [6,7]

$$\int_\Omega \{\nabla\delta T \cdot (\lambda\nabla T - \rho L D_{\theta v}\nabla\theta) + \rho c \delta T \frac{\partial T}{\partial t}\}d\Omega + \int_{S_2} \delta T \, \bar{h} \, dS = 0 \qquad (9)$$

and

$$\int_\Omega \{\nabla\delta\theta \cdot (D_T\nabla T + D_\theta\nabla\theta) + \frac{\partial}{\partial z}(\delta\theta) K_\theta n_z + \delta\theta\frac{\partial\theta}{\partial t}\} \, d\Omega + \int_{S_4} \delta\theta \, \bar{q} \, dS = 0 \qquad (10)$$

Following the usual procedures associated with the finite element method [6,7], we have developed an equation for a finite element from equations (9) and (10). Solution of the equation is achieved by using the time integration scheme proposed by Zienkiewicz [6] and the Gauss-Siedel iteration scheme [10].

4. AXISYMMETRIC MODEL OF THE BUFFER-CONTAINER ROCKMASS SYSTEM

In this section we shall apply the above numerical scheme to examine the coupled heat and moisture flow in a buffer-rockmass system due to the heating action of the embedded container. It should be observed that the borehole configuration proposed for the repository has a three-dimensional geometry (Figure 1). The model chosen for numerical analysis is however simplified axisymmetric configurations which is amenable to convenient numerical computation. In the model (Figure 2) a single borehole is considered to be located in a rockmass of dimensions (both lateral and vertical) sufficient enough to prevent heating effects at the boundary for the time scale of interest. The cylindrical boundary is considered to be insulated. In this model the backfill region is assumed to extend over the vault region as shown in Figure 2. The model also accounts for the rockmass region which is present above the apex level of the disposal vault. The temperature at the upper vertical boundary surface of the model is considered to be 15°C. The interfaces of buffer/container buffer/rock, buffer/backfill and backfill/rock are assumed to exhibit perfect thermal contact throughout the heating process.

The heat output to the waste container should be specified according to the decay time history given by Cameron [11]. In this study, however, we develop results for the case where the heat output from the container is maintained at 270 Watts without any decay.

In the present study we focus only on the moisture movement on the buffer region. The rockmass and backfill therefore act as purely heat conducting media. The buffer has an initial volumetric moisture content of approximately 28.4 percent, consistent with the optimum moisture content at which the buffer region is compacted. For water movement in the buffer regions, the boundary conditions at the buffer/container, buffer/rock and buffer/backfill are required. Because of the relatively-low hydraulic conductivity of the rock compared to the backfill and the imperviousness of the container, a no-moisture flux boundary condition is considered at the buffer/container and the buffer/rock-interfaces. For the buffer/backfill interface, the volumetric water content is expected to decrease from the original content with time. For demonstration purposes, the original volumetric water content of buffer is considered throughout the heating process.

The finite element discretizations used in the numerical analyses are shown in Figure 3. The thermophysical properties for the buffer material essentially vary with both temperature and the volumetric moisture content. In the absence of specific experimental data we shall assume that these thermophysical properties can be assigned a range of values that can be encountered in porous clay-sand mixtures. Typical values for the thermal properties of the buffer, backfill and rockmass regions are given by Radhakrishna [12], Tsui and Harris [13]. A summary of the thermophysical constants for the various materials are given in Table 1.

5. NUMERICAL RESULTS

The numerical results of primary interest pertain to the time-dependent variations of temperature and volumetric moisture content within the buffer region. The assessment of the buffer integrity can be directly related to the thermal gradients and the associated moisture losses from the buffer region. The results for three sets of material parameters of the buffer are shown in Figures 4 and 5. Figure 4 shows the variation of temperatures for the three cases (at the plane through the mid-section of the heater) at the buffer-container interface (T_c) and the buffer-rockmass interface (T_R) for the physical model of the repository under consideration (Figure 2). Figure 5 indicates the volumetric moisture content at the boundaries of the buffer region and at the mid-section of the heater. It is clearly evident that the diffusivity characteristics and the unsaturated hydraulic conductivity values have little influence on the temperature distribution. However, they exert asignificant influence on the variation of volumetric moisture content distribution within the buffer region. A change in the diffusivity characteristics can lead either to the complete drying of the buffer region or the relatively low loss of moisture during the heating process.

6. CONCLUSIONS

The numerical studies of heat and moisture flow within a porous geological material are motivated by the need for the examination of the thermal integrity of engineered geological barriers known as buffers which are used in high level nuclear waste disposal endeavours. An assessment of the thermal performance of the buffer is crucial to the evaluation of the reliability of the engineered multi-barrier concept. Results described in connection with this paper indicate that for the class of material parameters (i.e., heat conduction processes in the buffer, backfill and rockmass and moisture flow in the buffer region only) considered, the heat conduction processes stabilize relatively quickly. It is assumed that the moisture losses in turn do not substantially alter the heat conduction properties of the buffer region. The moisture movements due to thermal gradients can be studied quite conveniently by using the near steady state temperature distributions within the buffer region. In a general situation, however the effect of moisture movement and temperature on the thermophysical properties of the various media needs to be given due consideration.

8. REFERENCES

1. RUMMERY, T.E. and ROSINGER, E.L.J. The Canadian Nuclear Fuel Waste Management Program, Proc. Canadian Nuclear Society Int. Conf. on Radioactive Waste Management, 6, 1982. Also, Atomic Energy of Canada Limited Report AECL-7787, 1983.
2. RUMMERY, T.E. and ROSINGER, E.L.J. The Canadian Nuclear Fuel Waste Management Program, Proc. American Nuclear Society Int. Topical Meeting on Fuel Reprocessing and Waste Management, Jackson, Wyo, USA, Vol.1, 14, 1984.
3. BIRD, G.E. and CAMERON, D.J. Vault Sealing for the Canadian Nuclear Fuel Waste Management Program, Atomic Energy of Canada Limited Report, AECL Technical Record 145, 1982.*
4. LOPEZ, R.S. Disposal Vault Sealings, Proc. 18th Information Meeting of the Nuclear Fuel Waste Management Program, Atomic Energy of Canada Limited Tech. Record 320, 150-191, 1985.*
5. PHILIP, J.R. and DeVRIES, D.A. Moisture Movement in Porous Materials Under Temperature Gradients, Trans. Amer. Geophys. Union, 38, 222-232 (1957).
6. ZIENKIEWICZ, O.C. The Finite Element Method in Engineering Science, McGraw Hill, New York, 1977.
7. O'CONNOR, J.J. and BREBBIA, C.A. Finite Element Techniques for Fluid Flow, Butterworth Publ., London, 1976.
8. LEWIS, R.W., MORGAN, K. and ZIENKIEWICZ, O.C. (Eds.) Numerical Methods in Heat Transfer, John Wiley, New York, 1981.
9. LEWIS, R.W., MORGAN, K. and SCHREFLER, B.A. (Eds.) Numerical Methods in Heat Transfer, Vol.II, John Wiley, New York, 1983.

10. SELVADURAI, A.P.S. and AU, M.C. Coupled Heat and Moisture Flow in Buffer Materials Used in a Nuclear Waste Disposal Vault. Atomic Energy of Canada Limited Technical Record*(to be published).
11. CAMERON, D.J. Fuel Isolation Research for the Canadian Nuclear Fuel Waste Management Program, Atomic Energy of Canada Limited Report AECL-6834, 1982.
12. RADHAKRISHNA, H.S. Near Field Thermal Analysis of the Nuclear Waste Disposal Vault, Proc. 19th Information Meeting of the Nuclear Fuel Waste Management Program. Toronto, AECL Tech. Rec. 350, Vol.3, 608-618, 1985.
13. TSUI, K.K. and HARRIS, N.L. Comparison of Two- and Three-Dimensional Thermal Analyses for the Near Field of an Immobilized Waste Disposal Vault, Atomic Energy of Canada Limited Technical Record TR-266, 1984.

* Unrestricted, unpublished report available from SDDO, Atomic Energy of Canada Limited, Research Company, Chalk River, Ontario K0J 1J0.

Constants	Rock	Backfill	Buffer (1)	Buffer (2)	Buffer (3)
λ (J/m/s/°K)	3.0	2.0	2.0	2.0	2.0
D_θ (m^2/s)			7.5×10^{-6}	7.5×10^{-7}	7.5×10^{-8}
D_T (m^2/s/°K)			10^{-8}	10^{-9}	10^{-8}
$D_{\theta v}$ (m^2/s)			10^{-11}	10^{-11}	10^{-11}
ρc (J/m^3/°K)	2.24×10^6	1.6×10^6	3.0×10^6	3.0×10^6	3.0×10^6
K_θ (m/s)			10^{-10}	10^{-11}	10^{-10}

The Latent Heat of Vapourization L = 2.2×10^6 J/Kg.

Table 1: The Values for the Thermophysical Constants

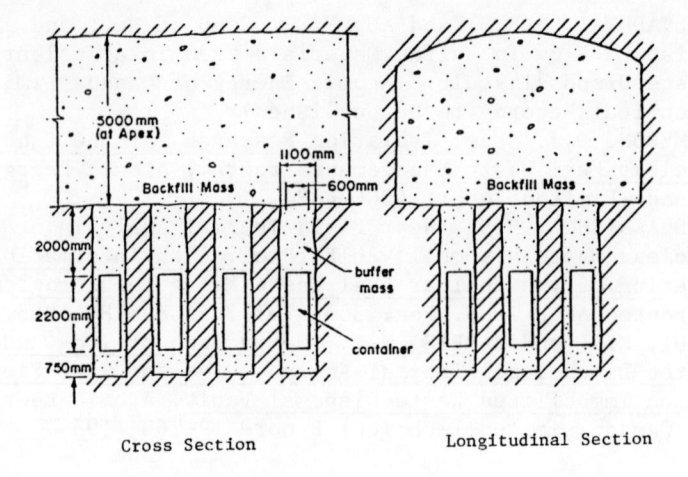

Figure 1: Typical Layout of a Room in a Disposal Vault

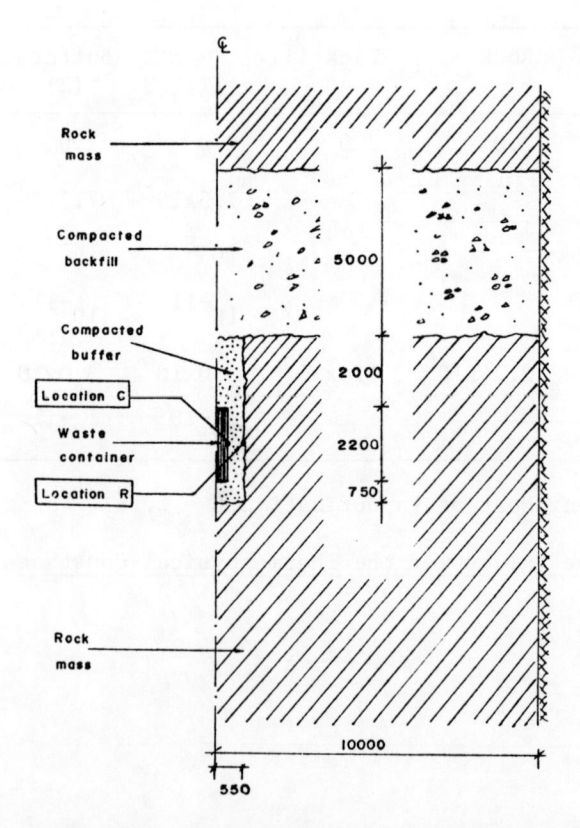

Figure 2: Modelling of the Container Buffer-Rockmass System (All dimensions in mm.)

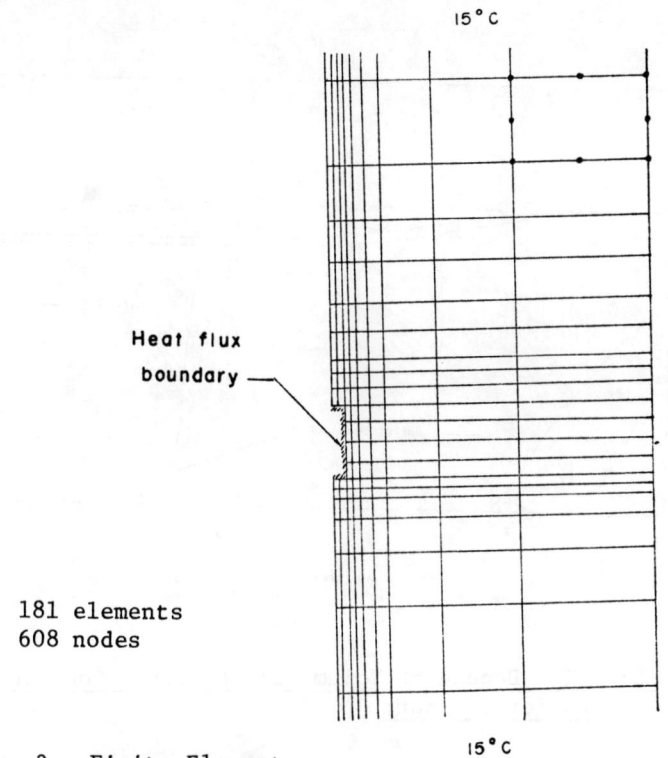

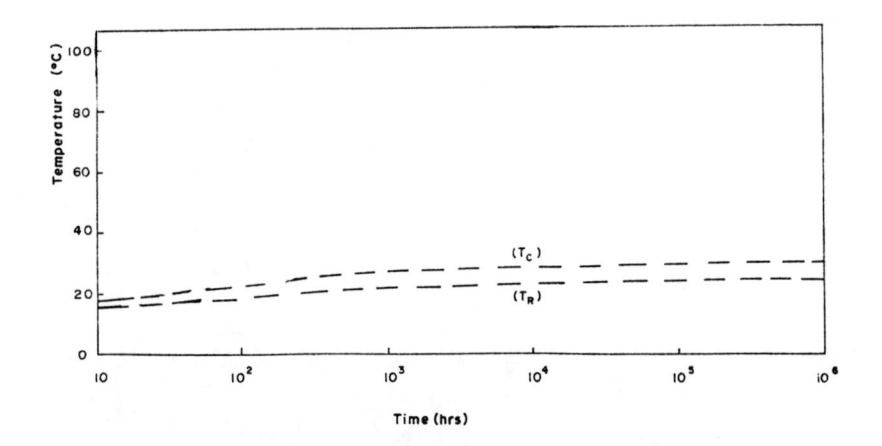

Figure 3: Finite Element Discretization of the Model

Figure 4: Time-Dependent Temperature at the Buffer Boundary

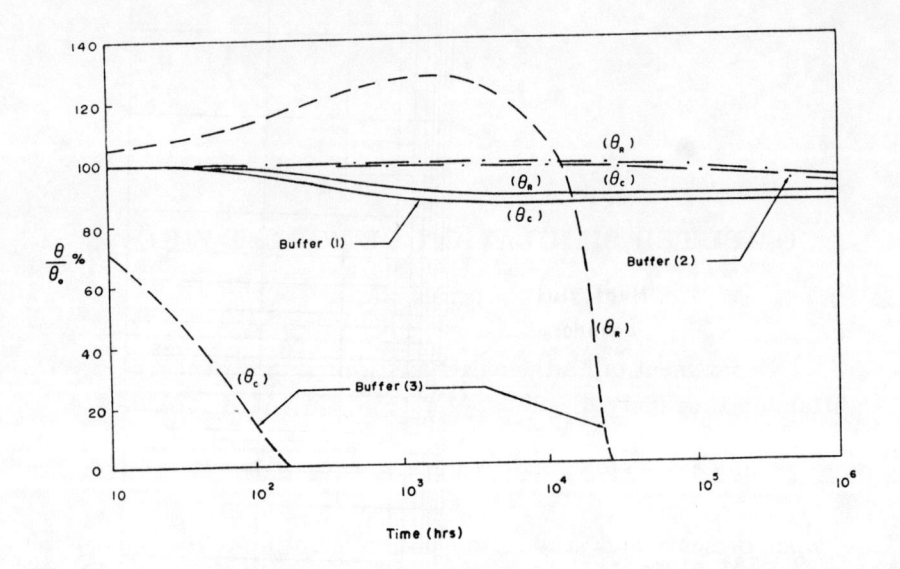

Figure 5: Time-Dependent Volumetric Moisture Content at the Buffer Boundary

COMPUTER SIMULATION OF FOREST FIRES

Gwynfor D. Richards

Department of Mathematics and Computer Science
Brandon University, Brandon, Manitoba, CANADA R7A 6A9

Summary

This paper presents and studies the properties of a general model to predict forest fire spread under different conditions. The model is based on n^{th} order reaction theory, and is sufficiently simple and flexible to conveniently deal with variations in wind velocity, forest properties, fire breaks and other phenomena of interest to the fire controller. The effects of changing various parameters in the equations are illustrated. Results for zero and constant wind velocity are given, together with the effect of abrupt wind changes and fire breaks.

Introduction

The forestry industry worldwide is a multi-billion dollar industry with millions of acres of valuable resources being lost each year to forest fires. However, little computer work seems to have been done to simulate this problem.[1,2] The ability to accurately model the behaviour of a forest fire would be useful to the forester in many ways. In developing optimal fire management strategies for different forest and weather conditions, and as an aid in deciding how best to control a fire which is already in progress. Long-term planning for an area, such as where to leave fire breaks when planting, would also be made easier.

The purpose of this paper is to present and study the properties of a general model for forest fires. This can simulate changes in wind velocity, forest properties, placement of fire breaks and other things of interest to the forester reasonably easily. The model is demonstrated under a variety of conditions.

Mathematical Model

Exact calculations of temperature and rates of combustion in a forest fire are impractical due to the large spatial and temporal variations of the parameters involved. Moreover, in the prediction of the behaviour of a fire, exact values are not required. The forest fire controller is merely interested in approximately how fast the fire is progressing, where the major areas of activity are and are likely to be, and in testing different strategies of control. The model used here is simple and computationally straight forward so as to conveniently give a useful prediction of the present and future state of a fire under different conditions.

Trees are assumed to be arranged in a grid pattern, each grid point being assigned integer coordinates (x,y). The model considers two variables, F, the amount of material in a state of combustion, and M, the amount of unburnt material remaining at a point. Using principles from n^{th} order reaction models, the rate of decrease of F is assumed to be proportional to the value of F at that point, raised to a power. For a single tree the rate of increase of F is proportional to the product of F and M raised to powers. However, burning trees effect neighbouring trees. It is assumed that a tree cannot effect another tree until F has reached a threshold value. The contribution a tree makes to the rate of increase of F at a neighbouring tree is taken to be proportional to the powered product of F at the effecting tree and M at the effected tree. At any point, the rate of increase of F is balanced by the rate of decrease of M. In differential form the model becomes

$$\frac{dF}{dt} = aM^{n-\alpha}(b_1 b_2) * F^\alpha - cF^\beta \qquad (1)$$

$$\frac{dM}{dt} = -aM^{n-\alpha}(b_1 b_2) * F^\alpha \qquad (2)$$

α, β and n are constants, all other variables being functions of x, y and time, t. The thresholded convolution $*$ being defined as

$$(b_1(x,y,t)b_2(x,y,t)) * F^\alpha(x,y,t)$$

$$= \sum_{u=-\infty}^{\infty} \sum_{v=-\infty}^{\infty} b_1(u,v,t) b_2(u,v,t) F^\alpha(x-u, y-v, t) \qquad (3)$$

subject to the threshold condition

$$b_1(u,v,t) b_2(u,v,t) F^\alpha(x-u, y-v, t) = 0$$

$$\text{if} \quad b_2(u,v,t) F^\alpha(x-u, y-v, t) < T(x,y,t) \qquad (4)$$

The initial conditions below complete the model

$$M(x,y,0) = M_0(x,y), \quad F(x,y,0) = F_0(x,y) \qquad (5)$$

The functions a, b_1, b_2, c and T are in general dependant on wind velocity and forest properties. b_2 and T control the threshold value F must reach at a neighbouring grid point before it has an effect, and the product $b_1 b_2$ controls its effect. The fire can be made to move predominantly in an upwind direction by biasing b_1 and b_2.

Numerical Method

The equations were solved simultaneously at each grid point using a 2nd order Runge-Kutta method, this technique was found to be adequate as results at small time intervals were required for video films. The wind velocity $\vec{v}$ was determined by the functions b_1 and b_2 which were defined at grid points for a wind direction of $\theta = 0$ to the x axis, for various wind speeds $|\vec{v}|$. For intermediate values of $|\vec{v}|$, b_1 and b_2 were calculated using linear interpolation. For other angles of θ, the functions were rotated by θ and values assigned to grid points by assuming bi-linear variation of b_1 and b_2 over grid cells.

Because, by nature, wind velocity is rarely constant, two probability distributions P_1(wind speed) and P_2(wind direction) can be used to randomly generate wind velocities. At each time step random numbers r and s between 0 and 1 are generated, then $|\vec{v}|$ and θ are chosen so as to satisfy

$$r = \int_0^{|\vec{v}|} P_1(w)dw, \quad s = \int_0^{\theta} P_2(w)dw \qquad (6)$$

Results and Discussion

In order to gain a feeling for the behaviour of the model, it was solved under various conditions. In all cases a square grid was used with T, a and c taken to be constant during a fire, in all experimemts $\alpha = 1, \beta = 1$ and $n = 2$. A unit of material (a tree) was placed at each grid point, with fires being started by igniting 10% of a single tree. Different values for b_1 and b_2 were used to model zero, constant and variable wind.

Zero Wind

For a uniform forest condition, zero wind infers that b_1 and b_2 should be symmetric about the x and y axis. For tests done here $b_1 = 0.5 b_2$ with b_2 only being non-zero at the origin and its eight surrounding points, with values defined as

$t = 200\text{s}$ $\qquad$ $t = 300\text{s}$

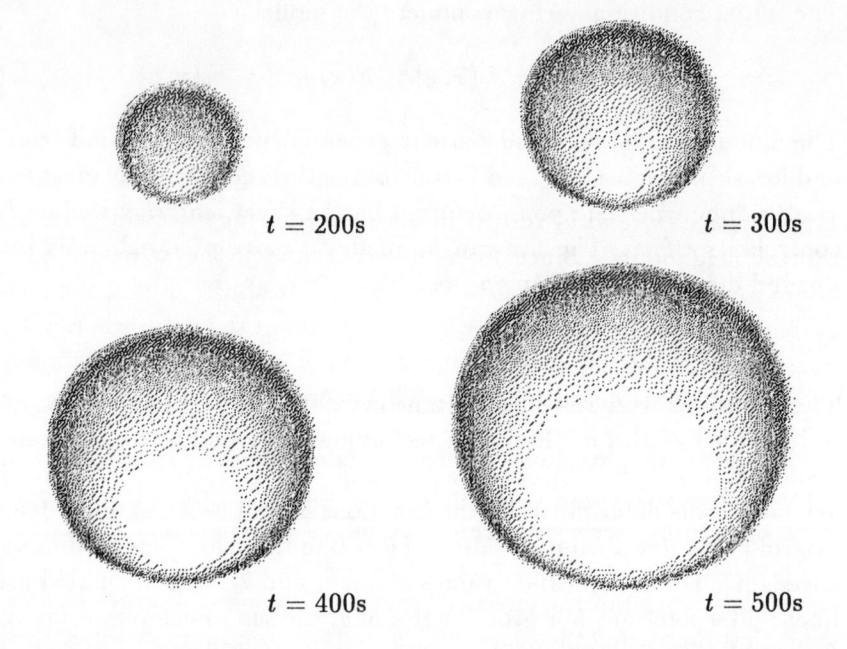

$t = 400\text{s}$ $\qquad$ $t = 500\text{s}$

Figure 2 Grey-scaled plots showing the amount of burning material (F) in a forest fire under the influence of a constant velocity wind. Model constants used are $T = 0.1, a = 0.1$ and $c = 0.01$.

The spread of wind-influenced forest fires is shown in Figures 2 to 4, these results have been grey scaled, white represents no combustion and jet black represents 100% of the original tree burning at a given time. To get an idea of the grey and spatial scale, the fire break in Figure 4 is 27 trees long and jet black.

Figure 2 illustrates the progress of a fire for a constant wind velocity. A closed band-like fire front forms which is stronger and faster upwind of the origin of the fire. The fire leaves behind it an expanding area of burnt material.

Figure 3 shows the same fire except that at $t = 320$s the wind direction is reversed, i.e. b_1 and b_2 are reflected in the x-axis. The major areas of the fire are blown back into the region of burnt material, and are reduced to an insignificant band of burning material progressing in the now downwind direction. The relatively insignificant regions of the fire that are now upwind of the origin of the fire are blown into regions of entirely unburnt material. This fire front becomes stronger and accelerates, and eventually the fire resembles an inverted version of the constant wind fire.

$\frac{1}{2\sqrt{2}}$	$\frac{1}{2}$	$\frac{1}{2\sqrt{2}}$
$\frac{1}{2}$	1	$\frac{1}{2}$
$\frac{1}{2\sqrt{2}}$	$\frac{1}{2}$	$\frac{1}{2\sqrt{2}}$

This choice of b_1 and b_2 results in a tree only being effected by its eight nearest neighbours. Since for non-zero x and y the values of b_1 and b_2 are inversely proportional to the distance from the origin, the results obtained are essentially radially symmetric about the point where the fire was started. In general, zero wind fires need not be radially symmetric as the problem is discrete as opposed to continuous.

In all the experiments done, an expanding ring-like fire front forms. This has a relatively constant cross-section and radial speed, and leaves behind it an expanding circle of burnt material. Figure 1 shows the values of F on a half cross-section along the x-axis for different times, t. These results are typical of the effects of varying T, a and c. Figures 1a and 1b show that increasing the threshold, T, reduces the speed of the fire front, as F at the leading edge must reach a higher value before the next tree is ignited. T seems to have relatively little effect on the cross-sectional shape of the front. The relative values of a and c primarily control the maximum value of F attained at a point and the time (after ignition) taken to reach this value. Figures 1a, 1c and 1d show that decreasing a or increasing c results in a slower moving fire front of lower height and width.

Non-Zero Wind

For a fire influenced by a wind, a burning tree has more effect on the trees upwind of it than those downwind. To simulate this, the functions b_1 and b_2 must be biased towards the downwind direction. For the results shown here $b_1 = 0.5b_2$ with b_2 only being non-zero at the origin and its eight neighbours, with their values defined as

0.15	0.2	0.15
0.5	1.0	0.5
0.5	0.8	0.5

This simulates a wind in the $+y$ direction. T, a and c were taken to be 0.1, 0.1 and 0.01 respectively.

a) $T = 0.1$ $a = 0.1$ $c = 0.02$

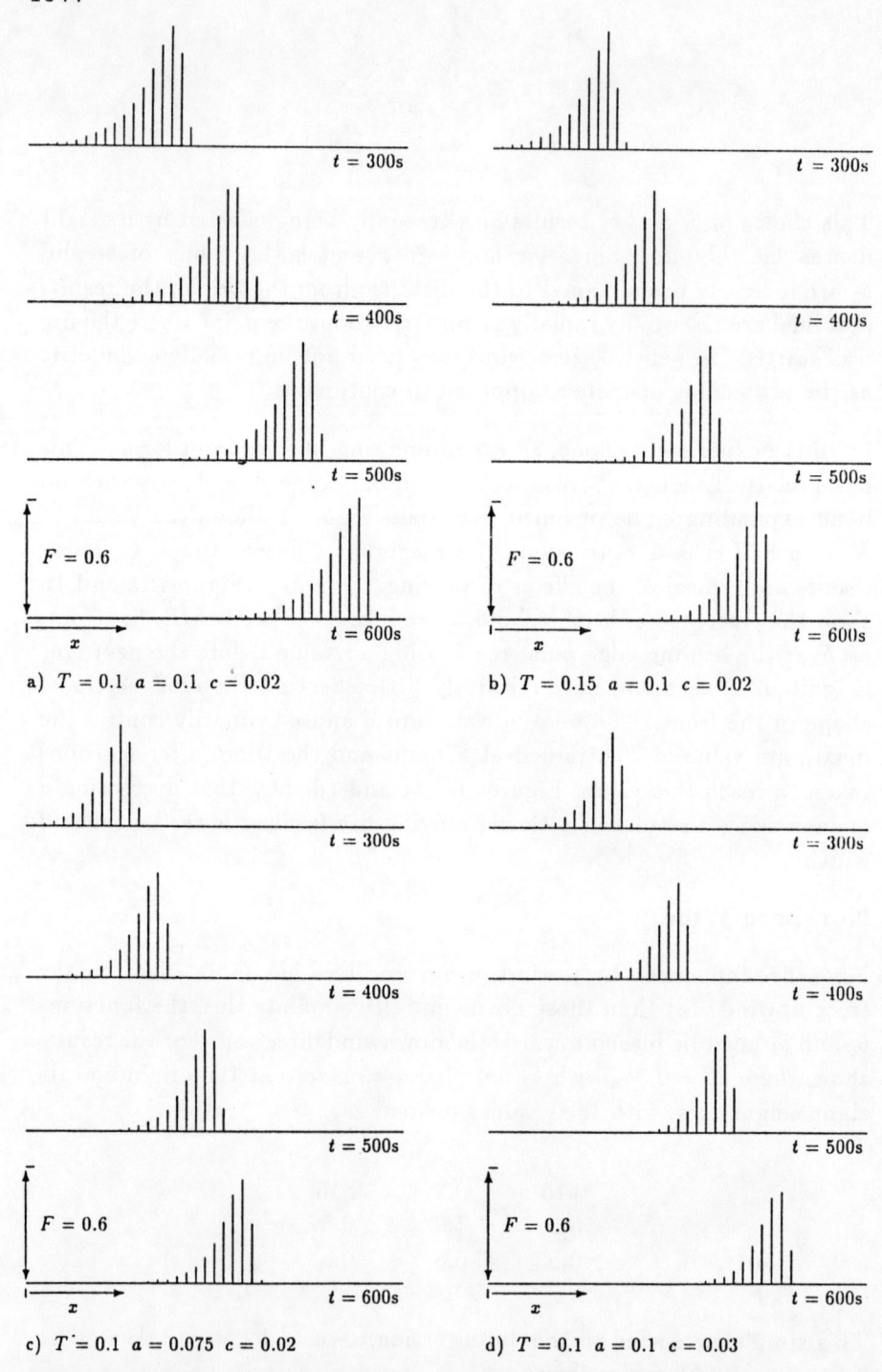

b) $T = 0.15$ $a = 0.1$ $c = 0.02$

c) $T = 0.1$ $a = 0.075$ $c = 0.02$

d) $T = 0.1$ $a = 0.1$ $c = 0.03$

Figure 1 Half cross-sectional plots of forest fire spread under zero wind conditions. Each bar represents a tree, the height gives the amount of burning material (F), x is the distance from the origin of the fire.

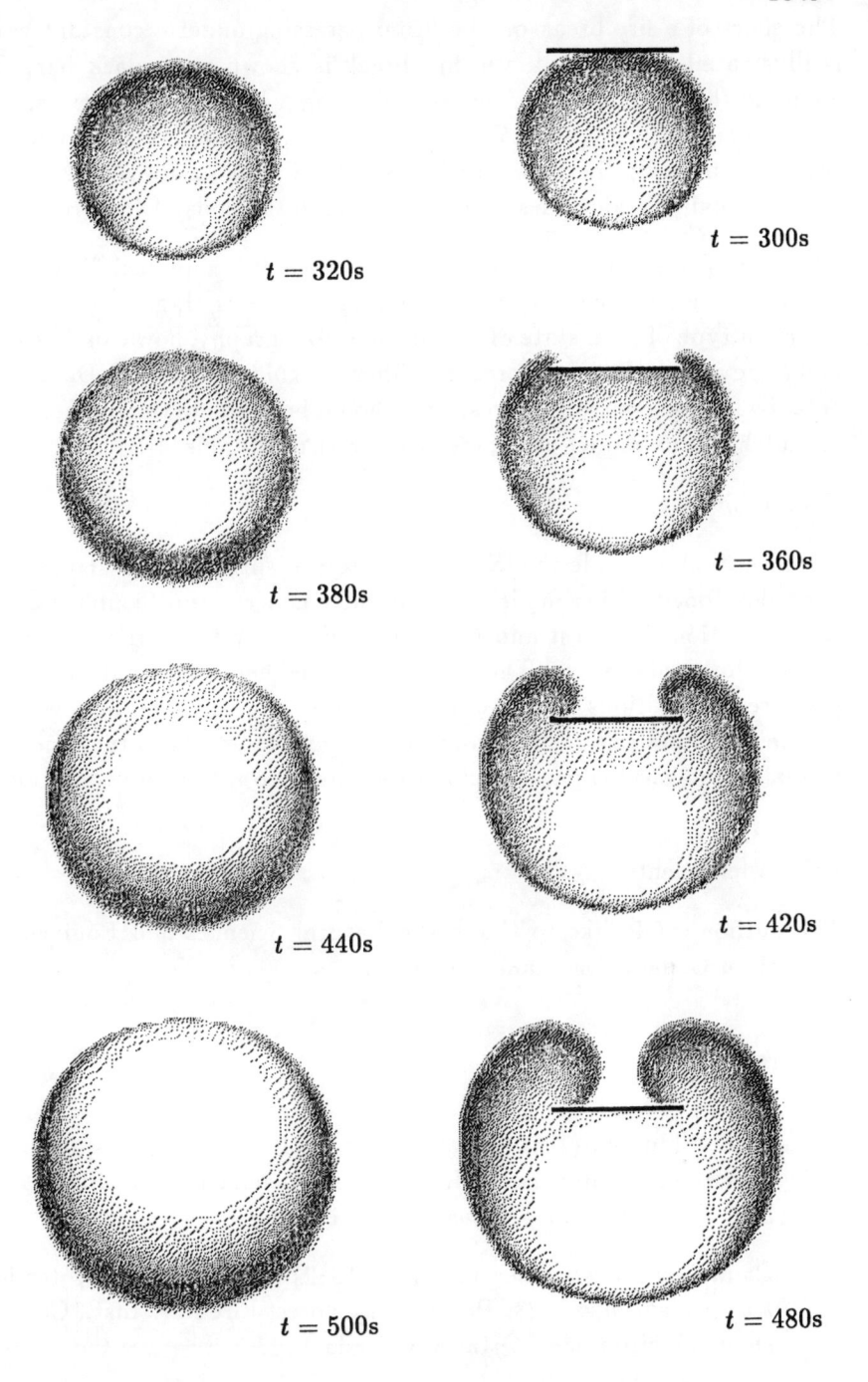

Figure 3 Grey scaled plots of burning material (F) under the same conditions as Figure 2, with the wind direction reversed at $t = 320$s.

Figure 4 Grey scaled plots of burning material (F) under the same conditions as Figure 2, with the inclusion of a fire break.

The effect of a fire break on the fire progressing under a constant wind is illustrated in Figure 4, the fire break is shown as a black bar. To simulate this, $M_0(x,y)$ was set to zero along a band of grid points, this was sufficiently wide so that the fire could not pass over it. The fire break temporarily checks the progress of the fire in an upwind direction, but eventually it continues by curling around the ends of the fire break.

The results given here demonstrate the applicability and flexibility of the numerical model used. However, the question arises of, given a forest of a certain type of tree, state of dryness and topography, how can T, a, c, b_1 and b_2 be assigned to that forest for different wind conditions? Data from actual wild and controlled forest fires should be useful in the further work of matching the model parameters to forest conditions.

Conclusions

A conceptually simple model to simulate the spread of forest fires has been developed. This model is sufficiently general and flexible to deal with variations in forest and weather conditions yet is computationally straight forward to solve. The results presented here show that the model produces simulations that could be of use in fire control. This work is preliminary, analysis of data from the field and forestry records is needed to match the various parameters in the model to actual fire conditions.

Acknowledgments

The author would like to thank the Natural Sciences and Engineering Council of Canada for financial support.

References

1. R. D. Shipman (1970) Abstracts of publications relating to forest fires research and use in Resource Management. Pa. State Agr. Sch. Forest Resour., Mimeo. pp 1-110.

2. Canadian Forestry Service (1950-1972). 'Forest Fire Protection Abstracts', Vols. 1-23. Prepared by Forest Fire Res. Inst., Canada Dept. Environment, Ottawa, Canada.

NONLINEAR VIBRATIONS OF A SKEW CYLINDRICAL PANEL AT ELEVATED TEMPERATURE

PARITOSH BISWAS

Department of Mathematics, P.D.Women's College
Jalpaiguri, West Bengal, India

INTRODUCTION

Skew panel structures find wide applications in modern aircraft and aerospace engineering. Thin and easily deformable elastic plates and shells have low flexural rigidity and can vibrate at large amplitude. Basic governing equations are nonlinear and the frequency of vibration depends on amplitude, thermal loading parameter, skew angle and aspect ratio.

Although extensive studies in the area of nonlinear free and forced vibrations of plates and shells without thermal loading have been carried out by different researchers, but the literature on skew or oblique panels are quite restricted when the structures are thermally stressed.

Nowinski [1] studied the nonlinear vibrations of isotropic skew plates with movable in-plane edges using von Karman field equations. Satyamoorthy and Ramlaxmi [2-5] also used von Karman equations for skew plates with movable in-plane edges and for immovable edge conditions, used Berger's approximation [6]. Analysis on anisotropic skew plates was carried out by Prathap and Varadan [7]. So far only a single paper on skew plate at large amplitudes including a thermal gradient has been published [8].

In this paper the von Karman equations for shallow cylindrical shell panel at elevated temperature have been transformed into oblique co-ordinates and solutions have been obtained for clamped movable and immovable edge conditions. Some interesting observations have been made after complete analysis of the problem.

Nonlinear vibrations of a skew cylindrical

BASIC GOVERNING EQUATIONS IN SKEW CO-ORDINATES

Cartesian co-ordinates (x,y) and skew co-ordinates (ξ, η) are related by [8]

$$x = \xi + \eta \sin\phi, \quad y = \eta \cos\phi \tag{1}$$

and

$$\nabla^2 = \frac{1}{\cos^2\phi}\left[\frac{\partial^2}{\partial \xi^2} - 2\sin\phi \frac{\partial^2}{\partial \xi \partial \eta} + \frac{\partial^2}{\partial \eta^2}\right] \tag{2}$$

In terms of oblique co-ordinates the basic equations can be expressed in the form

$$\nabla^4 F = Eh\left[\frac{1}{\cos^2\phi}\left\{\omega_{,\xi\eta}^2 - \omega_{,\xi\xi}\,\omega_{,\eta\eta}\right\} - \frac{1}{R_1 \cos^2\phi}\left\{\omega_{,\xi\xi} - 2\sin\phi\,\omega_{,\xi\eta} + \omega_{,\eta\eta}\right\} - \frac{1}{R_2}\omega_{,\xi\xi}\right] - \nabla^2 N_T \tag{3}$$

$$D\nabla^4 \omega + \rho h \frac{\partial^2 \omega}{\partial t^2} = \frac{1}{\cos^2\phi}\left[\omega_{,\xi\xi}\,\omega_{,\eta\eta} - 2\omega_{,\xi\eta}\,F_{,\xi\eta} + \omega_{,\eta\eta}\,F_{,\xi\xi} + \frac{1}{R_1 \cos^2\phi}\left(F_{,\eta\eta} - 2\sin\phi\,F_{,\xi\eta} + \sin^2\phi\,F_{,\eta\eta}\right)\right.$$

$$\left. + \frac{1}{R_2}F_{,\xi\xi}\right] - \frac{1}{1-\nu}\nabla^2 M_T \tag{4}$$

Nonlinear vibrations of a skew cylindrical

METHOD OF SOLUTION

For a skew cylindrical panel clamped along all the four edges and bounded by the periphery $\xi = \pm a$ and $\eta = \pm b$, a suitable mode shape that satisfies the boundary conditions

$$w = w_{,\xi} = 0 \quad \text{for} \quad \xi = \pm a \quad (5)$$
$$w = w_{,\eta} = 0 \quad \text{for} \quad \eta = \pm b \quad (6)$$

is

$$w = \frac{f(t)}{4}(1 + \cos\frac{\pi\xi}{a})(1 + \cos\frac{\pi\eta}{b}) \quad (7)$$

in which f is a function of time.

We consider a planar temperature distribution

$$T(\xi,\eta) = T_1 \sin\frac{\pi\xi}{a}\sin\frac{\pi\eta}{b} \quad (8)$$

T vanishes around the periphery of the panel and maximum at the centre such that

$$T_{max} = T_1 \quad \text{and} \quad M_T = 0 \quad (9)$$
$$N_T = T_1 h \sin\frac{\pi\xi}{a}\sin\frac{\pi\eta}{b} \quad (10)$$

Inserting required expressions into equation (3), one gets the particular integral in the form

$$F_p = b_1\cos\alpha + b_2\cos\beta + b_3\cos 2\alpha + b_4\cos 2\beta +$$
$$b_5\cos\alpha\cos\beta + b_6\sin\alpha\sin\beta + b_7\cos\alpha\cos 2\beta +$$
$$b_8\sin\alpha\sin 2\beta + b_9\cos 2\alpha\cos\beta + \sin 2\alpha\sin\beta$$
$$(11)$$

in which b_i ($i = 1,2,\ldots 10$) are known, $\alpha = \frac{\pi\xi}{a}, \beta = \frac{\pi\eta}{b}$

The complementary function of equation (3) is

$$F_c = \frac{1}{2}(L\eta^2 + M\xi^2) - N\xi\eta \quad (12)$$

where the constants L, M and N contribute directly to the in-plane stresses $N_{\xi\xi}$, $N_{\eta\eta}$ and $N_{\xi\eta}$ respectively and will be determined by using in-plane boundary conditions.

For the movable case, the boundary conditions are taken in the forms

$$\int_{-b}^{b} N_{\xi\xi}d\eta = 0 \quad \text{and} \quad \int_{-b}^{b} N_{\xi\eta} = 0 \quad \text{on the edges } \xi = \pm a \quad (13)$$

$$\int_{-a}^{a} N_{\xi\eta} = 0 \quad \text{and} \quad \int_{-a}^{a} N_{\eta\eta}d\xi = 0 \quad \text{on the edges } \eta = \pm b \quad (14)$$

Nonlinear vibrations of a skew cylindrical

For the immovable edges the conditions are $u = v = 0$ on all the four edges and are given by an averaging manner [7] as

$$u(a, \eta) - u(-a, \eta) = \int_{-a}^{a} u_{,\xi} d\xi = 0 \qquad (15)$$

$$v(a, \eta) - v(-a, \eta) = \int_{-a}^{a} v_{,\xi} d\xi = 0 \qquad (16)$$

$$v(\xi, b) - v(\xi, -b) = \int_{-b}^{b} v_{,\eta} d\eta = 0 \qquad (17)$$

$$u(\xi, b) - u(\xi, -b) = \int_{-b}^{b} u_{,\eta} d\eta = 0 \qquad (18)$$

Making use of the boundary conditions (13-14) one gets for movable edges

$$L = M = N = 0 \qquad (19)$$

and making use of the boundary conditions (15-18) one gets for immovable edges

$$L = \frac{3Ehf^2\pi^2}{128(1-\nu^2)}\left\{\left(\frac{1}{a^2}+\frac{\nu}{b^2}\right)-\left(\frac{\nu}{a^2}+\frac{1}{b^2}\right)\tan^2\phi\right\} +$$

$$+ \frac{(1-\nu)f}{4R_1}\cos 2\phi - \frac{f}{4R_2}\cos^2\phi \cdot (\sin^2\phi - \nu\cos^2\phi) \qquad (20)$$

$$M = \frac{3Ehf^2\pi^2}{128(1-\nu^2)}\left(\frac{\nu}{a^2}+\frac{1}{b^2}\right) - \frac{f}{4R_1(1-\nu)} -$$

$$- \frac{f\cos^2\phi}{4R_2(1-\nu^2)} \qquad (21)$$

$$N = 0$$

For rectangular plates $\phi = 0$, $R_1 = R_2 \infty$ and the constants L and M are given by

$$L = 3Ehf^2\pi^2(1/a^2 + \nu/b^2)/128(1-\nu^2) \qquad (22)$$

Nonlinear vibrations of a skew cylindrical

$$M = 3\,Ehf^2\pi^2(\nu/a^2 + 1/b^2)/128(1-\nu^2) \qquad (23)$$

Applying Galerkin procedure in equation (4) one gets after a lengthy but simple calculations the time-differential equation (Duffing's generalised equation) in the form

$$\ddot{f}(t) + (\alpha_1 - \alpha_2 N_T^*)f(t) - \beta f^2(t) + \gamma f^3(t) + \delta N_T^* = 0 \qquad (24)$$

in which $\alpha_1, \alpha_2, \gamma, \delta$ are known constants.

On the assumption that the large amplitude vibration possesses only one degree of freedom $f(t)$ can be represented as [10]

$$f(t) = A\,F(t) \qquad (25)$$

in which case the equation (24) is to be solved with the normalised conditions $F(0) = 1$, $\dot{F}(0) = 0$.

When the steady state vibrations have been established, then of the shell vibration can be assumed to that of pulsation of the thermal load and therefore one can express

$$N_T^* = \overline{N}_T^*\cos\omega t$$

$$f(t) = A\cos\omega t \qquad (26)$$

Applying Galerkin procedure in equation (24) one gets the ratio of the nonlinear and linear frequencies, i.e., the non-dimensional frequency as

$$\left(\frac{\omega_N}{\omega_L}\right)^2 = 1 + \frac{\gamma}{\alpha_1}(A/h)^2 + \frac{\delta}{\alpha_1}\overline{N}_T^* / (A/h)^2 \qquad (27)$$

Equation (27) is in the most general form and non-dimensional frequencies can be determined for variations of aspect ratio (a/b), non-dimensional **curvatures** as well as for variations of thermal loading parameter and non-dimensional amplitudes.

NUMERICAL RESULTS AND DISCUSSION

Table-1 shows the variations of non-dimensional frequencies for variations of thermal loading parameter and non-dimensional curvature considering the following set of values

$a/b = 1$, $\phi = 0$, $\lambda = (4/3)\overline{N}_T^*(a/b)^2 = 0$ & 1

$a^2/R_1 h = 10$ & 15, $R_2 = 0$

Nonlinear vibrations of a skew cylindrical

TABLE-I

$A/h = 0$	.2	.4	.6	.8	1.0
$\lambda=0$ $a^2/R_1h = 10$ 1	1.07	1.27	1.50	1.85	2.20
$\lambda=0$ $a^2R_1h = 15$ 1	1.04	1.14	1.29	1.47	1.63
$\lambda=1$ $a^2R_1h = 10$	1.23	1.34	1.58	1.00	2.22
$\lambda=1$ $a^2R_1h = 15$	1.15	1.19	1.32	1.49	1.70

Table-II shows the results for skew angle $\phi = 60°$

TABLE - II

	0	.2	.4	.6	.8	1.0
$\lambda=0$ $a^2R_1h = 10$	1	1.03	1.13	1.26	1.42	1.62
$\lambda=0$ $a^2R_1h = 15$	1	1.03	1.11	1.24	1.40	1.58
$\lambda=1$ $a^2R_1h = 10$		1.04	1.13	1.26	1.43	162
$\lambda=1$ $a^2R_1h = 15$		1.03	1.12	1.24	1.40	1.58

Nonlinear vibrations of a skew cylindrical

Interesting observations can be made from the results computed in the above two tables.

i) Non-linear-pistol frequencies are less in the absence of temperature than for those of shells under thermal loading.

ii) Effect of skew angle is to diminish non-dimensional frequencies.

iii) Effect of curvature is also to diminish the non-dimensional frequencies.

As thermal stresses and vibrations of plate and shell problems have wide applications in aeronautics and spacecraft structures so the present problem may be of use to the Designers and Engineers before proper designing of structural components in order to maintain desired frequencies.

ACKNOWLEDGEMENT

The present paper is a part of the advanced research project No. F. 8-15/83 (SR-III) supported by the University Grants Commission, New Delhi.

The author will remain grateful to the Department of Science and Technology (DST), Govt. of India and to the U.G.C. for possible financial grant to enable him to attend and present the paper at the International Conference on Thermal Problems to be held at Montreal, Quebec, Canada, June 29-July 3, 1987.

The author will also be thankful to the Governing Body of his Institution for considering the matter of leave for the purpose of attending the Conference.

REFERENCES

1. Nowinski, J.L. - Large amplitude oscillation of oblique panels with initial curvature, AIAA, $\underline{2}$, 1025-31 (1964).

2. Sathyamoorthy, M and Pandalai, K.A.V.- Nonlinear vibrations of elastic skew plates exhibiting rectilinear orthotropy, J. Franklin Institute, $\underline{296}$, 359-369 (1973).

3. _____ Nonlinear flexural vibrations of orthotropic skew plates, J.Sound Vib, $\underline{24}$, 115-120 (1972)

4. _____ Large amplitude flexural vibrations of simply-supported skew plates, AIAA, 11, 1279-82 (1973).

5. _____ Vibrations of simply-supported clamped skew plates at large amplitudes, J. Sound and Vibration, 27, 37-46 (1973).

6. Berger, H.M.- A new approach to the analysis of large deflections of plates, J. Appl. Mech., ASME, 22(4), 465-472 (1955).

7. Prathap, G. and Varadhan, T.K.- Nonlinear flexural vibrations of anisotropic skew plates, J. Sound and Vibr., 63,3, 315-23 (1979).

8. Dalamangas, A.- Vibrations of skew plates at large amplitudes including thermal gradient, Rev. Roum. Sci. Techn., Mech Appl., Tome 29,3, 263-69 (1984).

9. El. Zacuk, R.R. and Dym.- Nonlinear vibrations of orthotropic doubly-curved shallow shells, Jour. Sound and Vibration, 31,1, 89-103 (1973).

10. Nowinski, J.L. - Nonlinear vibrations of circular plates exhibiting rectilinear orthotropy, ZAMP, 14, 112-124 (1963).

LOCAL NON-SIMILARITY SOLUTIONS FOR FREE CONVERTION
PROBLEMS USING ACCELERATED SUCCESSIVE REPLACEMENT
METHOD

M.A. Muntasser
Associate Professor, Dept. of Mechanical Engineering
University of Al-Fateh, Tripoli, Libya
J.C. Mulligan
Professor, Dept. of Mechanical and Aerospace
Engineering, North Carolina State University
North Caroline, USA.

A laminar boundary layer analysis for steady state free convection on two-dimensional and axi symmetrical bodies has been developed in a general form with arbitrary surface temperature and variable transpiration velocities. A locally similar and locally non-similar boundary layer formulation has been presented for a horizontal cylinder and a sphere with and without suction and injection boundary conditions. Uniform temperature as well as suction and injection velocities through the wall of these geometries were treated. Heat transfer and flow data were reported at various locations on the surface of a horizontal cylinder and sphere for a moderate range of Prandtl numbers, and comparisons were made with the existing solutions in the literature.

The method of local non-similarity was found to be a simple, direct and accurate procedure in reducing the partial differential equations of streamwise non-similar free convection to a format of ordinary differnetial equations. When coupled with a computer routine for solving ordinary differential equations, such as the method of accelerated successive replacement, accurate and inexpensive boundary-layer computations are readily carried out.

The results of heat transfer with suction through an isothermal horizontal cylinder surface were also found to be of the same accuracy as that without suction. Again a two-equations model was found sufficient to obtain results within 1.0 per cent of the finite difference solution. The third level analysis was found necessary for the injection case for the isothermal horizontal cylinder in order to obtain results comparable to the finite difference solution. Similarly, for a sphere with suction at higher Prandtl numbers, the second and third level truncations gave results of the same order.

For the case of the isothermal sphere, it is found that the third level analysis is needed to provide an assessment of the accuracy of the first and the second level results. The third level computations were found to be sufficiently accurate and comparable to the series solutions and the experimental data available in the literature.

For the isothermal sphere with injection satisfactory results are obtained from the first and second levels of solutions at $x \_ 80^o$. The third level analysis is also found to be necessary for larger x to assess the accuracy of the first two levels of solutions in the local non-similarity analysis.

1. INTRODUCTION

The method of local nonsimilarity since it was developed by Sparrow and Co-workers [1,2] Several problems of natural convection heat transfer have been solved utilizing this method [3,4]. It was found (4) that for the case of the laminer, steady, free convection heat transfer from an isothermal cylinder that only two levels of computation gave results better than either integral metled or series solution and within one per cent of the finite difference solution. Since the anlysis for the horizontal cylinder outlined in reference [4], for the sake of brevity it will not be repeated here. In this paper a locally similar solution and locally non-similar boundary layer for laminar steady natural convection from two-dimensional and axisymmetric bodies with a constant surface temperature and constant transpiration velocity is developed and solved by the implicit numerical method of "accelerated successive replacement". Heat transfer results are given for a horizontal cylinder at various prandtl numbers. These results are then compared with the results of the existing finite difference solution [5]. The analysis is then extended to the sphere, for which results are reported here for the first time and presented in the same format as used for two-dimensional surface.

The laminar, steady free convection heat transfer from an isothermal sphere has been treated analytically in the literature on several occasions. Approximate similarity transformation (6,7) as well as approximate series representation (8,9) have been employed in obtaining solutions of the thermal boundary layer equation for this axisymmetric non-similar problem. The only numerical finite difference solution which seems to be available is the limiting case of an analysis of combined forced and free convection (10), in which data is only presented for the forward side of the surface.

The principal objective in this section is to present numerically accurate computations of the streamwise variation of the heat transfer on both forward and rearward surface of a sphere, computations which are not consistent and definitive in the literature and to demonstrate the utility of the method of "local non-similarity" in axisymmetric streamwise, non-similar convective heat transfer with suction and injection boundary conditions.

The set of equations corresponding to the first level (local similarity) for the isothermal spheres is

$$F''' + (3 + \frac{\cos x}{\phi_1^4})FF'' - [b(x) + 3](F')^2 + \Theta = 0 \quad (2.14)$$

$$\Theta'' + Pr(3 + \frac{\cos x}{\phi_1^4}) F \Theta' = 0 \quad (2.15)$$

with boundary conditions:

$$F'(x,0) = 0, \quad F(x,0) = -v_0/[3\phi_1(x) + \cos x/\phi_1^3],$$
$$\Theta(\,,0) = \Theta_w \quad (2.16)$$

$$F'(x,\infty) = 0, \quad \Theta(x,\infty) = 0. \quad (2.17)$$

The set of equations corresponding to the second level truncation (first approximation of a local non-similarity method) may be written as

$$F''' + (3 + \frac{\cos x}{\phi_1^4}) F F'' - [b(x) + 3](F')^2 + \Theta = \frac{\phi_2}{\phi_1}(F'E' - E F'') \quad (2.18)$$

2. ANALYSIS

The analysis of the two-dimensional free convection from a horizontal cylinder was previously considered in (4). The boundary conditions for injection and suction of the same general formulation is given below.

$$F'(X,0) = 0, \quad \phi_2 \frac{dF}{dx} + 3\phi_1 F = -V_0, \quad \Theta = 1 \text{ at } = 0 \quad (2.1)$$

$$F'(x, \infty) \quad \Theta = 0 \text{ as } \to \infty \quad (2.2)$$

The boundary conditions for injection and suction for the set of local similarity equations can be presented as

$$F'(X, 0) = 0, \quad F(X, 0) = -V_0/3\phi(x), \quad \Theta(X,0) = 1 \quad (2.3)$$

$$F(X, \infty) = 0 \qquad \qquad \Theta(X, 0) = 0 \quad (2.4)$$

Similarily, the boundary conditions corresponding to the second level truncation in the local non-similarity method for the same problem, is given as

$$F'(X,0) = 0, \phi_2(x) \frac{dF(X,0)}{dx} + 3\phi_1(X,0) = -V_0, \Theta(X,0) = 0 \quad (2.5)$$

$$F'(X,0) = H(X,0) \quad (2.6)$$

$$F'(X,\infty) = E(X,\infty) = \Theta(X,\infty) = H(X,\infty) = 0 \quad (2.7)$$

Finally, the set of six non-linear ordinary differential equations comprising the third level approximation can easily be obtained using the same method. The third level boundary conditions then become:

$$F'(X,0) = E'(X,0) = S'(X,0) = 0 \quad (2.8)$$

$$\phi_2(X) \frac{dF}{dx}(X,0) + 3\phi_1(X) F(X,0) = -V_0 \quad (2.9)$$

$$\phi_2 S(X,0) + 6\phi_1(X) E(X,0) + \frac{3b(X) \sin X}{\phi_1 \phi_2^2} F(X,0) = 0 \quad (2.10)$$

$$\Theta(X,0) = 1, \quad H(X,0) = 0, \quad Q(X,0) = 0 \quad (2.11)$$

$$F'(X,\infty) = E'(X,\infty) = S'(X,\infty) = 0 \quad (2.12)$$

$$\Theta(X,\infty) = H(X,\infty) = Q(X,\infty) = 0 \quad (2.13)$$

$$\Theta'' + \Pr(3 + \frac{\cos x}{\phi_1^4}) F \Theta' = \Pr \frac{\phi_2}{\phi_1} (F'H - \Theta'E) \quad (2.19)$$

$$E''' + (3 + \frac{\cos x}{\phi_1^4}) F E'' - (b(x) + 9) F'E' - \left[(b(x) - 6\right.$$

$$\left. - \frac{\cos x}{\phi_1^4}\right) FE'' - (4b(x) \cot x + \phi_2/\phi_1) F F'''$$

$$- b'(x) (F')^2 + H = 0 \quad (2.20)$$

$$H'' + \Pr(3 + \frac{\cos x}{\phi_1^4}) F H' + \Pr(b(x) - 3) F'H - \left[(b(x) - 6)\right.$$

$$\left. - \frac{\cos x}{\phi_1^4}\right] \Theta'E - (4b(x) \cot x + \phi_2/\phi_1) F \Theta' = 0 \quad (2.21)$$

With the boundary conditions

$$F'(x,0) = 0, \quad \phi_2(x) \frac{\partial F(x,0)}{\partial x} + \left[3\phi_1(x) + \cos x/\phi_1^3\right]$$
$$F(x,0) = -v_0 \quad (2.22)$$

$$E'(x,0) = 0, \quad H(x,0) = 0, \quad \Theta(x,0) = 1 \quad (2.23)$$

$$F'(x,\infty) = E'(x,\infty) = \Theta(x,\infty) = H(x,\infty) = 0. \quad (2.24)$$

The general set of equations comprising a Third level truncation (second level approximation of local non-similar method) for the isothermal sphere is

$$F''' + (3 + \frac{\cos x}{\phi_1^4}) F F'' - (b(x) + 3)(F')^2 + \Theta$$
$$= \frac{\phi_1}{\phi_2} (E'F' - E F'') \quad (2.25)$$

$$\Theta'' + \Pr(3 + \frac{\cos x}{\phi_1^4}) F \Theta' = \Pr \frac{\phi_2}{\phi_1} (F'H - \Theta'E) \quad (2.26)$$

$$E''' + (3 + \frac{\cos x}{\phi_1^4}) F E'' + (b(x) + 9) F'E' - \left[b(x) - 6\right.$$

$$\left. - \frac{\cos x}{\phi_1^4}\right] F''E - (4b(x) \cot x + \phi_2/\phi_1) FF''$$

$$-b'(x)(F)^2 + H = \frac{\phi_2}{\phi_1} \left[S' F' + (E)^2 - SF'' - EE'' \right] \quad (2.27)$$

$$H'' + \Pr\left(3 + \frac{\cos x}{\phi_1^4}\right) F H' + \Pr(b(x) - 3) F'H$$

$$-\left[b(x) - 6 - \frac{\cos x}{\phi_1^4}\right]\Theta'E - (\Pr)(4b(x)\cot x$$

$$+ \frac{\phi_2}{\phi_1}) F \Theta' = \Pr\frac{\phi_2}{\phi_1}(E'H + F'Q - H'E - \Theta'S) \quad (2.28)$$

$$S''' + \left(3 + \frac{\cos x}{\phi_1^4}\right)FS'' - 12F'S' - 2b(x) - 9 - \frac{\cos x}{\phi_1^4})F''S$$

$$-\left[b(x) - 6 - \frac{\cos x}{\phi_1^4}\right]EE'' - \left[\frac{\sin x}{\phi_1^4} + 8b(x)\cot x\right.$$

$$+ \frac{\phi_2}{\phi_1}\Big]FE'' - 3b'(x)F'E' - 12(E')^2 - \left[b'(x) + \frac{\sin x}{\phi_1^4}\right.$$

$$+ 8b(x)\cot x + \frac{\phi_2}{\phi_2}\Big]F''E - \left[4b'(x)\cot x + 3 - b(x)\right.$$

$$-\frac{4b(x)}{\sin^2 x}\Big]FF'' - b''(x)(F')^2 + Q = 0 \quad (2.29)$$

$$Q'' + \Pr\left(3 + \frac{\cos x}{\phi_1^4}\right)FQ' + 2\Pr\left[b(x) - 3\right]F'Q$$

$$- 2\Pr(b(x) - 6 - \frac{\cos x}{\phi_1^4})EH'$$

$$- \Pr\left[\frac{\sin x}{\phi_1^4} + \frac{\phi_2}{\phi_1} + 8b(x)\cot x\right]FH'$$

$$+ 2\Pr\left[b(x) - 3\right]E'H + \Pr b'(x) F'H$$

$$- \Pr\left[2b(x) - 9 - \frac{\cos x}{\phi_1^4}\right]S\Theta'$$

$$- \Pr\left[b'(x) + \frac{\sin x}{\phi_1^4} + \frac{\phi_2}{\phi_2} + 8b(x)\cot x\right]E\Theta'$$

$$- \Pr\left[4b'(x)\cot x + 3 - b(x) - 4b(x)/\sin^2 x\right]F\Theta'$$

$$= 0 \quad (2.30)$$

Where the boundary conditions are:

$$E'(x,0) = F'(x,0) = S'(x,0) = 0 \qquad (2.31)$$

$$\phi_2(x)\frac{\partial F(x,0)}{\partial x} + \left[3\phi_1(x) + \frac{\cos x}{\phi_1^4}\right]F(x,0) = -v_o \qquad (2.32)$$

$$\phi_2 S(x,0) + \left[6\phi_1(x) + \frac{\cos x}{\phi_1^3}\right]E(x,0)$$

$$+ \left[3\left(1 - \frac{\cos x}{\phi_1^4}\right)\frac{b(x)\sin x}{\phi_1 \phi_2} - \frac{\sin x}{\phi_1^3}\right]F(x,0)$$

$$= -\frac{\partial v_o}{\partial x} \qquad (2.33)$$

$$\Theta(x_1,0) = \Theta_w, \quad H(x,0) = \frac{\partial \Theta_w}{\partial x}, \quad Q(x,0) = \frac{\partial H(x,0)}{\partial x} = 0 \qquad (2.34)$$

$$F'(x,\infty) = E'(x,\infty) = S'(x,\infty) = 0 \qquad (2.35)$$

$$\Theta(x,\infty) = H(x,\infty) = Q(x,\infty) = 0 \qquad (2.36)$$

The term $\frac{\partial V_o}{\partial x} = \frac{\Theta_w}{x} = \frac{H}{x}$ and $\Theta_w = 1 = 0$ for the case of constant suction and injection.

3. RESULTS AND DISCUSSION

3.1. A Horizontal Cylinder

Computations are obtained from local **convective** heat transfer parameters defined by

$$Nu_x / |v_o| Gr_w 1/4 = -\phi_1(x) \Theta(x,0)/|v_o| . \qquad (3.1.)$$

Results are developed for the moderate range of Prandtl numbers 0.72, 1.0, 5.0, and 10.0, and for a range of wall velocities of $v_o = (0.5, 1.0, 2.0, -0.5, -1.0$ and -2.0. Figure (1) illustrates the data for $Pr = 1.0$, $v_o = -1.0$, for a horizontal cylinder with suction

The second level computations correct the virst level so as to provide results less than 1.0 percent from the finite difference (5) results, differenced in values getween the second level and the third level of computation is too small to be plotted in the figure.

Table 1 presents results for the local convective heat transfer parameter defined by equation (2.) for various suction velocities and Prandtl numbers of 0.72, 5.0 and 10.0. For large Prandtl numbers the heat transfer parameter varies slowly with x. It can also be seen in Table (1) that convective heat transfer increases with an increase of Prandtl number for the same suction velocities. The heat transfer for Pr = 10 is ten-fold higher than that at Pr = 0.72 and double that at Pr = 5.0 for v_0 = -2.0. Also, the local convective heat transfer parameter increases with increases in suction velocity.

Figure 2 shows the variation of the local convective heat transfer with x (radians) from a horizontal cylinder with moderate injection velocities v_0 = 1.0 and 2.0 for Pr = 1.0. In Figure 2 the results of the three levels of the present solutions are reported and compared with the finite difference solution (5). A discrepancy is expected at large x between the results of the first level and the finite difference solution. The deviation is seen to reach 19.0 per cent at x = 2.0 and v_0 = 1.0. The second level corrected this discrepancy to 11.0 per cent at the same condition. However, the second level result is improved appreciably as v_0 decreases. That is, the deviation becomes only 3.0 percent compared to the finite difference.

Table 1 Third level peripheral variation of the local convective heat transfer parameter $Nu_x/|v_0| Gr_w^{1/4}$ from a horizontal cylinder for various suction velocities and various Prandtl numbers

x^0	Pr = 0.72		Pr = 5.0		Pr = 10	
	v_0=-0.5	v_0=-1.0	v_0=-0.5	v_0=-1.0	v_0=-0.5	v_0=-1.0
0	1.1841	0.8559	5.2191	5.0228	10.0840	10.0350
30	1.1750	0.8513	5.2059	5.0195	10.0650	10.0340
60	1.1444	0.8360	5.1649	5.0091	10.0410	10.0320
90	1.0905	0.8094	5.0923	4.9906	10.0150	10.0310
120	1.0030	0.7670	4.9860	4.9582	9.9854	10.0290
135	0.9080	0.7342	4.9168	4.9317	9.9702	10.0250
150			4.8333	4.8870	9.9612	10.0170

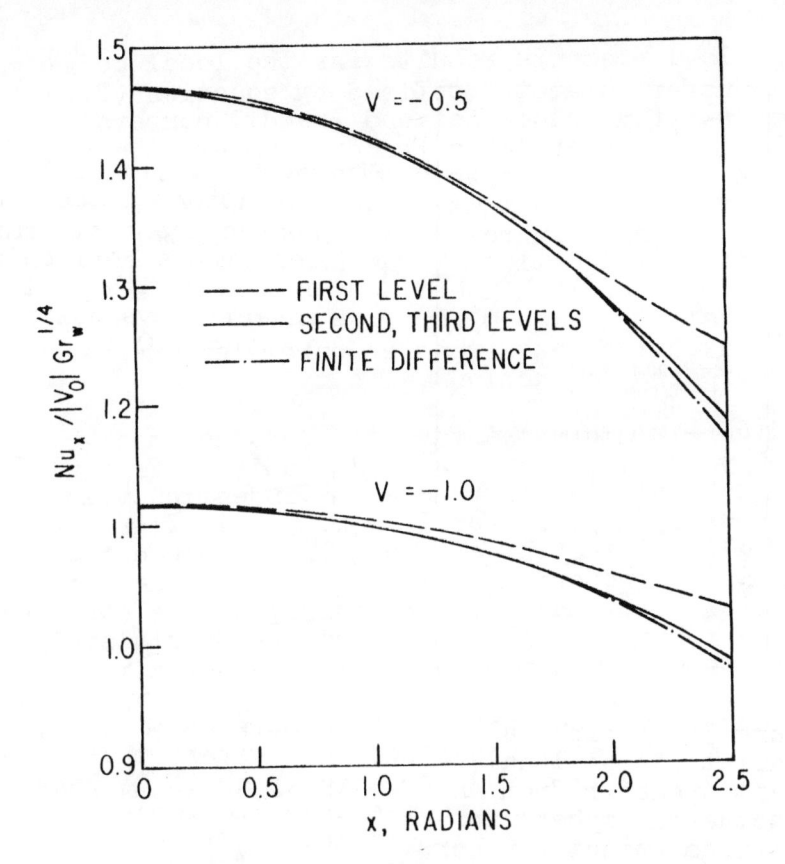

Figure 1. Peripheral variation of the local convective heat transfer from a horizontal cylinder with various suction velocities at Pr = 1.0

Result at the smaller $v_0 = 0.5$ for the same streamwise location $x = 2.0$. This discrepancy in the second level suggests that for the injection case a third level analysis is essential.

The third level computations are seen to overcorrect the second level results, as shown in Figure 2.

Table 2 shows the third level peripheral variation of the local convective heat transfer from a circular cylinder for various injection velocities, $v_0 = 1.0$ and 2.0, at Prandtl number 0.72, 5.0 and 10.0. The convective heat transfer decreases with Prandtl number, although increases for all injection velocities. The convective heat transfer tends to zero at Pr = 5.0 and $v_0 = 2.0$ and also at Pr = 10 for $v_0 = 1.0$ and 2.0. Here it is found that the temperature gradient is zero at = 0, even though it

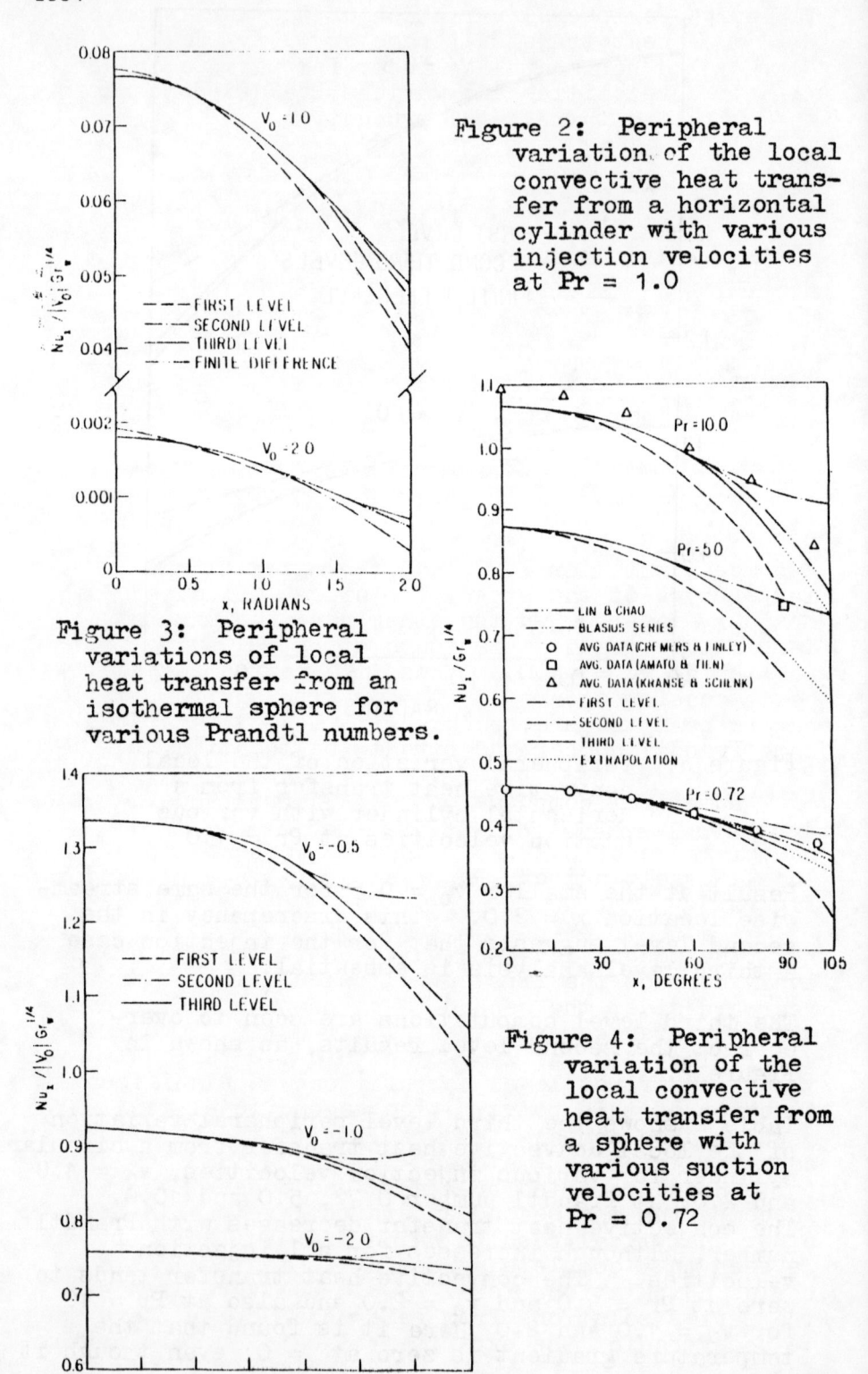

Figure 2: Peripheral variation of the local convective heat transfer from a horizontal cylinder with various injection velocities at Pr = 1.0

Figure 3: Peripheral variations of local heat transfer from an isothermal sphere for various Prandtl numbers.

Figure 4: Peripheral variation of the local convective heat transfer from a sphere with various suction velocities at Pr = 0.72

Table 2: Peripheral variation of the local convective heat transfer parameter $Nu_x/v_0\ Gr_W^{1/4}$ from a horizontal cylinder for various injection velocities and various Prandtl numbers for the third level truncation.

x^o	Pr = 0.72		Pr = 5.0		Pr = 10.0	
	$v_o=1.0$	$v_o=2.0$	$v_o=1.0$	$v_o=2.0$	$v_o=1.0$	$v_o=2.0$
0	0.0981	0.0052	0.0009	0.0000	0.0000	0.0000
30	0.0961	0.0050	0.0008	0.0000	0.0000	0.0000
60	0.0888	0.0042	0.0006	0.0000	0.0000	0.0000
90	0.0778	0.0032	0.0004	0.0000	0.0000	0.0000
120	0.0697	0.0030	0.0003	0.0000	0.0000	0.0000

becomes finite for 0, and the match is the temperature of the stagnant fluid at the edge of the boundary layer. On the other hand the maximum temperature gradient is found at Pr = 10 with suction at v_0 = -1.0. Thus, the maximum convective heat transfer occurs at Pr = 10 with suction, while a zero heat transfer occurs with injection at the same Prandtl number and streamwise location.

Numerical computations were carried out for the isothermal sphere for the intermediate Prandtl number range of 0.72, 5.0 and 10. The ratio of the local Nusselt number, Nu_x, and $Gr_W^{1/4}$ are plotted in Figure 3.

The second level computation is seen to effectively correct the first level calculation. On the rearward side, the second level computation agreed with the average experimental air data of Cremers and Finley [11] at x = 90° within 5.0 per cent, even though the first level computation was significantly in error at this point. The second level truncation results also compare favourably with the experimental data for water by Amato and Tien [12], and the data for benzene reported by Kranse and Schenk [13]. In comparing the second and third levels with the other analytical results which are available, it can be seen that complete agreement is achieved with series computations for x 90°. However, at x = 90° the series solutions from reference [8, 9], are seen to underpredict the experimental data as well as the third levels.

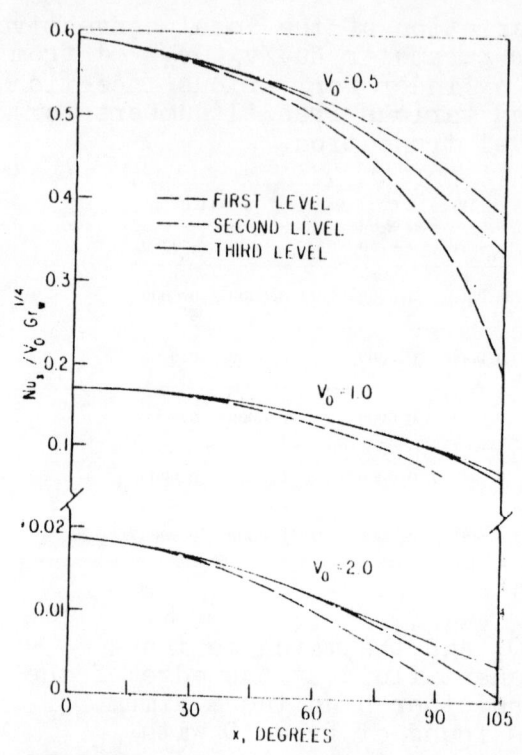

Figure 5: Peripheral variation of the local convective heat transfer from a sphere with various injection velocities at Pr = 0.72

Figure 4 shows the local convective heat transfer variation with respect to streamwise locations x from a sphere with various suction velocities at Pr = 0.72. The convective heat transfer results follow a similar trend as was found in [4] for an isothermal sphere. The results for the convective heat transfer from a sphere with various injection velocities through the wall at Pr = 0.72 using the three local similarity and local non-similarity models are shown in Figure 5.

The method of local non-similarity was found to be a simple, direct, and accurate procedure in reducing the partial differential equations of streamwise non-similar free convection to a format of ordinary differential equations. When coupled with a computer routine for solving ordinary differential equations, such as the method of accelerated successive replacement, accurate and inexpensive boundary-layer computations are readily carried out. Moreover, the technique preserves many of the advantages inherent in similarity analysis.

The third level computations were found to be sufficiently accurate and comparable to the series solutions and the experimental data available in the literature [6,7] for the isothermal sphere. The results of heat transfer with suchtion through an iso isothermal horizontal cylinder surface were also found to be of the same accuracy as that without suction. The third level analysis was found necessary for the injection case for the isothermal holizontal cylinder in order to obtain results comparable to the finite difference solution.

REFERENCES

1. SPARROW, E.M., QUAK, H. and BOERNER, C.J., "Local Non-Similar Boundary Layer Solutions", AIAA Journal, Vol. 8, 1936 - 42, 1970.

2. SPARROW, E.M. and YU, H.S., "Local Non-Similarity Thermal Boundary-Layer Solutions," J. of Heat Transfer, Trans ASME, Series C, Vol 93, 328 - 334 1971.

3. MINKOWYCZ, W.J. and SPARROW, E.M., "Local Non-similar Solutions for Natural Convection on a Vertical Cylinder," J. of Heat Transfer, Trans. ASME, Series C, Vol. 96, 178 - 183 1974.

4. MUNTASSER, M.A. and MULLIGAN, J.C., "A Local Non-similarity Analysis of Free Convection from a Horizontal Cylindrical Surface". ASME, J. of Heat Transfer, Vol. 100, No. 2, 165 - 167, 1978.

5. MERKIN, T.H., "The Effects of Blowing and Suction on Free Convection Boundary Layers", Int. J. Heat Mass Transfer, Vol. 18, 237 - 244, 1975.

6. MERK, H.J. and PRINS, J.A., "Thermal Convection in Laminar Boundary Layers, III", J. of Applied Sci. Res., Sec. A, Vol.4, 207 - 220, 1953-1954.

7. BRAUN, W.H., OSTRACH, S. and HEIGHWAY, J.E., "Free Convection Similarity Flows about Two-Dimensional and Axisymmetric Bodies with Closed Lower Ends, "Int. J. Heat Mass Transfer, Vol. 2 121 - 135, 1961.

8. LIN, F.N. and CHAO, B.T., "Laminar Free Convection over Two-Dimensional and Axisymmetric Bodies of Arbitrary Contour," J. of Heat Trans. Trans. ASME Series C, Vol. 96, 435 - 442, 1974.

9. CHIANG, T., OSSIN, A. and TIEN, C.L., "Laminar Free Convection from a Sphere," J. of Heat Trans. Trans. ASME, Series C, Vol. 86, 537 - 542, 1964

10. CHEN, T.S., and MUCOGLU, A., "Analysis of Mixed Forced and Free Convection about a Sphere", Int. J. Heat Mass Transfer, Vol. 20, 867 - 875, 1977.

11. CREMERS, C.J. and FINLEY, D.L., "Natural Convection about Isothermal Spheres", Paper NC 1.5, Fourth Int. Heat Trans. Conference, Paris - Versailles, Vol. IV, 1 - 11, 1970.

12. AMATO, W.S. and TIEN, C., "Free Convection Heat Transfer from Isothermal Spheres in Water", Int. J. Heat Mass Transfer, Vol. 15, 327 - 339, 1972.

TRANSIENT THERMAL BEHAVIOUR OF STEAM CHEST DURING TURBINE STARTUPS

G.SUBHASH, DR. SHAIK RAHAMATULLA, S.VASUDEVA RAO

BHARAT HEAVY ELECTRICALS LTD,
Corporate R & D Division, Vikasnagar,
HYDERABAD - 500 593, INDIA.

SUMMARY

Super heated steam supplied from the boiler is admitted to steam turbine through a steam chest which consists of Combined Isolating Emergency Stop (CIES) valves and governor valves. Its structural integrity against thermal transients due to various startup procedures has to be established by a rigorous thermal analysis.

The present paper deals with the details of analytical investigations carried out on the thermal behaviour of a typical steam chest for different cold, warm and hot startup procedures using finite element method. Dittus-Boilter equation[1] is used to determine the heat transfer coefficients at different zones of inner surface of the steam chest. At the areas of stagnant steam, Nusselt's formula for film condensation is used to calculate the heat transfer coefficients. With the above boundary conditions the temperature distributions and the resulting thermal stress are computed at different specified time steps. Results and conclusions are presented with the help of isotherms and isostress plots.

INTRODUCTION

Today numerical techniques are the accepted means for solving thermal problems particularly in realistic situations involving complex geometries. The computer based finite element method has emerged as a powerful analysis and design aid in diverse areas due to its versatility and adaptability. However, its use is often restricted because of 3-D modelling and computation. Consequently it is often recommended [3] to idealise a 3-D geometry as a 2-D model. In analysing components subjected to thermal and mechanical loading under complex transient conditions, one has to carefully judge which configuration and loading offers correct representation, yielding accurate temperature and stress distribution.

In this paper the idealisation of intricate steam chest of a steam turbine is carried out as a simplified axi-symmetric model for predicting the level of thermal stress at critical regions.

Low or high cycle fatigue arising out of transient thermal gradients is one of the most common phenomenon encountered in the case of thermal power plant components. The starting, operating and shut down conditions greatly influence the design of such components. Due to planned repairs, servicings or due to emergency situations turbines have to start and stop very frequently which will cause severe thermal gradients in valves, casings, rotors etc. The present work is taken up with a view to quantitatively asses the thermal stress levels in steam chest of 120MW steam turbine and locate the regions of critical stresses.

METHOD OF APPROACH

The stop valves and governor valve (fig. 1) are modelled as axi-symmetric bodies without pipe connections and internals. They are idealised into triangular ring elements as shown in fig.2. In computing thermal stresses, the temperature distribution is obtained a-priori.

Thermal analysis

The temperature distribution in the valves is obtained at different time intervals during the startup procedures assuming the following boundary conditions.

* The initial metal temperature is uniform

STOP VALVE   GOVERNOR VALVE

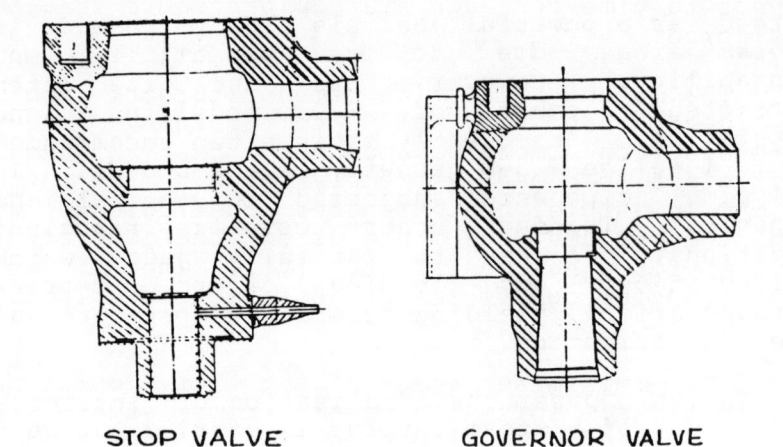

FIG.1: SECTIONAL VIEW OF THE VALVES.

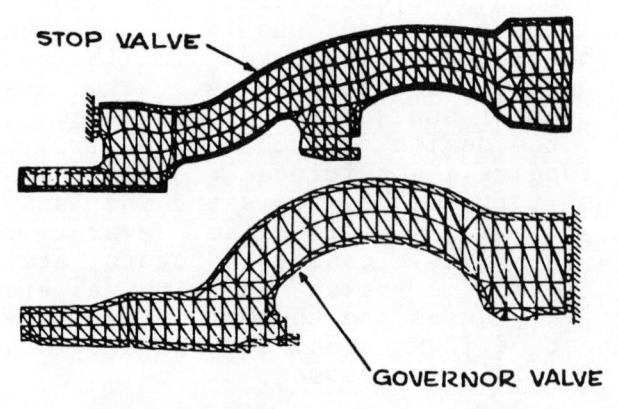

FIG.2: IDEALISATION OF VALVES

corresponding to the value shown by the cooling curves provided by manufacturing plant.

* Different operating parameters of steam such as pressure, temperature, mass flow etc., are varied with respect to time at different possible rates for each type of starting cycle which eventually define the starting curves.

* At the zones where the possibility of condensation exists, heat transfer coefficients due to film condensation are assumed and at other zones forced convection heat transfer coefficients calculated by Dittus-Boilter equation are used. Variation of heat transfer coefficients with respect to time for each startup procedure is also considered.

* The moment saturation temeperature of steam is higher than the wall temperature, steam condenses on metal and the heat transer coefficient becomes extremely high. As the minimum possible heat transfer coefficient, we used the values determined using the Nusselt's formula for a horizontal tube[2]. The minimum heat transfer coefficient thus selected is 10,000 W/Sq. m.

* To minimise the effect of solution perturbations, initially small time steps are taken and later, the time step size is increased.

Stress Analysis

The temperatures obtained from heat conduction analysis are used for evaluating the stress. The same idealisation which has been used for obtaining the temperature distribution is used for stress analysis also. The axial displacements in the stop valve are constrained at the base where it is bolted to the seating, giving due consideration to the bolt hole clearances which permit radial displacements only. In the case of governor valve also the axial displacements at the top surface where the valve is bolted to the flange are constrained (see fig. 2). Stresses are calculated at the specified time intervals in each start up. After obtaining the plane stresses, principal and vonMises stresses are also computed.

RESULTS AND DISCUSSIONS

Temperature distribution in the stop valve is obtained at various time intervals of starting procedures after 8hrs (hot) and 36 hrs (cold) shutdowns. Fig. 3 shows isotherms and isostess plots for stop valve for 36hrs startup at 21 minutes when peak stress occured during the startup. From the figure it can be observed that a maximum temperature gradient of 100°C is set up eventhough the metal temperature at this time is not very high. This gradient is due to lower initial metal temperature and higher steam temperature at the starting of the turbine.

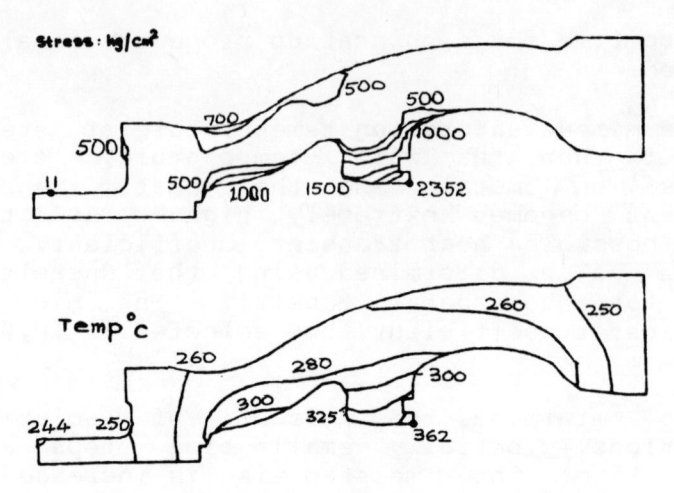

FIG. 3 : ISOTHERMS AND ISOSTRESS CURVES IN STOP VALVE AT 21 MIN. FOR 36 HRS. START UP.

Because of the large gradients, a maximum vonMises stress of 23.5 Kg/sq.mm is developed at the valve seating. The high stress concentration at the valve seat is due to sudden variation of curvature above and below the valve seating in addition to large temperature gradients.

Fig. 4 shows the isotherms and isostress plots in the stop valve for 8hrs startup corresponding to 37 minutes after the start of the turbine. Isotherms indicate that the maximum gradient is occuring at the valve seating as well as at the top portion where there is a variation in thickness. Because of these gradients and the pressure loading a maximum vonMises stress of 16.2 Kg/sq. mm is developed at 37 minutes near valve seating.

Variation of peak stress (which occured at the valve seating) with respect to time is given in fig. 5 for both the valves. It is to be observed that peak stress in 8hrs startup occurs at a later stage (at 37 minutes out of 50 minutes) than in 36 hrs startup (at 21 minutes out of 180 minutes). The reason is that in case of 8hrs startup, the initial mismatch between metal temperature and inlet steam temperature is very low; but subsequently as the rate of rise of inlet steam temperature imposed is more than the developed rate of rise of metal temperature, higher stresses are created at a later time.

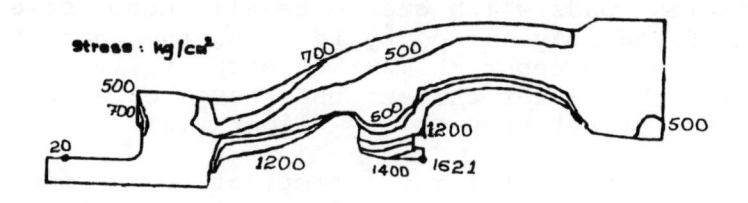

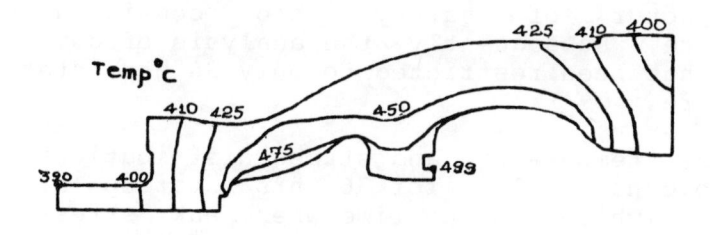

FIG. 4: ISOTHERMS AND ISOSTRESS CURVES IN STOP VALVE AT 37 MIN. FOR 8 HRS. STARTUP.

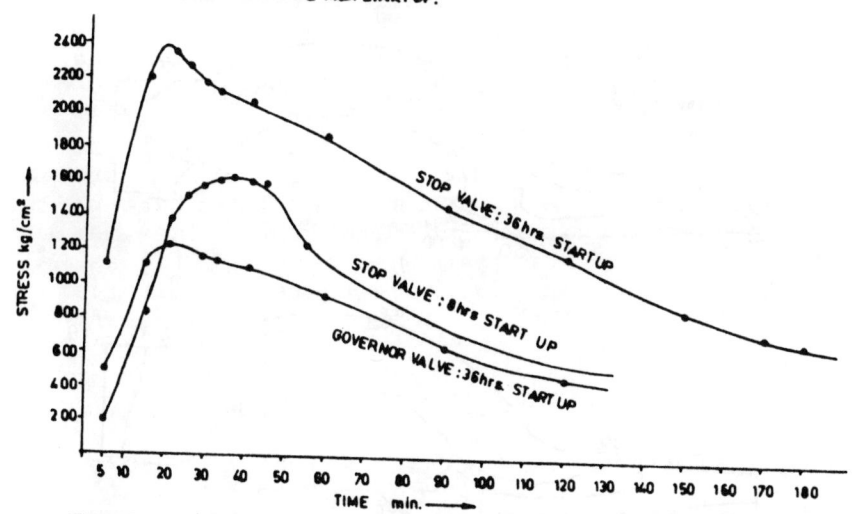

FIG. 5: VARIATION OF PEAK STRESS W.R.T TIME

Where as in the case of 36 hrs startup, though the mismatch of 120°C exists at zero time, higher stresses are developed a little later since the rate of rise of metal temperature lags behind the rate of rise of steam temperature upto some time after starting. This time lag is because of low steam parameters which cause less heat flux to the valve body at the starting of the turbine.

From the results it is also found that the pressure loads which create tensile hoop streses have favourably acted by reducing the net hoop stress in the upper chamber of stop valve, where as in the lower chamber they have effected in increase of net hoop stress slightly.

Even though the metal temperature is less in 36 hrs startup compared to 8 hrs startup when peak stress occurs, the former is more severe taking the factor of safety into consideration. Therefore, subsequently the analysis of governor valve has been restricted to only 36 hrs startup procedure.

The temperature and stress distributions in the governor valve for 36 hrs startup at 21 minutes (which is the time when peak stress is

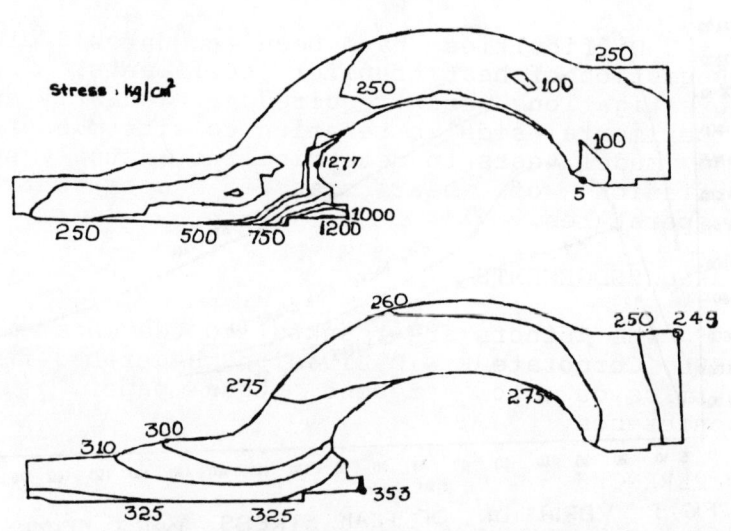

FIG.6: ISOTHERMS AND ISOSTRESS CURVES IN GOVERNOR VALVE AT 21 MIN. FOR 36 HRS. STARTUP

developed) is shown in fig. 6. In this valve also stress concentration is observed at the valve seating.

CONCLUSIONS

* From the analysis carriedout it is found that the stop valve is more stressed compared to governor valve.

* Results also showed that the cold start is more severe than the hot start.

* In both the valves, the zone near the valve seating is highly stressed.

* It is observed that the initial mismatch largely affects the level of thermal stresses developed. Therefore, it is recommended that the initial mismatch should be kept as low as possible with proper trade off to the losses involved in prolonging the starting cycle duration.

* Difficulties have been encountered in the assumption of heat transfer coeffients. Further investigations are required particularly on the experimental side in relation to site measurements and model tests to determine the actual spatial variation of heat transfer coeffients and temperatures.

ACKNOWLEDGEMENTS

The Authors are thankful to the management of BHEL, Corporate R & D Division, Hyderabad for kind permission to present this paper in the conference.

REFERENCES

1. Eckert, E.R.C. and R.M. Drake, - Heat and Mass transfer, McGraw Hill Book Company, Newyork, 1959.

2. Molchanov, E.I. and A.I.Aleshin, 'Temperature conditions and stress state of steam chest during startup', Thermal Engineering, 1969, 16(6), 13-16.

3. Sandsmark, N. and D.T. Saugerad, 'Effect of idealising three dimensional geometry in two dimensional model in temperature and stress analysis of engine components', Proc. Int. Conf. on Numerical Methods in Thermal Problems, Swansea, 1979.

NUMERICAL AND EXPERIMENTAL STUDY OF NATURAL CONVECTION IN PARALLELOGRAMMIC ENCLOSURES

G. Cesini, G. Lucarini, M. Paroncini, R. Ricci

Dipartimento di Energetica
Università degli Studi di Ancona - Italy

SUMMARY.

Natural convection in parallelogrammic cavities is investigated experimentally varying the inclination angle, the Rayleigh number and the aspect-ratio.
Finite differences numerical method is used to verify experimental data.
Numerical and experimental results are in good accord.

NOMENCLATURE

T — temperature
T_H — temperature of hot wall
T_C — temperature of cold wall
$T_0 = \dfrac{T_H + T_C}{2}$ reference temperature
g — gravity force
$Pr = \nu/\alpha$ Prandtl number
$Gr = \dfrac{g\beta D^3 (T_H - T_C)}{\nu^2}$ Grashof number
$Ra_D = Pr\,Gr$ Rayleigh number
D — width of cavity
H — height of cavity
L — lenght of cavity
t — time
$t' = \alpha t/D^2$ non-dimensional time
$V = V(u,v)$ velocity vector

$u_0 = \alpha/D$ reference speed
$u' = u/u_0$ non dimensional x-component of velocity
$v' = v/u_0$ non-dimensional y-component of velocity
α — Thermal Diffusivity
β — Coefficient of thermal expansion
γ — inclination angle
$\Theta = (T-T_C)/(T_H-T_C)$ non-dimensional temperature
ν — Kinematic Viscosity
ψ — Stream Function
$\psi' = \psi/\alpha$ non-dimensional stream function
ω — Vorticity
$\omega' = (\omega D^2)/\alpha$ non-dimensional Vorticity
∇ — Nabla operator
∂ — partial differential

1. INTRODUCTION

Numerical and experimental methods have been widely used in recent years to study free convective heat transfer in vertical and inclined rectangular enclosures.
On the contrary there is a great lack or information about the parallelogrammic geometry, although it can have many engineering applications in thermal insulation and heat storage.
In fact a parallelogrammic enclosure behaves as an oneway heat wall or "thermal diode" [1,4].
Chung and Trefethen [1] investigated numercially and experimentally the free convective heat transfer in inclined cavities filled with air or water.
Nakamura and Asako [2] carried out numerical and experimental investigations on parallelogrammic air filled enclosures.
Mackowa and Tanazowa [3] performed an experimental study on free convective heat transfer in liquids enclosed in parallelogrammic enclosures having several aspect-ratios and for tilt angles ranging between 0 and + or - 45 degrees. The experiments covered a Rayleigh number range between $3 \cdot 10^3$ and $2 \cdot 10^7$ and a Prandtl number range between 7 and 10^4.
Recently Seki et alii [4] carried out experimental measurements covering a range of Rayleigh numbers between $3.4 \cdot 10^4$ and $8.6 \cdot 10^7$ and Prandtl numbers between 0.70 and 480, for different tilt angles and an aspect-ratio of 1.44.
They also gave a correlation equation for Nusselt number as a function of tilt angle, Prandtl number and Rayleigh number.
In this work are reported the results of a numerical and experimental study on natural convective heat transfer in air enclosed in a parallelogrammic cavity.
A parallelogrammic enclosure as shown in Fig. 1 is considered.
Two walls, of height H, are vertical, and isothermal at different temperatures: TH (y=D) and TC (y=0) with TH>TC.
The other vertical walls are adiabatic and the remaining pair of walls are also adiabatic and are inclined by γ degree with respect to the horizontal plane.
We assume γ positive when it runs a clockwise, negative in the other case.
The governing differential equations which describe the motion of the fluid and the heat transfer are solved for a tilt angle of + 45 degrees.
An implicit alternating direction finite difference method is used to obtain the distributions of vorticity, stream func-

tion, velocities and temperature in the fluid as a function of Rayleigh number.
Temperature distribution is also experimentally measured by real time holographic interferometry technique.
Experimental results concern a Rayleigh number range between $4 \cdot 10^3$ and $1.4 \cdot 10^5$, for cavity aspect-ratios equal to 1 and 1.5 and for tilt angles equal to +45 and -45 degrees.
Heat transfer coefficients are obtained directly from the temperature distribution.
Overall correlations of heat transfer rate are evaluated using least-squares technique.

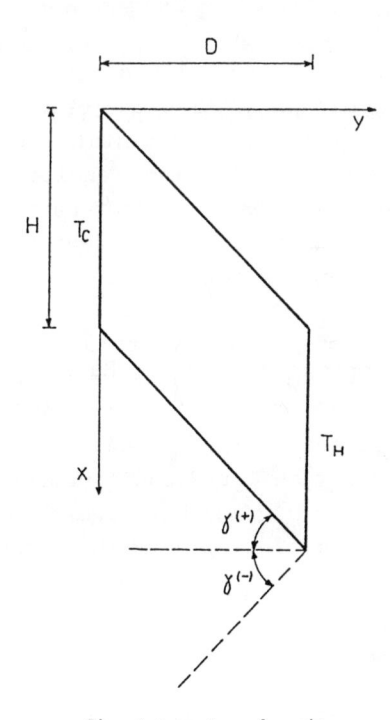

Fig. 1 Geometry of cavity

2. GOVERNING EQUATIONS AND NUMERICAL METHODS

Furthermore we consider a two-dimensional laminar flow in which the gravity force acts in the positive x direction and we assume the fluid to be incompressible with constant properties.
Then the governing differential equations, subjected to Boussinesque approximation are:

$$\frac{\partial T}{\partial t} + u \cdot \frac{\partial T}{\partial x} + v \cdot \frac{\partial T}{\partial y} = \alpha \cdot \left(\frac{\partial^2 T}{\partial x^2} + \frac{\partial^2 T}{\partial y^2}\right) \tag{1}$$

$$\frac{\partial \omega}{\partial t} + u \cdot \frac{\partial \omega}{\partial x} + v \cdot \frac{\partial \omega}{\partial y} = \nu \cdot \left(\frac{\partial^2 \omega}{\partial x^2} + \frac{\partial^2 \omega}{\partial y^2}\right) + g \cdot \beta \cdot \frac{\partial (T - T_o)}{\partial y} \tag{2}$$

$$\frac{\partial^2 \psi}{\partial x^2} + \frac{\partial^2 \psi}{\partial y^2} = -\omega \tag{3}$$

$$u = \frac{\partial \psi}{\partial y} \quad ; \quad v = -\frac{\partial \psi}{\partial x} \tag{4}$$

where ψ is a stream function and $\omega = \frac{\partial v}{\partial x} - \frac{\partial u}{\partial y}$ is the vorticity vector.

Introducing the dimensionless variables eqns (1) to (4) become:

$$\frac{\partial \theta}{\partial t'} + u' \cdot \frac{\partial \theta}{\partial x'} + v' \cdot \frac{\partial \theta}{\partial y'} = \frac{\partial^2 \theta}{\partial x'^2} + \frac{\partial^2 \theta}{\partial y'^2} \tag{5}$$

$$\frac{\partial \omega'}{\partial t'} + u' \cdot \frac{\partial \omega'}{\partial x'} + v' \cdot \frac{\partial \omega'}{\partial y'} = Pr \cdot \left(\frac{\partial^2 \omega'}{\partial x'^2} + \frac{\partial^2 \omega'}{\partial y'^2}\right) + Ra \cdot Pr \frac{\partial \theta}{\partial y'} \tag{6}$$

$$\frac{\partial^2 \psi'}{\partial x'^2} + \frac{\partial^2 \psi'}{\partial y'^2} = -\omega' \tag{7}$$

$$u' = \frac{\partial \psi'}{\partial y'} \quad ; \quad v' = -\frac{\partial \psi'}{\partial x'} \tag{8}$$

Equations (5) to (8) must be solved subject to the following boundary conditions:

$\theta = 0$ along the cold wall and
$\theta = 1$ along the hot wall
$\partial \theta / \partial n = 0$ along the adiabatic walls
$\psi' = u' = v' = 0$ along all the walls.

Furthermore we assume as initial condition

$\theta = 0.5$ and

$$\psi' = u' = v' = \omega' = 0$$

in the whole domain.

The parallelogrammic domain is overlayed with a grid having regular dimensions. The solutions for governing equation, with their boundary conditions, are then obtained at the grid points; equations (5) and (6) are solved by an A.D.I. method; the spatial derivates are approximated by central differences; the resulting tridiagonal set of algebric equations is then solved by using Thomas method [5]; the boundary vorticity values are obtained by using Tom formulation. Valuable improvements of the results were not obtained by using more accurate schemes of the second order like those proposed by Woods [6] or Kuskova [7]; this is probably due to the value of the velocity gradient and to the choice of the mesh.

The elliptic equation (7) has been solved at each time step by S.O.R. method.

This scheme gives:

$$\psi'^{p+1}(i,j) = \frac{\delta}{4} \cdot \left(\psi'^{p}(i+1,j) + \psi'^{p+1}(i-1,j) + \psi'^{p}(i,j+1) + \psi'^{p+1}(i,j-1) + \Delta x'^{2} \cdot \omega'(i,j) \right) + (1-\delta) \cdot \psi'^{p}(i,j)$$

(9)

where p is the iteration index and Δ the relaxation parameter. Numerical experimentation shows that the method is always convergent only using and underrelaxation procedure, (ranging between 0 and 1).

Equation (7) can also be solved by introducing a time dependent term, then an A.D.I. method is used to obtain the stationary solution of the "false transient" problem:

$$\frac{\psi'^{n+\frac{1}{2}}(i,j) - \psi'^{n}(i,j)}{\frac{\Delta \tau}{2}} = \frac{1}{\Delta x'^{2}} \cdot \left(\psi'^{n+\frac{1}{2}}(i+1,j) + \psi'^{n+\frac{1}{2}}(i-1,j) - 2\psi'^{n+\frac{1}{2}}(i,j) + \psi'^{n}(i,j+1) + \psi'^{n}(i,j-1) - 2\psi'^{n}(i,j) \right) + \omega'^{n}(i,j)$$

(10a)

$$\frac{\psi'^{n+1}(i,j) - \psi'^{n+\frac{1}{2}}(i,j)}{\frac{\Delta\tau}{2}} = \frac{1}{\Delta x'^2} \cdot \left(\psi'^{n+\frac{1}{2}}(i+1,j) + \psi'^{n+\frac{1}{2}}(i-1,j) \right.$$

$$- 2\psi'^{n+\frac{1}{2}}(i,j) + \psi'^{n+1}(i,j+1) + \psi'^{n+1}(i,j-1)$$

$$\left. - 2\psi'^{n+1}(i,j) \right) + \omega'^{n}(i,j) \qquad (10b)$$

where   is the false time step.

The accuracy of the two methods is practically the same but the second one is more time-consuming when the computational region is similar to ours; however it becomes convenient when using a very large number of nodes.

Local heat transfer coefficients at the isothermal walls are obtained from temperature distribution by using the relation:

$$Nu = \frac{hx}{H} = -\frac{\partial \theta}{\partial n} \qquad (11)$$

where n is the normal to the surface.
The average Nusselt number is then calculated from:

$$\overline{Nu} = \int_0^1 Nu \, dx' \qquad (12)$$

3. EXPERIMENTAL APPARATUS

It consists of a test cell and an holographic interferometer. The test cell is a parallelogrammic enclosure filled with air; two vertical metal plates are separated by two insulating walls inclined by 45° respect to the horizontal plane; two glass windows complete the cavity so that the light beam of the interferometer can passes through the fluid.

The height and thickness of the enclosure can be varied so that aspect-ratio equal to 1 and 1.5 can be obtained.

The dimension of the cavity in the direction parallel to the laser beam is as large that end effects are negligible; then the air layer may be considered as two dimensional phase object.

The temperature of each metal wall is controlled by water running through a jacket attached to the back surface.

One of the plates is kept at a uniform temperature of 20°C
with an error of 0.5%.
The temperature of the other plate can be varied between 20°C
and 50°C with an error of 0.5%.
The temperature distribution in the air layer is measured by
holographic interferometry technique.
The light source is a 5W Argon Laser with ethalon for the
514.5 nm wavelenght. The object beam has a maximum diameter
of 0.15 m. It is possible to use double-exposure and real
time techniques. The first technique is used for steady-state
measurements, the second one to study the temporal evolution
of transfer processes. The temperatures of different points
of the walls and the center of the air layer are measured
also with thermocouples.

4. RESULTS

A number of a computations with a 37x73 grid has been
carried out to verify the ability of the numerical code to
describe the convective heat process in the cavity with $\gamma = +45°$. Typical flow patterns and isothermal distributions are
shown in Figg. 2 and 3. Figures 2a and 3a show the
streamlines for Rayleigh numbers equal to $3 \cdot 10^4$ and $1.4 \cdot 10^5$.
The flow consists of a single counter-clockwise eddy. This
result is in a very good agreement with the qualitative
visual observation carried out in [4].
Figures 2b and 3b show the steady-state temperature
distribution for Rayleigh number equal to $3 \cdot 10^4$ and $1.4 \cdot 10^5$.
Isothermal lines are strongly affected by convection and show
the presence of two boundary layers, one growing up near the
hot wall and the other growing down near the cold wall. The
presence of a vertical temperature gradient in the central
region is also evident. Figures 2c and 3c are photographic
records of the experimental temperature distributions in a
cavity reproducing the same conditions previously
investigated by numerical methods. Since hologramms have been
obtained with the interferometer infinite fringe field, the
fringe pattern shows directly the distribution of the
isothermal lines.
The qualitative agreement between numerical and experimental
results is quite evident.
Figures 2d and 3d shcw the experimental temperature
distribution obtained one again for Rayleigh number equal to
$3 \cdot 10^4$ and $1.4 \cdot 10^5$ respectively but for a tilt angle equal to

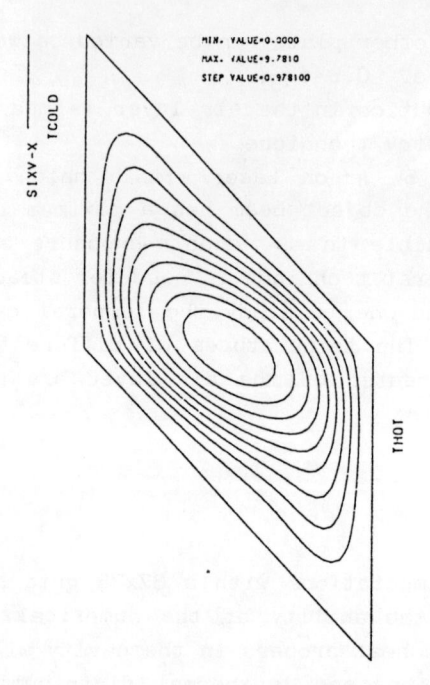

Fig. 2a Stream function distribution for Ra = $3 \cdot 10^4$

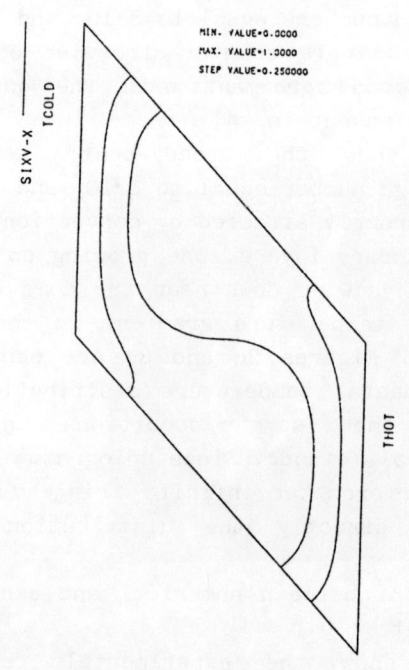

Fig. 2b Temperature distribution for Ra = $3 \cdot 10^4$

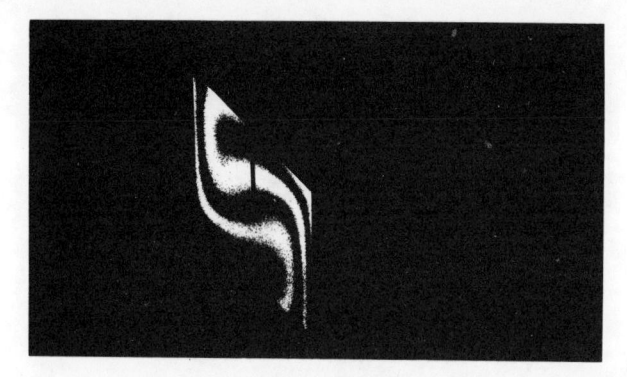

Fig. 2c Temperature distribution for Ra = $3 \cdot 10^4$
(γ = + 45°)

−45°. In this case the lower isothermal vertical plate is colder than the upper one. The interferogramm for Ra = $3 \cdot 10^4$ shows that the heat transfer in not greatly affected by convection.

The convective contribution is greather in the situation shown in Fig. 3d but temperature gradients at the isothermal walls are much less than in the case with γ = +45°.

Figure 4 shows the heat transfer coefficients obtained as a function of Rayleigh number by evaluating the experimental temperature distribution.

It is evident that the heat transfer process is strongly affected by the inclination angle. In fact, the average ratio $Nu_D(+45°)/Nu_D(-45°)$ is equal to 2.5 for aspect-ratio equal to 1.

Fig. 2d Temperature distribution for Ra = $3 \cdot 10^4$
(γ = − 45°)

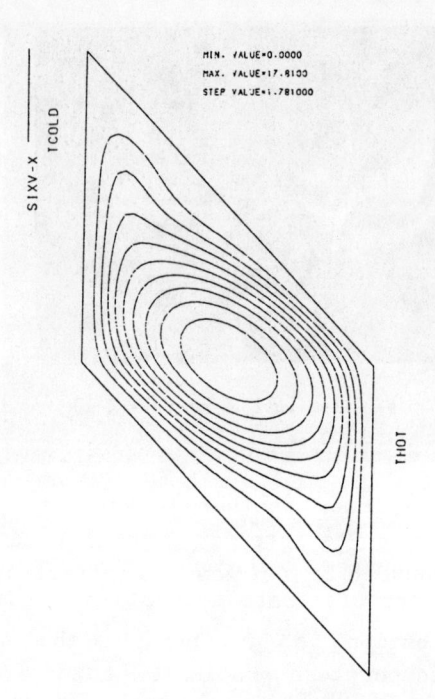

Fig. 3a Stream function distribution for Ra = $1.4 \cdot 10^5$

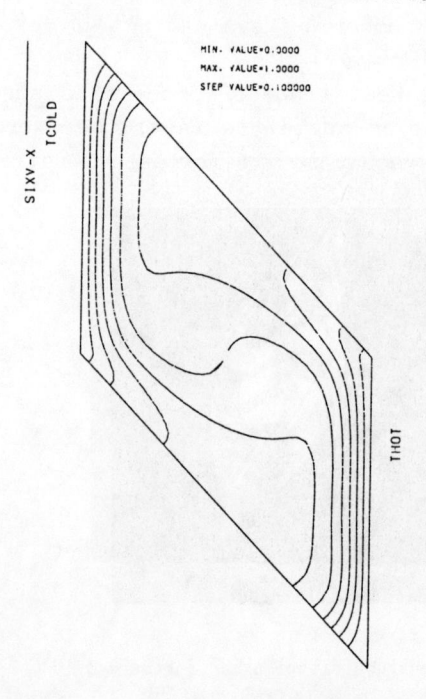

Fig. 3b Temperature distribution for Ra = $1.4 \cdot 10^5$

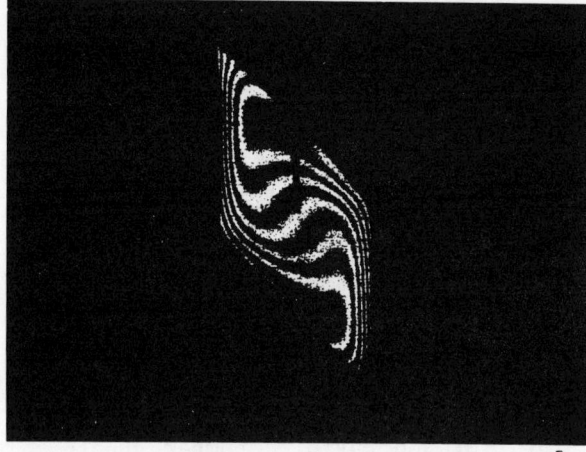

Fig. 3c Temperature distribution for Ra = $1.4 \cdot 10^5$
($\gamma = +45°$)

Employing the least-squares technique, the following correlation of the heat transfer rate are obtained from the data of figure 4:

$$\overline{Nu}_D = 0.228\ Ra^{0.252} \qquad (13)$$

$\gamma = +45°$, for A = 1 and 1.5, $4 \cdot 10^3 < Ra_D < 1.4 \cdot 10^5$ with a standard deviation of ±2%;

$$\overline{Nu}_D = 0.184\ Ra^{0.214} \qquad (14)$$

Fig. 3d Temperature distribution for Ra = $1.4 \cdot 10^5$
($\gamma = -45°$)

for $\gamma = -45°$, A = 1.5, $4 \cdot 10^3 < Ra_D < 2 \cdot 10^4$ with a standard deviation of ±5%;

$$\overline{Nu}_D = 0.050\, Ra^{0.305} \qquad (15)$$

for $\gamma = -45°$, $A = 1$, $2\cdot 10^4 < Ra_D < 1.4\cdot 10^5$ with a standard deviation of ±10%.

Finally the numerical results obtained for $\gamma = +45°$, $A = 1$, $4\cdot 10^3 < Ra_D < 1.4\cdot 10^5$ are correlated by the following equation

$$\overline{Nu}_D = 0.047\, Ra^{0.373} \qquad (16)$$

with a standard deviation of ±2%.

For comparison numerical and experimental correlation obtained for $\gamma = +45$ are reported in Figure 5 together with the correlation presented in [4], which is valid for γ ranging from 0 to 70, and Rayleigh number greather than $3.4\,10^4$ with a standard deviation of ±30%.

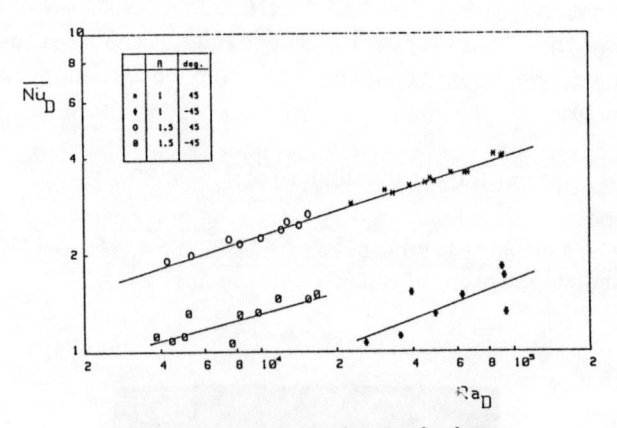

Fig. 4 Experimental correlations

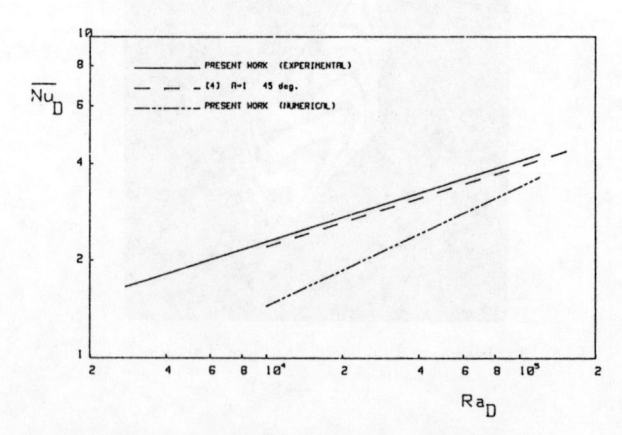

Fig. 5 Comparison between numerical and experimental correlations

5. REFERENCES

1. Chung K.C., Trefethen, Natural convection in vertical stack of inclined parallelogrammic cavities. Int. J. Heat Mass Transfer, 25, n. 2, pp. 277-284, 1982.
2. Nakamura H., Asako Y., Heat transfer in parallelogrammic enclosure. Trans. ASME, Vol. 46, pp. 471-481, 1980.
3. Maekawa T., Tamasawa I., Natural convection heat transfer in parallelogrammic enclosure. Proc. of 7th Int. Heat Transfer Conf., Munchen, Vol. 2, pp. 227-232, 1982.
4. Seki N., Fukusako S., Yamaguchi A., An experimental study of free convective heat transfer in parallelogrammic enclosure. J. Heat Transfer, Vol. 105, pp. 433-439, 1983.
5. Nogotov E.F., Applications of numerical Heat Transfer. McGraw-Hill, 1978.
6. Kuskova T.V., Chudov L.A.O., On approximate boundary condition for vortex in viscous incompressible flows. n. 11, pp. 27-31, Moscow, MGU, 1968.
7. Peaceman D.W., Rachford H. H., The numerical solution of parabolic and elliptic differential equations. J. Soc. Ind. Appl. Math., Vol. 3, pp. 28-41, 1955.

FINITE ELEMENT ANALYSIS OF HEAT TRANSPORT IN A HYDROTHERMAL ZONE

N. E. Bixler* and C. R. Carrigan**
*Fluid and Thermal Sciences Department 1510
**Geosciences Department 1540
Sandia National Laboratories
Albuquerque, New Mexico, 87185, U.S.A.

SUMMARY

Two-phase heat transport in the vicinity of a heated, subsurface zone is important for evaluation of nuclear waste repository design and estimation of geothermal energy recovery, as well as prediction of magma solidification rates. Finite element analyses of steady, two-phase, heat and mass transport have been performed to determine the relative importance of conduction and convection in a permeable medium adjacent to a hot, impermeable, vertical surface. The model includes the effects of liquid flow due to capillarity and buoyancy and vapor flow due to pressure gradients. Change of phase, with its associated latent heat effects, is also modeled. The mechanism of capillarity allows for the presence of two-phase zones, where both liquid and vapor can coexist, which has not been considered in previous investigations.

Our numerical method employs the standard Galerkin/finite element method, using eight-node, subparametric or isoparametric quadrilateral elements. In order to handle the extreme nonlinearities inherent in two-phase, nonisothermal, porous-flow problems, we compute steady-state results by integrating transients out to a long time (a method that is highly robust). In order to perform the time integrations efficiently, we have adopted a modified form of the implicit method of Gresho et al.[1], which includes an adaptive scheme for time-step selection.

Results, primarily isotherms, moisture distribution, and Nusselt number, are found to depend strongly on several parameters—permeability of the medium, temperature difference between the hot and cold boundaries, and magnitude of capillarity. In contrast to single-phase natural convection, Nusselt numbers reach a peak at intermediate temperature differences. The peak corresponds to maximum efficiency of countercurrent convection, which transports latent heat.

1. INTRODUCTION

Previous analyses of natural convection in two-dimensional porous media fall largely into two classes: (1) single-phase convection in saturated media, which has been investigated by both analytical and numerical means[2-4], and (2) two-phase convection where the phases are taken to be distinct, *i.e.*, where the medium is either locally fully saturated with liquid or with vapor, which has largely been investigated by approximate analytical techniques (see, *e.g.*, references 5-8). Another class of problems relevant to this work involves countercurrent convection in a porous heat pipe, which has been treated semianalytically by Udell[9,10], who shows that large Nusselt numbers can be achieved in a porous medium in the same way as in a standard heat pipe. A recent numerical study[11] also predicts the occurrence of enhanced heat transfer when capillarity is significant in the cooling of a volcanic dike intruded into permeable host rock. Drying of porous media, a topic that has been investigated extensively[12-15], is similar to flow in porous heat pipes, except that vapor transport is often dominated by diffusion rather than convection.

Here we use finite element analysis to model convective heat transfer resulting from motions of liquid and vapor phases in a square domain. In the limit as the fraction of water in the liquid phase approaches unity, the results would asymptote to those of single-phase natural convection in a porous medium. The presence of two phases significantly enhances overall heat transfer rates in the same way as in one-dimensional porous heat pipes.

This study is intended to be fundamental in nature. Nonetheless, the results are relevant to several problems of practical importance. An underground heat source, such as the nuclear material in a waste repository or a near surface magma intrusion, may lie within a partially saturated, hydrothermal zone. It is important to understand the nature of heat and mass transfer in such a zone in order to predict cooling rates, effects on species transport, and thermal stresses in the host rock. Another potential application is two-phase flow in a packed-bed reactor.

2. TWO-PHASE POROUS FLOW MODEL

The physical mechanisms used here to model two-phase heat transfer are as follows: (1) pressure- and buoyancy-driven convection of liquid and vapor and (2) change of phase, with associated latent heat effects. These mechanisms are a subset of those treated by Hadley[16] and by Bixler[17], which also include binary, Knudsen, and thermo-diffusion in a two-component gaseous mixture.

Conservation of liquid, vapor, and energy dictate the following relationships:

$$\rho_l \frac{\partial \Theta}{\partial t} = -\nabla \cdot \underline{j}_l - F_v \tag{1}$$

$$\frac{\partial (\Phi - \Theta)\rho_v}{\partial t} = -\nabla \cdot \underline{j}_v + F_v \tag{2}$$

$$[\rho C]_{ave} \frac{\partial T}{\partial t} = -(C_l \underline{j}_l + C_v \underline{j}_v) \cdot \nabla T + \nabla \cdot (\Lambda_{ave} \nabla T) - F_v L \tag{3}$$

Here, ρ is density, Θ is moisture content, Φ is porosity, t is time, $\underline{j}$ is mass flux vector, F_v is vaporization rate, C is heat capacity at constant pressure, T is temperature, Λ is thermal conductivity, and L is latent heat of vaporization. Subscripts l and v refer to liquid and vapor, respectively. The subscript, ave, refers to a volume weighted average of the solid, liquid, and gaseous phases. The fluxes in (1)–(3) are Darcian:

$$\underline{j}_l = -\frac{\rho_l k_l}{\mu_l} \nabla P_l \tag{4}$$

$$\underline{j}_v = -\frac{\rho_v k_v}{\mu_v} \nabla P_v \tag{5}$$

Here, k_l and k_v are the permeabilities of the porous material to flow of the liquid and vapor phases at local conditions; μ is dynamic viscosity; and P is effective pressure, i.e., $P = p + \rho g z$, where p is absolute pressure and g is the acceleration due to gravity. The density of the vapor phase is taken to be ideal and the Boussinesq approximation is made for the density of the liquid phase:

$$\rho_v = \frac{p_v M_v}{RT} \tag{6}$$

$$\rho_l = \rho_{l0}(1 - \beta \Delta T) \tag{7}$$

In keeping with the Boussinesq approximation, (7) is used only to evaluate density gradients; elsewhere, liquid density, ρ_l, is replaced by a constant reference value, ρ_{l0}. ΔT is the difference between the local temperature and the reference temperature corresponding to ρ_{l0}; β is the coefficient of volumetric expansion for water at local temperature.

HYDROTHERMAL ZONE HEAT TRANSPORT

The rate of vaporization, F_v, is taken to be proportional to the product of moisture content and the difference between equilibrium and local vapor pressures[17]:

$$F_v = c\Theta(p_v^{\sim} - p_v) \qquad (8)$$

where $p_v^{\sim}$ is the equilibrium vapor pressure at local conditions. When capillarity is important, the equilibrium vapor pressure is modified, as given by the Kelvin-Helmholtz equation:

$$p_v^{\sim}(T, p_c) = p_v^*(T)e^{-p_c M_v/\rho_l RT} \qquad (9)$$

Here, capillary pressure is $p_c = p_v - p_l$ and $p_v^{\sim}(T)$ is equilibrium vapor pressure as a function of temperature, i.e., steam table values. By setting the coefficient, c, to be large enough, (8) forces vapor pressure to be close to equilibrium everywhere except in locations where moisture content approaches zero. Thus, incorporating (8) into (1)-(3) gives a natural transition from enforcement of equilibrium to enforcement of conservation of mass and momentum for the vapor phase. (See Hadley[16], Bixler[17], and Bixler et al.[18] for further arguments and discussion.)

3. SOLUTION METHOD

We use the standard Galerkin/finite element method to handle spatial discretization of two-dimensional domains. We choose eight-node subparametric elements to represent the three dependent variables, P_l, p_v, and T, that appear in (1)-(3). The finite element mesh, shown in Figure 1, consists of 10 by 10 elements with refinement near the boundaries.

To achieve steady state results, we performed time integration by a second-order accurate predictor/corrector scheme, which is a modified form of the one proposed by Gresho et al.[1] In order to predict each of the dependent variables at time plane $n + 1$, time derivatives must be calculated at the two previous time planes, n and $n - 1$. In the original scheme proposed by Gresho et al., the following equation, which is accurate to second order, was used at both time planes to estimate rates of change:

$$\dot{y}_m = 2\frac{y_m - y_{m-1}}{\Delta t_{m-1}} - \dot{y}_{m-1} \qquad (10)$$

Here, y represents one of the coefficients in the finite element expansion of a dependent variable, dots represent derivatives with respect to time, and m is either n or $n - 1$. A problem with the exclusive use of (10) to calculate time derivatives is that errors at a time level are propagated to the next because of the $-\dot{y}_{m-1}$ term. This problem is most apparent

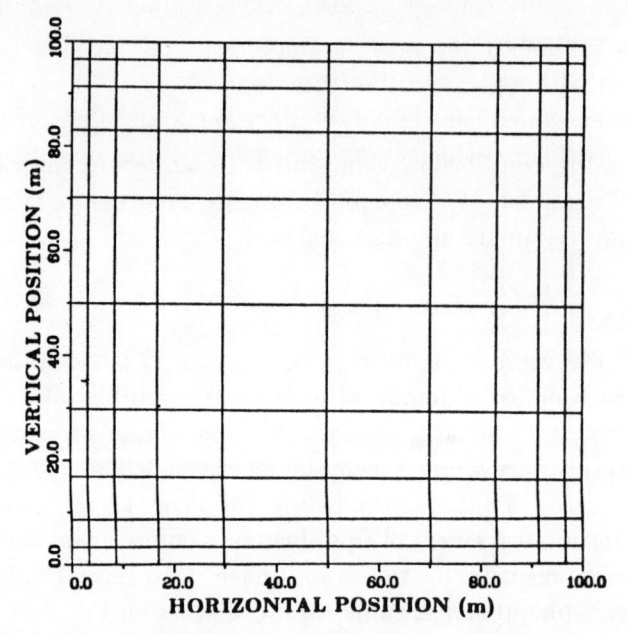

Figure 1: Finite Element Discretization of Domain.

as equilibrium is approached. Rather than settling down to the steady state, the predictor oscillates about it. The effect is to limit the size of the largest time step. A solution to this problem is to replace (10) with the following formula, which is also second-order accurate:

$$\dot{y}_m = \frac{\Delta t_{m-1}}{\Delta t_m + \Delta t_{m-1}} \frac{y_{m+1} - y_m}{\Delta t_m} + \frac{\Delta t_m}{\Delta t_m + \Delta t_{m-1}} \frac{y_m - y_{m-1}}{\Delta t_{m-1}} \quad (11)$$

Because (11) depends only on previous values of dependent variables, rather than on their derivatives, it settles down nicely as steady state is approached. Since evaluation of (11) for $m = n$ would require knowledge of y_{n+1}, we use (10) at time plane n and (11) at time plane $n - 1$.

A second difference between our time integration scheme and the one of Gresho et al.[1] is that we take multiple Newton iterations during a time step if the *rms* norm of the change in dependent variables is larger than a specified value, ϵ_2, up to a maximum of 5 per time step. In this problem we chose ϵ_2 to be 10^{-4}. Other parameters, defined in Gresho et al., were chosen as follows: ϵ, the target local time truncation error, equals 10^{-5} and α, the parameter that determines when a time step should be repeated, equals 0.9. Reference 17 describes the implementation of this time integration procedure in more detail.

HYDROTHERMAL ZONE HEAT TRANSPORT

4. PHYSICAL PARAMETERS AND BOUNDARY CONDITIONS

A large number of physical parameters are needed to define a particular problem, although several of those pertain to thermodynamic properties of water, which can be obtained from standard steam tables. The parameters used here, which characterize San Andres limestone[11] are tabulated below.

$c =$ 10^{-8} s/m^2
$C_l =$ 4220 J/kg/K
$C_s =$ 2600 J/kg/K
$C_v =$ 2000 J/kg/K
$e =$ 4 or 5
$g =$ 9.8 kg/s^2
$k_l =$ $1.0 \cdot 10^{-13}$, $3.5 \cdot 10^{-14}$, or $1.0 \cdot 10^{-14}$ m^2
$k_v =$ k_l
$L =$ interpolated steam table values in J/kg
$M_v =$ 18 kg/kg-mole
$p_v^* =$ interpolated steam table values in kg/m/s^2
$R =$ 8307 J/kg-mole/K
$\beta =$ interpolated steam table values in 1/K
$\Theta =$ $1.724 \cdot 10^e \Phi / p_c$ if $p_c \geq 1.74 \cdot 10^e$
     $[(1.724/1.74) + 1.8 \cdot 10^{-5-e}(1.74 \cdot 10^e - p_c)]\Phi$ if $p_c < 1.74 \cdot 10^e$
$\Lambda_l =$ 1.37 J/m/s/K
$\Lambda_v =$ 0
$\Lambda_s =$ 1.7 J/m/s/K
$\mu_l =$ $2.414 \cdot 10^{-5} \cdot 10^{247.8/(T-140)}$ kg/m/s
$\mu_v =$ $8.04 \cdot 10^{-6} + 4.2 \cdot 10^{-8}(T - 273)$ kg/m/s
$\rho_{l0} =$ 1000 kg/m^3
$\rho_s =$ 2600 kg/m^3
$\rho_v =$ Equation (6)
$\Phi =$ 0.1

No mass fluxes are permitted through any of the boundaries (Figure 1). The top and bottom boundaries are taken to be adiabatic; the side boundaries are isothermal, the right side being held at 273 K and the left side being hotter by 100 through 700 K in 100 K increments. Thus, for each of the cases described in the following paragraph, seven boundary conditions were used so that the dependence of Nusselt number on temperature difference could be demonstrated.

We consider four cases. In the first, we choose liquid- and vapor-phase permeabilities to be $3.5 \cdot 10^{-14}$ m^2 and the exponent, e, to be 4

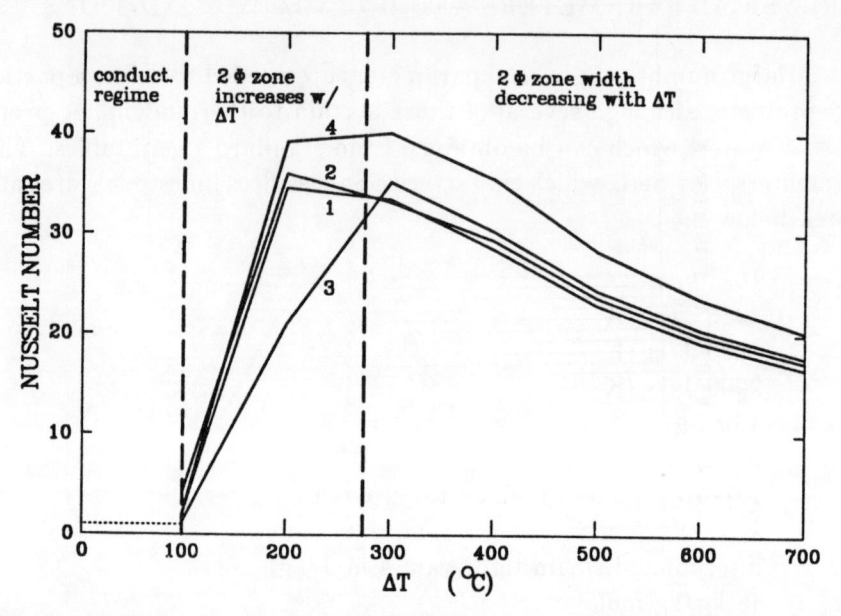

Figure 2: Dependence of Nusselt number on temperature difference, ΔT, for Cases 1–4.

in the formula for capillary pressure above. In the second and third cases, we choose permeabilities to be higher, 10^{-13} m^2, and lower, 10^{-14} m^2, respectively. In the fourth case, we choose capillary pressures to be higher by setting $e = 5$.

5. RESULTS

Figure 2 shows the dependence of Nusselt number, Nu, on the imposed temperature difference across the domain, ΔT, for all four cases, where Nu represents the ratio of the total heat flux to the conductive heat flux for a specified ΔT. Heat transfer is always most efficient at intermediate values of ΔT. Comparison of Case 2 with Case 3 at least suggests that increasing permeability has the two-fold effect of enhancing the maximum in Nu and of decreasing ΔT where the maximum occurs; however, more data points are needed to resolve the peaks, which is a future research goal. Increasing the parameter, e, which increase the magnitude of capillary pressure, both increases and broadens the peak in Nu (Case 4 compared with Case 1).

At small ΔT the moisture contours are nearly horizontal, as shown in Figure 3 for Case 1 when $\Delta T = 100$ K. Figure 4 shows the corresponding isotherms, which indicate that conduction dominates except near the hot boundary where horizontal circulation also contributes. At intermediate ΔT, moisture contours are diagonal owing to a competition between

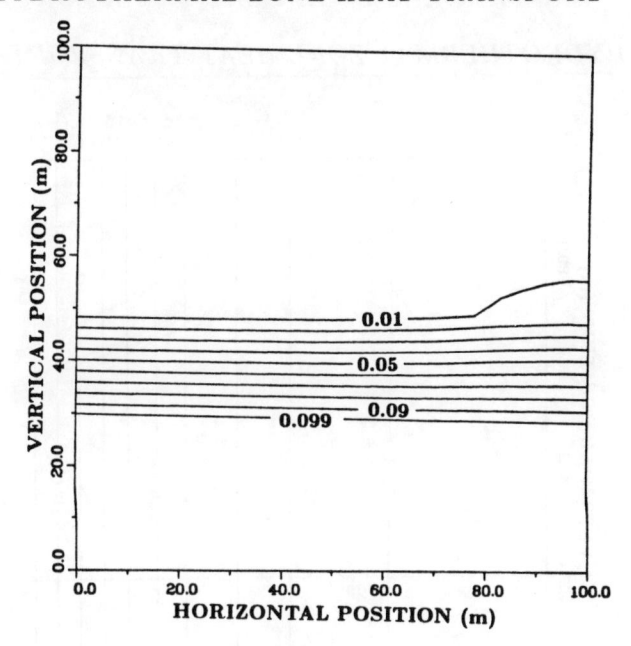

Figure 3: Moisture contours for Case 1 when $\Delta T = 100$ K.

gravity and evaporation near the heated boundary, as shown in Figure 5 for Case 1 when $\Delta T = 200$ K. At large ΔT the influence of gravity is overwhelmed by the process of evaporation near the hot boundary and condensation near the cold, which results in the nearly vertical contours shown in Figure 6 when $\Delta T = 400$ K. The Isotherms in Figure 7 indicate strong thermal boundary layers near both the hot and cold boundaries separated by a large region of horizontal convection.

6. DISCUSSION

Heat fluxes at small ΔT ($< 100°$C) are controlled by conduction, as indicated in Figure 2 by the dotted line. At larger ΔT (100 to 275°C) a two-phase convective zone forms and increases in width with increasing ΔT. However, because of the imposed thermal boundary conditions, the two-phase convective zone cannot occupy the entire width of the domain. Furthermore, above 275°C, the two-phase zone begins to shrink with increasing ΔT. This results from the formation and growth of a dry-out zone near the hot (left) boundary. In this regime of ΔT, Nu decreases with increasing ΔT. In fact heat transport is found to be flux limited so that the product of Nu and ΔT is approximately constant.

Within the two-phase zone, countercurrent convective transport of latent heat occurs, just as it does in the porous heat pipe[9,10]. Evaporation near the hot side raises vapor pressure and depletes moisture content

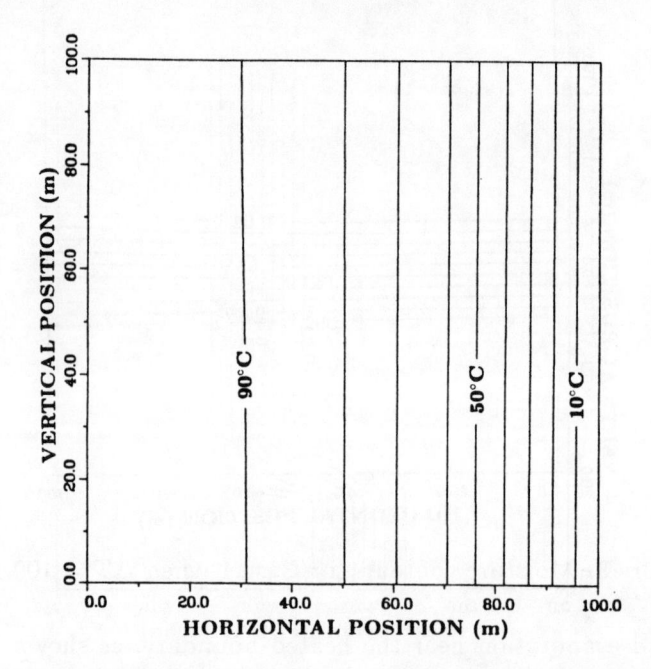

Figure 4: Isotherms for Case 1 when $\Delta T = 100$ K.

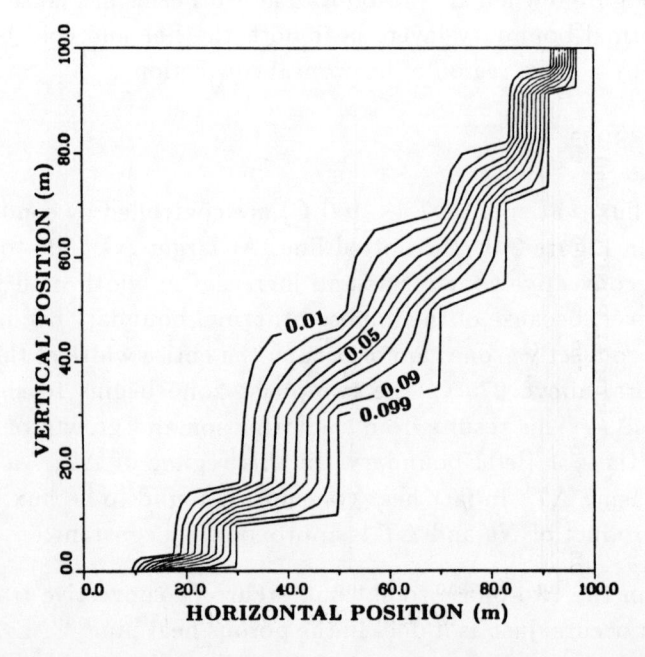

Figure 5: Moisture contours for Case 1 when $\Delta T = 200$ K.

HYDROTHERMAL ZONE HEAT TRANSPORT

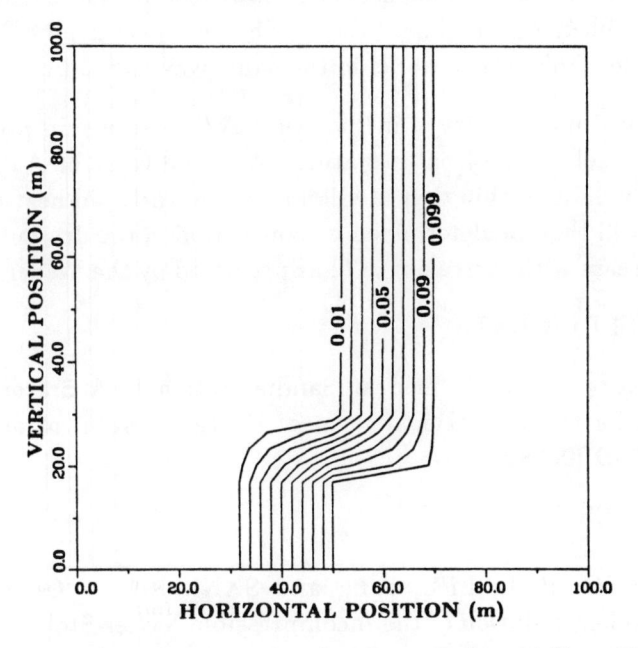

Figure 6: Moisture contours for Case 1 when $\Delta T = 400$ K.

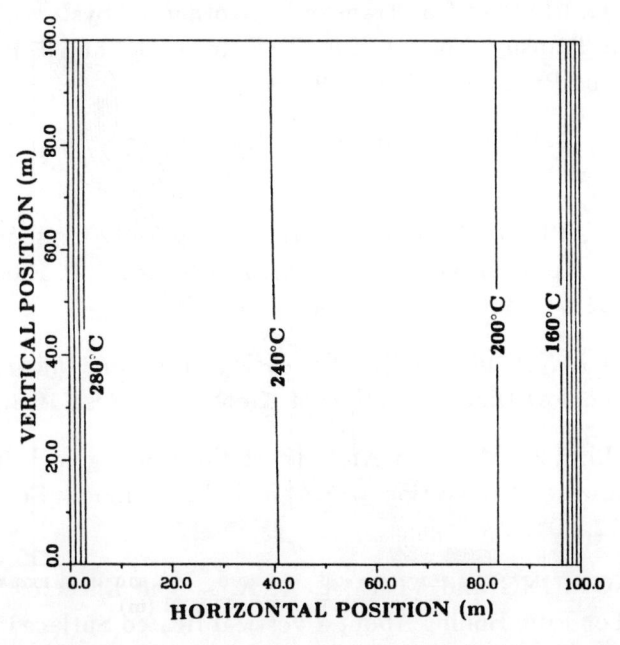

Figure 7: Isotherms for Case 1 when $\Delta T = 400$ K.

with respect to the cold side. This induces a net vapor flux toward the cold side (right) and a net liquid flux toward the hot side (left) due to capillary suction. Vapor condenses as it flows toward the cold side (right) and, thus, replenishes the liquid there. The net effect is an efficient heat transport mechanism that works in the same way as a porous heat pipe.

The maximum observed in the $Nu - \Delta T$ relationship presented in Figure 2 is likely to be local in nature. At some value of ΔT, the two-phase zone will be so thin that it will not significantly aid heat transport. Transport will then be determined by convection in the dry-out zone and Nu will increase with increasing ΔT, as predicted by the Weber solution[7].

ACKNOWLEDGMENTS

This work was performed at Sandia National Laboratories which is operated for the U. S. Department of Energy under contract number DE-AC04-76DP00789.

REFERENCES

1. GRESHO, P.M., LEE, R.L., and SANI, R.C. - 'On the Time-Dependent Solution of the Incompressible Navier-Stokes Equations in Two and Three Dimensions', Recent Advances in Numerical Methods in Fluids, Vol. 1, Eds. Taylor, C., and Morgan, K., Pineridge Press, Swansea, 1979, p. 27.

2. CHENG, PING - 'Heat Transfer in Geothermal Systems,' Advances in Heat Transfer, Vol. 14, Eds. T. F. Irvine Jr. and J. P. Hartnett, Academic Press, New York, 1978.

3. BEJAN, ADRIAN - Convective Heat Transfer, John Wiley and Sons, New York, 1984.

4. CHENG, PING - 'Geothermal Heat Transfer,' Handbook of Heat Transfer Applications, Eds. W. M. Rohsenow, J. P. Hartnett, and E. N. Ganic, McGraw-Hill, New York, 1985.

5. CARRIGAN, C. R. - 'A Two-Phase, Hydrothermal Infiltration Model for Shallow Dikes,' J. Volcanol. Geotherm. Res., 1986, 28, 175.

6. CATHLES, L. M. - 'An Analysis of the Cooling of Intrusives by Groundwater Convection which Includes Boiling,' Econ. Geol., 1977, 72, 804.

7. CHENG, PING and VERMA, A. K. - 'The Effect of Subcooled Liquid on Film Boiling About a Vertical Heated Surface in a Porous Medium,' Int. J. Heat Mass Transfer, 1981, 24, 1151.

8. PARMENTIER, E. M. - 'Two-Phase Natural Convection Adjacent to a Vertical Heated Surface in a Permeable Medium,' Int. J. Heat Mass Transfer, 1979, 22, 849.

9. UDELL, K. S. - 'Heat Transfer in Porous Media Heated from Above with Evaporation, Condensation, and Capillary Effects,' J. Heat Transfer, 1983, 105, 485.

10. UDELL, K. S. - 'Heat Transfer in Porous Media Considering Phase Change and Capillarity — The Heat Pipe Effect,' Int. J. Heat Mass Transfer, 1985, 28, 485.

11. BIXLER, N. E. and CARRIGAN, C. R. - 'Enhanced Heat Transfer in Partially Saturated Hydrothermal Systems,' Geophys. Res. Let., 1986, 13, 42.

12. CEAGLSKE, N. H. and HOUGEN, O. A. - 'Drying Granular Solids,' Ind. Engng. Chem., 1937, 29, 805.

13. WHITAKER, S. - 'Simultaneous Heat, Mass and Momentum Transfer in Porous Media: A Theory of Drying,' Advances in Heat Transfer, Vol. 13, Eds. T. F. Irvine Jr. and J. P. Hartnett, Academic Press, New York, 1973.

14. HADLEY, G. R. - 'Theoretical Treatment of Evaporation Front Drying,' Int. J. Heat Mass Transfer, 1982, 25, 1511.

15. WHITAKER, S. - 'Drying Granular Porous Media—Theory and Experiment,' Drying Technology, 1983, 1, 3.

16. HADLEY, G. R. - 'PETROS—A Program for Calculating Transport of Heat, Water, Water Vapor, and Air Through a Porous Material,' Sandia National Laboratories, Albuquerque, NM, SAND84-0878, 1985.

17. BIXLER, N. E. - 'NORIA—A Finite Element Computer Program for Analyzing Water, Vapor, Air, and Energy Transport in Porous Media,' Sandia National Laboratories, Albuquerque, NM, SAND84-2057, 1985.

18. BIXLER, N. E., EATON, R. R., AND RUSSO, A. J. - 'Drying Analysis of a Multiphase, Porous-Flow Experiment in Fractured Volcanic Tuff,' Fundamentals of Heat Transfer in Drying, Proceedings of the Second ASME/JSME Thermal Engineering Joint Conference, Honolulu, HW, 1987.

FURTHER ADVANCES IN THE USE OF BOUNDARY ELEMENTS FOR MODELLING
THERMAL PROBLEMS

C.A. Brebbia and L. C. Wrobel

Computational Mechanics Institute, Southampton, UK and
Federal University of Rio de Janeiro, Brazil.

1. INTRODUCTION

The boundary element method has been extensively applied
to solve steady state heat transfer problems [1]. Time dependent
temperature problems of the type governed by the diffusion
equations have been studied using a variety of formulations.
They include the finite difference type approximation of the
time derivative presented in [2] and applied in [3] amongst
others; the use of Laplace transforms [4]; applying the time
dependent fundamental solutions proposed by Chang [5] and
implemented numerically for a variety of problems by Wrobel
and Brebbia [6]. This formulation in terms of time dependent
fundamental solutions has also been used to solve axisymmetric [7] as well as three dimensional problems [8].

Each of the above formulations present certain disadvantages. The use of finite difference approximation for the time
derivative in combination with the boundary element approach
in space, means that the solution advances step by step in
time. This requires the computation of domain integrals to
calculate the initial conditions and the satisfaction of the
usual conditions for time step integration. The Laplace transform approach has the obvious attraction that the problem is
reduced to a time independent formulation in the domain of the
transform. The initial conditions become part of the governing
equation and after applying the appropriate fundamental solution
the problem reduces to a boundary only formulation. The disadvantage of this technique is mainly because of the difficulty
of defining a numerically efficient back transformation which
can take into account the type of arbitrary time dependent
boundary conditions and loads found in engineering practice.

The time and space dependent fundamental solution approach
has the advantage that the time stepping algorithms are not as
restrictive as when using a simple finite different scheme. In

addition the formulation can be applied using two different approaches, the first consists of dividing the domain into a series of cells and carrying out the domain integrals for the computation of the initial conditions from step to step. This approach has the obvious disadvantage that the method becomes a domain technique albeit with much larger time steps than those used when discretizing the time derivative using simple finite differences. To overcome this Wrobel and Brebbia [8] developed as a second approach an algorithm to reduce the solution on time using only boundary integrals. This technique, although easy to apply and mathematically elegant, has the disadvantage that it may require large amounts of computer time as at any given time the integral equation is always referred to the initial state.

Because of the disadvantages of the above methods for solving diffusion type problems, the authors investigated the use of the technique - now called Dual Reciprocity Method - first proposed by Brebbia and Nardini [9-10] for the solution of hyperbolic type problems. The method which is described in detail in reference [11] allows for the use of boundary integrals only and has the advantage that it works with the same type of integrals as the steady state case. It has been applied for the solution of transient heat transfer problems with linear material properties in reference [12].

The dual reciprocity approach can be extended to solve non-linear material properties using Kirchhoff's transformation and a modified time variable, which is needed when the diffusivity coefficient is temperature dependent. This time transformation which has been proposed by Kadambi and Dorri [13] results in a new time variable which is in itself function of position. The resulting system can be highly non-linear and need to be solved using a Newton Raphson type algorithm.

The approach has been implemented in the BEASY boundary element analysis package [14], and the resulting code is now being tested and used by several industries. The formulation relies on the accuracy of the approximations used in the dual reciprocity technique which - within certain guidelines - is satisfactory for engineering practice. The beauty of this technique is that it can be easily extended to solve a variety of other time dependent and non-linear problems in addition to describing arbitrary body, thermal forces, etc. which are difficult to take to the boundary using conventional methods.

2. BOUNDARY INTEGRAL FORMULATIONS

Consider the following temperature diffusion equation in which the material coefficients, (k, the conductivity, c, the specific heat and ρ the density) are all functions of temperature, T, i.e.

$$\frac{\partial}{\partial x_j}\left(k\,\frac{\partial T}{\partial x_j}\right) = \rho c\,\frac{\partial T}{\partial t} \tag{1}$$

One can first use the Kirchhoff's transform [15] to convert the above equation to a Laplacian form in the new variable u, where

$$du = k(T)\,dT \tag{2}$$

or

$$u = \int_{T_o}^{T} k(u)\,du \tag{3}$$

T_o: arbitrary reference value. The derivate with respect to time is

$$\frac{\partial u}{\partial t} = k(T)\,\frac{\partial T}{\partial t} \tag{4}$$

Hence equation (1) can now be written in a Laplacian form, i.e.

$$\nabla^2 u = \frac{1}{\kappa}\,\frac{\partial u}{\partial t} \tag{5}$$

where $\kappa = k/\rho c$ is a temperature dependent diffusivity potential. In the past [16] the variation of κ with temperature has been assumed to be negligible and a mean value of κ used to linearize (5), but this approximation is not accurate in all cases.

Based on reference [13] one can now define a new variable τ such that,

$$d\tau = \kappa\,dt \tag{6}$$

or

$$\tau = \int_{o}^{t} \kappa\,dt \tag{7}$$

By using this new variable, equation (5) can be written as a linear diffusion equation, i.e.

$$\nabla^2 u = \frac{\partial u}{\partial \tau} \tag{8}$$

This equation can now be solved using the Dual Reciprocity Method [10] which starts by considering the steady state operator only and proposes using the fundamental solution corresponding to

$$\nabla^2 u^* = -\Delta_i \tag{9}$$

where Δ_i is the Dirac delta function acting at a point 'i' and u* the corresponding fundamental solution of equation (9).

By applying reciprocity one can write,

$$\int_\Omega (u^*\nabla^2 u - u\nabla^2 u^*) d\Omega = \int_\Gamma (u^*q - uq^*) d\Gamma \tag{10}$$

where $q = \partial u/\partial n$ and $q^* = \partial u/\partial n$ are the normal fluxes. n is the normal to the Γ boundary and Ω indicates the domain.

Taking into account the character of u* (see equation (9)) one can write

$$\int_\Omega u^*\nabla^2 u \, d\Omega = -c_i u_i + \int_\Gamma (u^*q - uq^*) d\Gamma \tag{11}$$

c_i constant dependent on the value of the solid angle at 'i'. One can now replace the left hand side of (11) by the time derivatives on the right hand side of (8) and write,

$$\int_\Omega u^* \dot{u} \, d\Omega = -c_i u_i + \int_\Gamma (u^*q - q^*u) d\Gamma \tag{12}$$

where $\dot{u} = \partial u/\partial \tau$.

Integration of the left hand side terms in (12) would require subdivision of the Ω domain into cells. Alternatively one can apply the Dual Reciprocity technique which consists of two steps.

i) Defining an approximate function for $\dot{u}$ on the Ω domain in terms of a given set of functions.
ii) Finding the particular solutions for the set of functions when they are assumed to be on the right hand side of the governing equation.

Let us first assume that the $\dot{u}$ function can be represented by a set of N coordinate functions, f^j multiplied by an unknown function of time $\dot{\alpha}^j$, much the same as it is done over an element, in classical finite element theory, i.e.

$$\dot{u}(x,t) = \sum_{j=1}^{N} f^j(x) \, \dot{\alpha}^j(t) \tag{13}$$

The domain integral in equation (12) can now be written as,

$$\int_\Omega u^* \dot{u} \, d\Omega = \sum_{j=1}^{N} \dot{\alpha}^j \int_\Omega f^j u^* \, d\Omega \qquad (14)$$

The fundamental concept in dual reciprocity consists of assuming that for each of these f^j functions there exists a function $\hat{u}^j(x)$ such that it gives a particular solution to the governing equation - in this case Laplace's - i.e.

$$\nabla^2 \hat{u}^j = f^j \qquad (15)$$

This equation can now be substituted into (14) giving

$$\int_\Omega u^* \dot{u} \, d\Omega = \sum_{j=1}^{N} \left\{ \int (\nabla^2 \hat{u}^j) u^* \, d\Omega \right\} \dot{\alpha}^j \qquad (16)$$

Notice that as the above integral has a Laplacian form it is now simple to transform it into a boundary integral form by applying reciprocity (see equation (10)) between the $\hat{u}$, $\hat{q}$ and u^*, q^* fields, which gives,

$$\int_\Omega (\nabla^2 \hat{u}^j) u^* \, d\Omega = -c_i \hat{u}_i^j + \int_\Gamma (u^* \hat{q}^j - \hat{u}^j q^*) d\Gamma \qquad (17)$$

where $\hat{q} = \partial \hat{u}/\partial n$.

Substituting this equation into (17) one obtains the following expression involving only boundary integrals, i.e.

$$c_i u_i + \int_\Gamma q^* u \, d\Gamma - \int_\Gamma u^* q \, d\Gamma =$$

$$= \sum_{j=1}^{N} (c_i \hat{u}_i^j + \int_\Gamma q^* \hat{u}^j \, d\Gamma - \int_\Gamma u^* \hat{q}^j \, d\Gamma) \dot{\alpha}^j \qquad (18)$$

It is natural now to propose the same boundary interpolation functions for u, q as u^*, q^* which produces the same influence coefficients and avoids having to integrate again. Hence after discretization and integration equation (18) produces the following matrix system,

$$\underset{\sim}{H} \underset{\sim}{U} - \underset{\sim}{G} \underset{\sim}{Q} = [\underset{\sim}{H} \underset{\sim}{\hat{U}} - \underset{\sim}{G} \underset{\sim}{\hat{Q}}] \dot{\alpha} \qquad (19)$$

where $\hat{U}$ and $\hat{Q}$ are square matrices whose coefficients are given by the particular solution fields obtained from equation (15). The last step is to transform $\dot{\alpha}$ vector into the $\dot{U}$ unknown. This can be easily done by specializing (13) for the boundary points, which in matrix form gives

$$\dot{U} = F \dot{\alpha} \tag{20}$$

Upon inversion, $\dot{\alpha} = F^{-1} \dot{U} = E \dot{U}$ (21)

Equation (19) can now be written as,

$$C \dot{U} + H U = G Q \tag{22}$$

where C is a diffusion matrix, i.e.

$$C = - (H \hat{U} - G \hat{Q}) E \tag{23}$$

The matrix expressions in (22) are similar in form to those obtained using finite elements and the equations can be integrated in time using a standard scheme. The coefficients of H and G are found using the Laplacian fundamental solution and standard interpolation functions. The square matrices $\hat{U}$ and $\hat{Q}$ represent the values of $\hat{u}$ and $\hat{q}$ variables at different nodes and for different f^j functions.

The present version of BEASY uses a simple two-dimensional time integration scheme which employs a linear approximation for U and Q within each time step of the form,

$$U = (1-\theta)U^m + \theta U^{m+1}$$
$$Q = (1-\theta)Q^m + \theta Q^{m+1} \tag{24}$$
and $\dot{U} = \frac{1}{\Delta\tau}(U^{m+1} - U^m)$.

Substituting these approximations into (17) produces the following system,

$$\left(\frac{1}{\Delta\tau} C + \theta H\right)U^{m+1} - \theta G Q^{m+1} =$$
$$= \left(\frac{1}{\Delta\tau} C - (1-\theta)H\right)U^m + (1-\theta)G Q^m \tag{25}$$

The right hand side of equation (25) is now at time mΔt. After introducing the boundary conditions at time (m+1)Δt one can rearrange the left hand side of (25) and solve the resulting system of equations by using a direct solution procedure such as Gauss elimination.

For more complex non-linear problems, however, BEASY uses a Newton-Raphson type algorithm as presented by Wrobel and Azevedo [17].

For non-linear diffusion problems for instance the variable τ is dependent upon position and hence an iteration is required. At each node for instance

$$\Delta \tau_j = \kappa_j \Delta t \tag{26}$$

The solution starts by solving for the linear case for a particular time, hence $\Delta \tau = \Delta t$. Once the value of u has been found for all points, one can perform an inverse transformation to find values of T. Next we need to apply Newton Raphson, and if the material varies with respect to temperature, the values of k, ρ and c and their derivatives with respect to T need to be computed. These values are easy to calculate as the conductivity, density and specific heat dependency upon temperature are modelled in BEASY using piecewise linear functions. Values of $\Delta \tau$ can be calculated at all points as $\Delta_j = \kappa_j \Delta t$. With this information the tangent matrices required in Newton Raphson can be evaluated and error vectors computed. If the norm of the residuals is still too large new iterations will be carried out until convergence is deemed to have occurred. One can then proceed to the next time with the convergent solutions of U and Q.

The type of approximate functions selected for f^j determines the efficiency of the scheme. The ones usually recommended are simply the distance between the pole 'j' and any point in R, i.e.

$$f^j(\vec{x}) = R(\vec{x}, \vec{x}_j) \tag{27}$$

Notice that the set of functions are linearly independent provided that none of the poles coincide with another. Substituting this function in equation (15) and integrating gives very simple expression for the $\hat{u}$ and $\hat{q}$ field, i.e.,

$$\begin{aligned}\hat{u}^j &= \frac{1}{a} R^3 \\ \hat{q}^j &= \frac{3\,k}{a} R^2 \frac{\partial R}{\partial n}\end{aligned} \tag{28}$$

where a depends on the dimensions of the problem. (a = 9 for

2D and a = 12 for 3D cases).

In addition to the above mentioned functions it is generally necessary to include a function f = constant associated to an internal pole. In certain cases it is also necessary to introduce more internal poles if the problem does not possess enough degrees of freedom to approximate properly $\dot{u}$. These functions f for the internal points are the same as for the boundary points (i.e. given by (27)). It is important to point out that internal poles can also be used in cases for which the non-linearities are impossible to be taken to the boundary using other standard techniques.

3. APPLICATIONS

The diffusion analysis procedure described in this paper has been implemented into the BEASY analysis system [14] which can solve two dimensional, axisymmetric and three dimensional problems for homogeneous or piecewise homogeneous bodies. BEASY works with constant, linear or quadratic elements which can be either continuous or discontinuous. This paper presents the following applications:

i) One dimensional thermal shock.
ii) Semi-infinite slab with convection

i) One-dimensional Thermal Shock

The first problem studied was that of an infinite slab subject to a thermal shock. This simple linear problem was included to test convergence of the numerical solutions with refining discretizations and decreasing time step values. The problem is, in fact, one-dimensional but it has been modelled as two-dimensional with mixed boundary conditions, i.e. u = 1 prescribed along the faces $x_1 = \pm L$ and q = 0 along the faces $x_2 = \pm \ell$ of a rectangular region at zero initial temperature. The numerical values adopted for the geometrical dimensions and material properties were L = 5, ℓ = 4 and κ = 1. Results are presented for the temperature at the centre point, $x_1 = x_2 = 0$. All numerical values are assumed dimensionless.

Table 1 compares the analytical solution at several time levels[19] with numerical results obtained with discretizations of 7, 14 and 28 constant boundary elements over one quarter of the region. The BEASY system takes symmetric into account by a reflection and condensation process described in [1]. It can be seen from the table that the results converge to values which are, generally, within 3% of the corresponding analytical solutions. One should notice, however, that the present formulation applies the thermal shock linearly over the first time step (and not suddenly as in the analytical solution). The results can be further improved by smoothing the thermal shock following standard finite element procedures.

For the results in table 2, we keep the discretization of 14 constant boundary elements fixed and decrease the time step value. Clearly, the smaller the time step value the better the approximation of the thermal shock, and this is reflected in the accuracy of the results obtained.

Results in tables 1 and 2 have been obtained with 1 internal pole located at the centre of the region and a unit value of coefficient θ. Tables 3 and 4 present results for 14 constant boundary elements, $\Delta t = 0.5$, varying the number of internal poles and value of coefficient θ, respectively. They demonstrate the superiority of the fully implicit scheme ($\theta = 1$) and the need for one internal pole associated to a constant, as previously described.

Finally, table 5 presents a comparison of solutions obtained with the same number of constant, linear and quadratic boundary elements (14, with $\Delta t = 1$, $\theta = 1$, 1 internal pole). Once again, the results attest the consistency and efficiency of the present formulation. The linear and quadratic boundary element discretizations employed continuous elements with a partially discontinuous one at the corner, to account for the discontinuity of the normal vector at that point.

TABLE 1

TIME	ANALYTICAL	7 ELEM.	14 ELEM.	28 ELEM.
2	0.025	0.016	0.012	0.012
4	0.154	0.167	0.158	0.158
6	0.298	0.304	0.290	0.292
8	0.422	0.420	0.405	0.407
10	0.526	0.518	0.502	0.504
15	0.710	0.696	0.681	0.683
20	0.823	0.809	0.796	0.798
30	0.934	0.924	0.917	0.917

TABLE 2

TIME	ANALYTICAL	$\Delta t = 1.0$	$\Delta t = 0.5$	$\Delta t = 0.1$
2	0.025	0.012	0.012	0.011
4	0.154	0.158	0.162	0.165
6	0.298	0.290	0.298	0.304
8	0.422	0.405	0.414	0.422
10	0.526	0.502	0.512	0.520
15	0.710	0.681	0.691	0.698
20	0.823	0.796	0.804	0.811
30	0.934	0.917	0.921	0.925

TABLE 3

TIME	ANALYTICAL	NO POLE	1 POLE	2 POLES	4 POLES
2	0.025	1.006	0.012	0.033	-0.018
4	0.154	0.142	0.162	0.141	0.135
6	0.298	0.154	0.298	0.275	0.282
8	0.422	0.307	0.414	0.396	0.405
10	0.526	0.450	0.512	0.498	0.507
15	0.710	0.692	0.691	0.684	0.691
20	0.823	0.826	0.804	0.801	0.806
30	0.934	0.942	0.921	0.921	0.924

TABLE 4

TIME	ANALYTICAL	$\theta = 1$	$\theta = 2/3$	$\theta = 1/2$
2	0.025	0.012	-0.002	-0.009
4	0.154	0.162	0.151	0.146
6	0.298	0.298	0.292	0.289
8	0.422	0.414	0.411	0.410
10	0.526	0.512	0.511	0.510
15	0.710	0.691	0.692	0.693
20	0.823	0.804	0.807	0.807
30	0.934	0.921	0.923	0.924

TABLE 5

TIME	ANALYTICAL	CONSTANT	LINEAR	QUADRATIC
2	0.025	0.012	0.016	0.017
4	0.154	0.158	0.166	0.166
6	0.298	0.290	0.302	0.302
8	0.422	0.405	0.418	0.418
10	0.526	0.502	0.515	0.516
15	0.710	0.681	0.694	0.694
20	0.823	0.796	0.807	0.807
30	0.934	0.917	0.923	0.923

ii) <u>Semi-infinite Slab with Convection</u>

This example deals with a semi-infinite slab, initially at zero temperature, in contact with a medium which ambient temperature was suddenly raised to 600°C. Heat is then exchanged by convection with heat transfer coefficient h = 0.260 J/(cm^2 °C). The physical properties of the slab material are k = 0.520 (1 - 5 × 10^{-4}u) W/(cm °C) and ρc = 0.602 (1 + 2 × 10^{-4} u) J/(cm^3 °C), representative of aluminium for the range 0°C to 600°C.

This problem, although simple, presents one extra non-linearity associated to the definition of the convection boundary condition in the Kirchhoff transform space.

A slice of the slab, with cross-sectional dimensions 10cm × 2cm, has been analysed by discretizing one-quarter of it into 6 equal sized quadratic boundary elements. Results of the analysis are presented in table 6, along with results calculated on a constant property basis with the two extreme values of the temperature (0°C and 600°C), representing lower and upper bounds for the nonlinear solution. Also displayed in the table are results obtained with a simple, one-dimensional, finite difference calculation [11], for the first few time steps.

TABLE 6

TIME (s)	BEM $k = k_o$	BEM $k = k_{var}$	BEM $k = k_{600}$	FDM $k = k_{var}$
1	7	4	-9	0
2	41	37	22	30
3	71	66	51	75
4	97	91	77	-
5	119	113	99	-
6	139	132	119	-
7	157	149	137	-
8	171	164	152	-
9	185	177	167	-
10	197	189	180	-

4. CONCLUSIONS

The dual reciprocity technique is a powerful method in the solution of many boundary element problems, including time dependent and non-linear cases.

This paper demonstrates how the method can be used to solve thermal diffusion problems with temperature dependent non-linear material properties. The approach which has been implemented in a Boundary Element Analysis package [14] can be extended to solve a wide variety of other boundary element problems, which produce domain integrals difficult to transform into boundary integrals using conventional techniques [18].

REFERENCES

[1] BREBBIA, C.A., TELLES, J. and WROBEL, L. Boundary Element Techniques - Theory and Applications in Engineering Springer Verlag, Berlin, NY, 1984.

[2] BREBBIA, C.A. and WALKER, S. Boundary Element Techniques in Engineering Butterworths, London, 1981.

[3] CURRAN, D.A.S., CROSS, M., LEWIS, B.A. Solution of Parabolic Differential Equations by the Boundary Element Method using Discretization in Time Appl. Math. Modelling, Vol.4, pp.398-400, 1970.

[4] RIZZO, F.J. and SHIPPY, D.J. A Method of Solution for Certain Problems of Transient Heat Conduction AIAA Journal, Vol.8, pp.2004-2009, 1970.

[5] CHANG, Y.P., KANG, C.S., CHEN, D.J. "The Use of Fundamental Green's Function for the Solution of Problems of Heat Conduction in Anisotropic Media. Int. J. Heat Mass Transfer Vol.16, pp.1905-1918, 1973.

[6] BREBBIA, C.A. and WROBEL, L.C. The Boundary Element Method for Steady-State and Transient Heat Conduction, in Numerical Methods in Thermal Problems (R.W. lewis and K. Morgan, Eds), Pineridge Press, Swansea, 1979.

[7] PINA, H.L.G. and FERNANDES, J.L.M. Three-Dimensional Transient Heat Conduction by the Boundary Element Method, in Boundary elements V, (C.A. Brebbia et al., Eds) Springer-Verlag, Berlin, 1983.

[8] WROBEL, L.C. and BREBBIA, C.A. Time Dependent Potential Potential Problems, in Progress in Boundary Element Methods Vol.1, (C.A. Brebbia, Ed.) Pentech Press, London, 1981.

[9] NARDINI, D. and BREBBIA, C.A. A New Approach to Free Vibration Analysis using Boundary Elements Proceedings 4th Intern. Conf. on Boundary Element Methods Southampton University, September 1982. Springer Verlag, Berlin, 1982.

[10] BREBBIA, C.A. and NARDINI, D. Dynamic Analysis in Solid Mechanics by an Alternative Boundary Element Procedure Int. Jnl. Soil Dynamics and Earthquake Engineering Vol.2, pp.228-233, 1983.

[11] NARDINI, D. and BREBBIA, C.A. Boundary Integral Formulation of Mass Matrices for Dynamics Analysis in Topics in Boundary Element Research, Vol.2 (C.A. Brebbia, Ed.) Springer-Verlag, Berlin & NY, 1985.

[12] WROBEL, L.C., BREBBIA, C.A. and NARDINI, D. Analysis of Transient Thermal Problems in the BEASY System, in BETECH/86 (J.J. Connor & C.A. Brebbia, Eds), Computational Mechanics Publications, Southampton, UK, 1986.

[13] KADAMBI, V. and DORRI, B. Solution of Thermal Problems with Non-linear Material Properties by the Boundary Integral method, in BETECH/85 (C.A. Brebbia & J. Noye, Eds), Computational Mechanics Publications, Southampton, UK, 1985.

[14] NAGESWARAN, S. and BREBBIA, C.A. BEASY - Boundary Element Analysis System as a CIM Tool, CAE Systems Handbook, Vol.1, Computational Mechanics Publications, Southampton, UK, 1987.

[15] BIALECKI, R. and NOWAK, A.J. Boundary Value Problems in Heat Conduction with Non-linear Material and Non-linear Boundary Conditions Appl. Math. Mod., Vol.5, pp.417-421, 1981.

[16] SKERGET, P. and BREBBIA, C.A. Time Dependent Non-linear Potential Problems, in Topics in Boundary Element Research, Vol.2 (C.A. Brebbia, Ed.), Springer-Verlag, Berlin & NY, 1985.

[17] WROBEL, L.C. and AZEVEDO, J.P.S. A Boundary Element Analysis of Non-linear Heat Conduction, in Numerical Methods in Thermal Problems, Vol.4, (R.W. Lewis and K. Morgan, Eds), Pineridge Press, Swansea, 1987.

[18] NIKU, S.M. and BREBBIA, C.A. Dual Reciprocity Boundary Element Formulation for Potential Problems with Arbitrarily Distributed Sources. To be published in Eng. Analysis Jnl.

[19] CARSLAW, H.S. and JAEGGER, J.C. Conduction of Heat in Solids 2nd Edition, Clarendon Press, Oxford, 1959.